9th International Conference on Compressors and their Systems

AA003213

IOP Conference Series: Materials Science and Engineering
Volume 90

London, United Kingdom
7 - 9 September 2015

ISBN: 978-1-5108-0998-7
ISSN: 1757-8981

Printed from e-media with permission by:

Curran Associates, Inc.
57 Morehouse Lane
Red Hook, NY 12571

Some format issues inherent in the e-media version may also appear in this print version.

Copyright© (2015) by the Institute of Physics
All rights reserved. The material featured in this book is subject to IOP copyright protection, unless otherwise indicated.

Printed by Curran Associates, Inc. (2015)

For permission requests, please contact the Institute of Physics at the address below.

Institute of Physics
Dirac House, Temple Back
Bristol BS1 6BE UK

Phone: 44 1 17 929 7481
Fax: 44 1 17 920 0979

techtracking@iop.org

Additional copies of this publication are available from:

Curran Associates, Inc.
57 Morehouse Lane
Red Hook, NY 12571 USA
Phone: 845-758-0400
Fax: 845-758-2634
Email: curran@proceedings.com
Web: www.proceedings.com

Table of contents

Volume 90

9th International Conference on Compressors and their Systems
7–9 September 2015, London, UK

Accepted papers received: 6 July 2015
Published online: 30 July 2015

Preface

Papers

Key Notes

Screw Compressors

Reciprocating Compressors

Scroll Compressors

Vane Compressors

Turbocompressors

Novel Compressors

Expanders

Systems

Modelling

Technology Development

IOP Conf. Series: Materials Science and Engineering **90** (2015) 011001 doi:10.1088/1757-899X/90/1/011001

9th International conference on compressors and their systems

FOREWORD

The 9th International Conference on Compressors and their Systems will be held in London from 5th – 9th September 2015, and as its Chairman, it is my pleasure to welcome you.

This series of conferences started in 1999 organised by the Fluid Machinery Group of the Institution of Mechanical Engineers (IMechE) but since 2009 it has been managed by City University London in conjunction with the IMechE and the Institute of Refrigeration, both of which have been very proactive in promoting it. The Organising committee is grateful for their support and continued encouragement.

This year, after rigorous reviewing, we have accepted over 80 technical papers for publication, the highest number in the conference history. On behalf of the organising committee I would like to thank the reviewers for their hard work and assistance.

In addition to the main technical sessions, this year we have introduced a third day, specifically for Industry, to consider technology, business and market drivers on compressor developments. The traditional series of the short courses is this year continuing prior to the main event with the second short course/forum on Computational Fluid Dynamics in rotating positive displacement machines.

I would like to extend my special thanks to our main sponsors, Holroyd PTG, Howden and Kapp Niels for their continuing support for the conference. With their generous contributions we have managed to keep the conference fees at the same level as in 2013, despite extending it to 3 days and holding it outside the University this year.

The welcome reception on Sunday 6th September 2015 is dedicated to the celebration of the 20th anniversary of the Centre for Positive Displacement Compressors Technology which was formed at City University in 1995 with support from the Royal Academy of Engineering and Holroyd; its main aim being to assist British manufacturers of screw compressors. The Centre has since made a significant impact on the screw compressor world, far beyond the British Isles and continues to grow in its support for the screw machine industry. Recently renamed the Centre for Compressors Technology, its aim has been extended to help promote the development of all types of compressors and expanders.

On this occasion, I must thank all those who have helped to make this event happen. Special thanks go to Louise Gordon, Emma Leaver and Sham Rane who did most of the organisation work and to all members of the organising committee and the international liaison committee. I would like to commend Dr Kristina Soldatova who joined the international liaison committee two years ago and extensively promoted the Conference in Russia which brought a large number of papers to the Conference from that part of the world.

On behalf of the conference committee I wish you successful and enjoyable conference.

Professor Ahmed Kovacevic

Conference Chairman

Content from this work may be used under the terms of the Creative Commons Attribution 3.0 licence. Any further distribution of this work must maintain attribution to the author(s) and the title of the work, journal citation and DOI.
Published under licence by IOP Publishing Ltd

School of Mathematics, Computer Science and Engineering
City University London, United Kingdom

IOP Conf. Series: Materials Science and Engineering **90** (2015) 011002 doi:10.1088/1757-899X/90/1/011002

9th International conference on compressors and their systems

City University London would like to thank our sponsors and partners

Sponsored by

HOLROYD MACHINE TOOLS & PRECISION COMPONENTS — Platinum sponsor

Howden — Gold sponsor

KAPP NILES — Silver sponsor

Run in conjunction with

Institution of MECHANICAL ENGINEERS

IOR ior.org.uk

Welcome reception sponsorship provided by
Centre for Compressor Technology at City University London

Social reception sponsorship provided by
CPI Fluid Engineering

Exhibitors
PDM Analysis in association with Simerics and 8020 Engineering
CFX Berlin
VERT Rotors

Content from this work may be used under the terms of the Creative Commons Attribution 3.0 licence. Any further distribution of this work must maintain attribution to the author(s) and the title of the work, journal citation and DOI.
Published under licence by IOP Publishing Ltd

Organising Committee 2015

Ahmed Kovacevic (Chair)	City University London, UK
Graeme Cook	Howden Compressors, UK
Louise Gordon	City University London, UK
Chris Holmes	PTG Holroyd Advanced Developments Ltd, UK
Guy Hundy	Aleph Zero, UK
Amy Jacobs	IMechE, UK
Emma Leaver	City University London, UK
Sham Rane	City University London, UK
Ivor Rhodes	ACE Cranfield, UK
Melanie Rigby	Consultant, UK
Miriam Rodway	Institute of Refrigeration, UK
Ian Smith	City University London, UK
Nikola Stosic	City University London, UK

International Liaison Committee 2015

Ian Bennett	Shell Projects and Technology, The Netherlands
Cesar DeChamps	University of Santa Catarina, Brazil
Mitsushiro Fukuta	Shizuoka University, Japan
Kristina V. Soldatova	St Petersburg University, Russia
Amin Haghjoo	Ingersoll Rand, Germany
Liansheng Li	Hefei General Machinery Research Institute, China
Erich Machu	Consultant, Austria
Kim Tow Ooi	Nanyang Technology University, Singapore
Matthew Cambio	Ingersoll Rand/Trane, USA
Joost Brazs	Danfoss Turbocor Compressors, USA
Pega Hrnjak	University of Illinois, USA
Peter Duineveld	HOERBIGER BENELUX B.V., The Netherlands

Content from this work may be used under the terms of the Creative Commons Attribution 3.0 licence. Any further distribution of this work must maintain attribution to the author(s) and the title of the work, journal citation and DOI.
Published under licence by IOP Publishing Ltd

Reviewers 2015

Thank you to all those who took part in the reviewing process:

Adrian	Alford	Corac Energy Technologies	United Kingdom
Igor	Almeida	Federal Institute of Education, Science and Technology of Rio Grande do Norte	Brazil
Brian	Angel	Renuda	France
Bachir	Bachir	Emerson Climate Technologies	Germany
Indranil	Banerjee	Bhabha Atomic Research Center	India
Ian	Bell		Belgium
Aditya	Bhakta	General Electric	India
Nasir	Bilal	Purdue University	United States
Joost	Brasz	Danfoss Turbocor Compressors	United States
Andreas	Bruemmer	TU Dortmund	Germany
Xiaodan	Cai	United Technologies Research Center	United States
Kim	Cheolhwan	LG Electronics	South Korea
Roberto	Cipollone	University of L'Aquila	Italy
Graeme	Cook	Howden Compressors Ltd	United Kingdom
Diogo Londero	Da Silva	Federal University of Santa Catarina	Brazil
Joris	Degroote	UGent	Belgium
Cesar J	Deschamps	Federal University of Santa Catarina	Brazil
Ashvin	Dhunput	Dunham-Bush Ltd	United Kingdom
Hui	Ding	Simerics Inc.	United States
Marco	Diniz	POLO - UFSC	Brazil
Oleg	Dmitriev	VERT Rotors UK Limited	United Kingdom
Peter	Duineveld	Hoerbriger	Netherlands
Thiago	Dutra	POLO/ Federal University of Santa Catarina	Brazil
Ferran	Escanes	Huayi Compressor Barcelona, S.L.	Spain
Kenneth Ian	Fitzsimmonds		United Kingdom
Mitsuhiro	Fukuta	Shizuoka University	Japan
Yuri St	Galerkin	Peter the Great St.Petersburg Polytechnic University	Russia
Jos Luiz	Gasche	Unesp-Ilha Solteira	Brazil

Content from this work may be used under the terms of the Creative Commons Attribution 3.0 licence. Any further distribution of this work must maintain attribution to the author(s) and the title of the work, journal citation and DOI.
Published under licence by IOP Publishing Ltd

Adel	Ghenaiet	University of Science and Technology (USTHB), Algiers	Algeria
Francesco	Grasso	Università di Firenze - DINFO	Italy
Pengcheng	Guo	Xi'an University of Technology	China
Amin	Haghjoo	Ingersoll Rand	Germany
Martin	Heinrich	TU Bergakademie Freiberg	Germany
Manoj	Heiyanthuduwage	Howden Compressors Ltd	United Kingdom
Christopher S.	Holmes	PTG Holroyd	United Kingdom
Md. Anwar	Hossain	Daikin Mcquay	United States
Paul Xiubao	Huang	Hi-Bar MC Technologies, LLC.	United States
Bukac	Hubert	Lttle Dynamics, Inc.	United States
Clive	Hudson	Lontra Ltd	United Kingdom
Guy	Hundy	Aleph Zero Ltd	United Kingdom
Kirill	Ignatiev	Emerson Climate Technologies	United States
Harshad	Inamdar	Purdue University	United States
Prins	Jan	Danfoss	Denmark
Xiaoyun	Ji	Shanghai Hitachi Electrical Appliances Co., Ltd	China
Rigola	Joaquim	CTTC - UPC	Spain
Norbert	Kaemmer	Emerson Climate Technologies GmbH	Germany
Hirotaka	Kameya	Ishinomaki Senshu University	Japan
Joseph	Karnaz	CPI Fluid Engineering	United States
Fabio Henrique	Klein	Embraco	Italy
Ahmed	Kovacevic	City University London	United Kingdom
Rodrigo	Kremer	Embraco Compressors	Brazil
Venu Madhav	Kuppachi	Elgi Equipments Ltd	India
Gilbong	Lee	Korea Institute of Energy Research	South Korea
Vincent	Lemort	Thermodynamics Laboratory	Belgium
Yujun	Leng	Purdue University, West Lafayette	United States
Alex	Leyderman	EADS Fairchild Controls	United States
Daqing	Li	Carrier Corporation of United Technology Company	United States
Liansheng	Li	Hefei General Machinery Research Institute	China
Koen	Lodefier	Atlas Copco Airpower	Belgium
Qiao	Lu	B/E Aerospace Inc.	United States
Yang	Lu	Xi'an Jiaotong University	China
Erich Heinz	Machu		Austria

Cerdoun	Mahfoudh	Military Polytechnic School	Algeria
RK	Mandloi	M A National Institute of Technology, Bhopal	India
Matsui	Masaru	Panasonic	Japan
Mrzyglod	Matthias	BSH Hausgeräte GmbH	Germany
Duane	McCormick	United Technologies Research Center	United States
Elvedin	Mujic	BITZER Kühlmaschinenbau GmbH	Germany
Stefano	Murgia	Ing. Enea Mattei SpA	Italy
Timur	Mustafin	Kazan National Research Technological University	Russia
Kai	Nadler	Oerlikon Leybold Vacuum GmbH	Germany
Alexander	Nikolov	Technische Universität Dortmund	Germany
Tsutomu	Nozaki	Hitachi	Japan
Kim Tiow	Ooi	Nanyang Technological University	Singapore
Ahmet	Ozdemir	ARCELIK A.S.	Turkey
José Alberto Reis	Parise	Pontifícia Universidade Católica do Rio de Janeiro	Brazil
Maria	Pascu	Howden Compressors Ltd	United Kingdom
Andy	Pearson	Star Refrigeration Ltd	United Kingdom
Eduardo	Pereira	Tecumseh do Brasil	Brazil
Stefan	Posch	Graz University of Technology	Austria
Sham	Rane	City University London	United Kingdom
Matthew	Read	City University London	United Kingdom
Ramons	Reba	United Technologies Research Center	United States
Alexei F.	Rekstin	Saint Petersburg State Polytechnic University	Russia
Ivor	Rhodes	ACE Cranfield Ltd	United Kingdom
Guilherme	Ribeiro	Institute for Advanced Studies	Brazil
Melanie	Rigby	Consultant	United Kingdom
Link	Rodrigo	Embraco	China
Ronald	Sachs	Gardner Denver Schopfheim GmbH	Germany
Jack	Sauls	IngersollRand / Trane	United States
Abdulnaser	Sayma	City University London	United Kingdom
Ernane	Silva	POLO/UFSC	Brazil
Ian K	Smith	City University London	United Kingdom

IOP Conf. Series: Materials Science and Engineering **90** (2015) 011004 doi:10.1088/1757-899X/90/1/011004

Kristina	Soldatova	Peter the Great St.Petersburg Polytechnic University	Russia
Nikola	Stosic	City University London	United Kingdom
Alison	Subiantoro	TUM CREATE	Singapore
Kok Ming	Tan	SANDEN INTERNATIONAL SINGAPORE PTE LTD	Singapore
Yan	Tang	Kaishan Compressor	China
Kenji	Tojo	TOJO R&D Design Office / Waseda University	Japan
Luis Mendoza	Toledo	EPFL	Switzerland
Igor	Vaisman	Rolls-Royce Corporation	United States
Sergey	Vizgalov	Kazan National Research Technological University	Russia
James	Willie	Gardner Denver Schopfheim GmbH	Germany
Feng	Yinshan	United Technologies Research Center	United States
Xiaoling	Yu	Xi'an Jiaotong University	China
Dmyto	Zaytsev	GEA Refrigeration Germany GmbH	Germany
Ming	Zhang	Beijing University of Chemical Technology	China
Zaoxiao	Zhang	Xi'an Jiaotong University	China
yuanyang	zhao	Hefei General Machinery Research Institute	China
Zhao	Zhidan	Purdue University	United States
Davide	Ziviani	Ghent University	Belgium
Yang	Zou	Creative Thermal Solutions	United States
Huiming	Zou	Technical Institute of Physics and Chemistry, CAS	China

Peer review statement

All papers published in this volume of *IOP Conference Series: Materials Science and Engineering* have been peer reviewed through processes administered by the proceedings Editors. Reviews were conducted by expert referees to the professional and scientific standards expected of a proceedings journal published by IOP Publishing.

Content from this work may be used under the terms of the Creative Commons Attribution 3.0 licence. Any further distribution of this work must maintain attribution to the author(s) and the title of the work, journal citation and DOI.
Published under licence by IOP Publishing Ltd

IOP Conf. Series: Materials Science and Engineering **90** (2015) 012001 doi:10.1088/1757-899X/90/1/012001

Compressor Lubrication –the key to Performance

Sonny G.Sundaresan

CEO, Sundaresan Consulting Services, LLC. Sidney, Ohio 45365, USA; Retired from Emerson Climate Technologies, Sidney, Ohio 45365,USA

E-mail: sonnysundaresan@yahoo.com

ABSTRACT

Today, Low Global Warming Potential Refrigerants are being evaluated, optimized and applied. Higher-pressure refrigerants, refrigerant blends, and new chemical families of refrigerants are candidates. Compressor designs and system components are being evaluated to deliver the desired capacity, efficiency and reliable performance for intended applications.

Once the Refrigerant family is chosen based on Environmental and Heat transfer properties, System Chemistry between the Refrigerant and Lubricant families are scrutinized for optimum miscibility and solubility. These directly affect the Phase separation between the 2 liquid phases and reduction of viscosity. Over the entire Operating Envelope, the "oil return" and "viscosity reduction" problems are assessed. Compatibility aspects between chemicals (Refrigerants, Lubricants) and materials of construction (metallic, polymeric, elastomeric) are verified to avoid failures in use. System Components, compressor bearings, expansion devices and heat transfer coils are adversely affected due to contaminants circulating in systems. Rust, Copper plating, Formicary corrosion, elastomer swelling are example. Advances in Analytical Chemistry Techniques help the compressor and system manufacturers characterize the problems. Further, System Chemistry guides selection of lubricant additives.

This presentation explains how lubricant selection is accomplished using Tribology and System Chemistry aspects.

Key words: Compressor, Lubrication, Tribology, Stability,. Miscibility, Solubility, System Chemistry, Daniel Plot, Stribeck curve, Life Testing

Content from this work may be used under the terms of the Creative Commons Attribution 3.0 licence. Any further distribution of this work must maintain attribution to the author(s) and the title of the work, journal citation and DOI.
Published under licence by IOP Publishing Ltd

IOP Conf. Series: Materials Science and Engineering **90** (2015) 012001 doi:10.1088/1757-899X/90/1/012001

1. Introduction

- Refrigerant capacity and efficiency are the targets for each application category: car air conditioning, domestic refrigerators and freezers, air-conditioners, heat pumps and commercial refrigeration.
- Lubrication is the key to the Performance of a Reliable Air-conditioning and Refrigeration Compressor.
- Hermetic System Chemistry is a science needed from the beginning to assure thermal and chemical stability, compatibility with all materials of construction, miscibility/solubility during the selection of lubricant family, life testing with the selected formulated lubricants. Operating envelope and contaminants are parameters used in all chemistry evaluations. After acceptable system chemistry screening tests, Tribological evaluation tests are performed. Bench Lubricity Tests and Compressor Life Tests are the next step, [1,2]
- Lower GWP (Global Warming Potential) refrigerants such as HFO1234yf, HFO1234ze, HFC32, CO_2, Propane, Isobutane and their blends are of current interest. Lubricants such as Mineral Oil, Alkyl benzene ,Polyalphaolefin, Polyalkalyneglycol, Polyolester snd Polyvinyl ether are being evaluated to obtain desired net viscosity for specific applications.

2. Refrigerants

2.1 Refrigerant Properties

REFRIGERANT PROPERTIES

- **Thermodynamic Properties**
  - Critical Temperature and Pressure
  - Evaporating / Condensing Pressures
  - Boiling and Freezing Points
  - Specific Heat
  - Enthalpy of Vaporization
- **Chemical Properties**
  - Stability and inertness
  - Compatibility with polymers
  - Flammability
  - Toxicity
- **Environmental Impact**
  - Ozone Depletion Potential
  - Global Warming Impact
- **Transport Properties**
  - Thermal Conductivity
  - Viscosity
  - Surface Tension
- **Physical Properties**
  - Leakage
  - Miscibility
  - Odor
  - Dielectric Strength

2.2 Chemical Families, GWP,and Applications

Currently lower GWP refrigerants such as HFO1234yf, HFO1234ze, HFC32, CO_2 and Hydrocarbons are being evaluated. Refrigerant Blends with HFC, HFO, and Hydrocarbons with 2 to 7 components are being provided by Refrigerant Manufacturers as Low GWP alternatives. Applications cover a wide range, from single stage vapour compression cycle to multi stage in refrigeration, heat pump and air-conditioning systems. The temperature range spans from extra low evaporating temperatures in medical systems to high condensing temperature in high ambient countries. Selection of Refrigerant starts with the operating envelope of the end product and the lubricant selection follows.

3. Tribology

The Lubricant Selection and Optimization Step require the use of fundamentals of Tribology (Friction, Lubrication and Wear). It requires thorough understanding of contact surfaces, materials, types of motion, loads and speeds. Characterization of "wear modes" is the goal for various surface techniques from simple visual, optical microscopy to Scanning Electron Microscopy. The wear modes are adhesive wear, abrasive wear, fatigue wear, fretting wear, corrosive wear, and erosive wear. The very first step in understanding lubrication needs is to understand the role

IOP Conf. Series: Materials Science and Engineering **90** (2015) 012001 doi:10.1088/1757-899X/90/1/012001

of contacting surfaces and the materials. The analyst has a wide range of techniques depending on the need for depth of penetration, special resolution, and tribological applications. Light microscopy is useful for 200nm character and size of wear topography; and able to see surface film colour; SEM is good for 500 nm and microstructure of developing wear scar together with elemental analysis of transfer products as they are produced. SIMs provides 500nm resolution and surface chemistry analysis in vacuum. Atomic Force Spectroscopy has better resolution in the Nano range (0.1nm) and provides three-dimensional Scanning Electron Microscopy (SEM) and Atomic Force Microscopy (AFM). The compressor designer has to confirm the wear regimes of interest. The Stribeck Curve is the Road Map.

3.1 Surface Characterization and wear

Every application has Surfaces in Contact, and a Relative Motion. (e.g., sliding, rolling and impacting). The Surface is not simple , nor it is flat. All engineering surfaces have a roughness and this plays an important role in tribology. The Surface layer : on the top is the Lubricant film, then underneath is Adsorbed Contaminants, Oxide , "Disturbed Material" . Underneath these 4 is the Bulk Material . The film thickness of the lubricant is a few orders of magnitude over that of surface roughness (This ratio is called Lambda). The wear regimes can be based on the ratio; less than 1.2 is boundary lubrication; 1.2-3.0 is mixed or thin film lubrication; greater than 3 is full film hydrodynamic lubrication. [3.]

3.2 Analytical Tools

The following table provides all tools available in Surface Analysis. [4]

Analytical method	Maximum depth of penetration	Spatial resolution	Tribological applications
Auger	0.5–1.5 nm	1000 nm	Metal transfer, surface segregation Magnetic recording surface inspection Analysis in vacuum
XPS	1.5–7.5 nm	1–4 mm	Lubricant reaction products chemistry Analysis in vacuum
Ellipsometry	400–500 nm	1 mm	Transparent solid film thickness Analysis in air
Raman	400–500 nm	10 nm	Organic film thickness and chemistry Analysis in air
Rutherford backscattering	2 µm	10 nm	Surface film thickness composition
SIMS	Sputters 0.5 nm	500 nm	Surface chemistry Analysis in vacuum
In situ SEM wear test		500 nm	Microstructure of developing wear scar Elemental analysis of transfer products as they are produced Analysis in vacuum
Light microscopy		200 nm	Character and size of wear topography surface film color Nomarski texture
Fourier transform infrared spectroscopy		2000 nm	Chemistry of organic films: polymer transfer and boundary lubrication Analysis in air
Atomic force microscopy		0.1 nm	Friction forces on the atomic level Atomic level surface roughness
Profilometer		0.1–25 µm	Surface roughness by stylus Micro-topography of surface finish and wear

Surface Analytical Tools Used in Tribology [4]

3.3 Types of Wear (modes)

Adhesive wear occurs when two smooth bodies are slid over each other, and fragments are pulled of one surface to adhere to the other. These fragments may come off the surface on which they are formed and be transferred back to the original surface, or else form loose wear particle.

Abrasive wear occurs when a rough hard surface, or a soft surface containing hard particles, slides on a softer surface, and ploughs a series of grooves in it .This material from the grooves is displaced in the form of wear particles, generally loose ones.
Surface fatigue wear is observed during repeated sliding or rolling over a track. The repeated loading and unloading cycles to which the materials are exposed may induce the formation of surface or subsurface cracks, which eventually will result in the break-up of the surface with the formation of large fragments leaving large pits in the surface.
In each case, the root cause should be identified; for example, in many cases of abrasive wear, by re-examination of test samples, the origin of the wear particles needs to be identified; if these are contaminants that can be filtered, the designer has an option to select type and location of filter ahead of this component; however, if the abrasive particles are by-products from wear, for example, stiffeners used in plastic/elastomers, the design option may be different.
If fretting wear is observed, eliminating slip at the interface or use of solid film lubricants is helpful.
If corrosive wear is observed, the root cause of corrosion such as acidic or oxidative environment should be investigated and eliminated; for example, copper plating on critical bearing surfaces can be eliminated using an acid catcher in the lubricant.
The investigator's goal is to identify the wear mode and its mechanism; this is an iterative process and may take time and other resources.

4.Lubrication Regimes and Stribeck Curve

4.1 Stribeck Curve is a plot of Friction Coefficient versus Hydrodynamic parameter (fluid viscosity times relative speed divided by load). The various regimes covered are: Boundary Lubrication (BL), Mixed Lubrication (ML), Elastohydrodynamic Lubrication (EHL) and Hydrodynamic Lubrication [3,5]

4.2 Transition Diagram is a plot of Load Carrying Capacity F (N) versus Sliding Speed (m/s) .Three regions separated by two transition curves are shown. (These regions are governed by the measured friction-time characteristics). Region I/II represents Safe region with negligible wear rate. The lubricating conditions in this region are BL, ML and EHL. Region III is to be avoided due to incipient or micro scuffing. The third region III is an Immediate Failure Region by scuffing. [6]

4.3 Stribeck curve

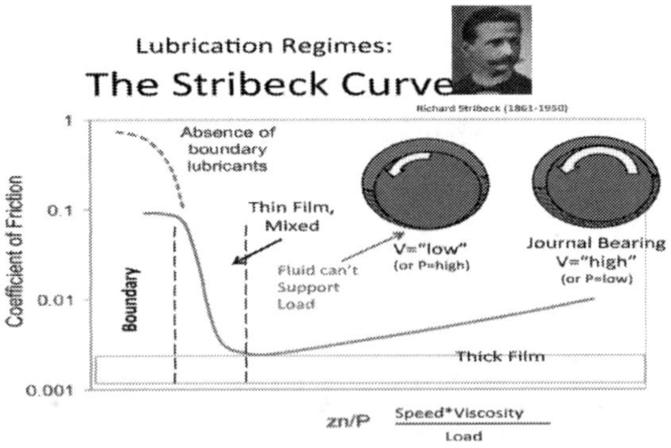

Source : S.Shaffer [3]

4.4 Scuffing

Temperature Rise and Scuffing

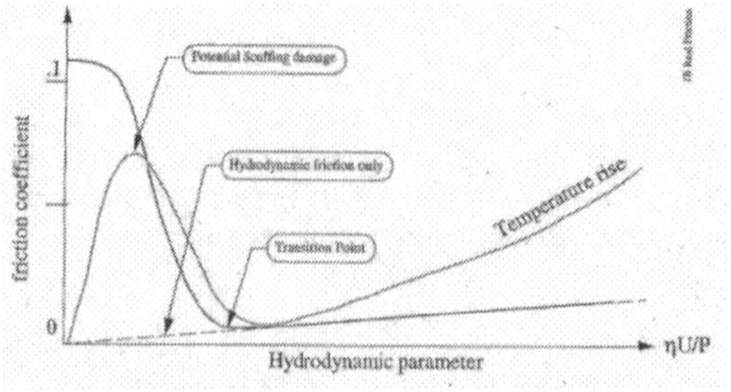

Source: J. L. Tevaarwerk[5]

4.5 Transition Diagram Combined with Stribeck Curve

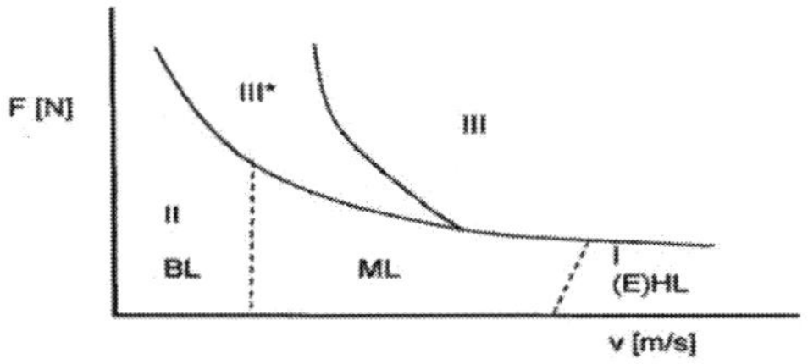

Source: [6]

4.6 Safe Operating Regime

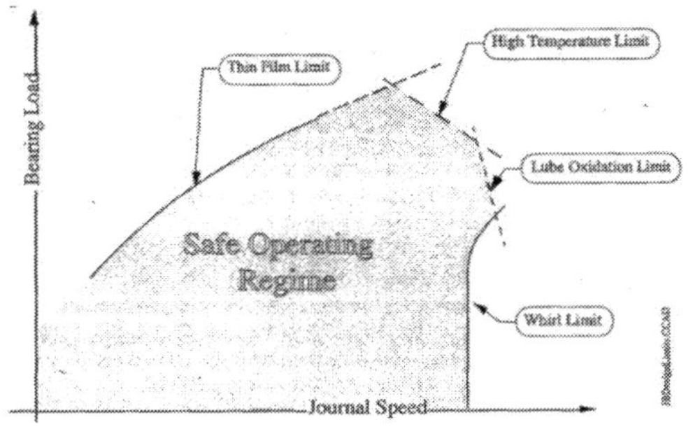

Source: J. L. Tevaarwerk[5]

5. LUBRICANTS and ADDITIVES

5.1 Lubricants

Lubricants are used in bearing designs to reduce friction and wear; other functions in a compressor application are sealing of clearances, noise control cooling and transport debris away fromInterface.heat transfer. Lubricants used in hermetic compressors are selected based on their viscosity and lubricity primarily; other relevant properties are chemical stability, thermal stability, low hygroscopicity, thermal conductivity, electrical resistivity, environmentally acceptable, compatible with elastomeric and polymeric components, soluble with refrigerant gas and non-deposit forming. The compressor manufacturer is ultimately responsible to determine acceptable values for the above list of properties.

In the Refrigeration & Air-Conditioning industry, the following is a list of unique needs of a lubricant: low temperature properties (low pour point, low wax), chemical stability, known partial miscibility with refrigerants and known solubility of refrigerants. The selection of lubricant is based on the three factors, miscibility, solubility and net viscosity.

5.1.1 Lubricant Families and The six chemical families of Lubricants in Air Conditioning and Refrigeration are: Naphthenic Mineral Oil , Alkyl benzene (Polyalphaolefin, Polyalkyleneglycol, Polyolester and Polyvinyl ether The available viscosity grade ranges from 10 cSt to 220 cSt (centi stokes). Viscosity Index(VI) is indicative of viscosity change over a temperature range .(40 Deg.C to 100 Deg.C). The pressure viscosity coefficient is a measure of effect of pressure on viscosity. This is relevant in elastohydrodynamic lubrication regime.

5.1.2 Lubricant Properties vary with the chemical family; the properties oil interests are viscosity, viscosity index, pour point and pressure viscosity coefficient. This figure compares 6 families belonging to 32 VG (viscosity grade).

IOP Conf. Series: Materials Science and Engineering **90** (2015) 012001 doi:10.1088/1757-899X/90/1/012001

WORKING FLUID COMBINATIONS

REFRIGERANTS	R11, 12, 22, 502	R123	R134a	R407C, R410A	R404A, R507A	NH_3	CO_2	HC Mixture
Lubricants Application	MO AB PAO	AB	PAG POE PVE AB	POE PVE AB (?)	POE PVE AB (?)	MO AB PAG	PAO MO POE PAG	MO AB

Note: Viscosity Grades range from 10 cSt to 100 cSt depending on family; refer to ISO 6743-3

5.1.3 Lubricity Additives are based on their functionality : Anti Wear (AW), Extreme Pressure (EP),and Friction Modifier(FM). AW usually is made from P chemistry (e.g. TCP, tricrysylphosphate and BTTP,butyl teraphalatephodphate);EP is made from P and Cl chemistry .[8]

5.2 Miscibility and Solubility

5.2.1An example of miscibility diagram for R410A and POE 22 is compared with R22 and two Mineral Oils.

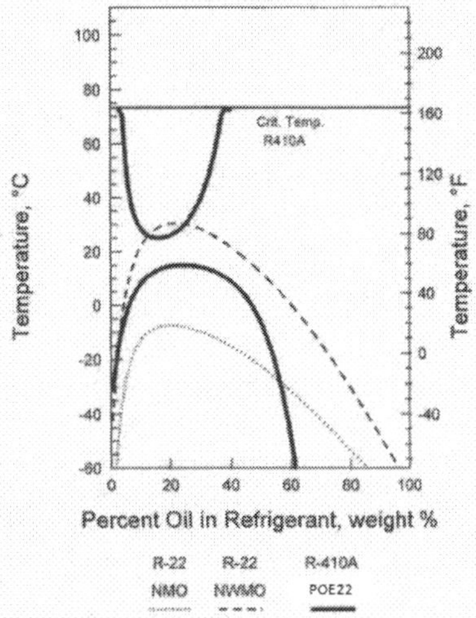

5.2.2 This miscibility/immiscibility has the following implications. At compressor, oil pick up location at start up and net viscosity (of the 2 phases mentioned above), foaming tendencies at start up decompression are affected. In suction line accumulators, location of oil return hole and its size is affected. The solubility affects net viscosity (reduction); this effect may help oil return but adversely affect lubrication. Also, solubility affects explosive decompression of elastomeric components and may result in foaming.

Viscosity selection depends upon minimum film thickness, maximum allowable viscous drag, minimum sealing, minimum flow rate and minimum variation within the operating envelope (temperature, pressure and refrigerant flood back). Viscosity index takes care of the change of viscosity alone with temperature. Solubility and Daniel Plot takes into account solubility, and percentage lubricant

6. Daniel Plots

Pressure versus Temperature is a measure of Solubility. Temperature versus % lubricant in refrigerant is another plot. Density versus temperature for pure Lubricant and with dilution of 5%,10%,15% and 20% refrigerant is also needed. Daniel Chart is Kinematic Viscosity versus Temperature from 100% lubricant, 95%,90% and to

IOP Conf. Series: Materials Science and Engineering **90** (2015) 012001 doi:10.1088/1757-899X/90/1/012001

85% .In Daniel Chart, constant pressure Isobaric) curves are incorporated to show the effect of pressure (typically 10bar, 15,25 and 35 bar).

The customer, for any operating point (e.g.: sump condition or at the bearing condition) can estimate the net kinematic viscosity in cSt. for comparison of different lubricant families and viscosity grades

6.1-6.4, The Daniel Plot of R410A with a 22 cSt POE is derived from refrigerant solubility data, density data and viscosity data versus temperatures. The combined PVXT data is called Daniel Plot and includes isobaric curves.[8]

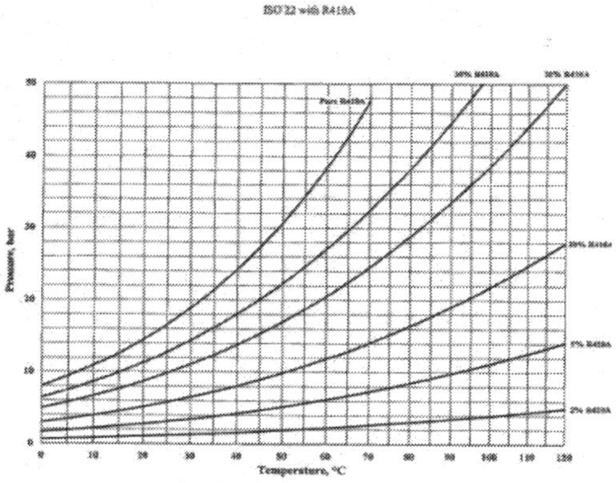

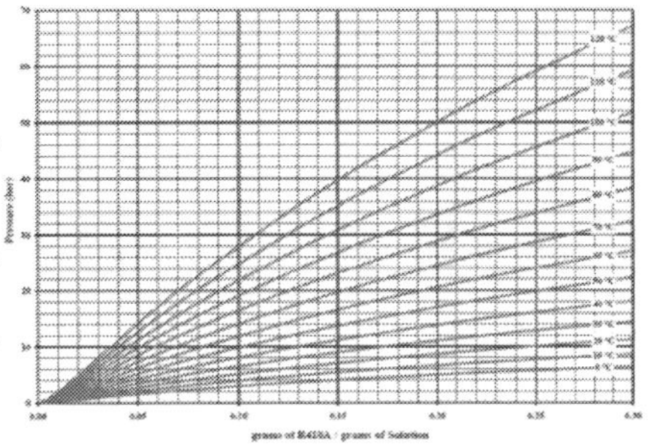

Solubility (Pressure versus Temperature) Pressure versus % Refrigerant Concentration

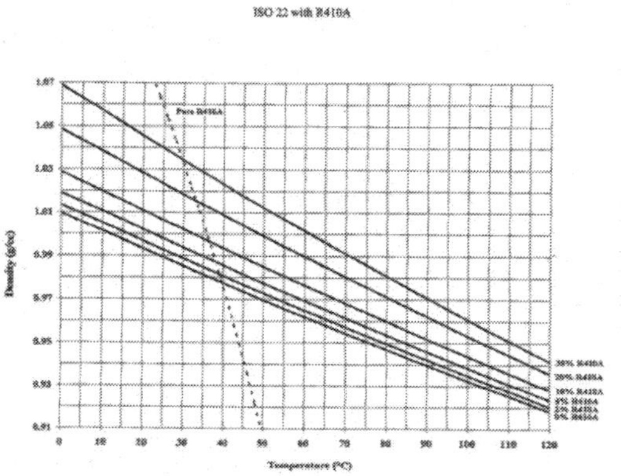

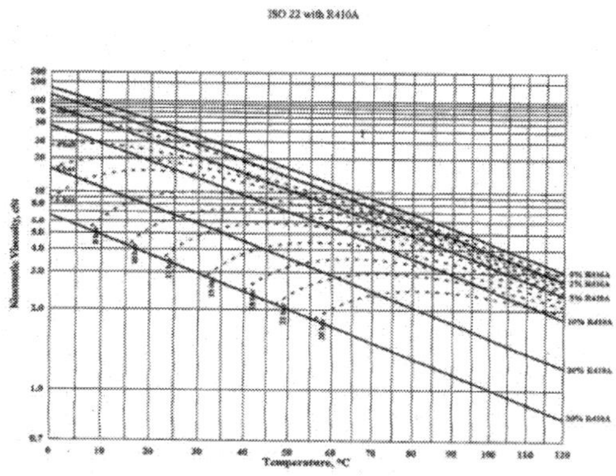

Density Data *Pressure-Viscosity-Composition-Temperature*

IOP Conf. Series: Materials Science and Engineering **90** (2015) 012001 doi:10.1088/1757-899X/90/1/012001

6.5 Determination of net viscosity using Daniel Plots

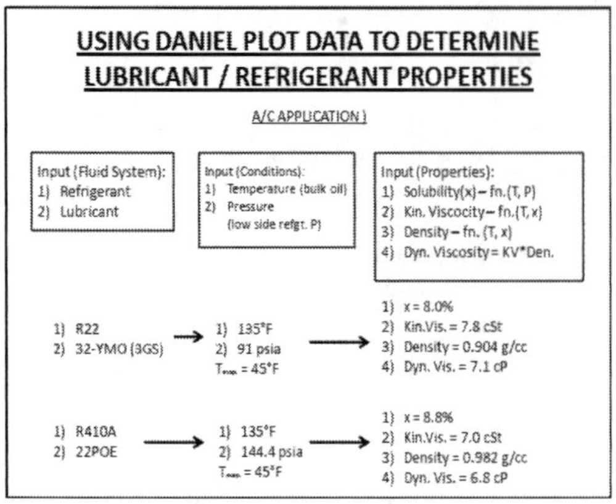

7.Quantifying Wear

The prerequisite to understanding how compressor components perform starts with detailed knowledge of Material, type of motion and extent of motion. Schematic of a Scroll Compressor and the key wear components are provided as an example. Measurements at contact points as a function of time are critically needed to understand Break-in wear and wear during operating life.

7.1 Schematic of a Scroll Compressor

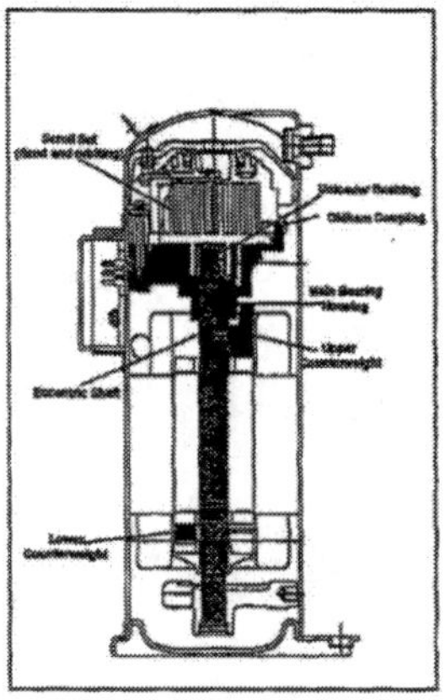

7.2. Key Wear Components in a Typical Scroll Compressor

Table 1. Key Wear Components In A Typical Scroll Compressor

Key Component	Contact Areas	Material	Mating Surface
Scrolls	Tips Bases Thrust Area (Orbiting scroll)	Cast Iron	Bases (other scroll) Tips (other scroll) Main Bearing Housing Thrust Surface
Oldham Coupling	Tabs (loaded side)	Diecast Aluminum based alloy	Scroll Key Slots
Drive Shaft	Flat	Heat Treated Carbon Steel	Unloader Bushing Flat
Unloader Bushing	Inner Surface Flat	Steel Powder Metal	Shaft Drive Flat
Drive Bearing	Inner Surface	Composite PTFE/Bronze liner with steel backing	Unloader Bushing OD
Floating Seal	Outer and Inner Lip	PTFE based	Fixed Scroll
Top Cap	Muffler Plate	Carbon Steel	Seal Face

8.Lubricity Bench Tests and compressor tests

8.1 Bench lubricity tests such as Falex Pin on V block, Block on Ring, Four Ball ,Cameron-Plint are performed depending upon the type of motion Materials of the wear couple can be changed. Lubricant manufacturers screen different base stocks and then different additives. They are inexpensive, relatively short duration and provide excellent value as a screening and ranking tool. The results are useful but may not be correlated to specific wear couple inside a compressor.

8.2 Recently, a High Pressure Tribometer with pin on disk method was developed to handle pressures up to 13.8Mpa and Transcrital CO_2 applications have been evaluated..[9]

9.Current Work on Lubricants for Low GWP Refrigerants including Propane and CO_2. :

Recent work for Low GWP candidates, R32 and CO_2 are best examples for development of new POE lubricants to meet their miscibility and net viscosity needs. [10,11,12,13)]. The discharge line temperature for R32 alone as a refrigerant was higher than that with R410A at equivalent compressor operating conditions. Compressor manufacturers to meet their reliability goals are still addressing this. Tran critical CO_2 applications operate at very high pressure and its effect on net viscosity is taken into account in improved lubricants.
With Propane and CO_2 refrigerants, and POE lubricant, solubility, miscibility and PVXT measurements (Daniel Plots) have been made and reported. [14]

9.1 Lubricants for Carbon Dioxide ,CO_2 (R744) [14]

POE lubricants showed good miscibility; however severe viscosity reduction possibility limits its uses in low temperature cascade systems.
PAG,PAO,MO, and AB are only partially soluble in CO_2. In Trans critical application, receivers,. Internal Heat Exchangers and compressor sump are affected.. However, the PAG type lubricant seems to give the best lubricity for trans critical applications based on the highest mixture lubricity. Also, high efficiency oil separators may make MO a viable lubricants in some cascade systems.

9.2 Lubricants for Propane,(R290) [14]

Miscibility, Solubility and Daniel Plots were determined for Five families of ISO 32 grade lubricants. They were compared to baseline POE with R134a in automotive air conditioning., the viscosity was compared at test condition (-6degC,Evaporating, 6 degree C Superheat, 10 degree C Suction Line HX and 20 deg C Compressor Heating). PAG was superior 50 cSt, better than POE (38 cost), MO (22 cost) and AB(20 cost). The baseline POE with 134a was 48 cSt). The details of the data on these candidates are provided in ref 14.

Summary

1. System Chemistry and Tribology play a vital role in Compressor Lubrication Selection, and Optimization in the Air Conditioning and Refrigeration Applications.
2. Studying contact surfaces using appropriate Surface Analysis Techniques and identifying the type and extent of wear mode is the first step.
3. Lubricant supply to where needed should be guaranteed in the inherent design.
4. Friction Co-Efficient Measurement and plotting Stribeck Curve is the next step.
5. Lubricant viscosity after refrigerant solubility should be taken into account. PVXT data (Daniel Plot) should be experimentally generated.
6. Lube additives, if any, should be thoroughly evaluated for adverse effects in the system.
7. Stribeck Curve helps designer optimize surface finish, lubricant viscosity and lubricity additive for boundary lubrication.
8. The selected final lubricant candidate should be validated in compressor life tests and system tests for Trouble Free Operation.

10.Future Trends and Needs

Changing Environmental Regulations.
Retrofits with new lubricants in existing equipment.
Refrigerants with poorer thermal stability.

Refrigerants with Higher Pressures
Refrigerant Blends with moderate to high glide
Compressor designs operating at higher speeds.
Compressors operating with High Discharge Line Temperature.
System designs requiring oil equalization between compressors.
Lubricants degradation at extreme contaminants level.
Removal of additives by filter driers.
Enhanced heat transfer tubing entrapping lubricants.
Oil Return to compressor sump.
Increased Oil Circulation rate outside the compressor.
Malfunction of expansion devices.
Modeling of interaction between Refrigerants and Lubricants.
Advanced Surface Analysis Tools.
Advanced Analytical Chemistry Tools (Separation and Identification)
Thesis in interdisciplinary sciences (Tribology and System Chemistry) Training (Workshops, Tutorials, Seminars and Symposiums)

11.Acknowledgements: Emerson Climate Technologies, Inc. Sidney, Ohio, USA for past experience and in particular Dr.Bill Finkenstadt, Aditi Mulay, Ron Watkins and Fadhel Zammouri for System Chemistry work on Refrigerants and Lubricants.

12.References

1 Sundaresan, S. G., Purdue 1998 Conference Proceedings. Paper 2170 "Evaluation of Lubricants for R410A/R407C Applications in Scroll Compressor".
2 Sundaresan, S. G., and Finkenstadt, W. R., "Polyalkylene Glycols and Polyolester Lubricants for use with HFC 134a in Reciprocating Compressors", *ASHRAE Transactions,* 92, 1992.
3.Shaffer,S., Bruker Nano Surface Division." Tribology 101 "2010.
4. W.A.Glaeser," Characterization of Tribological Materials", Butterworth-Heinemann, 1993
5..J.L. Tevaarwerk: "Notes on Dynamically Loaded Hydrodynamic Journal Bearing Design and Analysis," 1998.
6.Mark van Drogen, "The Transition to Adhesive Wear of Lubricated Concentrated Contacts". University Twene, PhD Thesis, 2005
7."Lubricant Additives Chemistry and Applications," edited by Leslie R. Rudnick, Book Publishing By Marcel Deker, Inc., New York. 10016. (ISBN>0-847-0857-1)
8. Hessel, E. Urrego, R. and Carr, D., PVXT data, on R410A and 22 ISO grade POE, Chemtura corporation, New Jersey, USA.2012
9.Demas, N.G and Polycarpou A.A.," Ultra High Pressure Tribometer for testing CO_2 Refrigerant at Realistic Compressor conditions."2006 Purdue Conference Proceedings, paper R072
10.Pham, H and Rajendran, R.; Emerson Climate, Sidney, Ohio, USA. Purdue 2012 Paper 2262, "R32 and HFOs as Low-GWP Refrigerants for Air-Conditioning"
11.Guo,W., Ji, G., Zhan, H., Wang, D., Emerson Climate, Suzhou, China, Purdue 2012 Paper 1337 " R32 Compressor Development for Air Conditioning Application in China".
12.Karnaz, J. 2014, "Lubricant Development to Meet Lower GWP Refrigerant Challenges."15th International Refrigeration and Air Conditioning Conference at Purdue.
13.Hessel, E. 2014 Ashrae TC 3.4,"Fundamentals of Lubrication and the Development of Lubricants Optimized for low GWP Refrigerant based applications.
14. Seeton, C. 2013 Ashrae TC 8.1/TC 3.4;" Solubility, Miscibility and Liquid Viscosity of Lubricants with CO_2 and Propane.

IOP Conf. Series: Materials Science and Engineering **90** (2015) 012002 doi:10.1088/1757-899X/90/1/012002

Trends in high performance compressors for petrochemical and natural gas industry in China

Yuanyang Zhao and Liansheng Li

State key laboratory of compressor technology, Hefei General Machinery Research Institute (HGMRI), Hefei, China

Email: yuanyangzhao@163.com

Abstract. Compressors are the key equipment in the petrochemical and natural gas industry system. The performance and reliability of them are very important for the process system. The application status of petrochemical & natural gas compressors in China is presented in this paper. The present status of design and operating technologies of compressors in China are mentioned in this paper. The turbo, reciprocating and twin screw compressors are discussed. The market demands for different structure compressors in process gas industries are analysed. This paper also introduces the research and developments for high performance compressors in China. The recent research results on efficiency improvement methods, stability improvement, online monitor and fault diagnosis will also be presented in details.

1. INTRODUCTION

Compressors are the key equipment of process systems in many industries. They are called the heart of process industry and have been used in many fields such as the petrochemical, metallurgy, energy, aerospace, refrigeration and so on. The main role of compressor is to transfer mechanical energy to the pressure or kinetic energy of gas.

The performance and reliability are very important for the compressors using in petrochemical and natural gas industry. For the compressors using in petrochemical and natural gas industry, the capacity and power is usually very large. Hence, the operate power and efficiency is very important for consumer. On the other hand, when the process system is operating, the compressors cannot shut down during several months, therefore the reliability of main parts of compressor is very important. The main parts are impeller of centrifugal compressor, crankshaft of piston compressor, rotors of screw compressor, and so on.

In the past decades, the compressor technologies in China are rapidly developed in structure design, large scale production, and the performances improvement with the technical progresses of the material, measuring, simulation, fault diagnosis control, and so on. Now, China is one of the largest counties for the compressor production and application. Figure 1 shows the sales of process gas compressor in China during the past years.

This paper summarizes the application trends and the technology researches for process gas compressor (Turbo compressor, screw compressor and reciprocating compressor), especially for compressor in the petrochemical and natural gas industry.

Content from this work may be used under the terms of the Creative Commons Attribution 3.0 licence. Any further distribution of this work must maintain attribution to the author(s) and the title of the work, journal citation and DOI.
Published under licence by IOP Publishing Ltd

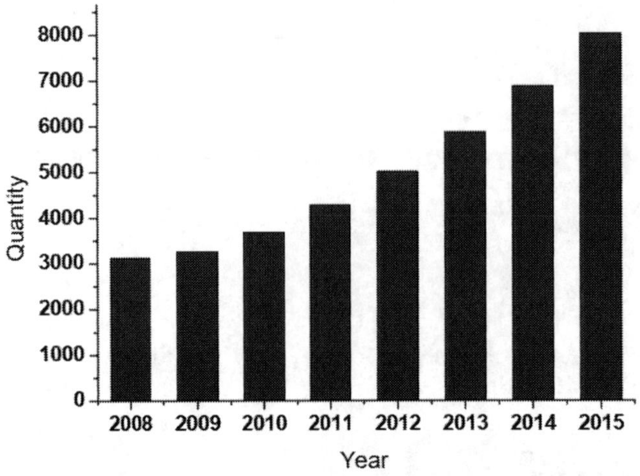

Figure 1. Sales of process gas compressor in China [1]

2. TURBO COMPRESSOR

Turbo compressors are used as the core unit in many process gas systems. As the size of process system increases, the capacity and power of turbo compressor is increasing. The compressors for 1×10^5 Nm^3/h air separation system are produced by Chinese companies in 2014. The capacity of the main compressor in this system is 5×10^5 Nm^3/h. For this kind of compressor unit, the technologies of aerodynamic design, structure design, rotor dynamic analysis, control system and online monitoring system are the key points for the perfect products.

2.1. Basic technology

Flow analysis

The most basic theory of turbo compressor is the internal flow theory. The unsteady flow in compressors results in some phenomenon, such as low efficiency, choke and surge, and so on. To research the flow mechanism of compressor, the simulation and experimental methods can be used. Nowadays, the CFD software is widely used during the design of turbo compressor to verify the design from model stage. Also, the CFD method can be used to predict the compressor performance under off-design and ultimate conditions.

At the same time, the PIV test method was also be used widely in China to research the internal flow. The flow in impeller and diffuser was tested and researched by PIV instrument.

Rotor dynamic analysis

As the scale increase of compressor unit and the rise of parameters, more and more turbo compressors in China are designed as the multi-shaft type. That means the speeds of every stage of compressor are different, every shaft has their critical speeds. The operation of compressor unit should keep away from all these critical speeds. Many Chinese researchers have investigated the rotor dynamic of multi-shaft type compressor carefully. Some new phenomenon of these systems was found.

2.2. Surge suppression technology

When the turbo compressors are operating at the low mass flow rate, the phenomenon of surge and/or rotating stall would occur. And this usually results in the serious accident when turbo compressor is operating at these working conditions long time.

Hence, the surge suppression of turbo compressor is very important. The main mechanism of surge suppression is to maintain the steady flow in compressor. The main flow separation in compressor can be restrained using the surge suppression methods.

There are many kinds of surge suppression methods. The traditional methods are exhaust bypass, inlet guide vane (IGV) method. The casing treatment and flexible blades are researched as the surge suppression methods in recent years in China.

For flexible blade method [2], the flexible material part is fixed on the local of blade. The location of flexible material part is at the flow separation position in impeller. When the compressor is operating, the shape of flexible material part is changed by the gas flow. This kind of shape change can be called self-excited vibration. This vibration of the flexible material part can change the flow patter in impeller. The flow separation can be restrained by this vibration. Hence, this method can be used as the anti-surge technology.

The casing treatment can also be used as the anti-surge technology. The caves on diffuser and shroud can improve the unsteady flow in compressor. The researches at this direction have been done in recent years in China.

Another casing treatment was proposed based on the active control method [3], which was called Active Control Casing Treatment (ACCT). Figure 2 shows the schematic diagram of ACCT. There are bleeding ports on the shroud near the blade tip. The injecting port is on the inlet pipe of compressor. There is the control device in the middle of the flow pass between injecting and bleeding ports. The control device can be control valve or small compressor. The flow parameters of the flow pass can be controlled by the control device.

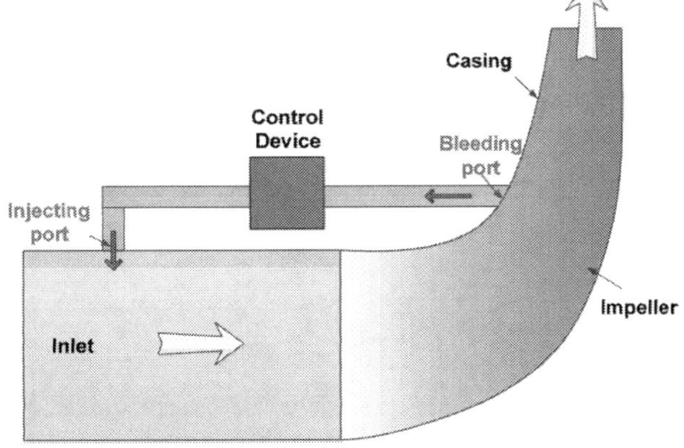

Figure 2. Schematic diagram of ACCT

ACCT can change the flow patter of impeller. Figure 3 shows the pressure distribution of impeller. Figure 3a shows the gas pressure in impeller when the ACCT is not used. Figure 3b shows the pressure when the ACCT is used. The area of low total pressure near the beginning end of split blade is decrease when using ACCT structure. The vortex in impeller is changed by the ACCT system. That means the flow loss is increasing and the flow is improved. The flow separation is restrained by the ACCT system.

Figure 4 shows the result of surge suppression by ACCT. From the results it can be seen that the pressure ratio increases as the increase of bleeding mass flow rate when the compressor operate at low mass flow rate condition. The ability of stability enhancement is increasing when the bleeding mass flow rate is increasing. When the bleeding mass flow rate is 0.64 kg/s, the ability of stability enhancement is 35%.

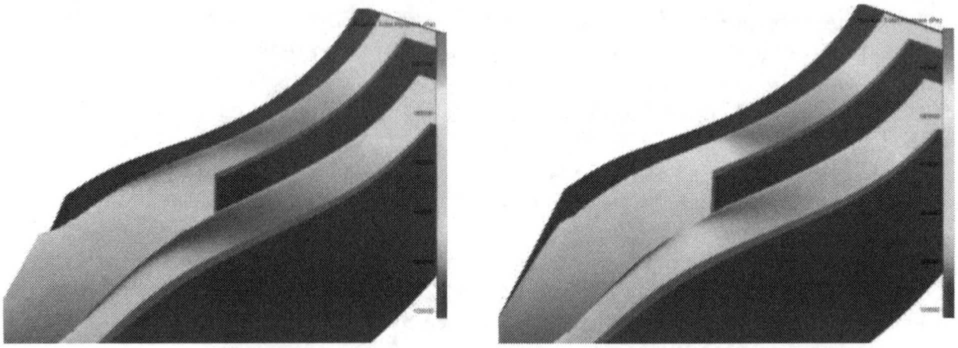

a. Original b. Use ACCT

Figure 3. Total pressure of impeller

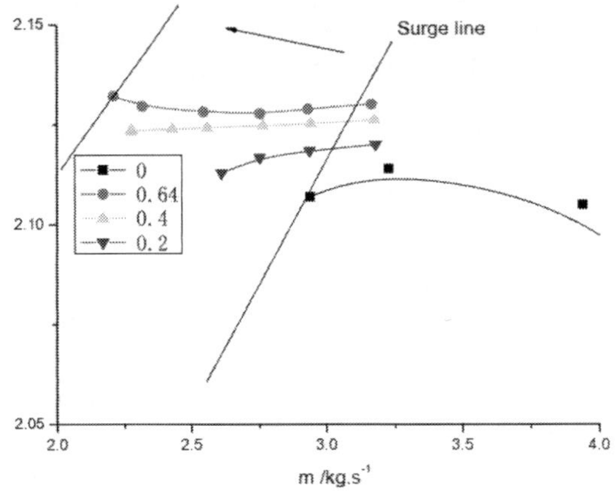

Figure 4. Performance of compressor

2.3. Reliability research

For the turbo compressor used in the petrochemical and natural gas industry, the main parts of compressor are affected by pressure, temperature, and chemical gas [4]. The reliability of compressor may be decreasing under these working conditions.

Figure 5 shows the fault mode of centrifugal compressor in China [5]. Almost 61% failure is owe to the problem of blade and seal.

Figure 6 shows an example of the centrifugal compressor failure. The failed impeller of recycle hydrogen centrifugal compressor is the closed type used for hydrorefining. The failure can be mainly attributed to sulfide stress cracking [4]. This is a typical chemical action which leads to the compressor failure.

To obtain the effect of pressure, temperature, and different chemical gases on the reliability of compressor impeller, the failure of impeller for different material ((such as AISI630/17-4PH and FV520B/14-5PH) was investigated in China.

The other way to maintain the reliability of turbo compressor is the technology of online monitoring and fault diagnosis. Nowadays, the online monitoring and fault diagnosis system is built in most of the turbo compressor used in petrochemical and natural gas industry. And many these kinds of system have the function of remote monitoring through internet. Therefore, the manufacture and customer can both get the situation of compressor unit in time. Some safety measure and action can be used before the occurrence of serious accident.

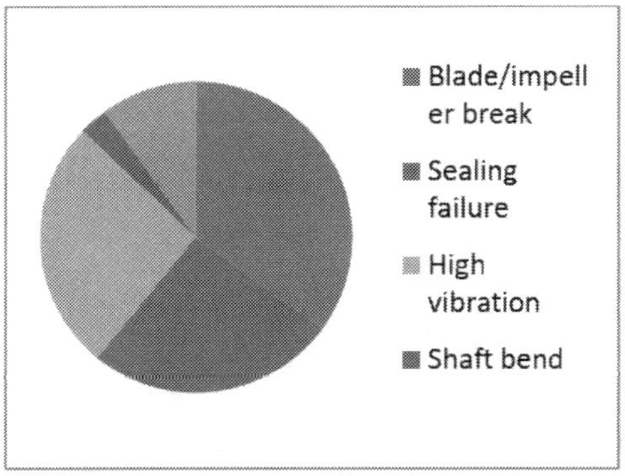

Figure 5. Fault mode of centrifugal compressor in China

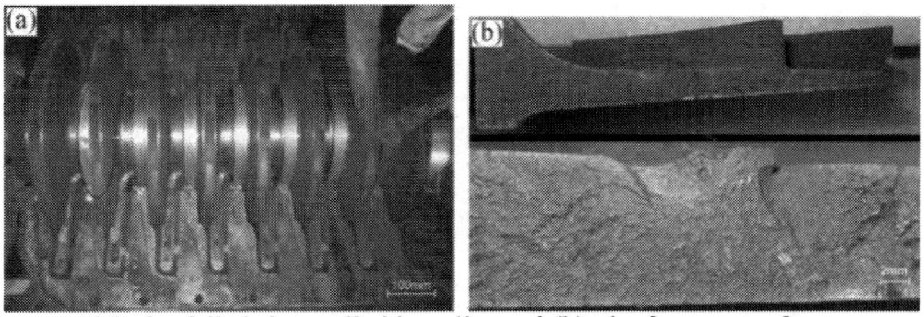

(a) the failed slot-welled impeller and (b) the fracture surfaces
Figure 6. Digital photographs [4]

3. RECIPROCATING COMPRESSOR

The reciprocating compressor is widely used in the petrochemical and natural gas industry. The higher gas pressure can be gotten through this type compressor. The biggest reciprocating compressor 4M150 was manufactured and used in the 1200 million tons/y oil refinery plant in China.

The high operating efficiency and reliability are the main development direction on reciprocating compressor. There are many reciprocating compressor is operating at the off-design working conditions. The real operating capacity of gas process system is usually less than the design capacity of compressor. Hence, the bypass system is usually used. The efficiency of whole system is very low. The capacity adjustment technology is very useful for this situation.

There are many failures of reciprocating compressor every year in China. Figure 7 shows the fault mode of reciprocating compressor in China [5]. Because there are many easily damaged parts for reciprocating compressor, the failure reasons are different.

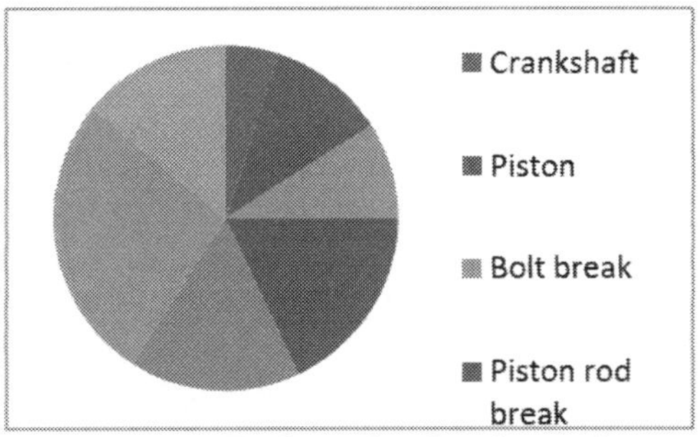

Figure 7. Fault mode of reciprocating compressor

Figure 8. p-V Diagram

3.1. Capacity adjustment technology

The method of Partial Stroke Press-off Inlet Valve (PSPIV) is the most energy conservation method to adjust compressor capacity [6]. The compressor working process is changed when the PSPIV system is used. In Figure 8, the area of the closed curve ABCDA expresses the compression power of reciprocating compressor with full load, and the area of the closed curve ABCEFA expresses that with part load. It shows that about half of power has been saved when the compressor operates with part capacity, which is described by area of the closed curve CDFEC. Figure 9 shows the schematic diagram of PSPIV system. The device of PSPIV was developed and applied in the petrochemical plant in China. The energy saving effect is very huge when this device was used. Figure 10 shows that the real power is decreasing with the real capacity approximate linearly.

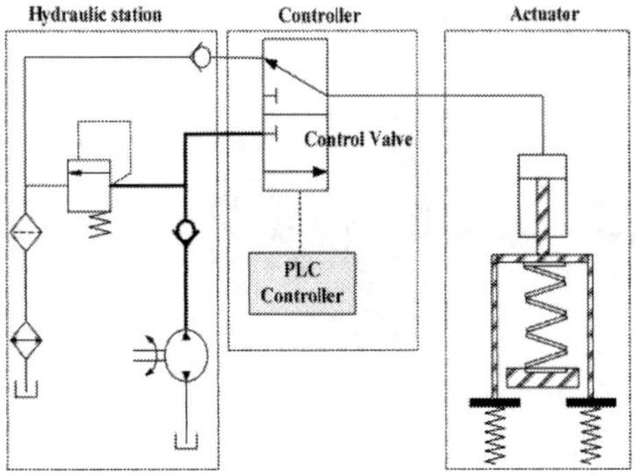

Figure 9. Schematic diagram of PSPIV

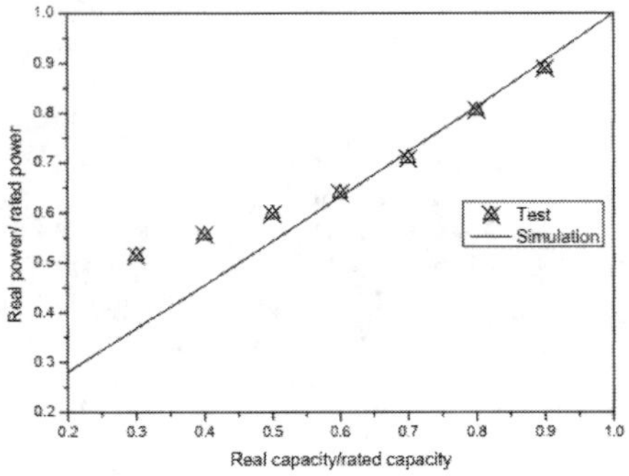

Figure 10. Compressor performance

3.2. Fault diagnosis technology

Base on the technologies of sensor, data acquisition, signal analysis, expert diagnosis, and internet, the advanced intelligence fault diagnosis system was developed and used in the process industry in China.

There is a relationship between compressor fault and vibration signal. Hence, the vibration signal is the key parameter of diagnosis system. According to the parameters of vibration, the possible fault position and model of compressor can be predicted. The enlargement of fault can be prevented by the instant repair or working condition adjustment.

Using this system on the reciprocating compressor unit, the thermodynamic and dynamic failures can be gotten before the serious accidents occur. The typical faults are the leakage of valve caused by the break of valve plate, the break of piston rod caused by the wear of piston ring, the crosshead and crank shaft fault by the wear of bearing.

4. SCREW COMPRESSOR

As a type of displacement compressor with high reliability, screw compressors are widely used in the process gas system, such as coal chemical, coal chemical, and natural gas industries. And many screw compressors were used to recovery the coalbed gas, flare gas, and oil associated gas in China.

The technical level of screw compressor for process gas was improved greatly. The profile theory, structure design, rotor manufacture, and gas seal are all researched in China. The biggest screw compressor was manufactured in China. The main parameters are as following:

Rotor diameter: ϕ 816 mm   Capacity: 650m^3/min
Power: 4500kW   Gas: CO_2

The sales of process gas screw compressor increased very fast during the past years (shown in Figure 11) [1].

Many experimental researches have been done on process gas screw compressor. Figure 12 shows a test system. The seal of shaft is dry gas seal, which can increase the reliability and service life of compressor. And the water injection on the performance of screw compressor was tested on this test rig.

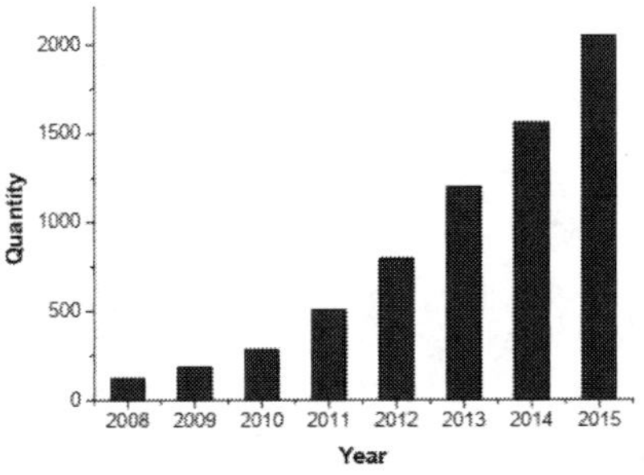

Figure 11. Sales of screw compressor

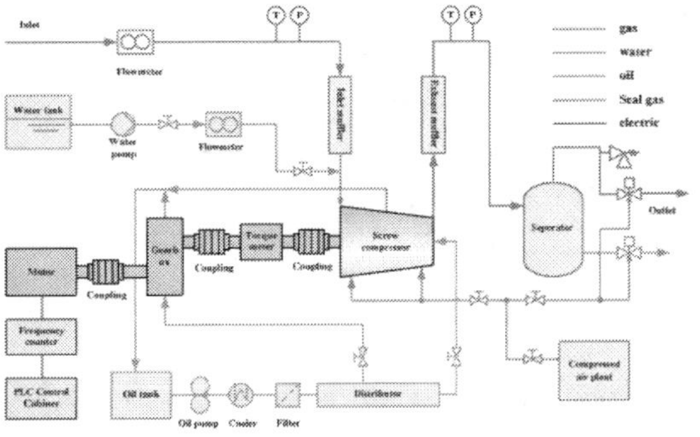

Figure 12. Test system of process gas screw compressor

Figure 13 shows some test results of screw compressor with water injection. When the amount of injected water increases from 0.12t/h to 3t/h, the discharge temperature decreases rapidly. The discharge temperature is determined by the amount of water injection. When the speed is 3000 r/min, the discharge temperature of compressor is 79℃ when the amount of injected water is 0.12 t/h. While the amount of injected water is rising to 6t/h, the discharge temperature is decreasing to 24.3℃ and approaches to the suction temperature. With the increasing of injected water, the capacity of compressor is rising quickly at the beginning. When the amount of injected water is increase to a certain value (about 1t/h), the capacity of compressor tends to be stable.

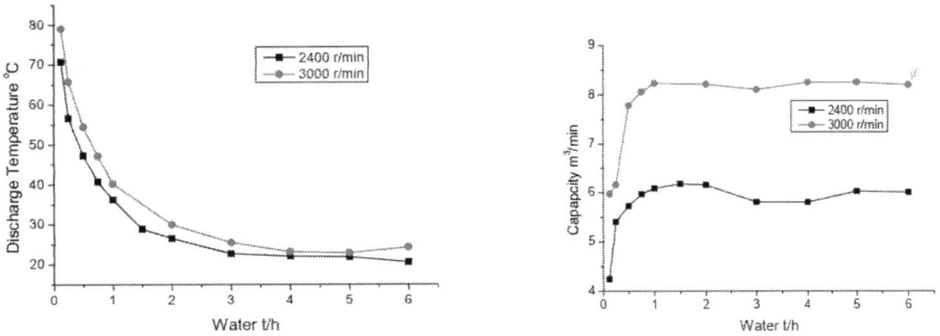

Figure 13. Effect of water injection on compressor performance

5. CONCLUSIONS

The application status of petrochemical & natural gas compressors in China is introduced. The turbo, reciprocating and twin screw compressors are discussed. This paper introduces some research and developments for high performance compressors in China. The recent research results on performance improvement, surge suppression, reliability improvement, online monitor and fault diagnosis is also presented in details.

ACKNOWLEDGEMENTS

The work was supported by the National Key Basic Research Program of China (2012CB026000), the Science and Technological Fund of Anhui Province for Outstanding Youth (1508085J05), and by the National Science & Technology Pillar Program (2013BAF01B02)

References

[1] Qian Jiaxiang. Industry status and industry status of process gas screw compressors. General machinery. 2013 (6): 42-45.

[2] Kang Wei, Zhang Jiazhong, LEI Pengfei, et al. Computation of unsteady viscous flow around a locally flexible airfoil at low Reynolds number. Journal of Fluid and Structure. 64, 42–58, 2014.

[3] Report of project "Fundamental Study on the High-end Compressor Units with High Efficiency, Reliability and Intelligence". National Key Basic Research Program of China. 2014.

[4] Defu Nie, Xuedong Chen, Zhichao Fan, Qiaoguo Wu. Failure analysis of a slot-welded impeller of recycle hydrogen centrifugal compressor. Engineering Failure Analysis, 2014, 42: 1-9.

[5] Lv Yunrong, Chen Xuedong, Gao Jinji, et al. Survey of Failure of Large Process Compressor in Chinese and Failure Prevention Countermeasure. Fluid machinery. 2013 (41): 14-20.

[6] Zhao Yuanyang, Tang Bin, Liu Guangbin, et al. Experimental research on dynamic response of capacity control system in reciprocating compressor. Journal of Mechanical Engineering Science. 2014, 228(2): 358–365.

Rotor clearance design and evaluation for an oil injected twin screw compressor

D Buckney[1], A Kovacevic[12], N Stosic[2]

[1]Howden Compressors Ltd., Research and Development, Glasgow, UK

[2]City University London, Centre for Positive Displacement, Compressor Technology, London, UK

Email: david.buckney@howden.com

Abstract. Designing twin screw compressors to safely operate at higher than normal temperatures poses a challenge as the compressor must accommodate larger peak thermal distortions while ideally maintaining efficiency at nominal operating conditions. This paper will present a case study of an oil injected compressor tested at elevated discharge temperatures with original and revised clearances. The local thermal distortions occurring within the compressor during operation were estimated using a procedure developed by the authors - thermodynamic results from a chamber model were used to approximate component temperature distributions that are then used to predict possible thermal distortions and the resulting affect on clearance gaps. The original and revised clearance designs are evaluated and performance penalties incurred due to the modifications are discussed.

1. Introduction

End users of twin screw compressors occasionally demand reliable compressor operation over a greater operating range than the original design envelope. If for example the allowable cooling oil temperature can be increased, this can reduce the cooling requirements and the overall plant cost. This poses a challenge as the compressor design needs to accommodate higher peak operating temperatures while ideally maintaining efficiency at nominal operating conditions.

In one sense the clearances in oil injected machines are easier to manage than in oil free machines as temperatures are usually kept within much lower limits however the lack of timing gears and the direct rotor to rotor drive does add more complexity to clearance analysis. Rotor contact is of course necessary for one rotor to drive the other; the objective is to control precisely where it is possible for contact to occur. Best practice for direct drive clearance design ensures only rolling contact at the pitch radius of each rotor as presented by Stosic et al [1].

Deviations from nominal design clearances, whether due to manufacturing and assembly variations or due to operational distortions can result in a shift in the relative rotation between the main and gate* rotors [3]: this further distorts the interlobe clearance distribution and must be considered in clearance evaluation for direct drive, oil injected compressors.

* The adopted rotor terminology is consistent with [1]. The main and gate rotors can also be referred to as the male and female rotors respectively, which is often the case.

Content from this work may be used under the terms of the Creative Commons Attribution 3.0 licence. Any further distribution of this work must maintain attribution to the author(s) and the title of the work, journal citation and DOI.
Published under licence by IOP Publishing Ltd

This paper will present a case study of an oil injected compressor that was tested at elevated discharge temperatures. The focus of this work is on potential thermal distortion and other factors affecting clearance distributions have been omitted from the current analysis. The local thermal distortions occurring within the compressor during operation will be estimated using a procedure developed by the authors [2] whereby thermodynamic results from a chamber model are used to approximate component temperature distributions based on available fluid boundary temperatures [4]. This is then used to predict possible thermal distortions and the resulting affect on clearance gaps.

2. Case study

The case study presented in this paper is based on historical data available in Howden Compressors Ltd. The compressor is a Howden WCVTA510/193/26 compressor, this is currently the largest Howden compressor with rotors just over 510mm in diameter. In this case the compressor has a built in volume ratio of 2.6.

In addition to air testing at contract pressure ratio and temperature the compressor was tested at an elevated discharge temperature by increasing the pressure ratio to 11 and then limiting injection oil cooling by increasing oil temperature and restricting flow. Due to power restrictions on the testing of this large compressor the speed was reduced from the contract speed of 1400rpm to 750rpm during this overload testing. After testing at up to a discharge temperature of 120°C (from suction at 20°C), tear down inspection revealed that rotor to rotor contact had occurred at the root of the main rotor as highlighted in Figure 1.

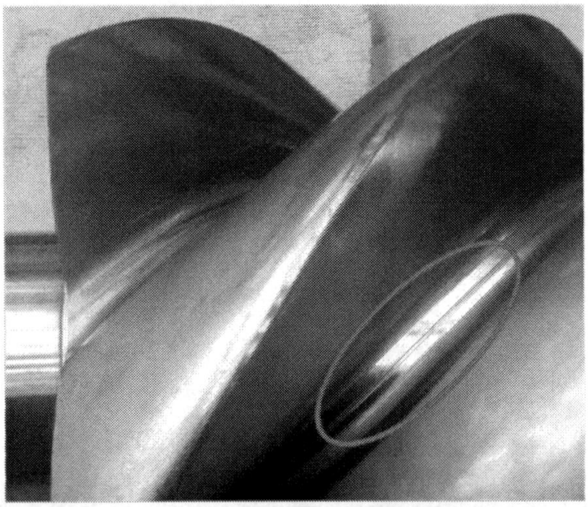

Figure 1. Main rotor with original interlobe clearance distribution showing rooting after testing at 120 degrees C.

The type of contact observed and the fact that this standard compressor had not presented any problems while operating within normal temperature limits points to thermal distortion of the rotors being the most likely cause of this rotor contact. Based on these test findings a revised interlobe clearance design was proposed. The revised clearance design was implemented on a different compressor at a later date and though teardown inspection revealed some very localised rotor contact, related to the tip sealing strip, there was no contact over the main part of the root and the clearance modification was approved for further use. The analysis performed in the remainder of this paper has been performed retrospectively in order to evaluate the original and revised clearance designs in more detail.

3. Clearance analysis procedure

3.1. Thermal analysis

The fluid temperature inside the compressor, assumed to be homogeneous throughout any given compression chamber, can be calculated with the use of well establish chamber models developed for twin screw compressors [1]. The overload test condition, with a discharge temperature of 120° C was modelled to determine how the temperature varied throughout the compression cycle for that particular duty. All modelling was done to replicate the test conditions where the overload discharge temperature was achieved by compressing air. It is worth noting that the intended gas for this compressor is actually a hydrocarbon mixture which has a lower isentropic index than air. The same discharge temperature is targeted regardless of the gas being used and for the time being it is assumed that the difference in temperature distributions on the rotors are not significantly different between these operating conditions.

Using a procedure developed by the authors [2], the fluid temperature from the compression cycle was mapped onto the surfaces of the rotors and the casing bores. These fluid boundary conditions at the surface of the compressor components can be used for further thermal analysis to determine the temperature distributions within the metal components of the compressor. Figure 2 shows the temperature of the fluid boundary on the main and gate rotor surfaces. The left image shows the instantaneous fluid temperature for a particular rotor position – this highlights how different compression chambers are exposed to different points in the compression cycle which are at different temperatures. In the right hand image these instantaneous temperatures has been averaged over one full rotation for each respective rotor in order to show the local average boundary temperature. In addition the temperature has also been averaged over each transverse plane of the rotors as this allows potential thermal distortions to be approximated analytically.

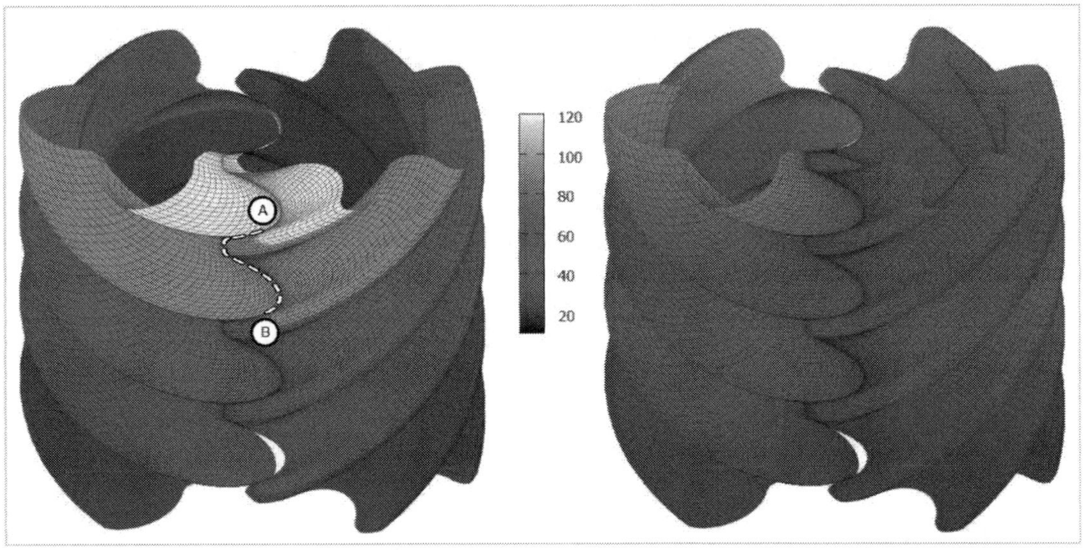

Figure 2. Instantaneous (left hand side) and averaged (right hand side) fluid boundary temperature mapped onto the surfaces of the main and gate rotors. Line AB highlights the interlobe sealing line for a single compression chamber.

Table 1. Average fluid boundary temperature across outlet plane.

Component Description	**Temperature (°C)**
Main rotor	86.8
Gate rotor	73.2
Casing	69.4

The peak temperatures occur at the outlet plane of the rotors therefore the analysis presented in this work will be based on the average fluid boundary temperatures presented in Table 1. It is clear that the averaged fluid boundary temperatures at critical locations on the compressor are actually much lower than the peak fluid discharge temperature of 120°C.

It is interesting to note that the average boundary temperature is higher on the main rotor than on the gate rotor, this is related to the number of lobes on each rotor. Consider the 4 compression chambers at the outlet end of the main rotor shown in the left hand figure. The gate rotor is also exposed to these 4 chambers but only over an angle of 240 degrees. For the other 120 degrees the gate rotor is exposed to 2 additional compression chambers at a lower temperature.

It is not possible to calculate the actual local temperature on any given component within the scope of this analysis however the average fluid boundary temperature gives a very good indication of the maximum temperature that the compressor components can reach. This results in a conservative approach that most likely overestimates actual thermal distortions due to heating by the compression fluid alone. Knowing the temperature of the rotors and casing, local clearances on a transverse section of the compressor can be evaluated. To generalise: rotor thermal distortions act to reduce the interlobe clearance while casing thermal distortions act to increase the interlobe clearance; if both rotors and casing were at the same temperature the net effect would be no change to the interlobe clearances.

3.2. Clearance presentation

Due to the 3D nature of the interlobe sealing line it can be difficult to present clearance information clearly so Figure 3 has been included in an attempt to make the clearances a bit more intuitive. This presents the same clearances mapped onto each of the rotors as vectors normal to the transverse rotor curve (actually showing the magnitude of clearances normal to the 3D rotor surface). This approach is very intuitive and clearly shows how clearances relate to difference areas on the rotor profiles. The difficulty with this clearance presentation is comparing small differences in similar clearance distributions. Magnifying the vectors quickly distorts the apparent clearance gap due to the curvature of the profiles. On the other hand, plotting local clearances on a linear axis such as ‘relative position on sealing line’ makes it difficult to determine where the clearances actually occur on the rotors.

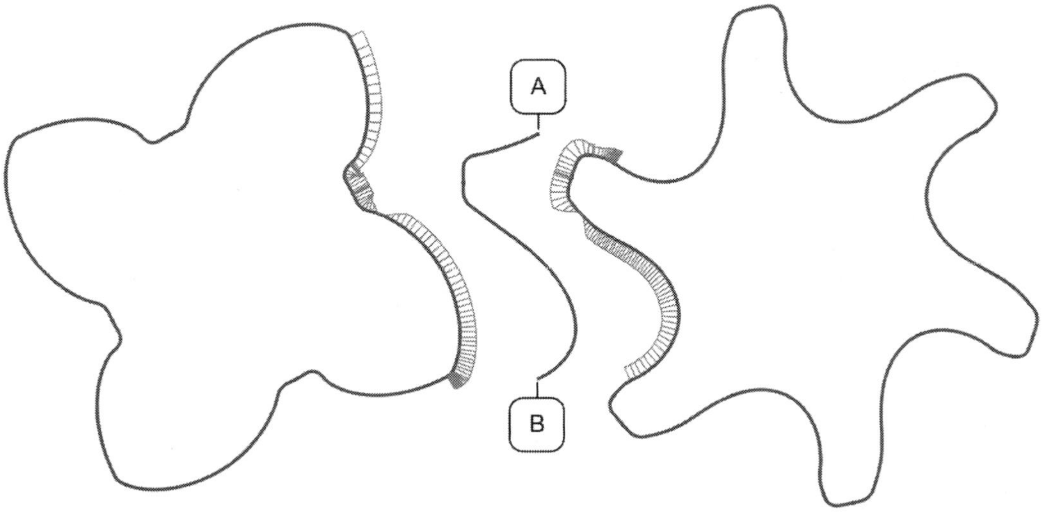

Figure 3. Transverse cross-section of main and gate rotors showing local interlobe clearances. The magnitude of the vectors represents the clearance *normal* to the 3D rotor surface. Line AB represents the conjugate rack, common to both rotors. The limits of this rack segment are the beginning and end of the sealing line for a single compression chamber, as previously highlighted in Figure 2.

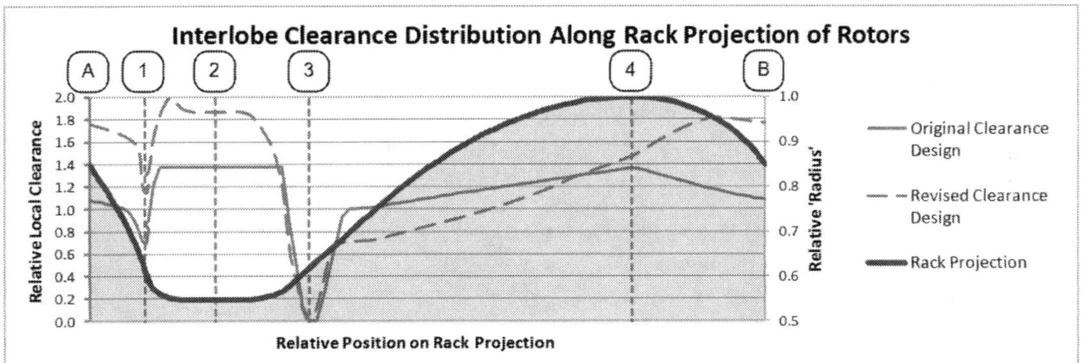

Figure 4. Interlobe clearance distribution along rack projection of rotors comparing original and revised clearance designs. Key clearance locations are highlighted along the common horizontal axis. The secondary axis shows the radius for the corresponding main rotor to give a better indication of where local clearances are located.

Table 2. Key clearance locations.

Location on rack	Description
A	Limit of sealing line for single chamber
1	Pitch radius on straight (undercutting) side of rotors
2	Root of main rotor / tip of gate rotor
3	Pitch radius on round (non-undercutting) side of rotors
4	Tip of main rotor / root of gate rotor
B	Limit of sealing line for single chamber

The compromise that has been taken is to plot the local clearances on the axis of a rack projection along which the rack would translate. Thus, the magnitude of the local clearances can be simply graphed and superimposed against the somewhat familiar shape of the rack as presented in Figure 4.

Rather than put a numerical value on the horizontal axis vertical lines have been added to identify key points along the length of this sealing line projection, details of which are provided in Table 2.

The original and revised clearance distributions are both presented in Figure 4. In all cases the relative rotation between the rotors is adjusted in order to provide zero clearance at location 3: the pitch radius on the round side of the rotors. This is where the driving force is transferred from the main to the gate rotor. In the revised clearance design the main difference is that the clearance in the main rotor root has been increased and to a lesser extent also at the main rotor tip. Notice that the clearance on the round flank of the rotors (between locations 3 and 4) has actually been decreased on the revised clearance design – this clearance will not significantly reduce during operation because the gap is maintained due to the nearby contact at the pitch point (location 3), therefore this clearance was deemed to be unnecessarily large. Conversely, on the straight flank (from 4 to B, and from A to 1) the gap is adversely affected by thermal distortions at location 3 which are transferred over to the straight flank. Therefore it was necessary to increase the clearances on the straight flank.

To quantify how these clearance changes impact the leakage area through the interlobe gap the local clearances where integrated over the length of the 3D sealing line. The total leakage area is 108.5mm^2 for the original design is and 142.7mm^2 for the revised design, an increase of 31.5%.

4. Results

4.1. Analysis of clearance distortions

The following results show the predicted operational clearance distributions for the original and revised designs respectively. As previously discussed, a similar temperature increase in both the rotors and casing results in little net change in the operational clearances, this case has been presented with the longer dashed lines on each figure, the trend is not far off the original clearance.

A second case has also been presented where the rotors do not move apart due to casing thermal expansion. This case is quite possible in the event that sudden temperature changes cause the rotors to heat up faster than the casing or alternatively the outlet end bearings might be situated remotely from the rotor outlet plane in a location where the casing temperature is much lower than estimated in Table 1. Unfortunately the distance between the two dashed curves for the different operational scenarios results in a large band of uncertainty but it does serve to highlight 1) where rotor contact is most likely to occur for a given clearance design, and 2) where clearances can be further reduced without compromising reliability.

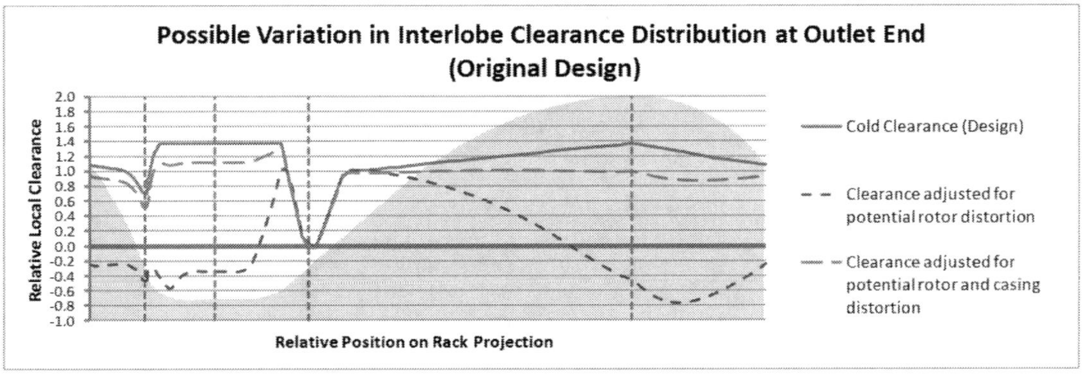

Figure 5. Possible variation in interlobe clearance distribution at outlet end (original design).

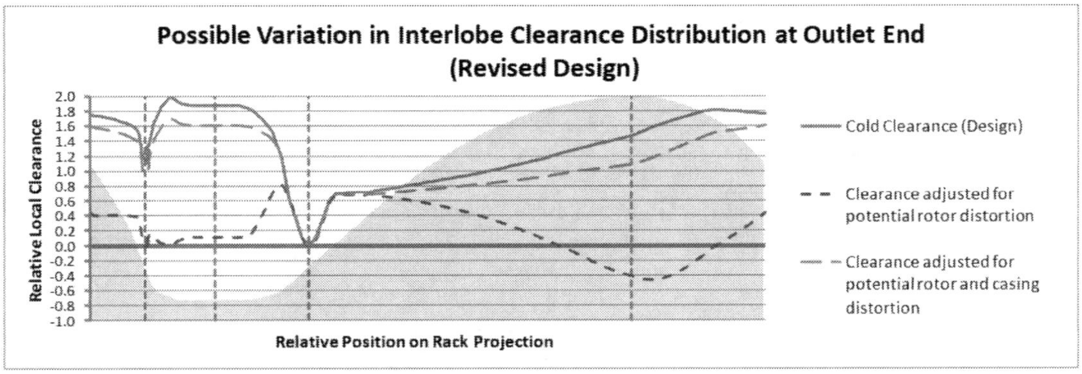

Figure 6. Possible variation in interlobe clearance distribution at outlet end (revised design).

This analysis supports the revised clearance in the root of the main rotor which shows contact cannot now occur there due to thermal distortion alone. The chance of contact at the tip of the main rotor has been significantly improved however the analysis suggests contact is still possible under extreme circumstances.

The most problematic clearance area is on the straight (undercutting) flank of the rotors – since distortions are transferred from the round contacting flank these clearances vary more than in other areas. What the figures do not show is that this represents a long part of the sealing line path due to the effect of undercutting, so this has a large effect on the overall leakage area. This area also experiences faster relative sliding motions between the rotor surfaces so contact should certainly be avoided. With these considerations in mind, there is probably still some scope to slightly reduce the clearances on both flanks just above the pitch locations.

4.2. Performance predictions

The models were also run at the nominal air test duty which is closest to the designed operating conditions: with a speed of 1400rpm, a pressure ratio of 5 and a discharge temperature maintained at approximately 90°C. The modelled deviation in the volume flow due to the revised design is compared with the deviation measured on test in Table 3.

Table 3. Performance penalty with revised interlobe clearance at pressure ratio of 5

	change in volumetric flow
Modelled	-0.5%
Measured on Test	-1.7%

The difference between the model result and the test result was higher than expected suggesting that the model is slightly underestimating the effect of leakages through the interlobe gap. Unfortunately as the test results were obtained using two completely different compressors the effect of other manufacturing and assembly tolerances can't be ruled out for the test results.

Importantly, the resulting change in flow on test was small enough that the compressor was still within normal operating tolerances in both cases. With the reduced flow there was a similar reduction in shaft power so there was very little change in overall efficiency. It is to be expected there would be some detrimental effect on performance at the nominal operating condition by increasing the rotor clearances to allow the compressor to operate at higher, non-standard, operating temperatures. This same design approach could be applied to investigate how far clearances could potentially be reduced for performance optimisation over a narrow operating band.

5. Conclusions

Results from a compressor model have been utilised to determine the fluid boundary temperature distribution on the surface of the rotors and casing. Estimated component temperatures were then used to evaluate how operational clearances are affected by thermal distortions.

Local interlobe clearance distributions have been superimposed against a rack projection of the rotors in order to compare small changes in local clearance while relating back to the location on the rotors.

Two interlobe clearance designs have been analysed to evaluate how suitable they are for operation at an elevated rotor discharge temperature of 120°C in an oil injected compressor with direct rotor drive. While there is a fairly large uncertainty on the predicted clearance deviations the results do confirm that the revised clearances should prevent rotor to rotor contact at the root of the male rotor. This is where contact was initially present on test, and later shown to be alleviated due to the revised clearance design.

Performance test results suggest that the increase in interlobe clearance has resulted in a 1.7% drop in flow however the overall flow was still within normal test tolerances. The change in flow predicted by the model was different from that measured on test.

References

[1] Stosic, N., Smith, I., Kovacevic, A., 2005, Screw Compressors, Mathematical Modelling and Performance Calculation, Springer.

[2] Buckney, D., Kovacevic, A., Stosic, N., 2014, Accounting for Local Thermal Distortions in a Chamber Model for Twin Screw Compressors, *International Engineering Compressor Conference at Purdue*.

[3] Sauls, J., Powell, G., Weathers, B., 2007, Thermal deformation effects on screw compressor rotor design, *International Conference on Compressors and their Systems*, London.

[4] Sauls, J., Powell, G., Weathers, B., 2006, Transient Thermal Analysis of Screw Compressors, Part I Use of Thermodynamic Simulation to Determine Boundary Conditions for Finite Element Analyses, *International Engineering Compressor Conference at Purdue*.

IOP Conf. Series: Materials Science and Engineering **90** (2015) 012004 doi:10.1088/1757-899X/90/1/012004

Analysis of the screw compressor rotors' non-uniform thermal field effect on transmission error

T.N. Mustafin, R.R. Yakupov, A.V. Burmistrov, M.S. Khamidullin and I.G.Khisameev

Kazan National Research Technological University, 68, K. Marx Str., Kazan, Russian Federation

E-mail: mustimur@rambler.ru

Abstract. The vibrational state of the screw compressor is largely dependent on the gearing of the rotors and on the possibility of angular backlash in the gears. The presence of the latter leads to a transmission error and is caused by the need for the downward bias of the actual profile in relation to the theoretical. The loss of contact between rotors and, as a consequence, the current value of the quantity, characterizing the transmission error, is affected by a large number of different factors. In particular, a major influence on the amount of possible movement in the gearing will be exerted by thermal deformations of the rotor and the housing parts in the working mode of the machine. The present work is devoted to the analysis of the thermal state in the operation of the screw oil-flooded compressor and its impact on the transmission error and the possibility of losing contact between them during the operating cycle.

1. Introduction

The technique that allows us to analyze the correctness of engagement during the compressor operation has been presented in the literature [1, 2]. Researchers have demonstrated the possibility of the use thereof during the engagement pre-analysis at the operating conditions of the compressor. At the same time, it is also important to note the fact that it is also suitable for a more detailed analysis of the influence of various factors on the engagement work of the screw compressor rotors in the event of some adaptation and refinement. Thermal deformations in these studies were calculated based on the assumption of uniformity of temperature fields in any cross section of the rotors, which was calculated on the basis of an empirical relationship generalizing the results of experimental studies. This approach is quite acceptable for preliminary analysis; however, it should be noted that, for a more detailed engagement analysis, it is also necessary to assess the influence of both operating and design features of the test compressor (rotor speed, design and size of the pressure and suction ports) that does not allow to making of the empirical formula in these studies. All of the above factors make it possible to formulate the problem of this study on the synthesis and improvement of methods of calculation of temperature fields of the rotors with subsequent analysis of their impact on the rotor engagement.

2. Dynamics of free rotors and their ability to lose direct contact with driving rotors

Changing the free rotor rotational correction angle causes it to rotate with a certain acceleration. This can be caused both by direct contact of the rotors and by gas forces torque. The gas forces torque

Content from this work may be used under the terms of the Creative Commons Attribution 3.0 licence. Any further distribution of this work must maintain attribution to the author(s) and the title of the work, journal citation and DOI.
Published under licence by IOP Publishing Ltd

is typically common for screw compressors operating at nominal conditions: rotational correction of the female rotor into contact with the male rotor in the direction of rotation of the rotors in the implementation of the male rotor drive. The value of this rotational correction angle can be calculated by the method described in the studies [1 and 2]. Therefore, for the continuation of the rotors being engaged, the possible acceleration of the rotor from gas forces should exceed the required acceleration associated with a change in the rotational correction angle.

Assuming the uniform rotation of the drive rotor in the first approximation that can be justified by its significantly higher moment of inertia thanks to the shells and partial compensation by the driving torque, while taking into account all the factors stated above, the condition of continuation being engaged can be written as follows:

$$\frac{d^2\theta_2}{d\tau^2} \geq \frac{d^2\beta_2}{d\theta_1^2}\cdot\omega_1^2, \tag{1}$$

where β_2 is a rotational correction angle of the female rotor into contact with the male rotor, ω_1 is the drive rotor angular velocity, θ_2 is the female rotor angle of rotation, τ is a time. The ">" sign in the inequality (1) would indicate that the free rotor can "try to twist" the drive rotor.

Approach to the definition of $\frac{d^2\theta_2}{d\tau^2}$ is considered in the studies [1, 2 and 3]. From the dynamic equation it can be written that:

$$\frac{d^2\theta_2}{d\tau^2} = \frac{M_{GF}+M_{AF}-\alpha_R\cdot M_R-\alpha_{F1}\cdot M_{F1}-\alpha_{F2}\cdot M_{F2}-M_{MEC}}{J_2}, \tag{2}$$

where J_2 is the free rotor moment of inertia; M_{GF} is the moment created by the gas forces, which is determined based on the indicator (P-V) diagram according to the sub-tense method [4]; M_{MEC} is the moment of resistance of mechanical compressor components, such as seals, dummy pistons and bearings (its value is determined largely by the design of mechanical components of the rotor); M_R, $M_{F1,2}$ are the moments of friction on the medium being compressed in radial and axial clearances, respectively; M_{AF} is the moment produced by the forces of adhesion; α_R, α_{F1} and α_{F2} are the correction factors that take into account the absence of radial and axial surfaces of the housing in the areas of suction and discharge ports execution (They were taken as the ratio of the surface areas on which there is a clearance, excluding the suction and discharge ports, to the area of the corresponding surface, including the suction and discharge ports area, i.e. to the complete cylindrical surface of the complete annular surface). It is not hard to show that the term "adhesion" in this case may be replaced with the more accurate term "cohesion". In this case, this moment will be proportional to the surface tension coefficient, the contact area and the contact angle; it should be noted at the same time that this moment will be considerably smaller than the other force moments, which in turn makes it possible to ignore it in subsequent calculations: $M_{AF}\approx 0$.

The moments of friction on the medium being compressed in end clearances depend on the operating mode [1, 2 and 4] according to the design diagram (Figure 1):

$$M_{F1,2} = \begin{cases} \dfrac{\pi\cdot\omega_2\cdot\eta_{MIX}\cdot R_{eq}^4}{2\cdot\delta_{F1,2}}\cdot\left(1-\dfrac{r_R^4}{R_{eq}^4}\right), & if \quad \mathrm{Re}_{F1,2}<10^4 \\ \dfrac{\rho_{MIX}\cdot\omega_2^2}{4}\cdot\left(c'_{M1,2}\cdot R_{eq}^5 - c''_{M1,2}\cdot r_R^5\right), & if \quad 10^6\geq \mathrm{Re}_{F1,2}\geq 10^4 \end{cases}, \tag{3}$$

where $c_{M1,2}=0{,}0277\cdot\mathrm{Re}_F^{-0,2}\cdot\left(\frac{\delta_{F1,2}}{R}\right)^{-0,2}$ is a coefficient of friction torque; r_R $(R=r_R)$ is used as a determining size to ascertain c''_M, and R_{eq} $(R=R_{eq})$ is used in all other cases; $\mathrm{Re}_F=\frac{R^2\cdot\omega_2\cdot\rho_{MIX}}{\eta_{MIX}}$ is the Reynolds number; $\omega_2 = \omega_1 \cdot \frac{z_1}{z_2}$ is the nominal angular velocity of the free rotor; ρ_{MIX}, η_{MIX} are, respectively, the density and kinematic viscosity of the gas-oil mixture in the working chamber of the compressor; δ_{F1}, δ_{F2} are the end clearances between the rotor and the housing on the suction and discharge side, respectively, z_1 и z_2 are the number of teeth of the male and female rotors.

Equivalent radius is defined as $R_{eq} = \sqrt[4]{\frac{1}{2\cdot\pi}\int_0^{2\cdot\pi}(r(t))^4 \cdot dt}$, where $r(t)$ is the function describing the radius vector of the rotor profile in the end section; t is the most convenient formal parameter selected by trial and error method (it is more convenient to use an angle in the polar coordinate system, but it is not always acceptable).

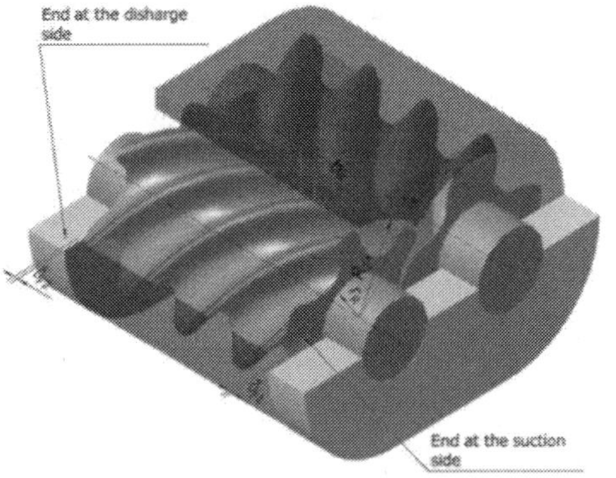

Figure 1: Design diagram of clearances to determine the moments of resistance

By analogy, the moments of friction on the medium being compressed in the radial clearance can be written as [1, 2 and 4]:

$$M_R = \begin{cases} \dfrac{2\cdot\pi\cdot\xi\cdot\omega_2\cdot\eta_{MIX}\cdot R_{2*}^3\cdot \tilde{z}_2\cdot b_R}{\delta_R}, & if \quad \mathrm{Re}_R < 2500 \\ \pi\cdot c_R\cdot\rho_{MIX}\cdot\omega_2^2\cdot \tilde{z}_2\cdot b_R\cdot R_{2*}^4, & if \quad 10^5 \geq \mathrm{Re}_R \geq 2500 \end{cases}, \tag{4}$$

where $c_R \approx 0{,}0076\cdot\mathrm{Re}_R^{-0{,}25}$ is a friction coefficient [1, 2, 4], $\mathrm{Re}_R = \frac{\omega_2\cdot R_{2*}\cdot\delta_R\cdot\rho_{MIX}}{\eta_{MIX}}$ is the Reynolds number, b_R is the rotor tooth thickness in the Z axis direction, R_{2*} is the outer radius of the rotor, δ_R is a radial clearance between the rotor and the compressor housing, $\xi = \frac{m_{OIL}}{m_{GAS}}$ is a gas-oil ratio, $\tilde{z}_2 = \frac{l_R}{l_S}\cdot z_2$ is the reduced number of the female screw teeth in the axial direction, l_R is the rotor length, l_S is the screw pitch in the axial direction.

Thermodynamic parameters of the mixture (kinematic viscosity, density, coefficient of cubical expansion and thermal conductivity, respectively) are determined by the following relationships [1, 2, 4 and 5]:

$$\eta_{MIX} = \left(\frac{1}{1+\xi}\right)\cdot\eta_{GAS} + \left(\frac{\xi}{1+\xi}\right)\cdot\eta_{OIL}, \quad \rho_{MIX} = (1+\xi)\cdot\rho_{GAS}, \quad \beta_{MIX} = \left(\frac{1}{1+\xi}\right)\cdot\beta_{GAS} + \left(\frac{\xi}{1+\xi}\right)\cdot\beta_{OIL} \qquad (5)\,(6)\,(7)$$

IOP Conf. Series: Materials Science and Engineering **90** (2015) 012004 doi:10.1088/1757-899X/90/1/012004

$$\lambda_{MIX} = \lambda_{GAS} + \left[0.72 \cdot \left(\frac{\xi}{1+\xi}\right)^2 + 0.28 \cdot \left(\frac{\xi}{1+\xi}\right)\right] \cdot (\lambda_{OIL} - \lambda_{GAS}), \quad (8)$$

In this case, the suction temperature for the clearance on the suction side and the discharge temperature for the clearance on the discharge side can be taken as a determining temperature in the preliminary calculations for the equation (3). The arithmetic mean temperature between the suction and discharge can be taken as a determining temperature for the equation (4). In addition, the ξ value can be taken as constant for all areas of the rotor in order to simplify the calculations.

3. Calculation of temperature fields of rotors and thermal deformation thereof

A raft of studies [6, 7, 8 and 9] is currently concerned with calculating the temperature fields of screw compressor rotors, which in varying degrees reveal the influence of various factors on the compressor rotors temperature field. The most comprehensive approach described for temperature field determination is presented in the studies [5 and 6]. However, at the same time, it should be noted that the methodologies in these studies require additional critical analysis and methodological justification.

The thermal condition of the rotors is determined by the heat conduction equation:

$$\frac{\partial T}{\partial \tau} = \alpha \cdot \nabla^2 T, \quad (9)$$

where α is the rotor material thermal diffusivity, T is a rotor temperature, ∇ is the nabla operator. The heat load on the compressor rotors is cyclical. However, the period of one cycle is short enough so that the time temperature field of the rotor remains practically unchanged. The latter thesis is confirmed by the research presented in the study [6]. This gives us the opportunity to assume that the pseudostationarity of thermal processes occurs in the rotor. Therefore, the equation (9) can be simplified to:

$$\nabla^2 T = 0. \quad (10)$$

Boundary conditions for the equation (10) are determined by the design diagram (Figure 2), on the basis of equality $\lambda_M \frac{\partial T}{\partial n} = q$, λ_M is the rotor material thermal conductivity, where $\frac{\partial T}{\partial n}$ is the normal derivative to the rotor surfaces upon which the heat flow q is acting. Thus, it is necessary to separate the heat gains from the gas-oil medium heat transfer with the rotor surfaces in the working chamber from the heat exchange of the rotor end surfaces with the gas-oil medium in the increasing spaces through the end surfaces of the suction and discharge ports, the heat leakage from the friction of rotors on the gas-oil medium at the end and radial clearances as well as the heat leakage throughout the shaft areas from the bearings.

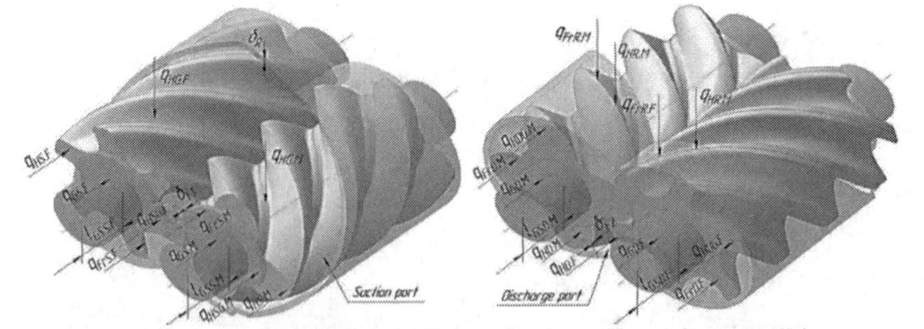

Figure 2: Design diagram for the determination of boundary conditions

The heat flow characterizing the heat transfer of the the gas-oil medium with the rotor surface in the working chamber is described by criteria equations. Thus, the equations [10, 11] were used to describe the heat transfer of the gas-oil medium with the rotor surface:

$$Nu_{GL,Re_{GZ}} = \left(49.371 + \left(1.685 \cdot \left(Re_{GZ} \cdot Pr_G\right)^{\frac{1}{3}} - 0.7\right)^3 + \left(\left(\frac{2}{1+22 \cdot Pr_G}\right)^{\frac{1}{6}} \cdot \left(Re_{GZ} \cdot Pr_G\right)^{\frac{1}{2}}\right)^3\right)^{\frac{1}{3}}, \quad Nu_{GT,Re_{GZ}} = \frac{\left(Re_{GZ} - 1000\right) \cdot \left(\frac{f}{2}\right) \cdot Pr_G}{\left(1 + 12.7 \cdot \left(Pr_G^{2/3} - 1\right) \cdot \left(\frac{f}{2}\right)^{0.5}\right)}, \quad (11)\ (12)$$

$$Nu_{G0} = \begin{cases} Nu_{GL,Re_{GZ}}, if \quad Re_{GZ} \le 2300 \\ Nu_{GT,Re_{GZ}}, if \quad Re_{GZ} \ge 10000 \\ Nu_{GL,2300} + \left[Nu_{GT,10000} - Nu_{GL,2300}\right] \cdot \frac{Re_{GZ} - 2300}{2700}, if \quad 10000 > Re_{GZ} > 2300 \end{cases}, \quad (13)$$

where $f = \left(1.58 \cdot \ln\left(Re_{GZ}\right) - 3.28\right)^{0.5}$ is a friction coefficient; Pr_G is the Prandtl number (a mean gas temperature in the working space in the given rotor section is taken as the determining temperature to calculate the Prandtl number); $Re_{GZ} = \frac{D_H \cdot \left(\frac{w_C}{2}\right) \cdot \rho_{MIX}}{\eta_{MIX}}$ is the Reynolds number; D_H is the hydraulic diameter of the corresponding rotor intertooth space end section; w_C is the determining velocity of the capacity from which a rotor's contact point velocity divided by 2 was taken. The equation (13) was obtained for the fixed channel, while the cavity of the screws is rotated around their axes. i.e. the heat exchange occurs in the gravitational (centrifugal) force field that intensifies the heat exchange somewhat. Correction for this fact is suggested in the study [12] and is slightly modified in the study [13]. According to the latter, the real Nusselt number is determined from the following relation:

$$\frac{Nu_G}{Nu_{G0}} = 0.262 \cdot \left(Gr_* \cdot \frac{Pr_G}{4}\right), \quad (14)$$

where $Gr_* = Gr \cdot Nu_{G0}$ is the modified Grashof number; $Gr = \frac{\omega^2 \cdot R_{MC} \cdot D_H^3 \cdot \rho_{MIX}^2}{\eta_{MIX}^2} \cdot \beta_{MIX} \cdot \Delta T$ is the rotating Grashof number; ω is an angular velocity of the corresponding rotor; R_{MC} is the corresponding rotor's intertooth space end section mass center radius vector; ΔT is the temperature difference between the rotor surface and gas in a working chamber. Then, the heat flows to the male and female rotors are respectively determined as follows (the second index in brackets refers to the female rotor):

$$q_{\substack{HG.M \\ [HG.F]}} = \frac{\alpha_{\substack{R1(M) \\ [R1(F)]}} \cdot Nu_{\substack{G(M) \\ [G(F)]}} \cdot \lambda_{MIX}}{D_{\substack{H(M) \\ [H(F)]}}} \cdot \left(T_g - T\right), \quad (15)$$

where the "(M)" index means that parameters of the male rotor are used as the geometrical parameters, while the "(F)" index means that geometrical parameters of the female rotor are used; T_g is the mean temperature of the gas-oil medium, with which the heat exchange takes place; T is the surface temperature, with which the heat exchange takes place; α_{R1} is a correction factor that takes into account the geometrical characteristics of the compressor (the ratio of the diameters of the rotor housing bores, the presence and size of ports on the radial part of the housing); these coefficients are equal to 1 for areas of rotors that do not form a radial clearance, and are equal to the relationship of the arc length whereon a clearance is formed to the circumference of the rotor housing bore for the sections forming a radial clearance.

Flowing of gas-oil fluid through the radial clearance is fairly complex. For the purposes of approximate description, this clearance may be represented as the portion of the rotary cylinder and the flow in it may be represented as the flow in the axial direction within the clearance between the rotary cylinder and the stationary housing. In this case, the flow is accompanied by the formation of Taylor vortices; the heat exchange in those conditions in the widest range of flow and rotation velocities as well as in the wide range of transport properties of the flowing medium is presented in the study [14]. It is proposed to use the following equation in order to describe the heat exchange therein:

$$Nu_R = 0.21 \cdot \left(Ta \cdot Pr_R\right)^{0.25}, \tag{16}$$

where $Ta = \dfrac{\left(\frac{w_C}{2}\right)^2 \cdot \delta_R^2 \cdot \rho_{MIX}^2}{\eta_{MIX}^2} \cdot \dfrac{\delta_R}{R_*}$ is the Taylor number; Pr_R is the Prandtl number (the mean temperature between the mean temperature of the gas-oil medium in the working chamber in the given rotor section and the rotor heat transfer surface temperature is taken as the determining temperature to calculate the thermal and transport properties of the medium being compressed); δ_R is a radial clearance; R_* is a rotor housing bore nominal radius. Then, the heat flows to the male and female rotor will be respectively defined as:

$$q_{\substack{HR.M \\ [HR.F]}} = \frac{\left(1-\alpha_{\substack{R1(M) \\ [R1(F)]}}\right) \cdot Nu_{\substack{R(M) \\ [R(F)]}} \cdot \lambda_{MIX}}{\delta_{\substack{R(M) \\ [R(F)]}}} \cdot \left(T_g - T\right), \tag{17}$$

The end surfaces heat exchange is determined by the following equation [15]:

$$Nu_{\substack{FG\_S \\ [FG\_D]}} = \begin{cases} 0.616 \cdot Re_W^{0.5} \cdot Pr^{0.435}, \text{ if } Re_W \le 2300 \\ 0.0267 \cdot Re_W^{0.8} \cdot Pr^{0.6}, \text{ if } Re_W \ge 10000 \\ 0.616 \cdot (2300)^{0.5} \cdot Pr^{0.435} + \left(0.0267 \cdot (10000)^{0.6} \cdot Pr^{0.6} - 0.616 \cdot (2300)^{0.5} \cdot Pr^{0.435}\right) \cdot \dfrac{Re_W - 2300}{7700} \\ \text{if } 10000 > Re_W > 2300 \end{cases}, \tag{18}$$

where $Re_W = \dfrac{R^2 \cdot \omega \cdot \rho_{MIX}^2}{\eta_{MIX}^2}$ is the rotating Reynolds number, R is a heat exchange surface point radius vector. The temperature in the port is the determining temperature here. The "$\_S$" index refers to the suction side, while the "$\_D$" index refers to the discharge side. Then, the heat flow to the rotor from the direction of the suction and discharge end face is respectively defined as:

$$q_{\substack{HS.M \\ [HS.F]}} = \frac{\alpha_{\substack{FS1(M) \\ [FS1(F)]}} \cdot Nu_{\substack{FG\_S(M) \\ [FG\_S(F)]}} \cdot \lambda_{MIX}}{R} \cdot \left(T_g - T\right), \quad q_{\substack{HD.M \\ [HD.F]}} = \frac{\alpha_{\substack{FD1(M) \\ [FS1(F)]}} \cdot Nu_{\substack{FG\_D(M) \\ [FG\_D(F)]}} \cdot \lambda_{MIX}}{R} \cdot \left(T_g - T\right), \tag{19) (20}$$

α_{FS} and α_{FD} are correction factors equal to the ratio of the port sectors area, which is intersected by the rotor during rotation on the end surface of the housing, to the area of the end section of the cylinder by the rotor housing bore radius.

The heat exchange at the end clearance can be defined in a similar fashion [15]:

$$Nu_{\substack{FGG\_S \\ [FGG\_D]}} = \begin{cases} 0.922 \cdot Re_W^{0.5} \cdot Pr^{0.435}, \ Re_W \le 2300 \\ 0.0251 \cdot Re_W^{0.8} \cdot Pr^{0.6}, \ Re_W \ge 10000 \\ 0.922 \cdot (2300)^{0.5} \cdot Pr^{0.435} + \left(0.0251 \cdot (10000)^{0.6} \cdot Pr^{0.6} - 0.922 \cdot (2300)^{0.5} \cdot Pr^{0.435}\right) \cdot \dfrac{Re_W - 2300}{7700} \\ 10000 > Re_W > 2300 \end{cases}, \tag{31}$$

$$q_{\substack{HSG.M \\ [HSG.F]}} = \frac{\left(1-\alpha_{\substack{FS1(M) \\ [FS1(F)]}}\right) \cdot Nu_{\substack{FGG\_S(M) \\ [FGG\_S(F)]}} \cdot \lambda_{MIX}}{R} \cdot \left(T_g - T\right), \quad q_{\substack{HDG.M \\ [HDG.F]}} = \frac{\alpha_{\substack{FD1(M) \\ [FS1(F)]}} \cdot Nu_{\substack{FG\_D(M) \\ [FG\_D(F)]}} \cdot \lambda_{MIX}}{R} \cdot \left(T_g - T\right), \tag{22) (23}$$

In this case, the mean temperature between the gas-oil medium temperature in the port and the rotor heat exchange surface temperature is taken as the determining temperature to calculate the thermal and transport properties of the medium being compressed.

The equations describing the friction in the clearances [4, 6] under the assumption of a uniform distribution of heat generated between the housing and the rotor takes the following form:

$$q_{\substack{FR.M\\[FR.F]}} = \frac{\left(1-\alpha_{\substack{R1(M)\\[R1(F)]}}\right)\cdot\eta_{MIX}\cdot R^2\cdot\omega^2_{\substack{1\\[2]}}}{2\cdot\delta_{\substack{R(M)\\[R(F)]}}}\cdot\left(T_g - T\right), \tag{24}$$

$$q_{\substack{FrG.M\\[FrG.F]}} = \frac{\left(1-\alpha_{\substack{FS1(M)\\[FS1(F)]}}\right)\cdot\eta_{MIX}\cdot R^2\cdot\omega^2_{\substack{1\\[2]}}}{2\cdot\delta_{\substack{F1(M)\\[F1(F)]}}}\cdot\left(T_g - T\right),\; q_{\substack{FrG.M\\[FrG.F]}} = \frac{\left(1-\alpha_{\substack{FD1(M)\\[FD1(F)]}}\right)\cdot\eta_{MIX}\cdot R^2\cdot\omega^2_{\substack{1\\[2]}}}{2\cdot\delta_{\substack{F2(M)\\[F2(F)]}}}\cdot\left(T_g - T\right). \tag{25) (26}$$

The parameters of the gas-oil medium in this case were selected using a method similar to that used in the calculation of the heat exchange in the relevant section. Calculation of the heat gain from the bearings through the rotor spike can be carried out according to the following dependences:

$$q_{\substack{GS.M\\[GS.F]}} = \frac{\lambda_R}{l_{\substack{GSS.M\\[GSS.F]}}}\cdot\left(T_{\substack{S.M\\[S.F]}} - T\right),\; q_{\substack{GD.M\\[GD.F]}} = \frac{\lambda_R}{l_{\substack{GSD.M\\[GSD.F]}}}\cdot\left(T_{\substack{D.M\\[D.F]}} - T\right), \tag{27) (28}$$

where λ_R is the thermal conductivity of the rotor material; l_{GSS}, l_{GSD} are the length of the rotor spikes on the suction and discharge side, T_D is the temperature in the section at the spike end; it may with a reasonable degree of reliability be assumed to be equal to the temperature of the oil discharged from the bearing.

The computational grid was structured and built by a method similar to that described in the study [16]. The gas temperature as well as the forces and moments acting on the rotors were determined based on the temperature and the indicator diagrams obtained using a mathematical model [4]. The gas-oil medium temperature was time-averaged for each section.

The equations describing the displacements and stresses ratio [17] were used to calculate the thermal deformations:

$$\left(\varepsilon_X - \beta_R\cdot\left(T - T_{St}\right)\right) = \frac{\sigma_X - \nu\cdot\left(\sigma_Y+\sigma_Z\right)}{E},\; \left(\varepsilon_X - \beta_R\cdot\left(T - T_{St}\right)\right) = \frac{\sigma_X - \nu\cdot\left(\sigma_Y+\sigma_Z\right)}{E}, \tag{29) (30}$$

$$\left(\varepsilon_X - \beta_R\cdot\left(T - T_{St}\right)\right) = \frac{\sigma_X - \nu\cdot\left(\sigma_Y+\sigma_Z\right)}{E},\; \gamma_{XY} = \frac{\tau_{XY}}{G},\; \gamma_{XZ} = \frac{\tau_{XZ}}{G},\; \gamma_{YZ} = \frac{\tau_{YZ}}{G}, \tag{31) (32) (33) (34}$$

where x, y, z are the arrangement of axes; β_R is the rotors material coefficient of thermal expansion, G is the rotors material modulus of elasticity in shear; E is the rotors material modulus of elasticity under tension; ν is the Poisson's ratio; σ, τ are the normal and shear stresses, respectively; ε, γ are the relative elongation and relative shear deformation, T_{St} is the standard rotor temperature that was assumed to be 293 K. In this case, the absence of axial movement of the mid-section of the rotor, and the absence of deformation of the rotors from external forces were taken as the boundary conditions (they were neglected as being small quantities [4]).

4. Calculation data

A standard screw compressor developed by CJSC "NIIturbokompressor" named after V.B. Shneppe" (successor of the Special Design Bureau for Compressor Engineering) was selected as the object of research with a rotor diameter of 250 mm, a length of the rotor profile section of 340 mm, a screw pitch of 440 mm and 660 mm for the male and female rotor, respectively, and a gear ratio of 4/6 similar to the profiles described in the studies [1, 2 and 4]. The following quantities were taken for the end clearance between the rotor and the housing: 0.5 mm on the suction side and 0.06 mm on the discharge side. The value of 0.2 mm was taken as the radial clearance. The following condition was selected as the study condition: the compressor's outdoor operating condition with compression from 1 to 9 at (pressure ratio of $\Pi = 9$), suction temperature of 25°C, discharge temperature of 90 °C, injected oil temperature of 60 °C and geometric compression ratio of 4.5. The gas-oil-ratio was assumed to be equal to 3.8. Rotor bearings center distance was taken as 200.11 mm on the suction side and 200.132 mm on the discharge side, respectively.

The results of the female rotor temperature fields' calculations, as well as the plots characterizing the engagement in the compressor operation process are shown in Figures 3 and 4. The calculations are given for two cases: when the compressor design provides for a suction port on the radial housing bore, and when it is not provided.

The results of temperature fields' calculations show that the radial rotor vertices are exposed to maximum warming. This is especially noticeable at the vertex of the female rotor, and can be explained by the significant frictional heat in the radial clearance. It should be noted here that the gas-oil medium temperature at the suction is lower, and hence its viscosity is considerably higher, and, as a consequence, the frictional heat in the clearance is also higher. This explains the higher rotor temperature in the housing without a suction port on the radial housing bore, as well as lower resistance of the latter to loss of contact in engagement in the course of operation because of the large frictional losses. Analysis of the results shows correctness of engagement during compressor operations in both considered variants. The gearing disconnection occurs only at the moment of the contact rerun between the rotor teeth. This moment is characterized by the discontinuity of the $\frac{d^2\beta_2}{d\theta_1^2}\cdot\omega_1^2$ function in the Figure 4. (At this moment the $\frac{d^2\beta_2}{d\theta_1^2}\cdot\omega_1^2$ function increases to infinity).

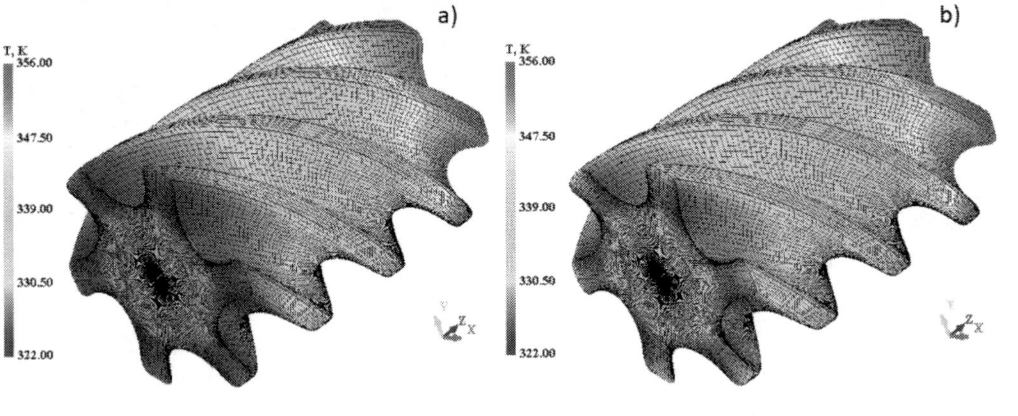

Figure 3: Temperature profiles of rotors:
a) with a suction port on the radial housing bore;
b) without a suction port on the radial housing bore

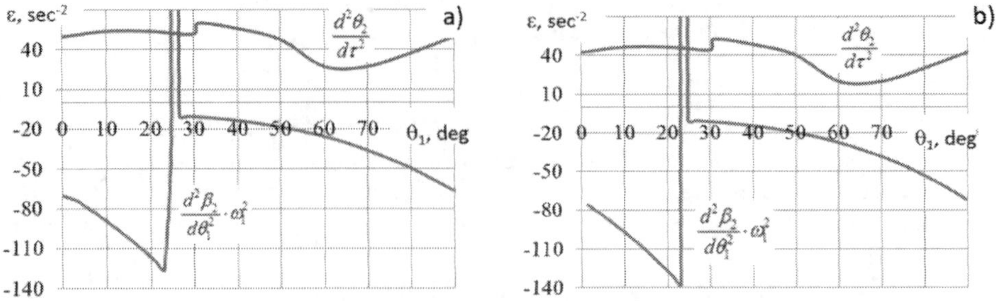

Figure 4: Acceleration of female rotor
(where θ_1 is the male rotor angle of rotation, ε is the female rotor angular acceleration):
a) with a suction port on the radial housing bore;
b) without a suction port on the radial housing bore

5. Conclusions

The developed computational model makes it possible to calculate the temperature fields of the rotors. At the same time, it allows us to estimate the impact of all factors, both individually and in their entirety, including not only the operating condition, but also the design features of the compressor.

The obtained results will also allow us to improve the accuracy of a screw compressor mathematical model, which can be used for the calculation of the gas parameters during the working process. In this case, the possibility exists to predict the sizes of the real clearances more accurately, which means the separation of the working chambers during the compressor working process. It should be noted that the additional heating of rotor teeth tips in the case of the absent ports on the radial surface of the housing may cause an increasing amount of back leakage gas temperature through the gaps, which are formed by the rotor teeth tips surface and the housing surface. Also, the compressing gas temperature may increase during the suction process due to the contact of the suction gas with hotter surfaces. These factors explain the reduction of the compressor capacity and increase of the power consumption in the case of the absent ports on the radial surface of the housing. The power consumption increases due to the increasing of the compressor indicated power and friction loss. This additional heating also exerts its influence on the rotor thermal field only in the boundary zones of the rotor teeth tips; therefore, the influence of this heating on the rotor thermal deformations of the contact zones and in consequence on the rotor meshing is marginal. In view of the above factors, it is recommended to manufacture the compressor ports or undercutting that repeat the port contours on the compressor housing radial surfaces.

The analysis of engagement during the operating conditions of the compressor was conducted on the basis of the obtained temperature field. This technique complements and builds upon the previously presented engagement analysis technique and can be the basis for further research of rotor engagement and vibration in the future.

Acknowledgments

This article was prepared at the Kazan National Research Technological University within the study guidelines and with the support of the implementation of the project "Establishment of high-tech production of oil-free scroll vacuum pumps for the industry of nanosystems and nanomaterials" of the open public tender for selection of organizations for the right to receive subsidies for the implementation of multiple-purpose projects aimed at the establishment of high-tech production according to the Decree of the Government of the Russian Federation dated April 9, 2010 N 218: "On measures of state support for the development of cooperation of Russian higher education institutions and organizations implementing multiple-purpose projects for the establishment of high-tech production".

References

[1] Yakupov R R, Mustafin T N, Nalimov V N, Khamidullin M S and Khisameev I G 2013 Discussion of actual profile clearances' calculation method in rotary compressors in the absence of rotor timing units *8th Int. Conf. on Compressors and their Systems* (London: City University) 209

[2] Yakupov R R, Mustafin T N, Nalimov V N, Khamidullin M S and Khisameev I G 2014 Analysis of transmission error depending on compressor working conditions *J of Proc. Mechanical Engineering* (Proc. of the Institution of Mechanical Engineers) Part E (London: SAGE Publications)

[3] Adams G P and Soedel W 1994 Dynamic Simulation of Rotor Contact Forces in Twin Screw Compressors *Proc. Int. Compressor Engineering Conference*, (West Lafayette: Purdue University)

[4] Khisameev I G and Maksimov V A 2000 *Dvukhrotornye vintovye I pryamozubye kompressory. Teoriya, raschetiproyektirovanie (Twin rotors screw and spur compressors. Theory, calculation and design (in Russian))* (Kazan: Fen)

[5] Poling B E, Prausnitz J M and O'Connell J P 2001 *The properties of gas and liquids Fifth Edition* (New York: McGraw-Hill)
[6] Hsieh S H, Shih Y C, Hsieh W H, Lin F Y and Tsai M J 2011 Calculation of temperature distributions in the rotors of oil-injected screw compressors *Int. J of Thermal Sciences* **50** 1271
[7] Hsieh S H, Hsieh W H, Huang C S and Huang Y H 2012 Numerical analysis of performance, rotor temperature distributions, and rotor thermal deformation of an R134a screw compressor, *Proc. Int. Compressor Engineering Conference*, (West Lafayette: Purdue University)
[8] Gao T Y, Yang D F, Cao F and Jiao J C 2011 Temperature and thermodynamic analysis of the rotors on a twin screw multiphase pump with high gas volume fraction *J of Zhejiang University (Applied Physics and Engineering)* **12** 720
[9] Sauls J, Powell G and Weathers B 2006 Transient thermal analysis of screw compressors, part 1: use of thermodynamic simulations to determine boundary conditions for fine element analyses *Proc. Int. Compressor Engineering Conference*, (West Lafayette: Purdue University)
[10] Kandlkar S G1998 Heat transfer characteristics in partial boiling, fully developed boiling, and significant void flow regions of subcooled flow boiling *J of Heat Transfer* **120** 395
[11] Gnielinski V 1995 A new calculation procedure for the heat transfer in the transition region between laminar and turbulent pipe flow (Ein neues Berechnungsverfahren für die Wärmeübertragung im Übergangsbereich zwischen laminarer und turbulenter Rohrströmung (in German)) *Forschung im Ingenieurwesen* **61** 240
[12] Wood J L and Morris W D 1980 A study of heat transfer in a rotating cylindrical tube Trans. *ASME J of Heat Transfer* **102** 612
[13] Shevchuk I V and Khalatov A A 1996 Heat Transfer and hydrodynamics in straight channels rotating about a parallel or inclined axis (Review). *J High Temperature* **34(3)** 455
[14] Tachibana F, Fukui S and Mitsumura H 1960 Heat transfer in an annulus with an inner rotating cylinder *Bulletin of JSME* **3** 119
[15] Dorfman L A 1960 *Hydrodynamic Resistance and Heat Loss of Rotating Solid. (Gidravlicheskoe soprotivlenie i teplootdacha vrashchayushchikhsya tel (in Russian))* (Moscow: Fizmatgiz)
[16] Smith I, Stosic N and Kovacevic 2007 *A Screw Compressors Three Dimensional Computational Fluid Dynamics and Solid Fluid Interaction* (Berlin: Springer-Verlag)
[17] Timoshenko S and Goodier J N 1970 *Theory of elasticity* (New York: McGraw-Hill)

IOP Conf. Series: Materials Science and Engineering **90** (2015) 012005 doi:10.1088/1757-899X/90/1/012005

Structural and torsional vibration analysis of a dry screw compressor

J. Willie, R. Sachs

Gardner Denver Schopfheim GmbH, Roggenbachstr. 58, Schopfheim, Germany

E-mail: `james.willie@gardnerdenver.com`

Abstract. This paper investigates torsional vibration and pulsating noise in a dry screw compressor. The compressor is designed at Gardner Denver (GD) and is oil free and use for mounting on highway trucks. They are driven using a Power Take-Off (PTO) transmission and gear box on a truck. Torque peak fluctuation and noise measurements are done and their sources are investigated and reported in this work. To accurately predict the torsional response (frequency and relative angular deflection and torque amplitude), the Holzer method is used. It is shown that the first torsional frequency is manifested as sidebands in the gear train meshing frequencies and this can lead to noise that is the result of amplitude modulation. Sensitivity analysis of the drive train identifies the weakest link in the drive train that limits the first torsional frequency to a low value. Finally, the significance of higher mode shapes on inter-lobe clearance distribution of the rotors is investigated.

1. Introduction

This paper is a state of the art presentation of how screw machines mounted on trucks can be modelled torsionally. The developed models presented successfully predict the first torsional mode in the dry screw compressors developed by Gardner Denver (GD). The modelled dry screw compressors are labelled generically as **A**, **B**, and **C**. Based on sensitivity analysis performed the weakest links in the drive trains are identified. The paper proceeds to show how the excitation of the first torsional mode and higher modes can be prevented. Driving through the male and female rotors is investigated and compared in **C** (See figures 1 and 2). In both **A** and **B**, the main drive is through the female rotor. The compressors modelled are mounted on to trucks using brackets suspended from their chassis rail. The connection to the truck drive is done using an input shaft to the compressor and the PTO transmission system that includes gears. A detailed description of the drive train is presented in section 2.

The prediction and measurement of torsional modes in screw compressors have been reported by the authors in [1]. The reported work included the use of a 90kW 4 pole asynchronous electric motor as the main drive. It also gave the sources of pulsation and vibration noise in screw compressors and investigated and found out that torque peak fluctuation in screw machines is driven by torsional resonance and not structural resonance. Crucially, the authors determined that at critical compressor input speeds, the driving torque is negative and this can lead to torque reversal in the drive train and to gear backlash and noise. In such a case, the compressor is attempting to drive the electric motor instead of the motor driving the compressor. The dependence of the torque excitation amplitude on prop-shaft angle is also investigated and reported.

Content from this work may be used under the terms of the Creative Commons Attribution 3.0 licence. Any further distribution of this work must maintain attribution to the author(s) and the title of the work, journal citation and DOI.
Published under licence by IOP Publishing Ltd

In the current work, the torsional models developed are use for a truck drive and include the PTO transmission system and gears. After the successful prediction of the first torsional modes and validation with measurement, the second torsional mode is considered. Most importantly, the effect of this mode on the inter-lobe clearance distribution of the rotors is investigated. This is followed by the presentation of the explanation of why beating-like or pulsating noise is possible in screw compressors. It is observed that during torsional resonance the first mode frequency is manifested as sidebands in the entire speed range of the compressor at the gear meshing frequencies (GMFs) and their harmonics. It is concluded that pulsating noise in screw compressors is due to amplitude modulation and not frequency modulation [2, 3]. The excitation of the first torsional mode is responsible for the torque peak fluctuation measured in the drive train of screw compressors and partly responsible for the pulsating noise inside them. That is, the excitation of the first torsional mode is a necessary but not a sufficient condition for generating pulsating noise in screw compressors. If this frequency is shifted, the torque peak fluctuation will be avoided, but the torque excitation from the engine remains the same. It implies that the modulation due to input torque fluctuation will still exist at the gear mesh and this may still drive pulsating noise in the machine. A sufficient condition will be to also consider the micro-gear geometry optimization, including gear backlash and tip-relief. The latter will ensure that there is only mid-flank contact at the gears. Misalignment in the drive train and rotor deflection must also be avoided by using tighter clearances and bigger rotors. This will ensure that transmission error is eliminated or minimized in the drive train.

The paper begins with the presentation of the development of torsional models in section 2. In section 3, the results and discussions are reported and finally, the conclusions are given in section 4 and this is followed by the acknowledgment.

1.1. Torsional vibration theory

In general, rotating elements driven by conventional drives (Electric motors, engines, etc) are susceptible to torsional resonance if any of the the natural torsional frequencies of the system is excited. It is therefore critical to determine the torsional frequencies of such systems and ensure that they do not lie within the input speed range of the machine and do not coincide with any excitation frequency. If they do, stiffness or inertia tuning can be used to shift the torsional frequencies of the system to the left or the right.

Two general drive trains are encountered in practice. In one, the torque is transmitted from the drive through all the rotating elements without branching (See figure 1). In the other case, two or more rotating elements are connected by parallel shafts connected to the same gear stage (See figure 2). In such a case the torque is divided in the shafts at the branched point.

To determine the system natural torsional frequencies, the structural dynamic equation

$$[I]\{\ddot{\theta}\} + [C]\{\dot{\theta}\} + [K]\{\theta\} = \{T(t)\} \tag{1}$$

is used, in which $T(t)$, the forcing torque or excitation, is set to zero, and C, which is the damping coefficient, is usually negligible in torsional systems and therefore is set to zero. I is the moment of inertia and K is the torsional stiffness. Lumped mass models, in which inertias are concentrated at gears and rotors are developed, with the shafts inertias as well as the stiffness of the gears neglected. Gears are assumed to be infinitely stiff. Each inertia is treated as a node, N, and the number of frequencies depends on N. For N nodes, we have N-1 natural torsional frequencies ω_1, ω_2,....ω_{N-1}, including rigid body modes. The corresponding relative angular deflection, θ, is a function of the angular frequencies. That is, $\theta_N = f(\omega_N)$.

To determine actual angular deflections, actual torques and hence the shear stress in the shaft elements, forced torsional models are required. In this case, $T(t) \neq 0$ and it is obtained by taking time series data of the torque from the drive and curve fitting it to obtain the analytical function, which is of the form $T(t) = T_0 sin(\omega t \pm \alpha) + \bar{T}$, where T_0 is the torque amplitude, $\bar{T}$ is the steady torque of the drive and α is the phase shift.

2. Torsional model development

This paper looks at the application of the Holzer method in the development of torsional models for screw machines [3, 4]. Other applicable methods include the transfer matrix method and the eigenvalue method. The Holzer method and transfer matrix method are applicable to both branched and unbranched drive trains whereas the eigenvalue method works for only unbranched drive trains. For drive trains with branching this method is not able to predict accurately the relative angular deflection. This is true because at the branched point, boundary conditions in torque and deflection must be met as reported in [1]. In the previously reported work, the models are developed for an electric motor drive and include both free and forced torsional models. The unforced (excluding motor input torque excitation) models are presented and used to determine the natural frequencies and mode shapes of the drive trains of the compressors. The amplification factors, which are an indication of the severity of the torque in the drive train at resonance, are determined and reported. Sensitivity analysis of the stiffness and inertia is done to determine the most critical elements in the drive that can be used in tuning the drive. This is followed by the development of forced torsional models (with the excitation torque included in the dynamic equation) that are used to determine the torque amplitude [6]. The effect of prop-shaft angle on torque amplitude is also investigated. In each case, validation is done with measurement data.

In the work reported in this paper, free torsional models are developed to include the drive of a truck. The models developed are for machines with generic drive trains **A**, **B** and **C**. Sensitivity analysis is used to determine the weakest link in the drive train outside the compressor and this can be compared with that in the electric motor drive train. The developed models are implemented in MathCAD.

Illustrations of the drive trains **A**, **B**, and **C** using line diagrams are shown in figures 1 and 2 for unbranched and branched drive trains, respectively. In figure 1, N denotes the nodes (concentration of mass or inertia) with N1=truck engine, N2=truck gear box, N3=PTO gear box, N4=first compressor stage gear assembly, N5= intermediate compressor stage gear assembly, N6=female rotor, N7=synchronization gear stage assembly and N8=male rotor. The same is true in figures 2(a) and 2(b), but in these cases N6 denotes the synchronization gears and N7 and N8 the female and male rotors, respectively. The stiffness is denoted by k in these diagrams and G denotes the gears. In the equivalent models, I and k denote the reduced inertia and stiffness, respectively. The reduction is done to the shaft between nodes 1 and 2 by multiplying the raw inertia and stiffness by the square of the overall transmission ratio.

In generic machine **B**, there is no intermediate gear stage and so node 5 is missing. Figures 2(a) and 2(b) depict the line diagrams of the generic machine **C** with drives through the female and male rotors, respectively. In this way, the influence driving through the male and female rotors has on torsional response can be investigated.

Figure 2(c) presents the reduced or equivalent models of generic machines **B** and **C**. The male rotor is taken as node 8 for consistency and comparison of the torsional response in all the models. The models developed neglect the stiffness of the gears, damping, and shafts inertias.

In figure 3, the details of the drive train outside the compressor are shown. The truck inertia (I1) is obtained from the manufacturer and the PTO transmission information is obtained by considering the sketch in figure 3, in which the PTO shafts and gear box are shown in addition to the truck gear box.

The stiffness, gear ratios and inertias of the drive train inside the compressors are known while for the PTO drive, the overall gear ratio, i_{14} is measured and G3 and G4 are known and so $i_{34} = \frac{G3}{G4}$ can be computed. It is therefore possible to compute i_{12} since $i_{14} = i_{12} \times i_{34}$.

IOP Conf. Series: Materials Science and Engineering **90** (2015) 012005 doi:10.1088/1757-899X/90/1/012005

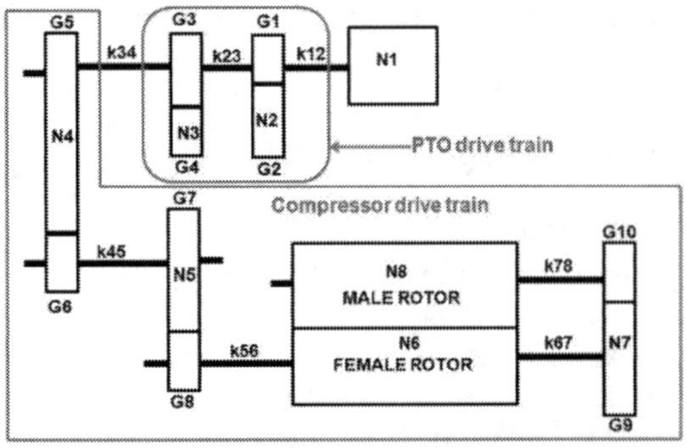

(a) Torsional model line diagram, unbranched drive train of generic machine **A**

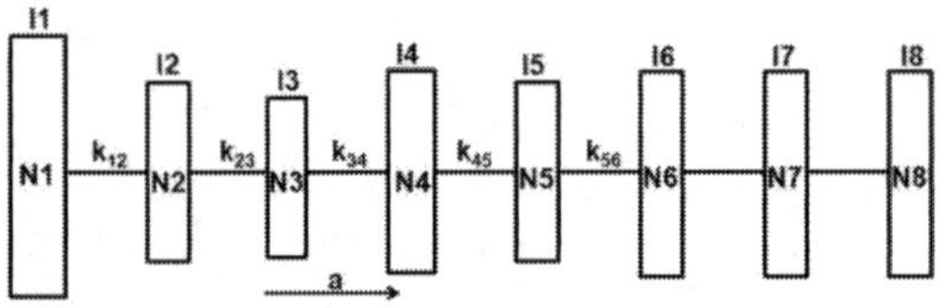

(b) Reduced torsional model (Generic machine **A**), unbranched

Figure 1. Torsional model line diagram and reduced model (Generic machine **A**)

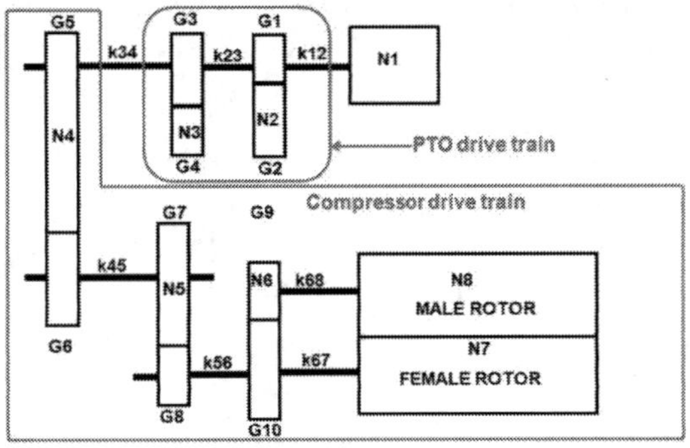

(a) Line diagram of generic machines **B** and **C**, female drive, branching

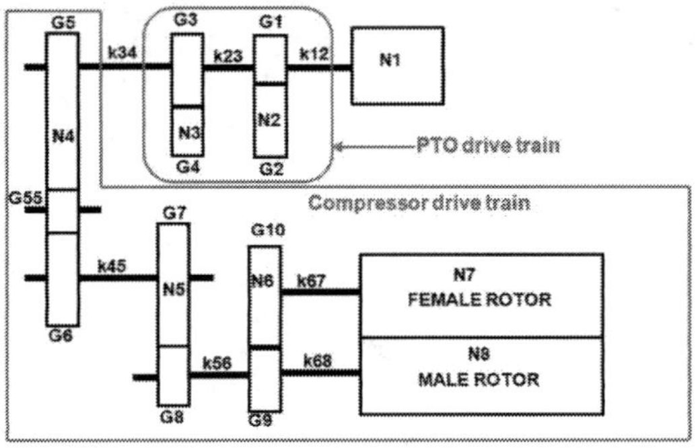

(b) Line diagram of generic machine **C**, male drive, branching

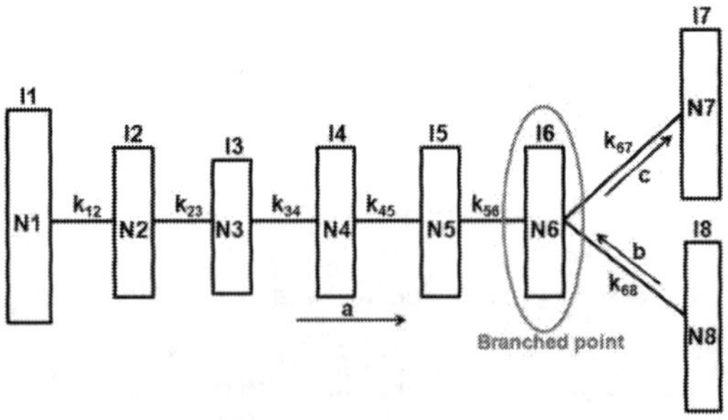

(c) Reduced (equivalent) torsional models (Generic machines **B** and **C**)

Figure 2. Line diagrams and reduced torsional models (Generic machines **B** and **C**)

Similarly, for the PTO drive train, the transmission shaft is assumed to be infinitely stiff and so its stiffness is neglected. The PTO shaft (thin) is known and so the stiffness is computed using $k = \frac{GJ}{L}$, where G is the shear modulus, J the polar moment of inertia about the principal axis ($=\frac{\pi D^4}{32}$ for a circular rotor and $=\frac{\pi(D_0^4 - D_i^4)}{32}$ with D_0 being the outer diameter and D_i the

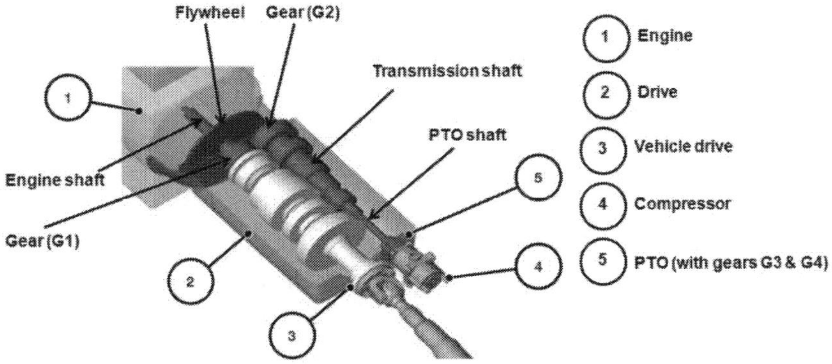

Figure 3. Schematic diagram of engine with PTO drive

Figure 4. Sectional view showing directional deformation of female rotor used in the determination of rotor stiffness from FEA

inner diameter, for a hollow rotor), and L is the rotor length. The PTO output shaft is also known and so its stiffness is computed. Unknown is the raw stiffness of the shaft between nodes 1 and 2 (engine shaft). The inertia of the flywheel is assumed to be included in that of the truck engine. To determine k12, an initial value equal to the raw or unreduced PTO shaft stiffness is assumed, and from measurements, the first torsional frequency measured is 10Hz on the truck (See figure 6). Because all other values in the model (Generic drive train **A**) are known, k12 is tuned till the first torsional frequency predicted by the torsional model is 10Hz. The k12 value obtained is used in the other torsional models (Generic drive trains **B** and **C**).

The rotors (male and female) are not uniform and so the method given above cannot be used to determine their stiffness. To do so, their equivalent diameters must first be determined and this is done using Finite Element Analysis (FEA) as shown in figure 4. From the maximum torque, T, applied and the maximum deflection, φ, obtained from FEA, the rotor stiffness, $k_r = \frac{T}{\varphi}$, is determined. From this, the rotor equivalent diameter, D_{eq}, can be determined using $D_{eq} = (\frac{32k_r L}{\pi G})^{\frac{1}{4}}$. This is especially true for the female rotor since the profile removes more material in it than in the male rotor for most applications. In the male, it might suffice to use the average of the tip diameter and the root diameter as the equivalent diameter.

2.1. Critical rotor speed determination

Conventionally, screw machines are designed to run below or above their critical speeds. This is done in order to prevent torsional resonance in the machine. The critical speed is determined by considering the sketch given in figure 5, in which the main drive is through the female rotor.

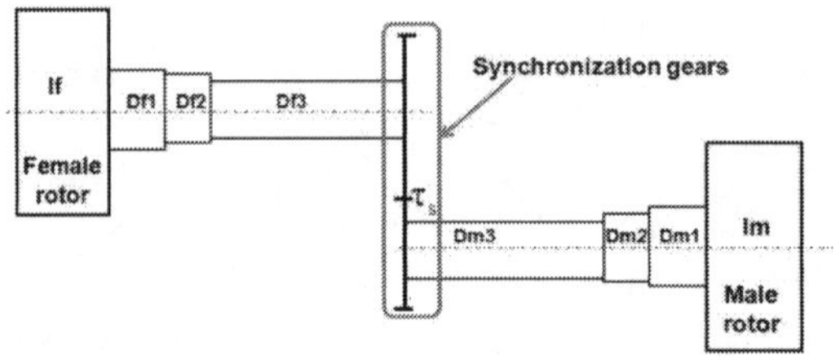

Figure 5. Determination of critical rotor speed

If I_f denote the raw female rotor inertia, I_m the raw male rotor inertia, D the shafts diameters (with subscripts f and m denoting female and male, respectively), and τ_s the gear ratio of the synchronization gears, then the critical female rotor speed is given by: $\omega_{cf} = [\frac{k_t(I_f+I_{mr})}{I_f I_{mr}}]^{\frac{1}{2}}$ in $\frac{rad}{s}$ or $n_{cf} = (\omega_{cf} \cdot \frac{60}{2\pi})$ in rpm. Here, I_{mr} is the reduced male rotor inertia and k_t is the equivalent stiffness of the female and male rotors shafts.

One of the findings in this paper indicates that a screw machine can be designed to run below or above critical speed and still experience torsional resonance when it is connected to a truck drive via a PTO. However, modelling this behaviour is not common in the literature and this phenomenon is difficult to identify in the design stage. The measurement work by Feese et al. [7] shows that this type of torsional resonance is possible in the drive train of compressors mounted on trucks.

3. Results and discussions

The first torsional frequencies and mode shapes of the generic machines **A**, **B** and **C** predicted by the torsional models are reported and validated with measurement data as shown in table 1 and figure 6.

Table 1. First torsional frequencies.

Generic machine type	Predicted first torsional freq. (Hz)	Measured first torsional freq. (Hz)
A	10.0	10.0
B	9.8	–
C, female	9.15	–
C, male	8.97	–

Torque is measured by mounting the torque flange onto the compressor input shaft. Brüel & Kjaer (B & K) microphones are used for the pulsed system in the torque measurement. In the truck test, the first torsional mode (10Hz or 600rpm) is occurring at $\frac{1}{2}\times$ critical input speed, whereas in the electric motor test, it was at $1\times$ critical input speed at 22Hz. Using sensitivity analysis, the weakest element outside the compressor is identified. The first torsional mode is therefore limited by this element. The system can be tuned by increasing or decreasing the stiffness of this element. The weakest element in the link is acting as a torque limiter to protect the truck gearbox.

The other tuning option is based on sensitivity analysis performed inside the compressor. In this case, the most sensitive element, based on inertia, is identified. To tune the system, the

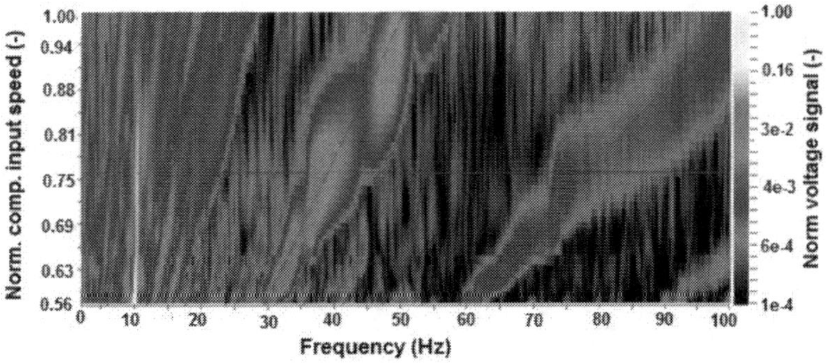

Figure 6. Measured torque run up spectra (Waterfall plot) of generic machine **A**

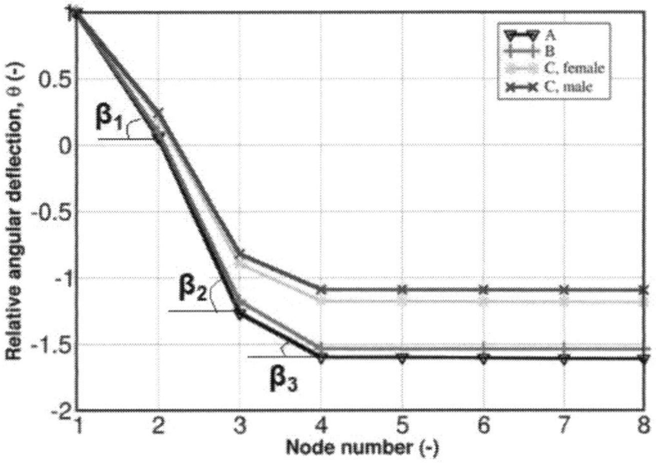

Figure 7. First mode shapes or torsional response of generic machines

inertia of the identified element can be increased or decreased. It has to be noted that because of the low first torsional frequency, shifting it will be very difficult, if not impossible, but the inertia effect of larger rotors can be used to damp and lower the torsional response.

The rotors used in the generic machines **A** and **B** have center distance (CD) of 93mm and an $\frac{L}{D}$ ratio of 1.6, while the CD of generic machine **C** is 105mm and it has an $\frac{L}{D}$ ratio of 1.6. The ratio of the discharge volume in machine **C** to that in **A** and **B** is 1.2. Similarly, the ratio of the male rotor speed in **A** and **B** to **C** is 1.3. The low speed in the latter (generic machine **C**) has noise advantage. Table 1 and figure 7 indicate that generic machine **C** has a lower first torsional frequency and response when compared to **A** and **B**. The slope, ($\tan\beta$), of the line connecting two nodal points is an indication of the severity of the torque in the shaft element [8]. Thus, the shaft element between nodes 2 and 3 has the largest slope and is therefore the most sensitive. This coincides with the position of the weakest element in the drive train as expected. Figure 7 also shows that the entire compressor is behaving as a rigid body and moving out of phase (anti-phase motion) and relative to the truck engine. The torsional response in generic machine **C** when we drive through the male rotor is better than when we drive through the female rotor.

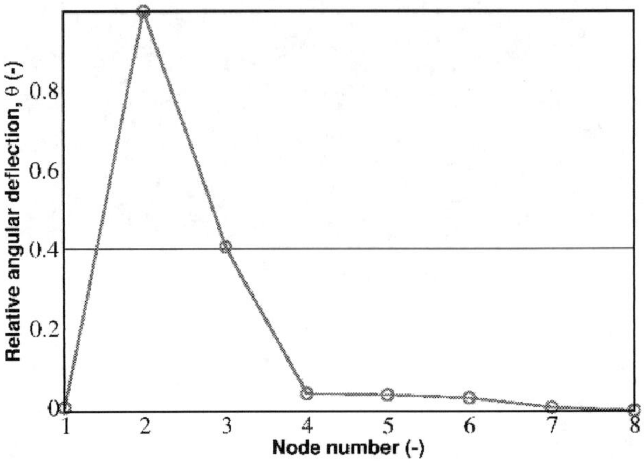

Figure 8. Second mode shape of generic machine **A**

Based on the outcome of the torsional response in figure 7, it is possible to determine the optimum position of the synchronization gears in the drive train. It ensures a stiffer and an optimum distribution of the torque in both the male and female rotors in the drive train.

3.1. Significance of higher mode shapes on inter-lobe clearance distribution

The effect of the first and second mode shapes on inter-lobe clearance is investigated and reported for generic machine **A**. It is found that in the first mode, there is likely to be no issues with clearance because in this case, there is no anti-phase motion (torsional wind-up) between the rotors (See figure 7). In the second mode and between nodes 6 and 7 and 7 and 8 (Figure 8), an anti-phase motion between the male and female rotors is present at the second modal frequency of 314Hz. In some applications, this frequency can be very close to the lobe-passing frequency of the rotors. If true, this mode can be excited, and this can lead to noise generation in the machine. Also, the relative clearance between the rotors will reduce making rotor-to-rotor contact possible (See figure 9). Using sensitivity analysis, the element that must be tuned in the drive train in order to shift this mode to the right is determined.

3.2. Pulsating noise

Pulsating noise is usually associated with screw compressors experiencing torsional resonance. This is especially true when there is torque reversal at the critical input speeds and when the input torque fluctuation is high and the gears are not optimized at the micro level. Transmission error in the drive train can also contribute to this effect. When there is torsional resonance, vibration signature (Fast Fourier Transform or FFT of time series vibration data) at the gears shows that this frequency is usually manifested as sidebands in the gear meshing frequencies and their harmonics of screw machines as depicted in figure 10. The noise shown in figure 11 is independent of the input speed. That is, irrespective of this speed, the 10Hz sideband is always present in the noise and vibration data at the gear meshing frequencies. The pulsating noise is therefore due to amplitude modulation and not frequency modulation [2, 3]. The noise data plot (Figure 11) shows that the amplitude is increasing and decreasing, but the 10Hz modulation

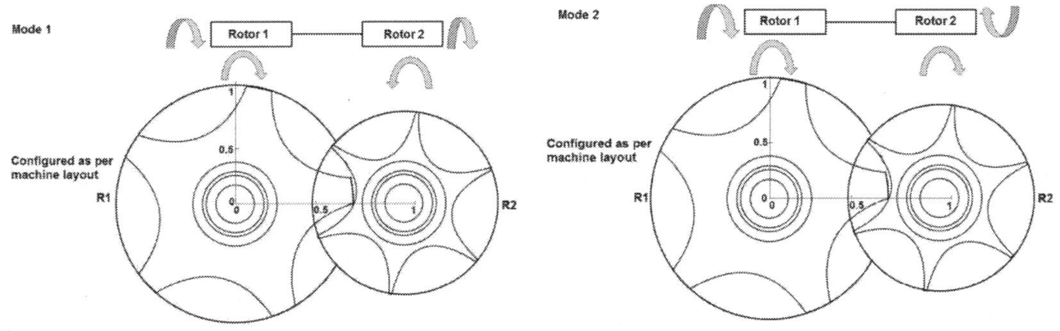

(a) Mode 1 (b) Mode 2

Figure 9. Effect of mode shapes on normalized rotors inter-lobe clearance distribution

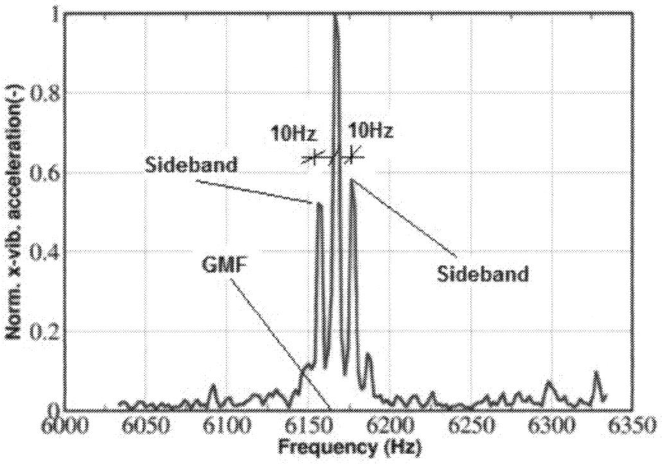

Figure 10. X-Vibration signature FFT at synchronization gears in a screw compressor

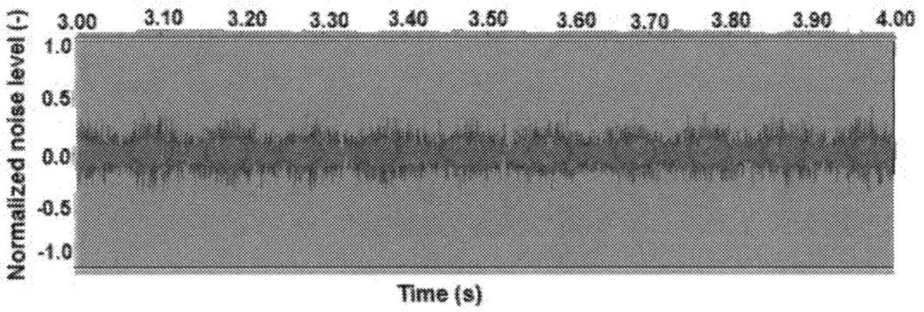

Figure 11. Pulsating or beating-like noise in a screw compressor

frequency is the same. The torque excitation coming from the engine is always present but the amplitude can be large or small depending on the prop-shaft angle in the drive train. The larger this angle, the larger the torque excitation. Tuning the drive train can prevent torsional resonance and torque reversal at this critical speed from occurring, but it may not be sufficient to eliminate the pulsating noise. A sufficient condition will be to optimize the micro-gear geometry in addition. Specifically, the tip-relief can be optimized in order to ensure mid-flank contact at the gears. An optimized backlash will prevent the fluctuating input torque from adversely affecting the movement of the gears and ensure that they move only in the drive direction. Fluctuation in torque will not lead to backlash at the gears in this case.

4. Conclusions

This paper presents the successful development of torsional models of screw compressors when they are driven by trucks. The models developed are an extension to those developed for an electric motor drive. It explains the effect of higher order modes on inter-lobe clearance between rotors and presents the source of the pulsating noise experienced in screw compressors. The following conclusions can be drawn: (i) Excitation of the first torsional mode can lead to torque peak fluctuation in screw compressors and to pulsating noise. (ii) Driving through the male rotor leads to lower torsional response. (iii) When the second torsional mode in screw compressors is excited it can lead to a reduction in the inter-lobe clearance and to rotor-to-rotor contact. (iv) Pulsating noise in screw compressors is due to amplitude modulation.

The following work is planned for the future: (a) Torque measurements in machines **B** and **C** and using the torsional frequency obtained to tune k12 and comparing it to the stiffness k12 obtained when torque measurement data of machine **A** is used as presented in this paper and (b) Presentation of the methodology that is used in the detailed calculation of the inter-lobe clearance distribution between rotors.

Acknowledgment

The authors would like to thank Steve Munt and Chris Witkowski, who carried out the measurements reported at GD DRUM, Bradford, UK. Their valuable contribution through discussions is also acknowledged.

References

[1] Willie J, Asal W and Sachs R 2014 *Analysis of the Noise and Vibration of a Dry Screw Compressor, Int. Conf. on Screw Machines 2014, TU Dortmund University, Dortmund, Germany, September 23-24 2014* (VDI-Berichte 2228) p 91
[2] Mobley R K 1999 *Vibration fundamentals* (Elsevier)
[3] Smith E H *Mechanical Engineers Reference Book, 12th Edition* (Elsevier)
[4] Den Hartog J P 1947 *Mechanical Vibrations 3rd Edition* (McGRAW-Hill Book Company, INC)
[5] Yuwen Y 2004 *Analysis of Multi-Branched Torsional Vibration for Design Optimization* Dissertation Submitted to the College of Engineering and Mineral Resources at West Virginia University in Partial fulfilment of the requirements for the degree of Doctor of Philosophy in Mechanical Engineering, 2004
[6] Glover G K 1977 *Mechanical Vibrations, 3rd Edition* (NEM CHAND and BROS ROORKEE (U.P.))
[7] Feese T and Maxfield R 2008 *Torsional vibration problem with motor/ID fan system due to PWM variable frequency drive, Proc. of the 37th Turbomachinery Symposium* p 45-56
[8] Leader M E and Kelm R D 2004 *Practical implementation of torsional analysis and field measurement, Vibration Inst. Annual Meeting, June 2004*

IOP Conf. Series: Materials Science and Engineering **90** (2015) 012006 doi:10.1088/1757-899X/90/1/012006

Numerical and Experimental Investigation of Pressure Losses at Suction of a Twin Screw Compressor

M Arjeneh, A Kovacevic, S Rane, M Manolis, N Stosic

Centre for Positive Displacement Compressor Technology, City University London, London, United Kingdom, EC1V 0HB

E-Mail: Mohammad.Arjeneh.1@city.ac.uk

Abstract. Rotary twin screw machines are used in the wide range of industrial applications and are capable of handling single and multiphase fluids as compressors, expanders and pumps. Concentration of liquid in the inlet flow can influence the performance of the machine significantly. Characteristics of the multiphase flow at the suction of a screw compressor depend on the local flow velocities and concentration. Local flow velocity measurements inside the screw compressors are difficult to obtain. However other flow properties such as local pressures are easier to attain. It is therefore useful to carry out experiments with local pressure variations in the suction which can be used to validate the 3D numerical Computational Fluid Dynamic (CFD) models that could help in studying the single and multiphase flow behaviour in screw compressors.

This paper presents experimental efforts to measure the local pressure losses inside the suction plenum of the screw compressor. Pressure variations are measured at 23 locations in the suction port at various operating conditions and compared with 3D CFD model. The grid generator SCORG^{TM} was used for generating numerical mesh of rotors. The flow calculations were carried out using commercial 3D solver ANSYS CFX. It was found that the local pressure changes predicted by the CFD model are in the good agreement with measured pressures. This validated the use of CFD for modelling of the single phase flows in suction of screw machines.

1. Introduction

Screw compressor developers, manufacturers and users are constantly challenged to broaden the scope of application of such machines and operate them more efficiently than ever. One of the objectives is use of screw compressors for handling of multiphase flows. Chamber models such as the one embedded in SCORPATH, as discussed in [1], provide very good accuracy in predicting standard oil injected and oil free compressor performance. However, despite of the good performance predictions with high gas volume fraction in a multiphase suction operation, it was found out that the prediction of the low gas volume fractions is less accurate [2]. The reason is that, although the thermodynamic models are computationally efficient and reasonably accurate they do not capture fully the characteristics of the fluid transport through the suction and discharge plenums. Therefore more detailed characterization of flow is required and it is believed that 3D CFD modelling can offer accurate prediction of the flow in the suction and discharge plenums of screw machines.

Content from this work may be used under the terms of the Creative Commons Attribution 3.0 licence. Any further distribution of this work must maintain attribution to the author(s) and the title of the work, journal citation and DOI.
Published under licence by IOP Publishing Ltd

A Number of reviewed research publications showed that CFD can effectively be used for analysis of the flow in the suction port of a screw compressor. Authors reported in [3] the efforts to visualize the flow by high speed camera and compare this with the CFD model results. Similar approach was used previously for predication of flow losses [4]. More examples on use of CFD in analysis of the flow in the suction plenum to determine potential loss areas [5], [6] showed possibilities to improve efficiency by the simple change in the design of the suction plenum.

The work presented in this paper is the part of the research program which aims to determine multiphase flow characteristics at the suction of a screw compressor as described in [7]. Modelling of multiphase flows in screw compressors using 3D CFD is still not part of a conventional design procedure due to the difficulties in obtaining solution and the speed of simulation. Previously, the authors [3] attempted to measure the multiphase flow characteristics (flow velocity) by means of Mie-Scattering technique. This technique proved to be very useful for evaluation of a single phase flow but posed challenges in measuring the multiphase flow due to formation of liquid film on the transparent test sections. Therefore, no quantitative and meaningful information were obtained for validation of CFD results. Local flow velocity measurements inside the screw compressors are difficult to obtain. However other flow properties such as local pressures can be measured with relative ease by using pressure measurement devices. For that reason, before the multiphase flow experiment is performed, it was proposed to carry out the measurements of pressure fluctuations with the single phase flow in order to obtain the pressure distribution in the suction chamber and validate the data obtained with CFD. This will demonstrate whether the local pressure measurement is a suitable method for validating the CFD results and determine suitability of the experimental setup for such measurements.

The screw compressor used for evaluation of pressure losses in the suction port was oil free compressor with the rack generated "N" rotor profiles and the internal two stage gearbox of 7.197:1 gear ratio. The main rotor has 3 lobes, 127.45 mm outer diameter and wrap angle of 285°. The gate rotor has 5 lobes, outer diameter 120.38. The compressors rotors are with length over diameter ratio of 1.6 and 93 mm centre distance. The manufactured interlobe and radial clearances are 180 microns. The compressor has axial suction port and combination of axial and radial discharge port, as shown in **Figure 1** Screw compressor configuration and sub-domains**Figure 1**.

In this paper the results obtained from the local pressure measurements of a single phase flow within the suction port at various locations and operating conditions are compared with the results obtained from the CFD model.

2. Grid generation and pre-processing for CFD calculation

Prerequisite for the successful full 3D CFD simulation of screw compressor is to obtain a numerical grid of rotating and stationary domains which will preserve conservativeness of space and equations and allow fast and reliable calculations. The decomposition in sub-domains and the topology of the numerical mesh used for CFD analysis of this oil free screw compressor research is shown in **Figure 1**.The mesh consists of the moving main and gate rotor sub-domains which are connected to the suction port and the discharge port mesh. The suction port is extended by a circular pipe. The rotor domains were meshed using SCORG^{TM} (Screw COmpressor Rotor Geometry Grid generator) [8], while the stationary parts were meshed using standard ANSYS mesh generators ICEM.

Two sets of meshes were generated for this research; one with the radial and interlobe clearances of 60 microns and the other with interlobe and radial clearances of 40 microns. Axial clearances were not taken in consideration. The reason for providing two sets of the mesh is to determine the most likely size of the leakage gaps in the operation of this machine. The size of mesh was just over 2 million cells. Table 1 shows breakdown of mesh size for all sub-domains.

Table 1 Size of the mesh for individual sub-domains

Component	No of cells in sub-domain
Main Rotor	451008
Gate Rotor	459360
Inlet Pipe	14650
Suction Port	568126
Discharge Port	590925
Total	2084069

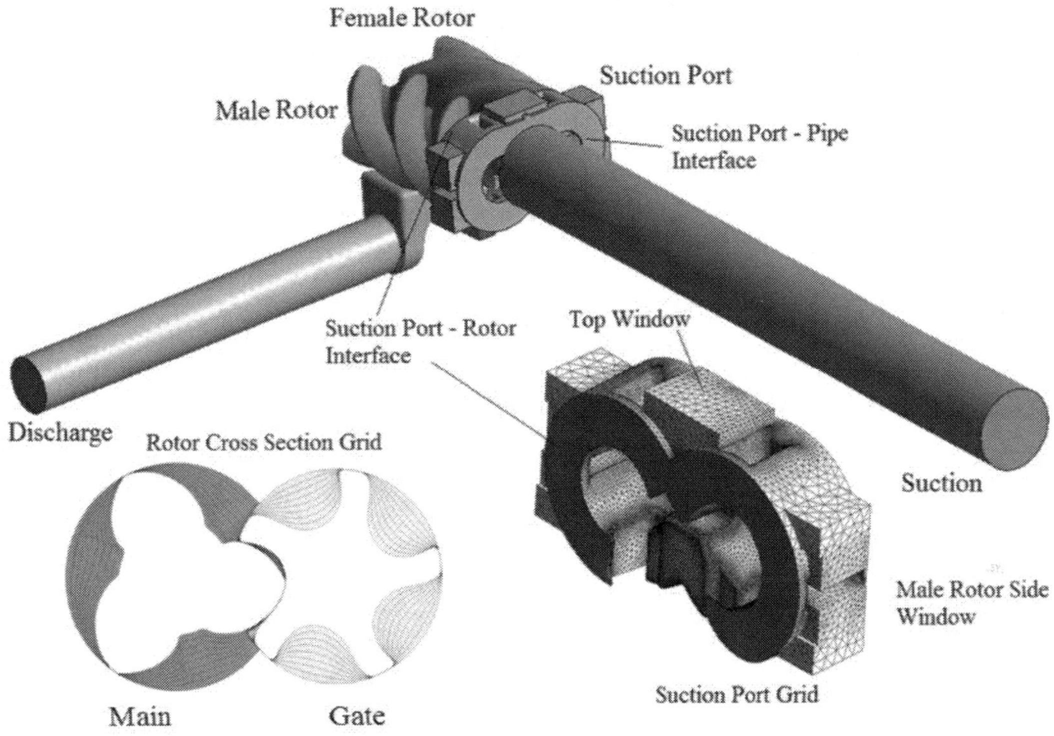

Figure 1 Screw compressor configuration and sub-domains

The CFD calculation was performed using commercial coupled pressure based solver ANSYS CFX. The working fluid is air with at 300 K suction. The suction pressure is 1.0 bar and discharge pressure is 2.0 bar. Convergence criteria were set at 1e-3RMS. The solver was allowed to run until satisfactory cyclic repetitions were achieved across all the monitored points. Selected turbulence model was SST K-Omega proved to be suitable for both main flow domains and in clearances.

3. Experimental Setup

Experimental investigation of the pressure fluctuations and performance of a test screw compressor was carried out in the air compressor test rig at City University London. **Figure 2** shows the layout of the measurement setup, the main test rig components and measurement points. In order to allow visual access into the suction port it was modified and Perspex windows were machined in it. Additionally, a transparent 1 meter long pipe was attached to the port in order to allow visualisation of the flow up stream of the suction plenum. The compressor is driven by a variable speed 75kW motor and has an internal synchronizing gear box with the gear ratio 7.197:1. The speed of the motor is adjusted with a speed controlled through the variable frequency converter. The torque meter is installed on the motor shaft while the digital encoder for the speed measurement is mounted on the main rotor shaft. The pressure and temperature of the gas are measured at the inlet, the discharge and upstream of the orifice plate. In addition, three pressure transducers are used for recording the interlobe pressures and are located in the working chamber through the compressor casing on the main rotor side. The air flow through the compressor is measured by use of an orifice plate installed in the discharge pipe system. The discharge pipe system contains the control valve for regulation of the discharge pressure.

Measured data from transducers are collected by a data acquisition system. It consists of a National Instrument Compact-RIO (CRIO-9022) Real-Time with an 8 slots chassis CRIO-9114. It has an industrial 533 MHz real-time processor for reliable real-time applications. Compact _RIO has 256 MB of DDR2 RAM and 2 GB of a non-volatile storage for holding programs and logging data. The unit also permits the acquisition of signals from the measurement transducers simultaneously during the experiment that are inputs to the unit itself and connected to a computer via an Ethernet cable.

The programming of the data acquisition system was by LabVIEW FPGA (Field Programmable Gate Array) which is suitable for high frequency data acquisition. The measurements obtained from the transducers are acquired on the cRIO (NI Compact RIO) and send to the PC where the latter instantaneously calculate values of the compressor power consumption, air flow rate, specific quantities and efficiency. All required information is displayed on a computer monitor in real time.

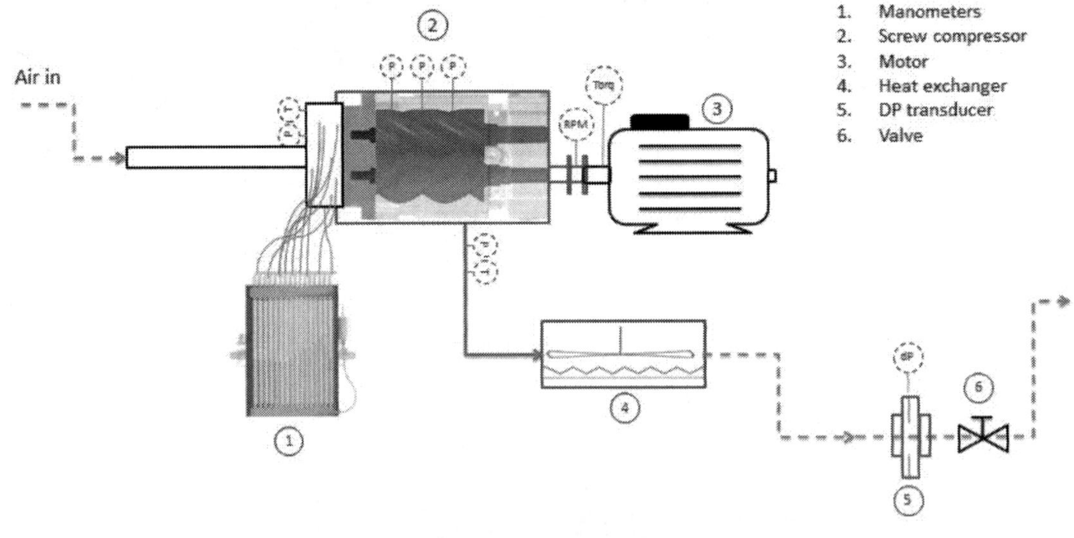

Figure 2 Experimental layout

Furthermore, the array of inclined manometers was used to map the pressure losses in the suction port. Pressures were measured in 23 places conveniently positioned within the suction plenum. Figure 3 shows the image of experimental setup while **Figure 4** shows location of the pressure probes. Measured pressures are averaged at each location over a number of revolutions.

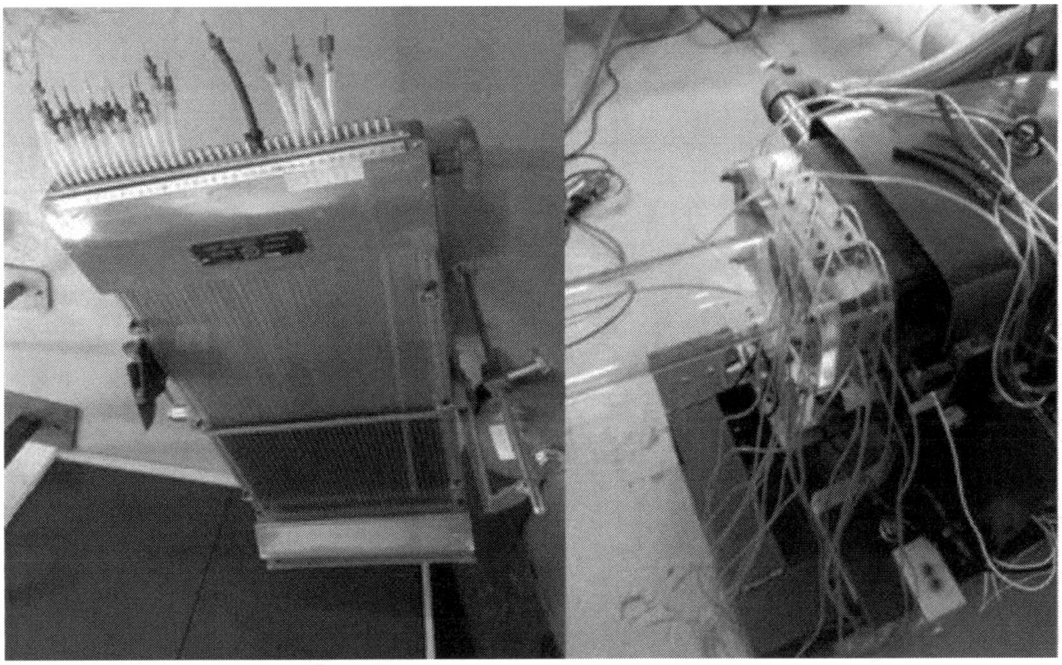

Figure 3 Suction pressure measurement setup

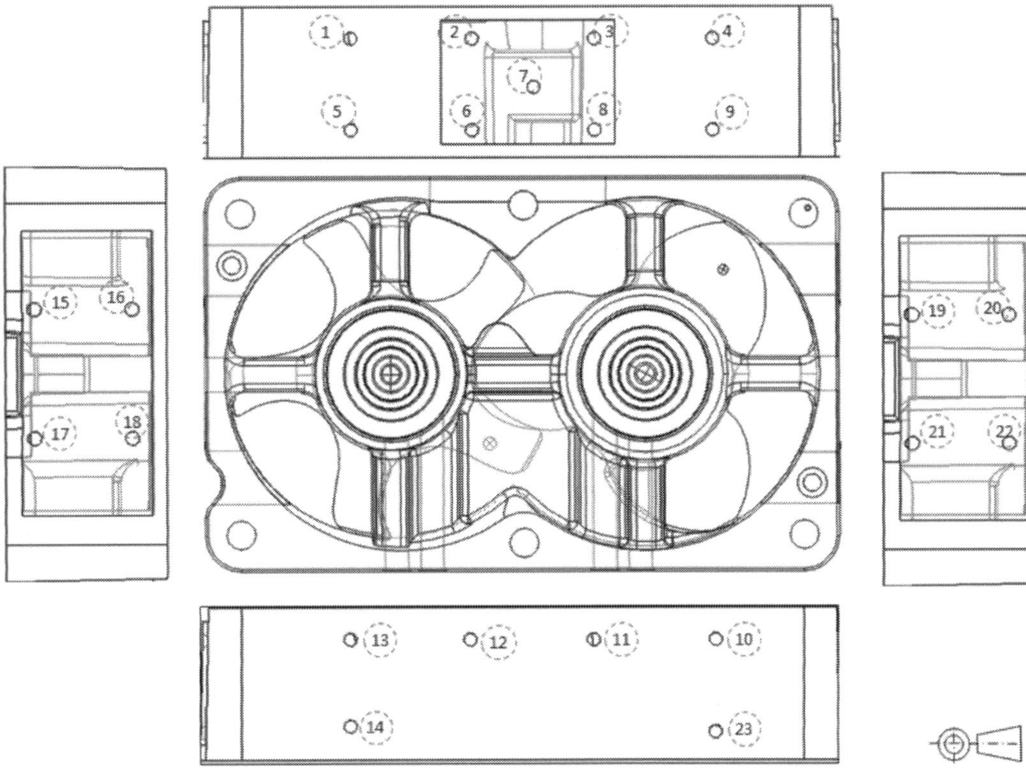

Figure 4 Location of pressure measurements in the suction plenum

IOP Conf. Series: Materials Science and Engineering **90** (2015) 012006 doi:10.1088/1757-899X/90/1/012006

4. Results and Discussion

The measured results are compared against 3D CFD simulations for the compressor main rotor speeds of 4000, 6000 and 8000 RPM. The CFD solver was allowed to run until the satisfactory repetitions of all the monitoring parameters were achieved in two consecutive rotor rotations. The CFD results presented in this paper are the averaged over twofull revolutions of the main rotor.

4.1. Internal Pressure and Compressor Performance

The comparisom of the compressor power and mass flow rate between the predictions and measurements is shown in Figure 5 for different compressor speeds. Two CFD results are shown in the diagram, one with 40μm and another with 60μm clearances. In case of 60μm clearance, the predicted mass flow is lower than the measured mass flow. However, for clearances of 40μm the predicted mass flow appeared to be almost identical to the emeasured flow rate. The predicted power is almost identical for two different clearance settings but it differs from the measured power.

The power measured in the test rig includes mechanical losses of the internal gearbox, bearings and seals while the power extracted from CFD calculations is the indicated power integrated from the pressure volume diagram which does not include any mechanical losses. The difference between the two increases with the increase in rotational speed as it would be expected.

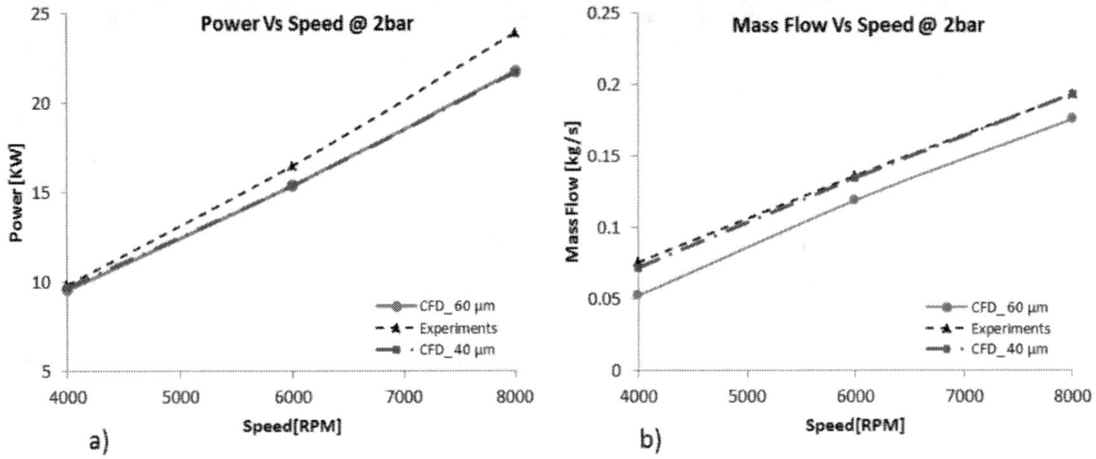

Figure 5 CFD and experiments comparison a) Power b) Mass flow rate

Pressure variations in the compression chamber as function of the main rotor rotational angle are shown in **Figure 6**Error! Reference source not found.. Measured pressure diagram matches well with pressure diagram extracted from CFD results at all speeds. The Distinctive difference between the measured and estimated internal pressure is observed in the peak pressure at low compressor speed of 4000 RPM. This could be due to the fact that at 4000 RPM the flow losses cause local temperatures to increase and the clearance changes affects mass flow rate which is marginally lower than measured mass flow rate at this speed. The peak pressure inside the working chamber increases as speed of the compressor increases. The peak pressures in both predicted and measured data are greater than the discharge pressure which indicates over compression.

IOP Conf. Series: Materials Science and Engineering **90** (2015) 012006 doi:10.1088/1757-899X/90/1/012006

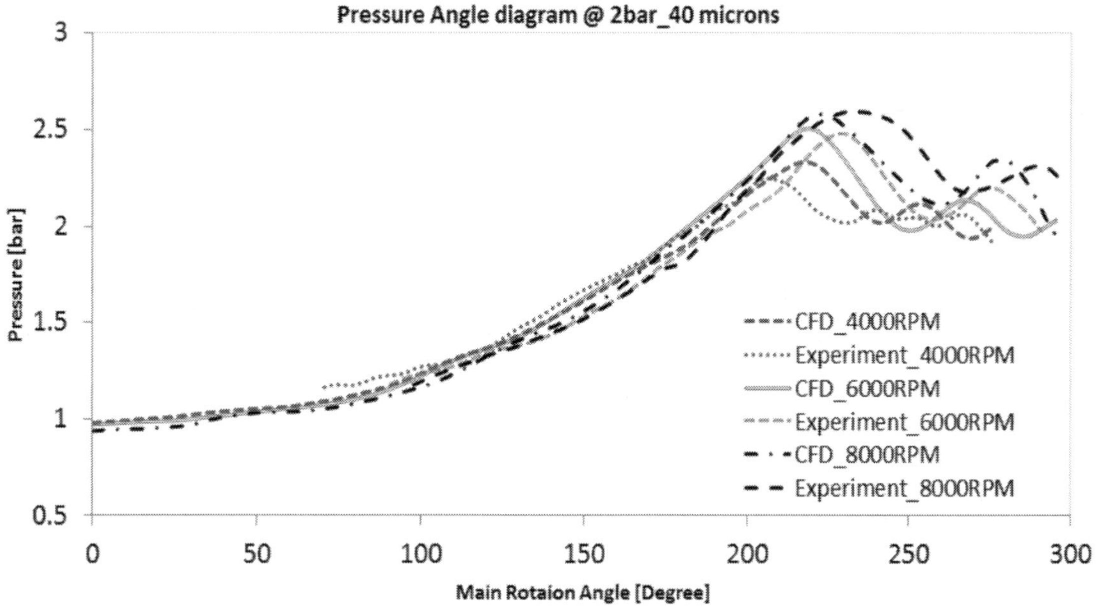

Figure 6 Pressure angle diagram with radial and interlobe clearances of 40 microns

4.2. Suction Pressure Map

Figure 7 shows the comparison of measured and calculated pressures at 23 locations in the suction plenum. The results are shown for three operating speeds, 4000, 6000 and 8000 rpm and two sets of clearances assumed in the numerical mesh of the compressor. The pressure loss at individual locations is the consequence of the change in local velocities. It can be seen from **Figure** 7a to **Figure** 7c that pattern of this pressure loss remains same at all operating speeds while only the amplitude of pressure losses changes. The lower the suction pressure as shown in diagrams, the higher the pressure loss. The highest pressure losses are observed in the bottom and top right corners in locations 4 and 23. This indicates that the higher amount of mass flows through these regions. This also indicates the position of the highest velocities within the suction plenum. Pressure loss in position 23 is higher than position 4 indicating that more mass flows towards the male rotor than the female rotor. Pressure loss at the manometer locations 12 and 11 is relatively low clearly reflecting the fact that the flow path is obstructed in this place.

It was previously reported in [9] how the mass flow depends of set clearances in CFD calculations. Since conjugate heat transfer and solid deformations are not accounted for in this CFD calculation, it is necessary to adjust clearances in the numerical mesh prior to the CFD calculation to account for thermal expansion of the compressor elements. The results presented in **Figure 4** show the effect of reducing clearances. The case with 40μm clearance gap predicts mass flows very close to the measured values as shown in Figure 5. Consequently the prediction of pressure losses in the suction plenum appears to be agreeing well with the measured values if the clearances are set to 40μm, as shown in **Figure 7**.

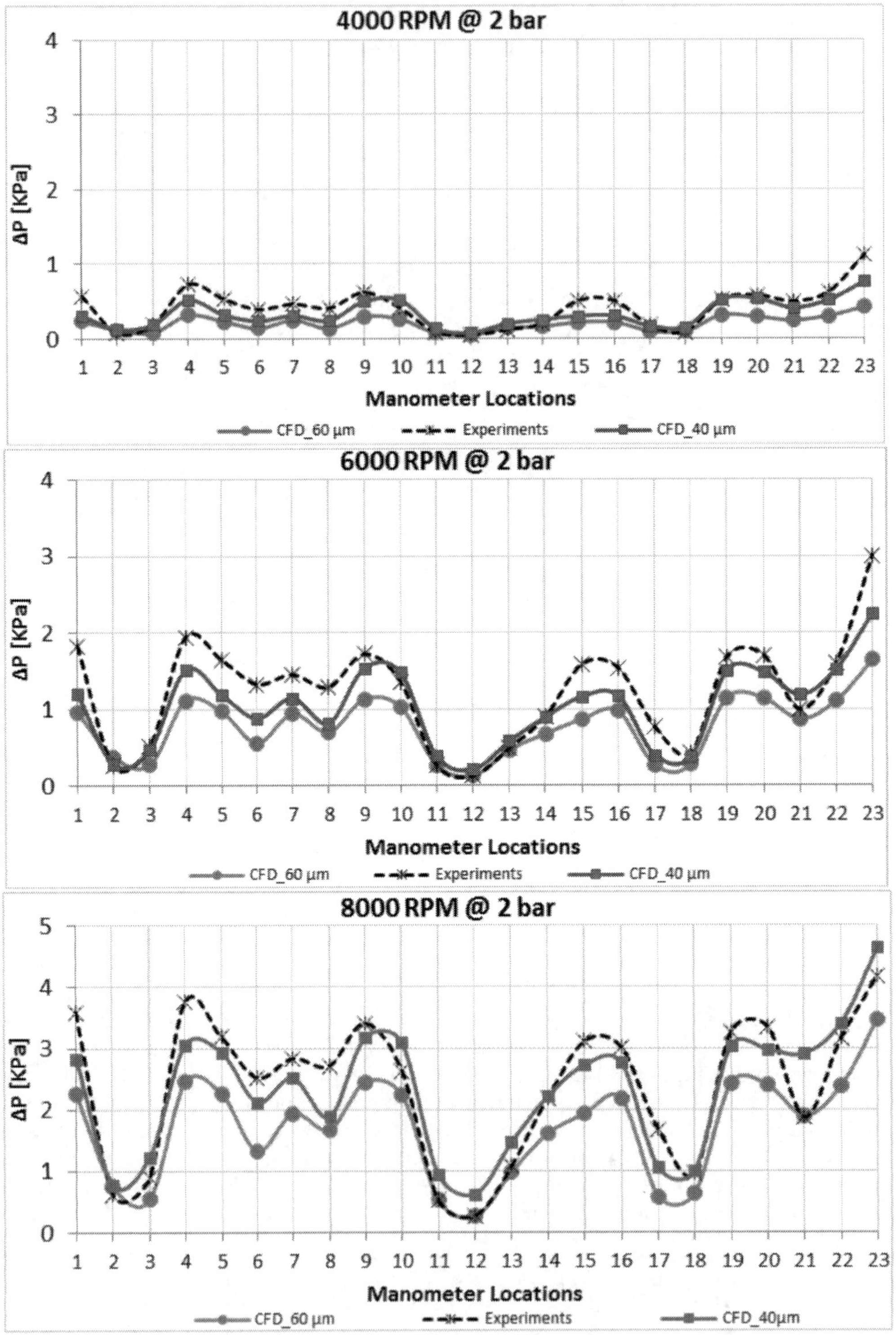

Figure 7 Pressure loss in the suction port at various locations

5. Conclusion

Measurements of pressure losses in the suction domain of an oil free screw compressor were carried out for the single phase flow of air by use of 23 inclined tube manometers. Alongside the power and flow measurements, the pressure variations within the compression chamber were recorded by use of three dynamic pressure transducers embedded in the casing on the main rotor side. The results are used to evaluate predictions of pressure losses in the suction plenum of the compressors derived from 3D CFD calculations. The following can be concluded:

- The predicted pressure fluctuations in the suction plenum have the same pattern as the experiment. At location 11 and 12 pressure losses are the lowest due to the obstruction in flow path. The highest pressure loss is measured at location 23 indicating the region with highest velocity in the bottom side of the main rotor.
- The measured and predicted chamber pressure matched well and the noticeable difference in the maximum pressure reduced substantially once the clearances of 40 μm were selected for simulations. This was due to better prediction of mass flow rate.
- The predicted values of suction pressure losses changed substantially with change in radial and interlobe clearances. The better prediction of losses coincided with more accurate prediction of mass flow rate at set clearance of 40 μm.
- It was found that CFD simulation the flow at the suction of a screw compressor using numerical grid generated by SCORG is able to predict the local flow features with high accuracy.

This study shows that 3D CFD can be used for study of the local flow features in the suction of screw compressors and can be used to optimise the shape of the suction plenum and port which is not usually considered in the current design practice. Furthermore it provides a good platform for studying effect of multiphase flows on flow characteristics at the inlet port of twin screw compressors. In the future the authors plan to continue experiments with multiphase flow at the suction of screw compressors,

6. References

[1] N. Stosic, A. Kovacevic, K. Hanjalic, and L. Milutinovic, "Mathematical Modelling of the Oil, Influence Upon the Working Cycle of Screw Compressors," 1988.
[2] F. Cao, T. Gao, S. Li, Z. Xing, and P. Shu, "Experimental analysis of pressure distribution in a twin screw compressor for multiphase duties," Exp. Therm. Fluid Sci., vol. 35, no. 1, pp. 219–225, Jan. 2011.
[3] A. Kovacevic, M. Arjeneh, S Rane, N Stosic, M. Gavaises, "Flow Visualisaton at Suction of a Twin Screw Compressor," in International Screw Compressor Conference at Dortmund, 2014.
[4] N. Stosic, A. Kovacevic, I. K. Smith "The CFD Analysis of a Screw Compressor Suction Flow," Int. Compress. Eng. Conf., p. Paper 1479, 2000.
[5] J Sauls, S Branch, Trane, "CFD analysis of refrigeration screw compressors," Int. Conf. Compress. their Syst., 2009.
[6] M Pascu, M Heiyanthuduwage, S Mounoury and U. Howden Compressors Ltd, "Influence of the suction arrangement and geometry of the inlet port on the performance of twin screw compressors.pdf," 8th Int. Conf. Compress. their Syst., 2013.
[7] M. Arjeneh, A. Kovacevic, M. Gavaises, and S. Rane, "Study of Multiphase Flow at the Suction of Screw Compressor," in 22nd International Compressor Engineering Conference at Purdue, 2014.
[8] A. Kovacevic, N. Stosic and I.K. Smith. Screw compressors - Three dimensional

computational fluid dynamics and solid fluid interaction, ISBN 3-540-36302-5, Springer-Verlag Berlin Heidelberg New York, 2007

[9] A. Kovacevic, S. Rane, N. Stosic, Y. Jiang, M. Furmanczyk, and S. Lowry, "Influence of approaches in CFD Solvers on Performance Prediction in Screw Compressors," in 22nd International Compressor Engineering Conference at Purdue, 2014, pp. 1–10.

IOP Conf. Series: Materials Science and Engineering **90** (2015) 012007 doi:10.1088/1757-899X/90/1/012007

Study on Design of Rotor Profile for the Twin Screw Vacuum Pump with Single Thread Tooth

Y Lu[1], B Guo[2] and M F[1] Geng
[1] Student, Xi'an Jiaotong University, Xi'an, China

[2]Professor, Xi'an Jiaotong University, Xi'an, China

E-mail: guobei@mail.xjtu.edu.cn

Abstract. The trapezoidal profile is always used as part of tooth profile in the screw vacuum pump. However, interferential phenomenon occurs when using trapezoidal tooth profile. The arc is added to the trapezoidal profile to obtain the self-conjugate curve. The geometric characters of new profiles are compared with the old ones from view point of area of leakage triangle, length of contacting line, area utility coefficient, and etc. The rotor with the self-conjugate curve has characteristics of the small axial leakage area. Radial leakage from the carryover was within the same volume and would not influence the mass flow rate of the vacuum pump. High ultimate vacuum can be reached by utilizing this profile.

1. Introduction

The dry vacuum pump is no mixing of the working fluid with oil and contact between the rotors is prevented by timing gears which mesh outside the working chamber. The tooth profile of the rotors is most important for a design of a vacuum screw rotor pump. Many studies on the theory and technology of the screw compressor could be now referred to D Dingguo and S Pengcheng on the generation of rotor profiles by the analytical method [1]; Stosic N, on the development of N type line [2]; Dmytro Z, on the modification and generation of meshing contour line [3]. Rotor profile is obtained through traditional trapezoid tooth and line correction using analytical method.

An efficient dry vacuum pump needs a rotor profile with a large flow cross section area, a short sealing line and a small blow-hole area. Stosic N describes details of a new lobe profile, which yields a larger cross-sectional area and shorter sealing lines resulting in higher delivery rates for the same tip speed [4]. In this paper, we proposed correction method of cycloidal meshing for a design of single screw pump using the traditional trapezoid tooth and designed a conjugate rotor for a screw vacuum pump. At the same time, we put forward a new profile and the shape of two rotors are exactly the same thus it could reduce rotor machining costs. We analysed and compared pump performance using the rotor profile improved from view point of contact line length, leakage triangle area and the area of utilization coefficient and carryover area of three profiles.

Content from this work may be used under the terms of the Creative Commons Attribution 3.0 licence. Any further distribution of this work must maintain attribution to the author(s) and the title of the work, journal citation and DOI.
Published under licence by IOP Publishing Ltd

Nomenclature

A center distance between male and female rotor
A_1 rotor area of male
A_2 rotor area of female
C_n Area of utilization coefficient
D rotor diameter of male and female rotor
τ rotational angle
t rotational angle
t_1 rotation angle of ab
t_2 rotation angle of bc
d diameter of dedendum arc
i transmission ratio, equal to 1
θ parameter of the generating profile
P Spiral characteristic number

2. Design and modification of profile

The specific requirements of profile design are as follows:

(1) In order to ensure there are no interferes with the rotor, rotor profiles are supposed to meet the theory of gearing [5].

(2) Tooth profiles are supposed to have good air tightness in the axis and horizontal direction in order to sealing between tooth volumes. Rotor contact line is a smooth continuous curve formed from addendum circle to dedendum one and minimize the length of contact line.

(3) To reduce leakage area of the triangle, the vertex of the contact line should reach the intersection of the female and male rotor's cylinder.

2.1. A subsection Mathematical model of the trapezoid tooth

Traditional Archimedes curve equation is obtained from rectangular (trapezoid) profile via the interference correction. This article selects the trapezoidal tooth. Rectangle tooth can be seen as a special case of the trapezoidal tooth. Axial cross section of Traditional trapezoid tooth is a trapezoid. The coordinate systems of the conjugate tooth are created as shown in figure 1.

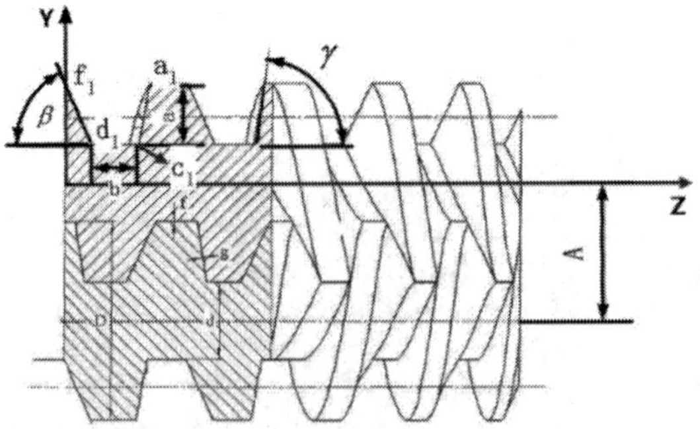

Figure 1. Axial section of trapezoidal tooth.

As shown in figure 1, Trapezoidal tooth can be regarded as the rotor helical motion under the trapezoidal tool cutting $a_1c_1d_1f_1$ [6].Therefore, tooth profile can be achieved through the spiral curve of four curve $a_1c_1d_1f_1$. Take c_1d_1 as example, calculating the corresponding tooth profile equation. Linear equation of c_1d_1 can be represented in the coordinate system as:

$$\begin{cases} X_u = 0 \\ Y_u = \dfrac{D}{2} - a \\ Z_u = t \end{cases} \qquad t \in [\frac{a}{\tan\beta}, \frac{a}{\tan\beta} + b] \tag{1}$$

Substituting equation (1) into equation (2);

$$\begin{cases} X = X_0(t)\cos(\tau)\, m\, Y_0(t)\sin(\tau) \\ Y = mX_0(t)\sin(\tau) - Y_0(t)\cos(\tau) \\ Z = Z_0(t) + \dfrac{P}{2\pi}\tau \end{cases} \tag{2}$$

The equation of the helix tooth space surface is obtained as following:

$$\vec{r}(u,\tau) = X\cdot\vec{i} + Y\cdot\vec{j} + Z\cdot\vec{k} \tag{3}$$

where

$$\begin{cases} X = -(\dfrac{D}{2} - a)\cos\tau \\ Y = (\dfrac{D}{2} - a)\sin\tau \\ Z = t + \dfrac{P}{2\pi}\tau \end{cases} \qquad \tau \in [\frac{a}{\tan\beta}, \frac{a}{\tan\beta} + b]$$

The screw tooth profile expresses the helix tooth space surface in the arbitrary OXY coordinate. Here, set Z=0 and we can obtain the axial curve segment c_1d_1. Substituting $t = -\frac{P}{2\pi}\tau$ into equation (3), it yields the radial equation as follows:

$$\vec{r} = -\frac{d}{2}\cos t\cdot\vec{i} + \frac{d}{2}\sin t\cdot\vec{j} \qquad t \in [\frac{-2\pi a}{P\tan\beta}, -\frac{2\pi}{P}(\frac{a}{\tan\beta} + b)] \tag{4}$$

As a result, radial equation is given by the axial cross section equation. Similarly, the remaining three sectional curves are calculated by the radial equations.

f_1d_1 section equation is:

$$\vec{r} = (-\tan\beta\frac{P}{2\pi}t - \frac{D}{2})\cos t\cdot\vec{i} + (-\tan\beta\frac{P}{2\pi}t + \frac{D}{2})\sin t\cdot\vec{j} \qquad t \in [\frac{-2\pi a}{P\tan\beta}, 0] \tag{5}$$

a_1f_1 section equation is

$$\vec{r} = -\frac{D}{2}\cos t\cdot\vec{i} + \frac{D}{2}\sin t\cdot\vec{j} \qquad t \in [-\frac{2\pi}{P}(\frac{a}{\tan\beta} + b + \frac{a}{\tan\gamma}), 0] \tag{6}$$

a_1c_1 section equation is

$$\vec{r} = -(\frac{d}{2} + \tan\gamma(-\frac{P}{2\pi}t - \frac{a}{\tan\beta} - b))\cos t\cdot\vec{i} + (\frac{d}{2} + \tan\gamma(-\frac{P}{2\pi}t - \frac{a}{\tan\beta} - b))\sin t\cdot\vec{j}$$
$$t \in [-\frac{2\pi}{P}(\frac{a}{\tan\beta} + b + \frac{a}{\tan\gamma}), -\frac{2\pi}{P}(\frac{a}{\tan\beta} + b)] \tag{7}$$

Basically, through substitution of β, γ, the common Archimedes curve and cycloid curves are obtained by the trapezoidal tooth correction.

2.2. Conjugate correction of profile

The equation of the trapezoidal tooth profile was obtained using method above. Generally β is given as 45 °. However, this kind of tooth profile not only does not meet the theory of gearing but also interfere may occur in segment a_1c_1. The theory of gearing was thus introduced for designing the vacuum pump rotor profile and reducing the design problem as shown in figure 2. The profile consists of cycloid (a b), epicycloid (b c), dedendum arc(c d), pseudo Archimedes curve (d e), Pseudo Archimedes conjugate envelope curve (e f), addendum arc (f a). Female rotor profile is composed of conjugate curves of male rotor profile and thus is the same with that of the male rotor and we can make the shape of the female rotor same with that of the male rotor. Through this correction, we can manufacture female and male rotors by the same processing, thus greatly reducing the cost of manufacture.

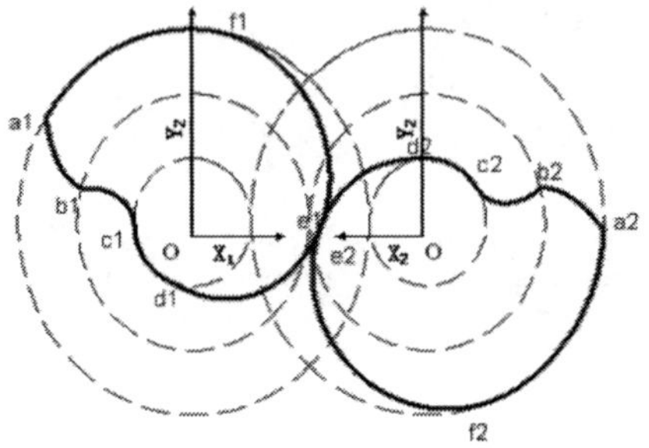

Figure 2. Conjugate correction of line.

The curve segment equations in the Cartesian coordinate are as follows:
Cycloid segment ab equation:

$$\begin{cases} x=A*\sin(2*t+t_1)-1/2*d*\sin(2*t+t_1)+A*\cos(t) \\ y=A*\cos(2*t+t_1)-1/2*d*\cos(2*t+t_1)-A*\sin(t) \end{cases} \quad (8)$$

Epicycloidal segment bc equation:

$$\begin{cases} x=A*\sin(2*t+t_2)-1/2*d*\sin(2*t+t_2)+A*\cos(t) \\ y=A*\cos(2*t+t_2)-1/2*d*\cos(2*t+t_2)-A*\sin(t) \end{cases} \quad (9)$$

Dedendum arc segment cd equation:

$$\begin{cases} x=-d*\cos(t)/2 \\ y=\ d*\sin(t)/2 \end{cases} \quad (10)$$

Equation for pseudo Archimedes curve segment de:

$$\begin{cases} x=-(a*\rho(t)+b)*\cos(t) \\ y=\ (a*\rho(t)+b)*\sin(t) \end{cases} \quad (11)$$

Archimedes curve ef is expressed by the rotation and translation of the conjugate curve segment de. The conjugate curve e_2f_2 is calculated from de segment equation:

$$\begin{cases} X_2=-x\cos(k\varphi)\mp y\sin(k\varphi)+A\cos(i\varphi) \\ Y_2=\mp x\sin(k\varphi)+y\cos(k\varphi)+A\sin(i\varphi) \\ \dfrac{\partial x}{\partial t}\times\dfrac{\partial y}{\partial\varphi}-\dfrac{\partial x}{\partial\varphi}\times\dfrac{\partial y}{\partial t}=0 \end{cases} \quad (12)$$

Their coordinate transformations are as follows, where $\mp$ is determined from the rotational directions of the two rotors:

$$\begin{cases} x=X_2*\cos\theta-Y_2*\sin\theta \\ y=X_2*\sin\theta+Y_2*\cos\theta \end{cases} \tag{13}$$

Addendum arc segment fa equation:

$$\begin{cases} x=-D*\cos(t)/2 \\ y= D*\sin(t)/2 \end{cases} \tag{14}$$

Female rotor profile is composed of its conjugate curves obtained by the envelope equation.

2.3. Smooth correction of Archimedes spiral

As shown in figure 3, Archimedes curve, the cusps between big arc and small arc, c, c', d, d' are not connected smoothly and it causes the abnormal junctions between Archimedes segment and arc segments.

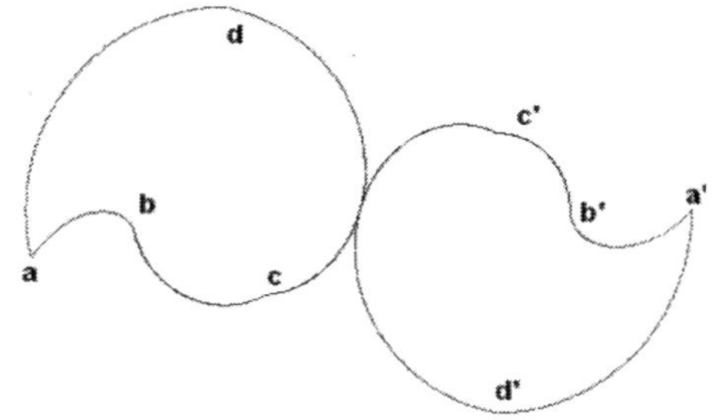

Figure 3. The traditional cycloid - Archimedes line.

Most of rotor profiles in the references and patents currently published have such as cusps and such a rotor profile is not conjugate each other. If the function f (x) in the interval (a, b) is of first order continuous differentiable, the graphic becomes a smooth tangent curve and it rotates continuously with the movement of the tangency points. The curve is called a smooth curve. However, this curve is not smooth, due to the first order discontinuous differentiable between the Archimedes curve and arc, as shown in figure 4 (from fig. 4 we could not read its cause, due to first order discontinuous differentiable or other.). We select a sine curve [7] for generating the spiral moving according to the spiral tooth surface model. Finally curve cd can be gotten as is shown in figure 5. In this way, the curves cd, ab, and bc, are connected smoothly and the cd curve, its conjugate curve $c'd'$, curve $a'd'$ and curve $b'c'$ form a continuous line. Not only their joints became smooth, but also two rotors satisfied the conjugate conditions.

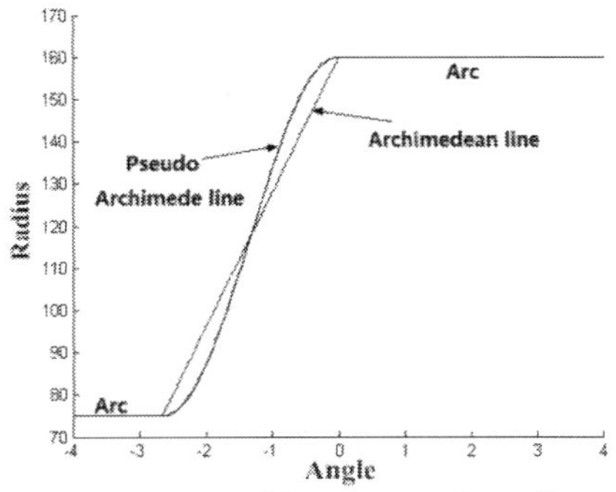

Figure 4. Archimedes angle-radius.

The Archimedes spiral equation is given by:

$$\rho = a\theta + b \tag{15}$$

Revised as following [7]:

$$\rho = a\sin(\theta) + b \tag{16}$$

The curve is corrected from Archimedes curve, so we call it "pseudo Archimedes curve". After correcting, the cusps in the line are eliminated and we can easily find out the conjugate curve as shown in figure 5.

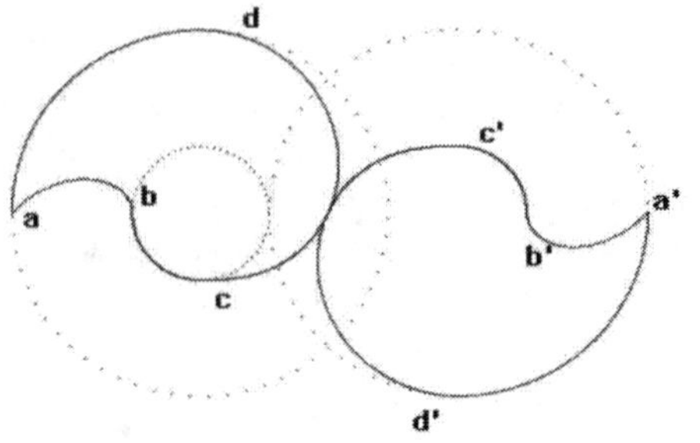

Figure 5. Smooth Archimedes spiral.

2.4. Profile of male rotor same with female one:

Even segments expressed by Archimedes conjugate curve are not the same each other, i.e., female rotor is not exactly the same with that of male. Two types of tools for hobbing, milling and grinding must be prepared in manufacturing screw vacuum pump, resulting in the cost increase. Therefore the Archimedes curve should be modified so as to be in coincidence with the conjugate curves.

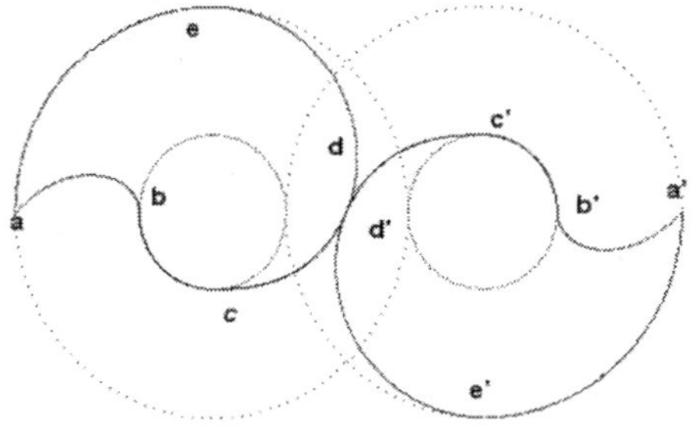

Figure 6. Rotor profile modified by line correction.

The original curve can be divided into two segments. One segment is the original Archimedes curve and the other segment is obtained from Pseudo Archimedes curve by the rotational and translational transformation. And the same rotors are obtained being conjugated with each other. The original curves are divided at point d and the conjugate curve d'e' is transformed by rotation and translation to the male rotor and replaced by de curve. In the same way, female rotor curve can be calculated. The line correction is validated as shown in figure 6. The female rotor given by its conjugate curve is the same as male rotor. Specific relation between curve and its conjugate can be seen in table 1.

Table 1. Corresponding relation between elements of the modified line.

Curve	Curve type	Conjugate	Conjugate curve type
bc	Arc	$a'e'$	Arc
cd	Pseudo Archimedes	$d'e'$	Pseudo Archimedes line envelope
de	Pseudo Archimedes line envelope	$d'c'$	Pseudo Archimedes
ea	Arc	$c'b'$	Arc
a	Point	$a'b'$	cycloid
ab	cycloid	a'	Point

2.5. Continuous correction of contact line

The conventional twin screw compressor is designed to form a sealing line between mated rotors. However, the sealing line is usually incomplete. We select a point in cycloidal curve and calculate its cycloid. Female and male rotors are calculated by the meshing line of the modified profile. The meshing line is shown in figure 7(a). As we can see from figure 7(a), meshing line is not closed so that vacuum tightness is broken [8]. But the vertex of the contact line reaches the intersection of the female and male rotor's cylinder. There is no leaking triangle between the contact line vertex and cylinder. By shifting the fixed point at the origin of meshing cycloid to the dedendum circle, the closed extent of meshing line is changeable. Eventually we can achieve the closed meshing line. The resulting meshing line is shown in figure 7(c).

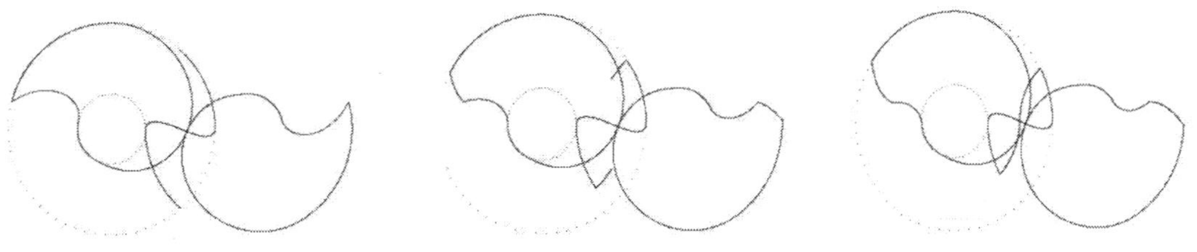

(a) (b) (c)

Figure 7. Rotor meshing line.

The meshed diagram of twin screw vacuum pump rotor is shown in figure 6. Male and Female rotors are single head screw. Two rotor tooth profiles are the same and mutually conjugated. The profile consists of cycloid, arc, Archimedes curve and its envelope curve.

3. Design examples and discussions of pump performance

As shown in figure 8, three types of rotor curve segments are modified according to the methods mentioned above. For comparing geometric features, let's take the diameter of the circle cd as 130 mm, the diameter of the circle ea 66 mm, and the polar Angle of curve de 0.8*pi. In figure 8(a), the cycloid ac consists of two different cycloids ab and bc. The corresponding meshing point of cycloid ab is point c and the corresponding meshing point of cycloid bc is point b; In figure 8(2), arc ac consists of ab and its conjugate curve bc; In figure 8(c), ac is a length of cycloid and corresponding meshing points is point c.

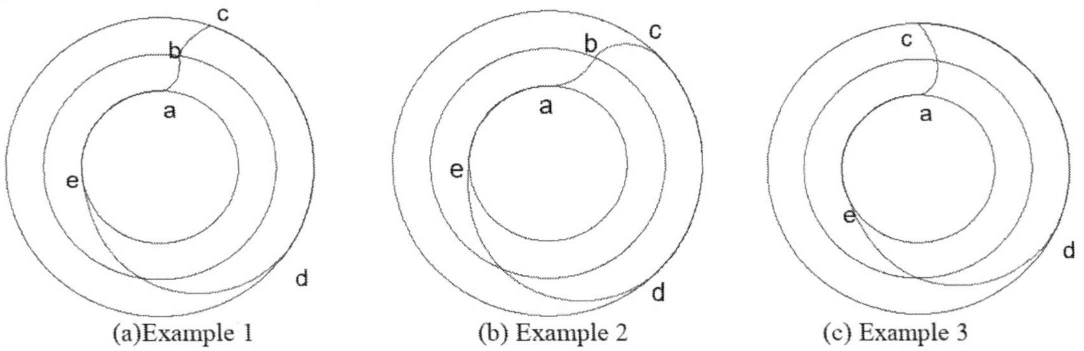

(a)Example 1 (b) Example 2 (c) Example 3

Figure 8. Rotor profiles.

3.1. Area of utilization coefficient

Let's select the lead as 72 mm for calculating the geometric parameters such as the rotor area, area between teeth and the area of utilization coefficient and the result is shown in table 2. The space area of utilization coefficient calculated by equation $C_n = \frac{(A_1 + A_2)}{D^2}$.

Table 2. Area of utilization coefficient.

	Example 1	Example 2	Example 3
rotor area /mm^2	8108.99	8118.31	7981.10
Tooth area/mm^2	5164.24	5154.92	5292.13
Area of utilization coefficient /%	61.12	61.00	62.63

As we can see from table 2, all kinds of rotor profiles are no major difference in the rotor area, tooth space area and the area of utilization coefficient, in which the maximum difference is within 2%.

3.2. Contact line

Meshing line or the contact line projected on the section is shown in figure 9.

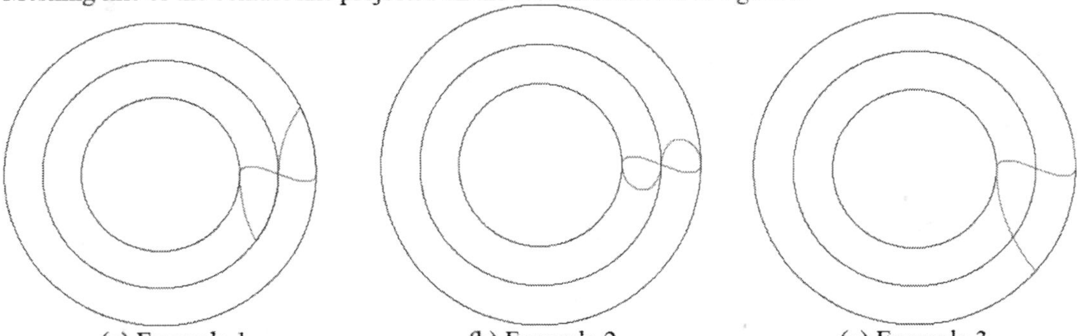

(a) Example 1 (b) Example 2 (c) Example 3

Figure 9. Contact line.

The contact line, L, is composed of three parts: the contact line length of cycloidal segment L1, the contact line length of pseudo Archimedes segment L2 and the contact line length of arc segment L3 and the contact line length and their partial contact line lengths are listed in table 3.

Table 3. Length of contact line.

	Example 1	Example 2	Example 3
L1/mm	123.80	58.12	94.52
L2/mm	46.32	46.32	46.32
L3/mm	39.34	33.79	43.20
L/mm	208.46	138.23	184.04

The result shows the epicycloid shortened can produce a shorter line of action. Thus, the pump efficiency could be increased.

3.3. Leakage triangle and carryover

As shown in figure 10, there is a carryover between two mated rotors. This means a larger gas may be carried from high–pressure port back to low-pressure port. This phenomenon will lead to larger leakage and reduce the pump performance [10].

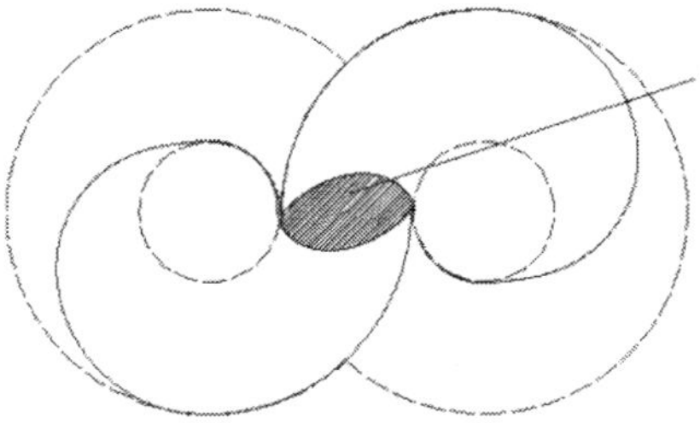

Figure 10. Carryover between two mated rotors.

Leaking triangle area S1 and carryover area S2 of three profiles are listed in table 4. The sealing performance of the rotor profiles, Example 3, is better than Example 2. In this paper, leaking triangle area of Example 3 is smaller than Example 1 by 27.26mm and by 87.18mm than Example 2. Further study of Example 3 founds that carryover connects the contiguous volume. Therefore, the rotor radial path of leakage does not affect sealing.

Table 4. Leakage triangle and carryover.

	Example 1	Example 2	Example 3
$S1/mm^2$	27.26	87.18	0
$S2/mm^2$	0	0	386.7

From view point of the air tightness, the contact line length of Example 3 is between Example 1 and Example 2 and no leakage triangle. The carryover of Example 3 connects contiguous volume and does not influence the air tightness. So the volumes of Example 3 is completely sealed. The profile of Example 3 can reach the higher vacuum and reduce the gas reflux.

4. Conclusion

The conjugate mathematical model was introduced to design three types of rotor profiles from which the conjugate tooth profile and its design constraints were derived. The following are our conclusions:

(1) We proposed a conjugated mathematical model for design of screw rotors for the single screw vacuum pump, by which it is enough to process two rotors with same shape in manufacturing the pump. This leads to the cost performance of the screw vacuum pump.

(2) We proposed a mathematical method to generate a conjugate curve to the original one to design two rotors conjugated each other using the rotational and translational transformation. a method of compositing new profile: curve-conjugate curve rotation and translation.

(3) We compared the pump performances depending on the different rotor profiles in terms of contact line length, leakage triangle area, the area of utilization coefficient and carryover. The sealing performance in case of Example 3 is better than Example 2.

Acknowledgements

The authors would appreciate the Department of Compressor Engineering of the Xi'an Jiaotong University for encouraging our works.

References

[1] D Dingguo and S Pengcheng 1989 *Rotary compressor* (Beijing)

[2] Stosic N On gearing of helical screw compressor rotors *Journal of Mechanical Engineering Science* **212** 587

[3] Dmytro Z and Xarlos A 2001 Screw compressor rotors line generation method based on pre-defined meshing line *International Journal of Refrigeration* **28** 744

[4] Stosic N, Smith I and Kovacevic A 2011 Geometry of screw compressor rotors and their tools, *Applied Physics & Engineering* **12** 310

[5] Litvin F 1989 *Theory of Gearing* (Washington DC :Nasa)
[6] Ohbayashi T, Sawada T and Hamaguchi M 2001 Study on the performance prediction of screw vacuum pump *Applied Surface Science* **169** 768
[7] Matsubara K, Riichi U and Muramatsu M 2007 Screw type vacuum pump US Patent: 7214036
[8] Kawamura T, Yanagisawa K and Nagata S Screw rotor and method of generating tooth line therefor 1998 US Patent: 5800151
[9] Ozawa O Gas Exhaust System and Pump Cleaning System for a Semiconductor Manufacturing Apparatus 1995 US Patent: 5443644
[10] H Chiu-Fan 2007 Study on the tooth profile for the screw claw-type pump *Mechanism and Machine Theory* **43** 812

IOP Conf. Series: Materials Science and Engineering **90** (2015) 012008 doi:10.1088/1757-899X/90/1/012008

Analytical Grid Generation for accurate representation of clearances in CFD for Screw Machines

S Rane, A Kovačević and N Stošić

City University London, Centre for Positive Displacement Compressor Technology, London, UK.
Email: sham.rane.1@city.ac.uk

Abstract. One of the major factors affecting the performance prediction of twin screw compressors by use of computational fluid dynamics (CFD) is the accuracy with which the leakage gaps are captured by the discretization methods. The accuracy of mapping leakage flows can be improved by increasing the number of grid points on the profile. However, this method faces limitations when it comes to the complex deforming domains of a twin screw compressor because the computational time increases tremendously. In order to address this problem, an analytical grid distribution procedure is formulated that can independently refine the region of high importance for leakage flows in the interlobe space. This paper describes the procedure of analytical grid generation with the refined mesh in the interlobe area and presents a test case to show the influence of the mesh refinement in that area on the performance prediction. It is shown that by using this method, the flow domains in the vicinity of the interlobe gap and the blow-hole area are refined which improves accuracy of leakage flow predictions.

1. Introduction

The CFD analysis of the working chamber of a screw machine is transient in nature and requires a grid which could accurately represent the deformation of the domain. In this respect, algebraic methods are convenient for quick adjustment and recalculation of the grid. *Kovačević, Stošić* and *Smith* [1, 2] have successfully used an algebraic grid generation method together with boundary adaptation and transfinite interpolation which has been implemented in program SCORG – Screw Compressor Rotor Grid Generator. *Sauls* and *Branch* [6] have applied SCORG to achieve results from CFD calculations in order to develop an improved one-dimensional thermodynamic model for refrigeration screw compressors by extracting calibration coefficients that influence the pressure variation during the discharge process. *Pascu et al.* [7] have reported optimization of the discharge port of a Twin screw compressor using SCORG grid generator. Optimization was based on the selection of the port geometry by relative comparison of flow field predicted by the CFD models. *Nouri et al.* [8] have carried out cycle-resolved velocity measurements within a screw compressor using LDV and these data have been used by *Kethidi et al.* [9] to compare with the results from 3D CFD models.

Voorde, Vierendeels and *Dick,* [10] have implemented a grid conversion algorithm for unstructured to block structured mesh from solution of Laplace equation for twin screw compressors and pumps using the differential methods. Based on this grid generation, flow in a double tooth compressor and a twin screw compressor has been analyzed and the results have been compared with experimental data over a range of discharge pressure and rotor speeds.

Recently CFX Berlin has introduced a software tool called TwinMesh to generate numerical mesh for ANSYS CFX solver. This tool generates grid for twin rotor machines on similar principles as that of

Content from this work may be used under the terms of the Creative Commons Attribution 3.0 licence. Any further distribution of this work must maintain attribution to the author(s) and the title of the work, journal citation and DOI.
Published under licence by IOP Publishing Ltd

the algebraic and differential approaches. *Hesse et al.* [11] have presented the application of such a deforming grid for the analysis of an oil free twin screw compressor.

Only limited studies are available on prediction of leakage flows using 3D CFD models. *Kovačević* and *Rane* have reported in [3] results from the analysis of a dry air expander and it was found that CFD models deviated in prediction of leakage flow much more under the operating condition of low rotor speed and higher pressure difference then at higher rotor speeds with low pressure difference. On the other hand, *Rane et al.* in [4] analyzed a twin screw compressor at one discharge pressure and one rotor speed and found that refinement of the rotor grid in circumferential direction has a direct influence on the prediction of the mass flow rate. However, pressure and power prediction showed lower sensitivity to grid refinement. In another study, *Kovačević et al.* [5] compared the predictions from two different CFD solution approaches and it was found that there is a big influence of the type of solver used for calculations on the prediction of leakage flows through the compressor. Again a low impact was found on the pressure and power prediction accuracy.

One of the major factors affecting the performance prediction of twin screw compressors by use of computational fluid dynamics is the accuracy with which the leakage gaps are captured by the discretization methods. In order to represent rotor profiles accurately, it is required to use point definition of the geometry but this definition gets compromised when a numerical grid is generated for flow calculations. One of the methods of improving the profile accuracy is by increasing the number of grid points on the profile. However, this method faces limitations when it comes to the complex deforming computational domain of the twin screw compressor because the grid quality deteriorates and computational time increases tremendously.

In order to address this problem, an analytical grid distribution procedure has been proposed in this paper that can independently refine the region of high importance i.e. the interlobe space. To achieve this, a new procedure is used for generation of the initial point distribution on compressor boundaries. This procedure uses equidistant distribution of points on the outer boundary of the domain which consists of the casing and rack to calculate distribution on rotors. By this means it is possible to refine mesh in the interlobe gap to allow for a more accurate prediction of leakage flows. The total grid size can be controlled by limiting the number of cells in the region not containing interlobe gap and blow-hole areas. The paper presents the principle of this analytical grid generation and a test case to demonstrate the influence of interlobe refinement on the performance prediction. A direct influence of interlobe grid refinement on leakage flow prediction has been verified from the study. This type of grid coupled with real time clearance variation procedure would be required in future to produce more reliable and accurate predictions from CFD analysis of screw machines.

2. Analytical grid generation for refinement of the interlobe numerical mesh

The rotor flow domain is divided into two subdomains belonging to two rotors using the rack plane which can be generated analyticaly or numerically using Willis meshing conditions [1]. Following that the main steps of mesh generation in each cross section are:

- Outer boundary in each block is defined as a combination of the rack segment and the casing circle segment. The rack segment stretches between the bottom CUSP point to the top CUSP point and is closed by the casing as shown in Figure 2.
- Rack segment is discretized using equidistant distribution. It is possible to use same distribution for both subdomains in order to maintain conformal interface if required.
- Casing segment in each subdomain is discretized using equidistant distribution which is usually different than for the rack segment. The distribution obtained on the outer boundaries of the two blocks is the reference for the rotor profile distribution.
- Nodes are distributed on the rotor profile with corresponding distribution available on the outer boundary. It is very likely that at first the initial cells will overlap for helical rotors, especially on the face of the gate rotor profile.
- Distribution on rotor profiles is regularized using background blocking.
- Interior nodes are distributed using transfinite interpolation [1].

IOP Conf. Series: Materials Science and Engineering **90** (2015) 012008 doi:10.1088/1757-899X/90/1/012008

- Orthogonalisation and smoothing iterations are performed to improve cell quality [1].

Most of the steps listed above are described earlier in [1]. Therefore only the background blocking is described in this paper. It can be considered as a uniform distribution of nodes on the rotor profile and on the outer circle as shown in Figure 1.

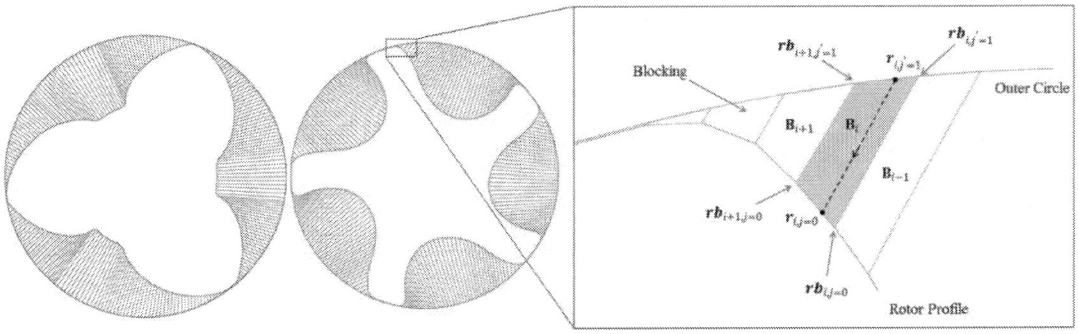

Figure 1. Background blocking for the main and gate rotors

The advantage of such blocks is that:

- Blocks do not have to be as refined as the final grid.
- Blocks can be used as reference for refinement in any required regions.
- Blocks have to be calculated only once and simply rotate for various rotor positions.

Suppose that the points distributed on the boundaries are represented in index notation with respect to the physical coordinate system as $\boldsymbol{r}_{i,j}$ (x,y). Points on the rotor profile are $\boldsymbol{r}_{i,j=0}$ (x,y), points on the outer boundary consisting of casing and rack curve are $\boldsymbol{r}_{i,j=1}$ (x,y) and the point distribution on outer full circle is $\boldsymbol{r}_{i,j'=1}$ (x,y). Each of the background blocks is identified by its index $\boldsymbol{B}_i$. The points on the inner boundary of the blocks which are the rotor profile nodes are $\boldsymbol{rb}_{i,j=0}$ (x,y) and the point distribution on outer full circle is $\boldsymbol{rb}_{i,j'=1}$ (x,y) as shown in Figure 1.

Starting from the bottom CUSP, nodes are distributed on the outer circle covering the rack part with required number of points i_{rack}. Nodes are then distributed on the outer circle covering the casing part with required number of points i_{casing}. At this stage the data are available for $\boldsymbol{rb}_{i,j=0}$ (x,y), $\boldsymbol{rb}_{i,j'=1}$ (x,y) and $\boldsymbol{r}_{i,j'=1}$ (x,y) and it is required to calculate $\boldsymbol{r}_{i,j=0}$ (x,y). This node distribution is based on equidistant spacing as given in equation 1.

$$\boldsymbol{r}_{i,j'=1}(x,y) = \boldsymbol{r}_{i-1,j'=1}(x,y) + S_i\, i \tag{1}$$

$$S_i = \frac{S_I}{I}$$

$$S_I = \boldsymbol{r}_{I,j'=1}(x,y) - \boldsymbol{r}_{i=0,j'=1}(x,y)$$

A scanning function is introduced that has the information of the background blocking. Starting from the bottom CUSP, the scanning function traces each node $\boldsymbol{r}_{i,j'=1}$ (x,y) and identifies the block $\boldsymbol{B}_i$ to which it belongs. There can be situation when a single block has multiple nodes present in it or there can be blocks with no nodes present. This is because the distribution on the rack curve can be refined in comparison to the blocking. Similarly the distribution on the casing can be coarse in comparison with the blocking. Once the nodes associated with each block are traced by the scanning function, an arc-length based projection is used to determine the nodes $\boldsymbol{r}_{i,j=0}$ (x,y) to be placed on the rotor profile. At the same time constraint is imposed on the node placement that they have to be bound in the same block $\mathbf{B}_i$ $\boldsymbol{B}_i$ as that of the outer circle nodes $\boldsymbol{r}_{i,j'=1}$ (x,y).

Figure 1 shows the projection of $\boldsymbol{r}_{i,j'=1}$ (x,y) on the inner boundary of the block in order to get $\boldsymbol{r}_{i,j=0}$ (x,y). This projection is based on arc length factor given by equation 2.

$$\boldsymbol{r}_{i,j=0}(x,y) = \boldsymbol{rb}_{i,j=0}(x,y) + \left(\boldsymbol{rb}_{i+1,j=0}(x,y) - \boldsymbol{rb}_{i,j=0}(x,y)\right)\frac{S_i}{S_I} \tag{2}$$

$$S_i = \boldsymbol{r}_{i,j'=1}(x,y) - \boldsymbol{rb}_{i,j'=1}(x,y)$$
$$S_I = \boldsymbol{rb}_{i+1,j'=1}(x,y) - \boldsymbol{rb}_{i,j'=1}(x,y)$$

The calculated positions $\boldsymbol{r}_{i,j=0}$ (x,y) of nodes ensure that they are guided by a regular rotor profile. Regularised distribution is superimposed onto the rack curve by finding the intersection points of the distribution lines and the rack curve. These intersection points are the new distributions $\boldsymbol{r}_{i,j=1}$ (x,y)on the rack curve as shown in Figure 2.

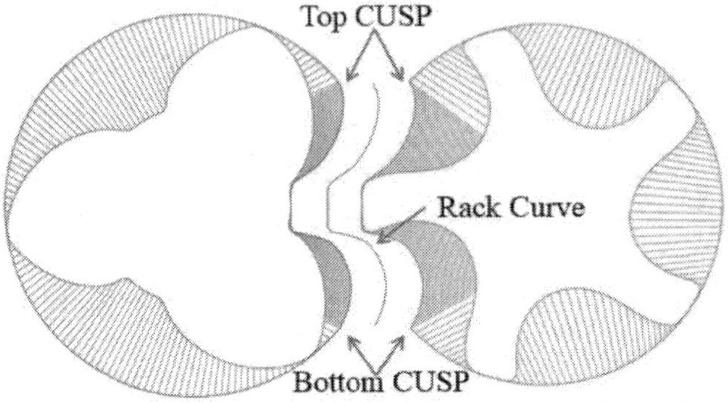

Figure 2. Refinement in the rack segment and superimposition of rack curve

Since the blocks on the main rotor side are different from the blocks on the gate rotor side, the intersection points obtained on the common rack curve from the two blocks are not identical. This results in a non-conformal map between the two rotor blocks. But this implementation gives an important feature to the grid to be able to construct refinement in the interlobe region. This region is important for capturing the rotor profile in the leakage gap and hence has strong influence on the leakage flow calculation from the solution. As an example, the cross section mesh with interface regions is highlighted and the three dimensional grid with helical lobes is shown in Figure 3. The mesh in the cross sections results in all quadrilateral cells with independent settings applicable for the number of nodes in the interlobe region. With the blocking approach the 3D grid is fully hexahedral and both the main and the gate rotor surfaces are smoothly captured. At the transition point from interlobe region to the casing region small non-aligned node movements are possible. However, these are positioned on the surface of the rotors and do not result in any irregular cells. The surface mesh on the casing is of the highest quality with regular quadrilateral cells. The surface mesh on the interlobe interface mostly follows axial grid lines with only small transverse movements in the vicinity of the top and bottom CUSP's which are cyclically repeating. These movements are on the surface of the interface and do not result in any irregular cells. The current implementation allows for a fully conformal interface with the equal index of the top and bottom CUSP points which ensures straight line in the axial direction. This accurately captures blow-hole area as shown in Figure 4.

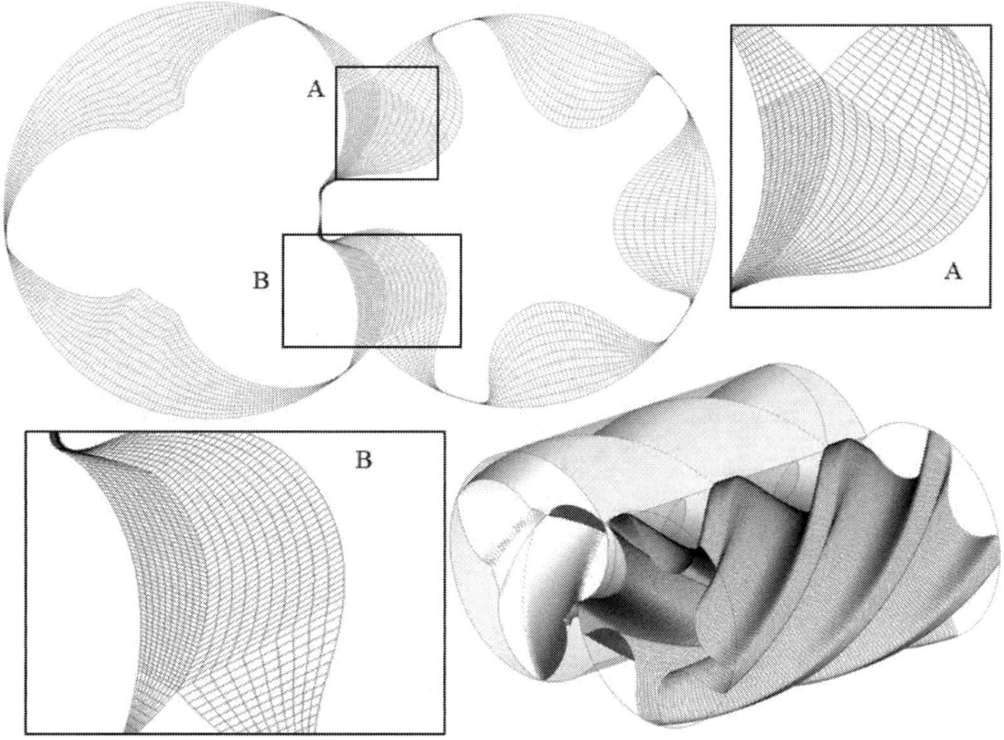

Figure 3. Grid generated with background blocking in screw compressor rotor

Modern rotor profiles usually feature small radii, as for example in the root of the main and the tip of the gate rotor. If the grid distribution starts from the rotor profile to the casing [1], the grid lines on the rotors follow the helix and capture these small radii accurately. However, when the grid distribution starts from the casing to the rotor profile, the grid lines in axial direction are connecting nodes with same circumferential indices in consecutive cross sections, hence they do not run along the helix of the rotors. This results in some faces becoming warped in these areas as shown in Figure 4. However, the cells remain regular and very low effect on the flow regime in this area is recorded.

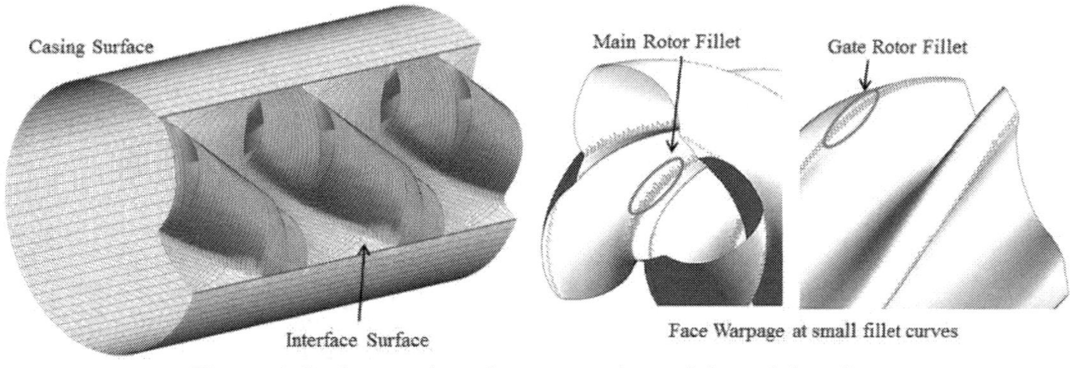

Figure 4. Surface mesh on the rotor, casing and the rack interface

3. Case study – Influence of interlobe grid refinement

An oil-free twin screw compressor with a 3/5 lobe arrangement of 'N' rotor profile rotors is modelled in this study. The male rotor diameter is 127.45 mm; the female rotor diameter is 120.02 mm while the centre distance between the two rotors is 93.00 mm. The length to diameter ratio of the rotors is 1.6 and the male rotor has a wrap angle 285.0 deg. The nominal interlobe, radial and end leakage gaps are 160 micro meters each. The built-in volume index of the compressor is 1.8.

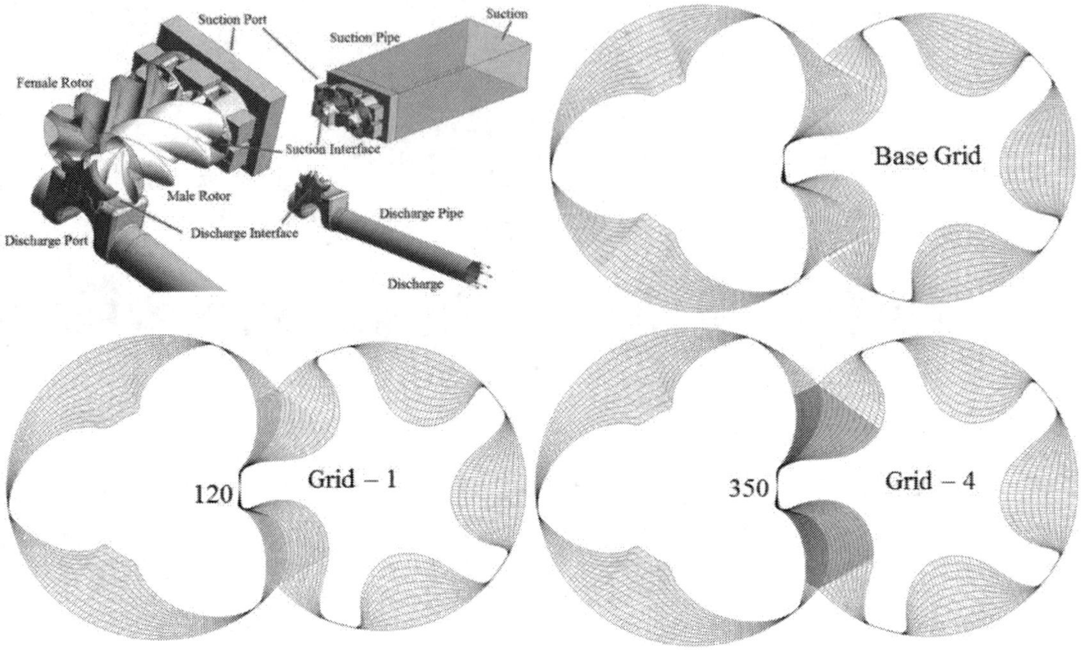

Figure 5. CFD Model Domain and Grid refinements in the interlobe clearance

Table 1. Grid levels and number of divisions.

Level	Circumference	Radial	Angular	Interlobe	Outer Circle	Mesh Size in Rotors
Base Grid	70	10	50		260	913682
Grid – 1	72	10	50	120	240	942480
Grid – 2	80	10	50	150	250	1049818
Grid – 3	90	10	50	200	250	1178100
Grid – 4	120	10	50	350	250	1570800

During operation, the rotors are subjected to thermal deformation which changes the clearance gap size. CFD models used in the study do not take into consideration such changes in clearances. Therefore, it is estimated that the clearances will reduce with the increase in temperature and the grids for CFD are generated with reduced uniform clearances of 60 micro meters in the interlobe and radial gaps in order to compensate for thermal distortions [5]. The axial end clearances are not included in the CFD model. Figure 5 shows the CFD model domain, boundaries and the sample grid in the cross section of the rotor domain. The working fluid is air with molar mass of 28.96 kg/kmol, Specific Heat Capacity $1.0044e^{03}$ J/kg K, Dynamic Viscosity $1.831e^{-05}$ kg/m s and Thermal Conductivity $2.61e^{-02}$ W/m K. A uniform pressure of 1.0bar was specified at the suction while the discharge pressure is 2.0bar. The main rotor

IOP Conf. Series: Materials Science and Engineering **90** (2015) 012008 doi:10.1088/1757-899X/90/1/012008

speed was 6000rpm and 8000rpm. High resolution scheme was used in the ANSYS CFX solver and cycle average mass balance between the rotor suction and discharge was achieved within 1% in all cases. Four levels of interlobe refinement have been generated with the new grid such that the first level has a refinement comparable to that of the base grid. The refinement is only on the profile in the interlobe space and on the casing side the number of divisions has been maintained equivalent to that of the base grid. Table 1 shows the different grid levels used for calculations and cross section mesh is shown in Figure 5. The grid in port domains was tetrahedral with 300736 nodes and constant in all the cases. On an Intel i7, 3.0GHz processor with 16Gb RAM each calculation took about 36hrs computational time for 5 cyclic repetitions of flow conditions. The experimental layout is shown in Figure 6.

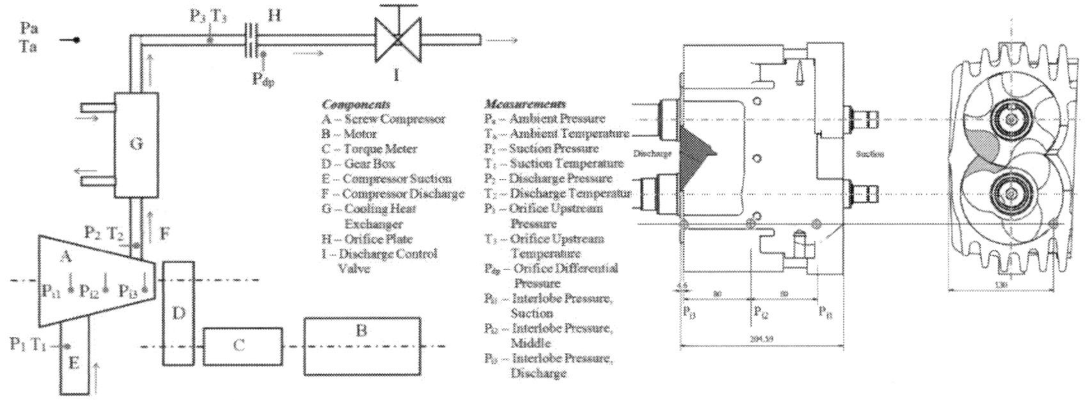

Figure 6. Layout of the test rig and location of the interlobe pressure transducers

3.1. *Results and Discussion*

The results from the analysis have been presented as a comparison of the internal pressure rise, indicated power (P_i), flow rate, volumetric efficiency (η_{vol}) and specific indicated power (P_{sp}) between the Base grid case and the various refinement levels of the new grid. Figure 7 is a plot of the pressure contour on the rotor surface with the two different grids. The surface of the rotor with Base grid is much regular as compared with the new grid but the pressure contour shows similar variation in both the cases.

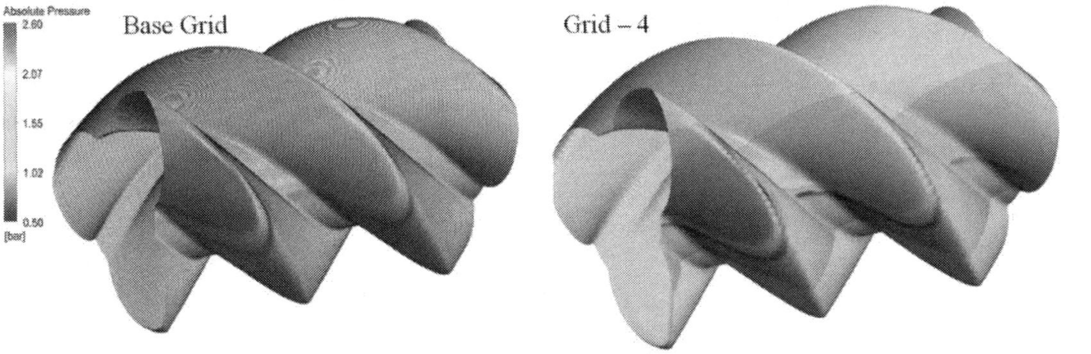

Figure 7. Pressure contours on the rotor surface with Base grid and Grid – 4, 8000rpm

3.2. *Pressure – Angle variation*

IOP Conf. Series: Materials Science and Engineering **90** (2015) 012008 doi:10.1088/1757-899X/90/1/012008

The interlobe pressure were measured using three pressure transducers mounted on the casing as shown in Figure 6. Pressure data from CFD results were obtained at the same coordinates as a cyclic average. Figure 8 shows the variation of pressure in the compression chamber at 6000rpm and 8000rpm. The plots from Base grid and new grids have been compared with experimental results. The internal pressure increase predicted by the base grid and new grids are very close to each other. Only a small increase in the compression rate is observed with the new Grid – 4. The pressure curves obtained by CFD are close to the experimentally obtained pressures, but the peak pressures from CFD models are slightly lower than experimental peak pressures at both 8000rpm and 6000rpm. Since the difference between the CFD and experimental peak pressures increases with speed, it is concluded that the thermal effects have big influence on the leakage gaps and further account of this phenomena will be necessary for better performance predictions. At the discharge pressure of 2.0bar and the current size of the discharge port, overcompression is observed with pulsations in the port which are accurately predicted by CFD.

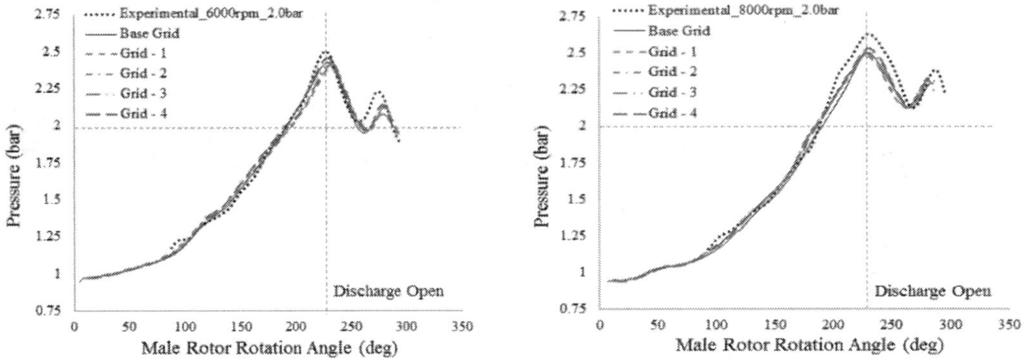

Figure 8. Pressure variation in the chamber with Male rotor rotation angle at 6000 and 8000rpm

3.3. Indicated Power.

Figure 9 shows the variation of the normalised indicates power (P_i) with speed for series of grid refinements. Predicted values and the experimental results of P_i have been normalized with respect to the indicated power obtained by the experiment. In order to extract the indicated power from measured power, it was assumed that the mechanical efficiency η_{mech} of the gearbox and bearings was constant at 95%. In reality it changes with speed and pressure ratio. The difference in P_i between CFD predictions on the base grid and the experiment is about –2.7% at 6000rpm and about –6.6% at 8000rpm. By using the new grid generation method, P_i is increasing at both speeds and all grid refinements getting closer to the measured values. The difference between CFD and experiment is about +1.4% at 6000rpm and about –2.8% at 8000 rpm with the finest Grid-4.

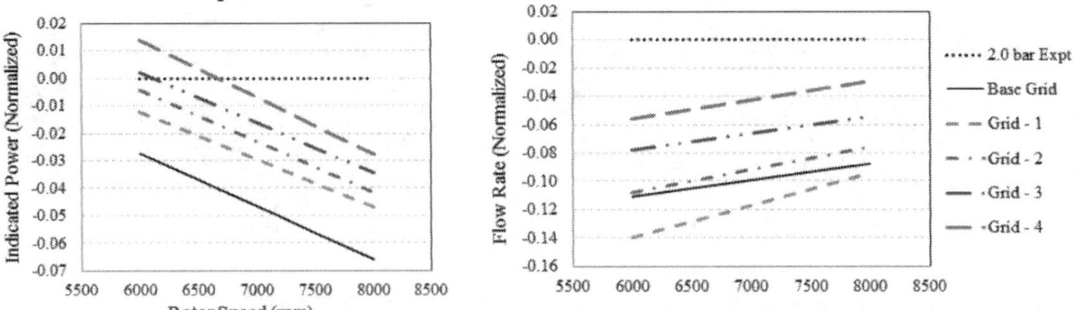

Figure 9. Comparison of Indicated power and Flow rate

3.4. *Flow rate*

Figure 9 shows the variation of the flow rate with speed with grid refinements. The predicted and measured flow rates are normalized with respect to the experimental results. The interlobe grid refinement has big influence on the flow rate. With the Base grid the difference between CFD prediction and experiment was about –11% at 6000rpm and –8.7% at 8000rpm. But this difference has strongly reduced with the Grid – 4 to –5.5% at 6000rpm and –2.9% at 8000rpm. The increase in flow rate also compares with the increase in P_i from Grid – 1 to Grid – 4.

The difference in flow rate prediction has reduced with interlobe refinement but it is still present. This is because the clearance gap of 60 micro meters has been assumed in the CFD model as the operating gap size. Due to thermal deformation and other operating uncertainties, the gap size is expected to vary. Few more iterations with reduced gap size could help in achieving further closer predictions from the model.

3.5. *Volumetric efficiency*

Figure 10 shows the variation of volumetric efficiency (η_{vol}) with speed for different grid refinements. The η_{vol} with Base grid is about 7% lower than the experiment. With new grid the difference is higher with Grid – 1 but reduces with further interlobe refinement. With Grid – 4 the difference in η_{vol} is just about 3% from the experiment. In line with the flow prediction, the difference is smaller at 8000rm as compared to that at 6000rpm.

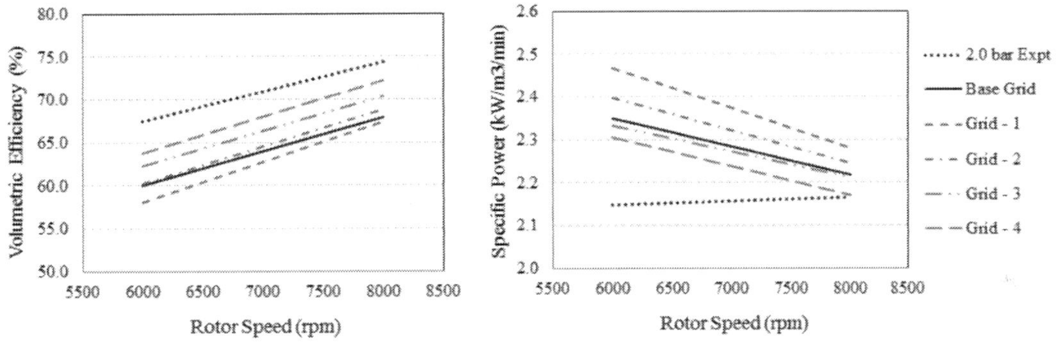

Figure 10. Comparison of Volumetric efficiency and Specific indicated power

3.6. *Specific Indicated Power.*

Figure 10 also shows the variation of specific indicated power (P_{sp}) with speed for different grid refinements. The P_{sp} is influenced by both the calculation of indicated power and the mass flow rate. With the Base grid specific indicated power is higher than experiment at both the speeds. The same trend is seen with the new grids because of the slight increase in indicated power as seen in Figure 9. From the level of Grid – 1 to Grid – 4 however this difference goes on reducing. At 6000rpm the difference between calculated and experimental P_{sp} is about 0.2 kW/m^3/min with Base grid and it has reduced to 0.15 kW/m^3/min with Grid – 4. At 8000rpm, the specific indicated power matches with the experiment with Grid – 4.

4. Conclusion

An algebraic grid generation procedure has been proposed for distribution of the boundary nodes in the grid generation of twin screw machines for CFD calculation. This distribution utilised the reference background blocking structure to identify nodes on the rotor profile based upon the nodes of the outer boundary. It allows the grid in the interlobe space to be refined and capture the rotor profile curvature more accurately. The test case was presented which compares the CFD results with series of refined grids with experimental results.

- The grids generated by this procedure contain only quadrilateral cells with improved orthogonality. The interface between the rotors is non-conformal with much improved alignment which improves with the interlobe grid refinement. The method of 'outer-to inner' boundary grid generation causes the cells located in the tip and the root radii of the rotors to have larger warp angle compared to the traditional rotor to rack generated mesh.
- The interlobe grid refinement significantly improves the curvature of the rotor profiles, quality of the mesh and consequently the prediction of the mass flow rate and leakage flows.
- The interlobe grid refinement showed insignificant change in pressure history capture showing that in both cases the internal pressure curve was close to the experimental results.
- Indicated power increased by a small amount from Grid – 1 to Grid – 4 and was close to experimental value at both the speeds.
- A difference between CFD predictions and experimental results still exists. This can be attributed to the consideration of right operational clearance gaps in the CFD model.

In order to obtain higher accuracy in CFD predictions, the change in leakage gap sizes due to the thermal deformation needs to be accounted for in the CFD models. This requires employment of fluid-solid interaction modelling or using experimental data. The new grid generation approach however provides a flexibility to improve the resolution of the rotor geometry and thereby reduce inaccuracy in leakage flow calculations significantly.

References

[1] Kovačević A., 2005. Boundary Adaptation in Grid Generation for CFD Analysis of Screw Compressors, *Int. J. Numer. Methods Eng.,* Vol. **64**: 401-426.

[2] Kovačević A., Stošić N. and Smith I. K., 2007. *Screw compressors - Three dimensional computational fluid dynamics and solid fluid interaction*, ISBN 3-540-36302-5, Springer-Verlag Berlin Heidelberg New York.

[3] Kovačević A. and Rane S., 2013. 3D CFD analysis of a twin screw expander, *8th International conference on compressors and their systems*, London, p. 417.

[4] Rane S., Kovačević A., Stošić N. and Kethidi M., 2014. Deforming grid generation and CFD analysis of variable geometry screw compressors, Computers and Fluids, 99, p. 124–141.

[5] Kovačević A., Rane S., Stošić N., Jiang Y., Furmanczyk M. and Lowry S., 2014. Influence of approaches in CFD Solvers on Performance Prediction in Screw Compressors, *Proc. Int. Compressor Conf. at Purdue*, Paper 1124.

[6] Sauls J. and Branch S., 2013. Use of CFD to develop improved one-dimensional thermodynamic analysis of refrigerant screw compressors. 8th Int conf on compressors and their systems, p. 591.

[7] Pascu M., Kovačević A. and Udo N., 2012. Performance Optimization of Screw Compressors Based on Numerical Investigation of the Flow Behaviour in the Discharge Chamber. Proc. Int. Compressor Conf. at Purdue. Paper 1145.

[8] Nouri J. M., Guerrato D., Stošić N., Kovačević A., and Arcoumanis, C., 2006, Cycle-resolved velocity measurements within a screw compressor. Proc. Int. Compressor Conf. at Purdue, Paper 1810.

[9] Kethidi M., Kovačević A., Stošić N., Smith I. K., 2011. Evaluation of various turbulence models in predicting screw compressor flow processes by CFD. 7th Int conf on compressors and their systems, p. 347.

[10] Voorde Vande J., Vierendeels J., 2005. A grid manipulation algorithm for ALE calculations in screw compressors. *17th AIAA Computational Fluid Dynamics Conference*, Canada, AIAA 2005-4701.

[11] Hesse J., Spille-Kohoff A., Hauser J., Schulze-Beckinghausen P., 2014. Structured meshes and reliable CFD simulations: TwinMesh for positive displacement machines. International screw compressor conference, TU Dortmund.

[12] Stošić N., Smith I.K. and Kovačević A., 2005. *Screw Compressors: Mathematical Modeling and Performance Calculation*, Monograph, Springer Verlag, Berlin, June 2005, ISBN: 3-540-24275-9.

CFD simulation of a screw compressor including leakage flows and rotor heating

Dr. Andreas Spille-Kohoff, Jan Hesse, Ahmed El Shorbagy
CFX Berlin Software GmbH, Karl-Marx-Allee 90A, 10243 Berlin, Germany

E-mail: andreas.spille@cfx-berlin.de

Abstract. Computational Fluid Dynamics (CFD) simulations have promising potential to become an important part in the development process of positive displacement (PD) machines. CFD delivers deep insights into the flow and thermodynamic behaviour of PD machines. However, the numerical simulation of such machines is more complex compared to dynamic pumps like turbines or fans. The fluid transport in size-changing chambers with very small clearances between the rotors, and between rotors and casing, demands complex meshes that change with each time step. Additionally, the losses due to leakage flows and the heat transfer to the rotors need high-quality meshes so that automatic remeshing is almost impossible.

In this paper, setup steps and results for the simulation of a dry screw compressor are shown. The rotating parts are meshed with TwinMesh, a special hexahedral meshing program for gear pumps, gerotors, lobe pumps and screw compressors. In particular, these meshes include axial and radial clearances between housing and rotors, and beside the fluid volume the rotor solids are also meshed.

The CFD simulation accounts for gas flow with compressibility and turbulence effects, heat transfer between gas and rotors, and leakage flows through the clearances. We show time-resolved results for torques, forces, interlobe pressure, mass flow, and heat flow between gas and rotors, as well as time- and space-resolved results for pressure, velocity, temperature etc. for different discharge ports and working points of the screw compressor. These results are also used as thermal loads for deformation simulations of the rotors.

1. Introduction

Nowadays, screw machines are essential devices in many industrial plants and processes. As compressors they provide pressurized process gas or near-vacuum conditions, as expanders they produce energy from high-pressure air or water vapour. Dry screw machines only work with a gaseous fluid, e.g. air, and need tight gaps between rotating parts and the casing to prevent bypass flow; the rotors are typically synchronized by gears. For higher pressure ratios between pressure and suction side, oil or water are used as sealing and lubricating fluids.

From a fluid dynamics point of view, screw machines are quite complex. The geometry of the lobes and the casing is quite complex and due to rotation the shape of the fluid regions is deforming each time step. Compressibility and therefore acoustic aspects play an important role, and oil or water in-

Content from this work may be used under the terms of the Creative Commons Attribution 3.0 licence. Any further distribution of this work must maintain attribution to the author(s) and the title of the work, journal citation and DOI.
Published under licence by IOP Publishing Ltd

IOP Conf. Series: Materials Science and Engineering **90** (2015) 012009 doi:10.1088/1757-899X/90/1/012009

jected machines require consideration of multi-phase effects. Though Computational Fluid Dynamics (CFD) is a broadly accepted tool, both in academia as well as in industry, to analyse fluid mechanics in complex systems in order to understand such systems, it is so far rarely used in industry for the examination and improvement of screw machines.

The numerical computation of screw machines requires accurate grid generation because the fluid flow is transient and depends on the rotor position. Furthermore, the grid needs a high resolution especially in the gaps and their vicinity. CFX Berlin has put its focus on the computation of positive displacement (PD) machines many years ago. Consequently, TwinMesh was developed by CFX Berlin with the aim to generate high quality grids in a short time. As a result, CFD can now be utilized on a day-to-day basis in industrial applications of PD machines.

The essential advantage of a numerical simulation is the visualization of the fluid flow inside the machine. Engineers commonly optimize PD machines based on experience, without knowing the detailed flow effects in particular. CFD results can give additional information and insights in local flow effects. This results in a better process understanding and thus, new ideas for product optimization.

In the following, the second chapter, a brief summary about available approaches to model PD machines will be given. In detail, three different methods will be presented and advantages and disadvantages will be discussed. The approaches realised within TwinMesh are shown in the third chapter. The fourth chapter shows workflow and setup for the simulation of a dry screw compressor. In the fifth chapter, simulation results are shown. Finally, a brief summary and outlook is given in the last chapter.

2. Methods for the Chamber Modelling

The fundamental challenge in PD machine modelling is the reproduction of the fluid domain depending on the rotor position. All physical processes involved depend on quite different length scales, starting at a few micro meters in the sealing gaps up to a couple of hundred millimetres in the working chamber. Thereby, local small scale effects can affect the macroscopic operating performance significantly. In such areas local discretisation in the fluid domain can be achieved utilising the following methods.

2.1. Immersed Solid Method

The immersed solid method is a comfortable method to investigate the rotor influence on the fluid flow inside a PD machine. Therefore, the fluid domain of the PD machine is meshed regardless the rotor geometry, i.e. the whole fluid domain is filled up with grid elements whereas the solid part of the rotor geometry is meshed independently. Commercial CFD codes, such as ANSYS CFX, then calculate the fluid flow inside the PD machine. During the simulation the region of overlap is determined each time step and momentum sources are applied to force the fluid there to follow the motion of the rotor. The advantage of this method is that the initial grids can be used for the whole simulation as remeshing or mesh deformation are avoided.

However, one of the main disadvantages is that this method only works for incompressible, single-phase flows and therefore, compressibility effects have to be neglected. Furthermore, the influence of the rotor surface on the flow is modelled insufficiently. For turbulent flows the results especially in the near wall region are rather loose. Very high numbers of elements are necessary to accurately resolve the near wall region. Since the gap position varies with time the grid also has to be refined circumferentially. CFX Berlin has investigated this method intensively in cooperation with the Technical University Berlin [1].

2.2. Remeshing

The remeshing method utilizes an algorithm that generates a new mesh during run-time depending on the rotor position and local mesh quality. At each time step the grid is deformed due to the new rotor positions, and a routine automatically decides whether this grid needs to be remeshed or not depending on the mesh quality. Thus, the user interaction is rather small. Compared to the immersed solid

IOP Conf. Series: Materials Science and Engineering **90** (2015) 012009 doi:10.1088/1757-899X/90/1/012009

method, the rotor surface is accounted for in a way that the near wall region can be treated more accurately by turbulence models. Also, multi-phase flows can be calculated.

However, the main disadvantage is that remeshing in small gaps is almost unavoidable for almost every angle increment. On the one hand, the time step size needs to be small in order to compute the mesh deformation accurately. On the other hand, every remeshing needs interpolated results from the previous time step, which can result in interpolation errors. Furthermore, the remeshing method must use an automatic grid generation program. Such programs usually use unstructured mesh elements that result in high elements numbers within the gaps or in poor mesh quality within the gaps.

2.3. Customised Grid Generation

In order to avoid the aforementioned disadvantages the grids must be customized by the engineer her/himself. Each new rotor position results in a change of the fluid domain and therefore, every position needs to be meshed individually. One popular software tool for this is ANSYS ICEM CFD Hexa which is capable of generating structured grids consisting only of hexahedral elements. Additionally, an O-grid type mesh can be utilized to mesh the rotor geometry, which is essential in accurately resolving flows in the gaps. If the topology, i.e. node number and element links, between the rotor positions remains the same, interpolation between the meshes is not necessary so that interpolation errors such as for the remeshing method are avoided.

The customized grid generation method is of high numerical quality and has been developed and utilized by CFX Berlin in cooperation with industrial partners [2, 3]. However, the time-consuming grid generation rules out that this method can be used efficiently for industrial purposes. Hence, TwinMesh has been developed by CFX Berlin to ensure high quality mesh generation at moderate operation times and user interaction.

3. Basics of TwinMesh

TwinMesh is a meshing software for PD machines with two axially parallel rotors, with complex rotor geometry, i.e. continuous (e.g. lobe pump) or discontinuous (e.g. screw compressor). Via IGES-format and alternatively point wise CSV-format, CAD-data of the rotor and casing can be imported that represents the fluid domain and is used for the structured grid generation. The topology usually contains two rotors, which are meshed using O-type grids and that are connected via an interface contour. Figure 1 shows the TwinMesh GUI in the case of a screw compressor.

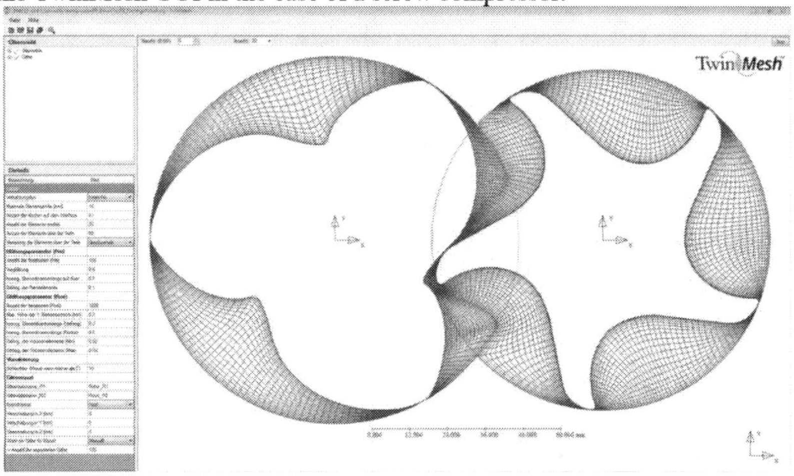

Figure 1: TwinMesh GUI for the mesh generation of a screw compressor.

After the import of the rotor and casing contours, TwinMesh generates 2D meshes for rotation angles of the rotors: For each 2D mesh, TwinMesh creates an interface line between both rotors and initial O-type grids around each rotor. The initial meshes are smoothed with an explicit and iterative method with several target criteria: Equal node distance and normal mesh lines at boundaries, and small aspect ratios and equal face angles in the rotor region. There are several meshing strategies concerning where the boundary nodes are fixed: On the rotor surface (leading to sliding nodes on the casing and the interface), on the casing (leading to sliding nodes on the rotor surface, but allowing 1:1 interfaces between rotors and to the stator meshes), or a mixed type. The number of elements is directly dependent on the node distribution along the geometry contour and is defined by the user. Figure 2 shows the high quality of the mesh lines in the male and female chambers.

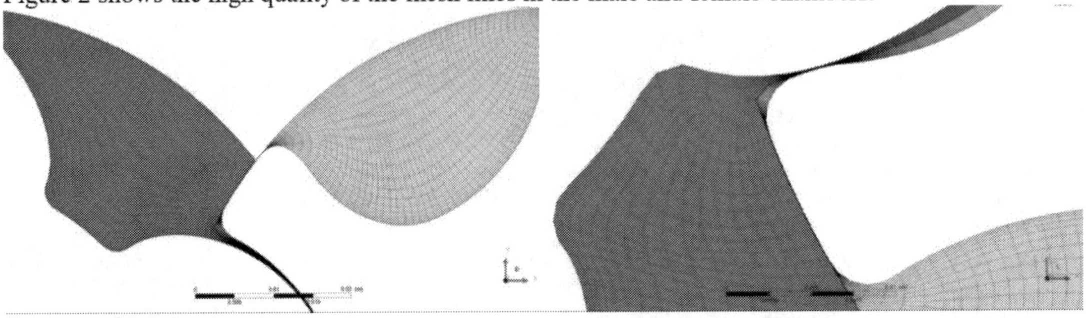

Figure 2: Mesh lines in a 2D cross section showing male (red) and female (green) chambers with 20 mesh lines in radial direction, refined boundary resolution, orthogonal mesh lines at boundaries and homogeneous resolution in gaps.

TwinMesh visualizes the mesh quality (minimum angle, determinant, aspect ratio, volume change) statistically for all meshes and as a coloured plot for a specified rotation angle. Once all 2D grids are generated, series of those grids are used to build appropriate 3D grids. Therefore, it is ensured before the simulation that all grids that will be used in the CFD solver have a high quality. In general, properties of a high quality mesh for PD machine simulations should fulfil certain criteria:

- Mesh should reflect the fluid domain as closely as possible.
- Interior angle > 18° and element size evenly distributed.
- Boundary layers should be adequately resolved with orthonormal elements.
- Small changes in node position between two subsequent angle increments.
- Number of meshes per revolution should be sufficient; preferably about 1° angle increments.

Mesh topology and node number remains the same for all 3D meshes so that the CFD solver needs no interpolation but directly uses the conservation laws for deforming meshes.

4. Workflow and Simulation Setup

ANSYS CFX was chosen as the solver as it provides a wide range of numerical models for the simulation of complex flows. For PD machine simulations the following models are of particular interest:

- Mesh deformation: mesh motion is included in the differential equations so that interpolation is not necessary; a FORTRAN routine is used that reads the TwinMesh data and calculates the mesh motion.
- Multi-phase models: cavitation in pumps, oil or water injection in compressors
- Complex fluid properties: Non-Newtonian or high-viscosity fluids in pumps, ideal or real gas models, dry and wet steam properties for compressors
- Heat transfer models: viscous heating, compressibility effects, heat transfer to/from solids
- Turbulence modelling: Reynolds-Averaged Navier-Stokes models and scale-resolving models

For the application of TwinMesh and ANSYS CFX on lobe pumps and gear pumps, see [4]. Geometry, setup and results for this screw compressor are based on a master thesis of one of the authors [5].

The simulation of a screw compressor starts with the geometric description of the rotating and stationary components of the machine, i.e. of both rotors, the casing and the fluid regions in between. In this simulation the rotors are of SRM type A with 4 lobes for the male (300° wrap angle) and 6 lobes for the female rotor (200° wrap angle), see Figure 3. The rotors have a length of 168.3 mm, an axis distance of 80 mm and rotation speed of 12 333 rev/min of the male and 8222 rev/min of the female rotor. The gap between the rotors is 100 μm, between rotors and casing the gap is 50 μm.

Suction and discharge ports were added to the rotors, see Figure 4, with a suction port of 200 mm length and a final diameter of 50 mm and a discharge port of 300 mm length and a final diameter of 55 mm. Three different control edges at the discharge side were designed to get expected pressure ratios of 3, 4, and 4.6, respectively, see Figure 5.

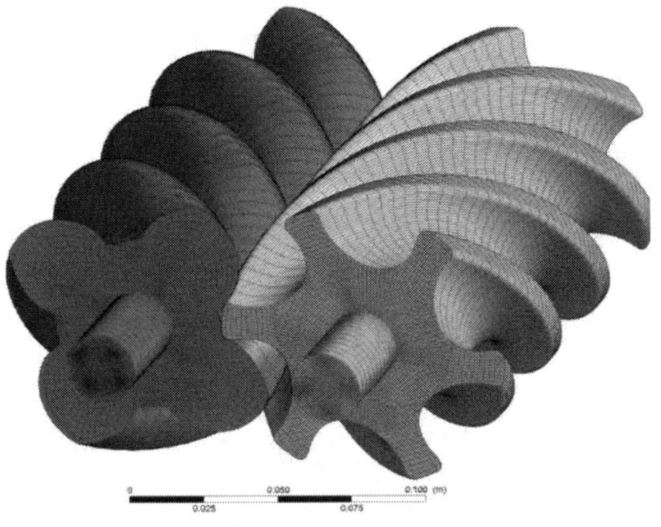

Figure 3: Geometry of male (left) and female (right) rotor with surface mesh.

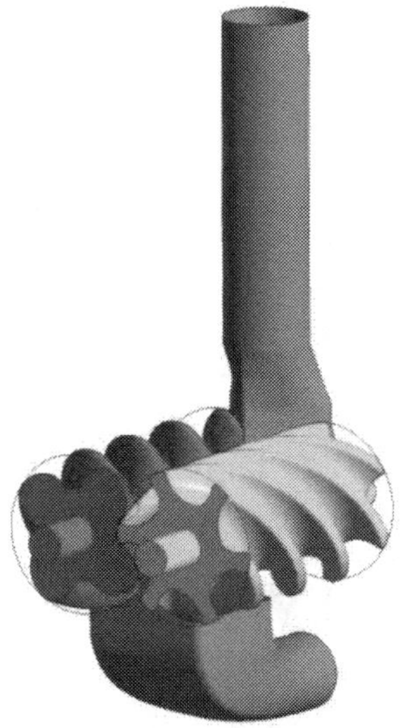

Figure 4: Geometry of suction (bottom) and discharge (top) port.

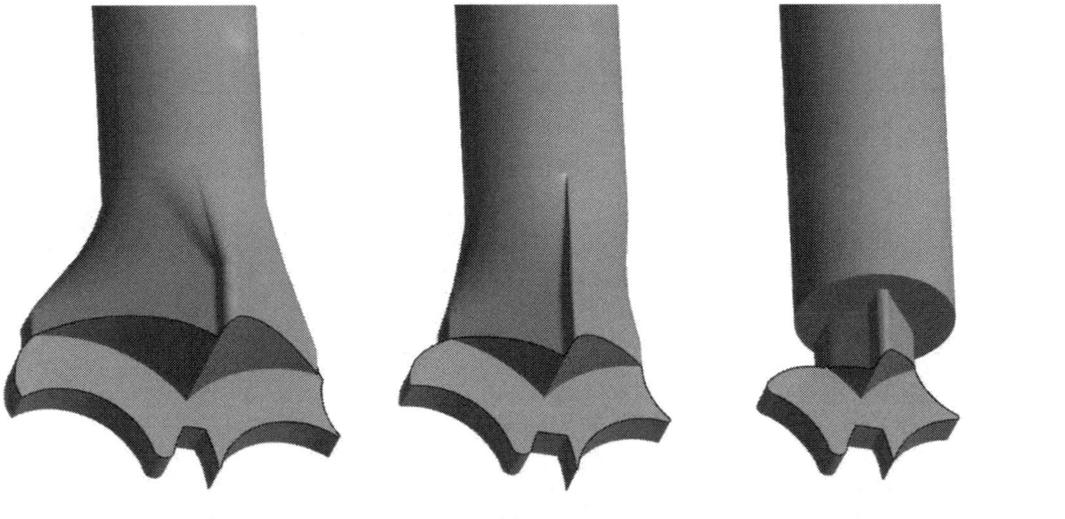

Figure 5: Three different control edges at the discharge side for expected pressure ratio 3 (left), 4 (middle) and 4.6 (right), i.e. for built-in volume ratios of 2.2, 2.7 and 3, respectively.

For the CFD simulation, the fluid region and, if a thermal simulation of the solids is desired, the solid regions have to be meshed. The resolution and the quality of the meshes have a great impact on both simulation time and result quality. With better and finer meshes, the results quality is improved, whereas coarse meshes may lead to a fast, but less meaningful simulation. For the fluid regions around the rotors, TwinMesh generates hexahedral meshes with fine resolution of gaps and smooth changes between small gaps (several microns) and larger chambers (several centimetres). In a grid sensitivity study [5], we found that a radial resolution of 20 elements, circumferential resolution of 300 to 400 elements, a spanwise resolution of 130 elements and a temporal resolution of 1° rotation angle ensure good results. With fewer elements in circumferential direction, the curvatures in the rotors cannot be resolved adequately; fewer elements in radial direction cause higher volume changes at interfaces leading to local errors.

The meshes for the stationary fluid regions, i.e. the supply pipes at suction and discharge sides, and the solid regions, are generated with common meshing tools as ANSYS ICEM CFD or ANSYS Meshing; here, structured hexahedral meshes are possible or unstructured hybrid meshes consisting of prisms for the adequate resolution of boundary layers and tetrahedrons for the inner parts. Hexahedrons have the advantages of a higher quality and less numerical effort. In contrast, the tetrahedrons and prisms are of more flexibility and less manual meshing effort.

In the setup shown in Figure 6, the meshes are combined and defined as fluid or solid regions:

- Meshes for rotating fluid regions around both rotors with axial and radial gaps from Twin-Mesh, rotated by reading mesh files via FORTRAN routine at run-time
- Mesh for stationary fluid region (pressure and suction side) from ANSYS Meshing
- Meshes for solid rotor regions from ANSYS Meshing, rotated by prescribed mesh deformation as rigid bodies

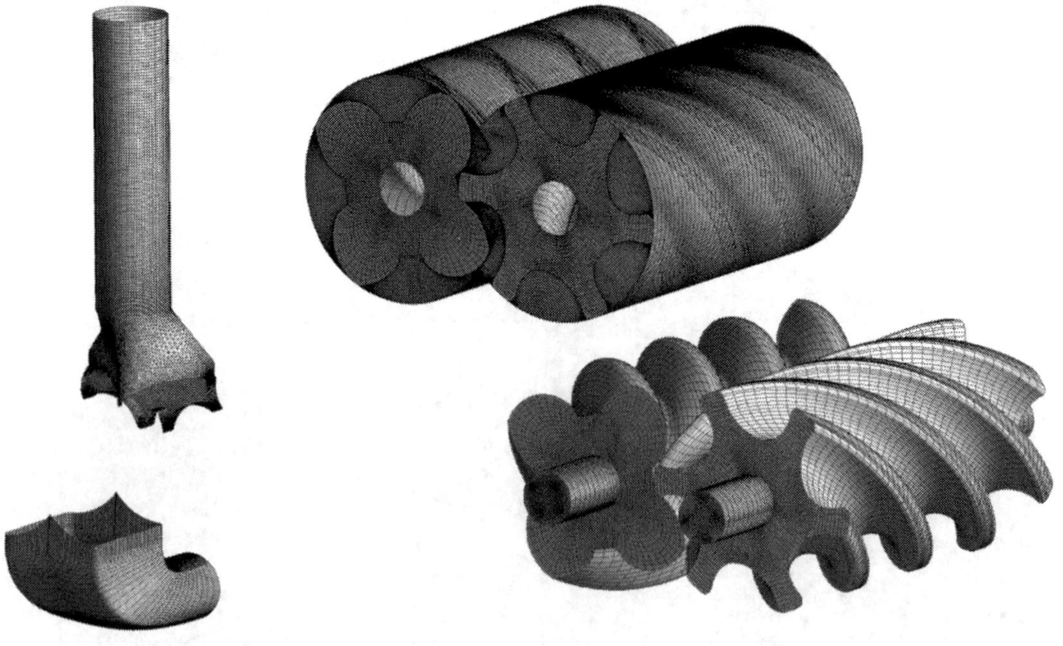

Figure 6: Stationary fluid domain (left) consisting of hexahedrons, tetrahedrons and prisms (approx. 1 mio nodes, 1.9 mio elements); both rotating fluid domains (top) consisting of hexahedrons (each approx. 550 000 nodes and elements); rotating solid domains (bottom) consisting of hexahedrons (each approx. 300 000 nodes and elements).

The different meshes are connected with interfaces so that flux of mass, momentum, turbulence and heat (for fluid-fluid interfaces) or heat (for fluid-solid interfaces) is conserved. Materials are assigned as air ideal gas for the fluid regions and steel as solid material. The user has to select appropriate physical and numerical models, e.g. whether viscous dissipation shall be taken into account or which turbulence model shall be used. Turbulence models are necessary to capture turbulent effects like increased friction and mixing without fully resolving turbulent structures spatially and temporally (direct numerical simulation). In these simulations, the shear-stress transport (SST) model is used, which is a Reynolds-Averaged Navier-Stokes (RANS) model so that turbulence is fully modelled. Hybrid models as Scale-Adaptive Simulation or Detached Eddy Simulation can resolve large turbulent structures while still modelling the smaller and near-wall scales. Furthermore, boundary conditions for all relevant numerical quantities are needed. For screw compressors, typically pressure and temperature at the inflow side (here 1 bar and 300 K) and pressure at the outflow side (here 3 bar) are prescribed; for the turbulent quantities, common conditions at the inlet are low or medium intensity.

In the simulation, time development is also discretized. With TwinMesh, rotor meshes for each time step have been generated and stored in this case for each 1° rotation angle of the male rotor. The time step size depends on this angle increment and the rotation speed and is 13.5 μs in this case. For each time step, the solver reads the corresponding rotor meshes and updates the connections between the parts and all quantities taking the mesh deformation and motion into account. Meshes were generated for a 90° rotation of the male rotor; then, simulation stops and a new simulation is initialized with the results of the previous one and run for the next 90 time steps, i.e. another 90° rotation. A typical initialization of the first simulation is zero velocity everywhere, and pressure and temperature at 1 bar and 300 K at the suction side and at 3 bar and 410 K (adiabatic compression) with linear interpolation at the rotors at the discharge side. Typically, several revolutions of the rotors are necessary to reach a periodic state.

The inlet and outlet boundaries are often positioned arbitrarily near the compressor to minimize computational effort. Prescribing the pressure at these artificial boundaries leads to numerical reflections of the pressure waves generated by the compressor, especially at the pressure side. The reflected waves and the outgoing waves superpose and form standing waves. So-called non-reflecting boundary conditions allow the waves to leave the computational domain. In ANSYS CFX, Navier-Stokes characteristic boundary conditions (NSCBC) can be used via Local One-Dimensional Inviscid (LODI) relations.

The result of the simulation includes temporally and spatially resolved fields of pressure, temperature, velocities, and turbulent quantities and derived quantities like density, shear strain rate, and eddy viscosity. Furthermore, integral quantities like moment on the rotors or force or heat flux on the walls are calculated within each time step. In order to consider heating of the solid rotors, they can be included in the simulation so that the heat flux from the compressed gas directly heats the solid material and vice versa. However, compared to the rotation of the rotors, a very slow process with several minutes in duration is needed to reach stationary temperature. Thus, several hundred thousand revolutions must be simulated. Alternatively, the simulation of the solid heating is decoupled from the fluid simulation by time-averaging the heat flows and imposing them on the solids for longer time scales of the order of minutes while using the solid temperatures in the fluid simulation as boundary conditions.

5. Simulation results

The aim of the screw compressor is to compress air from 1 bar to 3 bar; this can be illustrated by looking at the pressure in one chamber while the machine rotates, see Figure 7. In the beginning, the chamber is connected to the suction side at 1 bar and fills with air; after closing connection to the suction side, compression starts due to decreasing chamber volume until the control edge to the pressure side is reached so that the chamber is connected to the outlet. Depending on the geometry of the control edge (see Figure 5), it opens before the outlet pressure of 3 bar is reached (under-compression for the case with estimated pressure ratio 3 at approx. 2.7 bar) or at chamber pressures higher than the outlet pressure (over-compression for the cases with expected pressure ratio 4 and 4.6 at approx. 3.6

and 4.5 bar, respectively). In case of under-compression, gas initially flows from the outlet side into the chamber; in case of over-compression, gas immediately flows out of the chamber towards the outlet. Figure 8 shows pressure contours at a rotor angle of 26° (refers to 26°+2•90°=206° in Figure 7) for control edge designed for pressure ratio 4; for results with other pressure ratio, see [5]. The pressure peak has already left the compression chamber and moves (with speed of sound) towards the outlet. The strong pressure drop over the axial gap at the pressure side can be clearly seen here. Figure 9 shows pressure and temperature for this rotation angle in a cross section.

Figure 10 shows mass flow at inlet and outlet and torques on both rotors over the rotation angle. Mass flow at the inlet is almost constant at 0.133 kg/s, mass flow at the outlet fluctuates between 0.08 and 0.23 kg/s due to the periodic opening of the compression chamber at the control edge at 13° of the male rotor. Due to finite speed of sound, pressure and mass flow maximum reach the outlet later at approx. 70° of the male rotor. While the torque on the male rotor is quite high with an average of 18.2 Nm, torque on the female rotor is only 3.2 Nm in average. This leads to a power consumption of 23.3 kW on the male rotor and 2.8 kW on the female rotor; together 26.1 kW for the whole screw compressor.

Figure 11 shows leakage mass flows for the four male rotor gaps (sum of axial and radial part) and six female rotor gaps; gap mass flows up to 3% of total mass flow can be seen on male rotor and up to 2% on female rotor.

Figure 12 shows the temperature distribution on the rotor surface and thermal expansions of the rotor. The shaft ends are set to 70°C as a fixed boundary condition; the male rotor is heated up to 180°C and the female up to 160°C leading to a thermal expansion of over 300 µm axially and 90 µm (male) resp. 80 µm (female) in radial direction. However, this expansion of the rotors (and the casing) is not taken into account in this CFD simulation. The high temperatures in the rotors are due to high heat input of approx. 560 W at high pressure side while approx. 380 W leave the solids at low pressure side and heat the gas, and 180 W are leaving the solid through the shafts; with shorter shafts, this cooling effect would be larger.

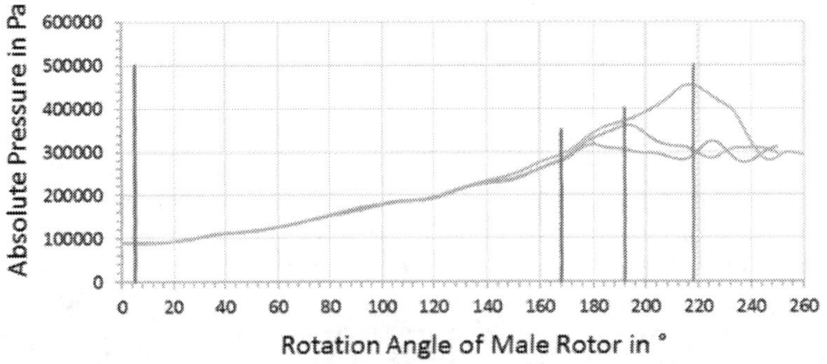

Figure 7: Chamber pressure over rotation angle of male rotor for all three control edges (see Figure 4); estimated pressure ratio 3 (blue line), 4 (orange line) and 4.6 (green line); vertical lines are for start of compression (red line, identical for all cases) and end of compression.

IOP Conf. Series: Materials Science and Engineering **90** (2015) 012009 doi:10.1088/1757-899X/90/1/012009

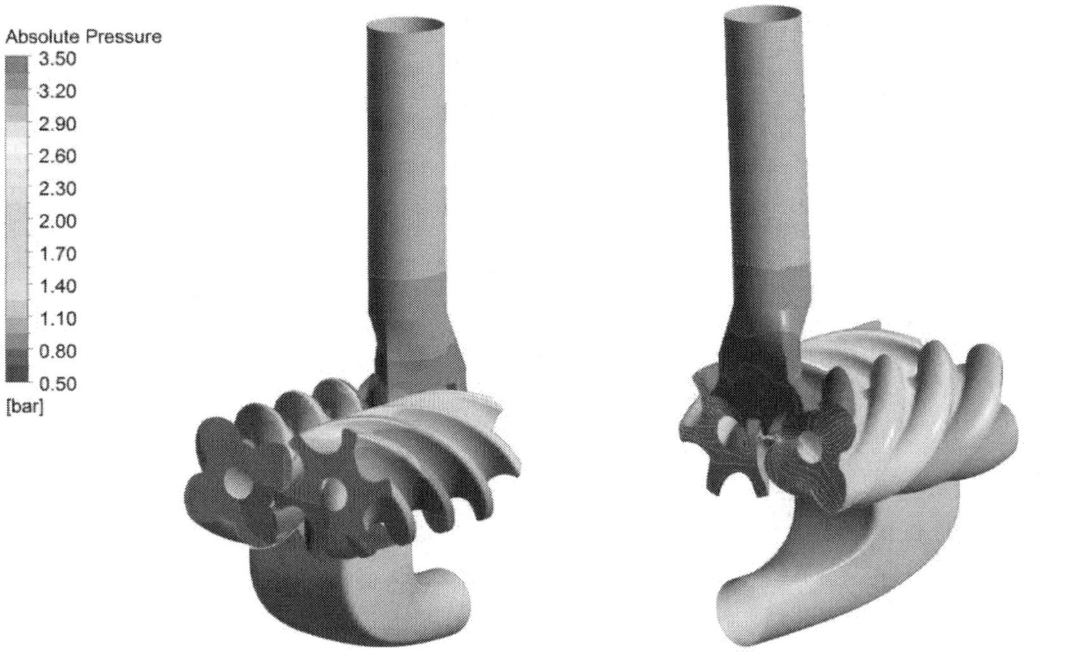

Figure 8: Pressure on casing wall and rotor surface at rotation angle 26°; control edge is reached at rotation angle 13°, control edge is for estimated pressure ratio 4.

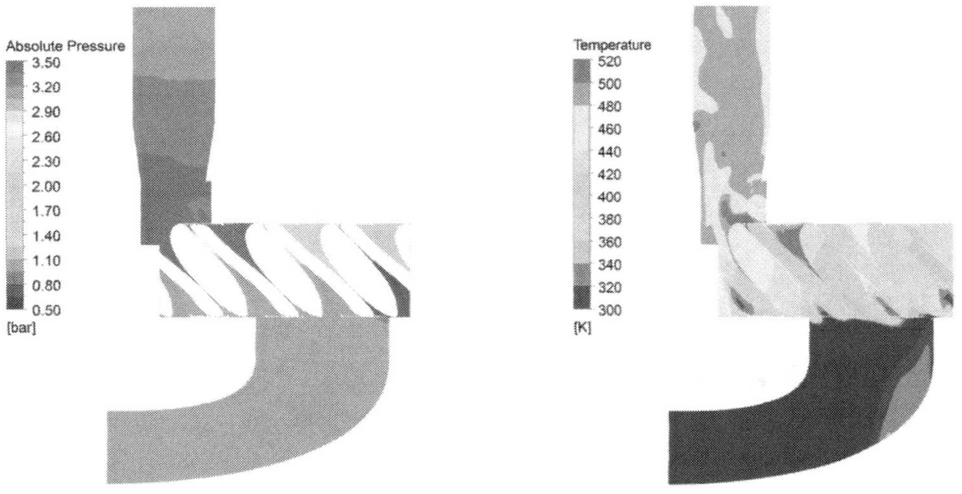

Figure 9: Pressure (left) and temperature (right) on mid cross section at rotation angle 26°.

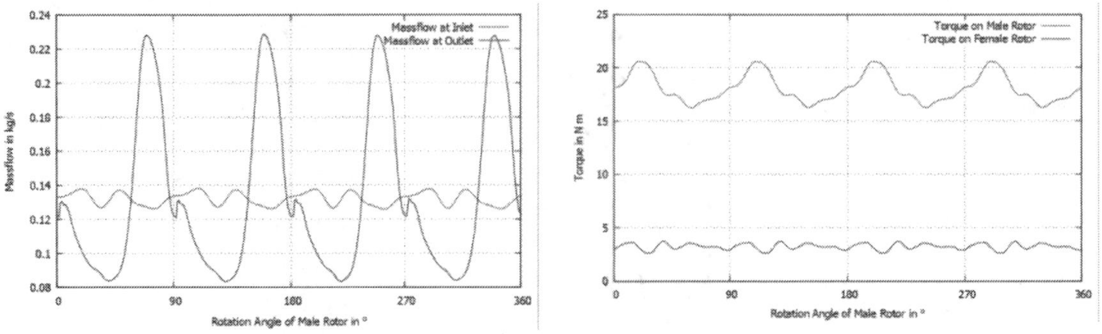

Figure 10: Mass flow (left) at inlet and outlet and torques (right) on male and female rotor over rotation angle of male rotor.

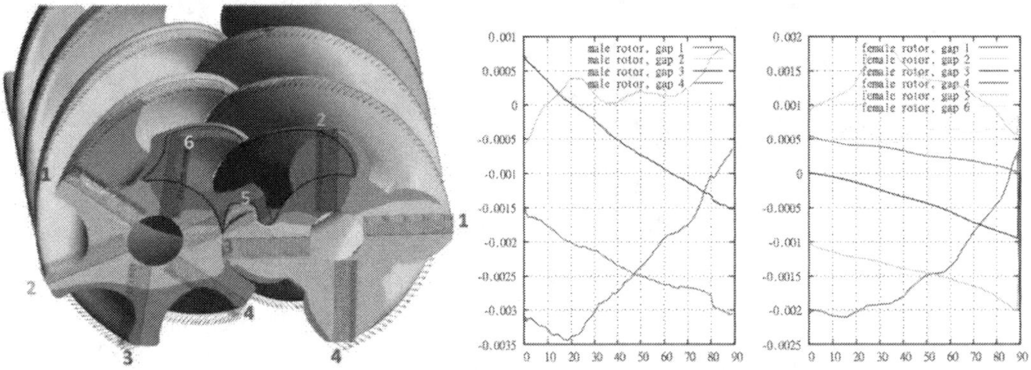

Figure 11: Position of axial and radial gap user surfaces at rotation angle 0° (left) with vectors showing rotation direction and pressure port as black line; mass flow through the gaps over rotation angle (right), positive means mass flow in rotation direction

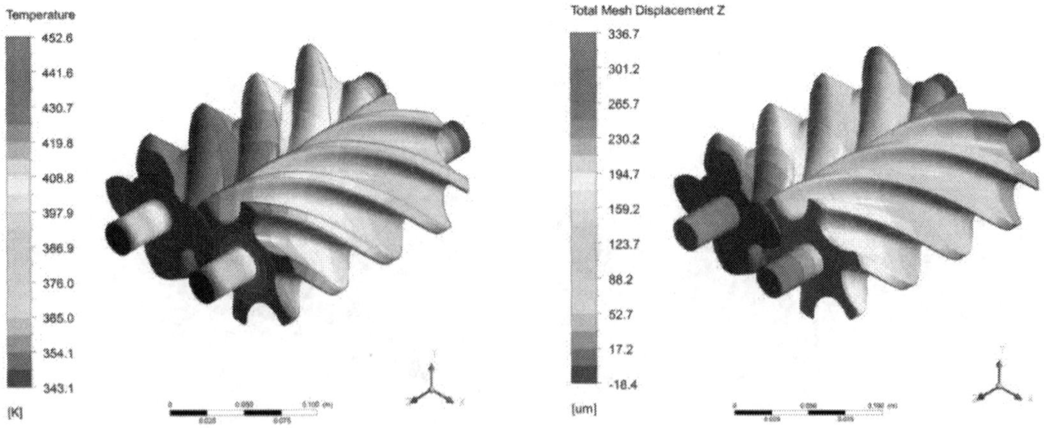

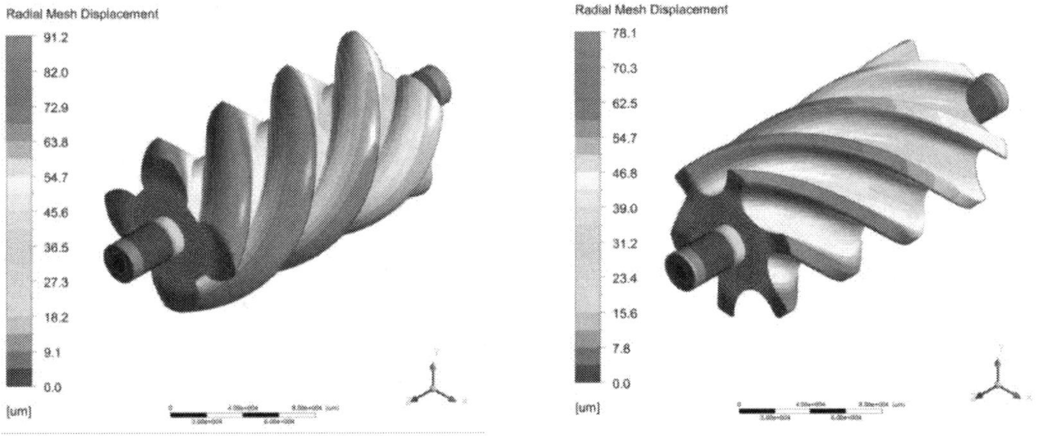

Figure 12: Results for the thermal expansion of the solid rotors; temperature on solid surface (top left) with shaft ends fixed to 70°C, axial expansion (top right) with right shaft end fixed, radial expansion of male (bottom left) and female (bottom right) rotor.

6. Summary and outlook

This paper shows simulation results for a dry screw compressor compressing air from 1 bar and 300 K to 3 bar and approx. 470 K. The simulations include axial and radial gaps and take the heating of the rotors into account. Meshes for the fluid regions around the rotors were generated with the special meshing program TwinMesh whereas those for stationary parts and the solid rotors were generated with ANSYS Meshing. TwinMesh creates hexahedral meshes for rotating PD machines with a high quality for each time step; these meshes are read by the CFD program during run-time. The transient simulations were performed with ANSYS CFX version 15.0.7 taking into account fluid flow, heat transfer including viscous dissipation and solids, and turbulence effects with the SST model.

We showed chamber pressure for three different control edges on the discharge side, one leading to under-compression, the other two leading to over-compression instead. For the case of moderate over-compression, pressure and temperature plots were shown for an arbitrary rotation angle. Here, the mass flow at the outlet showed a strong oscillation due to the periodic opening of the chambers. The main torque (on the male rotor) is approx. 18 Nm and the overall needed mechanical power is about 26 kW. Finally, thermal expansion results for both rotors in axial and radial direction were shown.

The usage of the thermal expansion results as a deformation of the rotors as well as the casing will be used in ongoing projects. Furthermore, additional work is needed to simulate oil or water injected screw compressors since multi-phase flows require very high mesh qualities for stable simulations.

References

[1] Schwotzer T 2009 *Simulation einer Drehkolbenpumpe mit der Immersed-Solid-Methode* Bachelor Thesis, Technical University Berlin, Berlin

[2] Ludwig A 2007 *Numerische Strömungsberechnung für Verdrängermaschinen mit Hilfe bewegter Berechnungsgitter* Master Thesis, Hochschule für Technik und Wirtschaft, Berlin

[3] Fuchs M 2010 *Numerische Simulation der instationären Strömung in einer Drehkolbenpumpe* Bachelor Thesis, Technical University Berlin, Berlin

[4] Hesse J, Spille-Kohoff A, Hauser J and Schulze-Beckinghausen P 2014 *Structured meshes and reliable CFD simulations: TwinMesh for positive displacement machines*, VDI-Berichte 2228 "International VDI Conference Screw Machines 2014"

[5] El Shorbagy A 2015 *Auslegung, Konstruktion und numerische Simulation eines trockenlaufenden Schraubenverdichters und Vergleich der Simulationsergebnisse mit den Entwurfsanforderungen* Master Thesis, Technical University Berlin, Berlin

IOP Conf. Series: Materials Science and Engineering **90** (2015) 012010 doi:10.1088/1757-899X/90/1/012010

Computation of Flow in Screw Compressors

Georgi Kalitzin[1], Xiaodan Cai, Ramons Reba, Gorazd Medic
United Technologies Research Center, 411 Silver Lane, East Hartford, CT 06118

E-mail: kalitzg@utrc.utc.com

Abstract. A CFD model enabling accurate and computationally affordable simulation of unsteady flow in screw compressors has been developed. This paper focuses on computational aspects, including real-gas CFD using hybrid structured/unstructured moving grids, and specifics of grid generation for moving rotors and their communication with the discharge plenum.

1. Introduction

A compressor is an essential part of the thermodynamic cycle in chiller systems. Rotary screw compressors are typically used in applications requiring more flow than is produced by reciprocating compressors but less than is produced by centrifugal compressors. They are produced in sizes that range from 10 to several 1000 cubic feet per minute. For chillers they are used for systems with the size ranging from 10 to about 500 tons of cooling capacity.

Screw compressors are positive displacement machines. The gas compression process is a continuous sweeping motion. Two or more helical screw rotors are intertwined in such a way that they create a closed volume between rotor lobes, casing and endwall. The rotation of the rotors decreases this volume continuously compressing the fluid. It is discharged through openings in the casing and endwall that are connected to a radial and axial port of the plenum, respectively. The ports open when a lobe passes one of these openings and releases flow into the discharge plenum. The flow discharge is very abrupt creating unsteady flow with large pressure pulsations. These create significant vibrations and noise issues.

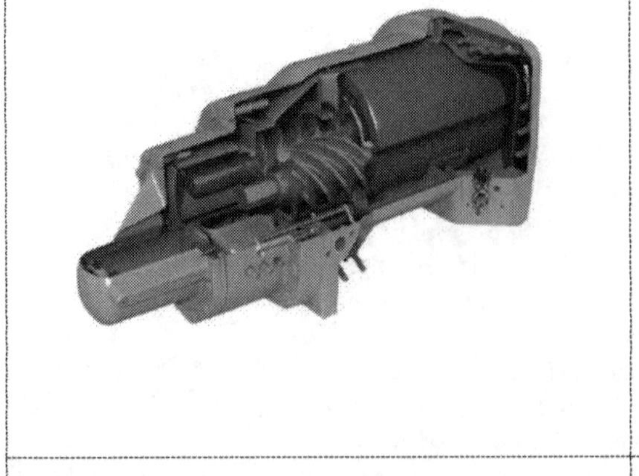

Figure 1. Typical industrial screw compressor.

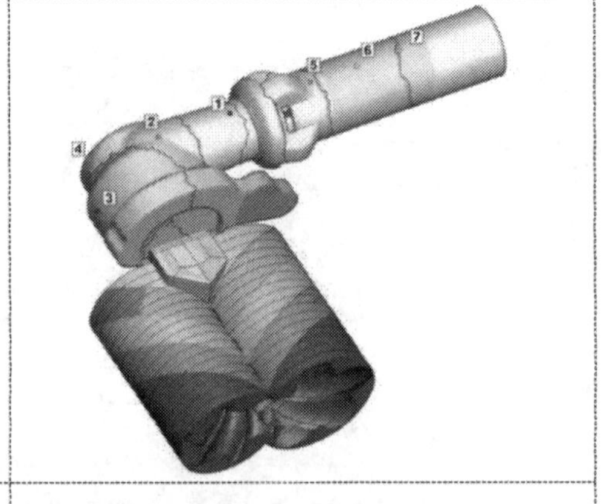

Figure 2. Computational model.

[1] To whom any correspondence should be addressed.

Content from this work may be used under the terms of the Creative Commons Attribution 3.0 licence. Any further distribution of this work must maintain attribution to the author(s) and the title of the work, journal citation and DOI.
Published under licence by IOP Publishing Ltd

Intensive research on gas pulsations in screw compressors have been carried out in both academia and industry. Fujiwara and Sakurai [1] measured gas pulsation, vibration, and noise in a screw compressor. Koai and Soedel [3] developed an acoustic model in which they analyzed flow pulsations in a twin screw compressor and related the pulsations to the compressor performance. Sangfors [4] and Huagen et al [2] developed mathematical models for the prediction of gas pulsations in screw compressor suction and discharge chambers. More recently, Kovačević and his coworkers [5], [6] developed a RANS-based three dimensional computational fluid dynamic model for the analysis and prediction of screw compressor performance.

In this paper, we present our three-dimensional CFD model for screw compressors used in chiller applications. We discuss the critical elements, which include rotor and plenum grid generation, non-matching interfaces, the real gas implementation and non-reflective boundary conditions for specifying average outflow pressure. The structured rotor grid generation method builds upon earlier methods such as the one presented in Kovacevic [10], while providing an alternative approach for controlling grid quality in the region between rotors.

The CFD model is validated for a typical screw compressor. Comparison of the computations with experimental data shows that the presented method accurately predicts the pressure spectra. The computational model provides insight into the three dimensional flow field and can be used to evaluate compressor design modifications for pulsation reduction.

2. Computational framework

The in-house CFD solver used in these simulations is based on a second-order-accurate (both in space and time), explicit-in-time numerical method for compressible flow equations [12], with the k-omega RANS turbulence model. For refrigerants, the solver has been extended to model real gas properties via look-up tables. Moving grids are handled using a standard methodology [13]. Both, structured and unstructured grids can be used. The structured solver provides higher speed and accuracy, while the unstructured solver provides higher flexibility in grid generation. Communication between the two solvers is realized with a non-matching interface as the grids are generally non-matching.

For real gas modeling, our in-house density-based compressible flow solver has been extended with NIST real gas property tables based on pressure and temperature. This approach avoids the construction of inverse tables for refrigerants. In order to analyze the screw compressor for a range of operating conditions, i.e. different discharge plenum pressures, a non-reflective outflow boundary condition that allows the specification of the mean pressure is required.

The investigated computational model consists of the two screw rotors, radial and axial port and plenum with check valve and exit pipe, shown in Figure 2. In this particular case, the radial and axial ports are gridded with a structured mesh and the plenum and check valve are gridded with an unstructured mesh. Inflow conditions of total pressure and temperature are imposed at the axial inlet of the screw rotors. A time averaged static pressure is imposed at the outlet of the pipe.

2.1 Rotor grids

The surface of the screw compressor rotors is of complex geometrical shape. However, the basic surface of each rotor is defined by very few parameters: the profile of one lobe, the axis location, the number of lobes, the casing radius and the skewing angle along rotation axis. Figure 3 shows the surfaces and an axial cut through the rotors of a typical screw compressor.

The difficulty in the rotor grid generation arises from the fact that the two rotors are located very close to each other. In fact, they are touching each other at a few points. In the present approach, a very small clearance is always assumed between the two rotors.

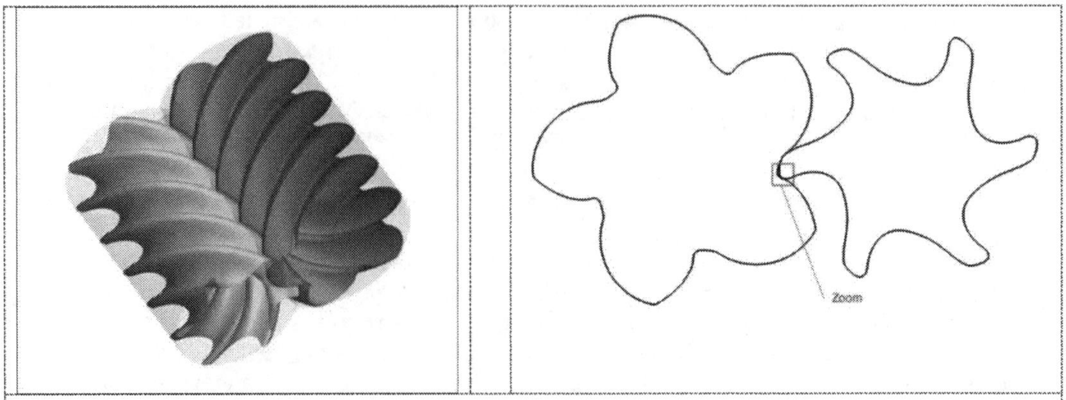

Figure 3. Screw compressor rotors (left) and axial cut through rotors (right).

The rotors are rotating and an unsteady computation requires grids for each time step. The difference of the coordinate location of each node between successive time steps determines the grid velocity. If the nodes move relative to each other the grid deforms. To reduce moving grid errors which are proportional to grid velocity and cell deformation, the grid sequence has to be smooth in time. A structured meshing approach seems to be most appropriate as it gives the best control upon the grid node distribution.

Although the rotor surface is three dimensional, the cross section cut through both rotors normal to the rotation axis in Figure 3 reveals a two dimensionality of the geometry. Moving this cut along the axis produces an apparent rotation of the two surface lines. The rotation repeats itself - the geometry contains a periodicity in the axial direction. Therefore, the general grid generation strategy is to create a sequence of two-dimensional structured O-grids around each rotor and assemble them along the rotation axis into a three-dimensional grid. In the computation, the rotation of the rotor grids is achieved by manipulating grid indices and moving grid planes along the rotation axis.

There are two ways to connect the two O-grids around the individual rotors to each other (see Figure 4). A one-to-one connection is desirable as it preserves the fluxes at the interface. Such a connection introduces a relationship between the grid points on the surface of the two rotors constraining the grid generation algorithm. The circumferential length of the two rotors differs and the grid nodes on the rotors need to move to maintain the one-to-one connection between them. This leads to very deformed grids in the gaps, potentially eliminating the advantage of flux conservation at the interface that the one-to-one approach provides. A higher grid count circumferentially and normal to the rotor surface smoothes the grid movement but increases the computational cost.

On the other hand, the grid generation simplifies if the two rotor grids are connected with a non-matching interface and the fluxes are interpolated. For this approach, the O-grids slide relative to each other at the interface and the deformation of the individual rotor grids is significantly reduced. In addition, a lower grid resolution is possible with this approach, although the high lobe curvature may lead to a local grid overlapping.

The grid generation process for the rotors is then as follows: a sequence of 2D Euler grids for the standalone rotors is created. Then an interface line is defined in the region where two rotors interact; this line is designed so that it changes smoothly in time. The standalone grids are then corrected in the interacting region by using interpolation along the wall normal gridlines. Finally, these 2D grids are assembled into 3D grids.

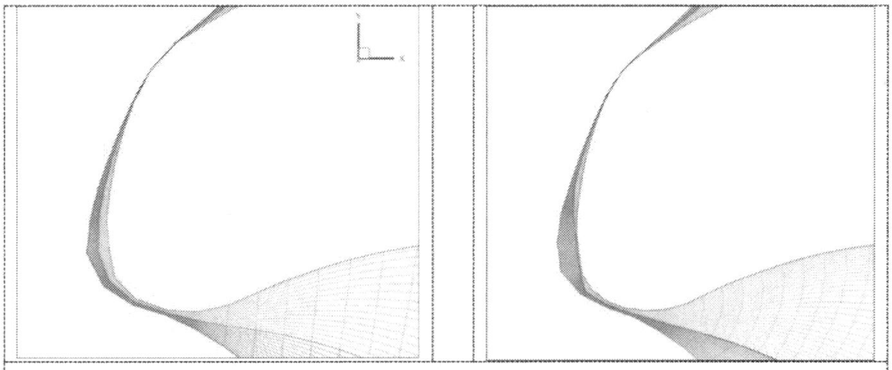

Figure 4. Grid between the rotors in the region highlighted in Figure 3 – for matching (left) and non-matching (right) interface.

2.1.1 2D grids for individual rotors

The 2D grid is generated from a grid for a single lobe defined by the area between the casing and the profile as shown in Figure 5, left. This area is gridded with 73x17 grid points where the points on the casing are spaced equidistantly and the points on the profile are spaced using a two sided exponential function. The first and last few points are aligned radially. Note that 73 points of circumferential resolution for one lobe is a good compromise between computational cost and geometry resolution, as can be seen in Figure 4. This one-lobe grid is then copied circumferentially. With 5 lobes for the male rotor there are 360 cells circumferentially. The grid rotates each time step 1 degree. A higher time resolution can be achieved by creating intermediate grids. In our computations we went up to 8 substeps for one degree of rotation.

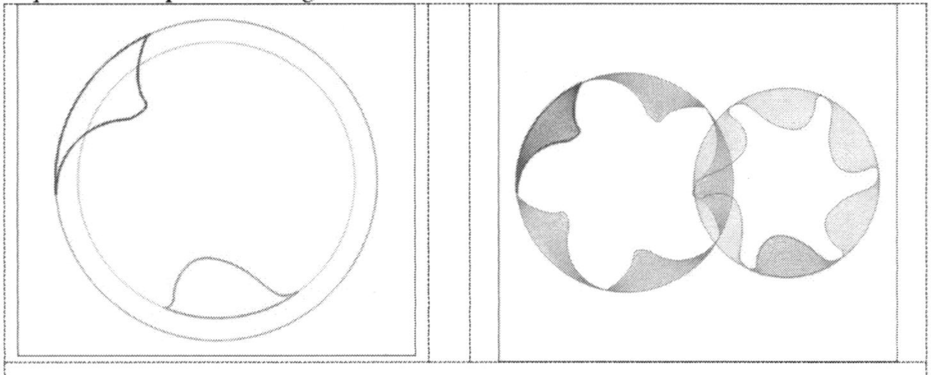

Figure 5. Lobe profiles and individual rotor grids.

2.1.2 2D grids for interacting rotors

Next step is to define a line that will separate the two interacting grids. It starts and ends at the stationary points where the rotor casings overlap (Figure 6). This interface line should not cross the rotor surfaces and should change smoothly when the rotors rotate.

The interface line is found through the construction of a high resolution one-to-one connected auxiliary domain. An example is shown in Figure 6. Part of the domain is bounded by the rotor surfaces on which von Neuman type boundary conditions are applied. An elliptic solver is used that moves grid points on the rotor surface creating a smooth Euler type of grid.

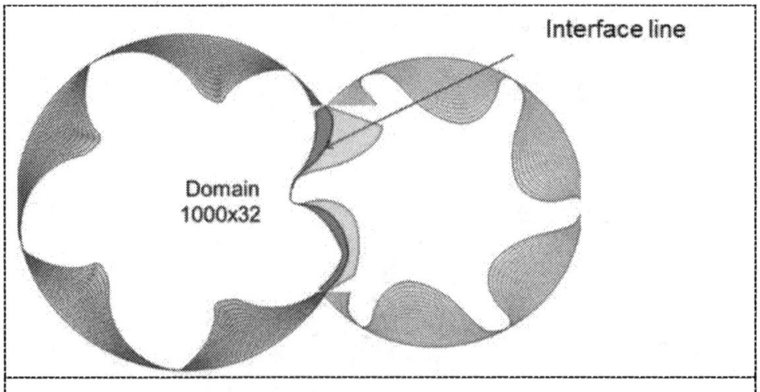

Figure 6. Auxiliary domain for overlap region.

An example of an resulting grid is shown on the left in Figure 7. Note, that for some positions of the rotors a large mismatch of 5 to 1 cells is observed. This is the result of the elliptic solver placing the interface line in the middle of the domain. An improved interface line can be obtained by constraining it with the rotor casing radius. This is shown on the right of Figure 7 where now only 2 cells interface 1.

The final interacting rotor grids are interpolated from the standalone rotors and the interface line. Points are interpolated only along the lines that originate from the rotor surface. Practically, the individual rotor grids are 'squeezed' normal to the rotor surface (Figure 8). First, the intersection of the grid line with the interface line is found. Then the length of the grid line from the intersection point toward the rotor surface is computed. Once this length is known points are distributed equidistanly or, if a viscous grid is need, using a two-sided exponential distribution for which a first cell height is specified on the rotor surface. A final 2D viscous grid with 73 x 33 gridpoints per lobe is shown in Figure 9.

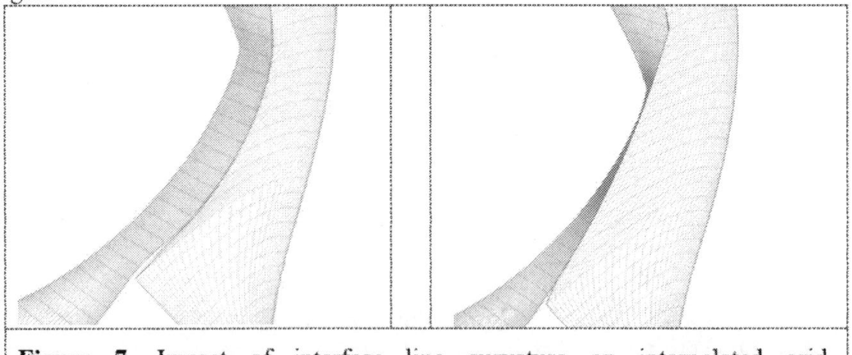

Figure 7. Impact of interface line curvature on interpolated grid. Communication of up to 5 to 1cells on the left and to 2 to 1 cells on the right.

2.1.3 3D rotor grids

In the final step, the 2D grids for interacting rotors are stacked in the axial direction to form a 3D grid. There are two ways of stacking the grids. When adding a 2D grid the circumferencial index can be kept constant or it can be rotated with the grids. The 'axial' gridlines are either parallel to the axis or follow the geometry features. Imagine moving the 2D grid plane axialy through the 3D grid: for the first method only the rotor surface rotates and the grid lines normal to the surface are 'squeezed' by the geometry features. For the second method, the grid rotates with the rotor and most of the grid only undergoes a solid body rotation. This is the reason why this stacking method has been chosen, as one can demonstrate that this reduces the numerical errors associated with moving grids. In addition, this method also allows easier enforcement of boundary conditions in the structured solver.

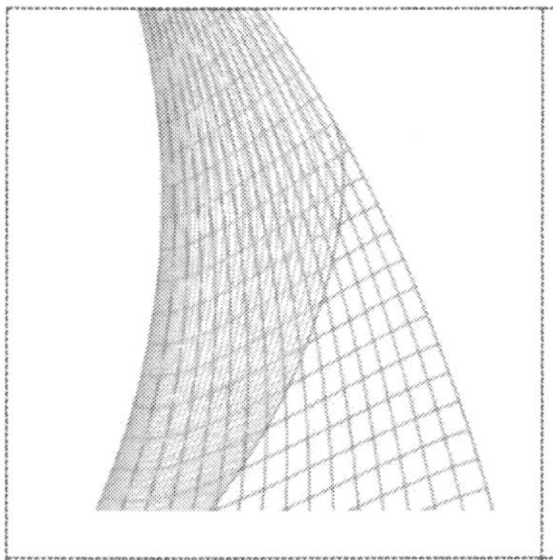

Figure 8. Interpolated grid for interacting rotors (in orange) overlayed over the individual (undistorted) rotor grid (in red)

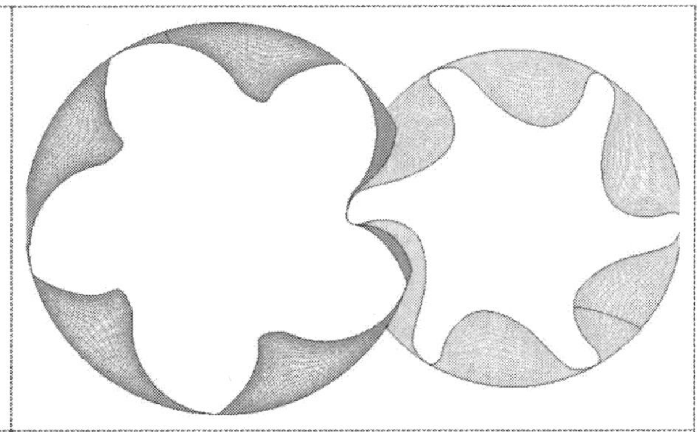

Figure 9. Final 2D grid for interacting rotors.

The number of cells used to resolve one lobe determines the periodicity in the axial direction. When running the code in parallel, we split the grid in axial slices and provide each processor with one periodic slice. Each processor gets all the information necessary to reconstruct the full rotation of the rotor. If a smaller time step is needed intermediate grid can be generated or interpolated from the current grid. Splitting the grid in slices is not only efficient for the reconstruction of the grid but also for code efficiency in general. For a structured code there is usually one 'inner' index for which the data is most compact alligned in the cache of the computer. Optimal speed can be achieved if this dimension is the largest which for the rotor grid is the circumferential direction.

2.2 Plenum grid

Downstream of the rotors we model the radial and axial ports, plenum, check valve and exit pipe. Some parts of the plenum are gridded separately and MMI interpolation is used to connect the parts. Generally, the port grid resolution should be comparable to the rotor grid resolution. However, the rotor grids slide above the ports and refining the grid in the ports in the direction of rotation increases the temporal resolution of the port opening and as such the accuracy of the interpolation in a region where large fluid jets and flow gradients appear. The plenum is gridded with an unstructured Euler grid with a general cell size such that at least the first two frequencies of the pressure spectra are resolved. A finer grid is used in the narrow channels of the check valve.

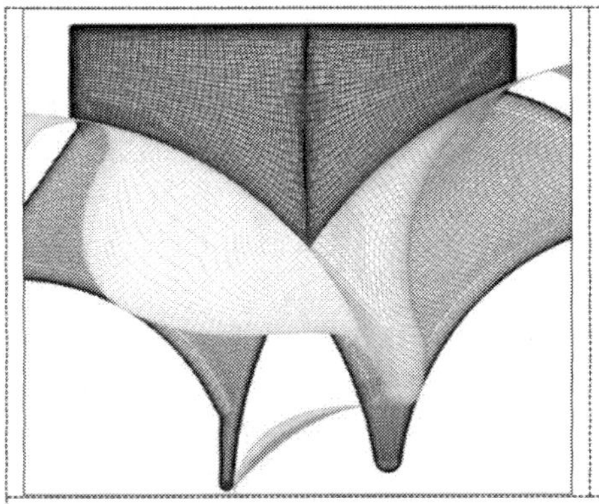

Figure 10. Rotor-Endwall (left) and Rotor to Rotor (right) interface.

3 Dynamic Mis-Matched Interface (MMI)

The flow solver exchanges data on a Mis-Matched Interface (MMI) by interpolating the flow variables. The additional complexity is that the boundaries of the two surfaces with mismatched grids that are interfacing with each other do not coincide and change in time. Thus the boundary of the interpolating region is difficult to specify. In our approach the interpolating region is defined by a flow-through criteria. A solid wall boundary condition is applied where the criteria is not fulfilled. This procedure is dynamic since at each time step the metric of the rotor grids is recalculated and the boundary of the interpolating region is re-determined.

Examples of MMI interfaces are shown in Figure 10. The axial port is shown on the left. The interface is flat but the grid on both patches is of very different density. The interface between the two rotors is shown on the right. Here the grid of both patches is of similar density but the interface is three dimensional with large gaps and overlaps. The challenge is to define an unique criteria that finds the correct flow through area and an accurate interpolation stencil for any of these interfaces. For moving interfaces, the physical location of the donor points as well as the interpolation stencil has to change in time smoothly. The detection criteria should not be binary; it should allow to distinguish between imperfect choices.

The user defines two surfaces which contain a flow through area. For every node from one surface, the MMI routine projects it onto a plane defined by the cell face of the other surface. After discarding the points for which the distance to the plane is above a certain threshold a detector is used to decides which cell face is closest. This is shown on the left of Figure 11 on the example of a triangular face. The projection P' of the point P creates with the nodes of the triangular face the areas S_a, S_b and S_c. The sum $S_a+S_b+S_c$ is equal to the triangle area S only if the projection P' is inside the triangle. For points outside, the ratio $(S_a+S_b+S_c)/S$ departs from unity quickly (Figure 11, right). The projection point for which the ratio is closest to unity is the best choice. The area ratio S_a/S also determines the weight with which the flow variable in point A is weighted for a linear interpolation.

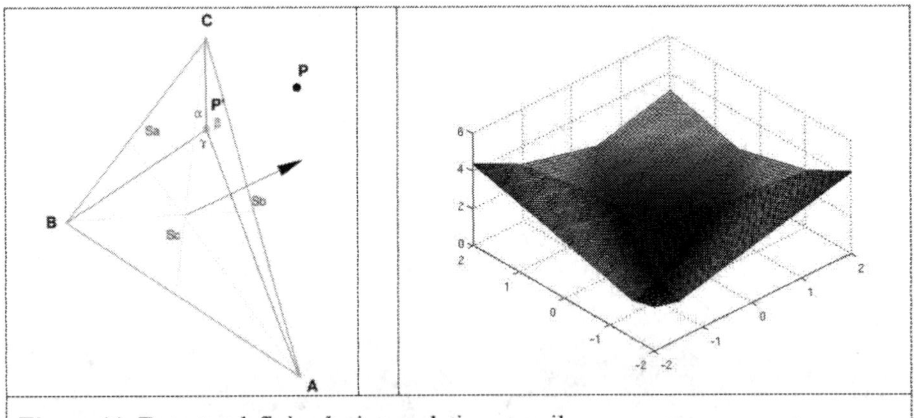

Figure 11. Detector defining he interpolation stencil.

4 Real gas formulation

Since the screw compressor is used for air conditioning applications the real gas properties of the refrigerant need to be considered. For this our in-house density based compressible flow solver has been extended with real gas property tables. The look-up tables are based on NIST data and all thermodynamic quantities are defined as functions of pressure and temperature. The essence of the implementation in our solver is that pressure and temperature are updated directly from the conservative variable changes and no inverse tables are used.

The governing flow equations

$$\frac{\partial Q}{\partial \tau} + \frac{\partial Q}{\partial t} + \frac{\partial F_i}{\partial x_i} - s = 0 \quad (1)$$

are solved with conservative variables to provide a solution change $\delta Q = (\delta\rho, \delta(\rho u_i), \delta(\rho E))$.

This solution change is then used to update the pressure and temperature by

$$\delta p = \left(\frac{\partial p}{\partial \rho}\right)\delta\rho + \left(\frac{\partial p}{\partial(\rho u_i)}\right)\delta(\rho u_i) + \left(\frac{\partial p}{\partial(\delta E)}\right)\delta(\rho E)$$

$$\delta T = \left(\frac{\partial T}{\partial \rho}\right)\delta\rho + \left(\frac{\partial T}{\partial(\rho u_i)}\right)\delta(\rho u_i) + \left(\frac{\partial T}{\partial(\delta E)}\right)\delta(\rho E)$$

Which can be written as

$$\delta p = \beta\{\delta(\rho E) - u_i\delta(\rho u_i) + (u^2/2 - \phi)\delta\rho\}$$

$$\delta T = \alpha \cdot \delta\rho - \frac{u_i}{\rho C_v}\delta(\rho u_i) + \frac{\delta(\rho E)}{\rho C_v}$$

With

$$\beta \equiv \frac{\left(\frac{\partial p}{\partial T}\right)_\rho}{\rho c_v}, \quad \phi \equiv (E - \frac{u^2}{2} + \frac{p}{\rho}) + \rho C_p\left(\frac{\partial T}{\partial \rho}\right)_p, \quad \alpha = \left(\frac{\partial T}{\partial \rho}\right)_p + \frac{1}{\rho C_V}\left(\frac{k^2}{2} - \phi\right)$$

The derivatives and fluid properties here are obtained from the looked up table using the pressure and temperature from the current iteration. Once the pressure and temperature is updated, the fluxes in equation (1) are computed with the thermodynamic properties obtained with the updated pressure and temperature.

5 Outflow boundary condition

Acoustic propagation is sensitive to boundary conditions. Non-reflective inflow and outflow conditions are essential for an accurate noise prediction. The non-reflective boundary condition treatment follows Hixon et al ([7]). The correction to the flow variables is decomposed into a mean flow correction to maintain the desired mean flow quantities and a non-reflective correction for the transient waves (Giles [8]):

$$(\delta Q_t)_{bc} = (\Delta\bar{Q}_t)_{MFBC} + (\delta Q_t)_{nrbc}$$

The correction to the mean quantities at the boundary is constructed using the 1-D characteristics of Thompson ([9]) which is based on the mean flow difference from the target values:

$$\Delta\bar{Q} = \bar{Q} - \bar{Q}^{target},$$

Where the mean flow variables $\bar{Q}$ are integrated over a given length of time, T, as

$$\bar{Q} = \frac{1}{T}\int_{t-T}^{t} Q dt$$

Let the variables $\Delta A_1, \Delta A_2, \Delta A_3, \Delta A_4$ represent the four characteristic variables:

$$\Delta A_1 = (\Delta\bar{P}) - C^2(\Delta\bar{\rho}),\ \Delta A_2 = (\Delta\bar{v}),\ \Delta A_3 = (\Delta\bar{P}) + \bar{\rho}C(\Delta\bar{u}),\ \Delta A_4 = (\Delta\bar{P}) - \bar{\rho}C(\Delta\bar{u})$$

which describe incoming and outgoing waves through the boundaries, and v is the tangential and u is the normal velocity. The outflow conditions presented here are non-reflective with the possibility to specify an average pressure at the outflow. At the outflow the three outgoing wave variables $\Delta A_1, \Delta A_2, \Delta A_3$ are set to zero, and the incoming-wave variable, ΔA_4, is corrected by

$$\Delta A_4 = (\Delta\bar{P}) - \bar{\rho}C(\Delta\bar{u})$$

whereby

$$\Delta\bar{P} = k \cdot \left(\bar{p} - \bar{p}^{target}\right)$$

$$\Delta\bar{u} = \left(\bar{u} - \bar{u}^{target}\right)$$

Hence, the corrected values for the mean flow variables are:

$$\Delta\overline{\rho}_{mfbc} = \Delta\overline{P} / C^2$$
$$\Delta\overline{u}_{mfbc} = -\Delta\overline{P} / \overline{\rho}C$$
$$\Delta\overline{v}_{mfbc} = 0$$
$$\Delta\overline{p}_{mfbc} = \Delta\overline{P}$$

6 Validation

The ability to predict noise in a screw compressor has been validated for a simplified model. The model consists of the main components: two screw rotors and the plenum with radial and axial ports, check valve and exhaust pipe. The working fluid is refrigerant R134a.

In the real machine the flow enters the screw rotors axially as well as radially through a suction inlet. This has been omitted and inflow conditions have been specified at the axial inlet to the screw rotors. We found that a crucial parameter for noise generation in a screw compressor is the start of discharge pressure, i.e. the pressure in the rotor lobes at the location of the ports just when the ports open. The pressure buildup in the rotors is significantly affected by the total conditions at the inlet. In the computation we adjusted these conditions such to match the start of discharge pressure. At the outflow, a time averaged pressure has been specified over a non-reflective boundary condition.

The rotor grids are viscous wall integration grids and adiabatic viscous wall boundary conditions are specified at the rotor casing. As described the rotor grids are constructed such that axial endwall of the screws rotors is not resolved. Therefore an Euler wall boundary conditions is specified on the screw rotor endwall. The ports also have wall integration viscous grids and the plenum is gridded with an Euler grid and slip wall conditions are applied.

A typical run for 10 screw rotor revolutions with computational grids with about 9.3 and 2 million structured grid points for the screw rotors and ports, respectively, and 0.120 million unstructured grid points for the plenum, requires about 2 days using 50 processors on a 2013 Linux cluster. This relatively low turnaround time enables CFD analyses of this type feasible within a product design cycle.

An instantaneous snapshot of the pressure field is shown in Figure 2. One can see the pressure buildup along the rotor lobes which are under an angle to the rotors axis. In the computation these pressure stripes move axially towards the plenum with an increase of pressure until they reach the radial port. Pressure waves bounce within the lobes as flow discharges.

Unsteady pressure history on the plenum wall has been stored for each time step at the locations labeled 1,2,3,4 in Figure 2. In a post-processing step these pressure histories have been decomposed in power spectra for the fundamental lobe passing frequency and its harmonics. The power spectra for the four locations are averaged and compared to corresponding experimental data in Figures 12-14. The dashed lines indicate a +/- 5dB range around the test data for reference. Repeatability of the test data is +/- 1dB. Figure 12 compares CFD results to test data for a nominal operating condition. It is seen that the fundamental and two subsequent harmonics are captured within measurement accuracy. Figures 13 and 14 show sensitivity to change in the back-pressure and rotor speed, respectively. It is seen that the CFD accurately captures sensitivity to these parameters. In particular, Figure 13 shows that the fundamental increases markedly with increased back pressure, and Figure 14 shows that higher harmonic content increases with speed.

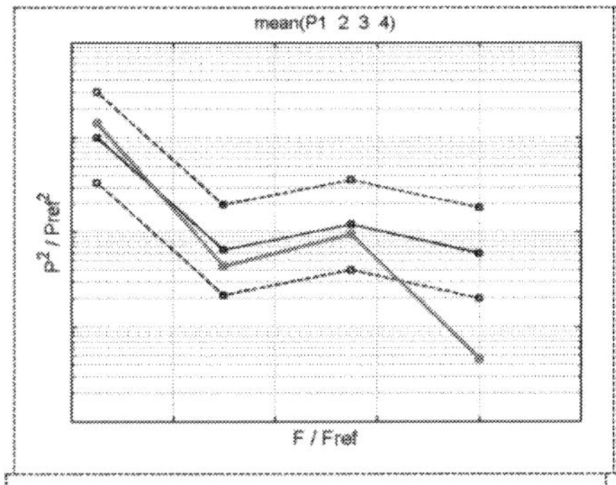

Figure 12. Computed and measured gas pulsation spectra at nominal conditions.

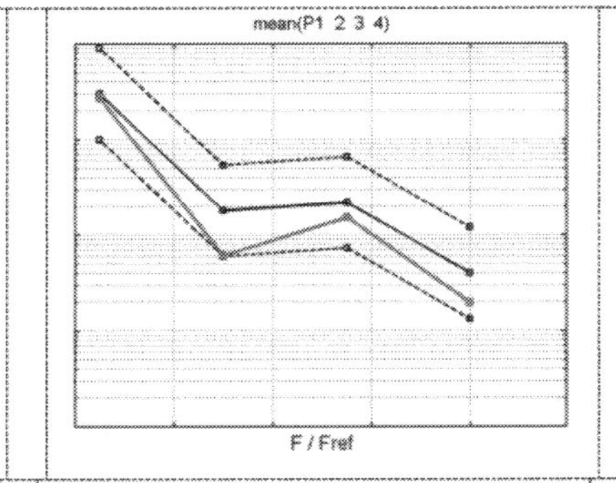

Figure 13. Computed and measured gas pulsation spectra at increased discharge pressure.

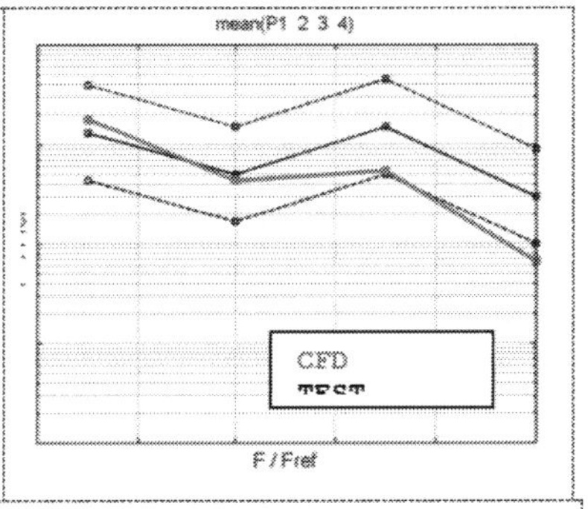

Figure 14. Computed and measured gas pulsation spectra at increased rotor speed.

7 Conclusions

A three-dimensional computational model of a screw compressor has been developed and validated for gas pulsation prediction.. It has been shown that the methodology accurately predicts the first three harmonics with computational cost suited for application in the product design cycle.

Several novel numerical items have been addressed and presented. These include the rotor and plenum grid generation, real gas property look up tables, non-reflective boundary conditions and mis-matching interface interpolation. The hybrid grid approach with structured grids for the screw rotors and unstructured for the plenum has proven to be an efficient and accurate way to compute screw compressor flow field in an industrial environment.

8 References

[1] Fujiwara A, Sakurai N 1986 *Experimental analysis of screw compressor noise and vibration*, International Compressor Engineering Conference, Purdue, USA.

[2] Huagen W, Ziwen X, Xueyuan P, Pengcheng S 2004 *Simulation of discharge pressure pulsation within twin screw compressors*, Proceedings of the Institution of Mechanical Engineers, Part A: Journal of Power and Energy, Volume 218.

[3] Koai K L, Soedel W, 1990 *Determination of port flow and interpretation of periodic volume source*, International Compressor Engineering Conference at Perdue.

[4] Sangfors B 1999 *Computer simulation of gas flow noise from twin screw compressors*, International Conference on Compressor and their Systems, London, UK.

[5] Kovacevic A, Stosic N, Smith I K 2006 *Screw Compressors Three Dimensional Computational Fluid Dynamics and Solid Fluid Interaction*, Springer.

[6] Kovacevic A, Mujic E, Stosic N, Smith I K 2007 *An integrated model for the performance calculation of Screw Compressors*, International Conference on Compressors and their Systems, London, UK.

[7] Hixon et al 2003 *Mean flow boundary conditions for computational aeroacoustics* (AIAA - 2003-3299)

[8] Giles M,B 1990 *Nonreflecting Boundary Conditions for Euler Equation Calculations* (AIAA Journal, Vol 28-10)

[9] Thompson K, W 1987 *Time Dependent Boundary Conditions for Hyperbolic Systems*

[10] Kovacevic A, *Boundary adaptation in grid generation for CFD analysis of screw compressors.*

[11] Mechels F, F 2004 *Formulas of Accoustics*

[12] Ni R-H 1982 *A multiple-grid scheme for solving the Euler equations* (AIAA Journal Vol 20-11)

[13] *CFL3d Tutorial*, http://cfl3d.larc.nasa.gov

Deformation Analysis of the Main Components in a Single Screw Compressor

Feilong Liu [1,3], Xueli Liao[2], Quanke Feng [1]*, Martijn Van Den Broek[3], Michel De Paepe[3]

[1] School of Energy and Power Engineering, Xi'an Jiaotong University, Xi'an, China
[2] Patent Examination Cooperation Center of the Patent Office, SIPO, Sichuan, China
[3] Ghent University, Department of Flow, Heat, and Combustion Mechanics, Gent, Belgium

qkfeng@mail.xjtu.edu.cn
*Corresponding Author

Abstract: The single screw compressor is used in many fields such as air compression, chemical industry and refrigeration. During operation, different gas pressures and temperatures applied on the components can cause different degrees of deformation, which leads to a difference between the thermally induced clearance and the designed clearance. However, limited research about clearance design is reported. In this paper, a temperature measurement instrument and a convective heat transfer model were described and used to establish the temperature of a single screw air compressor's casing, screw rotor and star wheel. 3-D models of these three main components were built. The gas force deformation, thermal-structure deformation and thermal-force coupling deformation were carried out by using a finite element simulation method. Results show that the clearance between the bottom of the groove and the top of star wheel is reduced by 0.066mm, the clearance between the side of groove and the star wheel is reduced by 0.015mm, and the clearance between the cylinder and the rotor is reduced by 0.01mm. It is suggested that these deformations should be taken into account during the design of these clearances.

Key words: single screw compressor, clearance, finite element, deformation

1. Introduction

The single screw compressor was invented by B. Zimmern based on a "pressure energy conversion device" in 1962[1]. Single screw compressors (both air and refrigeration) first appeared on the market with a rotating speed of 5000 rpm and a power of 10 HP [2]. In 1963, Peugeot managed to produce 20HP, 30HP, 40HP, and 70HP moving and stationary compressors [3]. Then, Daikin, Mitsui Seiki, Mitsubishi, Chicago Pneumatic, Hall and Grasso started to study and introduce the single screw compressor technology [4-5]. Now, the single screw compressor is widely used in many fields, including air compression, refrigeration, industrial processes, heat pumps etc.

Restricted by machining and assembly precision, certain clearances, which influence gas leakages and friction power consumption, are included in the design of the casing, screw rotor and star wheel. Oversized clearances increase the gas leakage, while clearances that are too small increase the frictional power of the lubricating oil, and can even lead to a jam of the compressor. During operation, different gas pressures and temperatures applied on the components can cause different degrees of deformation, which lead to a difference between the thermal cooperated clearance and designed clearance.

Content from this work may be used under the terms of the Creative Commons Attribution 3.0 licence. Any further distribution of this work must maintain attribution to the author(s) and the title of the work, journal citation and DOI.
Published under licence by IOP Publishing Ltd

In this paper, three-dimensional models of a single screw air compressor's casing (including the cylinder), screw rotor and star wheel are described. A finite-element method was used to analyze the gas force and thermal deformation of these three main components with ANSYS. According to the obtained results, during operation, these components were subject to different deformations, which caused a different decrease of the designed clearances and could lead to a performance reduction.

2. Modeling

The models were built with Unigraphics NX and were then imported into ANSYS Workbench to perform a finite element analysis. The models of the three components (the casing, screw rotor and star wheel) were all individual models without other assembly parts and they were built on the basis of the compressor's schedule drawing.

In this paper, the casing of the air compressor was integrally cast and the model was built based on geometric relationships and measurements. The result is reproduced in Fig. 1 (a). To render the consequent calculations more efficiently, the model was first simplified as follows, also shown in Fig. 1 (b):

1) Ignore the small characteristics of the suction and exhaust end, star wheels and exhaust chamber, flanges and oil path, including most of the small lug bosses, chamfers, round corners etc. ;
2) Ignore all the bolt holes on the star wheel side, the motor side, the suction and the exhaust side;

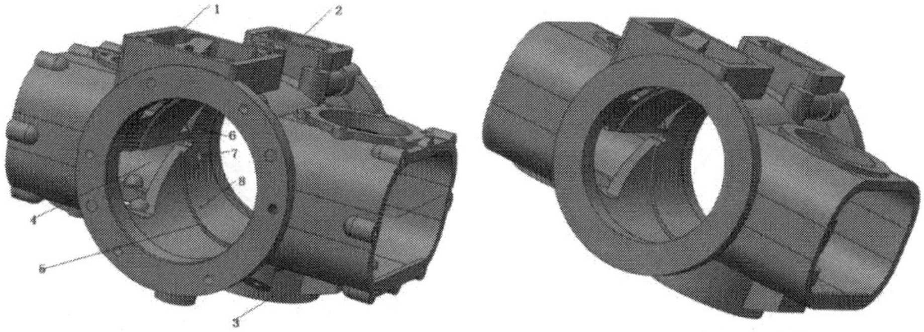

(a) Actual model (b) Simplified model

1—suction channel 2—oil path 3—exhaust channel 4—star wheel chamber 5—enclosed screw spiral 6—exhaust orifice 7—high pressure gas leak hole 8—oil injection hole

Figure 1. Actual model and simplified model of the single screw compressor.

The research presented in this paper is focused mainly on the deformation. Because the interference fit between the rotor and the axle has little influence on the deformation, the axle and the hole can be ignored. The geometries of the two star wheels are identical. The models of the rotor and star wheels built with Unigraphics NX are depicted in Fig. 2.

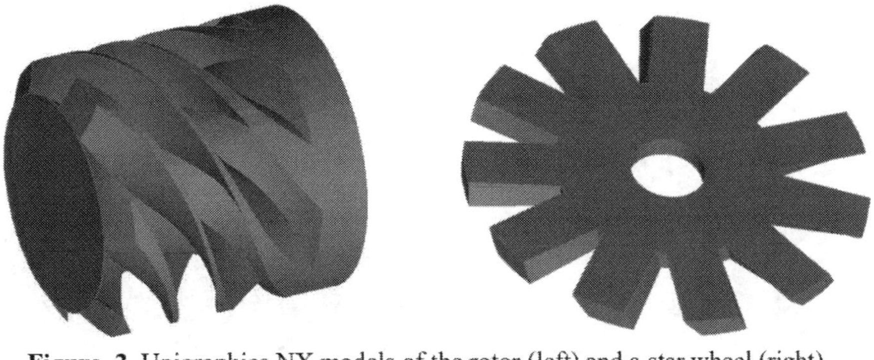

Figure 2. Unigraphics NX models of the rotor (left) and a star wheel (right).

IOP Conf. Series: Materials Science and Engineering **90** (2015) 012011 doi:10.1088/1757-899X/90/1/012011

3. Finite Element Analysis

3.1. Mesh generation

The ANSYS ICEM tetrahedral mesh (i.e. first generate the body mesh and then the surface) was used to mesh the models. In particular, a top-down mesh generation technique was used, which reduces the time to solve the geometry surface's repairing problems for some complicated geometries. Four mesh methods for a tetrahedral mesh are available in ANSYS ICEM:

1) Robust (Octree) method: a top-down mesh generation technology
2) Quick (Delaunay) method: adjust to Tetra/Mixed mesh type
3) Smooth (Advancing Front) method: adjust to Tetra/Mixed mesh type
4) TGrid method: adjust to Tetra/Mixed mesh type

Table 1. Material parameters of the single screw air compressor.

Components	Material	Poisson ratio	Elastic modulus GPa	Density kg/m^3	Linear expansion coefficient $10^{-6}/K$	Thermal conductivity $W/(m^2 \cdot K)$
Casing	HT250	0.25	120	7300	11.8	50
Rotor and star wheel	QT600-3	0.28	170	7300	12	35.5

The Robust (Octree) method was chosen for the mesh generation. After the material's parameters (Table 1) were set, the models were meshed, shown in Figs. 3-5. Local grid refinement and "mesh smoothing" were used for some of the key parts which had small local dimensions, to improve the mesh quality. The number of grid units and nodes for the casing were 505,172 and 89,148. The rotor had 136,137 grid units and 21,137 nodes. For the star wheel, these were 31,487 and 5739 respectively. After "meshing smoothing", the mesh quality meets the requirements for the three main parts.

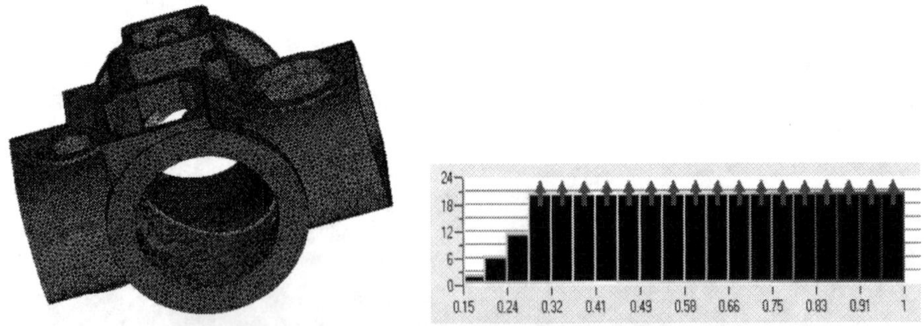

Figure 3. Mesh generation and quality of the casing.

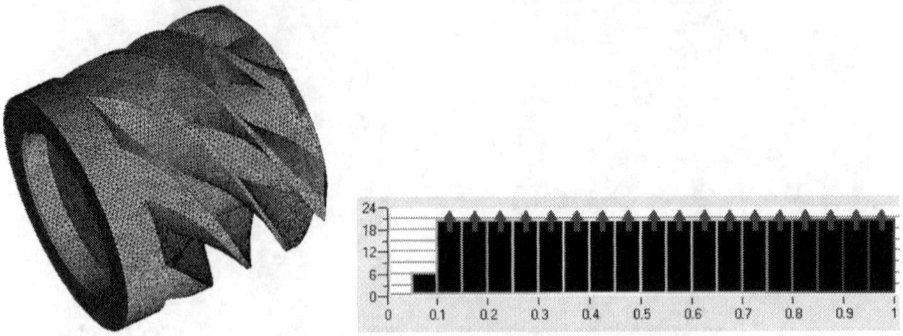

Figure 4. Mesh generation and quality of the rotor.

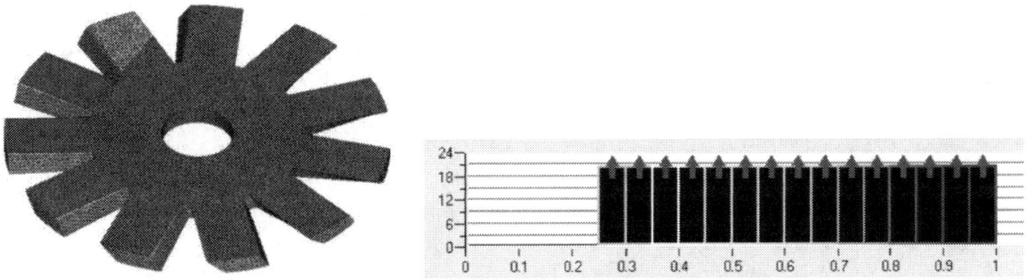

Figure 5. Mesh generation and quality of the star wheel.

3.2. *ANSYS finite element analysis*

The gas pressure varies during the operation of the compressor. In this paper, the time when the exhaust starts and the full-load condition (a gas flow rate of 100%) are taken as an example to perform the finite element analysis. The three continued grooves of the rotor undergo a complete compression cycle: suction-compression-exhaust. The boundary conditions imposed on the finite element model are based of the single screw compressor's operating conditions.

The pressure of the casing on the suction and exhaust ends were 0.096MPa, 0.7MPa (tested). The pressures in the three rotor grooves in the suction-exhaust direction were 0.096MPa, 0.4MPa, 0.7MPa. The pressure on the other parts out of the casing was 0.1MPa. The suction and exhaust temperature were 30℃ and 72℃.

In classical mechanics, the general dynamic equation is:

$$[M]\{\ddot{x}\}+[C]\{\dot{x}\}+[K]\{x\}=\{F(t)\} \tag{1}$$

Where [M] is the mass matrix, [C] is the damping matrix, [K] is stiffness coefficient matrix, {x} is the displacement matrix, {F(t)} is the force vector.

In Linear Static Structural Analysis, [K] should be a constant and continuous matrix, {F} should be a static force matrix which do not change with time loading on the model. Therefore, the equation is:

$$[K]\{x\}=\{F\} \tag{2}$$

This is used for calculating the structural deformation and displacement under a constant loading. The material's elastic modulus, passion ratio and linear expansion coefficient are all constant. The loading includes:

1. The external applied force or pressure.
2. Steady state inertia force, such as centrifugal force and gravity.
3. Displacement loading.
4. Temperature loading.

In this case, the pressures were used to do the Static Structural Analysis.

The three-dimensional steady-state heat conduction differential equation without inner heat source but constant property was taken as the numerical model for the temperature (t) distribution:

$$\frac{\partial^2 t}{\partial x^2}+\frac{\partial^2 t}{\partial y^2}+\frac{\partial^2 t}{\partial z^2}=0 \tag{3}$$

In a steady-state heat conduction problem, the boundary condition is the only definite condition, while the boundary conditions can be divided in three types [6]:

1) Dirichlet condition: the boundary temperature of the subject is given, t_w=constant.
2) Neumann condition: the boundary heat flux is given, q_w=constant.
3) Robin condition: $-\lambda(\frac{\partial t}{\partial n})_w = h(t_w - t_f)$,

Where t_w is the boundary temperature, t_f is the fluid's temperature and h is the heat transfer coefficient.

The surface temperature outside of the casing could be measured by a FLIR infrared thermal imager, shown in Fig. 6. The oil injection temperature was controlled around 60℃, which could certainly affect the temperature distribution in the suction chamber and the cylinder. The temperature inside the cylinder was not measured due to the oil injection (the temperature tested by the probe could be the temperature of the injection oil). The temperature of the suction and exhaust end is 56℃ and 70℃ respectively, which were enforced to the outside of the casing as the Dirichlet conditions. The temperature of the exhaust chamber was measured by the thermocouple, 72℃, which was also imposed to the inner surface of the exhaust chamber as the Dirichlet condition. The Robin condition was used for the inner surface of the casing. The convective heat transfer coefficients for the three chosen rotor grooves could be calculated, which were 360 $W \cdot m^{-2} \cdot ℃^{-1}$, 160 $W \cdot m^{-2} \cdot ℃^{-1}$, 80 $W \cdot m^{-2} \cdot ℃^{-1}$, and the related gas temperatures were 72℃, 50℃, 35℃.

Similar to the casing, the Dirichlet condition was also used for the rotor and the star wheel. According to reference [7], the temperature of the suction and exhaust end of the rotor were assumed to be 62℃ and 72℃. The temperature of the star wheel could be calculated, but details are not expatiated here. The surface and the tooth-top temperature were 64℃ and 66℃.

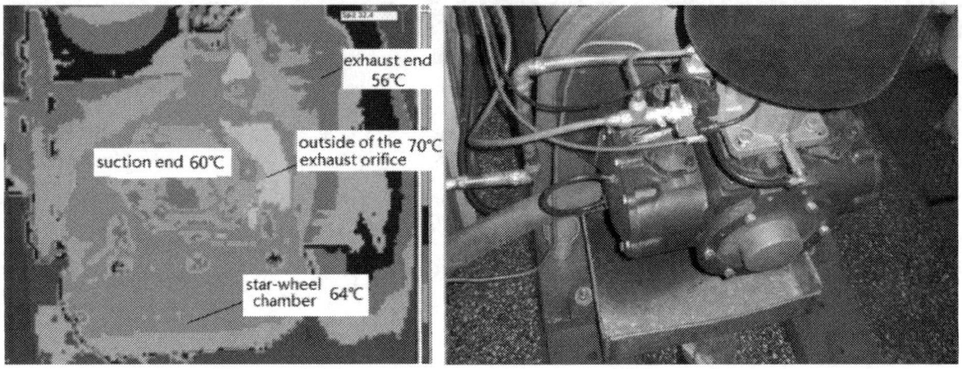

Figure 6. Temperature measurement of the casing.

First, ANSYS Workbench was used to do the static analysis and achieve the gas force deformation. Then, the temperature distribution achieved from the steady-state thermal analysis was put into the static structural analysis through the Workbench to realize the thermal-structural coupling analysis to get the thermal deformation.

Finally, the deformation caused by the combined action of the gas force and the temperature load is also acquired with ANSYS Workbench, through the built-in Thermal-Force Coupling Analysis method. The temperature distribution obtained from the Steady-State Thermal method was used as input for the Static Structural method, as well as the imposed gas pressure. The procedure is visualized in Fig. 7.

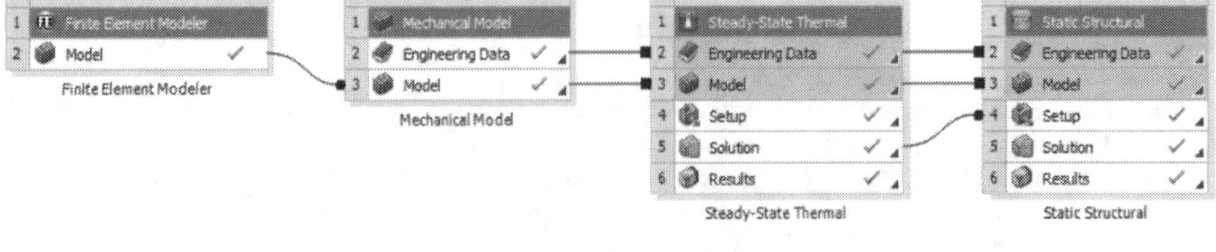

Figure 7. Procedure of the thermal-force coupling analysis.

3.3. Result and discussion

The computing results are shown in Table 2. It can be seen that the thermal deformation is much larger than that caused by the gas force, which means the gas force has little influence on the deformation. Since the actual deformation is the thermal-force coupling deformation, in the remainder of this paper we focus on this deformation.

Table 2. Deformation results.

	Gas force /mm	Thermal-structure/mm	Thermal-force/mm
Casing	9.82×10^{-3}	15.93×10^{-2}	15.98×10^{-2}
Rotor	2.49×10^{-3}	8.35×10^{-2}	8.36×10^{-2}
Star wheel	3.61×10^{-4}	4.27×10^{-2}	4.28×10^{-2}

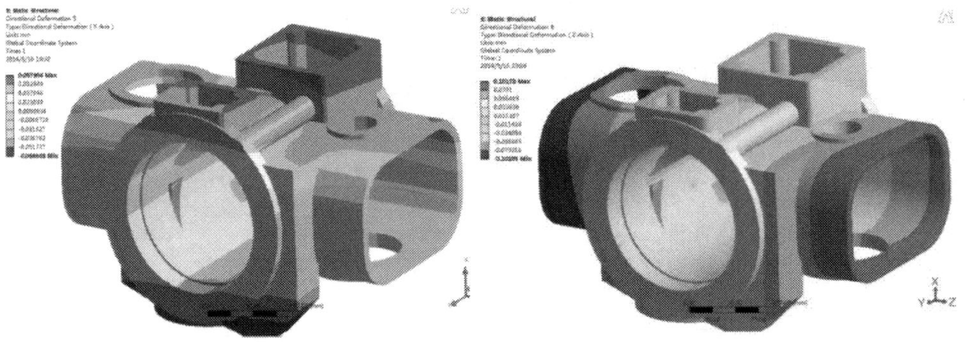

(a) Deformation in X-direction (b) Deformation in Z-direction

Figure 8. Deformation of the casing.

Fig. 8 shows the casing's deformation computing results through the direction of X and Z. In the X-direction, the maximum deformation takes place at the suction end and the minimum deformation happens at the exhaust end, which were 0.053mm and -0.067mm respectively. In the Z-direction the deformation concentrated in the star wheel chambers with a maximum deformation of 0.101mm and a minimum deformation of -0.102mm. Such large deformations could cause a change of the center distance between the star wheels and the rotor.

The deformation of the inner surface of the casing (the cylinder) is shown in Fig. 9. Obviously, the deformation increases from the suction end (minimum 0.009mm) to the exhaust end (maximum 0.047mm). The outward expanding of the cylinder may lead to a clearance change between the rotor and the cylinder.

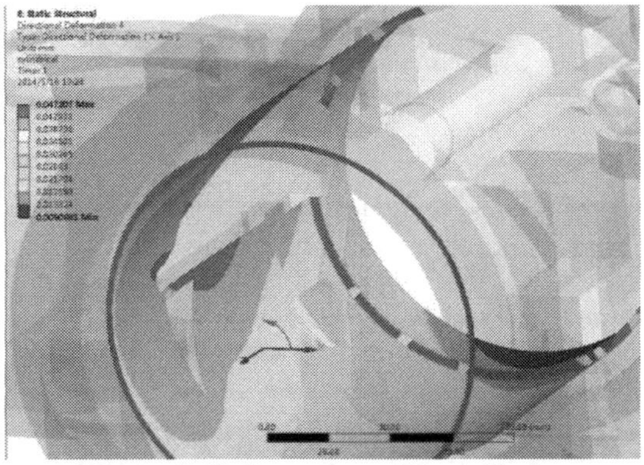

Figure 9. Deformation of the cylinder.

In order to get the accurate deformation of the cylinder in the radial direction, seven paths were delineated on the cylinder wall from the seal section to the suction end with a distance of 20mm, as shown in Fig. 10.

The computing results are given in Table 3, which shows that the largest deformation appears on the exhaust end, Path 2 and Path 3 (0.043mm).

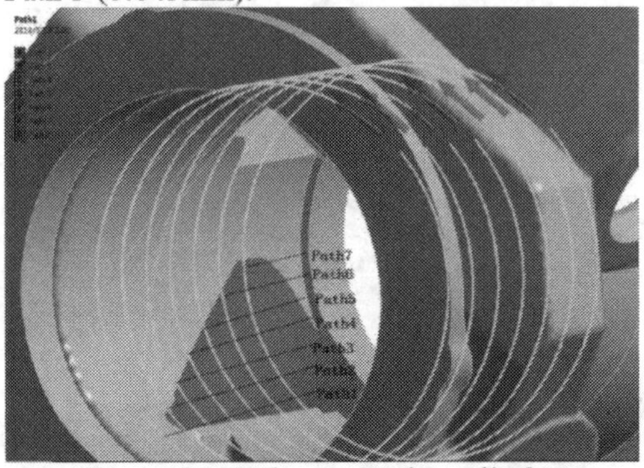

Figure 10. Paths set on the cylinder.

Table 3. Deformation of the cylinder through the radial direction.

Path	Path1	Path2	Path3	Path4	Path5	Path6	Path7
Deformation/mm	0.034	0.043	0.043	0.037	0.037	0.036	0.029

The deformation in the axial direction of the rotor was 0.082mm. Fig. 11 shows the whole deformation result in the radial (X) direction, which affects the clearance between the rotor and the cylinder. The largest deformation, which was 0.044mm, took place at the rotor's sealing edge, and the smallest deformation (0.021mm) was on the bottom of the middle groove.

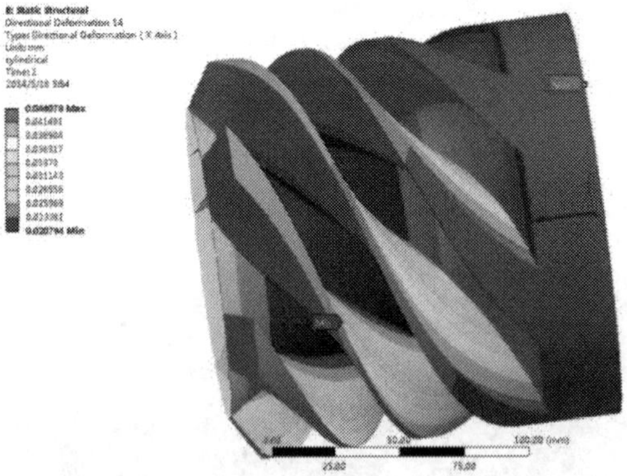

Figure 11. Deformation of the rotor in the radial direction.

Similar to the cylinder, for the radial direction, seven paths, with locations corresponding to those on the cylinder, were also set to get the accurate deformation, as depicted in Fig. 12. The result is shown in Table 4. The maximum deformation also happens at the exhaust end and decreases towards the suction end, from 0.044mm to 0.036mm.

Table 4. Deformation of the 7 paths on the rotor.

Path	Path1	Path2	Path3	Path4	Path5	Path6	Path7
Deformation/mm	0.044	0.043	0.043	0.041	0.040	0.038	0.036

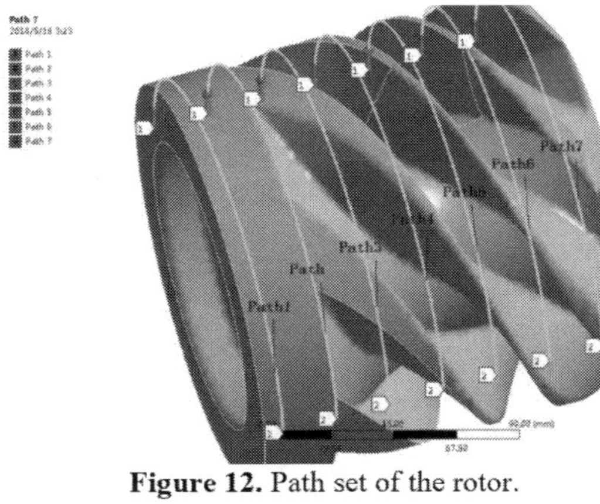

Figure 12. Path set of the rotor.

To research the clearance between the rotor groove and a star wheel tooth, the deformation on the bottom and both sides of the rotor groove in the normal direction also needs to be taken into account. Due to the rotor's complicated geometrical shape, it is hard to get the result directly from the calculation mentioned above. Since the groove bottom's projection is circular, a path is set as in Fig. 13 (a). Then the deformation in the path's normal direction coincides with the deformation of the groove's bottom. Similarly, the path for both sides of the groove can be set as in Fig. 13 (b). Thus, the required deformation corresponds to the deformation in the path's normal direction. The maximum of the bottom was 0.024mm (exhaust end) and the minimum was 0.021mm (suction end). For the groove sides, they were the same everywhere, 0.009mm.

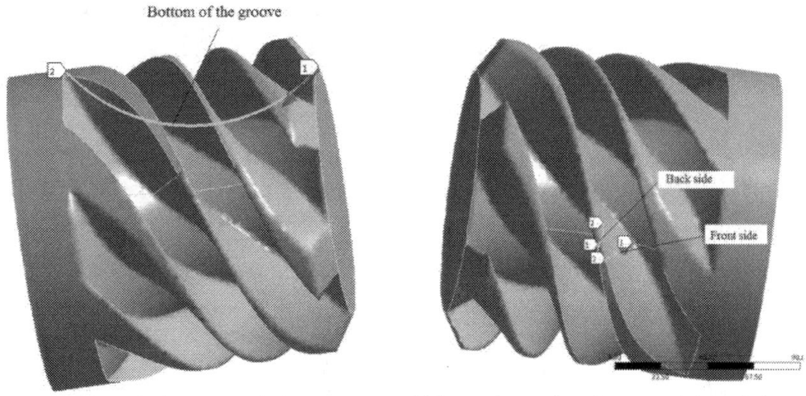

(a) Path set for the bottom of the groove (b) Path set for the two sides of the groove

Figure 13. Paths set on the groove's bottom and flanks.

The computing result of the star wheel is shown in Fig. 14. The star wheel extends from the central hole to the top of the teeth. The central deformation is 0.007mm, while on the top of a tooth the deformation is maximum, 0.042mm. Similarly to the rotor, paths were set to acquire information on the deformation of the tooth's edges, as shown in Fig. 15. The tangential deformation on the top of a tooth is 0.042mm, and it is 0.006mm for each side.

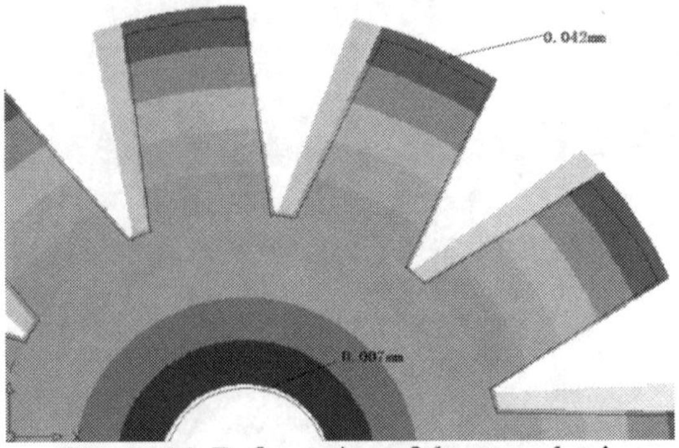

Figure 14. Deformation of the star wheel.

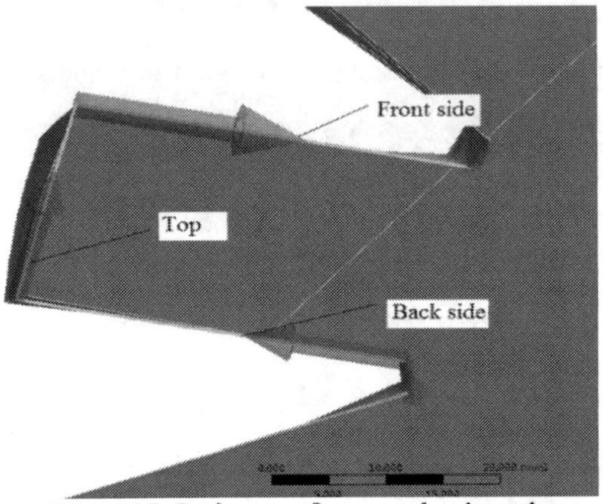

Figure 15. Paths set of a star wheel tooth.

The designed clearances of the compressor investigated (type of 9 $m^3 \cdot min^{-1}$) in this paper are given in Table 5, as well as some other types of compressors' designed clearances applying on the market. According to the computing result, during the operation, these clearances change due to the thermal-force coupling deformation.

Table 5. Designed clearances of the air compressors.

Components	Clearance/mm (different types of compressors)		
Types	9 $m^3 \cdot min^{-1}$	3 $m^3 \cdot min^{-1}$	12 $m^3 \cdot min^{-1}$
Cylinder and rotor	0.02	0.02	0.03
Top of the tooth and bottom of the groove	0.04—0.07	0.03—0.05	0.06—0.08
Flank of the tooth and the groove	0.03—0.05	0.02—0.04	0.05—0.07

For the rotor and the cylinder, they have the same deformation direction, but the rotor experiences a larger deformation than the cylinder, so the clearance decreases. Table 6 shows the deformation difference between the cylinder and the rotor for the seven paths. The largest change takes place at the rotor's sealing edge (0.10mm).

Table 6. Clearance change.

Path	Path1	Path2	Path3	Path4	Path5	Path6	Path7
Cylinder /mm	0.034	0.043	0.043	0.037	0.037	0.036	0.029
Rotor /mm	0.044	0.043	0.043	0.041	0.040	0.038	0.036
Difference/mm	0.01	0.000	0.000	0.004	0.003	0.002	0.007

The tooth top of the star wheel expands with 0.042mm, while the bottom of the rotor groove has a maximum deformation of 0.024mm, which is in the opposite direction of the tooth top. Thus, their clearance reduces by 0.066mm, close to the designed clearance, which would cause an elastic contact and lead to the wear of the star wheel.

Similarly, because of the opposite deformation direction, the clearance between the side of the rotor's groove and the side of the star wheel's tooth also reduced by 0.015mm, which was within the designed clearance.

4. CONCLUSIONS

1) A symmetric deformation could certainly happen to the casing and could lead to a change of the center distance between the star wheel and the rotor due to the thermal and gas force factors.

2) The largest deformation of the cylinder appears in the exhaust end, while that of the rotor happens in the rotor's sealing edge. Besides, the largest clearance change between the rotor and cylinder also takes place here, with a decrease of 0.10mm.

3) The decrease of the clearance between the top of the star wheel's tooth and the bottom of the rotor's groove is close to the designed clearance. As a result, a hard contact may turn up and finally lead to wear of the star wheel.

4) The clearance between the plank of a star wheel's tooth and the rotor's groove also reduces by 0.015mm, which is allowed by the designed clearance of 0.03mm-0.05mm.

5) It is suggested to do the thermal-force coupling deformation analysis for these three main parts of the air compressor, which could serve as a guide for the determination of the designed clearances.

References

[1] Zimmern B. Worm rotary compressors with liquid joints. U.S. Patent No. 3133695; 1965.

[2] Zimmern B. From Water to Refrigerant: Twenty Years to Develop the Oil Injection-free Single Screw Compressor [C]. Purdue, 1984: 513-518.

[3] Zimmern B, Patel GC. Design and Operating Characters of the Zimmern Single Screw Compressors[C]. Purdue, 1972: 96-99.

[4] Deren Wang, Development Situation of the Single Screw Compressor [J]. Electromechanical Device, 1994, (1): 2-6.

[5] Company C P T .A Revolution in Air Compressor Technology [J]. Transactions of the ASME, Journal of Applied Mechanics,1987: 217-224.

[6] J.H.Wu, G.X. Jing. Investigation on Suction Process of Oil-Flooded Rotary Positive Displacement Compressors [J]. Fluid Machinery, 1995, (5): 7-10.

[7] Chukanova E, Stosic N, Kovacevic A. Investigation of Start Up Process in Oil Flooded Twin Screw Compressors [C]. Purdue, 2012, 1138.

Economics of water injected air screw compressor systems

K. Venu Madhav[1] and A. Kovačević[2]

[1]ELGI Equipments Ltd, Coimbatore, India
[2]City University London, Centre for Compressor Technology, London, UK

Corresponding author e-mail address: a.kovacevic@city.ac.uk

Abstract. There is a growing need for compressed air free of entrained oil to be used in industry. In many cases it can be supplied by oil flooded screw compressors with multi stage filtration systems, or by oil free screw compressors. However, if water injected screw compressors can be made to operate reliably, they could be more efficient and therefore cheaper to operate. Unfortunately, to date, such machines have proved to be insufficiently reliable and not cost effective. This paper describes an investigation carried out to determine the current limitations of water injected screw compressor systems and how these could be overcome in the 15-315 kW power range and delivery pressures of 6-10 bar. Modern rotor profiles and approach to sealing and cooling allow reasonably inexpensive air end design. The prototype of the water injected screw compressor air system was built and tested for performance and reliability. The water injected compressor system was compared with the oil injected and oil free compressor systems of the equivalent size including the economic analysis based on the lifecycle costs. Based on the obtained results, it was concluded that water injected screw compressor systems could be designed to deliver clean air free of oil contamination with a better user value proposition than the oil injected or oil free screw compressor systems over the considered range of operations.

Key words: Water injected screw compressor, lifecycle costs, value proposition.

1. Introduction

ISO 8573-1 classifies the quality of air based on oil content in classes 0 to 4 [1]. Class 0 is completely free of oil while Class 1 will contain less than 0.01 mg oil per m^3 of air. Higher classes will allow one order of magnitude higher content of oil from each other. The pharmaceutical and food industries for products in direct contact with compressed air will require class 0 while industries dealing with food and beverage with non-direct contact with air, power product transport or film processing will allow use of air in class 1. It is expected that in future the legislative limits of oil contamination in air will lower. Typically, industrial air systems operate at pressures of 6-10 bar in power range of 15-300kW using packaged air compressors. The most widely used are rotary twin screw compressor systems due to their reliability, low maintenance and efficiency.

Oil injected screw compressor systems with multi filtration stages are associated with the high risk of oil contamination and at best can only meet air quality of Class 1 of ISO 8573-1. The presence of multiple filters decreases their efficiency and hence increases their power consumption, installation costs and associated maintenance costs.

A conventional oil free compressor system usually can offer a better quality air but it requires a 2-stage oil free compressor, inter cooler, special bearings to handle high temperatures and speeds, and an

Content from this work may be used under the terms of the Creative Commons Attribution 3.0 licence. Any further distribution of this work must maintain attribution to the author(s) and the title of the work, journal citation and DOI.
Published under licence by IOP Publishing Ltd

expensive sealing system to prevent mixing of oil and gas. This makes an oil free screw compressor system expensive, both in terms of initial and operating cost. It is usually more expensive than oil injected compressor in a package with multi filtration system.

Delivering compressed air to a manufacturing facility is an expensive operation as it requires expensive equipment that consumes significant amount of electricity and needs frequent maintenance. In spite of this, many users do not know how much their compressed air systems cost on an annual basis, or how much they could save by appropriate selection of the compressor type and system. In a situation, where various compressor types and sizes overlap, one may prove more energy efficient and controllable than the other for a particular duty. Equally, maintenance costs between the different types can vary considerably.

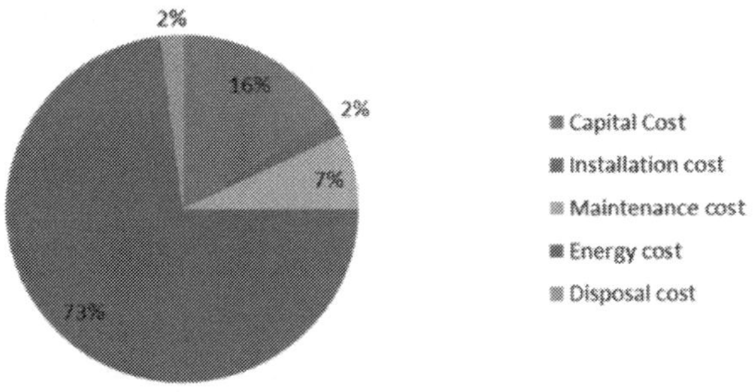

Figure 1 Lifecycle costs of a typical oil free air screw compressor system

A typical break-down of costs associated with a compressor utility over its assumed life cycle of 10 years is shown in Figure 1, as reported in [2].

This paper aims to compare the lifecycle costs of the newly designed water injected screw compressor system with traditional oil free compressor package and oil injected screw compressor in the multi-filtration system. As information for water injected compressors were not available for analysis, a family of water injected compressors was designed. A prototype water injected compressor system of 55-75 kW was built and used to check commercial viability of water injected screw compressor systems in the required range of operating conditions.

2. Design of the water injected screw compressor system

The water injected compressor systems consist of the specially designed water injected screw compressor driven by the constant speed electric motor. Other elements of the system include separator tank and filter, fan assisted air water cooler, air dryer and the control system. The layout of this new system is shown in Figure 2.

Table 1 Size of screw compressor blocks for water injected compressor systems

Power Range [kW]	Male rotor Diameter [mm]	Female rotor Diameter [mm]	L/D ratio
15-22	102	80	1.55
30-45	159	125	1.55
45-75	159	125	1.88
75-110	225	175	1.55
132-200	284	220	1.55
250-315	310	275	1.55

Six sizes of the water injected compressor systems were designed in the power range of 15 to 315 kW. The critical components of the system, the air compressors, are designed with "N" rotor profile [4], lobe combinations 4/5, length to diameter ratios1.55 and 1.88, built in volume ratio 4.5 and tip speed range between 10 and 32 m/sec. Some of previous designs using similar configurations are given in [5]. At a constant prime mover speed sets of gears with different gear ratios are used to vary the male rotor speed. The selected screw compressor sizes are shown in Table 1.

IOP Conf. Series: Materials Science and Engineering **90** (2015) 012012 doi:10.1088/1757-899X/90/1/012012

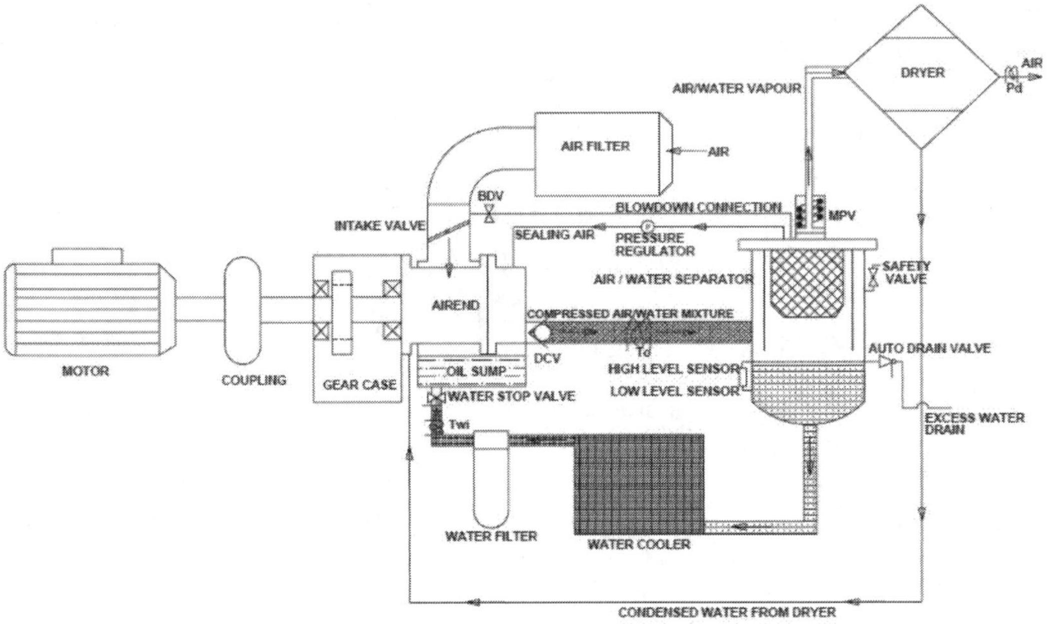

Figure 2 Water Injected compressor system lay-out

Other compressor system components were selected or designed considering the flow, power and performance requirements. The presence of water is associated with scaling & corrosion of materials and causes their deterioration. Hence, due consideration was given for their selection to prevent corrosion at the same time to avoid excessive costs. The special attention is given to water treatment. Water quality was maintained by the use of treated water in the loop . Corrosion was avoided by providing a special corrosion free coating on the parts that are exposed to water e.g rotors, housings tank, pipes etc. A stainless steel water filter was also provided in the water circuit for ensuring the purity of water entering the compression chamber.

The challenge of poor lubricity of water compared to oil for driving rotors in compression chamber is handled by using a time gear set that synchronizes the rotation of male and female rotors. An integrated oil sump was provided within the air end for lubricating the bearings and timing gears. The method of splash lubrication is used to lubricate the gears and bearings; this eliminates the need of a dedicated oil circuit. The cold water returning from the water cooler is used to remove the heat generated in the oil. A sealing system comprising of a combination of viscos-seals and floating carbon seals is used to prevent the mixing of water from the compression chamber with the oil from the gear and bearing chambers.

3. Analysis of lifecycle costs for different compressor systems

The lifecycle costs of the compressor system include initial capital expenditure, operating cost of energy used to drive the system, maintenance costs and disposal costs. The disposal costs are assumed equal for all evaluated compressors systems and therefore not elaborated further. The costs are based on modelling of the performance and independent audit reports on global manufacturing competitive index [6], [7], [8].

3.1. Initial costs

The initial cost of a compressor system is related to the required flow or power rating but is highly influenced by the type of compressor system. The logistics of the manufacturing location, transportation, installation etc. also influence the compressor initial cost.

The Initial costs (purchase prices) of oil free and oil injected compressors of various manufacturers were collected from the respective users of these compressors. The data was collected from users in seven countries including India, China, Indonesia, Germany, Italy, Brazil and the USA. The Initial costs (purchase price) of water injected compressors were estimated considering manufacturing costs of the designed compressor components made by the author and corrected to global manufacturing costs considering manufacturing index. Overheads, profits averaged by several compressor system manufacturers were considred for intial cost estimation..

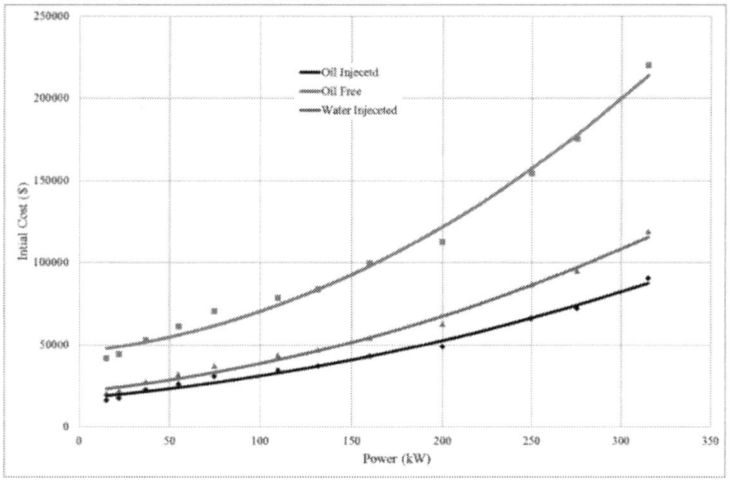

Figure 3 Comparison of average initial costs of three compressor system types

The average initial costs (purchase prices) of conventional oil free, oil injected compressors of various manufacturers at different parts of the World along with the estimated selling prices of water injected compressors in the power range of 15 to 315 kW were represented in graphical format in Figure 3. In order to show the relative comparison of costs for different compressor systems, costs for future analysis will be normalised to the relevant costs of the oil injected compressor system. The normalised initial costs for the range of considered compressor systems are shown in Figure 4.

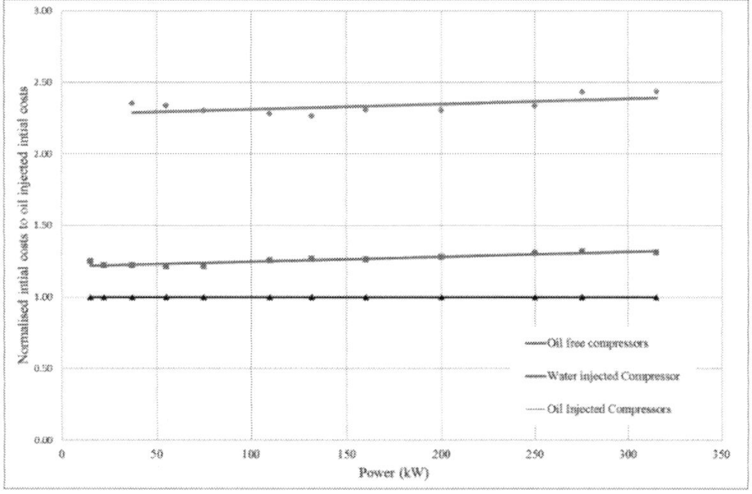

Figure 4 Normalised initial costs of the screw compressor systems.

The initial costs of water injected screw compressor system is 20-30% higher than the oil injected compressor system due to the additional costs associated with materials and surface coatings employed to prevent corrosion resistance. Oil free compressor systems have the highest initial costs due to the two stage configuration required for this pressure range, internal timing gears, gearbox, inter cooler and after cooler.

3.2. Maintenance costs

The cost of maintenance accounts for cost of parts which need to be changed at specified intervals and the cost of labour. It depends on the number of consumables, their service life and their cost. For oil injected and oil free compressors, consumable costs were based on the recommendations of several manufacturers for maintenance of their systems. For water injected compressors, the consumable cost is based on the expected life and the type of consumables. The cost of consumables is estimated by averaging quotes received from manufacturers of those parts.

Figure 5 shows the comparison of the maintenance costs in the power range between 15 to 315 kW for of all three types of compressor systems. The maintenance of oil injected compressor system will cost almost twice the maintenance of other two systems with water injected system being marginally more expensive than the oil free system.

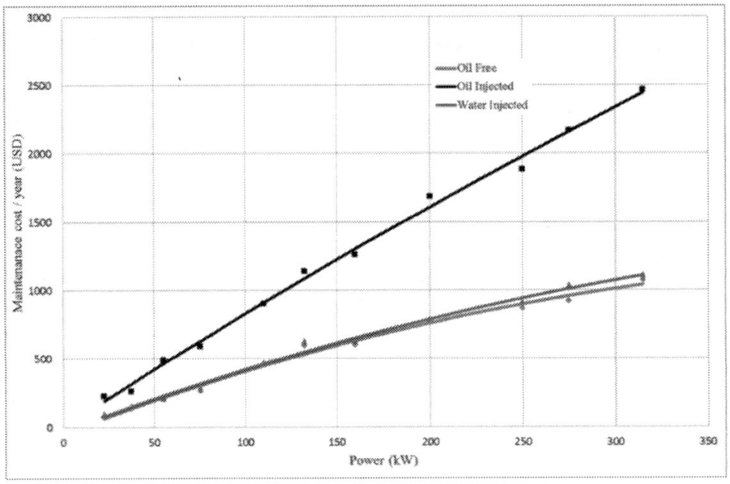

Figure 5 Estimated annual maintenance cost for different types of compressor systems

3.3. Operating costs

The operating cost of a screw compressor system is measured through the cost of electricity for operating the compressor in order to meet the user requirements. This includes the power required for the compressor and auxiliary devices inside a compressor system like the oil pump, air cooler fan, water cooler pump etc. The operating costs were calculated in the load and unload mode of operation and considering various cycles of load and unload which depend on the capacity of installed compressor system and the plant requirement for the compressed air.

The operating costs were calculated for three application cycles; 1) Full load, 2) Part load of 70% operation and 30% unload which is most commonly used in sizing of the system and 3) Part load of 30% operation and 70% unload which reflects emergency situations. These three modes are commonly accepted compressor application cycles in manufacturing industries [4]. The average number of hours in operation annually is assumed to be 5000. The assumed cost of electricity is 0.1 $/kW

The power consumption depends on the performance of the installed screw compressor system. The performance of the designed family of water injected compressor system in the power range of 15 to 315 kW was estimated using well established proprietary software suite for thermodynamic modelling of screw compressors, SCORPATH [4] developed at City University London. The performances aspects of the existing oil free and oil injected compressor systems in the same power range are established by collecting published data from various manufactures of these compressor systems and averaging them. The performance of all three considered compressor systems is shown in diagrams in Figure 6 and Figure 7.

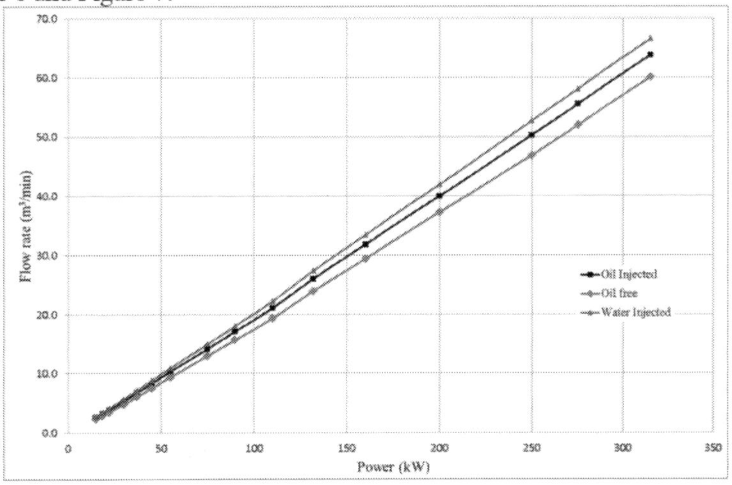

Figure 6 Average specific power of considered compressor systems at 7 barg discharge

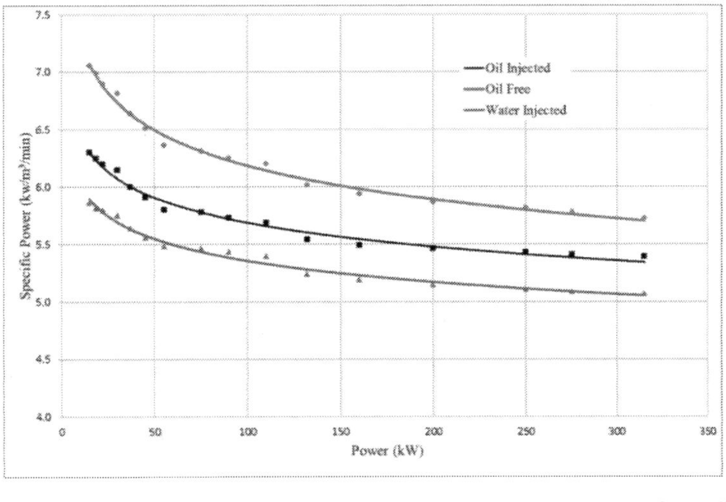

Figure 7 Averaged specific power of considered compressor systems at 7 barg discharge

The comparison of operating costs for the considered compressor systems at full load is shown in Figure 8. The operating power costs of water injected compressor systems are 5% and 15% lower than oil injected and oil free systems respectively in full load operation. The data indicates that the operating power cost trend follows the efficiency of the compressor type. As shown in the normalised operation cost diagram, Figure 9, this trend remains the same during the compressor operation in 70-30 mode. In the 30-70 application cycle, the operating costs of water injected compressor system are 10% lower than both, the oil injected and oil free compressor systems. This reduction is mainly due to the lower unloading power of the water injected compressor systems than that of the other two types.

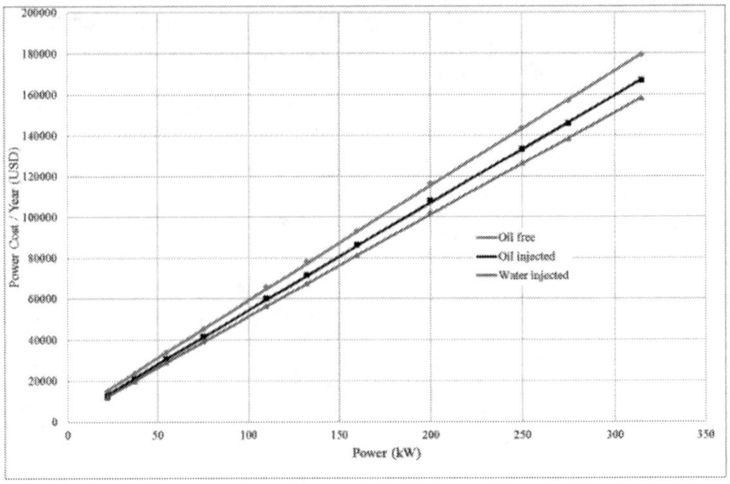

Figure 8 Operating annual power costs at full load and 7 barg discharge

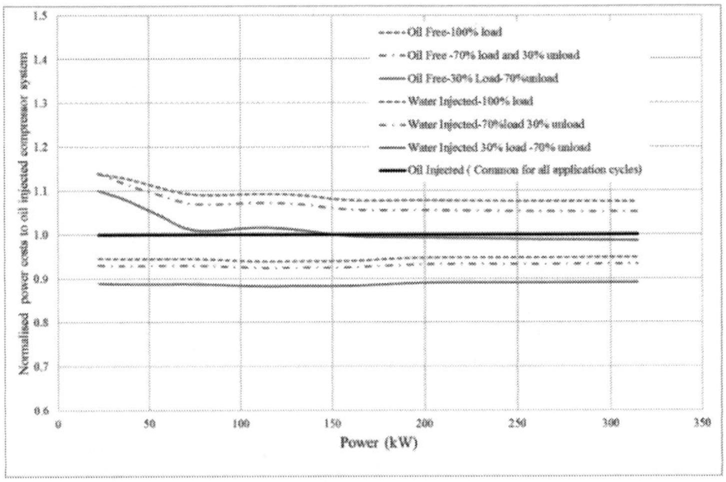

Figure 9 Operating costs of the considered compressor systems normalised to the oil injected system

3.4. Lifecycle costs

The lifecycle costs analysis takes into account the total fixed and variable costs of installing, maintaining and operating a compressed air system over its useful life. In this study, the lifecycle costs are calculated for 5 years operation and normalised to the cost of the oil injected screw compressor system. Full details on the analysis of the lifecycle costs are presented in [3].

The initial and lifecycle costs of water injected compressor system are significantly lower than the oil free compressor systems, Figure 10. For applications where oil injected compressor systems are not an option due to the high quality air requirements, water injected compressor systems would be better choice than oil free systems. The users get immediate benefit on cost during the purchase of water injected compressor system compared to oil free compressor system and even higher benefit on the long term use.

The initial cost of water injected compressor systems is higher than the oil injected compressor systems with multistage filtration; however, the life cycle cost of water injected compressors is lower up to 7%. The highest benefit is achieved for the 30-70 cycle. An average payback period for water

IOP Conf. Series: Materials Science and Engineering **90** (2015) 012012 doi:10.1088/1757-899X/90/1/012012

injected compressor systems will be less than 3 years in the power range of 55 to 315 kW, and between 4 to 5 years in the power range of 22 to 37 kW.

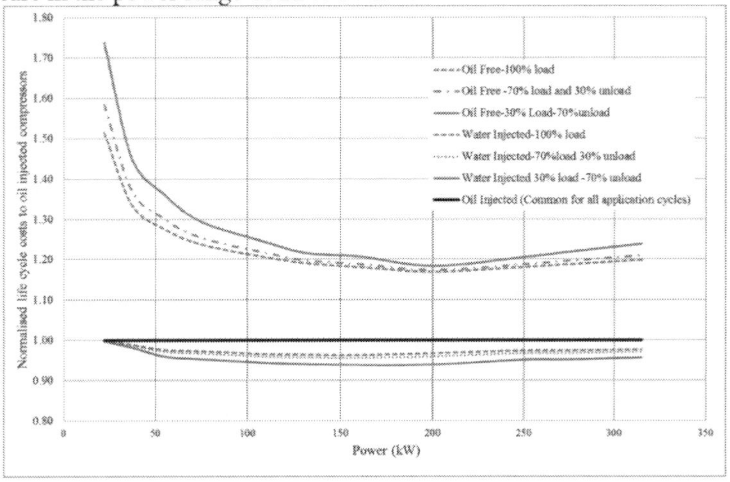

Figure 10 Normalized lifecycle costs of considered compressor systems to the oil injected system

4. Validation

4.1. *Performance measurement*

To validate the performance predictions, establish elements of the lifecycle costs and to examine reliability aspects of the material and components, a prototype of the water injected compressor in the power range of 55 to 75 kW was built. The prototype was tested over 1500 hours in order to obtain performance and to observe any potential performance deterioration and reliability issue over time, Figure 11.

The initial cost of the water injected prototype system was established and compared with the estimates made as part of the study.

The operating cost of a screw compressor is estimated through the cost of power which is related to performance. This also includes the power consumption of auxiliary devices inside a compressor plant like the oil pump, air cooler fan, water cooler pump etc.

Figure 11 Water injected compressor prototype under testing

Table 2 Measured performance and associated costs of operation at 7 barg discharge pressure

Compressor Power [kW]	Actual flow rate [m^3/min]	Total Power Full load [KW]	Total power Unloaded [kW]	Operating costs [$]		
				full load	load/unload 70-30%	load/unload 30-70%
55	9.52	56.1	11.2	28050	21315	12335
75	12.86	77.5	15.4	38750	29435	17015

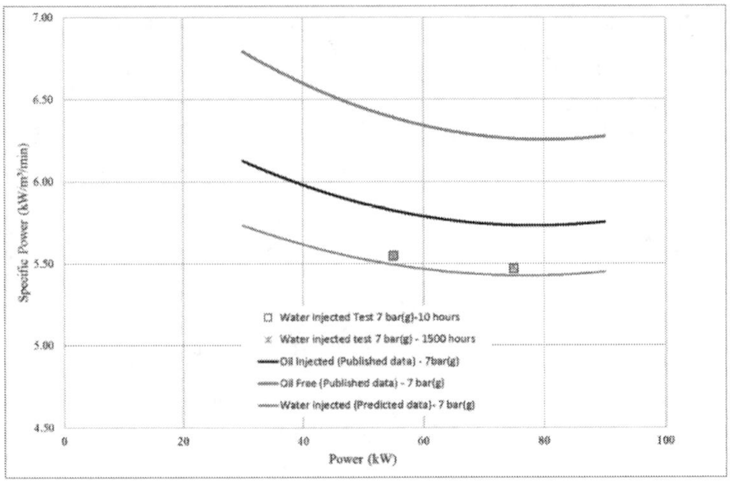

Figure 12 Comparison of measured and estimated specific power at 7 barg

The measurements showed that measured flow rates and powers for both compressors and both measured discharge pressures of 7 and 10 bar(g) did not change after 1500 hours of operation. Measured performance and associated costs are shown in Table 2 and Figure 12.

4.2. *Initial and operating costs*

The cost of building the prototype of water injected screw compressor system was compared with the estimates made earlier. The actual cost of the prototype was 9 % higher for the 55 kW and 11% higher that the estimated cost of the 75kW prototype. Since these two models were built on a common platform, the compressor components were specified such that all the components other than electric motor and drive gear sets are same for both 55kW and 75kW models. The production costs are recalculated based on the experience with building the prototype. The volume production will see the investment costs reduce 1-2% compared to the estimated value.

Since the performance of the prototype was within 2% of the estimated values and the operating costs of the measured compressor will be within the same deviation range to the estimated costs, r calculation model for operating costs is validated for the 55-75 kW water injected screw compressor system.

4.3. *Reliability Aspects*

The performance and reliability of the water injected compressor system was tested for 1500 hours of operation. The tests were performed in full load mode for 500 hours and 70% load and 30% unload mode for next 500 hours and then returned to full load mode and tested up to1500 hours.

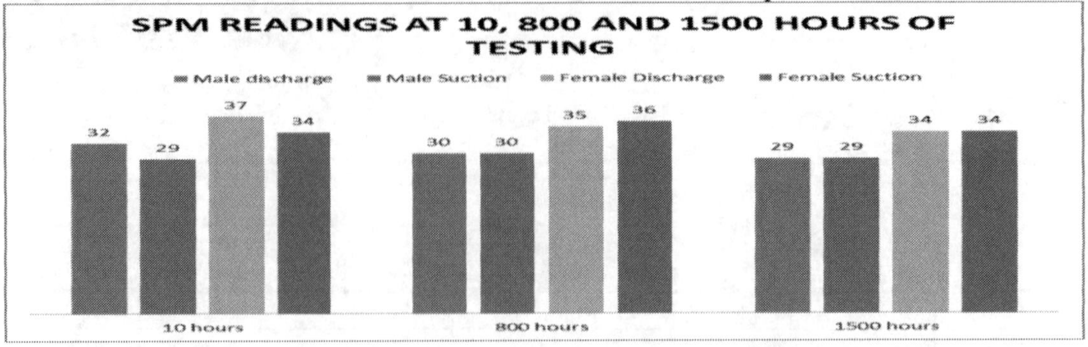

Figure 13 SPM (dBC) readings at 10, 800 and 1500 hours of operation

The compressor showed stable operation during this time. The measurements of bearing performance, seal performance and coating performance were taken at 10, 800 and 1500 hours of operations.

The bearing performance is normally determined by the shock pulse measurement (SPM). The increase in SPM by 15 units from the initial value is considered as the deterioration of bearings. The details of shock pulse meter (SPM) maximum readings (dBM) are shown in Figure 13. No visible increase in SPM was registered during the test.

The compressor housings, rotors, water separator tank and pipes, were coated with recommended coating and monitored during the endurance test of water injected system prototype. At the end of 1500 hours these were visually inspected. No visible signs of corrosion are noticed. The recommended coatings can provide reliable solution to prevent corrosion in the water injected compressor systems.

5. CONCLUSIONS

The study was carried out in order to identify the range of economically optimal water injected screw compressor systems considering performance, reliability and costs for industries requiring oil free air.

The performance and economic aspects including initial and lifecycle costs of oil free and oil injected screw compressors in the power range of 15 to 315 kW were established based on the data available from various manufacturers of these compressors. A family of water injected compressor system in the power range of 15 to 315 kW was designed. The performance and economic aspects of the newly designed water injected compressor system were estimated and compared with the established performance aspects of oil injected and oil free screw compressors.

Although the initial costs of the water injected screw compressors system are 20% higher than oil injected compressor, its operating costs over the period of 5 years are up to 11% lower depending upon the operation mode. Therefore lifecycle costs of the water injected compressor system are 5-7% lower than the oil injected compressor system and 25% lower than for the oil free system. The water injected system provides air completely free of oil.

Return on investment for water injected compressor systems in the power range of 55-315 kW is 2 to 3 years. Beyond this power, the break-even point is around 4 years.

The prototype of 55 to 75 kW power range was manufactured and tested over 1500 hours to validate the performance, reliability and economic aspects of water injected compressors. The testing confirmed estimated values.

Based on these findings it has been shown that water injected screw compressor systems are economically viable and reliable. These could become preferred option for delivering oil free air.

References

[1] ISO8173 standard – Compressed air quality, 2010
[2] "Compressed Air System Economics" by Office of Industrial Technologies, Department of Energy, USA 2012
[3] K. Venu Madhav, "The Commercial Viability of Water Injected Air Screw Compressor Systems", PhD Thesis, City University London, October 2014
[4] Stosic N, Smith I.K, Kovacevic. A, (2005): "Screw Compressors: Mathematical Modeling and Performance Calculation", Monograph, Springer Verlag, Berlin, 2005
[5] Venu Madhav K, Stosic N, Kovacevic A, Smith I. K, (200)1: "The Design of a Family of Screw Compressors for Oil–Flooded Operation", International Conference Compressors and Their Systems, London 2001
[6] "International comparison s of manufacturing productivity and labour cost trends", Bureau of Labour Statistics, US Department of Labourm, 2011
[7] "International Comparisons of Hourly Compensation Costs in Manufacturing Industries", Bureau of Labour Statistics, US Department of Labour, 2012
[8] "International Indexes of Consumer Prices 18 countries and areas, 1999-2013", Bureau of labour statistics, US department of Labour, 2013

IOP Conf. Series: Materials Science and Engineering **90** (2015) 012013 doi:10.1088/1757-899X/90/1/012013

The Characteristics and Causes of Push-off During Screw Rotor Manufacture, and its Compensation by Multi-axis Machine Path Calculations

X. Wang and C. S. Holmes
PTG Holroyd Advanced Developments, Holroyd, Harbour Lane North, Rochdale, OL16 3LQ, UK.

Email: Xianfa.wang@holroyd.com

Abstract. When grinding helical components, errors occur at the beginning and end of the contact path between the component and grinding wheel. This is due to the forces on the component changing as the grinding wheel comes into and out-of full contact with the component. In addition, shaft bending may add depth changes which vary along the length. This may result in an interrupted contact line and increased noise from the rotors. Using on-board scanning, software has been developed to calculate a compensated grinding path, which includes the adjustments of head angle, work rotation and infeed. This grinding path compensates not only lead errors, but also reduces the profile errors as well. The program has been tested in rotor production and the results are shown.

1. Introduction

Screw rotor pairs are designed to have a continuous line of contact along the entire length of their flanks, at or near the pitch line. Cutting or grinding tools exert forces on the rotor which might cause varying amounts of deflection along the length, thereby affecting the contact line.

This occurs when the grinding wheel comes into, and out of full contact with the component, as occurs at the beginning and end of each flute. In addition, bending of the rotor shafts can occur, resulting in depth changes which vary along the length. These behaviours are commonly known as 'push-off', and their effect is to cause intermittent contact along the rotor length as the rotors rotate, with undesirable consequences.

Davey [1] has described one effect of having positive metal at one end of the rotor body. As the rotors rotate, the raised section comes into and out of mesh with the mating rotor, creating gaps on the driving flank where contact should occur, and a consequent reduction in clearances on the non-driving flank. To allow for this effect, profiles may need to be modified to keep the non-driving side clearances within tolerance, but the overall effect is increased leakage and reduced compressor efficiency.

Holmes [2] has presented evidence that intermittent contact results in transmission errors which are a direct cause of increased levels of rotor noise during operation.

Fortunately, push-off behaviour is usually consistent from part to part, and Davey described a method of applying local accelerations and decelerations to the work rotation as the cutting tool enters and leaves the flute. This has the effect of offsetting profile errors at each end of the rotor body.

This paper presents a method of compensation which includes the continuous adjustment of head angle, work rotation and infeed, and using measurements from on-board scanning in a Holroyd Zenith machine. The on-board scans are in good agreement with CMM scans after calibration with a master rotor. Software has been developed to calculate a compensated grinding path. By measuring the lead at the pitch diameter of all flutes and root radius along the body length together with the 360° profile scans at 3 transverse sections (middle and each end), a networked computer can generate the compensated grinding path from this scan data. This grinding path compensates not only lead errors, but also reduces the profile errors.

Content from this work may be used under the terms of the Creative Commons Attribution 3.0 licence. Any further distribution of this work must maintain attribution to the author(s) and the title of the work, journal citation and DOI.
Published under licence by IOP Publishing Ltd

IOP Conf. Series: Materials Science and Engineering **90** (2015) 012013 doi:10.1088/1757-899X/90/1/012013

2. Calculation of compensation path

The software is developed based on Holroyd's Zenith machine [3], whose axis designations are shown in Figure 1.

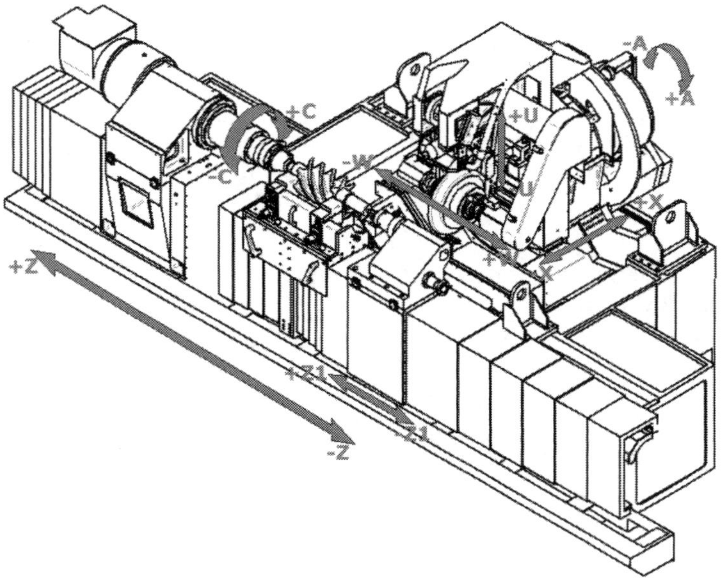

Figure 1. Zenith machine and its axes.

The push-off causes lead and profile errors which can be compensated by multi-axis adjustments as a function of work rotation, infeed, and head angle

$$\Delta w = \partial w/\partial C.\ \Delta C + \partial w/\partial X.\ \Delta X + \partial w/\partial A.\Delta A \qquad (1)$$

Where

W = Change in rotor contact position, mm

C = Work rotation, radians

X = Infeed, mm

A = Head angle, radians

$\partial w/\partial C$, $\partial w/\partial X$, and $\partial w/\partial A$ are calculated by Holroyd's HPMS program.

Formula (1) is used for both profile and lead.

The root scan is used to calculate ΔX within the rotor body. ΔX on the overrun section is calculated based on tests on rotor deflection (vertical and horizontal movements) under grinding force.

Three 360° profiles were scanned at the middle and 10 mm from each end of the body. The mean measured profile was calculated using the on-board software. The least squares method was used to calculate the head angle adjustments ΔA and infeed adjustments ΔX based on profile differences between the middle and end sections.

The lead scan results are used to calculated ΔW. As for the lead scan, $\partial w/\partial C$, $\partial w/\partial X$, and $\partial w/\partial A$ are constant coefficients at the pitch line. The head angle adjustment is applied linearly from the location where the push-off starts. ΔC is calculated along the grinding path from Formula (1). Further modification of the path may be needed if the first attempt does not achieve an acceptable result.

IOP Conf. Series: Materials Science and Engineering **90** (2015) 012013 doi:10.1088/1757-899X/90/1/012013

3. Grinding test results

For the purpose of testing, a male rotor was manufactured with the following geometry:

Flutes.	4
Nominal outside diameter	221.385 mm
Nominal lead	521.000 mm
Rotor body length	452.5 mm

A flute of the rotor was ground, then lead and profile were measured. A compensation path was created from the scan data, and applied to the full rotor grinding.

3.1 Profile scans

Figure 2 shows the profile differences between the middle and end sections before compensation. Figure 3(a) shows the predicted results of compensation calculated using the least squares method. It can be seen that the differences were mainly caused by depth change from the calculated result. Figure 3(b) shows the measured profile differences after compensation.

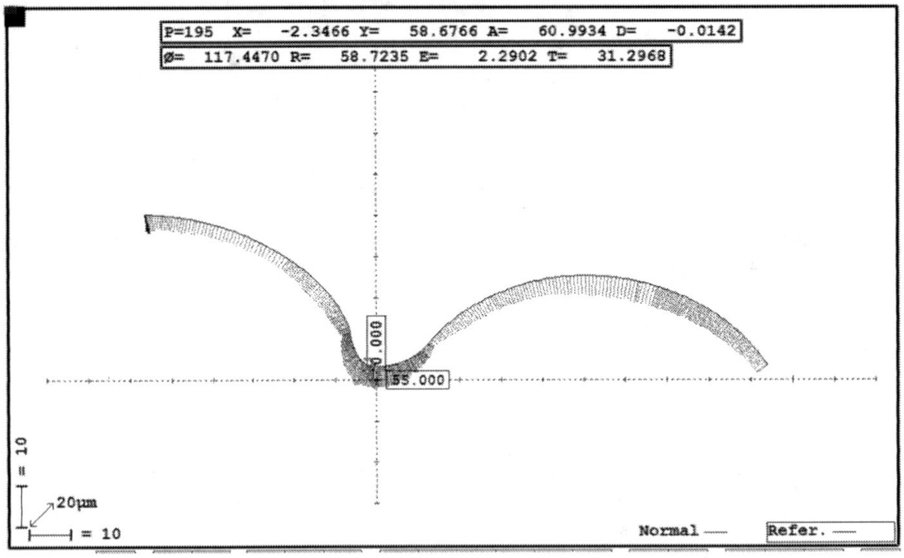

Figure 2. Profile differences between middle and end sections

IOP Conf. Series: Materials Science and Engineering **90** (2015) 012013 doi:10.1088/1757-899X/90/1/012013

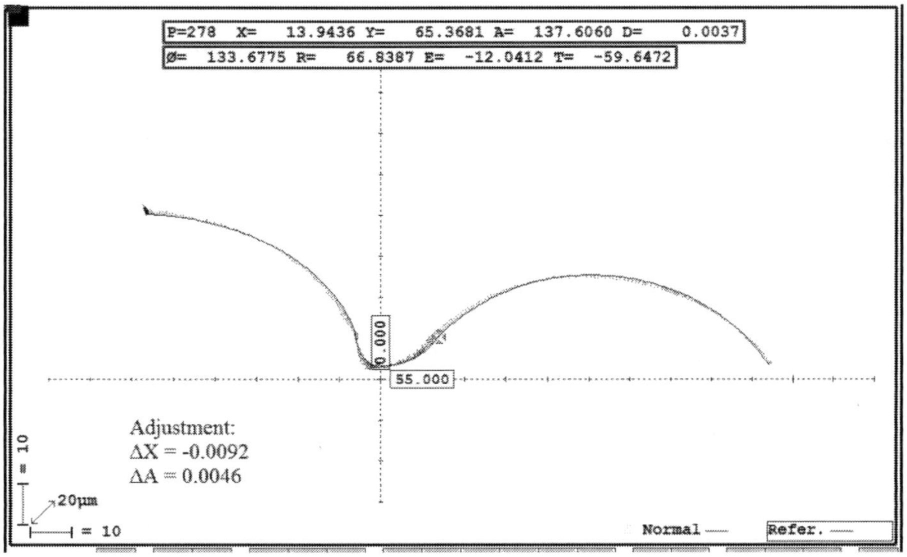

(a) Calculated results

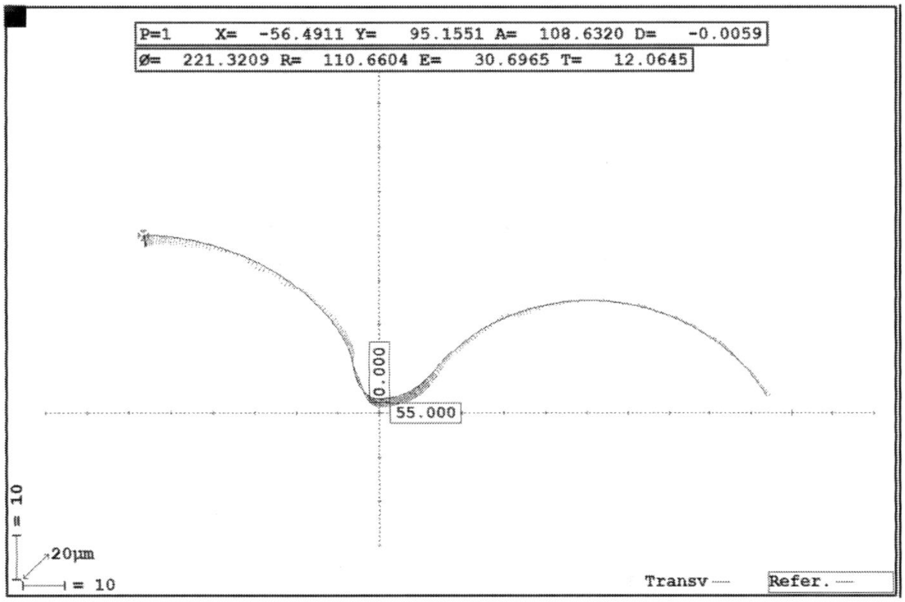

(b) Measured results

Figure 3. Profile after adjustment applied

IOP Conf. Series: Materials Science and Engineering **90** (2015) 012013 doi:10.1088/1757-899X/90/1/012013

3.2 Root depth

A typical root measurement scan is shown in Figure 4 (root depth versus traverse position). It can be seen that the rotor depth variation is 0.025 mm, and deepest at the end of the body.

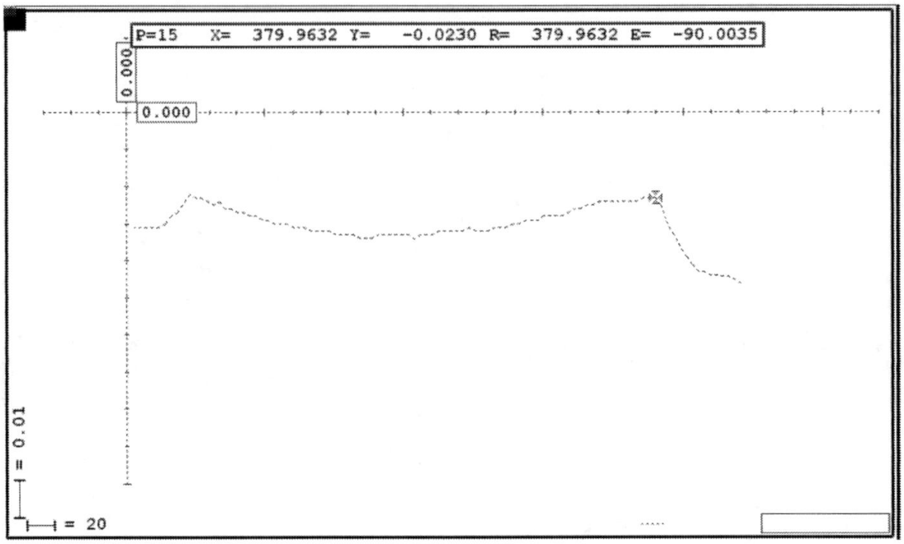

Figure 4. Root scan before compensation (vertical scale: 1 division = 0.01mm)

Figure 5 shows the root scan after the compensation. Note the change in vertical scale. It can be seen that the root variation has been reduced significantly to 0.006mm.

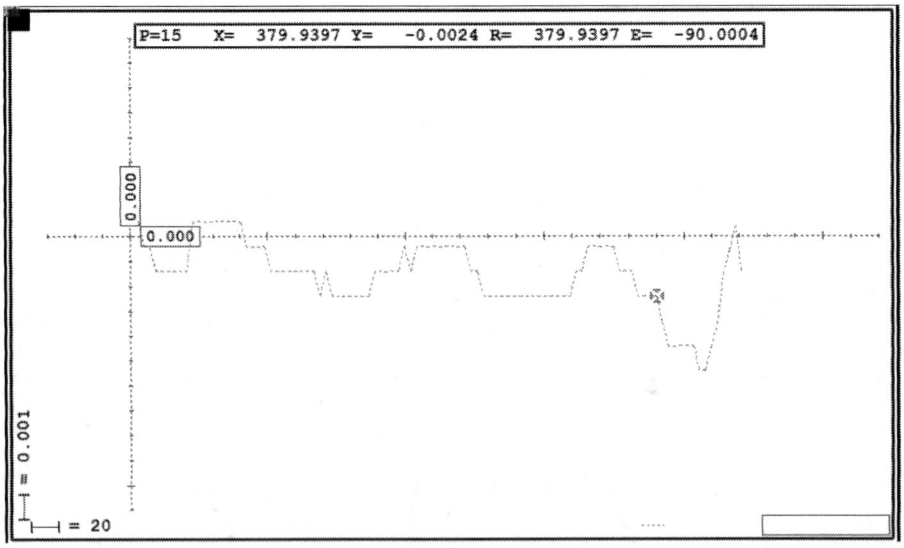

Figure 5. Root scan after compensation (vertical scale: 1 division = 0.001mm)

3.3 Lead results

The single flute lead of the rotor was scanned on Zenith and CMM for comparison. It appears that the scan results have good agreement as shown in Figure 6. The single flute lead scan results of the rotor ground with a compensation path are shown in Figure 7.

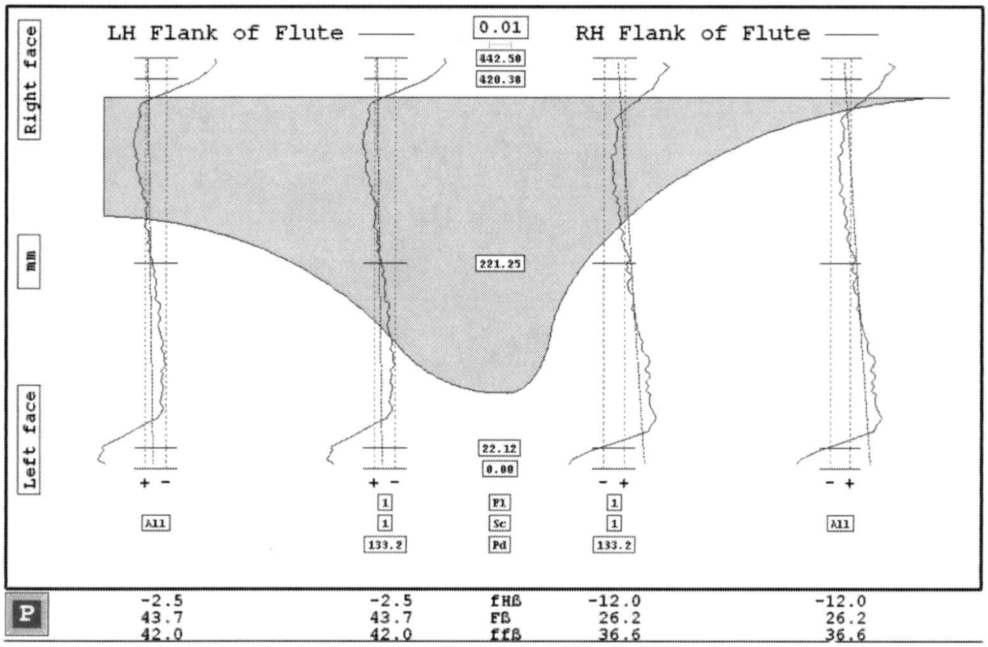

(a) Machine Lead Scan Results.

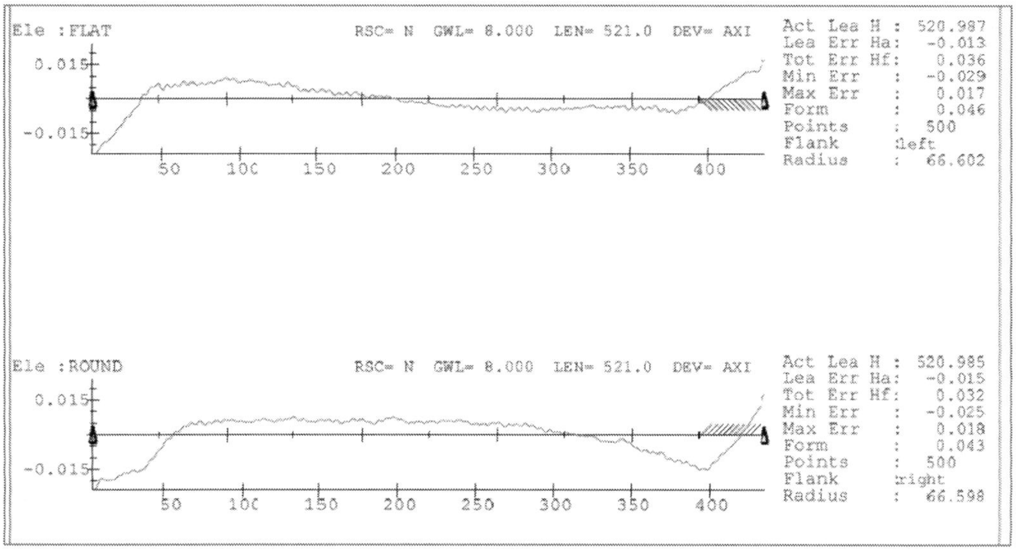

(b) CMM Lead Scan

Figure 6. Lead errors of a rotor ground without a compensation path

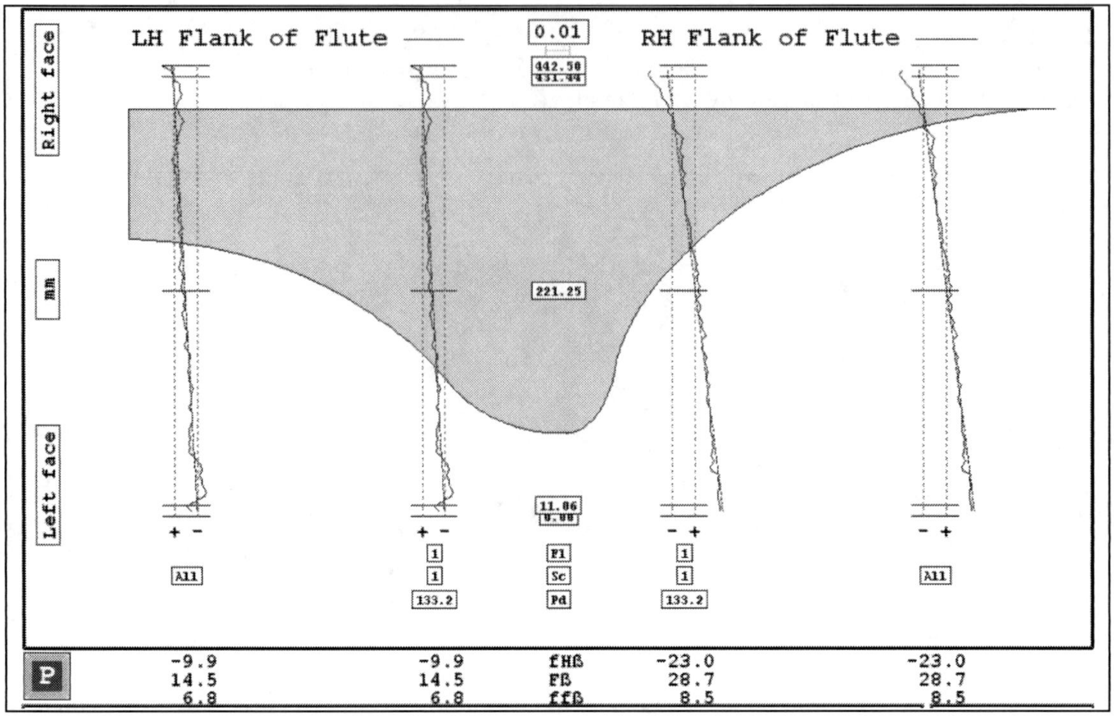

(a) Machine Lead Scan

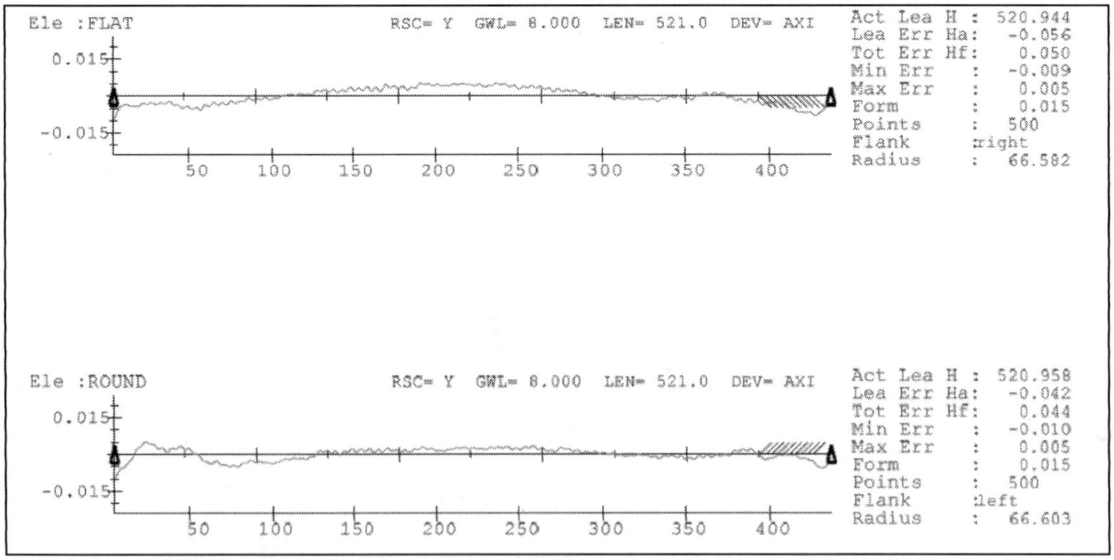

(b) CMM Lead Scan

Figure 7. Lead errors of a rotor ground with a compensation path

Figures 8 and 9 show full lead inspection of the rotor inspected on a customer's CMM. It can be seen that the leads have been improved significantly.

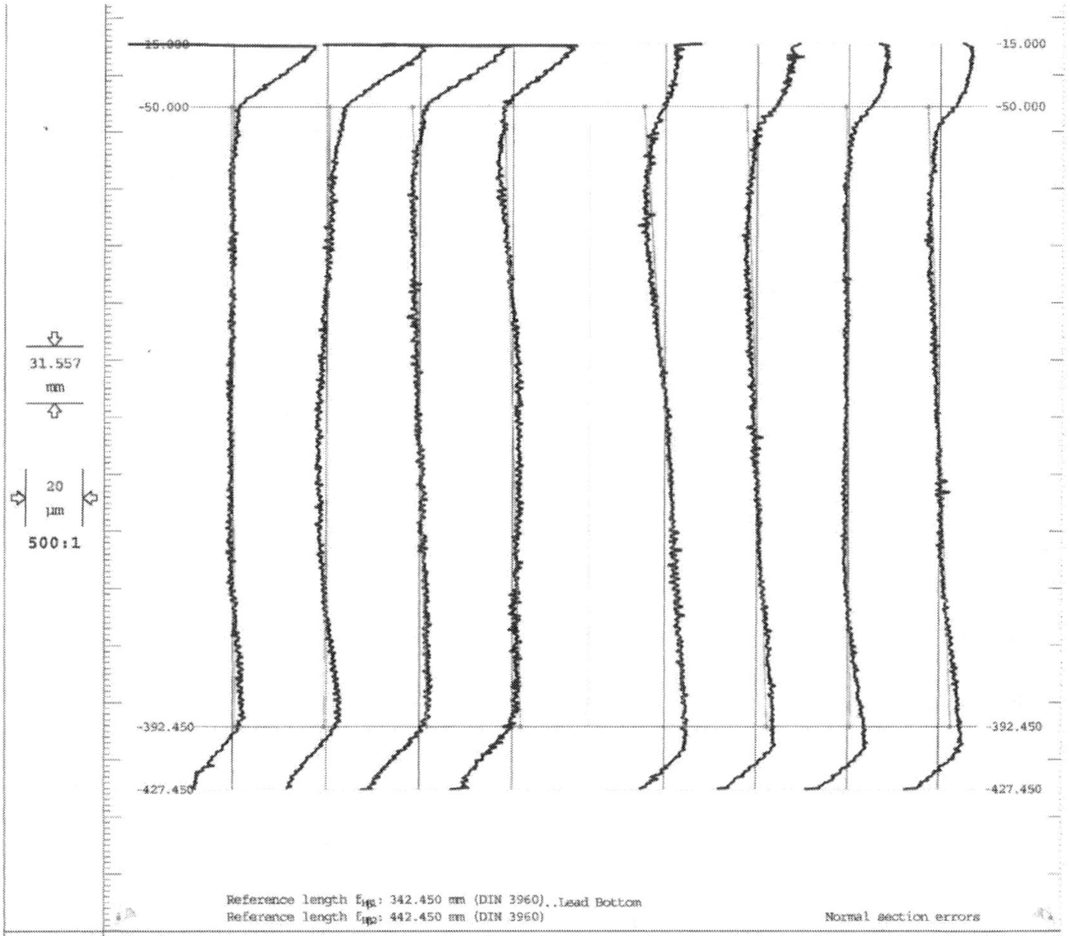

		[...]	Ø	4	3	2	1	1	2	3	4	Ø	[...]
F_{pz}	µm	±22	-8	-4	4	-15	-16	-36	-19	-5	-21	-20	±22
P_z	mm	521.022 / 520.978	520.992	520.996	521.004	520.985	520.984	520.964	520.981	520.995	520.979	520.980	521.022 / 520.978
F_β	µm	15	9	7	11	8	10	18	12	10	13	13	15
$f_{f\beta}$	µm	15	8	6	10	9	9	12	14	11	11	12	15
$f_{H\beta 1}$	µm	±12	3	2	-1	5	6	15	8	2	9	8	±12
$f_{H\beta 2}$	µm	±12	4	2	-2	7	8	19	10	2	11	11	±12

Figure 8. CMM inspection of a rotor ground without a compensation path

IOP Conf. Series: Materials Science and Engineering **90** (2015) 012013 doi:10.1088/1757-899X/90/1/012013

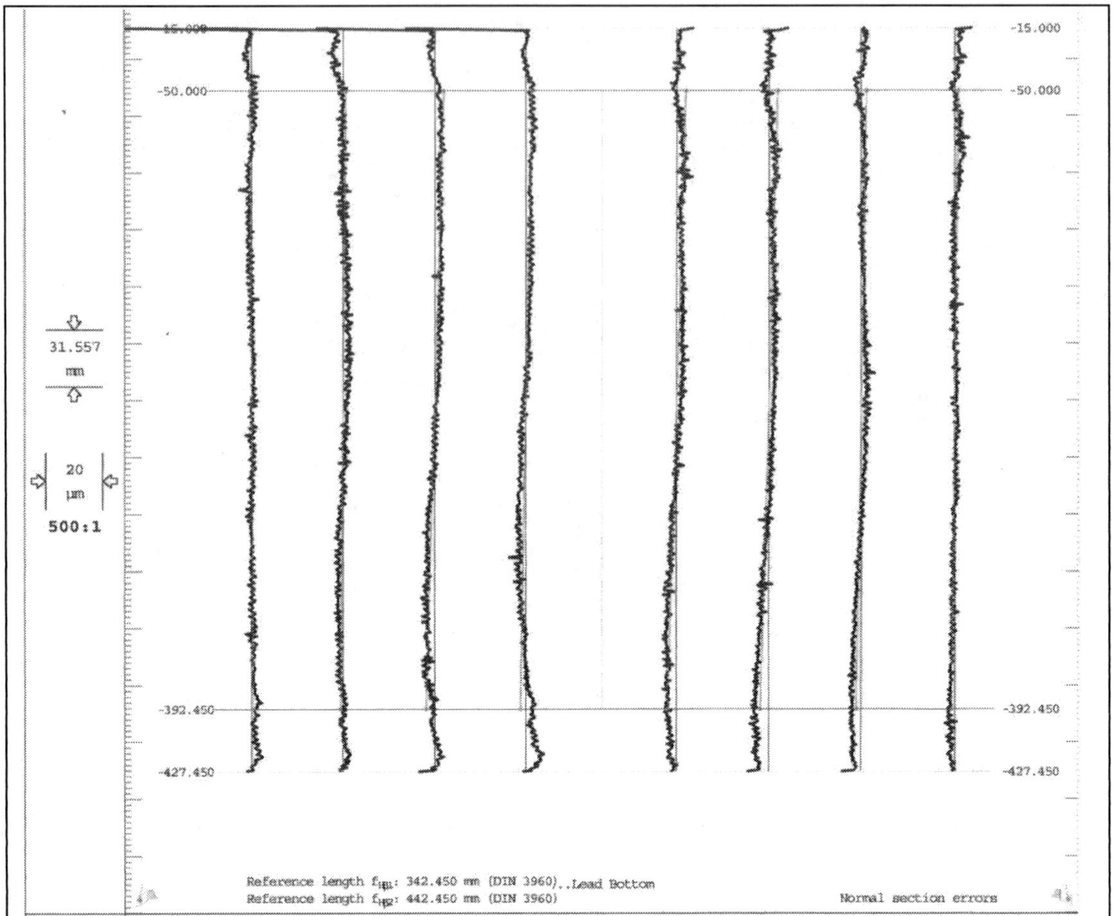

		[...]	Ø	4	3	2	1	1	2	3	4	Ø	[...]
F_{pz}	µm	±22	7	-3	5	17	11	15	14	9	5	11	±22
p_z	mm	521.022 520.978	521.007	520.997	521.005	521.017	521.011	521.015	521.014	521.009	521.005	521.011	521.022 520.978
F_β	µm	15	8	8	7	9	10	10	10	9	7	9	15
$f_{f\beta}$	µm	15	8	7	8	7	10	8	8	9	7	8	15
$f_{H\beta 1}$	µm	±12	-3	1	-2	-6	-4	-6	-6	-4	-2	-4	±12
$f_{H\beta 2}$	µm	±12	-4	1	-2	-8	-5	-8	-8	-5	-3	-6	±12

Figure 9. CMM inspection of a rotor ground with a compensation path

4. Conclusions

Using on-board scanning in the grinding machine, software has been developed to calculate a compensated grinding path, which includes the adjustments of head angle, work rotation and infeed. The on-board scans are in good agreement with CMM scans after calibration with a master rotor. By measuring the lead at the pitch diameter of all flutes and root radius along the body length together with the 360° profile scans at 3 sections, which are middle section plus two end sections, a networked computer can generate the compensated grinding path from this scan data. This grinding path compensates not only lead errors, but also reduces the profile errors as well. The program can also take into account the required lead modification in additional to push-off compensation.

The program has been tested in rotor production. The rotors produced using this method showed expected improvements on CMM lead scan at the pitch diameter. The profile errors at the beginning and end of the rotor body are also reduced. The quality of manufactured rotors is significantly improved.

References

[1] Davey, D. W., Keay, P, Yates, G. *Method for the correction of localised lead errors during screw rotor manufacture using CNC grinding machines*. International Conference on Compressors and Their Systems, City University London. IMechE, 2011.

[2] Holmes, C. S. *Noise reduction in screw compressors by the control of rotor transmission error*. International Compressor Engineering Conference at Purdue, C145, 2006.

[3] Holmes, C. S. *Developments in the productive grinding of screw rotors and other helical parts in response to user feedback*. International Conference on Compressors and Their Systems, City University London. IMechE, 2011.

[4] Holmes, C. S., Stephen, A. C. *Flexible profile grinding of screw compressor rotors*. International Conference on Compressors and Their Systems, City University London. IMechE, 1999.

IOP Conf. Series: Materials Science and Engineering **90** (2015) 012014 doi:10.1088/1757-899X/90/1/012014

Parallel operation of NH3 screw compressors - the optimum way

B Pijnenburg, J Ritmann

BITZER Kühlmaschinenbau GmbH, Jens Juuls vej 16, 8260 Viby J. Aarhus, DK

E-mail: bas.pijnenburg@bitzer.de, john.ritmann@bitzer.de

Abstract. The use of more smaller industrial NH3 screw compressors operating in parallel seems to offer the optimum way when it comes to fulfilling maximum part load efficiency, increased redundancy and other highly requested features in the industrial refrigeration industry today. Parallel operation in an optimum way can be selected to secure continuous operation and can in most applications be configured to ensure lower overall operating economy. New compressors are developed to meet requirements for flexibility in operation and are controlled in an intelligent way. The intelligent control system keeps focus on all external demands, but yet striving to offer always the lowest possible absorbed power, including in future scenarios with connection to smart grid.

1. Introduction

Compressor units for industrial ammonia refrigeration are typically designed for maximum operating conditions (highest temperature, load, ambient conditions, etc.). The typical system has excess of capacity in most operating conditions and in many cases it must adapt itself to continuously varying capacity demand. Different methods exist to vary the capacity of a compressor, whether it is a single compressor, several units in a rack, or several parallel compressor packages.

The choice of the right method is not always easy since each method can be better or worse with respect to the requirements to the capacity control system. The efficiency of the capacity control system is not only depending on the compressor construction. The driveline, especially in variable speed drive (VSD) applications plays an important role in the overall efficiency of the compressor unit.

The most common different capacity regulation methods are described first. A method to compare the total efficiency in the part load range of different units with either one or more compressors is developed. The method is then used to analyze the total efficiency for different number of compressors in a system and for two ways of capacity control.

2. Common methods and requirements for capacity control systems

Different methods exist to control the capacity of a compressor system. The quality of the capacity control systems can be measured on different parameters:

- The ability to accurately adapt the capacity to the cooling demand, i.e. how well a certain suction pressure set point can be followed.
- The influence on the efficiency of the compressor and the drive line, i.e. how much the COP of the system is affected.

Content from this work may be used under the terms of the Creative Commons Attribution 3.0 licence. Any further distribution of this work must maintain attribution to the author(s) and the title of the work, journal citation and DOI.
Published under licence by IOP Publishing Ltd

- The requirements to and load on the power supply.
- The possibility to cover a large capacity range between minimum and maximum load.
- Cost of the capacity control system.
- Operational reliability.
- Noise and vibration levels.

A thorough description of different methods can be found in the publication "Competence in capacity control" [1]. In this article only a short description is given of the most common methods for capacity control with open type ammonia screw compressors:

- On/off cycling
- Speed variation (variable speed drive)
- Slide valve regulation
- Control pistons

The first two methods are not incorporated in the compressor itself, but are about how to operate it. The last two methods require a specific construction inside the compressor. The first and last method offer step wise control, whereas the other two methods enable continuous variation of capacity. Each method is described briefly here, but a comparative study on speed and capacity slide control for screw compressors is described by Blumhardt (2006) [2].

2.1. On / off cycling
On/off switching is the most simple way of varying capacity. In simple systems it can lead to large variation in operating condition and high cycling rates. It can make sense on systems with small load variation or large system buffer capacity. The ability to regulate capacity precisely with this method improves with the amount of parallel mounted compressors.

2.2. Speed regulation
The flow through the compressor is varied with the rotational speed of the rotors. The speed regulated screw compressor requires a specific drive line with a VSD. In most cases the speed regulation does not require changes to the compressor itself, but it affects the way the compressor operates.

2.3. Slide valve regulation
A slide valve parallel to the rotor shaft can be moved to create an internal bypass. Just before the gas is compressed in the rotor cavities, it can bypass internally to the suction side. The slide valve system makes a fairly simple and robust system for both stepwise and step less variation of capacity. The slide valve regulation will also affect the volume ratio (VI) of the compressor. The influence on VI depends on the design of the compressor and the operating conditions. Some compressor designs are capable of independent regulation of capacity and VI.

2.4. Control pistons
One or more radially mounted pistons create an internal bypass for the gas. The control of the bypass is discrete, often with only one or very few steps. Operation with control pistons influences the volume ratio, similarly to slide valve regulation.

The described methods are used both individually but also in combination. It is also possible to regulate capacity on the plant side through bypass systems or suction throttling systems. These systems are not common in ammonia refrigeration, due to their very poor energetic efficiency and therefore not considered here.

In the further perspective of this article, focus will be on regulation with speed, slide valve and on/off cycling of compressors. All three methods are common engineering practice in ammonia refrigeration today.

The variation of capacity can influence both the volumetric and isentropic efficiency of the compressor. The amount in which efficiencies are affected differ for each method and also depend on the operating conditions and the construction of the compressor. Volumetric efficiency relates to actual swept volume. In this article the term part load index (λ) is used to indicate the ratio between actual volume flow (at a certain load) and the nominal swept volume at maximum capacity (full load). Variation of capacity will also influence the efficiency of the drive line (motor and or VSD).

3. Description of efficiency of capacity control methods

The efficiency of the compressor, the motor and the VSD can each be expressed as function of some form of capacity (not necessarily the system cooling capacity). This efficiency data, also called performance data, is typically expressed in the form of graphs, tables or polynomial functions. The total efficiency of the system can be obtained by coupling the performance data of different components in series and/or in parallel. The relation between cooling capacity and system efficiency cannot be obtained directly from the supplied performance data, but requires some calculation. The relation will also depend on the operation condition of the system (evaporating and condensing temperatures).

Typical dependencies of the compressor part load index (λ) and isentropic efficiency (η_c) for a typical chiller condition (T0/Tc = +5/+35) are shown in figure 1. The software to calculate compressor performance [3] contains data based on extensive measurements with different compressor models over the whole range of operating conditions, shaft speeds and capacity slide positions.

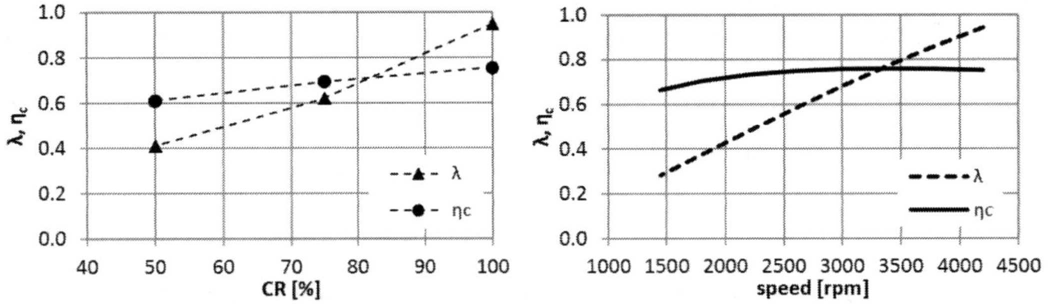

Figure 1. Example of compressor performance data curves.

A typical dependency of motor and drive efficiency, depending on shaft torque and shaft speed are shown in figure 2. Motor and drive data are calculated with software from a major global supplier of motors and variable speed drives.

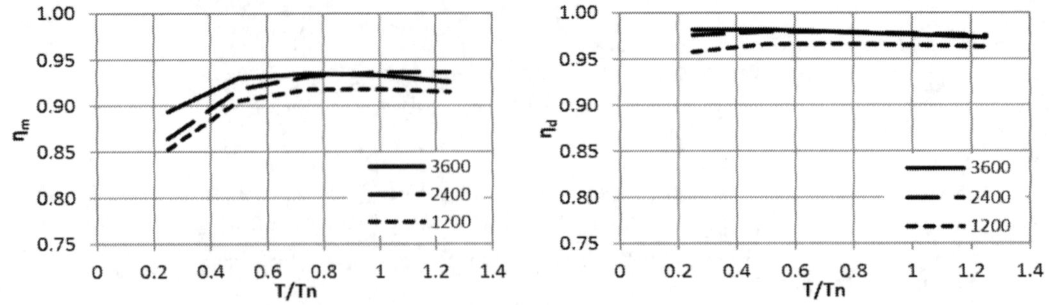

Figure 2. Example of motor and drive performance data curves.

The graphs show that cooling capacity is not mentioned directly. It must be derived from the performance data and the actual operating conditions of the system.

The capacity regulation systems can have other positive or negative effects on system performance, other than on energetic efficiency. Some of these effects that can be mentioned are:

- Potential for increased capacity with operation at speeds above synchronous with VSD.
- Built-in soft-start function in the VSD results in low motor and power supply load at start-up.
- Positive displacement compressors require practically a constant torque over the complete speed range. Therefor voltage vs. frequency ratio of the VSD must be constant. A VSD can normally not supply voltages above the supply voltage and this means that the motor will be supplied with "under-voltage" during operation above synchronous speed. This means it cannot supply the full torque, which can limit the possible operating conditions of the compressor in this speed range.
- Operation with economizer can be utilized for a large capacity range with speed regulation for most compressor designs. The economizer port is normally closed very early in the unloading process on compressors with slide valve regulation in combination with a fixed economizer port.
- Parallel compounding will require smaller compressors to deliver the same maximum capacity.
- Compressor, motor and drive efficiency increase to some degree with size.
- Parallel compounding automatically gives redundancy.
- Complexity of systems increases with the amount of compressors and combination of different regulation systems.

4. Method of part load efficiency evaluation

The part load efficiency of a compressor system can be evaluated based on the performance data of the components and the equations described in this section. The cooling capacity that a screw compressor can deliver is defined by equation (1).

$$\dot{Q}_0 = \dot{m} \cdot q_0 = \lambda \cdot V_S \cdot \rho \cdot q_o \tag{1}$$

The shaft power of a screw compressor is defined by equation (2).

$$P_c = \frac{P_{is}}{\eta_c} = \frac{\dot{m} \cdot p_{is}}{\eta_c} = \frac{\lambda \cdot V_S \cdot \rho \cdot p_{is}}{\eta_c} \tag{2}$$

The electrical power consumption of the screw compressor package is defined by equation (3).

$$P_{el} = \frac{P_c}{\eta_d \cdot \eta_m} = \frac{\lambda \cdot V_S \cdot \rho \cdot p_{is}}{\eta_d \cdot \eta_m \cdot \eta_c} \tag{3}$$

The coefficient of performance of the screw compressor package is defined by equation (4).

$$\mathrm{COP} = \frac{\dot{Q}_0}{P_{el}} = (\lambda \cdot V_S \cdot \rho \cdot q_o) / \left(\frac{\lambda \cdot V_S \cdot \rho \cdot p_{is}}{\eta_d \cdot \eta_m \cdot \eta_c}\right) = \frac{q_o}{p_{is}} \cdot \eta_d \cdot \eta_m \cdot \eta_c = \frac{q_o}{p_{is}} \cdot \eta \tag{4}$$

Plotting the COP as function of the cooling capacity ($\dot{Q}_0$) is the traditional way of evaluating the part load performance. One can see from the above equations, that for a certain operating condition and for a certain size of compressor system, there is a linear relation between the cooling capacity and the part load index (λ) and between the COP and the energetic efficiency (η). By showing the energetic efficiency as function of the part load index we have a way to show the part load performance on a dimensionless scale from 0 to 1, which makes it easy to compare for different operating conditions and different sizes of systems.

The total part load index and energetic efficiency for a compressor package with multiple compressors can be found from the weighted sum of the individual efficiencies according to equation (5) and equation (6).

$$\lambda = \frac{\sum_{i=1,n} \lambda_i \cdot V_{s,i}}{\sum_{i=1,n} V_{s,i}} \tag{5}$$

$$\eta = \frac{\sum_{i=1,n} \lambda_i \cdot V_{s,i}}{\sum_{i=1,n} \frac{\lambda_i \cdot V_{s,i}}{\eta_{d,i} \cdot \eta_{m,i} \cdot \eta_{c,i}}} \tag{6}$$

The performance data curves (figure 1) show that part load index (λ) and energetic efficiency (η_c) for a compressor can be controlled by two parameters: speed and capacity slide position. Operating conditions are assumed constant in this work, but they can also depend on capacity, i.e. evaporating temperature will rise as evaporator load decreases.

The energetic efficiency of the motor (η_m) and the VSD (η_d) depend on speed and torque (as seen in figure 2). The torque and in case of VSD also the speed will vary with capacity. The compressor shaft power relates to the torque and speed according to equation (7).

$$P_c = \frac{\lambda \cdot V_s \cdot p \cdot p_{is}}{\eta_c} = T \cdot \omega \tag{7}$$

The shaft speed is either constant or controlled actively, thus the shaft torque depends only on compressor part load index, energetic efficiency and speed. A motor is typically selected with 10% torque reserve at maximum (or nominal) load giving the torque at maximum system load to be 90% of the nominal motor torque ($T/T_n = 0.9$). The factor 0.9 is not constant and will differ based on the available motor sizes and application type. The part load torque of the motor relative to its nominal torque is defined by equation (8).

$$\frac{T}{T_n} = 0.9 \cdot \frac{\lambda}{\lambda_n} \cdot \frac{\eta_{c,n}}{\eta_c} \cdot \frac{\omega_n}{\omega} \tag{8}$$

The part load performance, described by the dependency between part load index and energetic efficiency can be derived quite easily from the compressor performance data. By combining equation (8) with interpolation of motor and drive performance data, it is also possible to derive the part load efficiency of the motor and drive as function of the part load index. An example of the relation between the part load index (λ) and the energetic efficiency (η) is shown in the figure 3.

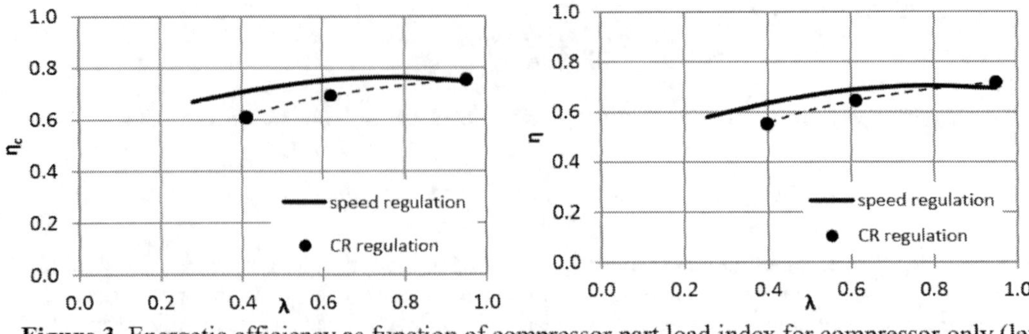

Figure 3. Energetic efficiency as function of compressor part load index for compressor only (left) and complete system (right).

5. Comparison of different solutions

The part load performance of different compressor package solutions is compared. The most basic compressor only regulates with a slide valve. Next step is to add a VSD and finally to use multiple compressors. Simulations of many different solutions were done to map the different combinations.

All comparisons are done for the same total maximum capacity. For easy comparison, the values for part load index were scaled (indicated by λ_s) to be exactly one at maximum capacity. The effect of size, where larger machines tend to have better efficiency is taken into account by using actual performance data from appropriate sizes of compressors, motors and drives. The effect of size and the effect of using a VSD becomes evident when looking at the nominal maximum load efficiency. Multiple smaller compressors will have a slightly lower maximum load efficiency when compared to fewer larger compressors. Solutions with VSD will have a slightly lower maximum load efficiency than comparable size compressors without. The losses in a VSD on multiple compressor systems naturally only affect the one compressor with VSD, which has a positive effect on the total losses.

Figure 4 shows a comparison of the total efficiency over the capacity range for a solution with one or two compressors. It shows that solutions without VSD only have an efficiency close to the maximum value in a limited range. The solutions with VSD show a more constantly high efficiency level. The solution with two compressors can keep the high efficiency down to approximately 28% capacity, whereas a single VSD compressor shows considerable drop in efficiency below 43% capacity. All solutions have high efficiency in the upper range above approximately 65% capacity.

The horizontal dash-dotted lines show a band width of 10% efficiency, in order to show the magnitude of the differences between the compared solutions. It can be noticed that compressor performance data according to EN12900 allow a tolerance of up to 10% on published compressor efficiency data (COP). Since 2014 this tolerance is applicable for the entire part load range.

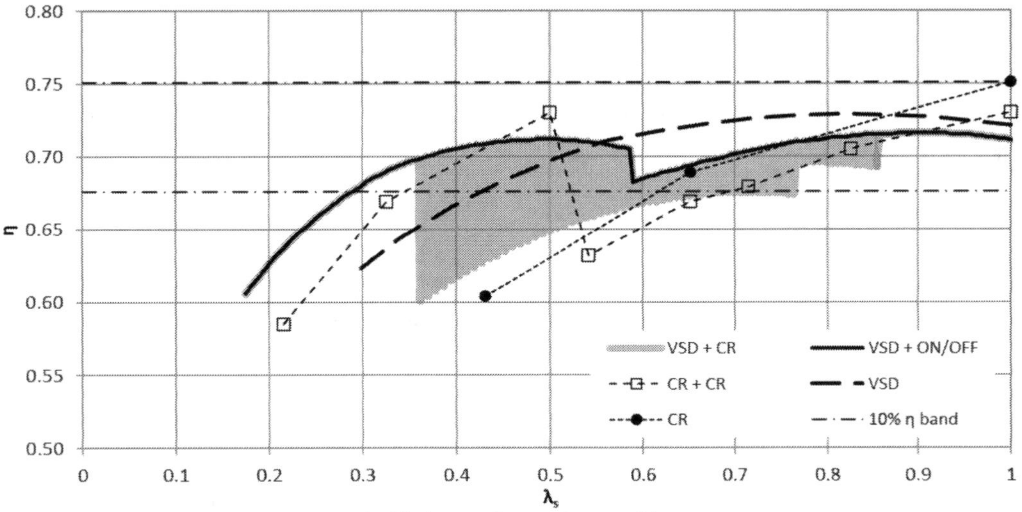

Figure 4. Comparison of part load efficiency for package with 1 or 2 compressors.

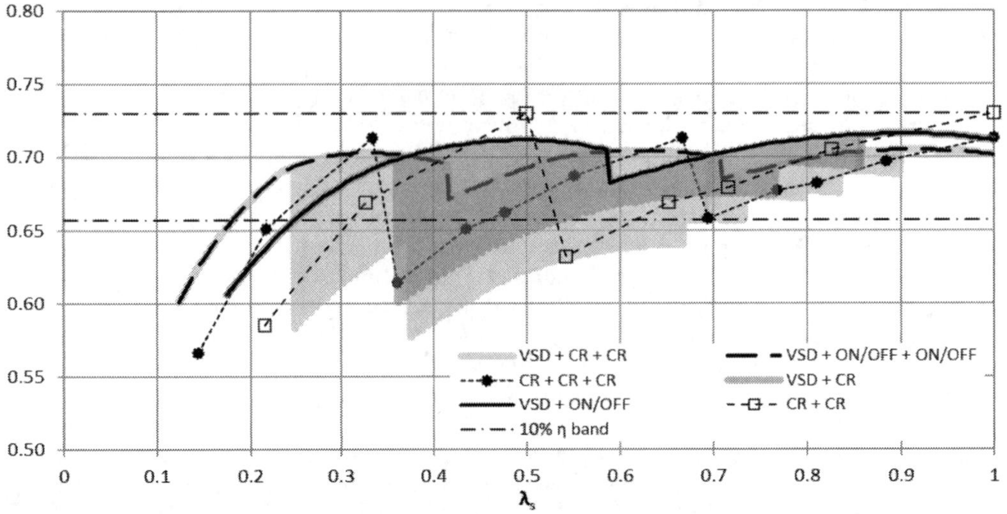

Figure 5. Comparison of part load efficiency for package with 2 or 3 compressors.

Figure 5 shows the comparison between a solution with either two or three compressor. Again it can be seen that solutions with VSD show a constant high efficiency. The solution with three compressors (one VSD) can even keep a very high efficiency down to approximately 17% capacity. It can also be noticed that a three compressor solution with only slide regulation (CR) has a high efficiency down to approximately 23% capacity, except from a minor load area around 37 - 43%.

The analysis of the different scenarios were done for operation at a fixed operating condition (T0/Tc = +5/+35°C). The results will change depending on the operating conditions and this should be considered when searching for the optimum solution. The results show that the flexibility to vary capacity increases with the amount of compressors and when adding minimum one VSD. Multiple compressors, combined with variable speed, increase the capacity range significantly in which the efficiency stays very close to the full load efficiency. The control system must off course be capable of handling this in order to take full advantage of this potential.

6. Conclusions

The optimum way of providing the highest possible part load efficiency over a widest possible range of capacities seems to be a compressor system with multiple compressors where one is fitted with VSD. The investigated systems can regulate capacity as far down as 17% (for a 3 compressor unit), while at the same time keeping the efficiency at a very high level (within 10% of the maximum).

If only operating in the upper capacity range (above approximately 65%), then all the presented solutions are within 10% of the maximum efficiency and the use of VSD may not even be justified economically.

Over a wide range of capacities, the differences in efficiency between the solutions are smaller than the allowed EN12900 COP-tolerance for the compressor only. Therefore a more detailed analysis is often needed to find the best solution. This analysis must take into account the variation of capacity demand, T0 and Tc over time. If this analysis is not available, then the solution with multiple compressor systems and VSD always adds flexibility in regulation and redundancy to the system.

7. Perspective

The importance of selecting the correct size of compressor, motor and drive suitable for the nominal load shall be mentioned. Very good part load efficiency should not be jeopardized by selecting excessively oversized equipment.

IOP Conf. Series: Materials Science and Engineering **90** (2015) 012014 doi:10.1088/1757-899X/90/1/012014

Control systems on the compressor package and other parts of the refrigeration system play a very important role to be able to take full advantage of the potential that multiple compressors and VSD offer. The maximum potential can only be exploited if the control system is able to control all parameters in an optimum way. Future control systems may also need to communicate with smart grid systems and production planning systems to be able to control capacity with some information of the expected future load. High flexibility and high efficiency in a large range of capacities is highly requested for these demands.

Nomenclature

COP	Coefficient Of Performance	[-]
CR	Capacity slide Regulation position	[%]
η	Total energetic efficiency of the compressor package	[-]
η_c	Isentropic efficiency of the compressor	[-]
η_d	Energetic efficiency of the variable speed drive	[-]
η_m	Energetic efficiency of the electrical motor	[-]
λ	Part load index, related to compressor nominal swept volume	[-]
λ_s	Scaled part load index, related to compressor nominal capacity	[-]
$\dot{m}$	Mass flow of refrigerant through compressor	[kg/s]
P_{el}	Electrical power consumption of the compressor package	[kW]
P_{is}	Isentropic power consumption of the compressor	[kW]
p_{is}	Specific isentropic compression work	[kJ/kg]
$\dot{Q}_0$	Cooling capacity of the compressor	[kW]
q_0	Specific cooling capacity	[kJ/kg]
ρ	Density of refrigerant at compressor inlet	[kg/m3]
T	Shaft torque	[Nm]
T0	Evaporating temperature	[°C]
Tc	Condensing temperature	[°C]
V_s	Nominal swept volume of the compressor at maximum load	[m3/s]
VSD	Variable Speed Drive	[-]
ω	Shaft rotational speed	[rad/s]
n	Subscript to indicate nominal values at maximum load	[-]

References

[1] BITZER publication A-600-5, 2014, Competence in capacity control
[2] Blumhardt R, 2006, Capacity control of screw compressors: speed or slide control – a comparitive study, BITZER publication SV-0402-GB
[3] Performance data for screw compressors: BITZER Software v. 6.4.2

IOP Conf. Series: Materials Science and Engineering **90** (2015) 012015 doi:10.1088/1757-899X/90/1/012015

New projects developed by COMOTI in gas industry

Marian Nitulescu
Dipl. Engineer COMOTI, *220 D Iuliu Maniu Bd., sector 6, cod 061126, OP 76, CP174Bucharest, Romania*
Valentin Silivestru
Scientific Researcher 1, Dr. Engineer President General Director , COMOTI, *220 D Iuliu Maniu Bd., sector 6, cod 061126, OP 76, CP174Bucharest, Romania*
Niculae Toma
Dr. Engineer COMOTI, *220 D Iuliu Maniu Bd., sector 6, cod 061126, OP 76, CP174Bucharest, Romania*
Cristian Slujitoru
Dipl. Engineer COMOTI, *220 D Iuliu Maniu Bd., sector 6, cod 061126, OP 76, CP174Bucharest, Romania*
Valentin Petrescu
Master. Engineer COMOTI, *220 D Iuliu Maniu Bd., sector 6, cod 061126, OP 76, CP174Bucharest, Romania*
Mihai Leahu
Director research and design department , *Transgaz, County Sibiu, Piața C. I. Motaș, Nr. 1, CityMediaș, Cod 551130*
Ciprian Oniga
Chief research dipl. engineer, *Transgaz, County Sibiu, Piața C. I. Motaș, Nr. 1, CityMediaș, Cod 551130*
Gheorghe Ulici
dipl. Engineer, *Transgaz, County Sibiu, Piața C. I. Motaș, Nr. 1 , CityMediaș, Cod 551130*

Abstract. The paper aims to present two new projects developed by the Romanian Research and Development Institute for Gas Turbines (COMOTI) in partnership with City University of London and GHH-Rand Germany, in the field of screw compressors/expanders. COMOTI passed, in recent years, from being a GHH-Rand licensed manufacturer for a range of oil-injected screw compressors, of CU type, to a new phase of range diversification, approaching screw compressors with a maximum discharge pressure of 45 bara. So, in cooperation with City University and GHH-Rand we design, manufacture and test, with air, in COMOTI test bench the new type of screw compressor named CHP 220. Also, the cooperation with GHH-Rand has resulted in the design, manufacture and air testing on the test bench, and then gas testing – in a gas compression station - for an electric generator driven by a screw expander. This paper presents how the tests were carried out, the experimental data and the interpretation of results

1. Introduction

From the beginning we want to emphasize that the long-term strategy of COMOTI is to build high efficiency, high returns and original solutions, technological and constructive, applied in engines construction, aviation systems and equipment - the Institute has a tradition of over 20 years - by creating and developing efficient products and technologies, especially requested by the oil and gas Companies. Trying to meet the demands / estimates on the evolution of the operating parameters of the gas stations, COMOTI developed two new projects:

-Screw compressors, which are positive displacement compressors and can find an increasing use in the gas industry and beyond, due to their advantages. So far, we have realized, under license from GHH-RAND, screw compressors with oil injection with a maximum flow rate of 3000 Nm^3/hour and a discharge pressure of maximum 30 bara. We emphasize that the good working relationships we had with the GHH-RAND company ended by allowing us to manufacture under license - since April

Content from this work may be used under the terms of the Creative Commons Attribution 3.0 licence. Any further distribution of this work must maintain attribution to the author(s) and the title of the work, journal citation and DOI.
Published under licence by IOP Publishing Ltd

2010 - the CU screw compressors family. Requests / signals from potential beneficiaries required addressing the need to design a new family of screw compressors, with oil injection, capable of developing a discharge pressure of 45 bar with flow rates up to 5000 Nm3/hour. This paper presents the experimental test phases – for the first screw compressor with discharge pressure up to 45 bara-type CHP220, realized by COMOTI in cooperation with City University –London – the basic rotor design, and GHH-RAND for manufacturing rotor pair. Continued research and development have allowed this type of screw-compressor to find a market position, diversifying its size range in the last decade, if we refer to other types of the compressors (centrifugal or reciprocating compressors).

To reduce the gas pressure in the regulating and measurement stations from the national piping network (with pressures up to 40 bara), lamination valves are used, usually heated to reduce the pressure from the national piping network to the local piping network. This leads to a significant loss of "green" energy that can be recovered through the efficient use of the differential pressure in the expander. The aim is to replace the rolling valves and regulators with a screw turbine which, through the adiabatic expansion of gases, produces a clean and usable energy. COMOTI developed an efficient equipment, a power conversion equipment that converts gas expansion in green electricity. This unique group works using the oil-injected screw technology. After years of research and experimentation, COMOTI has developed a new configuration of turbine screw expander with oil injection (an advantageous solution for relatively small flows) that will have all the screw compressors advantages: high efficiency, due to the compression evolution nearly isothermal, low maintenance costs thanks to very good reliability, fewer moving parts, and not in the least, competitive price. The solution has not been studied in Romania and is not applied to any plant or TRANSGAZ or ROMGAZ stations. Another advantage of using such equipment are, besides producing green electricity, the natural gas availability to other consumers, reducing the amount of CO2 released into the environment.

2. TEST with air, on COMOTI test bench for CHP220

We do not propose to analyze in detail the data related to the design, manufacture and completion of the compressor. But, in the case of the tests with a new type of compressor, conducted by COMOTI in cooperation with City University and GHH-Rand, we consider that it is necessary to provide data on the gas dynamics and the constructive parameters of the compressor CHP 220:

- oil flooded screw compressor that compresses a normalized flow of 5000 Nm3/h of natural gas from an admission pressure of 4,5 bara to a discharge pressure of 45 bara, when rotating at 1430 rpm,

-rotor design summary:

- 4/5-306 mm rotor diameter male/248,8 rotor diameter female, 413,1 mm length, L/D =1,35
- "N" rack generated Profile, UK Patent 9610289, PCT World 97/43550, US6,296,461, EP 0898 655
- Wrap angle 300 ^{0}
- Displacement 17,38 dm3/rot, meaning 15oo m3/h la 1430 rpm

The Performance chart of the 4/5-306 Compressor on gas, given by City University [7]

SUCTION PRESSURE 4.50 BAR
RESERVOIR PRESSURE 45.00 BAR

ROTOR TIP SPEED *(m/s)*	ROTATIONAL SPEED *(RPM)*	VOLUME FLOW RATE *(M3/MIN)*	VOLUMETRIC EFFICIENCY	INPUT POWER *(kW)*	SPECIFIC POWER	ADIABATIC EFFICIENCY	OIL FLOW *(kg/min)*
12.900	805.1	9.158	0.654	358.5	39.146	0.624	233.7
17.900	1117.2	14.350	0.739	490.4	34.177	0.715	235.3
22.900	1429.2	19.579	0.788	628.6	32.108	0.761	236.4
27.900	1741.3	24.790	0.819	772.3	31.155	0.784	237.4
32.900	2053.4	29.980	0.840	921.5	30.739	0.795	238.2
37.900	2365.4	35.154	0.855	1076.5	30.623	0.798	239.1
42.900	2677.5	40.328	0.866	1237.5	30.687	0.796	240.1

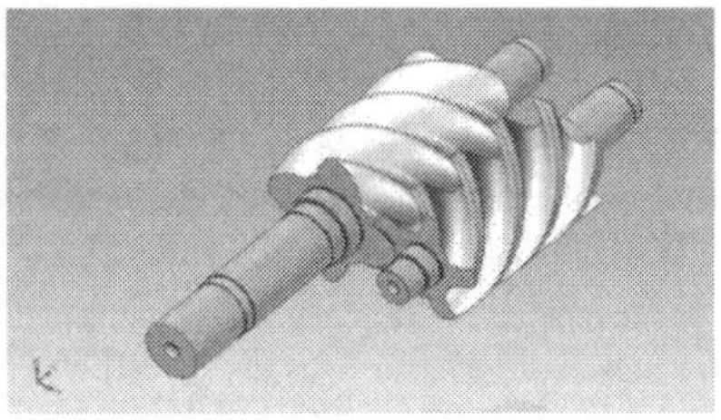

Figure 1. Rotors

To retrieve the very high load generated by pressure force on bearings, we choose the solution with balancing piston and radial/four point contact bearings , Explorer type – specially developed for application in screw compressors [8]. The rotors - see fig. 1 - were made of steel and were manufactured by grinding by GHH-Rand and therefore ensuring excellent precision and contact. The L/D, were L=length of rotors and D= male rotor diameter, has an optimal value - 1,35 [7], and so, the deflection is minimized providing compactness for the compressor. Because of the high discharge pressure – over 27,5 barg [4] – the housings were designed to be manufacture from steel. The structural integrity was verified by stress analysis, and after manufacture hydrostatic test was done at 76 bar. For safe and reliable operation a high quality dual seal was installed. After the manufacturing of the parts, during compressor assembly, each component was inspected to ensure that they have been cleaned and they are free of foreign material, corrosion products etc. After assembly, the compressor is ready for tests. The test was done on COMOTI's test bench, with air. Because this is a prototype, we intend to compare the theoretical - see fig. 2 - with measured parameters(for discharge pressure of 8.5 and 12 bara, City University have calculated the prediction parameter –[7]).

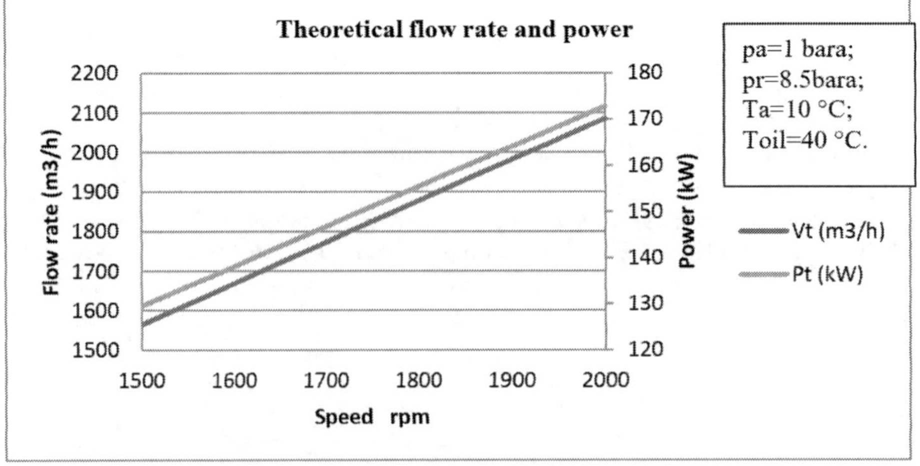

Figure 2. Theoretical/predicted performance on air

The instrumentation scheme of the test bench is shown in fig. 3, and respects the requirements stipulated in API619 and ISO1217. The parameters to be measured for the estimated performance of the compressor will be: compressor inlet/discharge temperature and pressure; speed of the male rotor; power of the electric motor driver; pressure/temperature of oil after the oil cooler; differential pressure, pressure and temperature across the venturi nozzle (designed according to ISO

5167-3); in/out temperature for water, in the oil cooler; - oil rates, in each oil point injection (seal, inlet bearings, injection, discharge bearings); pressure of oil in each oil injection point; atmospheric pressure and atmospheric temperature.

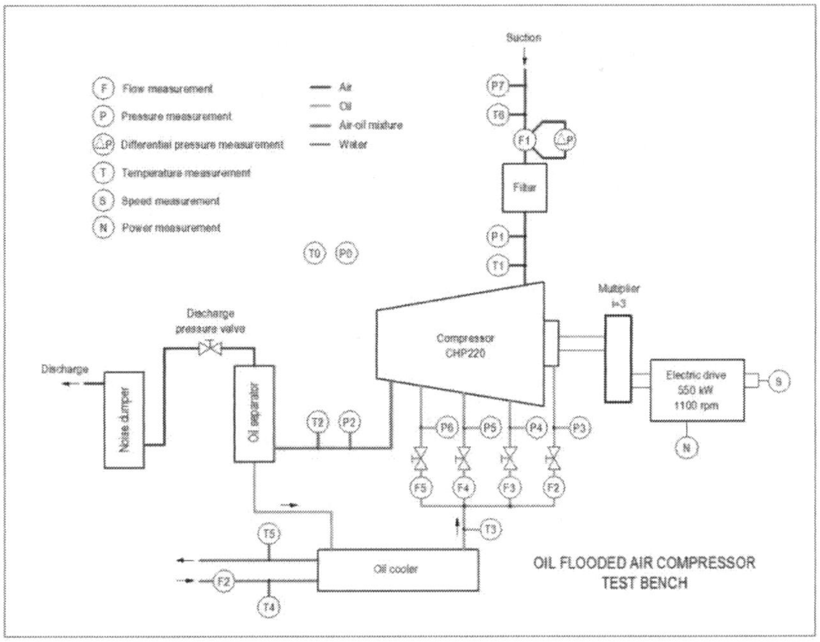

Figure 3. Instrumentation scheme of the test bench

All measurements were obtained from electrically generated signals derived directly, respectively, from the speed, pressure, temperature and flow rates measurements. All the installation parameters (4-20 mA analogical, 0-24 Vcc digital) are supervised by a Programmable Logic Controller. The values of parameters are recorded at a certain rate of time in computer memory (25 μs). The compressor process evolution was analyzed through parameter values data acquisition. Also, the parameters can be viewed in real time on the operating panel, which contains a series of screens that allow an operator to monitor the parameters.

In Fig. 4 – there is a photo of the test bench, with the compressor installed in it.

PLC oil cooler - oil filter suction/discharge

electric motor-multiplier separator compressor unit

Figure 4. Test bench, with the compressor installed

We can see: the electric motor driven; the multiplier; the compressor unit; the oil cooler/oil filter; the oil pipeline; the air pipeline suction and discharge; instrumentations for pressures, temperatures, flows, rotation; -air/oil separator etc.

During the course of the test, the temperatures of the compressor surfaces were measured automatically with a Fluke monitor - Fig. 5.

The test data were obtained, with atmospheric pressure at inlet and discharge pressures of 8.5, 12 and 15 bara. In Fig.6 we present print screens with the parameters at different moments of the tests. The maximum discharge pressure realized on the test bench was 36 barg (print-screen on the right side).

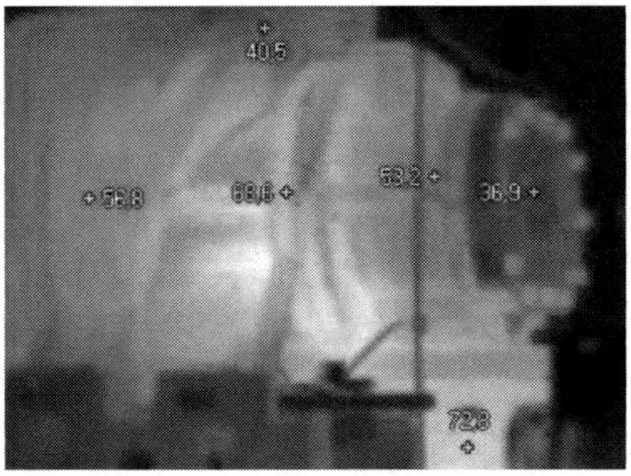

Figure 5. Temperatures of the compressor surfaces.

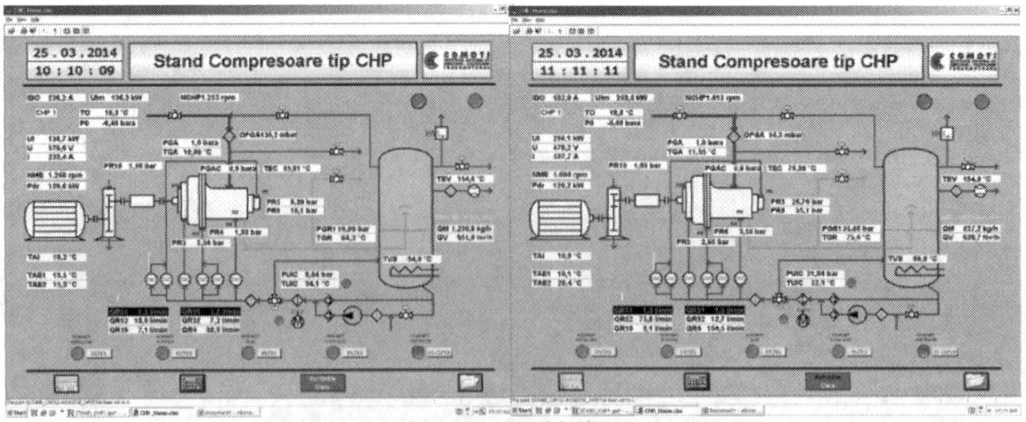

Figure 6. Print screens with the parameters

Using these parameters, and the relations mentioned :

k= adiabatic efficiency; R= gas constant [kJ/kgK]; Ta= suction temperature [K]; = compression ratio; G = mass flow [kg/s]; m_u = oil flow rate [kg/s]; = specific heat gas [kJ/kg K]; = specific heat oil [kJ/kg K]; = differential temperature between suction and discharge for gas [K]; = differential temperature between suction and discharge for oil [K]

IOP Conf. Series: Materials Science and Engineering **90** (2015) 012015 doi:10.1088/1757-899X/90/1/012015

η_v = volumetric efficiency = $Q_v^{masured}/Q_v^{theoretic}$

We elaborate the experimentally measured performance, as shown in Fig 7:

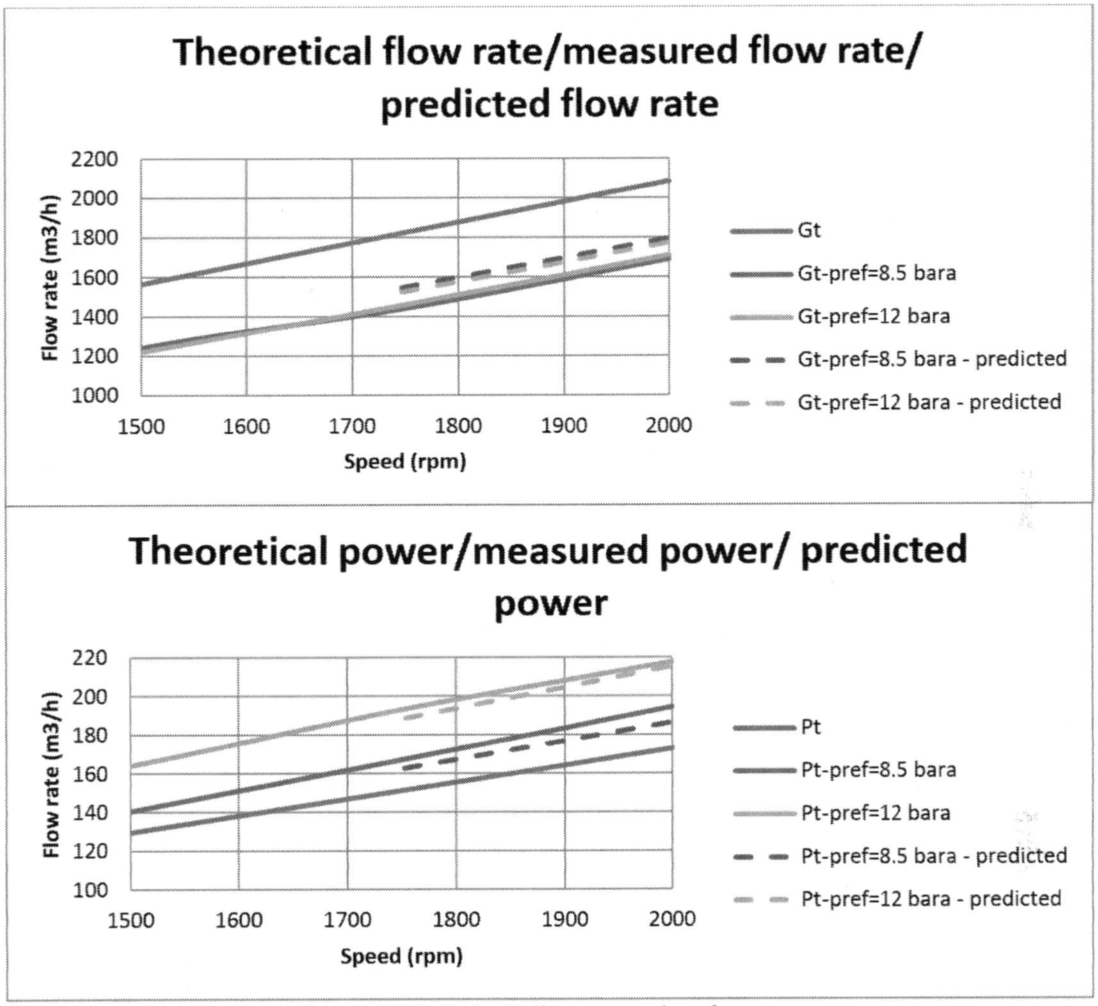

Figure 7. Experimentally measured performance

Given the conditions in which the experiments were conducted, the technical constraints that had to be taken into account, while comparing with recommendations of [5] - Annex C - we can say that the values of these parameters are as expected and in acceptable limits.

3. Test for screw expander

The screw expander - electrical generator is a viable and economical solution for recovery the lost gas expansion energy, in the gas pressure reduction station.

In order to reduce gas pressure in the gas pressure reduction station and the quantity on main pipelines, lamination valves are used, usually heated, which allow the reduction of pressure from the transport pressure to distribution pressure. This leads to a significant loss of energy "green" which can be recovered by decreasing pressure in the expander who drives the generator.

I.N.C.D.T. COMOTI has developed an efficient equipment for the conversion of the expansion energy of natural gas into electrical "green" energy. This group unique operates using screw technology (screw compressor), with oil injection, which has been used for a long time to screw compressors.

The conversion of screw compressor - type CU 128 G - in screw expander was made by GHH Rand, COMOTI performed tests, both air - in COMOTI's test bench - and gas - in a pressure reduction station.

The nominal system operating regime for the electric generator - screw expander corresponds to the following parameters:

- pressure of the natural gas at the entrance of the expander = 20 bara;
- discharge pressure of the expander = 5 bara;
- screw expander EU 128 G , Vi=2,6 , $n_{male\ rotor}$ =3838 rot/min, i=2,559;
- electric generator type GSIF EX dllBT4 250 kVA; 400 V, 1500 rpm (brushless synchronous generator).

Advantages:

- High efficiency of the product is due to detent almost isothermal - by the injection of hot oil, gas temperature;
- Does not pump, essential advantage in variation of parameters of gas at the entrance to expander;
- minimum structure requirements, due to the direction of continue rotation and the rotor dynamic levelling;
- Low maintenance, due constructive simplicity.

Screw Expander was tested on the COMOTI's test bench, see Fig. 8, in a configuration determined by the possibility of providing a source of air (air intake - 3 compressors CU128G).

All measurements were recorded digitally and then transferred to a computer used for online processing.

The command and control system has the tracking function of the parameters of thermal, mechanical and electrical processes, during normal or transitional working processes, adjusting and keeping under control, by the course of the operation in complete safety. The system for the expander-generator is based on a numerical system monitoring and management PLC type Versa Max produced by GE Fanuc. It connects with the components of execution by relays, contactors, transducers, signal adapters.

Figure 8. Expander on the COMOTI's test bench

Constructive solution of the of command and control system takes into account the working conditions of the assembly expander-generator, the nature of gas and noise area from the point of view of danger of explosion.

There are 3 automation lockers (1 locker for expander, 1 locker for energy monitoring and protection, 1 supply locker and generator protection) where the operator can follow the operating status of the assembly screw expander-generator (are located outside in areas Ex.).

For fixing the speed of expander at the level of 1500 rmp was put into the circuit of the generator a ballast resistor variable in three-phase. By the value of this resistor and the implementation of an excitation voltage has been obtained stabilization around the 1500 rmp (the operation was done in a insularized regime).

During the test the flow being discharged from the three compressors is up to 2426 kg / h and the expander inlet gas pressure reaching 6.75 bar. Oil was heated at 70 °C and gas at suction of the

IOP Conf. Series: Materials Science and Engineering **90** (2015) 012015 doi:10.1088/1757-899X/90/1/012015

expander at 60 °C. The maximum power obtained at generator was 51 kW (the expander power was 63.75 kW).

The tests confirmed the functionality of the equipment, tests on gas in gas reduction station will certify gas/generator parameters.

Experiments at Gas Reduction Station Onesti, see fig.9, were performed in the rpm range between 500 - 1500 rpm.

Figure 9. Expander-generator at Onesti

The technological scheme – see Fig. 10 - was designed and manufactured that:

- to ensure oil heating
- to ensure gas heating
- to ensure the integrity of the equipment in the station without affecting the normal operation;
- the control and PLC control equipment to ensure integrity in operation, process data collection and procesing

Tests were conducted at the Gas reduction Station in two stages:

-in the first phase gas flow varied between 1470..2087 Kg / h, suction pressure in expander between 11..13,7 bar, obtaining an output power (expander shaft) between 27..77 kW

-in the second phase gas flow varied between 1400..4500 Kg / h, suction pressure in expander between 10..15,7 bar see Fig. 11 obtaining an output power (expander shaft) between 25..111,25 kW (see fig . 12)

Experimental curves obtained at different pressures and flows- not the nominal parameters, because these was station available parameters -show a close evolution estimates by GHH-Rand Germany.

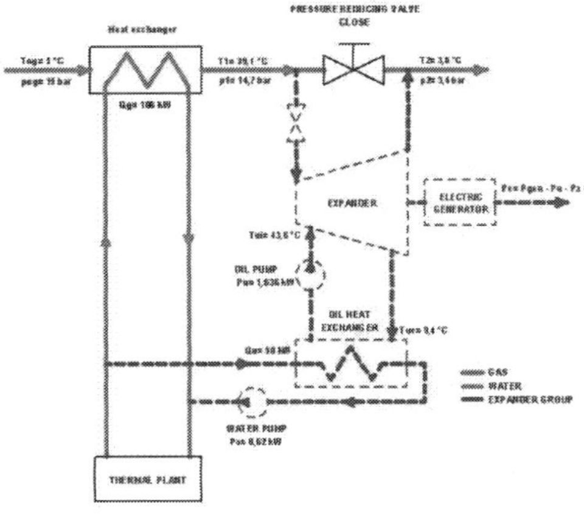

Figure 10. Technological scheme

Also, to stabilize the speed at 1500 rpm of the expander, a three-phase variable ballast resistance was placed in the circuit of the generator stator. The value of this resistance and an appropriate excitation voltage of the expander were stabilized at 1500 rpm, so the generator is ready for a possible coupling to the network.

IOP Conf. Series: Materials Science and Engineering **90** (2015) 012015 doi:10.1088/1757-899X/90/1/012015

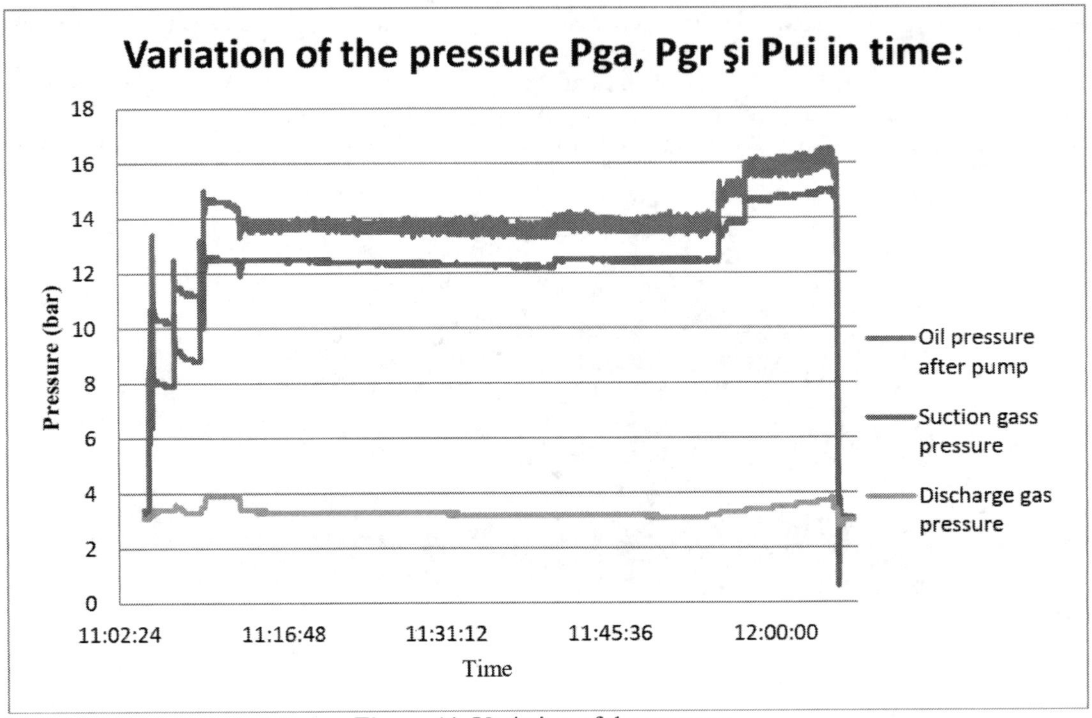

Figure 11. Variation of the pressure

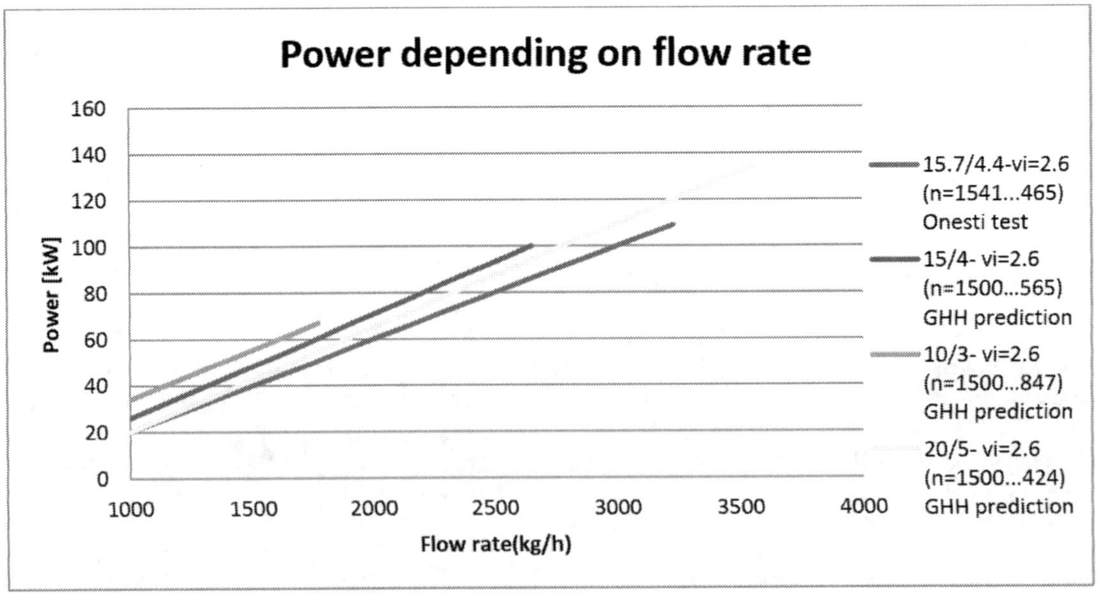

Figure 12. Power variation with flow rate

4. Conclusion

4.1 Screw compressor

The design of an oil injected screw compressor, with high discharge pressure - 45 bara - has passed the design and experimental phases. The compressor has undergone specific test production. Results can be considered normal for the test stage performed. The development of the tests stand for the compressors in the HP family allow the finalization of the experimental tests on air and doing tests on gas, in a gas station .The scope of this test was to validate the method for designing a compressor with high delivery pressure, to check how the compressors work. Given the conditions in which the experiments were conducted, the technical constraints that had to be taken into account, while comparing with information from technical literature, business prospects, we can say that the values of these parameters are as expected and in acceptable limits. So:

- volumetric efficiency between 0,72....0,82
- adiabatic efficiency between 0,77...0,92

We can see that the difference between prediction performance and measured performance are small.

4.2 Screw expander

It has been proven the functionality of equipment. Given the stand conditions (maximum flow 2400 kg/h at a pressure of 6,75 bar) the power at turbine shaft reaches 63,75KW (pGen = 51 kW), the isentropic efficiency 0,7. We obtain a high discharge temperature (above 20 ° C), which shows that we can achieve a diminished heating of the injected oil.

Tests were conducted at the Gas reduction Station in two stages:

-in the first phase gas flow varied between 1470..2087 Kg/h, suction pressure in expander between 11..13,7 bar, obtaining an output power (expander shaft) between 27..77 kW;

-in the second phase gas flow varied between 1400..4500 Kg/h, suction pressure in expander between 10..15,7 bar, obtaining an output power (expander shaft) between 25..111,25 kW.

Experimental curves obtained at different pressures and flows - not the nominal parameters - show a close evolution estimates by GHH - Rand Germany.

5. Bibliography

1. Smith, I.K., Stošić, N., Aldis, C. A. *Development of the trilateral flash cycle system. Part 3: the design of high-efficiency two-phase screw expanders.* Proceedings of the Institution of Mechanical Engineers, Vol. 210, 1996, 75-92.

2. Amosov, P. E., *et al. Vintovie kompressornie mashinii,* 1977 (Mashinostroenie, Leningrad).

3. Stošić. N., Milutinovic, L., Hanjalic, K. and **Kovacevic, A.** *Investigation of the influence of oil injection on the screw compressor working processes.* Int. J. Refrig., 1992, **15,** 206-220.

4. API 619/ISO10440.

5. ISO1217

6. N.Stosic, Ian K. Smith, A.Kovacevic *"Improving screw compressor performance"*

7. N.Stosic, Ian K. Smith, E. Mujic *"The design of a high efficiency oil flooded gas screw compressor"* - November 2009

8. SKF Explorer Catalogue

IOP Conf. Series: Materials Science and Engineering **90** (2015) 012016 doi:10.1088/1757-899X/90/1/012016

Thermal modelling and analysis of an oil-free linear compressor

M J Oliveira, M C Diniz and C J Deschamps[1]
POLO Research Labs for Emerging Technologies in Cooling and Thermophysics
Federal University of Santa Catarina, 88040-900, Florianopolis, SC, Brazil

E-mail: deschamps@polo.ufsc.br

Abstract. Gas superheating in the suction system may significantly reduce the volumetric and isentropic efficiencies of small reciprocating compressors adopted for household refrigeration. This paper reports a thermal modelling approach developed to predict superheating in an oil-free linear compressor. A simulation code based on the finite volume method was adopted to solve heat conduction in the solid components and gas flow inside the compressor shell. In order to reduce the computational cost, the compression cycle inside the cylinder was modelled with a transient lumped formulation, but in a coupled manner with the remainder of the solution domain. Comparisons between numerical and experimental results of temperature showed discrepancies in some solid components and in the gas path along the discharge system. However, the model was able to predict suction gas superheating in good agreement with measurements. A sensitivity analysis of the temperature distribution with respect to two design parameters was also carried out. The model is particularly useful for compressor design since no experimental calibration is required.

1. Introduction

Linear compressors work through the reciprocating motion of a piston provided by the combination of a linear motor and a spring [1]. This type of actuating mechanism significantly reduces the number of moving parts, increasing the mechanical efficiency and making the compressor suitable for miniaturization [2] and electronic cooling applications [3]. Most of current studies on compressor technology aim to increase the thermodynamic efficiency. An aspect that acts to reduce the volumetric and isentropic efficiencies of hermetic compressors is gas superheating due to heat transfer between internal components [4].

Heat transfer in reciprocating compressors has been extensively studied with different numerical approaches, following lumped, differential and hybrid formulations. Lumped models [5-8] are used when low computational cost is required in the first phases of compressor design. Differential models are necessary when full description of the phenomenon is needed [9-11]. Hybrid models [12-14] combine lumped and differential formulations and offer a trade-off between computational cost and accuracy.

Schreiner [15] carried out a comparative analysis between differential and hybrid models to predict the temperature distribution of a reciprocating compressor. The author observed that predictions of the hybrid model were in better agreement with experimental data. The discrepancies observed in the

[1] To whom any correspondence should be addressed.

Content from this work may be used under the terms of the Creative Commons Attribution 3.0 licence. Any further distribution of this work must maintain attribution to the author(s) and the title of the work, journal citation and DOI.
Published under licence by IOP Publishing Ltd

differential model were attributed to the flow of lubricating oil over the compressor components, which was not considered in the simulation. The hybrid model takes this into account through experimental calibration of thermal conductances.

This paper presents a model to predict the temperature field of an oil-free linear reciprocating compressor, with especial attention to the gas superheating during the suction process. The simulation approach consists of two coupled submodels: a transient lumped model for the compression cycle and a finite volume model for heat transfer in solids components and in the gas flow inside the compressor.

2. Simulation model

2.1. Compression cycle

The compression cycle is modelled with a transient lumped formulation [8] via four groups of equations: (i) equations to describe the variation of the compression chamber volume; (ii) equations for mass and energy balances inside the compression chamber; (iii) equations for valve dynamics; (iv) equations to estimate the mass flow rate through valves and leakage through the piston-cylinder gap. The model allows predictions for discharge temperature in the discharge orifice, compression power and mass flow rate.

The motion of the piston as a function of time is described by:

$$x_{pis}(t) = -\omega A \sin(2\pi f t) \tag{1}$$

where f is the operating frequency, A is the motion amplitude and ω is an amplitude factor that is adjusted between 0.85 and 1 to match the input data of power consumption. It should be noted that x_{pis} is referenced to the mid-stroke position, with $x_{pis}= -\omega A$ when the piston is at top dead centre and $x_{pis}= +\omega A$ when the piston is at bottom dead centre.

The instantaneous volume of the compression chamber is given by:

$$\forall_{cyl}(t) = \forall_{\delta,geo} + \{h_o + \omega A + x_{pis}(t)\} A_{pis} \tag{2}$$

where the first term is related to the geometric dead volume. In the second term, A_{pis} and h_o are the area of the head of the piston and minimum height of the compressor chamber, respectively.

The thermodynamic properties of the gas in the compression chamber are evaluated via a transient lumped formulation of the conservation equations. After some mathematical manipulations of the energy equation, the equation to evaluate the temperature of the gas, $T_G(t)$, is given by:

$$\frac{dT_G}{dt} = M - NT_G \tag{3}$$

where

$$M = \frac{1}{m_G c_v}\left[H_G A_w T_w - h_G \frac{dm_G}{dt} + (\dot{m}_{sv} h_{sv}) - (\dot{m}_{dv} h_{dv}) - (\dot{m}_{leak} h_{leak})\right] \tag{4}$$

$$N = \frac{1}{m_G c_v}\left[H_G A_w + \left.\frac{\partial p_G}{\partial T_G}\right|_v \frac{d\forall_G}{dt} - \left.\frac{\partial p_G}{\partial T_G}\right|_v v_G \frac{dm_G}{dt}\right] \tag{5}$$

In the above equations, H_G is the convective heat transfer coefficient required to estimate the instantaneous heat exchange between the gas and the cylinder wall, which was obtained from the correlation proposed by [16], c_v is the specific heat coefficient and v_G is the specific volume of the gas. The instantaneous volume of the compression chamber is obtained from equation (2) and the mass of gas inside the compression chamber, m_G, allows the determination of the gas density. The instantaneous gas pressure, p_G, is obtained by introducing the gas density and temperature in a state equation.

The instantaneous compression power is given by

$$\dot{W} = -f\int p_G d\forall_{cyl} \quad (6)$$

The simulation of the compression cycle requires estimates for the temperatures of the gas in the suction chamber ($T_{suc,o}$) and of the cylinder wall (T_w). The method used for this purpose will be explained shortly.

Linear compressors adopt reed type valves that open and close under changing pressure on each face. Valve dynamics is represented by a one degree of freedom mass-spring model:

$$M_v\ddot{x}_v + c_v\dot{x}_v + K_v x_v = F_v \quad (7)$$

where M_v, c_v and K_v are the reed equivalent mass, damping coefficient and stiffness, respectively. In equation (7), x, $\dot{x}$ and $\ddot{x}$ are the instantaneous reed lift, velocity and acceleration, respectively. The force F_v, resulting from the pressure load on the reed is obtained with reference to the effective force area, A_{ef}, which is determined numerically or experimentally from the pressure difference across the valve, Δ_{pv}, according to $A_{ef} = F_v/\Delta_{pv}$ [17]. The effective force area can be understood as a parameter related to how efficiently the pressure difference Δ_{pv} opens the valve. The differential equation for the valve dynamics, equation (7), was solved using an explicit Euler method.

The mass flow rate through the valve, $\dot{m}_v$, is evaluated with reference to the isentropic flow condition and the concept of effective flow area, A_{ee}, which accounts for viscous effects. Therefore,

$$\dot{m}_v = A_{ee}p_{up}\sqrt{\frac{2k}{(k-1)RT_{up}}\left[R_p^{2/k} - R_p^{(k+1)/k}\right]} \quad (8)$$

where p_{up} and T_{up} denote pressure and temperature upstream, k is the specific heat ratio, R is the refrigerant gas constant and R_p is the pressure ratio between the cylinder and discharge/suction chamber. Naturally, values for p_{up}, T_{up} and R_p vary along the compression cycle, whether the flow is critical or sub critical, and also with flow direction.

Gas leakage through the piston-cylinder gap was modelled employing a Couette-Poiseuille flow model [18].

2.2. *Fluid flow and heat transfer inside the compressor*

A commercial code (Ansys FLUENT v.14) based on the finite volume method was employed to predict heat transfer in solid components and along the gas path in the suction system, internal environment and discharge system. The system of governing equations is formed by the conservation equations for mass, momentum and thermal energy. The turbulent flow that prevails inside the compressor was solved with the concept of Reynolds-averaged quantities through the SST model.

A second-order upwind scheme was adopted to interpolate the flow quantities needed at the control volume faces. The coupling between the pressure and velocity fields was achieved with the SIMPLE algorithm [19]. The system of algebraic equations was solved with a segregated implicit algorithm.

The simulation model is prepared following three steps: (i) simplifications of the compressor geometry for mesh generation, (ii) specification of boundary conditions and (iii) simulation of fluid flow and heat transfer in a coupled manner with the lumped model for the compression chamber (section 2.1).

The solution domain of the differential model includes the suction and discharge systems, region inside the compressor shell and solid components. Especial care was taken to properly discretize regions of high gradients, such as near solid walls. The mesh for the solid and fluid solution domains had nearly 17 million tetrahedral and prismatic volumes. When it was imported to the solver, the number of volumes was reduced to around 7 million, with agglutination of tetrahedral volumes in the form of polyhedral volumes.

Boundary conditions are required to characterize the compressor operating condition and to ensure the coupling between the lumped and differential models. Figure 1 presents the compressor solution domain of the differential model, identifying different boundaries ($\mathcal{C}$). The volume ($\mathcal{V}_1$) of the linear motor coils was modelled as a heat source $\dot{Q}_{mot}$, which was estimated from the electrical motor efficiency. On the other hand, figure 2 illustrates the compression chamber where the lumped model is applied; hence there is no mesh inside it. In figure 2, it is also possible to identify the boundaries $\mathcal{C}_4$, $\mathcal{C}_5$ and $\mathcal{C}_6$ that exchange information between the lumped and differential models.

Evaporating and condensing pressures were imposed as boundary conditions for the inlet ($\mathcal{C}_1$) and outlet ($\mathcal{C}_2$), respectively. The inlet temperature was measured in the calorimeter (T_{in} = 32°C). The ambient temperature and the convective heat transfer coefficient on the compressor shell external surface were T_{amb} = 32°C and h_{amb} = 20W/m^{2}K, respectively.

Table 1 and table 2 present the variables related to each boundary ($\mathcal{C}$) and the volume ($\mathcal{V}_1$), as well as the source from which one is obtained: input data or from exchange of information between the lumped and differential models.

3. Solution procedure

The lumped and differential models are solved in a coupled manner until convergence is achieved, when the temperature field of the compressor is established. The lumped model is called inside the finite volume solver every four iterations.

The convergence of the solution procedure was verified by monitoring six global parameters: i) temperature of the gas at the suction orifice $\mathcal{C}_5$ ($T_{suc,o}$) ii) temperature at the cylinder wall $\mathcal{C}_4$ (T_w); iii) overall energy balance in the fluid flow domain; iv) heat transfer at the cylinder wall $\mathcal{C}_4$; v) mass flow rate at the compressor inlet $\mathcal{C}_1$ ($\dot{m}_{in}$); vi) mass flow rate at the compressor outlet $\mathcal{C}_2$ ($\dot{m}_{out}$). The convergence criterion for these global parameters is expressed as follows:

$$\frac{|\phi - \bar{\phi}|}{SF} < 10^{-3} \tag{9}$$

In equation (9), ϕ represents the global parameter being analysed and $\bar{\phi}$ is the average value of ϕ in the last ten most recent iterations. The denominator SF in equation (9) varies according to the global parameter ϕ of interest. Accordingly, $SF = T_{dis,o} - T_{in}$ when ϕ represents temperature; $SF = \dot{m}_{suc,o}$ when ϕ is mass flow rate and $SF = \dot{W}$ when ϕ is either energy balance or heat transfer.

In addition to monitoring the convergence of global parameters, the solution of the governing equations in the differential model was also verified for convergence by examining scaled residuals, as defined in [20]. The solution procedure of the governing equations for mass, momentum and turbulence was considered to be converged when their residuals were smaller than 10^{-3}, whereas a residual of 10^{-6} was used for the energy equation. The simulation was concluded when the convergence criteria for the six global parameters and differential model were simultaneously satisfied.

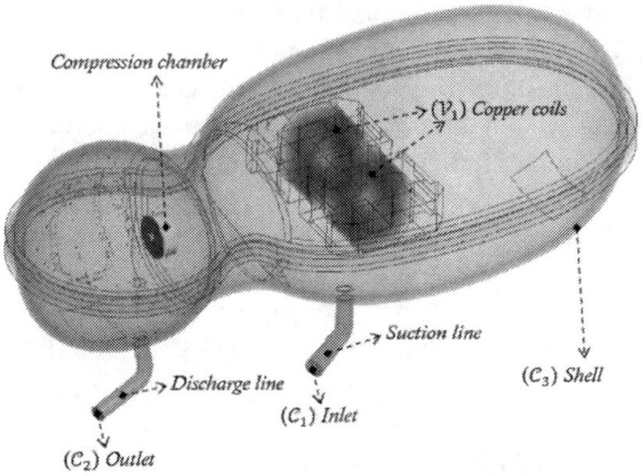

Figure 1. Boundaries of the compressor solution domain.

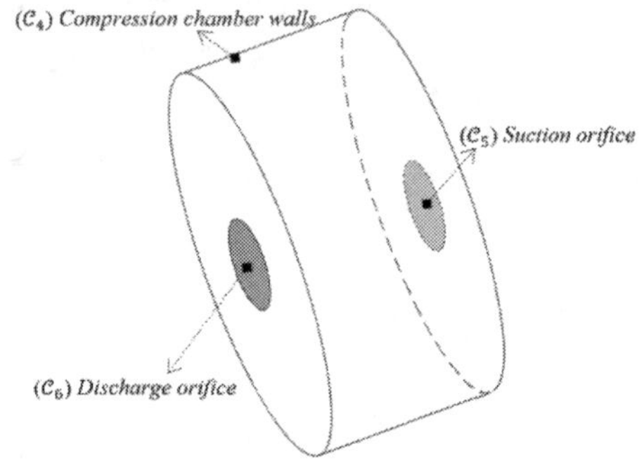

Figure 2. Boundaries of the compression chamber domain.

Table 1 and table 2 assigned each boundary ($\mathcal{C}$, $\mathcal{V}_1$), presented in figure 1 and figure 2, to a boundary condition necessary to create the coupling of lumped and differential models. In this way, an interactive loop is built (cycle) until the convergence, when the thermal profile of the linear compressor is found.

By following this solution procedure, about 1000 cycles were required to establish convergence, with a processing time of approximately 6 minutes per cycle on a computer with an Intel i7 3.2 GHz processor and 64 GB of RAM. As a comparison, the processing time of a complete simulation with hybrid models [13-15] is of order of minutes.

Table 1. Boundary and heat generation conditions of the differential model.

Boundary	Boundary conditions	Source
Inlet $\mathcal{C}_1$	p_{in}, T_{in}	Input data
Outlet $\mathcal{C}_2$	p_{out}	Input data
Compressor shell $\mathcal{C}_3$	h_{amb}, T_{amb}	Input data
Compression chamber walls $\mathcal{C}_4$	$\dot{Q}_w$	Lumped model
Suction orifice $\mathcal{C}_5$	$\dot{m}_{suc,o}$	Lumped model
Discharge orifice $\mathcal{C}_6$	$\dot{m}_{dis,o}$, $T_{dis,o}$	Lumped model
Copper coils $\mathcal{V}_1$	$\dot{Q}_{mot}$	Lumped model

Table 2. Boundary conditions of the lumped model.

Boundary	Boundary condition	Source
Compression chamber walls $\mathcal{C}_4$	T_w	Differential model
Suction orifice $\mathcal{C}_5$	$T_{suc,o}$	Differential model

4. Results

Simulations were carried out for a 120 Hz linear compressor, operating with R134a under evaporating temperature T_e = -23.3°C and condensing temperature T_c = 40.5°C. Accordingly, boundary conditions of pressure and temperature at the inlet were 115kPa and 305K, respectively. The compressor geometric dimensions are typical of small reciprocating compressors used for domestic refrigeration.

The numerical results were validated through comparisons with experimental data of temperature at different locations in the gas path (figure 3) and solid components (figure 4). Figure 5 shows the deviations between predictions and measurements of temperatures (ΔT) for points **A** to **J** identified in figure 3 along the gas path inside the compressor. As can be seen, there is reasonable agreement between both results, especially from point **A** to **F**, which represents the suction gas superheating. The only discrepancy occurs at point **H** located inside the discharge chamber, with a difference of approximately 17°C. Figure 6 presents the comparison between numerical and experimental results for the compressor solid components. As can be seen, there are considerable differences at points **R** and **S**, as well as at points **T** and **U** on the compressor shell surface.

The discrepancies observed in figures 5 and 6 may occur in part because the model neglects the motion of the mechanical kit inside the compressor. A dynamic mesh could be used to allow the modelling of this mechanism of convection intensification, but would result in a much more complex simulation model and higher computational cost. Since heat transfer inside the compressor is numerically solved, the difference between predictions and measurements may also be associated with truncation error due to inadequate grid discretization. This aspect will be further analysed in future work.

After validation, the model was applied to analyse the temperature distribution sensitivity in relation to two parameters: (i) convective heat transfer coefficient on the compressor external surface, h_{amb}, whose reference value is 20W/m^{2}K; (ii) thermal conductance of the insulating gasket (k/l), which is positioned immediately above positions **Q** and **R** (figure 4) to reduce heat transfer through conduction from the discharge system to the suction system. The conductance of the insulation gasket (k/l) was varied by specifying two thermal conductivities (k = 0.1 and 0.3 W/m K) different from the reference value (k = 0.2W/m K), but keeping its thickness (l) constant and considering h_{amb} = 15W/m^{2}K. Figures 7 and 8 show the temperature variations (ΔT) observed along the gas path due the aforementioned modifications in h_{amb} and k.

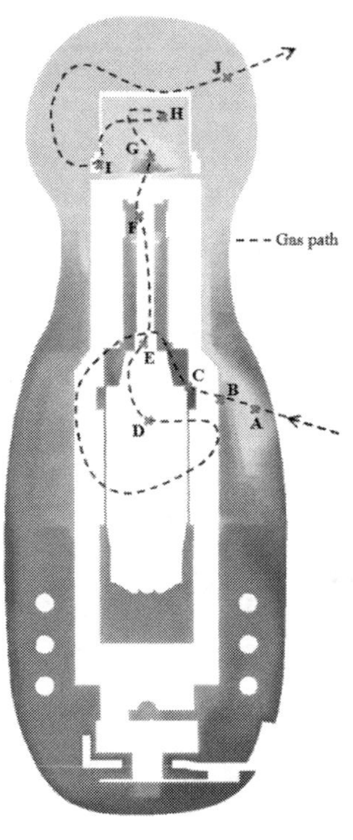

Figure 3. Temperature field of gas domain in the baseline condition (solid domain hidden).

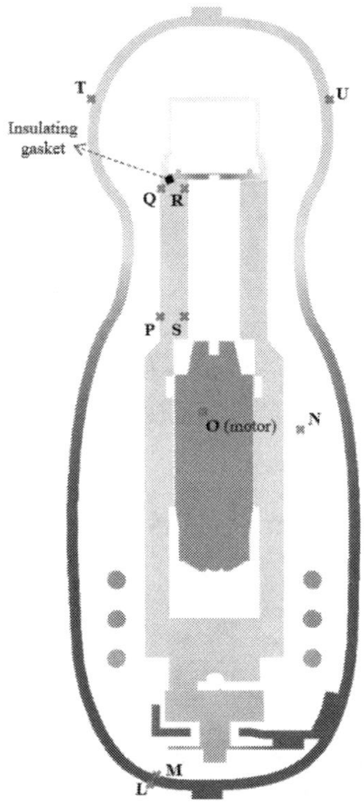

Figure 4. Temperature field of solid domain in the baseline condition (gas domain hidden).

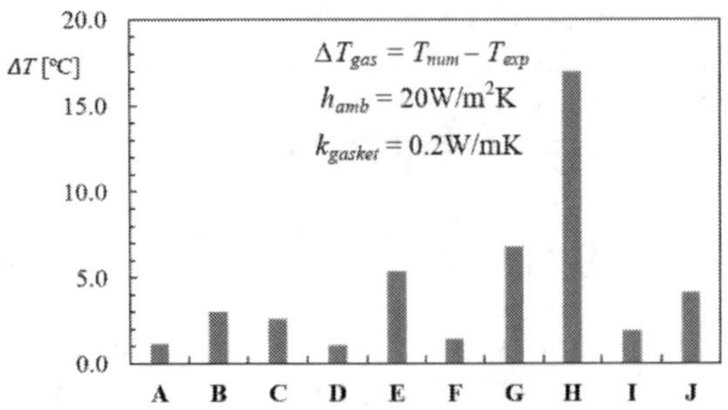

Figure 5. Differences between predictions and measurements of temperature along the gas path.

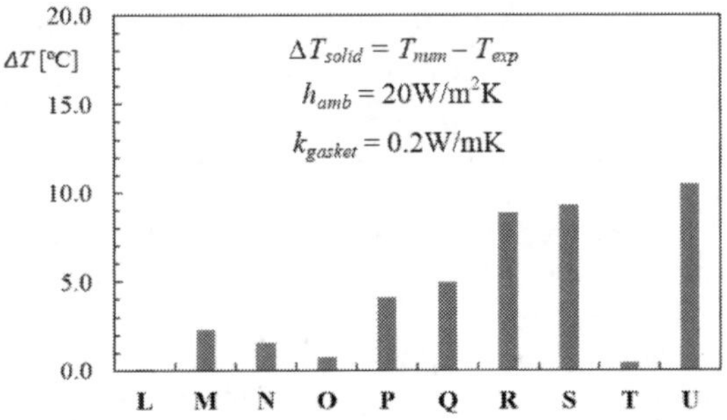

Figure 6. Differences between predictions and measurements of temperature in the solid components.

As expected, figure 7 shows a considerable increase of temperature when the convective heat transfer coefficient is reduced from its reference value (20W/m^{2}K). On the other hand, the compressor thermal profile is much less dependent on the insulating gasket conductance (figure 8). Nevertheless, point **F** (figure 3) corresponds to the position where the gas is suctioned into the compression chamber and, therefore, the temperature there affects the compressor isentropic and volumetric efficiencies. The model predicts an increase of 4°C when the thermal conductivity is varied from 0.1W/mK to 0.3W/mK.

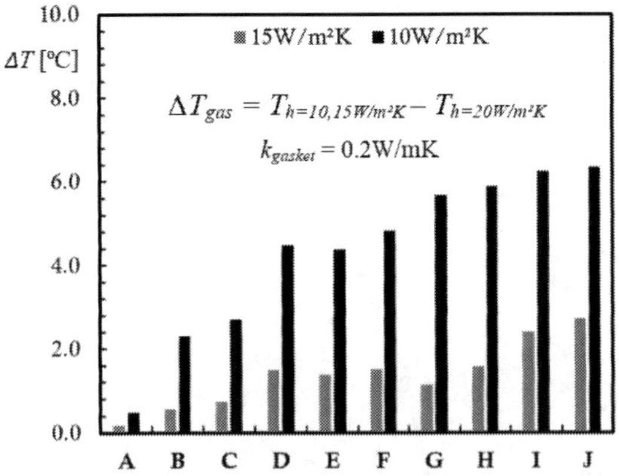

Figure 7. Effect of the external convective heat transfer coefficient on the temperature field (reference: h_{amb} = 20W/m^{2}K).

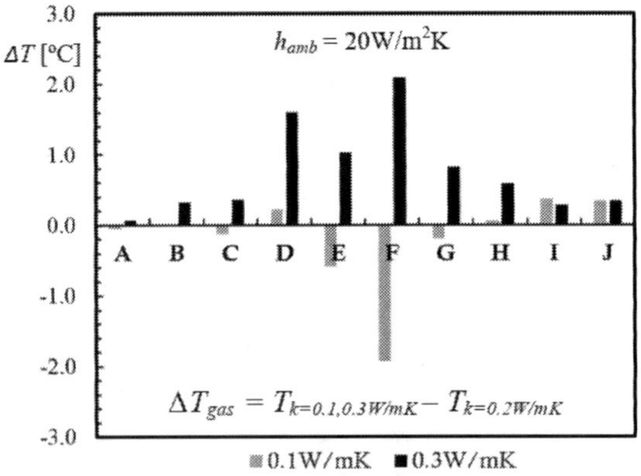

Figure 8. Effect of the insulating gasket thermal conductivity k on the temperature field (reference: k = 0.2W/mK).

5. Conclusions

A simulation model was developed and applied to predict the temperature distribution of an oil-free linear compressor designed for household refrigeration. The model adopts a lumped formulation for the compression cycle and a differential formulation for fluid flow and heat transfer in the remainder components of the compressor. Predictions and experimental data were found in reasonable agreement, with the exception of some locations in the gas path and solid components. The model was applied to analyse the influence of the external convection heat transfer and insulating gasket on suction superheating. The model is particularly useful for compressor design since it does not require experimental calibration.

Acknowledgments
The present study was developed as part of a technical-scientific cooperation program between the Federal University of Santa Catarina and EMBRACO. The authors also acknowledge the support provided by EMBRACO and CNPq (Brazilian Research Council) through Grant No. 573581/2008-8 (National institute of Science and Technology in Refrigeration and Thermophysics) and CAPES (Coordination for the Improvement of High Level Personnel).

References

[1] Van Der Walt N and Unger R 1994 Linear Compressors – A Maturing Technology *Proc. Int. Compressor Eng. Conf. at Purdue (West Lafayette, IN, USA)*
[2] Bradshaw C, Groll E and Garimella S 2011 A Comprehensive Model of a Miniature-Scale Linear Compressor for Electronics Cooling *Int. J. Refrigeration* **34** 63–73
[3] Unger, R L and Novotny S 2002 A High Performance Linear Compressor for CPU Cooling *Proc. Int. Compressor Eng. Conf. at Purdue (West Lafayette, IN, USA)*
[4] Prasad B G S 1998 Heat transfer in Reciprocating Compressors *Proc. Int. Compressor Eng. Conf. at Purdue (West Lafayette, IN, USA)*
[5] Meyer W and Thompson H 1988 An Analytical Model of Heat Transfer to the Suction Gas in a Low-Side Hermetic Refrigeration Compressor *Proc. Int. Compressor Eng. Conf. at Purdue (West Lafayette, IN, USA)*
[6] Ooi K T 2003 Heat Transfer Study of a Hermetic Refrigeration Compressor *J. Applied Thermal Engineering* **23** 1931–45
[7] Padhy S K 1992 Heat Transfer Model of a Rotary Compressor *Proc. Int. Compressor Eng. Conf. at Purdue (West Lafayette, IN, USA)*
[8] Todescat M L, Fagotti F, Ferreira R T S and Prata A T 1992 Thermal Energy Analysis in Reciprocating Hermetic Compressors *Proc. Int. Compressor Eng. Conf. at Purdue (West Lafayette, IN, USA)*
[9] Birari Y V, Gosavi S S and Jorwekar P P 2006 Use of CFD in Design and Development of R404a Reciprocating Compressor *Proc. Int. Compressor Eng. Conf. at Purdue (West Lafayette, IN, USA)*
[10] Chikurde C and Longanathan E 2002 Thermal Mapping of Hermetically Sealed Compressors Using Computational Fluid Dynamics Technique *Proc. Compressor Eng. Conf. at Purdue (West Lafayette, IN, USA)*
[11] Raja B, Seckar S J, Lal D M and Kalanidhi A 2003 A Numerical Model for Thermal Mapping in a Hermetically Sealed Reciprocating Compressor *Int. J. Refrigeration* **26** 652–58
[12] Almbauer R A, Burgstaller A, Abidin Z and Nagy D 2006 3-Dimensional Simulation for Obtaining the Heat Transfer Correlations of a Thermal Network Calculation for the Hermetic Reciprocating Compressors *Proc. Int. Compressor Eng. Conf. at Purdue (West Lafayette, IN, USA)*
[13] Ribas Jr F A 2007 Thermal Analysis of Reciprocating Compressors *Proc. Int. Conf. on Compressors and their Systems (London, UK)*
[14] Sanvezzo Jr J and Deschamps C J 2012 A Heat Transfer Model Combining Differential and Integral Formulations for Thermal Analysis of Reciprocating Compressors *Proc. Int. Compressor Eng. Conf. at Purdue (West Lafayette, IN, USA)*
[15] Schreiner J E 2008 Simulation Methodologies Development for Thermal Management Solution Analysis Applied to Refrigeration Reciprocating Compressor *MSc Dissertation (Federal University of Santa Catarina, Florianopolis, Brazil)*
[16] Annand W J D 1963 Heat Transfer in the Cylinders of the Reciprocating Internal Combustion Engines *Proc. Institution of Mechanical Engineers* **177** 973–96
[17] Deschamps C J, Prata A T and Ferreira R T S 2000 Modelling of Turbulent Flow Through Radial Diffusers *J. Brazilian Society of Mechanical Sciences* **22** 31–41

[18] Ferreira R T S and Lilie D E B 1984 Evaluation of the Leakage Through the Clearance Between Piston and Cylinder in Hermetic Compressor *Proc. Int. Compressor Eng. Conf. at Purdue (West Lafayette, IN, USA)*
[19] Patankar S V 1980 *Numeric Heat Transfer and Fluid Flow* (New York: Hemisphere Publishing)
[20] ANSYS Inc. 2014 *FLUENT User's Guide* (Lebanon/NH, USA)

IOP Conf. Series: Materials Science and Engineering **90** (2015) 012017 doi:10.1088/1757-899X/90/1/012017

Numerical Investigation of Flow Losses through Discharge Line of Household Type Refrigerator Compressors

I Yesilaydin[1] and L B Erbay[2]

[1] R&D Specialist, *Arcelik A.S. Compressor R&D Center,*
Compressor Plant, Organize Sanayi Bolgesi, 1. Cadde, Eskisehir, Turkey
[2] Professor, *Eskisehir Osmangazi University, Department of Mechanical Engineering*
Bati Meselik 26480, Eskisehir, Turkey

E-mail: ismail.yesilaydin@arcelik.com

Abstract. The technological developments, competition and energy policies are forcing the refrigeration market to increase the efficiency of their products as never before. The component that determines the performance and efficiency of the refrigeration system is the compressor, which is the major energy consumer in a compression refrigeration system. One of the essential elements of the total compressor efficiency is the discharge line flow efficiency that is subjected to this study. In this paper, the effects of design parameters on flow loses at discharge line of a hermetic reciprocating compressor were investigated with numerical flow analysis. At this study, three conceptual designs were modelled based on the discharge line design parameters such as line diameter, resonator volumes and line length. The pressure drop regions are determined by CFD analyses and they compared against the base model. Analyses are carried out by using commercial CFD software. Furthermore, the obtained numerical results were compared to experimental data and presented a good agreement in terms of pressure drop and discharge line flow efficiency.

1. Introduction

In recent years, the growing global need for energy and running out of the natural energy sources, prompt human being to find alternative sources and use the existing ones efficiently. By this way, energy saving practices and precautions will be reduced the effects of carbon dioxide emissions and global warming, which are serious threats to the future of the world.

According to electricity distribution and consumption statistics, 2.836.637 GWh of electrical energy is consumed by the countries of EU-27, 2010. The residential customers consumed 29.71% of this total electricity energy consumption. Only the refrigerators and freezers consume about 14.5% of the residential electric energy consumption. In this regard, energy consumption of residential refrigeration system is said to be 122.201 GWh approximately [1]. Total energy consumption by residential and commercial sectors in US is 21.7% and 17.8% respectively. Refrigerators and freezers consume 7.2% of total residential energy use and they are responsible for 11.9% of the residential electricity consumption [2]. As well as all over the word, regulations on energy saving are becoming increasingly stricter every year and also they have imposed on compressor manufacturers to design more efficient and environmentally friendly compressors.

Content from this work may be used under the terms of the Creative Commons Attribution 3.0 licence. Any further distribution of this work must maintain attribution to the author(s) and the title of the work, journal citation and DOI.
Published under licence by IOP Publishing Ltd

One of the most important parts of refrigeration system is a compressor which is the heart of refrigeration systems. Basically, compressors determine the performance and efficiency of the cooling system. This basic component of the cooling system consuming the electrical energy and converts it to energy stored in the form of pressure of the evaporated refrigerant. Although there are lots of different compressor types to achieving this goal but in the competitive environment as a result of globalization, the hermetic reciprocating compressor is the most commonly used type of compressors in household type refrigerators.

There are five fundamental requirements of a hermetic reciprocating compressor such as lifetime, acoustic noise level, compactness, efficiency and price. By the way, it has been known that compressor consumes the greater part of the electrical energy that input power of the refrigeration system. Even though all of the fundamental requirements are important for compressor manufacturers, but the compressor efficiency is getting more important because of the energy prices and environmental awareness [3].

Studies for improving the efficiency of compressors are grouped under three main headings: increasing the efficiency of electric motor, reducing the mechanical loses and also reducing the thermodynamic loses due to the irreversibilities in the suction, compression and discharge process. Several investigations have been reported concerning improving the performance of household reciprocating compressors. McGovern and Harte [4] presented a new approach in the field of compressor irreversibilities analysis and quantified the exergy destruction in particular regions. Birari et al. [5] investigated the existing compressor design with alternative refrigerant. With the CFD simulations, they checked for the heat balance across the compressor domain and suction and discharge path pressure drops. Ozdemir [6] examined discharge tube's and discharge muffler's heating effects to the compressor performance by using conceptual designs. This paper describes the use of CFD (Computational Fluid Dynamics) in the design and development of an efficient discharge line of reciprocating compressor. For this purpose, CFD simulations are carried out for numerically investigate effects of the structural changes in the discharge line of a hermetic reciprocating compressor to the flow loses that occur at that region.

Within the scope of this study, the effects of some of the structural changes to flow losses at the discharge line of a household type hermetic reciprocating compressor were investigated by numerical analysis. For this study, three different conceptual designs, which based on the discharge line design parameters such as orifice diameter, resonator volumes and line length, were modelled. The pressure drop regions are determined by CFD analyses and they compared against the base model. Analyses are carried out by using commercial CFD software. Furthermore, the obtained numerical results were compared to experimental results.

2. Numerical analyses of discharge line

In this section, modelling the discharge path of a hermetic reciprocating compressor, discretization of the generated model and boundary conditions are discussed.

2.1. Analysis

One of the major factors of total compressor efficiency is the discharge line flow efficiency. In order to determine the proper discharge line geometry for increase the performance of the compressor is intended. For this purpose, the effect of discharge lines which have different geometries on pressure drop due to the flow loses is examined with a commercial CFD software package ANSYS Fluent, which is based on the finite volume method. All the geometries in discharge line analysed that three dimensional and turbulent fluid flow pattern. Discharge process is assumed as steady state conditions and the thermal effects are neglected. Isobutane (R600a) is used as a refrigerant and in the simulations the fluid properties have been implemented as constant at mean temperature. At discharge line,

IOP Conf. Series: Materials Science and Engineering **90** (2015) 012017 doi:10.1088/1757-899X/90/1/012017

refrigerant gas flow is pulsating due to the opening and closing movements of the lead valves. In this study, refrigerant gas pulsation and also reed valve's movement are excluded.

2.2. Modeling

Four different exhaust lines are investigated and one of them is already being produced which can be seen in Figure 1.

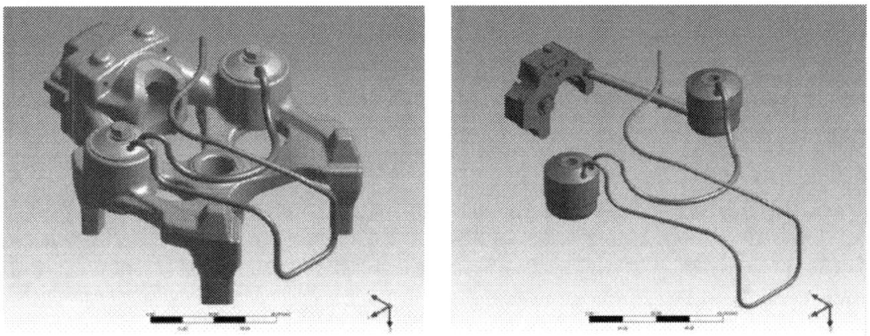

(a) (b)

Figure 1. Structure of the discharge line: CAD assembly (a) and flow path (b).

The other models were generated from the original model. Producibility is significant during modelling the alternative discharge lines. Model 2 designed to see the effects of widening the connection bore. At model 3, connection bore connected the fluid to the other resonator. Thus, the resonator volume is extended and discharge line length is reduced. Model 4 has only one resonator and discharge line length is also reduced. These models are presented in Figure 2.

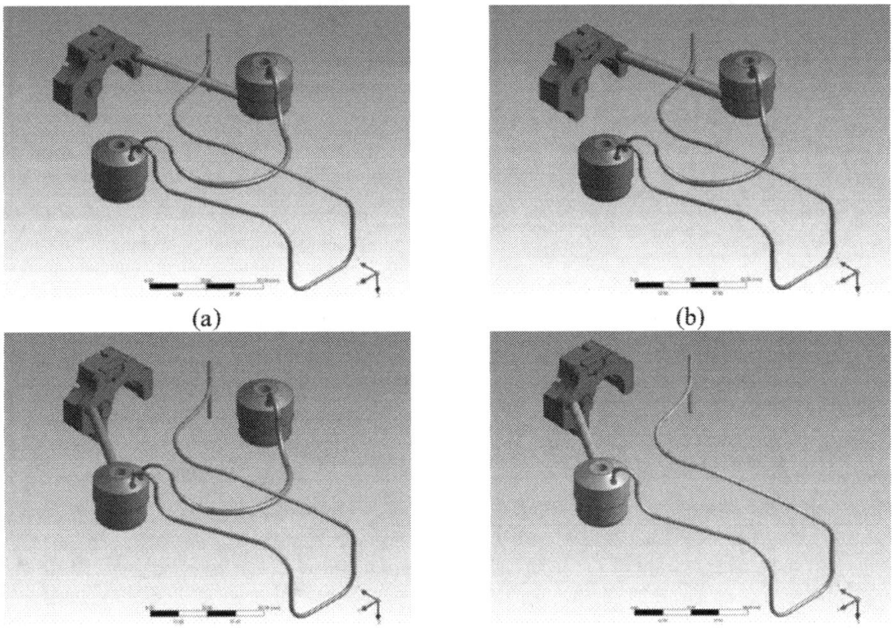

(a) (b) (c) (d)

Figure 2. Discharge line designs: Model 1 (a), Model 2 (b), Model 3 (c) and Model 4 (d).

A comparison between discharge line properties is presented in Table 1 where the distance of exhaust port between connection bore, diameter of connection bore and number of resonator volumes are given.

Table 1. Alternative dimensions of discharge lines.

Property	Unit	Model 1	Model 2	Model 3	Model 4
Distance of Exhaust Port between Connection Bore	δ	a	a	0.37 a	0.37 a
Diameter of Connection Bore	Ø	b	1.25 b	b	b
Number of Resonator		2	2	2	1

2.3. Discretization and boundary conditions

At discretization stage, tetrahedral mesh is generated for all domains of discharge line by using the ANSYS Mesh Platform. Figure 3 shows the applied mesh form. The smallest and the biggest element sizes are $1.9715 \cdot 10^{-5}$ m and $2 \cdot 10^{-3}$ m respectively. The total number of elements is approximately 7 million, which is enough for mesh independency.

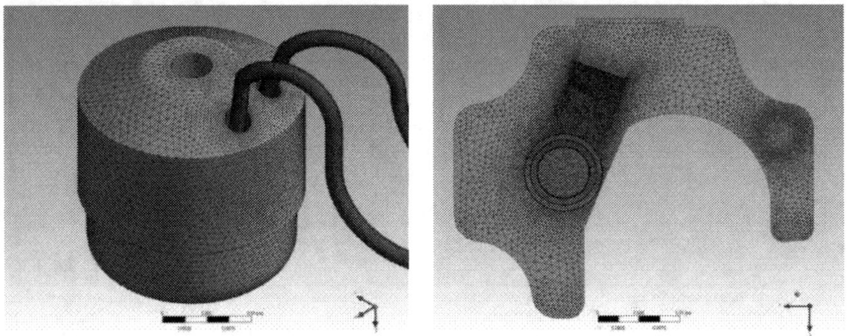

(a) (b)

Figure 3. Image of applied mesh: resonator (a), cylinder head group (b).

For the simulations, ASHRAE standard test conditions were used (54.4 °C condensing temperature, -23.3 °C evaporating temperature). Table 2 shows the refrigerant properties and boundary conditions used in the simulation.

Table 2. Properties and boundary conditions.

Refrigerant Properties (R600a)	Density (kg/m3)	10.992
	Dynamic Viscosity (kg/m-s)	1.0629e-05
Boundary Conditions	Mass Flow Inlet (kg/s)	6.256e-04
	Pressure Outlet (Pa)	0
	Operating Pressure (Pa)	761300

The behaviour of the turbulent flow has been considered using the Realizable k-Epsilon model with Enhanced Wall treatment. The turbulence parameters, such as turbulence intensity, were calculated and specified at appropriate zones. SIMPLE algorithm was used for the velocity pressure coupling.

3. Results and discussion

The pressure drop which is the main flow parameters, at the discharge path of the compressor is considered as the comparable result. The pressure drop through the discharge line which is extending from valve plate exhaust port to the discharge tube are compared according to inlet pressure of the discharge line in Figure 4. Due to the output pressure is zero; inlet pressure means the lost pressure.

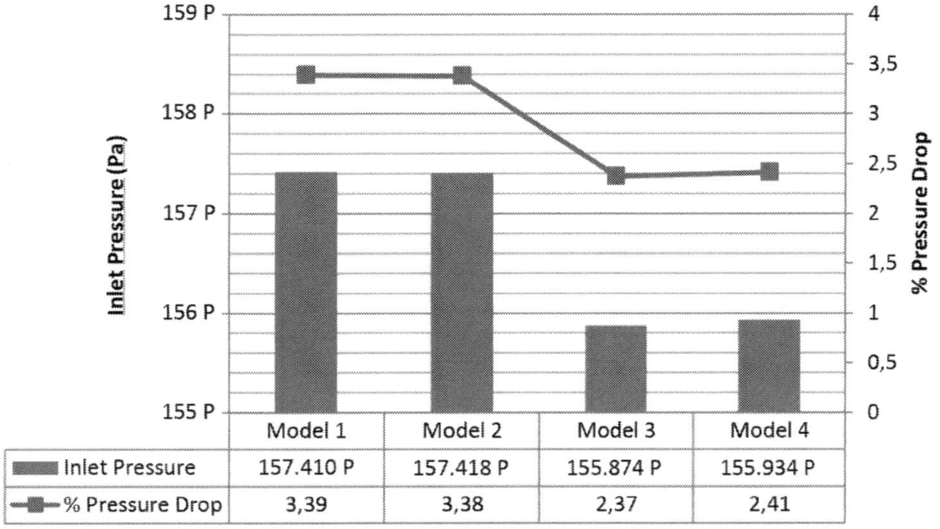

Figure 4. Inlet pressure and the pressure drop.

Figure 4 shows that the lost pressure of the Model 1 and 2 are greater than the others because of the flow across the single resonator and shortening the distance between the exhaust port and connection bore. Pressure drop of the both Model 1-2 and Model 3-4 are almost identical. Table 3 shows the comparison of the area weighted flow velocity at critical areas. It can be seen from the table, increasing the cross sectional area of the connection bore reduces the flow velocity about 0.19V m/s at this area. Aforesaid velocity reduction is reflected the outlet velocities.

Table 3. Area weighted flow velocities at critical areas.

Model	Inlet [m/s]	Connection Bore [m/s]	Outlet [m/s]
1	0.3433 V	0.5749 V	2.4957 V
2	0.3433 V	0.3815 V	2.3981 V
3	0.3433 V	0.5698 V	2.4745 V
4	0.3433 V	0.5381 V	2.4060 V

Figure 5 illustrates the streamlines of the discharge flow path. It is observed that there isn't any flow between the resonators at Model 3. Second resonator of the Model 3 pretends like it's excluded from the flow path. It is also observed that the streamlines between the exhaust port and connection bore is shortened at Model 3 and 4 with respect to Model 1 and 2.

IOP Conf. Series: Materials Science and Engineering **90** (2015) 012017 doi:10.1088/1757-899X/90/1/012017

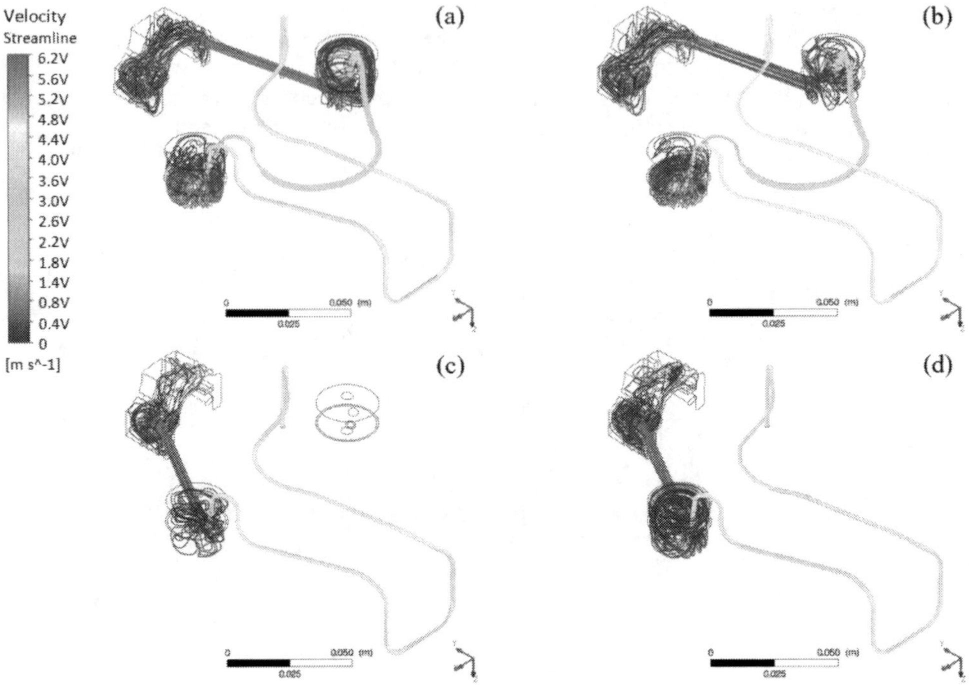

Figure 5. Streamlines: Model 1 (a), Model 2 (b), Model 3 (c) and Model 4 (d)..

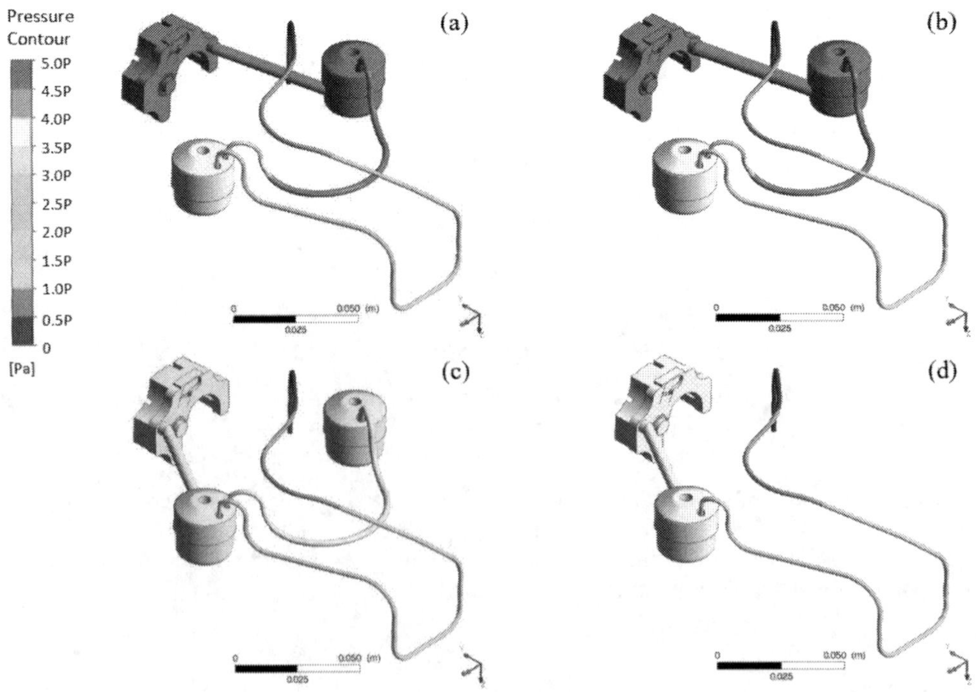

Figure 6. Pressure distribution plot: Model 1 (a), Model 2 (b), Model 3 (c) and Model 4 (d)..

Figure 6 shows the pressure distribution contour plot. It can be seen that there isn't a significant pressure drop between discharge port and first discharge muffler as against the shockloop area. For all models, the maximum pressure drop occurs in the shockloop flow path. Model 3 and 4 doesn't include tube between the resonator volumes on the direct flow path. For this reason the pressure drop that occurs in this area is prevented with these designs. It is also observed that the Model 3 has greater pressure drop against Model 4. It can be concluded that enlargement of the resonator volume with connection tube increased the pressure drop.

Only velocity streamlines or pressure distribution plots don't make sense for complex geometries like compressor flow path. For expanding on the numerical results, both pressure and velocity on centreline are illustrated in Figure 7 for Model 1.

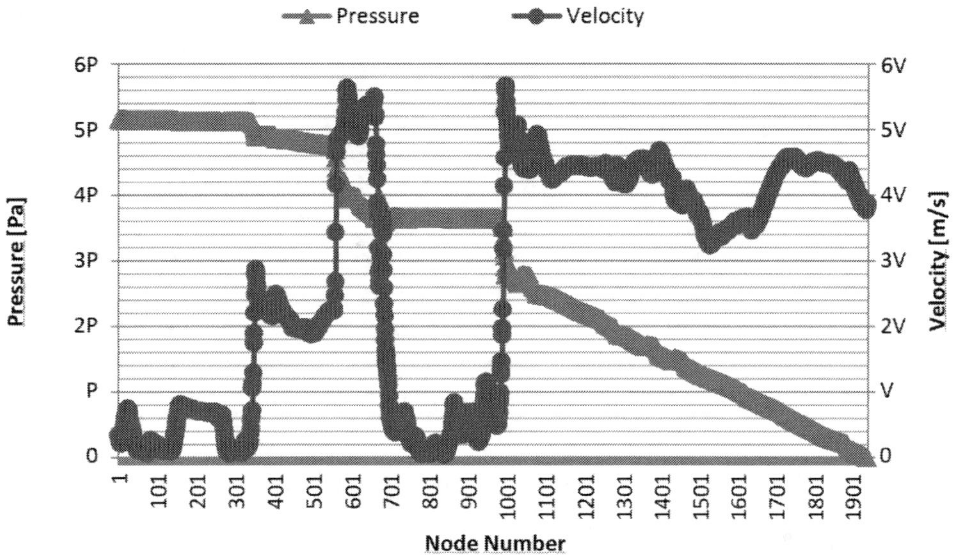

Figure 7. Pressure and velocity at centreline for Model 1.

In this figure, pressure drop areas can be recognised with the node numbers through the whole flow path. For this model, pressure value doesn't vary significant amount up to 319th node. In this section, refrigerant passes from first resonator to the connection tube. For this throttling effect velocity is increased up to 3V. Entrance of the connection bore at 158th node, velocity is increased up to 0.4 V. At the second throttling area, which is on the connection tube, is increased the velocity up to 5.6 V at 589th node. At the second resonator velocity is reduced to approximately V between the 661st and 994th nodes. After this region refrigerant passes to the shockloop and pressure reduces linearly with increased velocity.

Figure 8 shows the pressure and velocity at centreline for Model 2. It can be found that the characteristics of the pressure and velocity are the same with Model 1. For this model, it can be observed that the velocity is less than the Model 1 through the connection bore.

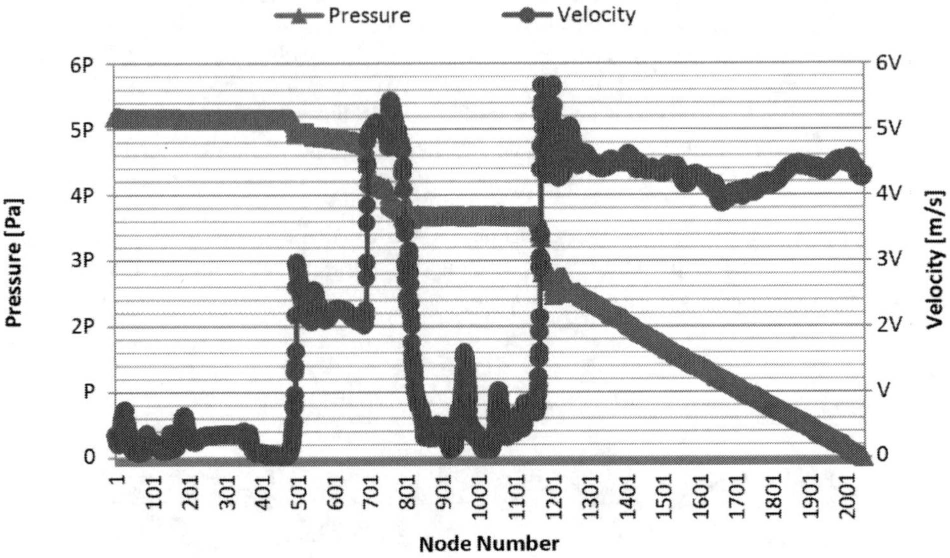

Figure 8. Pressure and velocity at centreline for Model 2.

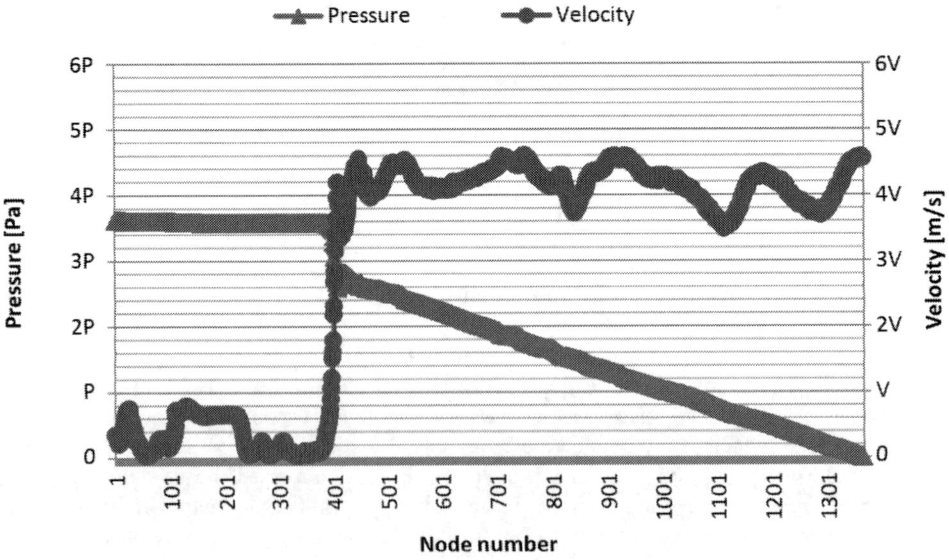

Figure 9. Pressure and velocity at centreline for Model 3.

In Figure 9 and 10 it can be observed that the flow path is simplified at these models. The drastic pressure drop is only occurred at the entrance of the shockloop. There isn't steady flow occurred between the resonators. For this reason it could not be obtained the effect of second resonator.

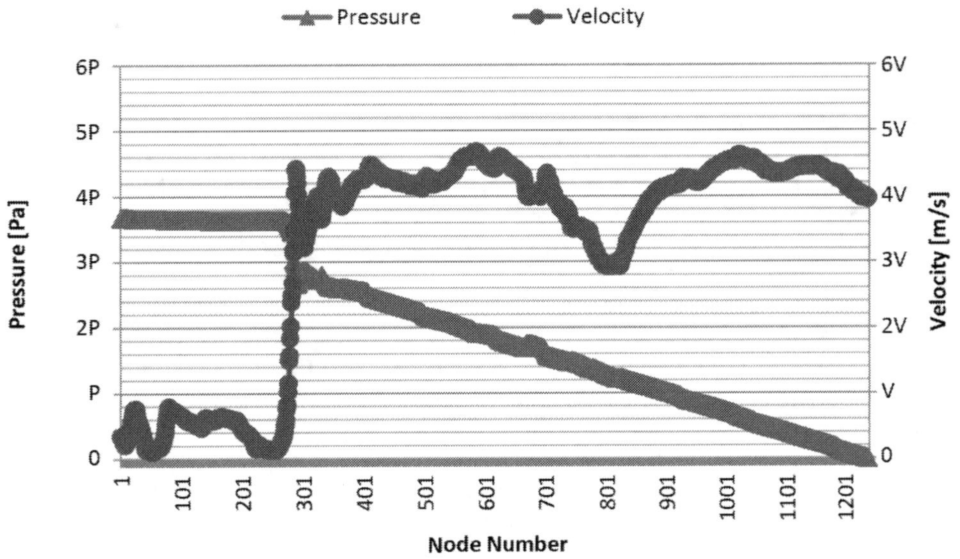

Figure 10. Pressure and velocity at centreline for Model 4.

In a comparison stage based on experimental study [7], it can be concluded that, the obtained numerical results have reasonable agreement in terms of pressure drop and discharge line flow efficiency. In Figure 11 comparisons of numerical and experimental results are presented. To improve the accuracy of the numerical results, pressure pulsations and thermal effect have to be taken into account.

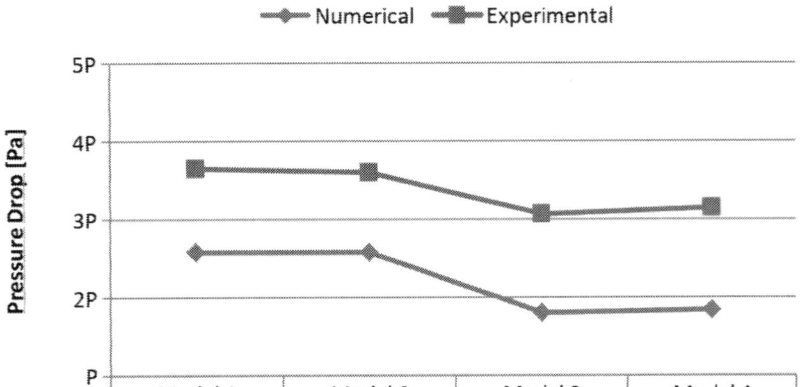

	Model 1	Model 2	Model 3	Model 4
Num.	0.2579 P	0.2575 P	0.1807 P	0.1837 P
Exp.	0.3648 P	0.3597 P	0.3064 P	0.3145 P

Figure 11. Comparison of numerical and experimental results.

4. Conclusion

In this study, numerical investigation of flow losses through discharge line and the behaviour of the pressure pulsation at the exhaust line of a reciprocating hermetic compressor were presented and the following conclusions were obtained:

- Reduction of the discharge flow path length reduces the pressure drop.
- The maximum pressure drop occurs in the shockloop flow path.
- Enlargement of the resonator volume with connection tube increased the pressure drop.
- Decreasing the discharge passage length might be affected the pressure pulsation value negatively. For this reason, decreasing the discharge passage length or resonator volume must also be reviewed in terms of acoustics.

Based on the results of this study it will be extended that:

- Investigation will be extended with the temperature and heat flux measurements.
- Pressure pulsation effect will be taken into account and correlated with acoustic measurements.

5. References

[1] Bertoldi P, Hirl B, Labanca N 2012 European commission joint research centre institute for energy and transport *Electricity Consumption And Efficiency Trends In The EU-27* (Ispra: European Union) pp 18-20.

[2] Energy Information Administration 2007 *Annual Energy Outlook 2007 with Projections to 2030* (Washington DC: Energy Information Administration) pp 142-3

[3] Rasmussen B D 1997, *Variable Speed Hermetic Reciprocating Compressor for Domestic Refrigerator* (Lyngby: Department of Energy Engineering Technical University of Denmark)

[4] McGovern J A, Harte S 1992 *Computer Simulation of Exergy Destruction within A Reciprocating Compressor* (West Lafayette: International Compressor Engineering Conference at Purdue) pp 310-5

[5] Birari Y V, Gosavi S S, Jorwekar P P 2006 *Use of CFD In Design And Development Of R404A Reciprocating Compressor* (West Lafayette: International Compressor Engineering Conference at Purdue) Paper 1727

[6] Ozdemir A R 2007 *An Investigation of the Heat transfer Characteristics among the Hermetic compressor Components* (Istanbul: Istanbul Technical University Graduate School of Science Engineering and Technology)

[7] Yesilaydin I, Erbay L B, Inan C 2014 *Investigation of Flow Losses Through Discharge Line of Household Type Refrigerator Compressors* (West Lafayette: International Compressor Engineering Conference at Purdue) Paper 2336

Acknowledgements

This study is part of a joint technical-scientific program between the Eskisehir Osmangazi University and Arçelik A.Ş. The authors would like to acknowledge the support of the management of the Research & Development Department, Arçelik A.Ş. Compressor Plant.

IOP Conf. Series: Materials Science and Engineering 90 (2015) 012018 doi:10.1088/1757-899X/90/1/012018

Modelling of fluid flow and heat transfer in a reciprocating compressor

J Tuhovcak[1], **J Hejcik and M Jicha**
Brno University of Technology, FME, Energy Institute, Technická 2, Brno 616 69, CZ

E-mail: tuhovcak@fme.vutbr.cz

Abstract. Efficiency of reciprocating compressor is strongly dependent on several parameters. The most important are valve behaviour and heat transfer. Valves affect the flow through the suction and discharge line. Heat flow from the walls to working fluid increases the work of the cycle. Understanding of these phenomena inside the compressor is a necessary step in the development process. Commercial CFD tools offer wide range of opportunities how to simulate the flow inside the reciprocating compressor nowadays, however they are too demanding in terms of computational time and mesh creation. Several approaches using various correlation equation exist to describe the heat transfer inside the cylinder, however none of them was validated by measurements due to the complicated settings. The goal of this paper is to show a comparison between these correlations using in-house code based on energy balance through the cycle.

1. Introduction

Reciprocating compressors with high efficiency and low production costs are employed in many technical fields, ranged from household appliances to large-scale compressors used in refineries and gas engineering. Reduction in compressor energy consumption by improving the efficiency is a relevant step in decreasing the global energy consumption. Analysis according to [1] shows that contribution of household compressors to overall electric energy consumption in US reaches up to 8 %. Statistics from industry sector are not known, however it is expected to be even higher. Overall efficiency of compressor consists of three main sub-efficiencies: electrical, mechanical and thermodynamic. Thermodynamic efficiency is the lowest (80 – 83 %), therefore there is a significant effort to improve compressor parts that cause decrease in efficiency: cylinder leakage, pressure losses in suction and discharge line and the superheating of the gas. High temperature causes material degradation (sealing, valves…), increases the risk of oil ignition inside the compressor and also increases the compression work. Therefore keeping the temperature of the gas as low as possible decrease the work input to a compressor and improve durability. The lowest compression work could be achieved with isothermal process, but refrigeration requires also high temperature on the compressor discharge to release heat in condensation unit. In this case the evaluation of thermodynamic efficiency is done by comparing isentropic and actual compressor work, see equation (1)

$$\eta_{isen} = \frac{w_{isentropic}}{w_{actual}} \tag{1}$$

[1] To whom any correspondence should be addressed.

Content from this work may be used under the terms of the Creative Commons Attribution 3.0 licence. Any further distribution of this work must maintain attribution to the author(s) and the title of the work, journal citation and DOI.
Published under licence by IOP Publishing Ltd

$$w_{isentropic} = \frac{\kappa R T_1}{\kappa - 1}\left[\left(\frac{p_2}{p_1}\right)^{(\kappa-1)/\kappa} - 1\right] \quad (2)$$

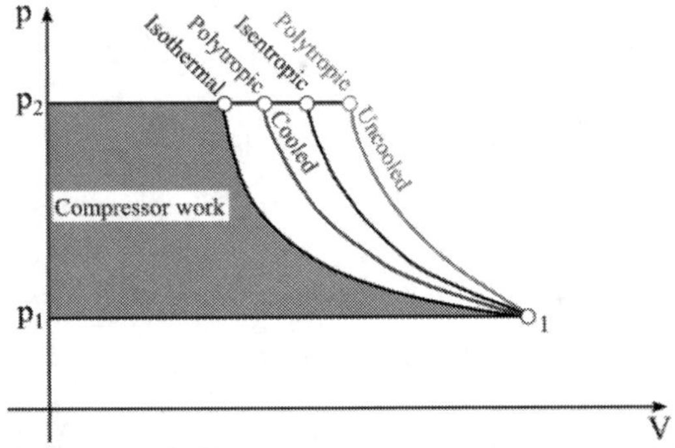

Figure 1. Processes inside the compressor cylinder.

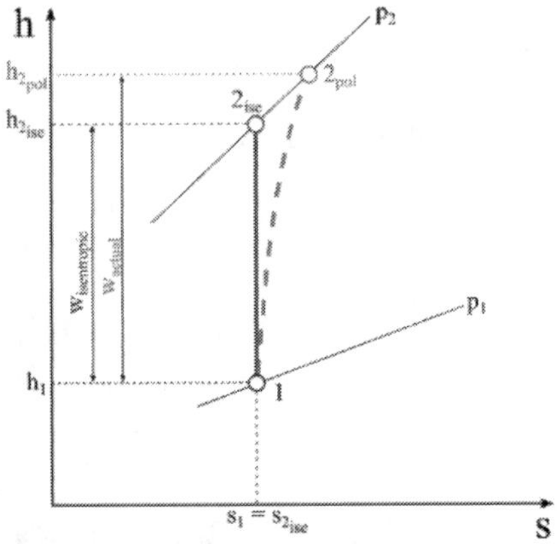

Figure 2. Comparison of isentropic and actual proces in the compressor.

Actual work of compression in a not cooled compressor is visible in the figure 1 and 2. In this paper actual work differs from isentropic (equation (2)) in heat transfer through the walls. Understanding heat and fluid flow inside the compressor during a working cycle is a fundamental step of development process aiming for better efficiency. As the compression of gas is the biggest source of heat, it is necessary to examine heat flow in the cylinder. It is possible to use numerical CFD tools, although integral models offer much faster results.

This paper presents a critical review of integral heat transfer models used for thermal analysis of compressor and their effect on the compressor efficiency.

2. Compressor modelling

Fluid flow modelling was done with 0D simulation tool using the energy balance over a control volume [2]. It has a benefit of fast calculation and could be easily combined with heat transfer model. Several approaches were used for heat transfer and they are presented in detail further in the paper. The 0D code was written in Matlab.

2.1. Fluid flow

The base of 0D model is an energy balance over a control volume using the 1st Law of Thermodynamics, equation (3). In this case the control volume is the whole volume of a cylinder. Flow inside the cylinder is neglected and both pressure and temperature are considered uniform in the cylinder.

$$dW + dQ + \sum_i dm_i \cdot h_i = dU \quad (3)$$

where Q stands for heat transfer from/to the cylinder, dm_ih_i is inflow or outflow through the valves, dU is the change of inner energy and dW is the piston work. The piston work is done by the change of a volume via piston motion

Mathematical description of valve movement was introduced by [3] as a mass-spring system, examined more deeply in the work of [4]. Description uses single degree of freedom equation (4) solved by using 4th order Runge-Kutta method.

$$m \cdot \ddot{x} + d \cdot \dot{x} + k \cdot x = \sum_i F_i \quad (4)$$

Properties of the spring system are formed by the mass of the valve *m*, stiffness of the spring *k* and damping constant *d,* which must be usually determined experimentally, as it is dependent on several parameters. The damping constant is usually very low, therefore it is often neglected in equation (4) [5]. Main forces acting on the valve are pressure difference over the valve, stiffness of the spring and adhesive forces due to oil between valve surface and seating. Mass flow through valves is calculated by Fliegner's equation (5):

$$\dot{m} = \varphi_{eff} \cdot \rho_1 \cdot \left(\frac{p_2}{p_1}\right)^{\frac{1}{\kappa}} \sqrt{\frac{2\kappa}{\kappa-1}\frac{p_1}{\rho_1}\left(1-\left(\frac{p_2}{p_1}\right)^{\frac{\kappa-1}{\kappa}}\right)} \tag{5}$$

where φ_{eff} is the effective flow area. It is determined from valve position and the local pressure loss coefficient ζ. p is the pressure and ρ is the density.

2.2. Heat transfer correlations

Improvement of the compressor efficiency could be done by optimization of thermal flows inside the compressor. The main source of the heat is the compression process inside the cylinder, nevertheless it is also important to analyze the heat transfer during the suction process. Hot walls of cylinder and suction manifold heat up the working fluid before the compression process and every increase of temperature by 1 K causes reduction of COP by 0.32 % [6]. Increase in temperature decreases the volumetric efficiency and increase in compression work, since the specific volume of a gas is proportional to temperature. This is particularly important during the suction process when the walls are superheating gas in suction line.

There are several approaches to quantify heat transfer coefficient, however the correct one has not been stated yet. Differences are significant especially during the suction or discharge process. Pereira [1] analyzed most important correlations for heat transfer and found the considerable disagreement with numerical simulation. All of these correlations are based on representation of Nusselt number (equation (6)) with different constants *a*, *b* and *c*, except for the approach used by Aigner [5], which uses Stanton number instead.

$$Nu = a\,\mathrm{Re}^b\,\mathrm{Pr}^c\,. \tag{6}$$

Correlations of Woschni (7) and Annand (8) were developed for internal combustion engines, in which a different flow could occur due to the different operation of the valves [7]. Equation (9) and (10) are used to calculate velocity in Woschni correlation.

$$Nu = 0.35\,\mathrm{Re}^{0.7} \tag{7}$$

$$Nu = 0.26\,\mathrm{Re}^{0.75} \tag{8}$$

$$u = 6.618 \cdot u_p \tag{9}$$

$$u = 2.28 \cdot u_p \tag{10}$$

Adair [7] proposed a new correlation (equation (11)) based on the investigation of reciprocating compressor. He used variable equivalent diameter *De*, equation (12) and (13) instead of cylinder diameter and the approximation of swirl velocity by angular speed of a crankshaft. Previous authors used a mean piston velocity in Reynolds number, which is much smaller compared to angular speed of crankshaft. According to Adair the compressor efficiency could be affected by heat transfer correlation by just 3 percent, however volumetric efficiency could decrease by 10 – 20 percent [7].

$$Nu = 0.053\,\mathrm{Re}^{0.8}\,\mathrm{Pr}^{0.6} \tag{11}$$

$$D_e = \frac{6 \cdot \textit{Cylinder volume}}{\textit{Cylinder area}} \tag{12}$$

$$\mathrm{Re} = \frac{\rho D_e u}{\mu} \tag{13}$$

The latest contribution to heat transfer modelling was done by Disconzi [8]. He divided the compressor cycle in 4 main processes: suction, compression, discharge and expansion. For each of them he developed new correlation. The definition of Nusselt number is similar to equation (6), however there are different constants for each process. Reynolds number is also different. Mean piston velocity is used during compression and expansion. The main innovation is in suction and discharge process, where the flow through the valve is taken into account, see table 1. The prediction of Disconzi's model was in good agreement with numerical simulation. He noted that the highest amount of heat is transferred during the suction because of the high temperature difference between sucking gas and cylinder walls. The heat transfer area is significantly larger compared to discharge, which increases the importance of a correct heat transfer prediction in cylinder.

Table 1. Reynolds number and constants for processes inside the cylinder according to Disconzi.

Process	**Reynolds number**	**Constants**		
		a	**b**	**c**
Compression	$\mathrm{Re} = \frac{\rho_{(t)} D u_p}{\mu_{(t)}}$	0.08	0.8	0.6
Discharge	$\mathrm{Re} = \frac{\rho_{(t)} D \left(u_p + u_p^{0.8} u_{c(t)}^{0.2}\right)}{\mu_{(t)}}$	0.08	0.8	0.6
Expansion	$\mathrm{Re} = \frac{\rho_{(t)} D u_p}{\mu_{(t)}}$	0.12	0.8	0.6
Suction	$\mathrm{Re} = \frac{\rho_{(t)} D \left(u_p + u_p^{-0.4} u_{c(t)}^{1.4}\right)}{\mu_{(t)}}$	0.08	0.9	0.6

$$u_c = \frac{\left|\dot{m}_{(t)}\right|}{\rho_{(t)} \cdot A_c} \tag{14}$$

where $\rho_{(t)}$ is the density of gas in the cylinder, A_c is the cross-sectional area of the cylinder and u_p stands for the piston velocity.

Aigner [5] also uses dimensionless numbers, but the correlation is built on Stanton number, which is the ratio of heat flux to the wall and energy flux in flow relative to wall, equation (15).

$$St = \frac{h}{\rho c_p u} = \frac{Nu}{\mathrm{Re}\,\mathrm{Pr}} \tag{15}$$

Stanton number could be determined as a function of skin friction coefficient C_f, which could be estimated from Reynolds number. In the work of Aigner, heat fluxes were reconstructed with the full three-dimensional calculation and for sufficient number of cases Aigner proposed more suitable Stanton numbers.

$$St = f\left(C_f\right) \tag{16}$$

The compressor cycle was also divided in main processes: inflow, outflow and compression with expansion together. Moreover he analyzed heat fluxes for individual surfaces in the compressor (piston, cylinder wall and cylinder head). It is worth noting that there is a difference between the cooling compressor and the compressor in Aigner's research. The latter is considerably bigger.

2.3. Compressor

Comparison of different heat transfer approaches was carried out with parameters of a real compressor. Dimension of the compressor and the properties of valves are shown in table 2. The working fluid differs in reality according to the application of the compressor. For the purposes of presented analysis an ideal air was used in order to simplify the comparison. Boundary conditions are shown in table 3.

Table 2. Compressor specification.

Parameter	Value	Unit
Cylinder bore	220	mm
Crank height	45	mm
Rod length	250	mm
Crankshaft speed	980	rpm
Clearance length	1.5	mm
Valve weight	0.08	kg
Spring stiffness	50000	N

Table 3. Boundary conditions used in the simulation tool.

Boundary condition	Value	Unit
Suction pressure	0.97	bar
Discharge pressure	5.0	bar
Inlet temperature	310	K
Wall temperature	350/450	K

3. Results and discussion

Numerical tools are used in the pre-development phase of the compressor to evaluate changes in design or working conditions. In comparison with experimental testing it is usually faster and less costly approach, nevertheless numerical results must be in reasonable agreement with experimental data. It has been proven that quasi-static simulation tool could be used for small reciprocating compressor with satisfying accuracy when examining the pressure behavior during the cycle. Evaluation of the heat transfer is however complicated task to do. Heat transfer rates change their direction and magnitude rapidly in fast running compressor and there are significant time lags in thermal processes. Mathematical description of the heat transfer is also not straightforward task as there are various approaches used in literature with no clear conclusion which one is most accurate. Literature shows the significant differences, but the effect of using different approaches on compressor efficiency was not discussed so far. Five correlations were examined in this paper and compared focusing on efficiency of the compressor. Evaluation of the efficiency of refrigerating compressor is convenient using isentropic efficiency, which is defined by equation (1) and (2).

The analysis of heat transfer models was carried out for different wall temperatures. In first case the wall temperature was set to 350 K. Results were compared with cylinder wall temperature set to 450 K. In the last case a balanced cycle temperature was used as cylinder wall temperature. This is the temperature, which is achieved when there is no heat transfer between cylinder and surrounded components or outside gas.

3.1. Heat transfer

Heat transferred during the crankshaft cycle is presented in the figure 3, where the negative values represent the heat flux from the cylinder to the walls. Positive values stand for heat going from walls to the gas. Each process inside the cylinder is demarcated by line. Interval A limits the compression, B is the discharge process, C is the expansion and D limits the suction process. The temperature rises as the gas is compressed and when it overcomes the wall temperature the heat changes the direction. This can be seen in the figure 3. Few discontinuities occur during the discharge process in the model of Woschni and Disconzi. The reason is the change of the characteristic velocity in Reynolds number. Woschni uses equation (9) during the discharge or suction and equation (10) for compression and expansion. Equations used by Disconzi are shown in table 1. The valve may bounce during the suction or discharge process closing up the cylinder. The discontinuities can be also seen in figure 3. Gas in the cylinder is heated up by surrounding walls during a part of expansion and the whole suction process, which is limited by discontinuities of Woschni's and Disconzi's model.

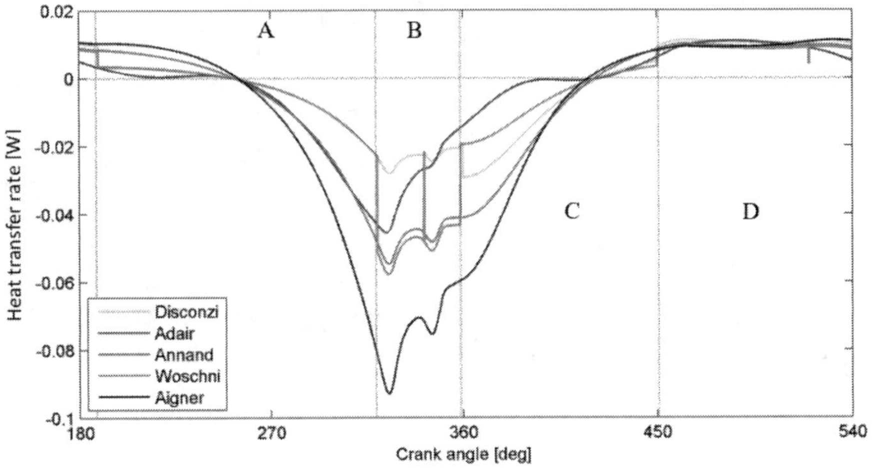

Figure 3. Heat transfer during the crankshaft cycle (T_{wall} = 350 K).

Heat transfer during the suction shows very similar behavior in all models. The total amount of transferred heat is very similar in table 4, except for the model of Disconzi and Aigner, which are slightly above all. Increase in the temperature causes reversed heat flow. The differences between models are rather significant during discharge. The models of Woschni and Annand are comparable and the values of total heat transfer rates are alike during the discharge. The reason could be in analogous expression of Nusselt number. The model of Aigner predicts the highest net heat flow during the cycle, however it must be pointed out that the Aigner adjusted Stanton numbers according to sufficient number of CFD analysis. In this work we used approach from [9] to calculate Stanton number.

The model used by Adair has different behavior during the cycle, especially after the suction and also the amount of heat transferred during discharge and re-expansion is very small. The reason could be also found in the definition of velocity and the characteristic dimension in Reynolds number. Whereas Woschni, Annand and Disconzi use mean piston velocity, Adair [7] uses the angular crankshaft speed as characteristic gas velocity.

Expression of characteristic gas velocity in all the models is simpler during the compression and expansion process, because authors usually use just piston mean speed. Nevertheless there are still significant differences in the amount of transferred heat caused by various expressions of Nusselt number or the velocity (Adair). However all models have analogous tendencies.

Heat transfer from cylinder outer surface is not known for the simulation purposes, therefore it was set as adiabatic to find balanced wall temperature. The balanced wall temperature will not be similar for all models as the net heat transfer is different. For this settings the highest amount of heat is transferred during the suction and discharge phase, see table 5 and figure 4. The total amount of heat transferred during the suction phase is very similar for all models in both cases (steady temperature and adiabatic wall). Other phases are significantly different. The differences between heat transfer models are expected since the authors use particular correlations. The effect of using various models on compressor efficiency was not examined in detail so far and this information could be beneficial for simulation tools to improve their accuracy.

Table 4. Heat transfer rate for particular models and wall temperature 350 K.

	Heat transfer rate [W]				
	Q_{cycle}	Q_{comp}	Q_{exp}	Q_{suc}	Q_{dis}
Isentropic	0	0	0	0	0
Disconzi	-13.92	-4.90	-9.31	9.73	-9.44
Adair	-17.46	-11.98	-1.77	8.09	-11.80
Annand	-30.17	-9.01	-11.93	9.10	-18.33
Woschni	-21.79	-4.66	-5.99	8.28	-19.41
Aigner	-47.26	-13.92	-14.44	9.82	-28.73

Table 5. Heat transfer rate for particular models and balanced wall temperature.

	Heat transfer rate [W]					
	T_{wall}	Q_{cycle}	Q_{comp}	Q_{exp}	Q_{suc}	Q_{dis}
Isentropic	-	0	0	0	0	0
Disconzi	383.33	0.00	-1.40	-6.09	15.51	-8.02
Adair	394.88	0.00	-5.34	-0.06	14.77	-9.37
Annand	404.69	0.00	2.78	-4.93	16.30	-14.15
Woschni	403.21	0.00	1.16	-2.40	15.79	-14.55
Aigner	421.27	0.00	7.20	-4.86	18.00	-20.33

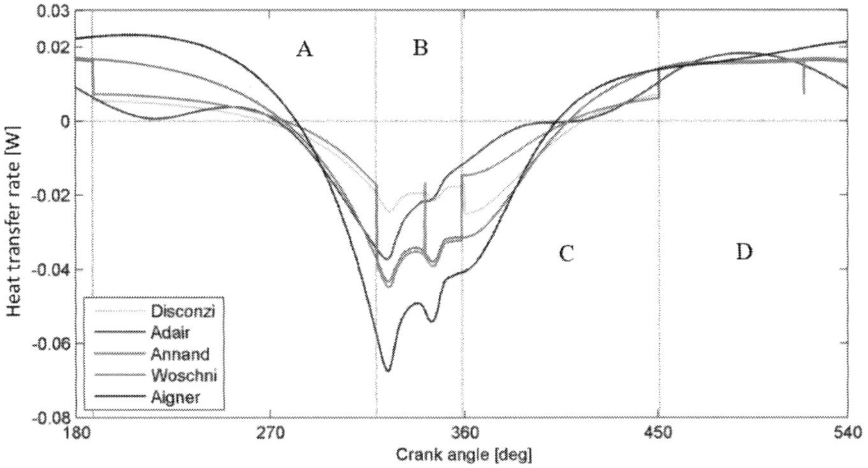

Figure 4. Heat transfer rate for particular models and balanced wall temperature.

3.2. Efficiency

The efficiency of compressor used in refrigeration is usually described with the isentropic efficiency, equation (1). Evaluation of the full crankshaft cycle shows that there is not necessarily analogy between isentropic efficiency and the amount of transferred heat. This is valid for all heat transfer models. Compression and expansion work corresponds slightly better with the bulk transferred heat. Overall the isentropic efficiency is not distinctive using different models and the differences between models are around 1 % (table 6). This is also hardly visible in indicated p-V diagram.

The initial temperature has a significant importance at the beginning of compression when determining the compressor efficiency. Higher wall temperature will cause increase in the initial gas temperature and decrease of the mass flowing into the cylinder. When the inlet temperature of the gas was set to 310 K and wall temperature was 350 K, almost all models predicted higher gas temperature at the beginning of compression by approximately 6 K, however all models were under the isentropic temperature. The model of Aigner predicted even lower temperature than it is in suction line, which is the consequence of gas re-expansion from a clearance volume. Gas overheating presented in table 6 is caused only by heat transfer from cylinder walls, the suction line is not taken into account. Decrease in the temperature at the suction helps to suck more mass inside the cylinder, which enhances the volumetric efficiency.

When the wall temperature was changed to 450 K, the mass inside the cylinder decreased by app. 6.5 %, see table 6. Hot wall heats up the gas and the specific volume raises, therefore the free volume of cylinder is filled up faster with smaller amount of mass. Change in the wall temperature by 100 K resulted in higher temperature by nearly 21 K at the beginning of compression.

Comparison of the cases with different wall temperature in table 6 showed also that the decrease in isentropic efficiency due to the raise of gas temperature at the beginning of compression is not significant, however the decrease in volumetric efficiency is more important. The difference of 21 K caused the lower mass load of the gas in the cylinder by approximately 0.31 % per 1 K. Important conclusion is the agreement of the models on this fact.

There are results of balanced cycle in the table 7. Balanced temperature is different for each model, nevertheless the isentropic efficiency remains very similar, ranging from 99.6 % to 96.9 %. Both the temperature and the mass in the cylinder after closing of suction valve are almost identical for all the models.

Table 6. Comparison of compressor with different wall temperatures.

	T_{wall} = 350			**T_{wall} = 450**		
	m_{comp_0}	T_{comp_0}	η_{isen}	m_{comp_0}	T_{comp_0}	η_{isen}
Isentropic	0.00446	318.59	100.00	0.00446	318.59	100.00
Disconzi	0.00448	317.11	98.91	0.00422	336.71	98.75
Adair	0.00449	316.60	99.99	0.00426	333.54	99.31
Annand	0.00455	312.37	98.41	0.00424	335.43	97.52
Woschni	0.00452	314.62	99.18	0.00428	332.53	98.78
Aigner	0.00462	307.80	98.03	0.00426	333.59	96.61

Table 7. Work and isentropic efficiency of compressor with balanced temperature.

	Work [J] and isentropic efficiency [%]			
	T_{wall}	m_{comp_0}	T_{comp_0}	η_{isen}
Isentropic		0.00446	318.59	100.0
Disconzi	384.80	0.00439	323.96	98.8
Adair	394.93	0.00438	324.24	99.6
Annand	404.71	0.00437	325.08	97.8
Woschni	403.21	0.00439	324.19	98.9
Aigner	421.28	0.00436	326.39	96.9

4. Conclusion

Overheating of the gas inside the compressor is the problem causing the decrease in the efficiency. In the presented paper several heat transfer models were examined to see their effect on the efficiency. It can be concluded from results that various heat transfer models do not affect compressor efficiency significantly, however the amount of transferred heat alters greatly. The results also show that overheating in the cylinder do not cause important drop in isentropic efficiency, but the difference in volumetric efficiency is rather more decisive. Further tests are still required to verify the statements for different compressors and conditions.

5. Nomenclature

A_c	[m²]	cross section of the cylinder	T_{wall}	[K]	wall temperature
c_p	[J/kg/K]	thermal capacity for const. pres.	u	[m/s]	velocity
C_f	[-]	friction coefficient	u_p	[m/s]	mean piston velocity
D	[m]	characteristic dimension	u_c	[m/s]	flow velocity through the valve
d	[-]	damping constant	U	[J]	internal energy
F	[N]	force acting on valve	x	[m]	valve position
h	[J/kg]	specific enthalpy	w	[J/kg]	specific work
k	[N/m]	spring stiffness	W	[J]	work
m	[kg]	mass	κ	[-]	adiabatic exponent
p	[Pa]	pressure	η_{isen}	[%]	isentropic efficiency
Q	[J]	rate of heat transfer	ρ	[kg/m³]	density
R	[J/kg/K]	gas constant	φ_{eff}	[m²]	effective flow area
t	[s]	time	μ	[m²/s]	kinematic viscosity
T	[K]	temperature			

Index	**Meaning**	**Index**	**Meaning**
cycle	full cycle	exp	expansion
comp	compression	suc	suction
comp_0	beginning of compression	dis	discharge

6. Acknowledgement

The presented research was supported by the project LO1202 NETME CENTRE PLUS with the financial support from the Ministry of Education, Youth and Sports of the Czech Republic under the "National Sustainability Programme I",", by the project DC components and grid of ENIAC Joint undertaking and by the Brno University of Technology project FSI-S-14-2355 and faculty project RP9052100712.

7. References

[1] Pereira E L, Deschamps C J and Ribas F A 2010 Numerical Analysis of Heat Transfer inside the Cylinder of Reciprocating Compressors in the Presence of Suction and Discharge Processes *Int. Compressor Engineering Conf.* Paper 1995.

[2] Lang W, Almbauer R, Burgstaller A and Nagy D 2008 Coupling of 0-, 1- and 3-d Tool for the Simulation of the Suction Line of a Hermetic Reciprocating Compressor *Int. Compressor Engineering Conf.* Paper 1881.

[3] Costagliola M 1950 The theory of spring-Loaded Valve for Reciprocating Compressor *Journal of Applied Mechanics*, vol. 17, pp. 415-20.

[4] Habing R 2005 Flow and Plate Motion in Compressor Valves: Doctoral Thesis.

[5] Aigner R and Steinrueck H 2007 Modelling Fluid Dynamics, Heat Transfer and Valve Dynamics in Reciprocating Compressor *5th Conference of the EFRC* 171 - 80.

[6] Almbauer R, Lang W, Nagy D and Berger E 2010 Application of Computational Fluid Dynamics for the Thermodynamic Development of a New Generation of Hermetic Reciprocating Compressor *Int. Compressor Egineering Conf.* Paper 1968.

[7] Adair R, Qvale E and Pearson J 1972 Instantenous Heat Transfer to the Cylinder Wall on Reciprocating Compressors *Purdue Compressor Technology Conf.*

[8] Disconzi F P, Deschamps C J and Pereira E L, 2012 Development of an In-Cylinder Heat Transfer Correlation for Reciprocating Compressor *Int. Compressor Engineering Conf.* Paper 2103.

[9] Bejan A 2013 *Convection heat transfer*, New Jersey, John Wiley & Sons, 4th ed. p. 658.

[10] Ribas F A, Deschamps C J, Fagotti F, Morriesen A and Dutra T 2008 Thermal Analysis of Reciprocating Compressors - A Critical Review *Int. Compressor Engineering Conf.* Paper 1907.

[11] Steinrueck H, Aigner R and Machu G 2008 Transversal waves in a reciprocating compressor *Acta Mechanica*, no. 201, pp. 231-47.

[12] Cengel Y A 2008 *Thermodynamics: an engineering approach*, New York: McGrawHill.

An Investigation on the Pin Bearings' Optimization of a Hermetic Reciprocating Compressor

A. R.Ozdemir[1] ; B.Hacioglu[2] ; E. Kasapoglu[2]
[1] Senior Specialist-R&D, Arcelik A.S. R&D Center, Istanbul, Turkey
[2] Senior Specialist-R&D, Arcelik A.S. Compressor Plant, Eskisehir, Turkey

E-mail: ahmetrefik.ozdemir@arcelik.com

Abstract. A hermetic reciprocating compressor is the most energy consuming component of the refrigerators. Therefore performance improvement studies of the compressor play an important role to reduce overall energy consumption of the refrigerators. Design of bearings is one of the major study areas influencing reciprocating compressor performance.

In this study crank pin and piston pin bearings in hermetic compressor applications are investigated and optimized. The effect of operating conditions, bearing offset between crank pin and piston pin, bearing clearance and bearing location along the shaft on the friction loss were investigated and optimal bearing designs were developed.

Efficiency measurements of the compressor showed that the improved crank pin and piston pin bearing designs provide up to 3.8% increase in the COP as a result of reduction in mechanical loss that is significantly influenced by the length of the bearings, bearing location along the shaft and operating conditions.

Keywords: Reciprocating compressor, Bearing, Mechanical Loss, Friction, Coefficient of performance (COP)

1. Introduction

Positive displacement piston type hermetic compressors play an important role in the energy efficiency of domestic refrigerators. Therefore the tightening energy consumption regulations for domestic refrigerators push the compressor manufacturers to design and produce more and more energy efficient hence environmentally friendly compressors. The most important development objectives for hermetically sealed compressors are increasing the Coefficient of Performance (COP); decreasing manufacturing and operating costs; lengthening operating life, reducing noise and vibration levels. Design of the hermetic compressor bearings is directly related with all these topics.

The performance of a reciprocating compressor is influenced by the mechanical loss. In hermetic reciprocating compressors, mechanical loss is the frictional loss at bearings, which are used to support moving and rotating parts. The main factors affecting the mechanical loss are bearing dimensions, type of bearings, lubrication characteristics, friction coefficient, loading conditions, surface roughness, operating temperature, material, bearing surface coating, relative bearing surface velocity, clearance between journal and bearing housing.

In this study crank pin and piston pin bearings in hermetic compressor applications are investigated and optimized. The effect of operating conditions, bearing offset between crank pin and piston pin, bearing clearance and bearing location along the shaft on the friction losses were investigated and the optimized bearing designs were developed. Detailed parametric numerical simulations were performed using commercial software. According to the simulation results performance measurements were carried out for the selected design parameters. The results of the numerical analysis have shown that

Content from this work may be used under the terms of the Creative Commons Attribution 3.0 licence. Any further distribution of this work must maintain attribution to the author(s) and the title of the work, journal citation and DOI.
Published under licence by IOP Publishing Ltd

IOP Conf. Series: Materials Science and Engineering **90** (2015) 012019 doi:10.1088/1757-899X/90/1/012019

the numerically calculated mechanical loss level is in correlation with the performance results measured in a calorimeter test system.
Bearing design influences the efficiency of reciprocating compressors. Appropriate bearing design, that provides sufficient oil film pressure/thickness while minimizing friction loss, increases COP level of the reciprocating compressor.

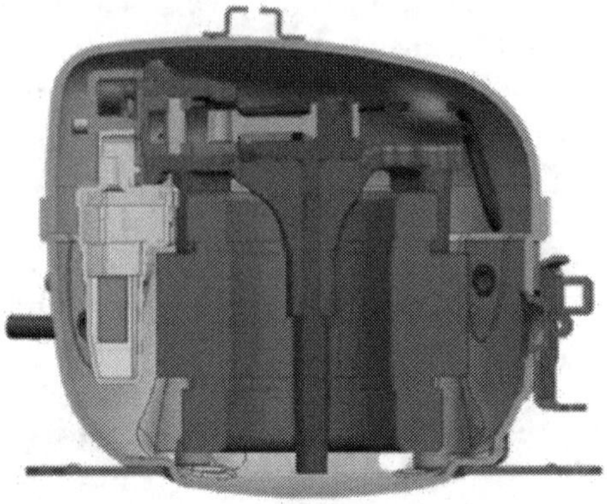

Figure 1. 3D CAD model of the studied reciprocating compressor

Numerical simulations with an engineering software coupled with experimental studies have become an essential way for improvement the performance of reciprocating compressors. A lot of research dealing with the modeling and simulating the compressors and their components are present in the literature. There are also some remarkable researches done in order to develop optimum compressor bearing designs.

M. Duyar [1] numerically solves the elastrohydrodynamic lubrication of a piston pin bearing of a reciprocating compressor. B. Hacioglu [2] presented a model that solves Reynolds' equation for piston cylinder sliding bearing for a compressor piston with oil feed groove. P.Mantri et al [3] modeled the piston-cylinder interaction inside a small hermetic compressor using the Reynolds equation which is solved using finite difference method. A.R. Ozdemir et al [4] investigated and optimized the crank shaft bearings in hermetic compressor applications. In this study the effect of crank shaft geometry, bearing clearance, lubricant viscosity and surface roughness on the friction losses were investigated and the new journal bearing designs were developed.

This study used the results of A.R. Ozdemir et al [4]'s main bearing designs and focused on the optimization of the other bearings (Crank Pin Bearing and piston pin bearing).

2. Experimental Studies

Calorimeter measurements and pV (pressure volume diagram) test results were the main experimental studies for determining the necessary boundary conditions for numerical calculations. Isobutene (R600a) refrigerant applied during the tests in the compressor calorimeter system. Measurement uncertainties of the calorimeter for different parameters are < %2. In pV set up piezo-resistive miniature pressure transducers were used to measure the pressure inside the cylinder for bearing force investigations. Reaction forces on crankshaft bearings were created by the gas force on piston. These reaction forces must be carried at the bearings by the oil film pressure generated. An optical encoder was placed on the shaft, for the determination of the cylinder volume and the piston position. pV measurements were conducted at ASHRAE test conditions to examine the pressure characteristics of the investigated compressor. The results of pV measurements of the reciprocating compressor are shown in Figure 2.

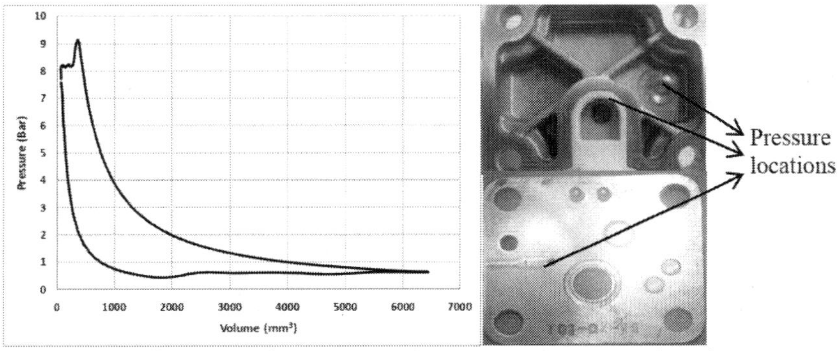

Figure 2. pV-diagram and temperature/pressure measurement locations

3. Numerical Analysis

The commercial bearing design engineering software, which provides different solution techniques for analyzing the dynamics of the bearings, used in this study. Determined conditions have been taken into account in the bearing analysis.

- Cylinder pressure variation with respect to the crank angle obtained from the ASHRAE conditioned pV measurements is used.
- Kinematic viscosity variation with respect to the temperature variation of the lubricant is calculated from Walther-ASTM equation.
- Boundary contact friction coefficient is selected as 0.08

3.1. Computational Model

Selected compressor bearings (crankshaft, piston, connecting rod and piston pin) are rigidly modeled. Calculation domain is divided into a number of cells by meshing process. In the computational model circular and axial mesh nodes were used according to the length (L.) of the bearing. Analyses properties and boundary conditions were given in Table 1. And the model of computational domain simulating the crankshaft and piston pin is shown at Figure 3.

Table 1. Analyses properties and boundary conditions

Upper Main Bearing Shaft Diameter	# of axial mesh nodes	# of circular mesh nodes
1,1	27	73
Lower Main Bearing Shaft Diameter	# of axial mesh nodes	# of circular mesh nodes
0,6	15	73
Crank Pin Bearing Shaft Diameter	# of axial mesh nodes	# of circular mesh nodes
0,9	23	73
Piston Pin Bearing Shaft Diameter	# of axial mesh nodes	# of circular mesh nodes
0,92	31	73
0,83	28	73
0,75	26	73
0,67	23	73

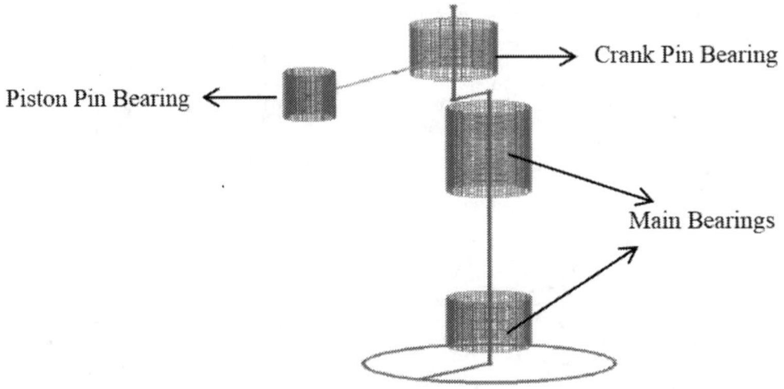

Figure 3. Model of bearings and computational domain

3.2. Parameters and Operating Conditions

Operating conditions and parameters for numerical analysis of bearings and the dynamic behavior analysis of the reciprocating compressor are given in Table 2. Finite difference discretization method which uses the Half Sommerfeld boundary condition to predict oil pressure for the given boundary conditions was used during the analysis. The ASHRAE conditions were used as the compressor test conditions *(54.4 °C condensing temperature, -23.3 °C evaporating temperature)*.

Table 2. Properties and operating conditions

Parameters and Operating Conditions	
Piston Pin Bearing/Shaft Diameter Ratio	0.8
Length of conrod/Shaft Diameter Ratio	3.2
Cylinder bore diameter/Shaft Diameter Ratio	1.8
Mass of conrod (gr)	19
Mass of piston pin (gr)	5.8
Mass of piston (gr)	18
Oil temperature (^{0}C)	80
Kinematic Viscosity (at 40^{0}C)	5 cSt
Nominal oil pressure (bar)	0.624
Working condition	ASHRAE
Maximum rated speed (rev/min)	2930

3.3. Solution Method

Cylinder pressure variation for one cycle of the compressor was used for kinematic analysis.

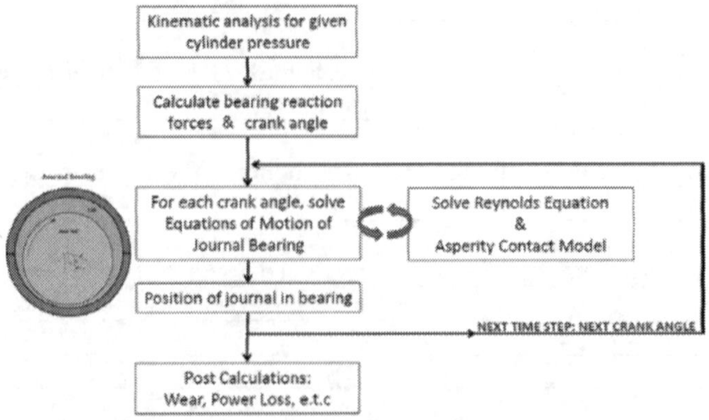

Figure 4. Flow chart of solution method

IOP Conf. Series: Materials Science and Engineering **90** (2015) 012019 doi:10.1088/1757-899X/90/1/012019

Solution methodology starts with the calculated bearing reaction forces for every crank angle. Equations of motion of journal bearings were solved according to the governing equations in which oil film pressure was calculated using Reynolds equation and asperity contact model for every crank angle in the second step of the program. Finally maximum film pressure, minimum film thickness, hydrodynamic power loss and boundary lubrication power loss were calculated.

4. Results and Discussions

Numerical calculations were realized with 4 cycle of computation for obtaining the cyclic convergence for the all cases. Results of numerical and experimental studies are summarized in this section.

4.1. Numerical Results:

144 different combinations of the parameters given in Table 3 were used for hydrodynamic bearing analysis in the numerical study. The load acting on the bearings was calculated under the ASHRAE operating conditions. The bearing analyses were realized for certain assumptions and results were compared with the baseline case of the model.

Table 3. Numerical Analysis Cases

Parameters	
Crank Pin B. Clearance/Shaft Diameter Ratio	1.9E-03/1.6E-03/ 1.3E-03
Piston Pin B. Clearance /Shaft Diameter Ratio	1.2E-03/4.2E-04/6.7E-04/9.2E-04
Piston Pin Bearing/Shaft Diameter Ratio	0.92/0.83/0.75/0.67
Bearing offset/Shaft Diameter Ratio	1.4E-01/ 8.3E-02/0
Upper Main Bearing/Shaft Diameter Ratio	1.1
Lower Main Bearing/Shaft Diameter Ratio	0.6
Crank Pin Bearing/Shaft Diameter Ratio	0.9
Kinematic Viscosity (at 40^0C)	5 cSt
Working condition	ASHRAE

Maximum film pressure, minimum film thickness and total power loss which include the sum of hydrodynamic power loss and boundary lubrication power loss were calculated for every case. Total power loss ratio variation with respect to the defined cases is given in Figure 4. Maximum film pressure, minimum film thickness and power loss ratio results for crank pin and piston pin bearings are given in Figure 5. and Figure 6. respectively.

According to the evaluation of the numerical results in defined analysis cases range; decreasing the length of the bearings reduces the total power loss, on the other hand minimizing the clearance between the crankshaft journal and its bearing increases the total power loss. The optimization of the bearing dimensions (diameter/length/clearance) is critical to reduce hydrodynamic, mixed lubrication and boundary contact power losses and wear in the compressor.

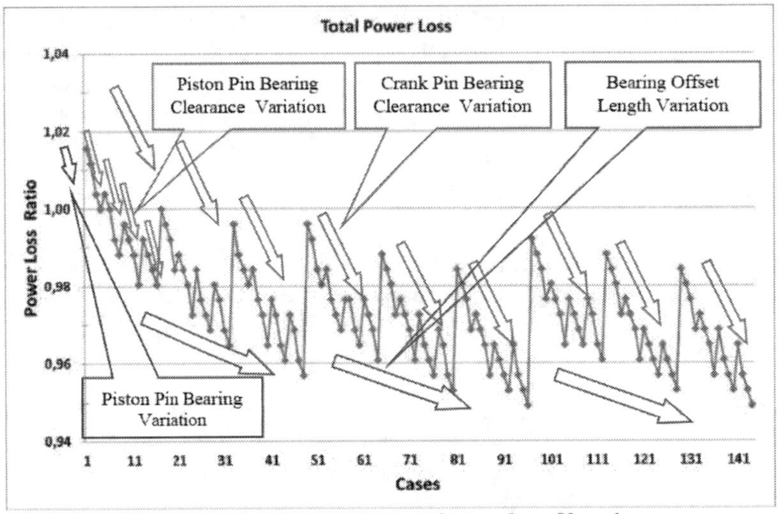

Figure 5. Total power loss ratio results of bearings

There is critical maximum film pressure limit for every application based on the experience. For this reason evaluation of the numerical analysis with experimental results is very crucial for the determination of the limits. On the other hand, critical lower limit of the minimum film thickness changes with the surface roughness characteristic of the bearing and journal.

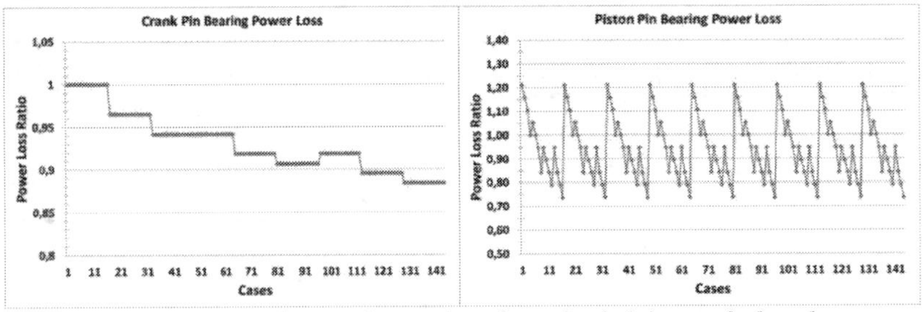

Figure 6. Power loss ratio results of crank pin/piston pin bearing

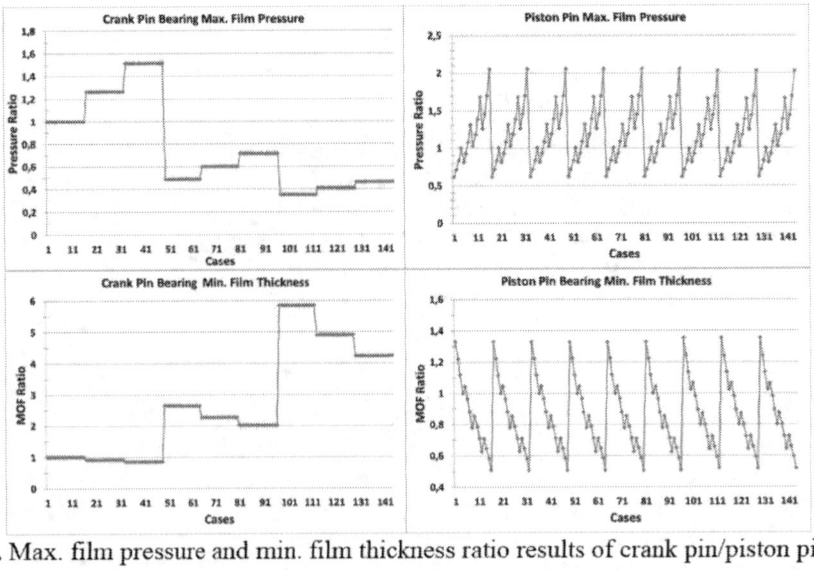

Figure 7. Max. film pressure and min. film thickness ratio results of crank pin/piston pin bearing

All the results of the numerical analysis were investigated in detail with statistical techniques. Some of the cases given in Table 4. were selected for summarizing the interaction between the parameters. Case4 represents the base design. Total power loss difference was evaluated with regard to base design. Variation of the maximum oil film pressure and minimum clearance with respect to the crank and bearing angle were also investigated in this study at different cases.

Table 4. Summary of the numerical analyses results

ASHRAE	Crank Pin Bearing (CPB)			Piston Pin Bearing (PPB)			TOTAL	Dimensions			
5 cSt	Max. film pres. Ratio	Min. film thickness Ratio	Power loss Ratio	Max. film pres. Ratio	Min. film thickness Ratio	Power loss Ratio	Power loss Ratio	CPB Clearance / Shaft Diameter	PPB Clearance / Shaft Diameter	Piston Pin Bearing / Shaft Diameter	Bearing offset / Shaft Diameter
Case1	1,00	1,00	1,00	0,62	1,33	1,21	1,02	1,3E-03	4,2E-04	9,2E-01	1,4E-01
Case5	1,00	1,00	1,00	0,81	1,04	1,05	1,01	1,3E-03	6,7E-04	9,2E-01	1,4E-01
Case4	**1,00**	**1,00**	**1,00**	**1,00**	**1,00**	**1,00**	**1,00**	1,3E-03	4,2E-04	6,7E-01	1,4E-01
Case16	1,00	1,00	1,00	2,05	0,51	0,74	0,98	1,3E-03	1,2E-03	6,7E-01	1,4E-01
Case36	1,51	0,86	0,94	1,00	1,00	1,00	0,98	1,9E-03	4,2E-04	6,7E-01	1,4E-01
Case52	0,49	2,64	0,81	0,94	1,00	1,00	0,98	1,3E-03	4,2E-04	6,7E-01	8,3E-02
Case48	1,51	0,86	0,81	2,05	0,51	0,74	0,96	1,9E-03	1,2E-03	6,7E-01	1,4E-01
Case144	0,46	4,23	0,76	2,03	0,52	0,74	0,95	1,9E-03	1,2E-03	6,7E-01	0

4.2. Experimental Results:

Parameters for the experimental studies given in Table 5. were defined according to the interaction between the numerical results of the cases. Components of the compressor were investigated with respect to the surface roughness, material, coating, form, clearance between journals and bearing housings before the assembling process of the compressor for eliminating the undefined influences. Same class components were selected for the assembling process.

Table 5. Defined experimental cases

Parameters	
Crank Pin B. Clearance/Shaft Diameter Ratio	1.9E-03 / 1.3E-03
Piston Pin B. Clearance /Shaft Diameter Ratio	1.2E-03 / 4.2E-04
Bearing offset/Shaft Diameter Ratio	1.4E-01 / 0
Piston Pin Bearing/Shaft Diameter Ratio	0.67
Upper Main Bearing/Shaft Diameter Ratio	1,1
Lower Main Bearing/Shaft Diameter Ratio	0,6
Crank Pin Bearing/Shaft Diameter Ratio	0.9
Piston Pin Bearing/Shaft Diameter Ratio	0.8
Kinematic Viscosity (at 40^{0}C)	5 cSt

Calorimeter measurements were conducted at ASHRAE test conditions to examine the performance characteristics of the investigated compressor samples. Repeatability and reproducibility of the measurement system evaluated during the tests. Minimum five samples with minimum three repetitions were tested in each configuration. The results of the performance measurements of the reciprocating compressor samples are shown in Table 6. According to the experimental results decreasing the bearing offset and increasing the clearance between the crank pin /piston pin journals and their bearings reduce the total power loss. Performance measurements of the compressor with the improved bearing design (Case 1) showed up to 3.8 % increase in the coefficient of performance (COP) with respect to the compressor with previous bearing design (Case 5).

Numerical and experimental studies for various operating conditions and oil viscosities will be useful for the overall bearing design of the compressor.

Table 6. Experimental test results of the samples

ASHRAE			Dimensions		
5 cSt	Average ΔCOP %	Comparison of Num.&Exp. Results ΔCOP %	C.P.B. Clearance Shaft D. Ratio	P.P.B. Clearance Shaft D. Ratio	Bearing offset Shaft D. Ratio
Case 1	**3.8**	2.8	1.9E-03	1.2E-03	0
Case 2	**2.5**	1.7	1.9E-03	4.2E-04	0
Case 3	**2.2**	1.6	1.3E-03	1.2E-03	0
Case 4	**0.8**	0.4	1.3E-03	4.2E-04	0
Case 5	**0.0**	0.0	1.3E-03	4.2E-04	1.4E-01

5. Conclusions

In this paper factors affecting the crank pin bearing and piston pin bearing optimizations are investigated:

- Decreasing the length of the bearing reduces the total power loss.
- Increasing the clearance between the crankshaft journal and its bearing decrease the total power loss.
- Optimization of the bearing dimensions (diameter/length/clearance) is critical to reduce hydrodynamic, mixed lubrication and boundary contact power losses and wear rate in the compressor.
- Statistical evaluation of the numerical analysis results show that the numerically calculated total power loss is closely related with efficiency gain measured at the experimental tests.
- Efficiency measurement of the compressor with the improved crank pin and piston pin bearing design showed up to 3.8 % increase in the coefficient of performance (COP) with respect to the compressor with base bearing design.
- This study shows that the mechanical loss characteristics are significantly influenced by the pin bearing offset and the pin bearing clearance.

Results of this study helps to characterize the optimum crank pin and piston pin bearing parameters which lead to improved overall mechanical efficiency of the compressor. Based on the results, this study will be extended by numerical analysis and experimental studies with respect to the parametric variation of operating conditions for various oil viscosities.

References

[1] M., Duyar, Z., Dursunkaya, 2002 "Design Improvement of a Compressor Bearing Using an Elastohydrodynamic Lubrication Model", *International Compressor Engineering Conference at Purdue,*

[2] B., Hacioglu, Z., Dursunkaya, 2008 "Effect of Oil Feed Groove on Compressor Piston Lubrication", *International Compressor Engineering Conference at Purdue,*

[3] P., Mantri, B., Tamma, B, Kachhia, A., Bhakta , 2014 "Parametric Study of Friction Model for a Reciprocating Compressor", *International Compressor Engineering Conference at Purdue,*

[4] A,R, Özdemir, E, Kasapoglu, B, Hacıoglu, 2014 "An Investigation on the Bearing Design and Friction Characteristics of a Hermetic Reciprocating Compressor", *International Compressor Engineering Conference at Purdue*

Acknowledgement

The authors would like to express their sincere appreciation to Arçelik A.Ş.

A thermal network model for induction motors of hermetic reciprocating compressors

T Dutra, C J Deschamps[1]
POLO Research Laboratories for Emerging Technologies in Cooling and Thermophysics, Federal University of Santa Catarina, Florianopolis, SC, Brazil

E-mail: deschamps@polo.ufsc.br

Abstract. This paper describes a simulation model for small reciprocating compressors with emphasis on the electrical motor modelling. Heat transfer is solved through algebraic equations derived from lumped thermal energy balances applied to the compressor components. Thermal conductances between the motor components are characterized via a thermal network model. The single-phase induction motor is modelled via an equivalent circuit, allowing predictions for the motor performance and distributed losses. The predicted temperature distribution is used to evaluate the stator and rotor windings resistances. The thermal and electric models are solved in a coupled manner with a model for the compression cycle. Predictions of temperature distribution, motor efficiency, as well as isentropic and volumetric efficiencies, are compared with experimental data at different operating conditions. The model is then applied to analyse the motor temperature as a function of input voltage and stator wire diameter.

1. Introduction

The development of simulation models is of major importance to design more efficient and reliable compressors, requiring modelling approaches to predict the compression cycle, temperature distribution and electrical motor performance. The compression cycle can be modelled based on polytropic processes [1], lumped-parameter models [2] or distributed-parameter models [3]. On the other hand, the temperature distribution inside the compressor has been usually predicted via lumped-parameter methods and thermal conductances to characterize heat transfer between the components. Several models [2] divide the compressor domain into a few control volumes and employ global conductances obtained from experimental data, whereas other models [4] use thermal network approaches (TNW). TNW models usually adopt a greater division of the compressor domain, allowing the prediction of a more detailed temperature distribution. Finally, full finite volume models [3] and hybrid models [5] have also been developed for compressor thermal simulation.

The electrical motor is usually taken into account in compressor simulations through experimental relationships between torque, efficiency and speed [3, 6]. Few works consider an electrical model for the driving motor [7-9]. He *et al.* [8] presented a simulation model for a semi-hermetic screw compressor combining a TNW approach with an electrical circuit for a three-phase induction motor. The authors validated their model and analysed the temperature distribution as a function of the motor geometrical parameters. Dutra and Deschamps [7] presented a simulation model for a hermetic reciprocating compressor employing lumped-parameter approaches for the thermal and electrical sub-

[1] To whom any correspondence should be addressed.

Content from this work may be used under the terms of the Creative Commons Attribution 3.0 licence. Any further distribution of this work must maintain attribution to the author(s) and the title of the work, journal citation and DOI.
Published under licence by IOP Publishing Ltd

models, respectively. Predictions were validated through comparisons with measurements of compressor efficiency and components temperatures at different operating conditions.

The present paper presents a simulation model similar to that of Dutra and Deschamps [7]. However, instead of a lumped formulation to predict a single temperature to characterize the electrical motor, a TNW approach is adopted to calculate the temperatures of different motor components. Therefore, a more detailed thermal characterization of the electrical motor is achieved with no significant increase of the computational cost. The proposed model is validated through comparisons with measurements and then a parametric study is carried out to illustrate its application, by analysing the effect of the stator main winding diameter on the motor temperature distribution.

2. Mathematical model

2.1. Compression cycle model

An unsteady-state lumped formulation was employed for the compression chamber modelling [10], resulting the following equations for the variations of mass and temperature in the compression chamber:

$$\frac{dm_c}{dt}=\dot{m}_s-\dot{m}_d-\dot{m}_l-\dot{m}_{bs}+\dot{m}_{bd} \tag{1}$$

$$\frac{dT_c}{dt}=\frac{1}{m_c c_v}\left\{\dot{Q}_w-h_c\frac{dm_c}{dt}-\sum\dot{m}h-\left[\frac{T_c}{\rho_c}\frac{\partial p_c}{\partial T_c}\right]_\rho\left(\rho_c\frac{dV_c}{dt}-\frac{dm_c}{dt}\right)\right\} \tag{2}$$

where t, T, m, c_v, $\dot{Q}_w$, V_c and p refer respectively to time, temperature, mass, specific heat at constant volume, heat transfer rate through cylinder walls, instantaneous volume of the compression chamber and pressure. Gas properties h and ρ denote specific enthalpy and density. Subscript c stands for compression chamber. Mass flow rate through suction and discharge valves are $\dot{m}_s$ and $\dot{m}_d$, whereas $\dot{m}_{bs}$ and $\dot{m}_{bd}$ correspond to backflows and $\dot{m}_l$ is leakage in the piston-cylinder gap. The aforementioned parameters are evaluated following Link and Deschamps [10]. Equations (1) and (2) are solved via the Runge-Kutta method.

The compression power is calculated from:

$$\dot{W}_c=-\frac{\omega}{2\pi}\oint p_c dV_c \tag{3}$$

where ω is the compressor speed. The required power to run the compressor is obtained by summing the total mechanical losses in the bearings, $\dot{W}_b$, and the compression power ($\dot{W}_r=\dot{W}_b+\dot{W}_c$). For the compressor considered in the present study, bearing losses are estimated from measurements to be approximately 9W.

2.2. Thermal model

2.2.1. General formulation. A steady-state energy balance applied to a generic lumped element i can be expressed as:

$$\dot{Q}_i-\dot{W}_i=\sum(\dot{m}h)_{i,in}-\sum(\dot{m}h)_{i,out} \tag{4}$$

where $\dot{Q}_i$ may represent the heat transfer rate between the element i and its surroundings, or any heat generation rate within this element. The rate of work associated with the element i is $\dot{W}_i$ and the energy advective transport terms into and from the element i are denoted by $\Sigma(\dot{m}h)_{i,in}$ and $\Sigma(\dot{m}h)_{i,out}$.

The reciprocating compressor was divided into eleven elements: suction muffler (1), compression chamber (2), discharge chamber (3), discharge muffler (4), discharge tube (5), internal environment (6), compressor housing (7), stator core (8), stator winding (9), rotor (10) and stator slots/teeth surface (11). A schematic view of the compressor elements and the thermal interaction among them is depicted in figure 1. Axisymmetric and circular sector schematic views concerning only the motor components are shown in figure 2.

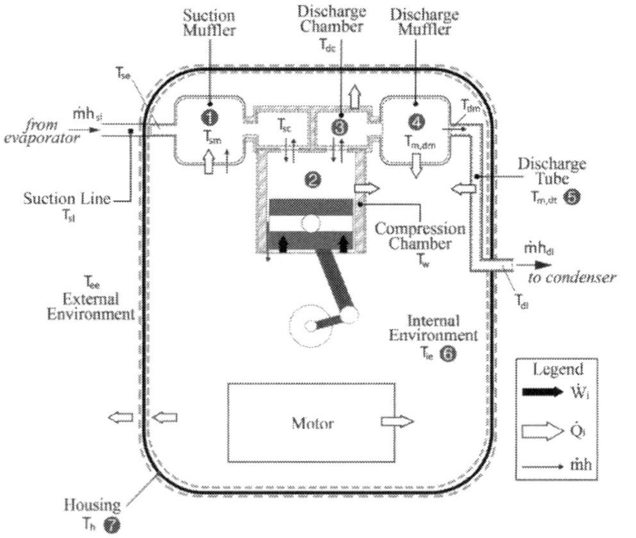

Figure 1. Compressor lumped elements.

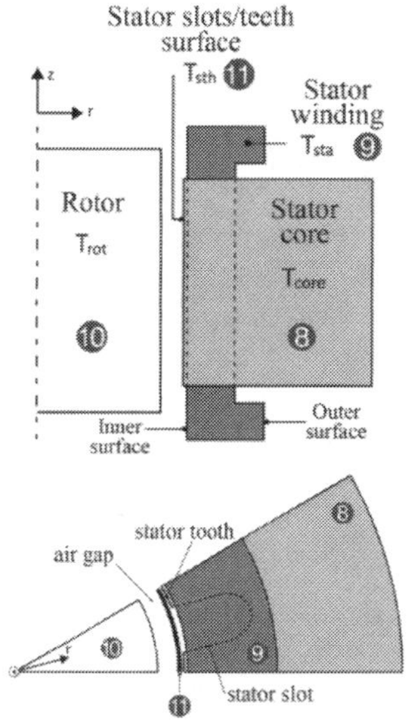

Figure 2. Motor elements.

Tables 1 and 2 show the energy balance terms $\dot{Q}$, $\dot{W}$, $\Sigma(\dot{m}h)_{in}$ and $\Sigma(\dot{m}h)_{out}$ associated with each one of the compressor elements defined in the thermal model. Several parameters, such as mass flow and backflow rates ($\dot{m}_s$, $\dot{m}_d$, $\dot{m}_{bs}$, $\dot{m}_{bd}$, $\dot{m}_l$) heat generation rates in components of the electrical motor ($\dot{Q}_{sta}$, $\dot{Q}_{rot}$, $\dot{Q}_{core}$), compression power ($\dot{W}_c$), mechanical losses in bearings ($\dot{W}_b$) and specific enthalpy in the compression chamber (h_c) are input data for the thermal simulation. The mass flow rates, compression power and specific enthalpy are obtained from the compression cycle model, whereas the mechanical and electrical losses are obtained from measurements and the electrical model, respectively.

Heat transfer between most of elements inside the compressor is taken into account by introducing global conductances, *UA*. These conductances are determined via temperature measurements at a single operating condition [2] and maintained fixed in thermal simulations under a wide range of operating conditions. Heat transfer between the motor components is modelled via local conductances, *H*, estimated from correlations available in literature for simple geometries. However, heat transfer between the motor components and the internal environment is evaluated by considering both global and local conductances. In such situations, heat transfer associated with global conductances represents an additional energy transfer rate that cannot be predicted by adopting only local conductances for simplified geometries.

The system of non-linear equations is solved via the Newton-Raphson method. The mean temperatures in the suction muffler, discharge muffler and discharge tube, denoted respectively by T_{sm}, $T_{m,dm}$ and $T_{m,dt}$, are given by:

$$T_{sm}=0.5(T_{sl}+T_{sc}); \qquad T_{m,dm}=0.5(T_{dc}+T_{dm}); \qquad T_{m,dt}=0.5(T_{dm}+T_{dl}) \tag{5}$$

Table 1. Heat transfer rate terms in the energy balances of the thermal model.

ID	$\dot{Q}$
1	$UA_{sm}(T_{sm}\text{-}T_{ie})$
2	$UA_{w}(T_{w}\text{-}T_{ie})$
3	$UA_{dc}(T_{dc}\text{-}T_{ie})$
4	$UA_{dm}(T_{m,dm}\text{-}T_{ie})$
5	$UA_{dt}(T_{m,dt}\text{-}T_{ie})$
6	$H_{rot,ie}(T_{ie}\text{-}T_{rot})+H_{sta,ie}(T_{ie}\text{-}T_{sta})+H_{core,ie}(T_{ie}\text{-}T_{core})+\sum UA_i(T_{ie}\text{-}T_i)$
7	$UA_{h,ee}(T_h\text{-}T_{ee})\text{-}UA_{h,ie}(T_{ie}\text{-}T_h)$
8	$(H_{core,ie}+UA_{core})(T_{core}\text{-}T_{ie})+H_{core,sth}(T_{core}\text{-}T_{sth})\text{-}\dot{Q}_{core}$
9	$(H_{sta,ie}+UA_{sta})(T_{sta}\text{-}T_{ie})+H_{sta,sth}(T_{sta}\text{-}T_{sth})\text{-}\dot{Q}_{sta}$
10	$(H_{rot,ie}+UA_{rot})(T_{rot}\text{-}T_{ie})+H_{rot,sth}(T_{rot}\text{-}T_{sth})\text{-}\dot{Q}_{rot}$
11	$H_{rot,sth}(T_{sth}\text{-}T_{rot})+H_{sta,sth}(T_{sth}\text{-}T_{sta})+H_{core,sth}(T_{sth}\text{-}T_{core})+UA_{sth}(T_{sth}\text{-}T_{ie})$

Table 2. Rate of work and energy advective terms in the energy balances of the thermal model.

ID	$\dot{W}$	$\sum \dot{m}h\|_{in}$	$\sum \dot{m}h\|_{out}$
1	-	$\dot{m}h_{sl}+\dot{m}_l h_{ie}+\dot{m}_{bs}h_c$	$\dot{m}_s h_{sc}$
2	$\dot{W}_c+\dot{W}_b$	$\dot{m}_s h_{sc}+\dot{m}_{bd}h_{dc}$	$\dot{m}_l h_c+\dot{m}_{bs}h_c+\dot{m}_d h_c$
3	-	$\dot{m}_d h_c$	$\dot{m}_{bd}h_{dc}+\dot{m}h_{dc}$
4	-	$\dot{m}h_{dc}$	$\dot{m}h_{dm}$
5	-	$\dot{m}h_{dm}$	$\dot{m}h_{dl}$
6	-	$\dot{m}_l h_c$	$\dot{m}_l h_{ie}$
7-11	-	-	-

2.2.2. TNW for motor components. Heat transfer between the motor components, i.e. rotor, stator winding and stator core, is solved by a TNW approach, commonly adopted to predict the temperature distribution of electrical machines [11, 12]. Local conductances between elements i and j are computed by combining theoretical estimates of conductive and convective thermal resistances:

$$H_{i,j}=\frac{1}{R_{i,j}};\quad R^{h}_{i,j}=\frac{1}{hA};\quad R^{k,z}_{i,j}=\frac{L}{kA};\quad R^{k,r}_{i,j}=\frac{ln(D_e/D_i)}{2\pi kL} \tag{6}$$

where h, A, L, k, D_e and D_i denote the convective heat transfer coefficient, contact surface area, characteristic length, thermal conductivity and outer and inner diameters of lumped elements, respectively. The superscripts h and k refer to convective and conductive resistances, whereas z and r represent axial and radial axes of the coordinate system, respectively. Heat transfer in circumferential direction is neglected. Figure 3 shows the resulting thermal circuit for the induction motor, which is similar to that used by Boglietti *et al.* [12]. The identification (ID) of the motor elements (tables 1 and 2) are used in the subscripts of the thermal resistances. It should be noted that thermal resistances between the components 8 (stator core) and 10 (rotor) and the internal environment are represented by conductive and convective resistances in series. The conductive resistances $R^{k,r}_{8,6}$, $R^{k,z}_{8,6}$ and $R^{k,z}_{10,6}$ represent the temperature drop within the motor element itself, i.e. between the centre of the element (node) and its border. The convective resistance is associated with heat transfer between the element border and the internal environment.

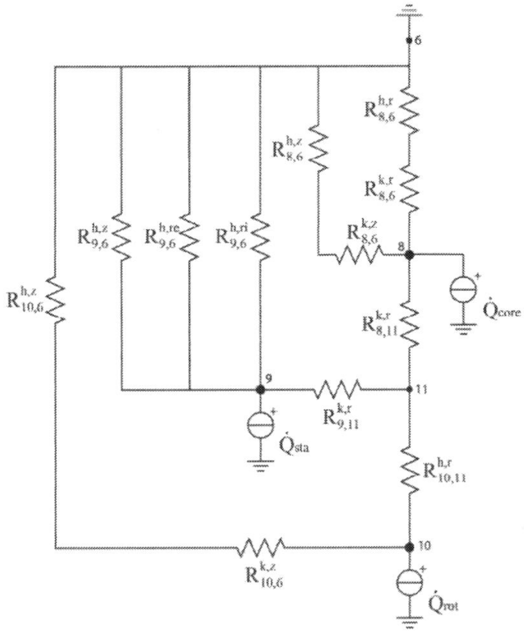

$R^{k,z}_{10,6}$ - Axial conductive resistance between rotor and ie.

$R^{h,z}_{10,6}$ - Axial convective resistance between rotor and ie.

$R^{h,r}_{10,11}$ - Radial convective resistance between rotor and stator slots/teeth surface (air gap).

$R^{h,z}_{8,6}$ - Axial convective resistance between stator core and ie.

$R^{h,r}_{8,6}$ - Radial convective resistance between stator core and ie.

$R^{k,z}_{8,6}$ - Axial conductive resistance between stator core and ie.

$R^{k,r}_{8,6}$ - Radial conductive resistance between stator core and ie.

$R^{k,r}_{8,11}$ - Radial conductive resistance between stator core and stator slots/teeth surface.

$R^{h,z}_{9,6}$– Axial convective resistance between stator winding and ie.

$R^{k,r}_{9,11}$ - Radial conductive resistance between stator winding and stator slots/teeth surface.

$R^{h,re}_{9,6}$ - Radial convective resistance between stator winding and ie (outer surface – figure 2).

$R^{h,ri}_{9,6}$ - Radial convective resistance between stator winding and ie (inner surface, facing the rotor – figure 2).

Figure 3. Motor TNW and corresponding thermal resistances label.

One of the most difficult tasks on adopting lumped thermal models based on local conductances is to determine characteristic lengths, contact surface areas and flow velocities for calculation of thermal resistances. Convective resistances are modelled with reference to local heat transfer coefficients computed from correlations available in literature. For instance, a correlation for turbulent flow over a flat plate is considered for the convective resistances $R^{h,z}_{9,6}$, $R^{h,re}_{9,6}$, $R^{h,ri}_{9,6}$, $R^{h,z}_{8,6}$ and $R^{h,r}_{8,6}$ between stator components and internal environment:

$$h=\frac{Nu_L k}{L}; \quad Nu_L=0.037Re_L^{0.8}Pr^{0.33}; \quad Re_L=\frac{\rho VL}{\mu}; \quad Pr=\frac{\mu c_p}{k} \tag{7}$$

where Nu_L, Re_L and Pr are the Nusselt number, Reynolds number and Prandtl number. Gas properties, such as density, ρ, specific heat at constant pressure, c_p, dynamic viscosity, μ, and thermal conductivity, k, are evaluated at the film temperature corresponding to the average between the temperatures of the motor component and internal environment. The velocity characteristic V is assumed to be a fraction of the crankshaft tangential velocity and L is the length in the flow direction, as proposed by Sanvezzo Jr. and Deschamps [5]. The remaining convective resistances $R^{h,z}_{10,6}$ and $R^{h,r}_{10,11}$ are calculated from correlations available for rotating discs [5] and flow in annular gaps [13], respectively.

2.3. Electrical model

The electrical motor operation was modelled with the equivalent circuit method. Figure 4 depicts the circuit of a single-phase induction motor operating at steady-state condition [14]. Such a circuit consists on a set of impedances for the stator, rotor and magnetization; Z_{sta}, Z_{rot}, Z_m, respectively. Impedances are given by combined electrical resistances, *R*, and reactances, *X*. In figure 4, the symbols (+) and (-) represent forward and backward loops, following the theory of rotating magnetic fields.

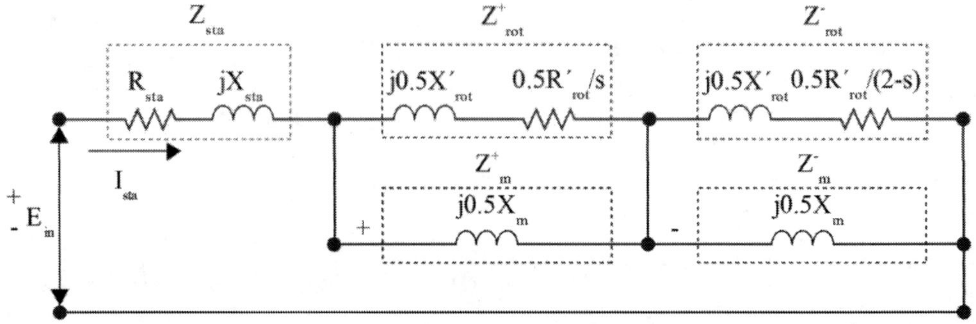

Figure 4. Equivalent circuit of a single-phase induction motor at steady state condition.

The solution of the motor equivalent circuit gives the electrical currents in all the branches [7], which allows the calculation of the torque supplied by the motor to the compressor shaft, τ_{sh}, as well as the joule-losses in the stator, $\dot{Q}_{sta}$, and in the rotor windings, $\dot{Q}_{rot}$:

$$\tau_{sh}=\left(\frac{P}{2}\right)\frac{\dot{W}_{sh}}{\omega};\qquad \dot{Q}_{sta}=I_{sta}^2R_{sta};\qquad \dot{Q}_{rot}=0.5\left({I_{rot}^{+}}^2+{I_{rot}^{-}}^2\right)R_{rot}^{'} \tag{8}$$

where superscript $'$ indicates the rotor resistance is referred to the stator side [14], *P* is the number of poles and I_{sta}, I_{rot}^{+} and I_{rot}^{-} are the electrical currents in the stator winding and rotor forward and backward branches, respectively. The shaft power, $\dot{W}_{sh}$, is evaluated by:

$$\dot{W}_{sh}=(1\text{-}s)\left({I_{rot}^{+}}^2\frac{0.5R_{rot}^{'}}{s}\text{-}{I_{rot}^{-}}^2\frac{0.5R_{rot}^{'}}{2\text{-}s}\right)\text{-}\dot{Q}_{core} \tag{9}$$

where the first and second terms on the right-hand side are the rotating magnetic field power [14] and the stator core loss, $\dot{Q}_{core}$, respectively. The latter is computed from estimates of magnetic flux densities in different regions of the stator core [15]. The solution of equation (9) depends on the slip ratio *s*, which is obtained by substituting equation (9) in the following equation:

$$\dot{W}_{sh}\text{-}\dot{W}_r=0 \tag{10}$$

where $\dot{W}_r=\dot{W}_b+\dot{W}_c$.

Equation (10) provides the coupling between the compression cycle model and the electrical model, expressing that the power required to run the compressor is equal to the motor shaft power. This equation is solved via an incremental search technique [16]. Once the shaft power and electrical losses are calculated, the compressor input power and motor efficiency are obtained:

$$\dot{W}=\dot{W}_{sh}+\dot{Q}_{sta}+\dot{Q}_{rot}+\dot{Q}_{core};\qquad \eta_{ele}=\frac{\dot{W}_{sh}}{\dot{W}} \tag{11}$$

The input data for the electrical model are input voltage, E_{in}, compressor speed, motor geometry and material properties. The latter two are employed to estimate reference values for electrical resistances and reactances through analytical relations [15]. The electrical resistances are continuously calculated during the simulation by using updated temperatures of motor components:

$$R=R_0+\beta R_0(T-T_0) \tag{12}$$

where subscript 0 refers to the reference state defined for motor temperature of 25°C and β is the temperature coefficient, which is a thermal property of the material. Details of the coupled solution procedure, encompassing compression cycle model, thermal model and electrical model are available in Dutra and Deschamps [7].

3. Results and discussions

A R-290 reciprocating compressor designed for household and light commercial applications was considered in this study. The compressor was instrumented with temperature sensors and was run at nine operating conditions, represented by the combination of evaporating temperatures (-10°C, -23.3°C, -35°C) and condensing temperatures (45°C, 54.4°C, 70°C). Each operating condition was tested four times in order to assess the repeatability uncertainty. Compressor external environment temperature, T_{ee}, and suction line temperature, T_{sl}, were set to (32 ±1)°C. Input voltage and nominal frequency were (228 ±1.2)V and 50Hz, respectively. Measurement uncertainties are ±1% for mass flow rate and compressor input power and ±2°C for temperatures, considering a 95% confidence interval.

The volumetric efficiency, η_v, is defined as the ratio between the actual mass flow rate and the ideal mass flow rate, whereas the isentropic efficiency, η_s, is given by the ratio between the isentropic compression power and the actual input power:

$$\eta_v=\frac{\dot{m}}{\dot{m}_{th}}; \qquad \eta_s=\frac{\dot{m}\Delta h_s}{\dot{W}} \tag{13}$$

Results of volumetric and isentropic efficiencies as a function of operating conditions are shown in figures 5 and 6. As can be seen, predictions follow the same trend of experimental data, presenting higher deviations when the pressure ratio is far from that adopted in the calibration condition (-23.3°C/54.4°C). Volumetric efficiency is overestimated and underestimated at -10°C and -35°C evaporating temperatures, respectively. Such an inaccuracy in the mass flow rate prediction affects the isentropic efficiency calculation, which partially explains the deviations found at the same operating conditions. Additionally, inaccuracy related to the input power prediction also increases deviations, especially at the evaporating temperature of -10°C. In this condition, viscous friction in the suction and discharge systems are greater, but the present approach has no specific model for the fluid flow through mufflers to account for this increase. Despite the underlined differences, predictions are considered satisfactory, with maximum deviations of 5.3% and 5.6% for volumetric and isentropic efficiencies.

Volumetric and isentropic efficiencies are decreased by the gas superheating along the compressor suction path. Therefore, the prediction of both parameters requires the calculation of temperatures in the suction chamber and at the cylinder wall. Figure 7 shows results of both temperatures as a function of the operating condition. Predictions become shifted in relation to the measurements when the evaporating temperature is varied. This is likely due to the adoption of fixed global conductances for the thermal model regardless the mass flow rate, which is strongly affected by the evaporating temperature. Despite this shortcoming, the maximum deviation between predictions and measurements is lower than 10°C and most deviations are within 5°C.

Temperatures of motor components are key parameters to be considered in compressor reliability analysis. Figure 8 presents stator winding and core temperature variations with operating conditions.

The maximum difference between predictions and experimental data is approximately 4°C at the operating condition (-10°C /45°C). Measurements indicate that the stator winding temperature is slightly higher (≅ 3°C) than the core temperature. The proposed model also indicates similar results, with smaller differences (<1.8°C) for high evaporating temperatures and greater differences (>4.5°C) for low evaporating temperatures, almost within measurement uncertainties.

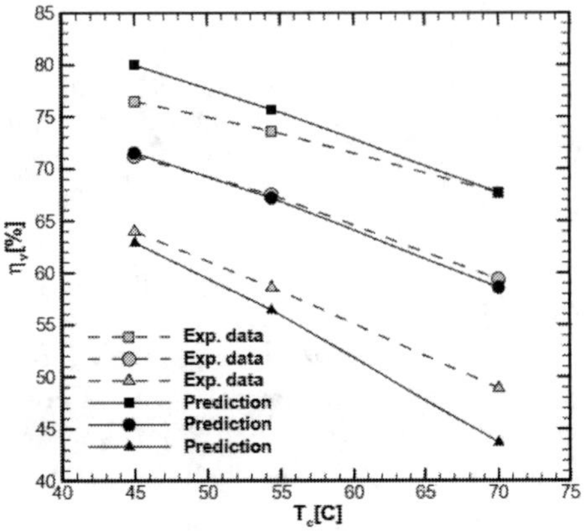

Figure 5. Influence of operating condition on volumetric efficiency: ■ T_e = -10°C; ●T_e = -23.3°C; ▲T_e = -35°C.

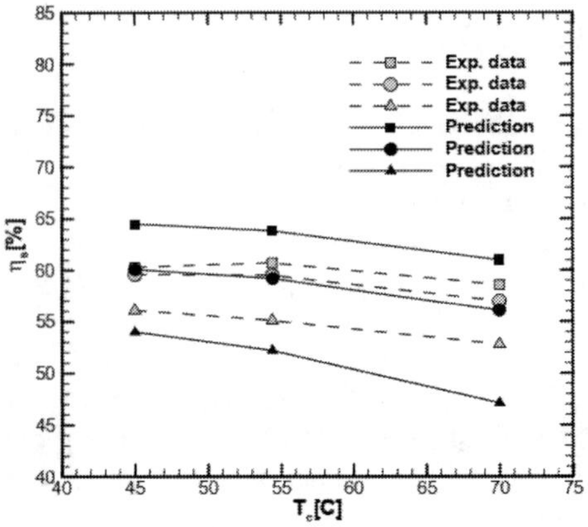

Figure 6. Influence of operating condition on isentropic efficiency: ■ T_e = -10°C; ●T_e = -23.3°C; ▲T_e = -35°C.

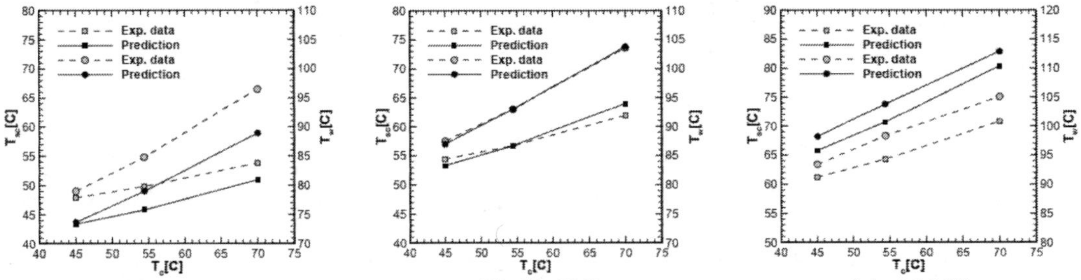

(a) -10°C (b) -23.3°C (c) -35.0°C

Figure 7. Suction chamber and cylinder wall temperatures: ■ suction chamber; ● cylinder wall.

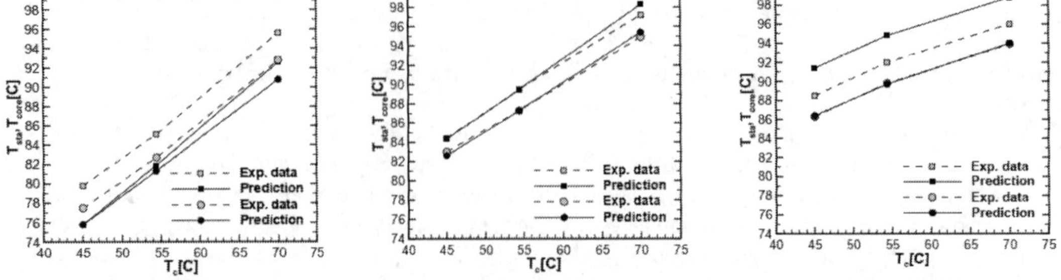

(a) -10°C (b) -23.3°C (c) -35.0°C

Figure 8. Motor components temperatures: ■ stator winding; ● stator core.

After its preliminary assessment, the model was applied for parametric analysis. First, the compressor was simulated for a low evaporating temperature condition (-35°C/45°C) and two input voltages (228V and 255V). Figure 9 shows that predictions and measurements are in reasonable agreement, with both results indicating an increase of temperature with the input voltage due to the increase of motor losses. Actually, stator joule-losses ($\dot{Q}_{sta}$) and core losses ($\dot{Q}_{core}$) are found to increase by 5% and 30%, respectively, decreasing the motor efficiency. For instance, figure 10 exhibits results of motor efficiency as a function of shaft torque for input voltages of 220V and 255V. Experimental results obtained with a dynamometer test are also plotted for the purpose of comparison. As can be clearly noted, the motor efficiency is reduced with input voltage increase at the low shaft torque conditions, e.g. -35°C/45°C (τ_{sh} = 0.28N.m).

Finally, the temperatures of the motor components were analysed as a function of the stator wire diameter for the operating condition (-35°C /45°C). The results depicted in figure 11 indicate the temperatures of the motor components rise when the wire diameter is reduced. This is an expected outcome, since the electrical resistance of the stator winding (R_{sta}) is increased, increasing the joule-losses. Hence, the temperature of the stator winding becomes much higher than temperature of the core as the wire diameter is reduced.

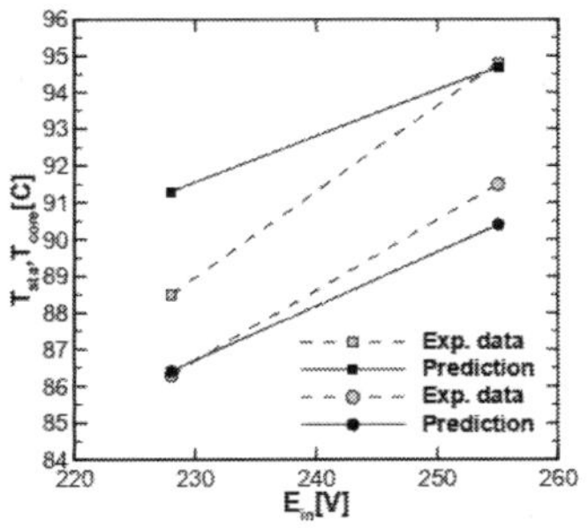

Figure 9. Temperature of the motor for different input voltages: ■ stator winding; ● stator core.

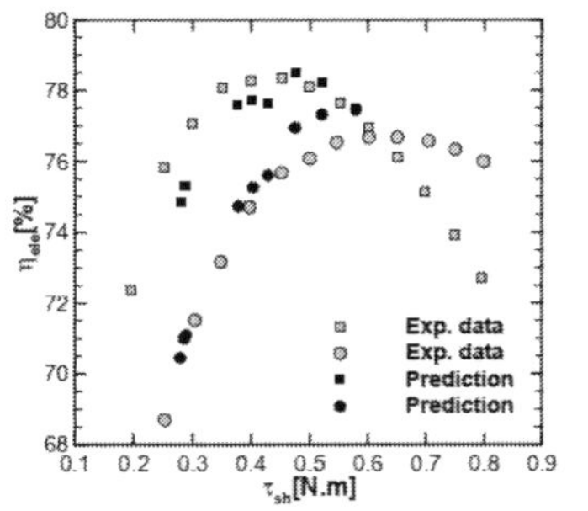

Figure 10. Electrical motor efficiency for different shaft torques: ■ 220V; ●255V.

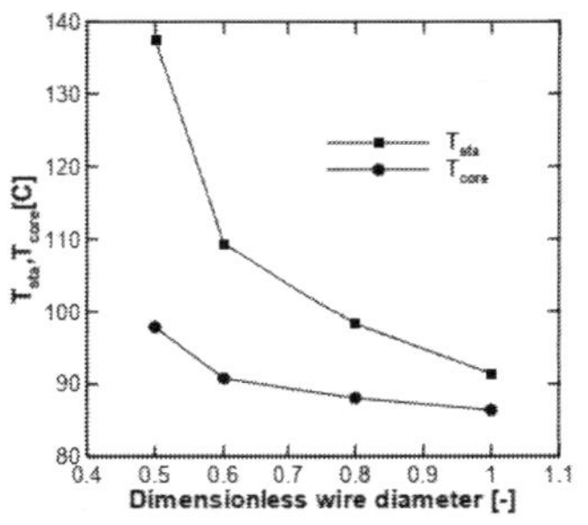

Figure 11. Temperature of the motor for different stator wire diameters.

4. Conclusions

This paper presented a multi-physic simulation model for hermetic reciprocating compressors adopted for refrigeration. In addition to models for the compression cycle and heat transfer between the compressor components, an approach based on the electrical equivalent circuit was employed for the induction motor. Heat transfer in the electrical motor was modelled via the thermal network concept, with thermal resistances being calculated from analytical and empirical relations. The comprehensive model provided satisfactory predictions for the compressor volumetric and isentropic efficiencies, as well as for the temperature distribution as a function of different operating conditions. The model was also applied in parametric analyses for the motor temperature as a function of input voltage and stator wire diameter.

Acknowledgments

This study was developed as part of a technical-scientific cooperation program between the Federal University of Santa Catarina and EMBRACO. The authors are also grateful to the Brazilian governmental agency CNPq (National Council of Research) for the grants 573581/2008-8 (National Institute of Science and Technology in Refrigeration and Thermophysics) and 141513/2012-8.

References

[1] Cavallini A, Doretti L, Longo G A, Rosseto L, Bella B and Zannerio A 1996 Thermal analysis of a hermetic reciprocating compressor *Proc. Int. Compress. Eng. Conference at Purdue*. West Lafayette, USA, 535- 540.

[2] Todescat M L, Fagotti F, Prata A T and Ferreira R T 1992 Thermal energy analysis in reciprocating hermetic compressors *Proc. Int. Compress. Eng. Conference at Purdue* West Lafayette, USA, 1419-28.

[3] Birari Y V, Gosavi S S and Jorwekar P P 2006 Use of CFD in design and development of R404a reciprocating compressor *Proc. Int. Compress. Eng. Conference at Purdue*. West Lafayette, USA, C072.

[4] Ooi K T 2003 Heat transfer study of a hermetic refrigeration compressor *Appl. Therm. Eng.* **23** 1931-45.

[5] Sanvezzo Jr. J and Deschamps C J 2012 A heat transfer model combining differential and integral formulations for thermal analysis of reciprocating compressors *Proc. Int. Compress. Eng. Conference at Purdue* West Lafayette, USA, 1-11.

[6] Rigola J, Pérez-Segarra C D and Oliva A 2000 Advanced numerical simulation model of hermetic reciprocating compressors. Parametric study and detailed experimental validation *Proc. Int. Compress. Eng. Conference at Purdue* West Lafayette, USA, 23-30.

[7] Dutra T and Deschamps C J 2014 Development of a lumped-parameter model for hermetic reciprocating compressor with thermal-electrical coupling *Proc. Int. Compress. Eng. Conference at Purdue* West Lafayette, USA, 1367.

[8] He Z, Xing Z, Chen W and Wang X 2013 Thermal and hydraulic analysis on the flow around the motor in semi-hermetic twin screw refrigeration compressors *Appl. Therm. Eng.* **58** 114-124.

[9] Mantri P, Bhakta A, Mallampalli S, Hahn G and Kusumba S 2014 Development and validation of integrated design framework for compressor system model *Proc. Int. Compress. Eng. Conference at Purdue* West Lafayette, USA, 1153.

[10] Link R and Deschamps C J 2011 Numerical modeling of startup transients in reciprocating compressors *Int. J. Refrigeration* **34** 1398-1414.

[11] Mellor P H, Roberts D and Turner D R 1991 Lumped parameter thermal model for electrical machines of TEFC design *IEEE Proc.-B* **138** 205-218.

[12] Boglietti A, Cavagnino A, Lazzari M and Pastorelli M 2003 A simplified thermal model for variable-speed self-cooled industrial induction motor *IEEE Transactions on Industrial Applications* 945-952

[13] Bouafia M, Ziouchi A, Bertin Y and Saulner J B 1999 Experimental and numerical study of heat transfer in an annular gap without axial flow with a rotating inner cylinder *Int. J. Thermal Sciences* **38** 547- 559.

[14] Fitzgerald A E, Kingsley C K and Umans S D 2003 *Electric Machinery* sixth ed. (McGraw-Hill Companies, Inc).

[15] Stepina J 2003 *Single-Phase Induction Motors: Construction, Theory and Calculation* (Motorsoft, Inc).

[16] Arora J S 2004 *Introduction to Optimum Design* second ed. (Elsevier Academic Press).

Torsional Vibration Analysis of Reciprocating Compressor Trains driven by Induction Motors

M Brunelli[1], A Fusi[2], F Grasso[3], F Pasteur, and A Ussi[2]
[1] *BSIM srl,* Corso Vittorio Emanuele II – 32, 10123 – Torino, IT
[2] C.S.T. Compression Service Technology Srl, V. Giovanni del Pian dei Carpini n.1, 50127 Firenze – IT
[3] Department of Information Engineering - DINFO, University of Firenze, via S. Marta, 3, 50139 Firenze, IT
[4] Siemens Industry Software S.r.l. Viale Piero e Alberto Pirelli, 10, 20126 Milano, IT

E-mail: francesco.grasso@unifi.it, andrea.fusi@cstfirenze.com

Abstract. The dynamic study of electric motor driven compressors, for Oil&Gas (O&G) applications, are traditionally performed in two steps separating the mechanical and the electrical systems. The packager conducts a Torsional Vibration Analysis (TVA) modeling the mechanical system with a lumped parameter scheme, without taking into account the electrical part. The electric motor supplier later performs a source current pulsation analysis on the electric motor system, based on the TVA results. The mechanical and the electrical systems are actually linked by the electromagnetic effect. The effect of the motor air-gap on TVA has only recently been taken into account by adding a spring and a damper between motor and ground in the model. This model is more accurate than the traditional one, but is applicable only to the steady-state condition and still fails to consider the reciprocal effects between the two parts of the system. In this paper the torsional natural frequencies calculated using both the traditional and the new model have been compared. Furthermore, simulation of the complete system has been achieved through the use of LMS AMESim, multi-physics, one-dimensional simulation software that simultaneously solves the shafts rotation and electric motor voltage equation. Finally, the transient phenomena that occur during start-up have been studied.

1. Introduction

Stresses acting upon mechanical parts can cause premature failure of both the driving electrical machine shafts and other mechanical parts due to fatigue phenomena, especially in the case of resonance, when the exciting frequency matches a torsional natural frequency. It is for this reason that the standard codes for reciprocating compressors recommend that the "torsional natural frequencies of the complete driver-compressor system (including couplings and any gear unit) shall not be within 10% of any operating shaft speed and within 5% of any multiple of operating shaft speed in the rotating system up to and including the tenth multiple" [8]. In fact, reciprocating compressor systems are more subject to these phenomena due to the great variations in torque, consequently the preventive analysis of torsional vibrations in reciprocating compressors is essential for meeting the American Petroleum Institute (API) requirements, thus preventing the failure of the crankshaft, coupling, engine dampers, and compressor oil pumps.

In order to predict the vibration behavior of any torsional system, the inertia and stiffness properties of all the components are required. The steady-state torque and power output of a polyphase induction

Content from this work may be used under the terms of the Creative Commons Attribution 3.0 licence. Any further distribution of this work must maintain attribution to the author(s) and the title of the work, journal citation and DOI.
Published under licence by IOP Publishing Ltd

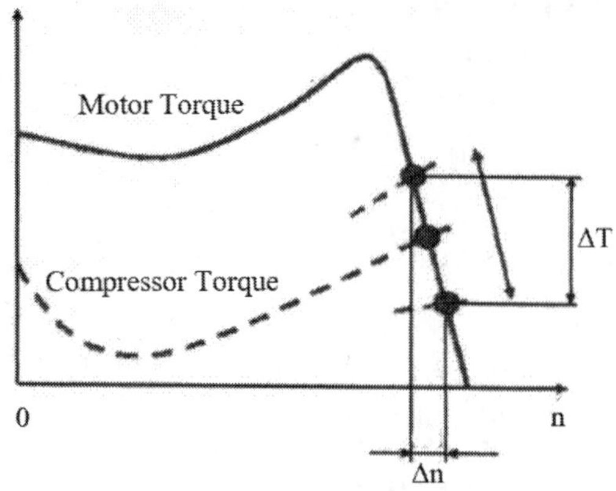

Figure 1. Induction motor torque reaction (ΔT) due to motor speed (n) variation.

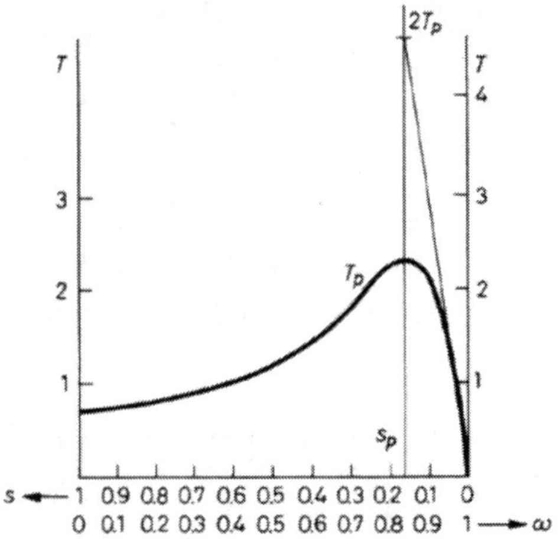

Figure 2. Static torque characteristic of an induction motor. T [Nm] ω [rad/s]

motor are the result of electromagnetic fields which act across the air-gap between the stator and rotor. If the rotor has torsional vibration superimposed over a steady rotation, the same electromagnetic fields across the air-gap can produce an additional torque. Traditionally, the electromechanical interaction in the air-gap is ignored in standard torsional vibration analysis, but it can be important when improved prediction accuracy of natural frequencies or damping is required. In steady-state conditions, this effect can be taken into account by adding to the model a torsional spring and damper between the motor and the ground, obtained through a linearization of the electric motor model.

In this paper, the influence of motor dynamics in reciprocating compressor trains has been studied by means of different simulation profiles and the effect of induction motor electromagnetic stiffness and damping on torsional vibration has been evaluated considering the influence of air-gap torque. When transient phenomena have to be simulated, the above linearization is no longer applicable; furthermore, the analytical expressions of transient torque, which are usually provided by electric motor manufacturers, do not represent the real behavior of the system. For this reason, the transient analysis has to be performed by means of multi-physics simulation software, which makes it possible to simultaneously solve the electrical and mechanical equations of the system.

The paper is structured as follows: Section 2 outlines the torque effort paying particular attention to the resistant torque produced by reciprocating compressors. Section 3 describes the torque calculation in induction motors. Section 4 presents the modeling of the induction motor air-gap that can be added to the equivalent shaft to develop a more accurate simulation. This model can be used in TVA and gives more realistic results. Section 5 presents the results of simulations in two case studies. In sections 6 and 7, the study of the steady-state and transient forced responses is discussed. To conclude, further possible developments in this field are indicated.

2. Considerations on Torque Effort

Reciprocating machinery and especially reciprocating compressors show high torque variations. Absorbed torque can be represented by a variable and a constant term, which is the mean absorbed torque C,

$$C = C_m + C(t) \tag{1}$$

where $C(t)$ represents a time variable function with a null mean value.

Reciprocating compressors are often driven by electric motors, which may be either synchronous or asynchronous. Electric motors react to these torque fluctuations with their own air-gap torque variations. These variations have an important effect on system torsional dynamics in terms of the natural frequencies of the system. This phenomenon can be intuitively understood by observing the induction motor torque-speed characteristic curve.

IOP Conf. Series: Materials Science and Engineering **90** (2015) 012021 doi:10.1088/1757-899X/90/1/012021

Figure 1 shows that the motor speed variation (mainly due to variable compressor torque) causes an adjustment of the developed motor torque (air-gap torque). As a result of this variation, there will also be an additional oscillation of the shaft speed.

The motion equation for an asynchronous (induction) motor (to which this study applies) is:

$$\frac{J}{p}\frac{d^2\varepsilon}{dt^2} + D\frac{d\varepsilon}{dt} = C(t) \tag{2}$$

where J is the motor inertia, p the number of pole pairs, D the damping, C(t) the torque fluctuation due to the compressor and ε the flux angle between the applied voltage and a generic rotor winding.

3. Air-gap Torque Estimation

For a better understanding of the considerations that follow, it helps to consider the steady-state operation of the motor. The simplest mathematical expressions of motor equations are obtained if the frame of reference is chosen so as to rotate with the voltage vector of the supply, because in that frame the voltage and current appear as DC quantities.

Let us consider the following electrical and mechanical equations of an induction machine:

$$u_s = i_s R_s + \frac{d\Psi_s}{dt} + j\omega_k \Psi_s \tag{3}$$

$$u_r = i_r R_r + \frac{d\Psi_r}{dt} + j(\omega_k - \omega)\Psi_r \tag{4}$$

$$J\frac{d\omega}{dt} + T_m = \frac{3}{2} p I_m \left(\Psi_s i_s^*\right) \tag{5}$$

where $\mathbf{u_x}$, $\mathbf{i_x}$ and $\mathbf{\Psi_x}$ are vectors of voltage, current and flux, of the stator (suffix s) and rotor (suffix r), $\boldsymbol{\omega_k}$ is the angular velocity of the chosen frame of reference, $\boldsymbol{\omega}$ is the mechanical angular velocity of the rotor, **p** is the number of pole pairs, **J** is the total moment of inertia, $\mathbf{T_m}$ is the air-gap torque, I_m represents the imaginary part of the quantity taken into account and * is the complex conjugate. R_s and R_r are the stator and the rotor resistance, respectively. At steady-state the flux vectors are of constant length, therefore in the chosen frame reference:

$$\frac{d\Psi_s}{dt} = \frac{d\Psi_r}{dt} = 0\,. \tag{6}$$

Thus we obtain:

$$u_s = i_s R_s + j\omega_k \Psi_s \tag{7}$$

$$u_r = i_r R_r + j(\omega_k - \omega)\Psi_r \tag{8}$$

$$T_m = \frac{3}{2} p I_m \left(\Psi_s i_s^*\right). \tag{9}$$

By considering the ratio of the magnetizing reactance and the stator reactance:

$$k = \frac{X_m}{X_s} \tag{10}$$

we obtain the motor torque curve as a function of the slip:

$$T_m = \frac{3}{2} p I_m \left[\frac{1}{X_s} + k^2 \frac{j}{\frac{R_r}{s} + jX_r^{'}}\right] = \frac{3}{2} p k^2 I_m \left[\frac{js}{R_r + jsX_r^{'}}\right] = \frac{3}{2} p k^2 \frac{sR_r}{R_r^2 + s^2 + X_r^{'2}} \tag{11}$$

The torque curve has a maximum torque known as pull-out torque (T_p) at s_p (pull-out slip):

$$T_p = \frac{k^2}{2X_r^{'}}; \qquad s_p = \frac{R_r}{X_r^{'}} \tag{12}$$

As can be seen in Figure 2, the slope for low slips can be approximated by 2 T_p/s_p. As a result the air-gap torque can be calculated by the following equation:

$$T_m = \frac{2T_p}{s_p} s \tag{13}$$

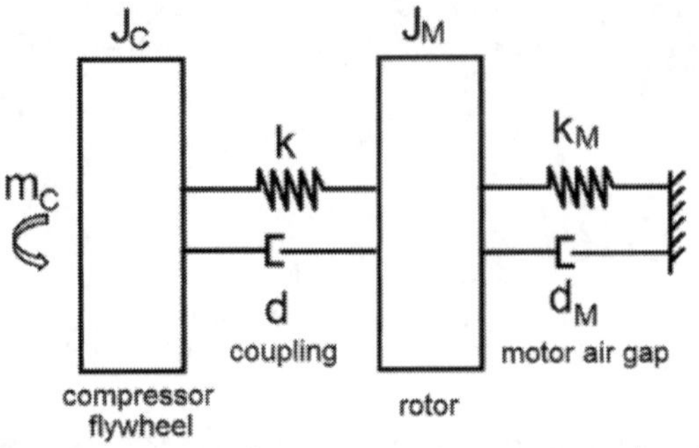

Figure 3. Motor air-gap torque representation by additional mechanical spring and viscous damper.

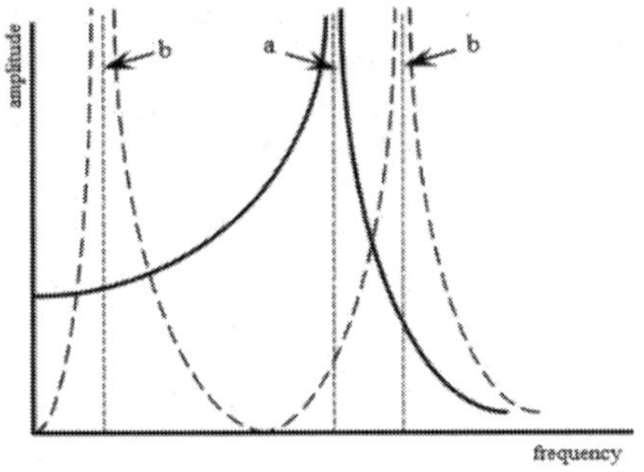

Figure 4. Electric motor effect on the natural frequencies of a system.

By introducing the pull-out slip, it is possible to rewrite the stator and rotor current equations as:

$$i_s = \frac{1}{jX_s} + \frac{k^2}{X_r'} \frac{s}{s_p + js} = \frac{1}{jX_s} + 2T_p \frac{s}{s_p + js} \tag{14}$$

$$i_r = 2T_p \frac{s}{s_p + js} \tag{15}$$

4. Motor Air-gap Influence on Torsional Vibration Analysis

Many references are available in technical literature on this subject, explaining the procedure for this kind of analysis [6]. The TVA is based on a lumped parameter model, approximating the behavior of the complete system, in which all the elements (compressor, flywheel, reducer, electric motors etc.) are represented by inertia disks linked by shaft pieces, modeled with springs and viscous dampers. API norms establish the rules and the simplification that can be done when making the lumped parameter model. Using the above mentioned model, the system is analyzed in a three-step procedure, calculating:

- Torsional Natural Frequencies and relevant Mode Shapes
- Steady-state Forced Response.
- Transient Forced Response.

During the last two stages, the excitation torques are applied to the system to evaluate the resulting torques on the shaft components and angular oscillations on the inertia disks.

According to API 618, the effect of the electric motor on the compression train is taken into account solely by ensuring that: “torsional natural frequencies are separated from the first and second multiples of the electrical power frequency by more than 10% and 5% respectively”. In fact, it is known that an electric motor produces an excitation couple at the frequency line and twice the frequency line. Recently, lumped parameter models of systems including induction motors, have been improved, taking into account the electric motor air-gap effect, in order to more accurately the system torsional natural behaviour [2][3].

This effect can be represented by an external stiffness and damping, like those shown in Figure 3, still provided that the induction motor model has been appropriately linearized, where J_C is the Compressor Inertia [kgm^2], J_M is the Electric Motor Rotor Inertia [kgm^2], k is the Coupling Stiffness [*Nm/rad*], d is the Coupling Damping [$Nm{\cdot}s/rad$], k_M is the Motor Air-gap Stiffness [*Nm/rad*], d_M is the Motor Air-gap Damping [$Nm{\cdot}s/rad$] and m_C is the Compressor Resistant Torque [*Nm*]. k_M and d_M depend only on electric motor parameters (like breakdown torque, rated torque, number of poles, rated slip, etc.) as shown in the next relations:

$$k_M = n_p T_B \frac{x^2}{1+x^2}; \qquad d_M = \frac{k_M}{\omega^2 T_L}; \qquad x = \omega T_L; \qquad T_L = \frac{1}{\omega_L} \frac{1}{2S_R} \frac{T_R}{T_B}; \tag{16}$$

where T_L is the Electric Motor Time Constant [*s*], T_R is the Rated Torque [*Nm*], T_B is the Breakdown Torque [Nm] and S_R is the Rated Slip.

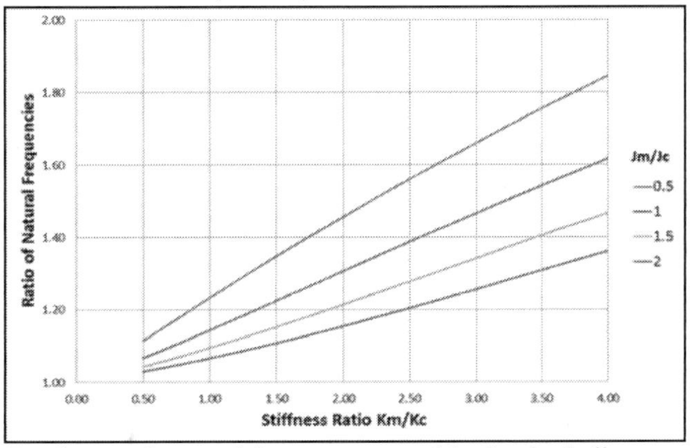

Figure 5. Fundamental natural frequency repositioning including motor air-gap influence

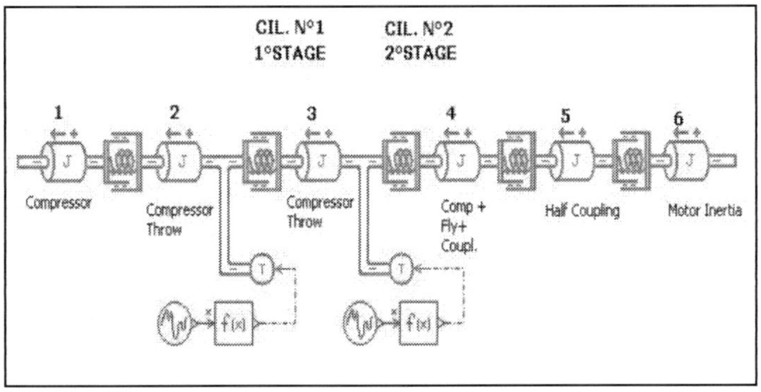

Figure 6. Torsional vibration analysis ignoring air-gap effect

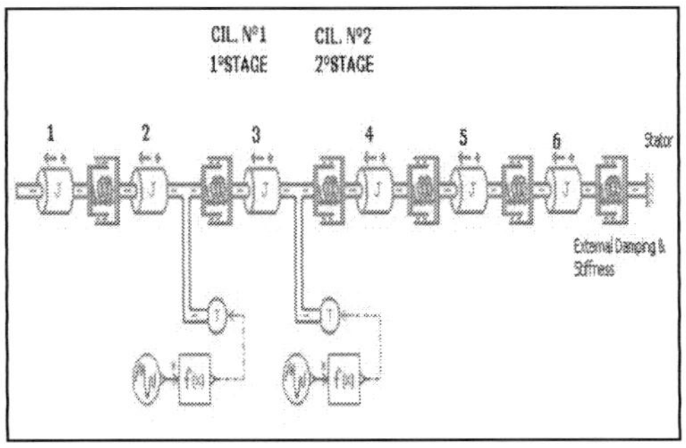

Figure 7. Torsional vibration analysis representing the motor air-gap torque by mechanical spring and viscous damper

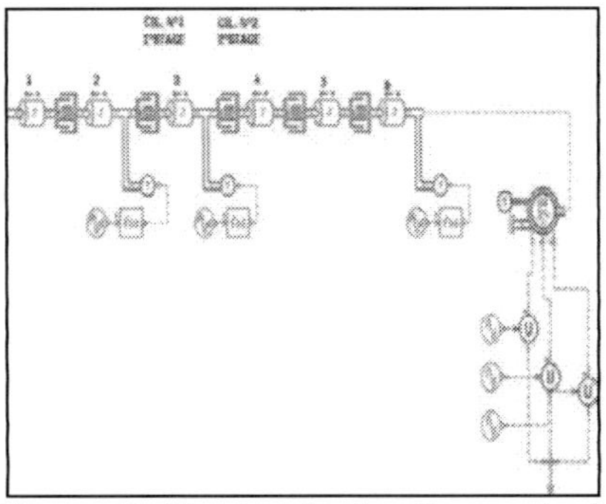

Figure 8. Torsional vibration analysis carried out using LMS AMESim induction motor model

Having considered the air-gap effect on the equivalent shaft has modified the original system. This has two major consequences on the system torsional natural frequencies, as shown in Figure 4, where letter *a* represent the fundamental natural frequency calculated ignoring the air-gap effect, while the new natural frequencies are represented with letter *b*.

- The previous first one, equal to zero (corresponding to the rigid mode of vibration), disappears and a new first one appears. The new frequency is very low effect, therefore, not considering it generally has no consequence.

IOP Conf. Series: Materials Science and Engineering **90** (2015) 012021 doi:10.1088/1757-899X/90/1/012021

- The previous fundamental natural frequency is increased by a certain percentage that depends on the electro-mechanical interaction of the system, i.e. on the inertia ratio between the motor and compressor (J_M/J_C) and on the stiffness ratio between the motor shaft and the coupling (k_M/k_C). The shift of the fundamental natural frequency can be quickly estimated by using the following graph, based on analytical expressions.

As can be seen from the graph, the error in the fundamental natural frequency estimation, due to neglecting the air-gap effect, can be very significant when the system includes soft coupling and/or low motor inertia.

Therefore, not taking into account the air-gap effect could lead to unacceptable errors in calculation of torsional natural frequencies and consequently in forced response analysis.

5. Simulation Results

The theoretical results obtained in the previous paragraph have been verified in two case studies, named case A and case B. This has been done using traditional mathematical methods and with the multi-physics simulation software LMS AMESim. The natural frequencies obtained through the models, that neglect and include the air-gap effect, have been compared. In the following figure, a lumped parameters model (case A) realized with LMS AMESim is shown. In this model, the air-gap effect is not included.

TABLE 1. SYSTEM PARAMETERS (CASE A)

Inertial masses			Shaft interval		
Item	Description	Moment of Inertia J kgm^2	Item	Description	Torsional Stiffness k Nm/rad
1	Compressor	0.059	[1-2]	compressor	1.11.E+07
2	Compressor throw	1.181	[2-3]	compressor	1.14.E+07
3	Compressor throw	1.171	[3-4]	compressor	6.13.E+06
4	Compr. + fly. + coupling	57.175	[4-5]	coupling	6.70.E+04
5	Half coupling	0.940	[5-6]	induction motor	2.78.E+06
6	Induction motor	52.570			

TABLE 2. UNDAMPED SYSTEM NATURAL FREQUENCIES IGNORING AIR-GAP EFFECT (CASE A)

1ST TNF	HZ	7.7
2ND TNF	HZ	240.4
3TH TNF	HZ	279.5
4TH TNF	HZ	746.9
5TH TNF	HZ	2253.1

TABLE 3. MOTOR AIR-GAP STIFFNESS AND DAMPING (CASE A)

External stiffness	k_m	Nm/rad	5.68E+04
External damping	d_m	Nm/rad	3.56E+01
Stiffness Ratio	k_m/k_c	/	0.85
Inertia Ratio	J_m/J_c	/	0.88

With the values of Table 1, the significant torsional natural frequencies (TNF) of the system are listed in Table 2.

Later, the air-gap effect was modeled with an external stiffness and damping obtained with equations (3) and (4), as explained in Paragraph 2 (see Table 3).

In this situation, the previous fundamental natural frequency results increased by 13% and, as explained in section 4, the natural frequency due to the rigid mode of vibration is no longer zero (see

Table 4). The results obtained with the scheme shown in fig. 7 have been confirmed also by using the induction motor model provided by LMS AMESim (fig. 8).

Adopting the same procedure used until now, the expected results have been obtained also for case B and the natural frequencies calculated by respectively including and ignoring the air-gap effect are summarized in table 6. Similar to case A, the results obtained with the scheme shown in figure 7 have been confirmed also by using the induction motor model provided by LMS AMESim (fig.8).

TABLE 4. TORSIONAL NATURAL FREQUENCIES INCLUDING MOTOR AIR-GAP EFFECT

1st TNF	Hz	3.1 (*new frequency)
2nd TNF	Hz	8.7 (+13%)
3th TNF	Hz	240.4
4th TNF	Hz	279.5
5th TNF	Hz	746.9
6th TNF	Hz	2253.1

TABLE 5. MOTOR AIR-GAP STIFFNESS AND DAMPING (CASE B)

External stiffness	k_m	Nm/rad	9.66E+04
External damping	d_m	Nm/rad	7.90E+01
Stiffness Ratio	k_m/k_c	/	0.92
Inertia Ratio	J_m/J_c	/	1.30

TABLE 6. UNDAMPED SYSTEM NATURAL FREQUENCIES IGNORING AIR-GAP EFFECT AND TORSIONAL NATURAL FREQUENCIES INCLUDING MOTOR AIR-GAP EFFECT (CASE B).

1st TNF	Hz	8.4	3.6 (*new frequency)
2nd TNF	Hz	240.4	9.2 (+10%)
3th TNF	Hz	245.5	240.4
4th TNF	Hz	746	245.5
5th TNF	Hz	2747.2	746
6th TNF	Hz		2747.2

6. Steady-state Response Analysis

According to API 618 the steady-state forced response is not required when the following conditions are verified:

- there are no torsional natural frequencies within 10% of operating shaft speed and within 5% of any multiple of operating shaft speed;
- torsional natural frequencies are separated from the first and second multiples of the electrical power frequency by more than 10% and 5% respectively.
-

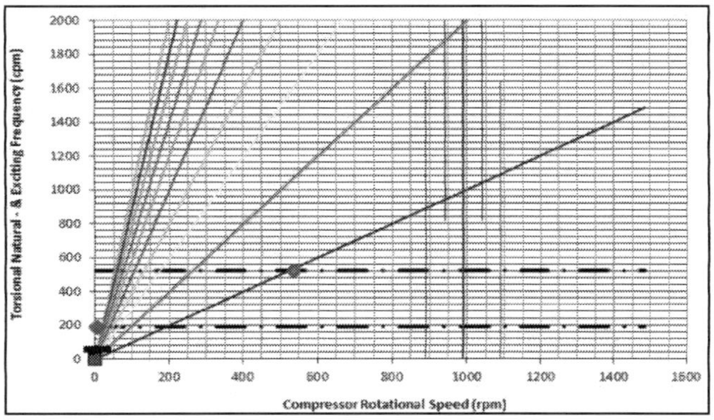

Figure 9. System A - Campbell Diagram

To evaluate whether or not these conditions are verified, a Campbell diagram is commonly used. In this diagram, the resonances are identified with the intersection between torsional natural frequencies (horizontal line) and excitations torque (oblique lines). The Campbell Diagram for case study A is shown in figure 9. It should be noted that, in this case, ignoring the electric motor air-gap effect, leads to a lower evaluation of the fundamental torsional natural frequency, consequently reducing the separation margin.

As can be seen from the Campbell diagram shown above, for case study A, according to API 618, a steady-state forced response is not required.

However it has been carried out to evaluate the resulting torque acting on the shaft intervals, using the model shown in figure 8.

Figure 10 shows the coupling torque calculated for case A, when the nominal load torque (4000 Nm) is applied: the torque oscillation has an amplitude that remains within acceptable values and, as expected, it has frequency components at about 16 Hz and 32 Hz, corresponding to the shaft nominal speed (994 rpm) and its second multiple.

7. Transient Response Analysis

API 618 establishes that transient analysis is not mandatory when induction motors without Variable Frequency Drive (VFD) are used, as in the examples presented in this work. However, when a soft coupling is used, also in these cases the start-up could be a matter of concern, since, starting from stand still, the system will certainly pass through one or more natural frequencies, and, as a consequence, the torque fluctuation can be amplified.

During the start-up phase, also the electric motor oscillating torque represents an excitation source that must be included in the analysis. Motor manufacturers usually provide analytical expressions for the torque as a function of the time for the main transient phenomena (start up, bi-phase and tri-phase short circuit, auto-reclosing). But these expressions are valid only with a locked rotor and cannot be used to simulate what happens in the transient periods. For example, in figure 11, the torque vs. time curve for the start-up of the motor used in case A, is shown. It can be seen that the torque oscillation has a time constant of about 2.5 s, therefore the machine would have appreciable oscillations for at least 10 seconds (4x time constant) without speeding up. This is in contrast with what common practice suggests, i.e. the starting of the whole system lasts much less time (usually 2 to 5 seconds, depending on the load torque). To obtain more realistic results for the analysis of the system start-up, the authors used the multi–physical software LMS AMESim that simultaneously solves the voltage equations of the motor and the mechanical motion equations of the shaft.

Different possibilities are offered to model an induction machine with LMS AMESim. In order to achieve a representative model to analyze the transient start-up of the machine, a dynamic double cage induction machine model has been set. This model has the advantage of being simple while ensuring an accurate torque-speed characteristic in all machine operating states (locked rotor, breakdown torque and rated speed).

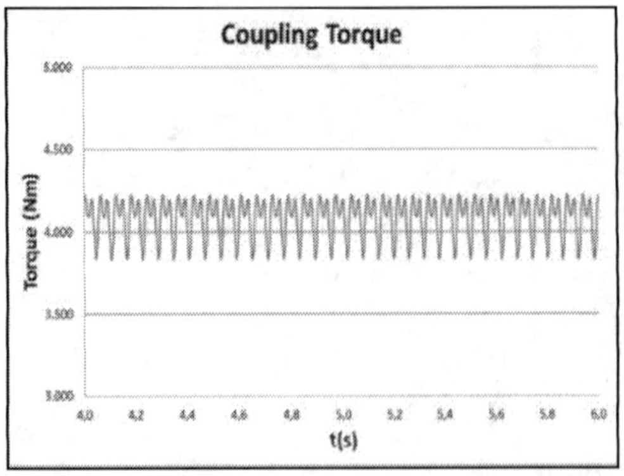

Figure 10. System A – Steady-state coupling torque

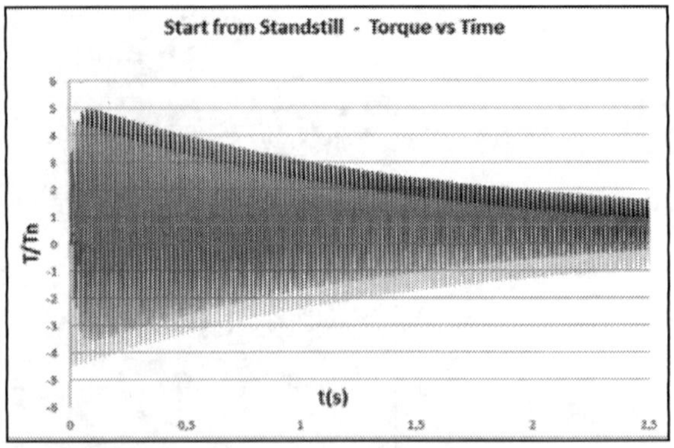

Figure 11. Induction motor Start-Up torque (provided by the manufacturer).

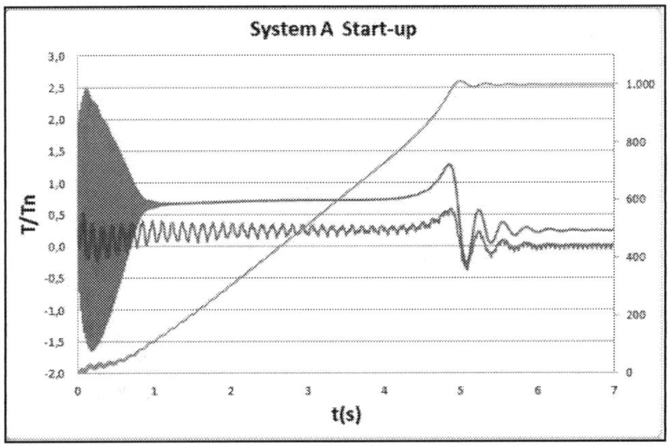

Figure 12. System A startup simulation performed by LMS AMESim

The voltage equations of the model expressed in the rotating reference frame linked to the electrical position of the rotor are the following:

$$U_{sd} = -\psi_{qs}\omega_{es} + I_{ds}R_s + \frac{d\psi_{ds}}{dt} \tag{17}$$

$$U_{sq} = -\psi_{ds}\omega_{es} + I_{qs}R_s + \frac{d\psi_{qs}}{dt} \tag{18}$$

As the rotor squirrel cage is short circuited:

$$U_{rd} = U_{rq} = 0 \tag{19}$$

$$0 = I'_{dr1}R'_{r1} + I'_{dr2}R'_{r2} + \frac{d\psi'_{dr1}}{dt} + \frac{d\psi'_{dr2}}{dt} \tag{20}$$

$$0 = I'_{qr1}R'_{r1} + I'_{qr2}R'_{r2} + \frac{d\psi'_{qr1}}{dt} + \frac{d\psi'_{qr2}}{dt} \tag{21}$$

The subscripts s and r refer respectively to the stator and to the rotor component of the quantity taken into account. The subscripts d and q refer to the direct and quadrature component of the Park's transformation reference frame. The prime subscript stands for rotor variables referred to stator windings. The 1 and 2 subscript refer to the cage 1 or cage 2 parameters. ψ is the magnetic flux, U is the voltage, I is the current, R is the resistance, L is the inductance and p the pole pair number.

The fluxes of the model are expressed as follows:

$$\psi_{ds} = L_s \cdot I_{ds} + L_m \cdot \left(I'_{dr1} + I'_{dr2}\right) \tag{22}$$

$$\psi_{qs} = L_s \cdot I_{qs} + L_m \cdot \left(I'_{qr1} + I'_{qr2}\right) \tag{23}$$

$$\psi'_{dr} = L_m \cdot I_{ds} + L'_{r1} \cdot I'_{dr1} + L'_{r2} \cdot I'_{dr2} \tag{24}$$

$$\psi'_{qr} = L_m \cdot I_{qs} + L'_{r1} \cdot I'_{qr1} + L'_{r2} \cdot I'_{qr2} \tag{25}$$

where L_m is the magnetizing inductance common to L_s and L'_{r1}, L'_{r2} are the stator, rotor cage 1 and rotor cage 2 inductances. The dynamic torque expression derives from the energy balance consideration. It is evaluated according to the following expression:

$$T_e = p \cdot L_m \cdot \left[\left(I_{qs} \cdot \left(I'_{dr1} + I'_{dr2}\right)\right) - \left(I_{ds} \cdot \left(I'_{qr1} + I'_{qr2}\right)\right)\right] \tag{26}$$

The parameters of this model are given by the manufacturer in the framework of a single cage model. Common model parameters have been set according to this manufacturer's datasheet information. We estimate the double cage parameters R'_{r1}, L'_{r1}, R'_{r2} and L'_{r2} thanks to an optimization process according to the torque speed characteristics provided by the manufacturer.

In figure 12, the results of the system A start-up simulation are shown. The simulation has been performed with a load torque of about 1/3 of the nominal torque, a value that approximates the effective resistant torque of a compressor started with valve unloaders on. The red line indicates the induction motor torque, the blue one the torque acting on the coupling, and the green one the rotor speed (starting from 0 rpm and reaching 994 rpm).

From the figure shown above some considerations can be made:

- Observing the torque oscillation of the electric motor, it can be deduced that the system electrical transient lasts only 1.2 s during the start-up (while the electric motor locked rotor transient lasts much more). However, the size of this oscillation (2.5 times the nominal torque) justifies a transient analysis although not required by the norms.
- The induction motor has oscillations that occur at line frequency (50 Hz) as expected. This justifies the API requirement to separate torsional natural frequencies from the first and second multiples of the electrical power supply frequency by more than 10% and 5% respectively.
- The coupling torque oscillations are not amplified when passing through resonance (516 rpm, 3.0 s). This is due to the presence of damping in the system and to the short time in which the compressor passes through the resonance region.
- The maximum torque acting on the coupling occurs at the electric motor breakdown torque.

8. Conclusions

In this work, the dynamic system of compression trains for O&G applications driven by electric motors has been studied. The study has been carried out considering the whole system, with the reciprocal effect that the mechanical and electrical parts have on each other. TVA has been performed in two case studies, comparing the results produced by a traditional lumped parameters model and a more recent model that represents the electric motor air-gap through a spring and a damper connected to earth. By using simulation software and traditional mathematical methods, it has been confirmed that the new model provides more accurate results in terms of system torsional natural frequencies.

However, the new model is obtained through a linearization of the non-linear induction motor model, therefore, it cannot be used to study the transient phenomena that occur, for example, during the motor start-up.

Furthermore, the use of the analytical expressions (usually provided by electric motor manufacturers) would lead to inaccurate results. In fact, these expressions are valid only in particular conditions (for example, with a locked rotor) and are obtained considering the electric motor separate from the mechanical part of the system, and consequently do not describe the real system behavior.

This limitation has been overcome by using the multi-physics simulation software LMS AMESim. This software makes it possible to consider the reciprocal effect of the electrical and mechanical part of the system by simultaneously solving the voltage equations of the electric motor and the mechanical equations of the shaft.

In this way, it has been possible to carry out the transient forced response of the system obtaining consistent results.

In further works, we will be extending our studies on the implications that the dynamic system of compression trains has on the supply line. To do this, we will be focusing more on the current pulsations that are generated on the line, due to the torque oscillations caused by the reciprocating compressors.

References

[1] Di Vito G 1943 Motori sincroni ed asincroni per azionamento di compressori *L'elettrotecnica*, **XXX-13** 3-18
[2] Knop G 2012 The Importance of Motor Dynamics in Reciprocating Compressor Drives *Proc. Int. Rotating Equipment Conf. EFRC (Düsseldorf, D, 27th–28th September 2012)* pp 277-86
[3] Hauptmann E Eckert B Howes B 2013 The Influence on Torsional Vibration Analysis of Electromagnetic Effect across an Induction Motor Air-gap *Gas Machinery Conf. 2013 (Albuquerque MN, 6th–9th October 2013)*
[4] Kern U Gamm M 2005 About transient torque impacts on turbo-compressor shafting driven by induction motors or synchronous motors *Int. Conf. on Compressors and Their Systems - IMechE, (London UK, 4th–7th September 2005)* pp 477-96
[5] Kovacs P K 1984 Transient phenomena in electrical machines (Amsterdam:Elsevier)
[6] Corbo M A Malanosky S B Practical Design 1996 Against Torsional Vibration Proc. of the Twenty-fifth Turbomachinery Symp (Houston, Texas, September 17 - 19, 1996) pp 189-222
[7] Krause P C Wasynczuk O Sudhoff S D Pekarek S 2013 *Analysis of electric machinery and drive systems* (HobokenWiley-IEEE Press)
[8] API STANDARD 618 2013 Reciprocating compressors for petroleum, chemical, and gas industry services 6th ed

Experimental study on performance of BOG compressor

Bin Zhao, Tao Wang, Xueyuan Peng*, Jianmei Feng

Xi'an Jiaotong University,No.28 West Xianning Road, Xi'an, Shaanxi, China
Corresponding auther*:xypeng@mai.xjtu.edu.cn

Abstract: The boil-off gas (BOG) compressor is widely used for recycling the excessive boil-off gas of liquefied natural gas (LNG), and the extra-low suction temperature brings about great challenges to design of the BOG compressor. In this paper, a test system was built to examine the effects of low suction temperature on the compressor performance, in which the lowest temperature reached -178℃ by means of a plate-fin heat exchanger with liquefied nitrogen. The test results showed that, as the suction temperature decreased from 20℃ to -150℃, the volumetric efficiency of the compressor dropped by 37.0%, and the power consumption decreased by 10.0%. The preheat of the gas by the pipe through the suction flange to suction valve was larger than 20℃ as the suction temperature was -150℃, and this value increased with the decreased suction temperature. The pressure loss through the suction valve at lower suction temperature was larger than that at ambient temperature while the volume flow rate was kept the same.

Keywords: BOG compressor, volumetric efficiency, power consumption, preheat

1. Introduction

Natural gas(NG) is one of the most promising clean energy sources and plays a more and more important role in our daily energy consumption. NG is usually stored and transported in the form of compressed natural gas(CNG) and liquefied natural gas(LNG). After liquefied, LNG is only 1/625 of NG in terms of volume at atmosphere pressure, and considering the physical nature of NG, it is obvious the most economical way for transportation of NG over long distances is in the form of LNG. While usually stored in containers at the low temperature of about -162℃,it is inevitable that heat transfers from surroundings to the cryogenic LNG, which results in continuous vaporizing BOG(0.05%/day, mass fraction)[1-3]. In the LNG terminals, BOG generation drastically increases when LNG ship unloads. The excessive BOG in the containers gradually increases the pressure of the container. Therefore, there must be effective way to deal with BOG [4]. The best two choices of recycling BOG are directly compressing to the distribution pipelines and liquefying, and the latter is

Content from this work may be used under the terms of the Creative Commons Attribution 3.0 licence. Any further distribution of this work must maintain attribution to the author(s) and the title of the work, journal citation and DOI.
Published under licence by IOP Publishing Ltd

more widely used to recover BOG due to its 30-60% higher energy-utilization efficiency[5]. A typical BOG recycle system in LNG terminal is shown in Fig.1[6], in which the BOG is liquefied in a BOG reliquefier after pressurized by the BOG compressor. BOG compressor is an essential part of dealing with the BOG in this system.

BOG compressors normally work at extra low suction temperature of about -162℃， and require operating at an oil-free condition. At present, the BOG compressors in LNG terminals are classified as either vertical labyrinth compressors [7] or horizontal piston ring compressors [8].The suction temperature of BOG compressor are between -162℃ and -100℃,and suction pressure between 0.003 and 0.01MPa(gauge pressure)[9]. The compressor usually has two or three stages, and the pressure ratio distribution is different with compressors working at ambient temperature, the pressure ratio of first stage is higher than the others. On this condition , the discharge temperature of the first stage of BOG compressor can arise up to -50℃,which requires less attention on the design of the rest stage.

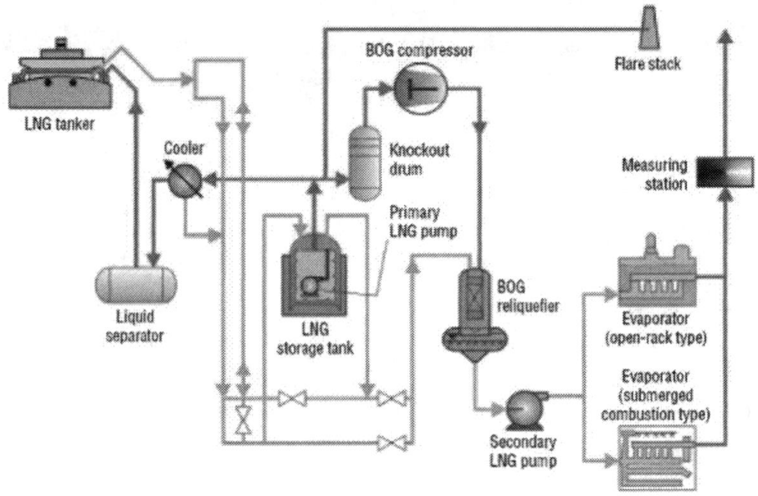

Figure 1. LNG system in a typical LNG terminal[6]

The temperature of BOG in the LNG system can be as low as -162℃ and the effects of such low temperatures must be taken into consideration when designing the BOG compressor. It was proposed that preheat was one of the most important considerations when designing BOG compressors in cryogenic service[6]. K. Murai et al.[10] measured a BOG compressor and provided a correlation to calculate preheat. Yoshida and Katsumi ShinyaKoji[11] researched the development of the suction and discharge valves using non-metallic valve plates for BOG compressors, Shen et al.[12] built a 3-D finite element model of BOG compressor to study the heat transfer in the cylinder of BOG compressor and the temperature distribution of the cylinder .The results of the simulation proved to be an effective way to acquire the temperature distribution of BOG compressor. Chen et al.[13] investigated the effect of low suction temperature on flow rate.

In the above investigations, preheat was always included. Because, at low temperatures the heat transfer from the cylinder to the gas affects the thermodynamic process in the compression chamber. This directly determines the design method and the parameters of BOG compressors. The large temperature difference between the low-temperature suction chamber and the surrounding environment leads to thermal stresses on the cylinder, influencing the structure and materials of the cylinder and the heat treatment process. As the piston is in direct contact with the freezing gas, the piston rings must be oil-free, and the self-lubricated material must have appropriate friction and wear characteristics. For these reasons, the heat transfer in the cylinder and the thermodynamic process in the compression chamber are critical issues in the design of BOG compressors. Rare research were done on the characteristics of BOG compressor,

As mentioned above, preheat is one of the most important design parameter of BOG compressor because it strongly affects the actual flow rate of BOG compressor. There is some research involving the experimental testing to measure preheat, however in those research either the structure of tested compressor is totally different with practical applied BOG compressor or the empirical correlation contains unknown parameters. In this paper, a BOG compressor was designed to investigate preheat and corresponded flow rate at different suction temperature and pressure, and the pressure-time diagram inside the cylinder was measured. To meet this demand, a testing system (shown in Fig.2) was set up to obtain different suction temperature and pressure. In this testing system the suction gas was firstly cooled by changing heat with liquid nitrogen through a plate-fin heat exchanger, and then it flowed through an evaporator and was accurately adjusted to designed temperature.

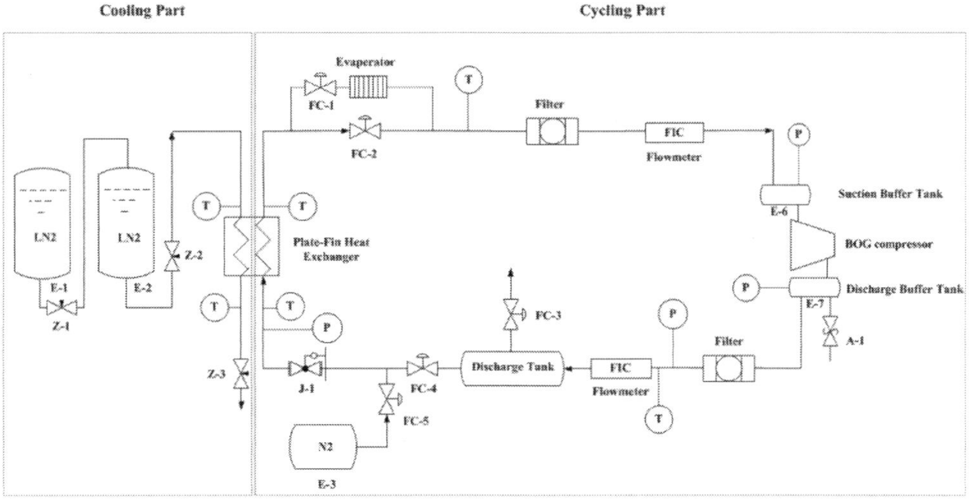

Figure 2. The Testing system of the BOG compressor

2. BOG compressor and testing system

2.1 BOG compressor

The tested BOG compressor, as shown in Fig. 3, was a L-type double-acting reciprocating unit with two cylinder running at 411rpm(6.8Hz).Only the cylinder of first stage was used in this experiment considering of the discharge temperature of the first stage was higher than -45℃. The working medium in the compressor is nitrogen. The diameter of first stage cylinder is 300mm, and the piston is 299mm.The nominal swept volume was 700.53m^3/h. The tested suction temperature is between -160 ℃ and -40 ℃. The suction pressure is between -0.03MPa to 0.11MPa and the discharge pressure is between 0.15MPa and 0.25MPa. The all connecting bolts around the cylinder were made of austenitic stainless steel.

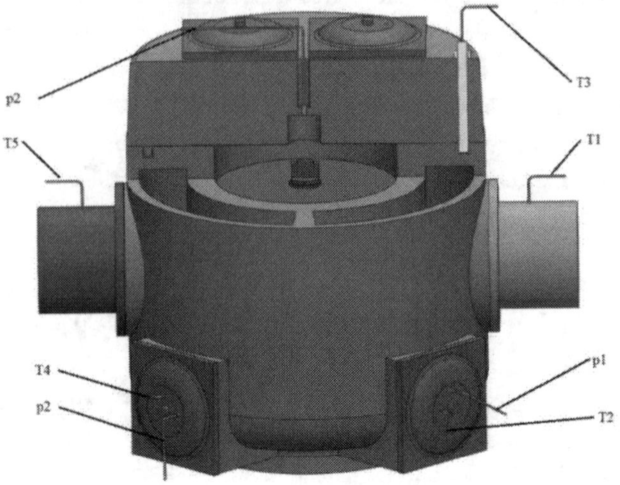

Figure 3. The BOG compressor studied

2.2 BOG compressor testing system

A testing system was set up to obtain different suction temperature and pressure, as shown in Fig2 .This system contains two parts: the cooling part and the cycling part. The cooling part mainly includes several liquid nitrogen containers which was connected to a plate-fin heat exchanger. The cycling part is made up of compressor, discharge buffer tank, discharge tank, flow meters, nitrogen tank, plate-fin heat exchanger, evaporator, filter and suction buffer tank.

In the testing system the nitrogen, after pressurized by BOG compressor, flows through the plate-fin heat exchanger and is cooled by liquid nitrogen flowing through the heat exchanger, then the cooled nitrogen flows through evaporator and is adjusted to different suction temperature. As the operating pressure of plate-fin heat exchanger is relatively low, pressure of nitrogen is decreased by a reducing valve. The nitrogen tank in the cycling part is used to replace the air inside the compressor, tanks and pipelines before start-up, and supply nitrogen when the discharge pressure increases after start-up. For the temperature of nitrogen is low toward -178℃ when flowing out of the heat exchanger, it is necessary that the pipelines, flow meter and suction buffer tank are covered with heat preservation cotton.

3. Experimental method

3.1 Experimental facilities

As shown in Fig 1, five thermocouples with an accuracy of ±0.2℃ were arranged to measure temperature around the cylinder. Four thermocouples were installed to measure the gas temperature around the cylinder, including the suction flange(T1), suction valve(T2, shown in Fig. 4), discharge valve(T4) and discharge flange(T5). The fifth thermocouple was installed to measure the temperature at the end of bolt that connected the cover head and the cylinder on suction side(T3). The thermocouple of T2 was used to measure the gas temperature before flowing through the suction valve. Three pressure sensors were installed in the cylinder of the compressor to measure the pressure of the suction chamber(p1), the pressure inside the cylinder(p2) and the pressure in the discharge chamber(p3). All the pressure sensors displayed the absolute pressure with an accuracy of ±0.25%. Two target flow meters were installed before suction buffer tank and after the discharge buffer tank. The accuracy of the flow meter was ±1%. The power was measured with a three-phase wattmeter with an accuracy of ±1%.

Figure 4. The pressure sensor and thermocouple installed on the suction side

3.2 p-v indicator diagram

Figure 5. The pressure sensor measuring the pressure inside the cylinder

An experimental study on recording the p-v diagram was conducted. In order to measure the transient pressure inside the cylinder changes with the rotation angle, a miniature pressure sensor was adopted. Considering the practical compressor structure, the pressure sensor was installed as shown in Fig. 5. The pressure sensor was embedded in the cover head and the signal wire was led out. The signal was collected and analyzed by the dynamic signal acquisition and analyzing system. The sensor had a

IOP Conf. Series: Materials Science and Engineering **90** (2015) 012022 doi:10.1088/1757-899X/90/1/012022

uncertainty of less than ±0.1% and its response frequency was 1400 kHz. The maximum working pressure was 0.7MPa with an output voltage of 100mv, and the range of working temperature was between -196℃ and 120℃. A top dead center sensor (TDCS) was also installed and the signal was diminished so that the maximum of output signal was a little higher than the pressure sensor's which allowed a much more convenient observe during the experiment..

3.3 Dynamic data acquisition system

The dynamic data acquisition system was based on an acquisition card with a high acquisition speed. The sampling frequency of the acquisition card was 1 MHz and the analog input was up to 16 simultaneous ways. The maximum sampling frequency of a single channel was 250 kHz. The signal of pressure sensors, TDCS and thermocouples were input to the conditioning device by means of 8 synchronous channels. After undergoing amplification and conditioning, both the pressure and temperature signals were collected by the high speed acquisition card and displayed on computer screen. The structure of the dynamic data acquisition system is shown as fig. 6. Fig. 7 showed the graphic display of acquisition system which was programmed by LabVIEW.

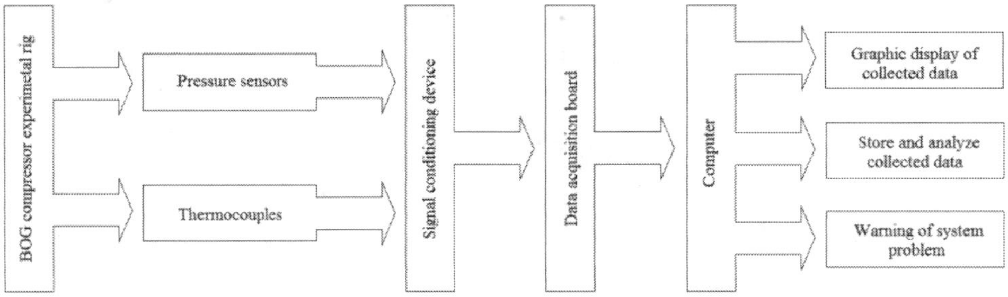

Figure 6. Dynamic data acquisition system

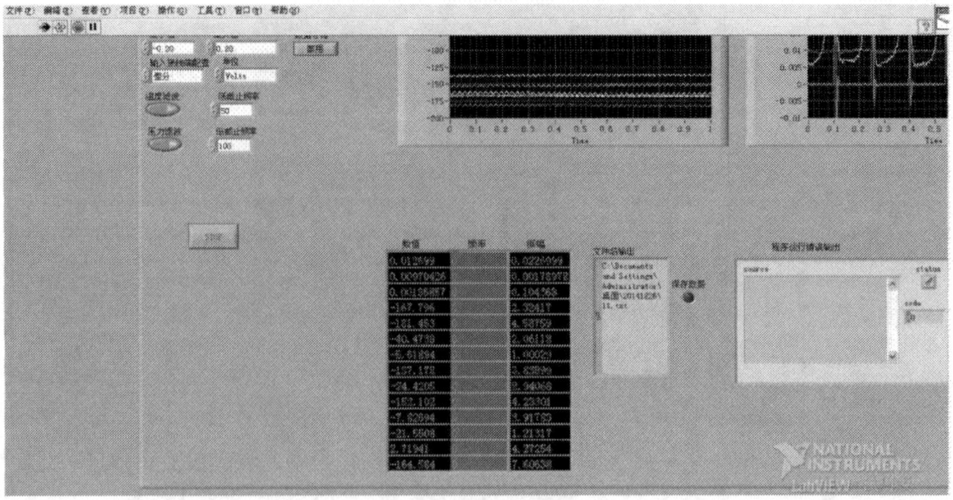

Figure 7. graphic display of acquisition system

4. Results and discussions

By adjusting the suction temperature, the temperature distributions of the cylinder were obtained, allowing for analyzing how greatly preheat influenced the flow rate of BOG compressor. Tests were performed using different suction pressures and pressure ratios. The pressure-time diagrams inside the cylinder at different suction temperatures were analyzed and were compared with that of ambient suction temperature in order to find out the difference of compression and expansion process between BOG compressor and ordinary compressor of ambient suction temperature.

According to the experimental results of the temperature distribution, the preheat increased with the decrease of suction temperature and increased with the increase of pressure ratio. The pressure-time diagrams at low suction temperature showed the suction pressure loss were higher than those of ambient suction temperature which resulted in the increase of the actual pressure ratio and power consumption and the decrease of flow rate. In addition, the flow rate and power consumption were recorded and the specific power consumption were calculated. The calculated results showed the power consumption of low suction temperature is less than that of ambient suction temperature.

4.1 Effects of different suction temperatures on BOG compressor

Suction pressure was kept constant at 0.1MPaA (Absolute pressure) and discharge pressure was kept 0.22MPaA, while the suction temperature was adjusted to -50℃, -70℃, -90℃, -110℃, -130℃ and -150℃. Fig.8 showed a picture of BOG compressor working steadily at a suction temperature of -150℃.The outer surface of the cylinder was covered with frost, and the frost of suction side was thicker than that of the discharge side.

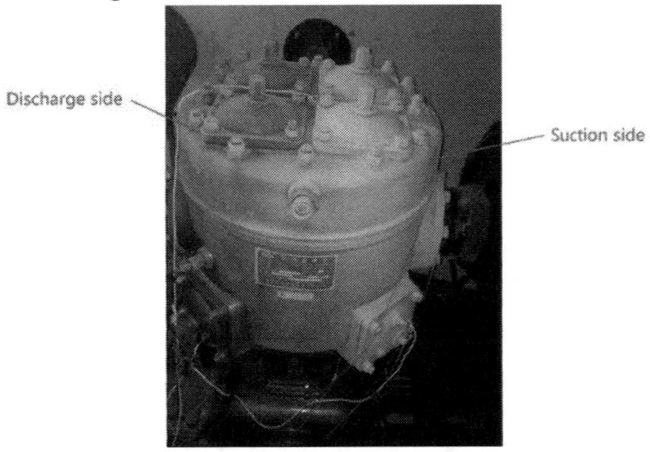

Figure 8. The cylinder of the BOG compressor at suction temperature of -150℃

Fig.9 illustrated the volumetric efficiencies and isentropic efficiencies of the BOG compressor under different suction temperature. The volumetric efficiencies decreased from 76% to 51%and the total isentropic efficiency decreased from 60% to 46% when the suction temperature decreased from -50℃ to -150℃, while these two values were 81% and 74% at suction temperature of 20℃. Fig.10 showed the temperature distributions of all measured points around cylinder and preheats at different suction temperatures. The temperature of T2, T3, T4 and T5 decreased with the decrease of suction temperature. When the suction temperature was -50℃, the temperature of T2, T3, T4 and T5 was

-45℃, 42℃, 39℃ and -1℃, respectively. The preheat, i.e. temperature difference between T1 and T2, was 5℃, and the value was 22℃ when the suction temperature is -150℃. When the suction temperature decreased from -50℃ to -70℃, the temperature diagrams of T4 and T5 shared a cross point. That was because heat transferred from cylinder to discharged gas when discharged temperature decreased rapidly. The temperature at end of bolt on the suction side was -39℃ while the suction temperature was -150℃, then the temperature of other end of bolt embedded into the cylinder could be between the suction temperature and temperature of suction valve, i.e. between -150℃ and -128℃, thus the temperature of the whole bolt was obtained, which contributes to the pretension design of bolts.

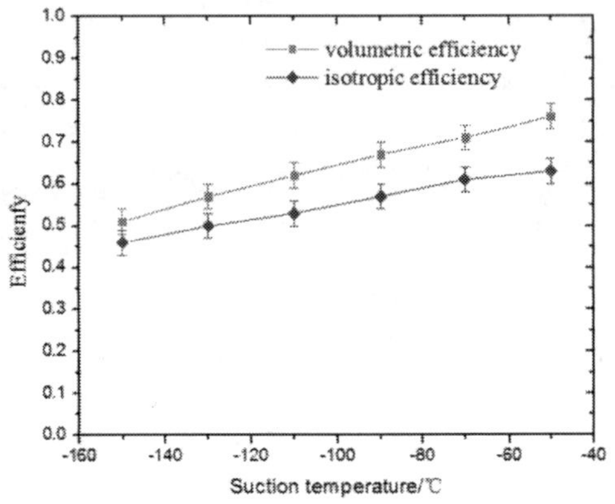

Figure 9. The volumetric efficiency and isentropic efficiency at different suction temperature

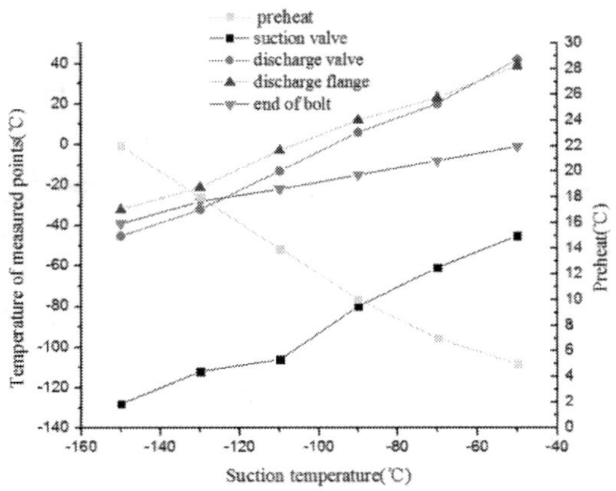

Figure 10. Temperature of measured points and preheat at different suction temperature

When the suction temperature was 20℃ and -150℃, the discharged flow rate was kept to 446.64Nm^3/h, while the power consumption was 17.06kW and 15.35kW, the power consumption decreased 10.02%. This meant it was more economical to pressurize at a lower suction temperature despite the decrease of volumetric efficiency and isentropic efficiency.

The p-v diagram was showed in fig.11 while the suction temperature was -150℃. The suction pressure loss consisted 17% of the power consumption, and the discharge pressure loss consisted 13%.

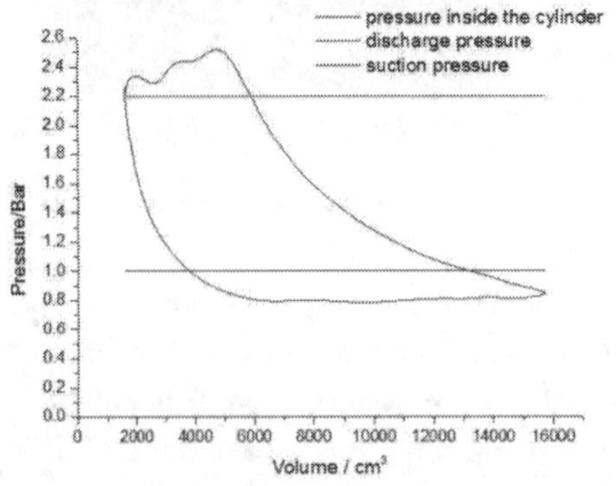

Figure 11. The p-v diagram under suction temperature of -150℃

4.2 Effects of different discharge pressure on BOG compressor

Suction pressure was kept constant at 0.1MPaA and suction temperature was kept -150℃, while the discharge pressure was adjusted to 0.15MPaA, 0.2MPaA, 0.25MPaA and 0.3MPaA. As shown in Fig. 12, volumetric efficiencies decrease from 72% to 48% and the total isentropic efficiencies decreased from 69% to 0.54% with the increase of discharge pressure from 0.15MPaA to 0.3MPaA. While the suction temperature was 20℃, the discharge pressure changed from 0.15MPaA to 0.3MPaA, the volumetric efficiency decreased from 78% to 61%, and isentropic efficiency decreased from 73% to 65%.

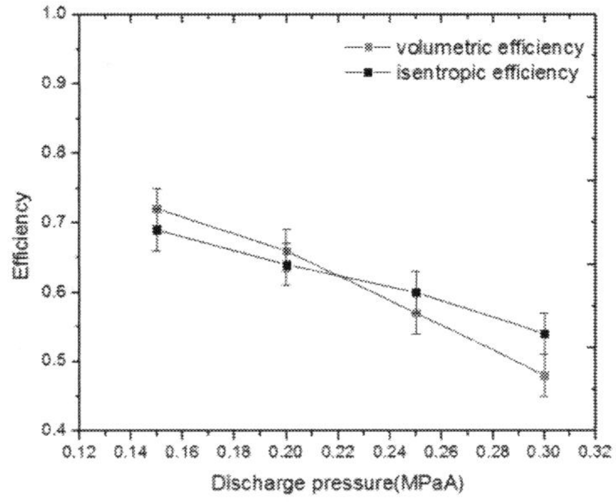

Figure 12. Volumetric efficiency and isentropic efficiency at different discharge pressure

5. Conclusions

A testing system was set up to measure the performance of BOG compressor, and the compressor worked steadily at the suction temperature low to -150℃. In this condition, the effects of low suction temperature on BOG compressor were examined.

When the compressor operated steadily with a suction temperature of -150℃,the temperature around the suction valve was about -128℃.Temperature around the discharge valve was relatively higher, with a temperature of -45℃.

The preheat was 5℃ with a suction temperature of -50℃,and it kept increasing to 22℃ while the suction temperature was -150℃.

The volumetric efficiencies decreased from 76% to 51% and the total isentropic efficiencies decreased from 6% to 46% when the suction temperature decreased from -50℃ to -150℃, while these two values were 81% and 74% at suction temperature of 20℃. The power consumption decreased 10% when the suction temperature decreased from 20℃ to -150℃.

ACKNOWLEDGEMENT

This work was supported by the Fundamental Research Funds for the Central Universities .

References

[1] W.S. Lin, N. Zhang, A.Z. Gu, LNG (liquefied natural gas): A necessary part in China's future energy infrastructure, Energy 35 (11) (2010) 4383–4391.

[2] G.H. Shi, Y.Y. Jing, S.L. Wang, X.T. Zhang, Development status of liquefied natural gas industry in China, Energy Policy 38 (11) (2010) 7457–7465.

[3] E. Querol, B.G. Regueral, J.G. Torrent, M.J.G. Martinez, Boil off gas (BOG) management in Spanish liquid natural gas (LNG) terminals, Appl. Energy 87 (11) (2010) 3384–3392.

[4] H. Sayyaadi, M. Babaelahi, Thermoeconomic optimization of a cryogenic refrigeration cycle for re-liquefaction of the LNG boil-off gas, Int. J. Refrig. 33 (6) (2010) ,p1197–1207.

[5] Mabuchi Nobuhiro (Osaka Gas Co. Ltd., Jpn), 2006. BOG reliquefaction with cold energy storage and utilization of LNG cold energy using cascade process. Energy Resour. 27, p334–336.

[6] Liquid gas transport and storage, Burckhardt Compression AG, Switzerland, <http://www.burckhardtcompression.com/webautor-data/196/21_laby_liquidgas_e-1.pdf>.

[7] P. Ernst, The LNG BOG labyrinth-piston compressor with flexible capacity control, Burckhardt Compression AG, Switzerland, <http://www.kgu.or.kr/ download.php?tb=bbs_017&fn=Ernst.pdf&rn=Ernst.pdf>.

[8] N. Akamo, Process critical compressors, Kobe Steel Ltd, Japan, 2008, <http://kobelcocompressors.com/_pdf/technical/Process%20critical%20compressors%20by%20akamo.pdf>.

[9] Sangeun Roh, Gihun Son, Gildal Song, Junghong Bae,Numerical study of transient natural convection in a pressurized LNG storage tank, Applied Thermal Engineering 52 (2013) 209-220

[10] K.Murai, K. Nagura, LNG boil-off gas reciprocating compressors, Kobe Steel Works Eng. Rep. 49 (1) (1999) 64–67.

[11] Yoshida, Katsumi ShinyaKoji, Iwawaki, Seiji, Development of suction and discharge valves using non-metallic valve plates for BOG compressors, 14th International Conference and Exhibition on Liquefied Natural Gas,p 883-888

[12] 3-D finite element simulation of the cylinder temperature distribution in boil-off gas (BOG) compressors, International Journal of Heat and Mass Transfer, v 55, n 23-24, p 7278-7286

[13] Chen, Jinfeng Wu, Zhimin,Feng, Jianmei, Peng, Xueyuan, Experimental investigation on flow rate of boil-off gas compressor at low suction temperature, Journal of Xi'an Jiaotong University 48(3),p68-71.

IOP Conf. Series: Materials Science and Engineering **90** (2015) 012023 doi:10.1088/1757-899X/90/1/012023

Modeling of an Oil-Free Carbon Dioxide Compressor Using Sanderson-Rocker Arm Motion (S-RAM) Mechanism

Bin Yang[1]*, Orkan Kurtulus[2], and Eckhard A. Groll[3]
Purdue University, School of Mechanical Engineering,
Ray W. Herrick Laboratories, West Lafayette, IN, USA

Email: [1]yang62@purdue.edu, [2]orkan@purdue.edu, [3]groll@purdue.edu

Abstract: A simulation model to predict the performance of a prototype CO_2 compressor is presented. This prototype compressor employs the Sanderson-Rocker Arm Motion (S-RAM) mechanism, which converts the rotary motion of the shaft into a linear reciprocating motion of the cylinders. The piston stroke can be variable by changing the incline angle between the connecting rod and compressor main shaft centerline. The compressor model is mainly composed of two main sub-models simulating the kinematics of the drive mechanism and the compression process. A valve sub-model is included in the compression process model.

1. Introduction

The prototype compressor is using the Sanderson Rocker Arm Motion (S-RAM) mechanism. This mechanism converts rotation of the compressor main shaft into the reciprocating motion of the piston. The stroke of the piston is variable by changing the incline angle of the connecting rod as shown in figure 1. The wobble plate is installed on the connecting rod. The ball joint type piston rod is employed to connect the wobble plate and piston similar to that in the conventional wobble plate compressor. The S-RAM mechanism is expected to have less frictional power loss compared to the traditional wobble plate compressor since there is no friction existing between the wobble plate and driving plate in the S-RAM mechanism. However, there is frictional power loss between the piston and cylinder wall, which is common for the wobble plate compressors using other mechanisms.

Some mathematical models for conventional swash/wobble plate compressor have been found in the literature. Below and David [1] analyzed the dynamics of a swash plate mechanism. A mathematical model was developed, which was essential for the evaluation of the bearing loads and counterweights. Ishii et al. [2] investigated the dynamic behavior of a wobble plate compressor, whose displacement was variable. The inertial forces and moments, constraint forces and friction forces were evaluated. It was found that the inertial force and friction forces

Content from this work may be used under the terms of the Creative Commons Attribution 3.0 licence. Any further distribution of this work must maintain attribution to the author(s) and the title of the work, journal citation and DOI.
Published under licence by IOP Publishing Ltd

IOP Conf. Series: Materials Science and Engineering **90** (2015) 012023 doi:10.1088/1757-899X/90/1/012023

had a significant influence on the compressor performance. Tojo et al. [3] made an analysis on the kinematics of a variable displacement wobble plate compressor. Tian et al. [4] developed a model for a variable displacement swash plate compressor. Analysis on the dynamics and compressor process were firstly conducted, and was validated with the experimental data consequently.

In this paper, a simulation model for the prototype compressor is presented to give a preliminary knowledge of the S-RAM mechanism. This model includes the model of the kinematics and a compression process model.

2. Geometry and Kinematics Model

A schematic of the geometry and kinematics of the compressor with S-RAM mechanism is shown in figure 1. The bottom dead center (BDC) is assigned as the starting point of the crank angle, θ. The piston displacement, x, and velocity, U, are expressed as a function of the crank angle as follows:

$$x = [R_{wb}^2 + R_{bj}^2 - (H - R_{cr}\sin\beta)^2]^{\frac{1}{2}} - [R_{wb}^2 + R_{bj}^2 - H^2 + 2HR_{cr}\sin\beta\cos\theta - (R_{cr}\sin\beta)^2]^{\frac{1}{2}} \quad (1)$$

$$U = HR_{cr}\omega\sin\beta\sin\theta[R_{wb}^2 + R_{bj}^2 - H^2 + 2HR_{cr}\sin\beta\cos\theta - (R_{cr}\sin\beta)^2]^{-\frac{1}{2}} \quad (2)$$

Where, H is the distance between cylinder centerline and compressor main shaft centerline. R_{wb} is the radius of wobble plate. R_{bj} is the piston rod length. R_{cr} is the length of connecting rod. β is the incline angle between the connecting rod and compressor main shaft centerline. ω is the main shaft rotation angular speed.

The instantaneous volume inside the cylinder is calculated by:

$$V_c = A_p \cdot x + V_0 \quad (3)$$

Where, V_0 is the clearance volume, defined by:

$$V_0 = \alpha \cdot A_p \cdot S \quad (4)$$

The relative clearance volume α is 0.1. A_p is the cross sectional area of the piston. S is the piston stroke.

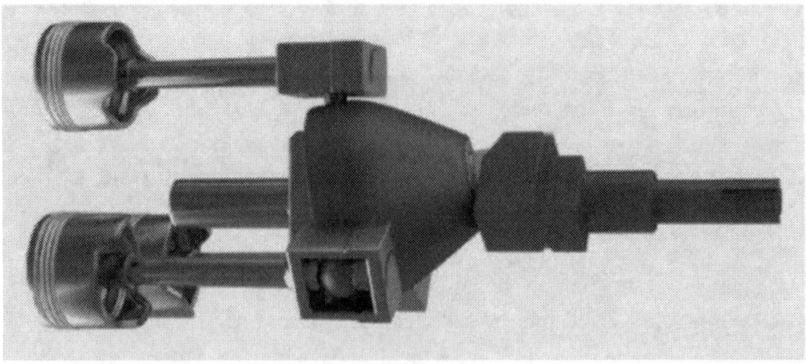

(a)

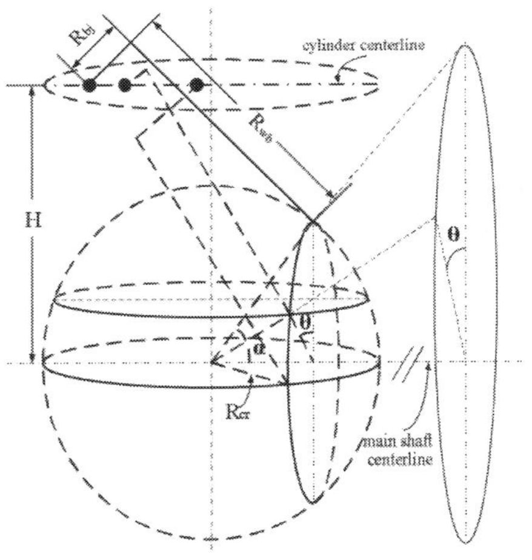

(b)

Figure 1: Schematic of geometry and kinematics of a compressor with S-RAM mechanism

3. Compression process model

3.1 Governing equationsb

The refrigerant inside the cylinder is assumed to be homogeneous, which means the thermal properties of the refrigerant are uniform anywhere in the cylinder. The refrigerant thermal properties are evaluated using REFPROP [5]. According the law of continuity, the mass change of refrigerant in the cylinder can be expressed by:

$$\frac{dm_c}{d\theta}=\frac{dm_{suc}}{d\theta}+\frac{dm_{li}}{d\theta}-\frac{dm_{dis}}{d\theta}-\frac{dm_{lo}}{d\theta} \tag{5}$$

Where, m_c refers to the refrigerant mass in the cylinder. The change of specific volume of refrigerant with respect to the rotation angle of the main shaft is:

$$\frac{dv_c}{d\theta}=\frac{1}{m_c}\frac{dV_c}{d\theta}-\frac{V_c}{m_c^2}\frac{dm_c}{d\theta} \tag{6}$$

Substituting equation (3) into equation (6) yields:

$$\frac{dv_c}{d\theta}=\frac{A_p}{m_c}\frac{dx}{d\theta}-\frac{A_p\cdot x+V_0}{m_c^2}\frac{dm_c}{d\theta} \tag{7}$$

The conservation equation of energy can be derived by employing the first law of thermodynamics:

$$\frac{dQ}{d\theta}+\frac{dm_{suc}}{d\theta}h_{suc}+\frac{dm_{li}}{d\theta}h_{li}=\frac{dW}{d\theta}+\frac{dm_{dis}}{d\theta}h_{dis}+\frac{dm_{lo}}{d\theta}h_{lo}+\frac{d(m_c u_c)}{d\theta} \tag{8}$$

Rearranging equation (8) and using equation (5) - equation (7) results in a relationship that describes the refrigerant temperature in the cylinder as a function of crank angle:

$$\frac{dT_c}{d\theta}=\frac{1}{[(\frac{\partial h_c}{\partial T_c})_v-v_c(\frac{\partial p_c}{\partial T_c})_v]}\{\frac{1}{m_c}[\frac{dm_{suc}}{d\theta}(h_{suc}-h_c)+\frac{dm_{li}}{d\theta}(h_{li}-h_c)+\frac{dQ}{d\theta}]+[-(\frac{\partial h_c}{\partial v_c})_T+v_c(\frac{\partial p_c}{\partial v_c})_T]\frac{dv_c}{d\theta}\} \tag{9}$$

The instantaneous heat transfer between the refrigerant and the cylinder wall is determined using the heat transfer coefficient, proposed by Annand [6] and later amended with a factor of three by Todescat et al. [7]. The heat transfer coefficient is defined as follows:

$$h_{heat}=3\cdot 0.7\cdot\frac{\tau_c}{D_p}(\frac{\rho_c U_{mean} D_p}{\mu_c})^{0.7} \tag{10}$$

Sequently, the heat transfer rate between refrigerant and cylinder wall is calculated:

$$\dot{Q}_{heat}=\frac{dQ}{dt}=h_{heat}A_{heat}(T_c-T_{wall}) \tag{11}$$

Adair et al. [8] reported that the cylinder wall temperature changed little in a reciprocating compresssor compared with its time averaged value in a cycle. Additionally, Recktenwald [9] found that the cylinder wall temperature had quite little effect on the overall performance of a reciprocating compressor. In this paper, the cylinder wall temperature is assumed to be constant, which is equal to the arithemetic mean value of the suction temperature and discharge temperature. It is worth noting that this assumption closes the energy balance equation for the control volume, but does not close the energy balance for the entire compressor. The analysis of the energy balance for the entire compressor requires the consideration of energy transfers between different components necessitating an exact temperature distribution of the cylinder wall, which will be the focus in future work.

3.2 *Valve sub-model*

The refrigerant flowing through the suction and discharge valves is shown in figure 2. The mass flow rates through the valves are determined by assuming an isentropic, compressible flow:

$$\dot{m}_{valve}=\frac{dm_{valve}}{dt}=C_{valve}A_{valve}(2\rho_{high}p_{high})^{\frac{1}{2}}\{\frac{\kappa}{\kappa-1}[(\frac{p_{low}}{p_{high}})^{\frac{2}{\kappa}}-(\frac{p_{low}}{p_{high}})^{\frac{\kappa+1}{\kappa}}]\}^{\frac{1}{2}} \tag{12}$$

The discharge coefficient C_{valve} of 0.58 is used for the suction valve and a coefficient of 0.6 is used for the discharge valve [10]. A_{valve} denotes the valve opening area for the suction and discharge valves. To calculate A_{valve} the displacement of the valve plate from the valve seat needs to be known. For a valve, there are two dominant regions, pressure dominant region and mass flux dominant region. In the pressure dominant region, the valve motion is dominated by the pressure difference across the valve. However, beyond the transition point the assumption that there is a significant pressure difference across the valve is no longer applicable and the valve motion is driven by the gas mass-flux. The transitional valve plate lift, below which the pressure force dominates, can be calculated by [11]:

$$h_{tr}=\frac{1}{4}\frac{D_{port}^2}{D_{valve}} \tag{13}$$

Where, D_{port} and D_{valve} are the diameter of valve port and valve plate, respectively. When the valve plate is located in the pressure dominant region, the valve flow area, A_{valve} can be determined by,

$$A_{valve} = \pi \cdot D_{valve} \cdot h_{valve} \tag{14}$$

Where, h_{valve} is the distance between valve plate and the valve seat. In the mass flow flux dominant region, the following equation is used instead,

$$A_{valve} = \frac{\pi}{4} D_{port}^2 \tag{15}$$

The motion of the valve plate can be modeled using a one degree-of-freedom model as shown by Kim et al.[11]:

In pressure dominant region:

$$m_{equivalent}\ddot{x} + k_{spring}x = (p_{high} - p_{low})A_{vp} + \frac{1}{2}C_D\rho_{port}V_{port}^2A_{vp} \tag{16}$$

In mass flux dominant region:

$$m_{equivalent}\ddot{x} + k_{spring}x = \rho_{port}(V_{port} - x)A_{port} + \frac{1}{2}C_D\rho_{port}V_{port}^2A_{vp} \tag{17}$$

Where, $m_{equivalent}$ is the equivalent mass of the valve plate spring; k_{spring} is the spring stiffness; C_D is the drag coefficient, which is assumed to be 1.17 as determined in Kim et al. [11]; and A_{vp} is the area of valve plate.

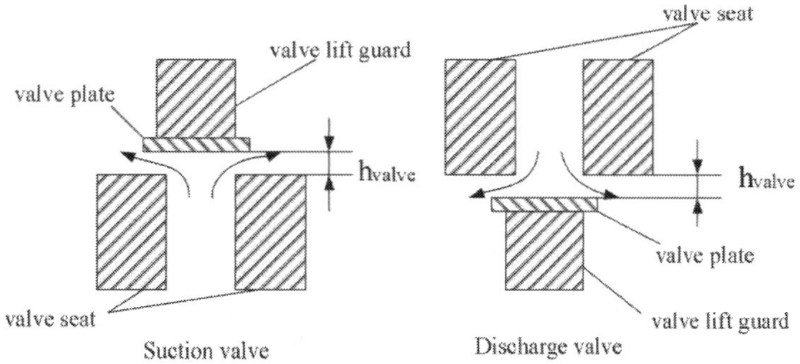

Figure 2: Mass flow at the suction and discharge valves.

3.3 Leakage sub-model

The gas leaking through the piston ring gap is modeled as an isentropic, compressible fluid flowing through an orifice. The mass flow rate of the gas leakage is calculated by:

$$\begin{cases} \dot{m}_{gap} = \dfrac{dm_{gap}}{dt} = C_{gap}A_{gap}p_u(\dfrac{2}{ZRT_u})^{\frac{1}{2}}\{\dfrac{\kappa}{\kappa-1}[(\dfrac{p_d}{p_u})^{\frac{2}{\kappa}} - (\dfrac{p_d}{p_u})^{\frac{\kappa+1}{\kappa}}]\}^{\frac{1}{2}}, \dfrac{p_d}{p_u} > 0.54 \\ \dot{m}_{gap} = \dfrac{dm_{gap}}{dt} = C_{gap}A_{gap}p_u(\dfrac{\kappa}{ZRT_u})^{\frac{1}{2}}[(\dfrac{2}{\kappa+1})^{\frac{\kappa+1}{\kappa-1}}]^{\frac{1}{2}}, \dfrac{p_d}{p_u} \le 0.54, choked \end{cases} \tag{18}$$

Where the discharge coefficient C_{gap} is assumed to be 0.86 (Furuhama et al., [12]) and the compressible factor Z is determined by using REFPROP.

IOP Conf. Series: Materials Science and Engineering **90** (2015) 012023 doi:10.1088/1757-899X/90/1/012023

4. Results and Discussion

The prototype compressor with S-RAM mechanism has five cylinders. Since the compression process of the refrigerant in each cylinder is the same, except for the compression process phase difference, the thermodynamic process in one of the cylinders is investigated. This compressor is assumed to be running at pressure ratio of 5 with the suction pressure of 1.8 MPa. The compressor motor is running at 900 RPM.

Figure 3 shows the change of piston displacement with respect to the main shaft rotation angle. The relationship between the piston velocity and main shaft rotation angle is shown in figure 4.

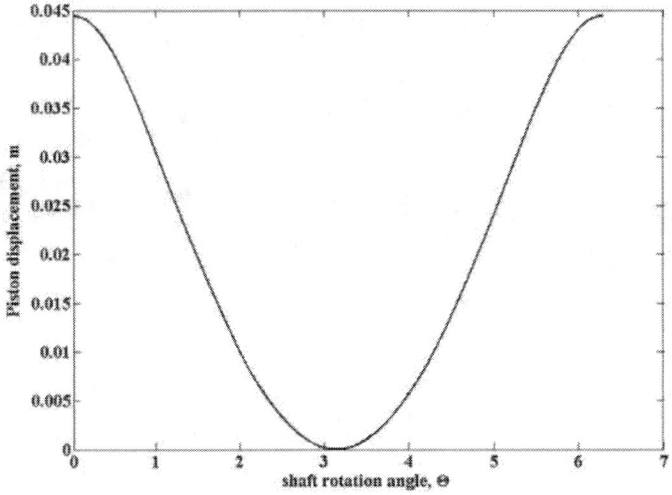

Figure 3: Piston displacement vs. main shaft rotation angle

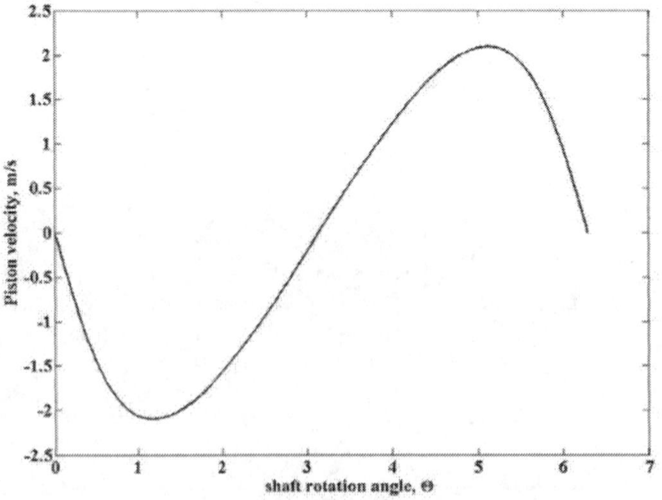

Figure 4: Piston velocity vs. main shaft rotation angle

Figure 5 shows the p-V indicated diagram for the prototype compressor from which the power losses occurring in the suction and discharge processes can be calculated. In the discharge process, the in-cylinder pressure reaches a value first higher than the pressure in the discharge chamber to overcome the valve spring force before it is open. As the discharge valve opens, some amount of refrigerant flows out of the cylinder decreasing the pressure inside the cylinder. The discharge valve starts to close when the in-cylinder pressure is not high enough to hold the resultant force of valve spring and discharge chamber pressure. Hence, the discharge valve moves back and forth during the discharge process as shown in figure 6. Similarly, at the beginning of the suction process, the pressure in the cylinder is lower than the pressure in the suction chamber, in order to create the necessary pressure difference across the suction valve to overcome the suction valve spring force. As the displacement of the suction valve increases, more and more refrigerant flows into the cylinder leading to the increment of in-cylinder pressure. It is worth noting that the pressure difference across the suction valve at the beginning of suction process is much lower than that when the discharge valve starts to open. The reason is that lower spring stiffness is used for the suction valve spring.

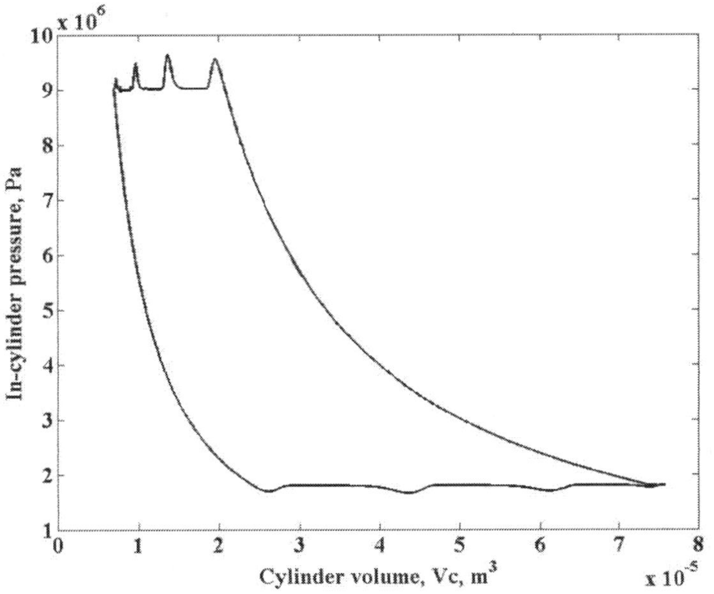

Figure 5: P-V indicated diagram of the compressor

Another interesting phenomenon is found in that the pressure fluctuation during the suction process is much smoother than that occurring in the discharge process although the vibrations of the valve plates are pretty similar for both suction and discharge valves as indicated in figure 6. The main reasons include that the in-cylinder volume is much bigger while the refrigerant temperature is much lower when the suction valve starts to open compared to the volume and temperature when the discharge valve begins to open.

Figure 6 shows the change of suction/discharge valve plate displacement with respect to the compressor main shaft rotation angle. It can be seen that the suction valve is still open when the compression process starts. The discharge valve also keeps open when the re-expansion process begins. Back flow occurs for both suction and discharge valves due to the late closing of the valve plate. The closing time of the valve plate can be adjusted by optimizing the valve spring stiffness and valve lift.

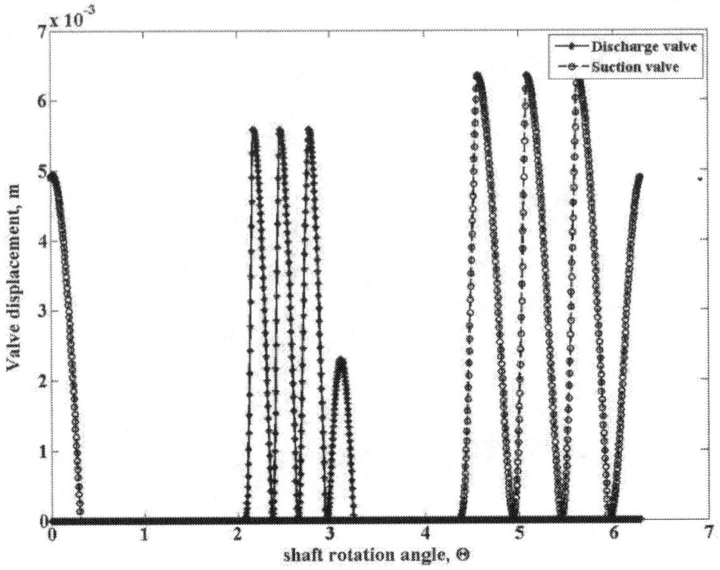

Figure 6: Suction/discharge valve plate displacements vs. main shaft rotation angle

5. Conclusions

A simulation model for a prototype CO_2 compressor using the S-RAM mechanism is presented. This initial model consists of geometry and kinematic model and compression process model. Valve, leakage and heat transfer model are included in the compression process model.

The in-cylinder pressure is investigated for one operating condition as an example. The suction/discharge valve displacements are studied. The fluctuation of the refrigerant pressure inside the cylinder is smoother for the suction process than for the discharge process. Both the suction valve and discharge valve are found to close later than expected, which may be improved by changing the valve spring stiffness and valve lift.

In this initial model, the dynamic analysis of the S-RAM mechanism is not included, which will be considered in future work. In addition, detailed experiments of the compressor performance and valve plate motion will be conducted to validate the model.

6. Nomenclature

A	area [m^2]	μ	dynamic viscosity
C	discharge coefficient [-]	ρ	density
D	diameter [m]	τ	thermal conductivity of refrigerant in control volume
Q	quantity of heat transferred to control volume through boundary from its ambient	ω	angular speed of the main crankshaft
R	length	*Subscripts*	
S	piston stroke	0	clearance volume
T	temperature	c	refrigerant in control volume

U	velocity	*cyl*	cylinder
V	volume	*d*	downstream
Z	compact factor of carbon dioxide	*dis*	discharge
h	unit enthalpy, or valve lift	*H*	hydrodynamic
m	mass	*high*	high pressure side
p	pressure	*li, lo*	leak in and out
u	unit internal energy	*low*	low pressure side
v	specific volume	*mean*	mean velocity of piston
x	piston displacement	*oil*	lubricating oil
Greek letters		*p*	piston
α	Relative clearance volume	*shaft*	crank shaft
β	incline angle between the connecting rod and compressor main shaft centerline	*suc*	suction
θ	crank angle	*u*	upstream
κ	specific heat ratio	*valve*	valve

7. References

[1] Below J F, Miloslavich D A 1984. Dynamics of the swash plate mechanism. *Proc. of the Int. Comp. Eng. Conf.* Purdue University, West Lafayette, IN USA, pp. 76-81.

[2] Ishii N, Abe Y, Taguchi T, Kitamura T 1990. Dynamic behavior of variable displacement wobble plate compressor for automotive air conditioners. *Proc. of the Int. Comp. Eng. Conf.* Purdue University, West Lafayette, IN USA, pp. 345-353.

[3] Tojo K, Takao K, Nakamura Y 1988. A study on the kinematics of a variable displacement compressor for automotive air conditioners. *Proc. of the Int. Comp. Eng. Conf.* Purdue University, West Lafayette, IN USA, pp. 496-504.

[4] Tian C, Liao Y, Li X 2006. A mathematical model of variable displacement swash plate compressor for automotive air conditioning system. *Int. J. Refrigeration* **29**, 270-280.

[5] NIST. 2010. *Reference fluid thermodynamic and transport properties (REFPROP) Version 9.0.*

[6] Annand W J D 1963. Heat transfer in the cylinder of reciprocating internal combustion engines. *Inst. of Mech. Eng.* **177**(36), 973-996.

[7] Todescat M L, Fagotti F, Prata A T, Ferreira R T S 1992. Thermal energy analysis in reciprocating hermetic compressors. *Proc. of the Int. Comp. Eng. Conf.* Purdue University, West Lafayette, IN USA, pp. 1419-1428.

[8] Adair R P, Qvale E R, Pearson J T 1972. Instantaneous heat transfer to the cylinder wall in reciprocating compressors. *Proc. of the Int. Comp. Eng. Conf.* Purdue University, West Lafayette, IN USA, pp. 521-526.

[9] Recktenwald G W 1989. *A study of heat transfer between the walls and gas inside the cylinder of a reciprocating compressor.* PhD. Thesis. University of Minnesota. 197-199.

[10] Lin M, Sun S Y 2006. Theory of the piston compressor. Xi'an Jiaotong University Press. Xi'an, China.

[11] Kim J H, Groll E A 2007. Feasibility study of a bowtie compressor with novel capacity modulation. *Int. J. Refrigeration* **30**, 1427-1438.

[12] Furuhama S, Tada T 1961. On the flow of the gas through the piston-rings (1st report, the discharge coefficient and temperature of leakage gas). *Bulletin of JSME* **4** (16), 684-690.

IOP Conf. Series: Materials Science and Engineering **90** (2015) 012024 doi:10.1088/1757-899X/90/1/012024

Systematic parametric design/calculation of the piston rod unit

V.Kacani
Leobersdorfer Maschinenfabrik GmbH, Vienna Austria

E-mail: vasillaq.kacani@lmf.at

Abstract. In this article a modern and economic method for the strength calculation of the piston rod unit and its components under different operating conditions will be presented. Herefore the commercial FEA - Software will be linked with the company-owned calculation tools. The parametric user input will be followed by an automatic Pre- and Postprocessing. Afterwards the strength calculation is processed on all critical points of the piston rod connection, assisted by an extra module, based on general standards and special codes for reciprocating compressors. In this process most arrangements of the piston rod unit as well as the special geometries of the single-components (piston, piston rod and piston nut) can be considered easily. In this article the modeling of the notches, especially on the piston rod, piston as well as the piston nut will be covered in detail.

1. Introduction

The paper investigates the parametric design and calculation of all components of piston rod unit considering the contact properties of all clamped parts and different piston design. The piston rod unit generally includes piston, piston nut, ring and piston rod. On crankcase side the piston rod is connected to the crosshead.

These components are under extremely high loads during the operation of the reciprocating compressor. The forces acting on the components are the pre-load force, the gas force, the thermal related forces as well as the mass force. During one revolution of the crankshaft the gas and mass forces are changing their value and the direction. Additionally for each crankshaft angle the forces are different on different sections of the piston rod unit. That means the strength calculation must be performed for each angular step with the actual pressure and acceleration at the local section and for each notch on all components of the piston rod unit.

Furthermore the boundary conditions such as geometry, embedding of the contact surfaces, materials, strength grade, tight techniques as well as different operating cases of the reciprocating compressor are into consideration.

In this paper a systematic design and calculation of the piston rod unit will be presented. This consists of three main steps: structure modelling, (preprocessing), definition of boundary condition, meshing, loads apply, calculation for each crank angle and, post-processing (results). All steps are realized automatically using manufacturer calculation/design tools, commercial programming software VB6, C++, FEA-Software[3], Scripting Language APDL[3] (ANSYS Parametric Design Language) as well as guideline and standards.

The strength verification will be performed according the different guidelines[1,2,4,5].

Content from this work may be used under the terms of the Creative Commons Attribution 3.0 licence. Any further distribution of this work must maintain attribution to the author(s) and the title of the work, journal citation and DOI.
Published under licence by IOP Publishing Ltd

A calculation software was developed to perform an automatic strength analysis of the components of the piston rod unit. The special features of the software are:

- Reduction of the simulation time
- Increasing the quality of calculation
- Calculation of stress curves at each notch along the notch angle α
- Systematic calculation/design and fault prevention through certified software.

2. Pre-processing

The software consists of several calculation modules and their interfaces between the components. The compressor data (like type, frame size, piston rod design, crosshead, connection rod etc) material-data of the main components (modulus of elasticity, ultimate tensile strength, yield strength, Poisson ratio, fatigue stress values, thermal expansion coefficient etc) can be generated automatically and internally from machine- and material databank.

The basic module is the manufacturer program for the design/calculation of the reciprocating compressors. The interface enables every user after the selection and design of a specific compressor, to use the software for the piston unit calculation.

The main data/parameters are a result of the previous compressor selection:

- Stroke, type of piston (one piece, two piece, welded piston etc), compressor speed..
- Actual pressure variation over the crankshaft angle, in each cylinder compartments (HE, CE, idle or active compartment, SV-Unloader .).
- Piston rod parameters: size, neck diameter, thread diameter, surface roughness, heat treatment..
- Piston nut data..

Data can be input/edited interactively by the user. Fig. 1 shows a screen shot of the user input form with the sketch of the two pieces piston and various input parameters.

- Notch geometry, radius, element type, element size or number of elements. Distance □s normal to the notch surface for the calculation of the stress gradient required for the fatigue analysis.

- Geometry of contact area between piston, piston rod and piston nut. Type and size of contact elements. Friction coefficient.
- Piston geometry (2D, 3D-modell)
- Welded piston with ribs
- Temperature distribution.
- The assembly preload force of the unit

Hence the geometry and the loads of the piston rod unit are determined. The data are written to a temporary file. The next steps are the generating the FE-model, the definition of boundary condition, applying of the loads, the calculation, as well as the stress evaluation. All this steps are effected automatically. A special interface for the FEA-simulation in the script language (APDL) was generated.

IOP Conf. Series: Materials Science and Engineering **90** (2015) 012024 doi:10.1088/1757-899X/90/1/012024

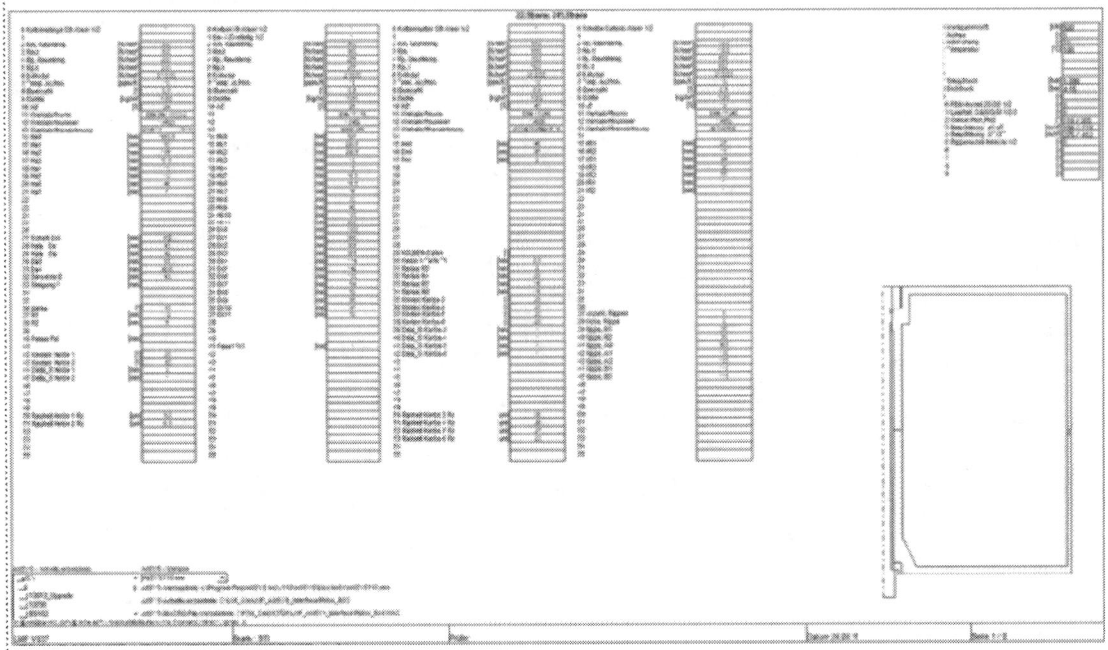

Figure 1. User Input Form

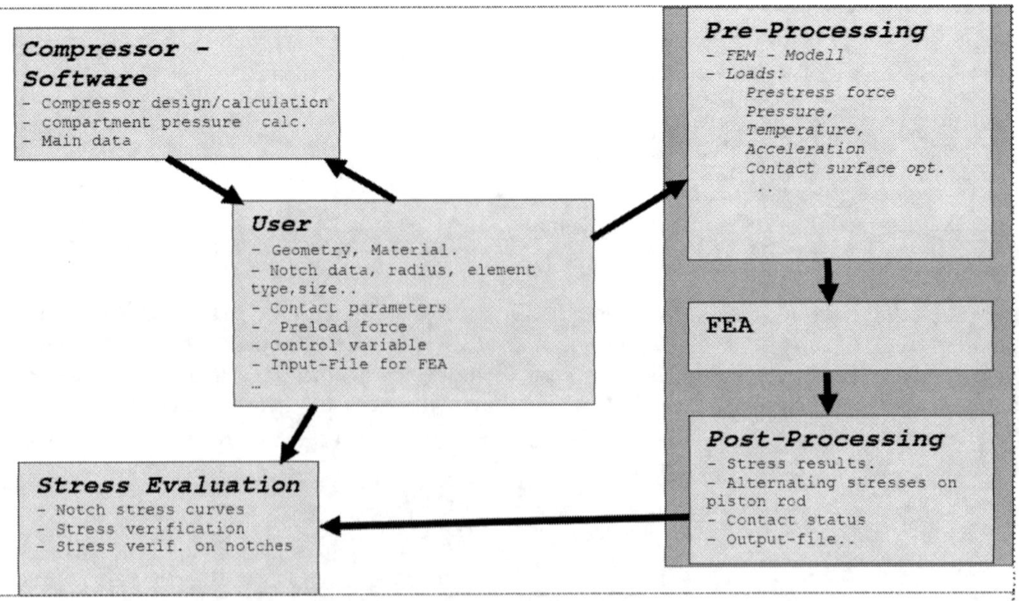

Figure 2. Work process flowchart

IOP Conf. Series: Materials Science and Engineering **90** (2015) 012024 doi:10.1088/1757-899X/90/1/012024

3. Critical stress zones and contact areas

Figure 3 shows the critical high stress zones 1 to 9 and the contact areas I, II, III,…VII.

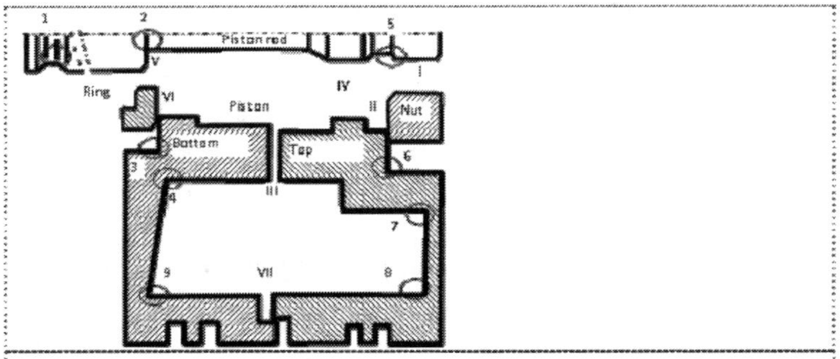

Figure 3. Stress zones and contact areas.

The critical notches 1, 2 and 5 are in the piston rod. Notch 1 on crosshead side of piston rod is a conical shoulder with radius. Notch 2 is a square shoulder with fillet (radius) in circular shaft. 5 represents a V-notch in a circular shaft (thread).

Notches 3, 4, 9 are located in bottom half of the piston 6, 7, 8 are in the top half of piston. These notches are representing changes on cross-section of piston with fillet of radius r. For welded piston the fillet radii shall be assumed to be 1mm (Radaj[5]).

4. Loads applied to the structure

The first load step is the preload force on the piston rod due to tightening of the piston nut. This load can be applied in ANSYS[3] with pretension elements. This is the basic load. All other loads are superimposed on the basic load. When specifying the minimum axial assembly forces the loss of the preload force[1] due to the plastic deformation of the contact surfaces must be taken into consideration. In API 618 the recommendation for the minimum preload is 1.5 times the maximum allowable continuous piston rod load. The preload force must be checked for each application under the actual operating loads: pressure, acceleration and temperature. Often the required preload force is much higher than the API recommendation. In The status of all contact areas, pressure and gap must be checked carefully for all operation condition. The prestress force has to prevent lifting or slide of all contact surfaces.

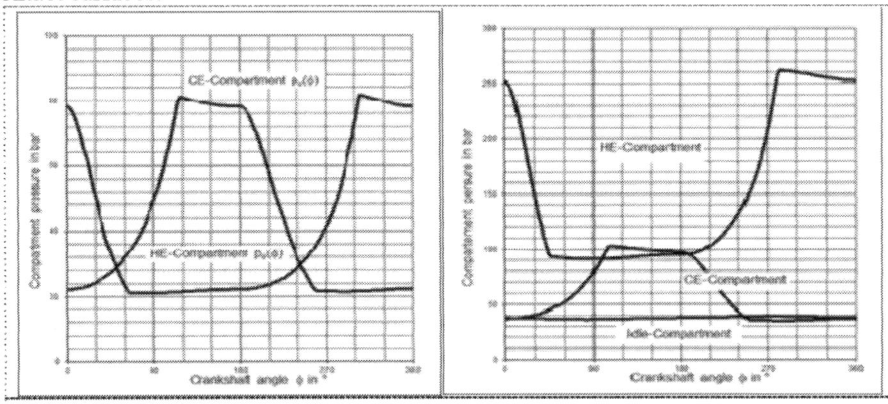

Figure 3, 4. Pressure versus crankshaft angle for double acting and stepped piston

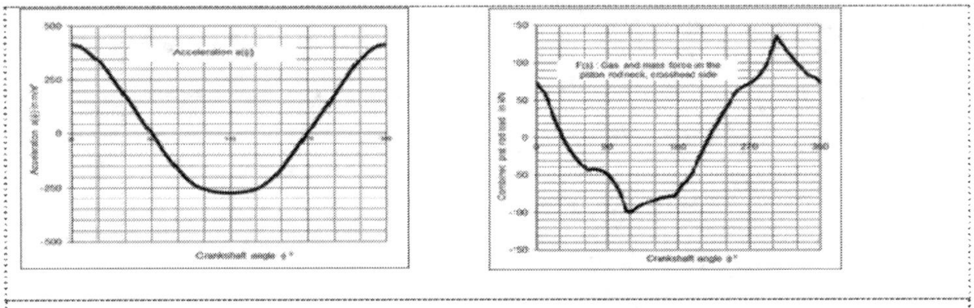

Figure 5, 6. Acceleration and combined rod load at the piston rod neck

Figure 3 and 4 shows the typical pressure curves $p_o(\phi)$, $p_u(\phi)$ on CE-, HE- and idle-compartment for double acting and for stepped piston. These loads are applied on the corresponding areas of the structures (Fig. 7). Figure 5 represents the acceleration and Figure 6 shows the resulting combined rod load at the piston rod neck crosshead side.

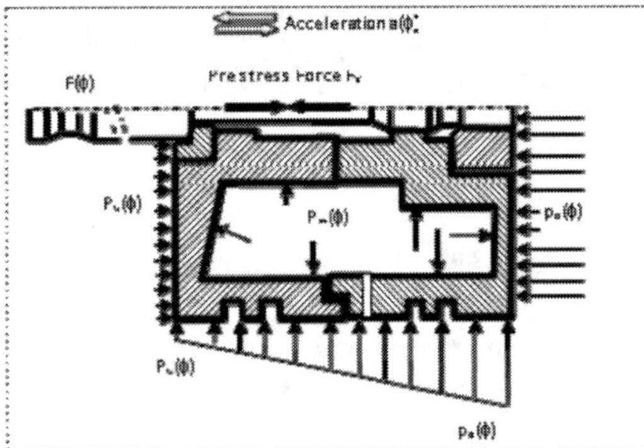

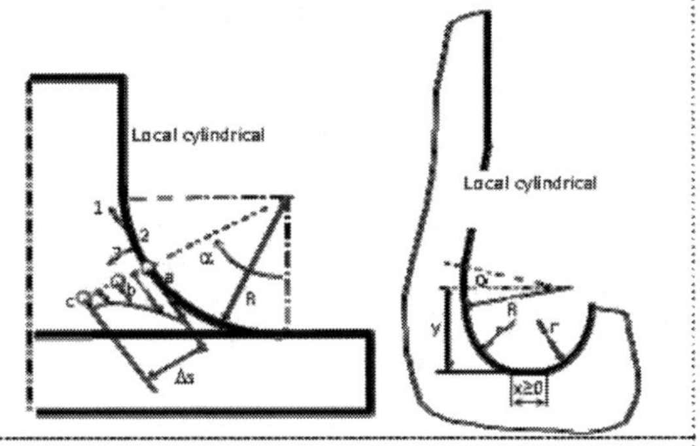

Figure 7. Loads on the piston rod unit: Preload Fv, pressure $p_u(\phi)$, $p_o(\phi)$ and acceleration $a(\phi)$

Figure 8. Typical notch shape and required parameters

5. Notch geometry, parameters and stress evaluation.

The main parameters of notches are (Fig.8):

- Shape of notch, local coordinate systems
- Geometry parameters: R, r, α, x, y
- Depth in vertical direction Δs
- Number of elements in circumferential direction 1
- Number of elements in radial direction 3

The principal stresses in the local coordinate system are calculated in three directions 1, 2, and 3 along the path (angle α) and for each step of crankshaft ($\Delta\phi$ acc. [1] is 5°). The direction 1 and 2 are always parallel to the free surface. The third direction is perpendicular to free surface of the component.

The stresses are calculated on three points: position a on the free surface $\sigma_a(\alpha,\phi)=(\sigma_{1a},\sigma_{2a},\sigma_{3a})$, position b in a depth of $\Delta s/2$ $\sigma_b(\alpha,\phi)=(\sigma_{1b},\sigma_{2b},\sigma_{3b})$, as well as position c in a depth of Δs $\sigma_c(\alpha,\phi)=(\sigma_{1c},\sigma_{2c},\sigma_{3c})$, (Fig.8). The stress value for σ_1 in perpendicular direction (Δs) to the free surface can be approximated as a polynom (second order): $\sigma_1=A_0 + A_1.s + A_2.\ S^2$. The coefficients A_0, A_1 and A_2 can be calculated easy from the three stress values (position a,b,c) established by the FEA-Analysis. Finally the stress gradient can be determined, as required in [2,5], for the strength verification. The formulas are listed in appendix. The equation for the gradient is:

$$\frac{d\sigma_1}{ds} = A_1 + 2 \cdot A_2 \cdot s$$

6. Meshing strategy of the structure

The meshing of the structure must be practice-oriented and the model should have a reasonable size. On the one hand the meshing of the components should be as coarse as possible to reduce the simulation time and on the other hand as fine as necessary with reference to the engineering tasks. For notches (Fig. 10) and the contact areas a fine controlled meshing is used to get the correct information for the required stresses, contact status as well as the contact pressure. For other areas of the structure a free coarse mesh can be used. The element type and size can be adjusted by user. The following figure shows the different piston design (Fig.9): stepped piston 2D-model, two pieces double acting piston 2D and 3D- model, double acting welded piston with ribs 3D-model. All these models can be calculated automatically with the computer program.

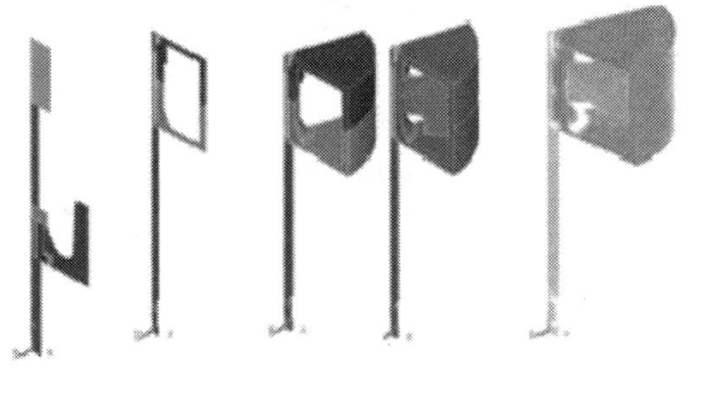

Figure 9. Piston designs

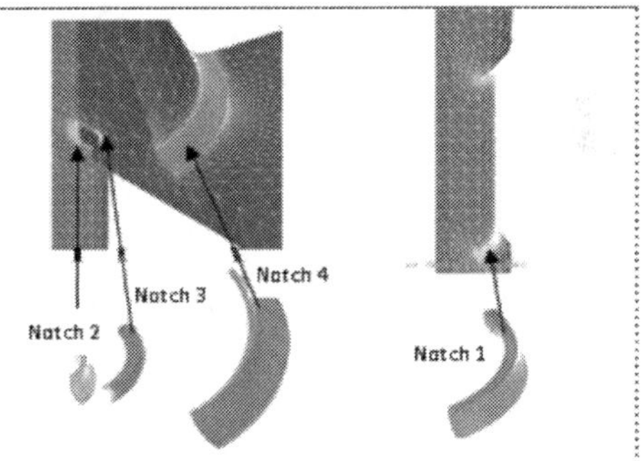

Figure 10 . Controlled fine and coarse mesh.

7. Finete Element Analysis and results

The first simulation is the calculation of the prestressed structure under the preload force F_v. The next calculations are successively under operating condition: pressure, temperature as well as acceleration for each crankshaft angle.

The main results of the Finite Element Analysis are the forces / stresses at the piston rod neck, thread (piston side and crosshead side) and the stresses on all notches.

The strength verification of pre-stressed connection piston rod, ring, piston and piston nut is performed according [1,4] guideline. The reduction of preload force of bolted joint due to the deformation of the contact surfaces is taken into consideration. From the FEA the actual preload force, the amplitude of the alternating force/stresses, maximum forces/stresses in the piston rod neck /thread as well as the resilience of the superimposed clamped parts and piston rod can be determined. From these values the safety factors are calculated. Furtherore the status and the pressure on all contact areas is checked.

The next step is the stress evaluation on the notches. From the FEA results the distribution (stress curves) of stresses on the free surface, on the depth $\Delta s/2$ as well as on the depth Δs along the angle α for each notch and crankshaft angle ϕ are calculated. Further the required alternating stress amplitude σ_a, average stress σ_m as well the stress gradient $d\sigma/ds$ for the fatigue strength verification according [2] are calculated.

Figure 11 to 14 shows the stress tangential to the free surface (direction 1) along the angle α in Notch 1, 2, 3 and 4 for two piston positions: Bottom-Dead-Center BDC and Top-Dead-Center TDC.

IOP Conf. Series: Materials Science and Engineering **90** (2015) 012024 doi:10.1088/1757-899X/90/1/012024

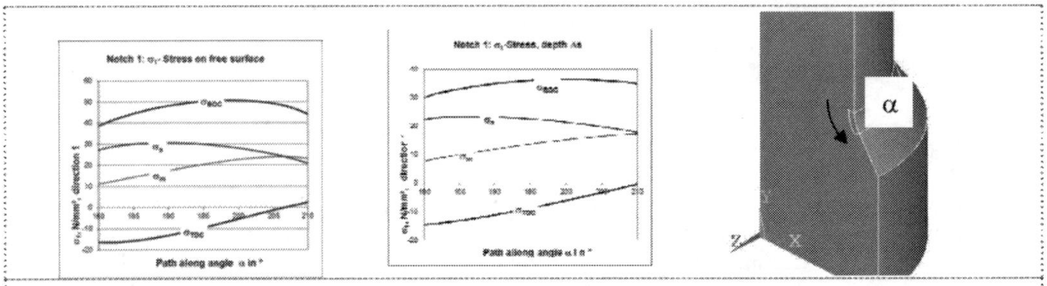

Figure 11. Notch 1, Tangential stress curves at BDC, TDC, amplitude and average stress on free surface and in depth Δs.

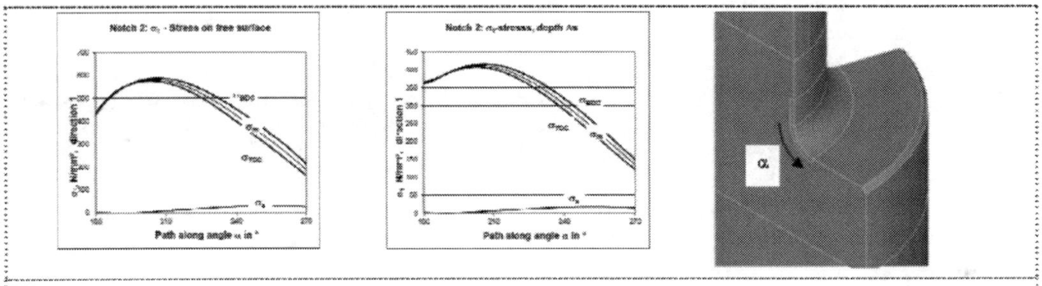

Figure 12. Notch 2, Tangential stress curves at BDC, TDC, amplitude and average stress on free surface and in depth Δs.

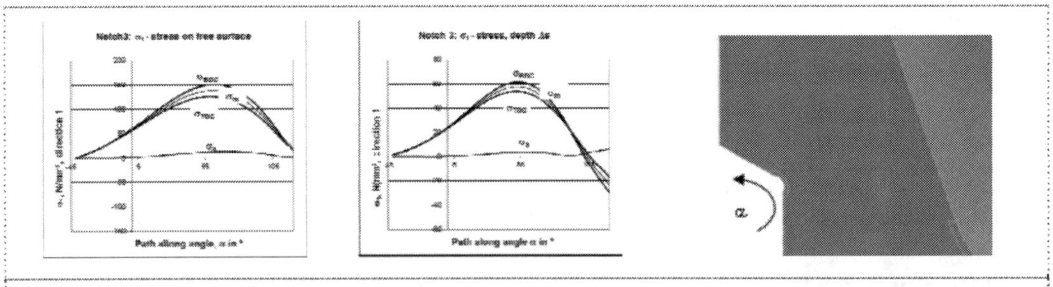

Figure 13. Notch 3, Tangential stress curves at BDC, TDC, amplitude and average stress on free surface and in depth Δs.

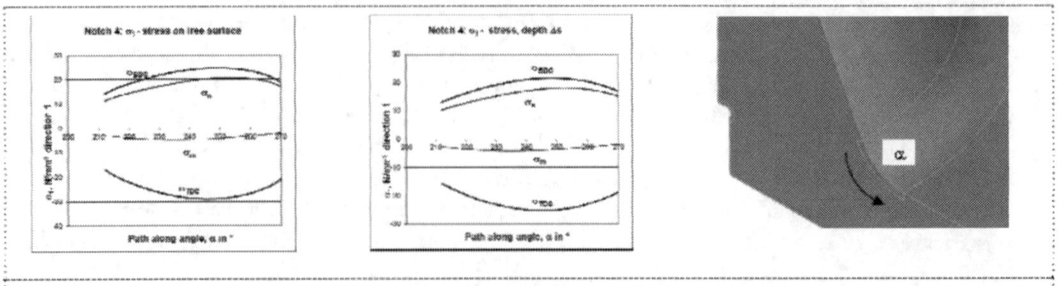

Figure 14. Notch 4, Tangential stress curves at BDC, TDC, amplitude and average stress on free surface and in depth Δs.

IOP Conf. Series: Materials Science and Engineering **90** (2015) 012024 doi:10.1088/1757-899X/90/1/012024

8. Summary

The components of the piston rod unit are under very high dynamic loads due to: pressure, temperature and acceleration. Depending on the operation mode of the reciprocating compressor, normal operation, SV-Unloaders, VSD, variable operation conditions the load profile of piston rod unit changes. The individual single calculation for each load step requires an enormous effort. The traceability and documentation of analysis would be also a big effort. The method presents a so-called expert system connects different tested and proved modules like company tools, standards, company owned software as well as commercial software together. The calculation is carried out automatically. The main steps of the calculation are:

- Compressor calculation
- User input data
- Modelling of the structure
- Finite Element Analysis
- Verification of the preload force
- Status of the contact surfaces
- Stress verification on the relevant critical notches.

Instead of the suggested factor 1.5 in API 618 5th ed., para.6.10.1, a minimum factor of 1.75 is recommended.

References

[1] VDI 2230 Richtlinien. Systematic calculation of high duty bolted joints. Joints with one cylindrical bolt.

[2] FKM-Richtlinie. Rechnerischer. Festigkeitsnachweis für Maschinenbauteile.

[3] ANSYS , Simulation Software.

[4] API Standard 618. Reciprocating Compressors for Petroleum, Chemical, and Gas Industry Services.

[5] Hobbacher. Empfehlungen zur Schwingfestigkeit geschweißter Verbindungen und Bauteile.

Appendix, Stress verification according FKM [2]

Static stress verification with local stresses

-Material Properties

Ultimate and yield strength of material from standard $R_{m,N},\ R_{p,N}$.

Technology Factor $K_{d,m}, K_{d,p}$

Constants $a_{d,m},\ a_{d,p}$

Anisotropy factor K_A

Effective diameter $d_{eff}, d_{eff,N,m}, d_{eff,N,p}$

Ultimate and yield strength of prefab. piece from the draw. $R_M = K_{d,m} \cdot K_A \cdot R_{m,N}, R_p = K_{d,p} \cdot K_A \cdot R_{m,p}$

Tension and shear stress factor $f_\sigma,\ f_\tau$

Compressive strength, shear strength $R_{c,m} = f_\sigma \cdot R_m,\ R_{c,p} = f_\sigma \cdot R_p\quad R_{s,m} = f_\tau \cdot R_m, R_{s,p} = f_\tau \cdot R_p$

-Characteristic construction values

Plastic stress concentration factor $K_{p,\sigma_1}, K_{p,\sigma_2}, K_{p,\tau}$

Plastic notch sensitivity factor $n_{pl,\sigma_1}, n_{pl,\sigma_2}, n_{pl,\tau}$

Stress concentration factor squer shoulder $K_{t,ZD}, K_{t,b}, K_{tt}$

Construction factors for block-shaped pieces $K_{SK,\sigma_1} = \frac{1}{n_{pl,\sigma_1} \cdot K_{NL}} \; K_{SK,\sigma_2} = \frac{1}{n_{pl,\sigma_2} \cdot K_{NL}}, K_{SK,\sigma_3} = 1$

-Component strength

Static local strength $\sigma_{SK,1} = \frac{f_\sigma \cdot R_m}{K_{SK,\sigma_1}}, \sigma_{SK,2} = \frac{f_\sigma \cdot R_m}{K_{SK,\sigma_2}}, \sigma_{SK,3} = \frac{f_\sigma \cdot R_m}{K_{SK,\sigma_3}}$

-Safety factors

Overall safety factor j_{Ges}

-Stress characteristics

The stress characteristics are result of FEA. The maximum extreme stresses will be evaluated from the stress curves in each notch Fig.11, 12, 13, 14 $\sigma_{Max,ex,1}, \sigma_{Max,ex,2}, \sigma_{Max,ex,3}$

-Stress verification

Static degree of utilization in three directions: $a_{SK,\sigma_1} = \left| \frac{\sigma_{max,ex,1}}{\sigma_{SK,1}/j_{ges}} \right| \le 1 \quad a_{SK,\sigma_2} = \left| \frac{\sigma_{Max,ex,2}}{\sigma_{SK,2}/j_{ges}} \right| \le 1 \quad a_{SK,\sigma_3} = \left| \frac{\sigma_{Max,ex,3}}{\sigma_{SK,3}/j_{ges}} \right| \le 1$

<u>Fatigue strength calculation with local stresses</u>

-Stress characteristics

The stress characteristics are result of FEA. The stress values: principal, maximum, minimum, mean as well as amplitude will be evaluated from the stress curves in each notch Fig.11, 12, 13, 14

Principal stresses in three directions $\sigma_{1,ZD} = \sigma_1, \sigma_{2,ZD} = \sigma_2, \sigma_{3,ZD} = \sigma_3$

Maximum and minimum stresses $\sigma_{min,1}, \sigma_{min,2}, \sigma_{min,3} \quad \sigma_{max,1}, \sigma_{max,2}, \sigma_{max,3}$

Stress amplitude, mean stresses in three directions, i=1,2,3 $\hat{\sigma}_{a,i} = \frac{\sigma_{max,i} - \sigma_{min,i}}{2} \quad \sigma_{m,i} = \frac{\sigma_{max,i} + \sigma_{min,i}}{2}$

Stress ratio $R_{\sigma,i} = \frac{\sigma_{max,i}}{\sigma_{min,i}}$

-Material characteristics

Tension, compressive and shear fatigue strength factor $f_{W,\sigma} \; f_{W,\tau} \; \tau_{W,Zd} = f_{W,\tau} \cdot \sigma_{W,Zd}$

$$\sigma_{W,Zd} = f_{W,\sigma} \cdot R_m$$

Polynomial function for the stress in direction i=1,2,3 $\sigma_i(s) = A_{0,i} + A_{1,i} \cdot s + A_{2,i} \cdot s^2$

-Construction characteristics

Stress gradient on free surface of the notch, direction i=1,2,3 $\overline{G}_{\sigma,i} = \frac{d\sigma_i}{ds} = A_{1,i}$

Constants, mean arithmetic roughness $a_G, b_G \quad A_{R,\sigma}, R_{m,N,Min} \quad R_Z \quad \widetilde{K}_f, K_{NL,E}$

Notch sensitivity factor

$$n_{\sigma,i} = \begin{cases} 1 + \overline{G}_{\sigma,i} \cdot 10^{-\left(a_G - 0.5 + \frac{R_m}{b_G}\right)} & \text{für} \quad \overline{G}_{\sigma,i} \le 0.1 \\ 1 + \sqrt{\overline{G}_{\sigma,i}} \cdot 10^{-\left(a_G + \frac{R_m}{b_G}\right)} & \text{für} \quad 0.1 \le \overline{G}_{\sigma,i} \le 1 \\ 1 + \sqrt[4]{\overline{G}_{\sigma,i}} \cdot 10^{-\left(a_G + \frac{R_m}{b_G}\right)} & \text{für} \quad 1 \le \overline{G}_{\sigma,i} \le 100 \end{cases}$$

Roughness factor for normal and shear stresses

$$K_{R,\sigma} = 1 - A_{R,\sigma} \cdot \log(R_Z) \cdot \log\left(\frac{2 \cdot R_m}{R_{m,N,\min}}\right)$$

$$K_{R,\tau} = 1 - f_{W,\tau} \cdot A_{R,\sigma} \cdot \log(R_Z) \cdot \log\left(\frac{2 \cdot R_m}{R_{m,N,\min}}\right)$$

Surface layer factor, coat factor construction factor K_V, K_S

$$K_{WK,\sigma_i} = \frac{1}{n_{\sigma_i}} \cdot \left[1 + \frac{1}{\widetilde{K}_f} \cdot \left(\frac{1}{K_{R,\sigma}} - 1\right)\right] \cdot \frac{1}{K_V \cdot K_S \cdot K_{NL,E}}$$

-Component strength

Alternate fatigue strength direction i =1, 2, 3

$$\sigma_{WK_i} = \frac{\sigma_{W,zd}}{K_{WK,\sigma_i}}$$

Residual stress factor, constants, mean stress sensitive factor $K_{E,\sigma} \quad a_M, b_M$

$$M_\sigma = a_M \cdot 10^{-3} \cdot R_m + b_M, \; M_\tau = f_{W,\tau} \cdot M_\sigma$$

Mean stress factor

$$K_{AK,\sigma_i} = \begin{cases} \frac{1}{1 - M_\sigma} & \text{Zone I } R_\sigma \succ 1 \\ \frac{1}{\left(1 + M_\sigma \cdot \frac{\sigma_{m_i}}{\sigma_{a_i}}\right)} & \text{Zone II } -\infty \le R_\sigma \le 0 \\ \frac{\left(\frac{1 + M_\sigma/3}{1 + M_\sigma}\right)}{\left(1 + \frac{M_\sigma}{3} \cdot \frac{\sigma_{m_i}}{\sigma_{a_i}}\right)} & \text{Zone III}, 0 \prec R_\sigma \prec 0.5 \\ \frac{3 + M_\sigma}{3 \cdot (1 + M_\sigma)^2} & \text{Zone IV } R_\sigma \succ 0.5 \end{cases}$$

Fatigue strength amplitude $\sigma_{AK_i} = K_{AK,\sigma_i} \cdot K_{E,\sigma} \cdot \sigma_{WK_i}$

Endurance strength factor $K_{BK,\sigma_i} \quad (K_{BK,\sigma} = 1 \text{ für } N = \infty)$

Endurance strength amplitude $\sigma_{BK_i} = K_{BK,\sigma_i} \cdot \sigma_{AK_i}$

-Safety factors

Safety factor j_D

- Fatigue stress verification

Degree of utilization in three directions i=1,2,3

$$a_{BK,\sigma_i} = \frac{\hat{\sigma}_{a_i}}{\left(\frac{\sigma_{BK_i}}{j_D}\right)} \le 1$$

A Diagnosis method of the small end fault on reciprocating compressor connecting rod

Zhinong Jiang [1] **Zhiwei Mao**[1] **Ziyun Yao**[2] **Jinjie Zhang**[3]

1. Diagnosis and Self-Recovery Engineering Research Center, Beijing University of Chemical Technology, Beijing, 100029, P. R. China;

2. Petro China Beijing Gas Pipeline Co., Ltd., Beijing, 100101, P. R. China

3. College of Information Science and Technology, Beijing University of Chemical Technology, Beijing, 100029, P. R. China;

E-mail: zjj87427@163.com

Abstract. The connecting rod is the key moving part of a reciprocating compressor, of which the stress state is extremely complicate and the wear fault of the small end is always a bottleneck problem in the field of fault monitoring and diagnosing. This paper is aimed to present a new method to diagnose the above wear fault. Firstly, a contact model of a clearance in the revolute joint of the small end of a connecting rod bearing (SECRB) was established and a multi-body simulation tool was utilized to simulate the slider-crank mechanism with a clearance, from which the dynamic influence of wear gap in SECRB of a slider-crank mechanism was obtained. Based on the study above, we extracted the characteristics of the wear fault of SECRB and then proposed a brand new approach to monitoring and diagnosing this wear fault by analyzing the angle domain of vibration signals. The availability was verified by conducting an experiment on a reciprocating compressor. And the experimental results show that this method can not only accurately diagnose the wear fault of SECRB but also approximately estimate its severity. This study laid a foundation for the online monitoring and early warning of this fault.

1. Introduction

Reciprocating compressor is one of indispensable key equipments in many sectors of the national economy, such as chemical process, gas transmission, power engineering, and other important occasions. However, due to the attributes such as complex structure, a lot of vibration sources, and many vulnerable parts of reciprocating compressors, serious accidents may happen frequently, which always cause irreversible consequences. The connecting rod as one of the key moving parts has complex motion states and poor working conditions, as a consequence of which, the wear faults often occur in its small end. In the early stage of the failure, dynamic behaviors of the unit will not be impacted seriously, but when the wear deteriorates to a certain degree, it may result in many malignant faults, such as cylinder knocking, scuffing, hitting, and even piston rod fracture. Therefore, researching the methods that can monitor and diagnose the wear faults of SECRB in the early stage on online has important theoretical significance and practical engineering application values.

Content from this work may be used under the terms of the Creative Commons Attribution 3.0 licence. Any further distribution of this work must maintain attribution to the author(s) and the title of the work, journal citation and DOI.
Published under licence by IOP Publishing Ltd

Nowadays, the studies, surrounding the diagnosis methods of wear faults of mechanical system, are mainly divided into two aspects[1]: oil analysis and vibration signal analysis,(both of which have some inherent drawbacks). the former can be valid in the detection of wear faults of the unit, but it can't locate where the wear occurs, making it difficult to guide the maintenance effectively, and apply it in the on-line monitoring system currently. In recent years, the latter has made great processes with the development of the dynamic theory of mechanical system with clearance joint, on which many researchers have done a lot of theoretical and experimental investigations. Plores et al. [2-4] studied and compared the models of dry friction and lubrication and the models of rigid body and flexible body, and did a parametric study on the dynamic response of planar multi-body systems with multiple clearance joints. Schwab et al[5]compared the revolute joint clearance models in the dynamic analysis of rigid and elastic mechanical systems, and presented a procedure to estimate the maximum contact force during collision in the impact model. Liu[6]presented a approximate model that provided a more direct and effective formula to describe the properties of the contact in the cylindrical joint with clearances based on the FEM analysis. Wenqing Lu [7] and Troy Feese et al [8] researched the fault prognosis methods for reciprocating machinery. However, most of those research works were aimed at the dynamic impact of different clearance models and the optimization of these models. The reports that apply the slider-crank mechanism with clearance joint in SECRB to actual reciprocating compressor, on whose piston exerted constantly changing gas force in working condition, and put forward an effective wear fault diagnosis method of SECRB, which can be applied to the engineering practice, have not been seen.

In this paper, a theoretical model of clearance joint in the SECRB was established based on the modern contact theory, and a practical model in a real reciprocating compressor was built in a multi-body dynamic simulation tool, in which many simulations were performed with the changing gas forces exerted on the piston. Based on the simulation results, from which the characteristics of the wear fault of SECRB were extracted, a new approach to monitor and diagnose the wear fault of SECRB through analyzing the angle domain of vibration signals has been put forward. In order to verify the validity and practicability of this method, an experiment was conducted on a real reciprocating compressor, from which the simulation entity model sizes were obtained. The experimental results reveal that this method is simple and effective to diagnose such wear faults, predicting strong practical value for engineering applications.

2 .Contact calculations

Due to the needs of the assembly and the relative rotation, there is a certain gap in the cylindrical joint, which is so small that it hardly makes any influence on the dynamic response on mechanical system with good lubrication conditions[9]. However, this gap will enlarge when wear become seriously caused by various reasons such as bad lubrication conditions. Meanwhile, the dynamic response on the mechanical system can be significantly enhanced[9]. Therefore, in order to study the increase impact, a cylindrical clearance joint contact model should be established.

2.1. Model of cylindrical clearance joint

Figure 1 shows the cylindrical clearance joint in real mechanism of the reciprocating compressor reciprocating compressor. The radial clearance can be defined as the difference between the radius of bearing and journal and given by:

$$c = R_b - R_j \tag{1}$$

where R_b is the radius of bearing and R_j is the radius of journal.

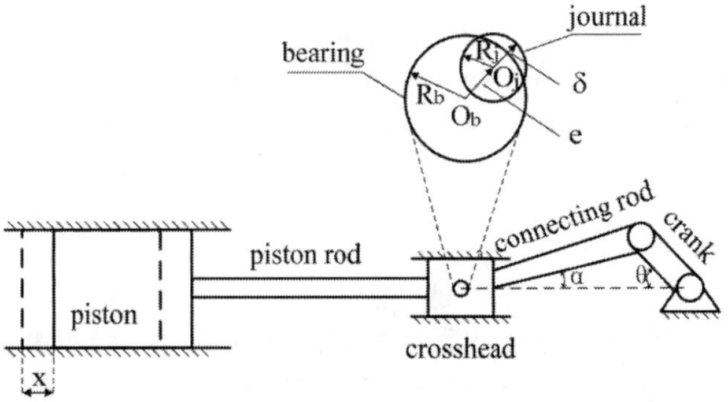

Figure 1. Cylindrical clearance joint in real mechanism of the reciprocating compressor

The existence of the gap makes two more degrees of freedom, that is, two less restrictions. The center of the journal and the center of the bearing are not coincident, expect (for) some special cases. Instead, bearing is limited in a certain area, namely, within the boundary of the bearing. The eccentricity ratio ε can represent the deviation degree of bearing center, which is a dimensionless physical quantity. And it can be given by:

$$\varepsilon = \frac{e}{c} \tag{2}$$

where e represent the eccentricity and c represent the radius clearance.

2.2 .Contact force model

Many kinds of contact models have been put forward in previous literatures. The most widely applied one was Hertz contact model, which, however, did not take energy dissipation into consideration in the process of collision, and was only applicable to completely elastic impact. Lankarani and Nikravesh[10]established a new contact model, which has been widely used in dynamic simulation of mechanical system with clearance joint by means of dividing the normal contact force into elastic part, represented by the first term in Eq.(3), and dissipative part, represented by the second term in Eq.(3). This model can be described as

$$F_n = K\delta^m + D\dot{\delta} \tag{3}$$

where K represents the contact stiffness, D is the hysteresis damping coefficient , δ is the depth of relative penetration, and $\dot{\delta}$ represents the relative impact velocity. The exponent m is usually chosen for 1.5 for metallic contact[10] . The contact stiffness can be given by:

IOP Conf. Series: Materials Science and Engineering **90** (2015) 012025 doi:10.1088/1757-899X/90/1/012025

$$K = \frac{4}{3\left(\sigma_b + \sigma_j\right)} \left[\frac{R_b R_j}{R_b + R_j} \right]^{\frac{1}{2}} \tag{4}$$

where σ_b and σ_j are the material properties and can be given by :

$$\sigma_z = \frac{1 - v_z^2}{E_z}, (z = i, j) \tag{5}$$

In Eq.(5) v_z is Poisson's coefficient and E_z is the Young's modulus.

Hysteresis damping coefficient can be expressed as:

$$D = \frac{3K(1 - c_e^2)}{4\dot{\delta}^{(-)}} \delta^m \tag{6}$$

where c_e represents the coefficient of restitution and $\dot{\delta}^{(-)}$ is the initial impact velocity. However, it is hard to determine these two parameters, so a modified approach[11]was used in this paper, given by:

$$D = d \frac{\dot{\delta}}{|\dot{\delta}|} |\dot{\delta}|^{m_2} \delta^{m_3} \tag{7}$$

where damping coefficient d can be chosen by the user.

2.3. Friction force model

Friction force acts when there is relative motion with two objects, along the tangent direction. Coulomb's friction law is the most simply and wildly used method to calculate the friction force in dry friction model. However, it is not continuous when the relative velocity equals to zero. In order to overcome this problem, a kind of representatively modified method[9]will be used to calculate the friction force in this paper, which is shown in Fig.2 and given by:

$$F_f = \begin{cases} -sign(v)\mu(v)\,|\,\mathrm{F_n}\,| \\ sign(F_f) \times \min(|\,F_n\,|, \mathrm{f_{max}}) \end{cases} \tag{8}$$

In Eq.(8), F_n represents normal contact forces, $\mathrm{f_{max}}$ is the maximum friction forces, and $\mu(v)$ is friction coefficient, which is a function of the relative speed in tangent direction, given by Eq.(9).

$$\mu(v) = \begin{cases} hav\sin(v, 0, 0, v_s, \mu_s) & if\; v < v_s \\ hav\sin(v, v_s, \mu_s, v_d, \mu_d) & if\; v \geq v_s \end{cases} \tag{9}$$

In this equation, (μ0, μ1) are Static and dynamic friction coefficient and (v0, v1) are threshold velocities.

3. Modeling and simulation results

A method for dynamic modeling and simulation was proposed and effectively verified by Gummer A and Sauer B in reference[11]. In this paper, the same method is adopted to establish model and do dynamic simulation for reciprocating compressor. The distinct differences are that a practical model of slider-crank mechanism with a clearance joint in a real reciprocating compressor is built and the simulations are performed with the changing gas forces, which can make a great influence on the dynamic behavior, exerted on the piston.

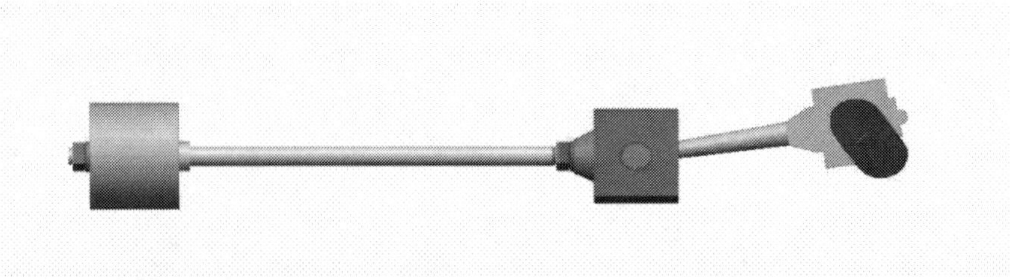

Figure 2. Entity model of slider-crank mechanism with clearance of a real reciprocating compressor

Table 1. The operation parameters of the reciprocating compressor

Operating parameters	Value	Operating parameters	Value
Working speed	495r/min	Crank radius	0.09m
Connect rod length	0.45m	Cylinder radius	0.125m
The suction pressure	100kPa	The discharge pressure	300kpa

The entity model set up in a multi-body dynamic simulation tool is shown in figure 2. The geometric dimensions of each component of the crank connecting rod mechanism and some physical parameters of a real reciprocating compressor are listed in table 1. In the simulation, contact forces are calculated according to Eq.(3), damping forces are calculated according to Eq.(7),and friction forces are calculated according to Eq.(8). All parameters required to set in the simulation are listed in table 2.

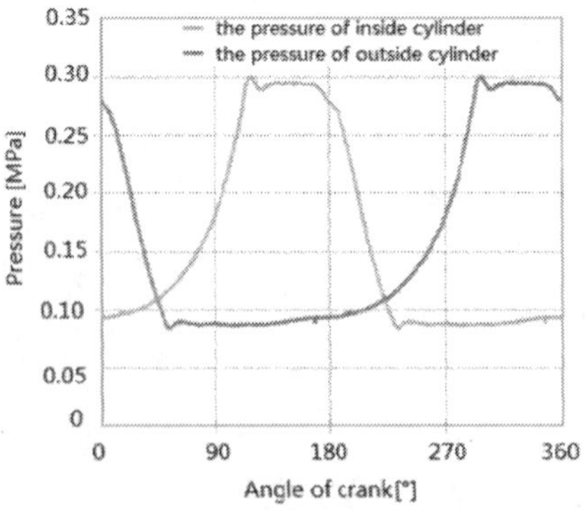

Figure 3. In-cylinder gas pressure curve

The changing gas forces acting on the piston of the reciprocating compressor can be got generally by two ways, that is, simulation and experimental test. This text used the latter to get gas forces. The relationship between gas pressure and crank angle is shown in figure 3.

In order to study the influence of wear clearance in SECRB of the reciprocating compressor on the mechanism dynamic behaviors, a lot of simulations of mechanism with different gap sizes joints or ideal joints are performed. Crosshead is attached directly to the small end of the connecting rod through the crosshead pin, for which the crosshead dynamic response that affected by the gap will be more significant. Therefore, the dynamic impact on the crosshead needs to be focused on.

Table 2. Parameters used in the dynamic simulation

Contact characteristics	Value	Contact characteristics	Value
Contact stiffness *k*	5603076/ N/mm	Damping exponent *m*2	0.0297
Stiffness exponent *m1*	1.5	Indentation exponent *m*3	1.5
Damping coefficient *d*	2003200	Coefficient of restitution *e*	0.9
Static and dynamic friction coefficient ($\mu 0,\mu 1$)	0.01	Threshold velocities (*v*0, *v*1)	1.0 mm/s

The simulate results of displacement and velocity of the crosshead are shown in figure 4, from which we can discover that the displacement and velocity in clearance joint conditions are same with ideal joint condition, respectively. Thus, it is explainable that the dynamic impacts on displacement and velocity by clearance are very little.

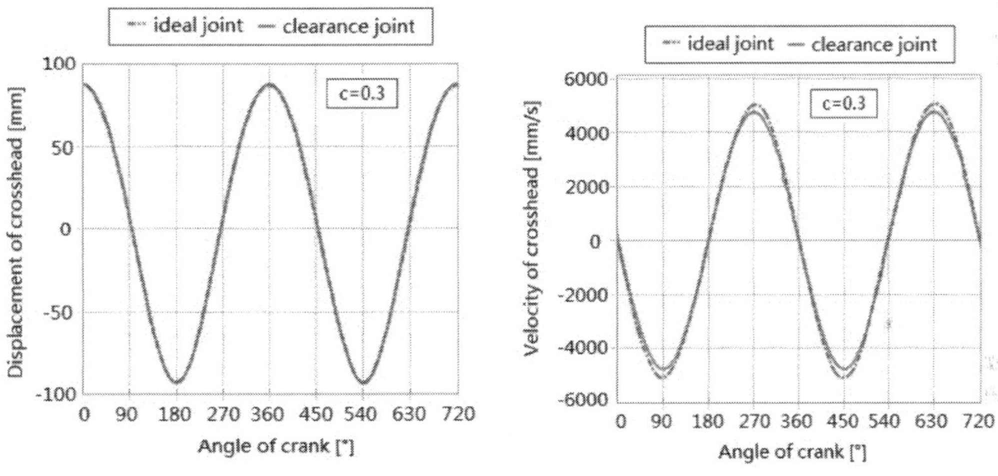

Figure 4. Simulation results of crosshead: (a) displacement; (b) velocity

However, as shown in figure 5, high acceleration amplitudes occur at the reversal points of crosshead, indicating high impact forces, when compared to ideal joint simulation results. What's more, the acceleration amplitudes become higher as the clearance size increasing, indicating higher impacts forces.

When wear clearance increases to a certain degree, forces and moment cannot be transmitted well, making the two connecting objects directly impact after crossing the gap. As a result, the impact forces will appear abnormal fluctuations and larger amplitudes. As shown in the simulation results, on one hand, it is on the reversal points of crosshead that forces transmit appear unstable and collisions happen between the two components, no matter how large the gap size is. On the other hand, the amplitudes of crosshead acceleration increase with the development of wear process.

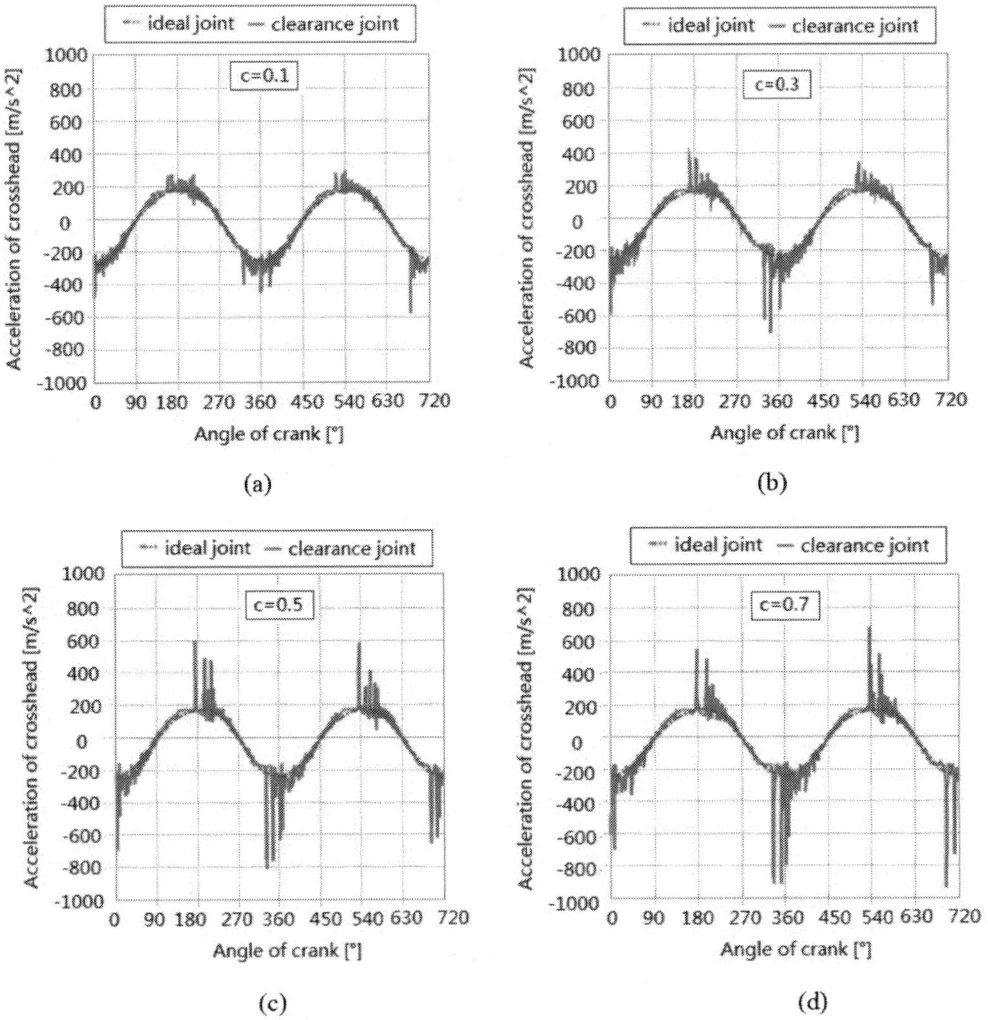

Figure 5. Simulated crosshead acceleration for different diametric clearances

4. Method presentations

It is known that mechanical parts are generally accompanied with high frequency vibrations when high frequency impacts are acting on. Thus, in wear conditions, crosshead will inevitably appear short-term high frequency vibrations, which can be transferred to the surface of the body though the directly connected slideways. Therefore, abnormal vibration signals can be obtained if a vibrating sensor has been installed on the surface right above of the crosshead.

Based on the above analysis, a new method for monitoring and diagnosing the wear fault of SECRB through analyzing the angle domain of vibration signals was put forward. First of all, the practical model of slider-crank mechanism with a clearance joint of a real reciprocating compressor should be established. Secondly, the in-cylinder gas forces should be obtained through experiments or simulations. Thirdly, dynamic simulations should be done in a multi-body dynamic simulation tool, the results of which should be analyzed to get the phase position where abnormal high acceleration amplitudes occur. Moreover, acceleration vibrating sensors should be installed on the surface right above the crossheads of the reciprocating compressor to monitor angle domain vibration signals. Finally, high frequency vibration signals with large amplitudes will appear at the same phase position got from dynamic simulations when wear of SECRB occurs.

5. Experimental verification

A reciprocating compressor on which an on-line monitoring system has installed is shown in figure 6. A lot of sensors were set up to capture various signals, and two kinds of them will be used to perform this verifiable experiment. One is the acceleration sensors installed on the right above surface of the crossheads and used to monitor their vibrations. The other one is the key sensor installed on the side of flywheel and used to determine the positions of the piston in the cylinder. Besides, the procedure we used to simulate the clearance joint in this text is as follows. Firstly, a crosshead pin 0.5 mm less than the standard one in diameter was employed to replace the original one. Then the reciprocating compressor was turned on to collect the experimental data for a period of time. Finally, the above steps were repeated again using a crosshead pin 0.7 mm less than the standard one in diameter.

Figure 6. Reciprocating compressor experimental platform

Vibration waveforms in normal operation and fault conditions are shown in figure 7, from which we can find that vibration waveforms appear obviously additional impacts at 0 and 180 degree in the fault conditions when compared to normal condition, and that the amplitudes of the vibration signals become larger with the gap size increasing. It is obviously that the fault characteristics gotten from experiment agree well with that extracted from simulation results, proving that the method put forward in this paper and based on vibration signal angle domain analysis is effective.

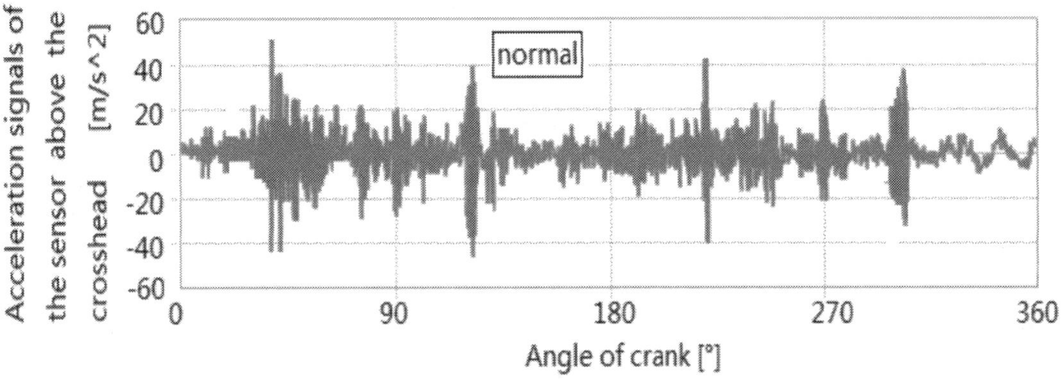

(a)

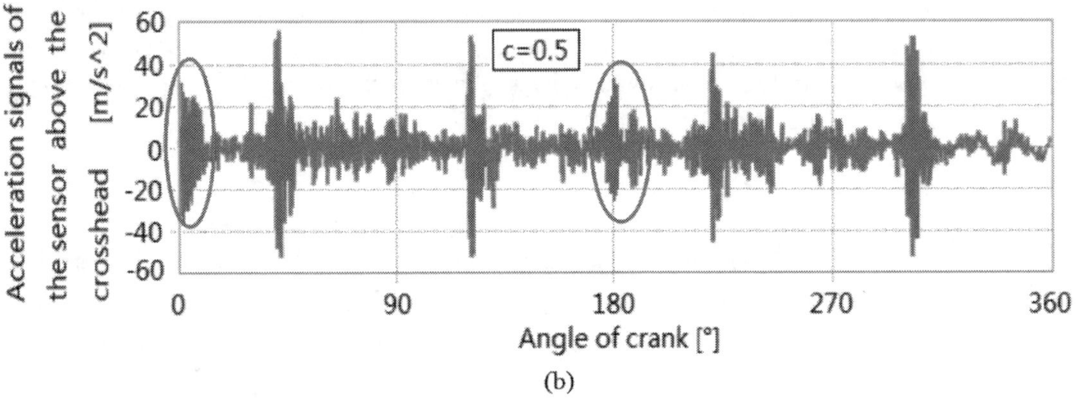

(b)

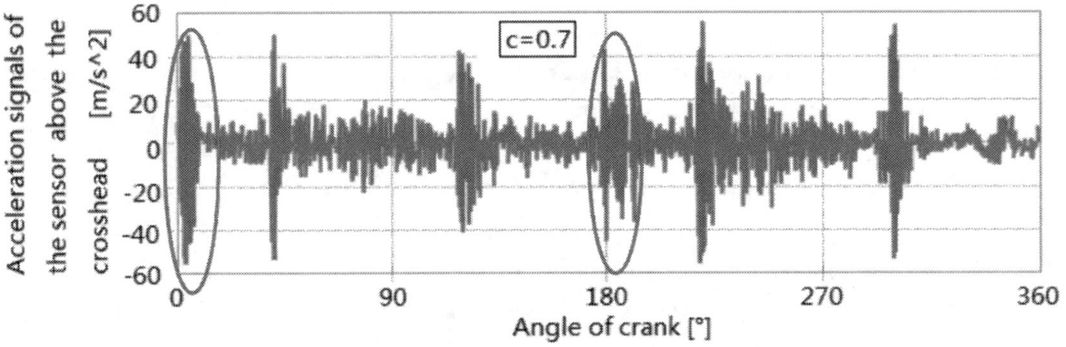

(c)

Figure 7. Acceleration vibration waveforms

6. Conclusion

A contact model of a clearance in the revolute joint of SECRB was established and a multi-body simulation method proved available was utilized to simulate the slider-crank mechanism with a clearance. A new method to diagnosis the wear fault of SECRB of reciprocating compressors effectively is presented, which can also be suitable in on-line monitoring system. This method can not only accurately diagnose the wear fault of SECRB but also approximately estimate its severity. Since calculation of wear is really difficult and not the purpose of this paper, the method about how to calculate the wear size from acceleration is not done. Experiments were conducted to demonstrate the availability of the approach. The method is universal and can be adapted to almost all reciprocating compressors.

Acknowledge

This work was supported by the National Basic Research Program of China (973 Program) under Grant No. 2012CB026000 and the National High Technology Research and Development Program of China (863 Program) under Grant No.2014AA041806.

Corresponding author

Jinjie Zhang, Email: zjj87427@163.com

References

[1] Peng, Z., N. Kessissoglou, and M. Cox 2005 A study of the effect of contaminant particles in lubricants using wear debris and vibration condition monitoring techniques *Wear*, **258** 1651.

[2] Tian, Q. and Flores, P. 2011 A new model for dry and lubricated cylindrical joints with clearance in spatial flexible multibody systems *Nonlinear Dynamics*, **64** 25.

[3] Flores, P. 2010 A parametric study on the dynamic response of planar multibody systems with multiple clearance joints *Nonlinear dynamics*,**61** 633.

[4] Erkaya, S. and İ. Uzmay 2010 Experimental investigation of joint clearance effects on the dynamics of a slider-crank mechanism *Multibody system dynamics*, **24** 81.

[5] Schwab, A., J. Meijaard, and P. Meijers 2002 A comparison of revolute joint clearance models in the dynamic analysis of rigid and elastic mechanical systems *Mechanism and machine theory*, **37** 895.

[6] Liu, C.-S., K. Zhang, and R. Yang 2007 The FEM analysis and approximate model for cylindrical joints with clearances *Mechanism and machine theory*, **42** 183.

[7] Lu, Wenqing, et al. 2010 Finite-Element Model-Based Fault Prognosis on Key Components of the Reciprocating Compressor. *8th International Pipeline Conference. American Society of Mechanical Engineers*, 221..

[8]Feese T, Hill C. 2009 Prevention of torsional vibration problems in reciprocating machinery *Proceedings of the thirty-eighth turbomachinery symposium, college station TX, Turbomachinery Laboratory, Texas A&M University*, 213.

[9] Flores, P. 2006 A study on dynamics of mechanical systems including joints with clearance and lubrication *Mechanism and Machine Theory*,**41** 247.

[10] Lankarani, H.M. and P.E. Nikravesh 1994 Continuous contact force models for impact analysis in multibody systems *Nonlinear Dynamics*, **5** 193.

[11] Gummer, A. and B. Sauer 2014 Modeling planar slider-crank mechanisms with clearance joints in RecurDyn *Multibody System Dynamics*, **31** 127.

IOP Conf. Series: Materials Science and Engineering **90** (2015) 012026 doi:10.1088/1757-899X/90/1/012026

Numerical analysis of the turbulent fluid flow through valves. Geometrical aspects influence at different positions.

J. Rigola[1], D. Aljure[1], O. Lehmkuhl[1,2], C.D. Pérez-Segarra[1] and A. Oliva[1]

[1]Heat and Mass Transfer Technological Center (CTTC) Universitat Politcnica de Catalunya BARCELONA Tech (UPC) C/ Colom, 11 08222 Terrassa (Barcelona), Spain

[2]Termo Fluids, S.L. C/ Magí Colet, 0 08204 Sabadell (Barcelona), Spain

E-mail: cttc@cttc.upc.edu

Abstract. The aim of this paper is to carry out a group of numerical experiments over the fluid flow through a valve reed, using the CFD&HT code TermoFluids, an unstructured and parallel object-oriented CFD code for accurate and reliable solving of industrial flows. Turbulent flow and its solution is a very complex problem due to there is a non-lineal interaction between viscous and inertial effects further complicated by their rotational nature, together with the three-dimensionality inherent in these types of flow and the non-steady state solutions. In this work, different meshes, geometrical conditions and LES turbulence models (WALE, VMS, QR and SIGMA) are tested and results compared. On the other hand, the fluid flow boundary conditions are obtained by means of the numerical simulation model of hermetic reciprocating compressors tool, NEST-compressor code. The numerical results presented are based on a specific geometry, where the valve gap opening percentage is 11% of hole diameter and Reynolds numbers given by the one-dimensional model is 4.22×10^5, with density meshes of approximately 8 million CVs. Geometrical aspects related with the orifice's shape and its influence on fluid flow behaviour and pressure drop are analysed in detail, furthermore, flow results for different valve openings are also studied.

1. Introduction

The radial outflow between two coaxial disks is technologically important for different applications (radial diffusers, air bearings, valve reeds, etc.). In all these cases the flow is quite complex, presenting a pressure gradient that may be either positive or negative depending on radial location. Furthermore, turbulent flow is often encountered, thus, turbulence models are critical on the numerical resolution of this technological relevant application.

The numerical results are obtained using the three dimensional, parallel, unstructured and object oriented code TermoFluids [1] using LES models, a high Reynolds number (close to real working conditions) and turbulent inlet flow. In fact, the present paper is an updated version of previous works [3] and [4] that used extruded meshes, low Reynolds numbers and did not use low Mach models[5].

Experimental ([6] and [7]) and numerical ([10] and [11]) have been performed setting a basis to understand the physics underlying in the flow. First attempts of numerical simulation of turbulent flows by means of RANS $k-\epsilon$ and $k-\omega$ models were carried out under low Re numbers ([12] and [13]). More recently, some papers presents numerical results based on LES turbulent models for low Reynolds number 2.5×10^4 [14], or RANS turbulent models for higher Reynolds numbers 1.6×10^6 [15]. Finally, two-dimensional numerical cases that were experimental validated under specific experimental apparatus, although for very low Reynolds numbers, were presented in [16].

From a compressor study viewpoint the fluid flow is composed by two different physics (entrance flow through a channel and a free jet through a surface)(see fig 1a). In that sense, the present paper is focused on the numerical simulation model of the fluid flow through the valve reeds, considering a simplified geometry of an axial hole plus a radial diffuser. The methodology proposed by [17], [18] and [19] is based on two parameters: i) effective flow area $(KA)_e$ that relates the actual mass flow

Content from this work may be used under the terms of the Creative Commons Attribution 3.0 licence. Any further distribution of this work must maintain attribution to the author(s) and the title of the work, journal citation and DOI.
Published under licence by IOP Publishing Ltd

IOP Conf. Series: Materials Science and Engineering **90** (2015) 012026 doi:10.1088/1757-899X/90/1/012026

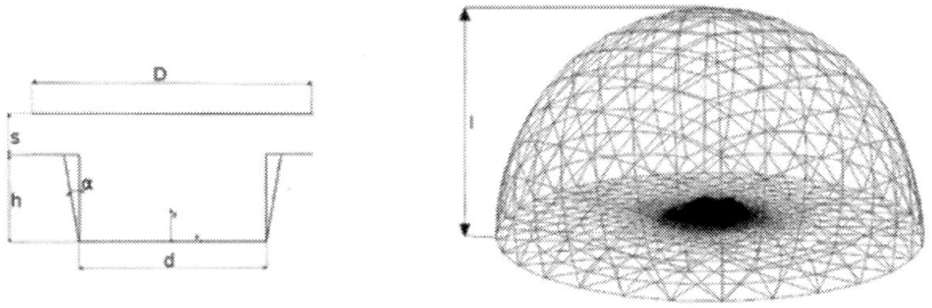

(a) (b)

Figure 1: (a) Outline and dimensions for the two valve geometries and (b) Outline of the computational domain

rate with an ideal mass flow rate per unit flow area (assuming isentropic contraction process) defined as $m = (KA)_e(\rho\nu)_{id} = (KA)_e Y\sqrt{2\rho_u(p_u - p_d)}$; and ii) the effective force area A_f defined as ratio of the actual net force on the valve and the force obtained assuming a constant pressure drop distribution, defined as $F = (p_u - p_d)A_f$, the most usual method for valve analysis and compressor design [2]

2. Definition of the case

The geometries to be considered are: geometry A (straight) where the hole is cylindrically shaped with constant diameter, and geometry B (cone) where the hole is drawn in a conical way linearly enlarging the diameter. Both cases are fitted with a top plate acting as the valve disc.

Fig. 1 shows the geometrical features and computational domain to be studied, where $D = 1.5d$, $s = 0.11d$, $h = 0.24d$, $\alpha = 20°$ and $l = 15d$. Flow inlet is done through the bottom boundary such that $Re = \rho W_{ref} d/\mu = 4.22 \times 10^5$, result obtained from NEST-compressor simulations. NEST analyses thermal, fluid-dynamic and aerodynamic behaviour in this type. Synthetic turbulence ([20], [21]) is used to create a turbulent inlet flow, using the Reynolds stresses values for a channel with a $Re_\tau = 6000$ ([22]. Flow outlet is the outer semi-sphere (see fig. 1b). The pressure based outflow condition uses a constant p chosen to fulfil the mass conservation condition at each time step, i.e. the mass inflow equals the mass outflow. In addition, a bufferzone at $r > 14d$ is used where the Reynolds number is decreased from 4.22×10^5 to 4.22×10^3 in order to avoid any influence from non-physical waves reflected by the artificial outflow boundary [23]. The orifice walls, the plate and bottom wall surfaces are modelled using a no-slip condition.

3. Mathematical formulation

3.1. Governing equations and numerical method

In order to study the flow, the filtered incompressible Navier-Stokes equations are solved:

$$\frac{\partial \overline{u_i}}{\partial t} + \frac{\partial(\overline{u_i u_j})}{\partial x_j} - \nu\frac{\partial^2 \overline{u_i}}{\partial x_j \partial x_j} + \rho^{-1}\frac{\partial \overline{p}}{\partial x_i} = -\frac{\partial \tau_{ij}}{\partial x_j} \quad (1)$$

$$\frac{\partial \overline{u_i}}{\partial x_i} = 0 \quad (2)$$

where $\overline{u_i}$ is the three-dimensional filtered velocity, $\overline{p}$ is the filtered pressure field, ν stands for the kinematic viscosity, ρ for the density of the fluid and τ_{ij} is the SGS stress tensor defined as $-2\nu_s\overline{S_{ij}} + 1/3\tau_{ij}\delta_{ij}$.

Governing equations are discretized on a collocated unstructured mesh by means of finite volume techniques. A second-order conservative scheme is used for the spatial discretization [24]. Such schemes preserve the symmetry properties of the continuous differential operators and ensure both, stability and conservation of the kinetic-energy balance. The velocity-pressure coupling is solved by means of a fractional-step algorithm. The temporal discretization for the convective, diffusive and derivative parts

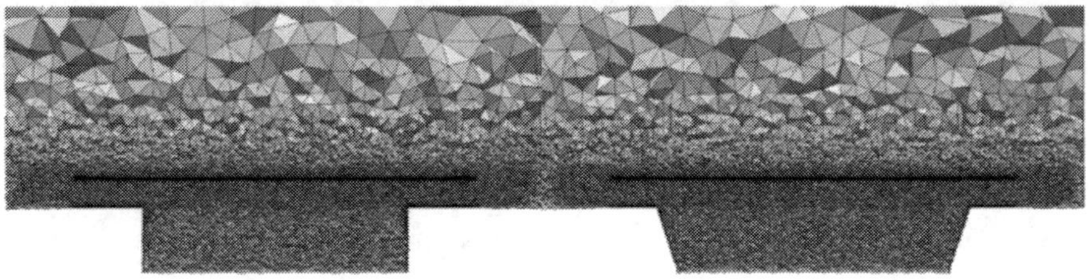

(a) (b)

Figure 2: Meshes along the xy cut plane, (a) straight geometry and (b) cone geometry

Geometry	Mesh	NCV	$y^+_{min}D$	$y^+_{avg}D$	$y^+_{min}O$	$y^+_{avg}O$
Straight	Mesh 1	2.13×10^5	2.98	111.97	9.24	106.28
	Mesh 2	1.11×10^6	0.13	18.44	2.27	18.66
	Mesh 3	7.83×10^6	0.04	8.51	0.34	9.63
Cone	Mesh 1	2.20×10^5	2.37	98.52	12.31	98.89
	Mesh 2	1.10×10^6	0.18	15.50	1.56	13.99
	Mesh 3	8.00×10^6	0.04	6.90	0.48	7.33

Table 1: Mesh parameters. NCV the total number of control volumes in the domain, $y^+_{min}D$ and $y^+_{avg}D$ minimum and averaged y^+ on the valve disc, and $y^+_{min}O$ and $y^+_{avg}O$ minimum and averaged y^+ on the discharge orifice surface.

of equation (1) was made using a second order self-adaptive scheme [27]; whereas a back-ward Euler scheme was used for the pressure gradient. For more details about the discretization the reader is referred to [28] and [29].

3.2. Large eddy simulation turbulence models

In LES the spatial filter applied to the Navier Stokes equations separates large and small scales. This operation introduces new variables into the system of equations, the sub grid stresses (SGS). In order to solve the equations this term must be properly modelled. Four closure models are used in the present paper, WALE, VMS, SIGMA and QR, presented in [30]. The WALE model was proposed by [31] based on the square of the velocity gradient tensor. This SGS model accounts for the effects of the strain and rotation rates, as well as, a y^3 near wall scaling for the eddy viscosity. The variational multiscale method was formulated for the Smagorinsky model on a spectral domain by [32]. Later, this method was extended to the classical filtering approach in LES. The VMS method consists on applying a spatial test filter to the already filtered scales to divide the resolved scales of motion into large and small. The later are used to model the turbulent viscosity. QR model was proposed by [33] responding to the question of damping sub filter scales properly. It is a SGS model based on the invariants of the rate-of-strain tensor. Finally, SIGMA model, proposed by [34], is a subgrid-scale model derived from the analysis of the singular values of the resolved velocity gradient tensor.

4. Numerical results

4.1. Mesh and time integration analysis

Hybrid tetra-prism unstructured meshes are constructed to discretize the equations. A fine near wall mesh is necessary to correctly solve the boundary layer and the physical phenomena associated with this zone. A prism layer is appropriate in this area due to the relative simplicity to place these control volumes (CV) close to the surface and the log-law grow ratio that can be applied. This elements are

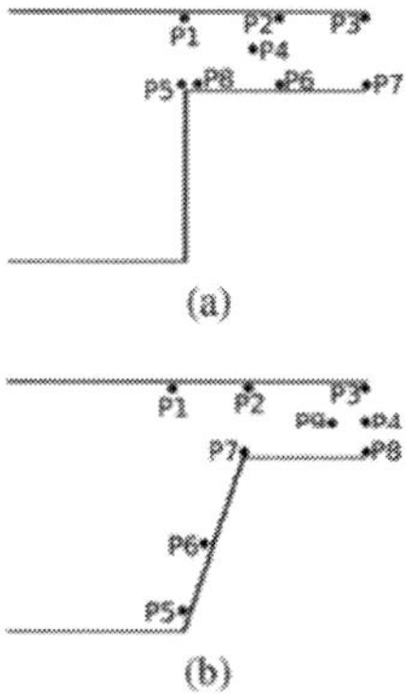

Figure 3: Numerical probes' location, (a) straight geometry, (b) cone geometry

Geometry	Straight	Cone
P1	$[0.5, 0.32]$	$[0.5, 0.32]$
P2	$[0.625, 0.32]$	$[0.625, 0.32]$
P3	$[0.75, 0.32]$	$[0.75, 0.32]$
P4	$[0.625, 0.28]$	$[0.75, 0.28]$
P5	$[0.5, 0.24]$	$[0.52, 0.1]$
P6	$[0.625, 0.24]$	$[0.56, 0.19]$
P7	$[0.75, 0.24]$	$[0.625, 0.24]$
P8	$[0.51, 0.24]$	$[0.75, 0.24]$
P9	-	$[0.68, 0.28]$

Table 2: Numerical probes' location

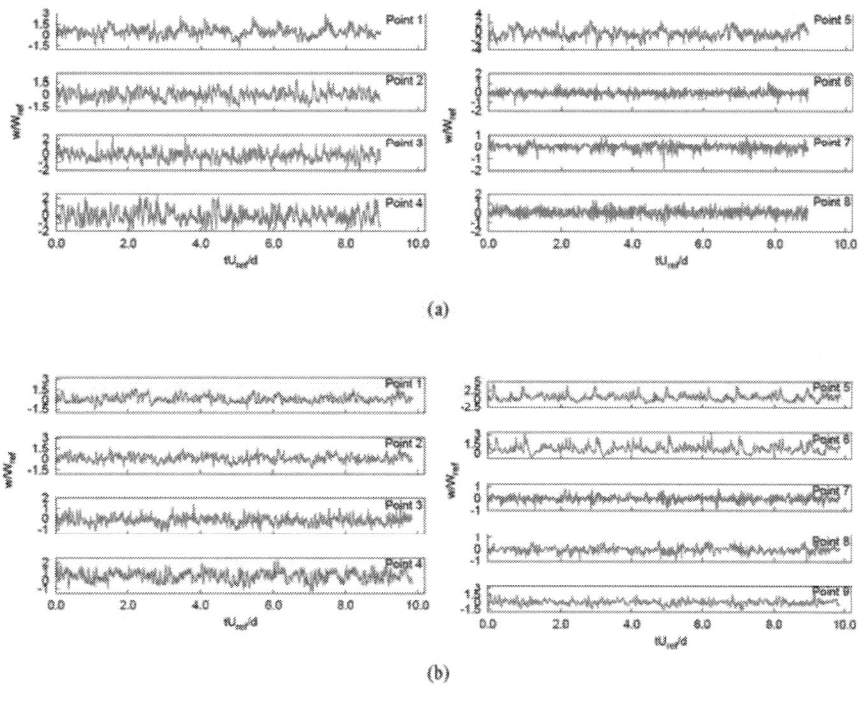

Figure 4: Streamwise velocity W time history for the different numerical probes in the (a) straight geometry and (b) cone geometry.

placed in the present case on the valve disc and discharge orifice geometry. Furthermore, the gap zone must be refined to correctly predict the flow discharge from the valve. As the flow moves away from the gap outlet the relative importance of the flow on the valve operation diminishes and so does the mesh resolution (see fig. 2). Three meshes are used with each valve model, see table 1 for details. Results in the present section have been obtained using the VMS turbulence model.

The fluid flows axially through the lower disk inlet orifice d and thickness h, after that, impacts on the upper solid disk of diameter D and finally, radially outflows through the gap between both coaxial disks. Present simulations use a turbulent inflow, which causes an unsteady turbulent behaviour characterized

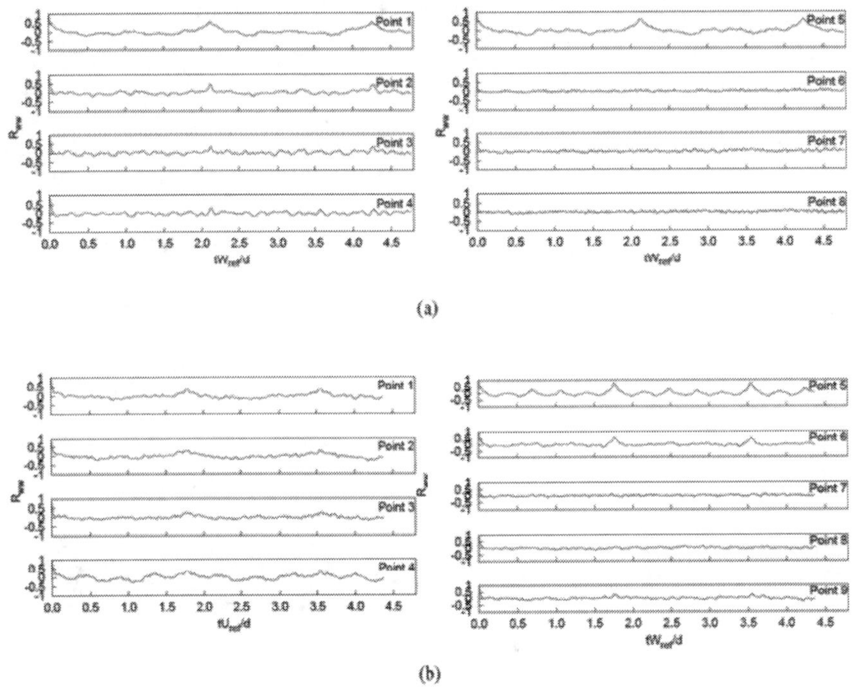

Figure 5: Z-component velocity autocorrelation function for the different numerical probes in the (a) straight geometry and (b) cone geometry.

by random-like velocity behaviour and large fluctuations. In order to perform a time integration analysis several numerical probes were placed within the valve domain. For the straight geometry points P1, P2 and P3 are located near the upper disc along the gap length, P4 in the center, both axially and radially, of the gap and points P5, P6, P7 and P8 are placed near the lower wall alongside the gap. For the cone geometry locations are: P1, P2 and P3 are placed near the top wall and along the gap in similar position as for the straight geometry. P4 is placed at the center of the gap near the exit, P5, P6 and P7 are located along the slanted wall, P8 in the lower wall, near the exit of the gap and P9 near the center of the gap. Locations are detailed in table 2 and fig. 3.

Simulations are started from a solution obtained using a laminar inflow flow and are advanced in time until the initial transient behaviour is washed out at around $0.5TU$ ($TU = tUref/d$). Afterwards statistics have been collected and averaged by integrating the instantaneous data. Streamwise velocity time history and autocorrelation functions for the z-component of velocity (W) at the different locations are shown in figs. 4 and 5. It is interesting to observe the behaviour for the straight case, where probes P5 and P8, although being very close to each other show quite a different behaviour. P5 is located near the valve seat corner and exhibits large velocity fluctuations. P8, located slightly downwind is located at the beginning of the recirculation region and, although fluctuation magnitude is less, its frequency is considerably higher P7 for the cone geometry is placed in a comparable location (just after the valve seat) and its turbulent behaviour is lower, i.e. lower fluctuation amplitude and frequency. This difference in flow behaviour is enhanced when observing probes P4 for the straight geometry and P9 for the cone geometry, both near the center of the gap. Flow for both geometries show some periodic like behaviour towards the valve outlet (see fig. 5a and 5b), however, as stated before, frequencies for the straight geometry seem higher than for the cone one.

Two important variables are shown in the numerical results presented. The first one is the fluid pressure distribution in the lower face of the upper disk, while the second one is the velocity magnitude (V) distribution along the gap between coaxial disks at different radial positions of $x = d/2$ (gap position 1), $x = d/4 + D/4$ (gap position 2) and $x = D/2$ (gap position 3) corresponding to the entrance, middle

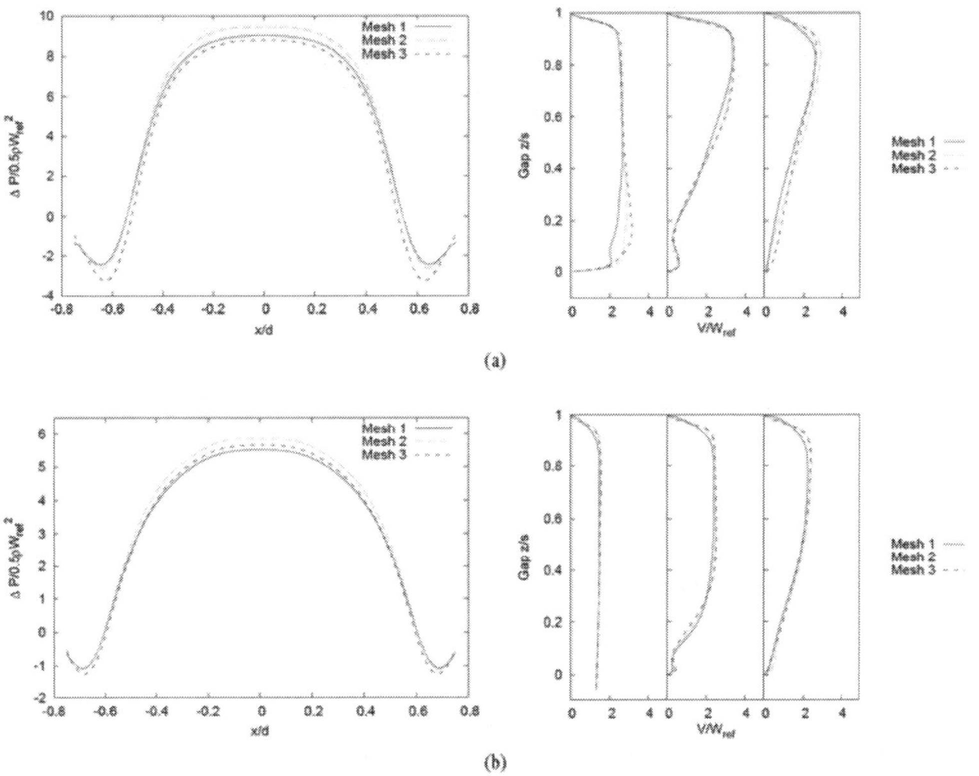

Figure 6: Pressure distribution over the lower face of the valve disc and velocity profiles over the gap at $x = d/2, x = d/4 + D/4$ and $x = D/2$ for (a) straight geometry, (b) cone geometry.

and exit of the gap, all shown in fig. 6. Mesh independence is achieved for the 8×10^6 CV meshes as both pressure and velocity profiles show almost no change for a further mesh refinement. Stagnation pressure is significantly higher for the straight geometry than for the cone one, this is due to the higher velocity the flow reaches when passing through the straight valve. In the cone geometry this shape acts as a diffuser lowering the velocity of the impinging jet and easing into the direction change, thus, the flow carries less energy when crashing onto the top plate. Effects of these flow modifications are also visible through the gap. For the straight geometry flow reaches a higher velocity, additionally, the recirculation zone created by the valve seat corner is larger for the straight geometry than for the cone geometry (see figs. 6a and 6b).

4.2. Turbulence model analysis

As exposed in section 3.2, four turbulence models were used to evaluate the results in the flow through the valves, VMS, WALE, QR and SIGMA. Results show great consistency when changing LES models, however, this is more evident in the cone geometry than in the conical one. Sudden direction change and the more pronounced valve seat corner create a more complex flow that poses a higher difficulty to solve.

WALE and QR models predict a slightly larger pressure distribution than the rest of the models for the straight geometry, whereas VMS and WALE do the same in the cone geometry. For the velocity profiles results take a different trend. In this zone velocities for the straight geometry are larger than for the conical one due to the sudden change of direction and area reduction. For the conical geometry this acceleration is more gradual resulting in lower final velocities. All models results are quite similar for the entrance of the gap, however the other location analysed show results difference. SIGMA and VMS models show almost identical result in the remaining gap locations. For the straight geometry large differences are seen in gap positions 2 and 3. QR and WALE models both show larger velocities

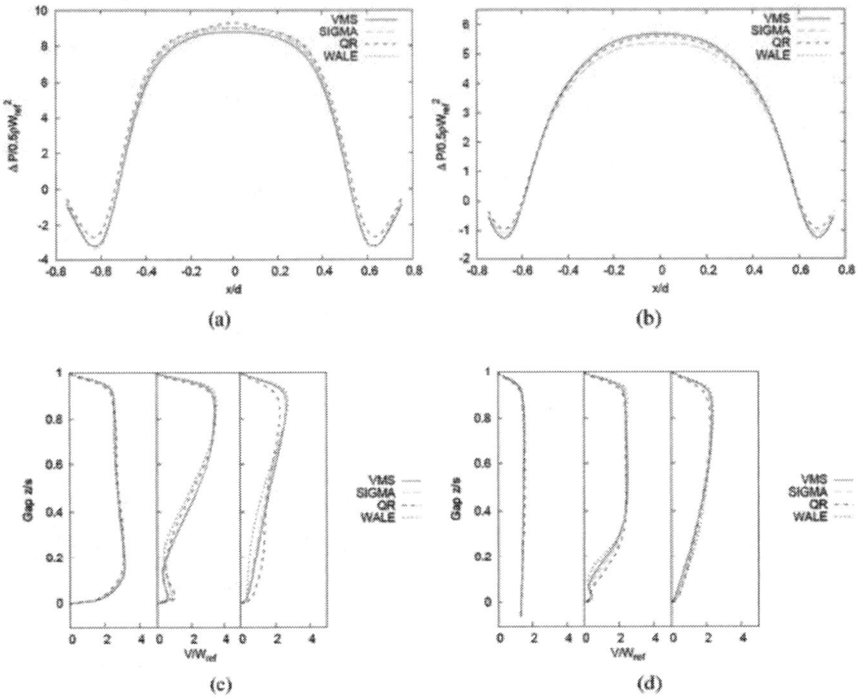

Figure 7: LES results for the pressure profile along the plate on (a) straight geometry, (b) cone geometry. LES results for the velocity profiles in the gap. (c) straight geometry, (d) cone geometry

in the recirculation region for both geometries. For the cone geometry these differences reduce for gap position 3, whereas for the straight geometry the differences become larger. VMS model shows results that fall in between the other models. much more evident for the velocity results than for the pressure ones. Previous experiences with turbulent flow suggest VMS results could be the best LES model to use for more detailed simulations.

4.3. Geometrical analysis

Changing the discharge orifice geometry greatly changes the velocity and pressure behaviour in the valve, as has been observed in the previous sections. To verify this behaviour different gap openings were studied as well, $s/d = 0.033$ and $s/d = 0.22$, using the same mesh size parameters than for

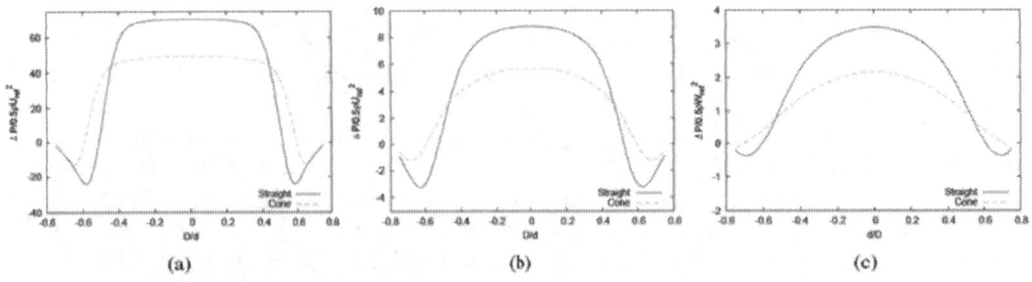

Figure 8: pressure distribution over the plate for both geometries, (a): gap opening $s/d = 0.033$, (b): gap opening $s/d = 0.11$ and (c): gap opening $s/d = 0.022$

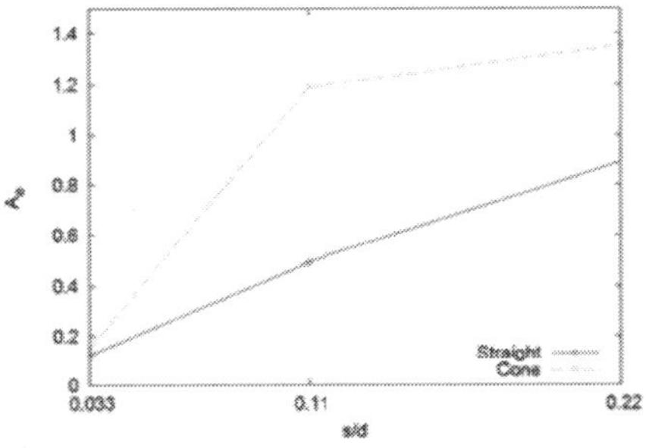

(a)

Figure 9: Effective flow area for the different valve openings.

the base case. Gradual expansion allows the flow to change its direction sooner and reduce the energy carried when it impinges into the valve disc. This results in a considerably lower pressure rise in this element, as observed in fig. 8. This lower pressure rise over the valve disc for the cone geometry results in a considerably higher effective flow area (see fig.9), where for the straight geometry increases from 0.124 to 0.89 as the valve opens and for the cone geometry from 0.152 to 1.35. Effective force area also increases for both geometries, from 0.569 to 1.52 for the straight geometry and from 0.74 to 2.45 in the cone one.

Furthermore, the recirculation region in the valve seat is much smaller for the conical geometry. The sudden right angle change in the flow's direction causes heavy flow separation. When the inlet channel geometry is changed to the cone the sudden direction change now possesses a smaller angle, thus, creating a smaller recirculation region. A second factor affecting this region is the velocity magnitude. As the flow is incompressible a reduction in flow area is accompanied by an increase in its velocity. Although, geometrically, the inlet and gap outlet areas are the same, effective flow areas are different due to the different velocity field and pressure drops present en each geometry (see fig.9). Observing fig. 10 there is a much larger velocity increment for the straight geometry than the conical one. As the valve opening increases from $s/d = 0.033$ to $s/d = 0.11$ both effective flow and force areas increase as well, however, as the gap opening further opens to $s/d = 0.22$ this trend changes. Effective flow area for the cone geometry reduces its rate of increase, whereas the effective force area decreases. In the straight geometry, although both effective areas increases, its rate of increment diminishes.

Finally non-dimensional pressure drop ($C_P = \Delta P/0.5\rho W_{ref}^2$) in the valve for the straight geometry and gap opening $s = 0.11d$ is $C_P = 4.11$, whereas for the cone geometry it is $C_P = 0.71$. This same behaviour remains when the valve opening is greater ($s/d = 0.22$) and smaller ($s/d = 0.033$), non-dimensional pressure drop for these valve openings and the straight geometry are $C_P = 1.26$ and $C_P = 65.32$ respectively, and for the cone geometry $C_P = 0.53$ and $C_P = 43.52$ respectively. Non-dimensional pressure drop reduces as the valve opens and is also significantly lower for the cone geometry, 57.9% for $s/d = 0.033$, 82.7% for $s = 0.11d$ and 76.4% for $s = 0.22d$.

5. Conclusions

Turbulent flow simulations through different valve geometries at an industrial type Re flow regime were carried out and their result compared to each other. Two mayor aspects were analysed and found to drastically change in shape and magnitude. Pressure profiles over the valve disc change from a nearly top-hat distribution to a smoother bell-shaped profile with the change of geometry. Additionally maximum pressure rise in this area is reduced nearly 40%. Velocity profiles in the gap show that for the straight geometry both, velocity magnitude and recirculation area, are larger then for the cone geometry. These effects greatly affect the valve operation, achieving a significant pressure drop reduction when using the cone geometry. Finally, it is important to note that the Mach number in some parts of the geometry surpass $Ma = 0.3$. In order to obtain better results compressible flow simulations are required.

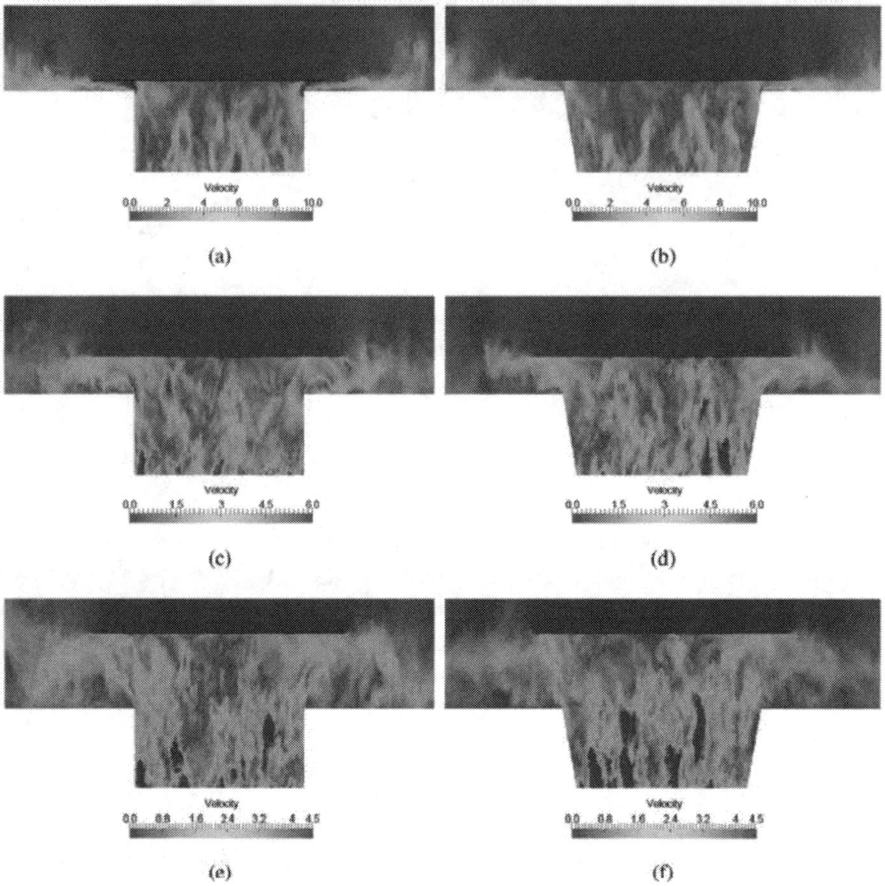

Figure 10: Instantaneous velocity field for both geometries, (a): gap opening $s/d = 0.033$, (b): gap opening $s/d = 0.11$ and (c): gap opening $s/d = 0.022$

Acknowledgements

This work has been financially supported by the "Ministerio de Economía y Competitividad, Secretaría de Estado de Investigación, Desarrollo e Innovación", Spain (reference ENE2010-17801), by the collaboration project between "Universitat Politècnica de Catalunya" and Termo Fluids S.L. and by the "Departamento Administrativo de Ciencia, Tecnología e Innovación - Colciencias" via their doctoral training program "Francisco Jose de Caldas".

References

[1] Lehmkuhl O, Borrell R, Pérez-Segarra C D, Soria M and Oliva A 2007 A new parallel unstructured CFD code for the simulation of turbulent industrial problems on low cost PC cluster *Parallel Computational Fluid Dynamics* Ankara.

[2] Damle R, Rigola J, Pérez-Segarra C D, Castro J and Oliva A 2011 Object-oriented simulation of reciprocating compressors: numerical verification and experimental comparison, *Int J Refrig* **34** **8** 1989-98.

[3] Rigola J, Lehmkuhl O, Pérez-Segarra C D and Colomer G 2009 Turbulent radial outflow between coaxial parallel disks *6th Int Symp of Turbulence, heat and Mass Transfer* Rome.

[4] Rigola J, Lehmkuhl O, Oliva A amd Pérez-Segarra C D 2009 Numerical simulation of the fluid flow through valves based on Large Eddy Simulation Models *Int Conf on Compressors and their Systems* London.

[5] Rigola J, Lehmkuhl O, Oliva, A, Ventosa J and Pérez-Segarra C D 2012 Numerical simulation of the fluid flow through valves based on Low Mach models *Int Compressor Engineering Conf at Purdue* Indiana.

[6] Wark C E and Foss J F 1984 Forces caused by the radial out-flow between parallel disks *J Fluid Eng* **106** 2927.

[7] Ervin J S, Suryanarayana N V and Hon Chai Ng 1989 Radial, turbulent flow of a fluid Between two coaxial disks *J Fluid Eng* **111** 37883.

[8] Piechna J R and Meier G E A 1986 Numerical investigation of steady and unsteady flow in the valve gap *Int Compressor Engineering Conf* Indiana pp. 1-14.

[9] Cyklis P 1994 CFD simulation of the flow through reciprocating compressor self-acting valves, *Int Compressor Engineering Conf* Indiana pp. 427-432.
[10] Satake S and Kunugi T 1998 Direct numerical simulation of an impinging jet into parallel disks *It J Numer Method H* **8** 76880.
[11] Lygren M and Andersson H I 1999 Turbulent flow between a rotating and a stationary disk. *J Fluid Mech* **426** 297326.
[12] Pérez-Segarra C D, Cadafalch J, Rigola J and Oliva, A 1999, Numerical study of turbulent fluid flow through valves, *Int Conf on Compressors and Their Systems* C542/021/99.
[13] Matos F F S, Prata A T and Deschamps C J 1999 Numerical analysis of the dynamic behavior of plate valves in reciprocating compressors *Int Conference on Compressors and Their Systems* C542/031/99.
[14] Rovaris J B and Deschamps C J 2006 Large eddy simulation applied to reciprocating compressors *J Braz Soc Mech Sci & Eng* **28 2** 208-15.
[15] Colaciti AK, Valdés L M, Aparecido H and Cabezas-Gomez L 2007 Numerical simulation of a radial diffuser turbulent air flow *Appl Math Comput* **189** 1491-1504.
[16] Mariani V C, Prata A T and Deschamps C J 2010 Numerical analysis of fluid flow through radial diffusers in the presence of a chamfer in the feeding orifice with a mixed Eulerian-Lagrangian method *Comput Fluids* **39** 1672-84.
[17] Schwerzler D D and Hamilton J F 1972 An analytical method for determining effective flow and force areas for refrigeration compressor valving systems, *Int Compressor Engineering Conf* Indiana pp. 30-36.
[18] Böswirth L 1982 Theoretical and experimental study on flow in valve channels, Parts I and II, *Int Compressor Engineering Conference* pp. 38-53.
[19] Soedel W Mechanics, Simulation and design of compressor valves, gas passages and pulsation mufflers, *Purdue University* 1992.
[20] Klein M, Sadiki A and Janicka J 2003 A digital filter based generation of inflow data for spatially develping direct numerical or large eddy simulations, *J. Comp Phys* **186** pp. 652-65.
[21] di Mare L, Klein, M, Jones, W P and Janicka J 2006 Synthetic turbulence inflow conditions for lare-eddy simulations *Phys Fluids* **18** p. 025107.
[22] Schultz M P and Flack K a 2013 Reynolds-number scaling of turbulent channel flow *Comput Fluids* **25** p. 025104.
[23] Verstappen R W C P and Veldman A E P 1997 Direct numerical simulation of turbulence at lower costs . *Eng. Math.* **32** pp143-59.
[24] Verstappen R W C P and Veldman A E P 2003 Symmetry-preserving discretization of turbulent flow *J Comput Phys* **187** pp. 343-68.
[25] Trias F X, Soria M, Pérez-Segarra C D and Oliva A 2006 Direct schur-fourier decomposition for the efficient solution of high-order poisson equations on loosely coupled parallel computers *Numer Linear Algebr* **13** pp. 303-26
[26] Yoshizawa A, Kobayashi K, Kobayashi T and Taniguchi N 2000 A non-equilibrium fixed parameter subgrid-scale model obeying the near wall asymptotic constraint *Phys Fluids* **12 9** 2338.
[27] Trias F and Lehmkuhl O 2011 A self-adaptive strategy for the time integration of the Navier-Stokes equations *Numer Heat Tr B-Fund* **60 2** pp. 116-34.
[28] Jofre L, Lehmkuhl O, Ventosa J, Trias, F X, and Oliva A 2014 Conservation properties of unstructured finite-volume mesh schemes for the Navier-Stokes equations *Numer Heat Tr B-Fund* **64 1** pp. 53-79
[29] Trias F X, Lehmkuhl O, Oliva A, Pérez-Segarra C D, and Verstappen R W C P 2014 Symmetry-preserving discretization of NavierStokes equations on collocated unstructured grids *J Comput Phys* **258** pp. 246–67.
[30] Aljure D E, Lehmkuhl O, Rodríguez I and A. Oliva 2014 Flow and turbulent structures around simplified car models *Comput Fluids* **96** 122-35.
[31] Nicoud F and Ducros F 1999 Subgrid-scale stress modelling based on the square of the velocity gradient tensor *Flow Turbul Combust* **62** 183200.
[32] Hughes T, Mazzei L and Jansen K 2000 Large eddy simulation and the variational multiscale method. *Comput Visual Sci* **3** 4759.
[33] Verstappen R 2011 When does eddy viscosity damp subfilter scales sufficiently? *J Sci Comput* **49 1** 94110.
[34] Nicoud F, Toda H, Cabrit O, Bose S and Lee J 2011 Using singular values to build a subgrid-scale model for large eddy simulations *Phys Fluids* **23** 085106

A parametric optimization procedure for the suction system of reciprocating compressors

W M Ferreira, E Silva, C J Deschamps[1]

POLO Research Laboratories, Federal University of Santa Catarina
Florianopolis, SC, Brazil

E-mail: deschamps@polo.ufsc.br

Abstract. The design of the suction system of compressors is of fundamental importance for efficiency and reliability. This paper reports a method developed to optimize the suction system of a reciprocating compressor, by using the genetic algorithm NSGA-II. The isentropic and volumetric efficiencies are used as objective functions, while the bending fatigue stress is used as a constraint to meet valve reliability. A simulation model of the compression cycle was coupled to the optimization procedure, with correlations for flow and force effective areas in terms of geometric parameters of the suction valve. Valve dynamics was numerically solved via the finite element method. The proposed optimization procedure was applied to a reciprocating compressor adopted for household refrigeration, identifying suction system geometries more efficient than the original design.

1. Introduction

There is a continuous effort to increase the efficiency of reciprocating compressors adopted in household refrigeration. According to [1], investments related to the development of such compressors in the period between 1980 and 2002 were responsible for an average increase of 60% in the COP. However, the performance of refrigeration systems is far from the theoretical limit and improvements in the compressor are still possible.

The performance of reciprocating compressors is commonly quantified through the isentropic efficiency, η_s, and volumetric efficiency, η_v, defined as

$$\eta_s = \frac{\dot{W}_s}{\dot{W}_i} \qquad\qquad \eta_v = \frac{\dot{m}}{\dot{m}_{sw}} \tag{1}$$

where $\dot{W}_s$ and $\dot{W}_i$ are the isentropic and indicated power, and $\dot{m}$ and $\dot{m}_{sw}$ are the measured and theoretical mass flow rates, respectively.

The isentropic efficiency is appropriate to analyze the efficiency of the compression cycle and is affected by inefficiencies in the processes of suction, discharge, compression, and expansion of the gas, not accounting for electrical and mechanical inefficiencies. The volumetric efficiency is reduced by gas re-expansion in the compression chamber, gas superheating, leakage, backflow in valves, and viscous friction in the flow through the suction and discharge systems.

[1] To whom any correspondence should be addressed.

Content from this work may be used under the terms of the Creative Commons Attribution 3.0 licence. Any further distribution of this work must maintain attribution to the author(s) and the title of the work, journal citation and DOI.
Published under licence by IOP Publishing Ltd

Much effort has been directed to improve the efficiency of the suction and discharge processes in compressors, with special attention to valve reliability because valve failure is a common cause of compressor shutdowns [2]. For instance, parametric optimization procedures have been used to identify optimum geometric configurations of compressor valves [3, 4], adopting the impact velocity of the valves against their seats as an optimization constraint. A similar procedure was adopted by [5], but using a genetic optimization algorithm and the impact velocity as part of the objective function rather than a constraint. Recently, Silva *et al.* [6] developed a genetic optimization algorithm for the optimization of valves in which the valve dynamics was solved via the finite element method in order to assess valve reliability for bending fatigue.

This paper presents a geometric optimization procedure for the suction system of reciprocating compressors. Correlations for effective area coefficients required to characterize valve performance were obtained in terms of geometric parameters of the suction system. The genetic algorithm NSGA-II for multi-objective optimization was employed and the isentropic and volumetric efficiencies of the compressor were set as independent objective functions. The bending fatigue stress was used as a constraint to meet valve reliability.

2. Compressor thermodynamic model

The numerical model used to predict the compressor performance is based on equations for the thermodynamic state of the gas in the compression chamber, valve dynamics, and flow in the suction and discharge mufflers [6].

The thermodynamic state of the gas is determined by applying the equations of conservation of mass and energy in the compression chamber in conjunction with the specification of an equation of state for the gas. The mass flow rates flowing in and out of the compression chamber are necessary in the equation of conservation of mass. Particularly, the mass flow rates through the suction and discharge valves are obtained with reference to the isentropic flow in a convergent nozzle and the introduction of the effective flow area coefficient, C_{ep}:

$$\dot{m} = C_{ep} A_o p_{up} \sqrt{\frac{2k}{RT_{up}(k-1)}} \sqrt{\Pi^{2/k} - \Pi^{(k+1)/k}} \tag{2}$$

where A_o is the area of the orifice, p_{up} is the upstream pressure, k is the ratio of specific heats, R is the specific gas constant, T_{up} is the upstream temperature, and Π is the pressure ratio. The leakage of gas through the piston-cylinder gap is obtained by assuming a fully developed Couette-Poiseuille laminar flow.

The dynamics of the suction valve was solved with the finite element model based on beam elements, since bending stress has to be evaluated in the optimization procedure. On the other hand, the dynamics of discharge valve is described as a rigid body moving perpendicularly to the seat, via a single degree-of-freedom model. In both cases, the resulting force acting on the valve may include the oil stiction force and preload stress imposed on the valve. The force due to the gas pressure load, F_v, is evaluated through the concept of effective force area coefficient, C_{ef}:

$$F_v = C_{ef} A_o \Delta p \tag{3}$$

where Δp is the pressure difference between both sides of the valve.

The pulsating compressible flow through the suction and discharge mufflers is modeled via the conservation equations for mass, energy, and momentum following a one-dimensional formulation. Such equations are solved using the finite volume method and taking into account heat transfer and viscous effects. Details of this modeling approach can be obtained in [7].

IOP Conf. Series: Materials Science and Engineering **90** (2015) 012027 doi:10.1088/1757-899X/90/1/012027

3. Correlations for effective flow and force areas

3.1. Geometric parametrization

The coefficients for effective flow area and effective force area are functions of the compressor geometry. Since the suction system is to be defined via an optimization procedure, correlations for these coefficients are required in terms of geometric parameters. Therefore, the geometry of the suction system was parametrized and CFD simulations were performed in order to obtain these correlations.

The solution domain of these simulations is presented in Figure 1. The parameters that describe the suction system geometry are: the valve opening, s, the valve opening angle, θ_v, the width of the orifice, w_p, the valve width, w_v, and the height and width of the seat, h_s and w_s, respectively (Figure 2). Some relations of dependence between these parameters can be specified. For instance, by assuming that no bending exists when the valve opens, s and θ_v can be easily related. Also, the valve width can be defined as

$$w_v = w_p + 2w_s + 2d_v \tag{4}$$

where d_v is the length of the valve that extends beyond the edge of the seat. If d_v is held constant, then C_{ep} and C_{ef} can be described in terms of four parameters: s, w_p $(= \mathrm{R_e} - \mathrm{R_i})$, w_s, and h_s.

The orifice geometry is completely described in terms of these parameters. For instance, the area of the suction orifice, A_p, is the sum of the areas A_1 and A_2 (Figure 3a), i.e.:

$$A_p = A_1 + A_2 \tag{5}$$

where,

$$A_1 = \frac{\pi}{8}(R_e - R_i)^2 \tag{6}$$

$$A_2 = \frac{\theta_o}{2}\left(R_e^2 - R_i^2\right) \tag{7}$$

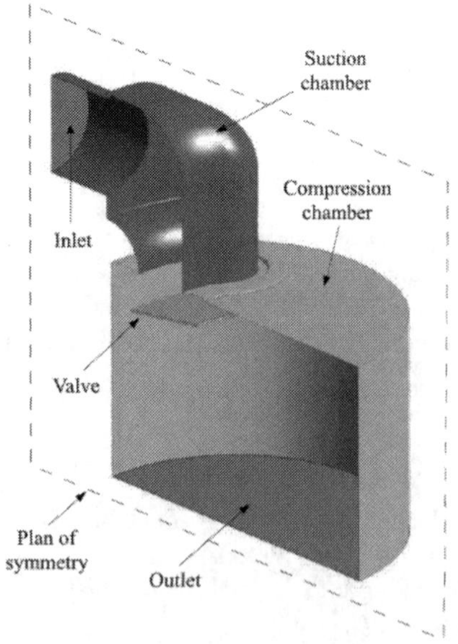

Figure 1. Domain of analysis.

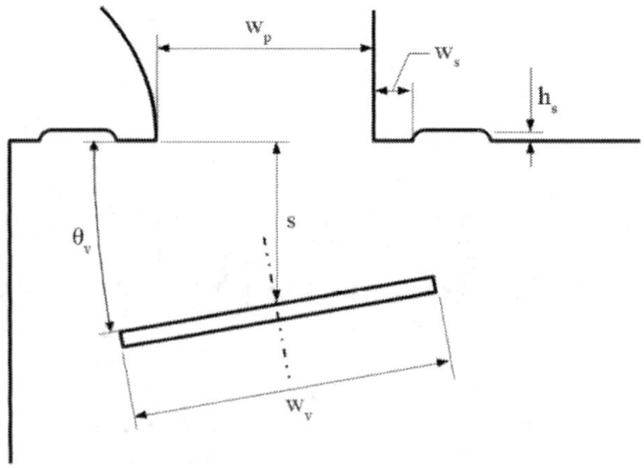

Figure 2. Geometric parametrization.

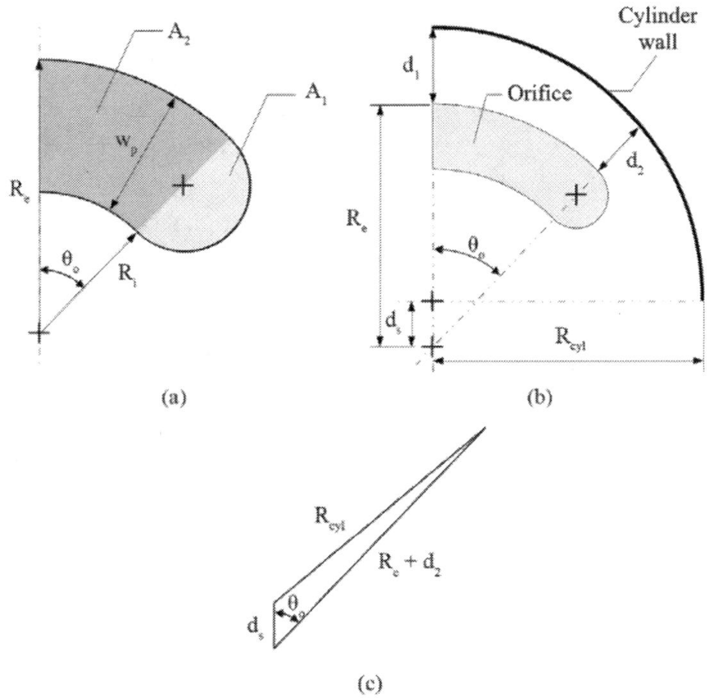

Figure 3. Compressor geometry: (a) suction orifice area, (b) suction orifice and cylinder profiles, and (c) trigonometric relation between geometric parameters.

In these equations, R_i and R_e are, respectively, the radius of the internal and external arcs used in the generation of the orifice profile and θ_o is their aperture angle.

From Figure 3b, the following relation can be established:

$$R_e = R_{cyl} + d_s - d_1 \tag{8}$$

where R_{cyl} is the cylinder radius, which has a fixed value for a particular compressor, d_s is the eccentricity between the axis of the cylinder and the origin of the arcs that define the orifice profile, and d_1 is the maximum distance between the external arc defining the orifice profile and the cylinder wall.

Also, from Figure 3c, a trigonometric relation can be obtained:

$$R_{cyl}^2 = d_s^2 + (R_e + d_2)^2 - 2d_s(R_e + d_2)\cos\theta_o \tag{9}$$

where d_2 stands for the minimum distance between the external arc defining the orifice profile and the cylinder wall. Therefore, if A_p, d_1, and d_2 are fixed, the orifice geometry is completely described solely by the parameter w_p. The seat and valve geometries are obtained directly from the orifice geometry and the parameters w_s and d_v.

3.2. Numerical simulation

The flow of gas in the suction system was solved with the commercial code ANSYS CFX 14.5. A steady solution was provided for each valve position of interest, considering the effect of gas compressibility. Based on the results and recommendations of [8, 9], the RNG k-ε turbulence model was adopted.

The boundary conditions of the solution domain (Figure 1) were based on pressure and temperature measurements carried out in the compressor. At the inlet section, pressure, temperature, and parameters related to the turbulence model were specified, while at the outlet only pressure was prescribed. The no-slip and temperature boundary conditions were adopted at the walls. A condition of flow symmetry was also adopted to reduce computational time, as indicated in Figure 1.

An unstructured computational mesh was adopted in the simulations, with prismatic elements near the boundaries and tetrahedral elements in the remainder of the domain. A mesh with approximately 700,000 elements was found adequate for the analyses. In order to reduce computational cost the isothermal condition was adopted, since the comparison of the effective area coefficients shows that they do not change significantly in comparison to the non-isothermal solution (Figure 4). This caused a reduction of almost 20% in the computational cost.

4. Optimization procedure

The variables of the optimization problem are the geometric parameters of the suction system, composed by the suction valve, valve plate, and suction muffler. As a result of the finite element method applied to predict the suction valve dynamics, the valve is represented by a group of beam elements. Therefore, the valve geometry is defined based on the valve thickness and the width of each beam element. In possession of correlations for the effective area coefficients, the orifice width, w_p, the height of the seat, h_s, and the width of the seat, w_s, are adopted as project variables related to the valve plate geometry. The diameter and length of the tube of the suction muffler connected to the suction chamber affect significantly the valve dynamics [10] and are included as variables of the optimization procedure.

As proposed by [6], the objective functions, η_s and η_v, are evaluated for the operating condition in which the highest performance is required. On the other hand, valve reliability is verified in the operating condition that result the highest stress levels, as in conditions where higher mass flow rates exist. Therefore, the compressor model must be run twice for each configuration to be evaluated. The Soderberg criterion provides a way to verify whether the failure limit is violated, being a constraint of the problem, together with the range limits of each of the variables considered. If this constraint is not satisfied, null values are set to the isentropic and volumetric efficiencies and the configuration is discarded from the optimization procedure.

The genetic algorithm NSGA-II (*Elitist Non-Dominated Sorting Genetic Algorithm*) developed by [11] was selected because it is commonly used in the solution of optimization problems with multiple objectives and real-valued variables. In this algorithm the ranking of the feasible solutions is based on the fitness and density of solutions in the objective space. In this case, there is no necessity to combine the efficiencies in a unique equation by specifying weighting factors, as described in [6]. The adopted stopping criterion is based on the number of generations of the algorithm. A simplified flowchart of the procedure is presented in Figure 5.

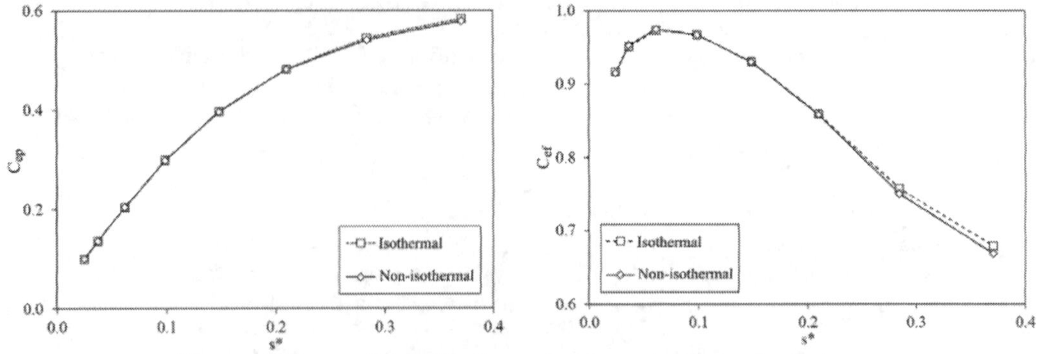

Figure 4. Comparison of isothermal and non-isothermal solutions.

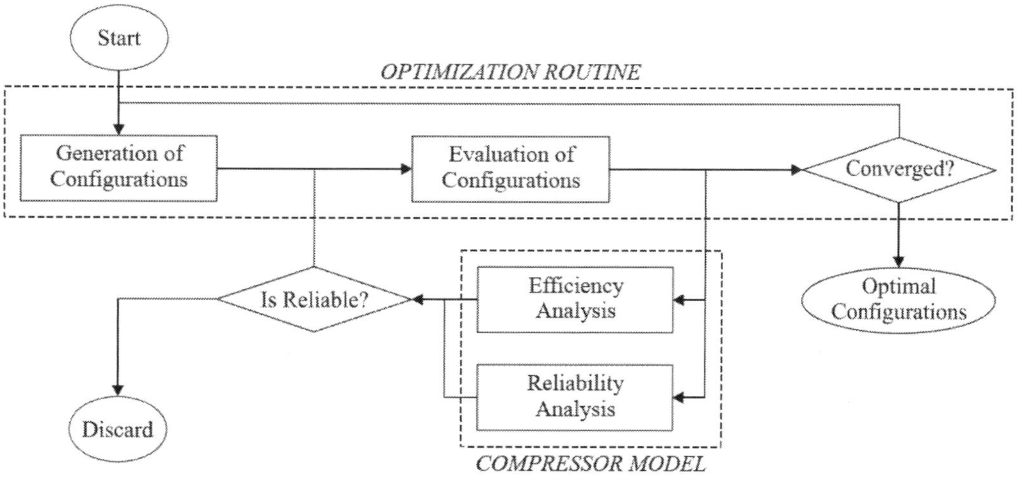

Figure 5. Schematic of the optimization procedure [6].

5. Results and Discussion

5.1. Correlations for effective flow areas

Correlations for coefficients of effective flow area and effective force area were obtained from CFD simulations, considering different geometric configurations of the suction system characterized by the variables h_s, w_s, w_p, and s. Three values were considered for each variable, except for the valve opening, s, which had eight levels, since it is considered the most influent parameter. Therefore 216 CFD simulations were run for a full factorial design of experiments.

Correlations for C_{ep} and C_{ef} were obtained, using the code Eureqa. From these results:

$$C_{ep} = C_1 \sin(C_2 s^*) + \frac{C_3 s^{*C_4} \cos(C_5 h_s^* + C_6 s^* + C_7 h_s^* w_s^*)}{C_8 w_p^*} \quad (10)$$

$$C_{ef} = C_9 + C_{10} w_s^* w_p^* s^* - s^* C_{11} C_{12}^{C_{13} s^*} e^{C_{14}(w_s^* - h_s^*)} + C_{15} s^{*2} + C_{16} s^* \sin(C_{17} s^*) \quad (11)$$

where the superscript "*" indicated non-dimensional variables, obtained by dividing the original variable by the equivalent diameter of the suction orifice. The values of the coefficients in these correlations are presented in Table 1.

Table 1. Coefficients of the correlations for C_{ep} and C_{ef}.

Coefficient	Value	Coefficient	Value	Coefficient	Value
C_1	7.33×10^{-1}	C_7	4.77×10^{-1}	C_{13}	8.09
C_2	3.48	C_8	8.1	C_{14}	8.09
C_3	7.76×10^{-1}	C_9	1	C_{15}	3.30
C_4	4.31×10^{-1}	C_{10}	1.44×10^{1}	C_{16}	5.19×10^{-1}
C_5	8.1	C_{11}	8.09	C_{17}	8.09
C_6	8.1	C_{12}	2.42×10^{-3}		

The values of C_{ep} and C_{ef} given by the correlations and CFD simulations were compared for two geometric configurations of the suction system not included to generate the correlations. The results in

Figure 6 show that the correlations give results in close agreement with the simulations, with a small deviation in the value for C_{ef} at large valve lifts s^*.

5.2. *Optimization of the suction system*

The optimization procedure was applied to redesign the suction system of an existing compressor for household refrigeration application. Two different methods were used to define the valve shape. In the first method (A), the width of each beam element in the finite element model was treated as an isolated variable of the optimization problem. In the second method (B), the widths of only a few elements were treated as isolated variables, and the others were obtained from a spline interpolation. Figure 7 presents the results given by both methods in terms of the valve thickness t_v commercially available. The results show that there is a relation between the region of the objective space explored by the algorithm and the valve thickness, with a trend of increasing the compressor efficiency as the valve thickness is reduced. However, thin valves are associated to higher stress levels, and hence are more susceptible to fatigue failure. Thus, the reliability constraint limits the search for more efficient valves.

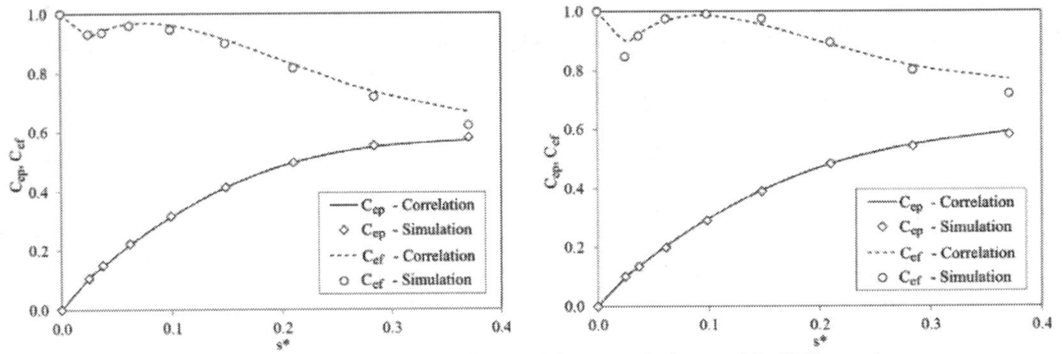

Figure 6. Comparative analysis of the correlations with CFD results.

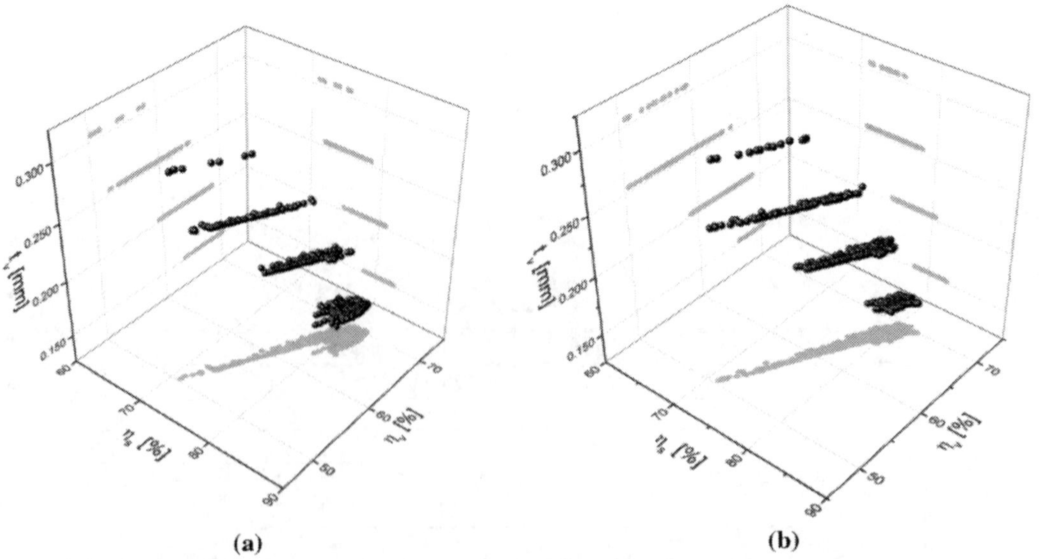

Figure 7. Solutions of the optimization procedure: (a) method A, and (b) method B.

Figure 8 presents the isentropic and volumetric efficiencies of the optimal configurations, i.e., the configurations located in the Pareto frontier found by both methods (A and B). The efficiencies predicted for the existing compressor design are presented for reference. It is clear that both methods found new design alternatives with greater performance. As expected, the reduced number of variables defining the valve geometry in method B produces fewer optimum solutions. However, method B simplifies the development of new valve concepts, since it avoids discontinuities between the widths of neighbor elements.

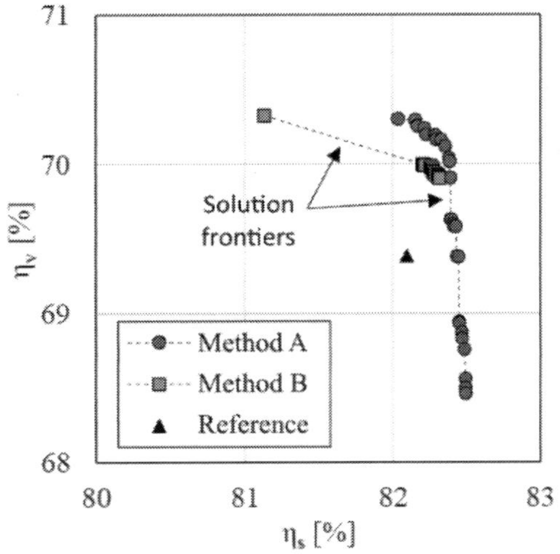

Figure 8. Efficiencies of configurations in the Pareto frontier.

6. Conclusions

This paper presented an optimization procedure for the suction system of reciprocating compressors. The isentropic and volumetric efficiencies were treated as individual objective functions to be maximized by the NSGA-II algorithm. A reliability constraint was used to prevent valve failure due to bending fatigue. Since the simulation model uses coefficients of effective flow area and effective force area to characterize valve performance, correlations for these coefficients were obtained in terms of geometric parameters of the suction system. The optimization procedure was applied to redesign the suction system of an existing compressor and found more efficient and yet reliable alternatives.

Acknowledgments

The present study was developed as part of a technical-scientific cooperation program between the Federal University of Santa Catarina and EMBRACO. The authors also acknowledge the support provided by EMBRACO and CNPq (Brazilian Research Council) through Grant No. 573581/2008-8 (National institute of Science and Technology in Refrigeration and Thermophysics) and CAPES (Coordination for the Improvement of High Level Personnel).

References

[1] Possamai F C and Todescat M L 2004 *Proc. Int. Compressor Engineering Conference (West Lafayette)* C067

[2] Hatch G and Woollatt D 2002 *Proc. Int. Compressor Engineering Conference (West Lafayette)* C13-4

[3] MacLaren J F T, Kerr S V, Hoare R G 1976 *Proc. Int. Compressor Engineering Conference (West Lafayette)* pp 389-396

MacLaren J F T, Kerr S V, Hoare R G 1978 *Proc. Int. Compressor Engineering Conference (West Lafayette)* pp 317-323

[4] Shu P C and Tramschek A B 1984 *Proc. Int. Compressor Engineering Conference (West*

Lafayette) pp 218-226
[5] Prins J, Ferreira C A I, Kalker-Kalkman C M 1996 *Proc. Int. Compressor Engineering Conference (West Lafayette)* pp 517-522
[6] Silva E, Deschamps C J, Fancello E A 2012 *Proc. Int. Compressor Engineering Conference (West Lafayette)* 1345
[7] Deschamps C J, Possamai F C, Pereira E L L 2002 *Proc. Int. Compressor Engineering Conference (West Lafayette)* C11-4
[8] Deschamps C J, Ferreira R T S, Prata A T 1996 *Proc. Int. Compressor Engineering Conference (West Lafayette)* pp 377-382
[9] Yakhot V and Orzag S A 1986. *J. Sci. Comp.* **1** 1-51
[10] Link R, Possamai F C, Pereira E L L 2004 *Proc. Int. Compressor Engineering Conference (West Lafayette)* C066
[11] Deb K, Pratap A, Agarwal S, Meyarivan T 2002 *IEEE Trans. Evol. Comput.* **6** 182-197

Fluid Flow in the Oil Pumping System of a Hermetic Compressor

S Posch[1], E Berger[1], M Heimel[1], J Hopfgartner[1], R Almbauer[1] and P Schöllauf[2]

[1] Graz University of Technology, 8010 Graz, Austria
[2] Secop Austria GmbH, 8280 Fürstenfeld, Austria

posch@ivt.tugraz.at

Abstract. This work deals with the investigation of the oil pumping system in hermetic compressors for refrigeration application. The oil pump which is used for this study consists of two pumping areas: the lower pumping area with a pick-up tube and an eccentric bore, and the upper pumping area with a helical groove. This study focuses on the helical groove in the upper pumping area. To analyse the fluid flow in the helical groove, a numerical approach is introduced. In this approach the Navier-Stokes equations are adapted to the problem and are solved by using the finite volume method. Compared to analytical models, this method is able to obtain the flow field in the cross section of the helical groove at higher resolution. The higher geometrical resolution also enables the analysis of the flow in the small gap between the rotating crankshaft and the stationary wall. The present method is used to quantify different operating parameters on the oil mass flow rate.

Nomenclature

Roman symbols

e	gap width
g_x	body force in *x*-direction
g_y	body force in *y*-direction
g_z	body force in *z*-direction
n	rotational speed
p	pressure
u	velocity component in *x*-direction
v	velocity component in *y*-direction
w	velocity component in *z*-direction
x	channel cross-section coordinate
y	channel cross-section coordinate
z	down channel coordinate

Greek symbols

α	helix angle
μ	dynamic viscosity
ρ	density

1. Introduction

The oil supply in hermetic reciprocating compressors for refrigeration application is usually provided by integrated pumping devices on the compressor crankshaft. Two different pumping systems are widely used in hermetic compressors: the helical pump and the centrifugal pump. These two pumping systems can be used separately or in combination. Based on the significant influence of the oil supply system on the reliability and the thermal behaviour of a compressor, a number of studies has been carried out and published which deals with the investigation of this system. In addition to experimental investigations, mathematical approaches are used to describe the fluid flow of oil pumping devices.

Content from this work may be used under the terms of the Creative Commons Attribution 3.0 licence. Any further distribution of this work must maintain attribution to the author(s) and the title of the work, journal citation and DOI.
Published under licence by IOP Publishing Ltd

These approaches can be divided into analytical (or semi-analytical) methods and methods based on computational fluid dynamics (CFD).

Numerical studies using commercial software (FLUENT) were carried out in [1] and [2]. The two-phase flow was simulated in three dimensions with the Volume of Fluid technique. The authors investigated characteristic parameters for the operation behaviour like the oil climbing time, the time to reach steady-state conditions or the oil mass flow rate. The focus in these studies was on the whole oil supply system which resulted in a significant large number of grid cells and a high computational time. Therefore, CFD simulations of the whole oil path are used to investigate existing models and they are not appropriate for the use in the design phase of hermetic compressors. A simplified uncoupled simulation model was presented in [3]. The authors split up the oil path and simulated only the helical pump using commercial software (CFX). The results of the calculated oil mass flow rate depending on the compressor speed and the immersion depth showed good agreement with experimental data in addition to a relatively low computational time of two hours.

Several analytical or semi-analytical models to describe the oil flow in hermetic reciprocating compressors were published in the open literature. In [4], a direct analogy between compressor oil flow and an electric circuit has been utilized. The authors analogized the individual lubrication elements by equivalent electric elements and formed an electric circuit which was solved by numerical methods. The calculated results of the oil flow rate showed good agreement with measured data. An analytical solution of an isothermal, Newtonian flow in a single screw extruder was developed in [5]. The authors adapted the Navier-Stokes equation to a finite channel with specified boundary conditions. To solve the resulting Poisson equation they applied the finite sine-transformation method. A comparison with experimental data from published literature showed satisfying accuracy of the presented solution. Based on the models presented in [5], [6] used a generalized integral transform technique (GITT) to solve the Poisson equation. The authors investigated the effects of screw geometry parameters on the oil mass flow rate. In [7], the same analytical solution was used to couple a screw channel pump and the shaft channel region. In this study the analytical model was compared on the one hand with experimental data to verify the calculated oil mass flow rate, and on the other hand with CFD data to verify the oil climbing time.

The base for the model developed in this study is the oil supply system of a hermetic reciprocating compressor for refrigeration application shown in Figure 1. The compressor operates in ON/OFF mode with constant speed. The oil supply system consists of two pumping areas: the lower pump is partially immersed in the oil sump and consists of a centrifugal pump, and the upper pump consists of a helical groove machined on the surface of the crankshaft.

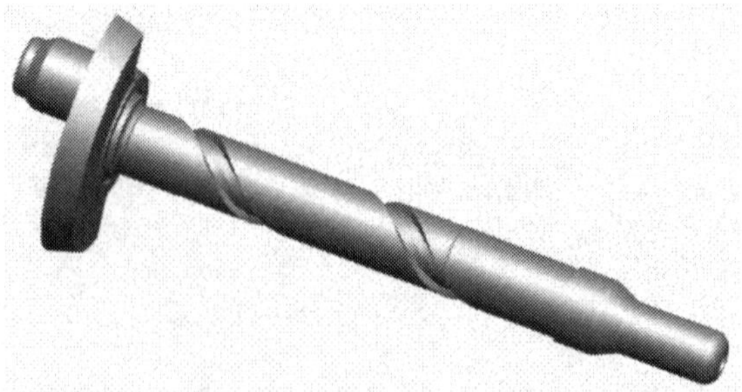

Figure 1. The crankshaft with the integrated oil pumping system.

The purpose of the present paper is the modeling of the flow in the helical groove of a hermetic compressor oil pump at higher resolution compared to analytical models. The relationship between the flow field in the cross section of the channel and the resulting oil mass flow rate is pointed out. Furthermore, the small gap between the crank shaft and the barrel wall is considered. The method is used to quantify the influence of different operating conditions (speed, oil viscosity) and geometric parameters (helix angle, gap width) on the mass flow rate.

2. Problem Formulation

The numerical modeling of the oil flow in the helical groove is based on the steady-state governing equations of motion in rectangular coordinates. Assuming that the fluid is isothermal, incompressible, Newtonian and has constant properties, the equations for laminar flow (the Reynolds-Number for this problem is approximately 250) are given by:

$$\rho\left(u\frac{\partial u}{\partial x}+v\frac{\partial u}{\partial y}+w\frac{\partial u}{\partial z}\right)=-\frac{\partial p}{\partial x}+\mu\left(\frac{\partial^2 u}{\partial x^2}+\frac{\partial^2 u}{\partial y^2}+\frac{\partial^2 u}{\partial z^2}\right)+\rho g_x \tag{1}$$

$$\rho\left(u\frac{\partial v}{\partial x}+v\frac{\partial v}{\partial y}+w\frac{\partial v}{\partial z}\right)=-\frac{\partial p}{\partial y}+\mu\left(\frac{\partial^2 v}{\partial x^2}+\frac{\partial^2 v}{\partial y^2}+\frac{\partial^2 v}{\partial z^2}\right)+\rho g_y \tag{2}$$

$$\rho\left(u\frac{\partial w}{\partial x}+v\frac{\partial w}{\partial y}+w\frac{\partial w}{\partial z}\right)=-\frac{\partial p}{\partial z}+\mu\left(\frac{\partial^2 w}{\partial x^2}+\frac{\partial^2 w}{\partial y^2}+\frac{\partial^2 w}{\partial z^2}\right)+\rho g_z \tag{3}$$

In order to simplify the problem, the cross section in the middle of the helical groove is investigated. The channel is considered to be unwrapped and the helical channel is assumed to make one full turn on the crankshaft, so the area of the cross sections expands half distance until a virtual neighboring channel. Figure 2 shows the volume of the channel with the position of the cross section and the unwrapped geometry with the coordinate system.

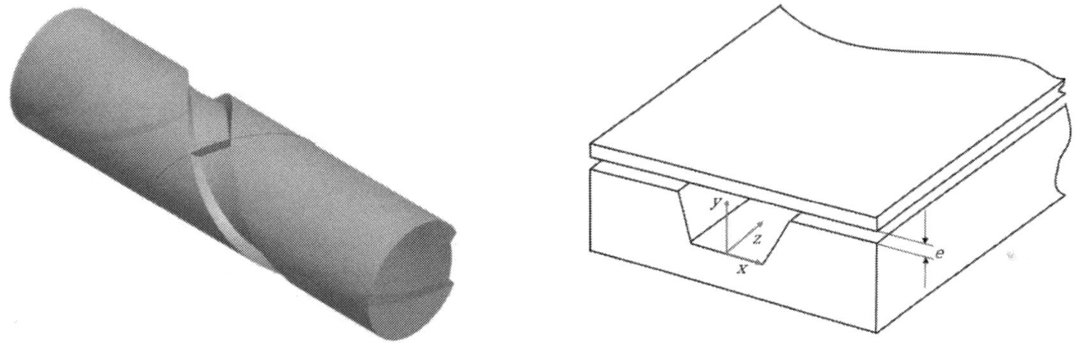

Figure 2. The position of the cross-section in the channel and the unwrapped geometry.

The following assumptions are made for the present problem: (i) the flow field is fully developed in *z*-direction and (ii) the body forces in *y*-direction are neglected. With these assumptions the above equations can be reduced to:

$$\rho\left(u\frac{\partial u}{\partial x}+v\frac{\partial u}{\partial y}\right)=-\frac{\partial p}{\partial x}+\mu\left(\frac{\partial^2 u}{\partial x^2}+\frac{\partial^2 u}{\partial y^2}\right)+\rho g_x \tag{4}$$

$$\rho\left(u\frac{\partial v}{\partial x}+v\frac{\partial v}{\partial y}\right)=-\frac{\partial p}{\partial y}+\mu\left(\frac{\partial^2 v}{\partial x^2}+\frac{\partial^2 v}{\partial y^2}\right) \tag{5}$$

$$\rho\left(u\frac{\partial w}{\partial x}+v\frac{\partial w}{\partial y}\right)=-\frac{\partial p}{\partial z}+\mu\left(\frac{\partial^2 w}{\partial x^2}+\frac{\partial^2 w}{\partial y^2}\right)+\rho g_z \tag{6}$$

Due to the assumption that the flow is fully developed in *z*-direction, the cross-section flow is independent of the transport velocity w and the equations (4) and (5) can be solved in a two-dimensional domain. The transport velocity w depends on the flow field in the domain and is treated like a passive scalar. The pressure term in the *z*-momentum equation can be neglected, if the pressure difference between the inlet and outlet of the channel is zero. The flow field in the cross-section is modeled with the finite volume method [8] using a grid with approximately 5000 cells. An evaluation concerning the mesh size has shown that there is no significant deviation between two grids with 5000 and 8000 cells, respectively. Below a grid size of 4000 cells significant deviations could be observed. The SIMPLE algorithm [9] applied on staggered grids in *x*- and *y*-direction is used for the pressure-velocity coupling. The discretization of the diffusion term is done by using the central differencing scheme and the discretization of the convective term is done by using the second order upwind scheme. Under-relaxation factors for pressure and velocity are implemented to get stable computations. The inclined walls in the cross-section are modeled with fine steps to get higher geometrical resolution. Figure 3 shows the computational domain and the set of boundary conditions which take the relative motion between the crankshaft and the barrel wall into account. The relative motion is modeled with the classical pumping model [10] where the screw channel walls are stationary and the barrel surface becomes a flat plate moving with constant velocity. The boundaries at both sides in the gap are modeled with periodic boundaries.

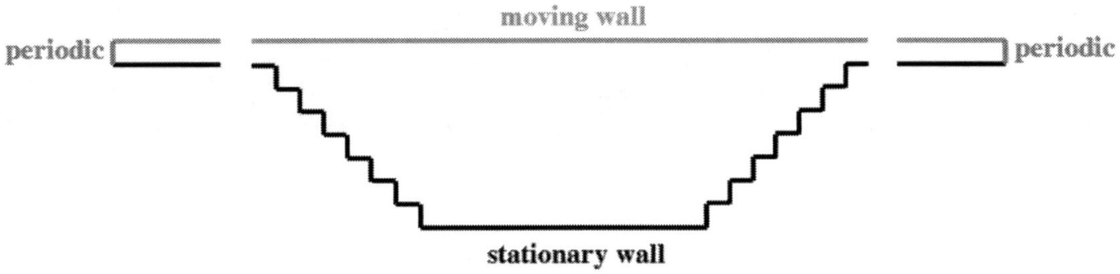

Figure 3. The computational domain with boundary conditions.

The net mass flow is calculated with the total mass flow in *z*-direction and the mass flow through the gap in *x*-direction.

3. Results

A fundamental set of input parameters is defined to verify the present model. The input parameters are related to the geometrical parameters of the investigated crankshaft and the operation parameters are adjusted to the conditions at an oil pump test bench. Table 1 shows the input parameters for the basic version. The test bench is able to measure the oil mass flow rate at different submersion depths of the crankshaft at constant rotational speed. An uncoupled investigation of the helical groove from the rest of the oil path is not possible, a fact which should be considered in the comparison between the calculated and the measured results. Due to the neglect of the pressure difference between inlet and outlet, the influence of submersion depth is not regarded in the numerical model and thus the validation with only one measuring point of the test bench is not meaningful. Two measuring points at different submersion depths are used to define a mass flow range. Table 2 shows the comparison between the calculated mass flow rate of the basic version and the measured mass flow rate. The considerable deviation between the calculation and the measurement can be explained by the neglect of the pressure term in *z*-direction. The pressure term would include the influence of the lower centrifugal pump which results in a higher pressure at the inlet compared to the outlet and an increase of the mass flow rate.

Table 1. Input parameters for the basic version.

Density [kg/m³]	832
Viscosity [cSt]	8
Rotational speed [rpm]	3000
Gap width [mm]	0.1
Helix angle [°]	49.3

Table 2. Comparison between calculated and measured mass flow rate (normalized on basic version).

Measured 7 mm submersion depth	1.51
Measured 2 mm submersion depth	1.23
Calculated basic version	1

The results of the simulated flow field are shown in Figure 4. The flow field in the gap between the crankshaft and the stationary wall appears as a simple Couette flow with gravitational influence. In the area of the channel the flow is diverted by the inclined walls and a vortex occurs. Due to the high dissipative impact of the oil on the flow the vortex is concentrated mostly on the top right side of the channel.

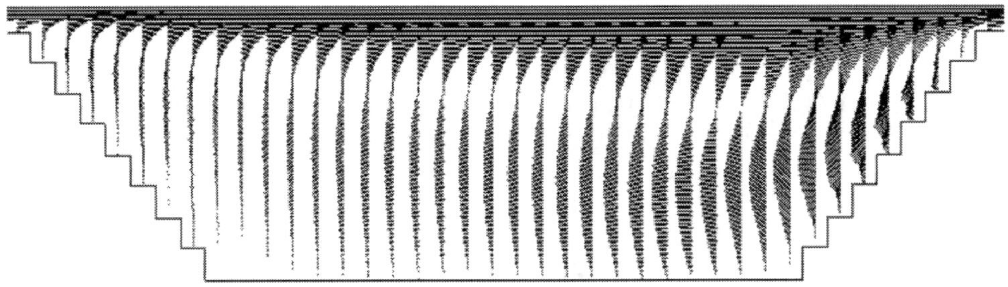

Figure 4. Velocity field in the channel cross-section.

After the simulation of the two-dimensional flow field, the calculation of the velocity in *z*-direction can be carried out. The result of the simulated *w*-field is shown in Figure 5. The classical analytical models ([5], [6] and [7]) solve only the Poisson equation of the velocity in channel direction which results in a symmetric solution over the channel cross-section. Compared to these models, the present model includes the influence of the two-dimensional flow field and a non-symmetric distribution of the *w*-velocity can be seen. The influence of the above mentioned vortex results in higher values of the *w*-velocity, concentrated at a small area at the top right of the domain. A comparison between the simulation with and without respect to the two-dimensional flow field and the gravitational forces results in a 3.15 times higher mass flow rate in the model without influence of the flow field. This shows that the cross-section flow field and the gravitational forces have a considerable influence on the results of the mass flow rate.

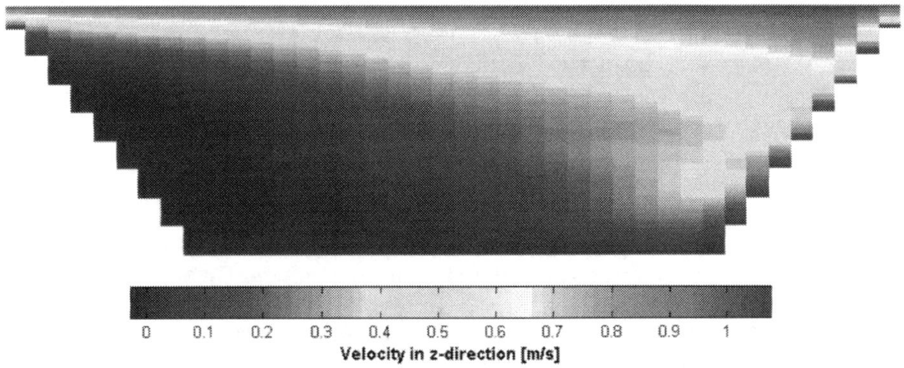

Figure 5. Distribution of the *w*-velocity.

The present model is used to consider the influence of the rotational speed, the oil viscosity, the gap width and the helix angle on the oil mass flow rate. The results are normalized to the oil mass flow rate of the basic version. The investigation of the influence of the several parameters on the friction behaviour of the oil pump is not included in the following variations.

3.1. Rotational speed

The influence of the rotational speed of the compressor on the operation behaviour of the oil pump is a very important detail in the design of variable speed compressors. Figure 6 shows the trend of the oil mass flow rate over the rotational speed. The variation of the speed shows an approximately linear dependency on the mass flow rate. Especially if the compressor runs at speeds below 2000 rpm for longer periods, the design of the oil pump should be improved to avoid lubrication failures.

3.2. Oil viscosity

As shown in Figure 7, the oil mass flow rate decreases significantly at lower viscosity. With oil viscosity over 8 cSt, which is used in the base calculation, an improvement in the efficiency of the oil pump regarding the mass flow rate could be observed. Due to the strong dependence on the temperature, the viscosity underlies significant variations during the compressor operation. Especially during so called 'Pull-Down' cycles in which the thermal load on the compressor is very high, the oil mass flow rate might decrease. The thermal behaviour of the compressor has a considerable impact on the choice of the oil.

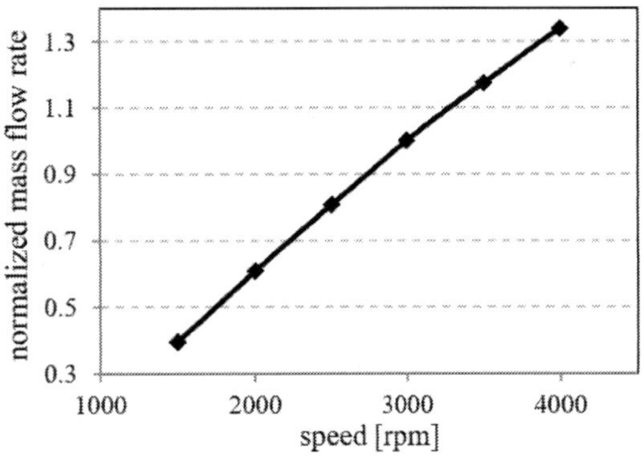

Figure 6. Influence of the rotational speed on the oil mass flow rate.

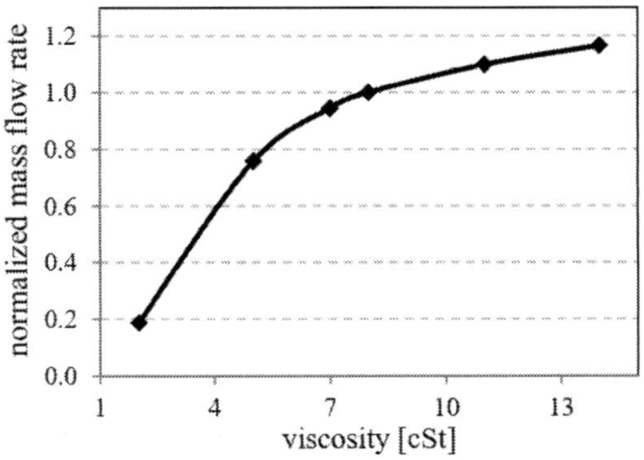

Figure 7. Influence of the oil viscosity on the oil mass flow rate.

3.3. Gap width

The width of the gap between the rotational crankshaft and the stationary barrel wall shows an approximately linear dependency on the oil mass flow rate (Figure 8). A smaller gap width results in a higher value of the mass flow rate due to a stronger influence of the velocity difference of the two parts on the velocity in channel direction. Also the higher mass flow rate in the gap at higher gap widths has a negative effect on the oil pump mass flow rate because as the mass flow rate in gap direction increases, the effective mass flow rate in channel direction decreases.

3.4. Helix angle

Another important influence factor on the pumping efficiency is the helix angle. The results of the helix angle variation are shown in Figure 9. Based on the vector components of the relative velocity between crankshaft and barrel wall, a smaller helix angle results in a higher velocity in channel direction which has a direct impact on the oil mass flow rate. An increase of the helix angle of 10° compared to the basic version has the same effect as a decrease of the rotational speed of about 2000 rpm.

IOP Conf. Series: Materials Science and Engineering **90** (2015) 012028 doi:10.1088/1757-899X/90/1/012028

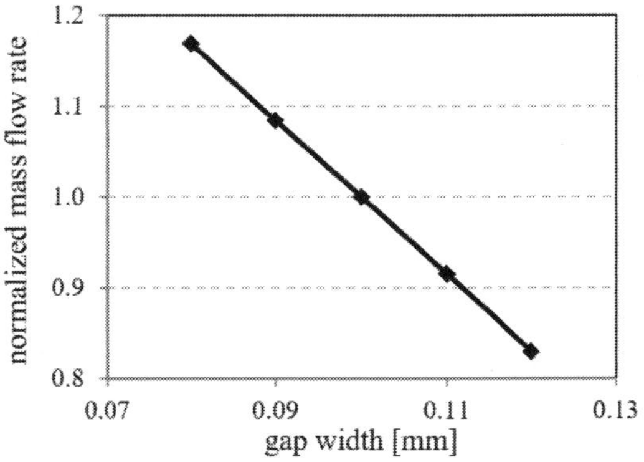

Figure 8. Influence of the gap width on the oil mass flow rate.

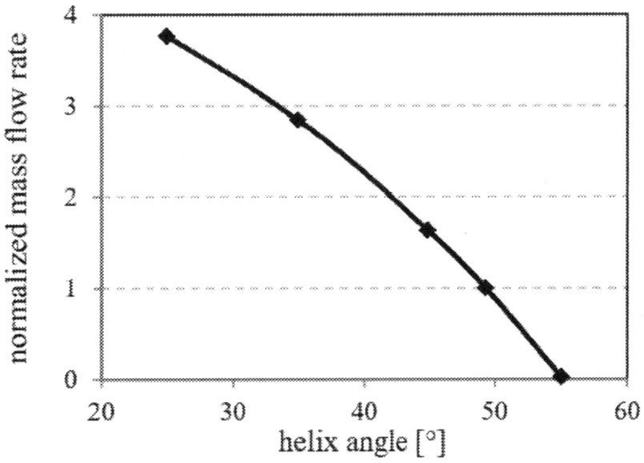

Figure 9. Influence of the helix angle on the oil mass flow rate.

3.5. Design variation
In order to show how to improve the pump efficiency for compressors which operate at low rotational speed, a calculation of a version with 1500 rpm, gap width of 0.08 mm and helix angle of 25° was carried out. Compared to the version without design variations, the mass flow rate can be increased significantly. The normalized mass flow rate with this configuration is 1.7. An additional investigation of the friction losses should be carried out to get a full evaluation of the design variation.

4. Conclusion
A numerical model for the investigation of the helical oil pump of a hermetic compressor for household appliances is presented in this paper. The model is based on the governing equations of motion which were adopted for the present problem. Compared to analytical models from published literature, the present model is able to calculate the flow field in the channel at higher geometrical resolution including the gap between the crank shaft and the barrel wall. The proposed model was used to carry out a parameter variation. The influence of the rotational speed, the oil viscosity, the gap width and the helix angle was investigated. An additional design variation was carried out to show the possibility to increase the mass flow rate at lower rotational speed with a decrease of the gap width and the helix angle.

The model can be used to rate design parameters of helical oil pumps at higher geometrical resolution compared to analytical models but less effort compared to three-dimensional CFD simulations with commercial tools as the time consuming meshing process is omitted. The influence of adjoining pumping devices on the boundary conditions of the model and a validation of the energy consumption of the pump will be explored in a future study.

Acknowledgements
This work has been carried out within the framework of ECO-COOL, a research project initiated and funded by the FFG (Austrian Research Promotion Agency). Furthermore the authors particularly acknowledge the technical support by Secop Austria GmbH, formerly ACC Austria GmbH and Liebherr-Hausgeräte Lienz GmbH.

References

[1] Lückmann A J, Alves M V C and Barbosa Jr J R 2009 Analysis of Oil Pumping in a Reciprocating Compressor *Appl. Thermal Eng.* **29** 3118-3123
[2] Kerpicci H, Alper Y and Onbasioglu S U 2013 Investigation of oil flow in a hermetic reciprocating compressor *Int. J. Refrig.* **36** 215-221
[3] Tada M P, Hoffmann T, Couto P R, Manke A L and de Bortoli G D 2014 Numerical and Experimental Examination for Oil Pump System Using a Simplified Uncoupled Simulation Model *Proc. 22nd Int. Compr. Eng. Conf. at Purdue* **1248** 1-7
[4] Kim H J, Lee T J, Kim K H and Bae Y J 2002 Numerical Simulation of Oil Supply System of Reciprocating Compressor for Household Refrigerators *Proc. 16th Int. Compr. Eng. Conf. at Purdue* **1533** 1-8
[5] Li Y and Hsieh F 1994 Modeling of Flow in a Single Screw Extruder *Journal of Food Engineering* **27** 353-375
[6] Alves M V C, Barbosa Jr J R and Prata A T 2009 Analytical solution of single screw extrusion applicable to intermediate values of screw channel aspect ratio *Journal of Food Engineering* **92** 152-156
[7] Alves M V C, Barbosa Jr J R, Prata A T and Ribas Jr F A 2010 Fluid Flow in a Screw Pump Oil Supply System for Reciprocating Compressors *Proc. 20th Int. Compr. Eng. Conf. at Purdue* **1326** 1-8
[8] Versteeg H K and Malalasekera W 2007 An Introduction to Computational Fluid Dynamics (Harlow: Pearson Prentice-Hall)
[9] Patankar S V and Spalding D B 1972 A Calculation Procedure for Heat, Mass and Momentum Transfer in Three-dimensional Parabolic Flows *Int. J. Heat Mass Transfer* **15** 1787
[10] Tadmor Z and Gogos C G 2006 Principles of Polymer Processing (Hoboken: John Wiley & Sons Inc.)

Performance Comparison of Capacity Control Methods for Reciprocating Compressors

L. Wang[1], G.B. Liu, Y.Y. Zhao, L.L. Li

National Key Laboratory of Compressor Technology, Hefei General Machinery Research Institute, Hefei, P.R.China

wangle4127@gmail.com

Abstract. Different capacity control methods are used for adjusting suction flow of reciprocating compressors to meet process need. Compared with recycle or bypass and suction throttling, three capacity control methods of speed control, clearance pockets and suction valve unloading are preferred due to their energy-saving at operating condition of partial load. The paper reviewed state of the art of the current capacity control technologies and their principles. A comprehensive mathematical model was developed to predict thermodynamic and dynamic performance of reciprocating compressors equipped with the capacity control systems of four above-mentioned methods. Comparison of shaft work and mechanical efficiency were conducted for different capacity control methods at the same condition. In addition, their influence on *p-v* diagram and valve motion were also studied, which is important for reliability and life of the reciprocating compressors. These results were helpful for selection of the capacity control systems by end-users and optimum design by manufacturers.

1. Introduction

Reciprocating compressors play a major role in the chemical, petrochemical, gas, and general industry processes. Unlike centrifugal pumps, the reciprocating compressor cannot self-regulate its capacity against a given discharge pressure, it will simply keep displacing gas until it is turned off. However, in the real processes of refineries, chemical plants, and gas transmission lines, we find that we have specific parameters within which to work, and that capacity is a unique quantity at any point in time. Thus, compressor capacity control methods are utilized to maintain a required delivery under variable process conditions.

The principle of several capacity control methods was presented early since 1930s [1]. In recent years, some innovation was introduced to the traditional capacity control methods. Automated, variable volume unloaders for clearance pocket capacity control provide the ability to smoothly load/unload reciprocating compressors to maintain ideal operations in ever-changing environments [2]. Stepless capacity regulation utilizes a very fast, digitally controlled, hydraulic solenoid valve to control suction valve opening time [3]. Some new concepts of capacity control were developed. Based on the pulse signal concept, a duty cycle regulation method for capacity control was introduced [4]. The Southwest Research Institute (SwRI) considered that the capacity control method which used linear motor to change length of piston stroke would be very promising [5].

2. Model formulation for capacity control methods

Content from this work may be used under the terms of the Creative Commons Attribution 3.0 licence. Any further distribution of this work must maintain attribution to the author(s) and the title of the work, journal citation and DOI.
Published under licence by IOP Publishing Ltd

A comprehensive mathematical model is proposed to evaluate the compression work and the friction losses for reciprocating compressors. The paper reviews the basic theories and the state of the art of the five capacity control technologies, namely recycle or bypass throttling, speed control, clearance pockets and suction valve unloading. Each of the capacity control methods is modeled respectively and the pressure-volume relationship and the valve displacement-angle relationship are obtained.

2.1. Modelling of compressor performance

The basic models describing the working process of reciprocating compressor include governing equations, internal leakage model and valve motion model [6]. The governing equation group is based on energy and mass conservation as well as equation of state as following:

$$\frac{dT}{d\theta}=\frac{1}{mC_v}\left\{-T\left(\frac{\partial P}{\partial T}\right)_v\left[\frac{dV}{d\theta}-\frac{v}{\omega}(\dot{m}_{in}-\dot{m}_{out})\right]-\sum\frac{\dot{m}_{in}}{\omega}(h-h_{in})+\frac{\dot{Q}}{\omega}\right\} \tag{1}$$

$$\frac{dm}{d\theta}=\sum\frac{\dot{m}_{in}}{\omega}-\sum\frac{\dot{m}_{out}}{\omega} \tag{2}$$

$$P=P(v,T) \tag{3}$$

By solving the governing equations considering the internal leakage and the valve motion simultaneously, some parameters such as the mass, temperature, pressure of working fluid and the valve displacement with respect to the orbiting angle at any working condition can be obtained. Thus, compression work P_c is calculated from p-V diagram by means of integrating the volume by the change of the pressure over the whole working process:

$$P_c=\oint V\mathrm{d}p \tag{3}$$

In addition, a novel friction model basing on journal bearing theory for reciprocating compressor is used to evaluate friction losses [7]. Hydrodynamic lubrication theory is applied to calculate the friction losses P_f of the journal bearings supporting the crankshaft and the connecting rod. Thus, mechanical efficiency η is ratio of the compression work to the shaft work P:

$$P=P_c+P_f$$
$$\eta=\frac{P_c}{P_c+P_f} \tag{5}$$

2.2. Recycle or bypass

One of the simplest methods of controlling capacity is to recycle, or bypass. To reduce the flow to process, one simply opens up the bypass line through some type of control valve and diverts the excess flow back to the compressor suction. This control scheme is extremely inefficient because the backflow gas is recompressed and the compression work is totally wasted. As a result this control technique is rarely used. Only in case of failure of other control methods, the control method of recycle or bypass is activated for safety. This control method does not make any changes to the working process of the reciprocating compressor.

2.3. Suction throttling

The gas suction pressure is reduced by throttling of a control valve into the compressor inlet. The density of the gas will be reduced at this lower pressure, thus helping to reduce the mass flow delivered. As a rule, the gas suction pressure is controlled as a function of either discharge pressure or flow. It is assumed that the process of the throttling is isenthalpic and the gas temperature is gotten by the equation of state.

Figure 1 and 2 show p-V and h-θ relationship under different suction pressure. As the suction pressure is reduced and the discharge pressure held constant, the compression ratio is increased. Suction throttling moves the characteristic curve of p-V to the left, indicating a lower volumetric efficiency and thus less flow. The valve motion is also altered by the increase of the compression ration. A sizable reduction in capacity means a dramatic increase of the compression ratio. This causes higher discharge temperatures. In the case, the discharge temperature is rising from 92℃ to 210℃ as the compression ratio rising from 2 to 5 for air.

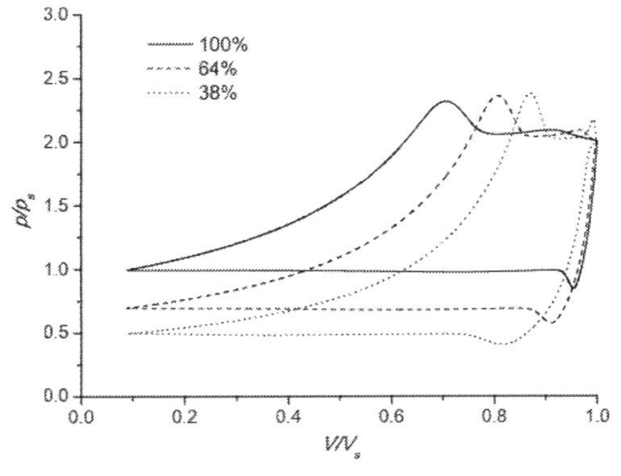

Figure 1. p-V diagram under different flow rates for suction throttling

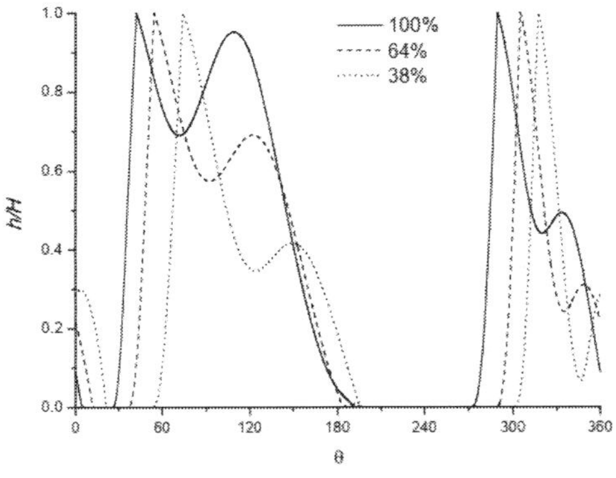

Figure 2. h-θ diagram under different flow rates for suction throttling

2.4. Speed control

Inverter-driven motors make it possible that the capacity is matched to the load by regulating the speed of the compressor. This method is widely used for refrigeration and air-conditioning system. However, it is rarely used in process compressor due to the problems of the cost and the reliability of the large-scale inverter-driven motors.

As shown in figure 3, the gas mass flowing into the compressor during one resolution is nearly equal for different orbiting speeds. Thus, the gas mass flow rate is approximately proportional to the orbiting speed. As shown in figure 4, the orbiting speed influences the valve motion. When the orbiting speed decreases, the status of both the suction and discharge valves gets worse.

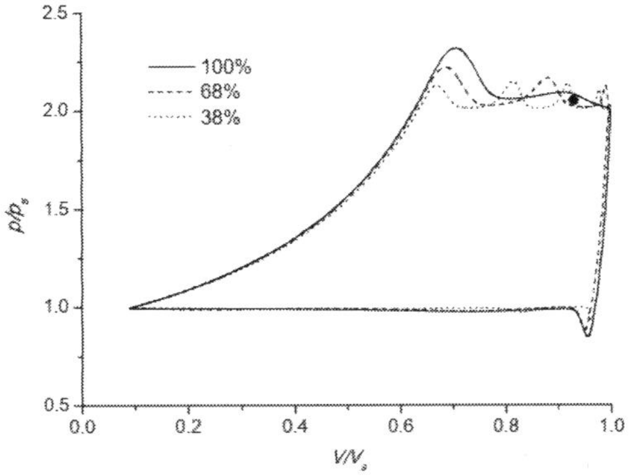

Figure 3. p-V diagram under different flow rates for speed control

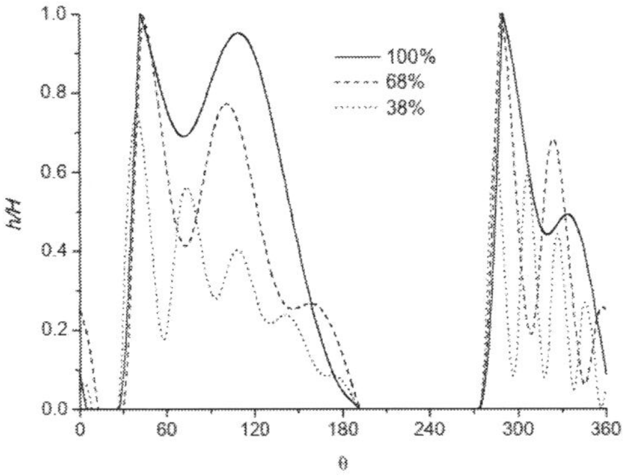

Figure 4. h-θ diagram under different flow rates for speed control

2.5. Clearance pockets

Volume pockets vary the compressor capacity by adding clearance to the piston area. Recently hydraulically controlled variable-volume clearance pocket instead of having fixed-steps is equipped in more and more process reciprocating compressor.

As clearance is increased, figure 5 illustrates that the effective suction volume (volumetric efficiency) decreases, thus reducing the amount of flow to process. Figure 6 illustrates that the time for the opening of both the suction and the discharge valve also decreases. It should be noted that the volumetric efficiency is not only affected by increased clearance but also is governed by the compression ratio. When the clearance volumetric coefficient is 0.8 for the compression ratio of 2.0, the flow rate is only 45% of the rated one. So, clearance pockets are ineffective and not useful with low compression ratios.

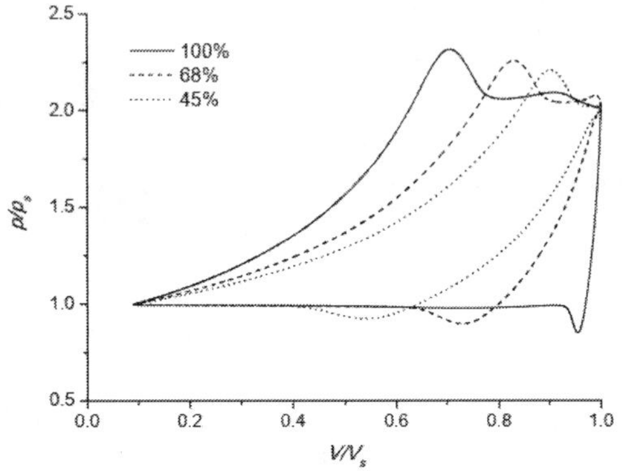

Figure 5. *p-V* diagram under different flow rates for clearance pocket

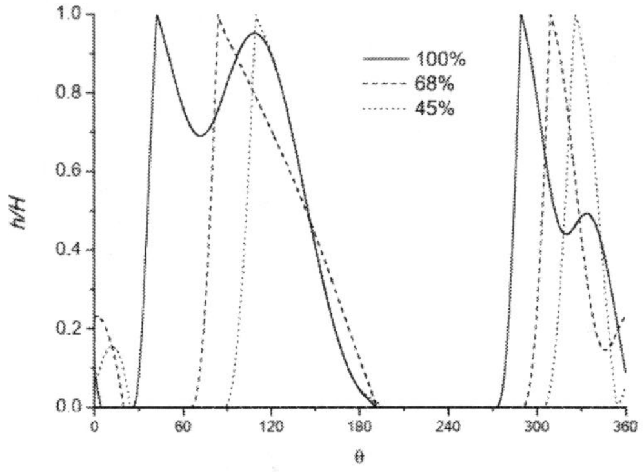

Figure 6. *h-θ* diagram under different flow rates for clearance pocket

2.6. suction valve unloading

Many researchers pay attention to the stepless capacity control system [7]. This system uses finger-type unloaders that are pneumatically actuated. It is to physically keep the cylinder from compressing gas by maintaining the suction valve opening for only a portion of one stroke. It is assumed that when suction valve unloading takes effect, the suction valve displacement h_s is equal to the valve stop H as follows:

$$\theta_0 \le \theta \le \theta_1 \quad h_s = H \tag{3}$$

θ_0 and θ_1 is the starting and ending orbiting angle for the suction valve opening respectively.

As shown in figure 7, suction valve unloading makes the curve of *p-V* move to the left, which means that the time for the suction process is extended and that for the compression process is shortened. As shown in figure 8, the displacement of the suction valve is set to the valve stop since the orbiting angle of 90 degree.

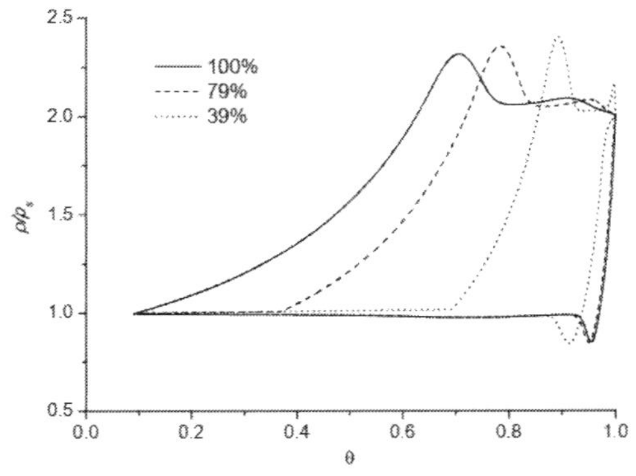

Figure 7. *p*-*V* diagram under different flow rates for suction valve unloading

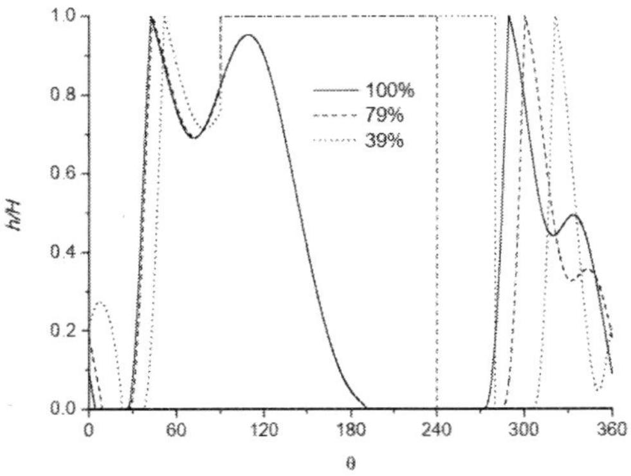

Figure 8. *h*-*θ* diagram under different flow rates for suction valve unloading

3. Performance comparison of different capacity control methods

Figure 9 plots the relationship between the shaft work (dimensionless) and the flow rate (dimensionless) for different capacity control methods. Except for the suction throttling, the shaft work is nearly linear to the flow rate for other three methods. The speed control is most efficient. The shaft work of the suction valve unloading is almost close to that of clearance pocket. The shaft work of the suction valve throttle is even a litter larger than the rated one at the rated flow rate of 60% to 100%.

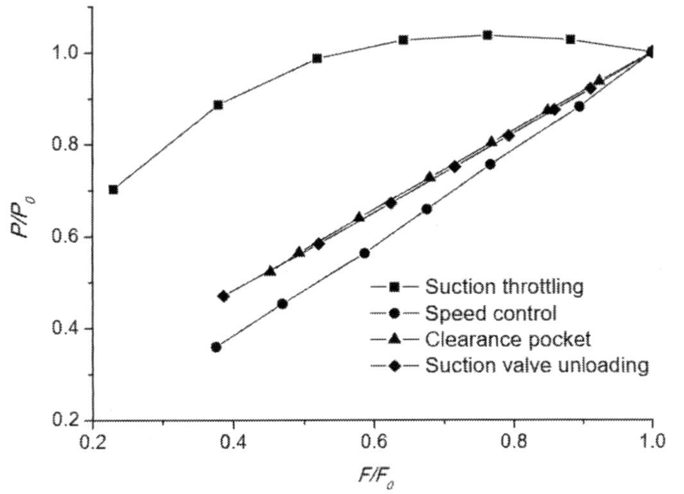

Figure 9. Comparison of shaft works for different capacity control methods

The mechanical efficiencies (dimensionless) in terms of the flow rate (dimensionless) for different capacity control methods are plotted in figure 10. The mechanical efficiency increases with the decrease of the rotation speed and is more than 1, which is attributed to reduction of friction losses. The mechanical efficiency for other three capacity control is less than 1, because the isentropic efficiency of the compression process decreases, which caused by the increase of the volumetric efficiency. Similar to figure 9, the mechanical efficiency for both the suction valve unloading and the clearance pocket shows the same behaviour.

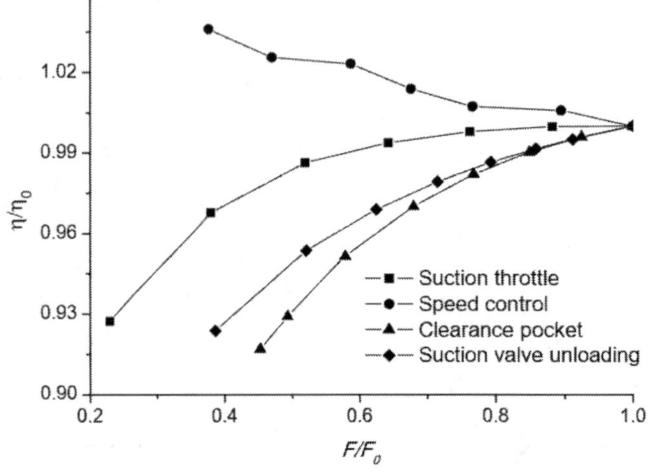

Figure 10. Comparison of mechanical efficiencies for different capacity control methods

4. Conclusion

Different capacity control methods are reviewed and simulated and their characteristics of p-V and h-θ curve are studied. The speed control is most efficient and its mechanical efficiency increases with the decrease of the rotation speed. The behaviour of the suction valve unloading and the clearance pocket for the shaft power and the mechanical efficiency are similar to each other.

Acknowledge

The authors gratefully acknowledge the financial support provided by the China National Basic Research Programme (Grant No. 2012CB026000), Anhui Provincial Natural Science Foundation (NO. 1408085MKL59).

Nomenclature

h	Enthalpy of working fluid, Valve displacement	P_f	Friction power
p	Pressure of working fluid	P_c	Shaft power
T	Temperature of working fluid	P	Power input
H	Valve stop	*Greek letters*	
m	Mass of working fluid	ρ	Density
$\dot{m}$	Working fluid mass flow rate	η	Efficiency
$\dot{Q}$	Heat flow rate into the control volume	θ	Orbiting angle
t	Time	ω	Angel speed
C_v	Specific heat capacity at constant volume	*Subscripts*	
v	Specific volume of working fluid	in	Flow in
V	Volume of the control volume	out	Flow out

References

[1] Heinz P 2006 *A Practical Guide to Compressor Technology 2nd* (New York: John Wiley &

Sons)
[2] Dwayne A, John S, Bruce M, Tom J 2002 *4th International Pipeline Conference* (Canada: ASME)
[3] Steinrueck P, Ottitsch F, Oberhuber A, Linskeseder M 1997 *Hycrocarb. Process* **76** 79-83
[4] Gu Z, Hou X, Wang Z, Feng S, Gao X, Li Y *Chinese Sci. Bull.* **56** 1967-74
[5] Danny M, Deffenbaugh A, Smalley K 2005 *Advanced Reciprocating Compression Technology* (USA: DOE)
[6] Wang L, Tang B, Li L 2011 The 11th Asian International Conference on Fluid Machinery (IIT Madras: AICFM)
[7] Wang L, Tang B, Zhao Y 2014 *Applied Mechanics and Materials* (Swizerland: Trans Tech Publications) vol 456 p320-23
[8] Tang B, Zhao Y, Li L, Liu G, Wang L, Yang Q, Xu H, Zhu F, Meng H 2013 *Appl. Therm. Eng.* **54** 380-86

IOP Conf. Series: Materials Science and Engineering **90** (2015) 012030 doi:10.1088/1757-899X/90/1/012030

PERFORMANCE ENHANCEMENT OF HERMETIC COMPRESSOR USING PHASE CHANGE MATERIALS

I M Mahmoud [a*], M A Rady [b], A S Huzayyin [a]

a Benha Faculty of Engineering, Benha University b Helwan Faculty of Engineering, Helwan University

Abstract

The present study is motivated by the need for the research of simple measures for increasing energy efficiency of hermetic compressor. The measure is the application of phase change materials for performance enhancement. The first experimental study should be guide for choice of PCM. It has been performed to investigate the effects of thermostat setting temperature on the performance of hermetic compressor. The effects of thermostat setting temperature with and without load on power consumption have been analyzed. Performance enhancement using phase change materials (PCMs) has been studied by employing a phase change material Rubitherm-42 (RT-42) on the top surface of compressor. Choice of PCM material is based on basic compressor performance measured in the first part of the present study. Experiments have been carried out for different load values and different quantities of PCM. The quantity and phase change characteristic of PCM are essential parameters that determine the percentage of performance enhancement in term of energy consumption. Reduction of energy consumption of about 10% has been achieved in the present study by using PCM. The present study shows that how to reduce the electrical power consumption to enhance compressor heat dissipation method to improve efficiency.

Keywords: Compressor, efficiency, Enhancement, Phase change material.

1. Introduction

Compressor is defined as component in any system to remove heat from any space. Compressor is using in small application such as refrigerator, large application such as chiller. The hermetic compressor is studied in this paper. Main components of a hermetic reciprocating compressor include a piston, cylinder, intake and discharge lines. The compressor also contains lubricant oil in the bottom. A schematic of such a compressor is shown in figure 1.

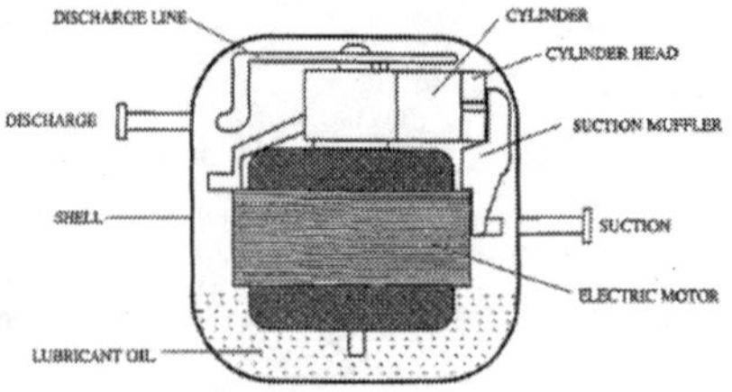

Figure 1. Schematic of a hermetic reciprocating compressor [Cavallini et al., 1996]

Content from this work may be used under the terms of the Creative Commons Attribution 3.0 licence. Any further distribution of this work must maintain attribution to the author(s) and the title of the work, journal citation and DOI.
Published under licence by IOP Publishing Ltd

Using hermetic reciprocating compressor is lead to design to satisfy the maximum load **[1]**. Variable Speed Compressor (VSC) and Variable Capacity Compressor (VCC) are efficient alternatives to control the refrigerating capacity, which continuously matches the compressor speed to the thermal load **[2]**. Embraco (2005) developed a VCC for household refrigerator using compressor. This test results show up to 45% energy saving by replacing a conventional on/off compressor by a VCC compressor technology **[3]**.

Compared the performance of different positive displacement compressors, including an open-type reciprocating, a semi-hermetic reciprocating and an open-type rotary vane compressor, for refrigeration applications are discussed. The results indicated that, all compressors showed an energy economy in the range of 12–24% when a speed control was used in comparison to the ones operating at constant speed **[4,5]**. Developed has been a numerical simulation of a refrigeration system with speed control. Simulations were carried out to verify the possibility of controlling the refrigeration system and the superheating of the refrigerant at the outlet of the evaporator by varying the compressor speed. The results suggested that the proposed model could be used to formulate an algorithm to control refrigeration systems **[2]**.

Studied the capacity control of a refrigeration system by injecting hot gas straight into the compressor suction side and demonstrated that the compressor discharge temperature significantly increases for this set-up **[6]**. On a later study carried out a comparative study of three different capacity control schemes of a vapor-compression refrigeration system. The comparative study was performed in terms of the system coefficient of performance and the operating temperatures, among other parameters **[7]**.

Bitzer compressors **[8,9]** are used inventor frequency with reciprocating compressors, hermetic and semi-hermetic. The main study used two frequencies and speed ranges are follows that: i) standard range of 30–60 Hz/850–1750 rpm, and (ii) extended range of 2570 Hz/ 700–2050 rpm.

Leducq et al. **[10]** mention that, although some commercial systems are already equipped with variable speed-compressors, the lack of knowledge about the dynamic behavior of these systems leads to poor control performance improvement. The industries where such systems have been implemented still have not been able to take full advantage of them.

In the present work temperature profile and power consumption have been measured for the different load of compressor. The effect of load on power consumption, efficiency has been analyzed.

2. Materials and methods

A compressor is used in the present work with the following characteristics:

- Compressor: Hermetic reciprocating.
- Model: HSL11Y-5, with mineral oil lubricant.
- Rated volt: 220-240 Volt.
- Frequency: 50 Hz.
- Rated current: 0.7 Ampere.
- Displacement: 4.51 cc/rev.
- Consumption power: 109 W.

- Cooling Capacity: 93 Watt.
- EER (Energy Efficiency Rate) =2.91 Btu/Wh.
- Motor type: Resistance Start Induction Run (RSIR).
- Cooling type: Static Cooling (ST).
- Net weight: 7 kg.
- Oil Qty.: 230 cc

Where Test Condition (ASHRAE) LBP (Low Back Pressure) applications:

1- Evaporating temperature : -23 °C (-10 F).
2- Condensing temperature : 54.4 °C (130 F).
3- Ambient temperature : 32.2 °C (90 F).
4- Gas superheated to : 32.2 °C (90 F).
5- Liquid subcooled to : 32.2 °C (90 F).

- Refrigerant: 550 g of Tetrafluoroethane (R134a).
- Lamp power: 15 Watt.
- Control: On/off by thermostat setting and self defrost. The thermostatic control of the original appliance is driven by a bulb placed in contact with the evaporator wall.
- Refrigerant: R134a.

Figure 2 shows the experimental set up. Temperatures at various locations (compressor inlet, compressor outlet and on the compressor surface) are measured with copper-constantan thermocouples type T, 0.5 x 10^{-3} m diameter. Compressor power consumption is measured by recording the instantaneous volt and current data entering to the computer. The measurement circuit includes current and voltage transformer. The uncertainty of measurements is ± 0.02% for temperature and ± 1% for power.

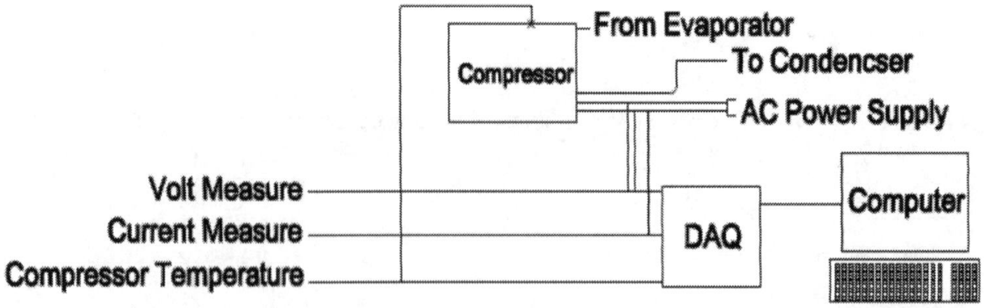

Figure 2. Experimental Setup

All signals output for thermocouple and power are connected to an NI DAQ system, consisting of a NI cDAQ-9178 chassis, NI 1913 thermocouple module input and an NI 9219 universal analog input module for measuring voltage and current.

All experiments were carried out in a temperature-controlled room (20 ± 0.2 °C). Stability of room temperature was first assured before starting any experiment. The hermetic reciprocating compressor is then allowed to work continuously at a selected thermostat setting temperature for at least 48 hours. Measurements of temperature and power consumption are recorded every 1 second. Data are stored in separate files for analysis.

3. Result and Discussion

In the following sections steady periodic results of the experimental measurements are anyalzed for two main cases. Namely, no load and load conditions.

The results are presented thermostat setting temperatures as shown in Table 1. It should be noted that the start and stop compressor are observed to be identical for a given thermostat setting at different load conditions.

Table 1. Thermostat setting temperature (Setting 4).

Thermostat Setting	**Setting Temperature (°C)**	
	Setting Value	**Range**
Setting 4	- 16.6	± 8

Table 2 shows total cycle time, compressor operation time and idle time in seconds at thermostat setting temperature -16.6 and no load with and without PCM.

Table 2. Total cycle time, operation time and idle time of compressor for no load at thermostat setting 4.

Thermostat Setting	**Cycle Times (s)**					
	Without PCM			**With PCM (10 g)**		
	On	**Off**	**Total Cycle**	**On**	**Off**	**Total Cycle**
Setting 4	166	445	611	178	430	608

The compressor temperatures corresponding to these setting are observed to be in the range of 40-46 °C. This range is compatible with the PCM phase change temperatures because the choice of PCMs is applicable. Thermostat settings 4 has been used to study the effect of PCM to match the phase change temperature range of RT42 (38-43 °C).

Table 3 shows the effect of PCM on refrigerator performance, summary of experiments.

Table 3. Effect of PCM on refrigerator performance, summary of experiments.

Exp. No.	Setting Temperature (°C)	Quantity of PCM (grams)	load condition (Watt)
1	- 16.6	0	0
2		10	0
3	- 16.6	0	5
4		5	5
5		10	5
6		15	5

Figure 3 illustrates variation of surface compressor temperature and power consumption for the case of no load at thermostat setting 4. As shown in Figure 3 below the compressor undergoes an initial transient period after which steady state periodic conditions are observed.

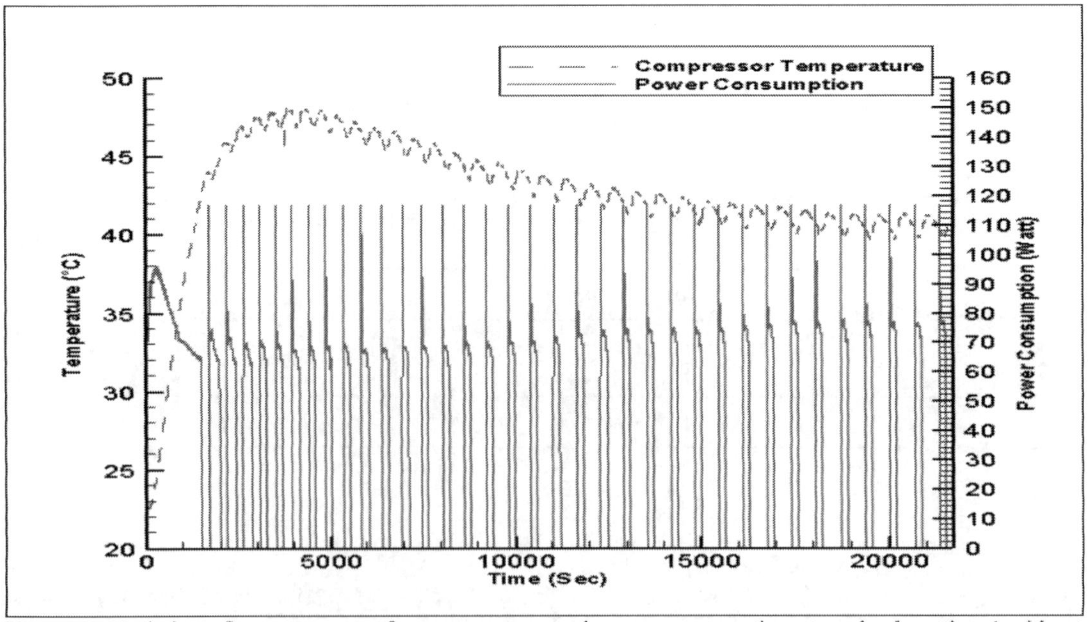

Figure 3. Variation of compressor surface temperature and power consumption at no load, setting 4 without PCM (transient state).

IOP Conf. Series: Materials Science and Engineering **90** (2015) 012030 doi:10.1088/1757-899X/90/1/012030

Performance comparisons at thermostat setting 4 with no load

Figure 4 illustrates the comparison of variation of temperature and power consumption of compressor at no load with PCM (10 g) and without PCM at thermostat setting 4 for a typical cycle.

It can be observed that the compressor is "on" for longer continuous periods while using 10 grams of PCM and thus resulting in the higher energy consumption as compared to the case without PCM at thermostat setting 4.

It has been observed during the experiment of using 10 grams of PCM that PCM does not undergo a phase change melting/solidification process at thermostat setting 4.

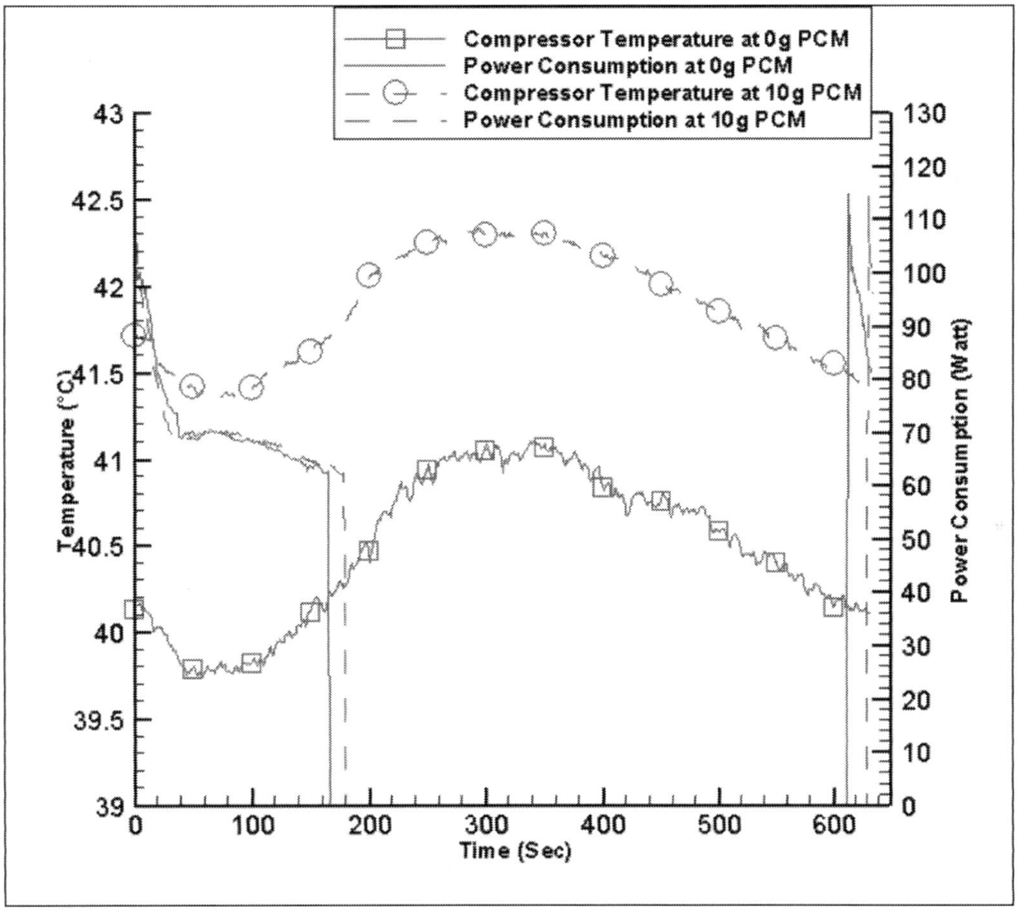

Figure 4. Variation of temperature and power consumption of compressor at no load with PCM (10 g) and without PCM at thermostat setting 4.

In this case, the PCM layer on the compressor top increase the thermal resistance between the compressor and the ambient and results in an adverse effect of an increase in compressor temperature.

Table 4 shows total cycle time, compressor operation time and idle time in seconds at thermostat setting temperature -16.6 and no load with and without PCM.

Table 4. Total cycle time, operation time and idle time of compressor for no load at thermostat setting 4.

Thermostat Setting	**Cycle Times (s)**					
	Without PCM			**With PCM (10 g)**		
	On	**Off**	**Total Cycle**	**On**	**Off**	**Total Cycle**
Setting 4	166	445	611	178	430	608

The total cycle time for thermostat settings 4 with and without PCM are 608 and 611 sec, respectively. The cycling time decreases with PCM applied on the compressor. The decrease of cycle time indicates larger daily operation time of the compressor and a consequent increase in overall energy consumption.

Table 6 shows the energy consumption in 1 hr. (kW.hr) and average compressor temperature at thermostat setting temperature -16.5 °C for no load with and without PCM.

Table 6. Energy consumption results at thermostat setting 4 for no load.

	Without PCM	**With PCM (10 grams)**
Energy Consumption in 1 hr (kW.hr)	0.0191	0.0209
Average Compressor Temperature (°C)	40.4	41.3

The compressor energy consumption for two cases has been calculated for one hour period of steady state operation by integration of the data shown in Figures 4.The results shown in Table 6 shows that an increase in energy consumption of compressor by about 9.44% is observed while using 10 grams of PCM of this thermostat setting.

Performance comparison at thermostat setting 4 at Different Load Conditions

Four experiments have been carried out to investigate the refrigerator performance at different load conditions with and without PCM. Namely, experiments with load value (5 Watt), three quantities of PCM (5, 10 and 15 grams) have been considered at thermostat setting 4.

The choice of above values of refrigerator load and thermostat setting have been done based on preliminary experiments. Present experiments parameters have been chosen to satisfy the first basic requirement of compressor temperature compatible with PCM phase change temperature. The second basic requirement related to the correct quantity of PCM is function of the temperature difference between compressor and ambient condition.

Performance comparisons at thermostat setting 4 for 5 watt load

In the following sections performance comparisons at thermostat setting 4 at load value of 5 watt with various quantities of PCM (0, 5, 10, 15 grams) are analyzed. The range of thermostat setting is -16.6 ± 8 °C.

Figure 5 illustrates the case of variation of temperature and power consumption of compressor at 5 watt load with various quantities of PCM (5, 10 and 15 g) and without PCM at thermostat setting 4.

For easy comparison, Figure 5 shows a typical periodic cycle of compressor operation at steady state for different values of PCM quantities. The start point for all the cases are aligned together for quantitative interpretation of results.

It can be observed from Figure 5 that, as compared to the base case without PCM, the compressor temperature using 5 grams of PCM is lower than without PCM. On the other hand, using 10 and 15 grams of PCM increases the compressor temperature. The compressor operation time for 5 grams is lower than no PCM case. However, using 10 and 15 grams of PCM increases the compressor operation time. Also, during the compressor on time the peak and average power consumption of compressor decreases by using 5 grams of PCM and increases by using 10 and 15 grams of PCM.

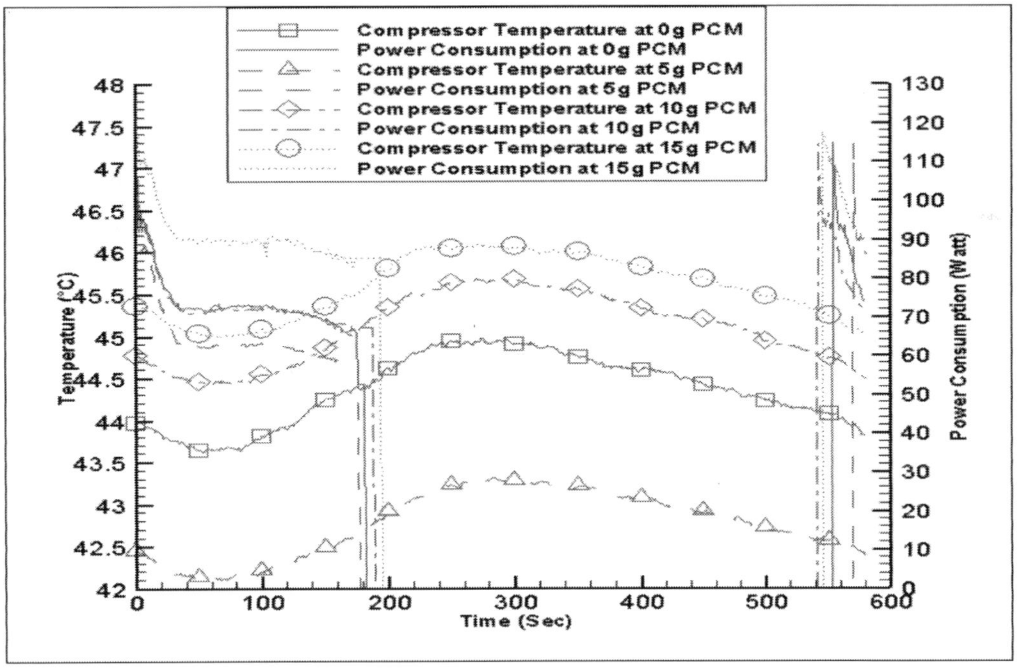

Figure 5. Variation of temperature and power consumption of compressor at 5 watt load with PCM (5, 10 and 15 g) and without PCM at thermostat setting 4.

Table 7 shows total cycle time, operation time and idle time in second at thermostat setting temperature -16.6 °C for different quantity of PCM with load 5 watt.

Table 7. Total cycle time, operation time and idle time of compressor for load 5 watt with different quantities of PCM at thermostat setting 4.

Thermostat Setting	Cycle Times (s)					
	Without PCM			With PCM (5 g)		
	On	Off	Total Cycle	On	Off	Total Cycle
Setting 4	181	371	552	175	392	567
Thermostat Setting	**Cycle Times (s)**					
	With PCM (10 g)			**With PCM (15 g)**		
	On	**Off**	**Total Cycle**	**On**	**Off**	**Total Cycle**
Setting 4	187	352	539	193	350	543

The total cycle time for thermostat settings 4 with and without PCM at different quantities are 567s (at 5g), 539s (at 10g), 543s (at 15g) and 552s (at no PCM) at load value 5 watt. As compared to no PCM, the cycle time decreases and on time increases with PCM applied on the compressor at 10 and 15 g. On the other hand, the cycle time increases and on time decreases with 5 g of PCM applied on the compressor. The increase of on time with decrease of cycle time compressor operation indicates larger daily operation time of the compressor and a consequent increase in overall energy consumption.

Table 8 shows comparison of the energy consumption and average compressor temperature at thermostat setting temperature -16.6 °C for 5 watt load with different PCM quantities. It can be observed that, the energy consumption of compressor decrease by about 9% while using 5 g of PCM as compared to no PCM case.

Table 8. Energy consumption results at thermostat setting 4 for load 5 watt with different quantities of PCM.

	Without PCM	With PCM (5 grams)	With PCM (10 grams)	With PCM (15 grams)
Energy Consumption in 1 hr (kW.hr)	0.0218	0.0198	0.02411	0.0274
Average Compressor Temperature (°C)	44.2	42.6	45.1	45.2

Table 9 shows the percentage increase of energy consumption at different thermostat setting temperature, quantities of PCM and load value condition.

Table 9. % Energy increase at thermostat setting temperature 4, different quantities of PCM and load value conditions.

Exp. No.	Thermostat Setting Temperature (°C)	Quantity of PCM (grams)	No/load Condition (Watt)	Energy Consumption	% Increase of Energy Consumption
1	-16.6	0	0	0.0191	-
2		10	0	0.0209	+ 9.44
3	-16.6	0	5	0.0218	-
4		5	5	0.0198	- 9
5		10	5	0.02411	+ 10
6		15	5	0.0274	+ 25

In general, as a guideline for reduction of energy consumption of hermetic compressor, proper choice of thermostat setting is a zero cost method. For high setting levels (low temperatures), the use of right quantity of PCM is lead to performance enhancement of about 10 %.

4. Conclusion of the Present Work

The major findings and contributions of the present work are summarized as follows:

1- The performance of hermetic compressor decreases with the increase of load value.

2- Performance enhancement methods should focus on enhanced heat dissipation methods for reduction of compressor temperature with the increase of load values.

3- Judicious selection of phase change temperature and quantity of phase change material applied on the compressor could reduce the compressor energy consumption.

4- The feasibility of using PCM applied on the compressor for performance enhancement of household refrigerators has been approved.

5- Enhancement of refrigerator performance is obtained at low values of thermostat setting temperature and using small quantities (5 g) of PCM. Reduction in energy consumption by about 10 % can be achieved.

6- High quantity of PCM may increase the power consumption of the hermetic compressor. This is due to higher compressor temperature.

In general, reduction in energy consumption of the hermetic compressor occurs when the PCM applied on the compressor undergoes a cyclic melting/solidification process. For cases when such processes do not occur the PCM acts as an insulating material for the compressor and worse the performance of refrigerator.

5. References

[1] Azzouz K, Leducq D, Gobin D. Enhancing the performance of household refrigerators with latent heat storage: An experimental investigation. International journal of refrigeration 2009;32:1634–1644.

[2] Koury R N N, Machado L, Ismail K A R, Numerical simulation of a variable speed refrigeration system. International Journal of Refrigeration 2001;24:192–200.

[3] Embraco, 2005. Annual report. Available from: http://www.embraco.com/.

[4] Buzelin L O S, Amico SC, Vargas J V C, Parise J A R. Experimental development of an intelligent refrigeration system 2004; 28: 165-175.

[5] Tassou S A, Qureshi T Q, Comparative performance evaluation of positive displacement compressors in variable-speed refrigeration applications. Intenational Journal Refrigeration 1998; 21:29–41.

[6] M. Yaqub, S.M. Zubair, J.U.R. Khan Performance evaluation of hot-gas by-pass capacity control schemes for refrigeration and air-conditioning systems, Energy 25 (6) (2000) 543–561.

[7] Yaqub M, Zubair S M, Capacity control for refrigeration and air-conditioning systems: a comparative study, J Energy Resour Technol-Trans ASME 123 (1) (2001) 92–99.

[8] BITZER Compressors, http://wwww.bitzer.de/; 2000.

[9] CAVALLINI A, DORETTI L, LONGO GA, ROSETTO L, BELLA B, ZANNERIO A. (1996): Thermal Analysis of a Hermetic Reciprocating Compressor, Proceedings of the 1996 International Compressor Conference, Purdue University, West Lafayette, Indiana, pp. 535-540.

[10] Leducq D, Guilpart J, Trystram G, Low order dynamic model of a vapor compression cycle for process control design, J Food Process Eng 26 (1) (2003) 67–91.

Development of an empirical model of a variable speed vapor injection compressor used in a Modelica-based dynamic model of a residential air source heat pump

Bertrand Dechesne[1], **Stephane Bertagnolio**[2], **Vincent Lemort**[1]

[1] University of Liege, Aerospace and Mechanical Engineering Department, Thermodynamics Laboratory, Chemin des chevreuils 7, Sart Tilman, Liege, Belgium
[2] Emerson Climate Technologies, Welkenraedt, Belgium

E-mail: bdechesne@ulg.ac.be

Abstract. This paper presents a steady-state model of a variable speed vapour injection scroll compressor. Two compressors were investigated. The developped empirical model is based on five dimensionless polynomials that were fitted using experimental data from a 2.7kW scroll compressor. A second set of data was used to show the prediction of the presented model for an other device which exhibits a different swept volume. In both cases, the suction and injection mass flow rate were respectively predicted with a coefficient of determination equal to 99.9 and 94.3% and for the consumed power , 98.4% and 95.6% . A Modelica based dynamic model is then presented. The steady-state validation of the main components models is performed. Finally the control of the cycle using two PID controllers is presented and commented.

1. Introduction

Computational time for dynamic modeling of refrigeration systems is continuously decreasing thanks to the ever-increasing power of computational packages and processors. Dynamic modeling allows engineers to simulate more accurately the control strategies which are of prime importance for system performances.

In this field, Rasmussen *et al.* provides a deep litterature review and very interesting simulation tutorial for the dynamic modeling of vapor compression cycles ([1] and [2]). Hongtao *et al.* ([3] and [4]) presented a model of flash tank injection heat pump developed in Modelica language. The studied compressor is a fixed speed scroll compressor. In [4], they presented results of the validation of their dynamic models for a system start-up and shut-down procedures. Jiazhen *et al.* [5] evaluated the transient performance of a heat pump working with different low-GWP refrigerants (R32 and D2Y60). In 2011, Li *et al.* [6] developed a dynamic model of a R134a automotive air-conditioning and showed the refrigerant mass migration during start-up and shut-down operations.

This paper presents a Modelica-based dynamic model of an air source heat pump (ASHP) using a variable speed vapor injection scroll compressor and R410A as working fluid. This work is based on the open-source and open-access ThermoCycle library [7] for the model components and CoolProp for the fluid thermodynamic properties [8]. An empirical model of the vapor

Content from this work may be used under the terms of the Creative Commons Attribution 3.0 licence. Any further distribution of this work must maintain attribution to the author(s) and the title of the work, journal citation and DOI.
Published under licence by IOP Publishing Ltd

injection scroll compressor is developed using empirical correlations for the volumetric efficiency, isentropic efficiency and the ratio between the injection and suction mass flow rate. A simple control strategy is implemented where the needle position of the injection and heating valves is controlled via PID controllers to obtain a target superheat on the injection line and at the exhaust of the evaporator. By using the simulation model, the performance and the control of the ASHP in the heating mode is presented.

2. System overview

The studied system is represented on figure 1.

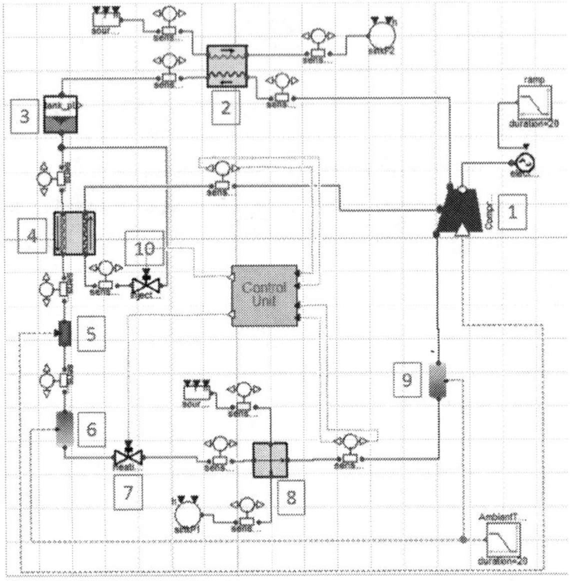

Cycle components:

1. Compressor
2. Condenser
3. Liquid receiver
4. Economizer
5. Drive cooler
6. Liquid line
7. Heating EEV
8. Evaporator
9. Vapour Line
10. Injection EEV

Figure 1. Heat pump Modelica model in heating mode.

It is a classical injection cycle with an economizer heat exchanger and two separated units. The outdoor unit is composed of the evaporator (8) and the heating electronic expansion valve (EEV, 7). Between these two units, split lines are present (vapour line 9 and liquid line 7). All the main components are presented in the following sections.

3. Compressor modeling

The proposed model uses a set of five dimensionless polynomials to predict the compressor behavior. The compressor model is treated as quasi steady-state, in fact, the time constants associated to the compressor dynamics are very small compared to those associated with heat exchangers and charge distribution [5]. The inputs of the compressor polynomials are the rotational speed (rpm) and the three pressure levels, i.e. the suction pressure (P_{su}), injection pressure (P_{inj}) and discharge pressure (P_{ex}) or combinations of these pressures, i.e.

$$\begin{cases} r_{p,tot} = \frac{P_{ex}}{P_{su}} & \text{the total pressure ratio} \\ r_{p,inj} = \frac{P_{inj}}{P_{su}} & \text{the injection pressure ratio} \end{cases}$$

The five polynomials are then used to compute the overall electrical power consumption, the discharge thermodynamic state and both injection and suction mass flow rates. The only parameter of the model is the compressor swept volume (V_s) which makes it easy to use for different size of devices.

These five polynomials are namely the volumetric efficiency (ε_v), the isentropic efficiency (η_s), the variable speed drive efficiency (η_{drive}), the injection ratio (X_{inj}), i.e. the ratio between the injection and the suction mass flow and finally the losses ratio (X_{loss}), i.e. the ratio between the compressor ambient losses and the electrical consumption of the compressor.

The volumetric efficiency is defined as the ratio between the actual mass flow and the theoretical mass flow determined via the swept volume, i.e.

$$\varepsilon_v = \frac{\dot{M}_{r,su}}{\rho_{r,su} \, . \, V_s \, . \, N_{rot}} \tag{1}$$

where $\dot{M}_{r,su}$ is the suction mass flow rate (kg/s), $\rho_{r,su}$ is the working fluid density defined at the suction port (kg/m^3) and N_{rot} is the rotational speed (rad/s).

The following formulation of the isentropic efficiency is implemented in order to take into account the injection process.

$$\eta_s = \frac{\dot{M}_{r,su} \, . \, (h_{r,ex,s} - h_{r,su}) + \dot{M}_{r,inj} \, . \, (h_{r,ex,inj,s} - h_{r,inj})}{\dot{W}_{in}} \tag{2}$$

where

$$\begin{cases} \dot{M}_{r,inj} & \text{is the injection mass flow rate, } kg/s \\ \dot{W}_{in} & \text{is the electrical power delivered to the compressor motor, } W \\ h_{r,su} & \text{is the refrigerant specific enthalpy at the suction port, } J/kg \\ h_{r,inj} & \text{is the refrigerant specific enthalpy at the injection port, } J/kg \\ h_{r,ex,s} & \text{is the specific enthalpy at the exhaust port for an isentropic compression} \\ & \text{from the suction to the exhaust pressure, } J/kg \\ h_{r,ex,inj,s} & \text{is the specific enthalpy at the exhaust port for an isentropic compression} \\ & \text{from the injection to the exhaust pressure, } J/kg \end{cases}$$

The isentropic compression power for an injection compressor can be interpreted as two separate volumes, a suction and an injection volume, performing separately an isentropic compression process respectively from the suction to the discharge pressure and from the injection to the discharge pressure.

The ratio between the electrical power privided to the compressor motor ($\dot{W}_{in}$) and the total electrical power delivered to the drive ($\dot{W}_{el}$) is defined as the drive efficiency. The injection mass flow rate is determined via the injection ratio X_{inj}.

By performing an energy balance on the compressor shell using the experimental data, a significant residual was calculated. Moreover, this residual had always the same sign, that is why it is generally considered as being linked to the compressor ambient losses. Thus, an additional dimensionless polynomial (X_{loss}) had to be added in order to predict the refrigerant temperature at the exhaust port accurately.

$$X_{loss} = \frac{\dot{Q}_{loss}}{\dot{W}_{el}} \tag{3}$$

Finally, the exhaust thermodynamic state can be computed via energy balance applied to the compressor shell, which is given by

$$(\dot{M}_{r,su} + \dot{M}_{r,inj}) \, . \, h_{r,ex} = \dot{W}_{in} + \dot{M}_{r,su} \, . \, h_{r,su} + \dot{M}_{inj} \, . \, h_{r,inj} - \dot{Q}_{loss} \tag{4}$$

IOP Conf. Series: Materials Science and Engineering **90** (2015) 012031 doi:10.1088/1757-899X/90/1/012031

In conclusion, we have:

$$\begin{pmatrix} \eta_s \\ \varepsilon_v \\ \eta_{drive} \\ X_{inj} \\ X_{loss} \end{pmatrix} = \mathbf{F}(rpm, P_{su}, P_{inj}, P_{ex})$$

and the outputs ($\dot{W}_{el}$, $\dot{M}_{r,su}$, $\dot{M}_{inj}$, $h_{r,ex}$) can all be computed using these polynomials and knowing the inlet and injection thermodynamic state as well as the rotational speed.

3.1. Mass flow rates determination

The validation process was made possible thanks to thirty experimental data points distributed all across the compressor operating map. As stated before, the rpm, the three pressure levels or the two pressure ratio were used as variables for the polynomials. The volumetric efficiency polynomial is a first order polynomial, i.e.

$$\varepsilon_v \quad = \quad a_0 + a_1 \,.\, rpm + a_2 \,.\, P_{ex} + a_3 \,.\, r_{p,tot} + a_4 \,.\, r_{p,inj} \tag{5}$$

Similarly, the injection ratio polynomial is given by:

$$X_{inj} \quad = \quad b_0 + b_1 \,.\, rpm + b_2 \,.\, P_{ex} + b_3 \,.\, P_{su} + b_4 \,.\, r_{p,inj} \tag{6}$$

Figure 2 presents the comparison between the predicted and measured values for the suction and injection mass flow rates. Both mass flow rates are given for two different compressors ($CP1$ and $CP2$). $CP1$ was used to fit the polynomials whereas the results presented for $CP2$ were obtained by using the same polynomials and by only updating the compressor swept volume.

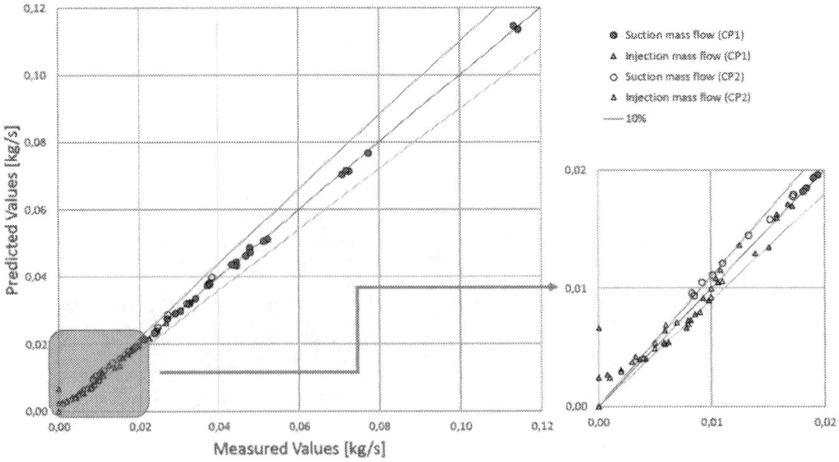

Figure 2. Predicted vs measured suction and injection mass flow rates for $CP1$ and $CP2$

IOP Conf. Series: Materials Science and Engineering **90** (2015) 012031 doi:10.1088/1757-899X/90/1/012031

3.2. Exhaust state and electrical consumption

Knowing the suction and injection states as well as both injection and suction mass flow rates and the discharge pressure, the isentropic efficiency polynomial gives the electrical power provided to the compressor motor. Via the losses ratio and the drive efficiency, it is then possible to determine the discharge temperature and the electrical power delivered to the variable speed drive. The drive efficiency is correlated to the rotational speed via a third order polynomials. A second order polynomial was chosen for the isentropic efficiency and a constant value for the losses ratio were selected, i.e.:

$$\eta_{drive} = c_0 + c_1 \cdot rpm + c_2 \cdot rpm^2 + c_3 \cdot rpm^3 \tag{7}$$

$$\varepsilon_s = d_0 + \sum_{i=1}^{2} (d_i \cdot rpm^i + d_{i+2} \cdot P_{ex}^i + d_{i+4} \cdot P_{inj}^i + d_{i+6} \cdot r_{p,tot}^i) \tag{8}$$

$$X_{loss} = e_0 \tag{9}$$

In fact, the correlation between the losses and the electrical power is pretty strong as it can be seen on Figure 3.

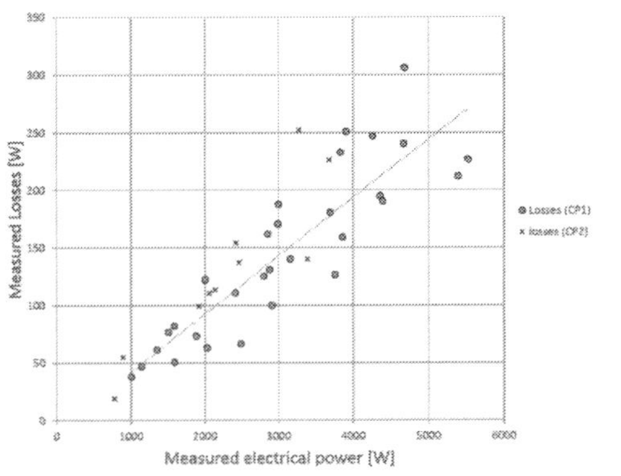

Figure 3. Compressor ambient losses versus electrical power for $CP1$ and $CP2$

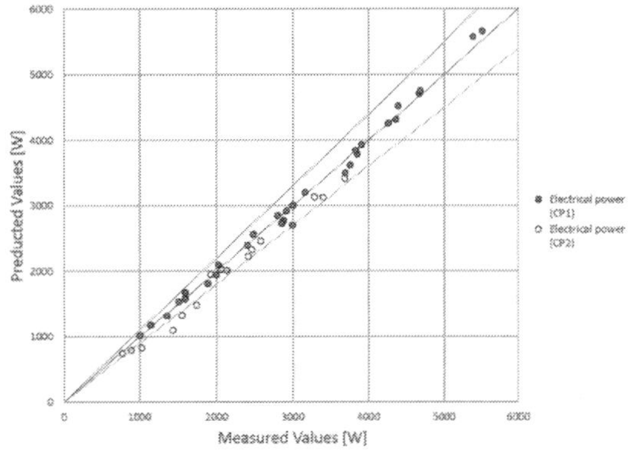

Figure 4. Predicted vs measured electrical power for $CP1$ and $CP2$

The discharge temperature is predicted with a RMSE of $2K$ for $CP1$ and $5K$ for $CP2$.

4. Heat exchangers

Dynamic heat exchanger models are based on the finite volume method where the energy and mass conservation equations are applied to several cells connected in series, more information can be found in [7]. The fluid cells are connected to a metal wall with no longitudinal conduction.

The main parameters of the dynamic heat exchangers models are the number of cells on each side, geometrical features such as the heat transfer area and the heat transfer coefficients (HTC) between both fluids and the metal wall. Different options are implemented to take into account the variation of the HTC (see [7]), in this case, a constant HTC in each zone (subcooled - two phase - superheated) was chosen.

Moving boundary models were developed in the "EES" software using the ε-NTU method in each zone for the steady-state validation of the models.

4.1. Condenser

Reviews and comparisons of heat transfer and pressure drop correlations for plate heat exchangers are widely available in the litterature, e.g. Garcia-Cascales *et al.* [16] and Ayub *et al.* [15]. In this study, Wanniarachchi [9] and Martin [12] correlations were used for single phase flows, Hsieh [10] correlations for evaporating and Yan [11] correlations for condensing flow were chosen.

The data set used to validate the models comes from a test rig similar to the cyle presented on Figure 1. In the condenser model, the secondary fluid inlet state is fixed, the working fluid inlet temperature and outlet subcooling are given as well. As the heat exchanger geometry is known, the main outputs of the models are the condensing temperature and the heat transfer rate once both mass flow rates are fixed.Results for the heat transfer rate are displayed on Figure 5. As the calculation of HTC correlations requires to compute transport properties of the fluid, which is particularly time consuming regardless of the properties calculation method, a constant averaged value per zone was chosen for the dynamic modeling of the heat exchangers. Moreover, a comparison using the HTC correlations and a constant averaged value of the HTC in each zone is performed. It can be concluded that using constant values in each zone does not affect greatly the results. This can be explained by the small pinch point that is present in this kind of heat exchanger.

The condensing temperature was determinated with a RMSE of $0.5K$ and the coefficient of determination for the heat transfer rate is equal to 98.9%.

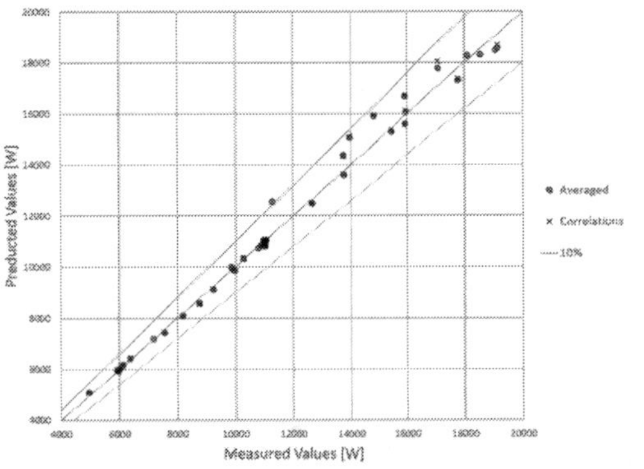

Figure 5. Predicted vs measured heat transfer rate at the condenser

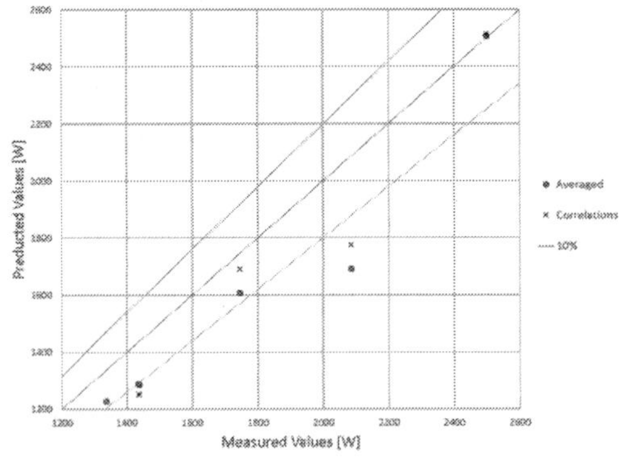

Figure 6. Predicted vs measured heat transfer rate at the economizer

4.2. Economizer

Similarly to the condenser analysis, the economizer results are given on figure 6. In this model, the considered inputs are both mass flow rates, on the high pressure side, the inlet state is fully determined and on the intermediate pressure side, the inlet temperature is considered as an input as well. No pressure drop is considered on the intermediate pressure side and the pressure on this line is therefore considered as the saturated liquid pressure at the inlet temperature. Hence, as the heat exchanger geometry is known, the main outputs of the models are the heat transfer rate, the high pressure side outlet temperature, the inlet quality and the injection superheat.

Once again, no noticeable difference is showed between the use of correlations or averaged values. However, the quality of the results is lower as the validation was performed on only seven experimental data points and the measurement of the injection mass flow rate is difficult.

4.3. Evaporator
The correlations used for the steady state validation of the evaporator model are developped by Wang ([13] and [14]). In order to determine the heat transfer resistance on the air side, an overall surface efficiency of 84% was determinated (Nellis and Klein, [17]). Results are given on Figures 7 and 8.

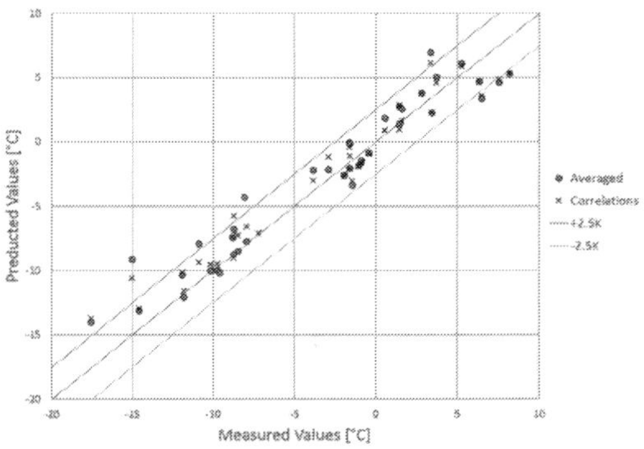

Figure 7. Evaporating temperature predictions vs measurements.

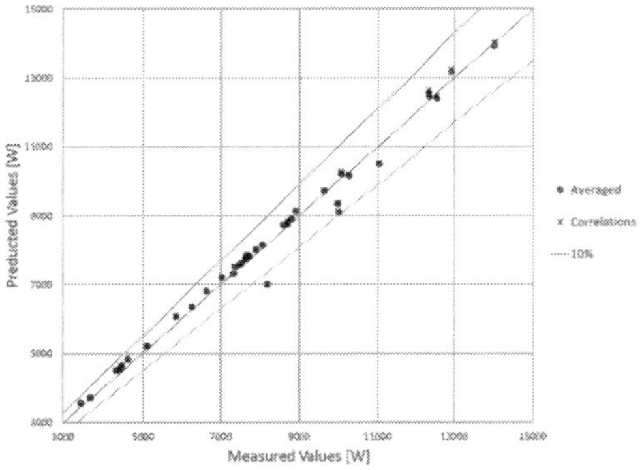

Figure 8. Evaporator heat transfer rate predictions vs measurements.

No significant discrepancy is found between results using the averaged heat transfer coefficients or the correlations.

5. Electronic Expansion valves
Several electronic expansion valve (EEV) models are available in the litterature. Jinghui Liu *et al.* [19] studied the thermodynamic behavior of choked flow in an EEV and gave the valve opening as a function of the pulse number. In an other study, Bach *et al.* [18] developped an adimensional law (8 PI group) for the valve mass flow using the Buckingham PI-theorem. They proposed laws for two refrigerants (R410A and R404A) valid for two-phase inlet conditions. Park *et al.* [20] also presented empirical laws for an EEV working with either R22 or R410a. The correlation yielded satisfactory predictions within a relative deviation of 15.0%. They developed a model based on a dimensionless correction coefficient C_d (see Eq.10) which is a function of the working conditions through six pi-groups. In the frame of this work, a constant coefficient of correction was chosen as the EEV behavior is not of prime concern.

$$\dot{m} \;=\; C_d \,.\, A \,.\, \sqrt{2 \,.\, \rho_{su,EEV} \,.\, (P_{su,EEV} - P_{su,EEV})} \qquad (10)$$

6. Split Lines
The split lines are present between the inside and ouside unit of the heat pump (c.f. Figure 1, component 6 and 9). They are modeled on the same principle than the heat exchangers. It is composed of several cells in series connected to a metal wall which can eventually exchange heat with the ambiance. In this case, the split lines have a maximum length of $25m$ and are under the surface of the ground, hence, the ambiance temperature was assumed to be constant at 10^oC. It is understandable that the liquid line is particularly important in the dynamic behavior of the cycle as a great part of the refrigerant charge is present in this pipe and because of its length.

IOP Conf. Series: Materials Science and Engineering **90** (2015) 012031 doi:10.1088/1757-899X/90/1/012031

An upwind scheme is chosen for the propagation of the fluid properties from cell to cell. The issue of numerical diffusion of the heat in the pipe has to be managed by choosing an appropriate number of cells in the split line.

7. Drive cooler

A drive cooler (DC) is present in the studied heat pump (c.f. Figure 1, component 5). This heat exchanger is used to cool down the drive by rejecting the heat on the high pressure side after the economizer which lowers the subcooling at the heating EEV inlet. In the Modelica implementation, this heat exchanger has been modeled assuming that all the losses from de drive, i.e.

$$\dot{Q}_{DC} = \dot{W}_{el} - \dot{W}_{in} = \dot{W}_{el} \cdot (1 - \eta_{drive}) \tag{11}$$

are injected at this point neglecting any dynamics from the heat echanger.

8. Liquid receiver and refrigerant charge

The liquid receiver is considered as one control volume and modeled using the energy and mass conservation principles assuming thermodynamic equilibrium at all times inside the control volume.

The overall refrigerant charge inside the cycle is calculated by adding the refrigerant mass from the different components. It has to be noted that no refrigerant mass is attributed to the compressor, drive cooler or EEVs as no control volume was considered for these components. The refrigerant charge is imposed by the initial conditions in the heat exchangers, split lines and tank. Hence, by varying the initial level in the liquid receiver, it is possible to adjust the total refrigerant charge in the cycle without changing the working conditions.

9. Control unit

The control unit is only composed of two PID controllers managing the injection and suction superheat by changing the orifice opening. Initialisation elements are added to disconnect the PIDs from the cycle during the first instants of the simulation which improves the robustness of the model without creating any time event.

10. Results

In this section, the cycle response to a compressor speed decrease from 7000 rpm to 3000 rpm is investigated. At the condenser, the water mass flow and inlet temperature are fixed to 0.6 kg/s and 50°C. At the evaporator, a dry air flow of 1.4 kg/s enters the coil at -4.2^oC. Results are shown on Figures 9, 10, 11 and 12.

At time $t = 50s$ the compressor speed ramps down from 7000 to 3000 rpm in $30s$. One can notice that the response of both EEV is not fast enough as the superheat drops significantly. As the EEVs openings decrease, both superheats increase until they stabilize at the set point values.

Figure 9 provides an interesting information on the refrigerant charge repartition, in fact, most of the fluid is present in the 15 meters long liquid split line, in the liquid receiver and in the condenser. Nevertheless, all the results presented in this section should be interpreted and used carefully as the dynamic model still has to be validated.

IOP Conf. Series: Materials Science and Engineering **90** (2015) 012031 doi:10.1088/1757-899X/90/1/012031

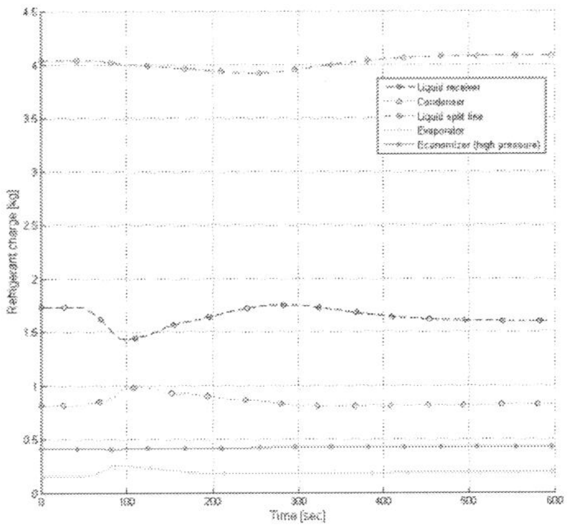

Figure 9. Refrigerant charge repartition

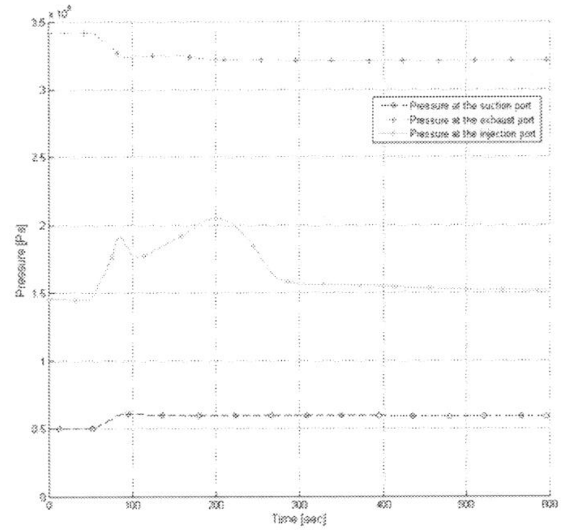

Figure 10. Evolution of the pressure levels

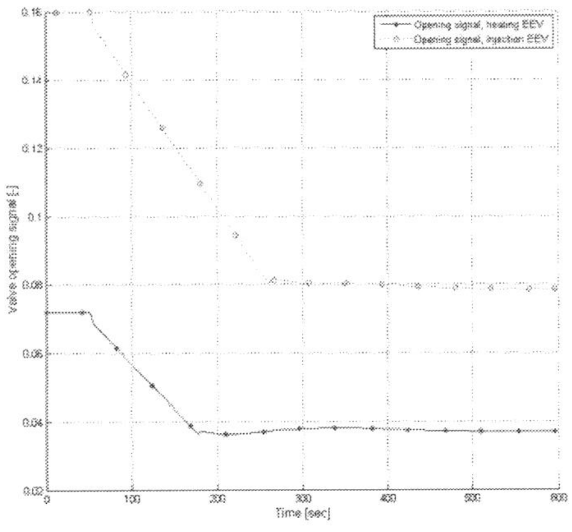

Figure 11. EEVs opening signal

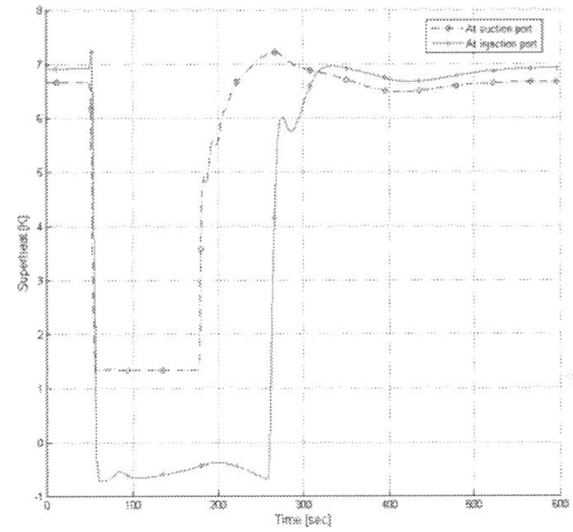

Figure 12. Evolution of the superheat at the suction and injection port

11. Conclusion and future work

In this study, a steady-state dimensionless model was developped for a variable speed injection scroll compressor. The only geometrical parameter of this compressor model is the swept volume. Five dimensionless polynomials (ε_v, η_s, η_{drive}, X_{inj} and X_{loss}) are needed to compute both mass flow rates (injection and suction) and the electrical consumption of the compressor.

Constant heat transfer coefficients were chosen for the heat exchanger to improve the model robustness and computational efficiency without damaging significantly the accuracy of the models. In fact, no representative difference was noticed between results when using either averaged heat transfer coefficients in each zone or correlations from the literature.

Finally an example is given with the dynamic response of the cycle to a compressor speed decrease.

In the future, the frosting and defrosting processes occuring on the air side of the evaporator will be taken into account. The dynamic validation of the present cycle model is also needed and will surely lead to adaptation of the proposed models. The reverse of the cycle, start-up and shut down simulations will also be considered.

References

[1] Rasmussen, Bryan P. "Dynamic modeling for vapor compression systems - Part I: Literature review." *HVAC& R Research* 18, no. 5 (2012): 934-955.

[2] Rasmussen, Bryan P., and Bhaskar Shenoy. "Dynamic modeling for vapor compression systemsPart II: Simulation tutorial." *HVAC& R Research* 18.5 (2012): 956-973.

[3] Qiao, Hongtao, Xing Xu, Vikrant Aute, and Reinhard Radermacher. "Transient modeling of a flash tank vapor injection heat pump systemPart II: Simulation results and experimental validation." *International Journal of Refrigeration 49* (2015): 183-194.

[4] Qiao, Hongtao, Xing Xu, Vikrant Aute, and Reinhard Radermacher. "Transient modeling of a flash tank vapor injection heat pump systemPart II: Simulation results and experimental validation." *International Journal of Refrigeration 49* (2015): 183-194.

[5] Ling, Jiazhen; Qiao, Hongtao; Alabdulkarem, Abdullah; Aute, Vikrant; and Radermacher, Reinhard, "Modelica-based Heat Pump Model for Transient and Steady-State Simulation Using Low-GWP Refrigerants" (2014). *International Refrigeration and Air Conditioning Conference.* Paper 1421. http://docs.lib.purdue.edu/iracc/1421

[6] Li, Bin, Steffen Peuker, Predrag S. Hrnjak, and Andrew G. Alleyne. "Refrigerant mass migration modeling and simulation for air conditioning systems." *Applied Thermal Engineering* 31, no. 10 (2011): 1770-1779.

[7] Quoilin, Sylvain; Desideri, Adriano; Wronski, Jorrit; Bell, Ian; and Lemort, Vincent, "ThermoCycle: A Modelica library for the simulation of thermodynamic systems" (2014). *Proceedings - 10th International Modelica Conference*; doi:10.3384/ECP14096683

[8] Bell, Ian H., Jorrit Wronski, Sylvain Quoilin, and Vincent Lemort. "Pure and pseudo-pure fluid thermophysical property evaluation and the open-source thermophysical property library Coolprop." *Industrial & engineering chemistry research* 53, no. 6 (2014): 2498-2508.

[9] Wanniarachchi, A. S., U. Ratnam, B. E. Tilton, and K. Dutta-Roy. *Approximate correlations for chevron-type plate heat exchangers.* No. CONF-950828–. American Society of Mechanical Engineers, New York, NY (United States), 1995.

[10] Hsieh Y. Y., Lin T. F. Evaporation Heat Transfer and Pressure Drop of Refrigerant R-410A Flow in a Vertical Plate Heat Exchanger *J. Heat Transfer* 125(5), 852-857 (2003) (6 pages); doi:10.1115/1.1518498

[11] Yan, Yi-Yie, Hsiang-Chao Lio, and Tsing-Fa Lin. "Condensation heat transfer and pressure drop of refrigerant R-134a in a plate heat exchanger." *International Journal of Heat and Mass Transfer* 42, no. 6 (1999): 993-1006.

[12] Martin, Holger. "A theoretical approach to predict the performance of chevron-type plate heat exchangers." *Chemical Engineering and Processing: Process Intensification* 35, no. 4 (1996): 301-310.

[13] Wang, Chi-Chuan, and Chi, Kuan-Yu. "Heat transfer and friction characteristics of plain fin-and-tube heat exchangers, Part I : new experimental data". *International Journal of Heat and Mass Transfer*, 1999.

[14] Wang, Chi-Chuan, and Chi, Kuan-Yu. "Heat transfer and friction characteristics of plain fin-and-tube heat exchangers, Part I : new experimental data". *International Journal of Heat and Mass Transfer*, 1999.

[15] Ayub, Zahid H. (2003), "Plate Heat Exchanger Literature Survey and New Heat Transfer and Pressure Drop Correlations for Refrigerant Evaporators",*Heat Transfer Engineering*,24:5,3 16; doi: 10.1080/01457630304056

[16] Garcia-Cascales, J. R., F. Vera-Garcia, J. M. Corberan-Salvador, and J. Gonzalvez-Macia. "Assessment of boiling and condensation heat transfer correlations in the modelling of plate heat exchangers." *International Journal of Refrigeration* 30, no. 6 (2007): 1029-1041.

[17] Nellis, G. and Klein, S. *Heat transfer.* Madison: Cambridge University Press, 2009. Print.

[18] Bach, Christian K.; Braun, James E.; and Groll, Eckhard A., "A Virtual EXV Mass Flow Sensor for Applications With Two-Phase Flow Inlet Conditions" (2012). *International Refrigeration and Air Conditioning Conference.* Paper 2122.

[19] Liu, Jinghui, Jiangping Chen, and Zhijiu Chen. "Investigation on the choking flow characteristics in electronic expansion valves." *International Journal of Thermal Sciences* 47, no. 5 (2008): 648-658.

[20] Park, Chasik, Honghyun Cho, Yongtaek Lee, and Yongchan Kim. "Mass flow characteristics and empirical modeling of R22 and R410A flowing through electronic expansion valves." *International Journal of Refrigeration* 30, no. 8 (2007): 1401-1407.

Calculation procedures for oil free scroll compressors based on mathematical modelling of working process

Y Paranin[1], A Burmistrov[2], S Salikeev[2], M Fomina[2]

[1] NIIturbocompressor n.a. V.B.Shnepp, 40 Sibirsky Trakt Str., Kazan, Russian Federation

[2] Kazan National Research Technological University, 68 K.Marx Str., Kazan, Russian Federation

Abstract

Basic propositions of calculation procedures for oil free scroll compressors characteristics are presented. It is shown that mathematical modelling of working process in a scroll compressor makes it possible to take into account such factors influencing the working process as heat and mass exchange, mechanical interaction in working chambers, leakage through slots, etc. The basic mathematical model may be supplemented by taking into account external heat exchange, elastic deformation of scrolls, inlet and outlet losses, etc. To evaluate the influence of procedure on scroll compressor characteristics calculations accuracy different calculations were carried out. Internal adiabatic efficiency was chosen as a comparative parameter which evaluates the perfection of internal thermodynamic and gas-dynamic compressor processes. Calculated characteristics are compared with experimental values obtained for the compressor pilot sample.

1. Introduction

Oil free scroll vacuum pumps and compressors hold much favour on the world market, which can be explained by their undeniable advantages such as low energy consumption, high reliability, counter balance and space saving. Scroll pumps and compressors are very science intensive type of production. The upgrading of scroll machines is carried out on the basis of physically valid and approbated mathematical models.

In this paper different approaches to modelling of working process of oil free scroll machines are considered.

The main characteristics of positive-displacement machine working as the compressor are volumetric efficiency (η_v), adiabatic efficiency (η_{ad}), and total adiabatic efficiency ($\eta_{ad.t}$).

2. Calculation procedure of working process taking into account heat and mass exchange in working chambers

Mathematical modelling of working process in a scroll machine makes it possible to take into account different factors influencing working process on each stage (heat and mass exchange, mechanical interaction in working chambers, leakage through slots, etc. [1-4]). The basic mathematical model may be supplemented by taking into account external heat exchange, elastic deformation of scrolls, inlet and outlet losses, etc., similar to [5-7], for example.

Working process in a scroll machine is performed with variable mass of working fluid. Mathematical model of a process is based on differential equations of working fluid parameters change which are widely used in compressors [1,2] and vacuum pumps [3,4] modelling. Differential equations describing the gas state in scroll machine chambers may be written in different forms. In this paper we'll use the form presented in [8-14]

$$dP_i = \frac{1}{V_i}\cdot\left[m_i R dT_i + RT_i dm_i - P_i dV_i\right] \quad (1)$$

$$dT_i = dT_{vi} + dT_{inleti} + dT_{mi} + dT_{Ti}\,. \quad (2)$$

The first of these expressions is obtained from the equation of state in differential form and takes into account the influence of mass change dm, temperature change dT and volume change dV on

Content from this work may be used under the terms of the Creative Commons Attribution 3.0 licence. Any further distribution of this work must maintain attribution to the author(s) and the title of the work, journal citation and DOI.
Published under licence by IOP Publishing Ltd

pressure change dP in compressor working chambers. The second equation takes into account the change of gas temperature dT_i in working chamber over an infinitely small period of time; dT_{vi} is the temperature change due to adiabatic volume change; dT_{inleti} is the temperature change in the chamber due to entering of gas into this chamber with some other temperature; dT_{mi} is the temperature change due to change of gas mass in the chamber; dT_{Ti} is the temperature change due to heat exchange of gas with the working chamber walls (dT_{Ti} may have both positive and negative sign depending on flow direction). The parameters of the chamber under study are given subscript "i"; V is the current volume value; R is the individual gas constant.

The following main assumptions are taken for the mathematical model:

- continuity and homogeneity of gas medium;
- compressible gas obeys ideal gas laws;
- when gas enters the working chamber from adjacent chambers parameters of state of the working gas change simultaneously over the whole volume of the working chamber;
- change of gas potential and kinetic energy is negligible;
- gas leakage through scrolls face seals is negligible;
- parameters of state in suction and discharge chambers are constant;
- gas-dynamic pressure loss at the suction is not taken into account;
- gas-dynamic friction loss between the gas and the chambers walls is not taken into account;
- working gas flow through design clearances is assumed to be isothermic;
- working gas flow through discharge port is assumed to be adiabatic;
- heat exchange between the gas and working chambers walls is convective and may be described by Newton-Richman law with a correction coefficient depending on torsion angle of the fixed scroll involute on all heat exchanging surfaces of the working chamber.

In figure 1 a design diagram of a scroll machine taking into account heat and mass exchange in working chambers is presented where Q_i, Q_{i+1}, Q_{i-1} is the heat supplied to (or removed from) the walls of working chambers i, $i+1$, $i-1$.

The total differential of mass change in a working chamber is given as

$$dm_i = dm_{i+1} - dm_{i-1} - dm_{Dii}, \tag{3}$$

where dm_{i+1} is the gas mass increase in the working chamber connected with gas flow from the foregoing working chamber with high pressure; dm_{i-1} is the gas mass decrease in the working chamber connected with gas flow into the working chamber with low pressure; dm_{Dii} is the gas mass increase or decrease in the working chamber connected with gas flow into the discharge port or from the discharge port.

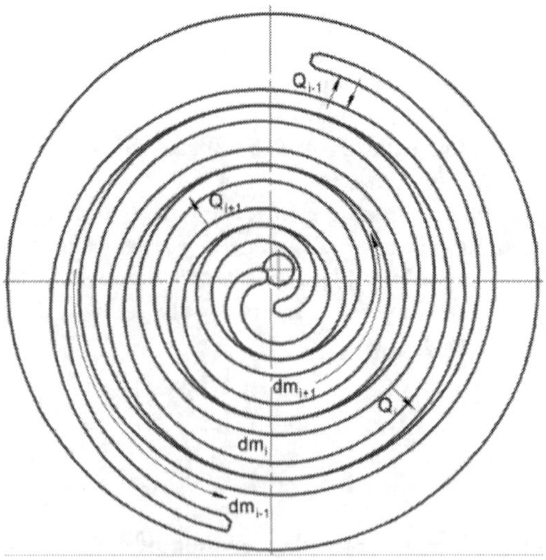

Figure 1. Design diagram of a scroll machine taking into account heat and mass exchange in working chambers.

Taking into account mass exchange between chambers of a scroll machine is necessary for calculation. The mass of gas flowing through the clearances (slots) of a working chamber formed by working elements of a scroll compressor or the discharge port is given as

$$dm_i = \phi_i F_i W_i \rho_i d\tau, \tag{4}$$

where ϕ_i is the gas flow rate coefficient through the slot or the discharge port; F_i is the slot or the discharge port cross-section area; W_i is the gas adiabatic exhaust velocity; ρ_i is the gas density in the chamber from which gas flows; $d\tau$ is the period of time over which exhaust occurs.

The mass of flowing gas is given [8] as

$$m = K_f F \left[\frac{1}{RT_2} \left(P_2^2 - P_1^2 \right) \right]^{1/2}, \tag{5}$$

where $K_f = \dfrac{1}{\sqrt{\ln\left(P_2/P_1\right)^2 + \xi + \theta\Sigma}}$

is the dimensionless experimental flow rate coefficient;
P_1, P_2 is the gas pressure before and after the slot; ξ is the coefficient taking into account local resistance at the inlet and the outlet of the slot; θ is the friction coefficient of gas flow in the slot; Σ is the slot form coefficient; $F=\delta_p h$ is the slot cross-section area; h is the scroll height; δ_p is the profile clearance (the clearance between the scroll profile surfaces).

K_f takes into account flow rate decrease due to losses when gas flows through the slot.

When a form and dimensions of the slot, physical properties and parameters of gas before and after the slot are known then determining of flow rate through the slot reduces to determining of the flow rate coefficient K_f.

The differential equations system of the working process is solved by the method of successive approximations. As a result, gas finite parameters in the chamber depending on shaft rotation angle are obtained including effective indicator and temperature charts of a compressor or a pump. In calculation the profile clearance δ_p is specified in initial data as the constant value not depending on rotation angle of the orbiting scroll.

Finally the main characteristics of the positive-displacement machines such as delivery rate, volumetric efficiency (η_v), adiabatic efficiency (η_{ad}) and total adiabatic efficiency ($\eta_{ad.t}$) are determined.

This calculation procedure is used for qualitative evaluation of main characteristics of scroll machines and for study and optimization of profile working elements geometry of such machines.

3. Calculation procedure of working process taking into account heat and mass exchange and elastic deformation of scrolls

Mathematical models which take into account change of operation clearances due to thermal and force deformations of scroll machine elements and heat exchange with external medium and heat exchange of gas with working chambers walls [6, 7, 13, 14] are used for quantitative evaluation of main characteristics of scroll machines.

The following additional assumptions are taken:

- heat from face seals friction is not supplied to gas;
- heat exchange on external surfaces of the orbiting and the fixed scrolls is convective;
- heat-transfer coefficient for the section of the heat exchange surface under study is constant;
- heat conductivity processes in scrolls are considered in steady-state conditions;
- scrolls material is uniform, isotropic, and elastic;
- heat conductivity of scrolls is constant and doesn't depend on temperature.

The design diagram of a scroll machine taking into account heat and mass exchange and elastic deformation of scrolls is presented in figure 2 where T_C, T_{cover}, T_{sh} are the temperatures of the casing, the cover and the shaft of the machine, respectively; T_a and T_w are the temperatures of cooling air and water; α_i, α_{cool1}, α_{cool2}, α_{air} are the heat exchange coefficients between the working fluid and working chamber walls, between cooling medium and the orbiting scroll surface, between cooling medium and the fixed scroll surface, between air and the orbiting scroll surface, respectively; E_{OSc} and E_{FSc} are the modulus of elasticity, μ_{OSc} and μ_{FSc} are the Poisson coefficients, G_{OSc} and G_{FSc} are the shear modulus, ψ_{OSc} and ψ_{FSc} are the coefficients of material linear expansion, λ_{OSc} and λ_{FSc} are the heat conductivity of the orbiting and the fixed scroll material, respectively.

For the fixed and the orbiting scrolls three-dimensional wedge cells mesh is built (figure 3).

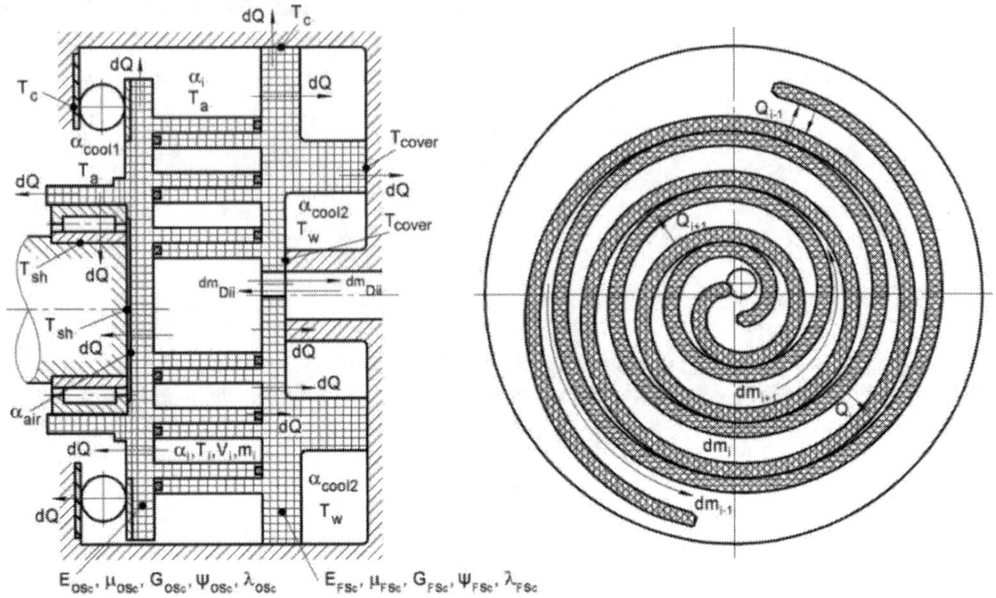

Figure 2. Design diagram of an air scroll machine taking into account heat and mass exchange and elastic deformation of scrolls (dQ are heat fluxes through structural elements).

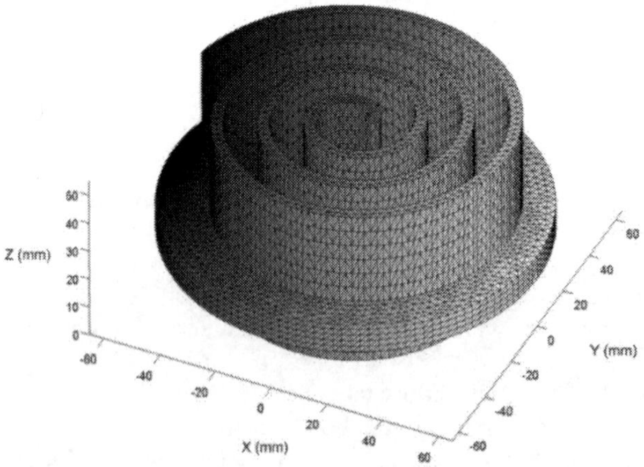

Figure 3. General view of three-dimensional mesh on the fixed scroll.

Scroll machines characteristics are determined by working fluid backward leakage through slots between scroll elements, and the leakage depends on clearances values.

There are two types of working clearances in a scroll machine: between scrolls profile surfaces (profile clearances) and between the top of one scroll and face disc of the other scroll (face clearances).

To minimize clearances is the most important line in efficiency increase of scroll machines working process. Moreover, for such machines to be failure proof it is necessary to set minimal clearances taking into account their change due to thermal and force deformations on the basis of non-contact motion of the orbiting scroll element under all regular operating conditions (in hot state).

Initial data for thermal and force deformations calculation are thermal fields and pressure diagrams of working elements.

It is necessary to add equations for calculation of profile clearances (6) and face clearances (7), equation of scrolls steady-state heat conductivity (8) with boundary conditions (9), (10) and equations of scrolls elasticity for linear and angular deformations taking into account temperature deformations (11) to the main equations of the scroll machine working process mathematical model (1-5).

Relationship for calculation of profile clearances taking into account thermal and force deformations of working elements at the rotation angle under study of the driving shaft is written in general form as

$$\delta_p(\varphi)=\delta_{pm}(\varphi)\pm\Delta\delta^T_{FSc}(\varphi)+\Delta\delta^T_{OSc}(\varphi)-\Delta\delta^T_e\pm\Delta\delta^F_{FSc}(\varphi)\pm\Delta\delta^F_{OSc}(\varphi)-\Delta\delta_r, \quad (6)$$

where $\delta_{pm}(\varphi)$ is the mounting profile clearance value at the angle φ; $\Delta\delta^T_{FSc}(\varphi)$, $\Delta\delta^T_{OSc}(\varphi)$, $\Delta\delta^T_e$ is the change of the profile clearance due to radial thermal deformations of the fixed and the orbiting scrolls at the angle φ and due to the eccentricity *"e"* increase of the eccentric shaft; $\Delta\delta^F_{FSc}(\varphi)$, $\Delta\delta^F_{OSc}(\varphi)$ is the change of the profile clearance due to radial force deformations of the fixed and the orbiting scrolls at the angle φ; $\Delta\delta_r$ is the change of the profile clearance due to radial clearance increase in the support bearing under the action of temperature difference between the shaft and the casing $\Delta\delta^T_r$ and under the action of the applied load $\Delta\delta^F_r$.

To calculate $\delta_p(\varphi)$ it is necessary to know the line of contact equation. A scroll machine with one-thread scrolls has got two lines of action: internal involute of the fixed scroll and the external involute of the fixed scroll.

The upper signs before the terms of the equation (6) connected with deformation are taken for the line of contact along the internal involute and the lower signs are taken for the line of contact along the external involute.

Scrolls with theoretical involute profile (without drive shaft eccentricity decrease) are most often used in scroll machines that is why in this paper method with drive shaft eccentricity decrease for the calculated value is used for obtaining of the profile clearance.

Face clearance in a scroll machine is

$$\delta_f=\delta_{fm}+\Delta\delta^T_C-\Delta\delta^T_{Sc}-\Delta\delta^T_B-\Delta\delta^T_{ball}+\Delta\delta^F_{Sc}+\Delta\delta^F_B-\Delta\delta^F_{ball}, \quad (7)$$

where δ_{fm} is the mounting face clearance value; $\Delta\delta^T_C$, $\Delta\delta^T_{Sc}$, $\Delta\delta^T_B$, $\Delta\delta^T_{ball}$ is the change of the face clearance due to axial thermal deformations of the casing, of one of the scrolls, of the adjoining base in the point with the maximal axial shift, and balls of the anti-rotation mechanism; $\Delta\delta^F_{Sc}$, $\Delta\delta^F_B$ is the change of the face clearance due to axial force deformations of one of the scrolls and adjoining base in the point with the minimal axial shift; $\Delta\delta^F_{ball}$ is the change of the face clearance due to elastic deformation in the point of contact of the most loaded ball with the rolling path.

In commercial machines, grooves are made at the top of both scrolls where seals are placed which nearly lean against the face disc providing the "tightness" increase of the face clearance. Face channel represents a plane rectangular slot the length of which is equal to the seal width along gas flow.

Delivery rate of the scroll compressor [9] is more sensitive to the change of face clearances than that of profile clearances (up to ten times).

Technological errors of manufacturing of scroll machine parts forming working chambers are not taken into consideration in equations (6) and (7) because they are manufactured with the help of high precision equipment.

Scrolls and casing thermal deformations due to high thermal load have the main effect on the change of working chambers clearances of oil free scroll compressors with pressure ratio π up to ten. Force deformations for the mentioned pressure ratio range are negligible.

Steady-state temperature distribution in a scroll is described by three-dimensional heat conductivity equation (Laplace's law)

$$\Delta T_i(x)=0, \quad x \in \Omega_i, i=1{,}2. \tag{8}$$

where Ω_i is the three-dimensional space occupied by a scroll; $x=(x_1,x_2,x_3)$ is the point of the three-dimensional space; $\Delta T(x)=\frac{\partial^2 T(x)}{\partial x_1^2}+\frac{\partial^2 T(x)}{\partial x_2^2}+\frac{\partial^2 T(x)}{\partial x_3^2}$ is the three-dimensional Laplace operator. i=1 corresponds to the fixed scroll, i=2 corresponds to the orbiting scroll, T is the scroll temperature.

The boundary B of the area Ω_i is divided into two parts $B=B_1+B_2$, where B_1 is the boundary part where the temperature is specified. The boundary condition is set for B_1

$$T(x)=T_d(x), \; x \in B_1, \tag{9}$$

where T_d is the specified function.

The boundary condition describing heat exchange with external medium with the specified temperature and the opposite scroll is set for B_2

$$k_i \frac{\partial T_i(x)}{\partial v}+\gamma(x)[T_i(x)-\gamma_1(x)T_0(x)-\gamma_2(x)T_{3-i}(x)]=q(x), \quad x \in B_2. \tag{10}$$

where k_i is the heat conductivity of the scroll material; $i=1{,}2$; v is the vector of the external normal to the scroll surface; $\gamma(x)=\alpha(x)+\beta(x)$; α is the heat exchange coefficient with external medium (air, cooling water, shaft, casing); $\beta=\lambda/l'$ is the heat exchange coefficient with the opposite scroll; λ is the air heat conductivity coefficient; l' is the distance between the corresponding sections of the orbiting and the fixed scrolls; $\gamma_1(x)=\alpha(x)/[\alpha(x)+\beta(x)]$; T_0 is the external medium temperature; q is the heat flux (due to bearings heat release and seal lip friction); T_{3-i} is the opposite scroll temperature.

The equation (8) with boundary conditions (9) and (10) was approximated by linear algebraic equations system according to finite elements method. Beforehand for the orbiting and the fixed scrolls three-dimensional mesh with triangular bases was built (figure 3). The mesh was adapted to the surfaces of the scroll rib, the base (discharge port is taken into consideration for the fixed scroll) and the boss. Isoparametric elements with nodes placed in the wedge vertices were used for approximation.

Elasticity equations for linear and angular deformations taking into account thermal deformations are

$$\begin{aligned}
\varepsilon_x &= \frac{1}{E}\left[\sigma_x-\mu\left(\sigma_y+\sigma_z\right)\right]+\alpha_1 \Delta T, \\
\varepsilon_y &= \frac{1}{E}\left[\sigma_y-\mu\left(\sigma_z+\sigma_x\right)\right]+\alpha_1 \Delta T, \\
\varepsilon_z &= \frac{1}{E}\left[\sigma_z-\mu\left(\sigma_x+\sigma_y\right)\right]+\alpha_1 \Delta T, \\
\gamma_{xy} &= \frac{1}{G}\tau_{xy}, \gamma_{yz}=\frac{1}{G}\tau_{yz}, \gamma_{zx}=\frac{1}{G}\tau_{zx}
\end{aligned} \tag{11}$$

where ε_x, ε_y, ε_z are linear deformations in the directions x, y, z; γ_{xy}, γ_{yz}, γ_{zx} are angular deformations; τ_{xy}, τ_{yz}, τ_{zx} are tangent stresses; σ_x, σ_y, σ_z are normal stresses in the directions x, y, z; α_1 is the coefficient of material linear expansion; ΔT is the temperature change on heating; E is the modulus of elasticity; μ is Poisson coefficient; G is the shear modulus. There are two subscripts in angular deformations and tangent stresses symbols: the first denotes the direction of the normal to the plane, the second denotes the direction of tangent stress vector. Equations (11) present the general form of the elasticity law (Hooke's law). The problem is solved in linear approach.

At present method of finite elements which is realized in software ANSYS [15] is widely used for calculation of stresses and deformations in design elements of intricate shape.

IOP Conf. Series: Materials Science and Engineering **90** (2015) 012032 doi:10.1088/1757-899X/90/1/012032

The obtained components of deformation vectors with the help of ANSYS are displayed in the form of isolines and fields.

Temperature fields and pressure diagrams of the orbiting and the fixed scrolls are obtained according to the calculation programme of thermal fields and in the form of a data file are sent to the initial data file of the deformation calculation programme.

Calculation results are presented in the form of radial thermal and force shifts distribution at the scrolls wrap middle height section (to determine $\Delta\delta_{FSc}(\varphi)$ and $\Delta\delta_{OSc}(\varphi)$) and in the form of axial thermal and force shifts distribution at the section over scrolls wrap face surface (to determine $\Delta\delta_{Sc}$ and $\Delta\delta_B$).

The calculation sequence of the compressor working process according to this method is the following. Equations (1)-(5) are numerically solved. Coincidence of indicator and temperature diagrams of two consecutive approximations with predetermined accuracy is obtained by successive approximations. Then coincidence of working chamber walls temperature curves is obtained by successive approximations. Temperature of working chambers walls and scrolls temperature fields are obtained according to equation (8) with boundary conditions (9), (10). Temperature fields are used in equations (11) for calculation of scrolls thermal and force shifts, and gas pressure diagrams in working chambers are calculated according to (1). To solve equations (11) the following boundary conditions are set: shifts following three spatial directions are lacking in the scroll and casing mating area.

Thermal and force shift values are used for profile and face clearances according to equations (6) and (7). These clearances are used for mass transfer calculation according to (5).

This method of calculation makes it possible to obtain information about influence of any scroll machine parameter on its characteristics and to predict the main machine characteristics with sufficiently high accuracy on the design development stage.

4. Effect of calculation procedure on exactness of characteristics evaluation

To evaluate the influence of procedure on oil free air scroll compressor characteristics calculation accuracy different calculation procedures were used: without taking into account heat exchange in the working chamber; taking into account heat exchange with one wall of the working chamber; taking into account external heat exchange, heat exchange with all walls of the working chamber and the change of the working clearances caused by elastic deformation of scroll compressor parts.

The compressor worked with water cooling (π is the pressure ratio; n is the rotation speed; V_w is the flow rate of water cooling the fixed scroll from the rear side). In this machine cast iron scrolls were used.

Internal adiabatic efficiency $\eta_{ad.int}$, was chosen as a comparative parameter. Some calculation characteristics are presented in figure 4 where solid lines correspond to calculated data and the dashed line corresponds to experimental data [12] (n=3000 rpm, V_w = 4 l/min).

The characteristics analysis makes it possible to make the following conclusions:

- heat exchange between gas and one wall of the working chamber (profile rib surface of the fixed scroll) results in inconsiderable decrease in $\eta_{ad.int}$ by 1.0 – 3.2% (characteristic 2) in comparison with characteristic 1 obtained without regard for heat exchange despite the fact that the fixed scroll is water-cooled. This can be explained by heating of the sucked gas caused by convective heat exchange with the hotter fixed scroll rib and therefore input energy increase. Experimental temperatures were used in this procedure. $\eta_{ad.int}$ may be significantly increased if more effective cooling of the fixed scroll is used but it results in design complication and considerable increase of the scroll compressor mass and bulk characteristics. For example, decrease of scroll rib temperature by 30 K results in $\eta_{ad.int}$ increase by 6 – 9% (characteristic 3) in comparison with characteristic 1. The maximal increase corresponds to the optimal value of π.

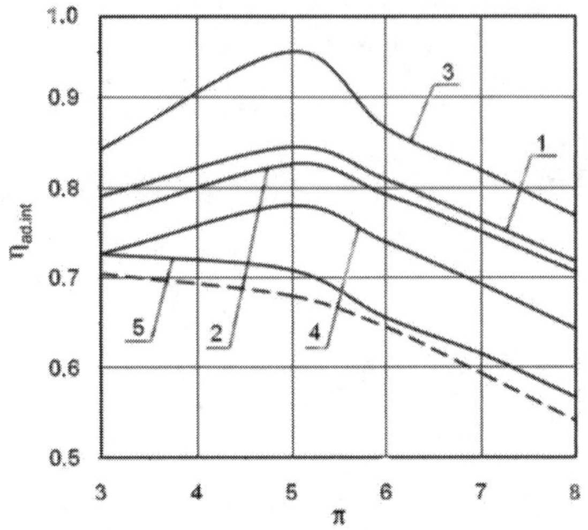

Figure 4. Effect of calculation procedure on exactness of internal adiabatic efficiency evaluation

1 - not taking into account heat exchange; 2, 3 - taking into account heat exchange with one wall of working chamber; 4 - taking into account external heat exchange and heat exchange in working chamber, 5 - taking into account external heat exchange, heat exchange in working chamber and elastic deformation of a scroll compressor elements.

- taking into account heat exchange with all walls of the working chamber (temperature of which was obtained by calculation according to the presented procedure) results in $\eta_{ad.int}$ decrease by 6.5–9.5% (characteristic 4) in comparison with characteristic 1. It can be explained by the sucked gas heating due to additional convective heat exchange with the orbiting scroll surfaces with temperatures higher than the fixed scroll temperatures. It should be noted that characteristics 1 – 4 were obtained with the constant clearance δ_p (which doesn't depend on φ and π) which is equal to the average calculation clearance value over the angle φ at π=7 of characteristic 5 (δ_p =0.035 mm);
- taking into account heat exchange with all walls of the working chamber and the working clearances δ_p change over the shaft rotation angle φ results in $\eta_{ad.int}$ decrease by 8.5–19.5% (characteristic 5) in comparison with characteristic 1, the larger decrease corresponding to larger π. It can be explained by the increase and redistribution of working clearances with increase of π caused by elastic deformation of the compressor parts. The clearances increase due to repeated compression caused by leakage in intermediate chambers results in input energy increase.

According to the figure 4 the most close to the experimental characteristic is characteristic 5 (the difference is no more than 3–5% over all experimental range).

It should be noted that an open type refrigeration scroll compressor (scrolls didn't have developed cooling surface) reconstructed as an oil free air scroll compressor is used as an object of the research. The feature of this compressor construction is the presence of the chamber at the discharge side which was intended for separation of oil and refrigerant. This chamber after the compressor reconstruction is used for water cooling of the fixed scroll, and the orbiting scroll is not cooled purposely. On some working regimes in the experiment the temperature in the centre of the orbiting scroll reached 140°C and at the periphery (at the inlet port) it reached 80°C. For the cooled fixed scroll these values were 70°C and 40°C respectively. Due to imperfection of the compressor cooling system the heating resulted in considerable decrease of efficiency in general.

Acknowledgements
This work was financially supported by the Russian Ministry of Education and Science according to Russian Federation governmental resolution no. 218 and performed within the integrated programme "Development of high-technological production of oil free scroll vacuum pumps for nanomaterials industry".

6. References

[1] Jand D and Jeond S 2006 *Int. J. Refrig.* **29** 744-53
[2] Uchikava N, Terrada H and Arata T 1987 *Hitachi Rev.* **36** 155

[3] ChenY, Halm N, Groll E and Braun J 2000 *Int. Compressor Engineering Conf.* (West Lafayette: Purdue University) pp 715-24
[4] Chen Y, Halm N, Groll E and Braun J 2000 *Int. Compressor Engineering Conf.* (West Lafayette: Purdue University) pp 725-34
[5] Kauder K and Unvert T 2003 *Compressornaya tehnika i pnevmatika* (*Compressors and pneumatics)* **1** 6-10 (in Russian)
[6] Lee G and Kim G 2001 *Int. Conf. on Compressors and their Systems* (London: City University) pp 123-32
[7] Peng B, Zhang H, Zhang L and Liu Z 2008 *Int. Compressor Engineering Conf.* (West Lafayette: Purdue University) pp 14-7
[8] Ibragimov E 2009 *Povishenie effektivnosti spiralnogo compressora suhogo szhatiya (Oil free scroll compressor efficiency increase)* (thesis) (Kazan: Kazan State Technological University) (in Russian)
[9] Paranin Yu, Yakupov R and Burmistrov A 2013 *Bulletin of Kazan Technological University* **16** (19) 267-71 (in Russian)
[10] Salikeev S, Burmistrov A and Raykov 2013 *Compressornaya tehnika i pnevmatika* (*Compressors and pneumatics)* **4** 37-42 (in Russian)
[11] Burmistrov A, Raykov A and Salikeev S 2012 *Bulletin of Kazan Technological University* **15** (8) 257-8 (in Russian)
[12] Paranin Y 2011 *Sovershenstvovanie metoda rascheta rabochego processa spiralnogo compressor suhogo szhatiya s ispolzovaniem rezultatov experimentalnyh issledovanii (Oil free scroll compressor working process calculaton method development using experimental results)* (Kazan: Kazan State Technological University) (in Russian)
[13] Paranin Yu and Khisameev I 2011 *Compressornaya tehnika i pnevmatika* (*Compressors and pneumatics* **)** **5** 16-23 (in Russian)
[14] Paranin Yu, Yakupov R and Burmistrov A 2014 *Bulletin of Kazan Technological University* **17**(1) 248-52 (in Russian)
[15] Structural Analysis Solutions 2014 *ANSYS Inc.* URL:http://www.ansys.com/Products/Simulation+Technology/ Structural +Analysis

Scroll Compressor Oil Pump Analysis

S Branch
Ingersoll Rand – Trane
Compressor Development, La Crosse, WI, USA

E-mail: sbranch@trane.com

Abstract. Scroll compressors utilize three journal bearings to absorb gas, friction and inertial loads exerted on the crankshaft. To function properly, these bearings must be lubricated with a certain amount of oil. The focus of this paper will be to discuss how computational fluid dynamics can be used to predict oil flow out of a single-stage oil pump. The effects of speed and lubricant viscosity on pump output will also be presented. The comparisons will look at mass flow rates, differences in pressure, and torque at various speeds and dynamic viscosities. The computational fluid dynamic analysis results will be compared with actual lab testing where a crankshaft bench tester was built.

1. Introduction

Trane scroll compressors utilize three journal bearings to absorb the radial gas, friction and inertia loads exerted on the crankshaft. To function properly, these bearings need to be adequately lubricated by a single-stage oil pump.

The purpose of this analysis is to evaluate a crankshaft with an integral oil pump that was designed at a fixed speed and simulated at variable speeds. A scroll compressor crankshaft using RL32H oil and R410a refrigerant was used for this comparison. The oil pump was simplified to not include the oil flow to the lower main bearing or the upper main bearing. The CFD simulation results will analyze the differences in pressure, torque, and flow rates at these different speeds. Finally, a comparison will be conducted to evaluate the differences in the CFD results with a crankshaft bench tester that was built and tested.

Literature review found two papers that performed similar CFD simulations; one from Bernardi (1) in 2000 and one from Cho (2) in 2002. The study conducted by Bernardi (1) performed a CFD analysis of an oil pump used for a scroll compressor and then compared the oil mass flow rates of the CFD results with actual lab testing. The oil mass flow rates differed between the actual lab testing and the CFD analysis by 17%. The study by Cho (2) in 2002 looked at an oil pumping system of a variable speed compressor using a swing pump. The predicted results from this study had a maximum deviation of 12.8%. Both of these studies were only conducted as a steady state analysis assuming laminar flow for the oil.

Content from this work may be used under the terms of the Creative Commons Attribution 3.0 licence. Any further distribution of this work must maintain attribution to the author(s) and the title of the work, journal citation and DOI.
Published under licence by IOP Publishing Ltd

2. Review of CFD Modelling Approach

The CFD simulations were conducted using CFX version 14.5.7. The simulation was built with two domains; one is stationary, and the inside one is rotating. The two domains are shown in Figure 1. Figure 2 shows the oil pump inside of the rotating domain along with the entrance and exit locations of the oil.

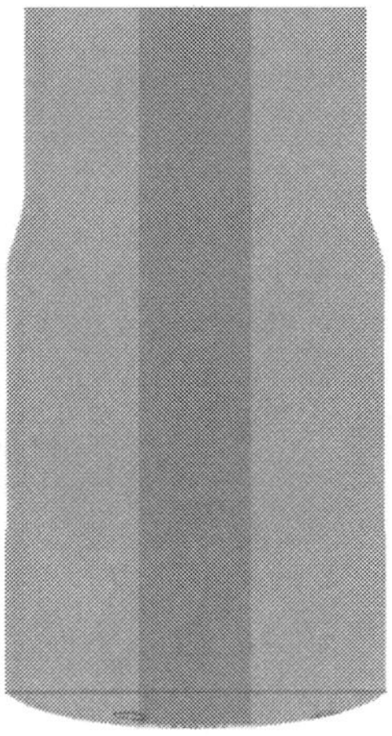

Figure 1. Stationary and Rotating Domains

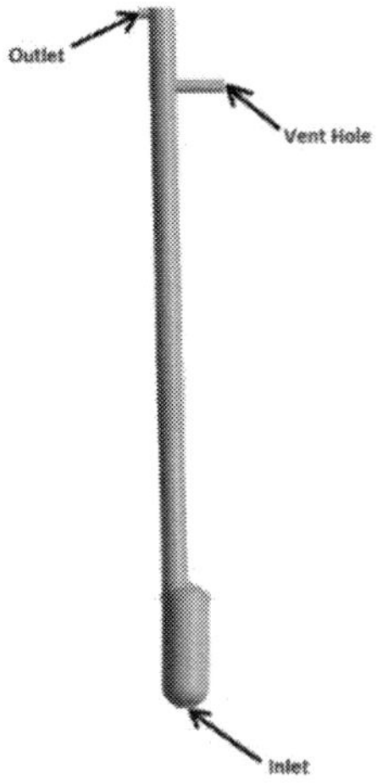

Figure 2. Oil Pump

The modelling approach used by Bernardi (1) and Cho (2) is similar to the approach used in this study. In both cases oil out-gassing was not considered. Oil out-gassing occurs when the refrigerant entrained in the oil is released or absorbed as its pressure or temperature changes with the oil. For this study the turbulence was modelled as a homogenous turbulence flow using a SST model because the average Reynolds number was greater than 20000. Bernardi (1) and Cho (2) modelled the oil as laminar, which is one of the differences in the two techniques. The inside section is a rotating domain which consists of the oil pump and the crankshaft. The rotating domain meshes are 2.67 million nodes. The size of outside stationary domain mesh is 1.68 million nodes. There is a general grid interface at the inlet, outlet, and vent locations in the rotating domain. There is also a frozen rotor or transient rotor stator interface between the rotating and stationary domains. All of the simulations used the same geometry and mesh, but different setup assumptions.

- ANSYS CFX Version 14.5.7 was used for the CFD Analysis
- The simulation was set up with two continuous fluids, R410 and POE oil (RL32H).
- There were no phase changes
- All the walls were adiabatic.
- RPM varied (1800 RPM, 3500 RPM, 6000 RPM)
- The oil start level is 42.8 mm above the bottom of the crankshaft
- Buoyancy was turned on and was in the negative z direction.
- Turbulence was set up as a homogenous SST model.
- The reference pressure is 86.3 psia.
- Heat transfer was set to fluid dependent.
    - R410 temperature was fixed at 48 °F

  - RL32H oil was fixed at 90 °F.
- The fluid pairs interphase transfer was set to a mixture model
  - Interphase length scale = 0.1 mm
- Drag coefficient = 0.44
- R410a refrigerant used an rgp table.
- The RL32H oil material properties:
- Molar Mass = 587 g/mol
- Density = 62.553 lbm/ft^3
- Specific Heat at Constant Pressure = 0.441 BTU/lbm F
- Oil Viscosity varied (2 cP, 25 cP, 50 cP)
- Thermal conductivity = 0.138 (BTU/ft^2 s (F/ft))

The simulation was started as a steady state analysis as was done by both Bernardi (1) and Cho (2). The steady state simulation ran until the mass and momentum converged to 1.0e-4. In addition, the imbalances were all less than 1%. Finally, the flows and pressures in to and out of the oil pump were not changing. After the steady state simulation converged, the results were used as the initial conditions for a transient rotor stator simulation.

The transient rotor stator simulation was simulated at 2 degrees of rotation per time step and ran until the solutions were converged. The convergence criteria were similar to the steady state method. Each time step needed to converge the mass and momentum to 1.0e-4, with imbalances less than 1% and stable flows and pressures in to and out of the oil pump.

3. CFD Results

Figure 3 shows the rotation of the crankshaft. Modelling the shaft rotation is important to predict the flow characteristics of the simulation. Figure 4 provides the streamlines to illustrate the oil flow path at one point in time. The oil travels from the inlet of the oil pump to the exit of the oil pump and then back to the oil sump. From Figure 3 where it shows the shaft rotating it can be observed in Figure 5 the vortex that is created with the oil; the lighter shade of grey is the oil which is at the bottom of the compressor shell, where the darker shade is the refrigerant. The transition region, shown in white, is a mixture of the oil and refrigerant.

Figure 3. Crankshaft Rotational Velocity.

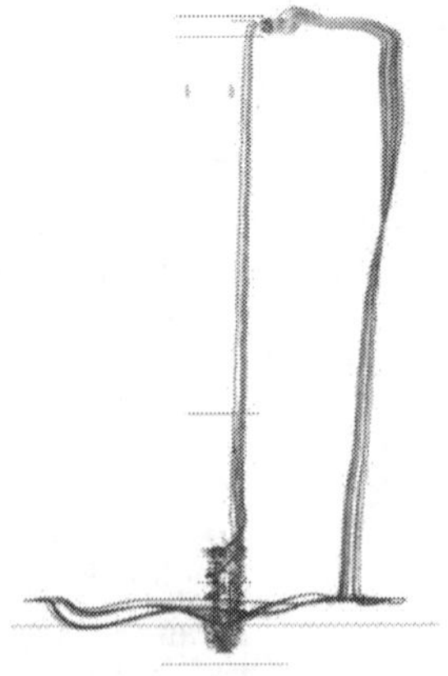

Figure 4. Oil Flow Streamline Velocities.

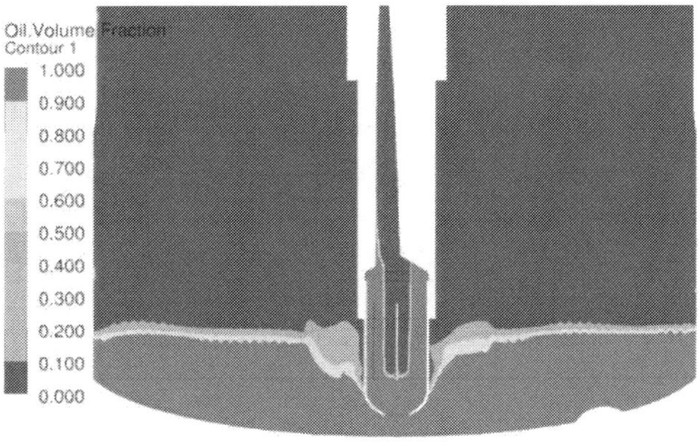

Figure 5. Oil Volume Fraction @ 3500 RPM and 50 cP

A potential concern with the oil gallery being full with oil was addressed with the 6000 RPM solution. At 6000 RPM the amount of oil in the ½ inch diameter oil gallery was only 30% full as shown in Figure 6 and Figure 7. The lighter shade from Figure 7 is the oil and the darker shade is the refrigerant.

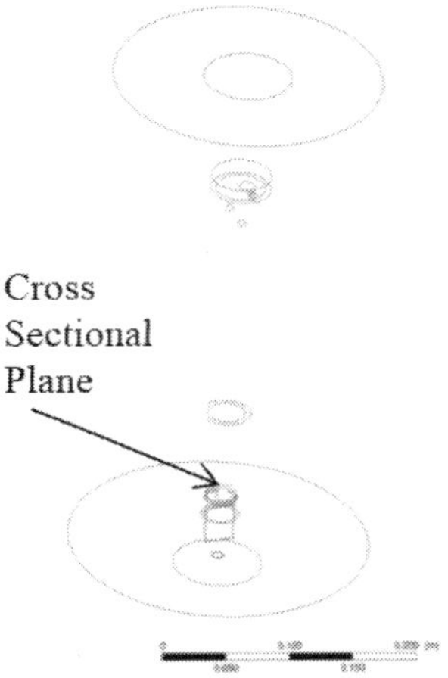

Figure 6. Cross Sectional Plane Location.

Figure 7. Oil Gallery Volume Fraction

IOP Conf. Series: Materials Science and Engineering **90** (2015) 012033 doi:10.1088/1757-899X/90/1/012033

Results of CFD simulations are shown in Table 1, including mass flow rates, pressure drop, and torque as a function of dynamic viscosity and various compressor speeds. The mass flow rate is an average value of the oil over a complete revolution. Torque is calculated from the oil and refrigerant of the crankshaft over one revolution. These results were used to compare with the bench tester. Figures 8, 9 and 10 come from the data in Table 1.

Table 1. Mass Flow Rate, Pressure Drop, and Torque Results.

MFR (kg/s)		RPM	
cP	1800	3500	6000
2	0.017	0.052	0.098
25	0.011	0.049	0.091
50	0.007	0.045	0.086

Pressure Drop (psia)		RPM	
cP	1800	3500	6000
2	0.084	0.168	0.393
25	0.078	0.136	0.270
50	0.074	0.115	0.251

Torque (Nm)		RPM	
cP	1800	3500	6000
2	0.002	0.012	0.033
25	0.002	0.011	0.033
50	0.001	0.010	0.032

Figure 8 shows the effects of mass flow rates. Figure 9 looks at the change in pressure from the inlet of the pump to the outlet of the pump. Figure 10 looks at the change in torque over the range of dynamic viscosities and compressor speeds.

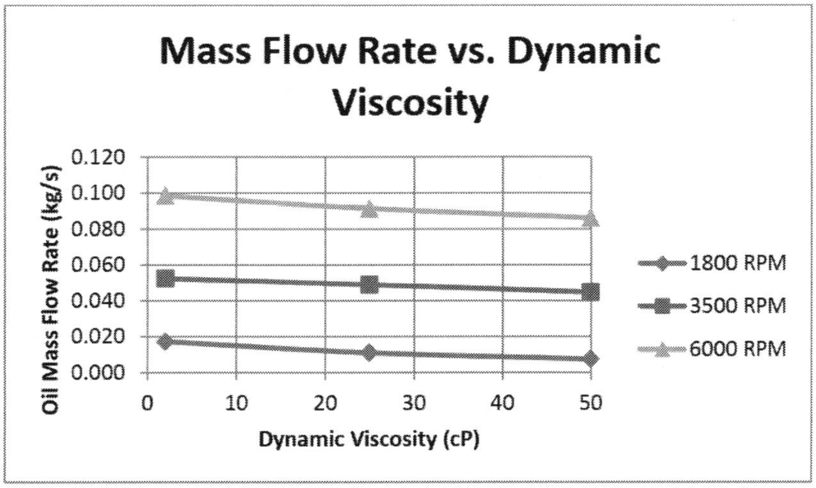

Figure 8. Mass Flow Rate vs. Dynamic Viscosity.

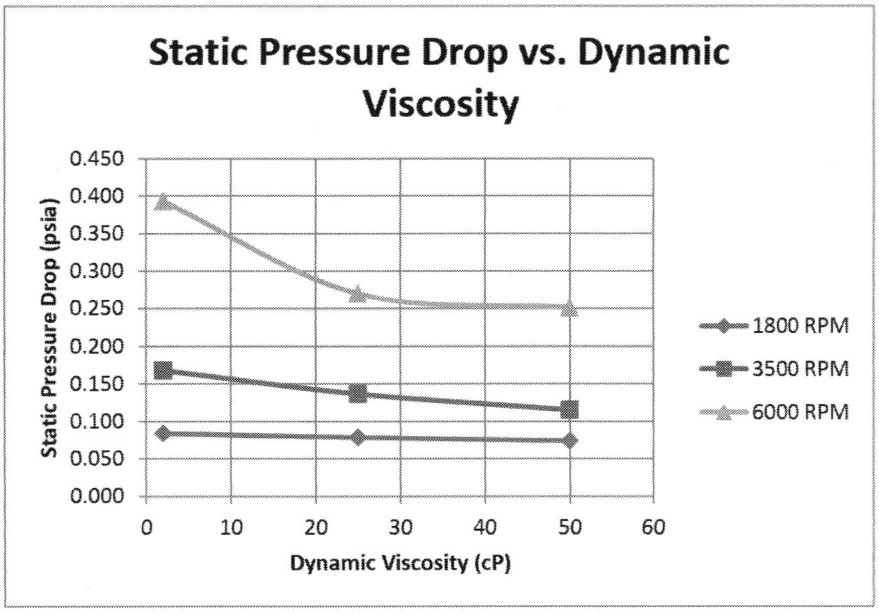

Figure 9. Static Pressure Drop vs. Dynamic Viscosity.

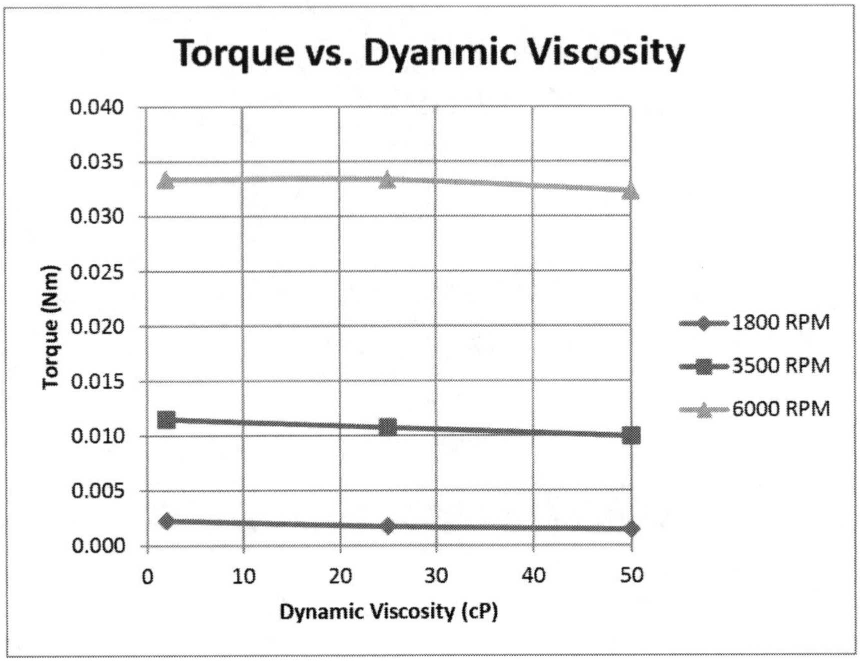

Figure 10. Torque vs. Dynamic Viscosity.

4. Comparison between the Bench Tester Results and the CFD Results

A bench tester was built to experimentally measure oil flow rates as a function of speed and lubricant viscosity, which is shown in Figure 11. The method that was used to measure the flow out of the crankshaft is shown on Figure 12. The flow exits the crankshaft and fills up the chamber. The change in pressure causes the oil to flow to the flow meter by flowing out of the chamber through three tubes, collect in the manifold and go through a flow meter. After the flow meter, the oil drains back into the bottom of the bench tester by gravity. The measurements were then compared to the CFD output to understand differences between the CFD model and the bench tester results. The bench tester consisted of an oil basin, the pump, a crankshaft to house the pump, and a chamber to collect oil discharged by the pump. A 2 hp electric motor provided torque to the assembly. A grade 68 cSt mineral oil was used for testing. Tests were conducted in air.

IOP Conf. Series: Materials Science and Engineering **90** (2015) 012033 doi:10.1088/1757-899X/90/1/012033

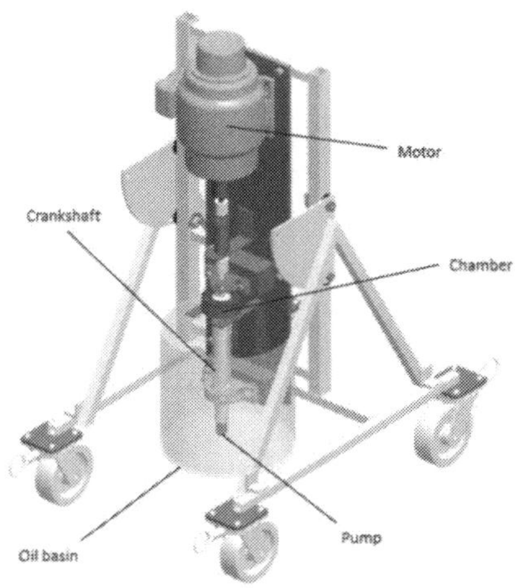

Figure 11. Bench Tester.

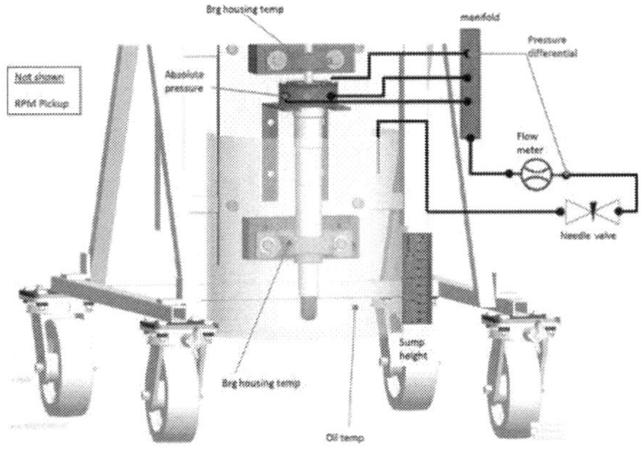

Figure 12. Bench Tester Flow Measurement Method.

Three areas were investigated to compare the CFD results with the bench tester. The first area looked at the vortex shape around the pump. The second area was the functionality of the gas vent at higher speeds. The third area was the oil flow rates at various speeds at a 50 cP viscosity.

Figure 13 shows the vortex around the pump. The vortex from Figure 13 does exist and is similar to what was predicted by the CFD on Figure 5.

Figure 13. Vortex of the Pump.

CFD predicted that the oil gallery was not full with oil at the 6000 RPM flow rates as shown in Figure 6 and Figure 7. When the bench tester ran at 6000 RPM it confirmed these results. The gas vent provides a method for the gas to leave the oil gallery without slowing the flow of the oil. The location of the gas vent is provided in Figure 14.

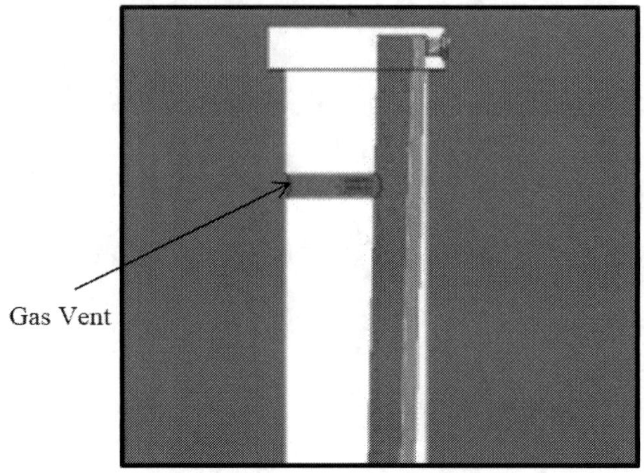

Figure 14. Gas Vent.

The bench tester was limited to a dynamic viscosity of 50 cP, therefore the flow rate comparisons were only conducted at that dynamic viscosity. Figure 15 shows the differences in the results from the CFD analysis and the bench tester. It is believed that a large portion of the error can be attributed to the measuring system of the bench tester putting extra pressure drop on the pump. If the CFD model had incorporated this extra pressure drop, there would have been less difference in the results.

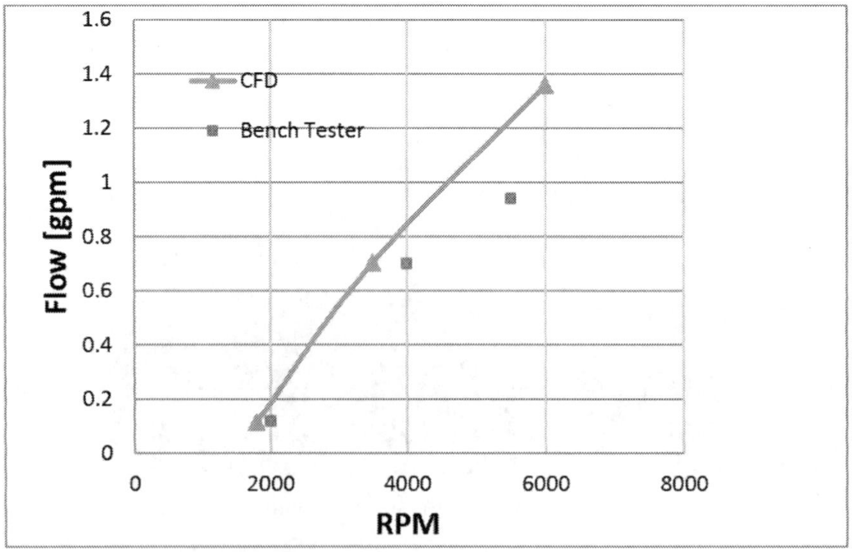

Figure 15. Model Validation Flow Rates at 50 cP.

5. Conclusion

- Although the crankshaft was designed for a fixed speed it performed adequately at variable speeds. The flow rates of the oil were sufficient to provide adequate lubrication to the bearings.
- The three areas that were compared between the CFD model and the bench tester showed the following.
  - In the CFD analysis the vortex shape appeared similar to the bench tester results. The main difference was in the CFD results it showed the transition region that was not visible from the bench tester.
  - The CFD simulation predicted that the gas vent would still function at 6000 RPM and was confirmed with the bench tester. This was verified by the percent of oil in the oil gallery in Figure 7.
  - The oil flow rates of the CFD analysis and the bench tester agreed well at low flow rates. As the flow rate increased the results deviated. It is believed that the differences are primarily due to the measuring system that was used for the bench tester which caused a larger pressure drop at higher flow rates.
- Further investigation would be to determine if there is a more appropriate turbulence model to better capture the behaviour of the oil and refrigerant.

Acknowledgement

The author acknowledges Derrick Lepak for his assistance and valuable insight in the bench tester set-up and operation details, as well as the contribution of Ingersoll Rand and the Trane Engineering organization in providing the environment in which work such as this is supported.

References

[1] Bernardi, J.de, 2000,"CFD Simulation of a Scroll Compressor Oil Pumping System", Proceedings of the International Compressor Engineering Conference at Purdue, West Lafayette, IN,1454

[2] Cho,H, Yoo,B.K, Kim,Y, 2002,"CFD Simulation on the Oil Pumping System of a Variable Speed Scroll Compressor", Proceedings of the International Compressor Engineering Conference at Purdue, West Lafayette, IN,1605

IOP Conf. Series: Materials Science and Engineering **90** (2015) 012034 doi:10.1088/1757-899X/90/1/012034

3D Transient CFD Simulation of Scroll Compressors with the Tip Seal

Haiyang Gao, Hui Ding and Yu Jiang
Simerics, Inc.,
1750 112th AVE NE Ste A203
Bellevue, WA 98004, USA

E-mail: hg@simerics.com

Abstract. A new template simulation tool is developed for scroll compressors/expanders capable of modelling tip seal leakages. This scroll template generates a high quality 3D multi-block structured mesh from user-input stationary and orbiting scroll surfaces. The mesh movement is then automatically calculated to account for every position of the orbiting scroll, maintaining good grid quality and smooth movement throughout the whole revolution. A state-of-the-art efficient CFD solver is used to solve Navier-Stokes equations, capable of simulation with both real gas and ideal gas. A case study is presented for a generic scroll compressor with refrigerant R410A. The case was run with and without the tip seal volumes. Comparisons are made to show the impact of tip seals on the compressor performance.

1. Introduction

Scroll compressors are widely used in many industries, including refrigeration, air-conditioning and automobile (as superchargers). It is believed that scroll compressors have the advantages of high efficiency, lower noise and vibration levels. The tip seal is a common sealing mechanism placed on top of involute wraps, in order to reduce axial leakage and maximize the efficiency.
Computational Fluid Dynamics (CFD) simulations can be a useful tool in understanding the flow field and help to reduce various flow losses. However, CFD simulation of a scroll compressor has proven to be a challenging task. Because the fluid domain of a scroll compressor contains gas pockets with large deformation and complicated leakage paths of tip seal and radial gaps. There are studies to solve it via simplified mathematical models [9], Two-Dimensional (2D) simulations [8] as well as a few unsteady, 3D numerical simulations in recent years [2, 4, 6]. However, there are few numerical studies in which the tip seal leakage volumes are included.

For most commercial packages, simulation of the scroll compressor with the tip seal leakages requires lengthy set up and running time. It would also require the user to have extensive CFD knowledge and reasonable familiarity of the tools on hand.

Gao and Jiang (2014) [3] successfully developed a fully automated template tool for scroll compressor simulation, which greatly simplified the simulation process and significantly reduced the turn-around time of scroll compressor/expender simulations.

In the current work, the tool is further extended to include the leakage flow of tip seals. The template can generate high-quality structure mesh from user-input outlines of stationary and orbiting rotors. The mesh for the tip seal is also built simultaneously based on a series of input parameters. The

Content from this work may be used under the terms of the Creative Commons Attribution 3.0 licence. Any further distribution of this work must maintain attribution to the author(s) and the title of the work, journal citation and DOI.
Published under licence by IOP Publishing Ltd

IOP Conf. Series: Materials Science and Engineering **90** (2015) 012034 doi:10.1088/1757-899X/90/1/012034

mesh movement is then calculated to account for every position of the orbiting scroll, maintaining good grid quality and smooth movement through the scroll revolution.

2. Scroll Rotor Template Mesher

To ensure mesh quality throughout the movement of the orbiting rotor, especially for areas with small gaps, a continuous high quality multi-block structured mesh is generated for the scroll compressor.

As the first step the rotor CAD surfaces (Figure 1) are imported into the software and the stationary and orbiting rotors are identified. These surfaces can be easily output from any CAD package as STL files. The mesh generator is designed in such a way that the resulted mesh movements are independent of the initial relative positions of the stationary and orbiting scrolls. To define the rotating motion, some geometric parameters, such as the center of both rotors and the rotation axis vector, are also provided as inputs. The parameters that define the shape of the tip seal volumes are also input by the user, more details to follow in the next section.

Figure 1: A typical CAD surfaces input by user, stationary rotor in grey, orbiting rotor in red.

To start the meshing process, a 2D outline is extracted from the imported 3D surfaces imported. Then the template mesher will search for several key geometry points, with which the computational domain is divided into several meshing partitions.

Figure 2 shows that six partitions are created for the scroll rotor domain. Partitions 1 through 3 are the two branches of the spiral fluid volumes and the outer boundary area. These narrow, long fluid volumes are meshed by lines that are normal to either rotor outline, with evenly distributed nodes between the rotors.

The discharge area is further divided into Partitions 4-6. Partitions 4 and 5 are established in such a way that at most one “contact” point is allowed in each partition. For a continuous mesh, the stationary and orbiting scroll cannot interfere with each other, but instead relatively small gaps are maintained. The template mesher uses lines that are normal to either solid wall to mesh the area close to the “contact”, to ensure high level of orthogonality, while uniformly distributed nodes are used to draw mesh lines for the rest of the volume. Special treatments are also employed to preserve the possible sharp corners in these partitions.

In Partition 6, there could be as many as two “contact” points. For meshing purposes, the partition is further divided into two smaller ones. The dividing line is defined such that it splits the domain into two symmetric ones for symmetric rotors, and it has the smallest angle from normal direction on both rotor walls (shown as the center dash line in Figure 3). After the division, the same procedure as in Partitions 4 and 5 is applied to mesh each of the two smaller partitions.

Ecliptic smoothing is then applied to the 2D mesh to improve the orthogonality in highly distorted areas. After the 2D mesh is built, it is stretched in the third direction to form a 3D grid. The meshing process is repeated for each time step when the orbiting rotor moves to a new position.

The meshing algorithm described above has been successfully applied to various types of scroll rotors, such as various shapes at the discharge, symmetric and non-symmetric scrolls, etc.

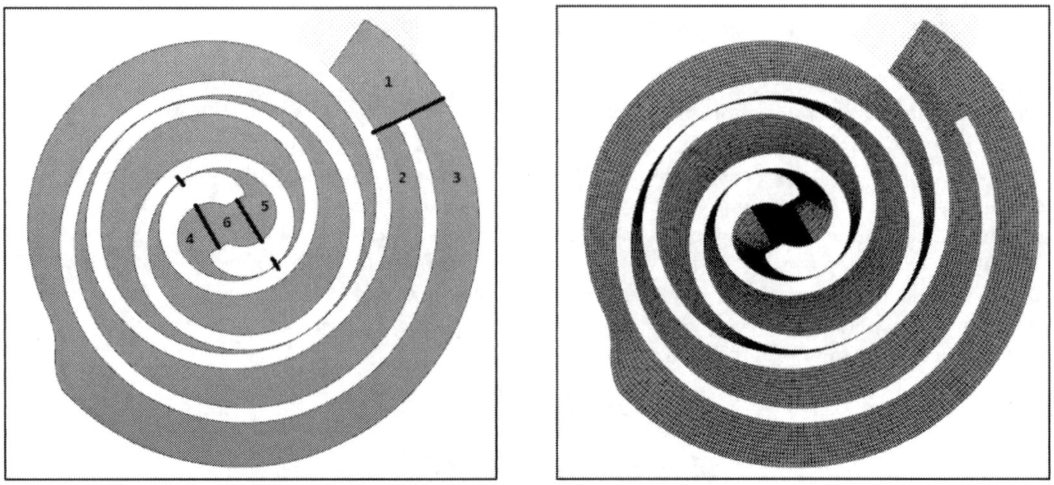

(a) (b)

Figure 2: Rotor domain partitions (a) and resulting mesh (b)

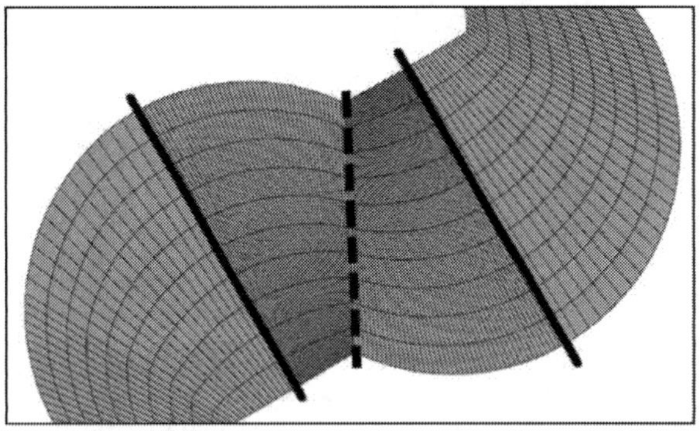

Figure 3: Further division of Partition 6.

3. Automated Meshing for the Tip Seal

A typical cross-section of the tip seal leakage path is shown in Figure 4. As demonstrated by Ancel et al. (2000) [1], the tip seal is pushed to the low pressure side of the groove. For a scroll compressor, we currently assume the tip seal is always attached to the outer side of the groove (Figure 4), considering the pressure is higher in the inner pocket most of the time. The resulting leakage path is the z-shaped fluid volume shown in Figure 5a. Once the user input all the five size parameters in Figure 4, the shape of the fluid domain is well defined. A multi-block structured mesh is then generated for the fluid volumes (Figure 5b).

From the top view of the scroll rotor in Figure 6, it can be seen that the tip seal leakage volume starts at the point where the width of the scroll becomes constant. The leakage volume ends when the spiral/involute shape stops. Both points are automatically detected by the template. Additional input is available for the user to control the starting and ending position of the leakage volume.

The scroll and the tip seal volumes generated in the last two sections are then connected via Mismatched Grid Interface (MGI) to form a single computational domain. Once the user input the necessary geometries and parameters, the whole procedure from the Sections 2 and 3 are fully automated.

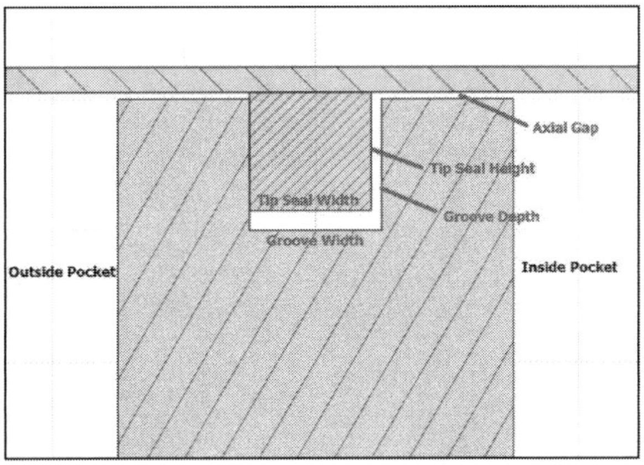

Figure 4: A typical cross section of tip seal leakage path.

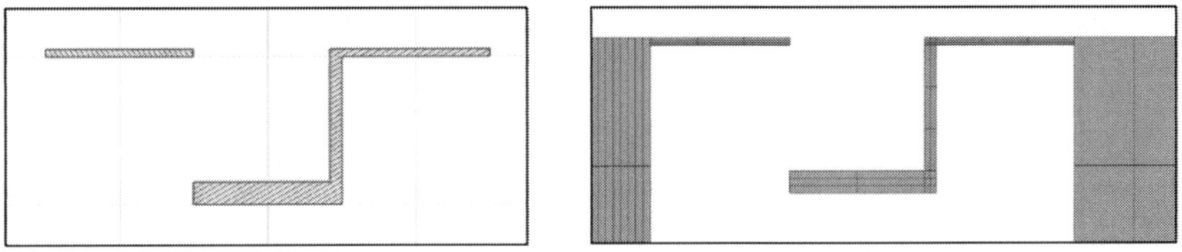

(a) (b)

Figure 5: Cross sections of tip seal fluid volume (a) and the mesh generated (b).

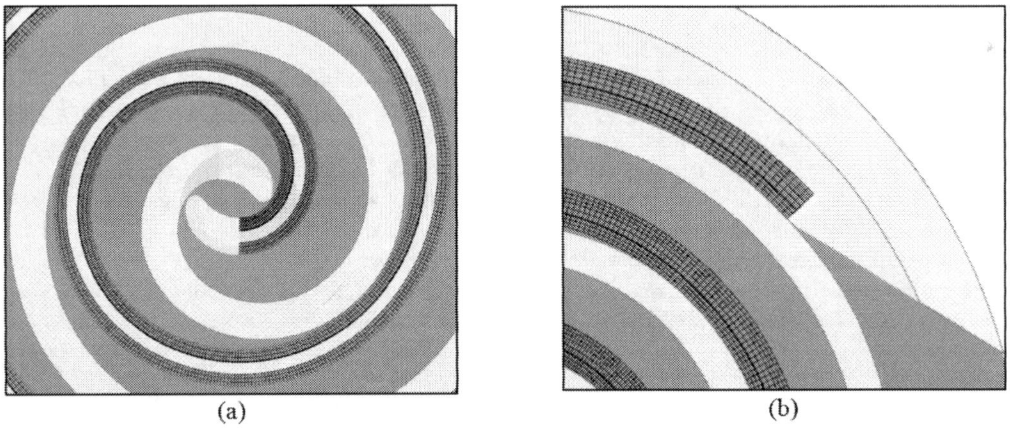

(a) (b)

Figure 6: Top view of the compressor showing the start (a) and end (b) of tip seal volumes.

4. A Case Study

Next, the scroll template tool is put to the test for a generic scroll compressor. The compressor works with refrigerant R410A, with inlet pressure at 10 bar and the outlet at 34 bar. The inlet temperature is set at 298K. The compressor has a diameter of 90mm and height of 20mm. The gap between stationary and orbiting scrolls is set to 18 μm. For the tip seal, the groove width is set to 2 mm, groove depth 1 mm, tip seal width 1.94 mm, tip seal height 0.9 mm and axial gap 50 μm.

A commercial CFD package, PumpLinx, is used for the simulation. A proprietary pressure-based algorithm [5, 7] is employed for solving the Navier-Stokes equations. The properties of R410A are read from property tables generated by the NIST database.

Static pressure boundary conditions are applied at inlet and outlet. For the energy equation, the inlet temperature is fixed at 298K while the fully developed temperature boundary [7] is assumed at the discharge.

Two simulations at 3000RPM have been performed: one includes the tip seal leakage (Figure 7b) while the other doesn't (Figure 7a).

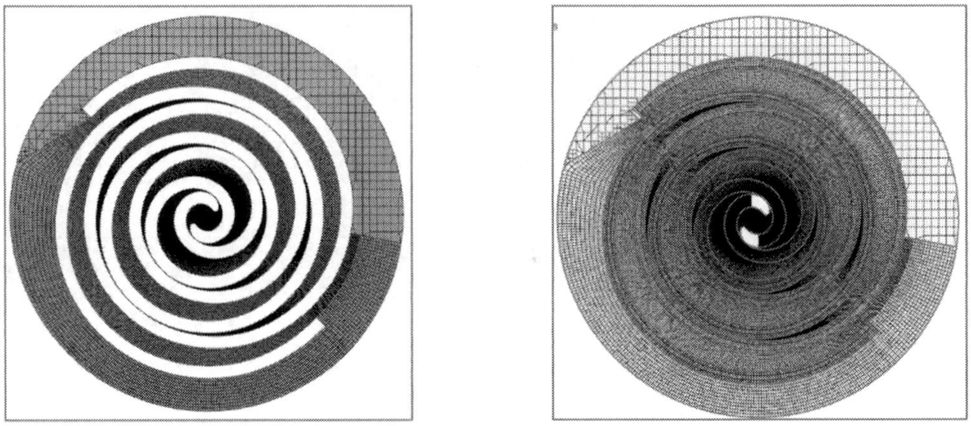

(a) (b)

Figure 7: Overview of computational domain: (a) without tip seal volumes (b) with tip seal volumes.

The meshes generated by the scroll template tool at four orbiting scroll positions are shown in Figure 8. It can be seen that for all positions, great mesh quality is maintained by the meshing algorithm.

Instantaneous pressure and temperature field is compared for both cases in Figure 9 and 10. From the pressure comparison, it can be observed that the tip seal volume results in substantial leakage between adjacent pockets, making an apparent difference in the pressure inside the pockets. From the temperature distribution, it is obvious the fluid inside the tip seal volume is heated up from the leakage flow, and this can lower the efficiency of the scroll compressor due to this extra energy loss.

Figure 11 and Figure 12 shows the comparison of discharge mass flow rate and temperature between two cases. Meaningful changes can be observed. Average mass flow rate is 12.1% lower when the tip seal leakage volumes are included.

The impact of the tip seal volumes on the compressor efficiency is also studied. The isentropic efficiency is defined as follows:

$$\eta = \frac{H_{2,isen} - H_1}{H_{2,act} - H_1}, \tag{1}$$

where H_1 is the enthalpy at the inlet, $H_{2,isen}$ is the enthalpy at the outlet from an isentropic process and $H_{2,act}$ is the actual enthalpy at the outlet. The leakage flow due to the tip seal volumes causes the heating of the fluid and thus lowers the efficiency. The average isentropic efficiency drops from 76.2% to 60.1% after including the tip seal volumes.

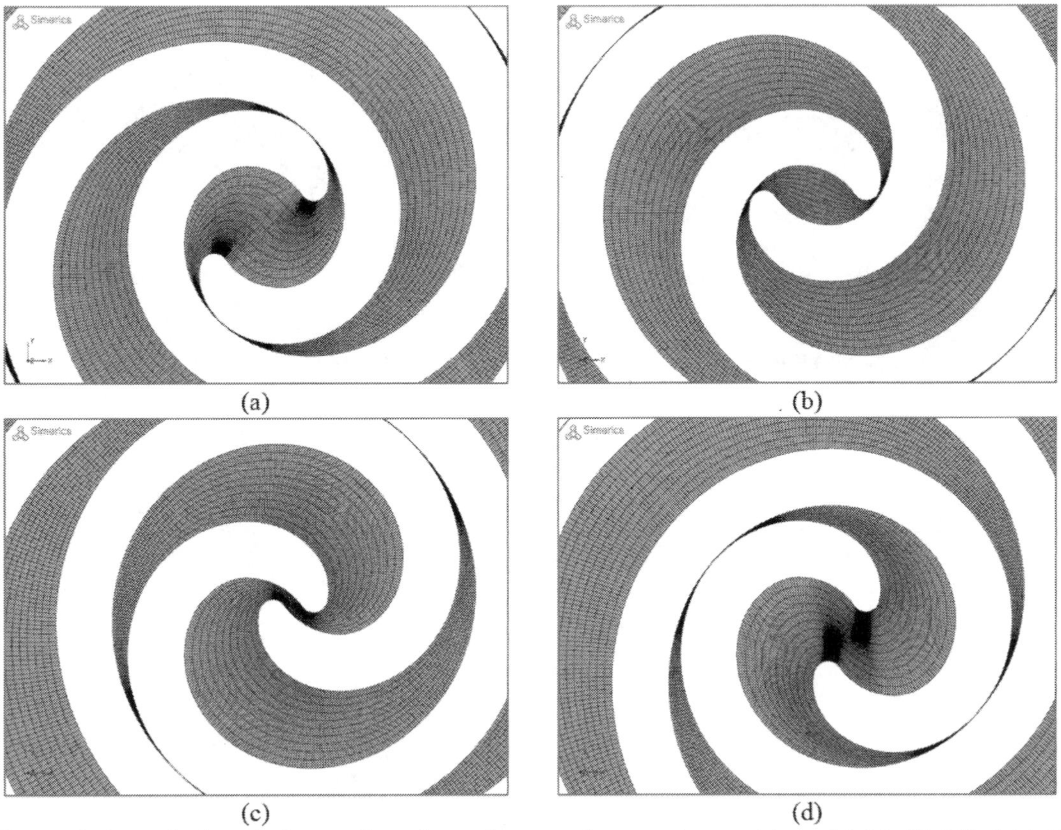

Figure 8: Smoothed mesh for scroll pocket volumes at four rotation positions: (a) 90, (b) 180, (c) 270 and (d) 360 degrees.

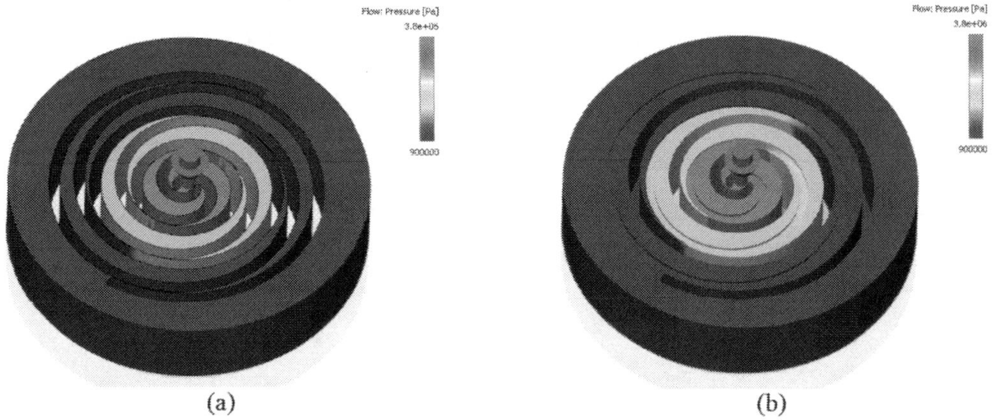

Figure 9: Instantaneous Pressure field at 180 degrees rotation position: (a) not including tip seal volumes (b) including tip seal volumes.

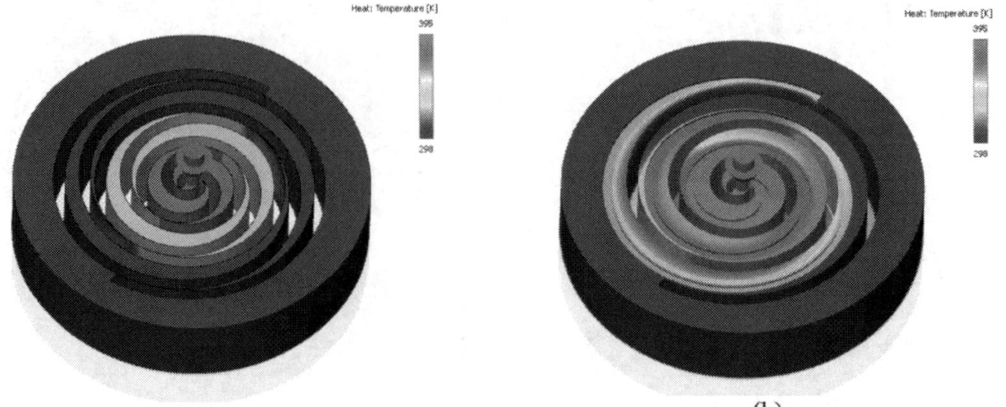

(a) (b)

Figure 10: Instantaneous Temperature field at 180 degrees rotation position (a) not including tip seal volumes (b) including tip seal volumes.

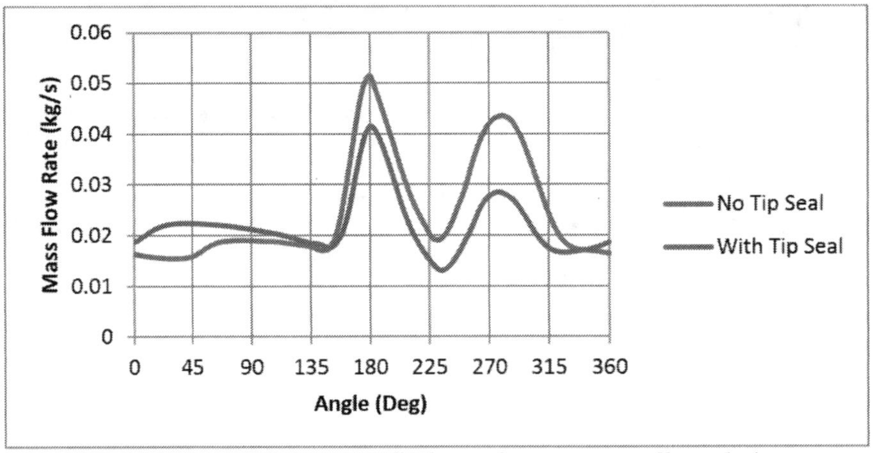

Figure 11: Mass flow rate at discharge during one scroll revolution.

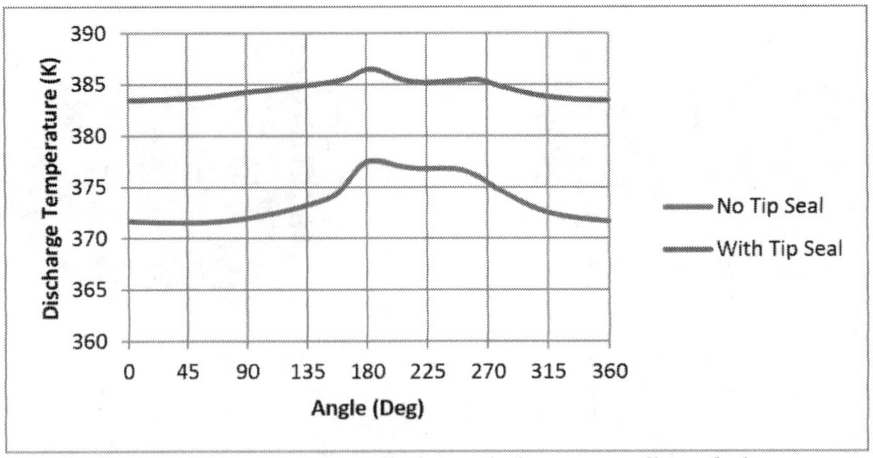

Figure 12: Temperature at discharge during one scroll revolution.

5. Conclusions

A new template tool that can streamline the simulation procedure of the scroll compressor and expander has been developed. The tool is also capable of simulating the tip seal leakage flow along with the scroll compressor. The new tool is tested on a generic scroll compressor, showing meaning impact by including the tip seal volumes.

With easy setup and a very efficient numerical solver, the unsteady flow field of scroll compressor along with the tip seal leakage path can be simulated in hours. The results of the simulation provide insightful flow information to guide the optimization and design of such systems, including pressure/velocity distribution, flow rate and temperature/efficiency of the compressor.

The ease of use and the short simulation turn-around time makes the proposed method an ideal candidate for simulations of scroll compressor systems.

References

[1] Ancel, C; Lamoine, P.; and Didier, F., “Tip Seal Behavior in Scroll Compressors” (2000). International Compressor Conference. Paper 1452

[2] Cui, M.M., “Numerical Study of Unsteady Flows in a Scroll Compressor”, 2006, *J. Fluids Eng.* Vol.128 no.5, p. 947-955.

[3] Gao, H. and Jiang, Y., "Numerical Simulation of Unsteady Flow in a Scroll Compressor" (2014). International Compressor Engineering Conference. Paper 2337.

[4] Feng, J., Qu, Z., and Lin, X., "Three-Dimension Numerical Simulation of Discharge Flow in a Scroll Air Compressor", 2004. *International Compressor Engineering Conference*. Paper 1647.

[5] Jiang Y., Furmanczyk M., Lowry S., Zhang D., and Perng C.Y., 2007, A Three-Dimensional Design Tool for Crescent Oil Pumps. SAE p. 2008-01-0003.

[6] Liu, S; Kang, X; Shan, C; and Hu, Y, "Numerical Simulation Of Three-dimension Unsteady Flow In The Compression Chambers Of A Scroll Compressor" (2014). International Compressor Engineering Conference. Paper 2277.

[7] PumpLinx 3.4, 2014, User help manual.

[8] Rovira, J., Rigola, Joaquim, Oliva, Asseni, Perez-Segarra and Carlos D., "Numerical Simulation and Experimental Comparison of Refrigeration and Air Conditioning Scroll Compressors. Application to Trans-Critical Carbon Dioxide Cycles" (2006). *International Compressor Engineering Conference*. Paper 1801.

[9] Wang, J., Yang L., Luo, X., Mangan, S.; Derby, J.W., "Mathematical Modeling Study of Scroll Air Motors and Energy Efficiency Analysis—Part I,", 2011, *Mechatronics, IEEE/ASME Transactions*, vol.16, no.1, p.112-121.

IOP Conf. Series: Materials Science and Engineering **90** (2015) 012035 doi:10.1088/1757-899X/90/1/012035

Model based design of an intercooled dual stage sliding vane rotary compressor system

Roberto Cipollone[1], Giuseppe Bianchi[1], Davide Di Battista[1], Giulio Contaldi[2], Stefano Murgia[2]

[1]Department of Industrial and Information Engineering and Economics, University of L'Aquila, L'Aquila, Italy

[2]Ing. Enea Mattei S.p.A, Vimodrone (Milan), Italy

E-mail: roberto.cipollone@univaq.it

Abstract. Energy saving is currently one of the most important driving factors for innovation all over the world. With reference to global electricity consumptions, electrical energy for compressed air production accounts for 4-5%. Among the rotary compressor technologies, Sliding Vane Rotary Compressors (SVRC) are characterized by noteworthy specific energy consumptions and demonstrated an unforeseen energy saving potential thanks to some intrinsic features specifically related to this kind of machines. The paper presents a further reduction strategy to lower energy consumptions in compressed air systems using SVRCs that relies on the combination of the recent use of a pressure swirled oil injection technology and a dual stage intercooling. The synergy between technologies already mature approaches to the lowest energy consumption and candidates SVRCs as superior machines in the energy context. The saving potential compared to the technology at the state of the art was evaluated thanks to a comprehensive mathematical modeling of the two compressor sections and the intercooling heat exchanger and fan. Results showed a reduction of the electrical power required to drive the compressor system up to 9.5%. The overall approach represents a model-based design for a new machine which is under development.

1. Introduction

Compressed air is an indispensable utility for industrial processes that is almost exclusively produced by electric energy and accounts for 10-15% of overall electricity consumption in the industrial sector [1, 2]. In particular, Figure 1 shows that the most energy consumer are machines in the range of 10-30 m^3/min [3]. Furthermore, if compressed air production for commercial and residential markets is taken into account, the overall consumptions would reach 20% [4]. Energy saving in Compressed Air Systems (CAS) might be accomplished through interventions upstream and downstream the compressor. Among them, compressor technology and operation belong to the first category and they account for a consumption share close to 10-20% [2].

In order to compare energy consumptions of current industrial air compressors and to outline suitable future improvement strategies, the Authors developed a procedure to define homogeneous operating points starting from energy consumptions at given operating conditions for different compression technologies and manufacturers provided in CAGI's datasheets [3]. Among the performance indicators, the global compressor efficiency, whose mathematical formulation is reported in Equation 1, not only depends on the compressor but also on its auxiliaries and electric motor [5].

Content from this work may be used under the terms of the Creative Commons Attribution 3.0 licence. Any further distribution of this work must maintain attribution to the author(s) and the title of the work, journal citation and DOI.
Published under licence by IOP Publishing Ltd

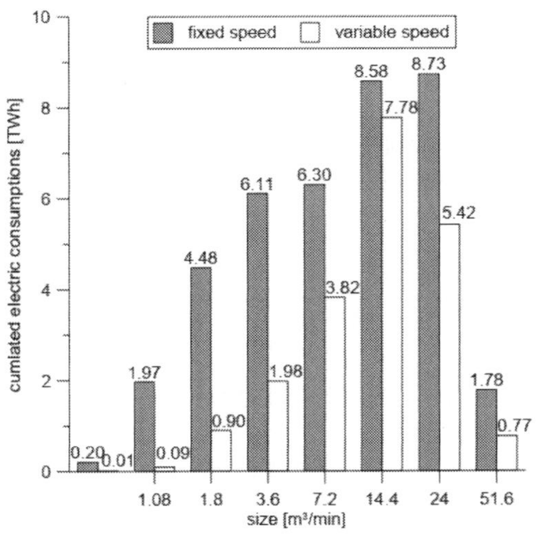

Figure 1: Cumulated electric energy consumption distribution per compressor size

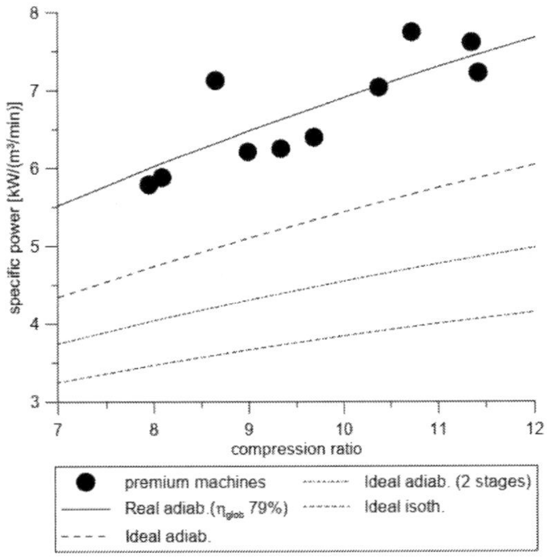

Figure 2: State of the art and goals for the energy saving in compressed air systems

$$\eta_{glob} = \eta_{vol}\, \eta_{ad-is}\, \eta_{mech}\, \eta_{org}\, \eta_{el} \tag{1}$$

The investigation carried out is summarized in Figure 2 and states that real performances of current premium machines closely fit a global compressor efficiency of 79%. Although improvements can still be accomplished, electric and mechanical technologies have reached a good level of development. Therefore, a great saving potential is mainly achievable on the thermodynamic side. The theoretical analysis reported in Figure 2 also reports ideal performances for adiabatic-isentropic and isothermal transformations. The latter trend could be achievable with infinite intercooled stages. However, almost half of the saving potential associated to a transition from an adiabatic-isentropic to an isothermal compression could be reached just with a dual stage compressor system. These general predictions were calculated disregarding the effects that mechanical and organic losses (fan consumption) have on the energy consumptions.

To address the energy saving strategy identified towards an industrial product, the paper presents a model based design of a dual stage intercooled compression system. The compressor technology taken as reference is the sliding vane rotary one. In literature, research was carried out through modeling and experimental approaches aiming at investigating the intrinsic features that belong to this kind of machines. These analyses provided relevant information concerned to the vane dynamics [6, 7], friction phenomena [8, 9], leakage between vanes [10, 11, 12] as well as methodologies that allowed to measure the vane pressure evolution in order to estimate the compressor performance [13, 14]. Even if the development of dual stage compressor prototypes was previously investigated for air conditioning applications [15, 16, 17], those machines were small scale ones ($< 1kW/(m^3/min)$). Hence, according to Figure 1, they would not lead to noticeable energy saving benefits in CAS in global terms. On the other hand, some researchers developed a model based design of a dual stage sliding vane rotary compressor system using a unique rotor [18]. Nevertheless, this theoretical concept might have drawbacks in terms of technical and economical feasibility of a real prototype because of sealing and stability issues for the compressor blades while sliding along stators with different curvature.

IOP Conf. Series: Materials Science and Engineering **90** (2015) 012035 doi:10.1088/1757-899X/90/1/012035

2. Design Methodology

In previous experimental studies on single stage sliding vane compressors, vane pressure was measured and allowed the reconstruction of the indicator (pressure-volume) diagram of the machine. These activities revealed that current sliding vane technology is characterized by a closed-volume compression transformation that follows an adiabatic trend [19]. Since minimum work occurs for an isothermal compression, to approach the lowest energy consumption, the innovative compressor design proposes to split the compression ratio in two stages, cooling the compressed air in between (Figure 3).

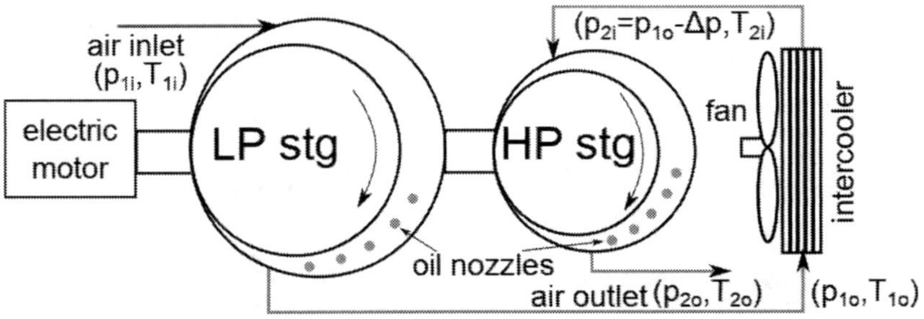

Figure 3: Dual compression stage with internal and intermediate air cooling

From technical and economic viewpoints, the best trade-off in terms of energy saving and feasibility of the technology resulted in a system layout with two compressors. In a dual stage compression, the compression ratio that minimizes the input work is ideally the square root of the total pressure ratio. However, in case of a real intercooler, a temperature difference between the compressed air at the heat exchanger outlet and the environment must exist. Hence, the intermediate pressure that minimizes the overall compression work can be analytically evaluated according to Equation 2.

$$p_{2i} = \sqrt{p_{1i}\, p_{2o} \left(\frac{T_{2i}}{T_{1i}}\right)^{\frac{\gamma}{\gamma-1}}} \tag{2}$$

In the current design approach, air is not only cooled during the passage from the Low Pressure (LP) stage to the High Pressure (HP) one, but also along both the compression processes thanks to an innovative oil sprayed injection technology whose features are presented in the following paragraph. Since both the machines were supposed to be driven by the same electric motor (i.e. same revolution speed), the HP compressor resulted smaller than the LP one. As concerns the air to air intercooler, a fan was needed to enhance the heat transfer. The energy consumption of this device was eventually discounted from the energy benefits achieved.

3. Compressor Model

The design of the dual stage sliding vane compressor was supported by a comprehensive simulation platform that is able to reproduce performances of real machines and to explore new design configurations. The model is composed of different modules that interact together according to the scheme reported in Figure 4. The mathematical model takes into account all the main physical phenomena occurring in sliding vane compressors [19, 20]. The vane filling and emptying processes were modeled through a one-dimensional quasi-propagatory approach. The parametric geometry modelling allows to investigate sliding vane machines (compressors and expanders) with any suction and discharge ports layout, both axial and radial. The blade dynamics module solves the Newton's 2nd of motion and the hydrodynamic lubrication at the blade tip which is the most severe location for friction losses.

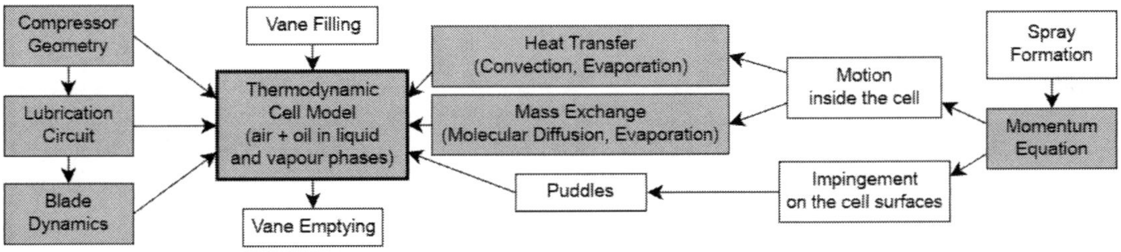

Figure 4: Compressor model block diagram

As concerns the lubrication circuit, two configurations may be studied: a conventional one, whereas oil is supplied through calibrated holes as liquid jets, and an innovative sprayed injection technique that relies upon atomizers to enhance the heat transfer between oil and air. The sprayed injection technique was modeled with a Lagrangian approach and applies conservation equations to oil particles after the spray break-up: downstream the break-up length, the Rosin Rammler oil droplet size distribution is discretized into parcels that, travelling along the compressor vane while it rotates, absorb heat from the compressing air until the impingement on the cell surfaces is reached and puddles build up. As shown in Figure 4, the core of the compressor model is the thermodynamic cell module that collects all the energy input and output fluxes to determine the pressure inside the compressor cells.

The mathematical model has been extensively validated both for the conventional and innovative oil injection techniques through experimental campaigns on industrial sliding vane compressors at different operating points (delivery pressure, revolution speed, etc.) [19, 21, 22]. The comparison between experimental and simulated indicator diagrams stated the accuracy and reliability of the model. Additional details on the modelling and experimental activities that led to the development of the compressor simulation platform can be found in the references [19, 20, 23, 24].

4. Intercooler Model

The benefits of splitting the overall compression ratio in two separate stages are strictly related to the chance of cooling the compressed air in between. To predict the air temperature at the inlet of the HP stage and pressure drops due to the intercooler, a mathematical model of the heat exchanger was developed relying on the know-how acquired on engine cooling circuit modeling [25]. Furthermore, an axial fan was considered to force the cooling air crossing the heat exchanger.

4.1. Heat Exchanger

Air-to-air heat exchangers that are commonly used in industrial applications are the tube and fin type because of high heat transfer rates and relatively low pressure drops. In these devices, the hot stream (compressed air) flows in circular tubes externally finned and it is cooled by ambient air pushed by a fan. After a preliminary design that estimated a suitable size and geometry of the intercooler, the heat exchanger was modeled through a one-dimensional approach, dividing it in a series of elementary ducts. For each of them, conservation equations are solved in a simplified way, in order to evaluate the thermal power exchanged, pressure drops and the resulting outlet flows conditions. To compute the thermal power exchanged in each elementary duct, an electrical analogy was used: as shown in Figure 5, a series of heat transfer resistances was considered. In particular a forced convective resistance was used between the compressed air and the internal tube wall while a conductive resistance was adopted between the metallic tube walls. The presence of fins on the external wall of the tube eventually led to an enhanced forced

IOP Conf. Series: Materials Science and Engineering **90** (2015) 012035 doi:10.1088/1757-899X/90/1/012035

convection with ambient air. Thermal resistances were evaluated considering the elementary heat exchanger geometry and the convective heat transfer coefficients according to Nusselt correlations. Mathematical formulations are reported in Equations 3 and 4 while symbols explanation are given in the nomenclature. The elementary thermal power exchanged was calculated considering the temperature difference between internal and external values according to Equation 5.

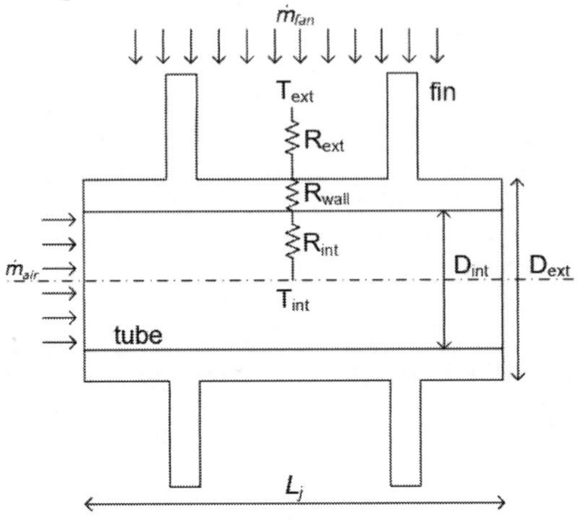

Figure 5: heat exchange model of an elementary duct

$$R_{int} = \frac{1}{h_{int}\, S_{int}} \qquad R_{wall} = \frac{ln(D_{ext}/D_{int})}{2\pi\, k\, L_j} \qquad R_{ext} = \frac{1}{h_{ext}\, S_{ext}} \tag{3}$$

$$Nu = C \left(\frac{\dot{m}}{\mu\, A} L_j \right)^n Pr^{1/3} \tag{4}$$

$$Q_j = \frac{T_{int} - T_{ext}}{R_{int} + R_{wall} + R_{ext}} \tag{5}$$

Pressure drops are evaluated considering friction between fluids and heat exchanger walls according to the corrected Darcy correlation reported in Equation 6:

$$\Delta p = f_m \frac{32\, \mu\, v}{D^2{}_{wet}} L_j \tag{6}$$

where the wet diameter D_{wet} corresponds to the inlet diameter at the inner tube side and to the diameter of the equivalent duct between two fins at the outer side. The correlation also takes into account a suitable friction multiplier f_m in both sides of the heat exchanger.

4.2. Fan Model

The mass flow rate of cooling air is calculated solving the model of an electrical axial fan. At a given revolution speed taken as reference, the fan operating curve (pressure increase vs. flow rate) was evaluated according to Equation 7.

$$\dot{V}(N_{ref}) = \dot{V}_{max} - \alpha \left(\frac{p_o}{p_i} - 1 \right)^2 \tag{7}$$

The latter expression required the knowledge of an operating point (e.g. design point) and the maximum flow rate $\dot{V}_{max}$ (which corresponds to a nil pressure increase). Operating curves at different revolution speeds were retrieved using a similarity law with respect to reference conditions (Equation 8).

$$\dot{V}(N) = \dot{V}(N_{ref}) \frac{N}{N_{ref}} \left(\frac{p_o}{p_i} - 1 \right) = \left(\frac{p_{o,ref}}{p_{i,ref}} - 1 \right) \left(\frac{N}{N_{ref}} \right)^2 \tag{8}$$

Electrical power required to drive the cooling air through the heat exchanger was evaluated using Equation 9 that assumed a product between mechanical and electrical efficiencies of 20%.

$$P_{el,fan} = \frac{\rho \dot{V} (p_o - p_i)}{0.2} \tag{9}$$

5. Results and discussion

The energy saving potential of a dual stage intercooled compression was assessed simulating different compressors system layouts and injection technologies. With reference to Figure 1, the analysis was carried out designing a compressor system in the most significant flow rate range for cumulated energy consumptions (23 m^3/min). Suction conditions were the ISO 1217 ones (1 atm and 20 $°C$) while the delivery pressure was set to 9 bar. A further parameter of investigation was the effectiveness of the intercooling process in terms of inlet temperature of the second stage (T_{2i}). In simulation cases A, B, C oil injection was performed in the conventional way, at only one angular location. Conversely, a series of five pressure-swirl nozzles were equally distributed along the compression process in cases D, E, F and supplied the same amount of lubricant of cases A, B, C respectively. Injection temperature was set to 40 $°C$ everywhere while injection pressure was equal to the discharge pressure of each compressor stage, as noticed experimentally [24].

Simulations data reported in Table 1 present a full energy breakdown of the compressor system: since sliding vane types are high-side compressors, according to Equation 10, in each machine mechanical power is not only needed to accomplish the compression process and to overcome friction losses but also to pressurize the oil that is injected in the compressor cells.

$$P_{mech} = P_{ind} + P_{fr} + P_{oil} \quad (10) \qquad \eta_{glob} = P_{ind}/P_{el} = \eta_{el}\, P_{ind}/P_{mech} = \eta_{el}\, \eta_{mech} \quad (11)$$

Compression power was eventually calculated from the area of the indicator diagram according to the methodology proposed in [19]. Package power, i.e. the electrical power needed to drive the compressor system and the intercooler fan (if present), was computed assuming an IE2 efficiency class for 4-poles electric motors operating at 50 Hz. This choice fixed the revolution speed of all compressors to 1500 RPM. However, since the efficiency of the electric motor increases with its size, the comparison was made in terms of mechanical power rather than the electrical one. Furthermore, since in SVRCs oil pressurization is accomplished without an auxiliary device, the mechanical efficiency reported in Equation 11 also takes into account organic losses. Figure 6 compares indicator diagrams of single stage compressor with conventional (case A) and innovative oil injection technology (case D). The internal air cooling accomplished by oil droplets would bring the adiabatic compression phase towards an isothermal trend, with a potential saving in the indicated power close to 3 kW. Since compressor geometry and operating parameters of the lubrication circuit (injection pressure and flow rate) were kept constant, friction losses and oil pressurization power would not change. This fact motivates the slight decrease of the mechanical efficiency in case D.

Figure 7 reports indicator diagrams related to a typical intercooled dual stage configuration. When the compression process is split in two steps, at the LP compressor outlet air is at high pressure. Furthermore, assuming that oil is separated from air downstream each stage, because of the heat transfer with oil and metallic surfaces of the machine, air temperature at the compressor outlet would be at a value much lower than the one it would have had inside the compressor cell before the discharge. With reference to previous experimental campaigns, assuming that at the compressor exit air and oil have the same temperature T_o, this outlet value can be calculated according to Equation 12 in which variations of the specific heat at constant pressure for air and oil have been neglected since they are lower than 1.5%.

$$\dot{m}_{air}\, c_{p,air}(T_o - T_{air,i}) + \dot{m}_{oil}\, c_{p,oil}(T_o - T_{oil,i}) = P_{mech} \tag{12}$$

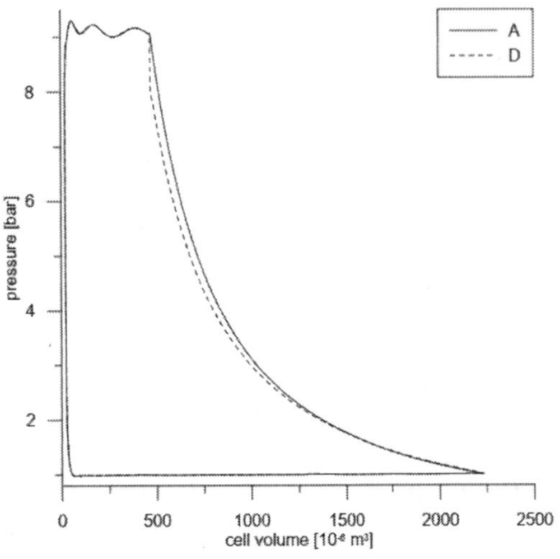

Figure 6: Single stage indicator diagram

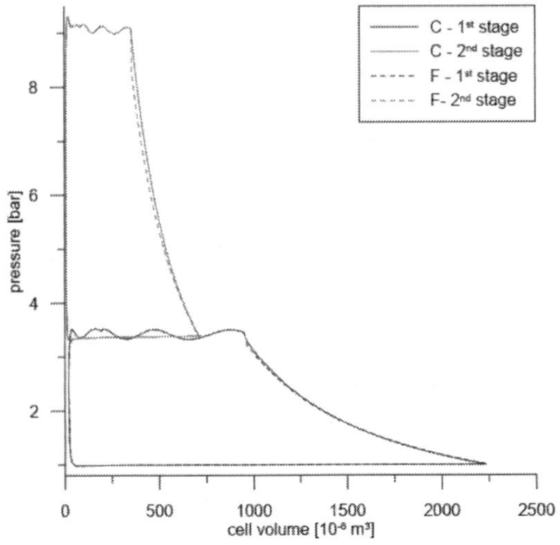

Figure 7: Dual stage indicator diagram

The resulting high fluid density would allow to reduce the volumetric capacity $\dot{V}_{air,2}$ of the HP compressor that would become more compact. In this way, friction losses could be reduced and the mechanical efficiency of the HP compressor would increase beyond 80%. Depending on the compressed air temperature at HP compressor inlet, the overall compression ratio was divided according to Equation 2 to achieve the optimal theoretical value of intermediate pressure. In simulation cases B and E, no intercooling was considered ($T_{2i} = T_{1o}$). On the other hand, in cases C and F air was further cooled up to 40 $°C$. Hence, the inlet temperature at the HP compressor got close to the ideal value ($T_{2i} = T_{1i}$) and reduced the difference between the compression ratios of the two stages. In order to keep the same machine geometry and its influence on friction losses between simulation cases B, C, E, F, optimal compression ratio was achieved tuning the angular displacement of the discharge ports. Simulations in Table 1 show that through a dual stage compression overall indicated power would decrease from 121.7 kW to 107.1 kW without intercooler and up to 104.4 kW with intercooler. These results are in agreement with the theoretical predictions reported in Figure 2 which compare specific compression powers of single and dual stage machines only from a pure thermodynamic point of view, i.e. without taking into account mechanical, organic and volumetric losses. However, since oil atomization depends on injection pressure, in case of dual stage compression the benefits of a sprayed injection technology would be less remarkable compared to the single stage process. This issue would be particularly critical for the first stage, whereas an injection pressure around 3.5 bar would not be able to properly atomize the oil jets. The resulting compression trend would be equal to an adiabatic one, as shown by the tendency of solid and dashed lines in Figure 7 to overlap. On the other hand, for the same lubrication circuit layout and overall oil flow rate, the compression split also divides the flow rate and injection pressure depending on the intermediate pressure value. Anyway, this results in a remarkable reduction of the pressurization power which accounts for 2.5 kW, as the energy saving that would be achived using the intercooling heat exchanger. Moreover, the first compression stage has the same geometry of the single stage compressor but accomplishes a lower indicated work to pressurize the air to 3 bar (instead of 9). Therefore, its mechanical efficiency results in very low values that consequently affect the mechanical efficiency of the whole compressor system. For these reasons, the 15% saving in terms of indicated power reduces to 9.5% with reference to electrical power.

Table 1: Simulations parameters and results
(p_{1i} 1 bar_a, T_{1i} 20.0°C, p_{2o} 9 bar_a, 1500 RPM)

case	A	B	C	D	E	F	
stages\spray\intercooling	1\N\N	2\N\N	2\N\Y	1\Y\N	2\Y\N	2\Y\Y	
$\dot{V}_{air,1}$	23.0	23.0	23.0	23.0	23.0	23.0	m^3/min
$\dot{V}_{air,2}$		7.1	7.4		7.1	7.4	m^3/min
$\dot{V}_{oil,1}$		148.0	129.4		148.0	129.4	L/min
$\dot{V}_{oil,2}$		148.1	170.9		148.1	170.9	L/min
$\dot{V}_{oil}$	303.9	296.1	300.3	303.9	296.1	300.3	L/min
$p_{1o} = p_{2i}$		3.7	3.4		3.7	3.4	bar_a
T_{1o}		55.5	56.3		55.5	56.3	°C
T_{2i}		55.5	40.0		55.5	40.0	°C
T_{2o}	81.0	46.1	45.6	81.0	46.1	45.6	°C
$P_{ind,1}$		62.6	57.0		62.3	56.8	kW
$P_{ind,2}$		44.5	47.4		43.3	46.2	kW
P_{ind}	121.7	107.1	104.4	118.7	105.6	103.0	kW
$P_{fr,1}$		26.3	26.3		26.3	26.3	kW
$P_{fr,2}$		7.3	7.4		7.3	7.4	kW
P_{fr}	27.3	33.6	33.7	27.3	33.6	33.7	kW
$P_{oil,1}$		0.9	0.9		0.9	0.9	kW
$P_{oil,2}$		1.6	1.9		1.6	1.9	kW
P_{oil}	5.1	2.5	2.8	5.1	2.5	2.8	kW
$P_{mech,1}$		89.8	84.2		89.5	83.9	kW
$P_{mech,2}$		53.4	56.7		52.2	55.5	kW
P_{mech}	154.0	143.3	140.9	151.1	141.7	139.4	kW
$\eta_{mech,1}$		69.7	59.2		69.6	67.6	%
$\eta_{mech,2}$		83.4	83.6		83.0	83.2	%
η_{mech}	79.0	74.8	74.1	78.6	74.5	73.8	%
P_{el}	162.3	151.0	148.5	159.2	149.4	146.9	kW

The intercooler design led to a heat exchanger made by aluminium with an internal heat exchange surface of 20 m^2 and a finned external one of 32 m^2. As shown in Figure 8, thermal performances are strictly related to mass flow rate of the cooling air which in turn depends on the fan revolution speed. In particular, cooling flow rate results from the equilibrium of the fan performance curves and the aeraulic pressure drop due to the heat exchanger (Figure 9). In cases C and F the thermal power exchanged in the intercooler is about 7.5 kW through an external air flow rate of 1200 m^3/h. At this operating point, fan would rotate at 1200 RPM and lead to a pressure rise of the cooling air around 30 Pa. On the other hand, pressure drops of the compressed air flow through the internal side of the intercooler would be close to 70 mbar. Fan electrical power absorbed would be eventually equal to 40 W. However, since the presence of this heat exchanger does not lead to significant efficiency improvements, in a future industrialization

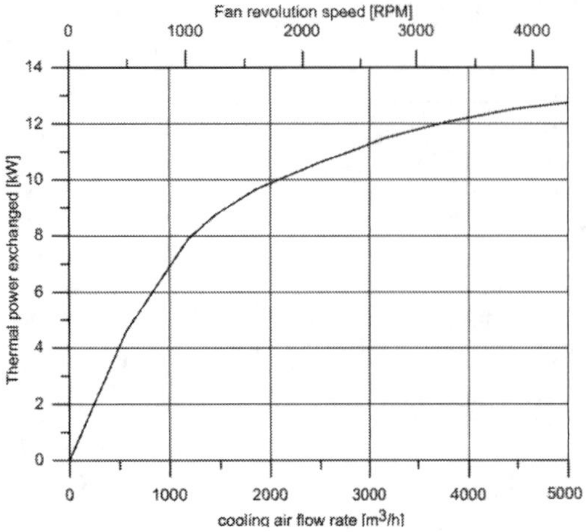

Figure 8: Influence of cooling mass flow rate (fan revolution speed) on the thermal power exchanged at the intercooler

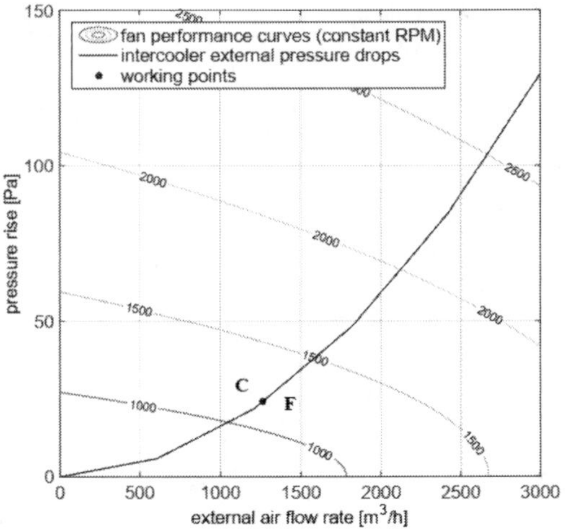

Figure 9: Fan performance map and operating points in simulation cases C and F

of the dual stage concept, the intercooler is not advisable.

6. Conclusions

The current research compared the performance of single and dual stage sliding vane compressors layout with conventional and innovative injection technologies. The split of the compression process in two phases revealed that:

- indicated power savings up to 15%;
- an intercooling heat exchanger would not provide additional benefits;
- less energy is required to pressurize the oil;
- oil spray injection would not be as effective as occurs in single stage compressors;
- lower mechanical efficiencies are expected because of greater overall friction losses.

If all these phenomena are taken into account, a dual stage compressor system might potentially lead to overall electrical savings up to 9.5%.

The economic feasibility of the solution outlined essentially depends on the utilization factor of the compressor system, on the cost of electricity and on the carbon tax. All these parameters have geographical spreads that do not allow to perform accurate predictions yet. However, in order to take into account also the additional maintenance costs due to a more complex compressor system, further investigations are going to be performed thanks to experiments on a real prototype.

7. Acknowledgment

The work has been done under the FP7 Project "Complete Vehicle Energy-Saving CONVENIENT" funded by the European Commission.

References

[1] Saidur R, Rahim N and Hasanuzzaman M 2010 *Renewable and Sustainable Energy Reviews* **14** 1135 – 1153 ISSN 1364-0321

[2] Radgen P 2001 *Compressed air systems in the European Union: energy, emissions, savings potential and policy actions* (LOG_X Verlag GmbH)
[3] European Association of Manufacturers of Compressors, Vacuum Pumps, Pneumatic Tools and Air and Condensate Treatment Equipment PNEUROP `www.pneurop.eu` last access: 2015-01-20
[4] Cipollone R 2014 *Proceedings of the Institution of Mechanical Engineers, Part E: Journal of Process Mechanical Engineering*
[5] Vittorini D, Bianchi G and Cipollone R in press *Energy Procedia*
[6] Badr O, O'Callaghan P and Probert S 1985 *Applied Energy* **19** 159 – 182 ISSN 0306-2619
[7] Wu J H, Zhang L, Huang L Q and Tang Y Y 2004 *Proceedings of the International Compressor Engineering Conference*
[8] Aradau D and Costiuc L 1996 *Proceedings of the International Compressor Engineering Conference*
[9] Badr O, Probert S and O'Callaghan P 1985 *Applied Energy* **20** 253 – 285 ISSN 0306-2619
[10] Reed W and Hamilton J 1980 *Proceedings of the International Compressor Engineering Conference*
[11] Fukuta M, Yanagisawa T and Shimizu T 1992 *Proceedings of the International Compressor Engineering Conference*
[12] Al-Hawaj O 2009 *International Journal of Refrigeration* **32** 1555 – 1562 ISSN 0140-7007
[13] Huang Y M and Yang S A 2008 *Measurement* **41** 835 – 841 ISSN 0263-2241
[14] Tramschek A and Mkumbwa M 1996 *Proceedings of the International Compressor Engineering Conference*
[15] Lee J, Lee U, Sung C, Han K and Kim E 2008 *Proceedings of the International Compressor Engineering Conference*
[16] Hirano K and Monasry J 2012 *Proceedings of the International Compressor Engineering Conference*
[17] Takashima K, Onoda I, Kitaichi S and Watanabe N 2004 *Proceedings of the International Compressor Engineering Conference*
[18] Valente R and Villante C 2008 *Proceedings of the 20th International Compressor Engineering Conference*
[19] Bianchi G and Cipollone R 2015 *Applied Energy* **142** 95 – 107 ISSN 0306-2619
[20] Bianchi G and Cipollone R 2015 *Applied Thermal Engineering* **84** 276 – 285 ISSN 1359-4311
[21] Cipollone R, Valenti G, Bianchi G, Murgia S, Contaldi G and Calvi T 2014 *Proceedings of the Institution of Mechanical Engineers, Part E: Journal of Process Mechanical Engineering*
[22] Bianchi G, Cipollone R, Murgia S and Contaldi G 2014 *Proceedings of the International Compressor Engineering Conference*
[23] Cipollone R, Bianchi G and Contaldi G 2012 *ASME 2012 International Mechanical Engineering Congress and Exposition* vol 6 pp 69 – 80 ISBN 978-0-7918-4522-6
[24] Bianchi G, Cipollone R, Murgia S and Contaldi G 2015 *International Journal of Refrigeration* **52** 11 – 20 ISSN 0140-7007
[25] Cipollone R, Battista D D and Gualtieri A 2013 *Energy Conversion and Management* **75** 581 – 592 ISSN 0196-8904

Nomenclature

α	fan coefficient $[m^3/s]$	C	turbulent coefficient $[-]$	$ad-is$	adiabatic-isentropic
γ	heat specific ratio $[-]$	D	diameter $[m]$	el	electrical
μ	dynamic viscosity $[Pa\,s]$	L	length $[m]$	ext	external
Δp	pressure drop $[Pa]$	N	revolution speed $[RPM]$	fr	friction
ρ	air density $[kg/m^3]$	Nu	Nusselt number $[-]$	$glob$	global
c_p	specific heat at const. p $[kJ/kg/K]$	P	power $[W]$	i	inlet
f_m	friction multiplier $[-]$	Pr	Prandtl number $[-]$	ind	indicated
h	heat transfer coefficient $[W/(m^2\,K)]$	Q	thermal power $[W]$	int	internal
k	thermal conductivity $[W/(m\,K)]$	R	thermal resistance $[K/W]$	j	discretization index
$\dot{m}$	mass flow rate $[kg/s]$	S	heat transfer surface $[m^2]$	$mech$	mechanical
n	Reynolds number exponent $[-]$	T	temperature $[K]$	o	outlet
p	pressure $[Pa]$	$\dot{V}$	volumetric flow rate $[m^3/s]$	org	organic
v	air flow speed $[m/s]$	1	first stage	ref	reference
A	cross sectional fluid area $[m^2]$	2	second stage	vol	volumetric

IOP Conf. Series: Materials Science and Engineering **90** (2015) 012036 doi:10.1088/1757-899X/90/1/012036

Design Evolution: From Rolling Piston to Revolving Vane to Cross-Vane Expander-compressor unit.

Ooi K T[1] and Yap K S[2]
[1,2]Nanyang Technological University, 50, Nanyang Avenue, Singapore 639798
[2] TUM CREATE (Technische Universität München - Campus for Research Excellence And Technological Enterprise), 1 CREATE Way, #10-02, CREATE Tower, Singapore 138602

Abstract

In this paper the design evolution from rolling piston compressor (RP) to revolving vane compressor (RV) and finally to cross-vane expander-compressor unit (CVEC) is presented and discussed. The details in the design philosophy which initiated this evolution will be presented and discussed.

It is estimated that more than 90% of the room air-conditioners uses RP in its compressor. This is because of its advantages: it has the few parts, it is simple geometrically and it is reliable. However, it is with no weaknesses. RP's weaknesses lie in three parts:
too many rubbing surfaces with high relative rubbing velocity with each other which give rise to high frictional losses;
components (eccentric and roller) are not rotating at their centres and resulted in unnecessary inherent vibration;
vane tip is constantly rubbing against the roller making it a weakest part in design.

To overcome RP's weaknesses, RV is introduced. As compared to RP, RV has fewer rubbing surfaces, and the relative velocities among these surfaces are reduced Components in RV rotate at their own centres and there is no inherent vibration; the rubbing at the vane tip has been eliminated completely.

However, like RP, RV also has a large rotor which occupied "useful" space and making the working chamber relatively small. To overcome this latter problem, CVEC is introduced. In this newly invented CVCE, not only parts are all concentric and rotate at their own centres, the unit also recovers expansion energy and hence significantly reduces energy required by the refrigeration systems. And, more importantly, all these are carried out not at the expense of the additional cost.

In this paper, details on these three compressors are explained, compared and their respective uniqueness are shown and discussed.

1. Introduction

Literature shows that significant research has been carried out to arrive at what we have today for a relatively energy efficient running refrigeration compressors. Among the most popular compressors in use today, there are rolling piston, reciprocating, scroll and screw. Rolling piston compressor (RP) is generally used for cooling capacity below 15 kW, though recently more and more RPs are used for cooling capacity of 15kW and above. In room air-conditioners applications, more than 90% of the

Content from this work may be used under the terms of the Creative Commons Attribution 3.0 licence. Any further distribution of this work must maintain attribution to the author(s) and the title of the work, journal citation and DOI.
Published under licence by IOP Publishing Ltd

compressor used RP. This is because of its simple geometry, few parts and relatively reliable. However, RP has some design weaknesses such as significant frictional losses due to a number of rubbing surfaces which have high relative rubbing velocity to each other; components (eccentric and roller) are not rotating at their centres and resulted in unnecessary inherent vibration; vane tip is constantly rubbing against the roller making it a weakest part in design.

To overcome RP's weaknesses, revolving vane compressor (RV) is introduced. RV has fewer rubbing surfaces, and the relative velocities among these surfaces are reduced. Components in RV rotate at their own centres and hence no inherent vibration, and there is no vane tip contact.

However, RP and RV have large rotors which occupied "useful" space. To overcome this, CVEC was introduced. In CVCE, not only parts are all concentric and rotate at their own centres, the unit also recovers expansion energy and hence significantly reduces energy required by the refrigeration systems. In this paper, details on these three compressors are explained, compared and their respective uniqueness are shown and discussed.

2. Rolling piston compressor

Rolling piston compressor is one of the most widely used compressors in room air-conditioners. It is mostly used in the applications where the cooling capacity is lower than 15 kW. Figure 1 shows a schematic diagram of a rolling piston compressor. In its basic form, it consists of five components: A cylinder (call stator), a roller, an eccentric, a vane and a discharge valve.

During operation, the roller rols in the cylinder but maintaining contact with the cylinder and the vane tip at all time, creatin two working chambers. These wrolking chambers are bounded by inner surface of the cylinder, out ter surface of the roller and the vane. While one chamber is expanding, hence suction proces, the other is compressing hence compression process. Once the gas pressure reachs that of the discharge pressure, the gas is discharged out f the working In this new designchamber through the discharge port.
There are six areas of frictional losses [2]:

1. Eccentric and the inner surface of the roller,
2. Roller face and the cylinder head face,
3. Eccentric face and the cylinder head face,
4. Vane tip and roller,
5. Vane sides and vane slot,
6. Outer roller surface and the inner cylinder surface.

For item (1) above, the rubbing between the inner surface of he roller and the eccentric is dependent on the forces acting on the roller and the viscous forces at the rubbing surfaces. For the items (2) to (6) the rubbing is between moving surfaces against staionary surfaces, hence the relative velocity reponssible for the friction is relatively large. Attention must also be paid to the rubbing at the vane tip and the roller. This latter rubbing region is a critical rubbing part, this is because a small region of the vane tip rubs against a relatively large outer surface of the roller, cuasing vane tip to wear easily, and in practice, the vane and the roller are harden to reduce wear.

It is noted that also, the main driving comoponent for the RP is the ecnetric. The motion of ecentric drives the roller and hence froming the comrepssion and the scution chambers mentioned earlier. This ecccentric inhenrently introduce "wobbling" motion which induces vibration.

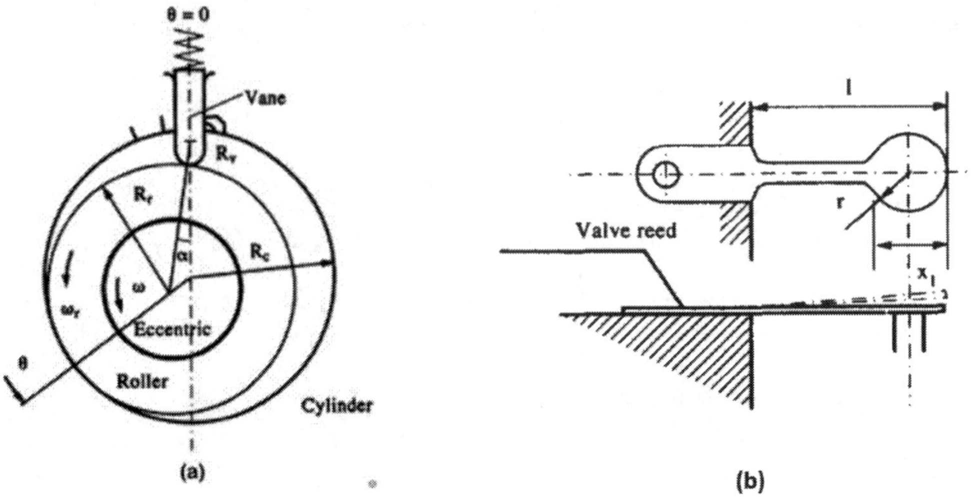

R_c	cylinder radius (m)
R_r	roller outer radius (m)
R_v	radius of vane tip (m)
R_s	shaft radius (m)
ω	angular velocity of crankshaft (rad/s)
ω_r	angular velocity of roller (rad/s)

Figure 1 (a) Schematic of a rolling pisont compressor [1], (b) Discharge valve [2]

3 Revolving vane compressor

RV compressor [3,4] is introduced to mitigate the disadvatges inherently in the RP. Figure 2 shows the schematics of the RV. In its basic form, RV compressor has 4 parts: a cylinder, a rotor, a vane and a discharge valve (no shown). Figure 3 shows the snapshots of the opration of the RV.

One of the distinguishing feature of the RV is that during the operation, the cylinder rotates together with the rotor. The rotating cylidner rotates the two end faces of the cylinder as well, hence making all the rubbing surfaces have lower relative velocity and hence reduces the fricitonal loss significantly. Refer to the six itms of frictional areas mentioned in section 2 for RP, in the design for RV, there is no eccentric, the frictional losses caused by items (1) and (3) are no longer presence. Since the cylinder is also rotating, items (2) and (6) in section 2, have significantly lower frcition. Vane tip in RV design is longer rubbng against any surface and hence item (4) is totally eliminated. Item (5) in section 2, has a very significant improvement. Since the vane is now rigid secured on to the cylinder, the vane slot which rubs agains the vane surfaces is now located at the rotor. It is this arrangement that predominantly caused the vane side frcition to be completely independet from the pressure differential across the vane! This has a huge significance especially when the compressor is operating under a high pressure difference condpition between the suction and the discharge, such as those when using CO2 as the refrigerant. Thus, fricitonal loss caused by item (5) has hugh implication in performnce improvement in the case for RV when compared to RP. The fricitonal loss caused by item (5) for Rv is only dependent on the rotating inertia of the rotor.

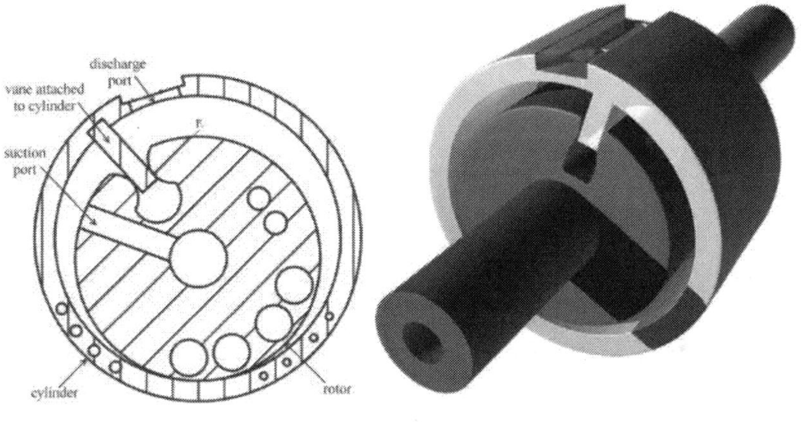

Figure 2 Schematics [3] and 3-D mark-up of the fixed-vane revolving vane compressor.

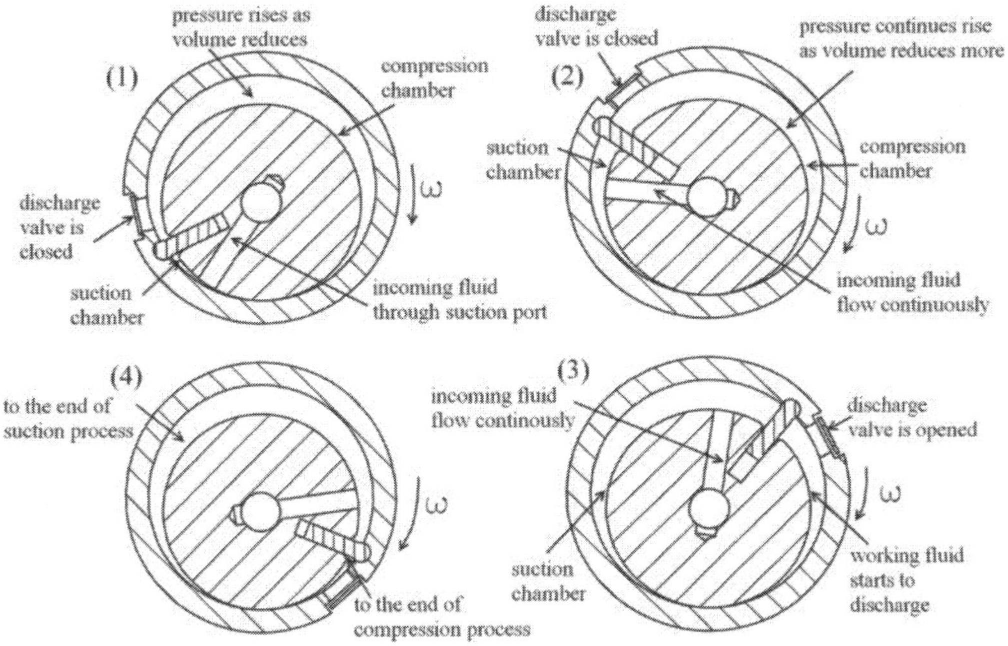

Figure 3 Snapshots of revolving vane comrpessor operation [4]

The revolving vane comrpessor can be dsigned with swiveling vane or fixed vane and the vane can be fixed either on the cylinder or on the rotor. Figure 5 shows the inertia torque variations for these different designs. It is advisable to fix the vane on the driving component, i.e. if the vane is fixed on the cylinder, the cylinder should be the driving component and vice-versa.

The revolving vane comrpessor can be dsigned with swiveling vane or fixed vane and the vane can be fixed either on the cylinder or on the rotor. Figure 5 shows the inertia torque variations for these different designs. It is advisable to fix the vane on the driving component, i.e. if the vane is fixed on the cylinder, the cylinder should be the driving component and vice-versa.

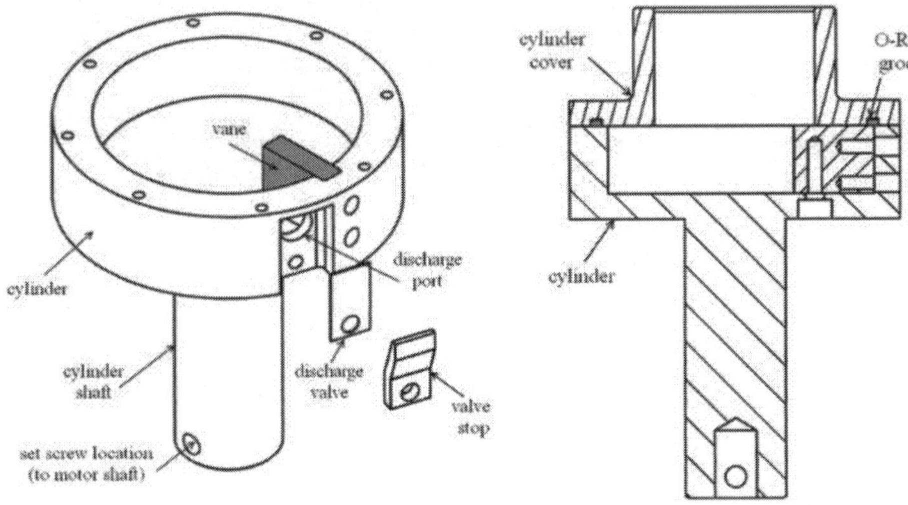

Figure 4 the cylinder assemby cmponents of revolving vane compressor [5].

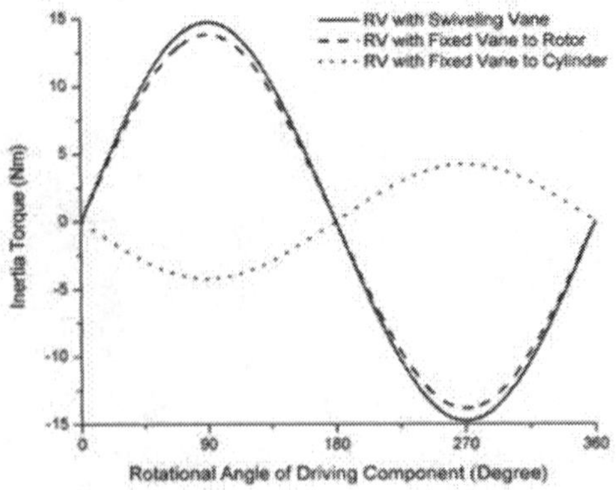

Figure 5 Variations of the inertia torque over three different revolving vane compressor designs [3]

4 Cross vane compressor-expander unit

Referring to RP and RV in figures 1(a) and 2 respectively, it can be seen that the available working spaces (working chamber) of the two compressors are relatively small as compared to the overall size of the compressor. This is because the useful space has been "occupied" by a relatively large rotor. It can be seen from the two figures that the rotors for both design are relatively large as compared to the cylinder, and the working chamber is comparatively small. The next question is that will we be able to make use of the "space" occupied by the rotor and makes the comrpessor or the refrigeration cycle works better. This thought has lead to the born of the cross-vane comrpessor expander unit (CVEC) [6,7]. Figure 6 shows the sschematics of the CVEC. It can be seen immediately that the "rotor" is no longer a "solid" cylinder but a "hollow" cylinder. In fact the space created by this "hollow" rotor is actuall an expander!

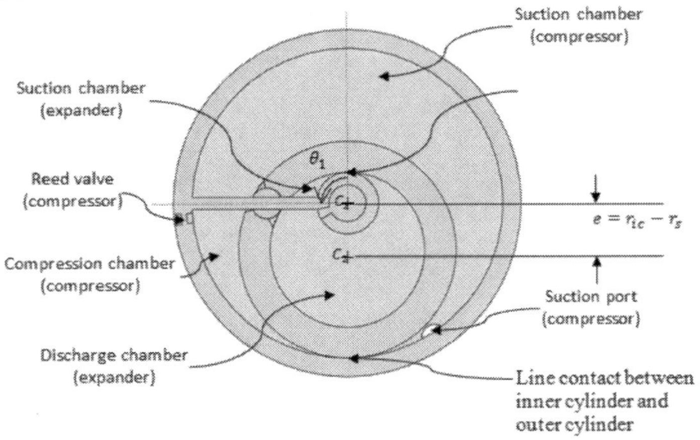

Figure 6 schematics of the cross-vane expander compressor unit.

During operation, the rotation of the shaft causes the vane to rotate which in turn rotates the cylinder for compressor and the expander. Figure 7 illustrates the snapshots of the operation of CVEC. The motion causes the volumes trapped within the suction and compression chamber of the compressor to vary, resulting in suction, compression and discharge of the working fluid. Similarly, the volumes trapped within the suction and discharge chamber of the expander varies too, resulting in suction, expansion, and discharge of the working fluid. Both the compressor and expander operate concurrently.

Such a machine is best to work with working fluids which operates on large pressures between suction and discharge of the compressor, such as CO_2. This large pressure difference will result in a significant power being recovered, which, otherwise wasted at the expansion valve, and hence increases significantly the overall COP of the CO2 cycle. Theoretical analysis (not shown) shows that 14-40% COP increase is possible.

Figure 8 shows schematics of a refrigeration cycle fitted with CVEC. It is noted that the expansion valve and the compressor is now replaced by CVEC.

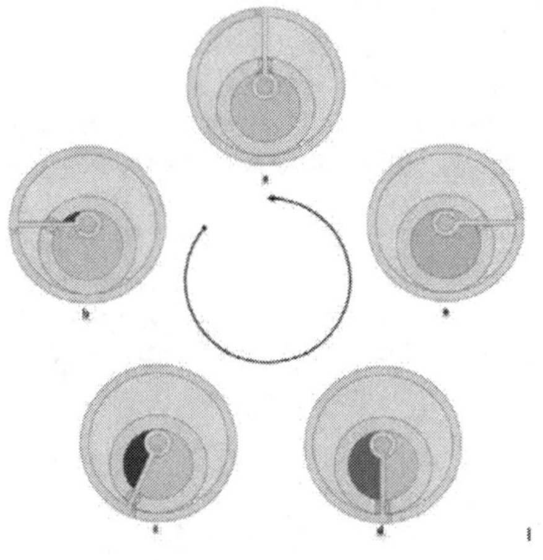

Figure 7 Operation snapshots of cross-vane expander comrpessor

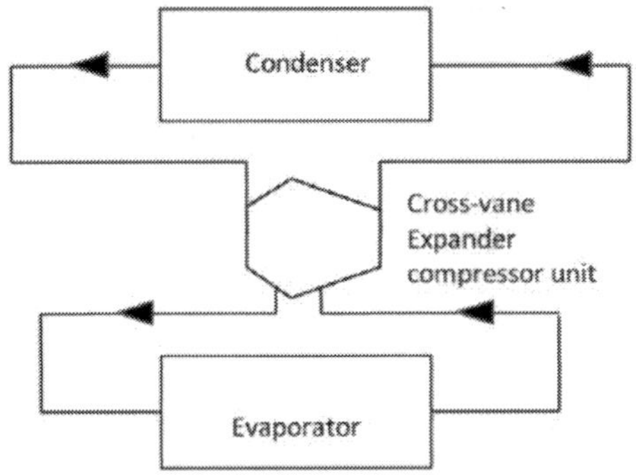

Figure 8 Schematics of a refrigeration cycle fitted with CVEC

To understand the behaviour of the CVEC in a greater detail, the components of the instantaneous power of the machine is shown in Figure 9 and these are: compression power, expansion power, inertial power, and total frictional power loss. The phase difference has resulted in the decrease of CVEC peak power of about 4.3%. This improvement is still relatively small, and ideally, the peak power of the compressor and expander should be matched at the same instance. This not only reduces the peak power of the CVEC further, but also ensures that all the recovered power from the expansion process is fully utilized throughout the cycle.

All in all, the average power of the compressor, expander, inertial effect, and total frictional loss are calculated to be 1181.6W, 213.2W, 0W, and 31.3W respectively. The average expansion power is 18.0% of the compression power, and when it is utilized to partially drive the compressor, an improvement of 21.3% in the COP of the refrigeration cycle could be achieved. In addition, the average total frictional loss is comparatively small too, and as a result, a mechanical efficiency of the CVEC of 96.9% could be obtained.

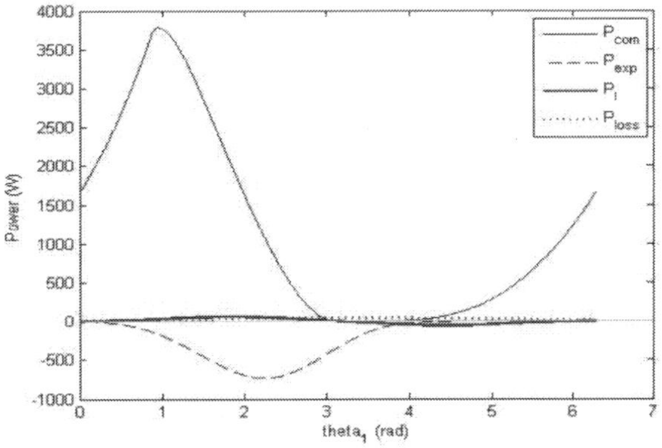

Figure 9 Instantaneous power constituents of cross-vane expander comrpessor

5. Conclusion

Rotary compressor has recorded an annual sale of more than 150 million units in 2013 [8]. With constant design improvement of the compessor, the energy efficiency of the refriegration cycles can be improved. This improvement will bring about significant energy and materials asving. In this paper, CVEC is useful not only it reduces the mechanical power loss but it also recoves expansion energy in the vapour compression refriegration cycle, which is otherwise lost, into useful eenrgy to supplement the power needed by the compressor. With its simple geometry and fewer parts, it is expected that there will not be any manufacturing cost increase when using this CVEC with the exisitng refrigeration cycle, but significant operational cost saving is expected. It is particularly true when it is employed in the CO2 refrigeration cycle.

References

[1] Ooi K T 2005 Design optimization of a rolling piston compressor for refrigerators *Applied Thermal Engineering*, Vol 25 Issues 5–6 pp 813 – 829
[2] Ooi K T and Wong TN, (1997), A computer simulation of a rotary compressor for house-hold refrigerators. *Applied Thermal Engineering*, 17(1) pp 65-78.
[3] Tan K M and Ooi K T 2011 A novel revolving vane compressor with a fixed-vane. *Int. J. Refrigeration* 34(8) pp 1980-1988.
[4] Tan K M and Ooi KT 2010 Journal bearings design for a novel revolving vane compressor, *Int. J. Refrigeration*, 34 (1), pp. 94–104
[5] Tan K M and Ooi K T 2014 Experimental study of fixed-vane revolving vane compressor. *Applied Thermal Engineering* 62(1) pp 207-214.
[6] Yap K S and Ooi K T and Chakraborty A 2014 Introduction of the Novel Cross Vane Expander-Compressor Unit for Vapour Compression Cycle *International Compressor Engineering Conference at Purdue* p 1129
[7] Yap K S, Ooi K T and Chakraborty A 2014 Modelling and Simulation of the Dynamics of Cross Vane Expander-Compressor Unit for Vapour Compression Cycle, *International Compressor Engineering Conference at Purdue*, p 1134
[8] Cover story, JARN, September 2013, Serial no 536, 45(9).

Study of torque matching of revolving vane compressor and expander

A Subiantoro[a1], **K T Ooi**[b]
[a]Group Leader, Technische Universität München - Campus for Research Excellence And Technological Enterprise (TUM CREATE), 1 CREATE Way #10-02, CREATE Tower, Singapore 138602
E-mail: alison.subiantoro@tum-create.edu.sg
[b]Professor, School of Mechanical and Aerospace Engineering, Nanyang Technological University, N3-02c-73, 50 Nanyang Avenue, Singapore 639798
E-mail: mktooi@ntu.edu.sg

Abstract. An investigation was carried out to find the most optimum configuration, particularly the torque matching characteristics, of an integrated Revolving Vane compressor-expander. To carry out the study, a mathematical model of the integrated compressor-expander was developed. An open cycle air refrigeration system was adopted. The controlled parameter was the angle shift between the compressor and the expander. The observed parameters were the peak torque requirement and the bearing load. The results show that when properly matched, the peak torque can be reduced by more than 65% while the bearing loads can be reduced by up to 25%, depending on operating conditions. Unfortunately, the optimum angle shifts for peak torque do not always coincide with those for bearing load. When the pressure and inertial components of the torques are comparable or when the inertial component is dominant, the optimum angle shifts for peak torque and bearing load are around 180° and 330°, respectively. When the pressure component is dominant, the optimum angle shift for peak torque is equal to the angle difference between the pressure peak torques of the compressor and the expander while for bearing load is around 150°.

1. Introduction

Revolving vane (RV) mechanism was introduced in 2006 [1] as a novel rotary mechanism that is simple and has low frictional losses. In its basic form, the mechanism has only 3 main components, i.e. a cylinder, a rotor and a vane. Unlike typical rotary mechanisms where the cylinders are stationary, the cylinder of a RV machine rotates together with the rotor. This results in reduced relative velocities between the rubbing surfaces and hence, the frictional losses are reduced too. The mechanism has been employed for compressor [2-6] and expander [7-10] applications. Moreover, various alternative designs have been introduced and studied [11-12]. The main components and the working principle as an expander of one of the variants, called "RV-I" elsewhere [11], are shown in figure 1.

[1] To whom any correspondence should be addressed.

Content from this work may be used under the terms of the Creative Commons Attribution 3.0 licence. Any further distribution of this work must maintain attribution to the author(s) and the title of the work, journal citation and DOI.
Published under licence by IOP Publishing Ltd

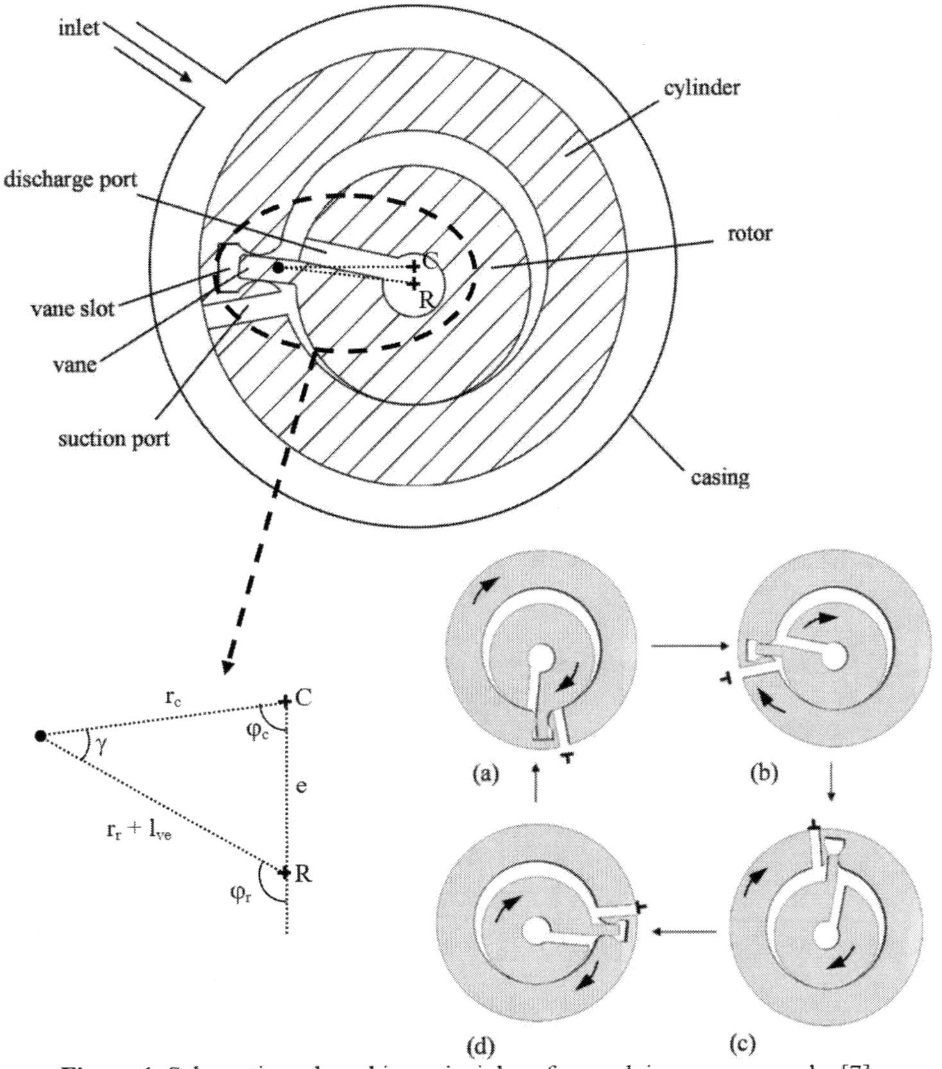

Figure 1. Schematic and working principles of a revolving vane expander [7].

As shown in the figure, the rotor and the cylinder rotate together eccentrically at their respective axes. A line contact is always maintained. Together with the vane, this line contact separates the space between the cylinder and the rotor into two working chambers. As the components rotate, one chamber decreases in volume while the other increases. In practice, the first is the discharge chamber and the latter is the suction chamber. It is important to note that to get a superior performance, the driving shaft must be connected to the component where the vane is attached to [11]. So, the driving shaft of the RV machine in figure 1 is at the rotor.

Expanders have been proposed in the recent years to increase the energy efficiency of refrigeration systems by recovering the usually wasted expansion energy in the expansion device. In conventional systems, the increase in Coefficient of Performance (COP) is more than 10% [13] while it is more than 50% in transcritical CO_2 refrigeration systems [14]. Expanders are also economically feasible and attractive once the technology is mature enough and mass production begins [15]. One of the remaining technological challenges left is in the integration of expander and compressor. This issue encompasses many aspects including design, manufacturing, performance, etc.

In this study, the torque matching aspect of an integrated RV expander and RV compressor for performance optimisation is studied. The RV design variant adopted was that shown in figure 1. It was assumed that the compressor and the expander shared a common shaft at their rotors. The controlled design parameter was the angle difference between the expander and the compressor. The performance was quantified by the peak torque requirement and the frictional loss. A low peak torque requirement is desirable as it allows for the use of a smaller motor. It also indirectly reduces the mechanical losses. Among all the frictional forces, the integration of RV compressor and RV expander affects particularly those at the shaft bearings due to the coupling of the loads. Therefore, the frictional loss analysis was focused on the bearing loads. A small average bearing load is desirable to minimise the bearing frictional loss.

2. Methodology

To carry out the study, a mathematical model of the integrated RV compressor-expander was developed. It combined the RV compressor and RV expander models developed previously [2-10], covering the geometrical, kinematics, dynamics and thermodynamics processes of the machine. The integrated compressor-expander machine was assumed to operate in an open cycle air refrigeration system where the compressor and the expander were linked with a shared shaft as illustrated in figure 2. In such a system, ambient air is first compressed by the compressor to increase its temperature and pressure. The discharged air is then channelled to a heat exchanger to reduce its temperature. This air is then expanded by the expander to reduce its temperature and pressure and is then used.

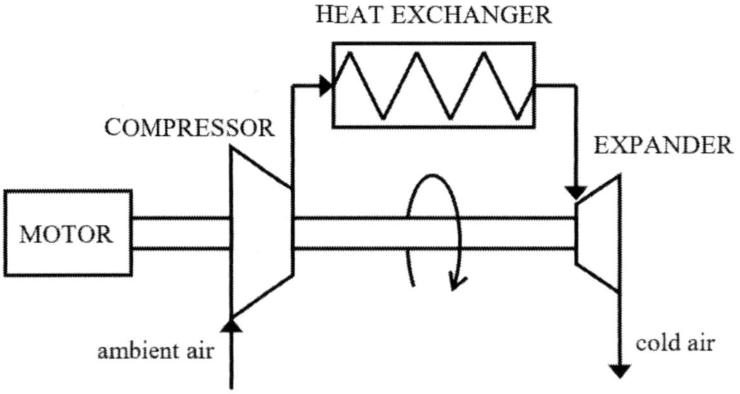

Figure 2. Schematic of an open cycle refrigeration system.

Assumptions were adopted to simplify the model. They include:

- Negligible pressure drop in the heat exchanger, so the compressor's outlet pressure was equal to the expander's inlet pressure
- Constant and equal operational speeds of compressor and expander
- Compressor and expander rotate in the same direction
- Negligible vane thickness, so the vane was always in contact with the driven component
- Negligible centrifugal forces
- Frictional losses were neglected in the analysis of peak torque
- Adiabatic compression and expansion processes
- Ambient air conditions were 30°C and 1 atm
- Compressor inlet was at ambient conditions
- Expander inlet temperature was 10°C hotter than ambient
- Expander outlet was at ambient pressure
- Negligible internal leakages in the compressor and the expander

- Negligible dead volume
- Ideal valves and ports
- Size of expander was calculated according to compressor capacity and operating conditions, expander's cylinder radius is constant while rotor radius is adjustable

Mathematical models of RV machines have been proposed and verified before [6,10]. The most important equations after the adoption of the assumptions above are reviewed briefly here. Detailed descriptions of the equations and terms are available in the relevant references.

The chamber volumes are shown in equations (1) and (2). The corresponding parameters are computed according to equations (3-5).

$$V_{scv} = \frac{l_{CV}}{2}\left(\varphi_c r_c^2 - e r_c \sin\varphi_c - \varphi_r r_r^2\right) \tag{1}$$

$$V_{dcv} = l_{CV}\pi\left(r_c^2 - r_r^2\right) - V_{scv} \tag{2}$$

$$e = r_c - r_r \tag{3}$$

$$l_{ve} = -e\cos\varphi_r - r_r + \sqrt{r_c^2 - e^2\sin^2\varphi_r} \tag{4}$$

$$\cos\varphi_c = \frac{e^2 + r_c^2 - \left(r_r + l_{ve}\right)^2}{2r_c e} \tag{5}$$

where e is eccentricity (m), l_{CV} is working chamber length (m), l_{ve} is exposed vane length (m), r_c and r_r are radii of cylinder and rotor, respectively (m), V_{dcv} and V_{scv} are volumes of discharge and suction chambers, respectively (m^3), φ_c and φ_r are angles of cylinder and rotor, respectively (rad).

The torque required by compressor generated by expander is calculated according to equation (6). The corresponding parameters are computed with equations (7-9). Equation (6) gives out a positive value for torque requirement of the compressor and a negative value for torque generated by the expander. Equation (6) also shows that the torque of a RV machine comprises three components, i.e. pressure, inertial and losses components. The first two are more dominant than the last. The nett torque of the integrated RV compressor-expander machine is the summation of the RV compressor's torque requirement and the RV expander's torque generated.

$$T = \left(p_{dcv} - p_{scv}\right) l_{ve} l_{CV}\left(r_r + \frac{l_{ve}}{2}\right) + \left(\frac{r_r + l_{ve}}{r_c\cos\gamma}\right) I_c\alpha_c \tag{6}$$

$$\cos\gamma = \frac{r_c^2 + \left(r_r + l_{ve}\right)^2 - e^2}{2r_c\left(r_r + l_{ve}\right)} \tag{7}$$

$$\alpha_c = \omega_r\frac{dl_{ve}}{dt}\left(\frac{4\left(r_r + l_{ve}\right)\left(r_c^2 - e^2\right)}{\left(r_c^2 + \left(r_r + l_{ve}\right)^2 - e^2\right)^2}\right) \tag{8}$$

$$\frac{dl_{ve}}{dt} = \omega_r e\sin\varphi_r\left(1 - \frac{e\cos\varphi_r}{\sqrt{r_c^2 - e^2\sin^2\varphi_r}}\right) \tag{9}$$

where I_c is the inertia of cylinder (kg·m^2), p_{scv} and p_{dcv} are pressures of suction and discharge chambers, respectively (Pa), T is torque (N·m), t is time (s), α_c is angular acceleration of cylinder (rad/s^2), ω_r is angular velocity of rotor (rad/s).

The shaft bearing load's components in the x and y directions are according to equations (10-11).

$$F_x = \left(\frac{I_c\alpha_c}{r_c\cos\gamma} - \left(p_{scv} - p_{dcv}\right) l_{CV}\left(r_r + l_{ve}\right)\right)\cos\varphi_r + \left(p_{scv} - p_{dcv}\right) l_{CV} r_r \tag{10}$$

$$F_y = \left(-\frac{I_c\alpha_c}{r_c\cos\gamma} + \left(p_{scv} - p_{dcv}\right)\left(r_r + l_{ve}\right) l_{CV}\right)\sin\varphi_r \tag{11}$$

where F_x and F_y are shaft bearing load components in the x and y axes, respectively (N)

A computer spreadsheet was developed to solve all the equations. The independent variable in the model was the rotor angle of the compressor, $\varphi_{r,compressor}$. It is important to note that some of the equations need inputs from other equations. Therefore, they had to be solved in order. The models were applied for the RV compressor and the RV expander. An "angle shift" was introduced between the compressor and the expander as expressed in equation (12). This shifted the torque profiles of the compressor and the expander, allowing for control of the matching of the corresponding torques. The shift will modify the amplitude and shape of the nett torque of the integrated machine. The average value of the nett torque was affected too, mainly through the change in the bearing loads. However, the average values of the pressure and inertia components of the torque were unaffected by this shift.

$$\varphi_{r,expander} = \varphi_{r,compressor} + \Delta\varphi \tag{12}$$

Major dimensions of the model and operating parameters are tabulated in Table 1. The dimensions were based on the RV expander prototype built and tested before [10]. The inertia values of the components were obtained with the SolidWorks design software.

Table 1. Major dimensions and operating parameters

Item	Value
Compressor	
Rotor radius	29 mm
Cylinder radius	35 mm
Length	25 mm
Capacity	30.2 cm^3
Cylinder inertia	0.00281 kg·m^2
Rotor inertia	0.00021 kg·m^2
Suction pressure	1 atm
Discharge pressure	8 - 80 bar (abs.)
Expander	
Cylinder radius	35 mm
Length	25 mm
Cylinder inertia	0.00281 kg·m^2
Discharge pressure	1 atm
Operating Parameters	
Working fluid	Air
Ambient temperature	30°C
Ambient pressure	1 atm
Polytropic constant	1.4
Operating speed	1000 - 3000 rev/min

3. Results and discussion

Results of the simulation are presented and discussed in this section. It is useful to start the discussion by observing the major components of a RV machine's torque, which are the pressure and the inertial components. Some of these are plotted in figure 3. Ratio between the pressure and the inertial components changed with different operating pressure and speed conditions. In the figure, it can be seen that when compressor discharge pressure was 8 bar and speed was 1000 rev/min, the pressure and inertial components were comparable. However, the total peak torque was practically following that of the inertial component. When pressure was 80 bar and speed was 1000 rev/min, the pressure component was dominant and hence, the total peak torque was dictated by the pressure component. When pressure was 8 bar and speed was 3000 rev/min, the inertial components was dominant. Consequently, the total peak torque followed that of the inertial component. It is also useful to observe that the profiles of the inertial components of RV compressor and RV expander torques were similar in shape. They were sinusoidal, positive in the first half of the cycle and negative in the second half. However, the trends of the pressure components were different. The peak of the pressure component of the RV compressor was in the second half of the cycle while that of the RV expander was in the first half.

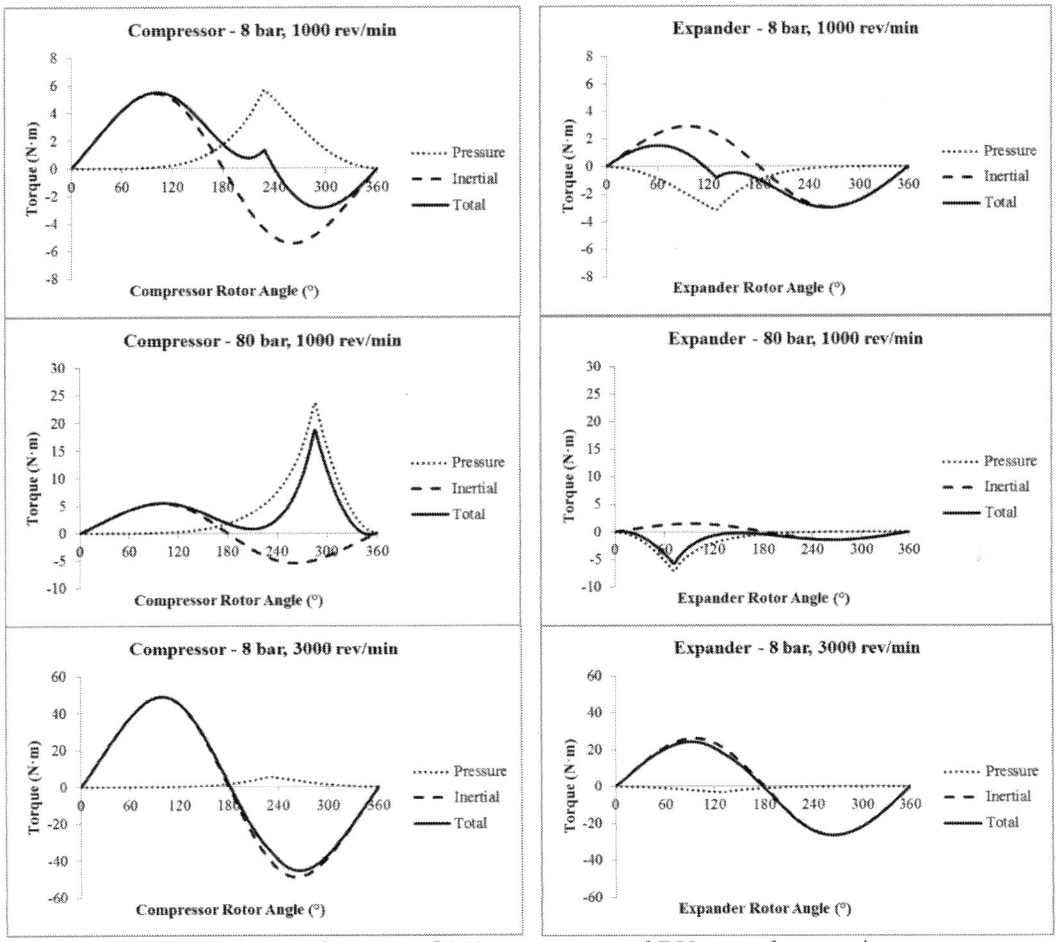

Figure 3. Components of torques of RV compressor and RV expander at various compressor discharge pressures and operating speeds.

The values of nett peak torque of the integrated RV compressor-expander machine at various angle shifts, compressor discharge pressures and operating speeds are plotted in figure 4. In general, it is found that it was not preferable to have an angle shift of 0°. When properly matched, the nett peak torque can be more than 65% smaller as compared to that with an angle shift of 0°, such as that in the case of 8 bar compressor discharge pressure and 3000 rev/min operating speed. On the other hand, when not properly matched, the nett peak torque can be up to 35% higher than when angle shift is 0°.

IOP Conf. Series: Materials Science and Engineering **90** (2015) 012037 doi:10.1088/1757-899X/90/1/012037

This change of nett peak torque was because of the modification in the nett torque profile as the angle shift changes as illustrated in figure 5. For example, when compressor discharge pressure was 8 bar, operating speed was 1000 rev/min and angle shift was 0°, the nett peak torque was more than 15% higher than that of the compressor alone. This was because when the compressor torque peaked, the expander was also positive. Hence, the nett torque is even higher than the compressor's peak torque. However, when the angle shift was 180°, the nett peak torque was more than 50% lower than the compressor peak torque because when the compressor torque peaked, the expander torque was negative. Hence, the nett torque was a smaller than the compressor's original peak torque.

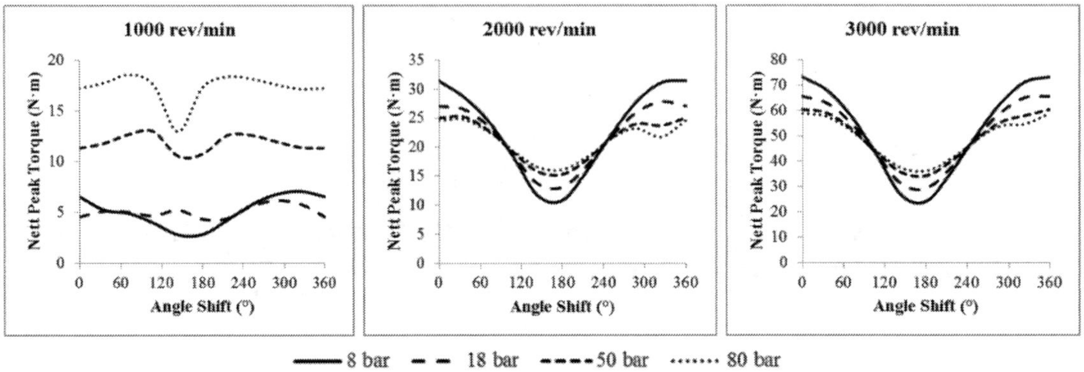

Figure 4. Variations of nett peak torque of the integrated RV compressor-expander with different angle shifts, operating speeds and compressor discharge pressure conditions.

The optimum angle shift depended on the operating speed and compressor discharge pressure. When the inertial and pressure components of torque were comparable, such as that in the case of 8 bar compressor discharge pressure and 1000 rev/min operating speed, or when the inertial component was dominant, such as when compressor discharge pressure was 8 bar and speed was 3000 rev/min, the optimum angle shift was in the vicinity of 180°. In such cases, the nett peak torque followed that of the inertial component as shown in figure 3 above. Moreover, the inertial components of RV compressor's and RV expander's torques were of similar profiles. Therefore, a shift of around 180° causes the two torque components to cancel each other as shown in figure 5. When the pressure component of the torque was dominant, such as when the compressor discharge pressure was 80 bar and operating speed was 1000 rev/min, the optimum angle shift was equal to the angle difference between the pressure peak torques of the compressor and the expander. These usually occur at the start of discharge in the compressor and at the end of suction in the expander, respectively.

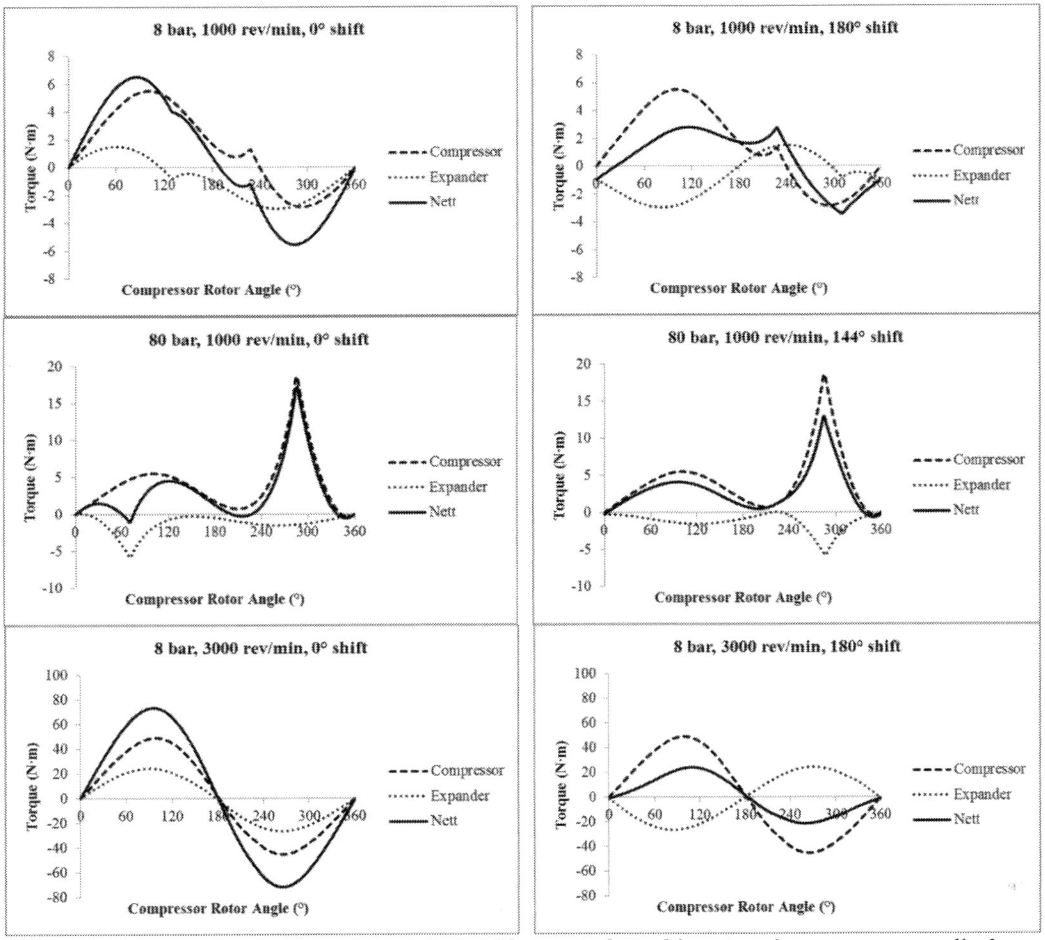

Figure 5. Torques of compressor, expander and integrated machine at various compressor discharge pressures, operating speeds and angle shifts.

The variations of average bearing load with angle shift, operating speed and compressor discharge pressure are shown in figure 6. When properly matched, the average bearing load can be up to 25% smaller than that with an angle shift of 0° (i.e. with 8 bar compressor discharge pressure and 1000 rev/min operating speed). On the other hand, when not properly matched, the average bearing load can be up to 50% higher than when the angle shift is 0° (i.e. at 8 bar compressor discharge pressure and 2000 rev/min speed) . In general, when the inertial component dominated (i.e. when speed is high or compressor discharge pressure is low), the optimum angle shift was in the vicinity of 330°. On the other hand, when the pressure component was dominant, the optimum angle shift was around 150°.

A note is necessary here to discuss the anomaly of the 8 bar, 3000 rev/min data in Figure 6. In general, bearing load is usually a weak function of the operational speed. This is true only when the pressure component of the load is significantly stronger then the inertial (speed) component. However, at 8 bar and 3000 rpm, the pressure component is smaller than the inertial component as mentioned in the previous sections. Hence, the load is dictated mostly by speed in such condition.

Comparing figures 4 and 6 shows, unfortunately, that the optimum angle shifts for bearing load reduction do not always coincide with those for peak torque reduction. Therefore, the designer will eventually have to decide which angle shift to use.

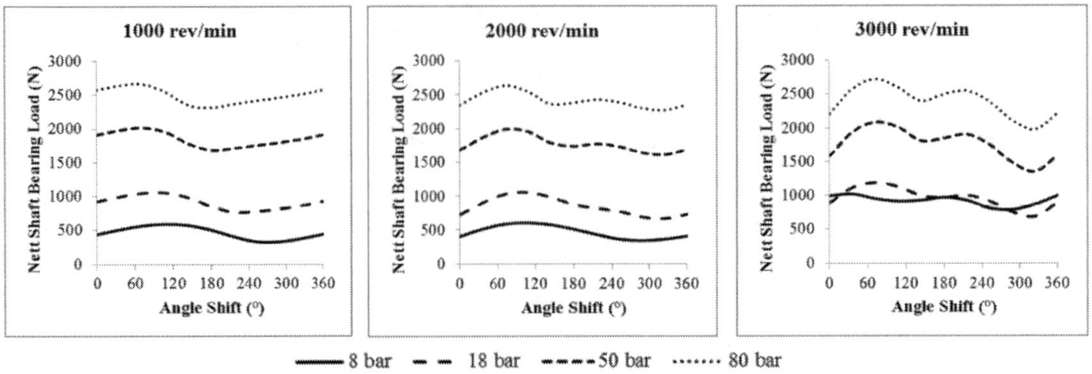

Figure 6. Variations of nett shaft bearing load with different angle shifts, operating speeds and compressor discharge pressure conditions.

4. Conclusions

Torque matching of an integrated revolving vane (RV) compressor-expander has been studied. The controlled design parameter was the angle shift between the expander and the compressor. The performance was quantified by the peak torque requirement and the shaft bearing load. To carry out the study, a mathematical model of the integrated RV compressor-expander was developed. The integrated compressor-expander machine was assumed to operate in an open cycle air refrigeration system. The low-side pressure was 1 atm, the high-side pressure was between 8-80 bar and the operating speed was between 1000-3000 rev/min. The compressor capacity was 30.2 cm^3.

From the study, the followings were observed:

- In general, it was not preferable to have no angle shift between the compressor and the expander as it may result in a high nett peak torque and/or a high average bearing load.
- When properly matched, the nett peak torque can be more than 65% smaller as compared to that with an angle shift of 0°. On the other hand, when not properly matched, the nett peak torque can be up to 35% higher than when angle shift is 0°.
- The optimum angle shift for nett peak torque depended on the operating speed and compressor discharge pressure. When the pressure component of the torque was dominant, i.e. when the pressure ratio was large, the optimum angle shift was equal to the angle difference between the pressure peak torques of the compressor and the expander. These usually occur at the start of discharge in the compressor and at the end of suction in the expander. In other cases, optimum angle shift was in the vicinity of 180°.
- When properly matched, the average bearing load can be up to 25% smaller than that with an angle shift of 0°. On the other hand, when not properly matched, the average bearing load can be up to 50% higher than when the angle shift is 0°.
- In general, when the inertial component was dominant (i.e. when speed is high or compressor discharge pressure is low), the optimum angle shift was in the vicinity of 330°. On the other hand, when the pressure component was dominant, the optimum angle shift was around 150°.
- The optimum angle shifts for bearing load reduction do not always coincide with those for peak torque reduction.

The study provides simple guidelines to help in the design process of an integrated RV compressor-expander. The method developed is simple and flexible enough to be used for further studies. However, more detailed studies and experimental validation will be conducted for better understanding of the phenomena.

5. Acknowledgment
This work was financially supported by the Singapore National Research Foundation under its Campus for Research Excellence and Technological Enterprise (CREATE) program. The author would also like to acknowledge the contribution of Mr. Chay Chinq Yee from the School of Mechanical and Aerospace Engineering, Nanyang Technological University, Singapore for the completion of this study.

6. References

[1] Teh Y L and Ooi K T 2006 Design and friction analysis of the revolving vane compressor *International Compressor Engineering Conference at Purdue* C046 pp 1-7
[2] Teh Y L and Ooi K T 2009 *Int. J. Refrig.* **32** 1092
[3] Teh Y L and Ooi K T 2009 *Int. J. Refrig.* **32** 1103
[4] Teh Y L and Ooi K T 2009 *Int. J. Refrig.* **32** 945
[5] Teh Y L and Ooi K T 2009 *Appl. Therm. Eng.* **29** 3235
[6] Tan K M and Ooi K T 2014 *Appl. Therm. Eng.* **62** 207
[7] Subiantoro A and Ooi K T 2009 *HVAC&R Res.* **15** 801
[8] Subiantoro A and Ooi K T 2012 *Int. J. Refrig.* **35** 1734
[9] Subiantoro A and Ooi K T 2012 *Int. J. Refrig.* **35** 1744
[10] Subiantoro A, Yap K S and Ooi K T 2012 *Appl. Therm. Eng.* **50** 393
[11] Subiantoro A and Ooi K T 2010 *Int. J. Refrig.* **33** 675
[12] Subiantoro A and Ooi K T 2014 *Energy* **78** 747
[13] Goncalves V A and Parise J A R 2008 A study on the reduction of throttling losses in automotive air conditioning systems through expansion work recovery *International Compressor Engineering Conference at Purdue* 2416 pp 1-8
[14] Fukuta M, Yanagisawa T, Nakaya S and Ogi Y 2006 Performance and characteristics of compressor/expander combination for CO_2 cycle *7th IIR Gustav Lorentzen Conference on Natural Working Fluids, Trondheim, Norway* pp 052 pp 1-8
[15] Subiantoro A and Ooi K T 2013 *Int. J. Refrig.* **36** 1472

Development of High Speed Inverter Rotary Compressor for the Air-conditioning System

Seoung-Min Kang, Eun-soo Yang, Jin-Ung Shin, Joon-Hong Park, Se-Dong Lee, Jong-Hun Ha, Young-Boo Son, Byeong-Chul Lee

*Energy Solution Research Lab., LG Electronics Inc., 327-23 Gasan-dong, Geumchun-gu, Seoul, 153-802, Korea

Tel: +82-10-9930-3195 Fax: +82-2-867-3941 E-mail: seoungmin.kang@lge.com (Seoung-Min Kang)

Abstract. In order to meet the various operating loads of an air-conditioning system, an inverter compressor with a wide operational range is necessary. One of the ways to achieve a wide operation range is to drive a small capacity compressor at high speed. Moreover, it is possible to maximize the efficiency in part-load operation condition close to actual operating conditions and to reduce the cost by compact design of a small capacity compressor. In addition, the shortage of maximum capacity, due to the small rated capacity, is covered through high speed operation. However, in general, if the compressor operates at high speed, problems occurs such as reduced efficiency due to friction, increased noise, increased amount of oil discharge and decreased durability of the main components. In order to solve these problems the following have been investigated: optimized dimension parameters of the compression chamber, enhanced shaft design and the structure for the reduction of oil discharge and noise at high speed operation. Finally the high speed inverter rotary compressor with high efficiency and more compact size has been developed as compared with the conventional rotary compressor.

Keywords: Rotary compressor, High-speed, Oil discharge, High efficiency

1. Introduction

In recent years, energy saving in the residential and industrial area has become a big issue in the world. In particular, the demand for energy-saving residential air conditioning systems, including heating and cooling, is increasing. The inverter rotary compressor has a key role in energy savings and reliability as the critical component accounts for 85% of the power consumption of air conditioning. In response to the various operational loads in a residential air conditioning system, Inverter technology to control the capacity by varying the rotation speed is important. Through inverter technology, we have developed a high speed inverter rotary compressor capable of controlling up to 150rps (rotor speed). In this development, we confirmed the problems of the rotary compressor at high speed operating conditions and maximized the efficiency in the operating conditions of the air conditioner through CAE analysis and experimental approaches. Also, it is possible to lower material costs through reducing the shell diameter from Φ112mm to Φ101mm.

Content from this work may be used under the terms of the Creative Commons Attribution 3.0 licence. Any further distribution of this work must maintain attribution to the author(s) and the title of the work, journal citation and DOI.
Published under licence by IOP Publishing Ltd

In this paper, the way to choose component specifications are introduced and the CAE(Computer Aided Engineering) / CFD(Computational Fluid Dynamics) analyses are performed for its reliability of developing a high speed rotary compressor.[1-4]

2. Overview of high speed rotary compressor

We have developed a rotary compressor with a displacement volume of 13.0cm^3, which is employed in a residential air conditioner. R410A and PVE oil are used as a refrigerant and lubrication oil. Figure 1 shows the structure of rotary compressor. The compressor consists of a cylinder, a reciprocating vane, roller, crankshaft with eccentric parts, main and sub bearings and brushless DC motor with rotor and stator. The lubrication oil in the sump is fed by the centrifugal force of the propeller, which is installed in the hollow hole of crankshaft. The compressor has two compression chambers which are operated alternately with a phase difference of 180 degree. The compression load generated by the compression chambers is supported by main and sub bearings positioned in the upper and lower parts of the compression chambers.

In a residential air conditioner, the high load occurs at the highest outdoor temperature in summer and at the lowest outdoor temperature in winter. In that case, the compressor must provide large capacity to fulfil the high load. In terms of a compact design compressor, it is necessary to achieve high speed operation with the high load.

In this paper, the optimized compression part design, the brushless DC motor design, the shaft reinforcement structure and oil discharge reduction structure to complete the high speed rotary compressor are introduced.

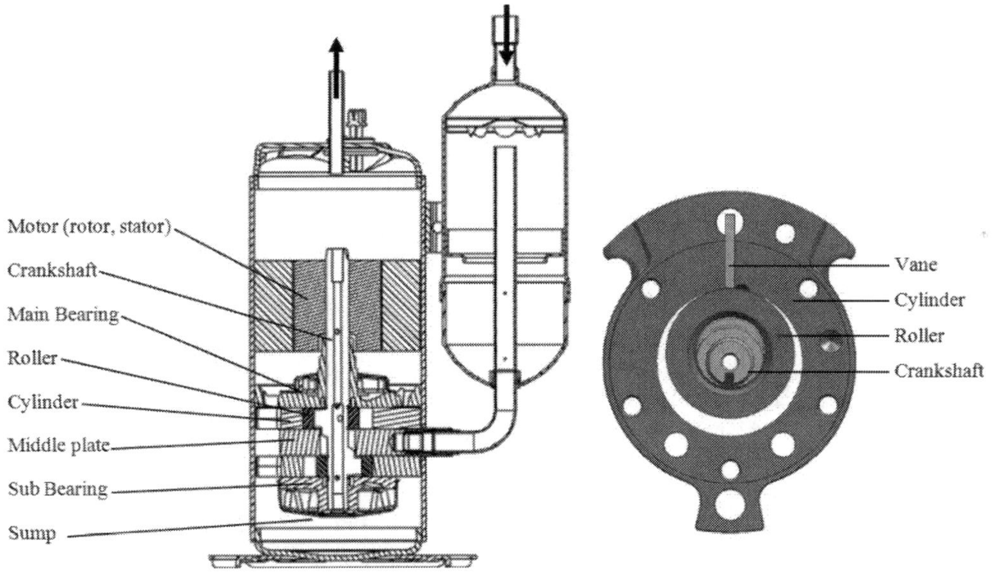

Figure 1 Structure of rotary compressor

3. OVERVIEW OF HIGH SPEED ROTARY COMPRESSOR

3.1. Optimization of cylinder dimension

To enhance efficiency at high speed operating conditions, we focused on reducing the leakage loss in the compression chambers and friction loss on the lubrication parts such as bearings. Figure 2 shows the design limitations of rotary compressor based on the cylinder inner diameter and height as the main specifications. The area A is suggested the optimized design area limited by geometry of compression chamber and reliability of main parts. These limitations are decided through implementation analyses and confirmation tests. As a result of adopting a lower cylinder height, the volumetric efficiency of developed model is improved by reducing the leakage loss from the radial clearance between the cylinder and roller.

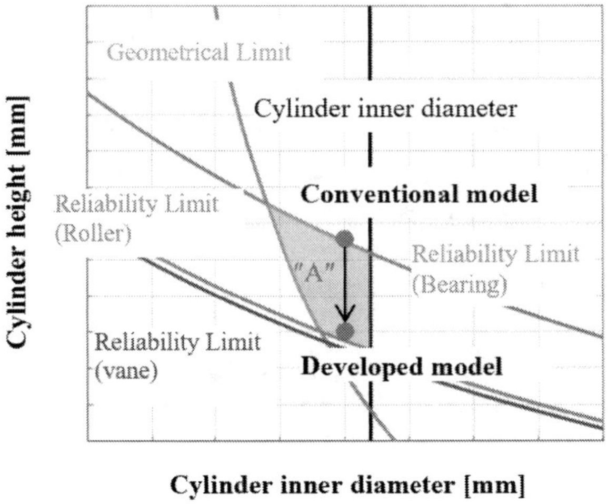

Figure 2 Design limitations of rotary compressor specifications

3.2. Reduction of crankshaft deformation and sliding surface pressure

The load acting on the crankshaft is composed of the sum of gas forces acting on the roller, vane force pushing the roller and centrifugal force of eccentric part(crank pin). The load deforms the parts at high speed operation under overload condition and is the maximum in the vicinity of crank angle when the discharge valve is opened. If a rotary compressor is operated with very high speed, the load is distributed in all areas by increasing the centrifugal force of eccentric part of crankshaft and its durability decreases by the high load acting on the sliding part.

Figure 3 shows the analytical result of crankshaft deformation between conventional and developed models. ANSYS was used for CAE analysis on the crankshaft deformation. From the analytical results, the shaft deformation was reduced by about 40% as compared to a conventional crankshaft at high speed. We confirmed that the maximum stress is located at the upper part of the crank pin. In order to verify the decrease of shaft deformation, we conducted an experiment on the basis of reliability standard and the result met with the criteria. Figure 4 shows the specifications for reducing shaft deformation. When we design the crankshaft considering the change to minimize crankshaft deformation, we should also consider the insertion of roller and middle plate. The shorter crank pin length is advantageous for shaft deformation and a certain length is required for a minimum of design in the assembly state. The pump part can be assembled by this bending length when the middle plate passes through the eccentric part. In other words, a middle plate can be assembled when this calculated bending length is less than the period overlapped between the eccentric part and middle plate.

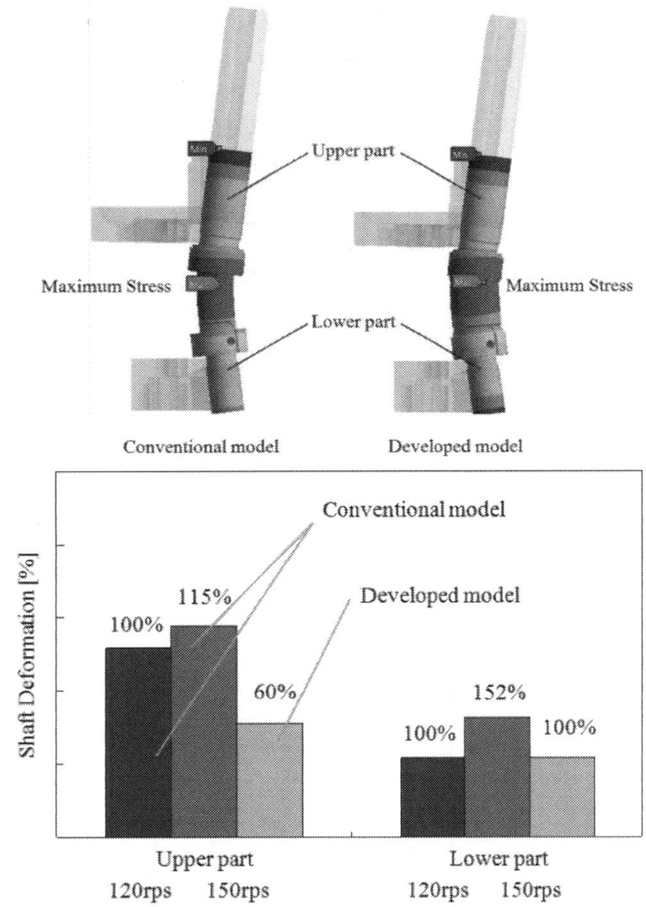

Figure 3 Comparison of crankshaft deformation at high speed

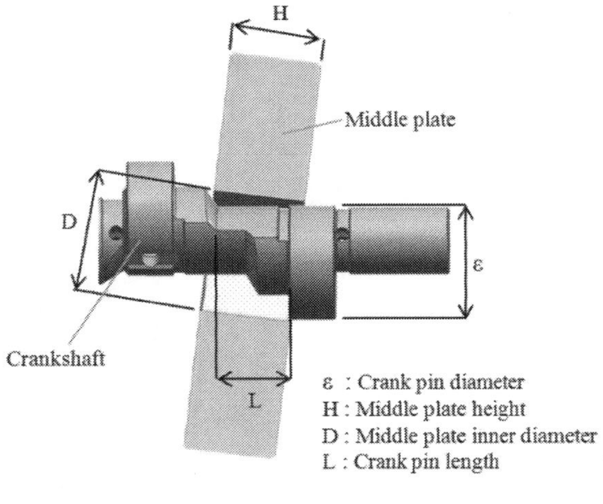

Design guide : $H \geq L \geq H - \sqrt{D^2 - \varepsilon^2}$

Figure 4 Specifications for reduction of shaft deformation

3.3. Optimization of Discharge Valve

When the pressure in the compression chamber is higher than the pressure outside the cylinder, the discharge valve is opened and then compressed gas is discharged through discharge port. The delayed opening of the valve influences the increase of over-compression during discharge process. When the rotary compressor operates in high speed, the over-compression becomes larger than normal speed. So the valve reliability problems can occur due to larger impact stress and velocity of discharge valve. For that reason, valve design is very important for reliability of compressor. The discharge valve design was determined by CAE analysis and experiments for specifically configured conditions of valve thickness and width, and L_mb to consider the operating and geometric conditions. L_mb is seated area of discharge valve at main bearing and it is key factor of valve reliability. Figure 5 shows the discharge valve structure and specifications. We performed the test with various valve specifications and compared them with the simulation results of valve stiffness

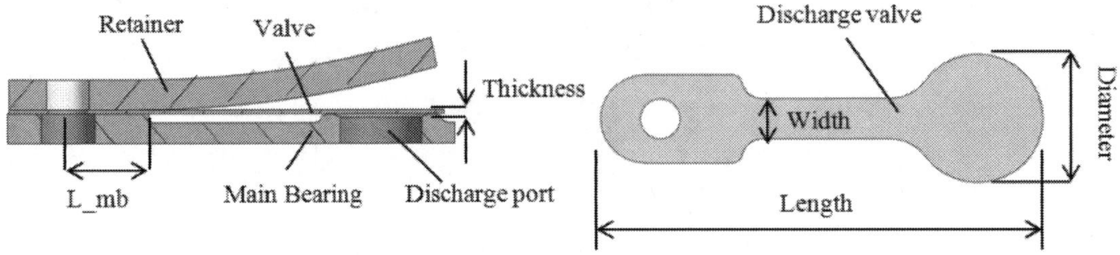

	Thickness [mm]	Width [mm]	L_mb [mm]	Stiffness [N/m]
Present	0.305	3.2	10	1413
Spec 1	0.254	3.2	10	816
Spec 2	0.254	5.0	7.6	850
Spec 3	0.254	5.0	6.5	682

Figure 5 Discharge valve structure and specifications

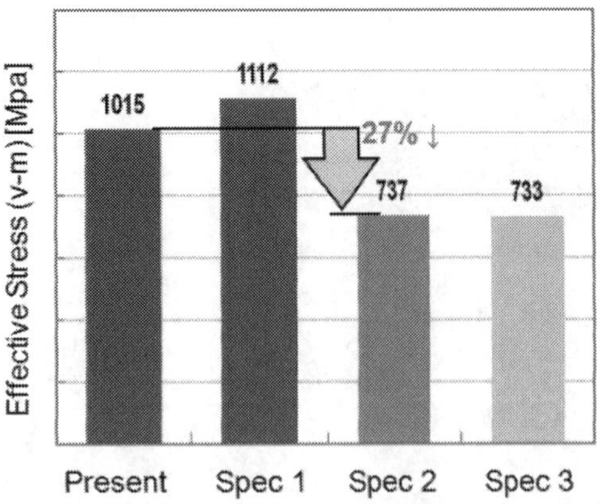

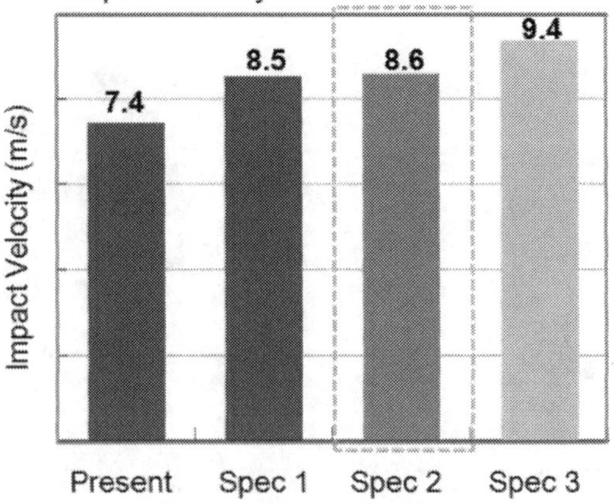

Figure 6 Simulation result of discharge valve

IOP Conf. Series: Materials Science and Engineering **90** (2015) 012038 doi:10.1088/1757-899X/90/1/012038

Figure 6 shows the effective stress and impact velocity of discharge valve according to the specifications. In the case of present and spec.1 with same valve width, if the discharge valve thickness becomes thin, the noise of the compressor increases and the reliability of compressor worsens due to increased stress and velocity. In spec.2 and spec.3, the effective stress of the discharge valve is reduced by about 27%. The spec.2 was slightly smaller impact velocity than spec.3. We selected spec.2 as the optimum design for reliability, noise and efficiency of the compressor.

3.4. Reduction of oil circulation ratio (OCR)

In order to keep a stable oil level in the compressor, it is necessary to optimize the circulation path of discharged oil. Most of the oil discharged with the refrigerant gas from the discharge muffler goes through the air gap between the motor stator and rotor and is separated by the centrifugal force caused by the rotation of the rotor. The oil moving to the upper space of the compressor is separated and dropped from the gas by the density difference. The separated oil is moved to the oil sump at the bottom of the compressor through the stator outer peripheral space. If the compressor operates at high speed, the oil circulation ratio (OCR : The amount of oil contained in the gas) increases due to lower separation capacity by high flow velocity. If the OCR increases, the reliability of the compressor worsens as the oil level decreases. Thus, even at the highest speed, OCR must be maintained at an appropriate level. In addition, if the OCR increases, the amount of oil withdrawn together with the refrigerant increases. For that reason, volumetric efficiency is decreased. In this study, we confirmed the flow pattern inside the compressor through a CFD analysis to improve the oil separation structure in the conventional compressor. Figure 7 shows the specifications for the reduction of OCR. The distance between the end of the discharge pipe and crankshaft was investigated from 5 to 10mm. The upper length and lower length of case were changed to decrease pressure difference between upper and lower spaces based on the motor as well.

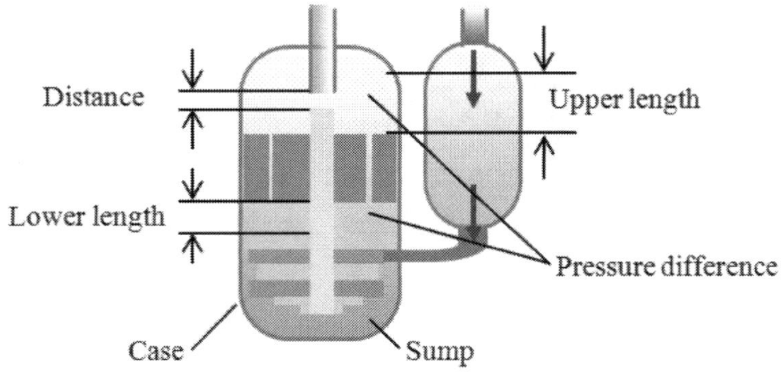

Figure 7 Design improvement of reduction of oil circulation ratio

Figure 8 shows the one example of CFD analysis result to understand the flow pattern inside the compressor and the experimental result of OCR according to the operating condition before and after changing the oil reduction specification. Finally, OCR could be reduced from 2.8 to 0.6wt% at150rps.

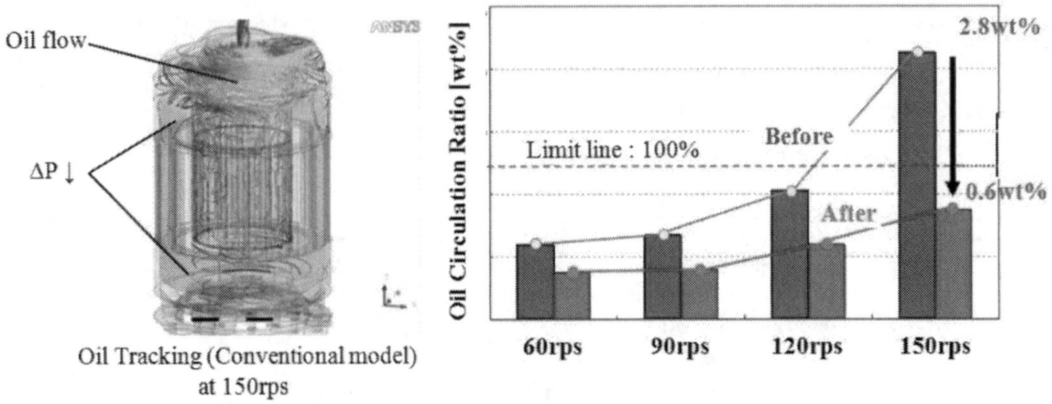

Figure 8 Oil circulation ratio experimental result

3.5. *High speed brushless DC motor*

Generally, the bridge of the rotor shows the weakness at high speed rotation. Thus, we reinforced the bridge to avoid rotor rib damage and prevent the escape of permanent magnet through CAE analysis. Figure 9 shows the bridge stress of conventional and developed models. The bridge stress is reduced from 314MPa to 79MPa with changing the rotor shape, and we acquired enough marginal safety ratio(Yield stress/Max. stress) in spite of high speed operation.

		Conventional model	Developed model
Stress at 150rps	Shape	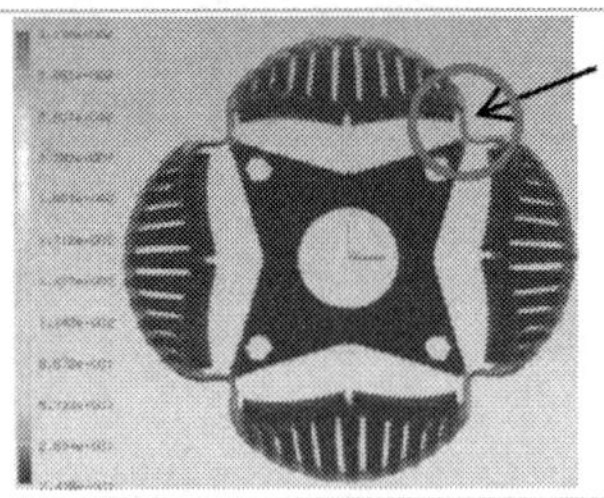 Bridge	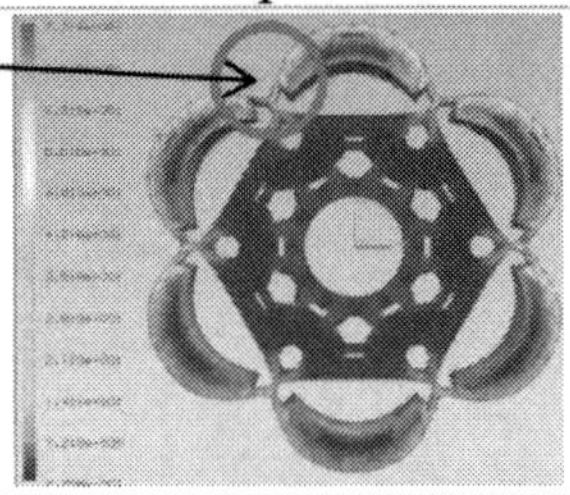
	Max [MPa]	314	79

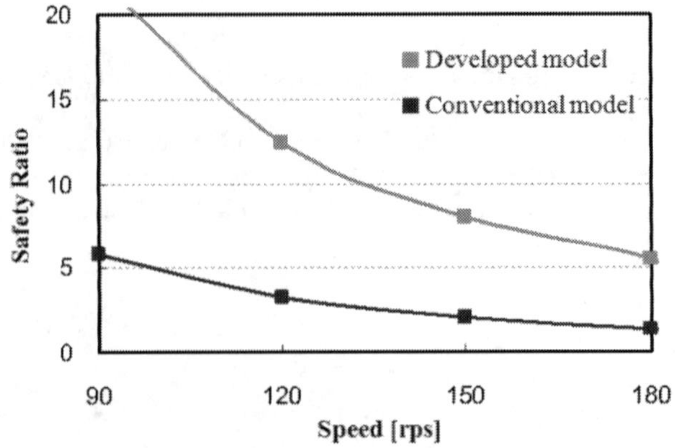

Figure 9 Maximum stress and safety ratio of bridge

IOP Conf. Series: Materials Science and Engineering **90** (2015) 012038 doi:10.1088/1757-899X/90/1/012038

3.6. *Noise reduction at high speed*

As the noise and vibration characteristics of the inverter rotary compressor are changed in accordance with operating conditions, The countermeasures of those are important to develop an the compressor. Especially, noise control at high speed operation is one of the most important factors because it plays an important role in technology competitiveness.

The increasing noise level at high speed is not related to a resonance but an excitation force and transfer path of the compressor. So, the study of noise characteristics and design optimization should be carried out for the related component parts. First, in this study, noise level is measured by using microphone. Then, frequency spectrum analysis is carried out to measure the signal to find the countermeasures of the noise problem. Finally, motor design and welding method are modified to improve the excitation force and the transfer path of the system, respectively.

The motor design which generates electromagnetic force is modified using computational simulation to reduce the noise level. Table 1 shows the comparison of motor specifications between conventional and developed motor. According to the simulation result, the excitation force of the developed motor decreased while motor efficiency is equivalent to the conventional design. In order to verify the computational analysis, the vibrations of conventional and developed models are measured.

	Conventional model	Developed model
Motor Stack [mm]	55	45
Coil Diameter [mm]	Φ0.95	Φ0.85
Turn No.	51	65
Slot fill factor [%]	80.3/81	82.4/83
Resistance [Ω]	0.828	1.078

Table 1. Specifications for developed motor

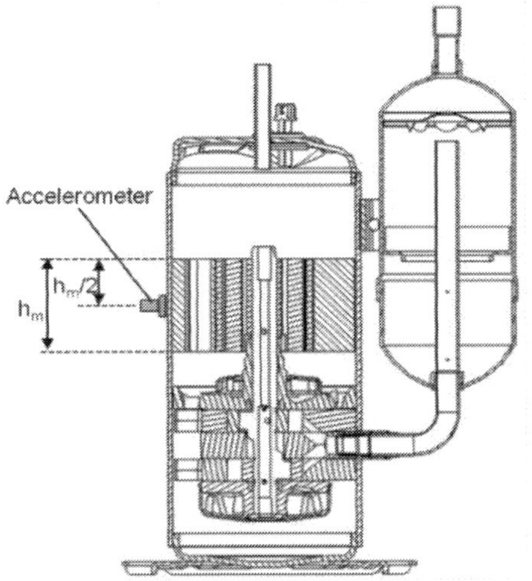

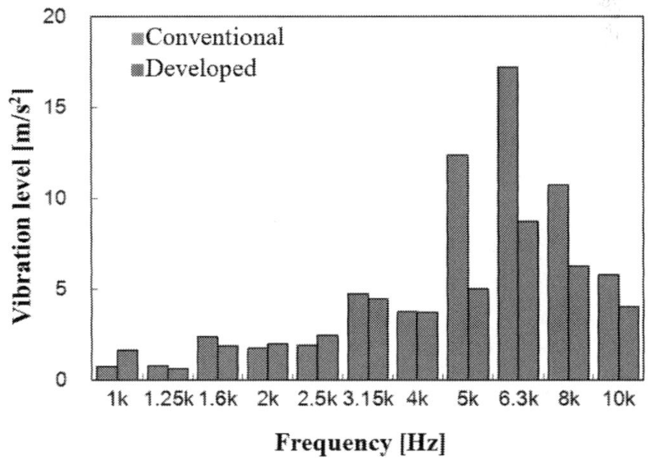

(a) Vibration measurement position with accelometer (b) 1/3 Octave band at 130rps

Figure 10. Comparison of vibration level between conventional and developed motors

Figure 10 shows the vibration measurement position and the experiment result using accelerometer. Radial vibration is measured at the half of motor stack height (hm) and 1/3 octave band frequency spectra of conventional and developed models are compared. As a result, the vibration level of the developed model is reduced by about 50%. Especially, 3~6 kHz frequency range is most effective because high order harmonic components of electromagnetic force are reduced. Since vibration of structure is the cause of noise radiation, reduction of vibration implies an improvement of noise level. Additionally, welding position of case and pump part was investigated to improve the vibration transfer path. In general, 3 points of main bearing are welded to the compressor case to support a pump part. However, the thickness of main bearing was too thin and the combined structure was too unstable to hold up the pump part. Thus, design modification was carried out by computational simulation and 3 points of sub bearing are additionally welded.

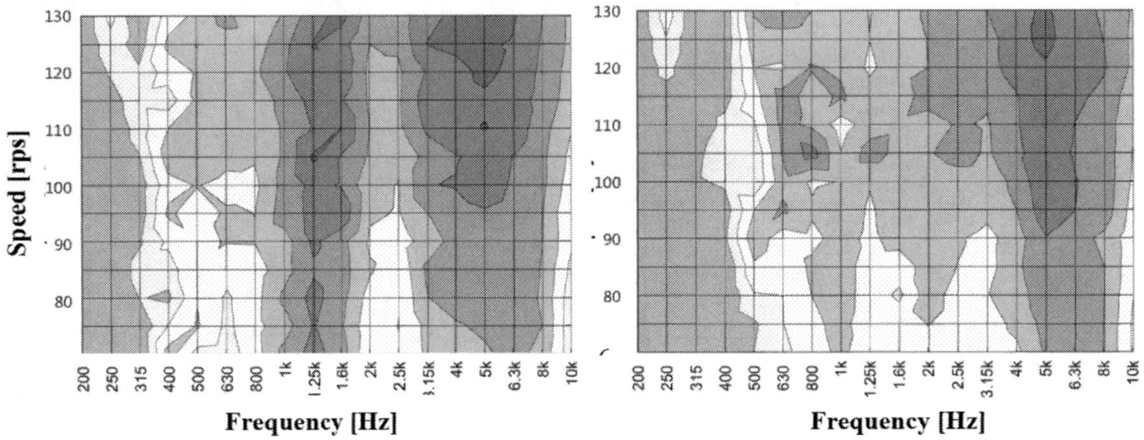

(a) Noise measurement result

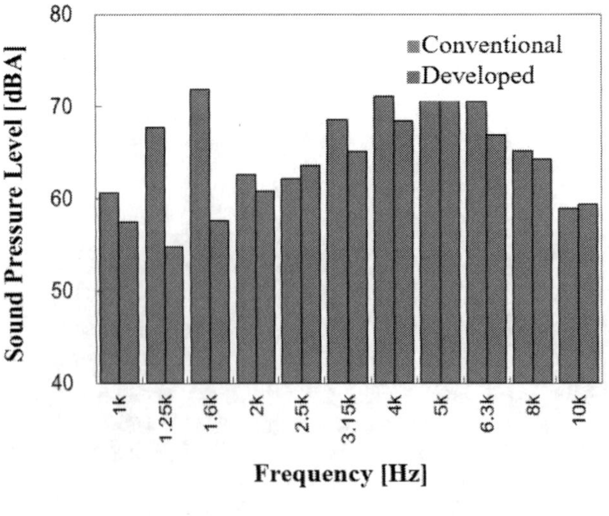

(b) Octave band at 130rps

Figure 11. Comparison of noise level between conventional and developed model

Figure 11 shows the noise reduction after design modification as compared with the conventional design. A microphone to measure the noise is placed at a distance of 30cm from the half the compressors body height and a noise signal is obtained after 15 seconds of stationary condition. Figure 11(a) is the color map of noise level with respect to operating speed. Both 1.25~1.6 kHz and 3~6 kHz frequency ranges are reduced due to the improvement of motor design and welding method, respectively. In order to check noise reduction at high speed, 1/3 octave band frequency spectra of conventional and developed designs are compared at 130rps operating condition as shown in Figure 11(b) Noise level in the 1.25~1.6 kHz frequency range is reduced by about 10dB and that of the 3~6 kHz frequency range is reduced by about 3dBA. Total noise level is reduced by about 3dBA.

4. Conclusion

This paper described the technology to solve the problems in the high speed operation of the compressor. For the better reliability and higher efficiency, the cylinder dimensions were compactly optimized to reduce the leakage loss from the radial clearance between the cylinder and rolling piston. The new crankshaft design concept of wide range application was confirmed and verified by CAE analysis and the experiment. The design guideline was also determined to assure the sufficient rigidity of the crankshaft. And blushless DC motor shape is modified for high speed operation and it is verified through rigorous testing. As a results, the high speed(150rps) inverter rotary compressor was developed with high efficiency and reliability for the air-conditioning system.

5. References

[1] Tsugio Itami, Masahiro Kubo, Makoto Sugiyama, Estimation of bearing load of Rolling piston type Rotary compressors under high speed operation, Proc. Of Intl. Comp. Eng. Conf., Purdue, Paper 547

[2] Mathias HARELAND, Anders HOEL, Stefan JONSSON, David LIANG, Guocai CHAI, 2014, SELECTION OF FLAPPER VALVE STEEL FOR HIGH EFFICIENT COMPRESSOR, Proc. Of Intl. Comp. Eng. Conf., Purdue, Paper 1408

[3] Jose Luiz GASCHE, Danilo Martins ARANTES, Thiago ANDREOTTI, 2014, Experimental Analysis of the Fluid Structure Interaction in a Suction Valve Model, Proc. Of Intl. Comp. Eng. Conf.,Purdue, Paper 1217

[4] Siva Rama Krishna Bolloju, Vamsi Tiruveedhula, Naveen Munnangi, Koteswara Rao Vaddadi, Prathap Reddy M, 2014, Efficiency Improvement of Rotary Compressor by Improving the Discharge Path through simulation, Proc. Of Intl. Comp. Eng. Conf.,Purdue, Paper 1543

[5] Jorg MAYER, Preben BJERRE, Fabian BRUNE, 2014, A Comparative Study of Different Numerical Models For Flapper Valve Motion, Proc. Of Intl. Comp. Eng. Conf.,Purdue, Paper 1315

[6] H. C. Kim, M. G. Cho, J. Kim, J. H. Park and J. Shim, Coherence technique for noise reduction in rotary compressor, 2012, Journal of Mechanical Science and Technology 26 (7) (2012) 2073~2076

Experimental investigation on materials and lubricants for sliding-vane air compressors

Stefano Murgia[1], **Gianluca Valenti**[2,3], **Giulio Contaldi**[1] **and Alessandro Valenti**[3]

[1] Ing. Enea Mattei S.p.a., Strada Padana Superiore, 307, 20090 Vimodrone (MI), Italy
[2] Politecnico di Milano, Dip. Energia, Via Lambruschini, 4A, 20156 Milano, Italy
[3] Valenti Energie S.r.l., Res. Mestieri Milano Due, 20090 Segrate (MI), Italy

E-mail: stefano_murgia@mattei.it

Abstract. Positive-displacement compressors and, among them, sliding-vane rotary machines are widely used in the compressed air sector. As in many other industrial fields, the efficient utilization of energy has become a major goal also in this sector. The aim of the present activity is the experimental investigation on the influence of two vanes materials (cast iron and aluminium with anodized surface) and of four commercial lubricants (characterized by different formulations and additives concentrations) on the performance of a mid-capacity sliding-vane rotary compressor in a number of operating pressures. The performance is identified by both the volume flow rate and the absorbed mechanical power, evaluated according to the international standard ISO 5167 and ISO 1217. The campaign indicates that the considered lubricants do not affect appreciably the volumetric flow rate. On the other hand, the specific lubricants determine a variation of about 1% of the mechanical power for both materials, while the specific material a variation between 0.9% and 2.6%. The best performance is achieved by aluminium vanes and a synthetic poly-α-olefin lubricant.

1. Introduction and test conditions

Positive-displacement air compressors are adopted widely in commercial and industrial applications. In these applications, compressed air may count for an appreciable fraction of the total energy use. Hence, increasing the efficiency of the compressors yields large benefits from many perspectives: energy, economic and environmental. With the scope of achieving such benefits, the present work investigates experimentally the influence of materials and of lubricants on the performances of sliding-vane rotary compressors (SVRC): the weight of vane and the lubricant properties affect directly the friction losses and, thus, the compressor efficiency. A commercial SVRC, whose features are reported in Table 1, is employed for the purpose and tested under diverse conditions.

Table 1. Properties of the sliding-vane rotary compressor.

Rated flow rate, l/min*	3500
Rated working pressure, bar-g	7.5
Rated power, kWe	22
Nominal rotor speed, RPM	1500

* Free air delivery as per ISO 1217:1996 annex "C"

Content from this work may be used under the terms of the Creative Commons Attribution 3.0 licence. Any further distribution of this work must maintain attribution to the author(s) and the title of the work, journal citation and DOI.
Published under licence by IOP Publishing Ltd

Each test condition is characterized by:

- the material of the vanes: 2 options, cast iron and aluminium with anodized surface;
- the lubricant: 4 market oils characterized by diverse formulations and additives concentrations;
- the delivery pressure: 5 options, from 6.5 to 8.5 bar-g by a 0.5-step;
- the operating temperature of about 85°C;
- the nominal rotor speed of 1500 RPM.

The combination of all characteristics generates 40 test conditions. All conditions are repeated 10 times in at least 2 different days to verify repeatability and increase the confidentiality of the results, yielding an experimental campaign of 400 tests in total. The sequence of the tests is randomized to cancel the effect of any correlation among them. The considered performances are: (i) the standard volume flow rate of air and (ii) the absorbed mechanical power of the packaged compressors calculated accordingly to standard ISO 1217 and ISO 5167.

The present activity is a continuation of an industrial-academic collaboration that has defined and proved already a number of technical enhancements to reach higher efficiencies in sliding-vane rotary compressors [1,2]. To the knowledge of the authors, this work is the first experimental investigation of materials and lubricants for positive-displacement compressors. The next sections will define in details the materials as well as the lubricants, the experimental setup, the results and the conclusions.

2. Materials and lubricants

The two materials options are cast iron vanes (CIV) and aluminium alloy vanes with anodized surface (AlV). The main features of the chosen materials are listed in Table 2. The differences between them are well known: cast iron offers better mechanical properties than aluminium, it is readily machined and, additionally, the machined surfaces are resistant to sliding wear. Furthermore, cast iron has a high thermal conductivity, low modulus of elasticity and the ability to withstand thermal shock. For these reasons this material is often used for the vanes. On the other hand, aluminium has a lower density and its superficial properties can be easily improved via a finishing superficial treatment, such as anodization. This treatment consists in the growth of an oxide thin coat on the material surface by way of an irreversible electrochemical process. The anodization allows improving the superficial mechanical characteristics and the corrosion resistance. Additionally, it improves the superficial hardness and increases the affinity with lubricants. The AlV used in the experimental investigation are characterized by an oxide coat with a thickness of about 20 μm. Both CIV and AlV have been manufactured with similar roughness indices (reported in Table 2) in order to evaluate only the effect of weight reduction on the performance of the SVRC.

Table 2. Properties of the chosen vane materials.

Properties	**cast iron**	**aluminium alloy**
Superficial treatment	None	Anodization (20 μm)
Density, kg/m^3	7200	2700
Specific heat*, J/kg/K	460	920
Thermal conductivity at 100°C, W/m/K	48.5	180
Coefficient of expansion*, μm/m/K	11.7	24
Modulus of Elasticity, GPa	120	70
Weight of a single vane	175	65
Roughness index, R_a	0.38	0.36

*Between 20°C and 100°C

IOP Conf. Series: Materials Science and Engineering **90** (2015) 012039 doi:10.1088/1757-899X/90/1/012039

Almost all positive-displacement air compressors require an oil to cool, seal and lubricate internal components. A correct lubrication ensures the reliability of the equipment, preventing corrosion and wear, protecting internal metal parts, and contributes to reduce the energy consumption. The formulations and properties of the four lubricants considered here are reported in Table 3: the performances of ISO-VG 68 diester-based synthetic lubricant (A), usually employed on the SVRC, are compared against those of an ISO-VG 100 Poly-α-olefin synthetic lubricant (B) and two different ISO-VG 100 diester-based synthetic lubricants (C, D). The lubricant viscosity affects the performances of air compressors in several working conditions: for example, viscosity determines the ease with which the equipment can be started in low-temperature environment or it can be kept running at high-temperature conditions. The four lubricants have been selected considering the operating temperature, the speed at which vanes are moving with respect to the rotor and the stator, and the load upon the compressor components of typical conditions. Although the considered lubricants have similar viscosity at higher temperatures, they differ appreciably in formulation and in additives concentration: these properties influence the tribology of the compressor and, ultimately, air volume flow rate and the absorbed mechanical power. In the present work, however, the lubricant effects on durability of lubricants are not evaluated.

Table 2. Properties of lubricants from datasheet.

Properties	Test Method	A	B	C	D
Type	-	Synthetic	Synthetic	Synthetic	Synthetic
Base	-	Diester	Poly-α-olefin	Diester	Diester
Additives concentration	-	high	low	medium	none
Viscosity, cSt @ 40 °C	ASTM D7042	68	91	96	95
Viscosity, cSt @ 100 °C	ASTM D7042	10	14.8	10.7	9.2
Viscosity index	ASTM D2270/ISO 2909	120	170	96	63
Total acid number, mg KOH/g	ASTM D664	0.17	0.03	0.18	0.11
Density, kg/m^3	ASTM D7042/ISO 12185	951	849	957	954

3. Experimental setup

The experimental rig, shown in Figure 1, employs the necessary instrumentation to measure: temperatures and pressures of air along the compression, the delivered volume flow rate, temperatures and pressures of lubricant along the process, shaft torque and rotational speed. The experimental setup is design to evaluate the compressor performance while varying the delivery pressure. In particular, the mechanical power is calculated as product of shaft torque and rotational speed measured using a flange torque meter installed between the compressor and the electric motor. A Kistler 4504B1KB1N1 torque meter (Full Scale of 1000 Nm, Accuracy ±0.5% FS) is controlled by the evaluation instrument Kistler CoMo Torque Type 4700A. The air volume flow rate is firstly determined considering the standard ISO 5167, than converted in the Free Air Delivery (FAD) conditions according to standard ISO 1217. The electronic data acquisition is performed thought the National Instruments cDAQ-9178 wired to a personal computer. LabView Signal Express 2011® is used to carry out the measurements. The test rig allows the easily replacement of vanes and lubricant to set the desired test condition. During every lubricant replacement, the oil circuit is accurately cleaned to prevent contaminations.

Figure 1. The test rig based on a mid-capacity sliding vane rotary compressor (top) and the acquisition system (bottom).

4. Results and discussion

As mentioned, the purpose of this study is the investigation on the effect of 2 vane materials and 4 commercial lubricants on the performance of a sliding vane rotary compressor at 5 delivery pressures, from 6.5 to 8.5 bar-g by a 0.5-step. The test conditions are performed at the operating temperature of about 85°C and operating speed of 1500 RPM. A single test condition is repeated 10 times to verify the repeatability of the measurements and increase the confidentiality of the results. The averaged values of the empirical data over the 10 repeated tests are evaluated to verify the variation of the performances. Figures 2 and 3 are representative of the experimental approach: they illustrate the average values of the standard volumetric flow rate and the mechanical power, calculated according to the international standards ISO 1217 and ISO 5167, for each lubricant and vane material configuration at the pressure of 7.5 bar-g. The standard relative uncertainty on flow rate and mechanical power are around 2% and 1%, respectively, for each test As anticipated, each test conditions is repeated over 10 different times to verify the data repeatability and increase the confidentiality. Therefore the averaged values are employed in the next comparisons.

The experimental results reported in Figures 2a and 2b show that it is not possible to establish clearly a correlation between the volumetric flow rate and the lubricant with the adopted instrumentation: lubricants do not affect appreciably the volumetric flow rate, and the measured variations fall within the range of uncertainty of measurement. On the other hand, as shown in Figures 3a and 3b, when the mechanical power is considered, the effects of lubricants on the SVRC performance are more significant. In particular, when the lubricant A is used, the lowest mechanical power is achieved. This is due not only to the lower viscosity of lubricant A but probably also to its formulation. At the considered operating temperature of about 85°C, indeed, the viscosities of all lubricants are comparable and cannot justify the differences in mechanical power, which are due probably to the different formulations and additives concentration. In this regard, the lower additives concentration of lubricant D justifies the worst performance of this lubricant. As shown in Figures 3a and 3b, when ISO-VG 100 lubricants B, C and D are employed, an increase of the absorbed mechanical power can be noted, especially in the case of CIV. However, results show that the lubricant replacements determine only a slight variation (up to 1%) of the mechanical power for both the considered materials.

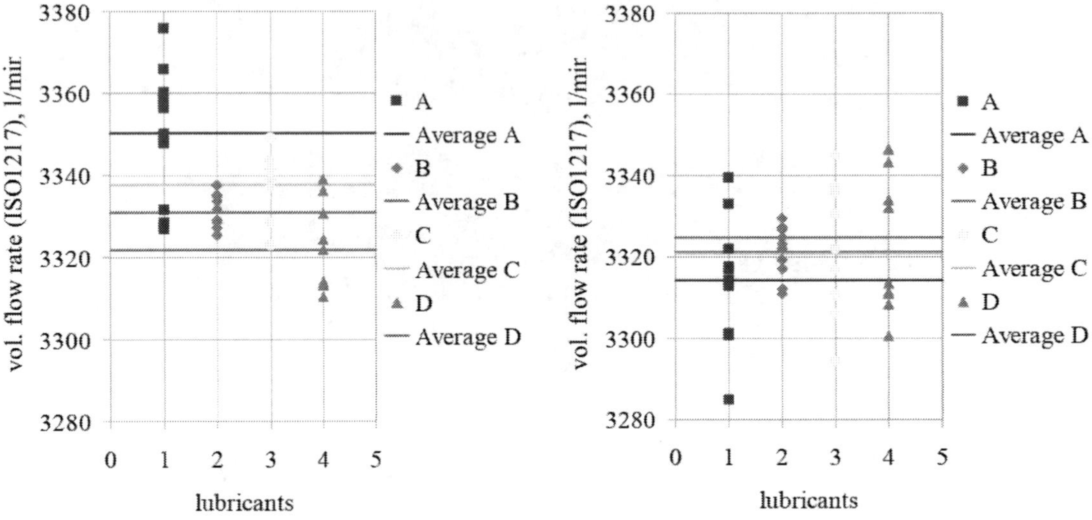

(a) Cast iron vane (CIV) **(b)** Cast iron vane (AlV).

Figure 2. Effect of lubricants on the volumetric flow rate of SVRC at 7.5 bar-g, 85 °C and 1500 RPM.

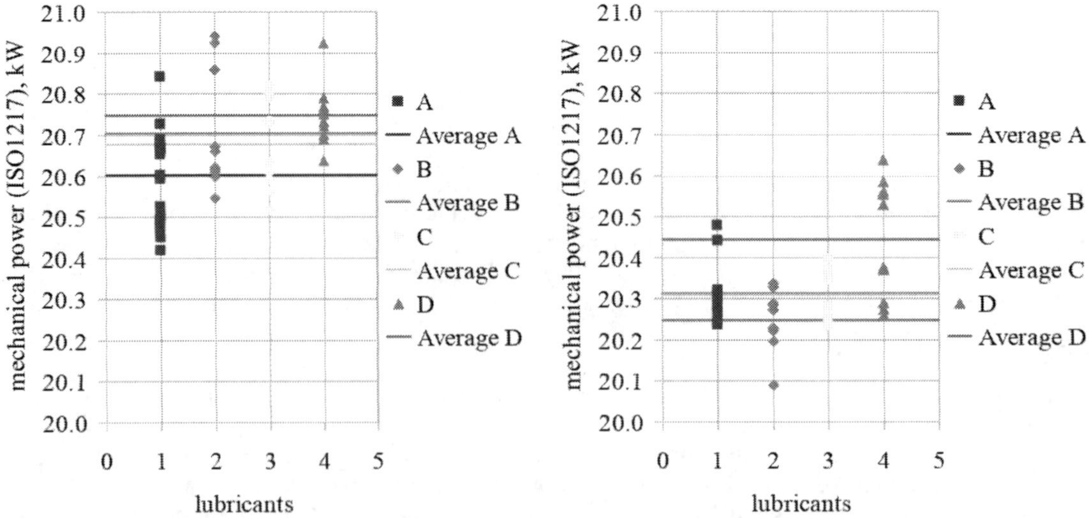

(a) Cast iron vane (CIV) **(b)** Cast iron vane (AlV).

Figure 3. Effect of lubricants on the mechanical power of SVRC at 7.5 bar-g, 85 °C and 1500 RPM.

The absorbed mechanical power decreases when AlV are used: the reduction of vane weight implies a reduction of the mechanical power on the order of 0.9-2.6%. The reduction of power consumption can be attributed not only to the lower density of the aluminium alloy but also to the improvement of its mechanical properties due to the superficial treatment. The chosen anodization allows to obtain a very high surface hardness, while the homogeneous surface finishing increases the affinity with the lubricant. Although the superficial treatment has a lower thickness comparing to conventional hard oxide coating, it resists very well to wear, improving the mechanical performance of AlV. As shown in the Figure 4, the performances of AlV are better than those of CIV in the range of the considered pressures: the reduction of the weight and the good surface finishing of the AlV determine an appreciable reduction of the adsorbed mechanical power. The maximum mechanical power reduction is achieved with the AlV and lubricant B. For both CIV and AlV the lubricant C is the only diester-based lubricant able to achieve the same performance of the Poly-α-olefin-based lubricant B. The good performances of the lubricant C are probably due to the use of some addictives (i.e. phosphorus, magnesium, etc.) in the formulation to improve the affinity between the lubricant and AlV surface. Results show that lubricant D offers the worst performance in the considered operating conditions: the lower viscosity index and, probably, the absence of additives justify the worse performances.

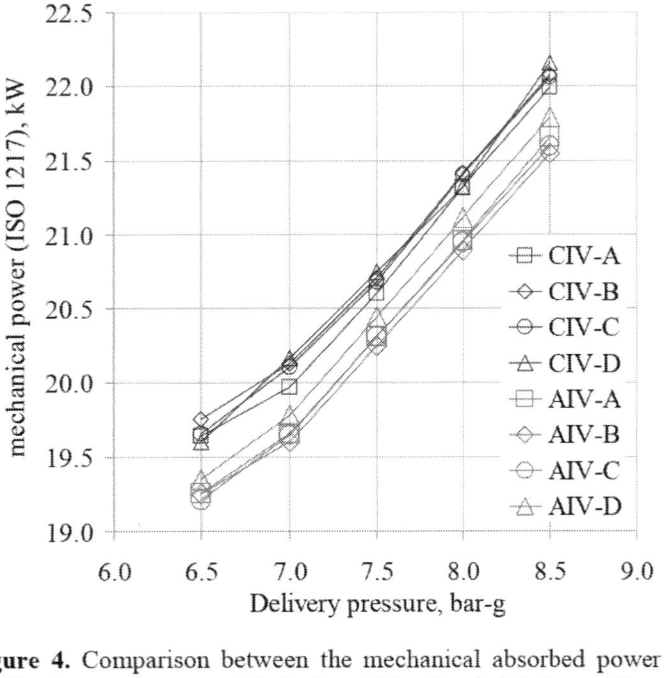

Figure 4. Comparison between the mechanical absorbed power profiles by varying materials (red and blue lines), lubricants (four markers) and delivery pressure at 85 °C and 1500 RPM.

Conclusions

This work investigates experimentally the performances of a commercial mid-capacity sliding-vane rotary compressor employing 2 kinds of materials for the vanes (cast iron referred to as CIV, and aluminium with anodized surface as AlV) and 4 different commercial oils (indicated as lubricant A to D) under 5 delivery pressures. All test conditions are replicated 10 times to verify the repeatability of the measurements and increase the confidentiality of the results. Volumetric flow rate and absorbed mechanical power for each test are calculated according to the international standards ISO 1217 and ISO 5167. The conclusions of the work are as follows.

- The considered vane materials and lubricants do not affect appreciably the volumetric flow rate, and the measured variations of performances fall within the range of the uncertainty of measurement. On the other hand, the effects of vane materials and lubricants on absorbed mechanical power are more significant: in the considered operating conditions, the lubricant replacement determines a slight variation of about 1% of the mechanical power for both the considered vanes materials.
- Lubricant A (ISO-VG 68) allows to reach the lowest mechanical power when cast iron vanes (CIV) are used. When lubricants B, C and D (all ISO-VG 100) are employed, an increase of the absorbed mechanical power can be noted, especially in the case of CIV. At the considered operating temperature of about 85°C the viscosities of all lubricants are comparable and cannot justify the slight differences in mechanical power. Performance are more probably influenced by the lubricant formulations and additive concentrations.
- The absorbed mechanical power decreases with aluminium vanes with anodized surface (AlV) by up to 2.6%. The reduction of power consumption can be attributed not only to the lower density of the aluminium alloy but also to the improvement of its mechanical properties due to the superficial treatment. The finishing surface of AlV allows to reach a good affinity with the lubricants. Results suggest that AlV is a proper solution for the energy optimization of SVRC.

The future work will focus on the evaluation of the effect of vane material and lubricants on a large-capacity 75 kW SVRC: in the future study the performance of AlV will be also investigated by varying the rotation speed of the compressor.

References

[1] Valenti G, Murgia S, Contaldi G and Valenti A 2014 *Case Studies in Thermal Engineering* Experimental evidence of the thermal effect of lubricating oil sprayed in sliding-vane air compressors **4** 113
[2] Valenti G, Colombo L, Murgia S, Lucchini A, Sampietro A, Capoferri A and Araneo L 2013 *Applied Thermal Engineering* Thermal effect of lubricating oil in positive-displacement air compressrs **51** 1055

Acknowledgements

The authors are very grateful to Daniele Colletta, Giacomo Ferrari, Lorenzo Marcellini and Maurizio Recalcati (listed in alphabetical order of last names) for conducting the tests during their internships.

IOP Conf. Series: Materials Science and Engineering **90** (2015) 012040 doi:10.1088/1757-899X/90/1/012040

NEW GENERATION OF UNIVERSAL MODELING FOR CENTRIFUGAL COMPRESSORS CALCULATION

Y. GALERKIN, A. DROZDOV
R&D Laboratory "Gas dynamics of turbo machines" Peter the Great St.Petersburg Polytechnic University, Polytechnical st. 29, St.Petersburg, Russia
Email:a_drozdi@mail.ru.

Abstract: The Universal Modeling method is in constant use from mid – 1990TH. Below is presented the newest 6th version of the Method. The flow path configuration of 3D impellers is presented in details. It is possible to optimize meridian configuration including hub/shroud curvatures, axial length, leading edge position, etc. The new model of vaned diffuser includes flow non-uniformity coefficient based on CFD calculations. The loss model was built from the results of 37 experiments with compressors stages of different flow rates and loading factors. One common set of empirical coefficients in the loss model guarantees the efficiency definition within an accuracy of 0.86 % at the design point and 1.22 % along the performance curve. The model verification was made. Four multistage compressors performances with vane and vaneless diffusers were calculated. As the model verification was made, four multistage compressors performances with vane and vaneless diffusers were calculated. Two of these compressors have quite unusual flow paths. The modeling results were quite satisfactory in spite of these peculiarities. One sample of the verification calculations is presented in the text. This 6th version of the developed computer program is being already applied successfully in the design practice.

Nomenclature

$\bar{b}$ - blade non-dimensional height
c – absolute velocity, m/s
c_u – tangential velocity, m/s
c' –velocity in view of a blade blockage factor, m/s
c'' – velocity in a blade row throat, m/s
D - diameter, m
D_2 - impeller diameter, m
$\bar{D}_h$ - hub non-dimensional diameter
$\bar{D}_0$ - impeller inlet non-dimensional diameter
i_3 - incidence angle
$\bar{l}_{bl}$ - non-dimensional length of a blade
$\bar{l}_{mbl}$ - non-dimensional length of a blade in meridian plane
$\bar{L}_m$ - non-dimensional axial length of impeller
$\bar{l}_{imp}$ - non-dimensional length in meridian section
$\frac{l}{t_{av}}$ - solidity of the vane cascade
M - Mach number
$M_u = \frac{u_2}{\sqrt{kRT_{t\,inl}}}$
$\bar{m}$ - mass flow rate, kg/s
Re - Reynolds number

Content from this work may be used under the terms of the Creative Commons Attribution 3.0 licence. Any further distribution of this work must maintain attribution to the author(s) and the title of the work, journal citation and DOI.
Published under licence by IOP Publishing Ltd

$\mathrm{Re}_u = \frac{u_2 D_2}{\mu_{inl}} \frac{p_{tinl}}{RT_{tinl}}$

$\bar{R}_h$ - hub non-dimensional radius of curvature

$\bar{R}_s$ - shroud non-dimensional radius of curvature

$\bar{S}_{bl}$ - non-dimensional blade surface area

$\bar{S}_o$ - non-dimensional average area of hub and shroud

$\bar{V}_{inl}$ - inlet flow rate, m^3/min

$X(i)$ - empirical coefficient in the math model

z - number of blades

α - absolute flow angle

α_v - vane angle (to tangential direction), ^{0}

β - relative flow angle

β_{bl} - impellers' blade angle to tangential direction, ^{0}

$\bar{\delta}_{bl}$ - non-dimensional blade's thickness

$\bar{\delta}_v$ - non-dimensional vane's thickness

φ - flow coefficient

φ_2 - angle of a shroud straight part

ε_3' - densities' ratio at inlets of VD and of an impeller

ε_4' - densities' ratio at outlet of VD and of an impeller inlet

λ_u - velocity coefficient

$$\lambda_u = \frac{u_2}{\sqrt{\frac{2k}{k+1} RT_{tinl}}}$$

$\pi = \frac{p_{out}}{p_{inl}}$ - pressure ratio

τ - blade blockage factor

$\psi_p = \frac{h_p}{u_2^2}$ - polytrophic head coefficient

ψ_i - work coefficient

$\psi_T = c_{u2} / u_2$ - loading factor

χ - angle of blade generatrix inclination to radial direction, ^{0}

$\eta = \frac{\psi_p}{\psi_i}$ - polytropic efficiency

Φ - flow rate coefficient

Subscripts

1 – impeller blade row inlet
2 – impeller outlet
3 - vaned diffuser inlet
4 - vaned diffuser outlet
av – average
bl – blade
calc – calculated
corr - corrected
cr - critical
des – design
exp – experiment

h – hub
init – initial
imp – impeller
inc - incidence
mean – mean
max – maximum
opt – optimum
out – outlet
s – shroud

1. INTRODUCTION

The complex of computer programs for gas dynamic design and flow path optimization named Universal Modeling method is in constant use from mid – 1990TH [1, 2]. The results of application and the process of the Method improvement are presented at conferences since 1995 [3]. Math model is the set of algebraic equations for calculation of head loss and mechanical work transmitted to gas. By means of 4th version math model several dozens of successful gas dynamic designs were created and realized by compressor manufacturers of Russia, Ukraine and Poland. The disadvantage of the model was that to calculate a stage performance with good accuracy an individual set of empirical coefficients must be applied. Different sets for stages with different design flow rate and loading factors were necessary. The problem was overcome in 5th version of the math model [4, 5]. The presented 6th version extends 5th version advantages on stages with 3D impellers and vaned diffusers (VD). Below is presented the main features of the 6th version of the Method.

2. 3D IMPELLERS CALCULATION

The new, 6th version expands the 5th version model achievements [4, 6, 7, 5, 8] on 3D impellers and vaned diffusers. Calculation algorithm used in the 6th version suggests more detailed calculation of head losses in 3D impellers. An important part is the more accurate definition of the hub/shroud surfaces and blades' surface area. Blades surface area is of the same order as hub/shroud surfaces' area – unlike 2D impellers. It is important to calculate blades' surface area precisely.
The previous, 4th version treated the problem in the very simplified way [9]. As it is shown at Figure 1 (left) only three geometry parameters specified a meridian shape: non-dimensional impeller inlet diameter $\bar{D}_0$, non-dimensional hub diameter $\bar{D}_h$,outlet height of blades $\bar{b}_2$. Not universal empirical formulas were applied to surface area calculations. Hub and shroud non-dimensional area in 4th model:

$$\bar{S}_{h+s} = \frac{S_{h+s}}{\frac{\pi}{4}D_2^2}\pi\left(1-\bar{D}_1^2\right)\left(\tau_1+\tau_2\right)/8 \tag{1}$$

Blades surface area is defined as the product of blade length $\bar{l}_{bl}$ and average blades height of an impeller:

$$\bar{S}_{bl} = 0,5\bar{l}_{bl}\left(\bar{b}_1+\bar{b}_2\right) \tag{2}$$

The empirical formula for a blade length:

$$\bar{l}_{bl} = \frac{1.15\pi}{4}\frac{\left(1+0.5\bar{D}_0-1.5\bar{D}_h\right)}{\sin\beta_{bl1}+\sin\beta_2} \tag{3}$$

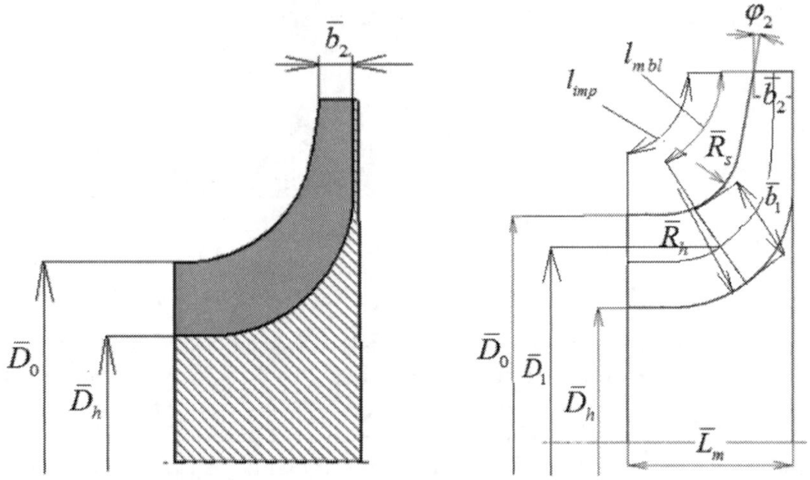

Figure 1. 3D impeller meridian section by the 4th program version (left) and the 6th version (right)

This approach does not allow estimate accurately friction losses on all surfaces of an impeller. It also does not give an opportunity to optimize geometric dimensions of the 3D impeller by candidates' comparison.

Figure 1 (right) demonstrates the set of geometry parameters that completely describe an impeller meridian shape in 6th version. This set is identical to one used for Q3D non-viscid calculations in the design process [1]. These parameters are: $\bar{D}_0$, $\bar{D}_h$, $\bar{b}_2$, a non-dimensional axial length of the impeller $\bar{L}_m$, hub and shroud curvature radius $\bar{R}_h$ and $\bar{R}_s$, the angle of a shroud straight part φ_2. Position of the blade leading edge is determined too by the parameter $l_{m\,bl}/l_{imp}$. The inlet blade height $\bar{b}_1$ and diameter $\bar{D}_1$ are calculated from the set of geometric quantities and is not user-defined.

The surface area of a blade in 6th version is:

$$\bar{S}_{bl} = \sum_{0}^{i} \frac{\Delta \bar{l}_{mi} \cdot \bar{b}_i}{\sin \beta_{bl\,i} \cos \chi} \tag{4}$$

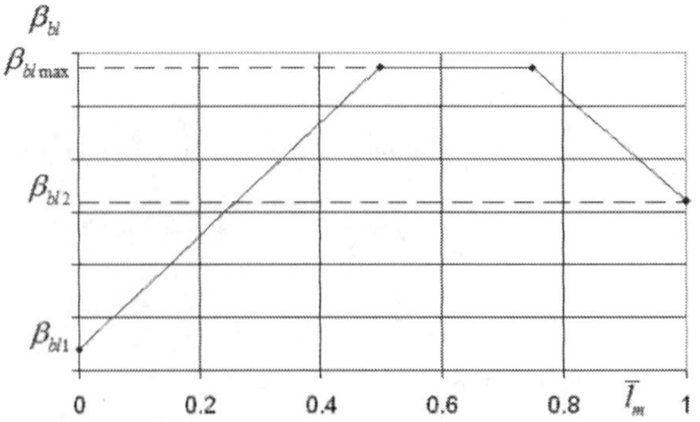

Figure 2. Schematization of blade angle along blade meridian length

To calculate blade area, it is necessary to know the function $\beta_{bl} = f(\bar{l}_m)$. The analysis of blade angle variation with the length for several versions of 3D impellers is made, and as a result the schematic linear approximation is proposed. The 3D impellers designed in accordance with [1], in most cases, have firstly an increasing blade angle, a middle part with practically constant angles and descending angles near the outlet:

- the initial portion $l_{m\,init}/l_m = 0 - 0.45$, assuming a linear angle increase from β_{bl1} to $\beta_{bl\max}$;
- middle portion from $l_{m\,mean}/l_m = 0.45$ - 0,75, assuming constant blade angle $\beta_{bl\max} = 1.35\,\beta_{bl2}$;
- outlet portion from $l_{m\,outl}/l_m = 0.75 - 1.0$, assuming a linear angle decrease from $\beta_{bl\max}$ to β_{bl2}.

The graphical view of such schematization is shown in Figure 2.

To calculate the blade surface area an impeller meridian length is divided into 40 sections. Figure 3 demonstrates position of a blade surface generatrix on a conical surface a-a.

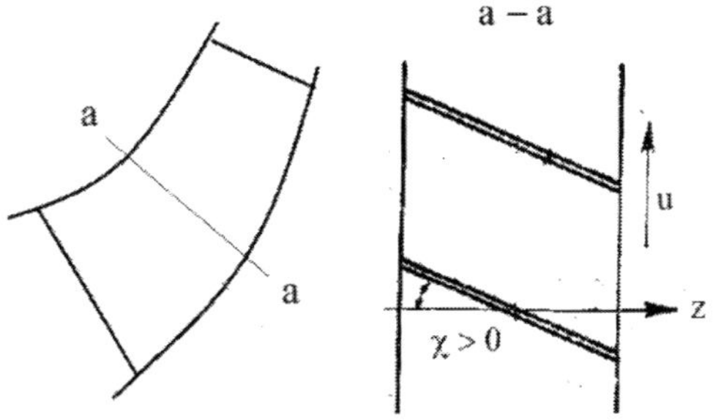

Figure 3. Blade meridian cross section (left) and blade generatrix position on a conical surface "a-a"

The angle χ of an inclination of a generatrix to a meridian plane on a surface "a-a" is not the independent design parameter. The angle varies along blade length and its value at a leading edge is equal 45 - 55^0 sometimes. Big χ increases blade surface area and a blade blockage factor. It is taken into account in a simplified way - the angle average value is assuming to be equal to 20^0 for all 3D impellers.

In accordance with loss model the friction force is proportional to average area of hub and shroud surfaces $S_0 = 0.5(\overline{S}_h + \overline{S}_{mean})$. The equation for calculations is:

$$\overline{S}_0 = \frac{S_o}{\frac{\pi}{4}D_2^2} = \overline{l}_{m\,bl} \cdot \left(2\left(1+\overline{D}_{av}\right) - z\frac{4\overline{\delta}_{bl}}{\pi \cdot \sin\beta_{bl\,av}}\right) \tag{5}$$

CFD calculations of stages with different 3D impellers have demonstrated that correct calculation of surface area is important. For instance, 3D impellers with slightly excessive number of blades are much less effective than could be expected.

3. VANED DIFFUSER LOSS MODEL

The sixth version uses a new algorithm for calculation the parameters in vaned diffusers. The vaneless part of greater or lesser extent is located before vaned diffuser. This is done to reduce the negative impact of uneven flow after impeller for vaned diffuser. Mechanical considerations are obvious. In this element, the approach presented in [4, 5, 6, 7, 8] is applied with some modifications in case of 3D impeller. Some equations of the mathematical modeling for the vane part of the diffuser are presented below.

Flow coefficient at vanes' inlet is:

$$\varphi_3' = \frac{\Phi}{4\varepsilon_3'\overline{D}_3\left(\overline{b}_3 - X(i)\right)\tau_3} \tag{6}$$

The math model empirical coefficient $X(i)$ takes into account a boundary layer blockage factor. The blade blockage factor is a geometry parameter:

$$\tau_3 = 1 - K_{\tau 3} \frac{z_{VD} \bar{\delta}_{vVD}}{\pi \bar{D}_3 \sin \alpha_{v3}} \tag{7}$$

The coefficient $K_{\tau 3}$ is different for vanes of different shape – airfoil type or other.

The non-incidence inlet condition and incidence losses are determined with taking into account vanes' load. The crucial factor is a direction of a streamline that goes to a critical point on a blade surface. Near a vane cascade a critical streamline deviates to a suction side of vanes where pressure is less. The scheme of this streamline behavior is presented in Figure 4.

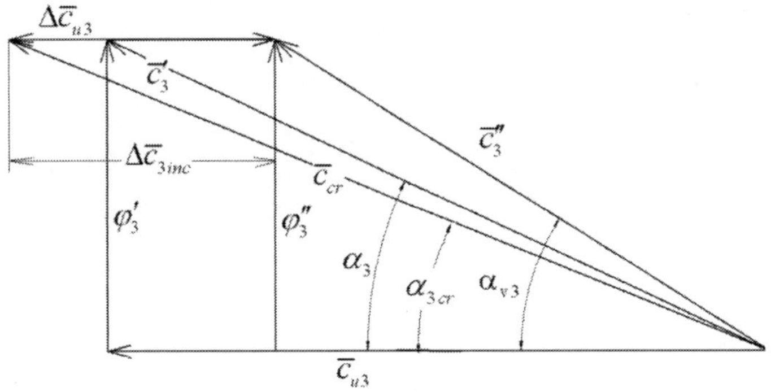

Figure 4. Velocity triangles at VD inlet

The critical streamline increases its velocity tangential component on $\bar{\Delta c}_{u3}$ under an influence of vanes' load. Non incidence condition is:

$$\alpha_{v3} = \alpha_{3cr} \tag{8}$$

The angle of a critical streamline is defined by the formula:

$$\alpha_{3cr} = arctg \frac{\varphi_3'}{\bar{c}_{u3} + \Delta\bar{c}_{u3}} \tag{9}$$

To calculate a critical streamline deviation the scheme from [1] is applied:

$$\Delta\bar{c}_{u3} = \frac{2\left(\bar{c}_{u3}\bar{D}_3 - \bar{c}_{u4}\bar{D}_4\right)}{z_{VD}(\bar{D}_4 - \bar{D}_3)} \tag{10}$$

If a critical streamline direction does not correspond to vane inlet angle then incidence losses take place. The analogy with sudden expansion losses is used. The losses are proportional to difference of vectors $\bar{c}_{3cr}$ and $\bar{c}_3''$ (velocity of inlet flow turned to direction of vanes) - Figure 4:

$$\Delta\bar{c}_{3inc} = \bar{c}_{u3} + \Delta\bar{c}_{u3} - \frac{\varphi_3'}{tg\alpha_{v3}} \tag{11}$$

Formula for calculating efficiency losses $\Delta\eta_{incVD}$ due to incidence inlet is similar to the corresponding formula in 4th version [10]:

$$\Delta\eta_{incVD} = X(i)(1 + X(i)(\lambda_{u2}\bar{c}_3'')^{X(i)})\frac{\Delta\bar{c}_{3inc}^2}{2\psi_T} \quad (12)$$

The exit angle α_4' of the flow differs from the blade angle α_{v4} on a deviation angle $\Delta\alpha_4$:

$$\alpha_4' = \alpha_{v4} - \Delta\alpha_4 \quad (13)$$

Deviation angle $\Delta\alpha_4$, according to [4, 11] is defined by the formula:

$$\Delta\alpha_4 = \frac{\alpha_{v4} - \alpha_{v3}}{\frac{\sqrt{l/t_{av}}}{0.346} + 1} \quad (14)$$

The optimum solidity of the vane cascade according to [12] is defined by the formula:

$$\left(\frac{l}{t_{av}}\right)_{opt} = z_{VD}\frac{\log\frac{\bar{D}_4}{\bar{D}_3}}{2.73\sin\frac{(\alpha_{v3} + \alpha_{v4})}{2}} \quad (15)$$

Measurements of flow structure at diffuser exit demonstrated sufficient non-uniformity. Kinetic energy of flow is bigger than by 1D calculation (1D representation corresponds to uniform flow). Kinetic energy influences head loss in the following element – return channel or scroll. Calculations by ANSYS CFX programs of a performance of the stator of the medium specific speed centrifugal stage have presented information on flow structure [9, 13]. The values of c_{43D} were averaged by energy $\bar{m}\cdot 0.5c_{43D}^2 = \sum d\bar{m}\cdot 0.5c_4^2$ at nine flow rates, i.e., at nine incidence angles $i_3 = \alpha_{v3} - \alpha_3$. The same element was calculated by 6th version computer program and c_{41D} values at the same conditions were defined. The ratio of the calculated by CFD velocity c_{43D} to velocity by 1D mode c_{41D} is presented in Figure 5 – black square. Coefficient K_{nu} takes into account flow non-uniformity at a vane cascade outlet:

$$K_{nu} = \frac{c_{43D}}{c_{41D}}. \quad (16)$$

Minimum of K_{nu} corresponds to zero incidence angle i_3.

The corrected value of non-dimensional velocity $\bar{c}_{4corr}$ is used in formula for head loss calculation in the element after vaned diffuser:

$$\bar{c}_{4corr} = (\varphi_4/\sin\alpha_4)\cdot K_{nu} \quad (17)$$

Presented in Figure 5 calculations were approximated by the formula:

$$K_{mu} = 1.16 + 20\sin^3\left(\left|i_3\right|\right) \quad (18)$$

The Eq. (18) approximates results of the numerical experiment data satisfactory in the practically important part of the performance.

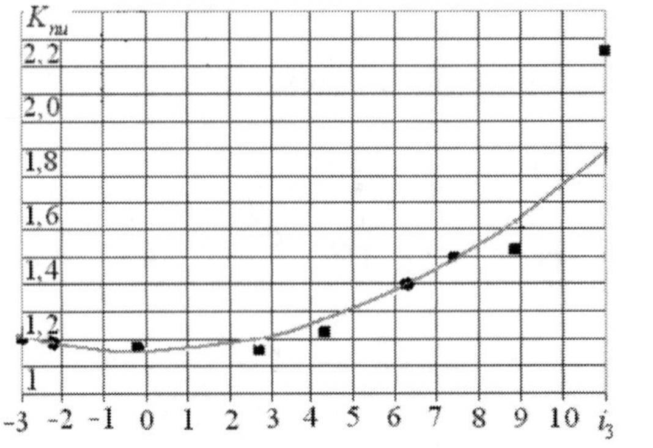

Figure 5. Non-uniformity coefficient versus the incidence angle i_3
Black square – CFD/6th version calculations, solid line – simulation by eq. (18)

4. MATHEMATICAL MODELLING VERIFICATION

Test results of 38 model stages were used for 6th version loss model identification (9 basic stages and their variants). The range of the main parameters of the model stages: $\Phi_{des} = 0.028 - 0.080$, $\psi_{T\,des} = 0.52 - 0.65$, $\bar{D}_h = 0.25 - 0.373$, $\bar{D}_4 = 1.428 - 1.615$, $M_u = 0.60 - 0.86$, $\mathrm{Re}_u = 4.8\times10^6 - 6.9\times10^6$. All stages consist of an impeller, vaneless or vaned diffuser and return channel. The basic stages were tested with varied hub ratio, vaneless diffusers (VLD) width, blade trailing edge configuration. Each test consisted of six different flow rate regimes. Thus, the identification process was made on the base of 228 empirical efficiency values.

The identification process consists of search of empirical coefficients values that satisfy the condition:

$$d\eta_{av} = \frac{\sum_{1}^{Z}\left|\eta_{exp} - \eta_{calc}\right|}{Z} \rightarrow 0 \quad (19)$$

After identifying the average error in calculating the maximum efficiency for all stages with vaned and vaneless diffusers which is equal to 0.007. Average calculation error for six points on efficiency performance curves is 0.028. Average calculation error for five points (except the highest flow rate regime) is 0.013. Several model stages test performance curves and results of their simulation by the 6th version model are presented at Figure 6.

Verification of a mathematical model was made by comparison of calculated efficiency performance curves with measured ones for several industrial compressors designed by the Universal modeling method and by other designer – design procedure is unknown. Test data are presented by manufacturers as results of official plant tests. There is one sample below.

The flow path scheme drawings of two pipeline booster compressors 16 MW with a delivery pressure of 7.45 MPa and pressure ratio $\pi = 1.7$. The designs of both compressors were made in due time by 4th version of the Method. Two-stage compressor flow path was places in the body of the standard pipeline compressor with $\pi = 1.44$. The elevated pressure ratio was achieved by increasing of impeller diameters and loading factor $\psi_{T\,des} = 0.82$. Low flow rate coefficient and big loading factor limited the efficiency by level of 80% - that is good for given non-dimensional parameters – Figure 7.

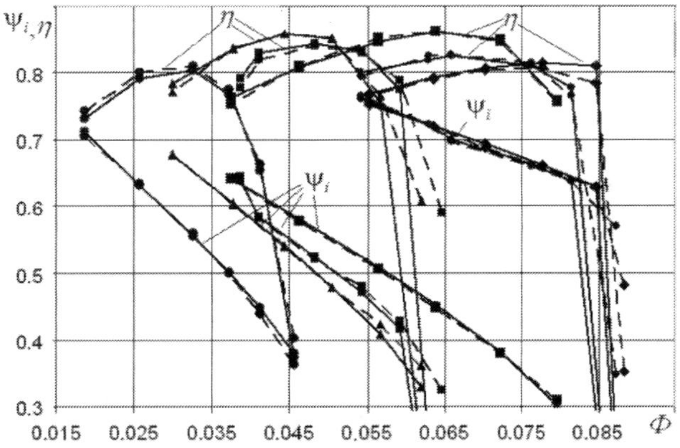

Figure 6. Five model stages test performance curves and results of their simulation by the 6th version model, dashed line – test, solid line - simulation

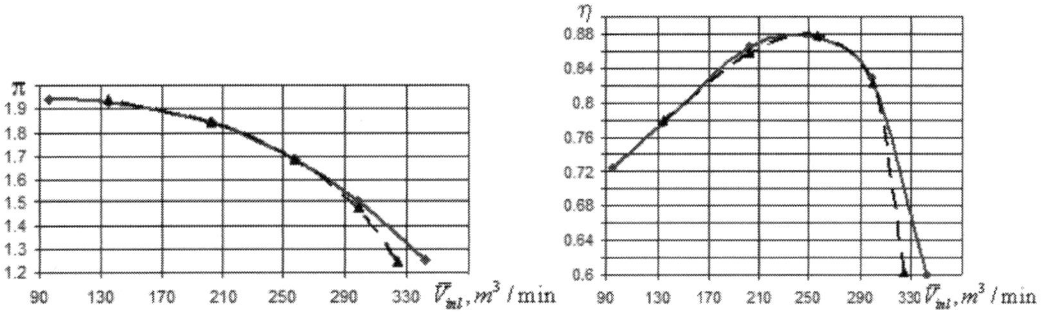

Figure 7. Performance curves of the 16 MWt four-stage pipeline booster compressor with VLD (air plant test). Solid line – test, dashed line – 6th version Math model calculation

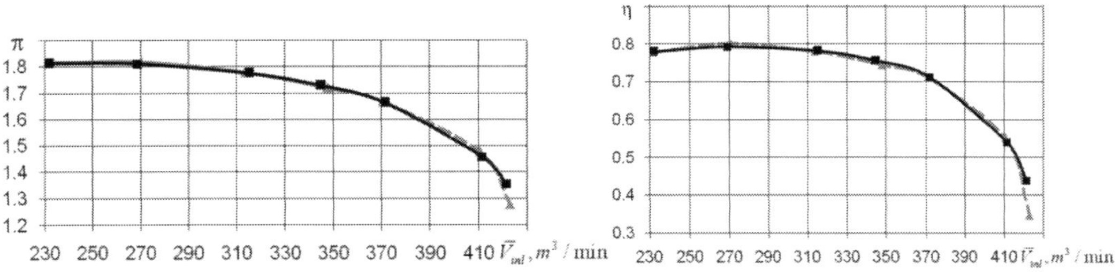

Figure 8. Performance curves of the 16 MWt two-stage pipeline booster compressor with VD (air plant test). Solid line – test, dashed line – 6th version Math model calculation

The four-stage compressor with the same dimensional parameters has optimal non-dimensional parameters. The loading factor at the design flow rate is slightly less than 0.5. The optimal design flow rate coefficient of stages with 2D impellers lies in range 0.050 – 0.080. Kinetic energy and surface area corresponds to minimum loss of a head in this case. Design flow rate coefficients of the compressor stages lie in this range. According to the official test maximum efficiency is 87.6%. Calculated and measured performance curves are compared at Figure 8.

Simulation results are good in main part of performances. Maximum flow rate corresponds to a compressor operation with low pressure ratio. Compressors do not operate at these regimes in practice. Poor simulation result at the maximum flow rate has no practical significance.

The analogy results were obtained for three more centrifugal compressors with significantly different parameters.

CONCLUSION

The new 6th version of the math model demonstrated ability to predict centrifugal stages and compressors gas dynamic performances with good accuracy. Precise calculations of stages and compressors with quite different design flow rate coefficients and loading factors are made by one common set of the empirical coefficients. Elevated power of modern PC opens way to further improvements of math modeling. One of evident ways is Q3D approach for modeling of 3D impellers that is inside visible future of the authors.

ACKNOWLEDGMENT

Work is performed with support of a Grant of the President of the Russian Federation for young PhD MK-7066.2015.8.

REFERENCE

[1] Galerkin Y. Turbo compressors. [text] / Galerkin Y. // LTD information and publishing center. - Moscow. – 2010 [Russian].
[2] Galerkin, Y. B., Soldatova, K.V. Operational process modeling of industrial centrifugal compressors. Scientific bases, development stages, current state. Monograph. [text] / Galerkin Y., Soldatova K.// Sankt-Peterburg. - SPbTU. – 2011. [Russian].
[3] Galerkin Y.B., Popova E.Y., Danilov K.A., Mitrofanov V.P. Quasi-3d Calculations in Centrifugal Impeller Design. VDI Berichte. № 1425. Hannover. – 1998.
[4] Galerkin, Y. B., Soldatova, K.V. Operational process modeling of industrial centrifugal compressors. Scientific bases, development stages, current state. Monograph. [text] / Galerkin Y., Soldatova K.// Sankt-Peterburg. - SPbTU. – 2011. [Russian].
[5] Galerkin, Y. B., Soldatova, K.V. The application of the Universal Modeling Method to development of centrifugal compressor model stages. [text] / Galerkin Y., Soldatova K.// International Conference "Compressors and their Systems".– London. – 2013. – P. – 477-487.
[6] Galerkin Y., Drozdov A., Gas dynamic performances simulation of centrifugal compressor stages with 3D impeller. [text] / Galerkin Y., Rekstin A. // Scientific and technical transactions of the TU SPb. – 2014. – No. 3. – Page – 45-53. [Russian].
[7] Y. Galerkin, A. Drozdov, K.Soldatova Centrifugal compressor efficiency types and rational application. [text] / Galerkin Y., Drozdov A., Soldatova K.// International Conference on Compressors and their systems. – London: City University, UK. – 2013.
[8] Galerkin, Y. B., Soldatova K.V., Drozdov A.A. New version of the Universal modeling for centrifugal compressor gas dynamic design. [text] / Galerkin Y., Drozdov A., Soldatova K.// Purdue Conference 2014. USA.
[9] Galerkin, Y. B., Marenina L.A. Research and improvement of centrifugal stages fixed elements by methods of computational dynamics. Part 1. / Galerkin Y., Marenina L.// Compressors and pneumatics, 2014. – №1. – P. 30-36. [Russian].
[10] Galerkin Y. Danilov K., Popova E. Design philosophy for industrial centrifugal compressor. [text] / Galerkin Y., Danilov K., Popova E. // International Conference on Compressors and their systems. – London: City University – 1999.
[11] Den G.N. Mechanic of a stream in centrifugal compressors. [text] / Den G. // – Leningrad. – Mechanical engineering. – 1973. [Russian].
[12] Ris V. Centrifugal compressors. [text] / Ris V. // - Leningrad. – 1981. [Russian].
[13] Galerkin, Y. B., Marenina L.A. Research and improvement of centrifugal stages fixed elements by methods of computational dynamics. Part 2. / Galerkin Y., Marenina L.// Compressors and pneumatics, 2014. – №2. – P. 10-15. [Russian].

SAMPLE OF CFD OPTIMIZATION OF A CENTRIFUGAL COMPRESSOR STAGE

Y. GALERKIN, A. DROZDOV
R&D Laboratory "Gas dynamics of turbo machines" Peter the Great St.Petersburg Polytechnic University, Polytechnical st. 29, St.Petersburg, Russia
Email:a_drozdi@mail.ru.

Abstract: Industrial centrifugal compressor stage is a complicated object for gas dynamic design when the goal is to achieve maximum efficiency. The Authors analyzed results of CFD performance modeling (NUMECA Fine Turbo calculations). Performance prediction in a whole was modest or poor in all known cases. Maximum efficiency prediction was quite satisfactory to the contrary. Flow structure in stator elements was in a good agreement with known data.
The intermediate type stage "3D impeller + vaneless diffuser+ return channel" was designed with principles well proven for stages with 2D impellers. CFD calculations of vaneless diffuser candidates demonstrated flow separation in VLD with constant width. The candidate with symmetrically tampered inlet part $b_3 / b_2 = 0{,}73$ appeared to be the best. Flow separation takes place in the crossover with standard configuration. The alternative variant was developed and numerically tested. The obtained experience was formulated as corrected design recommendations.
Several candidates of the impeller were compared by maximum efficiency of the stage. The variant with gas dynamic standard principles of blade cascade design appeared to be the best. Quasi – 3D non-viscid calculations were applied to optimize blade velocity diagrams – non-incidence inlet, control of the diffusion factor and of average blade load. "Geometric" principle of blade formation with linear change of blade angles along its length appeared to be less effective.
Candidates' with different geometry parameters were designed by 6th math model version and compared. The candidate with optimal parameters - number of blades, inlet diameter and leading edge meridian position – is 1% more effective than the stage of the initial design.

Nomenclature

$\bar{b}$ - blade relative width;
c_m – meridian velocity, m/s;
D - diameter, m;
$\bar{D}_h$ - hub relative diameter;
$\bar{D}_0$ - impeller inlet relative diameter;
$\bar{D}_{tmpr}$ - average relative tampering part diameter;
k - Isentropic coefficient;
$\bar{l}_m$ - relative length in meridian plane;
$\bar{L}_m$ - relative axial length of impeller;
M - Mach number;
$M_u = \dfrac{u_2}{\sqrt{kRT_{inl\,tot}}}$;
M_z - blade load momentum;
r, R - radius;
Re - Reynolds number;

Content from this work may be used under the terms of the Creative Commons Attribution 3.0 licence. Any further distribution of this work must maintain attribution to the author(s) and the title of the work, journal citation and DOI.
Published under licence by IOP Publishing Ltd

$\mathrm{Re}_u = \dfrac{u_2 D_2}{\mu_{inl}} \dfrac{p_{inl\,tot}}{RT_{inl\,tot}}$;

u_2 - impeller periphery speed, m/s;

$\bar{w} = \dfrac{w}{u_2}$ - relative velocity;

w' –velocity in view of a blade blockade factor, m/s;

z - number of blades;

α_2 - flow angle at an outlet of the impeller (to tangential direction), ^{0};

β_{bl} - blade angle to tangential direction, ^{0};

ρ - gas density, kg/m^3;

ζ – loss coefficient;

ζ_{fr} – friction loss coefficient;

ζ_{mix} – mixing loss coefficient;

ε_1' - densities' ratio at inlets of impeller and of an impeller;

τ - blade blockade factor;

$\psi = \dfrac{h_p}{u_2^2}$ - polytrophic head coefficient;

$\psi_T = c_{u2} / u_2$ - Euler work coefficient;

η^* - total efficiency;

Φ - flow rate coefficient;

Subscripts
0 – impeller inlet;
1 – impeller blade row inlet
2 – impeller outlet
3 - diffuser inlet
4 – return channel inlet;
des – design;
imp – impeller;
min – minimum;
p – pressure;
s – suction;
VLD – vaneless diffuser

1. SCOPE OF THE WORK

High specific speed stages have smaller radial dimensions. A high specific speed stage guarantees higher efficiency as the 1st stage of a multistage industrial compressor. If design flow rate coefficient Φ_{des} exceeds 0.07 – 0.075 well studied 2D impellers are not effective. Less studied and more complicated in flow path configuration 3D impellers with axial - radial blade disposition must be applied. Stator elements of high flow rate stages are wider that leads to specific flow behavior.

Industrial compressor stages with 2D impellers were objects of attention of the elder of authors for decades. Theoretic background, design principles and positive design practice are presented in [1]. 3D impellers have much more complex configuration. It is difficult to formalize and much more experiments are necessary to formulate design recommendations or obtain optimal solution of an exact impeller. CFD numerical experiments can be an effective instrument for gaining design experience. Universal modeling method – complex of computer programs for centrifugal compressors and stages performance curves calculations [2, 3, 4, 5, 6, 7] can be applied too.

The object of the optimization experiments is a stage "3D impeller + vaneless diffuser + return channel (crossover included)". The design parameters are $\Phi_{des} = 0.105$, $\psi_{T\,des} = 0.56$, $M_u = 0.7$, $\mathrm{Re}_u =$

5600000, k = 1.4. Design flow rate, Euler work coefficients and Mach number correspond to demand of modern pipeline compressors and alike. The similarity criteria Re_u and k correspond to air tests of a model stage. The design was made by means of Universal modeling method that is based on general principles formulated in [1] and that was presented at several conferences [3, 5, 6, 7].

The CFD numerical experiment was made by means of NUMECA Fine Turbo. The authors experience and publications [8, 9, 10, 11, 12], lead to the next conclusions about CFD calculations results:

- efficiency performance curve calculation for stages with radial impellers is more or less satisfactory;
- work coefficient prediction is overestimated on 6 - 7% at design flow rate for 2D and 3D impellers;
- 3D impeller calculated performance curves are shifted right;
- surge limit prediction of 3D impellers is unsatisfactory. Calculated surge limit is very close to the regime of maximum efficiency;
- calculated maximum efficiency of a stage with 3D impeller is very close to the measured value;
- loss coefficients of stator parts calculated by engineering Universal modeling method and CFD calculated are very close.

The authors keep in mind these considerations while planning and interpreting numerical experiments.

2. ANALYSIS OF STATOR STAGE ELEMENTS

Figure 1 demonstrates meridian configuration of the initial design and streamlines calculated by NUMECA Fine Turbo – design flow rate.

Absolute velocity angle at the VLD exit is 26.7^0. The recommendation for VLD with constant width of medium flow rate stages is $\alpha_{2des} \geq 25^0$ [1]. However CFD calculation points on flow separation at the very beginning of VLD at Figure1. The separation continues in the crossover. ANSYS CFX calculations of VLD with uniform flow at inlet and without crossover are presented in [11, 12]. For the VLD of the same configuration flow separation is expected if $\alpha_2 < 20$ - 22^0. The negative influence of flow non–uniformity at the impeller exit is evident.

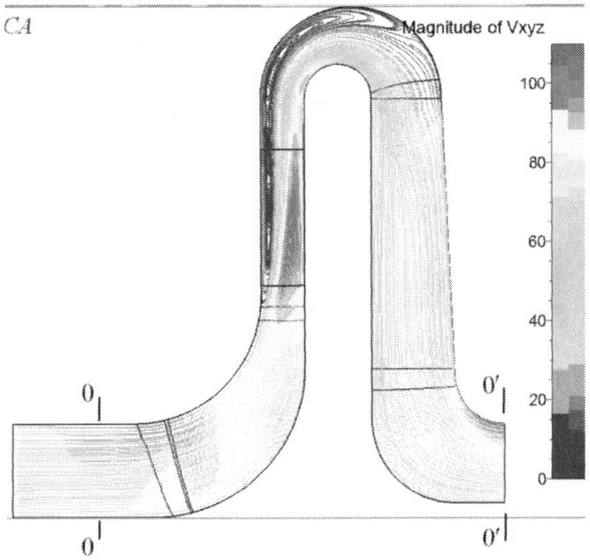

Figure 1. Initial design meridian configuration and streamlines (design flow rate, NUMECA Fine Turbo)

Tampering of VLD initial part is the known way to avoid flow separation for a long time [13].

Five candidates "impeller +VLD" were designed and compared. Geometry parameters of them are presented in the Table 1. The 1st candidate is the initial design. The others have initial parts with different tampering. The tampering is made from the hub side in three cases. This kind of tampering is recommended in [13] as effective gas dynamic solution. It helps also to diminish a stage axial length.

The candidates ## 4, 5 have longer tampering part. The candidate # 5 has symmetrically tampered initial part.

Table 1

Geometric parameters of VLD candidates

#	1	2	3	4	5
Tampering side	-	hub	hub	hub	symmetric
$\bar{b}_3$	0.0743	0.0628	0.0543	0.0543	0.0543
b_3/b_2	1	0.845	0.731	0.731	0.731
$\bar{D}_{tmpr}$	-	1.16	1.16	1.32	1.32

The candidates' ## 1, 2, 3 configuration and streamlines are shown at Figure 2.

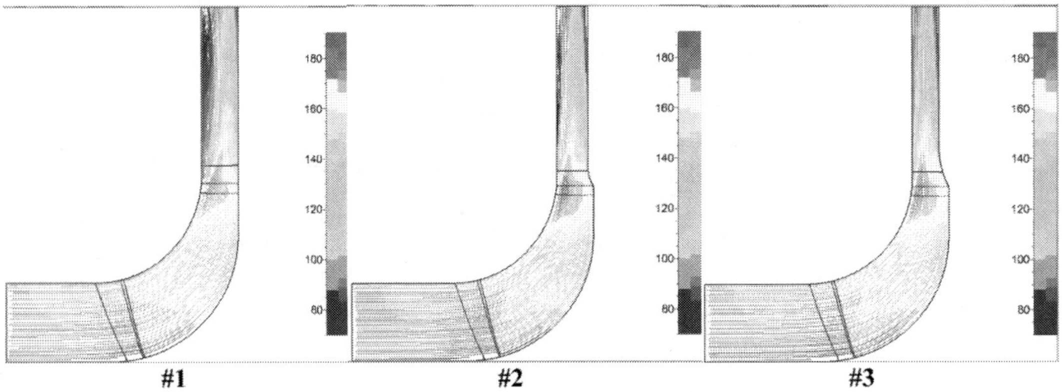

Figure 2. VLD candidates meridian configuration and streamlines (design flow rate, NUMECA Fine Turbo)

The flow separation in initial design is lesser than in the complete stage at Figure 1. Negative influence of the crossover is evident. Tampering ratio $b_3/b_2 = 0.845$ suppressed separation but has not eliminated it completely. The further tampering to $b_3/b_2 = 0.731$ eliminated flow separation. This tampering was accepted for further candidates.

The recommendation of [15] in tampering configuration was proven by calculation of two more candidates ## 4, 5. Candidate #4 has twice longer initial tampering part $\bar{D}_{tmpr}$ - 1.132. Candidate #5 has initial tampering part $\bar{D}_{tmpr}$ - 1.132 but the tampering was symmetric, i.e. was made from shroud and hub sides. The stages with different tampering were calculated by NUMECA Fine Turbo. The influence of tampering mode was noticeable. The comparison of three VLD loss coefficient performance curves is presented at Figure 3.

Symmetrical tampering (VLD #5) reduces the loss coefficient approximately to 10% in comparison with hub steep tampering (VLD #3). Rise of the stage efficiency is 0.4%.

To diminish negative influence of the crossover on a diffuser its configuration was modified. The initial design hub and shroud surfaces are formatted by arcs as it is shown at Figure 1. The modified configuration is presented at Figure 4 (right). The shroud surface is formatted by two arcs and a straight line. Streamlines are presented too (design flow rate, NUMECA Fine Turbo). The VLD tampering ratio is $b_3/b_2 = 0.731$ so there is no flow separation in this element. But there is the

IOP Conf. Series: Materials Science and Engineering **90** (2015) 012041 doi:10.1088/1757-899X/90/1/012041

separation zone at the shroud surface of the initial crossover. The modification of the shroud surface suppressed flow separation to the insignificant minimum.

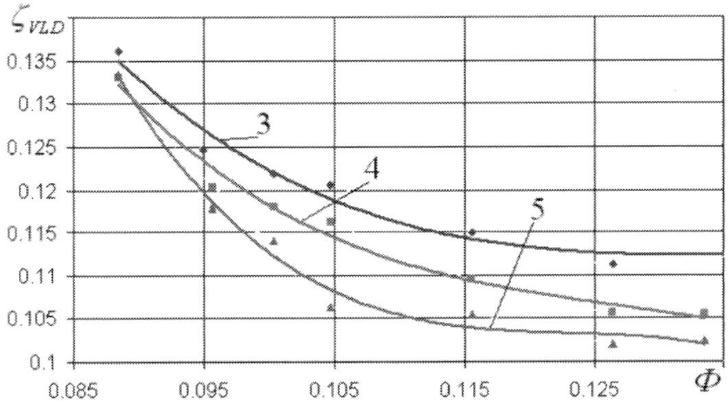

Figure 3. Loss coefficient performances of vaneless diffusers with different tampering

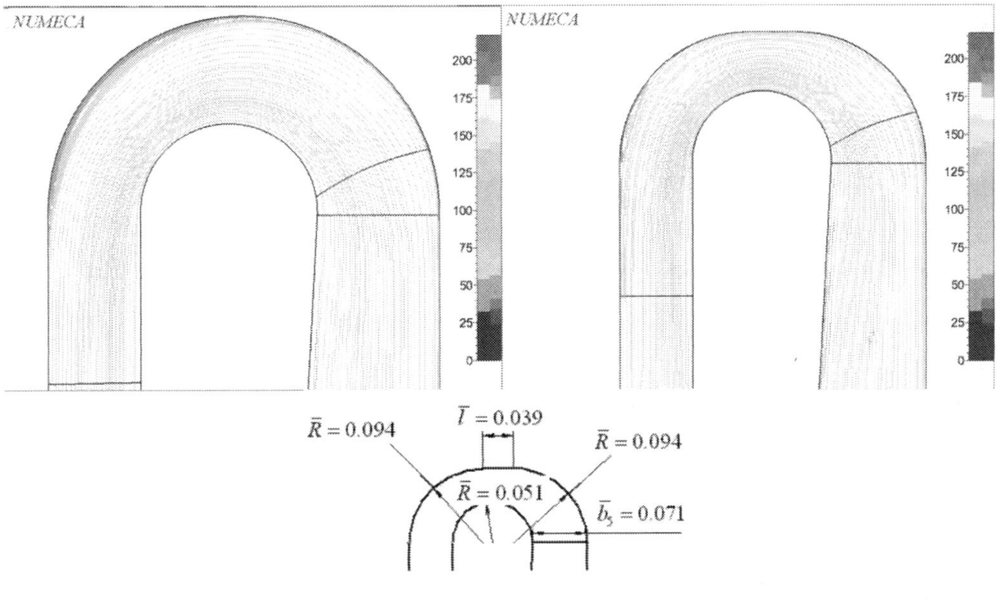

Figure 4. Left and center – initial and modified shroud surface and streamlines.
Right – modified shroud surface configuration

3. 3D IMPELLER BLADE CASCADE PROFILING AND OPTIMIZATION

Meridian dimensions and hub/shroud shape of the initial design were chosen as recommended in [1]. Blade cascade profiling is based on the choice of functions $\beta_{bl} = f(l_{mi}/l_m)$ on three blade to blade surfaces: shroud, middle, hub. An expert assessment of non-viscid Q3D velocity diagrams is applied. Figure 5 (left) demonstrates initial meridian configuration.

Four blade cascade candidates were designed. The 1st candidate is designed by the "geometry" principle that is known but is not applied in TU SPb practice. Three linear functions $\beta_{bl} = f(l_{mi}/l_m)$ of this candidate are shown at Figure 5. Inlet blade angles at three surfaces were chosen to minimize incidence losses along a leading edge height. The outlet angle was chosen in accordance with given Euler work coefficient. The chosen number of blades establishes an average blade load.

The TU SPb principle of profiling is realized in the candidate Imp-2. Function $\beta_{bl} = f(l_{mi}/l_m)$ at a shroud blade-to-blade surface has maximum. It helps to control velocity gradient along suction side of blades. There is an opinion that increase of the blade outlet angle from hub to shroud diminishes flow non-uniformity in the meridian plane.

The principle of the blade outlet angle increasing from hub to shroud is not proven so the candidate Imp-3 differs from Imp-2 by the constant outlet blade angle – Figure 5.

The principles of the candidate #4 is the same as for #2 but its relative length $\bar{L}_m$ is 0.35. The relative length of candidates ## 1 -3 is 0.30. View of blade cascades is shown on Figure 6. Velocity diagrams on hub and shroud surfaces for all candidates are shown on Figure 7.

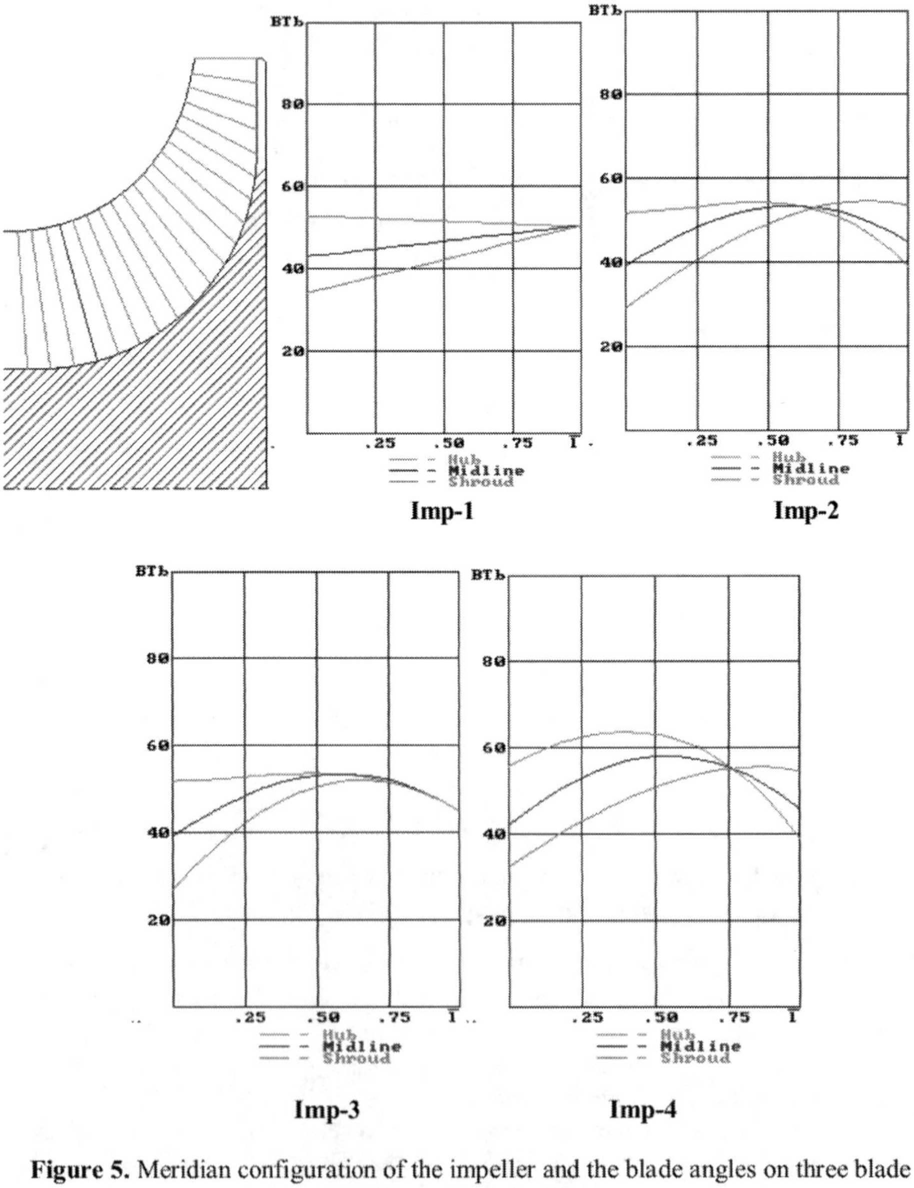

Figure 5. Meridian configuration of the impeller and the blade angles on three blade to blade surfaces – four candidates

IOP Conf. Series: Materials Science and Engineering **90** (2015) 012041 doi:10.1088/1757-899X/90/1/012041

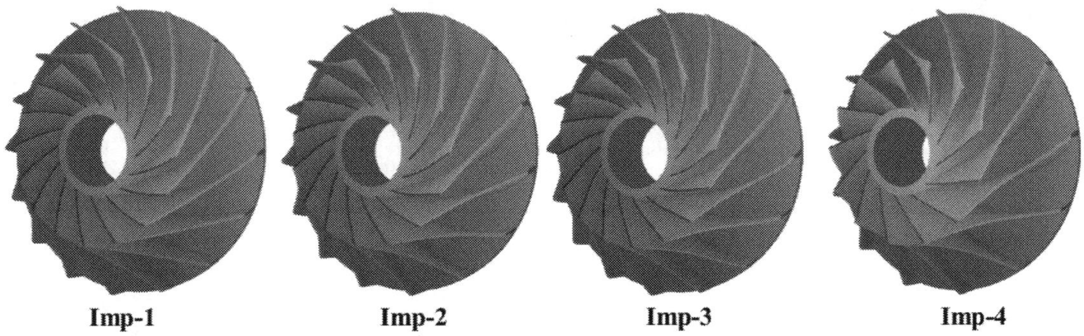

Figure 6. View of blade cascades – four candidates

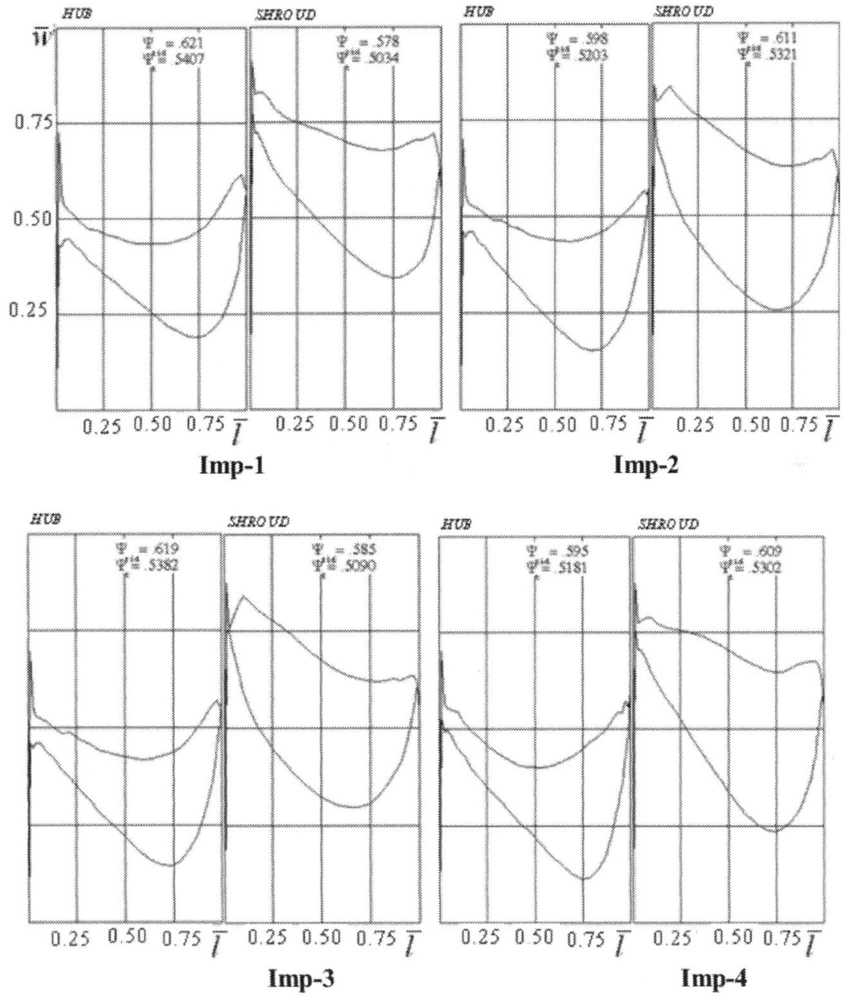

Figure 7. Velocity diagram on hub and shroud surfaces – four candidates

Calculations of performance curves of stages with four impellers – candidates lead to the next conclusions. In accordance with Q3D design program $\psi_{T_{des}} = 0.56$ of all four impellers – candidates.

CFD calculation predicts $\psi_{T\,des} \approx$ 0.63 for Imp-1. 2, 4 and 0.66 for Imp – 3. Mismatch with experiments of CFD calculation is a usual event. But the mismatch value 12-18% is too big and the problem deserves additional study.

Total efficiency performance curves – Eq. (1), (2) of the stages and impellers are presented at Figure 8.

$$\eta^* = \frac{\ln \frac{p_{0'}^*}{p_0^*}}{\ln \frac{T_{0'}^*}{T_0^*}} \frac{k-1}{k}; \quad (1)$$

$$\eta_{imp}^* = \frac{\ln \frac{p_2^*}{p_0^*}}{\ln \frac{T_2^*}{T_0^*}} \frac{k-1}{k}. \quad (2)$$

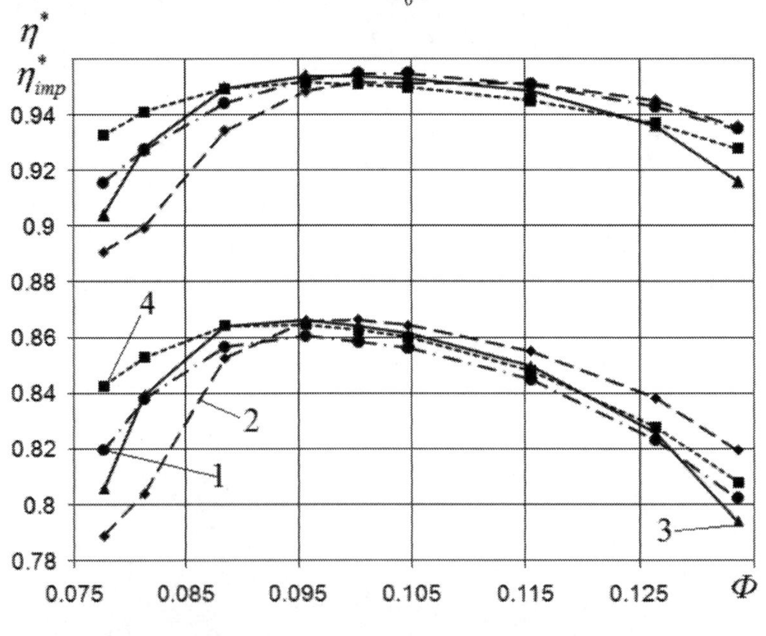

Figure 8. Total polytrophic efficiency of impellers and stages

It is well-known fact that an impeller total efficiency is not an objective parameter unlike stage efficiency. In the experiments and in CFD calculations averaged total pressure at an impeller outlet – section "2" - does not reflect impeller losses. Impeller's mixing losses do not occur in impeller, but in a following diffuser. The most effective Imp-3 demonstrated the most uniform outlet flow in the meridian plane. Imp-2 has less uniform outlet flow (the idea to diminish non-uniformity by variable outlet angle was not realized in this candidate). In spite of it the most effective is the stage with Imp-2 designed by principles well proven for 2D impellers [1].

The less effective is the stage with Imp-1 – "geometry" design.

The Imp-2 ($\bar{L}_m$ = 0.30) and #4 ($\bar{L}_m$ = 0.30) are designed on similar principles. Maximum efficiency of the stages differs little. Performance curves of the stage with Imp-4 shifted to left in comparison with the stage with Imp-2.

IOP Conf. Series: Materials Science and Engineering **90** (2015) 012041 doi:10.1088/1757-899X/90/1/012041

4. 3D IMPELLER OPTIMIZATION BY 6TH UNIVERSAL MODELIND VERSION

While CFD calculation is a proper mean to study details of a flow path configuration, the main simple geometry parameters can be studied by engineering calculation methods. The 6th math model and the proper version computer programs were applied for the optimization of three geometry parameters of the stage. All compared stage's candidates were designed by the standard method of the R&D Laboratory "Gas dynamics of turbo machines". The scheme at Figure 9 shows meridian dimensions of 3D impeller.

The principle of optimization is simple. The efficiency loss depends on inlet flow kinetic energy and loss coefficient:

$$\Delta\eta_{imp} = \frac{\zeta_{imp}}{\psi_T} 0,5\left(\frac{w'_{1s}}{u_2}\right)^2 = \frac{\zeta_{imp}}{\psi_T} 0,5\left(\frac{c'^2_{m1} + u^2_{1s}}{u^2_2}\right). \quad (3)$$

Euler work coefficient is represented in this equation but it is given for the stage. To minimize efficiency loss is necessary to minimize loss coefficient and inlet velocity. The standard procedure for an impeller inlet diameter choice is based on the principle of an inlet relative velocity minimization.

The more is an inlet diameter the more is u_{1s} and less c'_{m1s}. The minimum of the relative velocity corresponds to an inlet diameter by the equation that is known from 1930th:

$$\bar{D}_{0w\min} = \sqrt{\bar{D}^2_{\text{hub}} + 2^{\frac{1}{3}}\left(\frac{\Phi_{des}}{\varepsilon'_1\,\tau_1}\right)^{\frac{2}{3}}}. \quad (4)$$

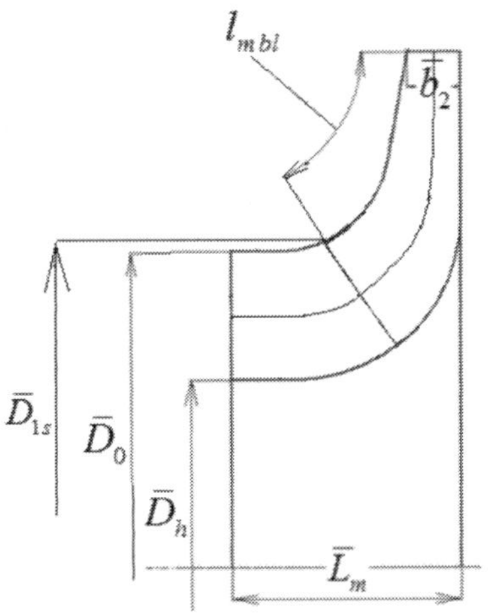

Figure 9. Meridian dimensions of 3D impeller

Different ζ_{imp} values correspond to different inlet diameters $\bar{D}_0$. The optimal inlet diameter $\bar{D}_0$ can be equal, less or more than $\bar{D}_{0w\min}$. The coefficient $A_d = \bar{D}_0 / \bar{D}_{0w\min}$ value is used in design practice as design recommendation based on experiments – physical or numerical.

The design experience has demonstrated that the loss coefficient of 2D impellers depends little of $\bar{D}_0$. The coefficient $A_{d\,opt}$ is equal to 1 in most cases.

The choice of $\bar{D}_0$ is not so simple in case of 3D impeller but firstly influence of blade number was investigated.

The initial stage for the blade number optimization has relative inlet diameter $D_0 = 0.60$, and the relative leading edge position $\bar{l}_{m\,bl}$ = 0.792. The result of blade number influence is presented in the Table 2.

Table 2

Stage's candidates with different blade number

z	β_{bl1}, ^{0}	β_{bl2}, ^{0}	w'_{1s}/u_2	ζ_{fr}	ζ_{mix}	η, %
10	27.9	67.9	0.717	0.0597	0.0335	84.8
12	28.7	57.9	0.717	0.0674	0.0225	85.2
15	29.7	51.1	0.719	0.0801	0.0127	85.3
18	30.5	48	0.721	0.0930	0.0073	85
22	31.4	46.2	0.724	0.1100	0.0043	84.5

The number of blades was varied in the wide range of $z = 10 - 22$. The gas dynamic flow parameters $\Phi_{des} = 0.105$, $\psi_{T\,des} = 0.56$ are given for all candidates. Therefore all impellers – candidates have different blade angles. The less is a blade number – the bigger is a blade outlet angle to obtain given $\psi_{T\,des}$. Inlet blade angles are bigger for candidates with bigger blade number due to blade blockade factor.

The maximum efficiency corresponds to 15 blades. The bigger blade load corresponding to smaller blade number leads to earlier flow separation and bigger mixing losses. Big blade numbers increases blades' surface area and friction losses increase as consequence.

Math model calculations demonstrate the next quantitative results. The candidate with 22 blades has friction losses on 1.84 times more in comparison with the candidate with 10 blades. Mixing loss coefficient 0.00425 is very little if z =22. If z = 10, the separation is sufficient and mixing loss coefficient is 0.0335 that is comparable with friction loss coefficient. Optimal number - fifteen blades well correlates with the design procedure recommendations.

The inlet diameter by eq. (4) for the analyzed impeller with 15 blades is equal to 0.613. The Eq. (4) estimates inlet relative velocity correctly if the leading edge is disposed at the section "0". Its position for an impellers analyzed below is determined by $\bar{l}_{m\,bl}$ = 0.792. The difference of $\bar{D}_0$ and $\bar{D}_{1s}$ is insignificant so the calculation results in the Table 3 are correct practically.

Candidates with $A_d < 1$ are more effective. The optimal value of $\bar{D}_0$ lies within 0.55 - 0.56, $A_{d\,opt} = 0.897 - 0.914$. The stage with optimum $\bar{D}_0$ has efficiency 0,861. The stage with $A_d = 1.0$ is 1.3 % less effective. The less is $\bar{D}_0$ if $A_d < 1$ the higher is kinetic energy at an impeller inlet. The efficiency becomes higher in spite of it. The reason is that blade surfaces are smaller. That diminishes friction loss coefficient in spite of higher average velocity w'_{1s}/u_2 that influence friction losses too. The positive factor is that flow deflection is less in impellers with smaller $\bar{D}_0$. It is important to note that the outlet blade angle of all candidates is the same, $\beta_{bl2} = 51^0$. The momentum that creates Euler work and is created by blade load is: $M_z = z\sum_0^2 \rho \cdot w \cdot (w_s - w_p) \cdot b \cdot dl_m \cdot r$. The more is $\bar{D}_0$ the more is a blade height b and less is an average flow velocity w. But that does not mean that their product $w \cdot b$ is the same for all candidates. The load $w_s - w_p$ depends on flow deflection, i.e. depends on blade angle gradient $\partial\beta_{bl}/\partial l_m \sim \frac{\beta_{bl2} - \beta_{bl1}}{l_m}$. The gradient is higher in candidates with bigger $\bar{D}_0$. It is not clear now if the constant value $\beta_{bl2} = 51^0$ is universal or is result of coincidence.

Table 3

Stage's candidates with different relative inlet diameters

$\overline{D}_0$	$A_d = \overline{D}_0 / \overline{D}_{0w\min}$	β_{bl1}, ^{0}	β_{bl2}, ^{0}	w'_{1s} / u_2	ζ_{fr}	ζ_{mix}	η, %
0.53	0.845	40.5	51	0.744	0.0553	0.0181	85.9
0.54	0.881	38.8	51	0.736	0.0578	0.0156	86
0.55	0.897	37.1	51	0.730	0.0607	0.0140	86.1
0.56	0.914	35.4	51	0.724	0.0640	0.0131	86.1
0.57	0.930	33.9	51	0.722	0.0673	0.0118	86
0.58	0.946	32.4	51	0.719	0.0713	0.0117	85.8
0.59	0.963	31	51	0.687	0.0756	0.0119	85.6
0.60	0.979	29.7	51	0.719	0.0801	0.0127	85.3
0.61	0.995	28.4	51	0.719	0.0857	0.0131	85
0.62	1.011	27.3	51	0.722	0.0913	0.0137	84.8
0.63	1.028	26.1	51	0.725	0.0977	0.0151	84.2

The optimal position of the blade leading edge $\overline{l}_{m\,bl}$ was determined by comparison of candidates presented in the Table 4.

Table 4

Stage's candidates with different position of a leading edge

$\overline{l}_{m\,bl}$	β_{bl1}, ^{0}	β_{bl2}, ^{0}	w'_1 / u_2	ζ_{fr}	ζ_{mix}	η, %
0.61	30.7	59.5	0.730	0.0522	0.0110	86
0.64	31.4	57.3	0.726	0.0542	0.0103	86.1
0.67	32.2	55.9	0.724	0.0560	0.0109	86.1
0.7	33	54.3	0.723	0.0579	0.0105	86.2
0.73	33.8	53	0.721	0.0600	0.0111	86.2
0.75	34.3	52	0.722	0.0615	0.0109	86.2
0.76	34.6	52	0.723	0.0619	0.0122	86.1
0.792	35.4	51	0.724	0.0640	0.0130	86.1
0.81	35.9	50.5	0.726	0.0651	0.0131	86
0.84	36.8	49.5	0.730	0.0673	0.0135	85.9

The optimal combination of w'_{1s} / u_2, ζ_{mix}, ζ_{fr} corresponds to the range of $\overline{l}_{m\,bl}$ = 0.7 - 0.75. The blade load is changing sufficiently in the studied range of $\overline{l}_{m\,bl}$ = 0.61 – 0.84. May be comparison of candidates with different blade number for $\overline{l}_{m\,bl}$ = 0.61 and 0.84 could correct the presented results to some extent.

ACKNOWLEDGMENT

Work is performed with support of the Grant of the President of the Russian Federation for young PhD MK-7066.2015.8.

CONCLUSION

The analysis of a typical high flow rate stage of an industrial compressor by CFD, Q3D and engineering-type computer program calculations demonstrates that some design recommendations well-proven on mean flow rate stages with 2D impellers are valid. The principle of 3D blades' design by Q3D non-viscid velocity diagrams optimization is valid. The new solutions and recommendations are proposed in other cases – symmetric sloping tampering of VLD is preferable, better crossover configuration is proposed, coefficient of an optimal inlet diameter $A_{d\,opt}$ appeared to be about 0.85 but not about of 1,0 as for 2D impellers. The principles of optimization can be applied to stages with other design parameters that will continue the process of design recommendations proof and correction.

REFERENCE

[1] Galerkin Y. Turbo compressors. [text] / Galerkin Y.// LTD information and publishing center. - Moscow. – 2010. [Russian].

[2] Galerkin Y. Pipeline Centrifugal Compressors – Principles of Gas Dynamic Design. [text] / Galerkin Y.// International Symposium «SYMKOM-05». Compressor & Turbine Flow Systems. Theory & Application Areas. Lodz. – № 128. Vol. 1. – 2005. – P.195-209.

[3] Galerkin Y., Danilov K., Popova E. Design philosophy for industrial centrifugal compressor. [text] / Galerkin Y., Danilov K., Popova E.// International Conference on Compressors and their systems. – London: City University, UK. – 1999.

[4] Galerkin Y.B., Danilov K.A., Popova E.Y. Universal Modelling for Centrifugal Compressors-Gas Dynamic Design and Optimization Consepts and Application. [text] / Galerkin Y., Danilov K., Popova E.// Yokohama International Gas Turbine Congress. – Yokohama. – 1995.

[5] Galerkin Y., Mitrofanov V., Geller M., Toews F. Experimental and numerical investigation of flow in industrial centrifugal impeller. [text] / Galerkin Y., Mitrofanov V., Geller M., Toews F.// International Conference on Compressors and their systems. – London: City University, UK. – 2001.

[6] Galerkin Y., Popova E. Industrial centrifugal compressors – gas dynamic calculation and optimization concepts. [text] / Galerkin Y., Popova E.// Texts of the Union of the German engineers. – Aachen. – Germany. – № 1109. – 1994.

[7] Galerkin Y.B., Popova E.Y., Danilov K.A., Mitrofanov V.P. Quasi-3d Calculations in Centrifugal Impeller Design. [text] / Galerkin Y., Popova E., Danilov K., Mitrofanov V.// VDI Berichte. № 1425. Hannover. – 1998.

[6] Galerkin Y.B., Drozdov A.A., Soldatova K.V. Centrifugal compressor efficiency types and rational application. [text] / Galerkin Y., Drozdov A., Soldatova K. // International Conference on Compressors and their systems. – London: City University, UK. – 2013.

[7] Galerkin, Y. B., Soldatova, K.V. Universal modeling method application for development of centrifugal compressor model stages. [text] / Galerkin Y., Soldatova K. // International Conference on Compressors and their systems. – London: City University. – UK. – 2013.

[8] Galerkin, Y. B., Marenina L.A. Research and improvement of centrifugal stages fixed elements by methods of computational dynamics. Part 1. [text] / Galerkin Y., Marenina L.// Compressors and pneumatics, 2014. – №1. – P. 30-36. [Russian].

[9] Galerkin, Y. B., Marenina L.A. Research and improvement of centrifugal stages fixed elements by methods of computational dynamics. Part 2. [text] / Galerkin Y., Marenina L.// Compressors and pneumatics, 2014. – №2. – P. 10-15. [Russian].

[10] Galerkin Y.B., Rekstin A.F., Soldatova K.V., Drozdov A.A.. High effective single stage pipe line compressor (gas dynamic design, model test results). [text] / Galerkin Y., Rekstin A ., Drozdov A., Soldatova K. // Compressors & Pneumatics. 2014. №8. P. 19-24.

[11] Galerkin, Y. B., Solovieva O.A., Improving the calculation methods vaneless diffuser centrifugal compressor stage based on computational experiments. Part 1. [text] / Galerkin Y., Solovieva O.// Compressors and pneumatics, 2014. – №3. – P. 35-41. [Russian].

[12] Galerkin, Y. B., Solovieva O.A., Improving the calculation methods vaneless diffuser centrifugal compressor stage based on computational experiments. Part 2. [text] / Galerkin Y., Solovieva O.//Compressors and pneumatics, 2014. – №4. – P. 15-21. [Russian].

[13] Seleznev, K.P., Galerkin, Y.B. Centrifugal compressors. [text] /K.P. Seleznev, Y.B.Galerkin // Leningrad. – 1982. – P. 271. [Russian].

IOP Conf. Series: Materials Science and Engineering **90** (2015) 012042 doi:10.1088/1757-899X/90/1/012042

Compressor map prediction tool

Arjun Ravi, Lukasz Sznajder and Ian Bennett.

Shell Projects and Technology
Lange Kleiweg 40
2288GC Rijswijk, The Netherlands

Abstract. Shell Global Solutions uses an in-house developed system for remote condition monitoring of centrifugal compressors. It requires field process data collected during operation to calculate and assess the machine's performance. Performance is assessed by comparing live results of polytropic head and efficiency versus design compressor curves provided by the Manufacturer. Typically, these design curves are given for specific suction conditions. The further these conditions on site deviate from those prescribed at design, the less accurate the health assessment of the compressor becomes. To address this specified problem, a compressor map prediction tool is proposed. The original performance curves of polytropic head against volumetric flow for varying rotational speeds are used as an input to define a range of Mach numbers within which the non-dimensional invariant performance curve of head and volume flow coefficient is generated. The new performance curves of polytropic head vs. flow for desired set of inlet conditions are then back calculated using the invariant non-dimensional curve. Within the range of Mach numbers calculated from design data, the proposed methodology can predict polytropic head curves at a new set of inlet conditions within an estimated 3% accuracy. The presented methodology does not require knowledge of detailed impeller geometry such as throat areas, blade number, blade angles, thicknesses nor other aspects of the aerodynamic design - diffusion levels, flow angles, etc. The only required mechanical design feature is the first impeller tip diameter. Described method makes centrifugal compressor surveillance activities more accurate, enabling precise problem isolation affecting machine's performance.

1. Introduction

In the oil and gas industry, the remote performance monitoring of critical machinery is essential to reduce downtime, maintenance costs and improve production rates of entire installations. A common machine that is central to most production and processing assets is the centrifugal compressor. Continuous performance assessment plays a vital part in evaluating the health of these machines. The performance of a centrifugal compressor is usually described by how the polytropic head varies with suction volume flow over a wide range of rotational speeds. Original equipment manufacturers (OEM) typically provide the design curves of polytropic head and efficiency, which serve as a benchmark in assessing the machine's performance - but for only a specific set of inlet conditions.

Ideally, a compressor would operate close to the OEM design conditions. In reality, the process conditions such as temperature, pressure, molecular weight and rotational speed deviate, sometimes substantially. This deviation impacts the accuracy of the performance assessment. There comes a point when the performance map provided by the manufacturer is no longer valid as a reference at new suction conditions. Accurate performance assessment of compressors requires a map that is valid for varying sets of suction conditions and rotational speeds. Unfortunately, obtaining new performance

Content from this work may be used under the terms of the Creative Commons Attribution 3.0 licence. Any further distribution of this work must maintain attribution to the author(s) and the title of the work, journal citation and DOI.
Published under licence by IOP Publishing Ltd

IOP Conf. Series: Materials Science and Engineering **90** (2015) 012042 doi:10.1088/1757-899X/90/1/012042

curves for an existing compressor from the vendor can be time consuming and resource intensive. Therefore, the methodology of predicting the compressor map at changing process conditions is proposed.

There are many approaches that can be adopted to obtain a compressor map. One of these is to create a numerical model of the centrifugal compressor stage and conduct CFD simulations. An alternative approach is to use empirical correlations for predicting performance maps. A detailed information of the stage geometry is needed; at least on a one dimensional (1D) basis. The detailed geometry is often unknown as this is propriety information of the vendor. Examples of 1D methods are given by [1], [2], [5].This paper presents a methodology that requires neither aerodynamic information (i.e. diffusion ratios, flow angles) nor the impeller geometry such as the throat areas, blade thickness and blade angles. It only requires the design performance curve from the OEM and the impeller tip diameter which is typically provided in the datasheet of the machine.

The following paper is structured as follows. Section 2 describes the methodology regarding how to correct the original performance map for a different set of suction conditions, limitations, validation and results. Conclusions are presented in the last section.

1.1. Nomenclature

D_{imp}	- Impeller tip diameter (m)
k	- Specific heat ratio (-)
Mu	- Tip speed Mach number (-)
MW	- Molecular weight (kg/kmol)
Q	- Volume flow rate (m3/s)
t	- Temperature (°C)
U	- Impeller blade tip speed (m/s)
R	- Universal gas constant [J/ (mol K)]
Z	- Compressibility (-)
H_p	- Polytropic Head (kJ/ kg)
N	- Rotational speed (rpm)
S	- Cubic spline polynomial
a, b, c, d	- Coefficients of spline polynomial (-)
σ	- Second derivative of cubic spline polynomial
h	- Equidistant spacing
m	- Number of intervals in the piecewise cubic spline polynomial
i	- Variable for iteration

Greek Symbols

Φ	- Volume flow coefficient (-)
Ψ	- Head coefficient (-)

Subscripts

OEM	- Original Equipment Manufacturer
design	- design inlet conditions
new	- new inlet conditions
suc	- suction conditions
p	- polytropic

2. General methodology and verification.

The process inlet conditions such as temperature, compressibility, specific heat ratio, molecular weight and circumferential speed all determine the compressor map. One non-dimensional parameter that takes all these factors into account is the tip speed Mach number. The key principle is that if the Mach numbers are the same for two different sets of suction conditions, then their corresponding non-dimensional curve of head coefficient vs. volume flow coefficient are equivalent. This approach follows similitude with equivalent Mach numbers, flow and head coefficient for the design and the new set of inlet conditions. An equivalent non-dimensional curve also ensures aerodynamic similarity which is important to simulate the equivalent flow of gas through the impellers. An equivalent flow coefficient for the design and the new set of suction conditions ensures the velocity triangles at the inlet of the stage to be the same. To visualize this concept a sample calculation has been done for two different set of suction conditions, for which the proximity of Mach numbers i.e. 0.517 to 0.5 was obtained. All the values are depicted in Table 1. Original polytropic head curves for these two set of inlet conditions provided by the OEM are shown in Figure 1. Corresponding non-dimensional curves obtained after conversion are displayed in Figure 2. Good agreement between these two curves validates the mentioned hypothesis. In both the graphs the red curve represents the design suction conditions and the black curve represents the new suction conditions.

Table 1 Set of inlet conditions and Mach numbers for corresponding performance curves depicted in Figure 1 and Figure 2.

	Temperature [°C]	Pressure [bara]	Molecular weight [kg/kmol]	Rotational speed [rpm]	Mach number[-]
Design inlet conditions	29.9	30	22.59	6778	0.517
New inlet conditions	40.9	30	17.24	7746	0.5

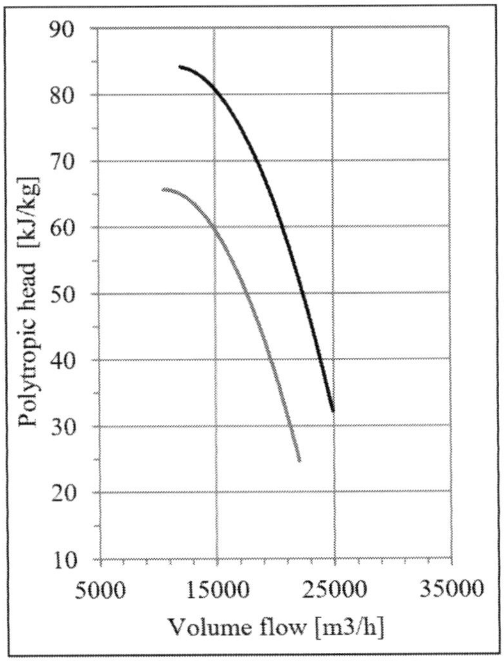

Figure 1 Polytropic head curves for different sets of inlet conditions.

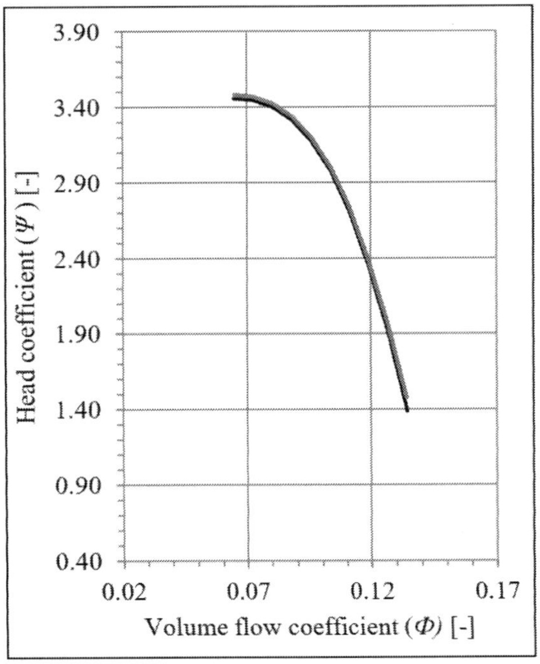

Figure 2 Equivalent non-dimensional curves for different sets of inlet conditions with the similar Mach number.

To present an example of the complete curve prediction methodology, data from a randomly chosen centrifugal compressor was used. The inputs are the design performance map of a variable speed compressor and its corresponding suction conditions i.e. pressure, temperature and gas composition for which the head curves were obtained. With the use of the presented methodology the new compressor map i.e. polytropic head (H_p) vs. volume flow rate (Q) relationships for a new set of suction conditions will be derived.

The coordinates of H_p and Q of original compressor map provided by the OEM at design suction conditions and varied N are shown in Table 2.

Table 2 Polytropic head (H_p) and volumetric flow (Q) for design suction conditions and varying rotational speeds (N) provided by the OEM. The rotational speeds are in % of the rated speed.

	N_1 100%		N_2 90%		N_3 80%		N_4 70%		N_5 55%	
	Q_1	H_{p1}	Q_2	H_{p2}	Q_3	H_{p3}	Q_4	H_{p4}	Q_5	H_{p5}
1	20171.2	137.7	15462.1	110.1	12505.4	89.4	10559.1	65.4	8152.6	41.3
2	22837.4	133.3	18499.0	110.2	15539.7	85.0	13594.6	63.2	10451.3	37.9
3	25593.6	125.4	21716.0	103.5	18113.8	80.5	16535.5	56.4	12289.8	34.6
4	27978.4	111.7	24562.2	92.2	20776.0	69.2	18923.0	47.4	14402.5	27.8
5	31183.5	84.4	28132.2	59.2	24535.3	45.3	21764.6	28.1	17246.1	11.9

2.1. *Main calculation procedure*

The design compressor map conversion to new suction conditions is done with the use of following non-dimensional parameters. Head coefficient (Ψ), volume flow coefficient (Φ) and Mach number (Mu) are calculated using outlined set of equations (2),(3) and (4) according to [3].

Mentioned equations require the conversion of N into impeller blade tip speed (U), which is done as follows:

$$U=\pi D_{imp}\frac{N}{60} \tag{1}$$

$$\psi=\frac{2H_p}{U^2} \tag{2}$$

$$\phi=\frac{4Q}{\pi D_{imp}^2 U} \tag{3}$$

$$Mu=\frac{U}{\sqrt{k_{suc}\,Z_{suc}\left(273.15+t_{suc}\right)R\,MW^{(-1)}}} \tag{4}$$

The polytropic head and volume flow curves presented in Table 2 are converted to non-dimensional curves in the form of Ψ and Φ for varied Mach numbers using the abovementioned equations. Mach number is varying as the rotational speed is changing. Obtained coordinates of converted curves, now in the form of non-dimensional curves are presented in Table 3 and visualized in Figure 3. Tip speed Mach number Mu_1 corresponds to the non-dimensional curve for the reference rotational speed i.e. N_1 as shown in Table 2. The same logic holds for Mu_2, Mu_3, Mu_4 *and* Mu_5.This non-dimensional map for varying Mach number is used as a baseline for the calculation of the compressor map for the new set of inlet conditions.

Table 3 The non-dimensional performance map values of head coefficient *(Ψ)* and volume flow coefficient (*Φ*) for varying Mach number *(Mu)* at design suction conditions.

	Mu_1 0.74		Mu_2 0.67		Mu_3 0.59		Mu_4 0.52		Mu_5 0.407	
	Φ [-]	*Ψ* [-]	*Φ* [-]	*Ψ* [-]	*Φ* [-]	*Ψ* [-]	*Φ* [-]	*Ψ* [-]	*Φ* [-]	*Ψ* [-]
1	0.087	3.595	0.074	3.554	0.067	3.662	0.065	3.479	0.064	3.567
2	0.097	3.516	0.087	3.524	0.082	3.533	0.080	3.424	0.080	3.369
3	0.108	3.338	0.100	3.363	0.096	3.276	0.096	3.201	0.096	3.007
4	0.119	3.016	0.114	2.995	0.111	2.853	0.111	2.744	0.111	2.420
5	0.134	2.168	0.134	1.892	0.132	1.813	0.134	1.478	0.135	0.996

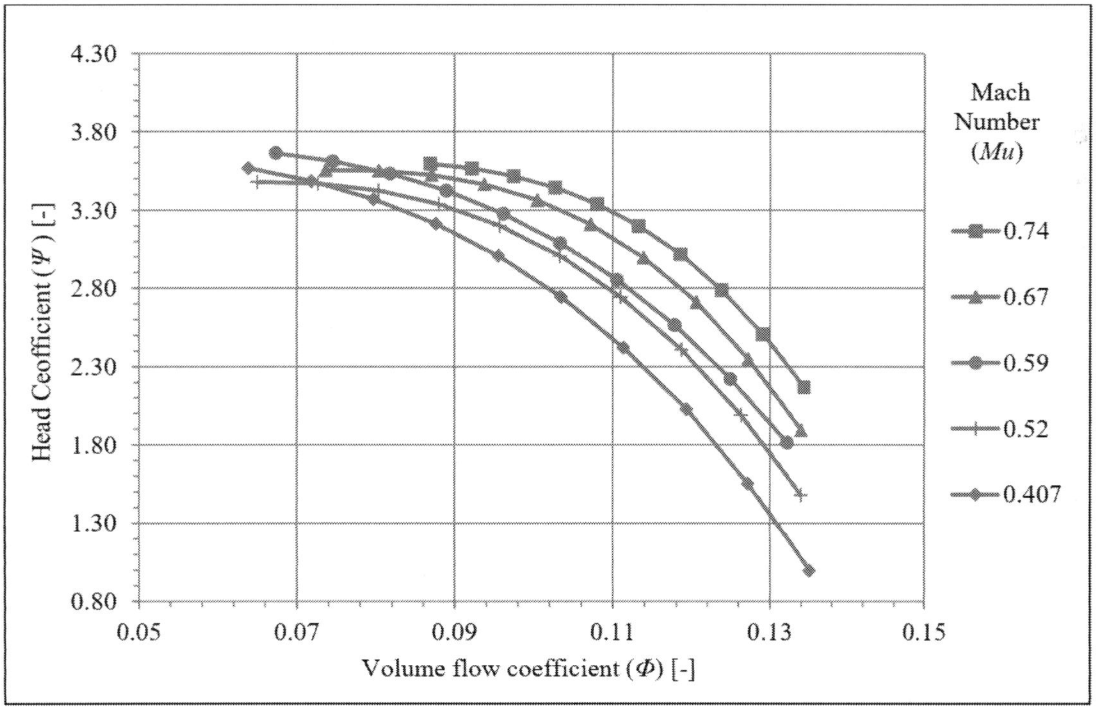

Figure 3 Invariant non-dimensional curves of head coefficient (*Ψ*) and volume flow coefficient (*Φ*) for different Mach numbers at design inlet conditions. The legend describes the impeller tip speed Mach number (*Mu*) for each curve.

At this point the Mach numbers for new set of inlet condition and shaft speeds are calculated (Mu_{new}). To draw the full non-dimensional curves for new conditions i.e. new head coefficient (Ψ_{new}) against new volume flow coefficient (Φ_{new}) the curve approximation method is used.

The mathematical method employed by the Authors was cubic spline interpolation technique, which is briefly presented in subchapter 2.2.1. It allowed obtaining coordinates of non-dimensional Ψ_{new} vs. Φ_{new} curves for desired Mu_{new} based on existing *Ψ* vs. *Φ* curves for different *Mu*.

In Table 4 the new coordinates of head coefficient (Ψ_{new})and volume flow coefficient (Φ_{new}) obtained in given example are presented The same is visualised in Figure 4

Table 4 Volume flow coefficient (Φ_{new}) and Head coefficient (Ψ_{new}) for new suction conditions.

	Mu_{new}= 0.65		Mu_{new}= 0.62		Mu_{new}= 0.56		Mu_{new}= 0.50		Mu_{new}= 0.44	
	Φ_{new}	Ψ_{new}	Φ_{new}	Ψ_{new}	Φ_{new}	Ψ_{new}	Φ_{new}	Ψ_{new}	Φ_{new}	Ψ_{new}
1	0.0717	3.5732	0.0689	3.6353	0.0662	3.6042	0.0646	3.4536	0.0640	3.5042
2	0.0854	3.5300	0.0831	3.5410	0.0810	3.4948	0.0802	3.4024	0.0798	3.3720
3	0.0992	3.3568	0.0972	3.3219	0.0958	3.2440	0.0957	3.1760	0.0956	3.0666
4	0.1130	2.9796	0.1114	2.9206	0.1105	2.8070	0.1112	2.7046	0.1114	2.5223
5	0.1336	1.8701	0.1326	1.8525	0.1326	1.7038	0.1344	1.3848	0.1351	1.1142

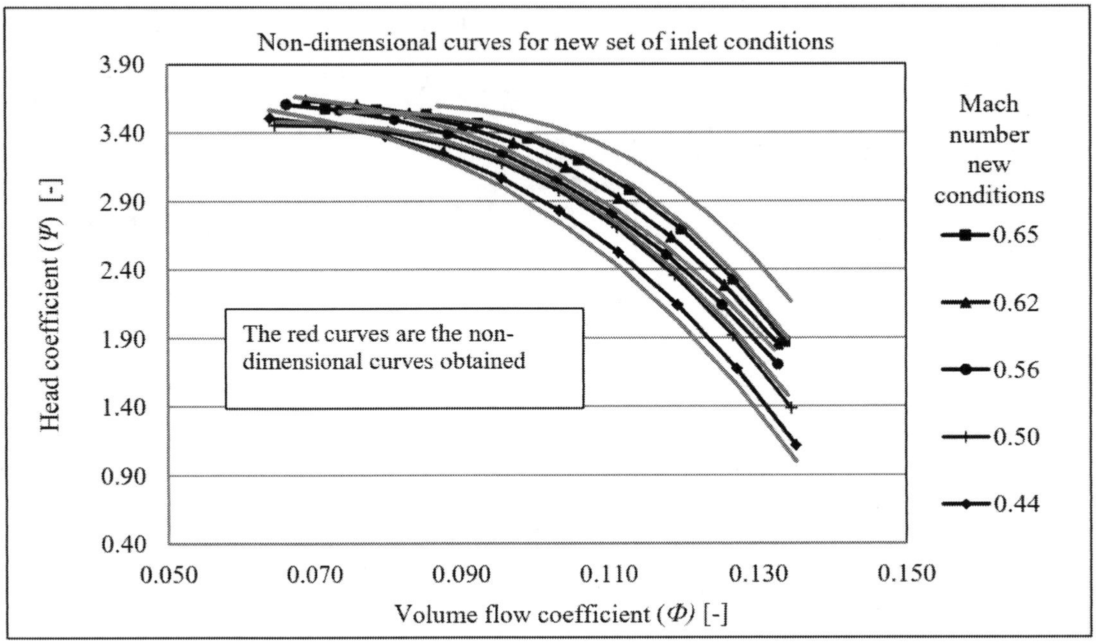

Figure 4 The interpolated non-dimensional curves of new head coefficient (Ψ_{new}) and volume flow coefficient (Φ_{new}) for varying Mach number at new suction conditions.

In Figure 4, the black curves with different markers represent the non-dimensional curves for varying Mach numbers for the new set of inlet conditions. The red curves are the non-dimensional curves for design suction conditions from Figure 3.In the last calculation step the coordinates of non-dimensional curves consisting of Ψ_{new} and Φ_{new} from Table 4 are recalculated back to obtain the actual polytropic head curves i.e. new polytropic head (H_{p_new}) vs. volumetric flow rate (Q_{new}) for desired set of process conditions and shaft speeds. It is done with the use of equations (5) and (6)

$$H_{p_{new}} = \psi_{new} \frac{U_{new}^2}{2} \quad (5)$$

$$Q_{new} = \phi_{new} U_{new} D_{imp}^2 \frac{\pi}{4} \quad (6)$$

To validate the methodology, the newly obtained compressor map was compared to the available map that was generated by the OEM at the same suction conditions during the machine's performance test. These curves are shown in Figure 5. The accuracy achieved in polytropic head prediction with the use of described methodology reaches c.a. 2-3% of error when predicted curve is compared to the original OEM's characteristic. The specific results regarding the accuracy of the method are depicted in Table 5. Such level of accuracy is regarded as satisfactory.

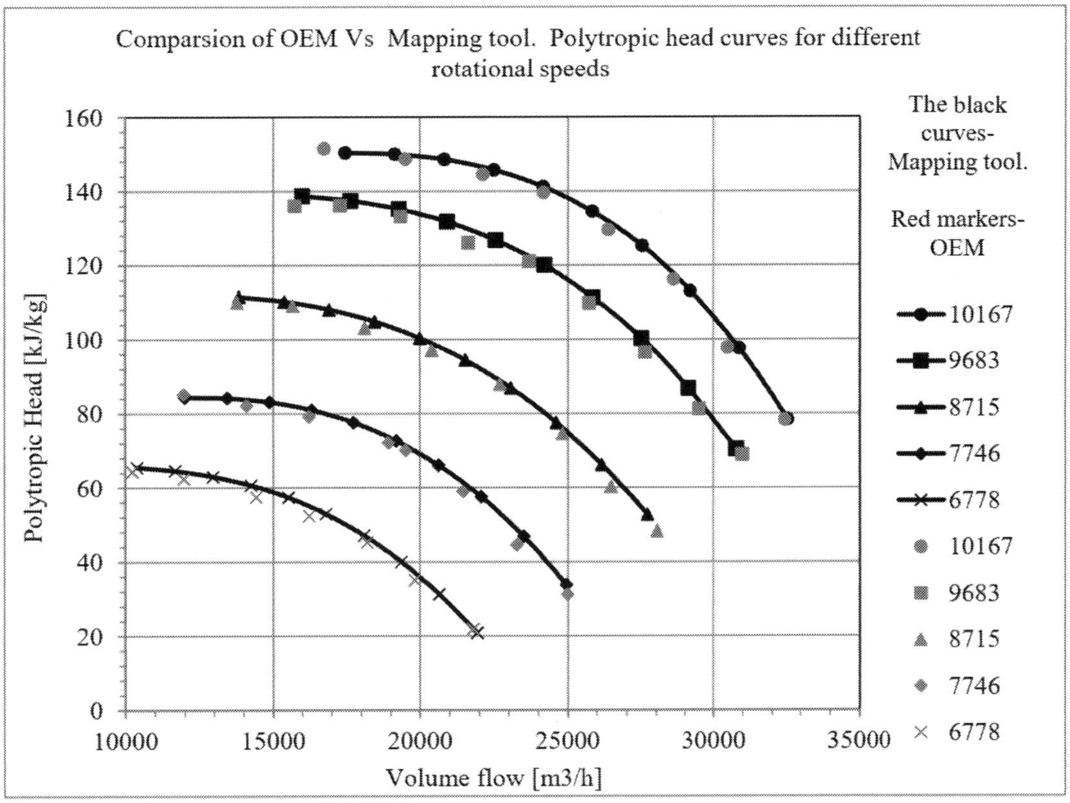

Figure 5 Comparison of compressor map generated by compressor mapping tool and map provided by the OEM. The red markers are the polytropic head curves from the OEM. The black curves are obtained from the compressor mapping tool.

Table 5 Validation of Compressor Mapping prediction tool with the values obtained from the OEM for the Rotational speed (N) = 9683 rpm.

	Flow [m3/h]	Polytropic head OEM [kJ/kg]	Polytropic head Mapping tool[kJ/kg]	Deviation in Polytropic head[%]
1	15991.0	136.0	138.7	2.0%
2	19277.3	133.3	135.2	1.4%
3	22563.7	124.1	126.7	2.2%
4	25850.0	109.1	111.3	2.0%
5	30779.5	71.7	70.4	1.8%

This method is only applicable when each stage of a centrifugal compressor is treated separately. Each stage refers to the number of impellers stacked together between the suction and discharge of a compressor.

2.2. Method limitations

One should remember that the accuracy of the method depends on the range of the operating envelope of the invariant non-dimensional curve varying with Mach number. Higher inaccuracies in curve approximation are seen when curve extrapolation is used.

The following convention is employed to define the upper and lower boundary of Mach numbers at design inlet conditions for which the presented methodology renders accurate results.

Mu_{low} - Mach number at design inlet conditions for lowest N;

Mu_{high} - Mach number at design inlet conditions for highest N;

Extrapolation limits the accuracy of presented method. Therefore the following constraints are imposed at newly calculated Mu_{new}. It is checked whether the Mu_{new} is within the range of Mach numbers calculated for the design conditions as shown below. There are three possible cases:

1. $Mu_{low}<Mu_{new}<Mu_{high}$ – the newly obtained Mach number is between boundary values of lowest and highest Mach numbers, therefore the interpolation method will be used to calculate coordinates of new head and volume flow coefficient curve. Experience has shown that interpolation allows to obtain non-dimensional curve approximation with high accuracy;
2. $Mu_{new}<Mu_{low}$ – the newly obtained Mach number is smaller than the lowest Mach number obtained for design inlet conditions, therefore the extrapolation method will be used to calculate coordinates of new head and volume flow coefficient curve;
3. $Mu_{new}>Mu_{high}$ – the newly obtained Mach number is greater than the highest Mach number obtained for design inlet conditions, therefore the extrapolation method will be used to calculate coordinates of new head and volume flow coefficient curve;

In 2nd and 3rd case the following condition must be met to ensure an acceptable accuracy of the curve approximation:

- Check if $\left(\dfrac{Mu_{low} - Mu_{new}}{Mu_{new}}\right) \leq 5\%$ (case 2)

- Check if $\left(\dfrac{Mu_{new} - Mu_{high}}{Mu_{new}}\right) \leq 5\%$ (case 3)

The three conditions are visualised in Figure 6. The red curves are the invariant non-dimensional curves for design inlet conditions. The dotted curves are the extrapolated curves corresponding to case 2 and 3. The dashed curve is the interpolated curve corresponding to case 1.

Extrapolation or interpolation of the new non-dimensional curve of Ψ_{new} vs. Φ_{new} for the new Mu_{new} is done depending on which of the three aforementioned conditions is met. The accuracy of the extrapolation largely deteriorates and the method is rejected for deviations in new Mach number greater than 5% from the highest or lowest design Mach number, as it was mentioned earlier.

It must be also mentioned that this method is only applicable when each stage of a centrifugal compressor is treated separately. Each stage refers to the number of impellers stacked together between the suction and discharge of a compressor.

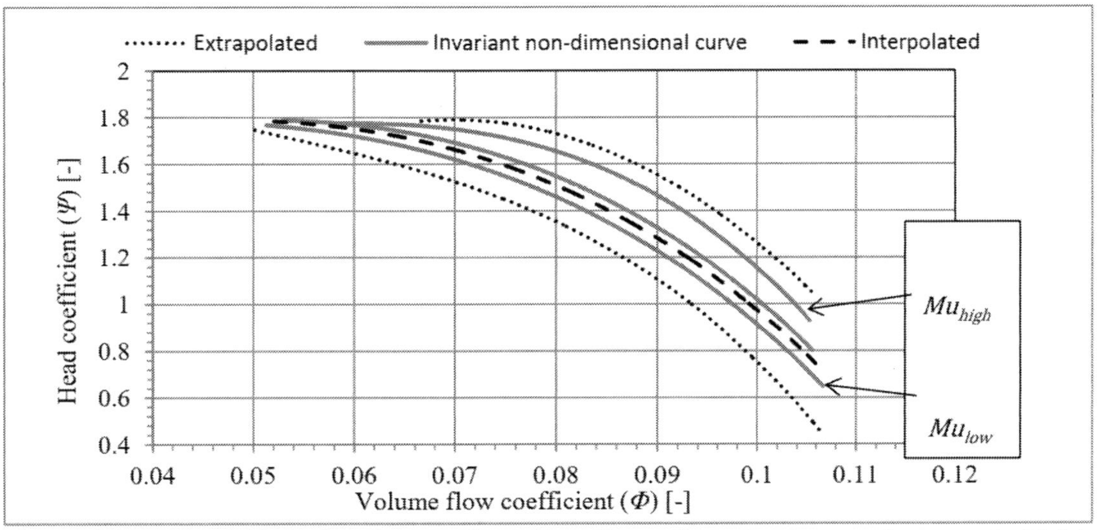

Figure 6 Interpolated (dashed) and extrapolated (dotted) non-dimensional curves for new set of inlet conditions from invariant non-dimensional curves

2.2.1. *Curve approximation method*

Mathematical method used to interpolate or extrapolate curves which was employed in the presented curve prediction tool is the Cubic Spline method. This technique provides smooth continuous approximated curves compared to piecewise linear interpolation. Cubic spline interpolation formula is continuous in the first and second derivatives and both within the intervals and interpolating nodes. It also provides an interpolating polynomial which has a smaller error compared to other interpolating polynomials such as Lagrange or Newton's A brief description of the original cubic spline interpolation method employed in the presented performance map prediction tool is provided hereunder.

$$f=CubicSpline(x_{input}, x, y)=S_i(x) \tag{7}$$

- x_{input} is the new Mach number for new set of inlet conditions;
- x is the dataset comprising known values of Φ obtained at design inlet conditions;
- y is the dataset comprising known values of Ψ obtained at design inlet conditions;

Spline interpolation is divided into small sub-interval and each of these sub-intervals is interpolated by using a third degree polynomial. According to [4], for a set of m+1 data points (x_i, y_i) and where no two x_i are the same on a given interval, a cubic spline (S_i) can be defined on each sub-interval (x_i, x_{i+1}) by

$$S_i(x)=a_i\ (x_{input}-x_i\ )^3+b_i\ (x_{input}-x_i\ )^2+c_i\ (x_{input}-x_i\ )+d_i \tag{8}$$

For m piecewise polynomial, four times m boundary conditions are needed to determine the coefficients a_i, b_i, c_i, and d_i. The boundary conditions are described in the equations outlined below:

$$S_i(x_i)=y_i \qquad for\ i=0:m-1 \tag{9}$$

$$S_{m-1}=y_m \tag{10}$$

$$S_i(x_{i+1})=S_{(i+1)}(x_{i+1})\ \ for\ i=0:\ m-2 \tag{11}$$

$$S_i'(x_{i+1}) = S_{(i+1)}'(x_{i+1}) \quad for\ i = 0:\ m-2 \tag{12}$$

$$S_i''(x_{i+1}) = S_{(i+1)}''(x_{i+1}) \quad for\ i = 0:\ m-2 \tag{13}$$

$$S_0''(x_0) = S_{(m-1)}''(x_m) = 0 \tag{14}$$

A new variable σ for the second derivative of the polynomial is introduced. Another variable h is defined for equidistant spacing.

$$\sigma_i = S''(x_i) \tag{15}$$

$$h = x_{i+1} - x_i$$

By using the equations (8), (9), (10), (11), (12), (13), (14) and (15)the coefficients a_i, b_i, c_i, d_i are obtained.

$$d_i = y_i; b_i = \frac{\sigma_i}{2}\ ; a_i = \frac{\sigma_{i+1} - \sigma_i}{6h}; c_i = \frac{y_{i+1} - y_i}{h} - h\frac{2\sigma_i + \sigma_{i+1}}{6} \tag{16}$$

3. Conclusion and Limitations

In this paper, a generic methodology is proposed, which enables correction of the original polytropic head curves for actual operating conditions based on the equivalence of non-dimensional parameters. The general procedure and validation of the method has been described. With minimal inputs of process conditions, first tip impeller diameter and a design performance map from the manufacturer, a compressor performance map for a new set of inlet conditions can be predicted within an accuracy of 2-3%.

The effective implementation of this method in remote condition monitoring system enables actual compressor performance assessment with the use of corrected polytropic head curves, which more accurately reflect the theoretical behaviour of a healthy machine at given conditions. Additionally, it can aid the performance review of the machine for new operating conditions and advise the engineers on site if the compressor can operate at the new conditions. Another key feature of this method is the accurate locations of the operating points on the compressor maps, which is essential for surge and choke protection. The presented feature can eventually facilitate isolation of undesirable compressor behaviour at an early stage and, therefore, help identify potential failure modes, thus reducing downtime and maintenance costs of the compressor and surrounding equipment. The method has been validated for 20 different centrifugal compressors, thus proving its readiness.

References

[1] Aungier R H 2000 *Centrifugal Compressor- A Strategy for Aerodynamic Design and Analysis* (New York, USA: ASME Press).

[2] Cumpsty N A 1989 *Compressor Aerodynamics* (Harlow, Essex, UK: Longman Scientific &Technical).

[3] Lüdtke K H 2004 *Process Centrifugal Compressors* (Berlin: Springer)

[4] Plato R. 2003 *Concise Numerical Mathematics* vol 57 (Providence R.I: American Mathematical Society).

[5] Swain E 2005 Improving a one-dimensional centrifugal compressor performance prediction method *Proc of the Institution of Mechanical Engineers Part A: Journal of Power and Energy* 219(8) pp.653-659.

NUMERICAL INVESTIGATION OF A CENTRIFUGAL COMPRESSOR WITH CIRCUMFERENTIAL GROOVES IN VANE DIFFUSER

X F CHEN[1], G L QIN[1], Z J AI[1, 2]

[1] School of Energy and Power Engineering, Xi'an Jiaotong University, Xi'an 710049, China
[2] School of Mechanical and Electronic Engineering, Sanming University, Sanming 365004, China

Email:glqin@mail.xjtu.edu.cn

Abstract. Enhancing stall and surge margin has a great importance for the development of turbo compressors. The application of casing treatment is an effective measure to expand the stall margin and stable operation range. Numerical investigations were conducted to predict the performance of a low flow rate centrifugal compressor with circumferential groove casing treatment in vane diffuser. Numerical cases with different radial location, radial width and axial depth of a circumferential single groove and different numbers of circumferential grooves were carried out to compare the results. The CFD analyses results show that the centrifugal compressor with circumferential grooves in diffuser can extend stable range by about 9% while the efficiency over the whole operating range decreases by 0.2 to 1.7%. The evaluation based on stall margin improvement showed the optimal position for the groove to be located was indicated to exist near the leading edge of the diffuser, and a combination of position, width, depth and numbers of circumferential grooves that will maximize both surge margin range and efficiency.

1. Introduction

The design of modern compressor is mainly focused on reducing stage numbers, which often leads to the problem of how to prevent flow instabilities such as surge and rotating stall. Additional to the ingenuity exercised in aerodynamic blade designing, various devices and methods have been tried out. Among them, casing treatment, small grooves or slots made on the endwall, has been proved to have a beneficial effect on increasing stable operating range of compressors.

Numerous configurations have been tested and studied since its development [1], and among them, large proportion of the recent research objectives seem to be devoted exclusively to two principal categories. One is the casing treatment which consists of circumferentially aligned axial slots [2, 3], and the other is the casing treatment with axially aligned circumferential grooves [4, 5]. Both have shown a sufficient capability in improving stall margin, but circumferential grooves seem to have an advantage in practical use from a perspective of efficiency drop which associates with the stabilizing effect and the cost-effectiveness in mechanical processing.

Recently, the mechanism through which circumferential casing groove enlarges the stable operating range in transonic compressor is becoming apparent through numerical and experimental investigations [5–7]. The interaction of passage shock and tip leakage vortex has often been pointed out as an important

Content from this work may be used under the terms of the Creative Commons Attribution 3.0 licence. Any further distribution of this work must maintain attribution to the author(s) and the title of the work, journal citation and DOI.
Published under licence by IOP Publishing Ltd

factor of stall inception in transonic compressors. The casing groove is considered to have an influence to this flow interaction [8]; thus contributes in delaying the onset of stall. However, this literature focus only on the multiple grooves, which cause a difficulty in assessing the effect of each groove and to evaluate their contribution to the stability improvement.

In general, the near casing flow field in compressor is highly complicated due to various effects of secondary flow [9, 10]. This is especially prominent in the case when shock wave is involved with the passage flow field, such as in transonic compressors. The flow structure changes drastically along the blade chord, and therefore, the circumferential grooves applied at the endwall will interact with diverse flow field according to their locations [11]. This indicates the necessity to investigate the effect of casing treatment individually corresponding to its geometric parameters in order to understand the stall suppression mechanism of casing treatment in detail. But up to now, there still do not seem to be a detailed research on centrifugal compressor.

The objective of the present study is to obtain fundamental knowledge on the effect of the circumferentially grooved casing treatment on near-casing flow and stall margin of centrifugal compressor. First, numerical analysis of centrifugal compressor was conducted on smooth wall condition with detailed examination of the flow field characteristics at near stall condition. Next, parametric studies were carried out with respect to the radial location, the radial width, the axial depth of the single casing groove. Finally, numerical cases with different numbers of casing grooves were carried out to compare the results. The effect of casing groove to the compressor performance and near-tip flow structure was evaluated for each case based on stall margin improvement.

2. Analysis Model

As previously explained, a centrifugal compressor rotor was adopted in the present work. The specifications of the impeller and diffuser are summarized in Table 1. The overall performance and flow passage data obtained by experiment were referred for the validation of numerical modeling, which will be presented below.

Table 1. Geometric parameters of impeller and diffuser

Impeller		Diffuser	
Blade inlet diameter (mm)	140	The first vane inlet diameter (mm)	251.7
Blade outlet diameter (mm)	240	The first vane outlet diameter (mm)	296
Tip clearance (mm)	0.21	The second vane inlet diameter (mm)	300
Blade outlet width (mm)	12	The second vane outlet diameter (mm)	380
Number of rotor blades	9/9	Number of stator vanes	17
Rotation speed (rpm)	30215	Vane outlet width (mm)	12

3. Numerical Method

3.1. *Numerical Scheme.*

Steady flow simulations were performed by solving compressible Navier-Stokes equations. The three-dimensional Reynolds-averaged Navier-Stokes equations were discretized in the computational domain by a cell-centered finite volume method and advanced the solution in time using explicit fourth-order Runger-Kutta method. The one-equation Spalart-Allmaras turbulence model was adopted to evaluate the eddy viscosity. In order to guarantee convergence, the artificial viscosity coefficient is added. The local time step, implicit residual smoothing, multi-grid method are used to improve the convergence speed.

The total temperature and total pressure was applied for the inlet boundary condition. As for the outlet boundary condition, a given mass flow was imposed, different point corresponding to different mass flow rate, initial assumptions and give pressure value at the same time. Blade and wheel set as rotating parts, rotating speed as the design speed 30,215 RPM; Wheel cover part set as stationary part.

In calculation of this paper, through the increase or decrease gradually to a certain flow approximation numerical boundary point (nearly choke point and near surge point), the final solution convergence before the numerical instability is defined as a numerical boundary point (choke point and surge point).

3.2. Computational Grid.

The calculation was done for the whole centrifugal compressor. The H&I topology structure is used in grid generation of impeller, proper encryption in the blade leading edge and trailing edge, the number of nodes is 1,223,952. Two series of vane diffuser using the default topology of NUMECA, the grid number is 202,223 and 198,695 respectively. The grids of circumferential groove is generated by ZR Effect function in Autogrid5. The grids of circumferential groove and blade channel are connected by full no-matching boundary (FNMB). The grid node of circumferential groove casing treatment in three directions is 41, 37, 37, respectively, the total number of nodes is 56,129. With a circumferential groove in the computational grid is shown in figure 1.

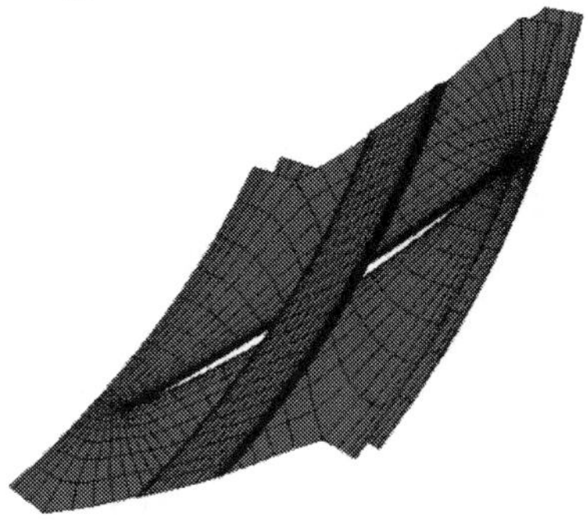

Figure 1. Computational grid with a circumferential groove

3.3. Validation.

The computational method was validated against the experimental datum. The overall characteristics are shown with the corresponding experimental data in figure 2. It can be seen from figure 2 that simulation gives a qualitatively good agreement with experimental results. In numerical simulations, the simple hypothesis for ideal gas, without considering the leakage loss, wheel resistance loss, constancy hypothesis do not tally with the actual situation and other factors, the efficiency and pressure ratio of the experiment are lower than the calculated result, but the overall trend is apparent, the computational method is able to predict the flow with sufficient reliability.

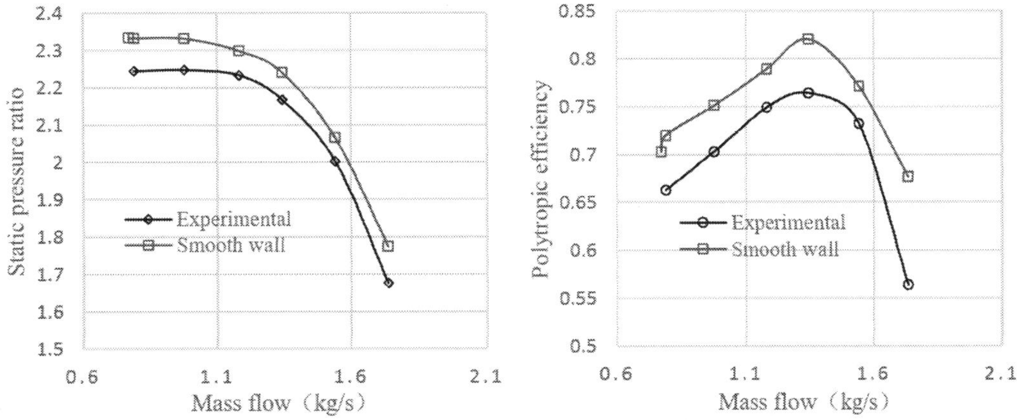

Figure 2. Characteristics plot of no-treated centrifugal compressor

4. Results and Discussion

4.1. Stall Characteristics and the Design of Circumferential Groove

4.1.1. Stall Characteristics in Diffuser. It is generally believed that the circumferential groove slot position depends on the magnitude of the location of the stall. Numerical simulation of the smooth wall found that in the near-stall condition, the deterioration of air flow in the impeller, lead to flow at a huge impact angle into diffuser leading edge, and made air separation and backflow on the no-working face of the first diffuser vanes. Figure 3 shows the Velocity streamlines at 95% span of diffuser in near-stall condition. Other than this, any notable vortex is not found among the passage. This implies that the flow which acts a dominant role in stall inception of the centrifugal compressor may range in the region near the casing.

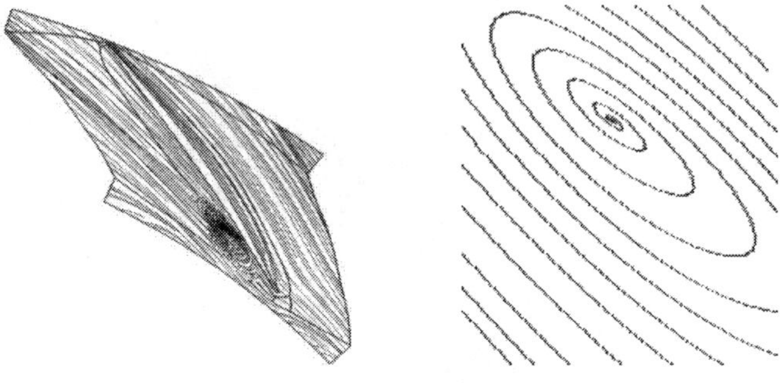

(a) Diffuser map (b) Partial enlargement of vortex

Figure 3. Velocity streamlines at 95% span of diffuser in near-stall condition

4.1.2. The design of Circumferential Groove. In this paper, the circumferential groove slot position designed in the first vane and the leading edge of the second vane of diffuser where stall occurred.

In order to determine the best slot position of the circumferential groove, a series of radial location of circumferential grooves were designed along the wheel side of the diffuser and numerical simulations were carried out. The circumferential groove with radial width of 6 mm and axial depth of 6 mm as initial. Figure 4 shows the slot location of circumferential grooves. For purpose of highlight the circumferential groove geometry characteristics, the circumferential groove naming rules would be designed as "axial depth _radial width_ the maximum absolute value of coordinates Y". Such as the number 2 circumferential groove in figure 4 is named as "6_6_139".

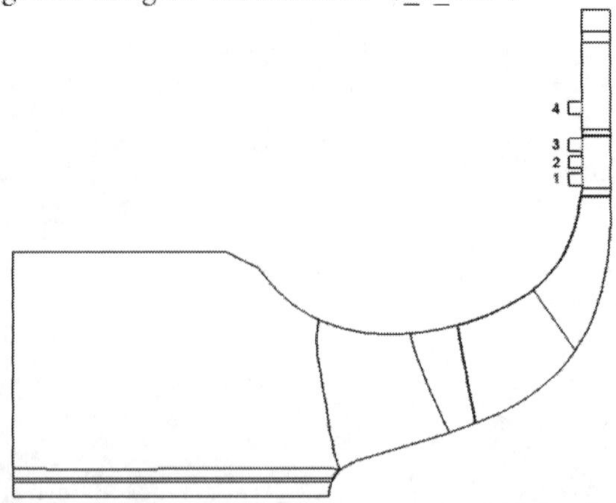

Figure 4. The slot location of circumferential grooves

For the sake of quantitative the improvement in stall margin and the influence of stage efficiency by circumferential groove, stable working range Φ, comprehensive improvement of stall margin ΔΦ, polytropic efficiency improvement at designed point $\Delta\eta_{pol}$ were evaluated in present work.

$$\Phi = \frac{Q_{m,des} - Q_{m,surge}}{Q_{m,des}} \times 100\%$$

$$\Delta\Phi = \frac{(Q_{m,surge})_{SW}}{(Q_{m,surge})_{CT}} \times \frac{(\varepsilon_{surge})_{CT}}{(\varepsilon_{surge})_{SW}} - 1 \qquad (1)$$

$$\Delta\eta_{pol} = \frac{(\eta_{pol,des})_{CT}}{(\eta_{pol,des})_{SW}} - 1$$

4.2. Effect of Applying Circumferential Groove

4.2.1. Different radial location of circumferential grooves. Table 2 lists the specific calculation results of different radial location of circumferential grooves as "6_6_131", "6_6_139", "6_6_147" and "6_6_159". Figure 5 shows characteristics plot of different radial location of circumferential grooves.

Compared with the results of the smooth wall, it can be found from Table 2 and figure 5 that the circumferential groove number 1 (6_6_131) reached the largest stable working range of 47.48% and the most comprehensive stall margin improvement of 9.10%, but also caused the efficiency of the compressor design points reduced by 0.28%. In the circumferential groove number 2 (6_6_139) had smaller efficiency to reduce by 0.23%, and stable working range is 47.11%, 8.37% stall margin improvement. In addition, it can be found from Table 2 that the circumferential grooves located in the leading edge of the first vane of diffuser can get a better expanding stability than the other circumferential grooves, it can be explained by the stall location which near the leading edge in diffuser.

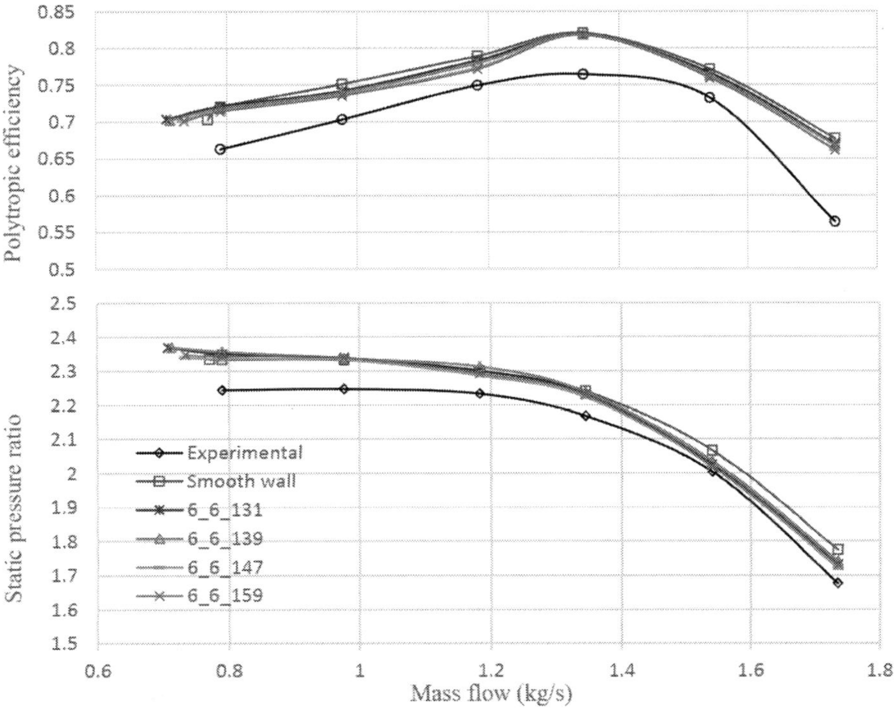

Figure 5. Characteristics plot of different radial location of circumferential grooves

Table 2. Results of different radial location of circumferential grooves

Circumferential groove	6_6_131	6_6_139	6_6_147	6_6_159	SW
Numbering scheme	1	2	3	4	-
$Q_{m,surge}$(kg/s)	0.708	0.713	0.733	0.735	0.771
ε_{surge}	2.3686	2.3694	2.3397	2.3465	2.3642
$\eta_{pol,des}$	0.8181	0.8185	0.8179	0.8177	0.8204
Φ	47.48%	47.11%	45.62%	45.47%	42.80%
$\Delta\Phi$	9.10%	8.37%	4.09%	4.11%	-
$\Delta\eta_{pol}$	-0.28%	-0.23%	-0.30%	-0.33%	-

4.2.2. Different radial width of circumferential grooves. Table 3 lists the specific calculation results of different radial width of circumferential grooves as "6_6_139", "6_10_139" and "6_14_139". Figure 6 shows characteristics plot of different radial width of circumferential grooves.

Compared with the results of the smooth wall, it can be found from Table 3 and figure 6 that the circumferential groove radial width increased from 6 mm to 10 mm, stable working range and comprehensive stall margin improvement enhanced with static pressure ratio fell and the efficiency decreases obviously; when the radial width increase from 10 mm to 14 mm, the static pressure continue to fall, stable working range and comprehensive stall margin improvement continue to increase, while the increasing amplitude is reduced, and the efficiency continue to decline.

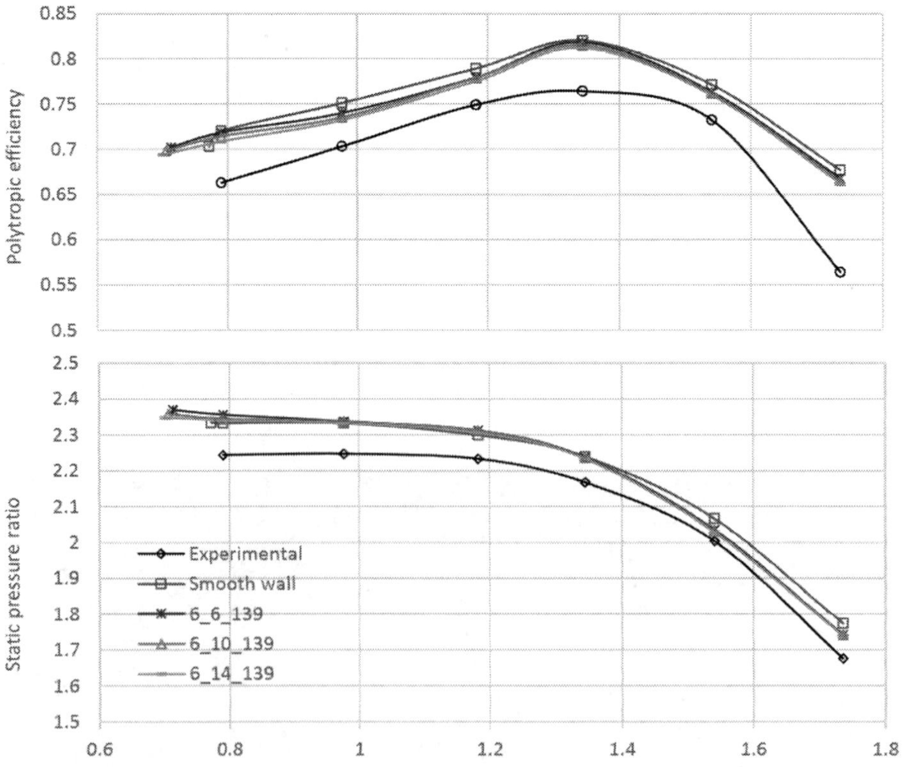

Figure 6. Characteristics plot of different radial width of circumferential grooves

Table 3. Results of different radial width of circumferential grooves

Circumferential groove	6_6_139	6_10_139	6_14_139	SW
$Q_{m,surge}$(kg/s)	0.713	0.706	0.701	0.771
ε_{surge}	2.3694	2.3583	2.3476	2.3642
$\eta_{pol,des}$	0.8185	0.8152	0.8123	0.8204
Φ	47.11%	47.63%	48.00%	42.80%
$\Delta\Phi$	8.37%	8.93%	9.21%	-
$\Delta\eta_{pol}$	-0.23%	-0.63%	-0.99%	-

4.2.3. Different axial depth of circumferential grooves. Table 4 lists the specific calculation results of different axial depth of circumferential grooves as "6_6_139", "10_6_139" and "14_6_139". Figure 7 shows characteristics plot of different axial depth of circumferential grooves.

Compared with the results of the smooth wall, it can be found from Table 4 and figure 7 that the circumferential groove axial depth increased from 6 mm to 10 mm, stable working range and comprehensive stall margin improvement enhanced with static pressure ratio fell and the efficiency decreases obviously; when the axial depth increase from 10 mm to 14 mm, stable working range and comprehensive stall margin improvement continue to increase, while the increasing amplitude is reduced, and the efficiency continue to decline.

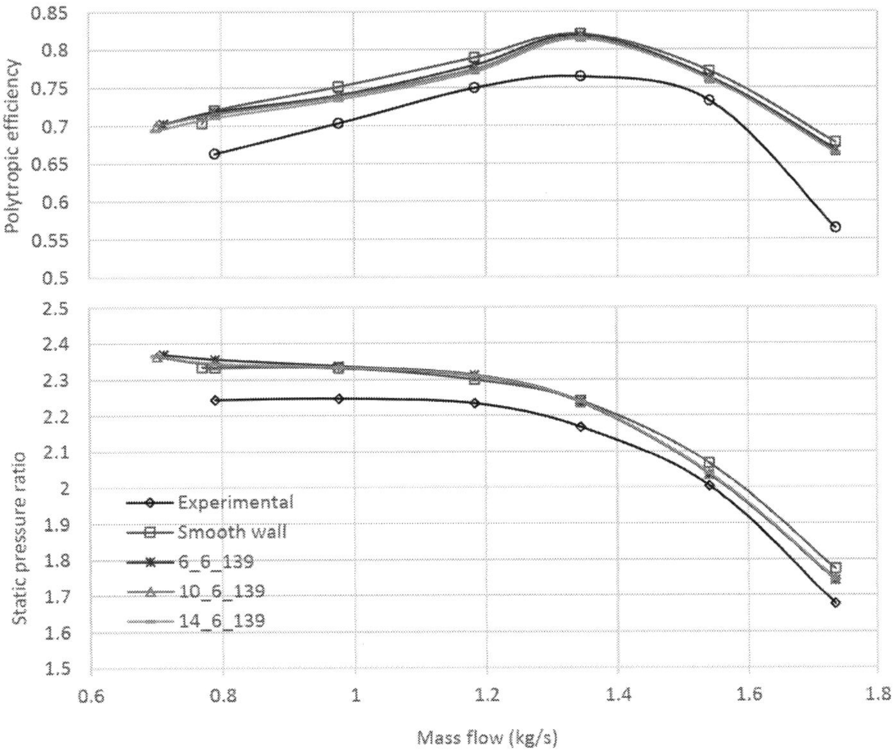

Figure 7. Characteristics plot of different axial depth of circumferential grooves

Table 4. Results of different axial depth of circumferential grooves

Circumferential groove	6_6_139	10_6_139	14_6_139	SW
$Q_{m,surge}$(kg/s)	0.713	0.703	0.701	0.771
ε_{surge}	2.3694	2.3647	2.3668	2.3642
$\eta_{pol,des}$	0.8185	0.8163	0.8136	0.8204
Φ	47.11%	47.85%	48.00%	42.80%
$\Delta\Phi$	8.37%	9.70%	10.11%	-
$\Delta\eta_{pol}$	-0.23%	-0.50%	-0.83%	-

4.2.4. Different numbers of circumferential grooves. Table 5 lists the specific calculation results of different numbers of circumferential grooves as "6_6_139_1" ("6_6_139"), "6_6_139_2" ("6_6_131" and "6_6_139") and "6_6_139_3" ("6_6_131", "6_6_139" and "6_6_147"). Figure 8 shows characteristics plot of different numbers of circumferential grooves.

Compared with the results of the smooth wall, it can be found from Table 5 and figure 8 that the circumferential groove number increased from 1 to 2, stable working range and comprehensive stall margin improvement enhanced with static pressure ratio fell and the efficiency decreases obviously; when the circumferential groove number increase from 2 to 3, stable working range and comprehensive stall margin improvement continue to increase, while the increasing amplitude is reduced, and the static pressure continue to decline, efficiency decreases more.

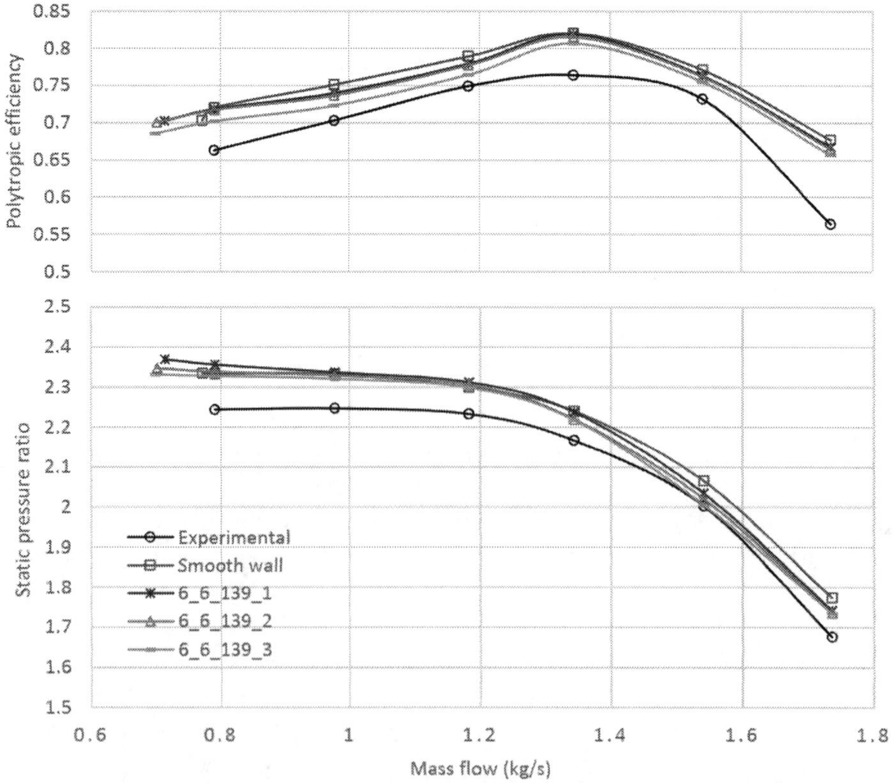

Figure 8. Characteristics plot of different numbers of circumferential grooves

Table 5. Results of different numbers of circumferential grooves

Circumferential groove	6_6_139_1	6_6_139_2	6_6_139_3	SW
$Q_{m,surge}$(kg/s)	0.713	0.701	0.699	0.771
ε_{surge}	2.3694	2.3468	2.3312	2.3642
$\eta_{pol,des}$	0.8185	0.8148	0.8063	0.8204
Φ	47.11%	48.00%	48.15%	42.80%
$\Delta\Phi$	8.37%	9.18%	8.76%	-
$\Delta\eta_{pol}$	-0.23%	-0.68%	-1.72%	-

5. Concluding Remarks

Steady three-dimensional Navier-Stokes flow simulations were conducted to investigate the effect of circumferential grooved casing treatment in vane diffuser of a centrifugal compressor. The conclusions are summarized as follows:

(1) In the centrifugal compressor, a tip leakage vortex was seen in near-stall condition. Numerical simulation of the smooth wall found that in the near-stall condition, the deterioration of air flow in the impeller, lead to flow at a huge impact angle into diffuser leading edge, and made flow separation and backflow on the suction face of the first diffuser vanes.

(2) When the groove is applied at the endwall of the diffuser, suitable position of circumferential groove can effectively expand the scope of the stable operation of the compressor, and efficiency has fallen a little at the design point.

(3) The width and depth of circumferential groove was increased, the stable working range and comprehensive stall margin improvements increase too, but the efficiency is decreased; the width and depth increasingly, stable working range and comprehensive stall margin improvements continue to increase, while the increasing amplitude is reduced, and efficiency is decreased more.

(4) The number of circumferential groove was increased, stable working range and comprehensive stall margin improvement enhanced with static pressure ratio fell and the efficiency decreases obviously; when the number of circumferential groove increasingly, stable working range and comprehensive stall margin improvement continue to increase, while the increasing amplitude is reduced, and the static pressure continue to decline, efficiency decreases more.

(5) The evaluation based on stall margin improvement showed the optimal position for the groove to be located was indicated to exist near the leading edge of the diffuser, and a combination of position, width, depth and numbers of circumferential grooves that will maximize both surge margin range and efficiency.

Nomenclature

Q = quantity of flow
ε = static pressure ratio
η = efficiency
Φ = stable working range
$\Delta\ \Phi$ = comprehensive improvement of stall margin
$\Delta\ \eta$ = efficiency improvement

Subscripts

m = mass
surge = surge point condition
pol = polytropic
des = design point

SW = smooth wall case
CT = casing treatment case

References

[1] Osborn W M, Lewis G W and Heidelberg L J 1971 Paper No. NASA TN D-6537
[2] Danner F C T, Kau H P, Muller M M, Schiffer H P and Brignole G A 2009 *ASME* Paper NO. GT2009-59647
[3] Lin F, Ning F and Liu H 2008 *ASME* Paper NO. GT2008-51541
[4] Houghton T and Day I 2009 *ASME* Paper NO. GT2009-59210
[5] Muller M W, Schiffer H P, Biela C and Hah C 2008 *ASME* Paper NO. GT2008-50135
[6] Chen H, Huang X, Shi K, Fu S, Bennington M A, Morris S C, Ross M, McNulty S and Wadia A 2010 *ASME* Paper NO. GT2010-23606
[7] Iyengar V and Lakshmi S 2005 *AIAA* 2005-0632
[8] Flisher F B 1989 *SAE transactions* **97** 1303-1310.
[9] Hunziker R, Dickmann H P, Emmrich R 2001 *Journal of Power and Energy* **215** 783-791.
[10] Xu W, Wang T and Gu CG 2011 *Science China Technological Sciences* **54(9)** 2483-2492.
[11] Zhu J and Chu W 2005 *AIAA* **809** 10-13.

IOP Conf. Series: Materials Science and Engineering **90** (2015) 012044 doi:10.1088/1757-899X/90/1/012044

GAS DYNAMIC DESIGN OF THE PIPE LINE COMPRESSOR WITH 90% EFFICIENCY. MODEL TEST APPROVAL

Y. GALERKIN, A. REKSTIN, K. SOLDATOVA
R&D Laboratory "Gas dynamics of turbo machines" Peter the Great St.Petersburg Polytechnic University, Polytechnical st. 29, St.Petersburg, Russia
Email:buck02@list.ru.

Abstract: Gas dynamic design of the pipe line compressor 32 MW was made for PAO SMPO (Sumy, Ukraine). The technical specification requires compressor efficiency of 90%. The customer offered favorable scheme - single-stage design with console impeller and axial inlet. The authors used the standard optimization methodology of 2D impellers. The original methodology of internal scroll profiling was used to minimize efficiency losses. Radically improved 5th version of the Universal modeling method computer programs was used for precise calculation of expected performances. The customer fulfilled model tests in a 1:2 scale. Tests confirmed the calculated parameters at the design point (maximum efficiency of 90%) and in the whole range of flow rates. As far as the authors know none of compressors have achieved such efficiency. The principles and methods of gas-dynamic design are presented below. The data of the 32 MW compressor presented by the customer in their report at the 16th International Compressor conference (September 2014, Saint-Petersburg) and later transferred to the authors.

Nomenclature

A_D - ratio of an impeller inlet diameter and diameter that corresponds to minimum of inlet velocity;

c_2 - absolute velocity at an impeller exit;

c_3 - flow velocity at vaneless diffuser exit;

c_m - meridian velocity;

c_{m1} - meridian velocity at a blade inlet;

D_1 - blade inlet diameter;

D_2 - impeller diameter;

D_3 - vaneless diffuser exit diameter;

D_{shaft} - shaft diameter of an impeller;

H_T - Euler work of a compressor;

h_T - Euler work of an impeller;

K_F - ratio of area at blade inlet and an impeller inlet;

K_n - specific speed;

$\overline{l}$ - meridian relative length of an impeller blade channel;

$\overline{m}$ - mass flow rate;

M_u - impeller Mach number;

n - RPM;

Content from this work may be used under the terms of the Creative Commons Attribution 3.0 licence. Any further distribution of this work must maintain attribution to the author(s) and the title of the work, journal citation and DOI.
Published under licence by IOP Publishing Ltd

$N_{i\,stage}$ - shaft power of a stage;
$\overline{V}_0$ - volumetric flow rate at a stage inlet;
u_2 - blade velocity;
w_1 - relative velocity at an impeller inlet;
z - number of stages;
α_2 - flow angle at an impeller exit;
Φ - flow rate coefficient;
Φ_{des} - design flow rate coefficient;
φ_0 - flow rate coefficient at an impeller inlet;
φ_1 - flow rate coefficient at a blade inlet;
φ_2 - flow rate coefficient at an impeller exit;
η_p - polytrophic efficiency;
$\Delta\eta_{imp}$ - loss of efficiency in an impeller;
$\Delta\eta_{stator}$ - loss of efficiency in a stator part of a stage;
ρ_0 - flow density at an impeller inlet;
ζ_{imp} - loss coefficient of an impeller;
ζ_{stator} - loss coefficient of a stator part of a stage;
ψ_p - polytrophic head coefficient;
ψ_T - loading factor;
$\psi_{T\,des}$ - design loading parameter.

1. OPTIMAL NON-DIMENSIONAL PARAMETERS

Pipe line centrifugal compressors consume huge amounts of energy. The highest possible efficiency of these machines is important economically and ecologically. A proper gas dynamic design is the first necessary step to the goal. There are several considerations that must be taken into account.

The general rule is that the specific speed of compressor stages must correspond to the optimum flow path dimensions. A specific speed of a compressor is [1]:

$$K_n = 3{,}545\frac{\overline{V}_0^{0,5}}{H_T^{0,75}}\frac{n}{60} \tag{1}$$

In most pipe line compressors all impellers are of the same diameter and design loading parameter. In this case the specific speed of a stage is:

$$K_n = 3{,}545\frac{\overline{V}_0^{0,5}}{\left(H_T / z\right)^{0,75}}\frac{n}{60} = \frac{\Phi^{0,5}}{\psi_T^{0,75}} \tag{2}$$

The most important design parameter of a stage is a flow rate coefficient at a design flow rate Φ_{des}. The bigger the coefficient, the smaller the radial dimensions of a compressor are. The maximum efficiency of stages with 2D impellers is in a range of Φ_{des} = 0.06 – 0.08. If Φ_{des} < 0.06, the surface friction losses in a flow path channels and losses in labyrinth seal and disc friction losses prevail. If Φ_{des} > 0,08 the high level of a flow kinetic energy at the impeller inlet diminishes efficiency. The eq. (2) demonstrates non dimensional kinetic energy $0,5(w_1/u_2)^2$ influence on the efficiency loss in an impeller:

$$\Delta\eta_{imp} = \frac{\zeta_{imp}}{\psi_T}\frac{(w_1/u_2)^2}{2} \tag{3}$$

The efficiency of stages with 3D impellers can be high at Φ_{des} = 0.11 – 0.12 due to better organization of a flow in wide blade channels.

The second important design coefficient $\psi_{T\,des}$ is present at the eq. (3) too. Formally the bigger ψ_T, the higher the efficiency. Its factual influence is quite opposite. The impeller loss coefficient ζ_{imp} is much greater for impellers with high design loading parameters. Losses in stator parts are bigger too, as the bigger is ψ_T the bigger is non dimensional kinetic energy after an impeller:

$$\Delta\eta_{stator} = \frac{\zeta_{stator}}{\psi_T}\frac{(c_2/u_2)^2}{2} = 0.5\zeta_{stator}\psi_T\left(1+tg^2\alpha_2\right) \tag{4}$$

The authors design practice demonstrates highest efficiency of stages with low values of $\psi_{T\,des}$ = 0.4 – 0.5.

It is important for GT-driven compressors to have maximum of consumed power N_i at a design flow rate. The power of a stage is proportional to two design coefficients product:

$$N_{i\,stage} \approx \bar{m}\cdot h_T = \rho_0 0.785 D_2^2 u_2^3 \cdot \Phi \cdot \psi_T \tag{5}$$

The maximum of $N_{i\,stage}$ at a design flow rate takes place if $\psi_{T\,des} \approx 0.5$.

One more advantage of low Euler coefficients is steeper loading parameter of performance $\psi_T = f(\Phi)$. The steeper the performance is, the smaller the flow rate coefficient is with highest delivery pressure that is a surge limit.

There are two most important constructive limitations that influence stage efficiency. The bigger is a hub ratio D_{shaft}/D_2 the bigger is an inlet blade diameter ratio D_1/D_2. It increases flow kinetic energy at an impeller inlet and diminishes efficiency of an impeller – eq. (3):

$$w_1/u_2 = \sqrt{(D_1/D_2)^2 + c_{m1}^2} \tag{6}$$

There is no problem to achieve necessary flow deceleration in a vane diffuser with limited radial dimension. Flow deceleration in vaneless diffuser is opposite to its radial length:

$$c_3/c_2 \approx 1/(D_3/D_2) \tag{7}$$

To obtain good efficiency the ratio D_3/D_2 must be not less than 1.60 and better about 1.70.

2. ADVANTAGES OF THE FLOW PATH SCHEME PROPOSED BY THE CUSTOMER. DESIGN MAIN FEATURES

The pressure ratio of the designed compressor is comparatively low and RPM of GT drive is high. It was possible to realize a single stage solution for this compressor. The necessary compressibility criteria $M_u \approx 0.7$ is not too high.

Advantages of the single stage design:

- console disposition of the impeller with zero hub ratio make possible to diminish flow kinetic energy at the impeller inlet,
- an axial inlet nozzle with minimal head losses can be applied.

The vaneless diffuser radial length of the compressor is close to optimum. The exit internal scroll was designed by especially formulated method that appeared to be rather effective.

Specific speed of the compressor – eq. (1), (2) - is quite good to apply traditional 2D impeller with $\Phi_{des} \approx 0.060$ and $\psi_{T\,des} \approx 0.52$. The new 5th version of the Universal modeling method computer program was applied to optimize the flow path and to calculate the expected performances. Basic information about the Method was presented at the conferences earlier [2, 3, 4, 5]. The new 5th version with better modeling abilities was presented at [6, 7, 8].

The designed impeller has no hub at all. The impeller inlet diameter is much smaller than ones of the tested analogs. Some geometry and flow parameters of the closest analog type 055 (upper line) and of designed impeller variants are presented in Table 1.

Table 1

Geometry and flow parameters of the closest analogs type 055 and of the designed impeller variants

Variants	K_F	A_D	D_1/D_2	φ_0	φ_1'	w_2'/w_1'	φ_2'	α_2'
055	0.825	1	0.592	0.300	0.285	0.729	0.250	24.9
1	0.80	1	0.435	0.318	0.302	0.941	0.153	15.0
2	0.825	1	0.445	0.316	0.309	0.921	0.153	15.0
3	0.85	1	0.444	0.306	0.308	0.924	0.153	15.0
4	0.90	1	0.452	0.295	0.313	0.908	0.153	15.0
5	0.95	1	0.460	0.285	0.318	0.892	0.153	15.0
6	0.825	0.975	0.436	0.328	0.323	0.920	0.153	15.0
7	0.825	1.025	0.458	0.299	0.296	0.915	0.153	15.0
8	0.825	1.05	0.469	0.284	0.282	0.911	0.153	15.0

The variants have different inlet dimensions. The other dimensions of the stage flow path - impeller and stator - were studied by the 5th generation program too. The final part of the impeller optimization is a qualitative one. The non viscid Q3D velocity diagrams were assessed by experts. The samples of the diagrams for the design flow rate are presented in Figure 1, 2.

The diagrams demonstrate several important features of the impeller design:

- non-incidence at inlet takes place along the leading edges height in spite of simple 2D blade configuration,
- flow deceleration along suction side is little. There is a hope that wake formation will not take place,
- velocity diagram area gives information about expected value of the Euler work coefficient.

3. MODEL TEST PROOF

The Universal modeling method has demonstrated reliability and the customers are realizing designs without any tests for two decades already. Because of new design features the customer of this design has decided to check the project by tests of the model at the 1:2 scale. Figure 3 demonstrates the model and the part of the test rig cross section.

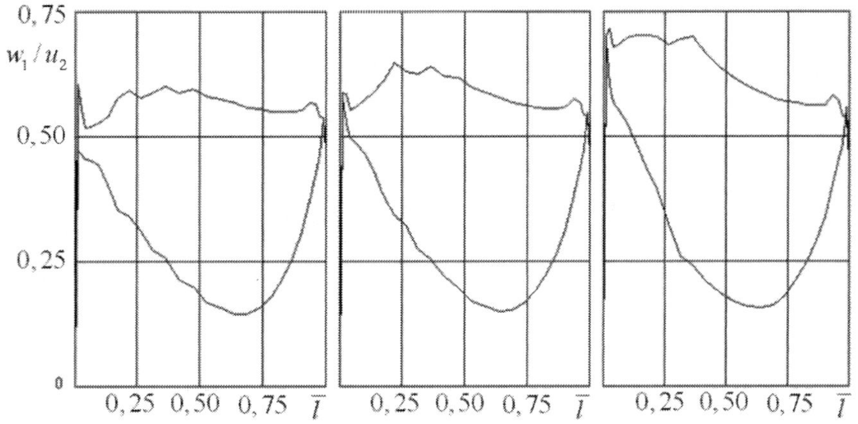

Figure 1. Velocity diagrams at three blade-to-blade surfaces (design flow rate). Left – hub, middle mean, right – shroud

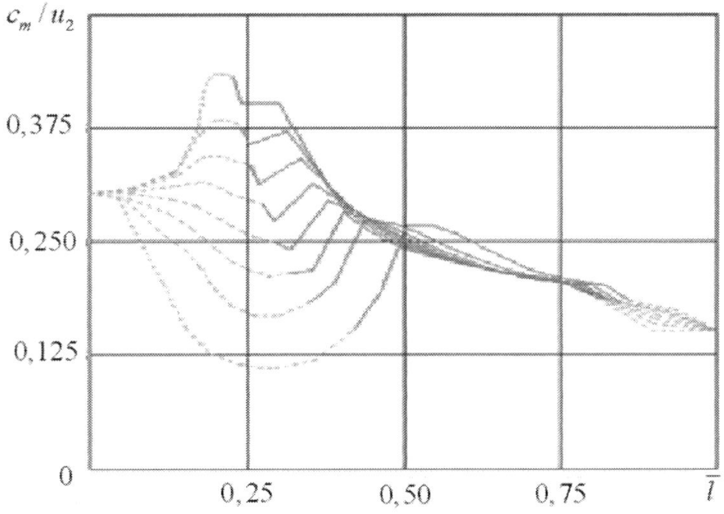

Figure 2. Meridian velocities along 7 blade to blade surfaces (design flow rate)

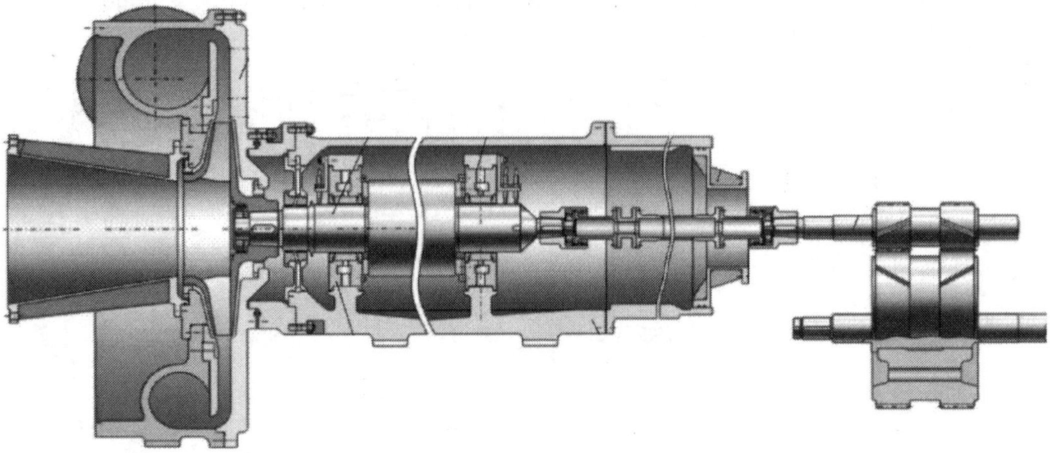

Figure 3. The model of the 32 MWt pipe line compressor at the test rig. Scale 1:2

The results of three tests in consequence are presented graphically in Figure 4.

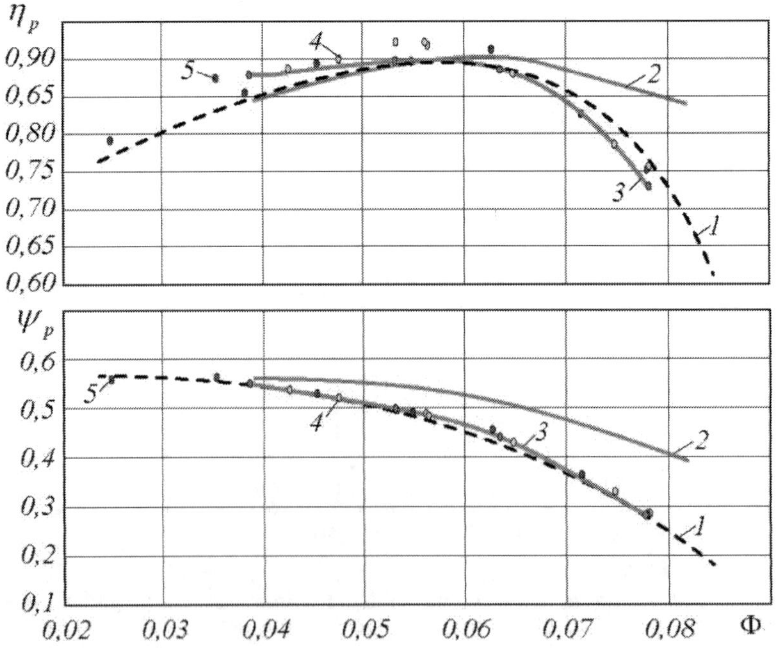

Figure 4. Compressor 32 MWt model test results
1- design performance, 2- ANSYS CFX, 3- $M_u = 0.705$, 4- $M_u = 0.710$, 4- $M_u = 0.700$

4. CONCLUSION

The result of ANSYS CFX modeling by the customer is presented in Fig. 4. The performance curves shift right from factual but the maximum efficiency is predicted well. The authors experience with different CFD programs and different impellers demonstrate the same results usually.

Design and test data comparison demonstrate two remarkable results. The polytrophic coefficient calculated performance is in very good conformity with all three model tests. The previous 4th version is very effective to predict design flow rate efficiency and head but not all performance. Some non-conformity at flow rate left and right of the design point are usual for that model. Fig. 4 demonstrates better level of modeling by the new 5th version. The second important fact is that the optimistic design assessment of the 90% efficiency is proven with high probability.

ACKNOWLEDGEMENT

Work is performed with support of a Grant of the President of the Russian Federation for young PhD MK-7066.2015.8.

References

[1] Galerkin, Y.B. Tubrocompressors. [text] /Y.B.Galerkin // SPbSTU. – SPb. – 2010. – P.650. [Russian].

[2] Galerkin, Y.B., Soldatova, K. V. Industrial centrifugal compressor model stages based on compressor tests results. [text] /Y.B.Galerkin, K.V. Soldatova// International symposium SYMKOM 2011. – Lodz. - Poland.

[3] Galerkin, Y.B., Danilov, K.A., Popova, E.Y.. Design philosophy for industrial centrifugal compressor. [text] /Y.B.Galerkin, K.A. Danilov, E.Y. Popova// Inernational Conference on Compressors and their Systems. – London: City University, UK. – 1999.

[4] Galerkin, Y.B., Soldatova, K.V., Drozdov, A.A. Specification of algorithm of calculation of parameters of a stream in the centrifugal compressor stage. [text] /Y.B.Galerkin, K.V. Soldatova, A. A. Drozdov // Scientific and technical transactions of the TU SPb. – 2010. – No. 4. – Page – 150-157. [Russian].

[5] Galerkin Y., Soldatova K.,Drozdov A. Modern state of the universal modeling for centrifugal compressors. [text] /Y.B.Galerkin, K.V. Soldatova, A. A. Drozdov// International Conference on Numerical Methods in Industrial Processes. World Academy of science, engineering and technology. Paris 2015 Conference. http://www.waset.org/Publications/?path=Publications. - Vol:9 No:01 2015. – № 242.

[6] Galerkin Y., Soldatova K., Drozdov A. New version of the universal modelling for centrifugal compressor gas dynamic design. [text] /Y.B.Galerkin, K.V. Soldatova, A. A. Drozdov // Purdue Conference 2014. USA.

[7] Galerkin Y., Soldatova K. The application of the Universal Modeling Method to development of centrifugal compressor model stages. [text] /Y.B.Galerkin, K.V. Soldatova// Conference "Compressors and their Systems". — London. – 2013. – P. 477-487.

[8] Soldatova, K.V. Universal modeling method application for development centrifugal compressor model stages. [text] / K.V.Soldatova //Conference «Compressors and their Systems».– London. – 2013. – P. 477-487.

Analysis of the gas-dynamic performance of a vaned diffuser with given velocity distribution along the vane's surfaces

M Kalinkevych[1], O Obukhov[2], O Obukhova[1], A Miroshnychenko[1]

1) Sumy State University (Sumy, Ukraine)
2) PJSC "Frunze NPO" (Sumy, Ukraine)

E-mail: obukhov_alexey@mail.ru

Abstract. Extension of the effective range of vaned diffusers is one of the promising ways to improve the centrifugal compressor's stages which are used in numerous fields of industry. The new method of profiling of the diffuser vanes has been developed using Stratford's results and boundary layer theory by Loytsanskiy. The developed method is based on the solution of the inverse task of gas-dynamic using given velocity distribution along the vane's surface. Comparison of the results of numerical simulations for different diffusers has shown that the performance of the diffuser designed with the resulting velocity distribution are better. Influence of the vane profile, number of the vanes, diffuser outlet diameter and the diffuser width on diffuser characteristics has been investigated. The results of the simulations have been used to formulate recommendations on the design of high-effectiveness vaned diffusers for centrifugal stages of different types.

NOMENCLATURE

$\bar{c}$	absolute velocity, m/s;
α	flow angle, °;
R_v	radius of mean camberline vane, m;
δ^{**}	momentum thickness;
τ	tangential stress
U_δ	velocity of the gas in a potential flow;
H, f	form-parameters;
ν	kinematical viscosity, m^2/s;
$\lambda = c/a$	coefficient of velocity;
a	critical sonic speed, m/s;
$\bar{l} = l/L$	relative length of the vane;
r	radius, m;
b	height of the vane, m;
z_v	number of vanes in diffuser;
D_2	tip diameter of the impeller;
δ	vane thickness, m;
ρ	density of the gas stream, , kg/m^3;
p	pressure, Pa;
T	temperature, K
R	gas constant, J/(kg·K);
$\pi(\lambda)$, $\varepsilon(\lambda)$	gas-dynamic function;
ζ	loss factor;
C_p	static pressure recovery coefficient;
F_0	flow coefficient;
$\overline{m}$	mass flow rate, kg/s;
V	volume flow rate;
U_2	impeller tip speed;
M	Mach number;
Subscripts	
in	diffuser inlet;
out	diffuser outlet;
ps	pressure surface of the vane;
ss	suction surface of the vane;
Superscript	
*	parameters of adiabatically stagnated flow.

Content from this work may be used under the terms of the Creative Commons Attribution 3.0 licence. Any further distribution of this work must maintain attribution to the author(s) and the title of the work, journal citation and DOI.
Published under licence by IOP Publishing Ltd

IOP Conf. Series: Materials Science and Engineering **90** (2015) 012045 doi:10.1088/1757-899X/90/1/012045

1. Introduction

Centrifugal compressors are widely used in numerous fields of industry including transportation of natural gas, petroleum and chemical industries. Power consumed by centrifugal compressors may reach *32MW*. Effectiveness of centrifugal compressors depends on aerodynamic flow quality of the elements of the stages.

The power of the compressor's drive is consumed to compress and to move the working gas and part of energy is consumed to overcome the aerodynamic resistance to the movement of gas. Losses of mechanical energy in centrifugal compressor are approximately uniformly shared between the impeller, vaned diffuser and return channel [1]. Therefore, the increase of efficiency of elements of the flow path of the centrifugal compressors is an essential task in engineering.

2. Vaned diffuser design procedure

Vaned diffusers with circular centrelines and symmetric vane's profile C-4 are widely used in centrifugal compressors [2]. These diffusers have simple geometry and are usually designed ignoring parameters of the compressed gas, real kinematics of the flow at the vane's inlet, vane-spacing ratio and geometrical parameters of the flow path between the vanes. As a result vaned diffusers with such geometry are characterized by flow separations from the pressure surface of the vanes at almost all conditions of operation. All these facts lead to narrow effective range of the stages with such diffusers (Figure 1).

Flow separation from the front surface of the vanes at all conditions of operation leads to an increase of the loss coefficient of vaned diffusers and decrease of the static pressure recovery coefficient.

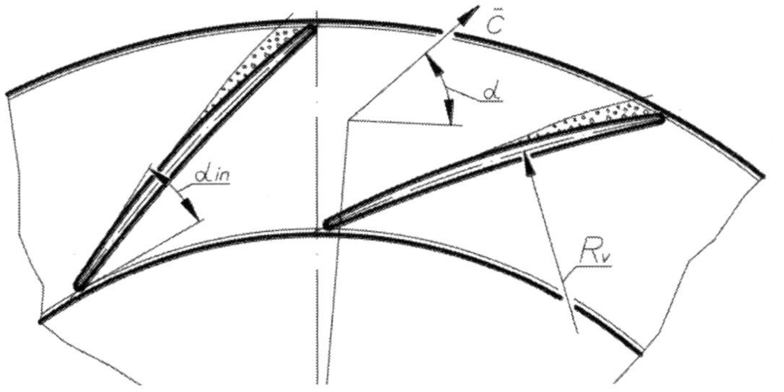

Figure 1 - flow separation from the pressure surface of the circular-arc mean camber line vane [3].

Based on the analysis of drawbacks of well-known design methods of vaned diffuser offered a method of profiling vanes of the diffuser with a given velocity distribution, based on Stratford's idea of profiling a flat diffuser with unseparated gas flow [4]. Unseparated gas flow provided by profiling the shape of the diffuser from the condition of zero tangential stresses on the walls of the diffuser. The boundary layer structure of gas flow on surfaces of the diffuser vanes was taken into account in the model suggested by Loitsianskiy [5]. Based on the above stated, the method of solving the inverse task of gas-dynamics for profiling high-performance vane diffuser was suggested.

Law of speed change of gas flow on the front surface of the vane is determined from the Karman's integral impulses ratio:

$$\frac{\delta^{**}}{dx} = \frac{1}{U_\delta}\frac{U_\delta}{dx}(2+H)\delta^{**} + \frac{\tau}{\rho U_\delta^2}, \tag{1}$$

According to the recommendations [5], the condition of unseparated flow to turbulent flat flow is as following:

$$-\frac{\delta^{**}}{U_\delta}\frac{dU_\delta}{dx}G(Re^{**}) \le f_s, \tag{2}$$

where f_s is the critical value of the form-parameter, $f_s = -0.02$; $G(Re^{**}) = (Re^{**})^{\frac{1}{6}}$.

Reynolds number for the momentum thickness is determined from the expression:

$$Re^{**} = \frac{\lambda \cdot a \cdot \delta^{**}}{\nu}, \tag{3}$$

By integrating the equation (1) we obtain:

$$\frac{d}{dx}\left(f\frac{U_\delta}{U_\delta^{`}}\right) = (1+m)\zeta - [2+m+(1+m)H]f\,, \quad (4)$$

where $f = -\frac{\delta^{**}U_\delta^{`}}{U_\delta}G(Re^{**})$; $m = \frac{Re^{**}G^{`}(Re^{**})}{G(Re^{**})}$; $\zeta = \frac{\tau}{\rho U_\delta^2}$.

The boundary layer will have the preseparation nature in the case when the form parameter f is equal to its critical value f_s, while $\zeta = 0$, $m = 0$ and $H = H_s$, where H_s is the critical value of the form parameter.

According to [5], the amount of turbulent and viscous friction depends solely on the form parameter, then for $f = -0.02$, equation (4) can be integrated in quadrature and written in dimensionless form [6]:

$$\lambda_{ps_i} = \lambda_{in}\cdot\left[1+\frac{(\bar{l}_i-\bar{l}_1)\cdot(2+H_s)\cdot(-f_s)}{\delta_{in}^{**}}\right]^{-\frac{1}{2+H_s}}, \quad (5)$$

In order to determine the shape of the vane profile with given by velocity distribution on its surface the equations system has been considered. It consisted of the equation of the moment of impulse, mass flow rate and state.

Momentum from the impulse equation for current streams is as following:

$$r_{in}\cdot\lambda_{in}\cdot cos(\alpha_{in}) - r_i\cdot\lambda_i\cdot cos(\alpha_i) = \frac{b\cdot z_v}{\bar{m}\cdot a}\cdot\int_{r_{in}}^{r_i}\Delta P\cdot r\cdot dr, \quad (6)$$

Mass flow rate equation using gas-dynamic functions is as following:

$$\bar{m} = \lambda\cdot\varepsilon(\lambda)\cdot a\cdot\rho_{in}^{*}\cdot\sigma\cdot b\cdot(2\cdot\pi\cdot r-\acute{\delta}\cdot z_v), \quad (7)$$

where ρ^* is density of the adiabatically stagnated flow; σ is function of total pressure loss; $\acute{\delta}$ is thickness of the vane ($\acute{\delta} = \delta_v + \delta^*$); $\delta^*$is displacement thickness.

As the equation of state the ideal gas law was used:

$$\frac{P}{\rho} = R\cdot T, \quad (8)$$

where R is gas constant of the working fluid; T is the temperature of the gas flow.

Nature of change in pressure between the surfaces of the vanes were accepted linear [7]:

$$\Delta P = P_{ps} - P_{ss} = P_{in}^{*}\cdot[\pi(\lambda_{ps}) - \pi(\lambda_{ss})], \quad (9)$$

The system of equations (6), (7) and (8) is produced by successive approximations. Definite integral in equation (6) is solved numerically by the central rectangles.

According to [5] region of the boundary layer is replaced by thickness repression, in which stands out thickness of the momentum of impulse losses.

Displacement thickness is determined from the expression:

$$\delta^* = H_s\cdot\delta^{**}. \quad (10)$$

The relative thickness of the loss of momentum of impulse is determined by the expression:

$$\overline{\delta^{**}} = 0.0159\cdot Re^{-1,15}\cdot\lambda^{-3,55}\cdot\left(\int_0^l\lambda^4\ l\right)^{0,85}, \quad (11)$$

Solving the inverse task of gas-dynamics is the distribution of the mean flow rate angle of flow in the space between vanes of diffuser that forms by the middle line of the vane.

3. Criterions of efficiency of the vaned diffuser

The main purpose of vaned diffuser as part of a centrifugal compressor stage is to convert the kinetic energy of the flow emerging from the impeller into static pressure with minimal losses in all conditions of operation. Criterions of efficiency for vaned diffuser are: loss coefficient and coefficient of recovery of static pressure depending on the nominal flowrate [8].

Loss coefficient is defined from the relation:

$$\zeta = \frac{P_3^*-P_4^*}{P_3^*-P_3}, \quad (12)$$

where P_3^* and P_4^* is the adiabatically stagnated pressure before and after the vane system of diffuser; P_3 is static pressure of gas flow before the vaned diffuser.

The coefficient of static pressure recovery is defined from the relation:

$$c_p = \frac{P_4-P_3}{P_3^*-P_3}, \quad (13)$$

where P_4 is static pressure of gas flow after vanes of the diffuser.

Flow coefficient, which characterizes the operation conditions of centrifugal compressor stage, is defined from the relation:

$$\mathrm{F_o} = \frac{4 \cdot V_0}{\pi \cdot D_2^2 \cdot U_2}. \tag{14}$$

4. Numerical modelling of gas flow in the vaned diffuser

To solve the direct task of gas-dynamics the software package ANSYS CFX was used.

Verification of numerical simulation technique was conducted by comparing the integrated gas-dynamic characteristics of centrifugal compressor stages derived from numerical and experimental investigation. The experiment was carried out at a test rig AD-400 of the PJSC "Frunze NPO" (Sumy, Ukraine). The value of impeller tip speed Mach number for experimental investigation was of 0.6. On the nominal operation mode (mass flow rate of 2 kg / s) stage provides a pressure ratio of 1.3. The results of comparison of gas dynamic characteristics of centrifugal compressor stages, obtained in numerical and experimental way, presented in [9].

Settings for modelling of process of gas flow in the flow path of the centrifugal compressor stages were cyclically symmetric and stationary.

At the inlet to the computational domain as the boundary conditions have been set parameters of adiabatically stagnated flow before of the impeller; at the outlet - proportion of the mass flow passing through the segment of the flow path.

The flow path of the centrifugal compressor stage has been replaced by a block-structured computational grid [10]. Computational grid was meshed near the walls for the application of SST turbulence model, which is able to predict the locations of occurrence of flow separation the most accurately. The mean value of y^+ on all test conditions of operation was not exceeding 2.

The choice of integration step was defined from the relation:

$$q = \frac{30}{\pi \cdot n}, \tag{15}$$

where n is impeller rotating speed, rev/min.

Fluctuations in the values of polytropic efficiency and coefficient of polytropic head were adopted as criteria for convergence of the numerical investigation. Their value for the centrifugal compressor stage is within the error of the model experiment. Processing of results of numerical simulation was performed in ANSYS CFX Post processor. It included determination of the parameters that are required to design the vaned diffuser and calculation of its criterions of efficiency.

Numerical simulation of gas flow passed in two stages. The purpose of the first stage of simulation was to determine the initial data for the design of the vaned diffuser. For this purpose two-element (impeller (D_2=460 mm, $\overline{b_2} = 0.0319$, $\beta_{v_1} = 30°$, $\beta_{v_2} = 104°$) and vaneless diffuser (D_4=759 mm, $b_3 = b_4 = 15.5\ mm$) centrifugal compressor stage has been calculated. Flow simulation in vaneless diffuser was due to the necessity of obtaining parameters of the gas flow behind the impeller without inverse influence of the diffuser vane system [2]. In the inlet of the vaneless diffuser were defined the following flow parameters at all study conditions of the centrifugal compressor stage: the absolute velocity of the air flow, the value of the static pressure and temperature, adiabatic stagnated pressure and temperature and density. These parameters of the vaneless diffuser are required for design of the vaned diffuser. At the second stage two-element (impeller and vaned diffuser) centrifugal compressor stage has been calculated. Geometric parameters are identical to the impeller parameters used in the first stage of the simulation. The second stage involved the determination of gas-dynamic characteristics of the investigated vaned diffusers.

5. Results of numerical modelling

For all stage models simulation of the air flow was carried out on at least 6 operating conditions, which differed in the value of mass flow rate. The values of Reynolds number and Mach number at the inlet of the vane system of the diffuser are in the ranges of $Re=(2.3\text{-}3.2)\cdot 10^5$ and $M=(0.28\text{-}0.51)$, respectively. Impeller tip speed Mach number for all numerical studies was $Mu_2=0.6$. The vane-spacing ratio, in the study of influence of the designing method the vaned diffuser, corresponded to the value $l/t = 1.5$. Symmetrical airfoil C-4 was used in the study of influence of a designing method of the vane diffuser, the vane-spacing ratio and the diffuser width.

The initial data for designing of the vaned diffuser with the given velocity distribution is shown in Table 1.

Table 1. Initial data for designing of the vaned diffuser

α_3, °	P_3, Pa	P_3^*, Pa	T_3, K	T_3^*, K	$c_3, \frac{m}{s}$	$\rho_3, \frac{kg}{m^3}$
21.6	120946	138973	309.72	322.2	156	1.360

5.1. The comparison of the efficiency of the diffuser with the given velocity distribution and the circular-arc mean camber line vaned diffusers

Variations in the loss coefficient and coefficient of recovery of static pressure for diffusers designed with the use of different methods are shown in Figure 2 and 3.

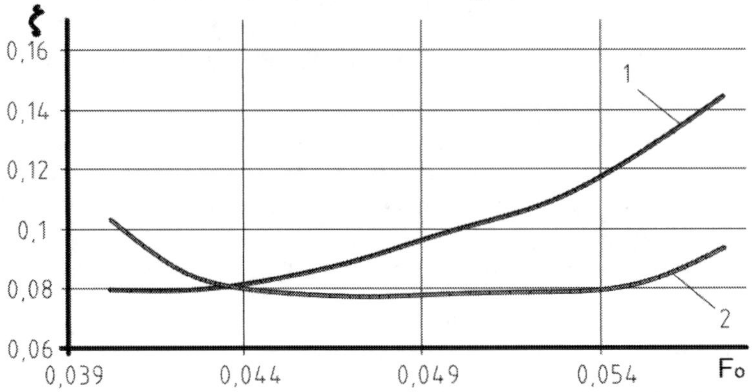

Figure 2. The loss coefficient of the vaned diffuser: *1* is the circular-arc mean camber line vaned diffusers; *2* is diffuser with given velocity distribution on the vane surface.

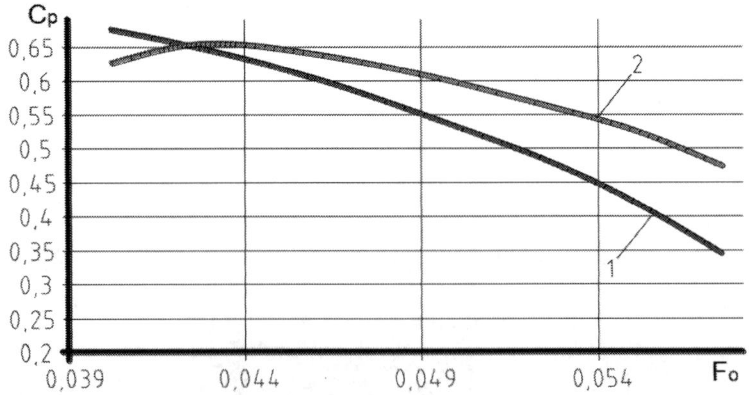

Figure 3. The coefficient of recovery of static pressure of the vaned diffuser: *1* is the circular-arc mean camber line vaned diffusers; *2* is diffuser with given velocity distribution on the vane surface.

Figure 2 shows that the vaned diffusers designed with taking into account the velocity distribution on the surfaces of vanes are more effective both on a design condition (by *28%*) when $Fo = 0.0502$ and on the maximum capacity operation condition (by *54%*) when $Fo = 0.0574$. On the near surge point of operation of the centrifugal compressor stage ($Fo = 0.0402$) there is slight deterioration in values of the loss coefficient of the diffuser (by *30%*), as a result of flow separation from the back surface of the vane. Separator of the air flow occurs due to impact inleakage on the diffuser vane with a positive angle of attack *(+5°)*. The coefficient of static pressure recovery of the vaned diffuser, which was designed by the proposed procedure, reaches higher values on design mode (by *9%*) and in the maximum capacity (by *21%*) relatively to the circular-arc mean camber line vaned diffusers (Figure 3). Whereas, there is a *8%* decrease in value of the static pressure recovery coefficient of the diffuser with given velocity distribution on near surge point of operation.

IOP Conf. Series: Materials Science and Engineering **90** (2015) 012045 doi:10.1088/1757-899X/90/1/012045

Thus, the analysis of integrated gas-dynamic characteristics of the vaned diffuser that are designed using different methods has shown a noticeable improvement in the performance of the diffuser with the given velocity distribution in the range of flow coefficient $Fo = 0{,}0435 \div 0{,}0574$, which is *80%* of the full range of stage operation.

5.2. Investigation of influence of the vanes shape on the efficiency of the vaned diffuser

Two types of profiles were compared. The first - a symmetrical airfoil C-4, the second - the profile of constant thickness ($\delta_v = 6\ mm$). Both diffusers had 19 vanes.

Figures 4 and 5 show a comparison of gas-dynamic characteristics of diffusers with different vanes profiles.

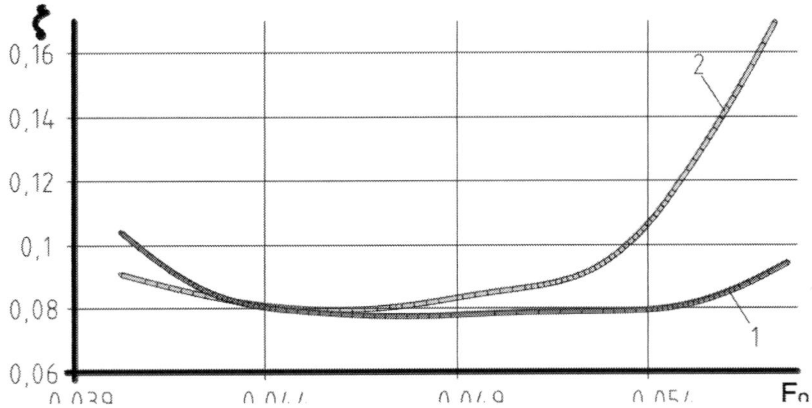

Figure 4. The loss coefficient of the diffusers with different vanes profiles: 1 – airfoil C-4; 2 – the profile of constant thickness

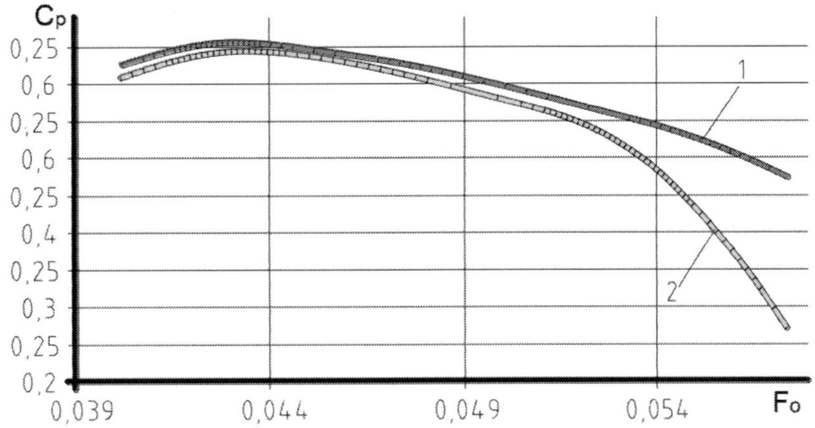

Figure 5. The coefficient of recovery of static pressure of the diffusers with different vanes profiles: 1 – airfoil C-4; 2 – the profile of constant thickness

From the analysis of gas-dynamic characteristics of the vaned diffuser there is an evident deterioration in diffuser's efficiency with a constant blade profile thickness in comparison with aerofoil C-4 vanes (Figure 4). Application of the airfoil C-4 in design of the vaned diffuser with a given velocity distribution allows to obtain the increase in values of the loss coefficient and the coefficient of static pressure recovery at the maximum flow coefficient ($Fo = 0.0574$) by 83% and 72%, respectively. A sharp constraint of the flow on the diffuser inlet is probable reason for the deterioration of gas-dynamic characteristics of the diffuser with a constant vane thickness. Although the influence of constraint coefficient can be reduced by making the diffuser's vane thinner and in such a way improving the flow around the inlet edge of the vane. Although the influence of constraint coefficient can be reduced by making the diffuser's vane thinner and in such a way improving the flow around the inlet edge of the vane, but this approach is impossible because of the requirements for strength and stiffness of the structure of the flow path elements of the centrifugal compressor.

5.3. Investigation of influence of the vane-spacing ratio on efficiency of the diffuser

Investigation of influence of the vane-spacing ratio (l/t) on efficiency of the diffuser designed with a given velocity distribution on the surface of the vane, was carried out for the following values: *1.0* (z_{34}=*12*); *1.25*(z_{34}=*16*); *1.5*(z_{34}=*19*); *1.75*(z_{34}=*23*); *2.0*(z_{34}=*27*).

Results of the comparison of gas-dynamic characteristics of the studied diffusers with different values of *l/t* are shown in Figures 6 and 7.

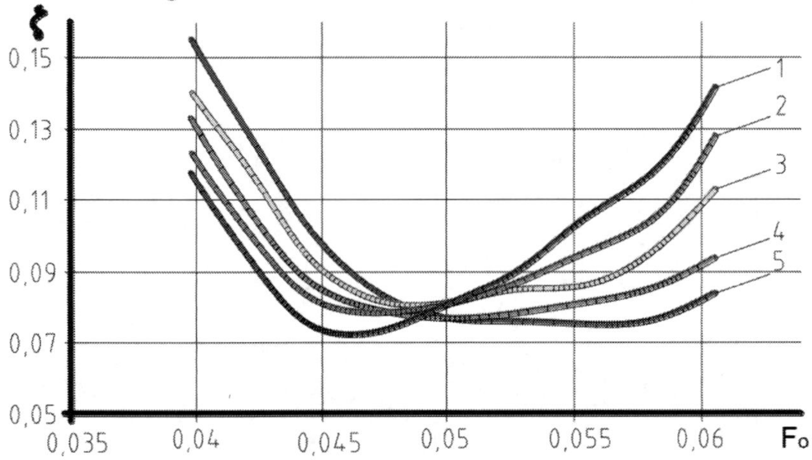

Figure 6. The loss coefficient of the diffusers with different vane-spacing ratio: *1 – l/t=1.0; 2 – l/t=1.25; 3 – l/t=1.5; 4 – l/t=1.75; 5 – l/t=2*

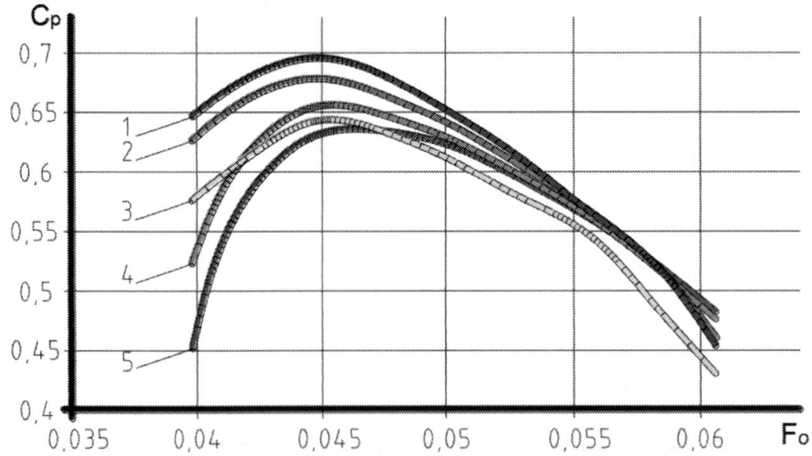

Figure 7. The coefficient of recovery of static pressure of the diffusers with different vane-spacing ratio: *1 – l/t=1.0; 2 – l/t=1.25; 3 – l/t=1.5; 4 – l/t=1.75; 5 – l/t=2*

The analysis of the gas-dynamic characteristics of the diffuser with different vane-spacing ratio showed that diffusers with values *l/t* equal to *1.75* and *2* have a pronounced peak of the loss coefficient, which is shifted towards lower flow coefficient. Diffusers with *l/t* = *1.5* have gentle changes in the values the loss coefficient and the coefficient of static pressure recovery.

5.4. Investigation of influence of the width of the diffuser on its efficiency

The angle of the gas flow depends on the cross-sectional area on the diffuser inlet. While studying the influence of the width of the diffuser on its efficiency an additional piece of research was conducted. It was aimed at identifying the initial data for profiling diffusers with the velocity distribution on the surfaces of the blades. Initial data for the design of the vaned diffuser with different widths is shown in Table 2.

Table 2. Initial data for the design of the vaned diffuser

b_3	$\alpha_3,°$	P_3, Pa	P_3^*, Pa	T_3, K	T_3^*, K	$c_3, \frac{m}{s}$	$\rho_3, \frac{kg}{m^3}$
0.9	19.9	120417	139514	310.3	323.6	162	1.351
1.0	20.1	121503	140187	310.6	323.5	160	1.363
1.1	20.6	121085	138830	310.9	323.3	157	1.357
1.2	19.1	121311	138533	311.1	323.1	154	1.358
1.3	17.6	121299	138430	311.2	323.1	154	1.358

The Investigation of the influence of the width of the diffuser on its efficiency was carried for the following values: $b_3 = 0.9 \cdot b_2$ (13.2 *mm*, z_{34}=20); $1,0 \cdot b_2$ (14.7 *mm*, z_{34}=18); $1.1 \cdot b_2$ (16.2 *mm*, z_{34}=18); $1,2 \cdot b_2$ (17.6 *mm*, z_{34}=16); $1,3 \cdot b_2$ (19.1 *mm*, z_{34}=16). Gas-dynamic characteristics of investigated diffusers with a given distribution of velocities and different value of the b_3 are presented in Figures 8 and 9.

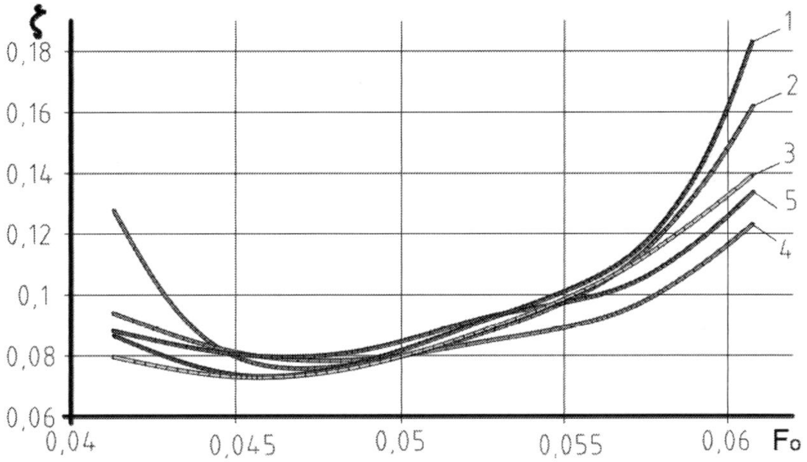

Figure 8. The loss coefficient of the diffusers with different value of the b_3: $1 - 0.9 \cdot b_2$ (13.2 mm); $2 - 1.0 \cdot b_2$ (14.7 mm); $3 - 1.1 \cdot b_2$ (16.2 mm); $4 - 1.2 \cdot b_2$ (17.6 mm); $5 - 1.3 \cdot b_2$ (19.1 mm).

Analysis of graphics presented in Figure 8 showed that the diffuser with height of the vane equal to $1.2\ b_2$ has the most gentle change in the value the loss coefficient. In the range of flow coefficient $Fo = 0.049 \div 0.0607$ vaned diffuser with $1.2\ b_2$has the minimal value of the loss coefficient among the investigated diffusers.

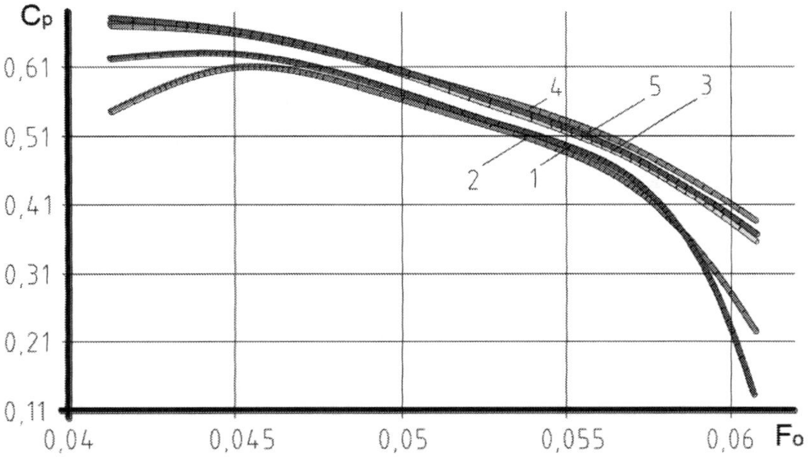

Figure 9. The coefficient of recovery of static pressure of the diffusers with different value of the b_3: $1 - 0.9 \cdot b_2$ (13.2 mm); $2 - 1.0 \cdot b_2$ (14.7 mm); $3 - 1.1 \cdot b_2$ (16,2 mm); $4 - 1.2 \cdot b_2$ (17.6 mm); $5 - 1.3 \cdot b_2$ (19.1 mm).

IOP Conf. Series: Materials Science and Engineering **90** (2015) 012045 doi:10.1088/1757-899X/90/1/012045

According to Figure 9 vaned diffuser with b_3=1.2 b_2 produces the most effective braking of the air flow in the entire range of values of the flow coefficient.

6. Comparison of the results numerical modeling and theoretical analysis

The suggested in 2nd Section method of solving the inverse task of gasdynamics allows to determine the following parameters of the flow in the vaned diffuser: pressure, density and temperature. The match in changes of the static pressure along the vane surface defined in different ways (using the proposed procedure and software package ANSYS CFX) is a confirmation of the correct choice of the mathematical model of the method of solving the inverse task of gasdynamics.

The values of static pressure along the surface of the vane in the proposed method are determined using gas-dynamic function:

$$P_i = P^* \cdot \left(1 - \frac{k-1}{k+1} \cdot \lambda_i^2\right)^{\frac{k}{k-1}}. \tag{16}$$

where the value of speed ratio for the pressure surface of the blade is determined from the equation (5), while for the suction surface - joint from the solution of equations (6) and (9).

Diagrams of the static pressure in the software package ANSYS CFX were determined in Post processor of solver along the line, located in the middle of the vane height.

Comparison of changes in static pressure along the surface of the blade (Figure 10) was carried out for the diffuser with given velocity distribution on the surface of the vanes, provided in Section 5.1. Pressure diagrams correspond to design condition of the investigated centrifugal compressor stage.

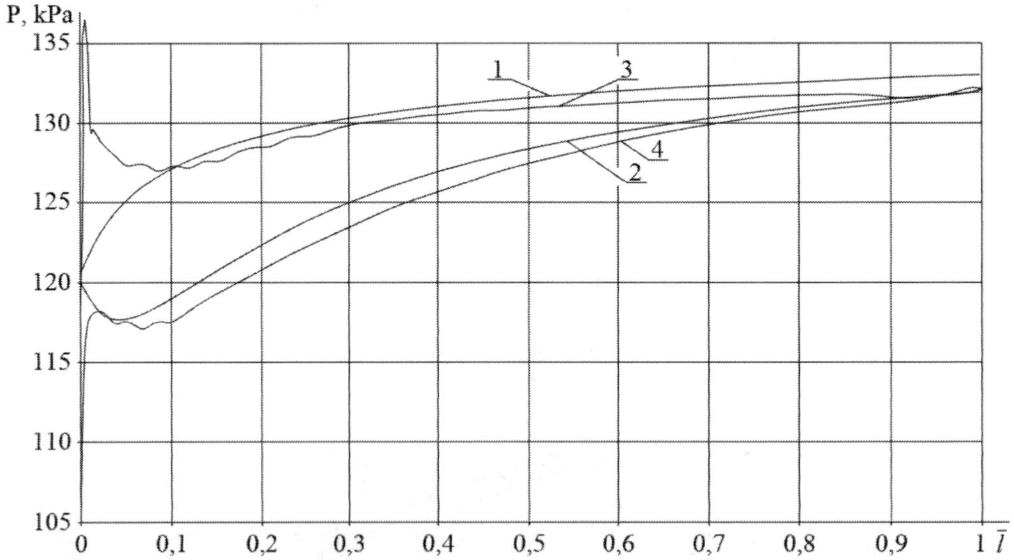

Figure 10. Static pressure distribution. *1* ,*2*– on the pressure and suction surfaces of the vane according to suggested method, respectively, *3*,*4* – on the pressure and suction surfaces of the vane according to ANSYS CFX, respectively

A comparison of distribution of static pressure along the vane surfaces, obtained by solving the inverse task of gasdynamics and numerical modeling, shows qualitatively satisfactory agreement. The maximum deviation of the compared values of the pressure does not exceed 3%, which indicates on the correct choice of the mathematical model of the proposed method for designing vaned diffusers. The discrepancy between the values of the static pressure distribution along the vane surface at its inlet section is explained by taking into account the constraint of flow in the transition from vaneless space between the impeller and vaned diffuser in the numerical modeling.

7. Conclusions

The suggested method of designing the diffuser vane with the given velocity distribution on the vane surfaces allows to extend the range of the effective flow deceleration up to 80% in comparison with the circular-arc mean camber line vaned diffusers. The analysis of variants with different design parameters of the vaned diffuser allowed to draw the following recommendations for the design of high-performance diffuser with a given velocity distribution on the surfaces of the vanes:

– in the case of application the airfoil C-4 vaned diffuser gas-dynamic characteristics are more gently sloping in comparison with the diffusers with a constant vane thickness;

– the vaned diffuser with a vane-spacing ratio $l/t=1.5$ has the lowest losses;

– the minimal values of the loss coefficient and the coefficient of static pressure recovery achieved for the width of the diffuser $b_3=1.2\ b_2$;

REFERENCES

[1] Djen G N 1990 *Centrifugal compressor flow path design* (Lenengrad, Maschinostroenie) p 232
[2] Gosteloy Dj 1987 *Aerodynamics of the turbomachine passages* (Moskow: Mir) p 391
[3] Seleznev K P 1982 *Centrifugal compressors* (Lenengrad, Maschinostroenie) p 272
[4] Stratford B S 1959 *The prediction of separation of the turbulent boundary layer* (J. Fluid Mech) pp 1-16
[5] Loicanskiy L G 1973 *Fluid mechanics (*Moskow: Nauka*)* p 847
[6] Kalinkevich N V 2009 *The design of vaned diffusers of centrifugal compressors based on the given velocity distribution* (Refrigeration technics and technology) pp 16-19
[7] Kalinkevich N V 2012 *Design method and numerical investigation of the channel diffuser of centrifugal compressor* (Compressor technics and pneumatics) pp 15-21
[8] Remizov A V 2013 *To the question of choice of gasdynamic effectiveness parameter of diffuser (*Engines and power plants aerospace aircrafts vol 10*)* pp 91-97
[9] Smirnov A V 2013 *To Application of the commercial CFD-products for the calculation of gas dynamic characteristics of centrifugal compressors* (Compressor and energy engineering vol 1(31)) pp 7-11
[10] Galerkin U B 2007 *The results of the calculation of viscous flow in stator elements of centrifugal compressors with the aid of ANSYS CFX* (Compressor technics and pneumatics vol 2) pp 42-47

Multi-objective optimization for impeller shroud contour, the width of vane diffuser and the number of blades of the centrifugal compressor stage based on the CFD calculation

A M Danilishin, Y V Kozhukhov, V K Yun

Department of Compressor, Vacuum and Refrigeration Engineering, Peter the Great St. Petersburg Polytechnic University, 195251 St. Petersburg, Russia, Politechnicheskaya St., 29

E-mail: lelliaa@mail.ru

Abstract. The study presents the results of multi-objective optimization for a high flow centrifugal compressor stage with impeller pressure ratio Π=1.3 and conditional flow rate coefficient Φ=0.108. The compressor stage includes an impeller with 3D backswept blades, a vane diffuser and an axial inlet. The goal of optimization is to increase polytropic efficiency and polytropic head ratio of the basic design. The CFD method is used to estimate compressor efficiency at rated duty. CAE optimization is applied based on the parametric optimization of impeller shroud contour and number of blades of the impeller and diffuser versus the original design. Parameterization of geometrics is also used to the width-ratio of vane diffuser in the range of b_3/D_2= 0.08-0.097. The study considers 52 cases of the optimization of impeller shroud contour and the number of blades in the search for the improved design. The optimization procedure uses the automatic generated computation grid and supplementary activation of solution to each design case. The numerical calculation for each case has been performed automatically by ANSYS CFX 14.5 soft application. The optimization allowed to obtain the improved design with total polytropic efficiency increase by 1.58% for the impeller and polytropic head coefficient increase by 0.58%. The polytropic efficiency and the polytropic head coefficient are calculated on the total parameters. The performance of the basic impeller has been exhaustibly validated by test records provided JSC "REPH ZAO". The resulted error range does not exceed 5% over the performance map, except near surge point.

Legend

b [-] dimensionless width
D [-] dimensionless diameter
h [$J \cdot kg^{-1}$] head
$\dot{m}$ [$kg \cdot s^{-1}$] mass flow
n [-] mode number
P [Pa] pressure
T [K] temperature
U [$m \cdot s^{-1}$] peripheral velocity
X [mm] driving dimension in axial direction
y^+ [-] wall distance in normal direction
Y [mm] driving dimension in radial direction
Z [-] number of blades
Π [-] total to total pressure ratio
β [-] loss coefficient
γ [-] performance parameter
ρ [$kg \cdot m^{-3}$] density
δ [-] relative error
η [-] efficiency coefficient
ψ[-] head coefficient
Φ[-] conditional flow rate coefficient

Subscripts

0, 1, 2,... - specified section
2,3 -blade element (impeller, diffuser)
p, d, i - polytropic, dynamic, internal
lk. , d.fr. - leak, disk friction
inl., out - inlet, outlet

Superscripts

*- total parameters

Content from this work may be used under the terms of the Creative Commons Attribution 3.0 licence. Any further distribution of this work must maintain attribution to the author(s) and the title of the work, journal citation and DOI.
Published under licence by IOP Publishing Ltd

IOP Conf. Series: Materials Science and Engineering **90** (2015) 012046 doi:10.1088/1757-899X/90/1/012046

1. Introduction

Centrifugal compressors are widespread in all main industries, and they are also used for production, processing, transportation and storage of natural gas. It is obvious that regardless of the service of the centrifugal compressor, the power consumption is estimated in thousands of kilowatts, which increases the requirements for the unit energy efficiency and encourages consumers and producers to find the ways to increase it.

Nowadays, the most commonly used methods for centrifugal compressor flow parts design are fluid dynamics methods (CFD). It is a standard way using for optimization of the flow part to increase its efficiency.

The scientific problem lies in the area of timesaving as the developing of simulation models for the CFD optimization is time-consuming due to the fact that the designer is required to calculate a few dozens or even hundreds variants of flow parts of centrifugal compressor stages to achieve the specified requirements of energy efficiency. This problem can be solved by the application of multi-objective optimization methods that automatically carry out a series of calculations with the use of supercomputing technology.

Multi-objective optimization allows to simultaneously search for cases of flow part by such important parameters (or criteria) as efficiency, pressure ratio, head, design constraints, etc ..

The above information confirms the need to develop the optimization procedure and to set forth the purposes of study:

- To validate CFD calculations of the basic design with experimental data.
- To define geometric parameters and energy efficiency criteria of the centrifugal compressor flow part for optimization.
- To develop the procedures for automatic optimization of the shroud contour, the width of vane diffuser and the number of blades for the centrifugal compressor flow part based on CFD.

2. Subject of research

The subject of research is a high flow centrifugal compressor stage model (Fig. 1), consisting of axial inlet nozzle, impeller with 3D backswept blades and vane diffuser designed by Nevsky Zavod (JSC "REPH ZAO") against advanced parameter of total to total pressure ratio $\Pi = 1.3$ in impeller for pipeline compressors of natural gas. The main dimensionless geometrical parameters of the flow part are provided in Table. 1.

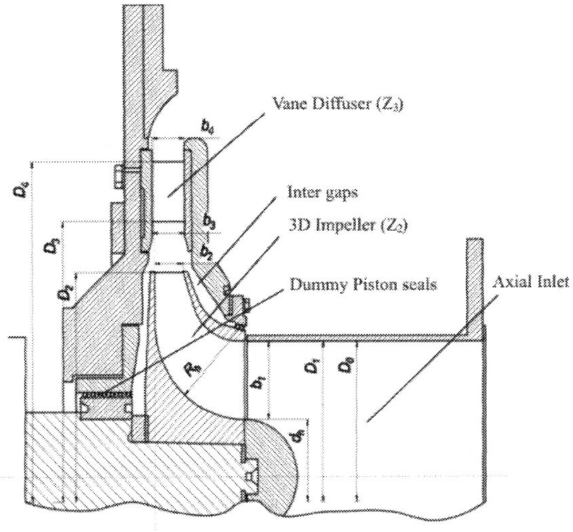

Fig. 1 Scheme of the object of study

Table 1. The basic dimensionless parameters of stage.

Parameter	Value
D_0	0.663
D_1	0.664
D_h	0.28
D_3	1.25
D_4	1.54
b_1	0.192
b_2	0.076
b_3	0.086
b_4	0.086
R_h	0.237
Z_2, pcs.	24
Z_3, pcs.	11

3. CFD model

CFD model has one vaned item with periodic boundary conditions. The collector chamber, inter gaps and dummy piston seals have not been simulated. The calculation is based on High-Re turbulence models considering the wall functions relevant to the flow near the wall at a relatively coarse computational grid, which allows to reduce the time period needed for simulation. Therefore, the 1st grid node should be in the logarithmic boundary layer ($y^+ = 30 \ldots 300$). A CFD Solver of Ansys CFX 14.5 was applied. The calculation was performed by averaging the Navier-Stokes equations RANS (Reynolds-averaged Navier-Stokes) for models of turbulence: Spalart-Allmaras (SA), RNG k-ε (RNG), Shear Stress Transport (SST). In this case SST turbulence model involves the use of a standard k-ε turbulence model, this is effected by the blending function depending on the wall distance. The service fluid is air (ideal gas).

The simulation of a large number of cases takes a lot of time hence, to shorten time necessary for calculations is one of the priorities. The main factor influencing the duration of each operation of calculation is the size of the computational grid, therefore, the first task was to find a computational grid with the minimum possible number of cells which adequately describes the physics of the flow. The outcome of the mesh independence studies provides the grid with 259275 cells.

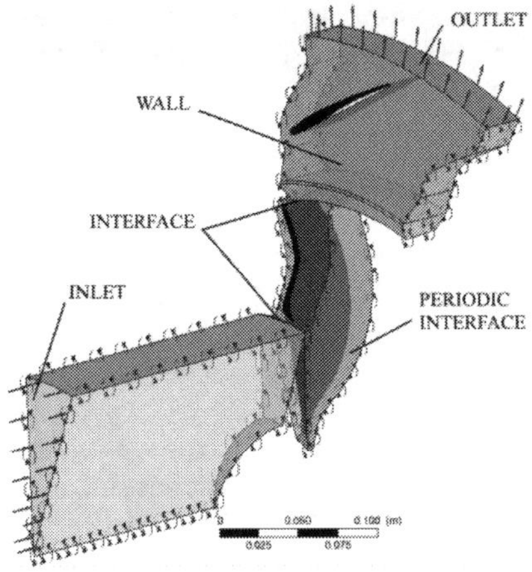

Fig. 2. CFD model and BL conditions of centrifugal compressor stage

4. Battery limits conditions

The inflow direction is assumed as normal to the axial inlet battery limits (BL) within grid computational domain. The simulations were run with P^*_{inl} and T^*_{inl} in inlet boundary and mass flow $\dot{m}_{out}$ at outlet boundary. The walls were assumed as adiabatic. The inlet and outlet BL conditions are listed in Table. 2. The Table also provides the conditional coefficient of flow rate (Φ), inlet total density (ρ^*_{inl}) and stage mass flow ($\dot{m}$). The BL conditions were determined by the outcomes of the full-scale experiment. An upwind discretization scheme of 2nd order was included into solver settings. The grid interface ('stage') is used for rotor-stator grid connection, which averaged flow parameters in the circumference direction.

Table 2. The parameters set on the inlet and outlet BL of the computational domain

Mode n	Φ	T^*_{inl}, *K*	P^*_{inl}, *Pa*	ρ^*_{inl}, $kg{\cdot}m^{-3}$	$\dot{m}$, $kg{\cdot}s^{-1}$	$\dot{m}_{out}$, $kg{\cdot}s^{-1}$
1	0.1491	291.53	90800	1.080	2.880	0.26181
2	0.1437	291.63	91500	1.088	2.796	0.25418
3	0.1369	291.73	92300	1.097	2.686	0.24418
4	0.1281	291.73	93200	1.108	2.539	0.23078
5	0.1195	291.87	94200	1.119	2.393	0.21754
6	0.1084	291.82	95300	1.132	2.197	0.19971
7	0.0946	292.06	96600	1.147	1.942	0.17655

5. Validation results

All modes developed the converging solutions (example of convergence history for mode 6 shown in Fig. 3), defined by the following parameters within the computational domain BL:

- the fall of the level of root-mean-square (RMS) residuals below 10^{-5},
- the relative error in mass imbalance is $(1\text{-}3)\cdot 10^{-3}$ %,
- the relative error in the energy imbalance is $(1\text{-}5)\cdot 10^{-3}$ %

The controlled values of integral parameters in the control sections do not change from iteration to iteration. The control of calculation accuracy is provided by means of validation with experimental data provided by JSC "REPH ZAO". Validation results are shown in Fig. 4. Hereinafter point №1 refers to the maximum flow mode, point №7 refers to the near surge point.

According to the experimental data, mode 7 has the maximum value of the sum of the coefficients of friction disk and leaks $\beta_{lk.} + \beta_{d.fr.} = 0.0165$; while mode №1 provides minimum sum $\beta_{lk.} + \beta_{d.fr.} = 0.0113$. These values are typical for high flow centrifugal compressors, which is described in detail in the literature [4]. The earlier studies [6] for middle flow 2D impeller with Φ range 0.06 - 0.09 and $\beta_{lk.} + \beta_{d.fr.}$ (0.0208-1.0146) showed that exclusion of inter gaps and seal from the computational domain increased the efficiency by 2.2 % average. Therefore, taking into account that for high flow centrifugal compressors stage the influence of such items is even less, we can assume that the efficiency increase in subject will not more than 2.2%.

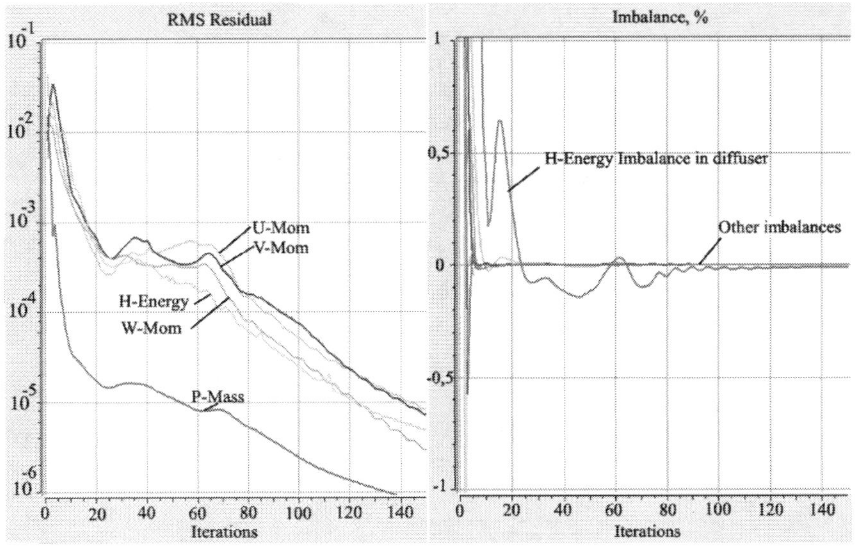

Fig. 3. Convergence history for RMS residuals mass and momentum, energy and imbalances within computations domains

Conditional flow rate coefficient is

$$\Phi = \frac{4\cdot \dot{m}}{\rho_{inl}^{*}\pi D_2^2 U_2} \tag{1}$$

Polytropic efficiency at full range of parameters is

$$\eta_p^* = {h_p^*}/{h_i}\,,\quad h_p^* = h_p + h_d. \tag{2}$$

Polytropic head coefficient at full range of parameters is

$$\psi_p^* = h_p^* / u_2^2 \tag{3}$$

Total to total pressure ratio coefficient

$$\Pi = P_{out}^* / P_{inl}^* \tag{4}$$

Mean relative error for n-modes throughout the performance

$$\delta_{mean} = \frac{1}{n}\sum_{i=1}^{n}\left|\frac{\gamma_i - \gamma_{i\,\text{experiment}}}{\gamma_{i\,\text{experiment}}}\right| \tag{5}$$

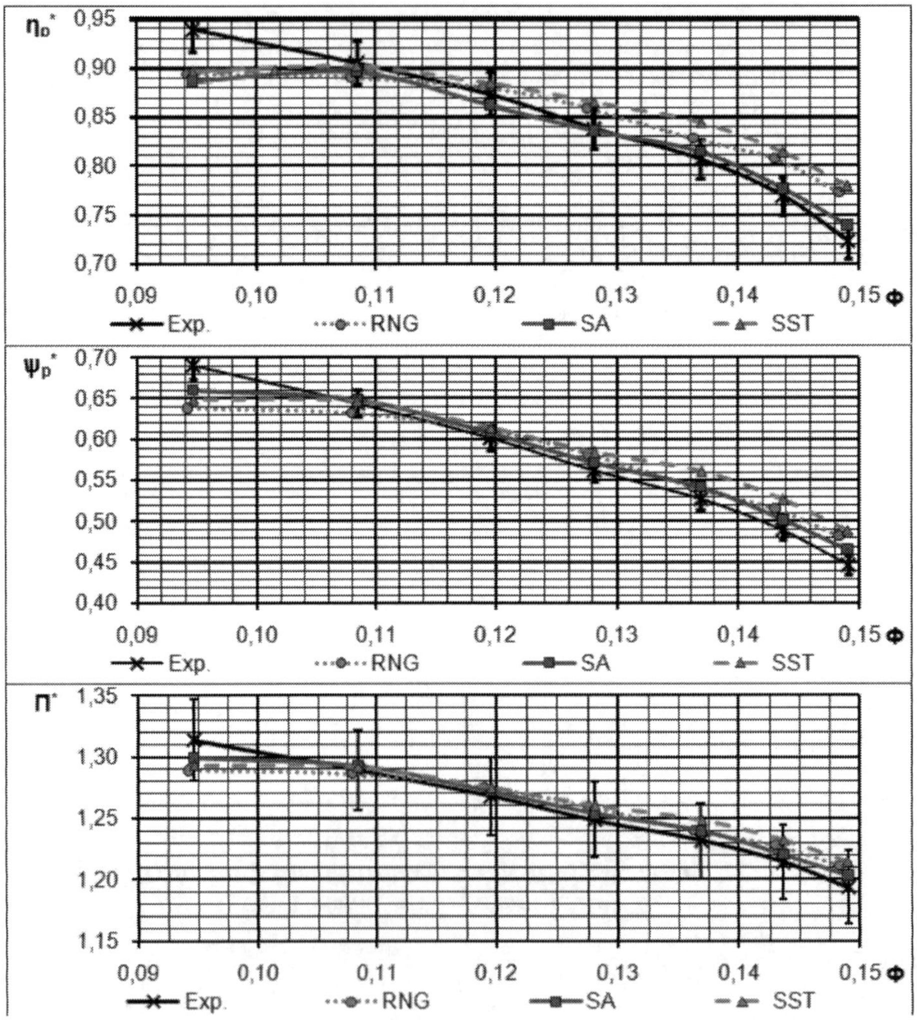

Fig. 4. Validation results for three turbulence models at the outlet of the impeller

Spalart-Allmaras turbulence model showed good accuracy with experiment and the received error does not exceed the level of engineering errors 5%, except near surge point that illustrated on fig. 4. Also it showed the best accuracy throughout the all performance parameters η_p^*, ψ_p^* and Π as shown in table 3. Using the formula (5), Table 3 shows the mean relative error δ_{mean} of 7- modes throughout the all performance graphs for each solution turbulence model. In the formula (5) is determined by the relative error performance parameters with respect to the experimental data, and then averaged by the number of the modes. Considering the said above, we selected the turbulence model Spalart-Allmaras (SA) for the further optimization.

Table 3. Mean relative error for 7- modes throughout the all performance graphs for each model turbulence solution.

	δ_{mean}		
Model turbulence:	η_p^*	ψ_p^*	Π
SST	3,9%	5,1%	1,1%
RNG	3,5%	4,4%	1,0%
SA	1,8%	2,4%	0,6%

6. The optimization procedure

The optimization procedure was developed by a soft tool Ansys Workbench 14.5 with the help of software Ansys Design Exploration v 14.5. the optimization was performed for the design mode №6. Figure 5 shows the optimization procedure carried out by Ansys Workbench 14.5 (left) and parameterization diagram (right) of the shroud contour centreline of the CC impeller. The same kind of parameterization was carried out for vane diffuser and inlet nozzle. All soft tools, including the grid generator and the solver were interfaced, which allows to adjust the computational domain automatically and to calculate the cases based on the input data.

Considering the design criteria one should remember that is better to minimize their number as it will lead to the increased amount of new cases and, consequently, solver actuations. It is also necessary to set optimization parameters adequately, that is why their values were discretely entered by hand. This approach allows to avoid a number of uncontrolled cases.

The optimization procedure divided into two steps. The first task is to optimize the shroud contour and width of vane diffuser. The shape is controlled by regulated dimensions in axial and radial direction, called X and Y (see fig. 5, right pic.).The hub contour is kept constant. The impeller shroud contour is adjusted so that it is remained lower than basic case. The dimensionless width of vane diffuser (b_3) is changed four times from $8\cdot10^{-2}$ to $9.7\cdot10^{-2}$. Regulated dimensions are set in a discrete mode for X and Y: 0, 4, 8. Eventually we come to three optimization parameters. The total number of cases is estimated as $(1+2^3)\cdot 4=36$. The pattern of variations for the impeller shroud contour is shown in fig. 6.

The second task is to optimize the number of blades. The number of blades is assumed as follows: Z_2=26, 24, 22, 20 for impeller blades and Z_3 = 11,13,15,17 for vane diffuser. Eventually we receive 2^4=16 cases in total.

The purpose of multi-objective optimization is to achieve the best efficiency for centrifugal compressor stage. The selected objective function of optimization for all tasks was the polytropic efficiency η_p^* and total to total pressure ratio Π on the impeller and the diffuser outlet sections. As soon as the optimization is initiated, the outputs are provided automatically (cases and grid generation).

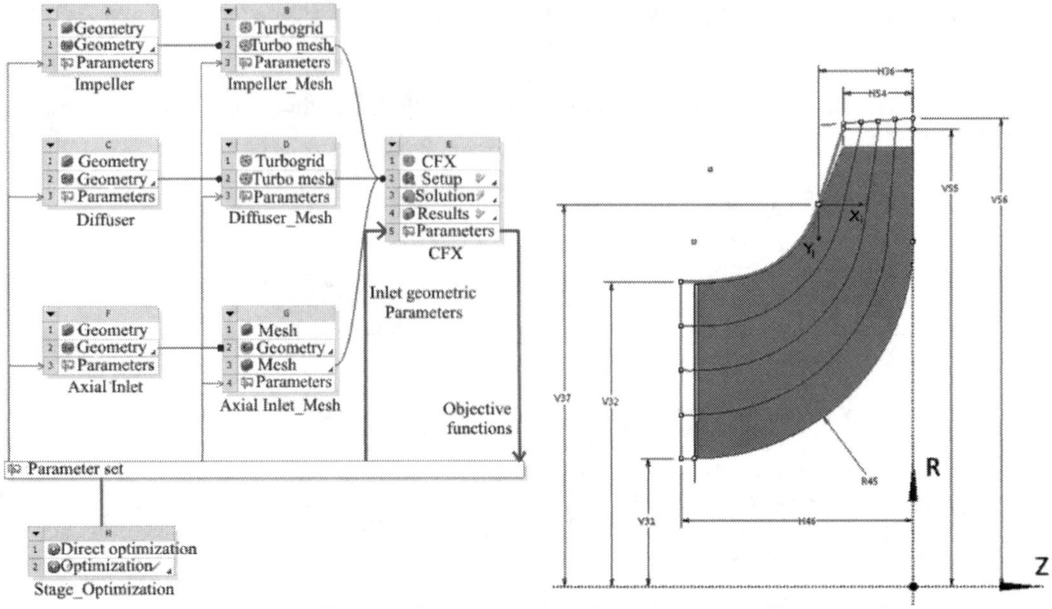

Fig. 5. The interface diagram between soft tools and the impeller shroud contour parameterization

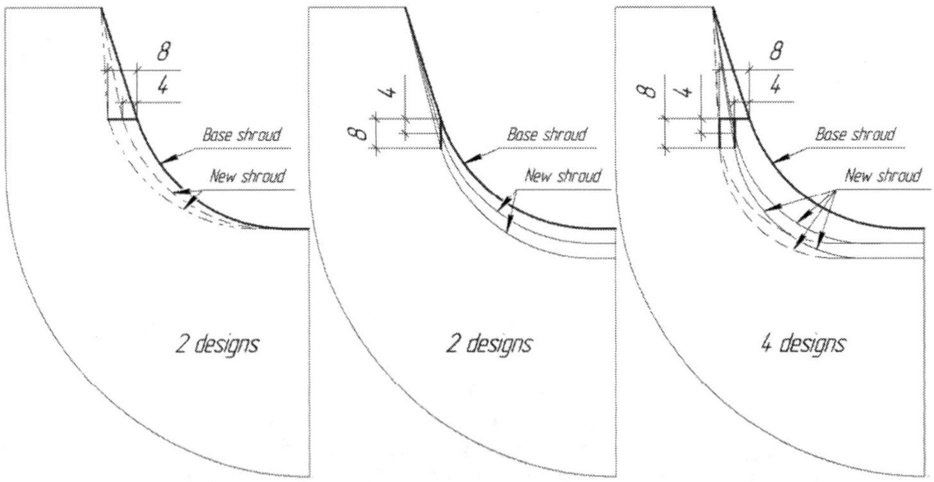

Fig. 6. The optimization cases for shroud contour of the impeller

7. Optimization results

Graphics (Fig 7 and 8) provide the results of the optimization. X-axis shows the cases for shroud contour of the impeller and number of vane diffusers. Y-axis shows η_p^* (left) and ψ_p^* (right). The review of these graphs (Fig 7,8) gives an idea of the efficiency increase of the stage versus adjustment of design parameters.

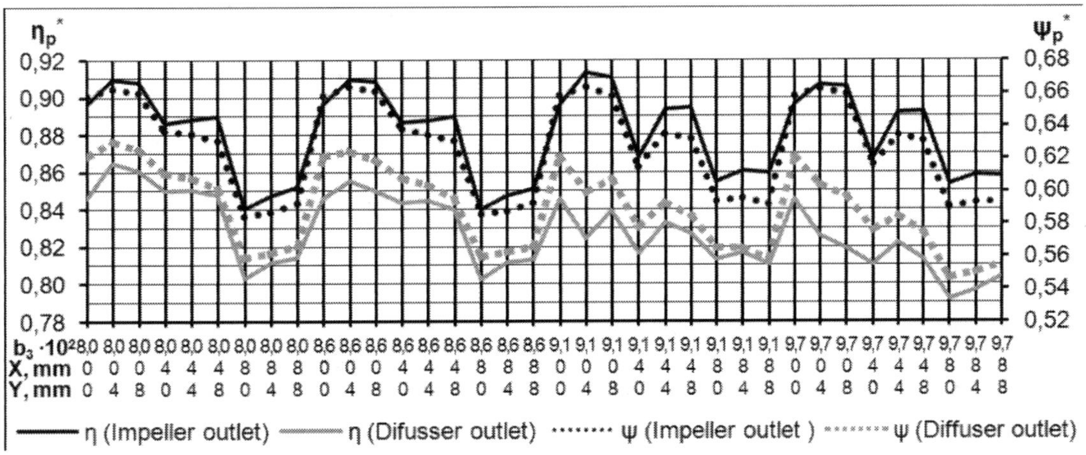

Fig. 7. The results of the optimization of the impeller shroud contour and the width of vane diffuser (b3) with controlled values X and Y.

The Optimization of impeller shroud counter showed a marked worsening of the efficiency and the head of the stage if controlled value of X is changed. The change of the width of the diffuser has little effect on the efficiency of the stage. The best result is achieved when controlled value of X = 4 or 8 mm and Y = 0, specifically η_p^* is increased by 1.33% and ψ_p by 0.57% at impeller outlet, and η_p^* by 0.96% and ψ_p^* by 0.32% at diffuser outlet, accordingly.

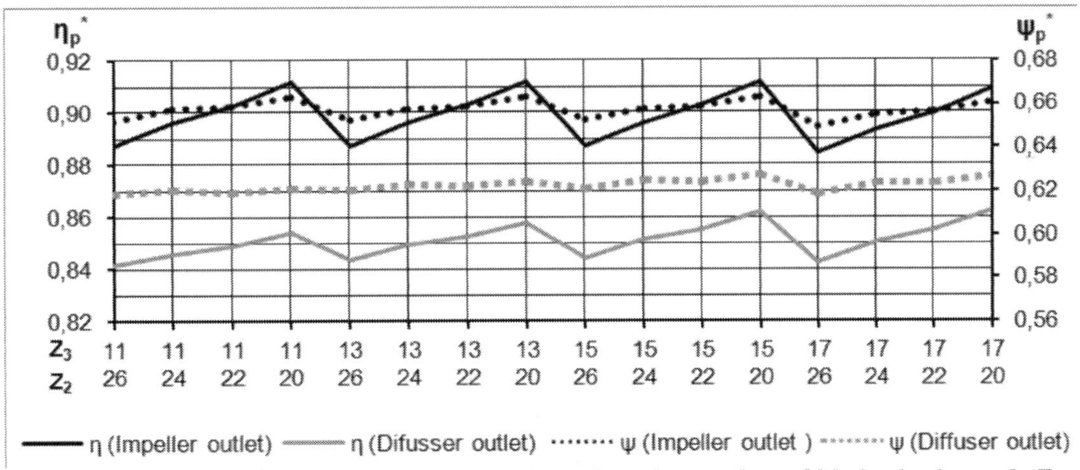

Fig. 8. The results of the basic case optimization when the number of blades is changed (Z_2 - impeller, Z_3 – diffuser)

Graph in fig. 8 shows that the reduction of the number of impeller blades from 24 to 20 increases η_p^* by 1.58% and ψ_p by 0.58%. It is justified by the less friction due to smaller work surface area. The simulation also shows that at Z_3 = 13 (number of vane diffuser blades) we have the increase of η_p^* by 1.66%.

Table 4 summarizes the best results for each of the two-step optimization (the best shroud counter optimization and the best number of blades optimization) versus the basic design. Here it is also presented the outcomes for the case where we try to combine the two best results (design parameters

X=4, Y=0, Z_2 = 20 and Z_3 = 13) of each step in attempt to obtain the best optimization. However, this idea failed to prove.

Table 4. The best optimization results v basic design.

Design	η_p^* Imp.	Change	η_p^* Dif.	Change	ψ_p^* Imp.	Change	ψ_p^* Dif.	Change
Base	0.896		0.846		0.658		0.621	
Shroud best	0.909	+1.33%	0.855	+0.96%	0.663	+0.57%	0.624	+0.32%
Z_2, Z_3 best	0.912	+1.58%	0.862	+1.66%	0.663	+0.58%	0.627	+0.65%
Both	0.911	+1.5	0.825	-2.1%	0.671	+1.3%	0.608	-1.3%

8. Conclusions

The soft tool Ansys 14.5 was used for the development of the multi-objected optimization. The subject of study for optimization is centrifugal compressor stage with impeller pressure ratio П=1,3 and conditional flow rate coefficient Ф=0.108. The turbulence model Spalart-Allmares was selected based on the preliminary validation with experimental data. % The deviations of the design parameters from the experimental data are within 5% over the performance map.

The optimization procedure was divided into two steps:

- to optimize the shroud contour based on two controlled values and four different values of the width of vane diffuser;
- to optimize the number of blades of impeller.

There were 52 design cases received. The optimization was carried out for the design mode №6.

The best result of Step 1 is the stage with X=0, Y=4 and increase of η_p^* by 1.33%, ψ_p^* by 0.57% at the impeller outlet and increase of η_p^* by 0.96%, ψ_p^* by 0.32% at the vane diffuser outlet. The best result of Step 2 is increase of η_p^* by 1.58%, ψ_p^* by 0.58% at the impeller outlet and increase of η_p^* by 1.66%, ψ_p^* by 0.65% at the vane diffuser outlet at Z_2 = 20 and Z_3 = 13. Then we developed the case which combined the best results of two steps, but the outcome was negative, in particular the decrease η_p* by 2.1%, ψ_p* by 1.3% at the vane diffuser outlet. This shows that for the best optimization it is necessary to consider the interaction of the stage units.

The developed optimization procedure gives broad opportunities for the fast adjustment of centrifugal compressor stages to the predetermined parameters of the power efficiency.

If such optimization tasks are paralleled using supercomputers, it is possible to achieve the wide range of cases by design parameters applying finer calculation grid. In addition the outcomes can be used to develop response surface. The mathematical methods of optimization can be applied to define local and global optimum, as well as Pareto set.

It is expected to further improve the possibilities of optimization by applying mathematical methods for the design parameters (criteria) analyses.

References

[1] Y. B. Galerkin, B. I. Zykov, K. P. Seleznev. Study of the shape of the meridional section of the impeller intermediate stage centrifugal compressor. Power engineering, 1965, No. 10, p. 33-39.

[2] H. P. Dickmann. Shroud contour optimization for a turbocharger centrifugal compressor trim family/ 10th European Turbomachinery Conference, Lappeenranta, Finland; 04/2013

[3] Florian Frese, Johannes Einzinger, Johannes Will. Design optimisation of an impeller with CFD and Meta-Model of optimal Prognosis (MoP) / 10th International conference on turbochargers and turbocharging, London, England; 2012

[4] K.P. Seleznev, Y. S. Podobuev YS, S.A. Anisimov. Theory and Design of turbocompressors:

Proc. manual for schools /; Ed. K.P. Seleznev - Leningrad: Mechanical Engineering, 1968. - 406p.

[5] ANSYS Inc. ANSYS CFX 14.5: Users Manual, 2012

[6] A.M. Danilishin, Y.V. Kozhukhov Analysis of the influence on the performance of centrifugal compressor stage of the model parameters in ANSYS CFX. Week of Science SPbSTU: Proceedings of the conference with international participation. Institute of Energy and Transport Systems SPbSTU. Part 2 - St. Petersburg Polytechnic University, 2014. - p.186-189

IOP Conf. Series: Materials Science and Engineering **90** (2015) 012047 doi:10.1088/1757-899X/90/1/012047

Numerical Investigation of Different Radial Inlet Forms for Centrifugal Compressor and Influence of the Deflectors Number by Means of Computational Fluid Dynamics Methods with Computational Model Validation

Y V Kozhukhov[1], V K Yun[2], L V Reshetnikova[3], M V Prokopovich[4]
[1] Deputy Head of the Compressor Department, Vacuum and Refrigeration Engineering. Saint-Petersburg Polytechnical University, Russia, Saint-Petersburg
[2] Professor of the Compressor Department, Vacuum and Refrigeration Engineering. Saint-Petersburg Polytechnical University, Russia, Saint-Petersburg
[3] Student of the Compressor Department, Vacuum and Refrigeration Engineering. Saint-Petersburg Polytechnical University, Russia, Saint-Petersburg
[4] Engineer of the Design department of special engineering, Russia, Saint-Petersburg

Email: lyubov5reshet@gmail.com

Abstract: The goal of this work is numerical experiments for five different types of the centrifugal compressor's inlet chambers with the help of CFD-methods and comparison of the computational results with the results of the real experiment which was held in the Nevskiy Lenin Plant in Saint-Petersburg. In the context of one of the chambers the influence of deflectors on its characteristics was investigated.
The objects of investigation are 5 inlet chambers of different types which differ from each other by deflectors' existence and by its number. The comparative analyze of the results of numerical and real experiments was held by means of comparison of relative velocity and static pressure coefficient distribution on hub and shroud region, and also by means of loss coefficient values change for all five chambers.
As a result of the numerical calculation the quantitative and qualitative departure of CFD-calculations results and real experiment were found out.
The investigation of the influence of the number of deflectors on flow parameters was carried out. The results of the study prove that the presence of the deflectors on flow path significantly increases the probability of the flow separations and reversed flows appearance on them. At the same time, the complete absence of the deflectors in the chamber significantly increases circumferential distortion of the flow; however the loss coefficient decreases anyway, the high values of which are caused by the shock flow existence. Thus, the profiling of the deflectors of the inlet chamber should be given a special attention.

1. Introduction

The radial inlet chamber is intended for a gas supply to the impeller of the first stage of the centrifugal compressor with the smallest energy loss and also a uniform flow profile into the inlet. From the quality of design, production and installation of the inlet chamber and, therefore, from aerodynamic perfection of its channels depends the value of the power consumed by the compressor on the commission of the work of gas compression to the demanded value of discharge pressure (P_{dis}). Wrong parametric choice of inlet device can become the reason for a decrease in efficiency of the centrifugal compressor of 3 … 5% and can considerable narrow down the range of its steady work. That is an essential negative factor, especially for compressors of high power [1].

Research into inlet devices made by the authors by means of CFD (Computational fluid dynamics) methods show that calculation in CFD programs are coordinated with the data of natural experiment. Both calculation and experiment confirm that changes of the design which will lead to the improvement of inlet device operation are possible. Based on this work and on the previous CFD calculations, the authors claim that similar programs can be used with confidence for the analysis, improvement and change of the radial inlet chambers. The unsteady flow phenomena which are

Content from this work may be used under the terms of the Creative Commons Attribution 3.0 licence. Any further distribution of this work must maintain attribution to the author(s) and the title of the work, journal citation and DOI.
Published under licence by IOP Publishing Ltd

observed at practical experiment show that CFD doesn't yield the full result at stationary problem definition RANS (Reynolds-averaged Navier–Stokes equations).

Also research shows that besides the traditional programs for a grid creation, such as ICEM CFD, Turbogrid, Solid Works, Fluent, the new programs can also be used. For example University of Florence uses Centaur Software and also TrueGrid for the hybrid grids creation of the radial inlet chambers [2].

Practical experiments for the inlet devices used in this work were held at the Nevsky plant as a part of a first model stage of a K4250-41-1 compressor in a single-cylinder performance [3]. This research was one of the sections of the design and research cycle of development works of the new type of K4250-41-1 compressor. This compressor is intended for the blasting pressure increase in the blast furnaces. The purpose of the practical experiment was clarification of influence of a form of the inlet chamber on the centrifugal compressor operation. Practical experiments were held for five various forms of the inlet chambers, including the option with axial intake, for the purpose of choice of the best variant.

Objects of this research in CFD programs are 5 computer models of the radial inlet chambers completely identical to the chambers tested at the Nevsky plant, further called RIC1, RIC2, RIC3, RIC4 and RIC5. CAD-models (Computer-aided design models) of investigated chambers are presented in figure 1.

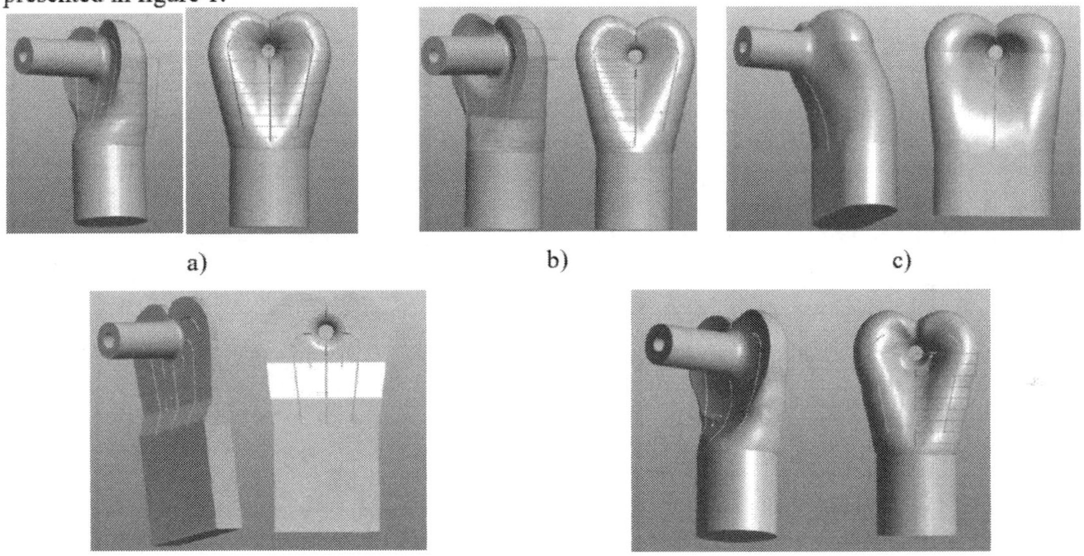

a) b) c)

d) e)

Figure 1. CAD-models of the investigated chambers a) RIC1, b) RIC2, c) RIC3, d) RIC4, e) RIC5

The purposes of the investigation are:

1) Performance of calculations and analysis of the flow at five inlet chambers of the centrifugal compressor by means of the software package ANSYS CFX, and also the structure definition of the flow pictures on various parts of computational space;

2) Calculation of the relative velocity $\overline{c}_2$, the static pressure coefficient K_P and the loss coefficient ζ for each chamber;

3) Comparison of received results of the modeling with the Nevsky plant experimental research data and conclusion about the possibility of the software package ANSYS CFX application for the flow calculation in the inlet chambers of centrifugal compressors;

4) Researches of the deflectors influence on one of the inlet chambers and the conclusions composition about the advisability or inadvisability of any changes introduction into the chamber design.

IOP Conf. Series: Materials Science and Engineering **90** (2015) 012047 doi:10.1088/1757-899X/90/1/012047

As a result of the carried-out work it is necessary to reveal an optimum construction of the inlet chamber among the investigated variants. It is also necessary to introduce the conclusion about the possibility of the software package ANSYS CFX application for the inlet chambers of the centrifugal compressors improvement.

2. Experimental investigations:

For carrying out experimental investigations of the single flow variant of the K4250-41-1 compressor the model of the first stage have been designed and manufactured. The model of the stage was made on a scale of $i = \frac{D_2}{D_2^*} = 5{,}41$ and was driven from the 125 kW electric motor, 380V, 2950 rpm through the increasing gearbox with transfer number of $i_p = 5{,}6$. The flow path of the model consisted of the impeller with the diameter of $D_2^* = 305\ mm$, the vaneless diffuser with return directing device and the outlet radial diffuser which imitated the flow entrance to the second stage. Air intake to the impeller was made from removable inlet chambers.

3. Analysis of the results of the experimental investigation:

Parameters on the inlet to the chamber: total temperature $T^* = 293\ K$, total pressure $P^* = 98100\ Pa$, mass flow $\overline{m}_p = 2{,}2$ kg/s, gas – air.

Based on the research the best characteristics were shown by the RIC1 chamber.

The RIC3 chamber which is characterized by the smoothest turns in the flow path is approximately equivalent to the RIC1 chamber. The flow inhomogeneity at an outlet from this chamber is slightly less, than behind the RIC1 chamber.

The next chamber by quality is the RIC5 chamber which has smooth turns in the flow path and is designed with higher acceleration of the flow in the middle part of the chamber than in the RIC1. The pressure loss ζ at this variant of the inlet chamber reaches 8,3% in comparison with an axial inlet. However, looking at the efficiency this chamber is close to the RIC1 and the RIC3 variants.

The RIC2 chamber is characterized by sharp turning of the chamber at the covering impeller disk. The model test of this chamber showed that loss of pressure increase in comparison with axial inlet on the design condition reaches 13,6%. Axial asymmetry of the flow and its inhomogenuity by radius reaches higher volume, than in the previous chambers.

The RIC4 chamber is small-sized, with the clamped sections and small curvature radiuses in flow path gives increase in pressure for 15% less at design mode, than at axial inlet and this chamber is the worst of all.

The loss coefficients ζ of all the studied chambers are compared on figure 2. By the loss coefficient the inlet chambers are distributed approximately in the same order, as on the basis of analysis of gasdynamic characteristics and uniformity of the flow at an outlet from the chambers. The RIC3 and RIC1 chambers have the smallest loss coefficient and RIC4 and RIC2 have the greatest. The loss coefficient for the chamber is determined by the formula: $\zeta = \frac{P_{1mid}^* - P_{2mid}^*}{\rho_{2mid} \cdot \frac{c_{2mid}^2}{2}}$

Characteristic feature of the investigated flow path of the stage is high $\frac{c_0}{u_2}$ value (ratio of the flow velocity before the impeller to the velocity on the impeller rim) which reached the value $\frac{c_0}{u_2} = 0{,}4$. This circumstance strengthened the role of the inlet chamber and its influence on the gasdynamic characteristics of the stage. The works which were carried out on the model with low ratio $\frac{c_0}{u_2} = 0{,}09$ showed that at such low $\frac{c_0}{u_2}$ values the role of the inlet chamber is much less, and it is possible to obtain gasdynamic characteristics of the stage with the inlet chamber installation, almost no different from characteristics with axial inlet.

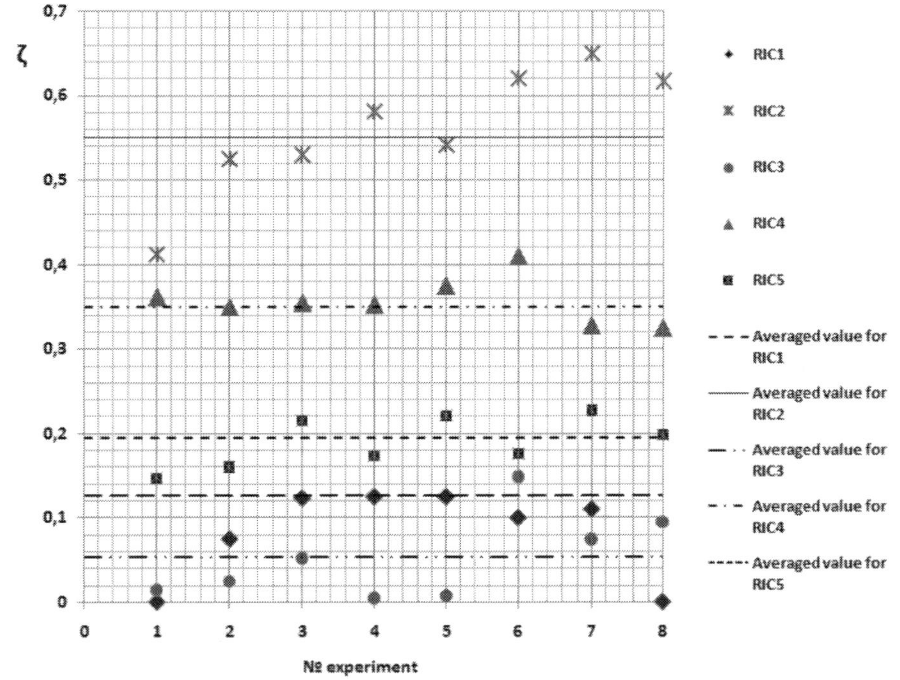

Figure 2. Experimental data of the loss coefficients of the investigated chambers with the values averaging

The following conclusions are drawn by results of the research:

1) The asymmetry of the flow in the output diffuser which distorted work of the stage took place. This occurred because the support plate was too close to the outlet confusor of the chamber.

2) Change of an operating mode of the stage by means of the replaceable diaphragms in the inlet, caused pressure decrease to 3000 mm w.c., which brought difficulties to the pressure measurement of the chambers.

3) Asymmetry of the outlet flow in the diffuser created an uneven temperature field that didn't allow correct estimation of the power consumption and the efficiency of the installation.

4) Measurement of total pressure heads at the outlet, from the return directing device, was performed only by one radius location.

5) Specified circumstances didn't allow research to the desired accuracy and detail. Nevertheless, the measured results allow a judgment to be made about the quality of the inlet chambers.

6) From five types of the inlet chambers which were exposed to these investigations, the best chamber by its qualities was the RIC1 chamber, the worst – the RIC2 chamber.

5. CFD- investigation of the inlet chambers of the centrifugal compressor:

During the numerical investigation 5 different models of the inlet chambers of the centrifugal compressor were under consideration. Their names are the same RIC1, RIC2, RIC3, RIC4, RIC5.

Preparation for the calculation of all radial inlet chambers was carried out in 3 steps:

1) Creation of the radial inlet chamber CAD-model in the program complex Pro/ENGINEER Wildfire 5.0. For creation of all 5 chambers the drawings were used which were provided by the Nevsky plant. The extended outlet section was created after the confusor for each chamber. This section was made for investigation of the flow unevenness which changes along its length after the confusor. CAD-models of investigated chambers are presented in figure 2.

2) Unstructured grid creation in the program ANSYS ICEM CFD 14.0 module with necessary concentration at walls. In the flow core the elements are tetrahedrons and 15 prismatic layers at walls.

3) In CFX-Pre module the boundary conditions for calculation were set. For obtaining the characteristics at different mass flows two types of boundary conditions were used:

- Total pressure and temperatures at the inlet and outlet of the chamber with a condition of two-way entry of the flow through the inlet and outlet (boundary condition "Opening");
- Total pressure and temperature at the inlet and boundary condition "Inlet", mass flow and boundary condition "Outlet" at the outlet.

Parameters at the inlet correspond to experimental conditions: total temperature $T^* = 293\,K$, total pressure $P^* = 98100\,Pa$, mass flow $\overline{m}_p = 2{,}2$ kg/s, gas – air.

Calculations were made with turbulence model SST as it is the most universal and is developed on the basis of the verified two-parametrical k-ε and k-ω models, uniting their best properties.

The mass flow $\overline{m}_p = 2{,}2\,kg/s$ is taken as the design mode of the investigated chamber which also corresponds to minimum loss coefficient at the chambers and to maximum efficiency of the stage. This mass flow also corresponds to design mass flow of the model chamber of the Nevsky plant. As during the experiment, all chambers were investigated at the test-stand with the same model stage, the design mass flow is identical at every chamber because the design mass flow of the stage doesn't change.

6. Analysis of the numerical experiment results:

As a result of numerical calculation when considering the loss coefficient value the best is the RIC3 chamber, the worst is RIC4. The others are intermediate between them.

The results analyses were carried out for each chamber. The maximum velocity appears on the peripheral surface of the confusor from the inlet when turning the flow. But, this chamber is characterized by smoother turns of the flow; therefore, there is less probability of the separation.

On figure 3 a) the placement of sections 1-1 and 2-2 is shown corresponding to the inlet and outlet sections of model chambers of the Nevsky plant. On figure 3 b) the scheme of static pressure tapping points in section 2 for model chambers of practical experiment and also a field of static pressure is presented in the corresponding plane of the computer model.

a) b)

Figure 3. Results of the CFD modeling for the RIC3 inlet device a) the placement of sections 1-1 and 2-2; b) the scheme of static pressure tapping points in section 2 for the model chamber of the Nevsky plant in the corresponding plane of the computer model; static pressure field

On Figure 4 the comparison of the relative velocity distribution $\bar{c}_2 = c_{2i}/c_{2mid}$ in section 2 is presented, where i – the static pressure tap point which was measured by holing the chamber walls.

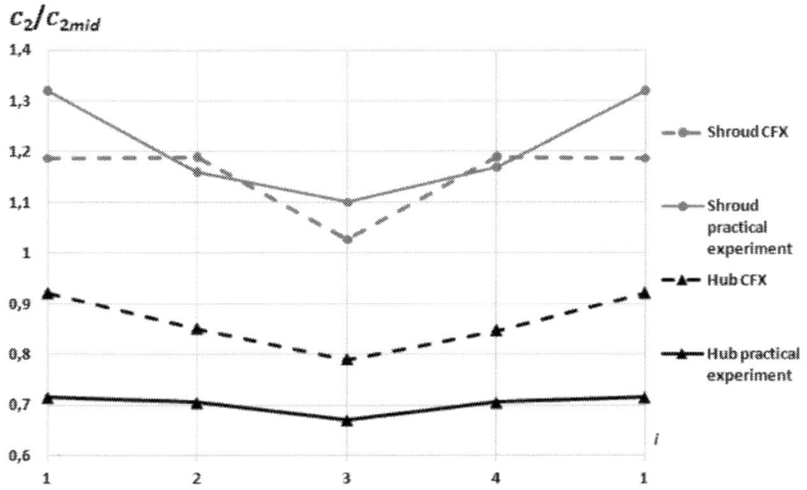

Figure 4. Comparison of the relative velocity distribution $\bar{c}_2 = c_{2i}/c_{2mid}$ in section 2 for the inlet device RIC3.

From figure 4 the qualitative coincidence of the velocity distribution is visible. The quantitative difference can be explained with the error of measurements at practical experiment and inaccuracy of iterative calculations by CFD methods.

On figure 5 the static pressure coefficient K_P distribution by circle in section 2-2 of the RIC1 inlet device is presented, where i – is a point of the static pressure tap. Loss coefficient is determined by the formula:

$$K_P = \frac{P_{1mid} - P_{2i}}{\rho_{2i} \cdot \frac{c_{2mid}^2}{2}}, \tag{1}$$

where P_{1mid}– static pressure on the inlet section in front of the inlet chamber, settling down behind the straight section of constant diameter; P_{2i} and ρ_{2i}–static pressure and density in section 2 in the point of pressure tap i respectively.

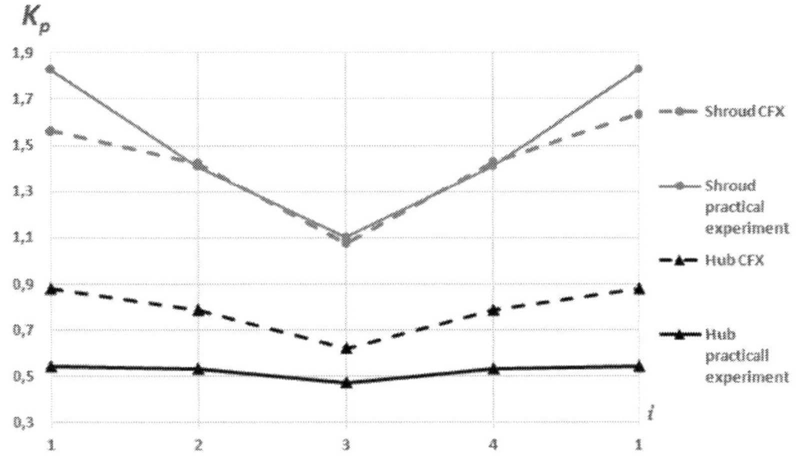

Figure 5. Comparison of the static pressure the K_P coefficient distribution by circle in section 2-2 of the RIC3 inlet device

The loss coefficient for the chamber is determined by the formula:

$$\zeta = \frac{P^*_{1mid} - P^*_{2mid}}{\rho_{2mid} \cdot \frac{c^2_{2mid}}{2}}, \qquad (2)$$

where P^* - total pressure.

By the results of the practical experimental investigations at the Nevsky plant the loss coefficient of the chamber on the design mode equals $\zeta_{exp_p} = 0{,}05$.

The loss coefficient of the chamber on the design mode by the results of calculation in ANSYS CFD:

$$\zeta_{CFD_p} = \frac{P^*_{1mid} - P^*_{2mid}}{\rho_{2mid} \cdot \frac{c^2_{2mid}}{2}} = \frac{98096{,}7 - 97812{,}375}{1{,}117 \cdot \frac{96{,}2^2}{2}} = 0{,}055 \qquad (3)$$

The difference of ζ calculation from the experimental value is about 9%, and $\zeta_{CFD_p} > \zeta_{exp_p}$. This error is admissible and shows good similarity of calculations for the inlet device.

7. General conclusions by the results of calculations for five chambers:

From the results of calculation for all five chambers it is possible to make the general conclusions for the results of calculation and their similarity to the results of practical experiment of the Nevsky plant. It is also possible to make the comparative analysis of all five chambers by the results of numerical calculation.

For example on figure 6 the summary plots are shown which are received for the comparison of relative velocity $\bar{c}_2$distribution on the shroud area. Also such plots were received for the hub area for the relative velocity and the static pressure coefficient K_p on the hub and shroud area. Data of the loss coefficient values ζ are summarized in table 1.

Table 1. Loss coefficient ζ for five different chambers

Chamber variant	Loss coefficient ζ	
	CFD results	Experimental
RIC1	0,13	0,11
RIC2	0,115	0,55
RIC3	0,055	0,05
RIC4	0,293	0,35
RIC5	0,092	0,19

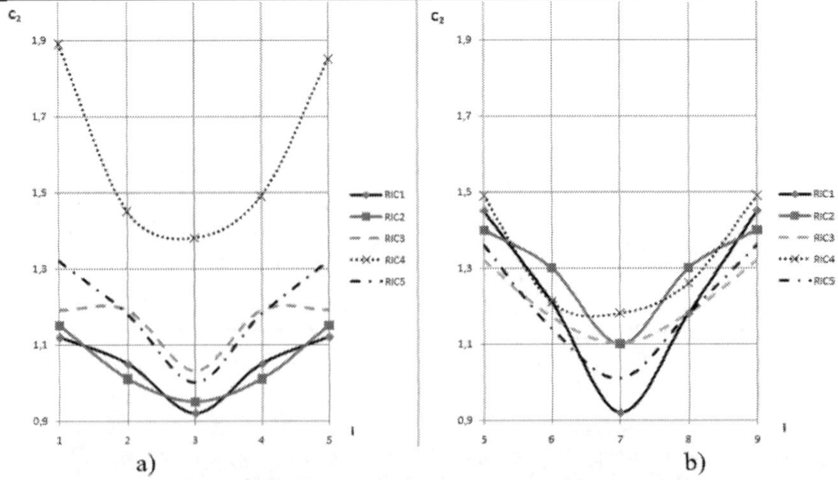

Figure 6. The relative velocity distribution $\bar{c}_2$ on the shroud area for five chambers: a) by the results of the CFD calculation, b) by the results of the practical experiment on the Nevsky plant

From the table 1 it is visible that the RIC1, RIC3 and RIC4 chambers correspond to experiment results of the numerical calculation best of all. The result received at the practical experiment for the RIC2 is in doubt as in comparison with chambers of the similar configuration the value of the loss coefficient is unreasonably overestimated to value $\zeta = 0{,}55$. Various negative factors which occurred while carrying out the experiment may be the reason.

As it was already told, as a result of numerical calculation by the loss coefficient value the best chamber is the RIC3 chamber, the worst is the RIC4. The others are intermediate between them.

From figure 6 it is visible that the nature of the calculated relative velocity distribution corresponds to the received values of the practical experiment. Thus, qualitative similarity between the results of calculation and experiment is available. Quantitative similarity is not observed at all chambers. The greatest divergence with the practical experiment is shown by the RIC2 and RIC4 chambers. One of the reasons for inaccuracy could be that at practical experiment the values on the hub region could cause difficulties and that could distort the results.

Estimating smoothness of the flow in the chambers, it is possible to notice that the best characteristics are at the RIC1 and RIC2, there is no flow separation in them after turning. Contraflows at the small central deflector could be observed at the RIC1 and RIC5 chambers. Separations and contraflows are especially accurately investigated at the RIC1 and RIC5, but also present at the RIC2 and RIC4.

In general, the results received from numerical experiment don't contradict the results of practical experiment and reflect the flow picture in chambers. Thus similar calculations are recommended for research and optimization of the inlet chambers of centrifugal compressors.

8. Research of the deflectors influence on the chamber's efficiency:

For research of the deflectors influence on the chamber's efficiency the RIC4 chamber was chosen since its design allows making the changes more evident.

Having analyzed the results of the numerical experiment, it was revealed that the best characteristics the RIC3 chamber possesses in which there are no side deflectors. Therefore the decision was made in the course of investigations to take away at first the part, and then the others side deflectors out of the RIC4 chamber.

On figure 7 the distinctions in the geometrical form between initial chamber variant and two changed chambers of the RIC4 are presented. Further the changed RIC4 chamber variants without short deflectors and completely without them are named RIC4-v1 and RIC4-v2 respectively.

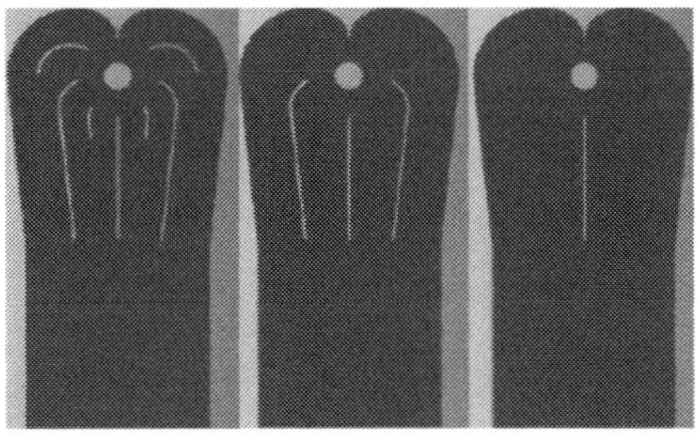

a) b) c)

Figure 7. CAD-models of the RIC4 chamber variants: a) RIC4, b) RIC4-v1, c) RIC4-v2

From the pattern of flow around the side deflectors which were received as a result of the calculations their shock flow was found out. The central deflector is flowed around in all cases without shocks.

For the changed chambers distributions of the relative velocity c_2/c_{2mid}, static pressure K_p and loss coefficient ζ were calculated. Values of loss coefficients calculated in CFD for three different variants are summarized in table 2. Values of relative velocity calculated in CFD and static pressure coefficient in comparison with the same sizes for an initial chamber RIC4 are given in figure 8 and figure 9

Table 2. Loss coefficient ζ for the RIC4, RIC4-v1 and RIC4-v2

Parameters	Chamber variants		
	RIC4	RIC4-v1	RIC4-v2
Loss coefficient ζ	0,293	0,291	0,229

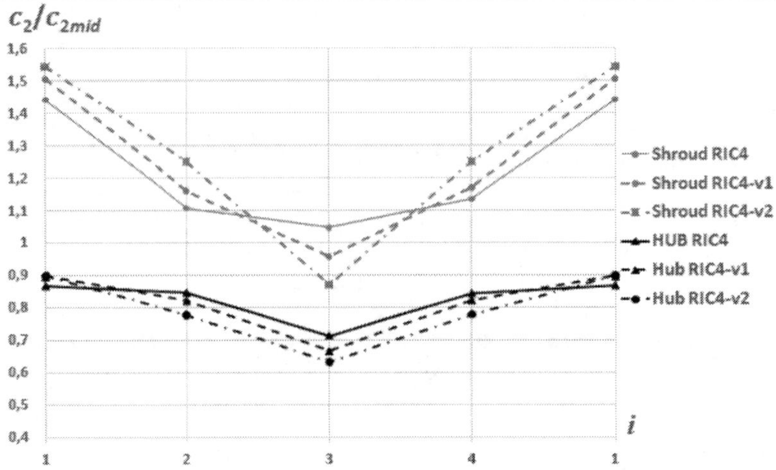

Figure 8. Comparison of the velocity distribution $\bar{c}_2 = c_{2i}/c_{2mid}$by the results of the CFX calculation.

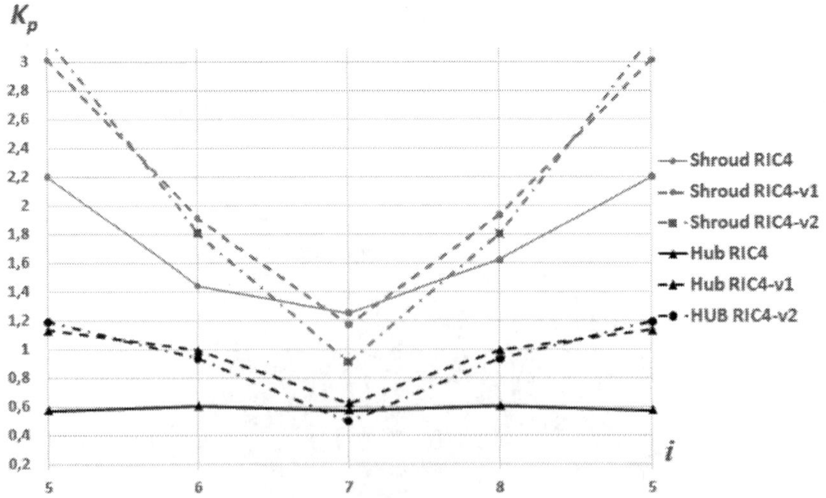

Figure 9. Comparison of the static pressure coefficient K_P distribution by the results of the CFX calculation.

The result when part of the deflectors was taken away from the RIC4 chamber, the loss coefficient did not change significantly while the flow became less uniform. Unevenness is explained by dividing deflectors being taken away from the chamber.

The chamber completely without deflectors shows better value of the loss coefficient due to the lack of loss arising at the flow while flowing around the deflectors. Unevenness of the flow at the chamber without deflectors has the greatest value in comparison with the variant with deflectors. Thus existence of the dividing deflectors in the chamber brings bigger loss, but does make the flow more uniform. Therefore, when designing the inlet chambers it is needed to pay special attention to the existence of the deflectors.

9. Conclusion:

The purpose of this work was to investigate five variants of the inlet radial chambers of the centrifugal compressor develop and research of the deflectors influence on the efficiency of one of the chambers tested at the Nevsky plant.

Received results were compared with the data of experimental investigations of the Nevsky plant by three parameters: relative velocity distributions $^{c_2}/_{c_{2mid}}$, static pressure coefficient $K_p$at the outlet section of the chamber and the loss coefficient ζ of the chamber. Research of influence of the side deflectors on parameters of the chamber's flow is made. The results of research prove that the presence of deflectors in the flow increases the probability of the flow separation and reverse flows on them. Full lack of deflectors in the chamber considerably increases the circular unevenness of the flow, thus nevertheless the loss coefficient from 0,293 to 0,229 where high values are caused by existence of the shock flow of deflectors. It is also possible to note the possibility of the analysis of the flow visualization when calculating in ANSYS CFD which allows defining the regions of the flow separation and contraflow in the chamber.

Results of the calculation show good compliance to the real flow and can be used for further researches and recommendations to use the ANSYS CFD for calculations of inlet radial chambers of centrifugal compressors. The possibility to increase the efficiency of inlet devices of turbocompressors is shown by the CFD methods.

At the same time it is necessary to specify that definition of pressure loss for the whole flow path of the centrifugal compressor or the stage has to be made for the whole compressor in general. It is connected with the fact that different elements of centrifugal compressor such as inlet chamber, impeller, diffuser etc. influence on each other in the flow. This fact has been confirmed in the practical experiment for the inlet chambers as a part of the compressor stage described in this work.

References:

[1] N A Karnauhova, Y V Gakerkin, Y V Kozhukhov. Research of the three dimentional flow at the radial inlet device of the centrifugal compressor with the use of the software package of ANSYS CFD, *the XL Week of science SPbSPU: materials of the international scientific and practical conference, SPb, Publishing house of Polytechnical university,* 2011.

[2] "Issues Surrounding Multiple Frames of Reference Models for Turbo Compressor Applications", *Z.Liu, D.L Hill, International Compressor Engineering Conference,* 2000.

[3] V K Yun. The analysis and assessment of the design methods of the flow path of centrifugal compressor machines//*Works of the XV international scientific and technical conference on compressor equipment. Part I. Kazan, 2011* pages 362-370.

IOP Conf. Series: Materials Science and Engineering **90** (2015) 012048 doi:10.1088/1757-899X/90/1/012048

NUMERICAL STUDY OF CENTRIFUGAL COMPRESSOR STAGE VANELESS DIFFUSERS

Y. GALERKIN, K. SOLDATOVA, O. SOLOVIEVA
R&D Laboratory "Gas dynamics of turbo machines" Peter the Great St.Petersburg Polytechnic University, Polytechnical st. 29, St.Petersburg, Russia
Email:buck02@list.ru.

Abstract: The authors analyzed CFD calculations of flow in vaneless diffusers with relative width in range from 0.014 to 0.100 at inlet flow angles in range from 10^0 to 45^0 with different inlet velocity coefficients, Reynolds numbers and surface roughness. The aim is to simulate calculated performances by simple algebraic equations. The friction coefficient that represents head losses as friction losses is proposed for simulation. The friction coefficient and loss coefficient are directly connected by simple equation. The advantage is that friction coefficient changes comparatively little in range of studied parameters. Simple equations for this coefficient are proposed by the authors. The simulation accuracy is sufficient for practical calculations. To create the complete algebraic model of the vaneless diffuser the authors plan to widen this method of modeling to diffusers with different relative length and for wider range of Reynolds numbers.

Nomenclature

b - width of diffuser;

c_2 - inlet velocity;

c_3 - outlet velocity;

D_2 - inlet diameter;

D_3 - outlet diameter;

h_w - lost head;

k - isentropic coefficient;

k_{rg} - surface roughness;

p_2 - static pressure at VLD inlet;

p_3 - static pressure at VLD outlet;

R – gas constant;

Re_{D2} - Reynolds number;

T_2 - static temperature at VLD inlet;

T_3 - static temperature at VLD outlet;

T_{t2} - total temperature at VLD inlet;

u_2 - blade velocity;

v_2 - kinetic viscosity at VLD inlet;

α_2 - flow angle at VLD inlet;

η - efficiency;

$\Delta\eta$ - loss of efficiency;

Content from this work may be used under the terms of the Creative Commons Attribution 3.0 licence. Any further distribution of this work must maintain attribution to the author(s) and the title of the work, journal citation and DOI.
Published under licence by IOP Publishing Ltd

λ_{c2} - velocity coefficient;
λ_{fr} - friction coefficient;
ξ - recovery coefficient;
ψ_T - loading factor;
ζ - loss coefficient;

Abbreviation
VLD – vaneless diffuser.

1. INTRODUCTION

Engineering methods of gas dynamic design and centrifugal compressor performances calculation are developing in many countries in spite of CFD methods wide application. There is no contradiction as engineering methods are and will be necessary as instruments of optimization of flow path main sizes and of quick overall analysis. Math models applied to calculate head losses in flow path elements step by step. The Universal modeling [1] principle is to identify models of all elements by comparison of calculated and measured efficiency of stages. This method is successfully applied in design practice since 1990th and is improved constantly [2, 3, 4, 5]. The authors propose simple algebraic equations instead of not the simple math model of a vaneless diffuser. The database for the equations is CFD calculations presented in [6, 7, 8, 9]. The authors of these works have compared different nets and turbulence models. Good correlation of CFD and math model performance curves of a stage stator parts (VLD included) is presented in [10]. This may be treated as an indirect proof of CFD calculation validity. The stage math model with the proposed equations will be identified by measured performances of stages. It will lead to necessary corrections.

2. BASIC INFORMATION ON VANELESS DIFFUSERS

Vaneless diffuser is the simplest element of a stage flow path. It is disposed between control planes $D_2 - D_3$ at Figure 1. But for a period of time the flow behavior in it was interpreted not properly. The empirical character of its performances is misleading [1]. It demonstrates Figure 2.

Loss coefficient has minimum if it is calculated on a base of pressures measured at a control plane disposed close to an impeller. The measurements at diameter 1.05 were traditional for Russian researchers [11]. The analogy with channel diffuser and its optimum divergence angle was the model to explain measured performances at Figure 2. Separation losses are absent in accordance with this model at low flow angles. The opposite situation must be if flow angles are big.

The study of flow structure, equilibrium equation analysis [12], experiments with diffusers separately of an impeller [13] demonstrated the real situation. The bigger is flow angle the lower are friction losses and less is flow separation possibility. The structure of flow based on measurements in an experimental stage is presented at Figure 3.

The difference of measured and calculated loss coefficients at Figure 2 is explained by flow structure at an impeller exit. Jet-wake mixing losses are result of impeller imperfectness. Mixing process is shown at Figure 4.

Mixing process ends at a distance 20% of an impeller diameter. It coincides with experimental data [12]. Diffuser performances calculated by measurements between $1{,}2D_2 - D_3$ do not depend of an impeller. It is also proven experimentally [12]. At Figure 2 mixing losses is the difference of measured and calculated values of the VLD loss coefficient.

The aim of this work is to propose simple algebraic model for loss coefficient of VLD. It could be used in computer programs of the Universal modeling method [4, 5, 11]. In the Method all impeller losses are calculated as if they start and end inside of it. Therefore the model of VLD loss coefficient must be simulated as it is shown at Figure 2 by solid line.

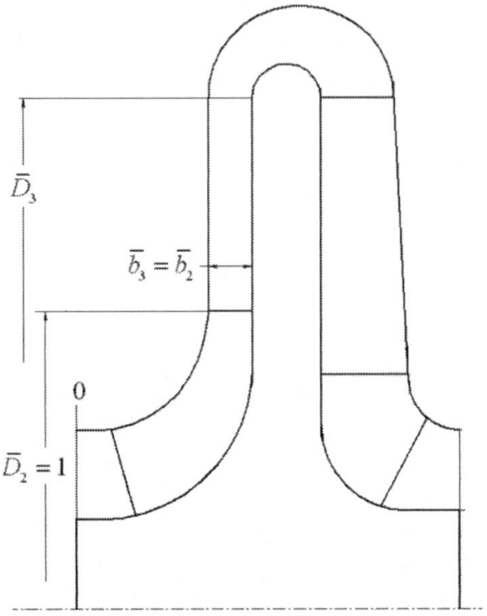

Figure 1. Vaneless diffuser in an industrial centrifugal compressor stage

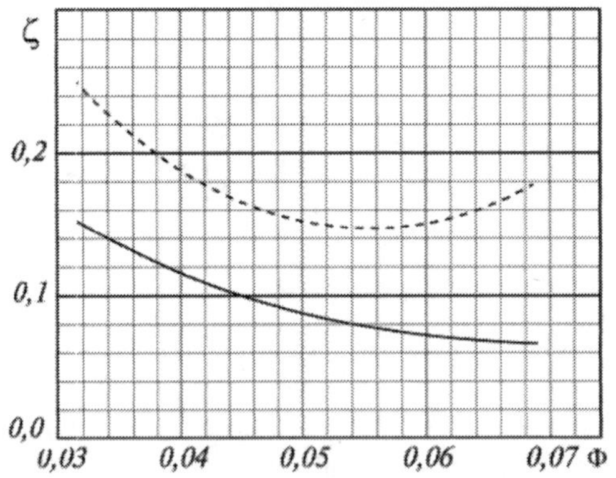

Figure 2. Measured and calculated vaneless diffuser loss coefficient performance curves. Stroke line – measurement, solid line – correct calculation [1]

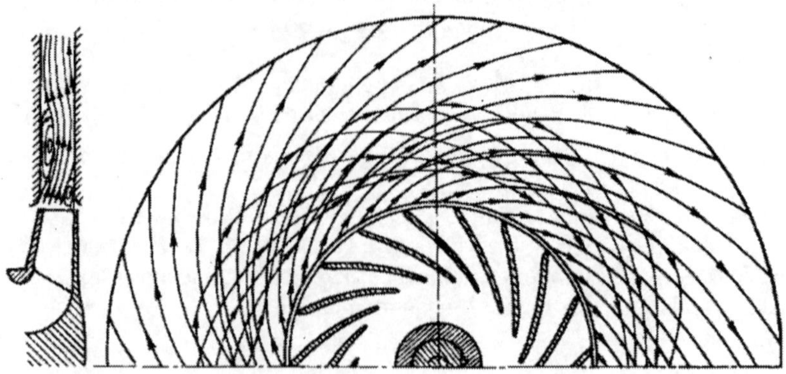

Figure 3. Flow separation in VLD at low flow angle [12]

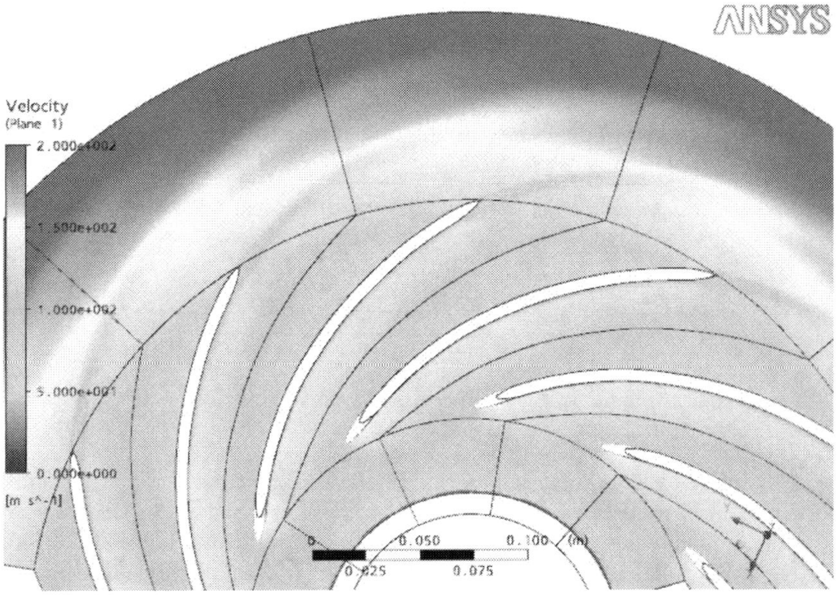

Figure 4. Relative velocity field in an impeller and VLD (design flow rate) [1]

3. OBJECTS AND METHODS

The objects for algebraic model creation were VLD with relative width in range $b/D_2 = 0.014 - 0.10$ and relative length $D_3/D_2 = 1.60$. The regime parameter is an inlet flow angle in range $\alpha_2 = 10 - 45^0$. Similarity criteria were varied in range of the velocity coefficient $\lambda_{c2} = c_2 / \left(\frac{2k}{k+1} RT_{t2}\right)^{0.5} = 0.23 - 0.82$ and of Reynolds number $Re_{D2} = \frac{c_2 D_2}{\nu_2}$ =3.8E6 – 10.3E6. Isentropic coefficient $k = 1.4$. Surfaces were hydraulically smooth. The surface roughness influence was studied by additional calculations in range of $\kappa_{rg} / 2b = 0 - 2.2 \cdot 10^{-3}$.

The program ANSYS CFX 14.0 and the net generator ICEM, H-type structured nets with 225216 elements, $k - \varepsilon$ turbulence model were applied. The information about flow behavior in these objects is presented in [6, 7, 8]. Some key facts are:

- flow separation takes place in diffusers with $b / D_2 \geq 0.029 - 0.10$ at flow angles < 10 – 25^0. The bigger width and velocity coefficient are, the bigger the critical angle is;
- there is no separation at any conditions in diffusers with relative width less than 0,029. High level of shear stresses due to walls' closeness prevents separation but leads to lower efficiency;
- non viscid core disappears in all studied VLD. The smaller relative width and flow inlet angle are, the earlier non viscid core disappears;

The equations for correct calculation of VLD efficiency and loss coefficient in a base of calculated flow parameters are:

$$\eta = \frac{\ln\left(\frac{p_3}{p_2}\right)}{\frac{k}{k-1}\ln\left(\frac{T_3}{T_2}\right)} \quad (1)$$

$$\zeta = \frac{h_w}{c_2^2/2} = \left[1-(c_3/c_2)^2\right](1-\eta) \tag{2}$$

4. DIFFUSER PERFORMANCE MODELING – SMOOTH SURFACE

In principle the performance $\zeta = f(\alpha_2)$ depends on:

- non-dimensional sizes $b/D_2, D_3/D_2$ and relative surface roughness $k_{rg}/2b$,
- flow inlet angle α_2,
- similarity criteria $\lambda_{c2}, \mathrm{Re}_{D2}, k$.

Therefore:

$$\zeta = f\left(b/D_2,\, D_3/D_2,\, k_{rg}/2b, \alpha_2,\, \lambda_{c2}, \mathrm{Re}_{D2},\, k\right) \tag{3}$$

The aim is to simulate CFD calculations by means of algebraic equations. The simplified analysis of VLD performance curves is presented in [12]. It is based on assumption that the constant value of a friction coefficient can be applied to estimate loss coefficient for diffusers of different configurations operating at different flow angles. The loss coefficient and the friction coefficient are connected by the equation:

$$\zeta = \lambda_{fr}\frac{1-\dfrac{1}{D_3/D_2}}{4\dfrac{b}{D_2}\sin\alpha_2} \quad \text{or} \quad \lambda_{fr} = \zeta\frac{4\dfrac{b}{D_2}\sin\alpha_2}{1-\dfrac{1}{D_3/D_2}} \tag{4}$$

In fact, the friction coefficient cannot be of constant value. The coefficient depends on the same parameters presented as arguments in eq. (3). From the other side, estimated in [12] diffuser performance curves are rather similar to expected ones. It shows that values of λ_{fr} are surely not constant but differ little. The result of simulation on the base of λ_{fr} is presented for the series of diffusers with most usual relative length $D_3/D_2 = 1{,}60$. Surfaces are hydraulically smooth. The graphic representation of λ_{fr} recalculation by eq. (4a) is presented at Figure 5 for λ_{c2}= 0.82, Re_{D2}= 10.3E6. The results for λ_{c2}= 0.64, Re_{D2}= 8.04E6 and λ_{c2}= 0.39, Re_{D2}= 4.9E6 are the same by character and differs not too much in Figures.

There are some facts that deserve special attention. At flow angle 45^0 the loss coefficients $\lambda_{fr\,45^0}$ are constant practically for VLD with different relative width at given Re_{D2}. The values of $\lambda_{fr\,45^0}$ for different Re_{D2} and λ_{c2} are presented in the Table 1.

Table 1

Friction loss coefficient $\lambda_{conv\,45^0}$ versus similarity criteria

Re_{D2}	4.9E6	8.04E6	10.3E6
λ_{c2}	0.39	0.64	0.82
$\lambda_{fr\,45^0}$	0.0238	0.0215	0.0209

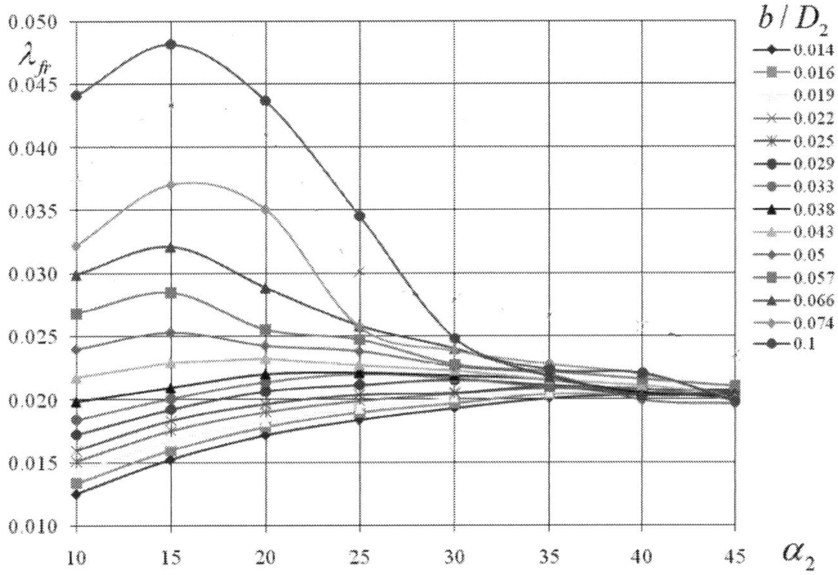

Figure 5. Friction coefficient λ_{fr} as function of inlet flow angle for series of VLD with b/D_2= 0.014 – 0.10. D_3/D_2= 1.60, λ_{c2}= 0.82, Re_{D2}= 10.3E6

Many authors use Reynolds criteria $\mathrm{Re}_b = \frac{2b \cdot c_2}{\nu_2}$ for vaneless diffusers. Evidently it does not work for the friction coefficient λ_{fr45^0}. Data at Figure 5 show that the friction coefficient λ_{fr45^0} is independent of b/D_2, i.e. of Re_b. From the other side data in the Table 1 shows that the bigger are λ_{c2} and Re_{D2} the lower is λ_{fr45^0}. It points on the fact that positive influence of higher Re_{D2} exists and prevails on negative influence of λ_{c2}.

The smaller is α_2 the lower is λ_{fr} for diffusers with $b/D_2 = 0.014 - 0.038$. Separation losses are absent or small in these diffusers. In diffusers with relative width $b/D_2 \geq 0.057$ higher friction coefficient value corresponds to inlet angles. Mixing losses prevail in wide diffusers.

The diffusers with b/D_2= 0.043 – 0.050 have practically constant values of the friction coefficient. Not monotonous part of the function $\zeta = f(\alpha_2)$ corresponds to regimes with developed separation zones. Real stages don't operate at these regimes due to dangerous unsteady processes.

Data analysis has shown that at inlet flow angle 10^0 friction coefficient λ_{fr10^0} is a linear function of a diffuser relative width. The values are independent of λ_{c2} and Re_{D2}.

The above mentioned facts have made possible to offer simple set of Eq. (5-7) to simulate results of numerical experiment:

$$\lambda_{fr} = \lambda_{fr45^0} + \left(\lambda_{fr10^0} - \lambda_{fr45^0}\right)\left(\frac{\alpha_{2\max} - \alpha}{\alpha_{2\max} - \alpha_{2\min}}\right)^{X_1} = \lambda_{fr45^0} + \left(\lambda_{fr10^0} - \lambda_{fr45^0}\right)\left(\frac{45^0 - \alpha_2}{35^0}\right)^{X_1} \quad (5)$$

$$\lambda_{conv45^0} = X_2 \frac{0,316}{\mathrm{Re}_{D2}^{0,25}} \tag{6}$$

$$\lambda_{fr10^0} = \lambda_{fr10^0 b\min} + \frac{\lambda_{fr10^0 b\max} - \lambda_{fr10^0 b\min}}{b_{\max}/D_2 - b_{\min}/D_2}\left(b/D_2 - b_{\min}/D_2\right) =$$

$$= 0,012 + 0,314\left(b/D_2 - 0,014\right). \tag{7}$$

Coefficients X_1, X_2 are empirical coefficients proposed by the authors. Eq. (6) looks as known Blasius formula for a friction coefficient of a pipe. Reynolds number in the eq. (6) is proportional to an inlet diameter of a diffuser. Reynolds number in Blasius formula is proportional to a diameter of a pipe. Therefore the value of X_2 is far from 1.

Graphics at Figure 6 demonstrate results of λ_{fr} simulation for several diffusers.

Simulation doesn't match CFD results in zones of flow separation in wide diffusers. It is not important practically. Average error for all VLD at all three λ_{c2} and Re_{D2} is less than 1,2% - separation zones included. Average error for non-separated zones is 0.5%. The error of stage efficiency estimation is much less. Loss of efficiency in a diffuser is [1]:

$$\Delta\eta = \frac{\zeta}{\psi_T} 0.5\left(\frac{c_2}{u_2}\right)^2. \tag{8}$$

Calculated by CFD and simulated performances $\zeta = f(\alpha_2)$ are presented at Figure 7.

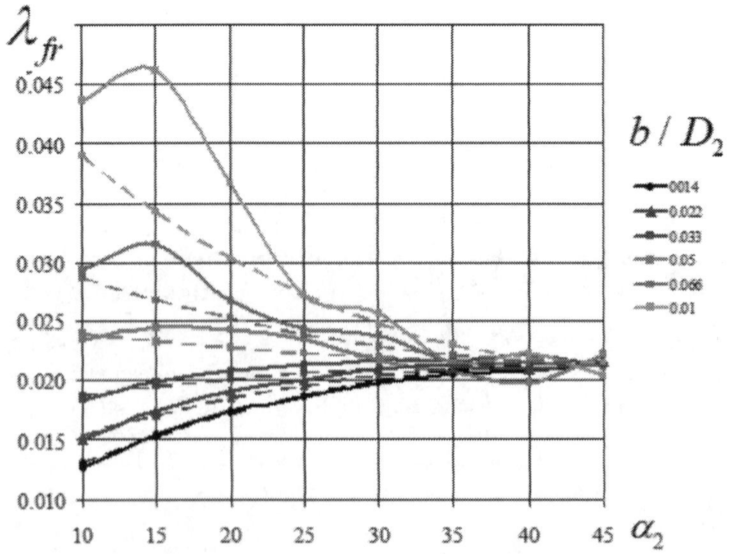

Figure 6. Friction coefficient λ_{fr} as function of inlet flow angle for series of VLD with $b/D_2 =$ 0.014 – 0.10. $D_3/D_2 = 1.60$, $\lambda_{c2} = 0.64$, $\mathrm{Re}_{D2} = 9.2 \cdot 10^6$
Solid – CFD calculation, stroke – simulation by eq. (5 -7)

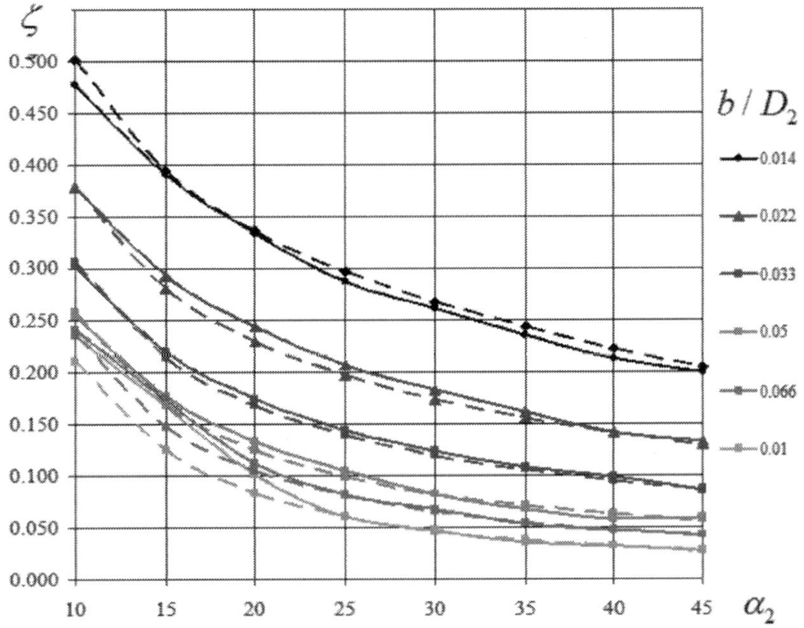

Figure 7. Loss coefficient as function of inlet flow angle for series of VLD with $b / D_2 =$ 0.014 – 0.10. $D_3 / D_2 = 1.60$, $\lambda_{c2} = 0.64$ и $\mathrm{Re}_{D2} = 9.2 \cdot 10^6$. Solid – CFD calculation, stroke – simulation by eq. (5 -7)

5. SURFACE ROUGHNESS INFLUENCE

The sample of surface roughness influence is presented at Figure 8. The range of relative roughness covers the zone of practical interest. For instance, in case of an impeller with diameter 0.85m the width of the VLD that is presented at Figure 8 is $0,8 \cdot 0,057 \approx 0,046\,m$. The maximum absolute roughness is $(0,046 \cdot 2 \cdot 2,2 / 10^3) \cdot 10^6 \approx 200\,\mu m$, i.e. 0.2 mm. Accordingly, the minimal relative roughness 0.055E-3 corresponds to $5\,\mu m$.

Calculation analysis has shown that friction coefficient of rough surfaces $\lambda_{fr\,rg} = f(\alpha_2)$ has the same character as in case of hydraulically smooth surface. I.e. the eq. (5) can be applied in principle. Therefore for the "rough" friction coefficient simulation is sufficient to find function $\lambda_{fr\,rg\,45^0} = f\left(\frac{k_{rg}}{2b}\right)$ and $\lambda_{fr\,rg\,10^0} = f\left(\frac{k_{rg}}{2b}\right)$.

The ratios $(\lambda_{fr\,rg} / \lambda_{fr})_{45^0}$ and $(\lambda_{fr\,rg} / \lambda_{fr})_{10^0}$ appeared to be the linear functions of the relative roughness logarithm presented as it is shown at Figure 9.

$$\left(\frac{\lambda_{fr\,rg}}{\lambda_{fr}}\right)_{45^0} = 1 + 4,37 \cdot \left(\frac{1}{-\lg\left(\frac{k_{rg}}{2b}\right)} - \frac{1}{-\lg\left(0,031 \cdot 10^{-3}\right)} \right). \tag{9}$$

$$\left(\frac{\lambda_{fr\,rg}}{\lambda_{fr}}\right)_{10^0}=1+2,62\cdot\left(\frac{1}{-\lg\left(\frac{k_{rg}}{2b}\right)}-\frac{1}{-\lg\left(0,031\cdot10^{-3}\right)}\right). \quad (10)$$

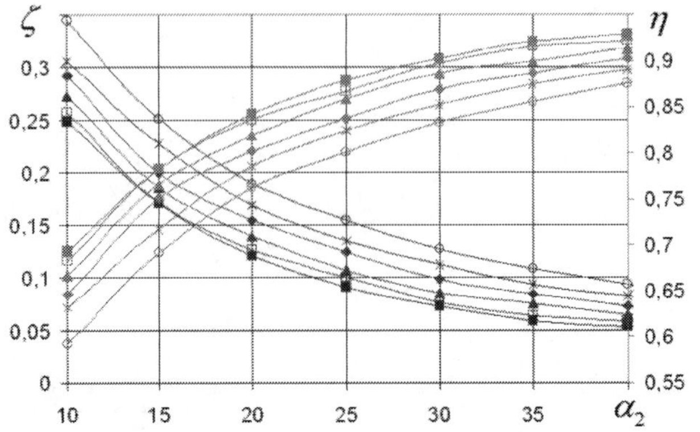

Figure 8. VLD performanes with different relative roughness. b/D_2=0.057, λ_{c2} = 0.64, Re_{D2}=8.04E6. $\kappa_{rg}/2b = 0.055\cdot10^{-3}$, $0.11\cdot10^{-3}$, $0.55\cdot10^{-3}$, $1.1\cdot10^{-3}$, $2.2\cdot10^{-3}$

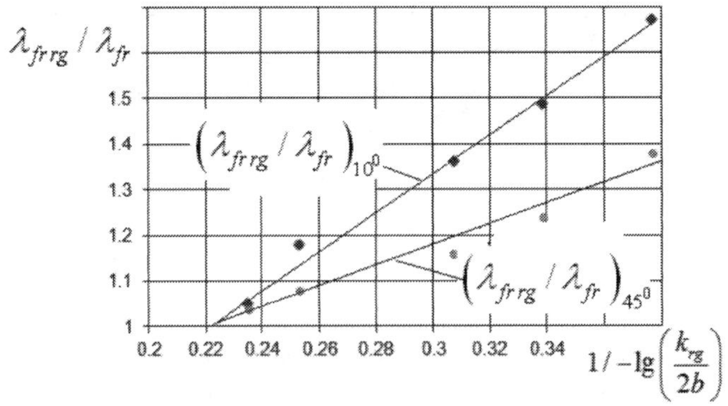

Figure 9. Relation of rough and smooth surfaces friction coefficient. The linear function are presented by eq. (9) and (10)

The ratio $\lambda_{fr\,rg}/\lambda_{fr}$ is equal to one when the relative roughness is $k_{rg}/2b = 0,031\cdot10^{-3}$. It means that the surface is hydraulically smooth when $k_{rg}/2b \le 0,031\cdot10^{-3}$. In case of rough surface a friction coefficient can be calculated by Eq. (5, 6, 7) meaning that:

$$\lambda_{fr\,rg\,45^0}=\lambda_{fr\,45^0}\left(\frac{\lambda_{fr\,rg}}{\lambda_{fr}}\right)_{45^0},\quad \lambda_{fr\,rg\,10^0}=\lambda_{fr\,10^0}\left(\frac{\lambda_{fr\,rg}}{\lambda_{fr}}\right)_{10^0}. \quad (11)$$

CONCLUSION

The presented way to simulate CFD – calculated VLD loss coefficient performance by algebraic equations is simple and precise. The analysis of VLD performances with different relative length and more profound study of Reynolds numbers and relative roughness will solve the problem of algebraic equations model of VLD for engineering centrifugal stage design methods.

ACKNOWLEDGEMENT

Work is performed with support of a Grant of the President of the Russian Federation for young PhD MK-7066.2015.

References

[1] Galerkin, Y.B. Tubrocompressors. [text] /Y.B.Galerkin// LTD information and publishing center. – Moscow. – KHT. -2010. – P. 596. [Russian].
[2] Galerkin Y., Soldatova K., Drozdov A. Modern state of the universal modeling for centrifugal compressors. [text] /Y.B.Galerkin, K. V.Soldatova , A.A. Drozdov// International Conference on Numerical Methods in Industrial Processes. World Academy of science, engineering and technology. Paris 2015 Conference. http://www.waset.org/Publications/?path=Publications. - Vol:9 No:01 2015. – № 242.
[3] Galerkin Y., Soldatova K., Drozdov A. New version of the universal modelling for centrifugal compressor gas dynamic design. [text] /Y.B.Galerkin, K. V.Soldatova , A.A. Drozdov// Purdue Conference 2014. USA.
[4] Soldatova, K.V. Universal modeling method application for development centrifugal compressor model stages. [text] / K.V.Soldatova //Conference «Compressors and their Systems».– London. – 2013. – P. 477-487.
[5] Galerkin, Y.B., Soldatova, K. V. Industrial centrifugal compressor model stages based on compressor tests results. [text] /Y.B.Galerkin, K.V. Soldatova// International symposium SYMKOM 2011. – Lodz. - Poland.
[6] Galerkin Y.B, Solovieva O.A. Flow behavior and performances of centrifugal compressor stage vaneless diffusers. [text] /Galerkin, Y.B., Solovieva, O.A. // International Conference on Numerical Methods in Industrial Processes. World Academy of science, engineering and technology. Paris 2015 Conference.
[7] Y.B. Galerkin, O.A. Solovieva. Improvement of vaneless diffusers calculation based on CFD experiments. Part 1 [text] /Galerkin, Y.B., Solovieva, O.A. // Compressors and pneumatics. - 2014. – № 3. – P. 35-41. [Russian].
[8] Y.B. Galerkin, O.A. Solovieva. Improvement of vaneless diffusers calculation based on CFD experiments. Part 2 [text] /Galerkin, Y.B., Solovieva, O.A. // Compressors and pneumatics. - 2014. – № 4. – P. 15-21. [Russian].
[9] Galerkin Y., Rekstin A., Soldatova K. Aerodynamic designing of supersonic centrifugal compressor stages. [text] /Galerkin, Y., Rekstin A., Soldatova K.// International Conference on Numerical Methods in Industrial Processes. World Academy of science, engineering and technology. Paris 2015 Conference. http://www.waset.org/Publications/?path=Publications. - Vol:9 No:01 2015. – № 216.
[10] Galerkin Y., Marenina L. Investigation and perfection of centrifugal compressor stages by CFD methods. [text] /Galerkin, Y, Marenina L.// International Conference on Numerical Methods in Industrial Processes. World Academy of science, engineering and technology. Paris 2015 Conference. http://www.waset.org/Publications/?path=Publications. - Vol:9 No:01 2015. – №. 249.
[11] Galerkin, Y.B., Rekstin A.F. Methods of research of centrifugal compressors. [text] /Y.B.Galerkin, A.F. Rekstin//. Leningrad. – 1968. [Russian]
[12] Seleznev, K.P., Galerkin, Y.B. Centrifugal compressors. [text] /K.P. Seleznev, Y.B.Galerkin // Leningrad. – 1982. – P. 271. [Russian].
[13] Jansen W. Quasi-unstady flow in a radial vaneless diffuser MIT. Report № 60. - October. 1960.
[14] Galerkin, Y.B., Danilov, K.A., Popova, E.Y.. Design philosophy for industrial centrifugal compressor. [text] /Y.B.Galerkin, K.A. Danilov, E.Y. Popova// Inernational Conference on Compressors and their Systems. – London: City University, UK. – 1999.

IOP Conf. Series: Materials Science and Engineering **90** (2015) 012049 doi:10.1088/1757-899X/90/1/012049

Spline for blade grids design

Andrei Korshunov*

Vladimir Shershnev*

Ksenia Korshunova**

*St. Petersburg Polytechnic University, Russia
**TU Ilmenau, Germany
E-mail: korshunov@spbstu.ru

Abstract. Methods of designing blades grids of power machines, such as equal thickness shape built on middle-line arc, or methods based on target stress spreading were invented long time ago, well described and still in use. Science and technology has moved far from that time and laboriousness of experimental research, which were involving unique equipment, requires development of new robust and flexible methods of design, which will determine the optimal geometry of flow passage.This investigation provides simple and universal method of designing blades, which, in comparison to the currently used methods, requires significantly less input data but still provides accurate results. The described method is purely analytical for both concave and convex sides of the blade, and therefore lets to describe the curve behavior down the flow path at any point. Compared with the blade grid designs currently used in industry, geometric parameters of the designs constructed with this method show the maximum deviation below 0.4%.

1. Introduction

The Since conversion of kinetic energy to potential energy happens mainly on the diffuser and the wheel of compressor stage, the lack of attention had been paid to the designing of return channel blades. However, it had been established, that by low-consuming compressor stages energy conversion is being performed on every element including return channel blades. Their importance can also be judged from the increasing amount of recent works [1,2] dealing with influence of the meridional section geometry on the blade efficiency. In this work the detailed step-by-step designing of the return channel blades has been described.

2. General explanations

By low-consuming low volume flow rate compressor stages the b_4/b_5 ratio can reach rather big values (up to 10)(Fig.1). As a consequence, the meridional width of the return channel increases and the value of angle α_5 (in Russian notation angles are defined between vector of flow and tangent to a circle) decreases to 15-20 degrees (by ultra-high-pressure compressor stages in low volume flow rate mode it can decrease even more).

Content from this work may be used under the terms of the Creative Commons Attribution 3.0 licence. Any further distribution of this work must maintain attribution to the author(s) and the title of the work, journal citation and DOI.
Published under licence by IOP Publishing Ltd

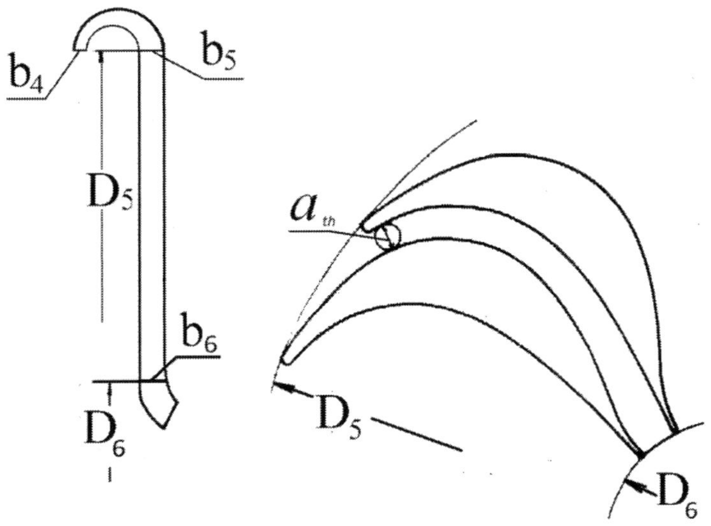

Figure.1 Meridional and radial view of return channel.

Since the gas flow has to have no twist on the next stage entrance (that is, the velocity component C_u=0 (tangential projection of flow velocity)), the return channel blade must be able to turn the gas flow almost 90 degrees. The length of the channel increases with the increasing of the turning angle of the flow, therefore the surface friction must be taken into account during the construction of the return channel blade. To decrease surface friction it is recommended to design channel segment with a square section which would minimize the surface of the channel.

The blade grid as a diffuser is an important part of the low volume flow rate compressor stage. It has to be designed in such a way, that the flow without separation would be possible on any compressor operating mode. To accomplish that, the blade width has to be increased (this kind of blade is called body blade).

One aim of the present work was to reduce the number of control parameters without sacrificing the accuracy of the blade geometry under given conditions. As a result of the current work the number of parameters has been reduced to 12 (Tab.1), which simplifies the equations and allows to solve the system of equations analytically.

Due to the fact, that return channel blade grid has complex geometry, it would be unpractical to set a middle line of the blade like it is commonly done for blades with uniform width. In this case, a concave side of the blade, i.e. its pressure side, was chosen as a base line for blade design. A cubic spline-function was used to design the convex side of the blade because of its controllability – since the cubic spline-function is constructed of piecewise third-order polynomials, it is a twice differentiable function, which allows one to control its behavior on a given interval effectively.

The convex side of the blade consists of two adjoining areas:

Segment I - entrance region (from leading edge to the section of the throat, i.e., to the beginning of the channel segment of the blade grid)

Segment II – channel region (from the section of the throat to the exit of the blade grid)

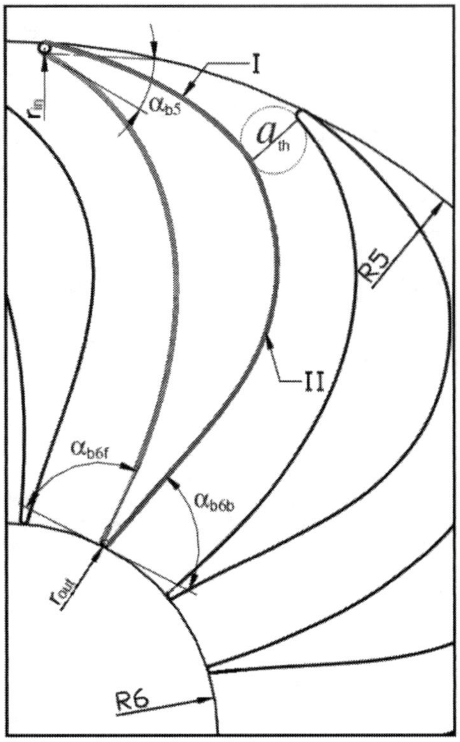

Figure. 2 – Description of blade grid parameters.

Control parameters	
Inlet angle on concave side	α_{b5}
Exit angle on concave side	α_{b6f}
Exit angle on convex side	α_{b6b}
Inlet radius	R_5
Exit radius	R_6
Inlet channel depth	b_5
Exit channel depth	b_6
Leading edge radius	r_{in}
Exit edge radius	r_{out}
Number of blades	Z
Absolute value of throttle size	a_{th}
Parameter	D

Table 1: Control parameters for blade greed design

Both segments I and II have equal first derivative values at the joining point, that is, the angle of the end of the entrance segment I (beveled cut area) is equal to the angle of the beginning of the segment II (channel part of the convex side of the blade), which allows to get a smooth, kink-free curve of the convex side of the blade.

Due to the rotational symmetry of the centrifugal compressors spline-function equations were transformed into cylindrical coordinates.

3. Concave side of the blade

In order to design a blade grid one can use flow parameters on inlet of blades, which can be obtained either experimentally or numerically. In his work [3], Carl Pfleiderer proposed a following method for designing of the blade middle line for radial flow compressors, which can be represented as:

$$\frac{d\varphi}{dr}=\frac{1}{r\,tg(\alpha)} \tag{1}$$

The alpha-parameter in Pfleiderer's work is defined as a constant. In this case, the solution of the equation (3.1.2.) would be a logarithmic spiral, i.e., the path of the undisturbed flow of the ideal liquid without taking friction into account. Later, in [4] the following equation was proposed:

$$\frac{d\varphi}{dr}=\frac{1}{r\,tg(\alpha(r))} \tag{2}$$

In this case, the angle α becomes a function of r (where r is a radial coordinate). The function $\alpha(r)$ has the following form:

$$\alpha(r)=\cos\left(Ar^2+Br+C\right) \tag{3}$$

The equation (3) was proposed in the work [4].
From physical point of view, this curve represents angular deflection from the undisturbed flow (where alpha is a constant) according to the parabolic law.

$$\cos(\alpha_{b5}) = AR_5^2 + BR_5 + C$$
$$\cos(\alpha_{b6f}) = AR_6^2 + BR_6 + C$$
$$D = 2AR_6 + B \qquad (3)$$

In order to find the coefficients in the equation (3) it is necessary to build an equation system (4) with boundary conditions and solve it for the said coefficients. As the boundary conditions, the geometrical parameters of the blade being designed are taken, such as R_5 (inlet radius of the blade grid), R_6 (outlet radius of the blade grid), $\cos(\alpha_{b5})$ and $\cos(\alpha_{b6f})$ (blade grid inlet and outlet angles, respectively).

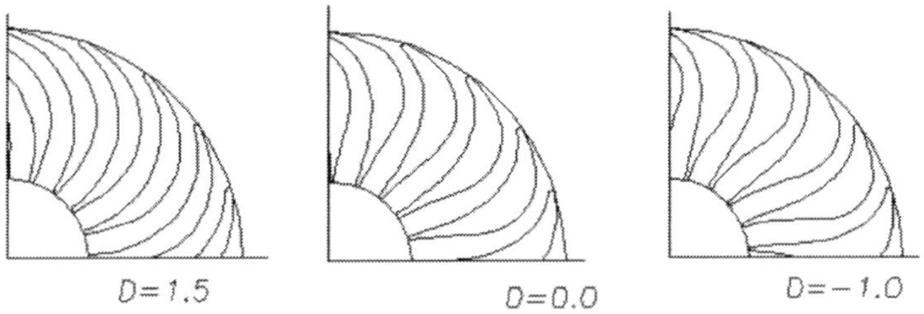

Figure. 3 – Influence of D-parameter on blade shape down to negative D

Since there are only two boundary conditions for three required coefficients, it is necessary to add the third equation to the system, which would be the first derivative of the boundary condition for R_6. That, in its turn, leads to the adding of the supplementary parameter D. This parameter defines the sign and the velocity of the tangential angle change along the length of the blade (Fig. 3 and 4).
If the parameter D is positive, the angular length of the blade increases; if negative, an inflection point appears on the concave side of the blade. Its exact position on the blade is also defined by this parameter – the smaller the value of D is, the closer the inflection point gets to leading edge. In case of D=0 the inflection point sits exactly on the radius R_6.
Coefficients A, B and C can be found by solving the system of equations (4), the rest of the system members are boundary conditions for blade designing.

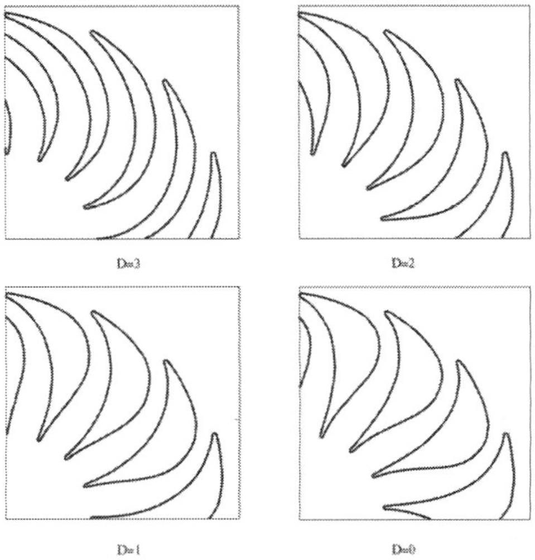

Figure. 4 – Influence of the parameter D on the angular length and the form of the body blades)

4. Convex side of the blade

In order to design the convex side of the blade one can use spline-function in polar coordinates, defined by two third-order polynomials:

$$\phi = A\,r^3 + B\,r^2 + C\,r + D \tag{4}$$

Coefficients of the cubic equation (4) can be defined by the use of first-type boundary conditions [5]. As a first step, coordinates of both ends of the curve must be set, then the values of the first derivative of the desired function on the ends of the given interval. This leads to the smooth connection between the inlet and the channel segments of the spline. As a result, the equation system for defining coefficients A,B,C and D of the cubic equation (4) has following form:

$$\begin{aligned} \phi_0 &= Ar_0^3 + Br_0^2 + Cr_0 + D \\ \phi_1 &= Ar_1^3 + Br_1^2 + Cr_1 + D \\ \frac{d\phi_0}{dr_0} &= 3\,Ar_0^2 + 2\,Br_0 + C \\ \frac{d\phi_1}{dr_1} &= 3\,Ar_1^2 + 2\,Br_1 + C \end{aligned} \tag{5}$$

To define the position of the inflection point on the inlet segment of the convex side of the blade (Segment I on the Fig. 2) it is necessary to add a new condition: the second derivative of the equation (4) with respect to r equals zero. The following relation describes position of the inflection point:

$$\begin{aligned} \frac{d\,\phi}{dr} &= 6\,Ar + 2\,B \\ r &= \frac{B}{3\,A} \end{aligned} \tag{6}$$

Best results were obtained when the inflection point sits on the inlet radius of the blade grid R_5. The channel segment of the blade (Segment I on the Fig.2) agrees in coordinates and in values of the first derivatives with the inlet segment, described with spline-function.

Conclusion

As a result of the current studies a designing method for return blade grid had been developed. It allows to design blades with constant thickness as well as body blades and can be applied for both axial and radial kind of blades. Another significant advantage of this method is its simplicity, since the number of control parameters was significantly reduced. The comparison of the geometrical parameters of the existing blade designs with the parameters calculated by the use of the proposed method revealed a very low (>0,4%) deviation. This leads to a conclusion, that, despite of its simplicity, the proposed method does not sacrifice the accuracy of the blade geometry and its physical adequacy.

References

[1] Ludtke, K. H. 2004. Process Centrifugal compressor. Basics, Function, Operation, Design, Application. Berlin Heidelberg: Springer-verlag, – 328p.

[2] S.Pazzi, F.Martelli, V.Michelassi, M. 2002. Giachi The use of artificial neural networks for performance prediction of return channels for industrial centrifugal compressors. Amsterdam – ASME TURBO EXPO – 2002. p. 273-287.

[3] C. Pfleiderer, H. Petermann. 2005. Stromungsmaschinen. Springer-Verlag Berlin Heidelberg. 589 p.

[4] Стрижак, Л.Я. Исследование влияния формы межлопаточных каналов центробежного компрессорного колеса на его характеристики. Дисс. канд. тех. наук / Л.Я. Стрижак – Л – ЛПИ – 1968 – 235 с.

IOP Conf. Series: Materials Science and Engineering **90** (2015) 012049 doi:10.1088/1757-899X/90/1/012049

[5] Шикин, Е.В. Кривые и поверхности на экране компьютера. Руководство по сплайнам для пользователей. Е.В.Шикин, А.И. Плис – М.: Диалог-МИФИ , 1996. – 237 с.

IOP Conf. Series: Materials Science and Engineering **90** (2015) 012050 doi:10.1088/1757-899X/90/1/012050

Performance Measurements of a Low Specific Speed TurboClaw® Compressor

J. Parra, R. Cattell, S. Etemad
Dynamic Boosting Systems Ltd, London

K R Pullen
City University London, School of Mathematics, Computer Science and Engineering

Abstract

Low specific speed compressors have been historically based on positive displacement machines. Attempts to bring advantages of turbomachinery such as oil free, low parts counts, low cost of manufacture, and reliability to low flow rate applications have not been sparse, but the principle difficulty has always been that the conventional turbomachine design operates at ultra-high speed to deliver low volume flow rates. This is synonymous with low efficiency due to higher losses (windage, surface finish, and tip clearances). The innovative TurboClaw® design is a low specific speed turbomachinery with forward swept impeller geometry. It owes its high efficiency and operational stability to careful design of its nearly tangential forward swept blading and diffuser geometry.

The present contribution describes the design and development of a high speed test rig to accurately measure TurboClaw efficiency using shaft torque. The test rig employs a permanent magnet high speed motor designed and developed for this purpose. The motor shaft drives a single stage TurboClaw® compressor via a high speed torque sensor. The results are presented and compared to data obtained through an energy balance method well established for this geometry since its first innovative development at Imperial College in 2003. The accurate performance measurements have been used to design and test a two stage compressor to recover energy from waste steam. This is also briefly described.

1. Compressor systems

Compressors are fluid machinery devices that convert mechanical work into enthalpy rise of the fluid increasing its pressure and temperature such that the fluid is compressed. The compressed air can be stored or used for a variety of applications, usually by utilizing the stored energy as it converts to kinetic energy as it is depressurized. Compressors group into two main categories, positive or negative displacement machines but the latter group is more commonly known as dynamic machines, see Figure 1. Positive displacement machines work by intermittently trapping the fluid in a volume then reducing this volume before expelling to a high pressure. Common types include piston which pumps up a chamber and using controlled valves to draw in and discharge the working fluid, rotary screw with helical screws with reducing volume, and vane compressors with slotted rotor with varied blade

Content from this work may be used under the terms of the Creative Commons Attribution 3.0 licence. Any further distribution of this work must maintain attribution to the author(s) and the title of the work, journal citation and DOI.
Published under licence by IOP Publishing Ltd

placement to guide working fluid into a chamber and compress the volume. It is noted that for positive displacement compressors sealing on moving surfaces are required, invariably requiring lower operational speeds, oil lubrication and/or costly tight tolerances. These characteristics increase wear, emissions and cost and reduce durability and reliability. For a given volume flow rate, positive displacement machines are larger and heavier than dynamic machines due to the combined effect of lower operational speeds and intermittency.

Dynamic compressors describe machines that transfer energy between a rotor and the working fluid and include centrifugal compressors, see Figure 1. These use a spinning impeller to accelerate the gas tangentially to high velocity and usually create pressure rise in the impeller by decelerating the fluid in the relative frame of the rotor. The exit fluid is then decelerated in stationary diffusers to convert kinetic energy to pressure. These machines are governed by Newton's second Law of Motion and Euler's turbomachinery equation. Turbomachines are inherently oil-free, but are driven by shafts supported by bearings, mostly oil lubricated but also grease packed, as well as oil-free air or magnetic bearings. It is noted that very high rotor speeds in the order of 100's m/s are necessary in order to obtain pressure rises of similar magnitude to atmospheric pressure. To illustrate this point, ignoring compressibility effects, the velocity needed for air at atmospheric pressure to have a dynamic pressure of 1 bar is 408 m/s based on Bernoulli principle. For a 50 mm diameter compressor impeller a rotational speed of 156,000 rpm would therefore be required. Two other important features of turbomachines are their continuous compression characteristics and high flow rates.

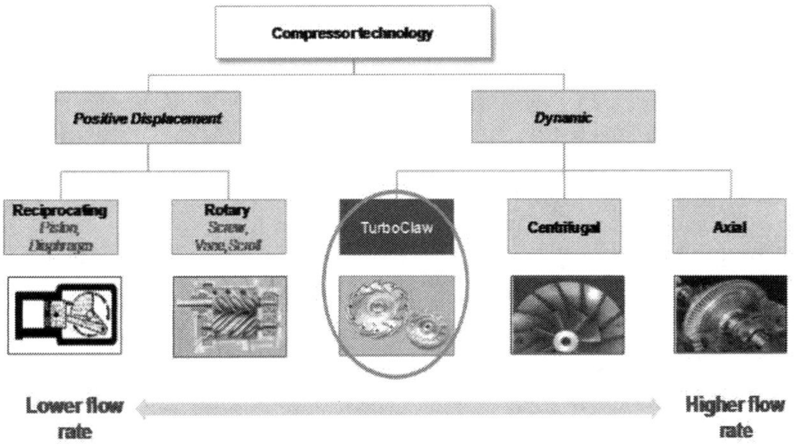

Figure 1: Compressor systems

The present contribution is based on research directed towards low flow rate turbomachine technologies and their development. The innovative TurboClaw is described and its performance characterised using a newly developed test rig. Geometric and performance scaling is briefly described as well as an important TurboClaw application, namely Mechanical Vapour Recompression (MVR).

2. Turbomachines with low flow rates

To determine the choice of machine to use for a given volume flow rate and pressure rise, a non-dimensional specific speed, N_s, is commonly used in preliminary design:

$$N_s = \omega \dot{Q}^{0.5} \Big/ \Delta h^{0.75} \qquad \text{Equation 1}$$

where ω, $\dot{Q}$, and h are speed, volumetric flow rate and enthalpy respectively. Equation 1 is a particularly useful tool for choice of machine type and also for gauging performance of turbomachinery operating speed. The performance of some of the different types of machine is depicted in Figure 2 from [1] and shows turbocompressors are very efficient for large flow rates.

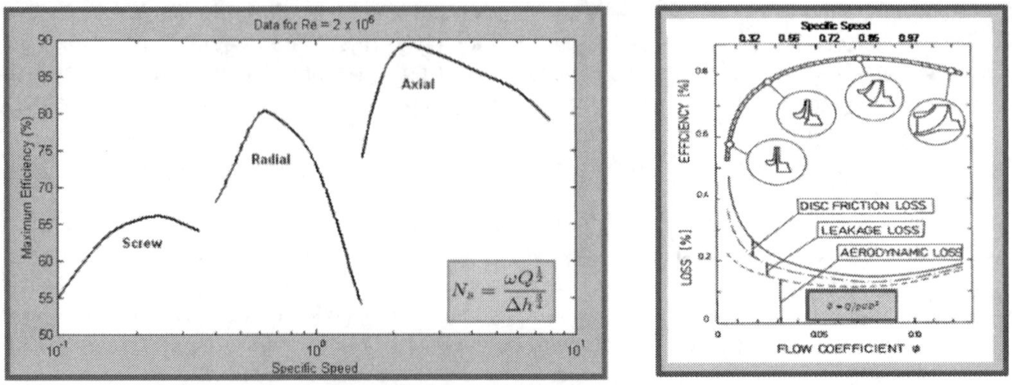

Figure 2: Left: Efficiency and specific speed [1]; Right: Low flow rate losses [2]

However these dynamic machines suffer from loss of efficiency when the specific speed is not optimal. It is immediately apparently from Equation 1 that for a given enthalpy rise, a low flow rate will necessitate either ultra-high shaft speeds or a suboptimal specific speed and efficiency will suffer. High shaft speeds may be impractically high that cost is unaffordable and life inadequate for the bearings and motors needed for support and drive. Shaft speed may also be restricted if the rotor is one of a number of stages or limited by the needs of other machinery. The reasons for poor efficiency under suboptimal specific speed are due to lower blade height to impeller diameter ratio with larger aerodynamic leakage and windage losses; see Figure 2 from [2].

To overcome the low flow rate turbomachines shortcomings of low efficiency and ultra-high speeds, designs including partial admission or partial emission in which the flow at impeller inlet or outlet is restricted, regenerative compressors in which the flow re-circulates in a single impeller, wedge- type with thick blading, and Barske with restricted volute have been devised. Table 1 presents comparative data of low specific speed technology demonstrating persistent shortcomings in both lower efficiency as well as high operational speeds.

Table 1 Comparative data for low specific speed machines

	Casey [3]	Wedge [4]	Barske [5]	Regenerative [6]
Impeller Diameter (m)	0.250	0.334	0.265	0.282
Tip speed (m/s)	200	347	483	74
Volume flow (m3/s)	0.0500	0.1878	0.0512	0.2600
Ns	0.248	0.411	0.073	0.100
Polytropic Efficiency %	51.3	57.9	39.8	45

3. Innovation

The turbocompressor pressure ratio p_{out}/p_{in} is governed by the thermodynamics relationship, Equation 2, derived from [7];

$$\frac{p_{out}}{p_{in}} = \left(\frac{\eta_i \Delta h_o}{c_p T_{o,in}} + 1 \right)^{\gamma/\gamma-1} \qquad \text{Equation 2}$$

Where T is temperature, η is efficiency, c_p is the specific heat at constant pressure, and γ is the ratio of specific heats. The Euler Turbomachinery equation, Equation 3;

$$\Delta h_o = U_{out} C_{\Theta out} - U_{in} C_{\Theta in} \qquad \text{Equation 3}$$

where Δh_o is the work input to the gas and U and C_Θ , the blade speed and tangential velocities respectively. Subscripts are given to denote flow in and out of the impeller. Turbocompressors can be axial, radial or mixed flow but for the lowest volume flow-rate and for a particular pressure head, radial flow is the best, see Figure 1. For reasons of efficiency and widest operating map, it is normal practice to make the blade outlet angle slope away from the direction of rotation, creating what is termed backswept.

The use of forward sweep in order to increase Euler work, (Equation 3) and to reduce blade speed has been understood for many years. However, previous attempts found that efficiency deteriorated and the flow range of the compressor reduced, hence seldom used. The present innovation, trademarked TurboClaw, addresses this shortcoming by increasing forward sweep so substantially that the outlet blade angle becomes almost tangential. The design which ensues from this approach is shown in Figure 3 [8] along with the velocity triangle for the rotor outlet. It is also described in [9] and [10].

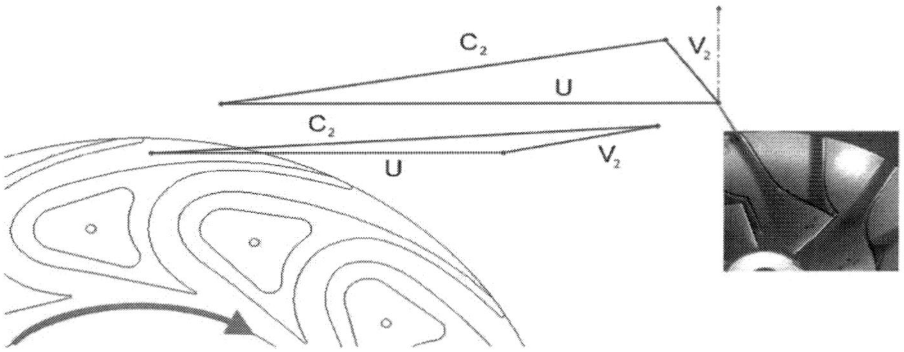

Figure 3: Impeller exit velocity triangle for TurboClaw (left) vs backswept turbocompressors (right)

Due to the effect of very thick blades which reduce the flow, it was found that the radial velocity can be extremely low relative to the tangential velocity without causing instability problems. The ratio $C_{\theta 2}$ to C_{m2} can be up to 25:1 which is an order of magnitude above the limits for conventional designs. Since the radial velocity is so low, the increase in tangential velocity which arises due to increased flow rate is also very low. This means that the work input does not increase markedly with flow (see Equation 2) hence the slope of a constant speed line on the map is only marginally positive. In reality, the increase in losses with increased flow tends to mask this effect in any case leading to a compressor map not too dissimilar to a conventional backswept machine.

4. Newly developed test rig

Performance measurements of TurboClaw based on energy balance was reported in [11]. Compressor efficiency was measured using the ratio of ideal power to real power. Ideal power was evaluated by

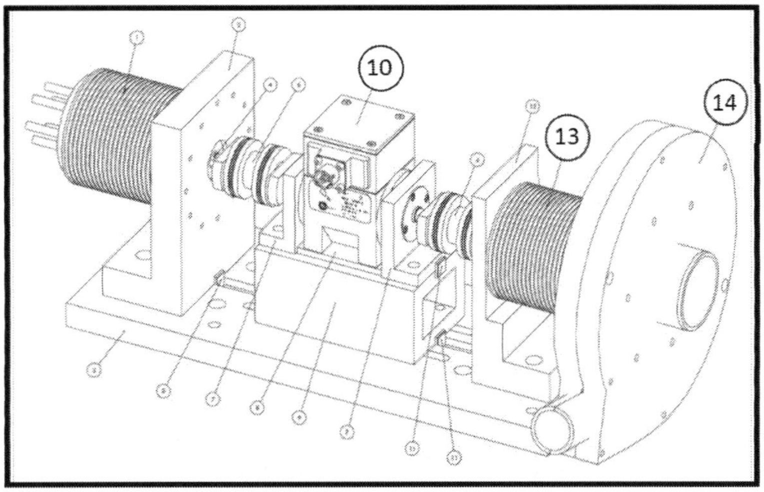

Figure 4: Newly developed test rig for compressor performance measurements

$\dot{m}c_pT_0$ $(PR^{(\gamma-1)/\gamma}$ -1) where $\dot{m}$ is the mass flow rate. The real power was determined based on measurement of heat to air and estimation of heat loss. The present contribution reports the development of a new test rig for high speed operation and shaft torque measurements, τ in Nm, where power is calculated as the product of τ and ω. Referring to Figure 4, a high speed permanent magnet motor (item 1) is coupled via a torque sensor (item 10) to a second bearing-cartridge (item 13) supporting the compressor (item 14). The motor and bearing cartridge shafts are aligned accurately and connected to the torque sensor shaft using two diaphragm couplers (item 6).

The inline torque measuring system is based on measurements from a torque-proportional transformer coupling. This consists of two concentric shaft cylinders, together with corresponding concentric stationary coils, on each side of the shaft's deformation zone. A high frequency alternating current is used in the primary coil and a torque-proportional voltage is detected on the secondary coil when angular deformation is measured. Speed measurements are made using an optical sensor reading a toothed path machined directly on the shaft.

5. Performance Measurement

To assess compressor performance accurately, the test rig is operated without the compressor to establish a base case measuring losses such as bearings, rotor windage, and the windage of the diaphragm coupler (item 6). A regression analysis was performed for torque (τ in Nm) vs shaft speed (ω in rpm) data to establish equation 4.

$$\tau = 1.547x\ 10^{-18}\omega^3 + 1.067\ x10^{-11}\omega^2 + 6.485\ x10^{-8}\omega + 1.6x\ 10^{-3} \qquad \text{Equation 4}$$

The correlation coefficient was 0.988. It is noted that windage and bearing losses were just 45% of the total loss with the diaphragm coupler (item 6) making up the difference of 56%.

The pressure ratio versus mass flow rate is depicted in Figure 5 for an 85 mm TurboClaw impeller developed as an electric supercharger [12] with Ns of 0.27 as calculated from Equation 1. With the compressor in position, tests were performed to accurately measure pressure ratio and mass flow rates as well as shaft torque that were reduced by Equation 4 to determine compressor torque. Compressor efficiency was then calculated based on the change of enthalpy and torque measurement, namely:

$$\eta_c = \frac{(h_{2s}-h_1)\,\dot{m}}{\omega\tau_c} \qquad \text{Equation 5}$$

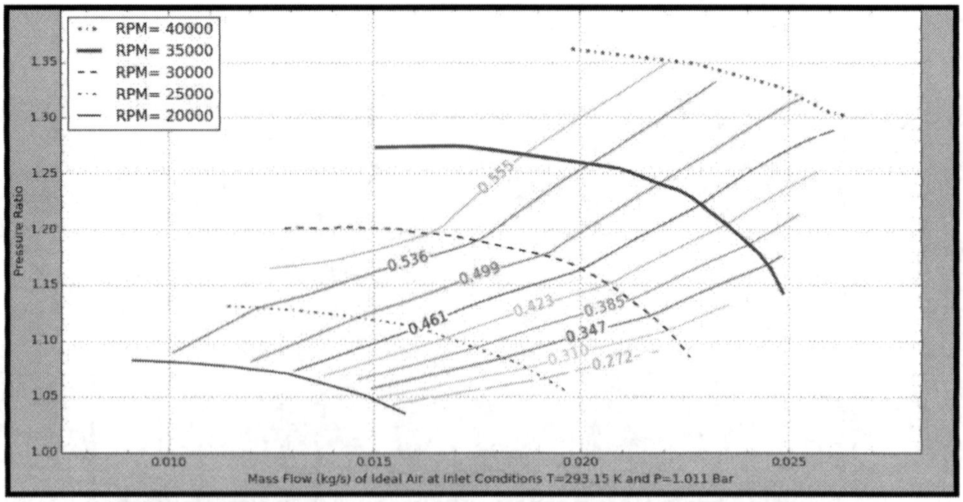

Figure 5: Performance measurements using test rig depicted in Figure 4

IOP Conf. Series: Materials Science and Engineering **90** (2015) 012050 doi:10.1088/1757-899X/90/1/012050

where η_c is compressor efficiency, ω is shaft speed, h is enthalpy with subscripts 1 for inlet and 2s for delivery, τ_c is compressor torque, and $\dot{m}$ is mass flow rate. It should be noted that this is not a true isentropic efficiency since the real compression process is not fully adiabatic and heat transfer from the compressor will slightly reduce the actual work input as compared to a machine fully insulated. However, this efficiency is more meaningful as the ratio of isentropic work to actual work and is a correct measure of how much power is required to drive the compressor to produce the desired pressure rise.

The data in Figure 5 shows the familiar rising PR for lower mass flow rates to surge. Only five constant speeds lines are shown for brevity. The peak efficiency measured was at 59.4% that far exceeds the measurement of 52% of earlier designs reported in [11 & 12] for a larger impeller with *Ns* of 0.2 based on the less accurate and non-conservative energy balance method described earlier. The efficiency data is also shown to be superior to the data presented in Table 1 for other known low specific speed machines and particularly so when compared to data from [3] at similar specific speeds.

6. Scaling effects

Efficiencies approaching 60% are highly competitive when measured against what can be achieved, if even possible, for oil free dry positive displacement compressors of similar low volume flow rates. It is known that turbomachines also greatly benefit from scaling effects and this aspect has been investigated. Turbomachinery can readily be scaled using Buckingham's similitude principles [13]. The scaling can be geometric (linear dimension), Kinematic (flow coefficient- same velocity triangles), Dynamic (same loading coefficient), and Energetic (same power coefficient). Scaling efficiency is however problematic due to difficulties in scaling surface finish, leakage, and matching fluid dynamics characteristics. The non-dimensional Reynolds number (ratio of inertial forces to viscous forces) is conventionally used to ensure dynamic similitude to scale efficiency based on the pioneering work of Pampreen [14], see Figure 6.

In practice, this is not always sufficient since very small changes to shape and surface roughness can result in very different flows. Nevertheless, Reynolds number is an important tool widely used to scale efficiency.

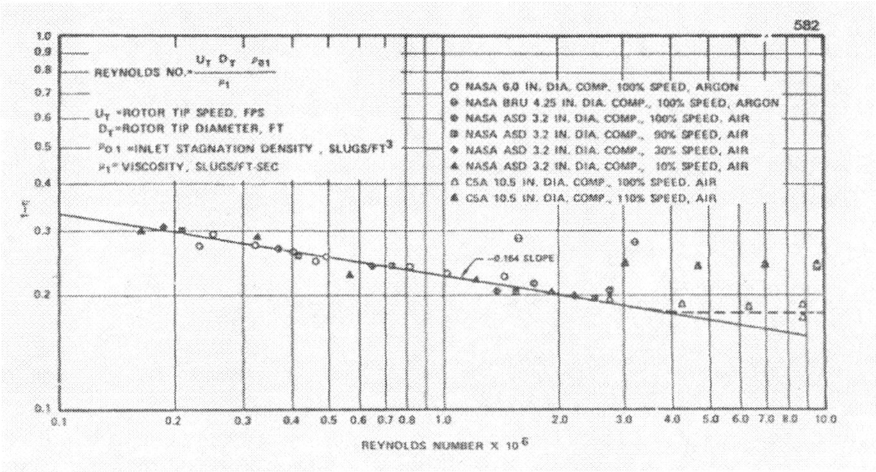

Figure 6: Reynold number scaling (data by Pampreen [14])

It is noted that the formulation for Reynolds number as in Equation 6 is based on a characteristic dimension stated as diameter here, but scaling blade height is also widely used.

$$R_e = {\emptyset V \rho}/{\mu} \qquad \text{Equation 6}$$

Where $\emptyset$ is diameter, V is impeller tip speed ($r\omega$), ρ is density, and μ is kinematic viscosity.

The performance superiority of TurboClaw has been further investigated by re-plotting the data resented in Figure 2 from [1], see Figure 7. This shows conventionally designed radial turbomachinery with a Reynolds number of $2x10^6$ to drop to 50% efficiency at specific speed of 0.2. At lower Reynolds number the conventional designs suffers even further, see curve at Reynolds number 60,000 generated by scaling Moody friction factor and using the methodology prescribed by [14] and presented in Figure 6. TurboClaw data presented in Figure 5 are at specific speed of 0.27 and with measured efficiency of 59.4%. This is shown to have a superior performance at low specific speed for corresponding Reynolds numbers, see Figure 7. This is being further assessed but from Figure 7 the potential is substantial based on the difference between the green and blue curves for conventional machines.

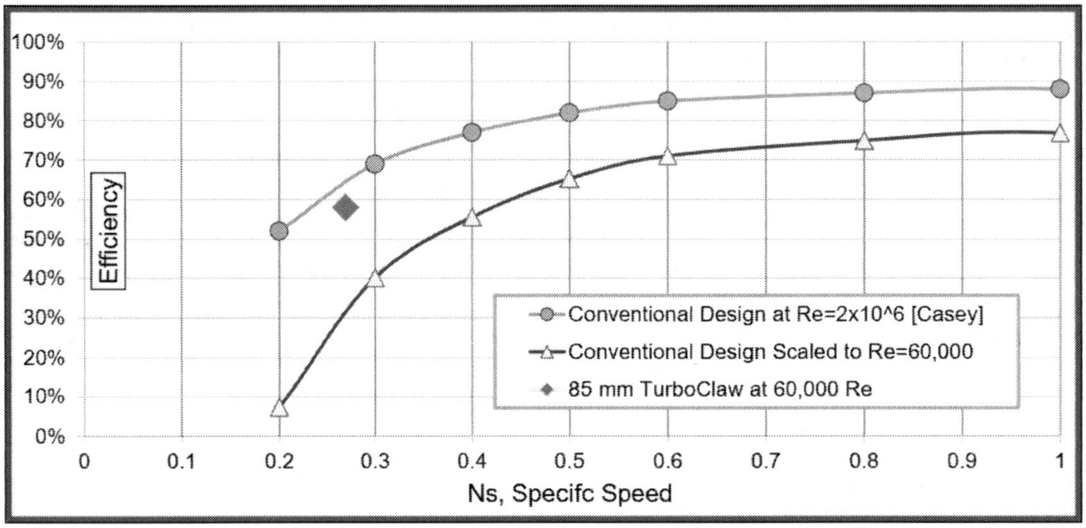

Figure 7: Comparison of measured TurboClaw data with corresponding data from the radial machine

For ideal gases, to predict the performance of a new compressor based on results from a different geometry, constant specific work is assumed even at different temperature and pressure inlet conditions using the corrected mass flow rate and speed. The assumption effectively means that the gas velocity is constant, hence compressor diameter and blade height become the main variables to keep specific work constant. This allows the designer to scale compressor geometries, but does not allow for change of working fluid or changes in tip speed when temperature changes.

For real gases, the changes to specific work can still stay constant, rather than the Mach number; provided the efficiency of the scaled geometry is corrected for changes in Reynolds number, see [14]. However, it is noted that for these cases the pressure ratio will be different, even for the same efficiency.

To assist application engineering, scaling software has been devised based on principles of similitude and Reynolds number scaling. The software uses a measured compressor map to predict candidate compressor performance and is interfaced with the data base from National Institute of Standards and Technology (NIST) to account for changes in the working fluid, [15]. To demonstrate, a two stage compressor for energy recovery from waste steam is described based on the Mechanical Vapour Recompression (MVR) principle. There are a multitude of applications for MVR including food and drinks processing. The main benefit of MVR is to recover waste energy in steam by compressing it to a higher pressure (hence temperature) and subsequently recover this heat in heat exchangers for re-use. System effectiveness could readily be gauged by working out the coefficient of performance (CoP) that is defined as heat recovery to drive electrical power ratio. The so called spark spread (price of gas versus electricity) can also be employed to work out the economic benefit in terms of pay-back time.

The two stage compressor was sized for a pressure ratio of around 1.5 with an electric motor of 30 kW. This enabled a range of pressure inlets and mass flow rates. Figure 8 depicts Coefficient of Performance (CoP) and pay-back period for the compressor against the mass flow rate that corresponds to inlet pressure. The two stage MVR compressor has been evaluated at pressure inlets of atmospheric, 3 bar, and 5 bar. Data show close correspondence to the predicted results depicted in Figure 8. Further tests are in progress.

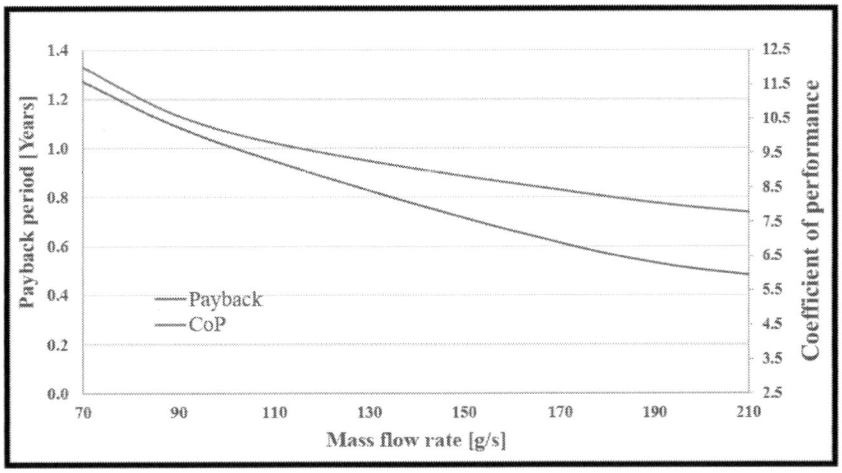

Figure 8: Two stage MVR with PR 1.5

7. Conclusions

Accurate efficiency measurements of forward swept low specific speed TurboClaw has been reported showing superior performance to conventional centrifugal designs at low specific speeds. This data has been used in proprietary software to design a compressor system for the mechanical vapour recompression application. The results show suitably high coefficient of performance to enable to quick pay-back period.

References:

[1] Casey, M.V. AGARD Lecture Series 195
[2] Dalbert et al Journal of Mech Eng Science, v213, n1, 1999, pp71-83
[3] Casey, M.V., Dalbert, P., Schurter, E. "Radial Compressor Stages for Low Flow Coefficients" IMechE paper C403/004 1990.
[4] Rusak, V. "Development and Performance of the Wedge Type Low Specific Speed Compressor Wheel" ASME 27th International Gas Turbine Conference and Exhibition 1982
[5] Barske, V.M. "Development of Some Unconventional Centrifugal Pumps" Proc. IMechE Vol 174 (1960) 437-461 129
[6] Sixsmith, H., Altmann H. "A Regenerative Compressor" Trans ASME J. Engng for Industry, August 1977, 99, 637-647.
[7] Sayers, A. T., "Hydraulic and compressible flow turbomachines" McGraw Hill 1990
[8] Pullen, K. R. Etemad, S. and Cattell R., Proceedings of ASME Turbo Expo 2012, June 2012, Copenhagen, Denmark (GT2012-69074)
[9] Patent GB2.366.333A, dated 06-03-2002.
[10] PCT/GB04/003752, dated 9-03-2005.
[11] Vine, A., PhD Thesis 'The Aerodynamic Design and Evaluation of a Low Specific Speed Turbo Compressor', Imperial College London 2005.
[12] Electric supercharging for small downsized engines by Pullen, K., and Etemad, S., presented at Nlett Turbocharging Seminar Sept 2014 Tainjin, China
[13] Buckingham, E. The principle of similitude. Nature 96, 396-397 (1915).
[14] Pampreen, R. C., J. Eng. Gas Turbines Power 95(3), 251-256
[15] http://www.nist.gov version 9.1

IOP Conf. Series: Materials Science and Engineering **90** (2015) 012051 doi:10.1088/1757-899X/90/1/012051

A Stability Enhancement Method for Centrifugal Compressors using Active Control Casing Treatment System

Yuanyang Zhao*, Jun Xiao, Liansheng Li, Qichao Yang, Guangbin Liu, Le Wang

State key laboratory of compressor technology, Hefei General Machinery Research Institute, Hefei, China

E-mail: yuanyangzhao@163.com

Abstract. The centrifugal compressors are widely used in many fields. When the centrifugal compressors operate at the edge of the surge line, the compressor will be unstable. In addition, if the centrifugal compressor runs at this situation long time, the damage will be occurred on compressor. There are some kinds of method to improve and enlarge the range of the centrifugal compressors, such as inlet guide vane, and casing treatment. For casing treatment method, some structures have been researched, such as holed recirculation, basic slot casing treatment and groove casing treatment. All these researches are the passive methods. This paper present a new stability enhancement method based Active Control Casing Treatment (ACCT). All parts of this new method are introduced in detail. The control strategy of the system is mentioned in the paper. As a research sample, a centrifugal compressor having this system is researched using CFD method. The study focuses on the effect of the active control system on the impeller flow. The vortex in impeller is changed by the active control system. And this leads to the suppression of the extension of vortex blockage in impeller and to contribute to the enhancement of the compressor operating range.

1. INTRODUCTION

The centrifugal compressors are widely used in many fields. The centrifugal compressors have a very important character, which is that there is a stable working condition range. When the centrifugal compressor operates at/pass the edge of the surge line, the compressor will be unstable. Hence, in the industry systems, there is the anti-surge control system to make the compressors working at the stable condition.

There are some kinds of method to improve and enlarge the stable range of the centrifugal compressors, such as inlet guide vane [1-3] and casing treatment [4-8]. For casing treatment method, some researches have been done, such as holed recirculation, basic slot casing treatment and groove casing treatment. All these researches are the passive methods.

This paper present a new stability enhancement method based Active Control Casing Treatment (ACCT). All parts of this system and the control method are introduced in detail in this paper.

As a research sample, a centrifugal compressor having this system is researched using CFD method. The study focuses on the effect of the active control system on the impeller flow. The vortex in impeller is changed by the active control system, which leads to the suppression of the extension of

Content from this work may be used under the terms of the Creative Commons Attribution 3.0 licence. Any further distribution of this work must maintain attribution to the author(s) and the title of the work, journal citation and DOI.
Published under licence by IOP Publishing Ltd

vortex blockage in impeller. Hence, the stability of compressor enhanced and the operating range enlarged.

2. STABILITY ENHANCEMENT METHODS FOR COMPRESSORS

Surge suppression of centrifugal compressor is very useful for centrifugal compressor when it is operating at the small mass flow rate. The main mechanism of surge suppression is to maintain the steady flow in impeller. The main flow separation in compressor is avoided using these surge suppression methods.

There are many kinds of methods, which can enhance the stability of the centrifugal compressor. The details of these methods are presented as follows.

2.1. Inlet guide vane (IGV)

The inlet guide vanes can significantly extend the stable operating range of centrifugal compressors. The inlet guide vanes can be turned to induce a controlled prewhirl in the front of the compressor impeller. This method is widely used to regulate the pressure ratio and the mass flow rate at constant rotational speed centrifugal compressors in industry system.

Wallace studied the theory and experiment of IGV at 1975[10]. The experimental results indicated that the stable operation could be extended to low mass flow rates by positive prewhirl using IGV. Figure 1 shows the Geometry of IGV and the performance of compressor at different setting angles.

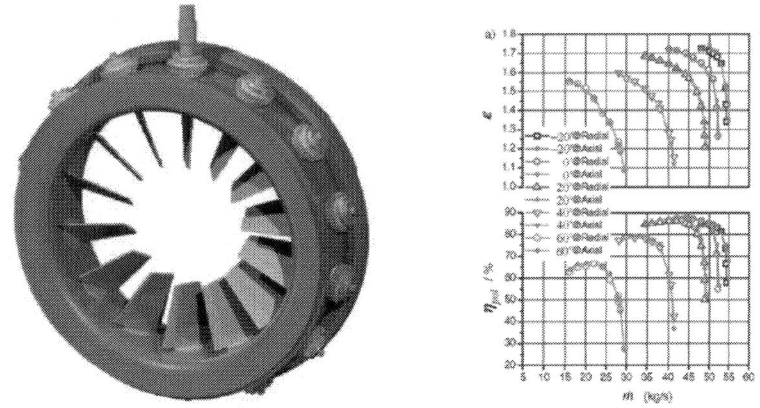

Figure 1. Geometry of IGV and compressor performance [9]

2.2. Traditional casing treatment

The casing treatment was proposed and researched in axial compressor firstly. Many research have been done based the cave/cavity on the diffuser and casing of compressor. Fisher researched a recirculation casing treatment in a turbocharger compressor [4]. It was called ported shroud housing (shown in Figure 2a). Wei Xu proposed and researched a recirculation casing treatment structure in recent years based on the research of Fisher [5]. Figure 2b shows the main structure of it.

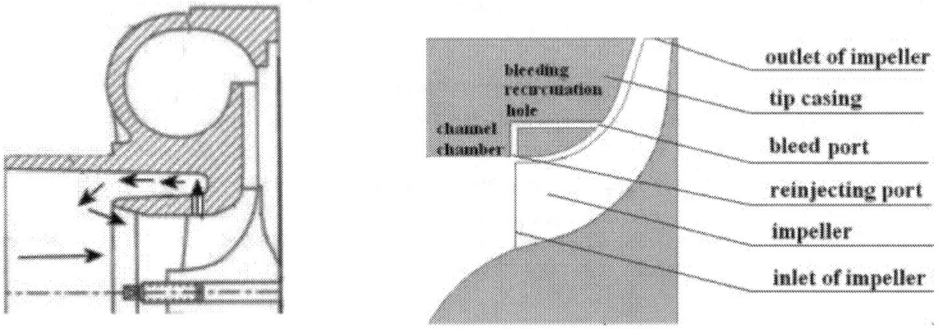

a. Ported shroud housing [4] b. Holed casing treatment [5]

Figure 2. Structure of casing treatment.

For this traditional casing treatment, there is a flow path from the blade tip to the inlet of impeller on the shroud of compressor. When the compressor is running, some gas flows in this channel. Usually, the gas bleeds from the tip of blade and then inject into the front of impeller.

The flow resistance in the flow path can be utilized to balance the static pressure difference between the bleed port and injecting one [5]. That means the flow rate is decided by the static pressure difference of this two ports. At the low mass flow rate conditions, the static pressure near the bleed port is high enough to drive an injected flow from the bleed port to the injecting one. The flow field at the tip of impeller is disturbed by the flows. In addition, the unsteady flow in the impeller is changed by it. The steady operating range of centrifugal compressors is enlarged by this method.

2.3. New casing treatment

For the traditional casing treatment, all of them are passive methods. The casing treatment is working at all the time when the compressor is running. People cannot change the situation of casing treatment. That means when the compressors operate at the design point, the traditional casing treatment are still working. That leads to the decrease of the efficiency of compressor.

Because the static pressure difference between the two ports is small, the flow rate in this flow pass is small, which can not change the unsteady flow in the impeller when the compressor flow rate is very low.

The Active Control Casing Treatment (ACCT) is presented in this paper basing on the traditional passive casing treatment methods. Figure 3 shows the basic structure of ACCT. there are bleeding ports on the casing near the tip of blades. The injecting port is on the inlet pipe of compressor. There is the control device in the middle of the flow pass between injecting and bleeding ports. Here the control device can be control valve or small compressor.

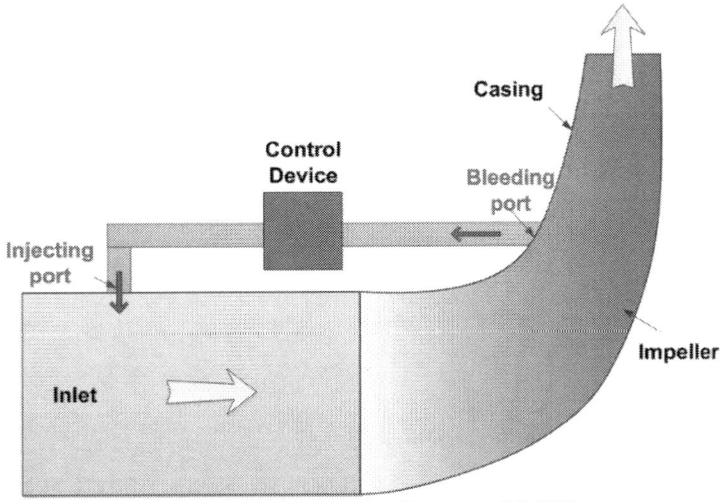

Figure 3. Schematic diagram of ACCT

Figure 4 shows the control flow chart of ACCT. The difference between ACCT and anti-surging systems also shows in Figure 4.

The yellow part in Figure 4 is the anti-surging system. In the anti-surging system, there are the pressures or other kinds of sensors, which can measure the stability of centrifugal compressors. When the compressor is on unsteady working condition, the controller of anti-surging system calculates the value of the control signal and put this signal to the anti-surge valve. Then some high-pressure gas flows back to the inlet of compressor through the anti-surge valve. Here the gas with the discharge pressure is put back to the inlet. The power which was used to compress this part gas is waste.

The blue part in Figure 4 shows the ACCT system. In the ACCT system, the sensors and controller are like with the anti-surging system. The main difference between these two systems is the state of gas that flows back to the inlet of compressor. In the ACCT system, the static pressure is just a little higher than the suction pressure of the compressor, which means that only a very small power was waste in this system.

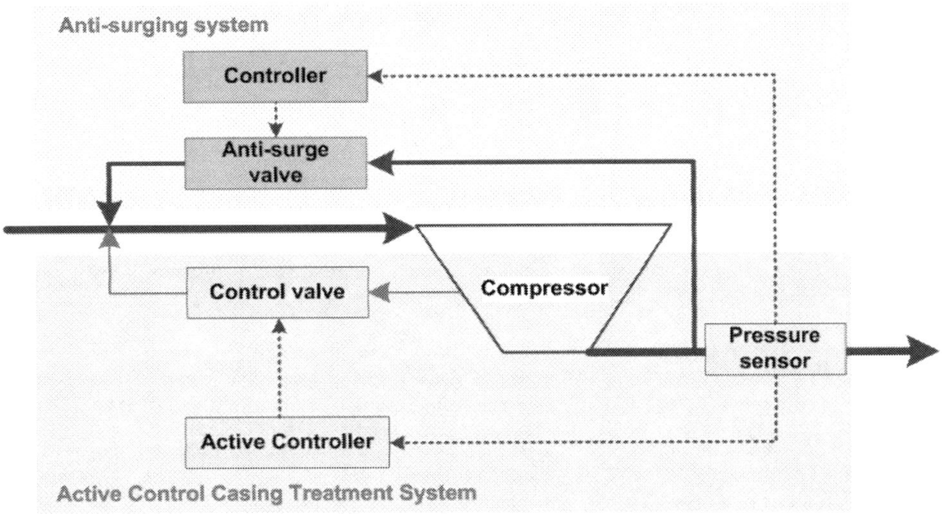

Figure 4. Schematic diagram of ACCT

The difference between ACCT and traditional casing treatment systems is the control ability of the system. In the traditional casing treatment system (Figure 2), there are always some compressed gas flowing from the impeller to the inlet of compressor, whatever the compressor is at the stability working conditions or not. That will lead to the low performance when the compressor operates at the normal working conditions (design working conditions). However, in the ACCT system, when the compressor operates on the normal working conditions, the control valve can be closed by the controller. When the compressor operates on the unstable working condition, the controller can decide the lift of control valve based on the strength of instability of compressor.

Using the above control method, the ACCT system can make the compressor maintain at the stable condition like the anti-surging system. At the same time, the efficiency of compressor is not decreasing largely on the low mass flow rate working conditions.

3. PERFORMANCES SIMULATION

As a sample, a centrifugal compressor is simulated by CFD method to verify the reasonable of ACCT system.

3.1. CFD model

Figure 5 shows the simulation mesh using in the CFD calculation. The centrifugal compressor has a radial impeller. The number of normal blade and splitter blade is 11 respectively. The outlet diameter of impeller D is 300 mm. The cycle speed of impeller is 300 m/s.

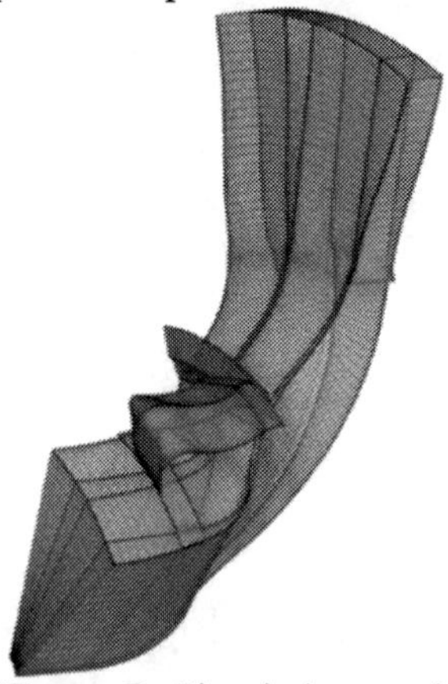

Figure 5. Simulation mesh

3.2. Define of stability

It is very important to evaluate the stability enhancement of centrifugal compressors. Here the ability of the stability enhancement of centrifugal compressors is defined as the following equation.

$$\phi = (m_{sta,ori} - m_{sta,CT}) / (m_{cho,ori} - m_{sta,ori}) \tag{1}$$

In Figure 6, the red line is the normal compressor performance curve. The pink short dash line is the extension part of performance curve when the ACCT system is used. The blue dot dash line is the boundary of unstable working conditions. The move of blue line from right to left (low mass flow rate) means the enhancement of stability of centrifugal compressor.

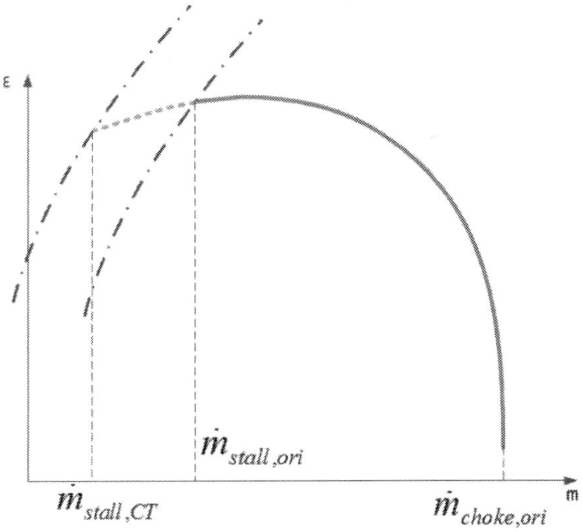

Figure 6. Define of the ability of the stability enhancement

4. RESULTS AND DISCUSSIONS

4.1. *Flow analysis*

Figures 7 and 8 show the details of flow in the impeller of centrifugal compressor. For the original structure compressor, the flow speed is low at the near of split blade (shown in Figure 7a). When using the ACCT structure, there is some gas bleeding at the split blade and that leads to the increase of flow speed at this area. This effect has the benefit to the stability of centrifugal compressor.

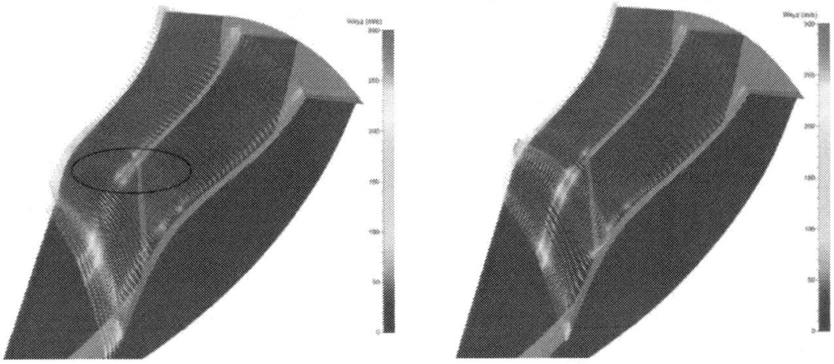

a. Original structure b. ACCT structure

Figure 7. Velocity profile of compressor near the blade tip

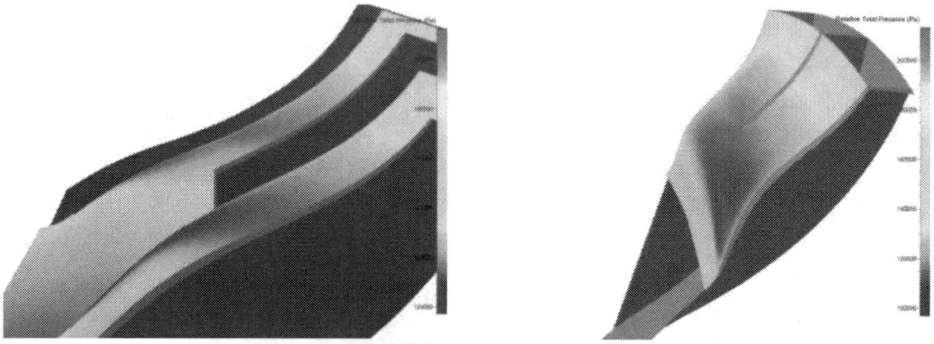

a. Original structure

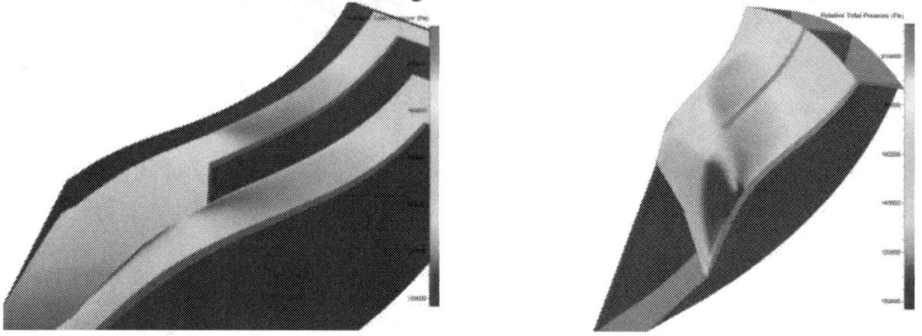

b. ACCT structure

Figure 8. Total pressure of impeller

Figure 8 shows the pressure distribution of impeller. Figure 8a shows the pressure of gas in impeller at two different directions when the original structure is used. Figure 8b shows the pressure when the ACCT structure is used. The difference of these two structures can be gotten from the CFD simulation results. The area of low total pressure near the beginning end of split blade is decrease when using ACCT structure. The vortex in impeller is changed by the ACCT system. That means the flow loss is increasing and the flow is improved.

4.2. Ability of Stability Enhancement

The performance and efficiency of compressor under different ACCT parameters is simulated using the CFD methods.

Figure 9 shows the efficiency of compressor when the bleeding mass flow rate is changed (0, 0.2, 0.4, and 0.64 kg/s). From the curves of Figure 9 it can be seen that the efficiency of compressor decreases as the bleeding flow rate increases when the compressor operate at the low mass flow rate working conditions. It should be mentioned that the compressor is already at the instable working condition for original design when the mass flow rate is less than 2.95 kg/s.

Figure 10 shows the relationship between pressure ratio and mass flow rate of centrifugal compressor. From the results it can be seen that the pressure ratio increases as the increase of bleeding mass flow rate when the compressor operate at low mass flow rate condition. The ability of stability enhancement is increasing when the bleeding mass flow rate is increasing. When the bleeding mass flow rate is 0.64 kg/s, the ability of stability enhancement is 35%.

Because the focus of this paper is the stability of centrifugal compressor at the small flow rate, the performance of compressor at higher mass flow rate is not simulated.

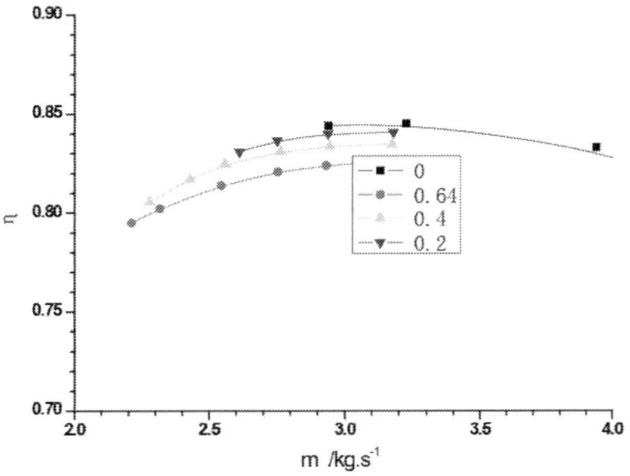

Figure 9. Efficiency of compressor

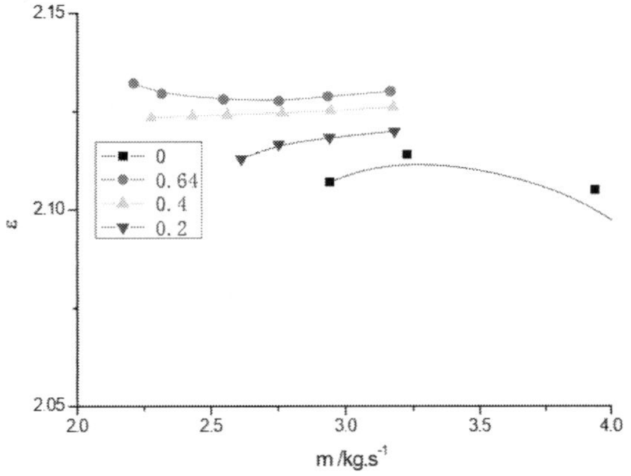

Figure 10. Performance of compressor

The test system of ACCT is building in the research lab. The test results will be gotten during the following period.

5. CONCLUSION

The main stable enhancement methods for centrifugal compressor are introduced in this paper. A new stable enhancement method based Active Control Casing Treatment is proposed in this paper. A centrifugal compressor having this system is researched using CFD method. The study focuses on the effect of the active control system on the impeller flow. The vortex in impeller is changed by ACCT system. And this leads to the suppression of the extension of vortex blockage in impeller and to contribute to the enhancement of the compressor operating range.

The pressure ratio increases with the increase of the bleeding flow rate. The compressor efficiency decreases with the increase of bleeding flow rate. The ability of stable enhancement increases when the bleeding flow rate is increasing. In addition, it is 35% when the bleeding flow rate is 0.64 kg/s.

Using ACCT system, the moment and the ability of stable enhancement can be gotten by the controller. That lead to the compressor can operate at the high efficiency and stability when the mass flow rate is low.

ACKNOWLEDGEMENTS
The work was supported by the National Key Basic Research Program of China (2012CB026000) and by the Science and Technological Fund of Anhui Province for Outstanding Youth (1508085J05).

References

[1] Y Zhang, D Qi, and Y Mao. Experimental investigation and improvement of the inlet guide vane with plate vane in a centrifugal fan. J. Power and Energy, 2009(223): 401-413.
[2] M Coppinger, E Swain. Performance prediction of an industrial centrifugal compressor inlet guide vane system. Journal of Power and Energy, 2000:153-164.
[3] M M Cui. Unsteady flow around suction elbow and inlet guide vanes in a centrifugal compressor. J. Aerospace Engineering, 2006:11-28
[4] Fisher, F. B. Application of Map Width Enhancement Devices to Turbocharger Compressor Stages. 1988, SAE paper No. 880794.
[5] Wei Xu, Tong Wang, Chuangang Gu, et al. A study on the influence of hole's diameter with holed casing treatment. Proceedings of ASME Turbo Expo 2011. GT2011-46167.
[6] Hideaki Tamaki. Effect of recirculation device on performance of high pressure ratio centrifugal compressor. Proceedings of ASME Turbo Expo 2010: Power for Land, Sea and Air. GT2010-22570.
[7] Liang Ding, Tong Wang, Bo Yang, et al. Experimental investigation of the casing treatment effects on steady and transient characteristics in an industrial centrifugal compressor. Experimental Thermal and Fluid Science, 2013(45): 136-145.
[8] Xinqian Zheng, Chuanjie Lan. Improvement in the performance of a high-pressure-ratio turbocharger centrifugal compressor by blade bowing and self-recirculation casing Treatment. J Automobile Engineering, 2014(228): 73–84.
[9] J Tan, X Wang, D Qi, and R Wang. The effects of radial inlet with splitters on the performance of variable inlet guide vanes in a centrifugal compressor stage. J. Mechanical Engineering Science. 2011(225):2089-2105
[10] Wallace, F. J., Whitfield, A., and Atkey, R. C. Experimental and theoretical performance of a radial flow turbocharger compressor with inlet prewhirl. Proceedings of the Institution of Mechanical Engineers. 1975(189): 177–186.

IOP Conf. Series: Materials Science and Engineering **90** (2015) 012052 doi:10.1088/1757-899X/90/1/012052

Centrifugal compressor surge detecting method based on wavelet analysis of unsteady pressure fluctuations in typical stages.

R Izmaylov[1], A Lebedev[2]

[1]Prof., Dr.Sci. (Tech), St. Petersburg State Polytechnic University
St. Petersburg, Russia
[2]Cand. Sci. (Tech), St. Petersburg State Polytechnic University
St. Petersburg, Russia
E-mail: LAA7777@yandex.ru

Abstract. Centrifugal compressors are complex energy equipment. Automotive control and protection system should meet the requirements: of operation reliability and durability. In turbocompressors there are at least two dangerous areas: surge and rotating stall. Antisurge protecting systems usually use parametric or feature methods. As a rule industrial system are parametric. The main disadvantages of anti-surge parametric systems are difficulties in mass flow measurements in natural gas pipeline compressor. The principal idea of feature method is based on the experimental fact: as a rule just before the onset of surge rotating or precursor stall established in compressor. In this case the problem consists in detecting of unsteady pressure or velocity fluctuations characteristic signals. Wavelet analysis is the best method for detecting onset of rotating stall in spite of high level of spurious signals (rotating wakes, turbulence, etc.). This method is compatible with state of the art DSP systems of industrial control. Examples of wavelet analysis application for detecting onset of rotating stall in typical stages centrifugal compressor are presented. Experimental investigations include unsteady pressure measurement and sophisticated data acquisition system. Wavelet transforms used biorthogonal wavelets in Mathlab systems.

1. Introduction

Centrifugal compressors are difficult power equipment. The following requirements are for control, automation and protection systems: firstly, operating reliability and operation algorithms; secondly, durability. The special feature of the dynamic action compressors is the presence of bands in the compressor map where work is not permitted- a zone of surge fig. 1. For reference, the surge is a complete loss of stability, which is an unacceptable phenomenon for centrifugal and axial compressors, as the compressor surge control system must function at its operation. On the other hand, surging is an auto-oscillatory process in the "compressor-network" system. The surge limit is determined by the characteristics of the compressor and the network. As a rule, it is located within the expenditure area, which is lower in comparison with the boundary of rotating stall and is significantly less than optimal. Eventually, the surge is usually preceded by an intense rotating stall. There are strong fluctuations, generated in addition to circumferential distortion aerodynamic measures and a rotating stall during the surge, which are formed by a contra flow of the gas through flow channel, which creates high pressure on the compressor and the machine alone.

Content from this work may be used under the terms of the Creative Commons Attribution 3.0 licence. Any further distribution of this work must maintain attribution to the author(s) and the title of the work, journal citation and DOI.
Published under licence by IOP Publishing Ltd

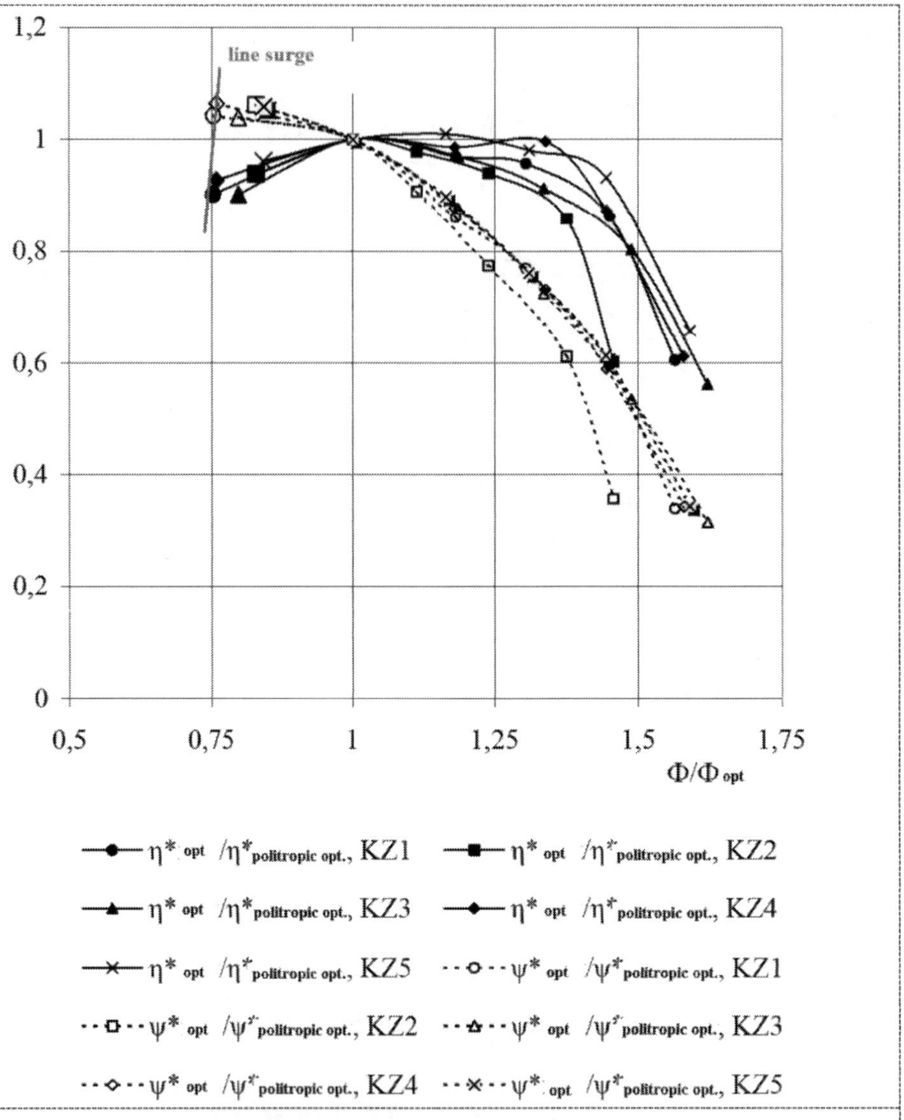

Figure 1. Description "pressure expense" of the KZ stage, formed from impeller of the model stage of supercharger and immobile elements.

2. Experimental stand

For research of the non-stationary phenomena with the purpose of selection of descriptions and properties of stream in area of the before-surge state. The experimental stand of ECC-1M of Compressor, vacuum and refrigeration technique department of SPbPU (before SPbSPU and LPI) was used. Vast cycles of researches of transients are executed in the model stages of centrifugal compressor [3] with measuring of quick changing processes both in impellers and in immobile elements (diffusers) by means of heat loss anemometer and fast-response pressure transmitter. Pneumometric measuring were executed on traditional methodologies. Results got in the wide range of working description - from a maximal expense to the border of surge. The chart of stand is brought around to a fig. 2-3 [1].

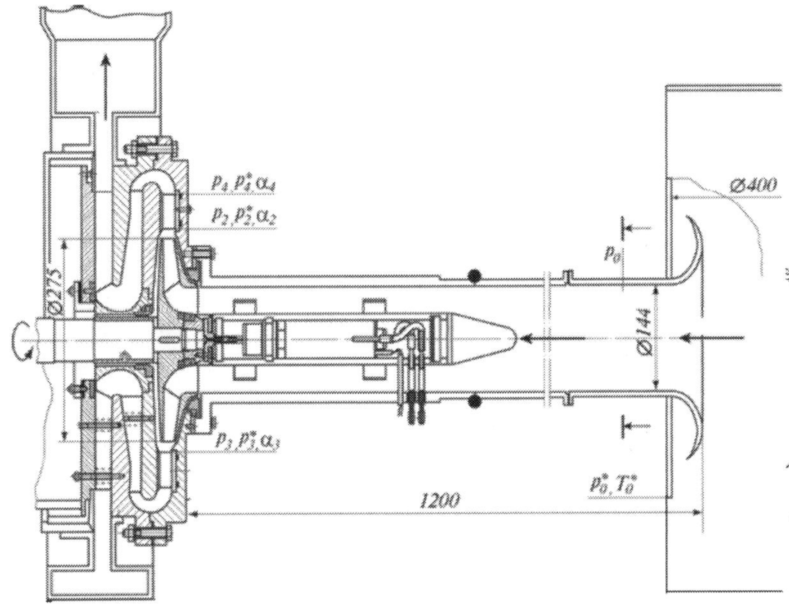

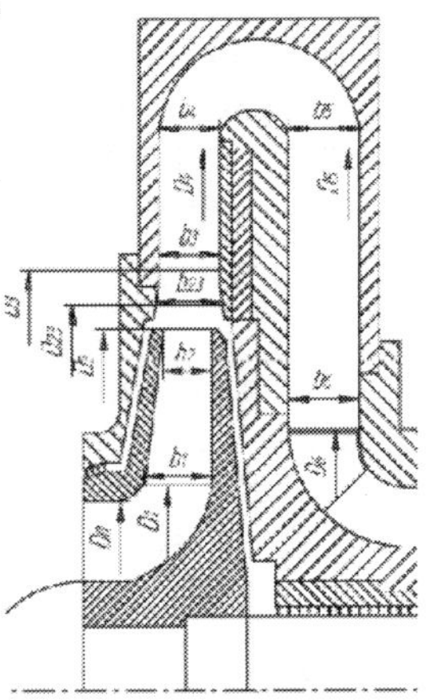

Figure 2. Experimental stand of ECC-1M [4].

Figure 3. Running part of the model stage.

In the experimental unit ECC-1M nonsteady process investigated in the intermediate stage formed by the closed type impeller and combinations with a vaned and vaneless diffuser with parallel walls. Quick-response transducers of pressure (fig. 4-5) are used in an experiment. Information processing in real time is performed by specific information and measurement system. Total expenses of preparation time and realization of experiment about 1 year , presentation of data about 2 years, analysis of the got signals more than 6 years.

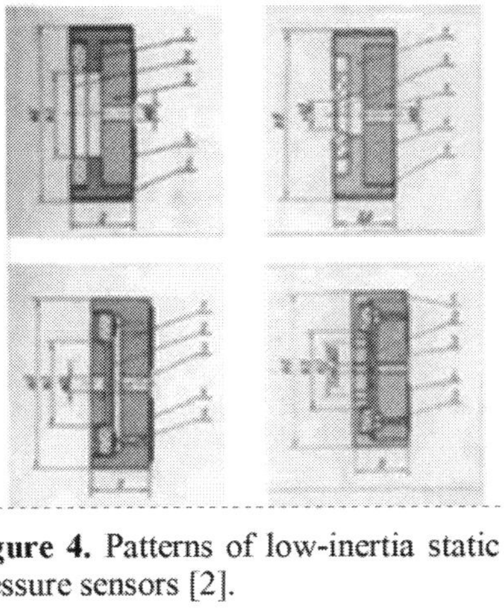

Figure 4. Patterns of low-inertia static pressure sensors [2].

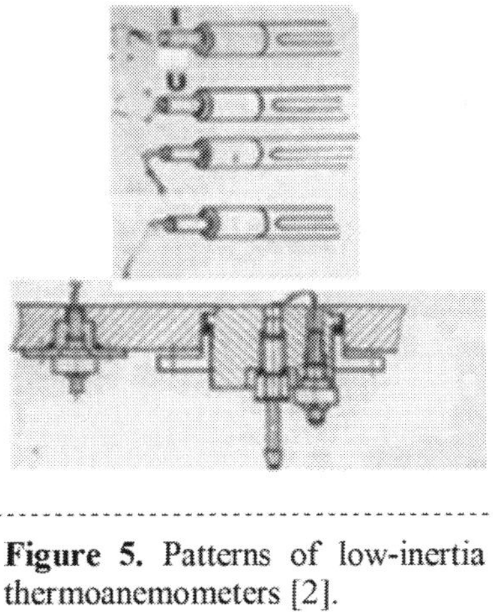

Figure 5. Patterns of low-inertia thermoanemometers [2].

3. The object of study

A few model stages of industrial natural gas superchargers were used as objects of research. Series of experiments [2, 4, 7] code-named "KZ" were conducted on the model of the first stage of supercharger. In all described cases an operating environment is air, suction was produced from an atmosphere through filtration chamber. The levels of equivalent circumferential speeds correspond to the u2 level of natural gas superchargers (to 266 m/s).

On calculated gas-dynamic factors, the prospected stage of "KZ" (fig. 3) belonged to middle flowrate ($\Phi p \approx 0,047$) and low-head type. The closed type impeller ($b2/D2=0,07$ - width of impeller, $\beta 2= 22,50°$- outlet angle of impeller blade, $D1/D2=0.564$- inlet diameter of impeller blades/impeller diameter, $b1/D2=0.096$ - width of the impeller at the inlet, $\beta 1=260°$- inlet angle of impeller blade, $z2=12$- number of impeller blades) has cylindrical permanent thickness blades, circumferences outlined by arcs. Leading edge of blades are symmetrically rounded. In a meridional plane leading edge lines and the planes of basic disk are perpendicular. The immobile elements of the model stage formed by vaned diffuser $z3 = 18$- number vaned diffuser blades, $D3/D2 = 1,163$- relative diameter of vaned diffuser inlet, $D4/D2 = 1,5$- relative diameter of vaned diffuser outlet, $b3 = b4$-width vaned diffuser, angle of inlet- $\alpha 3=150°$, angle of outlet- $\alpha 4 = 300°$. The swinging elbow of the stage is executed fluently on the arcs of circumference with small diffusion airfoil $b5/b4=1,135$. Return channel blades ($\alpha 5=330°$-angle inlet, $\alpha 6=90°$ outlet, $b6/b5 = 1$ -width) have a symmetric falcate form and outlined by the arcs of circumferences. Leading edge of a blade and trailing edge of a blade have identical large radiuses of rounding. In a meridional plane return channel made with straight parallel walls.

4. Method of experimental research

Research of unsteady process is one of the most difficult divisions of experimental aerodynamics. Complications are explained by the wide range of distribution of frequencies and near intensity at the prospected processes, and also by absence of clear borders of processes.

The characteristic periods of unsteady process in a centrifugal compressor are presented on a fig. 6. Transition process have a characteristic time around seconds and higher. The result of influence of such processes is noticeable on the testimonies of traditional measuring devices, for example, of manometers. These periods are not in multiple correlation in course of time of one turn of rotor.

The periods of pulsations of pressure at surging make from hundreds of milliseconds to seconds and tens of seconds (in big units with the ramified pipelines). The periods of rotating stall in absolute motion change from tens of milliseconds to tens of seconds, in relative motion from hundreds of microseconds to tens of milliseconds. As a rule, the periods of these processes (both in absolute and relative motion) are not multiple to the period of rotation of rotor.

Periods of the processes caused by circumferential distortion of distribution of aerodynamic sizes, to hundreds microseconds change from hundreds of milliseconds; they are multiple to the period of rotation of rotor.

Characteristic times of processes of type of coherent structures are calculated by values from 10 microseconds a to 100 milliseconds. In general case, the period of these processes is aliquant to the period of rotation of rotor.

Turbulence is a broadband casual process the spectrum of that embraces an area from the stakes of Hz to hundred KHz (accordingly temporal scales from tens of microseconds to tens of seconds) [4].

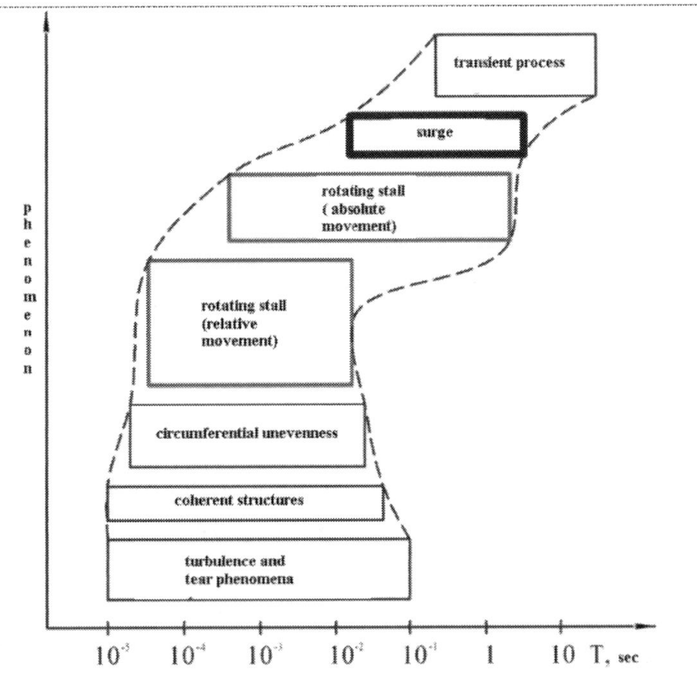

Figure 6. Characteristic temporal scales of aerodynamic unsteady process [4].

On the basis of foregoing it is possible to come to the following: for "complete" research of unsteady process devices are needed, possessing a high fast-acting and wide stripe of the reproduced frequencies: from the stakes of Hz to ten of KHz (taking into account nonsinusoidal character of fluctuations to hundred KHz). Naturally, at research of a limit class of the phenomena (surging, running around derangement in absolute motion) the top limit of frequency range of measuring device can be lowered, however, existence of pulsations of type of running around tracks results in corresponding complication of measuring and necessity of account of possibility of appearance (parasite) of resonances, especially at measuring, for example, of pulsations of pressure with the use of long tricking into channels.

A no less meaningful task is a choice of amount of measuring points on spatial coordinates (as the field methods measuring as a rule, based on optical effects, both in an experiment and in external environments create the problems related to dustiness of gas [12]). For the decision of this task it is necessary to know spatial lengths of waves of the prospected phenomena, and also places of appearance of the sought after effects.

5. Layout of measuring points chart

Location of measuring points of pulsations static pressures, is chosen with a condition, to provide testing with the purpose of determination of moment of origin and parameters of the beforesurging state. The sensors p1 and p2 are located on the area vaneless diffuser before inlet of a diffuser, and sensors of p3 and p4 on an exit from a diffuser (fig. 7). In the series of tests of the stage of KZ the sensors of p5 and p6 are located on the walls of swinging elbow of the stage (fig. 7). The sensors not located in a diffuser allow to prospect the nonstationarity of processes on length of channel of flow channel of centrifugal compressor.

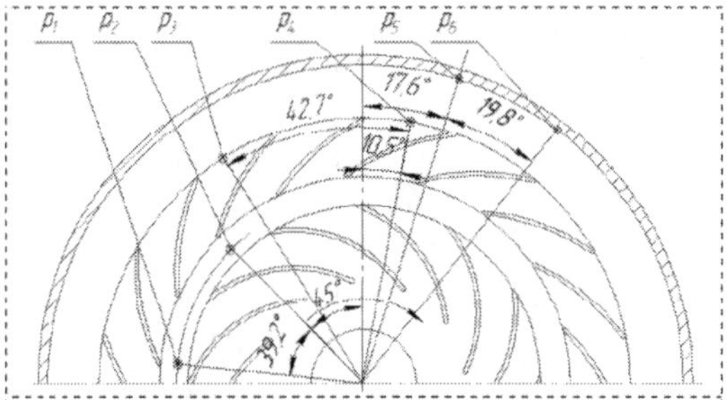

Figure 7. Layout of measuring points chart is in the series of tests of KZ[7].

6. Defence of compressors from surging

Presently there are many methods of defence of centrifugal compressors from surging. They can be divided into two groups: self-reactance and sign methods [3]

Now all systems of anti-surging defence are based on a self-reactance method. The lacks of self-reactance method is a presence of intensive pulsations of pressure, existing in the place of measuring of overfall of pressure in the intake pipe coupling on near to surging mode of operations of compressor, on that the productivity, expense, is calculated, methodology of indirect determination of gas ($\overline{V}_0$= f(ΔP))[4]. The anti-surging system, that this methodology is the basis of, can not to a full degree provide safe work of compressor on the modes of small charges, because will have large errors of measuring.

The self-reactance method of defence of centrifugal compressors from surging firstly is not reliable, secondly not economical, for some types of centrifugal compressors, it is also proved by that in researches the world conducted on defence of compressors by other method by a sign.

Methodology of sign method consists in the following: on the basis of informative criteria it is necessary in good time to find out the origin of rotating stall (or other characteristic phenomenon, antecedent of surging) in future to make decision about approaching to the dangerous area (surging). The choice of criterion is conducted taking into account the characteristic features of behavior of gas stream in flow channel of turbo-compressor before appearance of surging. Presently use different signs for realization of this method. One of methods of diagnostics of the before-surging state, offer J.Z. Gyzelbaev [1].,and another way, offer Joule. Mackey (McKee) of and other, is base on researches of countercurrent through a front end play between the walls of cover disk of impeller and corps[11].

7. System of analysis of signals of nonsteady process and algorithm of defence of centrifugal compressor from surging.

On the Compressor, vacuum and refrigeration technique department of SPbPU, institute of EaTS, is conducted numeral experiments on the analysis of signals of unsteady process in centrifugal compressors. The methods of exposure of periodicity are presently tested: spectrology, cross-correlation analysis [7],[10], and also methods of the signal of filtrations processing and wavelet decomposition fig.8 [7],[8].

Results of application of method wavelet decomposition evidently demonstrated on fig.9,11, experiment on that the records of the mode were produced pre-stall, as a result on the pictures of closenesses of power spectrum of periods of signal lances of fig.10 are visible and with advancement on the channel of fig 12 they become clearer, on the spoiled signal the peak of fig.7 is also noticeable.

The total number of modes of data on which the algorithm was checked is 162, inclusive of 6 sensors and 3 interested levels and other analysis methods total number of analyzed sensors is 972, and the total number of analyzed graphs more than 5000. Entries have the typical character and for presenting a new concept of a method of protection against surging the most revealing and characteristic mode preceding surge was selected. Wavelet analysis is used for a long time (more than 20 years),in this work the standard procedure of signal processing software package for mathematical analysis is applied[7], [8].More detailed information on the parent function algorithm of wavelet decomposition is a commercial secret.

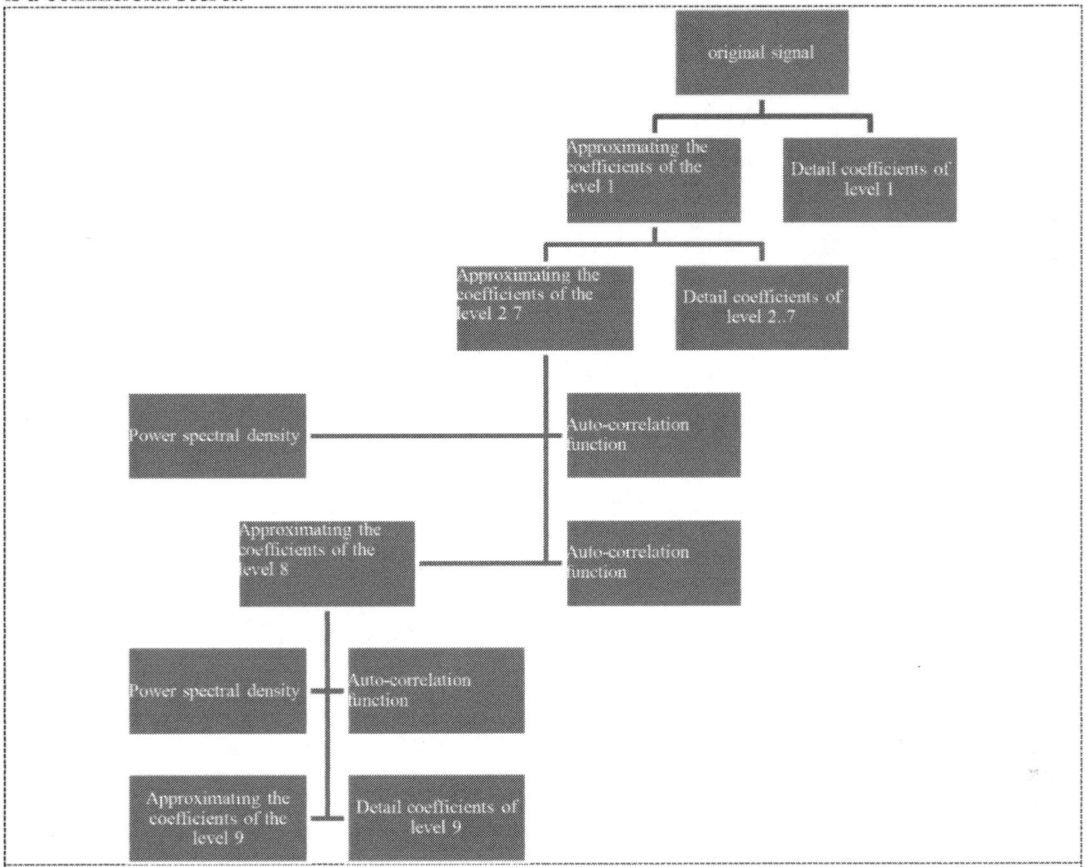

Figure 8. The procedure of data processing by the combined method of detecting hidden periodicities.

An algorithm for protecting of centrifugal compressor from surging based on timely exposure rotating stall. Feature of algorithm of fig. 13 consists of that at treatment of signals wavelet is used transformation, and after a regenerate signal is exposed to two parallel methods of exposure of period of pulsations : to the spectrology and autocorrelation analysis. Procured periods and their powers are compared, amplitude of pulsations, offer in-process Ngyen Mine Hai Finding fault with, is also analysed [10], such sequence allows to avoid the false system actuation. At the task of borders of periods of system actuation, it is also necessary to take into account that the clearness of the got values can depend on length of temporal interval on that their determination is produced. Time of interval is also related to the that condition, that time is required the system of defence on execution of the decision got in an algorithm, and this time must not exceed a minimum time domain for that the state of intensive running around derangement passes to the mode of surging

For the decline of dependence of interval length from total time of algorithm work and time expended by the executive mechanisms of defence, a next chart is used with the 90% ceiling. The chosen record of signal from a sensor in a certain temporal interval is exposed to the first analysis, at a next analysis from this record the first ten percents of record retire and new ten percents are added in the end, in the total it allows freely to vary length of area for an analysis. For optimization of algorithm and exception of value of the surplus ceiling him it is necessary to set for the certain stage and setting, because reduction of ceiling can result in the increase of amount of analysable intervals and, as a result, height of overall sizes of the system of analysis and cost. The choice of condition for a decision-making on the values of period is done so that the calculated periods for the different intervals of time can take on unequal critical values, for the account of this fact final value taken on an on the amount of intervals with the critical values of period for a few intervals on that determination of period was produced.

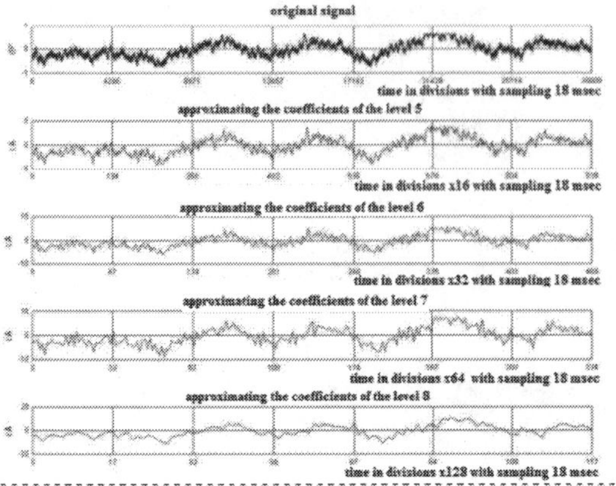

Figure 9. A typical example of stage of KZ1, mode pre-stall. Initial and decomposed signals approximating coefficients from 5 to 8 levels for the sensor of P_{4I}.

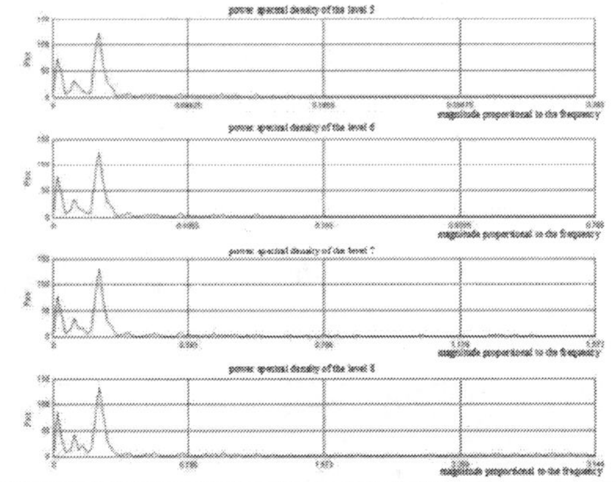

Figure 10. A typical example of Power spectral density for approximating the coefficients of expansion from 5 to 8 levels, for the sensor of P_{4I}.

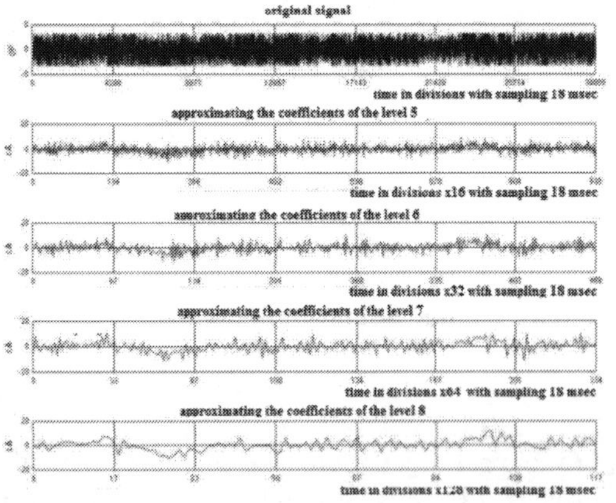

Figure 11. A typical example of stage of KZ1, mode pre-stall. Initial and decomposed signals approximating coefficients from 5 to 8 levels for the sensor of P_{4II},

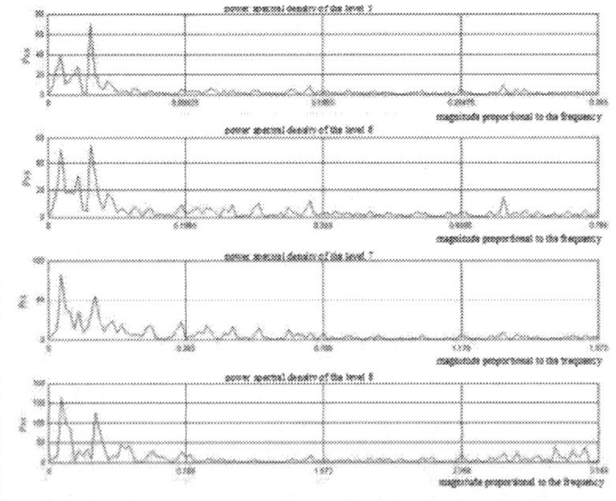

Figure 12. A typical example of power spectral density for approximating the coefficients of expansion from 5 to 8 levels for the sensor of P_{4II},

It is explained by that for some types of the stages at stall the period of stall is more stable, and a period at the developed derangement can take on a few values on the different intervals of time. For determination of type of stall - steady or pre-stall - two values equal to the relation to the number of areas with a critical period to the amount of all areas can be got. The number of all areas gets out proportionally to the interval of time necessary for establishment of fact of existence of the developed running around stall. At being in the zone of pre-stall an amount of areas will be less than, and, consequently, their relative value will diminish, and at the developed stall an amount of areas with a critical level will be more, and a relation will be bigger.

The considered algorithm can work in the next chart of anti-surge defence of fig.14. System consists of four blocks. The block of measuring channels provides work of four pressductors and gives out the analog signal of pulsations of pressure. A block ADC will transform analog signal in digital, produces treatment, analysis and decision-making on the basis of algorithm. Both industrial computers can be set in this block and DSP (digital programmable processors). The amount of pieces of equipment depends on the type of compressor and necessary reliability indexes. The block of register is counted on the record of the values educed during switching on the system of defence and casual extrass of signal, and also for reserve storage of the program of algorithm of defence and frontier values. A block - terminal and control stand destroy current values about the state of the system, and also has the opportunity to change value of borders at presence of necessary access level.

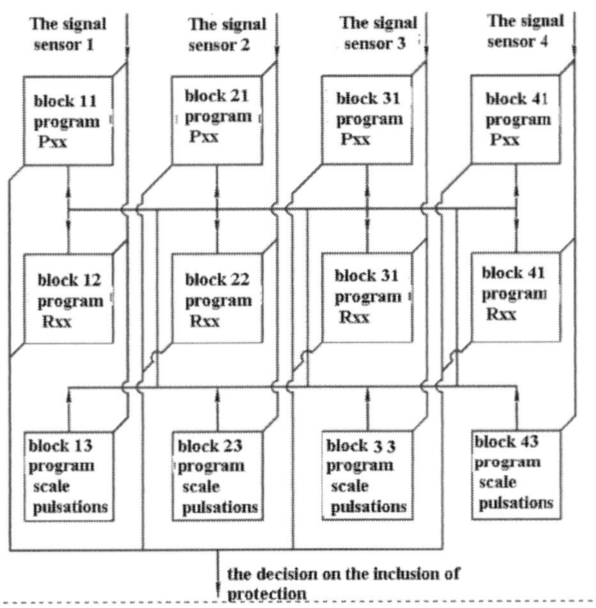

Figure 13. Concept chart of anti-surge algorithm

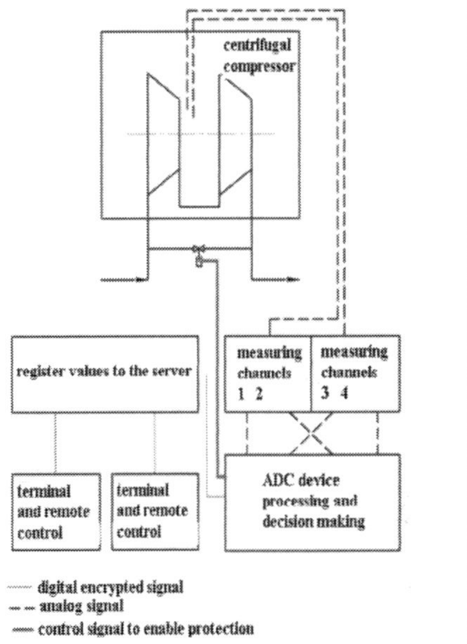

Figure 14. Concept chart of the system of anti-surge defence of centrifugal compressor.

8. Methods of prognostication and numerical simulation of nonsteady process in compressors

The periods got at the numeral design of CFD can serve for the task of primary border terms of wearing-out of the system of defence, for certain flow channel of centrifugal compressor. These calculations are conducted by Lopulalan H, development of virtual stand is in-process considered for research of transients in running part of compressor, that allows to find out the origin of running around stall (called pre-stall) in area $\varphi 0$опт $< \varphi 0 \leq \varphi 0$r.s., corresponding to experimental data[6]. But in researches differences were got from an experiment, namely "Parameters of running around derangement in impeller in order of size correspond to experimental data, the calculated frequency of pulsations in a vaneless diffuser substantially differs from experiment. It is related to the features of calculations with the use of interface "transient rotor – stator", transmitter data from a running around rotor in an immobile vaneless diffuser. As at running around derangement of zone of separation move on a circumference with a low rotation speed (ωз << ωрот), synchronization on frequency of rotation, carried out by an interface, results in disparity of numeral estimations of running around stall, looked after in an experiment". These educed features it is necessary to take into account at the task of border periods of wearing-out of the system of protecting from surging

Specialty designing and adjusting of the system of the anti-surging defence based on the timely exposure of signs of appearance of running around stall.

Designing plan and adjusting of the anti-surging system, based on the considered algorithm, consists of the next consistently executable stages.

1) To develop the chart of placing no less than two littleinertia pressductors on the circumference of one diameter on inlet of diffuser of the stage and on an exit from it, and originally to set in an amount four units

2) To Choose frequency of discretisation from minimum values for the period of stall, being base on non-stationary numeral design of flow channel.

3) To Check the error of the measuring system of rotation frequency, frequency of discretisation of measuring of rotation period must be not less frequency of discretisation of pressurometer.

4) To Conduct the tests of setting in place of making for adjustment of frontier values for a period at stall for present in the algorithm of management borders certain on a virtual stand, and to confirm final values in place of permanent exploitation, and also at a measuring seating capacity, exceeding an amount channels to choose most suitable.

References

[1] Gyzelbaev J Gas-dynamical nonstationarity of the processes in centrifugal compressor. The surge and methods of its detection 2000 *KGTU* **144** *Russia Kazan*

[2] Izmailov R Unsteady flow in centrifugal compressor numerical modelling and experimental investigation 2012 *ISUAAAT 13 S7-1 Tokyo Japan*

[3] Izmailov R Development of diagnostic systems prepumping condition of the centrifugal compressor of Chemical and petroleum engineering 1998 *HOM №11* **31-34** *Russia Saint Petersburg*

[4] Izmailov R Unsteady aerodynamic processes in centrifugal compressors. 1987 *LPI* **527** *USSR Leningrad*

[5] Izmailov R, Akylshin Y and Krytikov T System diagnostics prepumping condition of the centrifugal compressor 2004 *Turbines and compressors№3,41* **15** *Russia Saint Petersburg*

[6] Izmailov R The creation of the prototype of anti-surge system control PNG (unit pumping natural gas) 1995 *SPbSTU* №306415/803-226 **52** *Russia Saint Petersburg*

[7] Krytikov T The diagnosis of surge in a centrifugal compressor 2003 *SPbSTU* Ref 05.04.06 **16** *Russia Saint Petersburg*

[8] Lebedev A Wavelet analysis of nonstationary processes in centrifugal compressor 2009 *SPbSTU* Dis 05.04.06 **300** *Russia Saint Petersburg*

[9] Lopulalan H A virtual test facility for the investigation of non-stationary processes in stage centrifugal compressor 2014 *SPbSTU* Dis 05.04.06 **200** *Russia Saint Petersburg*

[10] Ngyen Min Hai Informative criteria for detection of a surge condition of the centrifugal compressor 2007 *SPbSTU* Dis 05.04.06 **190** *Russia Saint Petersburg*

[11] McKee R J, Deffenbaugh DM Increased flexibility of turbo-compressor in natural gas transmission though direct surge control *Southwest Research Institute®, mechanical and Materials Engineering Division*, Final report October 2001 - February 2005. SwRi® Project No. 18.04990 DOE Award No. DE-FC26-01NT41163. -60p.

[12] Wernet M Bright M and Skoch G An Investigation of Surge in a High-Speed Centrifugal Compressor Using Digital PIV *NASA Glenn Research Center*, December 2002, **23** p. (http://gltrs.grc.nasa.gov)

[13] Daubechies I, Ten Lectures on Wavlets *CBMS-NSF Reg. Conf series in Appl. Math. 61*, Soc Lnd, Appl. Math., 1992 *Philadelphia USA*

[14] Chui C, An introduction to Wavelets 1991 *Department of Mathematics Taxas A&M University College Station Texas USA*

IOP Conf. Series: Materials Science and Engineering **90** (2015) 012053 doi:10.1088/1757-899X/90/1/012053

Optimization of the axial compressor flow passage to reduce the circumferential distortion

G Popov[1], D Kolmakova[2], A Shklovets[3], A Ermakov[4]

[1] Junior Researcher, Samara State Aerospace University (SSAU), Samara, Russia
[2] Junior Researcher, Samara State Aerospace University (SSAU), Samara, Russia
[3] Junior Researcher, Samara State Aerospace University (SSAU), Samara, Russia
[4] Professor, Samara State Aerospace University (SSAU), Samara, Russia

E-mail: grishatty@gmail.com

Abstract. This work is motivated by the necessity to reduce the effects of the flow circumferential distortion in the flow passage of the aircraft gas turbine engine (GTE). In previous research, the authors have proposed the approaches to decrease of the flow circumferential distortion arising from the mid-support racks of GTE compressor and having a negative impact on the blade rows, located upstream. In particular, the idea of introducing the circumferentially non-uniform blade pitch and profile stagger angle of guide vanes located in front of the support was contributed in order to redistribute the flow and decrease the dynamic stresses in the rotor wheel of the same stage. During the research presented in this paper, another principal of reduction of the flow circumferential distortion was chosen. Firstly, the variants of upgrading the existing support racks were found. Secondly, the new design of support was offered. Both the first and the second version of the support design variation took into account the availability of technological and structural limitations associated with the location of oil pipes, springs and others elements in the support racks. Investigations of modified design showed that the support with altered racks provides a reduction of dynamic stresses by 20% at resonance with the most dangerous harmonic, and the new design of support can give the decrease of 30%.

1. Introduction

Many problems that are currently faced by researchers and engineers cannot be solved analytically or require huge costs for the experimental realization. Often, the only possibility to carry out the express analysis of engineering problems is a computer and mathematical simulation. Progress in the development of numerical methods significantly increased the number of tasks available for analysis [1-3]. The obtained results based on these methods are used in almost all fields of science and technology.

The finite element method finds its most important application in the design analysis. At the same time, bridges, buildings, marine hulls, aircraft components, machine parts, pistons, tools, i.e. any engineering construction are understood under "construction" in the design analysis.

Turbomachinery impellers are very critical parts of gas turbine engines (GTE). The reliable operation of the engine and flight safety of the aircraft depend on their reliable operation of turbomachinery impellers to a large extent.

While aircraft gas turbine engine operation, the static, dynamic, and thermal loads act on the rotor blades and disc, causing the complex stress pattern in the details.

Content from this work may be used under the terms of the Creative Commons Attribution 3.0 licence. Any further distribution of this work must maintain attribution to the author(s) and the title of the work, journal citation and DOI.
Published under licence by IOP Publishing Ltd

The possibility of appearance of variable dynamic stresses in the impellers associated with the action of the changing in time loads on it. One of the main types of oscillations are forced, which can appear as dangerous.

Forced oscillations are due to the impact of external forces on the object, time-varying, which do not depend on the dynamic behavior of a vibrating object and do not change under the assumption of object non-deformable. In turbomachines, first of all, these forces associated with circumferential uniform of flow incident on the rotating impeller. The circumferential distortion, even being in a constant rotating coordinate system relative to the rotating impeller, is transformed into power load that varies in time.

2. The calculation method of forced oscillations.

Circumferential nonuniformity of the gas stream flowing around the blades is the main source of vibration excitation of the gas turbine engine rotor wheels (RW). It appears in the form of nonuniformity of the velocity and pressure fields in the stream before and after the RW. Nonuniformity leads to the fact that the gas load intensity in the circumferential and radial directions is inconstant around the circumference of the flow passage. As a result, the integrated gas-dynamic force Q_g changeable in value acts on any of the blades during RW rotation.

Since the gas-dynamic force Q_g is a periodic value, i.e. $Q_g(\alpha)=Q_g(\alpha+2\pi)$, it can be expanded in a Fourier series:

$$Q_g=\sum_{m_e}^{\infty}Q_m\cdot\cos(m_e\alpha-\gamma_m)=\sum_{m_e}^{\infty}Q_{gm} \tag{1}$$

where Q_m – the amplitude of the component harmonic; m_e – harmonic number; α – centre angle, γ_m – phase shift along the circumference.

Expansion (1) allows presenting the gas load having a complex distribution along the circumference as a sum of component harmonics. Each of them represents a succession of load waves that fit around the circumference of the flow passage.

For rotating impeller, any of the components in (1) is a exciting harmonic, which is the backward running waves succession. The load rotates with an angular velocity ω, which equal to the angular velocity of RW. Thus, the circumferential gas flow nonuniformity for rotor wheels is equivalent to an effects of infinite set of exciting harmonics, each of which represents a succession m_e of backward running waves of load oscillates in time with frequency $f_e=m_e n_s$, where $n_s=\dfrac{\omega}{2\pi}$.

To determine the dynamic stresses in the blade airfoil, Ansys software was used. Gas dynamic load force on the blades is determined in static CFD-calculation. Using APDL programming language, the program was written that imports the load distribution from finite element model of blade row in CFX to finite element model of the blade row in Ansys Mechanical APDL. Then, the load is decomposed in Fourier series at congruent nodes of blades and is represented as a backward running wave. Campbell diagram is used to determine the most dangerous engine modes, and correspondingly, the most dangerous harmonics. Dynamic stress analysis is carried out only with the most dangerous harmonic. Schematically the method of calculation is shown in Figure 1.

Damping in the system is defined on the basis of experimental data as a viscous in material. Thus we cannot state that this calculation method allows evaluating the quantitative dynamic stresses in the blade at the design stage. However, this method is acceptable for qualitative assessment of changes of dynamic stresses during the optimization calculations.

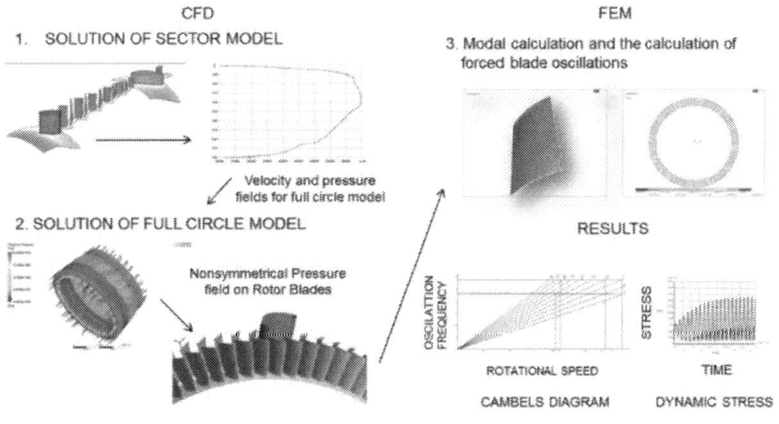

Figure 1. Scheme of method for calculation compressor rotor blades forced oscillations.

3. Practical implementation of the method

Practical implementation of the method is discussed in [4]. The object of the investigation is the rotor wheel blade of fifth stage of five-stage medium pressure compressor (IPC) of gas turbine engine (GTE) (Figure 2). Downstream the fifth stage of the IPC, the middle GTE support is located.

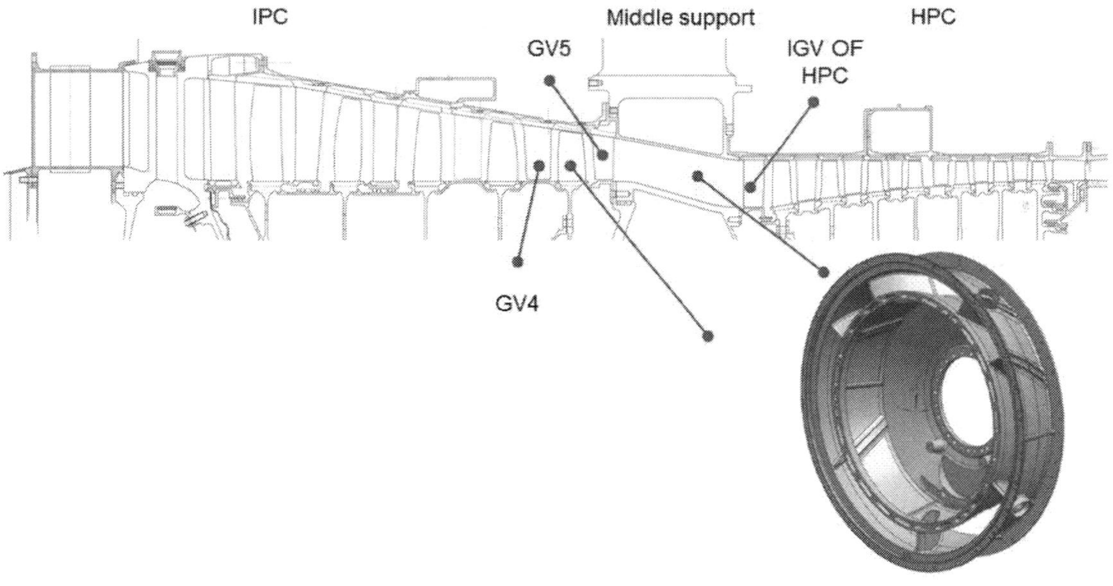

Figure 2. The diagram of the investigated compressor.

Seven racks of different cross-sections are unevenly distributed in the flow passage of support casing (Figure 2). These racks cause the circumferential variation of the gas flow in gas-turbine engine flow passage, which leads to increased dynamic stresses in the fifth rotor wheel blades, as a consequence, to its breakage (Figure 3).

As the number of support racks is 7, the blade was detuned from the dangerous seventh harmonic at the design stage. The natural frequency of the RW blade vibrations is selected so that the resonance is possible with the eighth harmonic and higher. Seventh harmonic is beyond the engine operating conditions. Experimental research revealed the destruction of the rotor blade of compressor stage at resonance with the 12th harmonic.

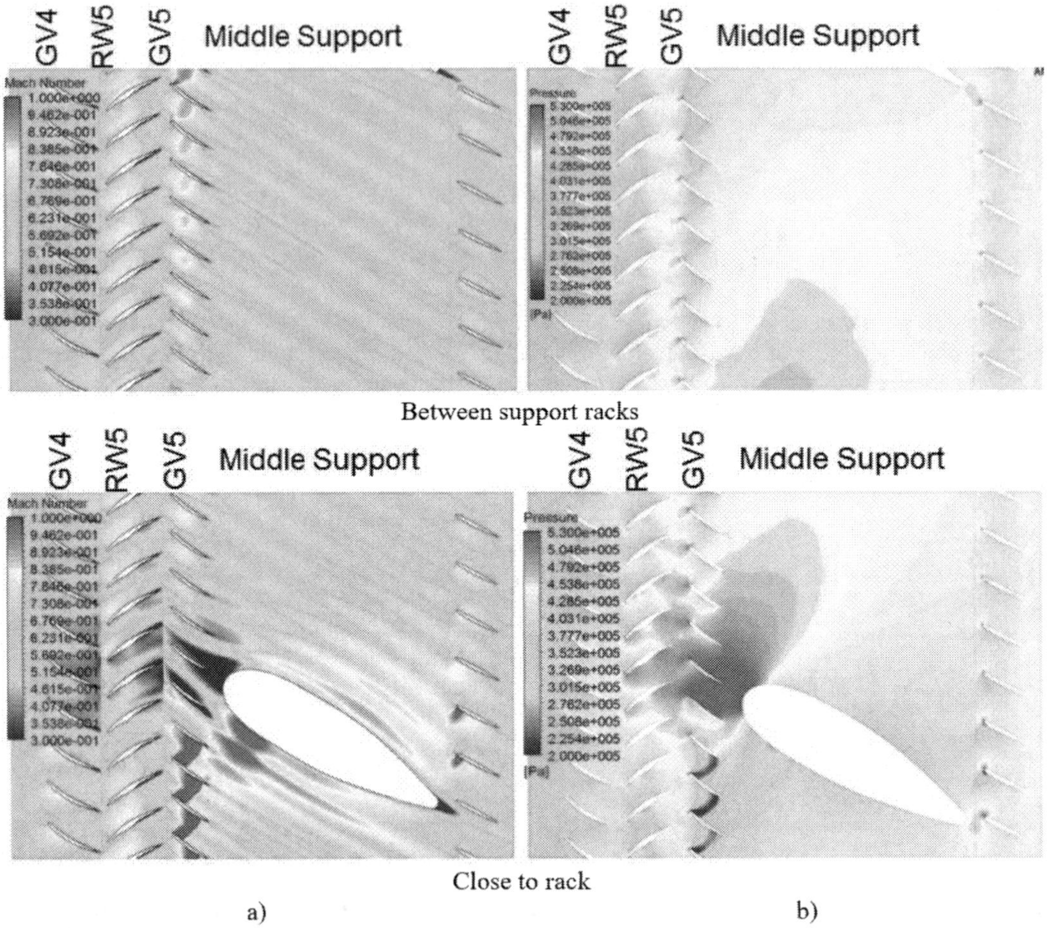

Figure 3. Mach number (a) and pressure (b) fields between support rack and close to racks.

On the basis of the above-described method of calculation of blade forced oscillations, the maximum alternating stresses were defined in the fifth stage rotor blade of IPC reference design operating in the conditions of gas flow circumferential distortion.

To reduce the circumferential distortion of the flow, stagger angles and pitch of fifth stage guide vanes were changed to non-uniform around the circumference of the rotor wheel (Figure 4). There were created 11 variants with different stagger angles and pitch. This allowed the redistribution of the flow and the discharge of GV channels located opposite the support racks. All this has led to a reduction of dynamic stresses in the RW blades.

As a result, the optimal variant of the design was chosen, in which the amplitude of the dangerous 12th harmonic was reduced by 2 times, while the number of changeable guide vane (GV) blades was equal to 14 (total blade number is 76). Figure 5 shows the value of the mean amplitude of the exciting harmonics for the reference compressor design, for the variant with 42 variable GV blades (variant 4) and the final compressor variant (variant 9) with 14 variable GV blades.

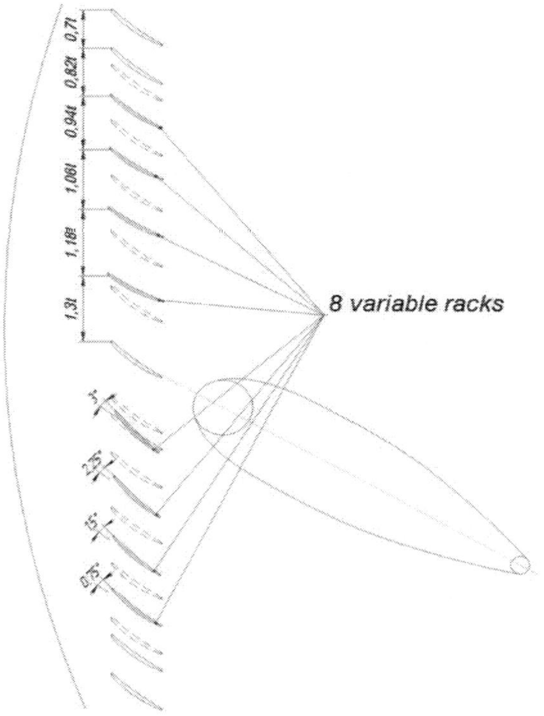

Figure 4. The algorithm of introduction of different stagger angle and pitch.

4. Effect of the support design on circumferential distortion of gas flow

4.1 Changing the angular location of the support racks

Technologically, changing the stagger angle of the guide vane blade is challenging, so it was decided to change the configuration of the support rack.

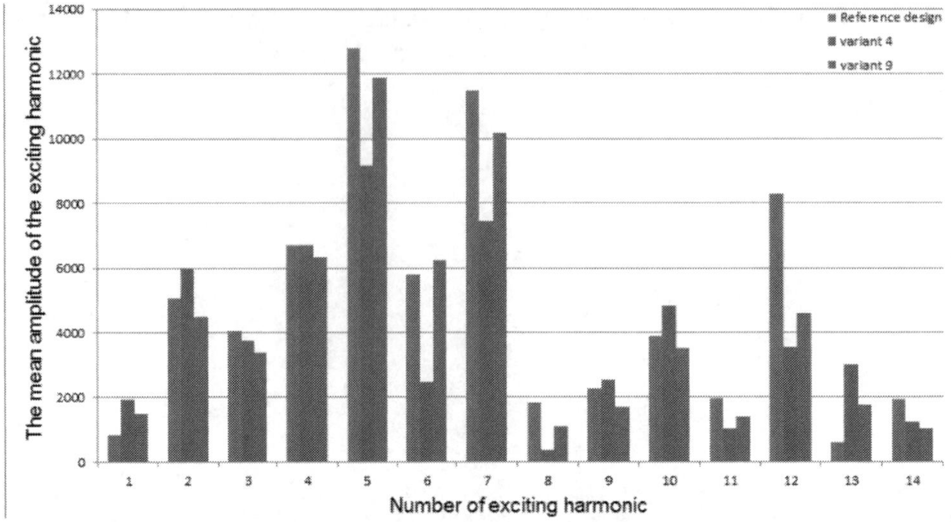

Figure 5. The level of the exciting harmonic amplitudes affecting the fifth rotor wheel blades.

Therefore, the experimental variant of support with 13 racks instead of standard support with seven racks was considered at the next stage of the research (Figure 6). The experimental tests were conducted for this support design by JSC "Kuznetsov" (Samara, Russia) –the enterprise of aviation and space propulsion engineering. [5].

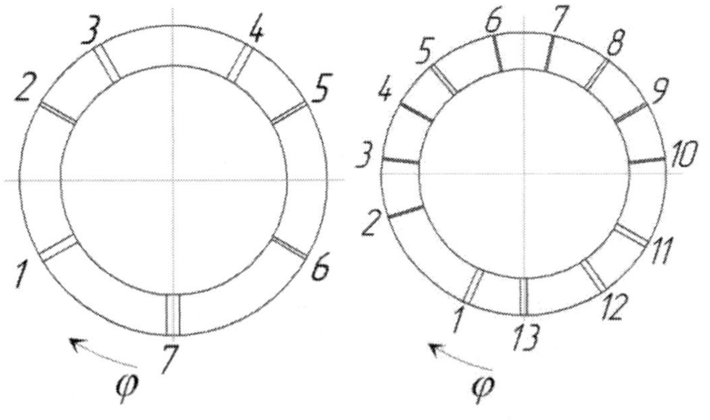

Figure 6. Scheme of standard 7-racks support and modified 13th-racks support.

It was revealed that the mean amplitude of dangerous 12th harmonic decreased by 2 times in the computational studies (experimental data confirm this). The coincidence of obtained results with experimental data confirms the adequacy of computational models and techniques. The figure 7 shows that there is a significant decrease in the amplitudes of all dangerous harmonics.

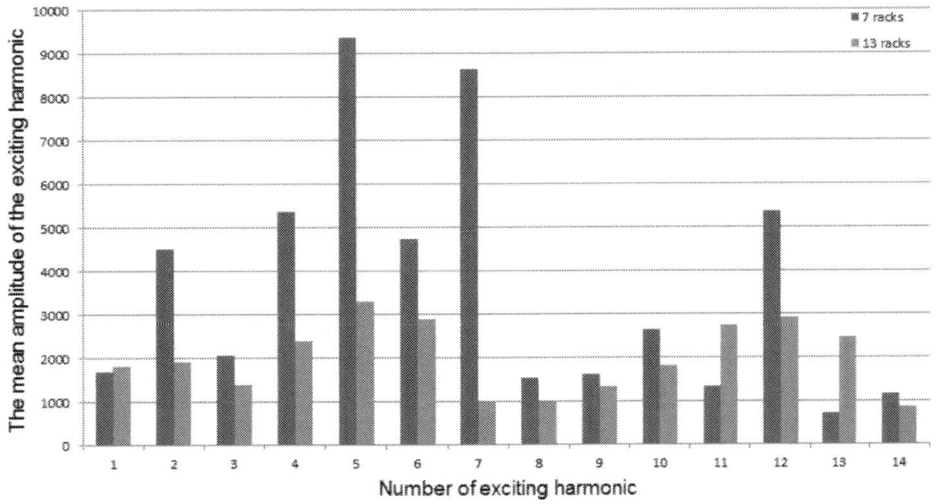

Figure 7. Mean amplitudes of exciting harmonic of support variants with seven and 13 racks.

The rack thickness, in which the engine systems are located, and their angular disposition were changed in the 13-racks experimental support. Therefore, such support cannot be applied at modernized engine. Consequently, one of the conditions was to keep unchanged the racks 1, 3, 4, and 7 of standard 7-racks the support (Figure 6) when optimizing the angular arrangement of racks.

For this, the optimization technique of the angular position of support racks was developed. Angular position of racks was represented as a periodic function. Function was provided in the form of discrete data array containing the nominal pressure before each rack, which was equal to one. The array contained 180 values; each value corresponded to the angle of the circumferential location of the rack. Then the function value was 0 if there was no the rack and the value was 1 if there was rack is. Discrete function for seven-rack support is shown in Figure 8.

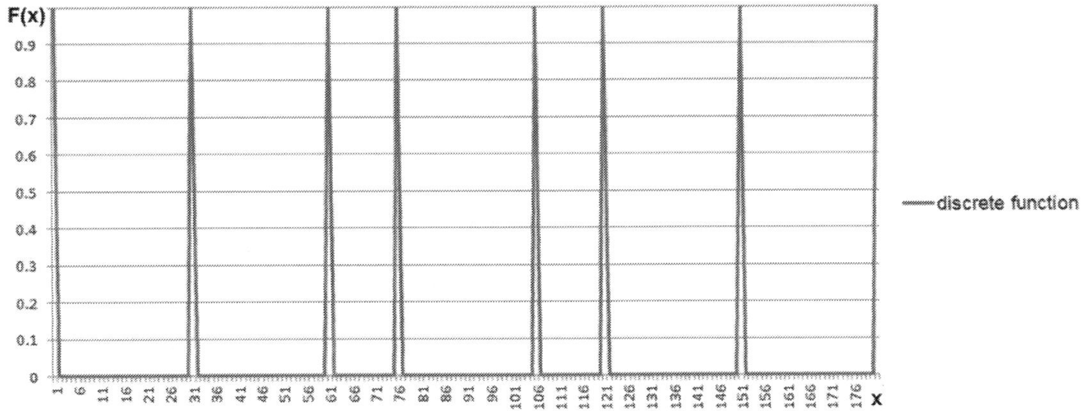

Figure 8. Discrete function for seven-rack support.

Changing the angular arrangement of free racks, decrease of dangerous harmonics amplitude can be achieved.

Reduction in the amplitude of the 12th harmonic was performed using optimization methods implemented in the software package IOSO [6]. The goal of optimization was to decrease the amplitudes of 10th and 12th harmonics. Consequently, the optimization criterion was the amplitudes of 10th and 12th harmonics.

Angles of the 1, 3, 4, 6, 7, 9-12 racks were used as variable parameters. In formulation of optimization problem, the restrictions on the amplitude values of 8, 9, 11 and 13 harmonics were imposed. These restrictions were assigned so that the amplitude values of these harmonics obviously did not lead to the destruction of the blades.

During the optimization, the Pareto front was obtained - the set of unimprovable solutions, a compromise between the decrease in the amplitude of 10th and 12th harmonics (Figure 9). A unique array of angular location of 1, 3, 4, 6, 7, 9-12 racks corresponded to each point of the Pareto set.

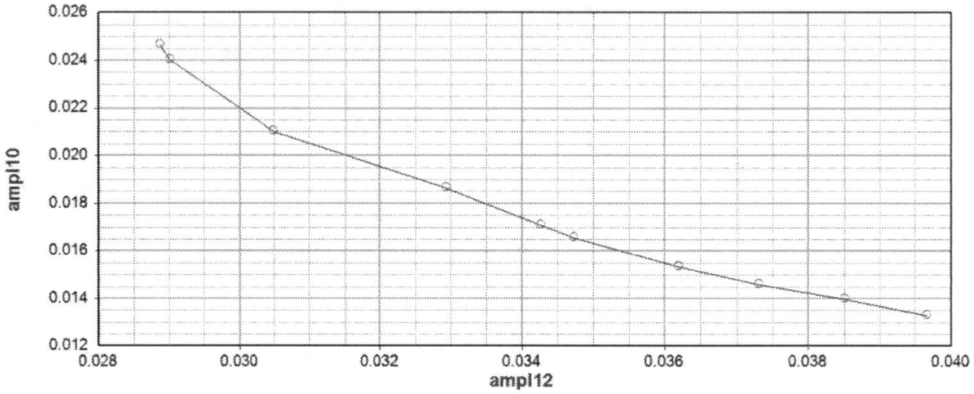

Figure 9. The Pareto set of multicriteria optimization.

For further analysis, the leftmost point of Pareto front was selected (Figure 9). This point corresponds to the maximum possible reduction in the amplitude of the 12th harmonic, as the most dangerous. Location of the racks, shown in Figure 10, corresponds to this point of the Pareto set.

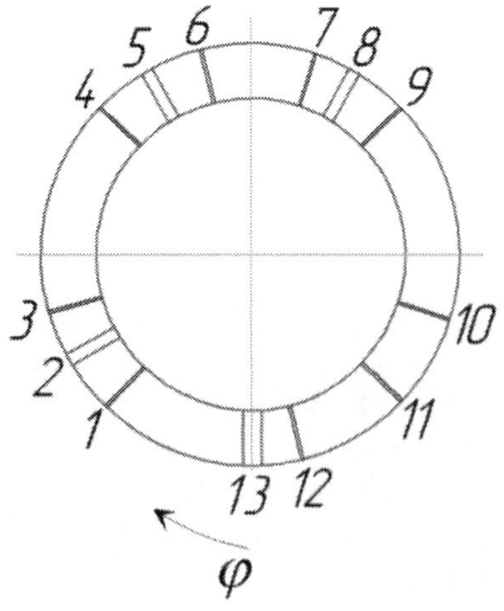

Figure 10. Location of support racks, providing the minimum value of the 12th harmonic, while maintaining the amplitude of the 10th harmonic at an acceptable level.

4.2 Changing the position of the support racks relative to the blades of the upstream located guide vane

There are experimental studies that prove that the lattice, which is located in front of the cylinder, streamlined by the flow, enhances the upstream transmission of the high pressure zone [7].

The results of that research allow us to offer an additional way to reduce the dynamic stresses. It represents a change in the form of support racks due to remoteness of the leading edge of support racks (cylinder) from the trailing edges of GV blades of fifth stage (lattice). At the same time, the internal cavities of racks remain unchanged for the placement of engine systems (Figure 11).

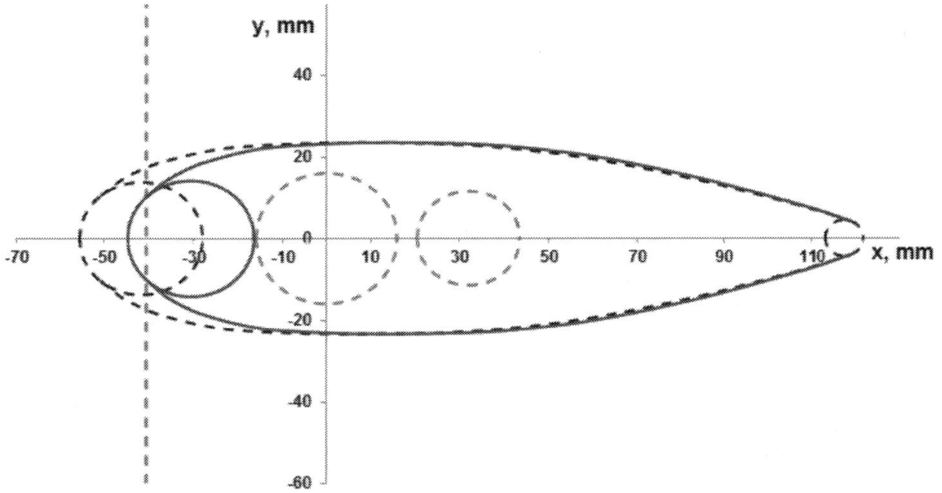

Figure 11. The algorithm of shift of the rack leading edge

Figure 12 shows a comparison of the harmonic amplitudes of the optimized 13-racks support without shifting the leading edges and with shifted leading edges of racks. Figure 13 shows a comparison of harmonic amplitudes of the 7-racks support, the reference design of 13-racks support (by JSC "Kuznetsov") and optimized 13-racks support with shifted leading edges (by SSAU).

Thus, the implementation of optimized 13-racks support with shifted leading edges reduces the amplitude of the most dangerous harmonic in 1.8 times, while the position of four the support racks with engine systems remains unchanged.

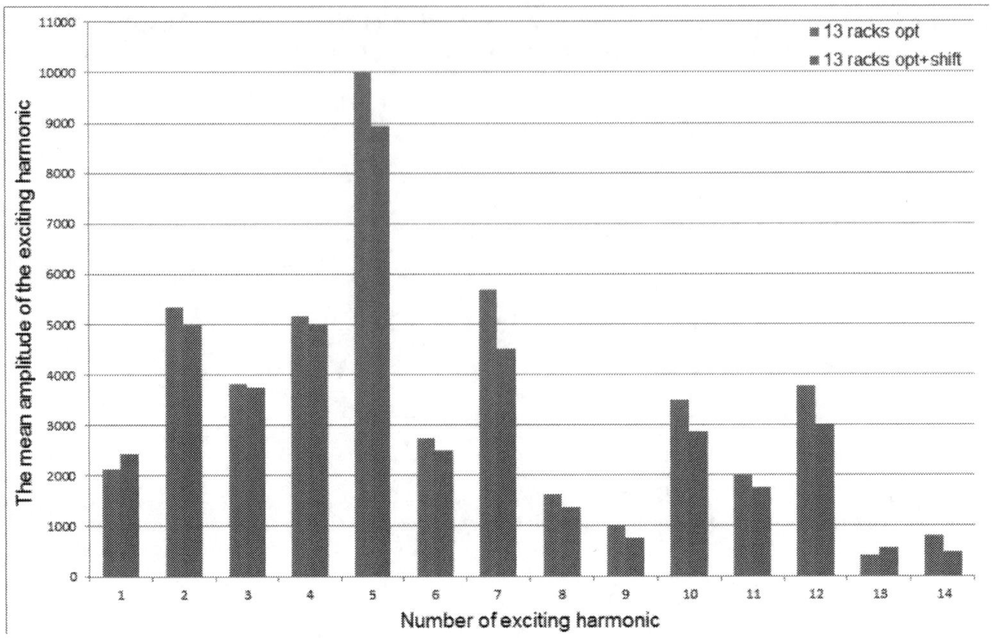

Figure 12. Mean amplitude of the exciting harmonics of the 13-racks support with and without shifted leading edges.

Conclusion

The high level of circumferential distortion of the gas flow in aircraft engine is caused by many factors. One of the major is the presence of support racks in the flow passage (e.g., after the compressor).

Change in the angular position of support racks allows redistributing and reducing the amplitude of the exciting harmonics.

During the research, the optimization technology of angular arrangement of support racks was developed by the software IOSO due to the discrete representation of a response from the rack.

This technology allowed obtaining the optimum racks location of 13-racks support, providing decrease in the amplitude of 12th harmonic in 2 times.

Furthermore, the variant of variable dynamic stresses reduction was proposed due to shape correction of support racks (shift of the rack leading edges relative to the trailing edges of GV blades located in front of the support).

This method provided a decrease in the amplitude of the 12th harmonic in 1.8 times.

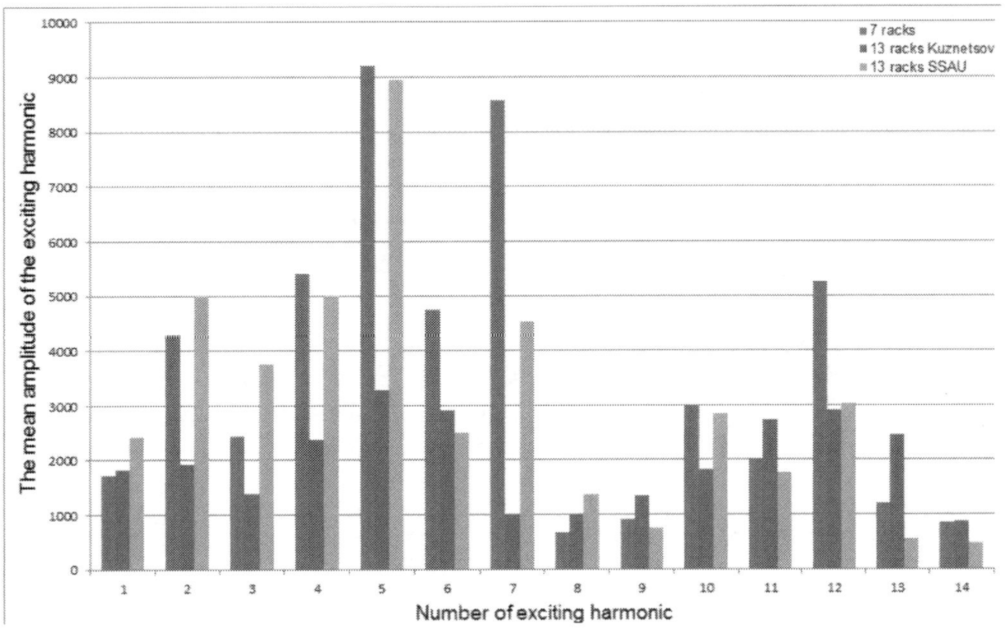

Figure 13. Mean amplitude of the exciting harmonics of the 7-racks support, reference 13-racks support (Kuznetsov) and optimized 13-racks support with shifted leading edges (SSAU).

Acknowledgments
This work was supported by the Ministry of Education and Science of the Russian Federation in execution of the order №218 from 09.04.2010 (theme code *2013-218-04-4777*)

References

[1] Popov G, Baturin O, Kolmakova D and Krivcov A Improvement Results of TK-32 turbocompressor turbine with gas-dynamics and strength CAE-systems *International J. of Engineering and Technology* **6** 2297-03
[2] Saigakov A, Gorshkalev A, Kayukov S and Blagin E 2014 Strength analysis of the internal combustion engine elements by using CAD/CAE-systems *Research J.l of Applied Sciences* **9** 669-673
[3] Zubanov V, Shabliy L and Krivcov A 2014 Centrifugal Kerosene Pump CFD-Modeling *Research J.l of Applied Sciences* **9** 629-634
[4] Ermakov A, Shklovets A, Popov G and Kolmakova D 2014 Investigation of the effect of the gas turbine compressor supports on gas flow circumferential nonuniformity *Research J.l of Applied Sciences* **9** 684-690
[5] JSC "Kuznetsov", Accessed February 10, 2014. http://www.kuznetsov-motors.ru/en
[6] Egorov I, Kretinin G, Leshchenko I, Kuptzov S 2002 IOSO Optimisation Toolkit - Novel Software to Create Better Design *Proc. 9th AIAA/ISSMO Symposium on Multidisciplinary Analysis and Optimisation (Atlanta, Georgia, 04 - 06 September 2002)*
[7] Saren V 1984 Flow around irregular lattice of plates placed in front of the cylinder *Technical Report* (Moscow: Central Institute of Aviation Motors) p 36.

IOP Conf. Series: Materials Science and Engineering **90** (2015) 012054 doi:10.1088/1757-899X/90/1/012054

Simulation of erosion by a particulate airflow through a ventilator

A Ghenaiet

Faculty of Mechanical and Process Engineering, University of Sciences and Technology, USTHB, BP32 El-Alia, Bab-ezzouar, 16111, Algiers, Algeria

E-mail: ag1964@yahoo.com

Abstract. Particulate flows are a serious problem in air ventilation systems, leading to erosion of rotor blades and aerodynamic performance degradation. This paper presents the numerical results of sand particle trajectories and erosion patterns in an axial ventilator and the subsequent blade deterioration. The flow field was solved separately by using the code CFX-TASCflow. The Lagrangian approach for the solid particles tracking implemented in our in-house code considers particle and eddy interaction, particle size distribution, particle rebounds and near walls effects. The assessment of erosion wear is based on the impact frequency and local values of erosion rate. Particle trajectories and erosion simulation revealed distinctive zones of impacts with high rates of erosion mainly on the blade pressure side, whereas the suction side is eroded around the leading edge.

1. Introduction

The ingestion of dust particles by air compressors and ventilation systems is inevitable and results in drastic degradations of their aerodynamic performance and structural integrity. The quartz is the most abundant erosive constituent and its abrasive impacts provoke a considerable erosion wear near the leading edge and tip of axial blades [1]. Measurements by Balan and Tabakoff [2] in an isolated axial compressor that followed erosion cycles indicated a reduction in compressor adiabatic efficiency and stage loading and a severe erosion of blade leading edge and pressure side with increased roughness. Ghenaiet et al. [3] characterized the increase in tip clearance and the reduction of blade chord of a high speed conta-whirl axial fan rotating at 11000 rpm, and noticed 7.1 % and 9.5 % drop in efficiency and pressure rise coefficient, respectively, after 9 hours of sand ingestion. Experimental procedures to assess erosion and performance degradation remain very expensive and reveal limited information because of different operating conditions of a turbomachine; subsequently the numerical simulations are a good alternative. Hussein and Tabakoff [4] pioneered in the particle trajectory simulations through the axial flow turbomachines and the use of experimentally based particle rebound factors. Ghenaiet et al. [5] presented a complete numerical procedure for simulating particle trajectory, erosion and the subsequent aerodynamic performance degradation, which was validated upon a contra-whirl axial fan stage consisting of seven IGVs and ten twisted rotor blades operating at a speed of 11000 rpm. In other simulations carried out by Ghenaiet [6] for a larger axial fan stage operating at a speed of 4650 rpm, results have shown that for different positions of rotor blade the main impacted areas are seen over the blade leading edge corner and the pressure side towards the tip, in addition to a strip over

Content from this work may be used under the terms of the Creative Commons Attribution 3.0 licence. Any further distribution of this work must maintain attribution to the author(s) and the title of the work, journal citation and DOI.
Published under licence by IOP Publishing Ltd

the suction side. In radial compressors the computations of trajectories carried out by Ghenaiet [7], have shown that are consistently different from those in axial turbomachinery, because of flow structure and direction of Coriolis and centrifugal force, and there are distinctive zones of high rates of erosion and the blade pressure side is more eroded. The prediction of erosion in turbomachinery is still a very difficult task due to several involved factors including; initial location of rotor blade, flow conditions, random particle size distribution and uncertainties in initial particle locations and impacting conditions. This paper presents the results of particle trajectories through an axial ventilator and the subsequent blade erosion. The trajectory and erosion computations used our developed Lagrangian tracking code which has been validated and used in previous studies [3, 5-7]. The studied ventilator is manufactured by Woods and rotates at a fixed speed of 1440 rpm. The rotor made from aluminium (figure 1) has seven highly staggered (root stagger 73 deg and tip stagger 82 deg) C4 blades of a maximum chord length 117.5 mm. The tip diameter is 479.6 mm and the hub diameter is 156.5 mm. A test-rig was used to measure the characteristics (figure 2) that depicts the stable and instable branches of pressure rise and the maximum efficiency occurring near full opening. It was not possible going beyond a flow rate of 1 m^3/s due to throttling.

Figure 1. Axial fan Woods

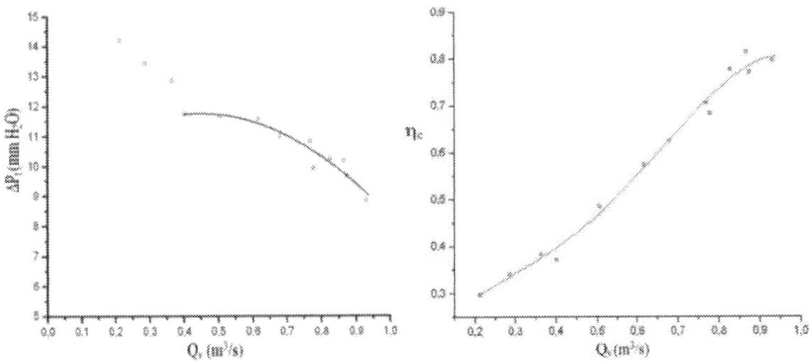

Figure 2. Measured characteristics, (left) Pressure rise, (right) Isentropic efficiency

2. Gas path computations

The flow-field was computed separately from solid phase by means of the time averaging Navier-Stokes solver TASCflow. The computational domain in addition to the rotor blade contains an extended hub housing the motor, inlet and outlet sector with a step transition, as shown by figure 3.

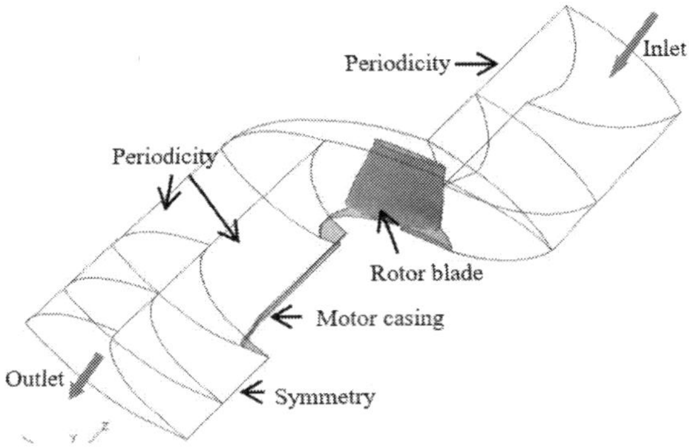

Figure 3. Computational domain

The computational mesh is of an H type as shown for the hub to shroud and for blade to blade (figure 4). A fine grid is attached to the tip-clearance region and a fine clustering is used near the walls. The six sub-grids are attached using stages interfaces.
The boundary conditions used are: a constant total pressure at the inlet and a mass flow at the outlet. Periodic boundaries are applied at one pitch of the rotor blades, at the hub and the extended domains which are joining on the axis of rotation, where a symmetry boundary condition is applied. In these CFD simulations the blade root stagger angles was set at 73°, and the mass flow rate corresponded to maximum opening.

Figure 4: Rotor grid: **a)** Hub to shroud **b)** hub, mid-span and tip

Figure 5 shows the flow velocity vectors plotted near the hub, mid span and in tip clearance for. Upstream and around the blade leading edge the relative flow velocities are relatively high and increasing with the blade height. At the trailing edge of fan blade there is evidence of wake formation, which is also clear from total relative pressure (plot not shown). The most complex flow features occur in the tip clearance region, where the leakage around the blade tip develops into a vortex and mixes out with the annulus wall boundary layer, thus increasing losses. The flow at rotor exit is strongly influenced by the back-pressure at the rotor step, where a strong diffusion occurs, and the amount of flow rotation tends to reduce after nearly two diameters distance.

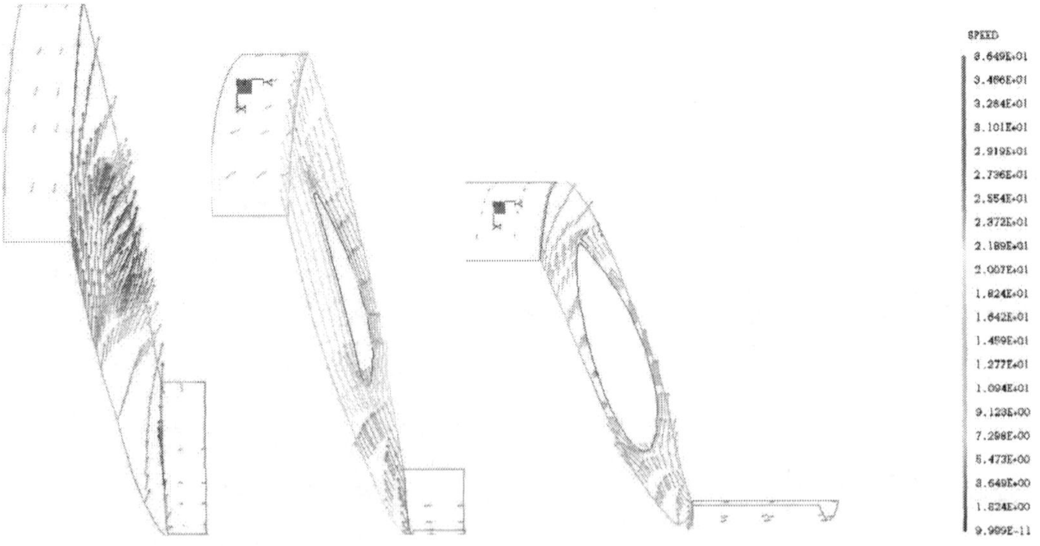

Figure 5. Velocity vectors near hub, mid and tip, at maximum opening

3. Particle trajectories and erosion computation

The Lagrangian approach considers the tracking of individual particles from different starting positions, thus giving an ability to handle in more details some physical aspects, such as the interaction with walls. This approach is preferred since the effect of particle phase on flow solution is neglected as the volume fraction is less than the upper limit of 10^{-6} [8].

The particle trajectory equations are derived from the superposition of different involved forces, but in most particulate air flows the drag force is dominant. The general expression for the drag coefficient is given as below according to Haider and Levenspiel [9], for Reynolds number from 0.01 to $2.6\text{x}10^5$. For small Reynolds numbers Re<0.5 the viscous effect is dominating and this is referred as the Stokes regime C_D=24/Re and this value is used. The constants A, B, C and D depend on the particle shape.

$$\vec{F}_D = \frac{\pi}{8} d_p^2 \rho_f C_D \left(\vec{V_f} - \vec{V_p}\right) \left\| \vec{V}_f \right| - \left| \vec{V}_p \right\| \quad , \quad C_D = \frac{24}{Re_p}\left(1 + A Re_p^B\right) + C/\left(1 + D/Re_p\right) \quad , \quad Re_p = \frac{\rho_f}{\mu_f} d_p \left|\vec{V}_f - \vec{V}_p\right| \tag{1}$$

If a particle is sufficiently large and there is a large velocity gradient, there will be a particle lifting force called Saffman force due to fluid shearing forces, which depends on the particle based Reynolds number and the shear flow Reynolds number [8] as follows.

$$\vec{F}_S = \frac{\rho_f \pi}{8} d_p^3 C_{LS} \left(\vec{V}_f - \vec{V}_p\right) \times \nabla \vec{V}_f \quad , \quad C_{LS} = \frac{4.1126}{\sqrt{Re_s}} f\left(Re_p, Re_s\right) \quad , \quad Re_s = \frac{\rho_f}{\mu_f} d_p^2 \left|\vec{\omega}_f\right| \tag{2}$$

The particle inertia forces namely; the centrifugal and Coriolis forces are derived from the second derivative of the vector position. Across a rotating frame the absolute velocity of a particle is changed to the relative velocity. By considering the drag and Saffman force as the main external forces, the following set of second order non-linear differential equations is derived.

$$\begin{cases} \dfrac{\partial^2 r_p}{\partial t^2} = \dfrac{3}{4d_p} \dfrac{\rho_f}{\rho_p} C_D \sqrt{\left|V_{fr} - V_{pr}\right|^2 + \left|V_{f\theta} - V_{p\theta}\right|^2 + \left|V_{fz} - V_{pz}\right|^2} \left(V_{rf} - V_{rp}\right) + r_p \left(\omega + \dfrac{V_{\theta p}}{r_p}\right)^2 + \dfrac{F_{Sr}}{m_p} \\ \dfrac{\partial^2 \theta_p}{\partial t^2} = \dfrac{3}{4d_p} \dfrac{\rho_f}{\rho_p} C_D \sqrt{\left|V_{fr} - V_{pr}\right|^2 + \left|V_{f\theta} - V_{p\theta}\right|^2 + \left|V_{fz} - V_{pz}\right|^2} \dfrac{\left(V_{\theta f} - V_{\theta p}\right)}{r_p} - 2\dfrac{V_{rp}}{r_p}\left(\omega + \dfrac{V_{\theta p}}{r_p}\right) + \dfrac{F_{S\theta}}{r_p m_p} \\ \dfrac{\partial^2 z_p}{\partial t^2} = \dfrac{3}{4d_p} \dfrac{\rho_f}{\rho_p} C_D \sqrt{\left|V_{fr} - V_{pr}\right|^2 + \left|V_{f\theta} - V_{p\theta}\right|^2 + \left|V_{fz} - V_{pz}\right|^2} \left(V_{zf} - V_{zp}\right) + \dfrac{F_{SZ}}{m_p} \end{cases} \tag{3}$$

The simulation of a particle motion in a turbulent flow requires the determination of the mean and fluctuating flow components which corresponds to a particular eddy and calculated from the local turbulence properties. The turbulence effect is assumed to prevail as long as the particle-eddy interaction time is less than the eddy lifetime and its displacement relative to an eddy is less than the eddy length. The eddy length and lifetime are calculated based on the local turbulence properties of the flow, according to Gosman and Ioannides [10].

Particle tracking through the computational cells requires transforming each particle position into local co-ordinates by solving a set of non-linear equations and updating its position if exceeding the value unity. The flow field data at the representative mesh points are used to interpolate the flow properties during the integration of particle trajectories by Runge-Kutta-Fehlberg seventh order technique. The integration time step is estimated based on the sizes of computational cells and flow velocities, but if a particle interacts with an eddy, the interaction time is considered as the effective time step. Near a boundary condition a more accurate time step is re-evaluated to get an impact within half diameter distance. At a point of impact, the rebound velocity and angle are computed statistically based on the values of mean and standard deviation of the rebound factors derived experimentally by Tabakoff et al. [11] given as polynomial regressions.

$$V_{P2}/V_{P1} = \sum_{i=0}^{4} a_i \beta_1^i \quad , \quad \beta_{P2}/\beta_{P1} = \sum_{i=0}^{4} b_i \beta_1^i \tag{4}$$

The particles are released randomly between 0 to 1000 μm (MIL-E5007E distribution) but confirmed to a specified concentration profile. Iterations on particles size and number were repeated till reaching the convergence in mass. Particle concentrations were varied from a low (50 mg/m^3) to a high concentration (1000 mg/m^3), and the fan operated at a speed of 1440 rpm for the maximum flow rate.

The intensity and pattern of erosion depend on the physical properties of sand, size, concentration, velocity and angle of impingement, in addition to the geometry and material. The most successful erosion prediction equation directly applied to turbomachinery is that due to Grant and Tabakoff [12] who first generated an erosion model based on aluminium alloy and silica sand as abrasive particles.

$$\varepsilon = K_1 f(\beta_1)(V_{1\theta}^2 - V_{2\theta}^2) + f(V_{1N}) \quad (5)$$

The erosion rate is expressed as the amount of material removed (milligrams) per unit of mass of impacting particles (grams), which depends on the local impact conditions, material and geometry of blades. The values of mass erosion (mg) are calculated from the local erosion rates, and then cumulated and used to compute the equivalent erosion levels in mg/g/cm^2.

$$\varepsilon_0 = \frac{1}{A_e \sum_1^N m_{p_i}} \sum_1^N m_{p_i} \varepsilon_i \quad (6)$$

4. Results and discussions

Small size particles such as 10 micron (figure 6) tend to follow the flow path and are strongly influenced by the turbulence, secondary flows and flow leakage tip, and are entrained by the large eddy downstream the rotor. For particles of size 100 micron (figure 7) the inertia becomes important, and this why they deviate considerably from flow streamlines through rotor blade and outlet duct. Due to centrifugal forces compared to drag force, large size particles 1000 micron (figure 8) move upwards towards the shroud and keep closer to the outlet wall.

The spread of impacts over the pressure side and on the shroud is shown by figure 9, revealing also the different size of impacting particles and their local velocities and rates of erosion, for which high values are seen upward and over the suction side from leading edge. Large particles tend to impact around the leading edge, whereas the impacts due to small and mid sizes particles are spreading over the pressure side. As revealed the local rates of erosion are low due to low impact velocities which do not exceed 36 m/s. There are several impacts found on the shroud due to the centrifugation. Many particles hitting around the leading edge rebound to reach the blade pressure towards the trailing edge.

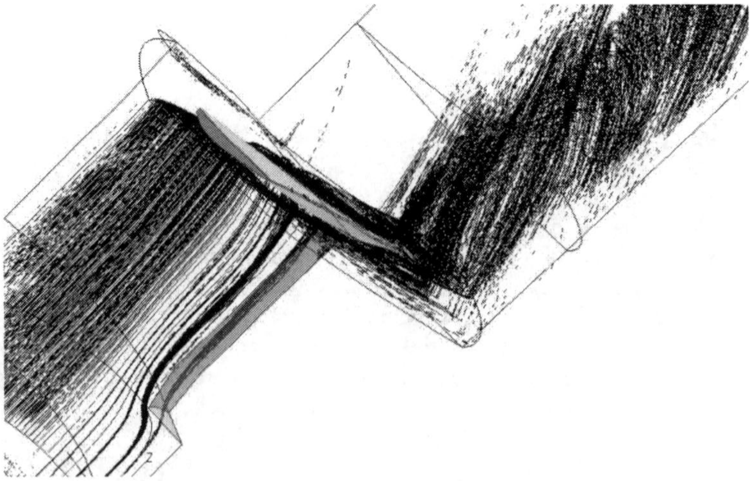

Figure 6. Trajectories of 10 micron sand particles

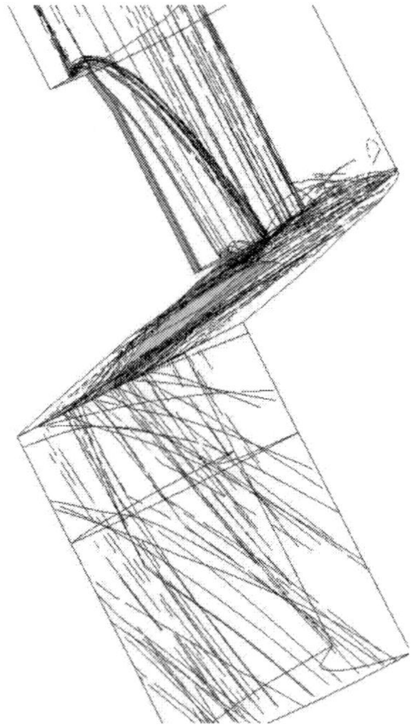

Figure 7. Trajectories of 100 micron particles

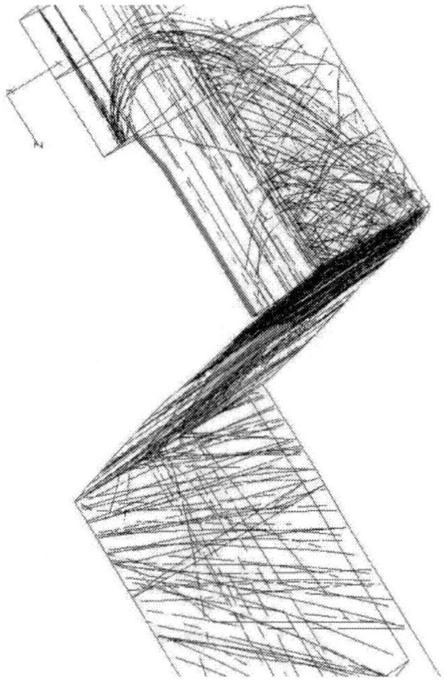

Figure 8. Trajectories of 1000 micron particles

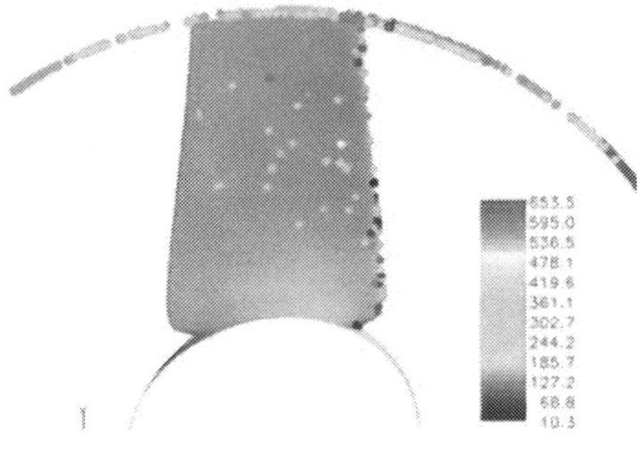

(a)

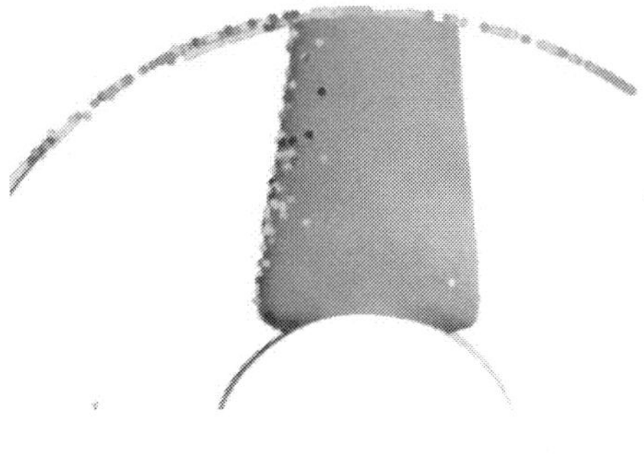

(b)

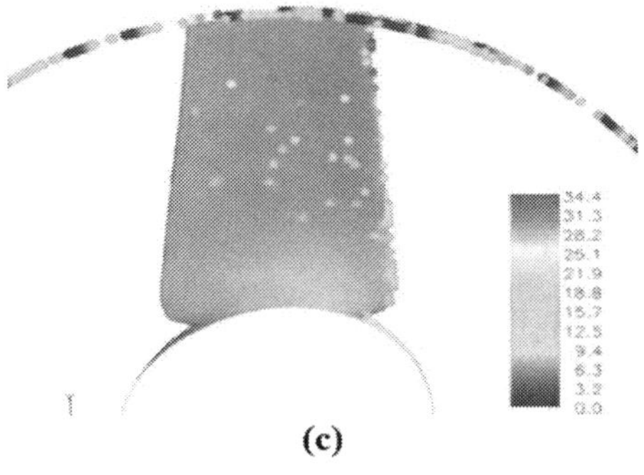

(c)

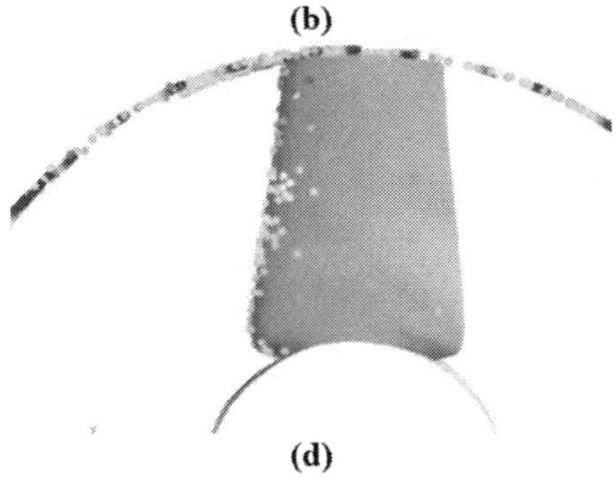

(d)

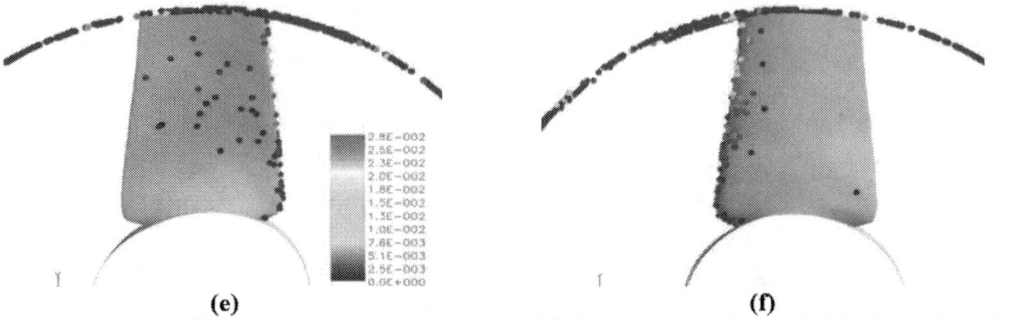

Figure 9. Sample of impacts with MIL-E5007E sand, left is pressure side and right is suction side, **a-b)** Impact diameter (μm), **c-d)** Impact velocity (m/s), **e-f)** Local erosion (mg/g)

Figure 10 and figure 11 are depicting the predicted erosion patterns corresponding to low concentration and very high concentration. It should be noticed as expected; that these rates of erosion are low for this fan as compared with high speed compressors, because of the low rotational speed 1440 rpm. The predicted erosion patterns are shown to depend largely on the concentration. There is a noticeable erosion of the blade leading edge from root to tip, and over the blade tip and the casing. On the suction side, the erosion is located around the leading edge and expanding towards the tip in a triangular shape. The critical erosion wear is at the top corner from the blade leading edge. On the pressure side, the erosion areas tend to spread, and at the extreme concentration the erosion pattern depicts a rounding of the blade from this side due to material removal. On the shroud the erosion is shown to spread over the tip and downstream trailing edge in scattered areas owing to large impacts encountered, resulting from centrifugal forces and rebound from the rotor. On the hub there are few impacts, mainly near the leading edge.

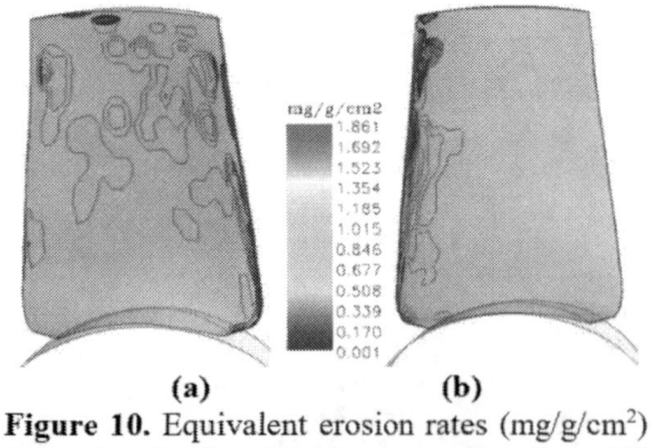

Figure 10. Equivalent erosion rates (mg/g/cm^2) over the blade, due low concentration (50 mg/m^3) of MIL-E5007E sand, **a)** pressure side, **b)** suction side

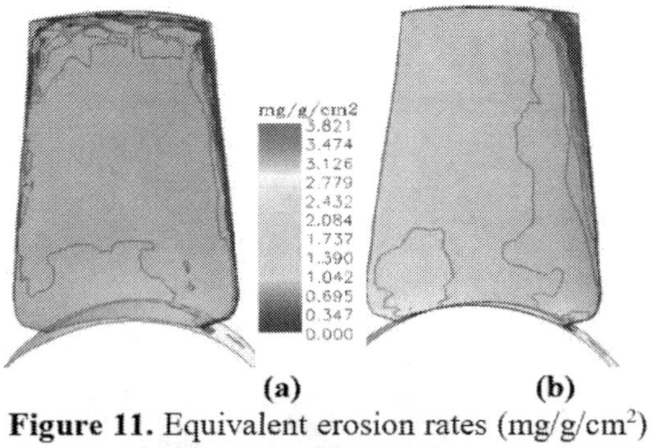

Figure 11. Equivalent erosion rates (mg/g/cm^2) over the blade, due high concentration (1000 mg/m^3) of MIL-E5007E sand, **a)** pressure side, **b)** suction side

The results of erosion parameters and blade geometry deterioration after one day of sand particle ingestion are presented in table 1. It is clear that the material removal and geometry changes are related to the mass of particles impacting the blade, concentration and duration of exposure. From the simulations for different operating conditions, one may conclude that the mass erosion from the rotor blade and the shroud increase with the concentration in between 50-1000 mg/m^3. The assessed parameters are the mass erosion, overall erosion rate and geometry deterioration estimated as percentages of chord and thickness reduction and plotted in figure 12. It is clear that the material removal and the geometry changes are related to the particle size and concentration and duration of exposure. Accordingly the mass erosion of one blade increases with particle concentration from 2.889

mg to 180.482 mg, which is well illustrated by figure 12(a). Also, it is noticed that the overall erosion rate is almost constant (table 1) and has an average value of $2x10^{-3}$ mg/g. The mass erosion of the shroud (figure 12(b)) is so important and varies with concentration from 4.3 mg to 71.626 mg, whereas that of the hub is extremely lower i.e. from $1.145x10^{-6}$ mg to $1.802x10^{-3}$ mg. The average reduction in blade chord is from $4.175x10^{-4}$ % to $4.054x10^{-3}$ % (table 1 and figure 12(c)), and the thickness from $4.241x10^{-6}$ % to $4.541x10^{-5}$ %, as seen from table 1 and figure 12(d). Also the tip clearance increased from $1.826x10^{-3}$ % to $3.670x10^{-2}$ % according to figure 12(e).

Table 1. Erosion and geometry deterioration due to sand ingestion after one day

Erosion parameters	Concentration (mg/m³)										
	50	100	200	300	400	500	600	700	800	900	1000
Particle rate (mg/s)	10.776	1.8.0667	36.642	57.070	69.451	90.789	111.3919	128.525	142.427	166.943	180.482
Erosion (mg)	2.889	3.580	6.298	9.253	11.190	15.3015	18.983	20.849	24.727	29.777	30.848
Reduction in mass (%)	7.272E-04	9.009E-04	1.584E-03	2.328E-03	2.815E-03	3.850E-03	4.777E-03	5.246E-03	6.222E-03	7.493E-03	7.762E-03
Overall erosion rate (mg/g)	3.104E-03	2.293E-03	1.989E-03	1.675E-03	1.725E-03	1.950E-03	1.972E-03	1.877E-03	2.009E-03	2.064E-03	1.978-03
Blade length reduction (%)	4.175E-04	4.276E-04	9.007E-04	1.477E-03	1.739E-03	2.341E-03	2.784E-03	3.062E-03	3.428E-03	3.934E-03	4.054E-03
Thickness reduction (%)	4.241E-06	5.103E-06	9.436E-06	1.564E-05	1.693E-05	2.375E-05	2.841E-05	3.127E-05	3.663E-05	4.284E-05	4.541E-05
Tip clearance Increase (%)	1.826E-03	3.708E-03	7.842E-03	9.449E-03	1.217E-02	1.326E-02	1.618E-02	1.902E-02	2.082E-02	2.783E-02	3.670E-02
Erosion of hub (mg)	1.145E-06	7.648E-03	1.680E-05	5.467E-04	4.077E-04	3.250E-04	7.185E-04	8.292E-04	6.442E-03	6.035E-04	1.802-03
Erosion of shroud (mg)	4.300	6.965	12.261	20.748	25.706	33.198	38.317	48.342	51.565	63.423	71.626

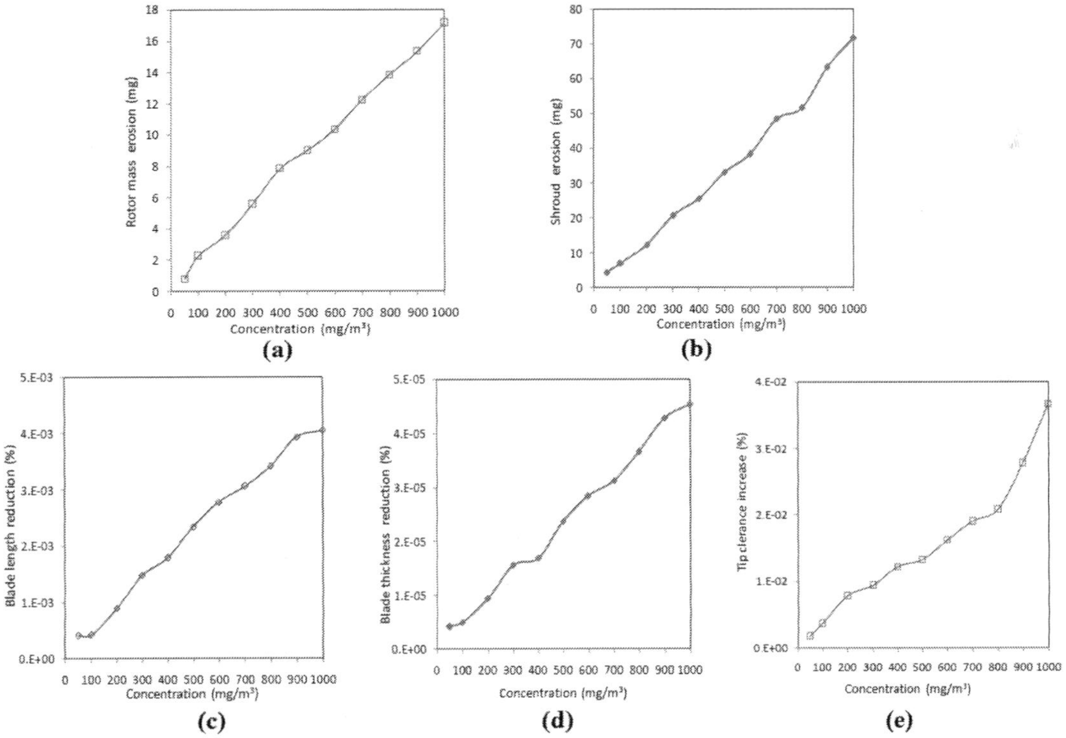

Figure 12. Erosion degradations after one week of sand ingestion: **a)** Blade erosion, **b)** shroud erosion, **c)** Blade length reduction, **d)** Blade thickness reduction, **e)** Tip clearance increase

IOP Conf. Series: Materials Science and Engineering **90** (2015) 012054 doi:10.1088/1757-899X/90/1/012054

As exhibited by figure 13, the erosion phenomenon is shown to distort the leading edge and reduce the chord and thickness of blade, progressively from mid span to tip. The process of erosion is more likely to cut the fore of tip section and the thickness is changing. In addition, the distortion at leading edge affects the flow circulation and moves the stagnation point, which should alter the static pressure distribution and its aerodynamic quality.

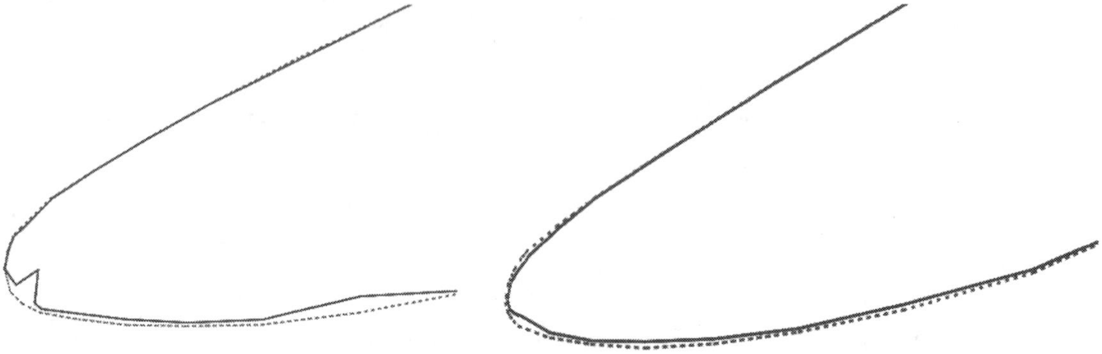

(a) (b)

Figure 13. Eroded profiles, **a)** Tip profile **b)** Mid-span profile

5. Conclusion

High impacts and erosion rates are predicted around the leading edge of the rotor blade due to its direct exposure to incoming sand particles. Impacts on the suction side are related to particles entering the rotor blades at high negative angle of direction. Impacts on the remainder of the pressure side are due to flow condition at the rotor entrance and rebounds around the leading edge. On the shroud, the impacts and erosion are connected to the particles acquiring centrifugal forces. On the hub there are few impacts. In general, the obtained rates of erosion are lower compared to high speed compressors. The consequent amount of predicted erosion data may help assessing fans of similar design operating in polluted environments, as well as envisaging adequate coating for the critical eroded area to extend the time of life.

Nomenclature

A	Area (m^2)
C	Concentration (mg/m^3)
C_D	Drag coefficient
C_L	Lift coefficient
d	Diameter (m)
K_1	Material constant
m	Mass (kg)
P	Pressure (Pa)
r	Radius, radial co-ordinate (m)
t	Time (s)
U	Peripheral velocity (m/s)
V	Absolute velocity (m/s)
W	Relative velocity (m/s)
z	Axial co-ordinate (m)

Greek letters

β	Impact angle (deg)
ε	Erosion rate (mg/g)
ε_p	Equivalent erosion rate ($mg/g/cm^2$)

ρ	Density (kg/m^3)
ω	Speed of rotation rad/s
θ	Angular position (rad)

Subscript

D	Drag
f	Fluid
N	Normal
p	Particle, pressure
r	Radial
θ	Tangential component
S	Saffman
z	Axial direction
1,2	At impact and rebound from a surface

References

[1] Tabakoff W 1984 Review of Turbomachinery performance Deterioration Exposed to Solid Particulate Environment, *ASME Journal of Fluids Engineering*, Vol 106, June 1984 , pp125–134

[2] Balan C and Tabakoff W 1984 Axial Flow Compressor Performance Deterioration, *AIAA paper 84-1208*, June 1984

[3] Ghenaiet A, Tan S C and Elder R L 2004 Experimental Investigation of an Axial Fan Erosion and Performance Degradation, *IMechE Journal of Power and Energy*, September 2004 Vol. 218 N. A6, ISSN : 0957-6509

[4] Hussein M F and Tabakoff W 1974 Computation and Plotting of Solid Particle Flow in Rotating Cascades, *J Computers and Fluids*, Vol. 2, No. 1, 1974, pp. 1–15.

[5] Ghenaiet A, Tan, S C and Elder, R L 2005 Prediction of an Axial Turbomachine Performance Degradation *IMechE Journal of Power and Energy*, June 2005 Vol. 219 N. A 4, ISSN: 0957-6509

[6] Ghenaiet A 2009 Simulation of Sand Particles Ingestion Through a Cooling Fan, IMechE, *Compressors and Their Systems 2009 paper L20/C682/056*, Chandos Publishing Oxford 2009, ISBN 978-1-84334-581-7

[7] Ghenaiet A 2005 Numerical Simulations of Flow and Particle Dynamics within a Centrifugal Turbomachine, *Compressors and Their Systems 2005, IMechE paper C639_52*, ISBN: 0-470-02576-X Wiley publisher

[8] Sommerfeld M 2000 Theoretical and Experimental Modelling of Particulate Flow - Overview and Fundamentals, *VKI Lecture series* 2000–06, April 3-7 2000, Von Karman Institute, Belgium

[9] Haider A and Levenspiel O 1989 Drag Coefficient and Terminal Velocity of Spherical and non Spherical Particles, *J Powder Technology*, 58 (1989) 63.

[10] Gosman A D and Ionnides E 1981 Aspects of Computer Simulation of Liquid Fuelled Combustors, *AIAA 19th Aerospace Science Meeting St Louis Missouri*, paper AIAA 81-0323

[11] Tabakoff W, Hamed A and Murugan D M 1996 Effect of Target Materials on the Particle Restitution Characteristics for Turbomachinery Application, *Journal of propulsion and power* Vol. 12 No2, March-April 1996

[12] Grant G and Tabakoff W 1975 Erosion Prediction in Turbomachinery Resulting from Environmental Particles, *AIAA Journal of Aircraft*, Vol. 12, No. 5, 1975, pp. 471–478.

IOP Conf. Series: Materials Science and Engineering **90** (2015) 012055 doi:10.1088/1757-899X/90/1/012055

A miniature Rotary Compressor with a 1:10 compression ratio

Olly Dmitriev[a], Eugene Tabota[a], Professor Ian Arbon EurIng, CEng, FIMechE [a]

[a]VERT Rotors, 12 Miller Row, Edinburgh, EH4 3BQ

olly@VERTrotors.com

Abstract. Micro compressors have applications in medical devices, robotics and "nano-satellites". The problem of active cooling for photo detectors in "nano-satellites" becomes more important because the majority of space missions target Earth observation, and passive cooling does not provide the required temperatures to achieve the desired SNR levels. Reciprocating compressors used in cryocoolers cause vibrations. VERT Rotors has built an ultralow-vibration rotary compressor with 40mm-long screws, and our prototype delivered 1:10 compression ratio. This "nano" compressor is a non-conventional conical type consisting of an Inner conical screw rotor revolving inside an Outer screw rotor.

1. The need to develop a very small rotary compressor with high compression ratio

Heat removal and temperature control has become a highly important issue with "nano-satellites" (100x100x100mm – scale, so-called "CubeSats") and small satellite platforms (1m x 1m x 1m - scale). With increasing power budgets on CubeSats, increasing radio power requirement and higher data rates, an active heat rejection technology is required. The tasks for such a system include moving tens of watts away from a heat source to radiate at an external face in order to keep radio units and optical lenses cool, or for transferring that heat to another subsystem on orbits where there are long eclipse periods (for example keeping the battery at operational temperature without using a wasteful heater).

Thermal control of CubeSats is usually passive with heat sinks and optical tape on the outer structure keeping the structure, with the exception of the solar panels, in a range of approximately 258 K to 313 K for sun synchronous orbits [1]. This is not sufficient for cooling down the photo detectors to 180-200K, which becomes a major requirement when Earth observation and remote sensing is estimated to account for 52% of small satellite missions in 2014-2016 [2,3]. The signal-to-noise ratio (SNR) of photo detectors is affected by temperature because dark noise from thermal excitation increases with temperature, as it follows the temperature dependence of the concentration of electrons in the conduction band, shown in the equation below:

$$n_0 = k_3 T^{-1.5} e^{-(k_4/T)} \tag{1}$$

where n_0 is the concentration of electrons in the conduction band, T is the temperature, and k_3 and k_4 are constants that depend on technology. Based on this equation, in order to obtain an SNR of 100 a temperature of about 187K is required [4]. While this is just a particular example, it is generally true that, in order to achieve reasonable SNR, cooling down of the sensor becomes a necessity.

In space, excess heat can only be removed from satellites by radiation from black panels. Due to the limited outer surface area of CubeSats, the effectiveness of passive cooling by radiation is limited. An active system would be significantly more effective with the compressor heating the refrigerant gas through compression. Specifically, the Stefan–Boltzmann law states that while the total energy radiated per unit surface area of a black body is linearly dependent on the surface area of the radiating panels; it depends on the fourth power of the black body's thermodynamic temperature T:

Content from this work may be used under the terms of the Creative Commons Attribution 3.0 licence. Any further distribution of this work must maintain attribution to the author(s) and the title of the work, journal citation and DOI.
Published under licence by IOP Publishing Ltd

$$P = A\,\varepsilon\,\sigma\,T_{Rad}^4 \qquad (2)$$

where P is the power radiated to space in W, A is the surface area of the radiator in m^2, $\boldsymbol{\varepsilon}$ is the emissivity of the black panels, and $\boldsymbol{\sigma}$ is the Stefan–Boltzmann constant (5.6704×10^{-8} Watt / $m^2 \cdot K^4$) [5].

An active cooling system elevating the temperature of gas through compression and enhancing radiation of heat through black panels would be the most feasible solution. A few key requirements of a compressor in an active system are:

- To provide a compression ratio of at least 1:3 to heat the gas.
- To support removal of 10-20W of heat.
- Very small footprint within 100mm x 100mm x 40mm.
- Low mass, below 100g, because of the high cost of delivering the payload to space.
- Most importantly, the compressor should produce as little vibration as possible, because vibration is unfavourable for optics in Earth observation satellites.

The other small compressors that are available are well-suited for certain applications, but for Earth observation missions the level of vibrations must be improved to reduce the impact on image sensors. Existing reciprocating-type micro-compressors naturally have inherent vibrations. One such compressor built by Lockheed Martin has mass of 190g; the net cooling capacity of this machine is 650mW at 150K, corresponding to specific power of 15.4 W/W [6]. More advanced active techniques, such as the space proven Stirling and Joule–Thomson cryocoolers surpass volume and mass capabilities of CubeSats. Examples of these include the 4.3 kg Oxford cryocooler employed on UARS with cooling capability of 0.8W at 80 K [1].

It was concluded that a rotary design would be the most appropriate for a low-vibration very small compressor. The challenge of the oil-free rotary designs is that they do not provide high compression ratios over 1:1.5 when implemented in very small sizes. For this reason it is typical that piston-type compressors are used in very small applications.

2. Designing a "nano" conical rotary compressor

2.1. Results of the design project

A nano rotary conical compressor with a 1:4 volume ratio (codename "MK03") was designed. The screw length is 40mm, and it fits onto a 100mm x 100mm PCB. MK03 is a rotary compressor which ensures very low vibration.

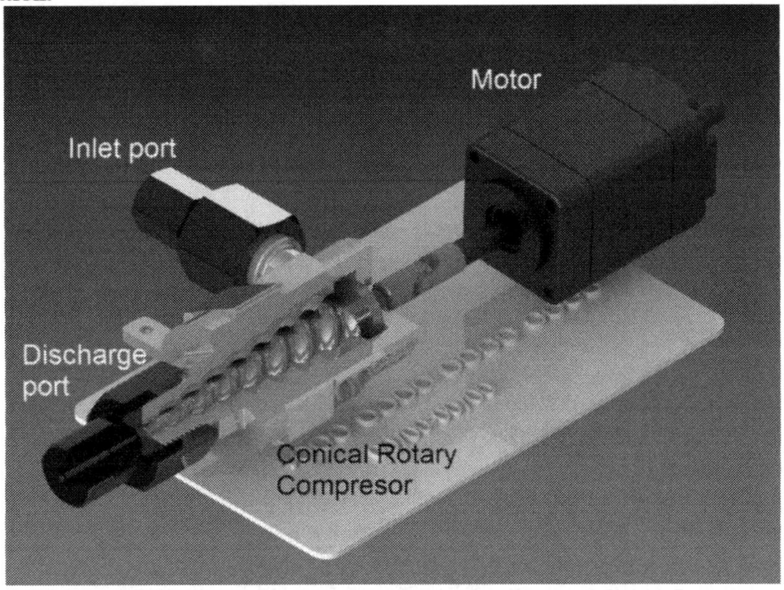

Figure 1. MK03 in comparison with a credit card

The word "nano" here is not literal, it suggests that this compressor is designed for 1-10 kg satellites that are usually referred to as "nanosatellites" [2].

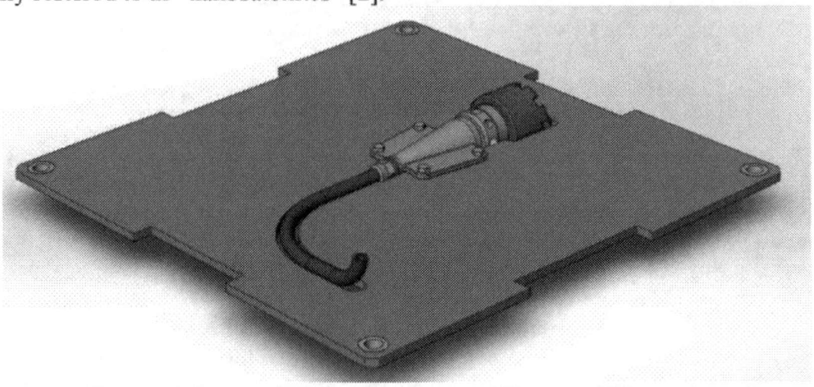

Figure 2. Nano CRC mounted on a 100mm x 100 mm board

MK03 is a Conical Rotary Compressor (CRC) which is a novel design consisting of one Inner conical screw rotor revolving inside an Outer conical screw rotor. In comparison, conventional twin-screw compressors consist of three elements, two screws revolving inside the housing.

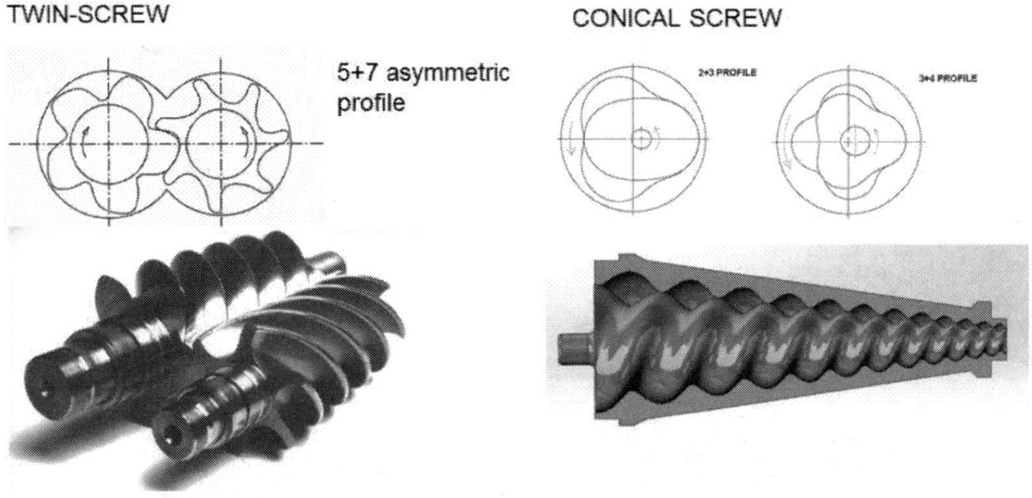

Figure 3. Comparison of twin-screw and conical screw profiles

To create the MK03, a mathematical formula and algorithm for generating mating conical screw rotors was developed. 3D CAD software that outputs screw profiles in 3D format, suitable for further CAD-CAM processing and machining was written. After several years of optimisation, the generation of screw profiles was fully automated and the process duration has been reduced to a few hours.

2.2. Operation of Conical Rotary Compressor

The compression chamber is formed by the volume trapped in between the Inner and the Outer rotors. The Inner rotor axis is angularly offset from the Outer rotor axis and revolves inside it as the rotor itself rotates.

In operation, a compressible gas is drawn into the assembly at the large end of the cone. The chamber coloured in blue is in the position that the flow will be cut off from the suction port [7].

As the inner rotor and outer rotor rotate, each of the closed chambers reduces in size as it travels from the large end to the small end of the cone, thereby compressing the gas. High-pressure gas discharges from the assembly at the small end of the cone.

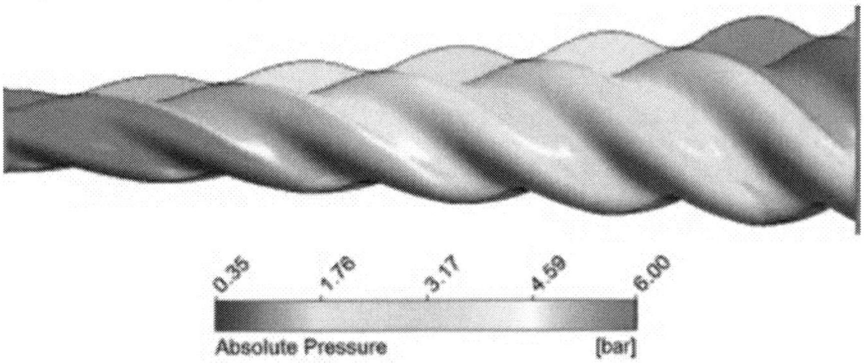

Figure 4. Gradual increase in pressure in CRC

2.3 Comparison of conical screw and twin screw designs

Perhaps the most important difference is that, within the CRC, the Inner rotor makes a rolling motion inside the Outer rotor thereby using the full volume inside the Outer rotors, therefore the CRC does not have a "blowhole", which is a typical feature of twin-screw compressors [8]. Leakage flow is the main cause of reduced efficiency in twin-screw compressors [9]. The "Blowhole" is a gap where two screws and a housing meet, and compressed gas from the high-pressure area leaks back to the low-pressure area. To overcome this problem, oil-free twin-screw compressors run at high rotational speeds. The CRC, not having a "blowhole", can potentially run at a higher efficiency, but it can be impaired by higher friction. To realise its potential the CRC should be made from low friction materials.

Secondly, due to the small discharge area and the gradual reduction of the swept volume, the internal pressure ratio of the CRC is "self-adjusting" to the external pressures, i.e. the required pressures of the system in which the compressor operates, which allows the compressor to work to the optimal discharge pressures. Therefore, the CRC does not suffer from over-compression or under-compression [7].

The efficiency of a positive displacement compressor (piston, scroll, screw, vane or any other) greatly depends on the capability of the machine to adjust the internal pressure ratio to the "external", or system, pressure ratio.

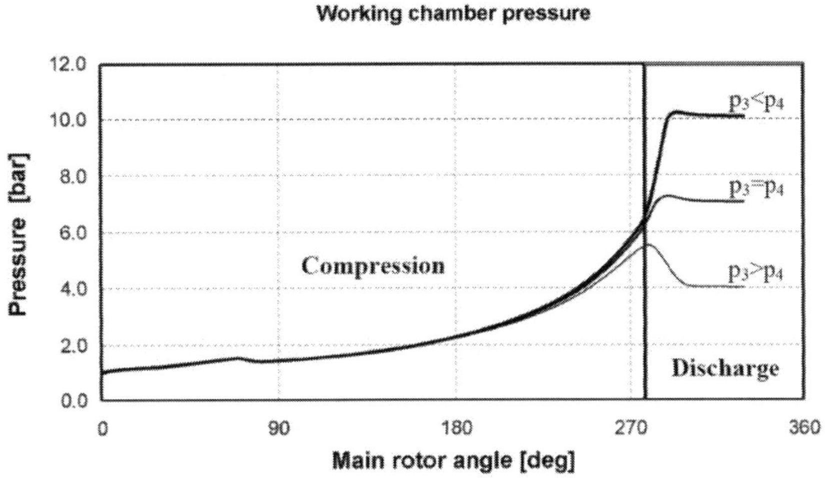

Figure 5. Pressure angle diagram of twin-screw compressors

If the "external", or system, pressure is higher than the internally-achieved pressure ($p3<p4$), this condition is called under-compression and requires additional work to push the fluid out of the compressor, by which process its pressure will increase to the external discharge pressure. If the external pressure is lower than the internally achieved pressure ($p3>p4$), this condition is called over-compression and it indicates that unnecessary work has been used in the compressor to compress the gas to a higher pressure level than is required. Either case is undesirable, as both cause a higher power consumption than necessary.

If the volume ratio [V_i] is constant, the pressure increase during the compression process (called internal compression) will be more or less constant and independent of the external, or system, pressure ratio. It can be calculated as:

$$p_3 = p_2 V_i^{\gamma} \tag{3}$$

where γ is the isentropic exponent of gas.

As shown in Figure 5, if the internal pressure p_3 matches the external pressure p_4 the compression process is optimal and no additional losses are generated.

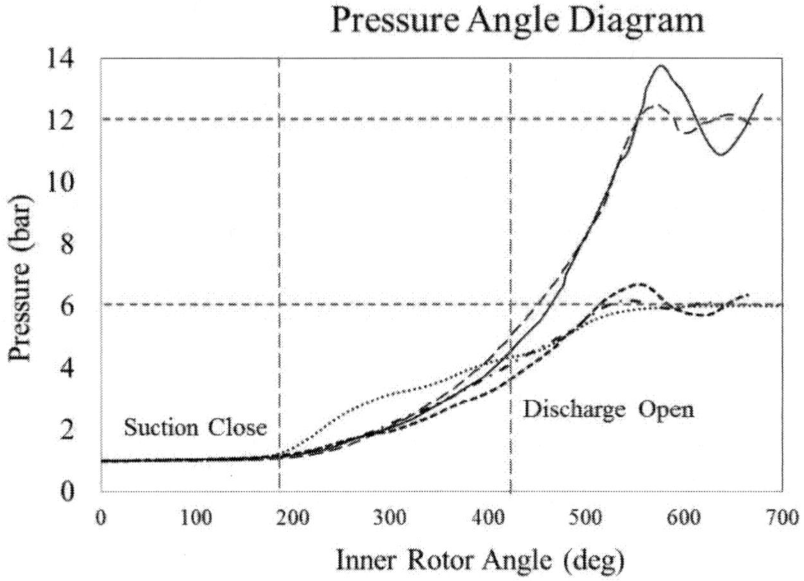

Figure 6. Pressure angle diagram of CRC

Figure 6 shows the calculated indicator diagram for a CRC. As can be seen in the diagram, this machine in all calculated cases achieves matching internal and external (system) pressures irrespective of speed or pressure ratio. It can be seen, therefore, that the CRC has a "self-adjusting" internal pressure ratio.

The working chamber which is connected to the discharge port has a very long slender shape, while the area through which it connects to the discharge pipe is small. This is a favourable feature, as it causes the compressor's "internal" pressure to gradually match the "external", or system, pressure at the rate at which this happens during internal compression.

In a typical twin-screw compressor this is different since, in these machines, the discharge port normally opens much wider, causing much faster equalisation of internal and external pressures which, although reducing the dynamic flow losses, still increases the required power.

The relatively small size of the discharge port in a CRC and its gradual change additionally helps to reduce pressure oscillations in the discharge port. Therefore, as seen in Figure 6, pressure pulsations exist but are not large at any pressure. In a typical twin screw compressor, these oscillations will be very small when the internal and external discharge pressures are matching but could be excessive if these are very different.

Thirdly, an axial discharge arrangement also improves energy-efficiency. As the report by the Centre for Positive Displacement Compressors concludes: "The conical design of the compressor and the axial discharge arrangement … is a unique advantage of this compressor."[7] In a typical twin-screw compressor, the suction and discharge ports are tangential to the screw axis. This creates additional resistance and leads to energy losses.

IOP Conf. Series: Materials Science and Engineering **90** (2015) 012055 doi:10.1088/1757-899X/90/1/012055

In the CRC, the suction and discharge both lie on the same axis as the housing, which ensures optimal gas flow without resistance.

2.4 The most compact rotary machine compressing air at 11 bar(g) in a single stage
The experimental studies carried out by our team in Edinburgh showed that MK03 was compressing air at 11 bar(g) in a single stage at 6,000 r/min shaft speed. This confirms that even very small CRCs can provide high compression ratios at standard shaft speeds thanks to progressive compression of gas inside the Outer rotor while it travels from the suction end to the discharge end. Effectively the CRC is the smallest rotary screw design on the market today capable of providing such compression ratio. Today's screw compressors are big industrial machines with a lower limit of displacement standing at approximately $20m^3/h$ at 3000 r/min [10], but typically at $80m^3/h$ [11].

In the oil-injected applications MK03 can be used to compress air to 8-10 bar(g), while in the oil-free operation the reasonable discharge pressures should be limited to 3 bar (g) to avoid overheating.

2.5 Thermodynamic model
To adjust the performance of the CRC, a numerical model for cooling a CubeStat with RF boards generating 20-30W of heat (80W of heat in total including contribution of the solar radiation and the compressor itself) was developed.

CubeSat is cooled from any temperature to which it was heated by the RF panels and optics and converge and stabilise at the equilibrium temperature.

2.5.1. Assumptions
Refrigerant gas: Helium

Refrigerant Flow Rate: Flow = 3 l/min

Compression ratio 1:3

The size of the black radiation panels for CubeSat is $0.12m^2$ (two 300mm x 100mm x 100 mm "cards" on the shadow side of a CubeSat).

Emissivity of black panels $\varepsilon = 0.96$

2.5.2. The model. The gas discharged from the Compressor needs to be heated to the temperature that will be sufficient to radiate 80W of heat through the black panels of the Condenser. The required temperature is derived according to the Stefan-Boltzmann law, formula (2) above.

The required flow of the coolant = 8.923×10^{-6} kg/s

Temperature of gas leaving the compressor: $T_1 = 368K$

Energy that is radiated by the black panels from the heated refrigerant is calculated using the formula (2), in this case it is 119.4W

After the gas has left the "black panel" condenser, its temperature drops to $T_2 = 239K$

The pressure in the condenser decreases to $p_2 = 1.9$ bar (a)

When the gas passes the Expansion valve, its pressure drops to 1 bar (a).

In the Evaporator the gas expands and its temperature goes down to $T_3 = 183K$

The refrigerant cools down the payload. Its temperature rises to $T_0 = 237K$, which will be the equilibrium temperature of the CubeSat.

3. Production and experimental validation

Several working prototypes of CRC were built to validate the concept. A prototype "MK3" unit with 167 mm-long conical rotors with a 1:10 volume ratio provided air compression ratio of 1:22 in a single stage in oil-free operation, at 830 r/min. Another machine codenamed "MK7" with 220mm-long conical rotors with a 1:2 volume ratio was extensively tested for oil-injected air compression over a period of 3 months. It was producing 130 l/min ($7.8 m^3/h$) at 8 bar (g) load at 52.5Hz. MK7 has a remarkably low noise level even without an acoustic enclosure, allowing two people to have a conversation while standing next to the working compressor. These prototypes validated the concept of conical rotors.

Figure 7. MK3 machine test showing 22 bar(g) in a single stage

Based on the ground work that had been done with larger machines, a working prototype "MK03" with 40mm-long conical rotors with a 1:4 volume ratio was built and tested This was produced by rapid prototyping, and during initial tests this machine compressed air to 10 bar (g) in a single stage.

Figure 8. MK03 prototype

Test results demonstrate that CRC is the smallest operational rotary compressor that runs at the 1:10 pressure ratio.

4. Next steps: Developing a nano-cryocooling kit

For the next step, the authors aim to build and test the ultra-compact and low-vibration cooling kit based on the MK03 compressor. The end result will be a fully operational 1:1 experimental model mounted onto a 100mm x 100mm PBC, suitable for installing into a CubeSat (see Figure 9 below).

IOP Conf. Series: Materials Science and Engineering **90** (2015) 012055 doi:10.1088/1757-899X/90/1/012055

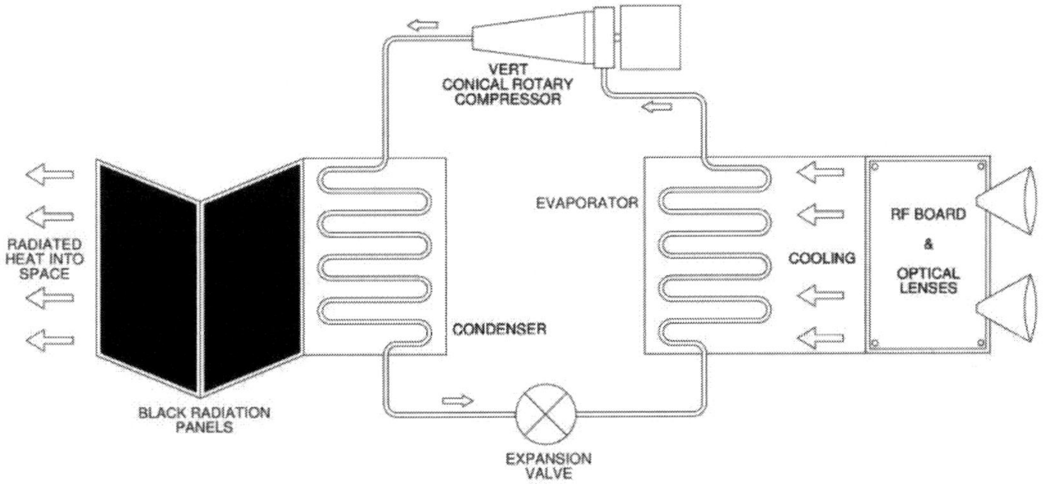

Figure 9. Diagram of the cooler based on the MK03 compressor

References

[1] A survey and assessment of the capabilities of Cubesats for Earth observation, D.Selva, D.Krejci, Acta Astronautica 74 (2012) 50-68

[2] 2014 Nano/Microsatellite Market Assessment, SpaceWorks Enterprises Inc.

[3] Small Satellite Missions, National Aeronautic and Space Administration, www.nasa.gov

[4] G.H. Rieke, Detection of Light: From the Ultraviolet to the Submillimeter, Cambridge University Press, Cambridge, UK, 2003.

[5] Stefan-Boltzmann Law, Encyclopaedia Brittanica, www.brittanica.com

[6] Microcryocooler for Tactical and Space Applications. J.R. Olson , P. Champagne, E. Roth, T. Nast, E. Saito, V. Loung, A.C. Kenton and C.L. Dobbins. AIP Conference Proceedings, 2014.

[7] Kovacevic, A., Rane, S. (2103), CFD Analysis of VERT Labs Helical Rotary Compressor HRC-1, City University London, UK

[8] The Design and Application of Rotary Twin-Shaft Compressors in the Oil & Gas Process Industry, EurIng Ian M. Arbon CEng, FIMechE, Mechanical Engineering Publications, 1994.

[9] Buckney, D., Kovacevic, A., Stosic, N.,(2011), Consideration of clearances in the design of screw compressor rotors, 7th International Conference on Compressors and their Systems, City University London

[10] http://www.atlascopco.co.uk

[11] Kevin Glass, Compressors: Small screw drivers, Bitzer UK http://www.acr-news.com/compressors-small-screw-drivers

IOP Conf. Series: Materials Science and Engineering **90** (2015) 012056 doi:10.1088/1757-899X/90/1/012056

Development of a Gas Dynamic and Thermodynamic Simulation Model of the Lontra Blade Compressor™

Jerome Karlovsky

Senior Analysis Engineer, Lontra Ltd, Unit 7, Folly Lane, Napton on the Hill, Warwickshire CV47 8NZ

Email: jerome.karlovsky@lontra.net

Abstract. The Lontra Blade Compressor™ is a patented double acting, internally compressing, positive displacement rotary compressor of innovative design. The Blade Compressor is in production for waste-water treatment, and will soon be launched for a range of applications at higher pressure ratios.

In order to aid the design and development process, a thermodynamic and gas dynamic simulation program has been written in house. The software has been successfully used to optimise geometries and running conditions of current designs, and is also being used to evaluate future designs for different applications and markets.

The simulation code has three main elements. A positive displacement chamber model, a leakage model and a gas dynamic model to simulate gas flow through ports and to track pressure waves in the inlet and outlet pipes. All three of these models are interlinked in order to track mass and energy flows within the system.

A correlation study has been carried out to verify the software. The main correlation markers used were mass flow, chamber pressure, pressure wave tracking in the outlet pipe, and volumetric efficiency. It will be shown that excellent correlation has been achieved between measured and simulated data. Mass flow predictions were to within 2% of measured data, and the timings and magnitudes of all major gas dynamic effects were well replicated.

The simulation will be further developed in the near future to help with the optimisation of exhaust and inlet silencers.

1. Introduction

After the filing of an initial patent, Lontra was founded in 2004 by Steve Lindsey, the inventor of the blade compressor. After favourable concept and market studies carried out at Southampton University, Imperial College, and Cosworth Technology, it became apparent that in order to evaluate the design further, it would be necessary to develop a parametric model of the compressor. At first this took the form of a simple geometry model driven by a small number of basic dimensions, which quickly enabled

Content from this work may be used under the terms of the Creative Commons Attribution 3.0 licence. Any further distribution of this work must maintain attribution to the author(s) and the title of the work, journal citation and DOI.
Published under licence by IOP Publishing Ltd

the calculation of chamber volume with shaft angle. This was useful for estimating the necessary speed and size of machine for a given application. The geometry model was a springboard for the development of a basic time marching thermodynamic model using simple port geometry with isentropic port flow, and calculation of chamber properties by using conservation of mass and energy. This gave useful estimates of flow, work and temperature that could be modified by imposing assumed values of volumetric and isentropic efficiency.

It became obvious that the next refinement should be the inclusion of a leakage model. To date 21 separately modelled leak paths have been identified. The leakage calculations are based on a method for calculating viscid flow of air in a narrow slot set out by Shires (1950). This method, which calculates the pressure drop down a slot due to viscous drag, is combined with an isentropic nozzle calculation which finds the pressure drop needed to accelerate the fluid to the speed at the entry of the leak path. The two are solved simultaneously to find the pressure at the start of the slot.

Lontra is constantly improving its testing and instrumentation capabilities. Pipe pressure measurements using high speed transducers have shown that there were measurable pulsations in the ducting of the compressor, and that the chamber pressure was being influenced by wave effects in the outlet pipe. These waves if timed correctly, can be used to increase efficiency, but if timed badly, can have a detrimental effect. This effect was simulated using a one dimensional unsteady gas dynamic simulation, a technique which is used extensively in the automotive industry to tune engine manifolds. The method used by Lontra is that set out by Blair (1999), but also uses some elements of the non homentropic method of characteristics described by Benson (1982). Comparison of measured and predicted pressure traces have shown that the timing and magnitude of the major pressure fluctuations are predicted well. It is planned that this technique will be used in the future to help with the design of silencers.

The simulation has been used successfully to optimize the geometry and port timing of a 1.5 bar compressor with a displacement of 11 litres running at 3000 rpm. This machine has performed very successfully in site trials at Severn Trent's Worcester sewage works where it is used to aerate sewage sludge. It has shown to be up to 20% more efficient than the existing compressors in this application.

2. Core design

The basic design uses a ring shaped chamber with a continually open inlet port and an outlet port valved by the action of the rotor. The principal components are a bladed rotor which rotates within a housing, and a rotating disc. The blade is shaped so that it can pass through a slot in the disc (fig.1a). The machine is double acting, drawing in gas behind the blade and compressing it ahead of the blade. The compression chamber is formed as the blade, rotating within the housing, works against the disc (fig 1b). The compressed gas exits though a port in the housing (fig 1c). This simple, elegant design creates a basic compressor. An animation of the design is available at www.lontra.co.uk

2.1 Geometry

The geometry of the machine is derived from 5 key parameters. In order to quickly evaluate different sizes and machine proportions, volumes and surface areas, including port areas are generally calculated analytically using double and triple integrals within the simulation code. This information is stored in arrays, and during the running of the simulation, angle based geometry data is found using interpolation of the array data. Unusual or complicated geometries can also be catered for by entering CAD derived angle based data as an input to the program.

(a) (b) (c)

Fig. 1 (a) the blade can pass through a slot in the disk, (b) anticlockwise rotation sucks in air behind the blade and compresses it in front of the blade, and (c) air exits through a port in the rotor.

3. Model of the working chamber

The working chamber is modelled using principles of conservation of mass and conservation of energy. The assumption is made that the chamber properties only vary with time and do not vary with position. This is commonly known as a zero dimensional model.

Conservation of mass over a time step dt

$$m_{t+dt} = m_t + dm_{in} - dm_{out} \tag{1}$$

Conservation of energy over a time step dt

$$U_{t+\mathrm{dt}} = U_t + dH_{in} - dH_{out} + Q - W \tag{2}$$

where

$$W = \frac{(p_{t+dt} + p_t)}{2}(v_{t+dt} - v_t) \tag{3}$$

The method of calculating chamber properties at the end of a time step is to calculate the specific internal energy from knowledge of mass and internal energy at the end of a time step. Temperature can then be found directly from internal energy, and the pressure calculated using the ideal gas equation. It can be seen that the work term uses values of pressure from the beginning and end of a time step. This means that the calculation of properties at the end of a time step needs to be found iteratively, by repeating the calculation procedure until values of P and T are found to within acceptable limits.

In equations 1 and 2 mass and enthalpy flows are due to leakage flows which will be described in section 2, and port flows, which will be described in section 3.

Heat transfer within the chamber is accounted for using the Dittus Boelter relationship, where

$$Nu = 0.023Re^{0.8}Pr^{0.4} \tag{4}$$

This formula is meant for fully developed turbulent flow in pipes, but it is the closest that the author has found to the situation within the blade compressor. For calculation of the Reynolds number, the gas is assumed to be flowing at the average speed of the blade.

4. Leakage calculations

Like many other oil free compressors, the Lontra blade compressor relies on close running non contacting surfaces for sealing. The mass flow through a narrow slot increases with width and clearance and decreases with length. In most designs, the clearance seals are formed by non conforming surfaces, which results in a short effective length of clearance seal (Fig. 2a). However the major clearance seals in the blade compressor are made of conforming surfaces that allows for seals that are much longer and offer increased resistance to flow (Fig. 2b).

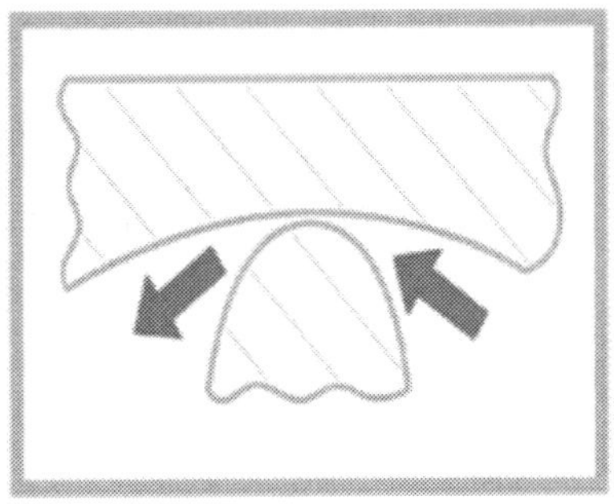
(a)

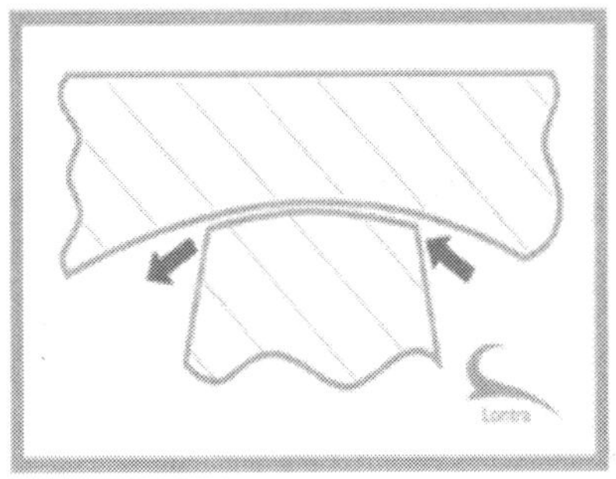
(b)

Fig 2 (a) Non conforming seals have a short effective seal length, (b) conforming surfaces offer a greater seal length.

Looking at figure 3, for steady flow, with force on the element acting to the right, the balance of forces on a fluid element is described by equation 5.

$$-\frac{\partial p}{\partial x} = -\frac{\partial s}{\partial y} + \frac{\partial(\rho u^2)}{\partial x} \tag{5}$$

Or by using the definition of viscosity $s = \mu\frac{\partial u}{\partial y}$

$$-\frac{\partial p}{\partial x} = -\mu\frac{\partial^2 u}{\partial y^2} + \frac{\partial(\rho u^2)}{\partial x} \tag{6}$$

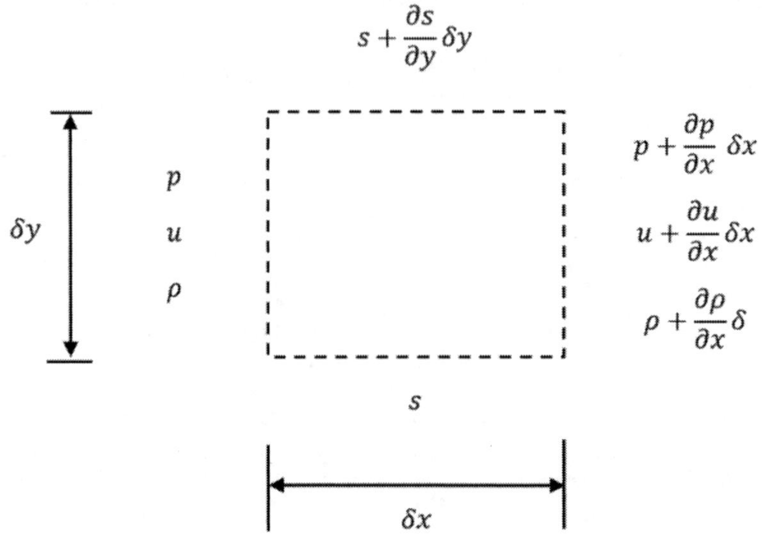

Fig 3. Volume of fluid

The term on the left hand side of equation 6 is the force due to the pressure gradient, the first term on the right is the force due to viscous shear, and the second term is the change in momentum. The methodology used to calculate mass flow through the clearance seals is based on an approach proposed by Shires (1950) which has become a standard approach in the design of air bearings. It assumes that for laminar flow in a narrow slot, the viscous forces are far higher than the inertia forces. Therefore the momentum term can be neglected and the equation simplifies to equation 7.

$$\frac{\partial p}{\partial x}=\mu\frac{\partial^2 u}{\partial y^2} \qquad (7)$$

Integrating between limits and rearranging yields

$$p_1{}^2-p_2{}^2=\frac{24\mu RT\dot{m}}{bc^3}(x_2-x_1) \qquad (8)$$

An extension to this theory, suggested by Shires, that uses the concept of a resistance coefficient to extend the useful range of application of this approach to include turbulent flow has been included. The unknowns are p_1, the pressure at the entry of the slot, and mass flow. The length of the slot is (x_2 – x_1), and p_2 is the pressure at the exit of the slot which is assumed to be at the pressure of the reservoir that gas is exiting into. Equation 8 has been combined with equation 9 for an isentropic nozzle, which is used to model the pressure drop needed to accelerate the fluid from rest in the upstream reservoir (subscript 0) to the speed of the fluid at the start of the slot. The two are solved simultaneously to find a value of p1 for which equations 8 and 9 give the same mass flow rate. The speed of the gas at the slot exit is monitored, and if this exceeds the speed of sound, then the routine is modified so that the speed of the gas at the slot exit is limited to the speed of sound.

IOP Conf. Series: Materials Science and Engineering **90** (2015) 012056 doi:10.1088/1757-899X/90/1/012056

$$\dot{m} = A\rho_0 \sqrt{\left\{2\left(\frac{\gamma}{\gamma-1}\right)\frac{p_0}{\rho_0}\left(\frac{p_1}{p_0}\right)^{2/\gamma}\left(1-\left(\frac{p_1}{p_0}\right)^{\left((\gamma-1)/\gamma\right)}\right)\right\}} \tag{9}$$

5. Gas dynamic calculations

Computer simulation of pressure wave propagation in ducts is a science that is over 40 years old. The technique is used extensively in the automotive industry to simulate unsteady flow in internal combustion engine ducting to help realise the benefits of pressure wave tuning. There are numerous books on the subject, the most comprehensive of which are by Benson (1982), Blair (1999), and Pearson and Winterbone (2000). There have also been a small number of researchers who have simulated unsteady gas dynamics in the ducts of positive displacement compressors for example (Stosic N. and Hanjalic K. (1978). In order to achieve accurate correlation, and also to investigate pressure wave tuning, it was decided that simulation of unsteady gas dynamics in the ducting of the blade compressor was important to Lontra

The routine used at Lontra is based on the method set out by Blair with elements of Benson's approach, and includes, plenums, junctions, tapered pipes, non isentropic inflow and outflow through ports, friction and heat transfer. The port flow coefficients chosen are those published by Blair for two stroke ports, as these are considered the most appropriate for the ports in the blade compressor. The model is object oriented, meaning that it is simple to join together any number of elements quickly and easily to produce the desired system. The details of the wave tracking method will not be given in this paper, as the three previous references give a very thorough explanation of the various techniques. Figure 4 shows simulated port and chamber pressure, with and without pressure wave tracking in the pipe system that is currently used on the test bed. It shows that gas dynamics accounts for significant pressure fluctuations in the outlet port and the working chamber, and that wave tracking is necessary for accurate correlation and simulation.

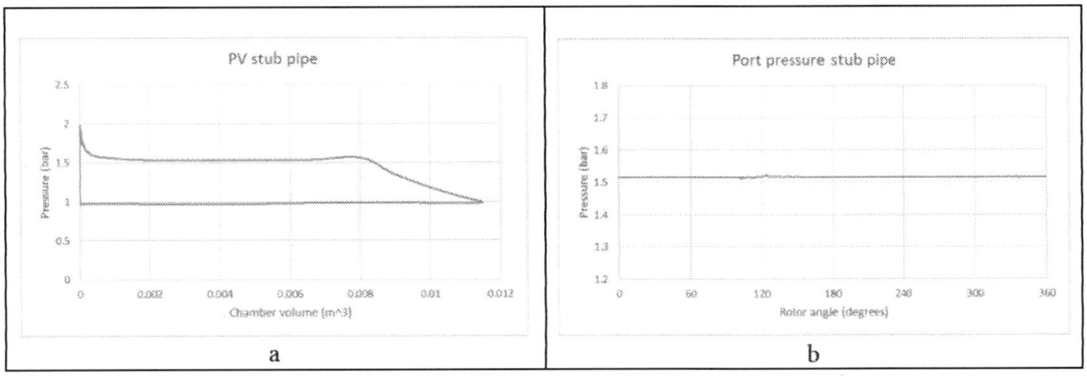

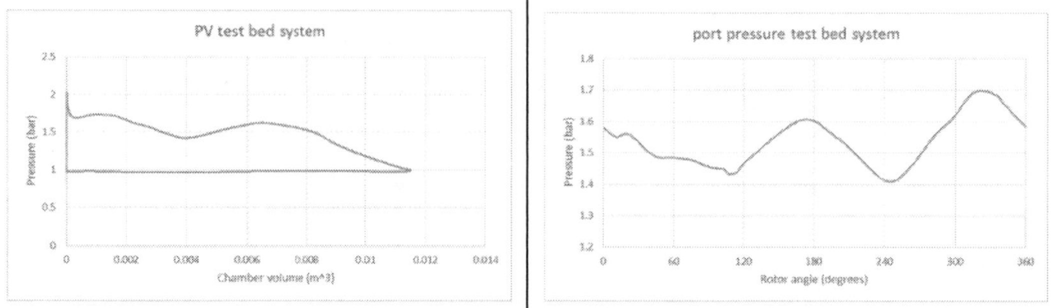

c d

Fig. 4. Simulated outputs at 0.5 bar outlet pressure, (a) PV diagram with no pipe, (b) port pressure with no pipe, (c) PV diagram with the test bed outlet system, (d) port pressure with the test bed outlet system

6. Test data and correlation.

6.1 The test rig

An instrumented test rig has been built to evaluate and develop the blade compressor design (figure. 5). The working chamber is instrumented with 10 Kulite high speed pressure transducers. Inlet and outlet pipe pressures are also measured with Kulite high speed pressure transducers of the appropriate pressure range, and temperatures are measured with PT100 class 1 PRTs. Torque is measured with an HBM T40 torque transducer. Angular position is determined using two encoder wheels in an offset configuration. Flow rate is measured using an orifice plate flow meter designed in house to ISO 5167. All data is acquired at 25 kHz, using an NI 9237data acquisition card and the test bed is controlled using bespoke software programmed in Labview. Knowledge of volume change with angle allows indicated work and power to be calculated directly from the pressure-volume history. Software has been written in house to regress the test data.

Figure 5. The Lontra test bed and user interface.

IOP Conf. Series: Materials Science and Engineering **90** (2015) 012056 doi:10.1088/1757-899X/90/1/012056

6.2 Correlation

In order to correlate the model over a wide range of operating conditions, measured data was acquired at 25 test points. The outlet pressure range chosen was from 0.2 to 0.6 bar in 0.1 bar increments and the speed range was from 1000rpm to 3000rpm in 500rpm increments. Geometric inputs into the simulation have been found either from physical measurements, or from CAD models. Calculated data such as volumes and surface areas have also been verified by CAD measurements. The dimensions that are most difficult to measure are the seal clearances. 21 separate clearance seals have been identified, some of which can be measured as the machine is being assembled, but others can only be estimated from machining tolerances. Comparisons of measured and simulated mass flow vs speed for the lowest and highest outlet pressures using the best estimates of clearances are shown in figure 6 and 7.

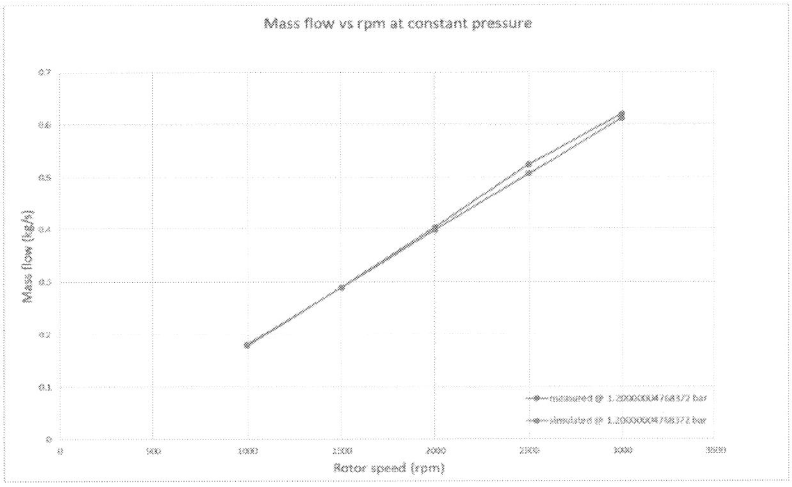

Figure 6. Predicted and simulated mass flow vs speed with a 0.2 bar outlet.

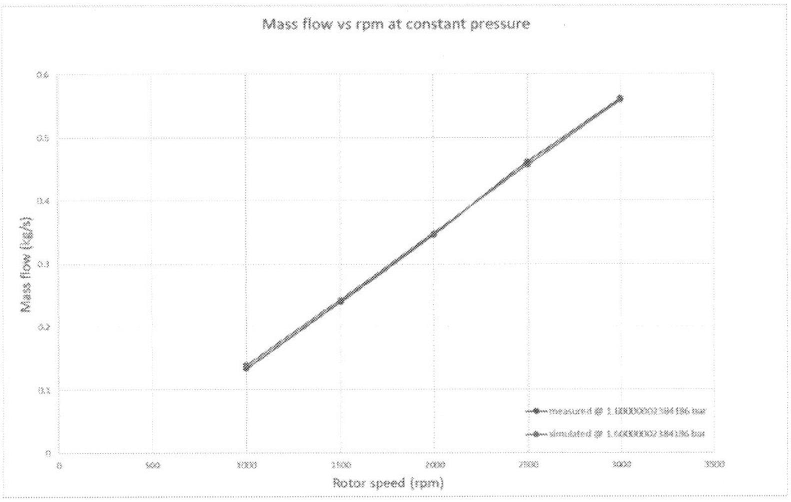

Figure 7. Predicted and simulated mass flow vs speed with a 0.6 bar outlet

IOP Conf. Series: Materials Science and Engineering **90** (2015) 012056 doi:10.1088/1757-899X/90/1/012056

Graphs at the intermediate pressure points show similar agreement between measured and predicted flow rates. The maximum error in mass flow prediction across all 25 test points was 4.2%, and the average mass flow error was 1.53%.

Graphs of predicted vs measured chamber pressure and outlet pipe pressures are shown in figures 8 and 9.

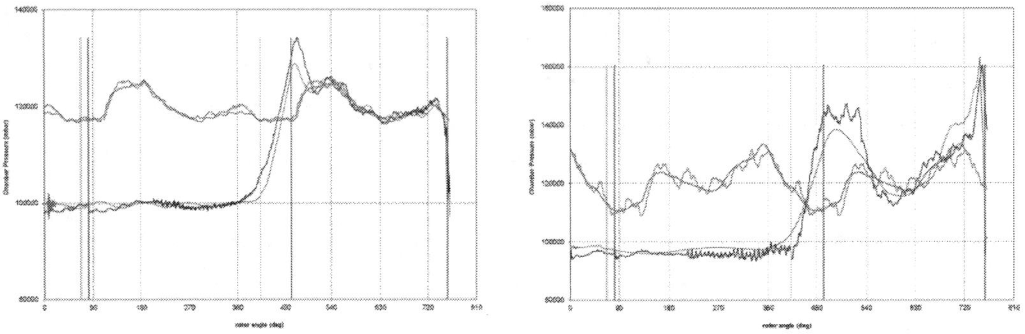

(a) 1000rpm (b) 3000rpm

Figure 8. Predicted and measured pipe pressure (pink) and chamber pressure (blue) at 0.2 bar back pressure. Smoother line in both cases is the simulated result.

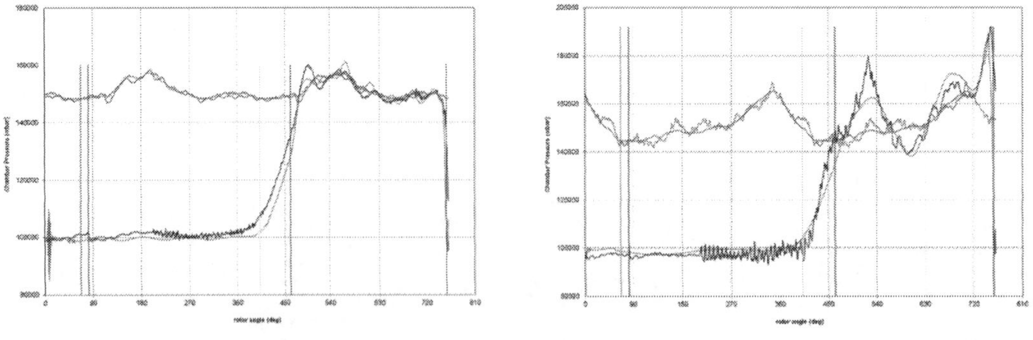

(a) 1000rpm (b) 3000rpm

Figure 9. Predicted and measured pipe pressure (pink) and chamber pressure (blue) at 0.5 bar back pressure. Smoother line in both cases is the simulated result.

It can be seen that the gas dynamic simulation tracks the major oscillations in the outlet pipe and their effect on the chamber pressure. In some cases the simulation under predicts chamber pressure during compression and outlet opening, and this is thought to be due to spatial variation of pressure within the working chamber that the simulation cannot model. Lontra is also working on implementing silencer modelling which could help to improve the accuracy of the prediction of intake pressure waves.

7. Conclusions

The Lontra blade compressor is an innovative design of positive displacement compressor. As with all new designs, it is necessary to understand the operation of the device in detail in order to improve and develop the product effectively. The simulation software that has been developed at Lontra is able to provide insight into the physical processes involved during compressor operation and can effectively inform design decisions. It is also possible to accurately predict mass flow for a given set of clearances, and accurate pressure predictions allow trends in indicated torque and efficiency to be evaluated. With new products being developed at Lontra, the simulation software has become an invaluable tool in evaluating new concepts while at the same time speeding up the development process and saving valuable resources that would otherwise have to be spent on a greater number of physical prototypes. It has also allowed Lontra to help our licensees with "what if" simulations, to quickly assess the effect of production changes (casting and machining tolerances for example) on machine operation and efficiency forming a very valuable tool for production cost management.

Nomenclature

m = mass (kg)
U = internal energy (J)
H = Enthalpy (J)
Q = heat transfer (J)
W = work (J)
p = pressure (Pa)
v = volume
Nu = Nusselt number
Re = Reynolds number
Pr = Prandtl number
ρ = density (kg/m^3)
s = shear stress (Pa)
u = velocity (m/s)
μ = viscosity (Pa.s)
R = specific gas constant (J/kg.K)
T = temperature (K)
b = breadth of slot (m)
c = slot clearance (m)
γ = ratio of specific heats
A = area (m^2)
$\dot{m}$ = mass flow rate (kg/s)

References

[1] Shires G. L. (1950) The Viscid Flow of Air in a Narrow Slot, Aeronautical Research Council paper 13
[2] Blair G. P. (1999) The Design and Simulation of Four Stroke Engines (Society of Automotive Engineers)
[3] Benson R. S. (1982) The Thermodynamics and Gas Dynamics of Internal Combustion Engines (Oxford University Press)
[4] Winterbone D. E. and Pearson R. J. (2000) Theory of Engine Manifold Design (Professional engineering publishing)
[5] Stosic N. and Hanjalic K. Computer Simulation of Two Stage Reciprocating Compressors (1978) International Compressor Engineering Conference, Paper 281.

IOP Conf. Series: Materials Science and Engineering **90** (2015) 012058 doi:10.1088/1757-899X/90/1/012058

Optimisation of two-stage screw expanders for waste heat recovery applications

M.G. Read, I.K. Smith and N. Stosic

Centre for Positive Displacement Compressor Technology, City University London, Northampton Square, London, UK

m.read@city.ac.uk

Abstract. It has previously been shown that the use of two-phase screw expanders in power generation cycles can achieve an increase in the utilisation of available energy from a low temperature heat source when compared with more conventional single-phase turbines. However, screw expander efficiencies are more sensitive to expansion volume ratio than turbines, and this increases as the expander inlet vapour dryness fraction decreases. For single-stage screw machines with low inlet dryness, this can lead to under expansion of the working fluid and low isentropic efficiency for the expansion process. The performance of the cycle can potentially be improved by using a two-stage expander, consisting of a low pressure machine and a smaller high pressure machine connected in series. By expanding the working fluid over two stages, the built-in volume ratios of the two machines can be selected to provide a better match with the overall expansion process, thereby increasing efficiency for particular inlet and discharge conditions. The mass flow rate though both stages must however be matched, and the compromise between increasing efficiency and maximising power output must also be considered. This research uses a rigorous thermodynamic screw machine model to compare the performance of single and two-stage expanders over a range of operating conditions. The model allows optimisation of the required intermediate pressure in the two-stage expander, along with the rotational speed and built-in volume ratio of both screw machine stages. The results allow the two-stage machine to be fully specified in order to achieve maximum efficiency for a required power output.

1. Introduction

The Organic Rankine Cycle (ORC) provides a means of recovering useful energy from low temperature heat sources. In comparison with conventional high temperature steam Rankine cycles, the low temperature of these heat sources means that the attainable cycle efficiency is much lower, while the required surface area of the heat exchangers per unit power output is much higher. The lower latent heat of evaporation of organic fluids relative to steam also means that the feed pump work required in ORCs is a significantly higher proportion of the gross power output.

For low source temperatures, the power generation cycle normally considered is that where the working fluid enters the expander as dry saturated vapour, as shown in Figure 1. However, in most cases, this leads to the working fluid leaving the expander with some superheat, which must be removed before condensation begins.

Content from this work may be used under the terms of the Creative Commons Attribution 3.0 licence. Any further distribution of this work must maintain attribution to the author(s) and the title of the work, journal citation and DOI.
Published under licence by IOP Publishing Ltd

IOP Conf. Series: Materials Science and Engineering **90** (2015) 012058 doi:10.1088/1757-899X/90/1/012058

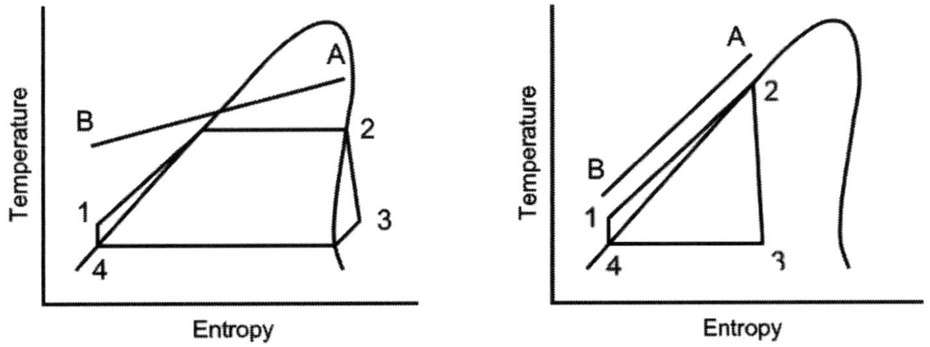

Figure 1: Illustrative T-s diagrams showing conventional ORC with dry saturated vapour at the expander inlet, and TFC with wet saturated vapour at the expander inlet

Maximising net power output from the cycle is a compromise between increasing the mean temperature of heat addition (which, in accordance with Carnot's principle, can increase cycle efficiency) and increasing the amount of heat extracted from the source, which requires a lower evaporation temperature.

This can be achieved in a Trilateral Flash Cycle (TFC) which expands the working fluid from a saturated liquid state as shown in Figure 1. Although this system has been considered for many years, to date, no large scale demonstration unit of it is known to have been built. This is because of the lack of suitable two-phase expanders with adiabatic efficiencies approaching those of dry vapour turbines.

By the use of a screw expander, instead of the more conventional turbine, it is possible to admit the working fluid to the expander as wet vapour and thereby eliminate both the need to desuperheat the vapour after expansion and simultaneously to raise the evaporation temperature, thus improving the cycle efficiency. The potential cost and performance benefits of using screw expanders in ORC systems have been extensively studied for geothermal applications [1-3].

In the fields of geothermal and waste heat recovery systems, there is growing interest in generating power from heat sources with initial temperatures in the 170-200°C range. At these temperatures, simple ORC systems are less attractive, as the method recommended for their use under these circumstances is to operate two such systems in a cascade arrangement. The first would operate over a higher temperature range and the condenser of this unit would act as the evaporator of the second unit with different working fluids in each closed loop. Alternatively, Kalina type systems, which require at least three heat exchangers, may be suitable. In the light of the relative complexities of these systems, the authors re-examined the possibility, first considered some thirty years ago, of using a TFC system for power recovery from higher temperature resources.

At resource temperatures in the 170-200°C range, a suitable working fluid for a TFC system is pure n-pentane. With such a working fluid, the expansion process involves a volume ratio of expansion of the order of 160:1, which, using twin-screw machines, requires a two stage expander to achieve efficient expansion. The problem, therefore, lies in the design of the first and second stage expanders, to admit saturated liquid and wet vapour respectively. While previous work has studied the performance of combined twin-screw and turbine systems [3], recent progress in developing and validating a computational twin-screw model has allowed the performance of single and two-stage systems to be investigated in greater detail. This model allows the optimisation of the expander parameters for a particular application, and can be incorporated with other detailed component models to allow multi-variable optimisation of low temperature heat recovery systems.

2. Twin-screw expander model

A full thermodynamic model of the expander has been created for investigating the performance of two-stage expanders. This is based on the quasi one dimensional analysis of twin-screw machines as described by Stosic and Hanjalic [4,5], which has been extensively validated for compressors for a wide range of working fluids and operating conditions. For expanders, the model has been validated for expansion of low dryness fluid (including saturated liquid) using the refrigerant R113 [6], and more recently for the expansion of high dryness wet steam [7]. Using this procedure, machine geometry and rotor profiles have been optimised for a particular set of operating conditions representative of those considered in this paper, and have been fixed for the purposes of the current study. The City University 'N' rotor profile described in Figure 2 has been used in the current analysis, as this geometry is known to have benefits including greater throughput and a stiffer gate rotor than is possible using alternative profiles with similar blow-hole area and sealing line lengths [8].

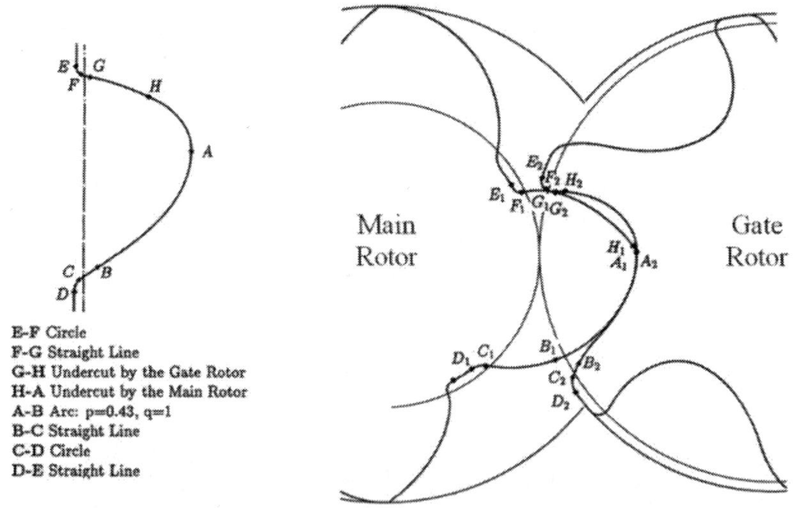

Figure 2: Description of the City University 'N' rotor profile for positive displacement screw machines

In principle, the screw machine geometry optimisation can be integrated with the expander analysis described here to ensure the best profile is used for the required operating conditions, but this would be very computationally intensive and is not expected to significantly affect results. Furthermore, from a manufacturing perspective it would be prohibitively expensive to produce an optimised machine for every different application. For a specified geometry, the characteristics of the twin-screw machine such as the curve of working chamber volume against angular position (illustrated in Figure 3), sealing line lengths, blowhole area and axial/radial clearances between the rotors and the casing are defined as fixed inputs for the expander model. An important machine parameter in the built-in volume ratio, BIVR, defined as the ratio of working chamber volumes at discharge port opening and suction port closing. Figure 3 illustrates how increasing the BIVR for a particular machine increases the volume of working fluid admitted through the suction port per revolution. For a particular rotational speed of the machine, the volumetric and mass flow rates can be determined.

IOP Conf. Series: Materials Science and Engineering **90** (2015) 012058 doi:10.1088/1757-899X/90/1/012058

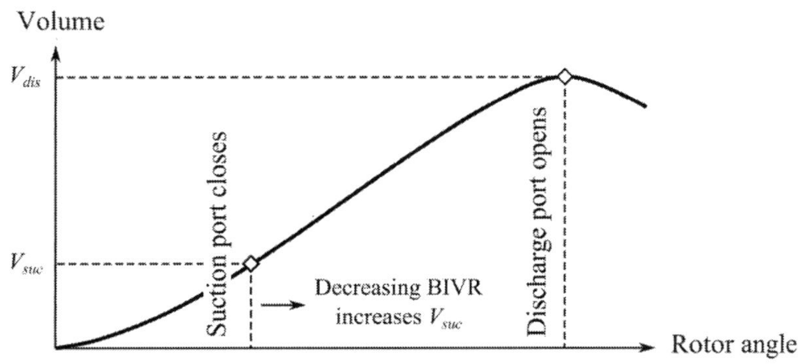

Figure 3: Schematic diagram of working chamber volume as a function of rotor angle

For single stage screw machines, the inlet dryness fraction and the pressures at the inlet and discharge are defined by the requirements of the cycle. The variable input parameters required for the expander model are then limited to the expander size, main rotor speed and BIVR. Two approaches can be taken to match the machine operation to the required cycle conditions:

i. The BIVR is specified and iterations are performed to find the rotor speed required to match the mass flow rate of expander to that of the working fluid required in the cycle – no limits are imposed on rotor speed, which in some cases can become impractically high.
ii. The rotor speed is fixed and iterations are performed to find the value of the BIVR required to match the mass flow rate – if the BIVR is greater than the limit for the chosen screw machine geometry then the expander cannot meet the requirements of the cycle conditions.

For two-stage machines, the intermediate pressure between the two stages is an additional input parameter. There is also an additional constraint, as the mass flow rate through both the high pressure (HP) and low pressure (LP) stages must be the same. While the mass flow rate of the HP stage is largely dependent on the inlet conditions and the size and BIVR of the HP machine, it is also dependent on the intermediate pressure, as this affects leakage flows in the machine. To characterise the performance of a two-stage expander for particular conditions, the following iterative approach is therefore required:

i. Specify the size, speed and BIVR of both stages.
ii. Estimate the intermediate pressure, and calculate the mass flow rates of the HP and LP stages.
iii. While the difference between the HP and LP mass flow rates is greater than an allowable error, repeat step ii.
iv. While the difference between the converged and required mass flow rates is greater than an allowable error, repeat steps i-iii, fixing either the speed or BIVR of the two stages as required.

The single or two-stage expander efficiency calculated using these approaches can be used in a thermodynamic cycle model to calculate overall cycle performance for specific operating conditions. It is possible to then apply an iterative numerical procedure to identify the optimum operating conditions for the cycle. However, the focus of this paper is to illustrate how the optimum expander parameters can be selected for specified cycle conditions.

IOP Conf. Series: Materials Science and Engineering **90** (2015) 012058 doi:10.1088/1757-899X/90/1/012058

3. Modelling of single and two-stage expanders for TFC application

The analysis presented in this paper has been performed for a simple heat recovery application from a single phase source fluid, defined as follows;

Assumptions:	
Heat Source Inlet Temperature	190°C
Heater Pinch Point Temperature Difference	5°C
Available Cooling Water Temperature	20°C
Cooling Water Temperature Rise in Condenser	5°C
Condenser Pinch Point Temperature Difference	5°C

For the proposed TFC system, in view of the fact that expansion must begin from the saturated liquid condition, a suitable working fluid for this case is n-pentane, which has a critical temperature of 196.6°C. The pressure of the working fluid in the condenser was constrained to be greater than or equal to atmospheric, so as to prevent air leaking into system. An initial cycle analysis program was used to identify suitable operating conditions for the expander;

Expander Inlet dryness fraction	0 (i.e. saturated liquid)
Expander Inlet Temperature	175°C
Condensing Temperature	36°C

Rather than specify the heat input to the cycle, and thereby determine a required mass flow rate for the working fluid, it is useful to characterise the performance of a range of single and two-stage expanders as a function of mass flow rate at these conditions. Standard twin-screw machine sizes, with main rotor diameters ranging from 145-408mm, have been analysed in order to illustrate what is achievable with practical single and two-stage expanders. To identify the maximum mass flow rates possible with these machines, and to ensure high efficiency, the performance has been considered at maximum allowable rotational speeds corresponding to a main rotor tip speed of 60m/s.

4. Results of TFC expander modelling

For the application specified above, a two-stage machine requires a relatively small HP machine in comparison with the size of the LP machine, due to the much higher density of the fluid at the HP inlet. A combination of a 145mm HP machine with a 408mm LP machine has been found to achieve good overall performance with well-matched expansion in both stages. The mass flow rate, overall adiabatic efficiency, required intermediate pressure and total shaft power are all dependent on the BIVRs of the HP and LP machines. The overall performance of the two-stage machine has therefore been calculated over a range of BIVR values, and contour maps of the key results are shown in Figures 4 and 5.

The results in Figure 4 show that mass flow rate is, as expected, very strongly dependent on the BIVR of the HP stage. Figure 5 shows that the maximum overall adiabatic efficiency occurs at BIVR values of 3.4 and 3.6 for the HP and LP stages respectively. This corresponds to an intermediate pressure of 6.9 bar(abs), mass flow rate of 9.9 kg/s and total shaft power of 520 kW. For a fixed HP BIVR, it can be seen that the required intermediate pressure increases as the LP BIVR increases. The maximum efficiency point corresponds to the case when the BIVRs of both stages are well matched to the expansion, but at lower values of LP BIVR, the intermediate pressure falls, leading to under expansion for the HP stage and over expansion for the LP stage. Conversely, at higher values of LP BIVR the rise in intermediate pressure leads to over-expansion for the HP stage and under-expansion for the LP stage. In conclusion, the circular efficiency contours are a result of over and/or under expansion in one or both of the expander stages.

Power output can be increased by moving away from the maximum efficiency point, but it is important to choose the BIVR values so as to ensure that efficiency is maximised for a particular power output. Figure 5 shows that maximum power for any HP BIVR occurs at a constant LP BIVR value of around 3.6, and that this closely corresponds to the maximum efficiency possible for a particular value of HP BIVR. It is therefore possible to plot curves showing the maximum values of shaft power and adiabatic efficiency as functions of the mass flow rate. These are shown in Figure 6 along with the corresponding performance of single stage expander for the same application. In all cases, the curves show the full range of performance achievable within the practical range of BIVR values.

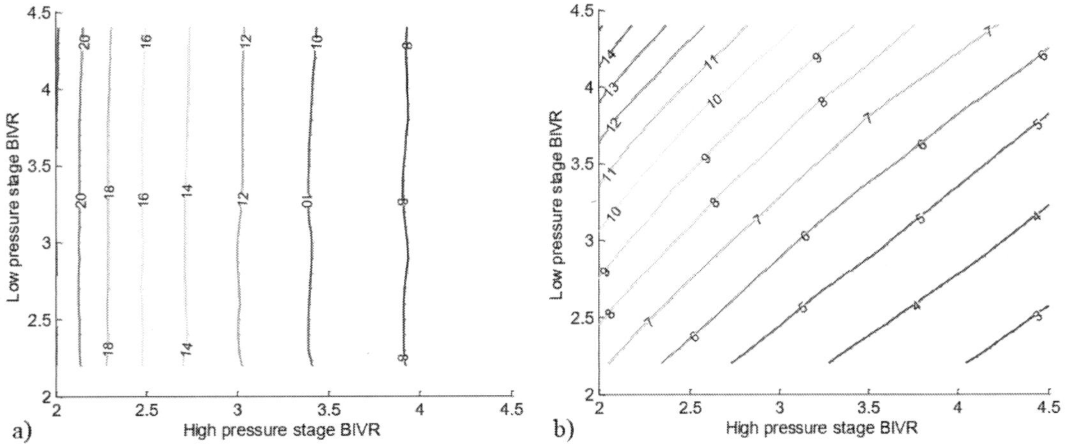

Figure 4: Contour maps showing a) mass flow rate (kg/s) and b) intermediate pressure (bar) of two-stage expander

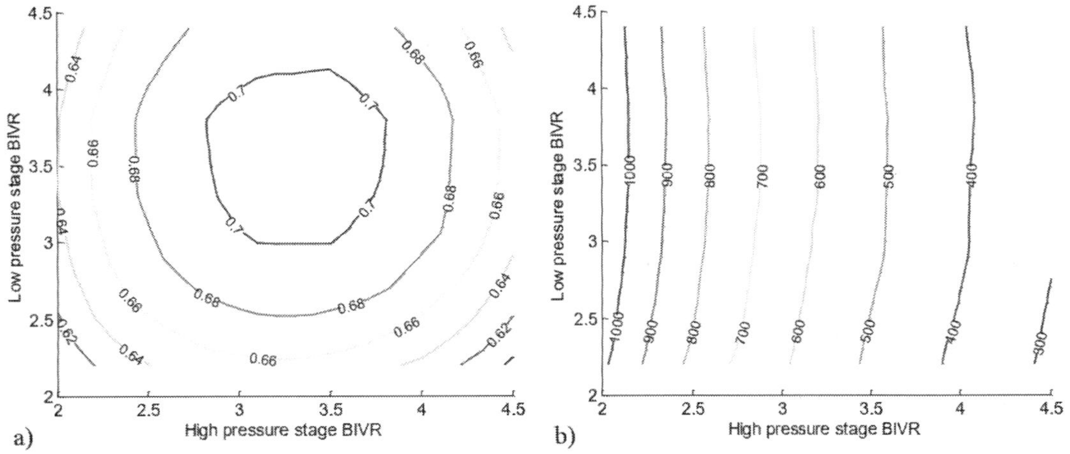

Figure 5: Contour maps showing a) adiabatic efficiency and b) total shaft power (kW) of two-stage expander

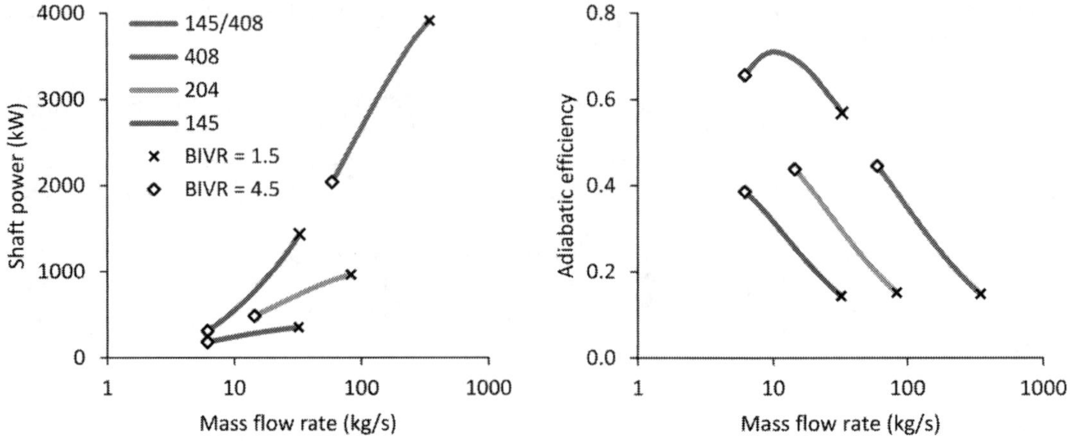

Figure 6: Performance comparison of single and two-stage expanders

The results in Figure 6 show that for single stage expanders, increasing the BIVR increases the efficiency, while reducing the mass flow rate through the expander. For all of the single stage expanders considered, the maximum efficiency is well below 50% due to the large degree of under expansion. This is a result of the limited BIVR being much lower than the actual specific volume ratio of the working fluid over the expansion process.

The two-stage machine achieves a much greater combined BIVR, and is therefore able to better match the overall expansion. The peak of the efficiency curve in Figure 6 shows the point where the expansion in the two-stage machine is best matched to the operating conditions; at higher mass flow rates the efficiency falls due to over expansion of the working fluid in the HP stage, while at lower mass flow rates it falls due to over expansion in the LP stage. Interestingly, the results in Figure 6 suggest that a two-stage machine may be viewed as equivalent to the LP machine operating as a single stage but with a BIVR higher than the practical limit; this is illustrated by the fact that the shaft power and efficiency curves for the two-stage 145/408 machine are essentially extensions of the performance curves for the single 408 machine, covering a lower range of mass flow rates. It is also worth noting that, as the mass flow rate of the two-stage machine is largely a function of the BIVR of the HP stage, this range of achievable mass flow rate is very close to that of the single 145 machine. In summary, compared to the LP stage operating alone, the addition of the HP stage can be seen to increase efficiency, but only by reducing mass flow rate and hence power output.

5. Effect of expander selection on TFC performance

The efficiency and mass flow rate of the expander affect the required power input and net power output of the TFC system. Two important measures of the overall system performance are the conversion efficiency, defined as the net power output divided by the *available* heat input, and the cycle efficiency, defined as net output power divided by the *actual* heat input. The following component efficiencies have been used in order to estimate the performance of both the TFC and a conventional saturated vapour ORC (see Figure 1).

Assumed component efficiency:

Feed pump	0.7
Motor	0.9
Generator	0.95
Turbine	0.82

For the TFC system, net power output is dependent on the mass flow rate, as this affects both the expander efficiency and work done in the feed pumps. Using the above component efficiencies, the net electrical power output of the system is shown in Figure 7, and the resulting cycle and conversion efficiencies are shown in Figure 8.

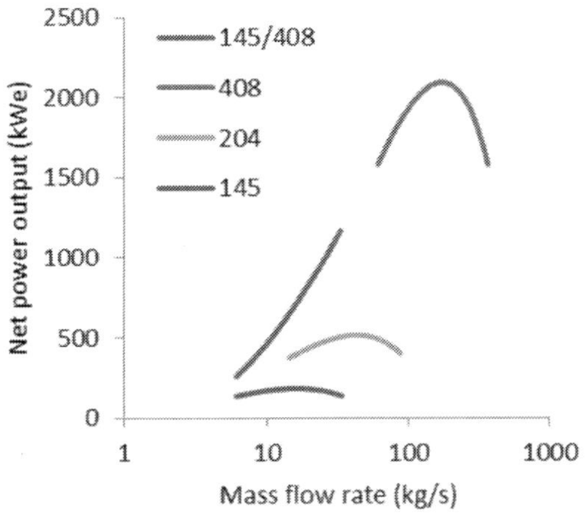

Figure 7: Net power output from TFC system using single and two-stage expanders

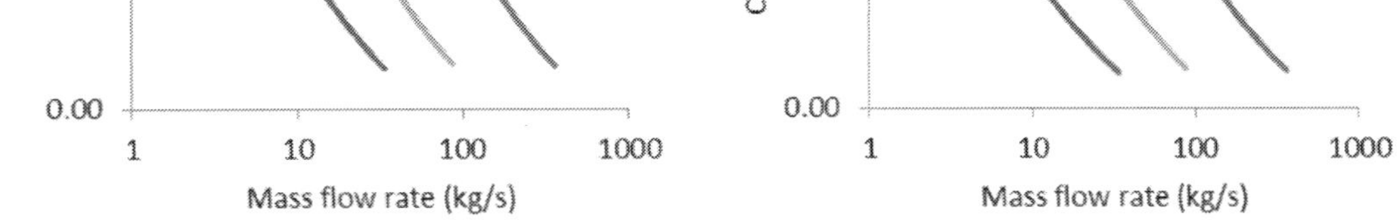

Figure 8: Achievable TFC cycle and conversion efficiency using single and two-stage expanders

Using the same operating conditions and component efficiencies in a thermodynamic cycle model, a simple saturated vapour ORC without recuperative feed heating, operating under the same conditions, was optimized in order to achieve maximum net power output. This was found to occur with cycle and conversion efficiencies of 12.5% and 8.8% respectively. The results from Figure 8 suggest that in the range of system sizes covered by the TFC with two-stage expander (250-1200kWe), the conversion efficiency is always greater than that which can be achieved in a simple saturated vapour ORC, largely due to the greater recovery of available heat from the source fluid.

6. Conclusions

The study presented in this paper shows that two-stage screw expanders can match the required volume ratio for the expansion of saturated liquid in waste heat recovery applications, and achieve high overall adiabatic efficiency. The design parameters for the two-stage machine can be optimised in order to maximize shaft power output for a given mass flow rate, and the possible range of operation of the two-stage machine has been mapped out. This allows a direct comparison of the performance of different single and two-stage machines operating under the same conditions. In the application discussed in this paper, the TFC using a two-stage expander is predicted to achieve a higher overall conversion efficiency than a conventional saturated vapour ORC.

References

[1] Smith, I.K., Stosic, N., Kovacevic, A., 2001, Power recovery from low cost two-phase expanders, *Transactions of Geothermal Resource Council*, p. 601-606.

[2] Smith, I.K., N Stosic, N., Kovacevic, A., 2005, Screw expanders increase output and decrease the cost of geothermal binary power plant systems, *Transactions of Geothermal Resource Council*, p. 25-28.

[3] Smith, I.K., Stosic, N., Kovacevic, A., 2004, An improved System for Power Recovery from Higher Enthalpy Liquid Dominated Fields, *GRC Annual Meeting, Indian Wells*.

[4] Hanjalic, K., Stosic, N., 1997a, Development and Optimization of Screw Machines with a Simulation Model - Part II: Thermodynamic Performance Simulation and Design Optimization, *Journal of Fluids Engineering*, 119(3): 664-670.

[5] Stosic, N., Hanjalic, K., 1997b, Development and Optimization of Screw Machines with a Simulation Model - Part I: Profile Generation, *Journal of Fluids Engineering*, 119(3): 659-663.

[6] Smith, I.K., Stosic, N., Aldis, C.A., 1996, Development of the Trilateral Flash Cycle System: Part 3: The Design of High-Efficiency Two-Phase Screw Expanders, *Proceedings of the Institution of Mechanical Engineers, Part A: Journal of Power and Energy*, 210(1): 75-93.

[7] Read, M., Stosic, N., & Smith, I. K., 2014, Optimization of Screw Expanders for Power Recovery From Low-Grade Heat Sources, *Energy Technology & Policy*, *1*(1), 131-142.

[8] Smith, I.K., Stosic, N., Kovacevic, A., 2014, *Power Recovery from Low Grade Heat by Means of Screw Expanders*, Elsevier

IOP Conf. Series: Materials Science and Engineering **90** (2015) 012059 doi:10.1088/1757-899X/90/1/012059

Numerical simulation of a twin screw expander for performance prediction

Iva Papes,Joris Degroote,Jan Vierendeels
Department of Flow, Heat and Combustion Mechanics,Ghent University, Belgium
E-mail: iva.papes@ugent.be

Abstract.
With the increasing use of twin screw expanders in waste heat recovery applications, the performance prediction of these machines plays an important role. This paper presents a mathematical model for calculating the performance of a twin screw expander. From the mass and energy conservation laws, differential equations are derived which are then solved together with the appropriate Equation of State in the instantaneous control volumes. Different flow processes that occur inside the screw expander such as filling (accompanied by a substantial pressure loss) and leakage flows through the clearances are accounted for in the model. The mathematical model employs all geometrical parameters such as chamber volume, suction and leakage areas. With R245fa as working fluid, the Aungier Redlich-Kwong Equation of State has been used in order to include real gas effects. To calculate the mass flow rates through the leakage paths formed inside the screw expander, flow coefficients are considered as constant and they are derived from 3D Computational Fluid Dynamic calculations at given working conditions and applied to all other working conditions. The outcome of the mathematical model is the P-V indicator diagram which is compared to CFD results of the same twin screw expander. Since CFD calculations require significant computational time, developed mathematical model can be used for the faster performance prediction .

1. Introduction

For more than half a century, screw machines have extensively been used as compressors. In recent years, screw machines as expanders play an important role in waste heat recovery applications. One of the leading technologies in these applications is the Organic Rankine Cycle (ORC). Low-grade enthalpy should be converted to useful work in ORC systems. Given its power range, efficiency and price, the twin screw expander is highly suitable [1]. The optimization of a screw expander's performance can be covered from different points of view such as efficiency, power output, cost, noise and vibrations, etc.

Analytical procedures for the compressors performance prediction have been extensively reported by numerous authors among which Stosic, Fleming, Fujiwara and Hanjalic contributed [2, 3, 4, 5]. Surprisingly, only few publications on the analysis of screw expanders can be found in the literature. The first analytical model for dry screw expanders was presented in [6]. More recently, the numerical and experimental study of an oil injected twin screw expander for both air and R113 has been presented in [7]. The mathematical model was verified with the experimental study and flow coefficients used in the leakage models were evaluated as constant. However, the limitation of the experiments is that they cannot accurately measure flow properties in the

Content from this work may be used under the terms of the Creative Commons Attribution 3.0 licence. Any further distribution of this work must maintain attribution to the author(s) and the title of the work, journal citation and DOI.
Published under licence by IOP Publishing Ltd

IOP Conf. Series: Materials Science and Engineering **90** (2015) 012059 doi:10.1088/1757-899X/90/1/012059

clearance area or during the filling. As it is shown in [8], with CFD calculations it is possible to overcome these limits. Therefore, in this model results of the 3D CFD analysis are used in order to better predict the leakage flows and pressure drop during the filling.

Modelling of the leakage flows presents an important and one of the most challenging parts of mentioned mathematical models. A typical leakage model used when modelling scroll or screw types positive displacement machines, is the isentropic converging nozzle [6, 9]. The drawback of this model is that it does not take friction into account. Another possibility in modelling leakage flows is to use the compressible adiabatic flow with friction, known as Fanno Flow, as presented in [11]. In [10], Bell et al. presented an empirical frictional correction factor for the isentropic nozzle model which is derived by calculating the mass flow through variable area leakage paths with real gas properties and then it is correlated to the prediction of an isentropic nozzle model. But this model can be used only when the flow is not choked. From the analysis of 3D CFD calculations it could be seen that in the leakage paths inside the twin screw expander and for the working conditions presented in this paper, flow is usually chocked. Therefore, authors used the isentropic converging nozzle model in this paper.

The aim of this paper is to present a mathematical model of a twin screw expander and to validate it with Computational Fluid Dynamic (CFD) results. Moreover, the goal is to extract the coefficients used in the isentropic converging nozzle leakage model from the 3D CFD analysis. The CFD analysis of the twin screw expander has been presented by the authors in [8, 12]. However, additional calculations and analysis were done in these CFD simulations in order to get enough data for accurate model comparison. In the end, performance prediction from the proposed mathematical model is validated with CFD results.

2. Twin Screw Analysis

The geometry of the twin screw expander analysed in this paper is shown in figure 1a. It is a standard air compressor used in opposite sense of rotation together with R245fa instead of air. The configuration of the rotor lobes is 4/6 (male/female). The outer diameter of the male and female rotors is approximately $70mm$ with the L/D ratio of 1.9.

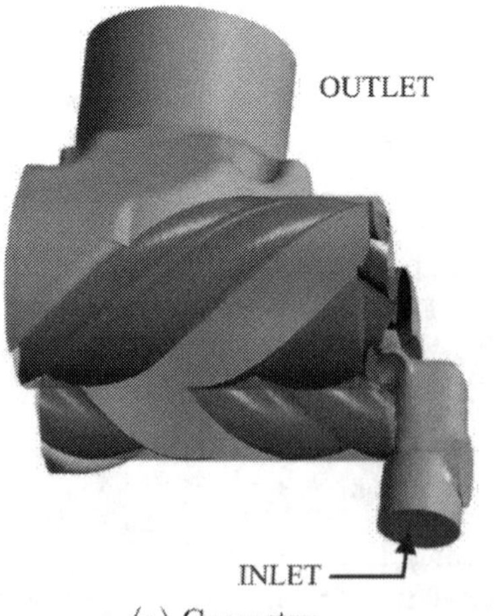

(a) Geometry

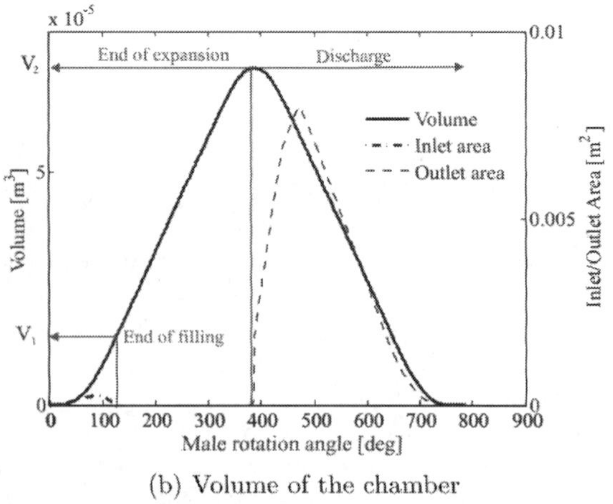

(b) Volume of the chamber

Figure 1: Geometry of the twin screw expander with the analytical description of the geometrical parameters used in the mathematical model

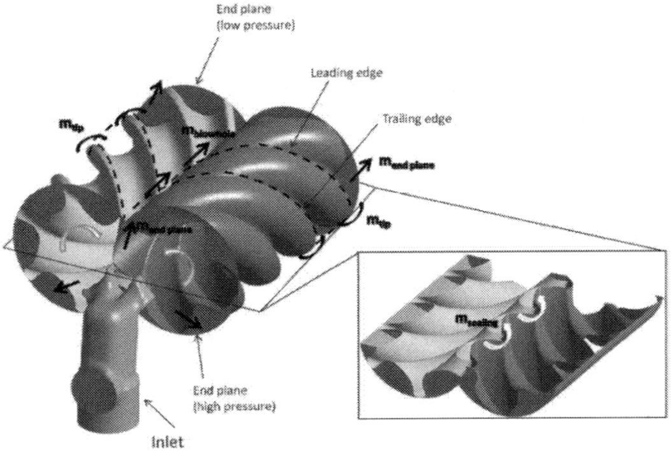

Figure 2: Different leakage types inside the twin screw expander

The volume curve of the screw expander is shown in figure 1b. The formation of a chamber starts at $\theta = 0°$ in figure 1b. When $\theta = 7°$, the chamber is in connection with the inlet port. As the rotors rotate, the volume of the chamber is rising together with the increase in the inlet surface area, and the chamber is filled with the working fluid. When the inlet area starts to decrease, the volume of the chamber is still increasing. This can already cause the **pre-expansion** of the working fluid. The filling ends at $\theta = 126°$ after which the working fluid expands with increasing volume of the chamber. At $\theta = 387°$, the working chamber is connected to the outlet and the working fluid is discharged through the outlet port.

Within twin screw expanders, it is possible to identify four types of leakage paths as shown in figure 2. All leakages are characterized by the length of the leakage path and the area of the clearance.

3. Mathematical model of a twin screw expander

Since the fluid enters through the boundaries of the working chamber, the zero dimensional mathematical model described in this paper represents an open thermodynamic system. The mathematical model employs the mass and energy conservation equations, accompanied by the geometrical model which describes the change in volume with the time or angular position, as well as change in the inlet and leakage path areas.

When analysing the flow within the screw expander, the following assumptions have been made:

- Potential and kinetic energy of the working fluid are negligible
- The flow through the leakage paths and inlet port is assumed to pass through an isentropic nozzle
- The heat transfer between working fluid and the rotor or between the casing and the ambient are not included in the model (they are also neglected in CFD simulations)
- Mechanical losses of the screw expander are not included in the model
- Working fluid of the screw expander is R245fa which is described by Aungier Redlich-Kwong Equation of State (ARK EoS)
- Leakage flows through the end planes are not included in the model since they are not modelled in the CFD calculations

- Once the chamber is connected to the outlet port, the pressure in the working chamber is equal to discharge pressure.

The conservation of mass for the chamber which represents the control volume can be expressed as follows:

$$\frac{dm_{ch}}{dt} = \sum_i \dot{m}_i \tag{1}$$

Energy conservation law applied to the control volume is defined as:

$$\frac{dE}{dt} = -p\frac{dV}{dt} + \sum_i \dot{m}h_i + \dot{Q} \tag{2}$$

If the process is assumed to be adiabatic with conservation laws converted into derivatives of temperature with respect to time, the rate of temperature changes is as follows:

$$\frac{dT}{dt} = \frac{-T\left(\frac{\partial p}{\partial T}\right)_v \left[\frac{dV}{dt} - v\frac{dm_{ch}}{dt}\right] - h\frac{dm_{ch}}{dt} + \sum_i \dot{m}_i h}{m_{ch}c_v} \tag{3}$$

The properties of the working fluid are calculated by using the same EoS as used in CFD calculations [8, 13]. The system of differential equations is solved by using the forward Euler method applied simultaneously on all chambers. Once the geometrical inputs (volume change and leakage areas) have been provided to the model, the pressure and the temperature are initialized with guess values. After that, for each iteration step, the mass flows going in or out of the chamber are calculated and the temperature in the next step is obtained. Since the mass, the volume and the temperature are then known, the pressure and the density can be updated.

The mass flow through the leakage paths and the inlet port is calculated using the isentropic nozzle model [14]:

$$\dot{m}_{nozzle} = CA\sqrt{p_{up}\rho_{up}}\sqrt{\frac{2k}{k-1}\left(p_{ratio}^{2/k} - p_{ratio}^{(k+1)/k}\right)} \tag{4}$$

Where the function of pressure ratio is defined as:

$$p_{ratio} = \begin{cases} \left(1 + \frac{k-1}{2}\right)^{k/(1-k)} & p_{down}/p_{up} \leq \left(1 + \frac{(k-1)}{2}\right)^{k/(1-k)} \\ p_{down}/p_{up} & p_{down}/p_{up} > \left(1 + \frac{(k-1)}{2}\right)^{k/(1-k)} \end{cases}$$

Flow coefficients C are constant in time and are obtained from CFD calculations. The indicated work of a twin screw expander can be expressed as the area of the P-V indicator:

$$W_{ind,cycle} = \int_{cycle} V dp \tag{5}$$

The power of the twin screw expander can be then calculated as:

$$P_{ind} = \frac{W_{ind,cycle}zn}{60} \tag{6}$$

with z the number of lobes and n the rotational speed.

4. 3D CFD analysis of twin screw expander

In this paper, 3D CFD (Computational Fluid Dynamics) calculations of a twin screw expander are used for a validation of the developed mathematical model.The flow calculations inside a screw expander are performed using Ansys Fluent with the use of User Defined Functions (UDFs) to handle the grid movement and real gas model as presented in [8]. The mathematical model consists of a set of momentum, energy and mass conservation equations, which are accompanied by the Aungier Redlich-Kwong (ARK) EoS and $k-\varepsilon$ turbulence model. The spatial discretization is second order upwind. Both CFD and mathematical model presented in this paper, used the same geometry from figure 1b.

To obtain good comparison between models following constraints are satisfied:

- The same EoS is used in both the 3D CFD calculations and the proposed mathematical model
- To obtain the flow coefficient for different leakage types, results of CFD calculations were used for every angular position. The mass flow rates through the corresponding leakage path were correlated with the pressure ratio between chambers that are forming that leakage path and it's area
- Pulsations in the pipe before the inlet port are obtained from the results of CFD calculations and are used in the developed mathematical model

By analysing the results of CFD calculations, it was seen that pulsations in the pipe before the inlet port play an important role in the pressure difference which influences the mass flow during the filling. These pulsations can be captured with 3D CFD calculations or with a one dimensional model. Since the aim of this study was to develop the mathematical model of the working chamber of the screw expander, the conditions of the inlet port were obtained from the results of CFD calculations as shown in figure 3. The results of pressure pulsations were calculated through the reference plane as shown in figure 3.

5. Results

The results of the CFD analysis and the developed model have been compared for a few parameters. The first parameter is the performance of the twin screw expander. This is evaluated through the P-V indicator diagram for both models shown in figure 4a. It can be seen that the developed model predicts the pressure in the working chamber well. Using the P-V diagram, power has been calculated using eq. 5 and eq. 6. The difference in power prediction is around 2%.

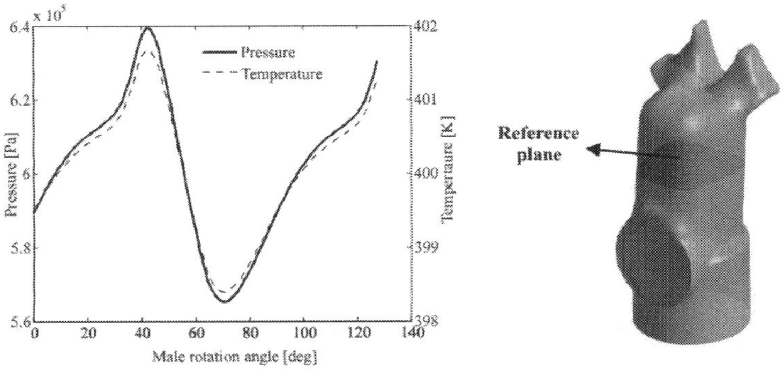

Figure 3: Pressure pulsations in the reference plane of the inlet port (results from CFD analysis)

The second parameter is the mass in the working chamber. The comparison between the models is shown in figure 4b. This parameter is very important because it shows if the mass in the chamber after the filling is approximately the same compared to the CFD results. But also, it shows if the leakage flows are reducing the mass in the chamber in the same way as in CFD results.

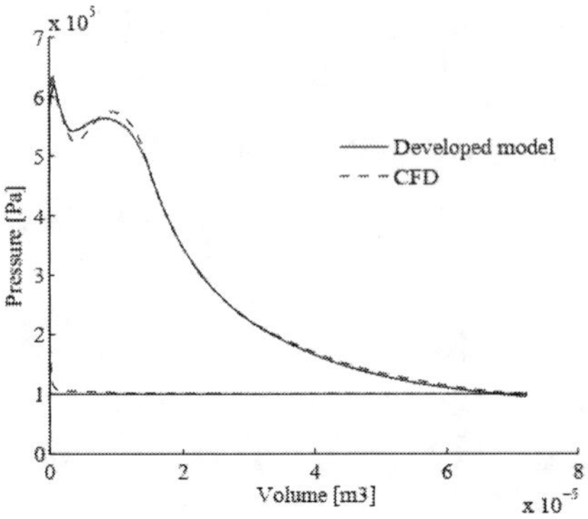

(a) P-V indicator diagram

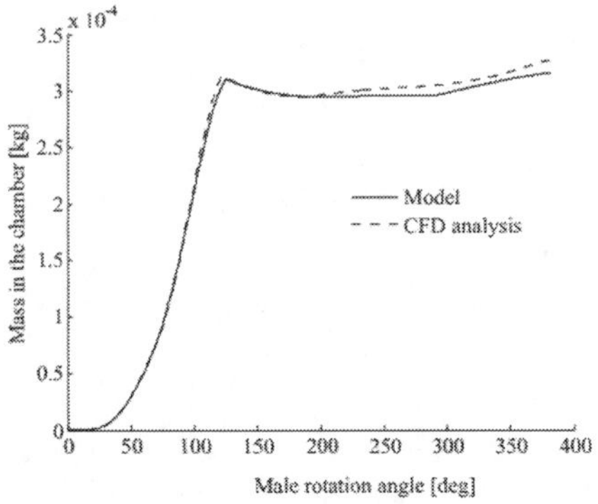

(b) Mass in the working chamber

Figure 4: Model predicted and CFD calculated results in the screw expander with pressure ratio 6 and rotational speed of 6000rpm

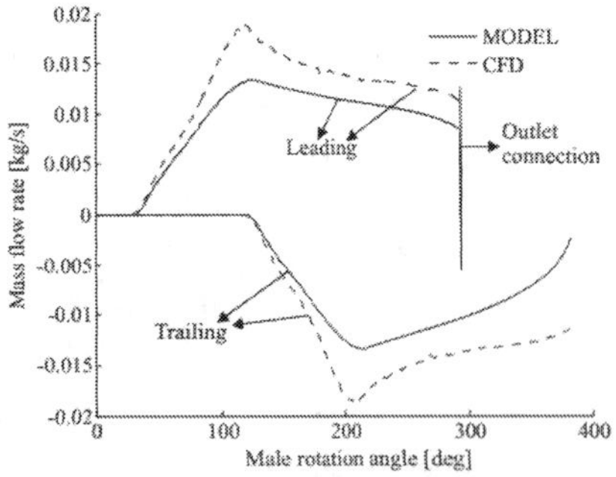

(a) Tip Leakages

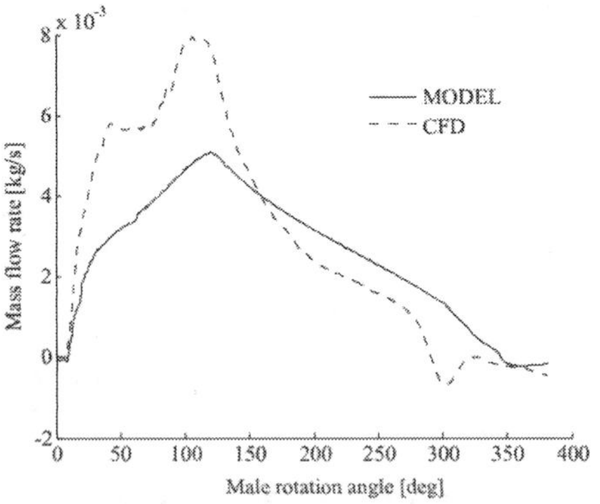

(b) Sealing leakage

Figure 5: Model predicted and CFD calculated results in the screw expander with pressure ratio 6 and rotational speed of 6000rpm

The last two parameters are the mass flow rates through the tip and sealing leakage path. Here it is very important to see if using the constant flow coefficient derived from the CFD calculations can estimate these flows correctly. It can be clearly seen that the trend of these curves is matching well with CFD results. It is important to note that the flow rates are small compared to the flow rate through the machine.

5.1. Evaluation of the model for different pressure ratio and rotational speed

Additionally, comparison has been made with a different pressure ratio and rotational speed as shown in figure 6a and figure 6b. The difference in power and mass flow rates between the CFD

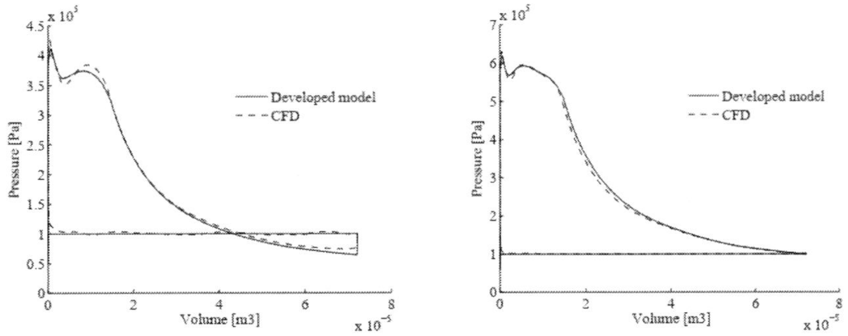

(a) Pressure ratio 4 and 6000 rpm (b) Pressure ratio 6 and 4000rpm

Pressure ratio	Speed	CFD		MODEL	
		Flow rate [kg/s]	Power [kW]	Flow rate [kg/s]	Power [kW]
6	6000	0.1469	4.72	0.1415	4.61
4	6000	0.0955	2.18	0.0922	2.05
6	4000	0.1055	3.18	0.1028	3.23

Table 1: Results for mass flow rates and power outputs for the developed model and the CFD analysis

and the developed model are presented in Table 1. The biggest difference is for the rotational speed of 4000rpm which is to be expected since the influence of the leakages is higher at lower speeds. It is important to note that coefficients of the model have not been adjusted for these changes in pressure ratio and speed.

6. Conclusion

A mathematical model for calculating the performance of a twin screw expander has been developed. As the input for the developed mathematical model, geometrical parameters such as volume, inlet and leakage areas are provided to the model. Additionally, the pressure and the temperature in the inlet port as a function of rotational angle should be provided to the model. These data can be obtained from the CFD calculations or a 1D model. The model shows good agreement compared to the CFD results. Further improvement will be done in the inlet model which in this moment shows the biggest difference compared to the CFD results.

7. Acknowledgement

This study is part of the ORCNext project with financial support of IWT Flanders.

Nomenclature

Symbols

$\dot{m}$	Mass flow rate (kg/s)
$\dot{Q}$	Heat transfer rate (kW)
ρ	Density (kg/m^3)
A	Area of the leakage path/inlet area (m^2)
C	Flow coefficient $(-)$
E	Energy (kJ/kg)
h	Specific enthalpy (kJ/kg)
k	Specific heat ratio$(-)$
m	Mass (kg)
n	Rotational speed (rpm)
P	Power (W)
T	Temperature (K)
V	Volume (m^3)
W	Work (J)
z	Number of lobes $(-)$

Subscripts

ch	Chamber
$down$	Downstream
i	Number of boundaries of the working chamber
in	Indicated
up	Upstream

Acronyms

CFD	Computational Fluid Dynamics
EoS	Equation of State
ORC	Organic Rankine Cycle
RKA	Redlich-Kwong Aungier

References

[1] Lemort V, Guillaume L, Legros A, Declaye S and Quoilin S 2013 A comparison of piston, screw and scroll expanders for small scale Rankine cycle systems *Proc. 3rd Int. Conf. on Microgeneration and Related Technologies.*

[2] Stosic N, Milutinovic Lj, Hanjalic K and Kovacevic A 1992 Investigation of the Influence of Oil Injection upon the Screw Compressor Working Process *Int. J. Refrig.* **15(4)** 206.

[3] Fleming J S, Tang Y and Cook G 1998 The Twin Helical Screw Compressor Part 2, A Mathematical Model of the Working process *J. Mech. Engineering Science* **212** 369.

[4] Fujiwara M and Osada Y 1995 Performance Analysis of Oil Injected Screw Compressors and their Application *Int. J. Refrig* **18** 4.

[5] Hanjalic K and Stosic N 1997 Development and Optimization of Screw Machines with a Simulation Model Part II Thermodynamic Performance Simulation and Design *ASME Transactions J. Fluids Engineering* **119** 664.

[6] Margolis D L 1978 Analytical Modelling of Helical Screw Turbines for Performance Prediction *ASME J. Engr. for Power* **100** 482.

[7] Wang Z, Zhang Y, Sun T and Wei L 2010 Numerical simulation and experimental study on the performance of screw expander *ASHRAE Transactions* part 2 **116** 218-225.

[8] Papes I, Degroote J and Vierendeels J. 2014 Analysis of a Twin Screw Expander for ORC Systems using Computational Fluid Dynamics with a Real Gas Model *Int. Compressor Engineering Conf. at Purdue University.*

[9] Chen Y, Halm N, Groll E and Braun J 2002. Mathematical Modeling of Scroll Compressor, Part I- Compression Process Modeling *Int. J. Refrigeration* **25** 731-750.

[10] Bell I H, Groll E A, Braun J E and Horton W T 2013 A computationally efficient hybrid leakage model for positive displacement compressors and expanders *Int. J. Refrigeration* **36** 1965–1973.

[11] Kang D J, Kim J W and Sohn C B 2002. Effects of leakage flow model on the thermodynamic performance of a scroll compressor *Int. Compressor Engineering Conf. at Purdue University.*

[12] Papes I, Degroote J and Vierendeels J. 2013 3D CFD Analysis of an Oil Injected Twin Screw Expander, *ASME International Mechanical Engineering Congress and Exposition.*

[13] Aungier R H 1995 A fast, accurate real gas equation of state for fluid dynamic analysis applications *J. of Fluids Engineering* **117** 277-281.

[14] Bell I 2011 Theoretical and Experimental Analysis of Liquid Flooded Compression in Scroll Compressors *PhD Thesis.*

IOP Conf. Series: Materials Science and Engineering **90** (2015) 012060 doi:10.1088/1757-899X/90/1/012060

Influence of liquid in clearances on the operational behaviour of twin screw expanders

Melanie Gräßer and Andreas Brümmer

TU Dortmund University, Chair of Fluidics, Leonhard-Euler-Straße 5, 44227 Dortmund, Germany

E-mail: `melanie.graesser@tu-dortmund.de`

Abstract. A lot of effort has been expended on understanding the influences of an injected auxiliary liquid on a twin screw expander's performance. Sealed clearances improve performance on the one hand, but involve considerable frictional losses on the other hand. This paper contributes to an evaluation of these opposing effects with regard to the efficiency of screw expanders. First, thermodynamic analyses using the multi-chamber model-based simulation tool *KaSim*, developed at the Chair of Fluidics, are presented for a test screw expander in order to show the maximum potential of clearance sealing. This analysis involves thermodynamic simulations for sealed and unsealed clearances and leads to an order of priorities for different clearance types. Second, hydraulic losses within front and housing clearances are calculated, applying an analytical model of incompressible one-phase clearance flow. Subsequently dry and wet screw expanders are evaluated while both clearance sealing and frictional losses are considered for the simulation of a liquid-injected machine.

1. Introduction

Rotary positive displacement machines are designed to be either dry or wet running. While dry running twin screw machines can be calculated well by chamber model based simulations, no general model for liquid injected machines, including all significant auxiliary liquid effects, exists. Consequently simulating these machine types to produce a satisfactory match with experimental results is not possible at present. Initial investigations in understanding the operational performance of wet running screw machines are presented by Kauder [1], who deduced the "Oil-Surge-Theory" by experiments with an oil injected screw compressor. According to this approach, the injected oil leads, besides other effects, to sealed clearances but also to higher frictional losses. The ratio of these two effects mainly depends on the volume of oil. Kauder and Deipenwisch [2] point out that oil has a considerable influence within the working chambers of screw compressors: cooling the working fluid, sealing clearances, rotor lubrication and noise reduction. However, there are also dissipative processes such as friction and liquid acceleration. An analytical model of dissipation is developed, including oil acceleration during injection and within clearances, as well as frictional losses due to clearance flows. The results show clearance priority, and indicate that more than 90% of power losses within the screw compressor investigated result from front and housing clearances. An initial model for oil distribution covering droplet trajectory within working chambers, film flow on rotor lobes and clearance flows is introduced by Harling [3]. Harling succeeded in verifying his model partially by visual experiments; however, also unconsidered issues such as oil surge or gas bubbles within the liquid

Content from this work may be used under the terms of the Creative Commons Attribution 3.0 licence. Any further distribution of this work must maintain attribution to the author(s) and the title of the work, journal citation and DOI.
Published under licence by IOP Publishing Ltd

flow are apparent. Oil distribution within working chambers is extremely inhomogeneous, as confirmed by Rinder and Moser [4] who took samples of an oil-gas-mixture to determine the amount of oil. They relate irregular spreading to insufficient distribution of the injected oil jet. Previously described research work deals with screw compressors exclusively and only a few research papers focus on screw expanders. Zellermann [5] provides a first method for the optimal design of liquid-injected machines. His theoretical approach for calculating oil distribution within working chambers is similar to Harling's [3]. Using analytical relations, droplet trajectory, film flow and clearance flow are calculated based on oil injection. Experimental approaches also show inhomogeneous oil distribution. Moreover, the considerable influence of liquid injection is demonstrated by varying injection location, auxiliary liquid (oil and water), auxiliary liquid volume flow and operating parameters, such as rotational speed and pressure ratio. Further investigations presented by Nikolov [6], demonstrate measured deviations of effective isentropic efficiency in dry and wet running screw expanders. In these machines, isentropic change of state in unsaturated humid air is shown as a function of circumferential speed. At a low circumferential speed, the efficiency of a wet running machine rises due to the sealing of clearances; however, with rising circumferential speed frictional losses within clearances become predominant and therefore efficiency decreases. In addition, effective isentropic efficiency is analysed, based on an isentropic change of state of a humid air-water mixture. The advantages of water injection at low circumferential speeds are not very significant. For a small water volume flow, water injected machines are even less efficient than dry running expanders. Differences in isentropic efficiency compared with unsaturated humid air result from continual interaction between water and air during expansion. The research already illustrates the main effects of liquid within screw machines. In order to provide further understanding of liquid-injected machines, this paper focuses on comparing the potential of clearance-sealing with frictional losses in an oil-injected screw expander.

2. Test screw expander

In this paper a liquid injected, unsynchronised test screw expander is simulated [1]. The expander has an inner volume ratio of 3.5 and is operated at rotational speeds up to $16000\,\mathrm{min}^{-1}$.

[1] The tests were performed at an external industry test rig. Due to privacy restrictions no exact absolute values are specified.

Table 1. geometrical data and fluids of investigated screw expander

	male	female
number of lobes	5	7
wrap angle	300°	214°
length / diameter ratio	1.6	1.74
inner volume ratio	3.5	
distance between shafts	$\approx 60\,\mathrm{mm}$	
normalized housing clearance height	3	4.67
normalized profile clearance height (mean height)	1.67	
normalized front clearance height (high pressure side)	1	
normalized front clearance height (low pressure side)	5	
working fluid	ethanol	
injected auxiliary liquid	oil	

Geometrical data is given in **Table 1**. The distance between the shafts is approximately 60 mm. All clearance heights are declared in relation to minimum clearance height, which arises at high pressure front clearance. In this paper clearance height is defined by two solid boundaries (rotors or housing) whereas clearance length is given by direction of flow. Clearance width indicates the third dimension which is normal to flow direction. **Figure 1** illustrates the different clearance types. The expander is investigated with organic fluid ethanol which is superheated during the whole process for all operating points. For simulation the injected auxiliary liquid (oil) is assumed to have a constant temperature of 95° C. The fluid parameters used to calculate clearance losses are dynamic viscosity $\mu = 0.0265\,\mathrm{Pa{\cdot}s}$ and density $\rho = 791\,\mathrm{kg{\cdot}m}^{-3}$. All simulations are carried out for the same operation point. The inlet pressure of the expander is set to $15 \cdot 10^5$ Pa and the working fluid is superheated by 30° C.

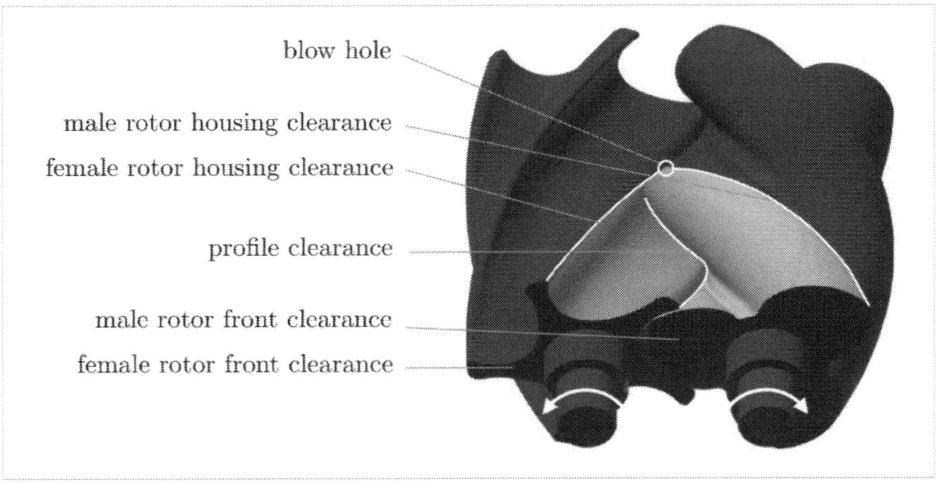

Figure 1. clearance types

3. Potential of clearance sealing regarding the expander's efficiency

In this section the influence of different clearance types on the operational behaviour of the test screw expander is investigated. In order to determine internal isentropic efficiency

$$\eta_{\mathrm{i,s}} = \frac{w_{\mathrm{i}}}{w_{\mathrm{s}}} = \frac{P_{\mathrm{i}}}{w_{\mathrm{s}}\dot{m}} = \frac{P_{\mathrm{i}}}{(h_{\mathrm{I}} - h_{\mathrm{O,s}})\dot{m}} \tag{1}$$

mass flow $\dot{m}$ and internal power P_{i} must be calculated taking all internal loss mechanisms such as inlet throttling, non-isentropic change of state and clearance mass flow of the working fluid into account. These parameters are computed with the multi-chamber simulation program *KaSim* using the main characteristics of positive displacement machines, which are cyclically changing working chambers [7]. Homogeneous fluid states within each chamber are determined for discrete points in time. The extensive state variables, i.e mass, internal energy and the specific volume of the working fluid are calculated based on the chamber volume curve and conservation of mass and energy. Subsequently, intensive parameters, such as pressure and temperature, are determined according to the defined thermodynamic state. In order to compute these state properties with maximum accuracy, *KaSim* accesses the *NIST Reference Fluid Thermodynamic and Transport Properties Database (REFPROP)*, providing equations of state for various fluids [8]. Equations used for ethanol are given in [9]. The injection of an auxiliary liquid is not simulated explicitly;

however, the effect of clearance sealing can be included in this modelling. The assumed changes of state for the reference cycle under consideration are isobar filling, isentropic expansion and isobar discharge. Thus, the specific efficiency of this ideal process w_s is determined directly by the difference between inlet (I) and isentropic outlet (O,s) enthalpy.

Simulated differences in internal, isentropic efficiency $\Delta\eta_{i,s}$ between a dry running expander with unsealed clearances and completely sealed clearances (overall) are shown in **Figure 2**. Furthermore, different clearance types are closed separately in order to show their individual effect on operational behaviour (housing clearance, blow holes, profile clearance, front clearance (high and low pressure side)). The values presented demonstrate an increase in internal isentropic efficiency due to clearance sealing. They are therefore interpreted as representing the maximum potential. Apparently this potential is largest for small rotational speeds. If all clearances are closed, internal isentropic efficiency increases by 40 % at $U_{mr} = 10\,m{\cdot}s^{-1}$ which is remarkable. Decreasing potential at higher rotational speed can be traced back to the declining influence of clearances. At high rotational speed working cycles are considerably shorter, leading to less mass transportation through clearances. Due to this effect the internal isentropic efficiency of machines with unsealed clearances converges with that for machines with sealed clearances and therefore potentially decreases; however, for the investigated screw expander, even at a high circumferential speed of $90\,m{\cdot}s^{-1}$ a significant potential of 10 % exists.

As far as clearance priority is concerned, the influences of housing and profile clearances are significant, whereas front clearances (high and low pressure side) and blow holes have a negligible effect on internal isentropic efficiency. The almost non-existent impact of high pressure front clearances is explained by the relatively small clearance height and the resulting small cross-section area in the direction of flow. At the male rotor, the high pressure front clearance cross-section area is only 4.5% of the mean cross-section area of the housing clearance. This approach applies also for the small blow hole cross-section areas of this expander. At the low pressure side, front clearances generally show small mass flows due to almost vanishing pressure differences between clearance inlet and outlet. Thus neither blow holes nor high and low pressure side front clearance sealing lead to significant changes in the operational behaviour of the investigated expander, as even for unsealed clearances mass flows are negligible. In contrast, high pressure differences combined with large cross-section areas lead to significant influences of housing and profile clearances.

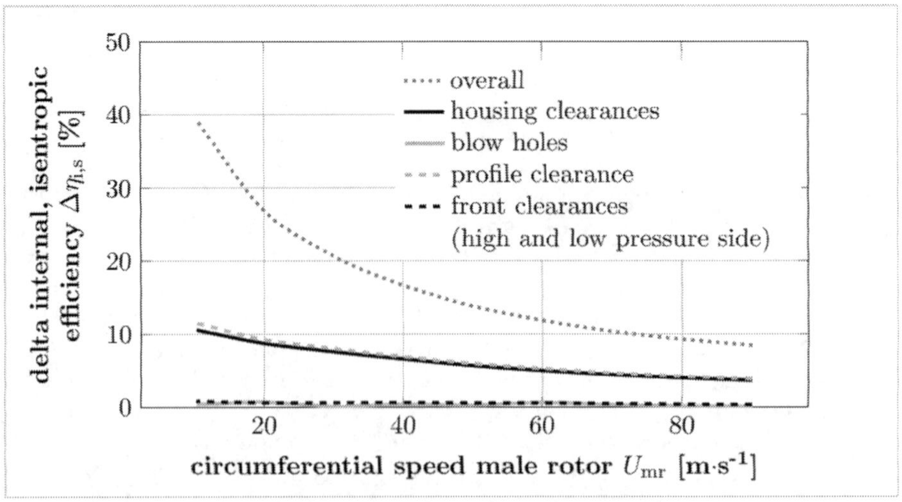

Figure 2. potential of clearance sealing

Note that the delta internal isentropic efficiencies $\Delta\eta_{i,s}$ of each clearance type cannot be added together for an overall result, as there are interdependencies. Sealing a single clearance type affects chamber pressures, which leads to different mass flows in all other clearances. Therefore, linear combination is not possible and each clearance configuration has to be simulated separately. Furthermore, the potential for clearance sealing might change in the screw expander under investigation, as shown, for example, by Dreißig [10], who points out significant influences of high pressure front clearances and blow holes. Nevertheless, two main tendencies which are a major influence on profile and housing clearance and a minor influence on the other two types concerning the efficiency of screw expanders correspond to Figure 2. Generally speaking, the major potential of clearance sealing is demonstrated; however, liquid injection in order to take advantage of this potential always involves frictional losses which will be analysed for high pressure front and housing clearances in the next section.

4. Frictional losses within front and housing clearances of liquid injected screw expanders

In this section frictional losses within front and housing clearances are calculated based on the analytical model presented in [11]. Assuming completely sealed clearances, a steady one-phase flow of auxiliary liquid is calculated based on the superposition of Couette and Poiseuille flow. In both cases either laminar or turbulent flow type is considered, depending on boundary conditions. Front clearance geometry is modelled by finite elements which are connected in parallel and do not interact with each other. Calculation is done separately for each element and the solution is deduced explicitly. In contrast, finite elements of housing clearance are arranged in series leading to additional boundary conditions between these elements. Thus, the resulting equations must be solved iteratively. Moreover, flow direction has to be taken into account due to the clearance geometry corresponding to nozzle and diffuser respectively, **Figure 3**. During simulation, calculations are performed for both flow directions; however, if the flow is defined, only one valid solution exists. If more or even no solutions exist, the flow is undefined and cannot be obtained by this model. This usually occurs with an almost vanishing mass flow. Therefore, mass flow is set to zero for all undefined states of flow, whereas hydraulic power is interpolated by limiting valid solutions. Furthermore it can be shown that the large opening angle of the diffuser geometry leads to flow separation. Therefore, the clearance outlet will always correspond with the minimum clearance height. For more details see [11]. Chamber pressure is determined for each time step by *KaSim*. Thus, pressure differences attached to all clearances are given, assuming chamber pressure at clearance in- and outlet. Using these boundary conditions, quasi-steady flow is calculated for each time step. Before analysing integral hydraulic losses depending on the expander's circumferential speed, hydraulic power and clearance mass flow during the working cycle are analysed as an example for male rotor housing clearance.

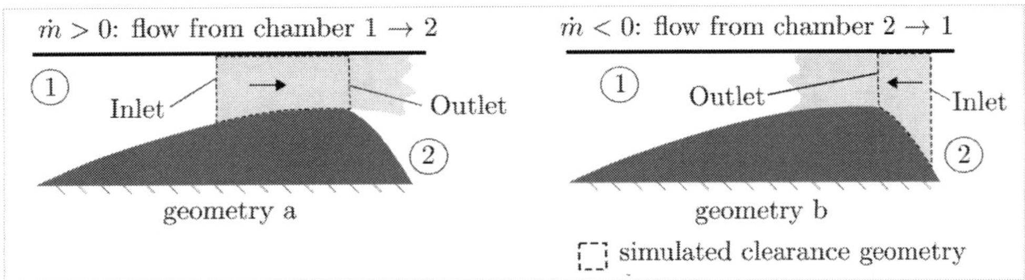

Figure 3. simulated geometry at male rotor housing clearance according to [11]

Figure 4 presents parameters of male rotor housing clearance as functions of normalized phase angle ϕ, defined as

$$\phi = \frac{\alpha_{\mathrm{mr}} z_{\mathrm{mr}}}{360°} \tag{2}$$

with rotational angle α and lobe number z. Thus, the thermodynamic state of the expander is equal at the start and end of each phase. All parameters are normalized to their maximum absolute values which are

$$\Delta\bar{p} = \frac{\Delta p}{\Delta p_{\mathrm{max}}},\ \bar{w}_{\mathrm{c}} = \frac{w_{\mathrm{c}}}{w_{\mathrm{c,max}}},\ \bar{\dot{m}}_{\mathrm{c}} = \frac{\dot{m}_{\mathrm{c}}}{\dot{m}_{\mathrm{c,max}}} \text{ and } \bar{P}_{\mathrm{c}} = \frac{P_{\mathrm{c}}}{P_{\mathrm{c,max}}}. \tag{3}$$

The upper diagram of Figure 4 shows normalized pressure difference $\Delta\bar{p}(\phi)$ at housing clearance, which is calculated from the pressure difference between two consecutive chambers. Furthermore, normalized housing clearance width $\bar{w}_{\mathrm{c}}(\phi)$ is given, resulting from the screw expander's geometry. Both parameters are used to compute clearance mass flow $\bar{\dot{m}}_{\mathrm{c}}$ and

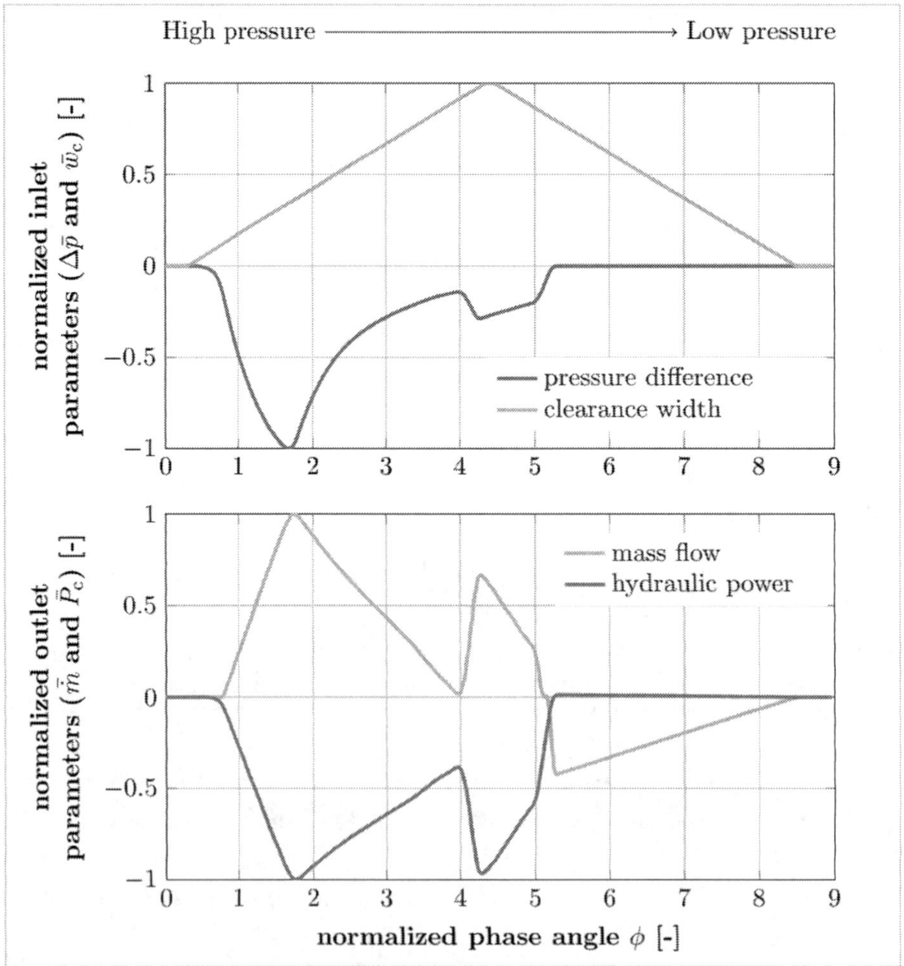

Figure 4. normalized housing clearance parameters during working cycle of screw expander

hydraulic power $\bar{P}_c$, as indicated in the lower diagram. As presented in [11], negative pressure differences correspond to lower pressures in the subsequent chamber (chamber 2 in Figure 3). As there are no positive pressure differences shown in Figure 4, overexpansion, that is pressure at the end of expansion below low pressure, does not occur at the investigated operating point. During charge, the working chamber under consideration (chamber 1 in Figure 3) increases, while the subsequent chamber is already expanding. Thus pressure difference at the housing clearance increases first ($0.4 < \phi < 1.7$) and decreases with the expansion of the working chamber ($1.7 < \phi < 4.0$). Underexpansion of the working fluid leads to a sudden pressure drop within the subsequent chamber at $\phi = 4.0$. Thus the pressure difference between the working chamber and the subsequent chamber shows an abrupt rise. Finally, pressure in both chambers corresponds to the outlet pressure ($\phi > 5.3$) resulting in vanishing pressure difference at the housing clearance during discharge. Clearance width shows a linear increase up to maximum value, and then decreases.

In the following, resulting clearance mass flow and hydraulic power are analysed. As shown in Figure 3 positive mass flow corresponds to mass transported from considered (1) to subsequent (2) chamber whereas a negative sign indicates the opposite direction. First ($0.4 < \phi < 0.8$) no mass flow arises due to an undefined state of flow that is explained above. For this set up (geometry, fluid and rotational speed) normalized pressure differences between -0.05 and -0.14 lead to clearance flow which cannot be calculated explicitly with the described analytical model. Subsequently, mass flow increases due to rising absolute values of pressure difference and clearance width ($0.8 < \phi < 1.7$). With decreasing pressure difference, mass flow drops although clearance width is still increasing ($1.7 < \phi < 4.0$). Increases and decreases at ($4.0 < \phi < 5.1$) result from changing pressure difference. Again, clearance width also has an impact on gradients. At $5.1 < \phi < 5.2$ the flow is undefined, again. For vanishing pressure difference negative mass flow results ($5.2 < \phi$). At a constant pressure difference the transported mass decreases linearly due to decrease in clearance width.

Hydraulic power can be either negative or positive. Frictional loss is indicated by a positive sign, while negative power corresponds to driving power. If driving power is transferred to the rotors, Poiseuille flow dominates over Couette flow, which always causes frictional loss. These simulations are computed at a moderate rotational speed, leading to low Couette flow. Therefore normalized pressure differences below -0.007 result in additional power. Increase and decrease in hydraulic power ($0.4 < \phi < 5.1$) can be traced back to pressure difference and clearance width, analogous to mass flow. For undefined states of flow (($0.4 < \phi < 0.8$) and ($5.1 < \phi < 5.2$)) indirect calculation by linear interpolation seems to be reasonable. For pressure differences above -0.02, that occur at $\phi > 5.2$, infinitesimal frictional loss arises which decreases linearly due to declining clearance width. Local maximums at $\phi = 1.7$ and $\phi = 4.3$ are nearly equal although pressure differences vary by 75%. This points out a dramatic influence of clearance width on hydraulic power. These results indicate that incompressible clearance flow might lead to additional power averaged over the working cycle. The behaviour of mass flow and power within front clearances is not analysed during the working cycle as their clearance geometry is kept constant and pressure difference is the only changing parameter. Furthermore, undefined states of flow do not exist for front clearances. Therefore qualitative functions of mass flow and power always follow the change in pressure difference.

In order to determine the averaged hydraulic power, incompressible flow is simulated for all high pressure front and housing clearances during one phase. Simulation of front clearances is only possible if just two chambers are connected. **Figure 5** shows a sample male rotor front clearance connecting chambers a, b and c. At present, calculation of this flow situation is not straightforward with the analytical model. Furthermore, inlet area has to be considered as front clearances exist only partially at this point. Thus, hydraulic loss of high pressure front clearances is calculated for the rotational angles α_{mr} and α_{fr}, respectively. Calculations start

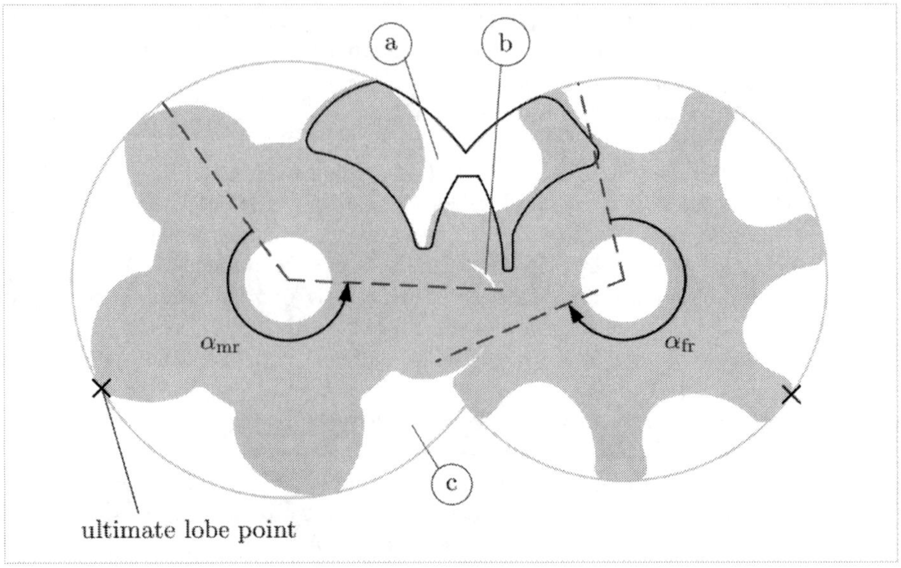

Figure 5. rotational angle at the sample rotor profile and the inlet area for calculation of hydraulic power at high pressure front clearance

when the ultimate lobe point passes the first dashed line and ends at the second line. Thus rotational angles around the inlet area are neglected.

Normalized hydraulic power of the housing and high pressure front clearances is presented in **Figure 6** as a function of the expander's rotational speed. The power of both clearance types is normalized, with maximum power occurring at high pressure front clearances in order to illustrate proportions. Furthermore, overall power is split between the male and female rotor. Simulations with *KaSim* are carried out assuming the clearance type to be sealed, whereas all other clearances are unsealed. First of all, a different indication of power at the high pressure front and housing clearance is apparent. While the frictional effects of oil cause significant losses within front clearance, flow within the housing clearance leads to additional power. With rising circumferential speed, driving power at the housing clearance actually increases, which is unexpected at first due to greater Couette flow; however, Figure 2 already indicates significant influence of circumferential speed on the performance of this screw expander. As only the housing clearance is sealed, rising rotational speed affects transported mass in all other clearances, which is explained in section 3. Therefore chamber pressures are influenced and pressure difference at the housing clearance changes, which affects Poiseuille flow. Rotors are driven more effectively as long as Couette flow is dominated by opposed Poiseuille flow. Thus the impact of changing pressure difference exceeds the increase in Couette flow. However, additional power arising from flow in housing clearances is negligible compared to hydraulic losses within high pressure front clearances. Obviously, even for small circumferential speeds, Couette flow is not dominated by Poiseuille flow as pressure differences are too small. This effect increases with the circumferential speed of the expander. In particular, high frictional losses within the male rotor front clearance are remarkable. They can be traced back to a large transfer area in the direction of flow. At the male rotor this area is approximately double in size compared with the female rotor.

Generally speaking, the analysis of frictional losses within the front and housing clearances confirms the predominant influence of front clearances. At the operation points investigated, housing clearance flow leads to additional power that is, however, negligible compared to front

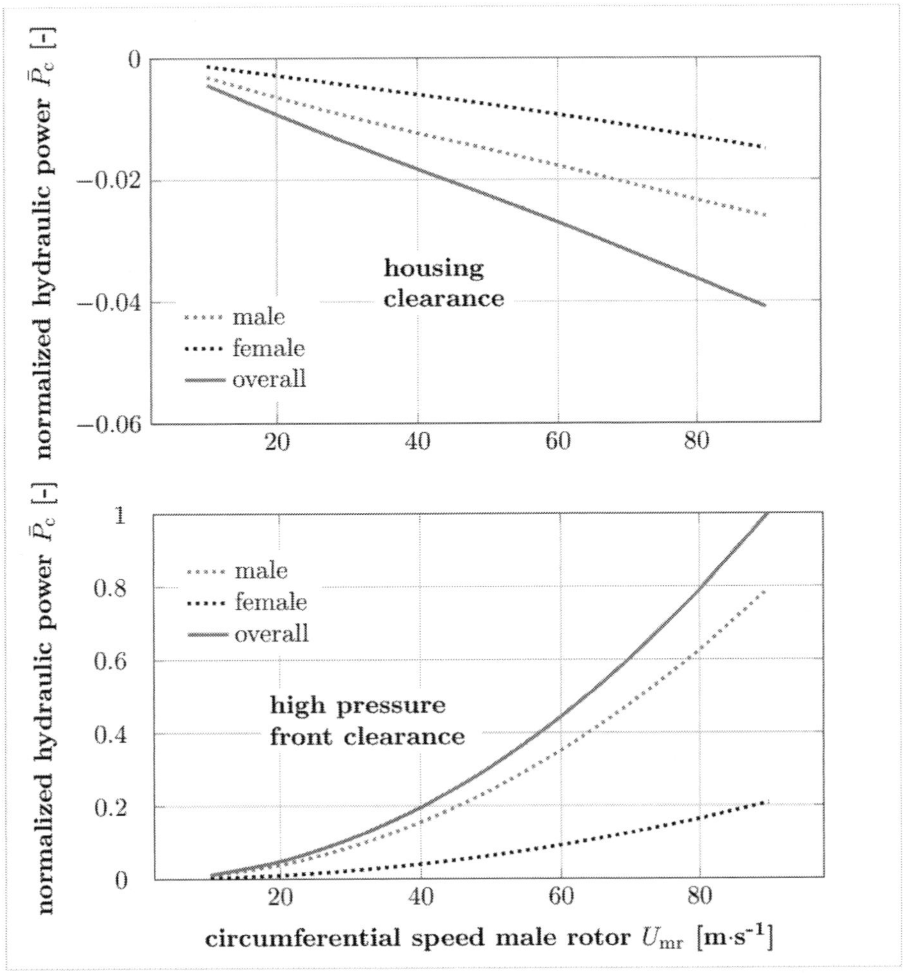

Figure 6. normalizd hydraulic power within housing and high pressure front clearances

clearances. Section 3 has already illustrated the thermodynamic potential of clearance sealing while not considering losses. In order to evaluate the influence of liquid on the operational behaviour of screw expanders within clearances, these two influences have to be contrasted, as outlined in the next section.

5. Comparison of potential by clearance sealing and frictional losses

In previous sections, the potential of clearance sealing and hydraulic losses has been analysed. In order to investigate the overall effect of the mechanisms described, the expander is simulated for both, dry and wet running conditions. Internal isentropic efficiency for dry screw expanders is determined according to equation (1). Liquid injected machines are modelled with sealed high pressure front and housing clearances. Profile clearance and blow holes are kept unsealed as no incompressible flow model exists at present. Therefore, hydraulic isentropic efficiency is defined as

$$\eta_{\mathrm{hyd,s}} = \frac{|P_{\mathrm{i}}| - P_{\mathrm{c}}}{w_{\mathrm{s}}\dot{m}} \tag{4}$$

IOP Conf. Series: Materials Science and Engineering **90** (2015) 012060 doi:10.1088/1757-899X/90/1/012060

where P_i is the internal power of an expander with sealed clearances and P_c is the hydraulic power of these sealed clearances (high pressure front clearances and housing clearances).

Figure 7 shows the isentropic hydraulic efficiency for wet running expanders with two different clearance configurations. While scaling factor f = 1 corresponds to relative clearance heights according to Table 1, high pressure front and housing clearance are scaled by 0.5 for the second configuration. All other clearance types are kept constant. Apparently, both types of hydraulic efficiency presented in Figure 7 at first increase with circumferential speed. At low circumferential speeds hydraulic losses are small compared to the expander's internal power. Improved efficiency can be traced back to sealing effects within the profile clearance, which are described in section 3. With rising circumferential speeds these sealing effects decrease while the hydraulic power of high pressure front clearances increases. According to Figure 6 friction within housing clearances is negligible. Therefore, the gradients of both types of hydraulic efficiency decline and even reverse their signs. Due to higher frictional losses, the hydraulic efficiency of the expander with reduced clearance heights is smaller compared to the unscaled machine for all operating points. Furthermore, the positive gradient of the scaled machine is smaller and the maximum hydraulic efficiency occurs at a lower circumferential speed ($U_{\mathrm{mr}} = 50\,\mathrm{m{\cdot}s^{-1}}$).

Moreover, isentropic efficiency for dry running expanders is indicated in Figure 7 in order to compare both applications. For small circumferential speeds liquid injection leads to higher efficiency. On the one hand, the potential of sealed high pressure front and housing clearances is remarkable. Note that the potential of these combined clearance types in not shown in Figure 2 and cannot be calculated from a linear combination of the values presented. On the other hand, frictional losses are small. Thus improved thermodynamics due to reduced clearance mass flow dominate frictional losses. With increasing circumferential speed, friction in high pressure front clearances rises, which leads to smaller positive gradients. At the same time, the internal isentropic efficiency of a dry running machine rises due to the sealing effects of all clearance types, which were described in section 3. Therefore, at high circumferential speeds, clearance friction dominates thermodynamic advantages and liquid injection is unfavourable.

It is clear that especially the hydraulic efficiency of liquid injected machines depends substantially on clearance heights. With decreasing clearance heights, frictional losses

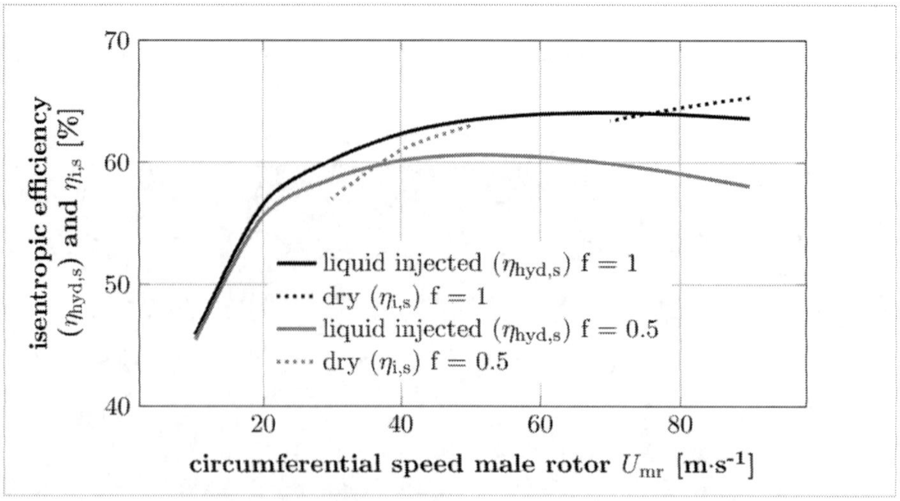

Figure 7. internal / hydraulic isentropic efficieny of dry / wet running screw expander for different high pressure front and housing clearance heights (scaled with factor f)

IOP Conf. Series: Materials Science and Engineering **90** (2015) 012060 doi:10.1088/1757-899X/90/1/012060

increase rapidly, leading to reduced hydraulic efficiency. The different intersection points at ($U_{mr} = 37\,m{\cdot}s^{-1}$) for f = 0.5 and ($U_{mr} = 76\,m{\cdot}s^{-1}$) for f = 1 point out the necessity for precisely defined hot clearance heights, as shown for example by Nikolov et al. [12].

6. Conclusion

This paper contributes to the understanding of liquid within twin screw expanders, focusing on the opposed effects of clearance sealing and frictional losses. First the maximum potential of clearance sealing is demonstrated. The predominant influence of housing and profile clearances on the thermodynamic behaviour of the investigated geometry is demonstrated. Especially at low circumferential speeds, the sealing of clearances leads to higher internal isentropic efficiency. Besides sealing effects, frictional effects within high pressure front and housing clearances were analysed. It was demonstrated that additional power is generated within housing clearances. However, this is negligible compared with losses within high pressure front clearances. Subsequently, dry and wet running screw expanders were analysed, taking both effects into account. At low circumferential speeds clearance sealing outweighs frictional losses, leading to higher hydraulic isentropic efficiency compared to the internal isentropic efficiency of a dry running expander. With increasing circumferential speed this order changes and dry running expanders are favourable in terms of efficiency. These results illustrate the influence of liquid injection in principle.

An analysis of the thermodynamic potential of clearance sealing shows significant effects of profile clearance. Thus a model for incompressible flow for this clearance type has to be developed in order to determine the intersection point of internal and hydraulic isentropic efficiency more precisely. Moreover, further effects, such as oil surges or cooling of the working fluid, have to be investigated. Nevertheless, two basic effects have been demonstrated. On the one hand, sealing the housing clearance involves remarkable potential for operational behaviour of investigated screw expander; however, friction within this clearance type is infinitesimal. On the other hand, high pressure front clearance shows considerable hydraulic losses while thermodynamic advantages are small. Both effects apply for a widen range of screw expanders. Certainly, the influence of different clearance types changes with varying geometry, for example clearance heights, as demonstrated in this paper; however, for standard applications basic relations between flow cross-section areas and frictional areas in flow direction do not change.

Notation

symbol	dimensions	meaning
α	[°]	rotational angle
Δ	[-]	difference
h	[N·m·kg^{-1}]	specific enthalpy
η	[-]	efficiency
$\dot{m}$	[kg·s^{-1}]	mass flow
μ	[Pa·s]	kinematic viscosity
p	[N·m^{-2}]	pressure
P	[N·m·s^{-1}]	power
ρ	[kg·m^{-3}]	density
U	[m·s^{-1}]	circumferential speed
ϕ	[-]	normalized phase angle
w	[N·m·kg^{-1}]	specific work
w_c	[m]	clearance width

index	meaning
c	clearance
hyd	hydraulic
i	internal
I	inlet
max	maximum
mr	male rotor
O	outlet
s	isentropic
$\bar{}$	normalized

Acknowledgments
Supported by: Federal Ministry for Economic Affairs and Energy on the basis of a decision by the German Bundestag

References

[1] Kauder K 1979 Energiewandlung in öleingespritzten Schraubenverdichtern *Technische Mitteilungen* **72(6)** 404-10

[2] Kauder K and Deipenwisch R 1999 Oil as a design parameter in screw-type compressors: oil distribution and power losses caused by oil in the working chamber of a screw-type compressor *Int. Conf. on Compressors and their Systems* IMechE conference transactions **1999-6** 49-58

[3] Harling B 1993 *Untersuchung zur Ölverteilung in Schraubenkompressoren mit Schmiermitteleinspritzung* Dissertation University Dortmund

[4] Rinder L and Moser I 1990 Untersuchung der Ölverteilung in den Arbeitsräumen naßlaufender Schraubenkompressoren *Schraubenmaschinen.../'90* VDI-Bericht **859** 1-14

[5] Zellermann B 1996 *Optimierung von Schraubenmotoren mit Flüssigkeitseinspritzung* Dissertation University Dortmund

[6] Nikolov A and Brümmer A 2014 Influence of Water Injection on the Operation Behaviour of Screw Expanders *9th Int. Conf. on Screw Machines 2014* VDI Bericht **2228** 43-60

[7] Janicki M 2007 *Modellierung und Simulation von Rotationsverdrängermaschinen* Dissertation University Dortmund

[8] Lemmon E W, Huber M L and McLinden M O 2013 NIST Standard Reference Database 23: Reference Fluid Thermodynamic and Transport Properties-REFPROP, Version **9.1**, National Institute of Standards and Technology, Standard Reference Data Program

[9] Schroeder J A 2011 *A New Fundamental Equation of Ethanol* University of Idaho, Master's Thesis

[10] Dreißig B 1989 *Ein Beitrag zur Auslegung von trockenlaufenden Schraubenmotoren* Dissertation University Dortmund

[11] Gräßer M and Brümmer A 2014 An analytic model of the incompressible one-phase clearance flow in liquid injected screw expanders *9th Int. Conf. on Screw Machines 2014* VDI Bericht **2228** 71-89

[12] Nikolov A, Huck C and Brümmer A 2012 Influence of thermal deformation on the characteristic diagram of a screw expander in automotive application of exhaust heat recovery. *21st Int. compressor engineering conf. at Purdue* paper **1478**

IOP Conf. Series: Materials Science and Engineering **90** (2015) 012061 doi:10.1088/1757-899X/90/1/012061

Low-order models of a single-screw expander for organic Rankine cycle applications

D. Ziviani[1], **A. Desideri**[2], **V. Lemort**[2], **M. De Paepe**[1], **M. van den Broek**[1]

[1]Department of Flow Heat and Combustion Mechanics - Ghent University - UGent, Sint-Pietersnieuwstraat 41, 9000 Gent, Belgium

[2]Aerospace and Mechanical Engineering Department, University of Liège, Campus du Sart Tilman B49, B-4000 Liège, Belgium

E-mail: davide.ziviani@ugent.be, adesideri@ulg.ac.be, martijn.vandenbroek@ugent.be, michel.depaepe@ugent.be, vincent.lemort@ulg.ac.be

Abstract. Screw-type volumetric expanders have been demonstrated to be a suitable technology for organic Rankine cycle (ORC) systems because of higher overall effectiveness and good part-load behaviour over other positive displacement machines. An 11 kWe single-screw expander (SSE) adapted from an air compressor has been tested in an ORC test-rig operating with R245fa as working fluid. A total of 60 steady-steady points have been obtained at four different rotational speeds of the expander in the range between 2000 rpm and 3300 rpm. The maximum electrical power output and overall isentropic effectiveness measured were 7.3 kW and 51.9%, respectively. In this paper, a comparison between two low-order models is proposed in terms of accuracy of the predictions, the robustness of the model and the computational time. The first model is the Pacejka equation-based model and the second is a semi-empirical model derived from a well-known scroll expander model and modified to include the geometric aspects of a single screw expander. The models have been calibrated with the available steady-state measurement points by identifying the proper parameters.

1. Introduction

A combination of concern for the environment, current energy policies, and favorable economics continues to push energy research in two key areas: further development of alternative energy sources and better utilization of traditional energy sources. With reference to the latter, waste heat recovery is gaining attention and organic Rankine cycle (ORC) technology has been widely demonstrated to be viable to harness low-grade heat sources. The operating principle is based on exploiting the thermal energy from the heat source by means of an evaporator and producing electricity by reducing the high-enthalpy content working fluid through an expander coupled with an electric generator. Besides turbines [1] which often are too expensive even though they present high efficiency, researchers are focusing on finding suitable expansion devices for different applications and power ranges. Furthermore, especially in the low power range, dynamic expanders become less cost-effective. The production of useful work from an ORC relies on the expander behavior and especially at lower capacity range it is challenging to find a device with efficient performance. Volumetric type expanders have been suggested by several authors as a suitable

Content from this work may be used under the terms of the Creative Commons Attribution 3.0 licence. Any further distribution of this work must maintain attribution to the author(s) and the title of the work, journal citation and DOI.
Published under licence by IOP Publishing Ltd

Nomenclature

A	Area (m^2)
AU	Overall heat transfer conductance (kW/K)
D	Diameter (m)
h	Specific enthalpy (kJ/kg)
i	Transmission Ratio (-)
L	Length (m)
m	Mass (kg)
$\dot{m}$	Mass flow rate (kg/s)
N	Rotational speed (-)
p	Pressure (kPa)
$\dot{Q}$	Heat rate (kW)
r_v	Volume ratio (-)
r_p	pressure ratio (-)
T	Temperature (K)
u	Uncertainty (-)
v	Specific volume (m^3/kg)
V	Volume (m^3)
$\dot{V}$	Volumetric flow rate (m^3/s)
w	Specific work (kJ/kg)
$\dot{W}$	Power (W)
z	Number of grooves or teeth (-)
ϕ_{FF}	Filling factor (-)
ε	Effectiveness (-)
η	Efficiency (-)

Subscript

amb	ambient
crit	critic
ex	exhaust
exp	expander
el	electric
d	discharge
eff	effective
g	groove
int	internal
is	isentropic
l	liquid
m	mechanical
meas	measured
nom	nominal
oa	over-all
pp	pump
su	supply
sh	shaft
sr	screw rotor
sw	starwheel
r	refrigerant
ref	reference
th	theoretical
v	volume

solution because of their higher expansion ratio and higher efficiency in the small scale and their tolerance to two-phase flow conditions. In many cases, the volumetric expanders are obtained by reversing and optimizing the corresponding compressors available on the market, e.g. scroll-type [2]. Among them, screw-type expanders are characterized by large volume ratios (from 2 to 5.3) and isentropic efficiency up to 85% [3]. In this paper, the performance of a single-screw expander is characterized by two low-order models based on experimental data obtained from an 11 kWe ORC running with R245fa. Insights on a deterministic model of the same machine are presented in a companion paper [4].

2. Single-screw expander and ORC test rig

In the present study, a single-screw expander (SSE) is considered. The SSE is based on the meshing between a grooved central rotor with two toothed gaterotors (or starwheels), as shown in Fig. 1(a). In particular, the rotor has 6 grooves and each of the starwheel disks have 11 teeth. The meshing isolates a flute on each side of the rotor leading to a symmetrical expansion process with respect the rotation axis of the rotor. As a consequence, the machine is characterized by balanced loading, low vibrations and long service life. The SSE tested, shown in Fig.1(a), is an adapted oil-injected air-compressors coupled with a 11 kWe asynchronous generator and the main geometric parameters are listed in Tab. 1. The SSE is installed in a small-scale industrial-based ORC. The layout of the system and the location of the sensors are illustrated in Fig.1(b). The experimental setup includes a 14-stage centrifugal pump with a maximum operating pressure of 14 bar and 2.2 kWe nominal power, three identical brazed plate heat exchangers, acting as evaporator, condenser and internal heat exchanger and a liquid receiver. Additional information regarding the installation can be found in [5]. The pump motor and

IOP Conf. Series: Materials Science and Engineering **90** (2015) 012061 doi:10.1088/1757-899X/90/1/012061

Table 1. Geometric parameters of single-screw expander.

i	[-]	11/6
D_{sr}	[mm]	122
$V_{g,max}$	[cm^3]	57.39
$r_{v,built-in}$	[-]	4.7
w	[m]	18.5
L_{rotor}	[m]	121

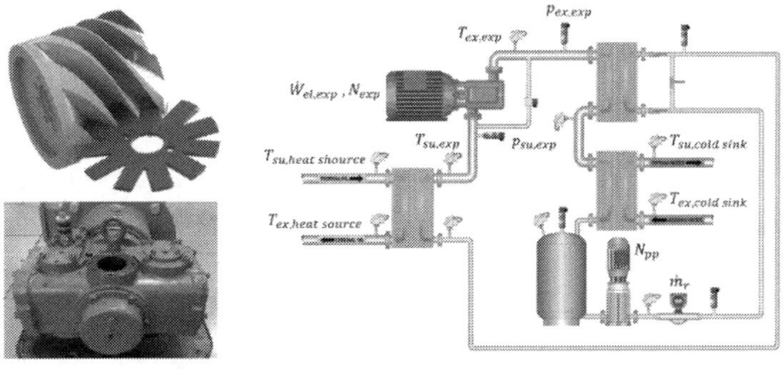

(a) (b)

Figure 1. (*a*) 3D CAD view of the meshing of the main rotor and the gaterotor of a single-screw expander (top); picture of the single-screw expander installed in the ORC system (bottom); (*b*) schematic of the ORC installation.

Table 2. Minimum and maximum values of the measured and derived variables.

	Measured independent variables			Measured dependent variables				Performance indicators	
	$p_{su,exp}$ (kPa)	$p_{ex,exp}$ (kPa)	$T_{su,exp}$ (°C)	ΔT_{sh} (K)	$\dot{m}$ (kg/s)	$\dot{W}_{el,exp}$ (W)	$T_{ex,exp}$ (°C)	$\varepsilon_{is,oa,exp}$ (-)	ϕ_{FF} (-)
Min	566	120	106.7	26	0.12	1283	75.5	0.2058	1.038
Max	1230	232	124.9	50	0.37	7364	104.6	0.5191	1.331

the expander electrical generator are controlled by two inverters. The working fluid selected is R245fa, which replaces the SES36 previously investigated by the authors [5]. The refrigerant is mixed with approximately 3% vol. of lubricant oil and a bleeding line after the pump ensures the lubrication of the expander. Thermodynamic calculations are performed in real time with the CoolProp wrapper for LabView [6].

3. Experimental data discussion

A total of 60 steady-state points have been determined from the experimental data. The maximum and minimum values of the measured and calculated variables are listed in Tab. 2. The tests have been performed by maintaining fixed the thermal oil supply temperature at 125°C as well as the cooling loop flow rate (water + 30% vol. glycol) at 4.166 kg/s. The mass flow rate of the working fluid has been controlled by setting the pump frequency between 25 Hz and 40 Hz. Four different rotational speeds of the expander have been considered, i.e. 2000 rpm, 2500 rpm, 3000 rpm, 3300 rpm. For all the collected measurements the fluid was entering the expander in vapor phase/ superheated condition. Due to the fact that the cold sink relies on the external conditions (air cooled loop), the condensing temperature and pressure of the ORC were not constant. This situation is shown in Fig. 2(a) where the power output is plotted as a function of the condensing pressure with a contour representing the condensing temperature. The tests have been carried out by considering two rotational speeds in each testing day and the measured points are clearly grouped. The external temperature was varying not only during the daytime, as normally occurs, but also different weather conditions have been encountered in different days. By analyzing Fig. 2(a), it is clear that the data are influenced by the air-cooled condenser. The test conducted at 2000 rpm and 3000 rpm will be considered as a reference, although the models will be validated by considering all the rotational speeds. In Fig. 2(b), the data shows the range of electric power out and pressure ratios reached during the testing. Generally, the performance of a positive-displacement expander can be expressed in terms of isentropic efficiency and volumetric efficiency. The latter is normally defined by the filling factor which can assume values greater than one for an expander. The adiabatic definition of the isentropic efficiency cannot be directly applied in the case of volumetric expander [2]. Mathematically, the filling factor is given by:

$$\phi_{FF} = \frac{\dot{m}_{meas} v_{su,exp}}{\dot{V}_{swept,th}} = \frac{\dot{m}_{meas} v_{su,exp}}{2\, z_{sr}\, V_{g,su}\, (N_{rot,exp}/60)} \tag{1}$$

where the volume of the groove at suction closure, $V_{g,su}$, is determined by the geometric model described in details in the companion paper [4]. Because of the absence of a torque meter installed at the expander shaft and temperature measurements on the expander shell, an overall isentropic effectiveness of the expander and generator is adopted by relating the measured electrical power output at the inverter and the isentropic expansion power, i.e.:

$$\varepsilon_{is,oa,meas} = \frac{\dot{W}_{el,meas}}{\dot{W}_{is,exp}} = \frac{\dot{W}_{el,meas}}{\dot{m}_{meas}\, (h_{su,exp} - h_{ex,is,exp})} \tag{2}$$

At the nominal speed of 3000 rpm, the maximum electrical power output and overall isentropic effectiveness achieved were 7364 W and 51.91%, respectively. Additional considerations can be done on the volumetric performance of the machine. In fact, in Fig. 2(c), the filling factor (1) at different rotational speeds is plotted as function of the expander inlet pressure. It becomes clear that the filling factor of this machine mainly depends on the rotational speed and it is less sensitive to the supply pressure variations. In particular, the higher the rotational speed the closer the filling factor reaches unity, meaning the leakages are reduced. Thus, since 3000 rpm is the nominal speed, the contour shows the maximum overall isentropic effectiveness around $\phi_{FF} \cong 1.05$. After the tests, an inspection has been conducted to assess the status of the expander. The tooth flank sealing profiles were worn out, which could have played a major role in the leakage. This fact may partially explain the value of the filling factor greater than one. Additional issues have been identified for the lubrication systems. In fact, most of the orifices that ensure the injection of the oil, were occluded. Most likely, the expander was running without proper lubricant injection for several working hours, affecting negatively the performance. The only oil entering the expander was the amount in solution with the refrigerant. This fact may explain large leakages at lower rotational speeds, i.e. higher values of the filling factor. Due to the variability of the test conditions, the overall isentropic effectiveness exhibits different trends by varying the rotational speed which it makes difficult to draw general conclusions on the performance of the machine, as shown in Fig. 2(d). However, it is possible to notice that the overall isentropic effectiveness increases with the increase of the pressure ratio until the internal limit of the SSE is reached, i.e. fixed built-in volume ratio. At higher pressure ratios, the power output still increases, but the overall isentropic effectiveness is penalized. The measurements for rotational speeds 2500 rpm and 3300 rpm do not provide any further information. These data points present a similar trend to 2000 rpm and 3000 rpm at least between pressure ratio, $r_{p,exp}$, 3 and 4. Additionally, by adopting the correlations obtained by [7] for the inverter electrical effectiveness and the generator efficiency, the expander isentropic efficiency at the shaft, $\eta_{is,sh}$, has been estimated and included in Fig. 2(d).

4. Low-order models

The characterization of a volumetric expander is based on a set of typical parameters such as pressure ratio, actual volume ratio and rotational speed as well as performance indexes, typically

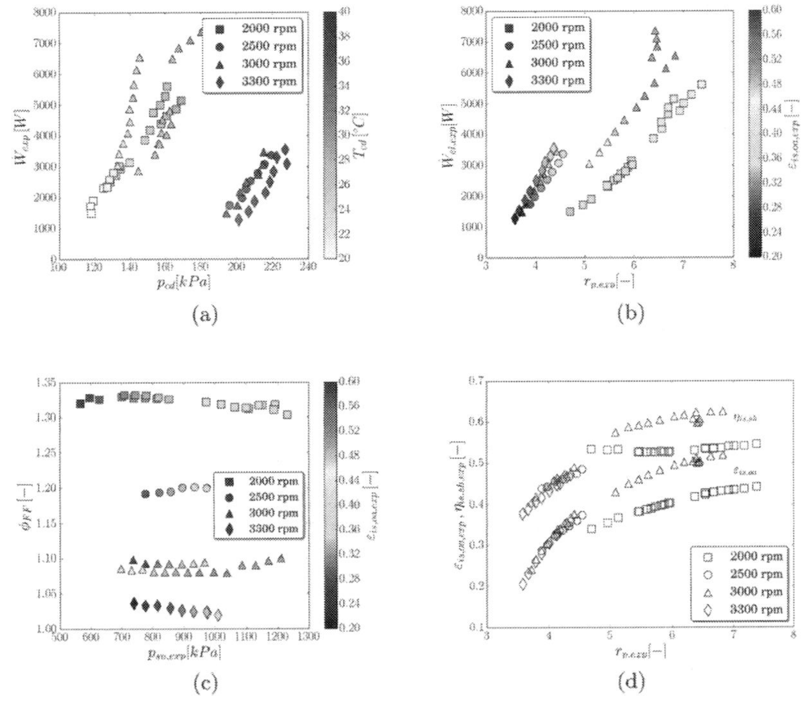

Figure 2. (a) Influence of the condensing temperature and pressure on the electrical power output due to the air condenser;(b) plot of the expander electrical power output as function or the pressure ratio at different rotational of the 60 steady-state points obtained. The contour is representative of the overall isentropic effectiveness ; (c) influence of the supply pressure of the expander on the filling factor; (d)Overall isentropic effectiveness, $\varepsilon_{is,oa}$, and isentropic efficiency, $\varepsilon_{is,sh}$, at the shaft as a function of pressure ratio at different rotational speed.

isentropic efficiency and filling factor. Empirical and semi-empirical models, usually referred as low-order models, are useful to describe the behavior of the expander with low computational effort with respect to more detailed models. Furthermore, these types of models can be easily integrated in cycle simulations to evaluate the impact of the expander behavior at full or part-load conditions. In this paper, two methodologies are proposed to describe the performance of the single screw expander. It should also be noted that the analyses and considerations within this paper are valid for R245fa. Desideri et al. [8] analyzed the same expander by using SES36 as working fluid.

4.1. Pressure ratio based equation

The inlet pressure, $p_{su,exp}$, the pressure ratio over the expander, r_p, and the rotational speed, $N_{rot,exp}$, are the three working condition variables selected to represent the expander behavior. In this paper, a similar nomenclature has been adopted. The aforementioned variables are expressed in terms of non-dimensional form by considering a set of reference conditions:

$$r_p^* = \frac{r_p - r_{p,ref}}{r_{p,ref}}\,;\,N_{exp}^* = \frac{N_{exp} - N_{exp,ref}}{N_{exp,ref}}\,;\,p^* = \frac{p_{su,exp} - p_{su,exp,ref}}{p_{su,exp,ref}} \tag{3}$$

IOP Conf. Series: Materials Science and Engineering **90** (2015) 012061 doi:10.1088/1757-899X/90/1/012061

where $r_{p,ref} = 6$, $N_{exp,ref} = 3000$ and $p_{su,exp,ref} = 10$. A generic non-dimensional expression similar to Pacejka's equation is used to calculate the overall isentropic effectiveness as a function of the previously defined variables and additional parameters. The complete mathematical formulation can be found in [2].

4.2. Semi-empirical model

The semi-empirical model adopted is based on the model described by Lemort et al. [9] which retains the main physical aspects of the expansion device. The entire expansion process from suction to discharge is decomposed into a number of consecutive steps: an adiabatic supply pressure drop, an isobaric supply cooling-down, an isentropic expansion to the pressure imposed by the built-in volume ratio of the expander, an adiabatic expansion at a constant machine volume, an adiabatic mixing between supply and leakage flows and an isobaric exhaust cooling-down or heating-up. The model estimates the discharge temperature, the mass flow rate, the overall isentropic effectiveness, the expander work and power losses. Due to the fact that the model was tailored mainly on scroll-type machines, an extension is proposed hereafter to include the particular features of a SSE. A graphical representation of the model is shown in Fig. 3. The expansion process in the SSE occurs simultaneously on both sides of the rotor and the two parts are not in communication unless unforeseen leakages occur. Therefore, the inlet flow (su) enters the expander and it is subject to a pressure loss ($su \rightarrow su_{in}$) due to the suction port, modeled with a converging-diverging nozzle characterized by a throat cross-section, $A_{c,su}$. The total mass flow rate entering the expander, $\dot{m}_r$, splits to reach the two suction ports, $\dot{m}_r = \dot{m}_r^{'} + \dot{m}_r^{''}$. Theoretically, the mass flow is equally distributed inside the expander. However, a slight pressure difference may occur due to the difference in shape and length of the internal channels in the housing of the expander. Two lumped pressure drops account for the internal dimension of the channels ($su_{in,1}^{'} \rightarrow su_{in,2}^{''}$). Heat losses during the supply process are considered by defining a fictitious isothermal wall representing the metallic housing of the expander with associated temperature T_{wall}. The cooling down due to temperature difference between working fluid and the wall is modeled by the ε-NTU method characterized by an overall heat transfer $AU_{su}^{'}$ [9] and the total heat loss $\dot{Q}_{su}$ is the sum of the two contributions $\dot{Q}_{su} = \dot{Q}_{su}^{'} + \dot{Q}_{su}^{''}$ In a single-screw machine 9 leakage paths can be distinguished [10]. In this model, the leakages are lumped together by considering an equivalent isentropic flow of a perfect gas through a simply convergent nozzle, where the throat pressure is limited by choked flow conditions. The nozzle is characterized by a fictitious area A_{leak} [9]. Since the leakage model depends on the current pressure and temperature conditions, each side of the rotor may have different values of the leakage cross-section area. The expansion process is divided in two steps accounting for an adiabatic and reversible expansion, $w_{int,s=const}^{'}$, followed by an adiabatic expansion at constant volume,$w_{int,v=const}^{'}$. The work associated with both steps can be expressed by:

$$\dot{W}_{int}^{'} = \dot{m}_{r,in}^{'} \left[h_{r,su2}^{'}\left(p_{r,su2}^{'}, T_{r,su2}^{'}\right) - h_{r,int}^{'}\left(s_{r,int}^{'}, v_{r,int}^{'}\right) + v_{r,int}^{'}\left(p_{r,int}^{'} - p_{r,ex2}^{'}\right)\right] \quad (4)$$

The expanding working fluid on both sides of the rotor contributes to generate mechanical work through the shaft, i.e. $\dot{W}_{int} = \dot{W}_{int}^{'} + \dot{W}_{int}^{''}$. Therefore, the mechanical losses due to friction between the starwheel teeth and the rotor as well as in the bearings can be lumped to a fictitious torque, τ_{loss}, and a term proportional to $\dot{W}_{int}$:

$$\dot{W}_{loss} = \dot{W}_{loss,0} + 2\,\pi\,\tau_{loss}\,\frac{N_{exp}}{60} \quad (5)$$

Once the expansion process is completed, the internal refrigerant flow is mixed with the leakage flow. An additional mixing occurs prior discharge between the flows from each side of the rotor.

The exhaust heat transfer, $\dot{Q}_{ex}$, and the outlet pressure drop are derived similarly as on the supply ones [9]. To be noted is that the inlet pressure drops have a key role in the case of an expander. Finally, an overall energy balance with the fictitious isothermal wall is minimized by the solver to guarantee convergence of the model:

$$\begin{aligned} \dot{W}_{loss} + \dot{Q}_{su} - \dot{Q}_{ex} - \dot{Q}_{amb} = 0 \\ \dot{Q}_{amb} = AU_{amb}\left(T_{wall} - T_{amb}\right) \end{aligned} \tag{6}$$

Given the supply conditions and the discharge pressure, a set of parameters have to be identified in order to run the model. The model inputs and outputs a well as the set of parameters are listed in Fig. 3.

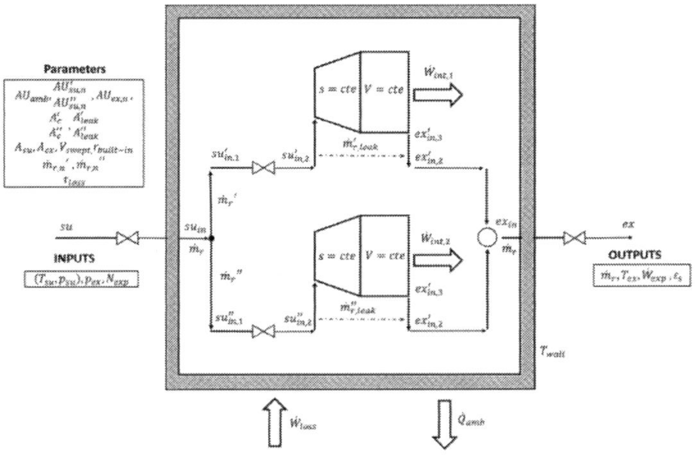

Table 3. Identified parameters of the semi-empirical model.

Parameter	Unit	Value
AU_{amb}	W/K	8.63
$AU'_{su,n}$	W/K	11.75
$AU''_{su,n}$	W/K	13.75
$AU_{ex,n}$	W/K	6.44
$\dot{m}'_{r,n}$	kg/s	0.21
$\dot{m}''_{r,n}$	kg/s	0.27
A'_{leak}	m^2	4.61e-6
A''_{leak}	m^2	6.61e-6
A_{su}	m^2	8.36e-5
A_{ex}	m^2	1.45e-3
α_{loss}	-	0.28
τ_{loss}	Nm	1.25
$\dot{W}_{loss,0}$	W	60.8

Figure 3. Semi-empirical model of the single screw expander.

5. Parameters identification

The two models described include a number of parameters that have to be identified in order to accurately predict the performance of the SSE. The set steady-state points are used to calibrate the models. The empirical parameters, $\mathbf{p} = (a_0, ..., a_6)$ of the Pacejka equation [2] are obtained by minimizing an objective function defined as the sum of the quadratic deviation of the measured and calculated values of the overall isentropic effectiveness scaled by the data uncertainties :

$$r_{Pacejka} = \sum_i \frac{\left(\varepsilon_{is,oa,exp,meas,i} - \varepsilon_{is,oa,exp,calc,i}(\mathbf{p})\right)^2}{u_i^2} \tag{7}$$

Since the the overall isentropic effectiveness measured values, shown in Fig. 2(d), have not a unique trend, a constrained Sequential Least SQuares Programming (SLSQP) minimization algorithm is used to drive to zero the objective function. In particular, the parameters which have physical meaning, i.e. $r_{p,0,nom}$, $r_{p,max,nom}$, $\varepsilon_{is,oa,max}$, $(d\varepsilon_{is,oa,max}/dr_{p,0})_{nom}$, $N_{exp,nom}$, have been determined in accordance to [2]. Instead, for the empirical parameters $a_0, ..., a_6$ and ξ a set of boundaries have been determined.

Table 4. Identified Pacejka equation parameters.

Parameter	Value	Empirical parameters:
$r_{p,0,nom}$	3.07600000e+0	a_0 =-1.93529023e-3
$r_{p,max,nom}$	5.99558088e+0	a_1 =-1.25256170e-2
$\varepsilon_{is,oa,max}$	5.19000000e-1	a_2 =-1.17292542e-1
$(d\varepsilon_{is,oa,max}/dr_{p,0})_{nom}$	7.08651674e-1	a_3 =-4.73874764e-2
ξ	1.23877317e+0	a_4 =7.80410436e+0
$N_{exp,nom}$	3.54700000e+3	a_5 =6.36530295e-4
		a_6 =5.78187315e-1

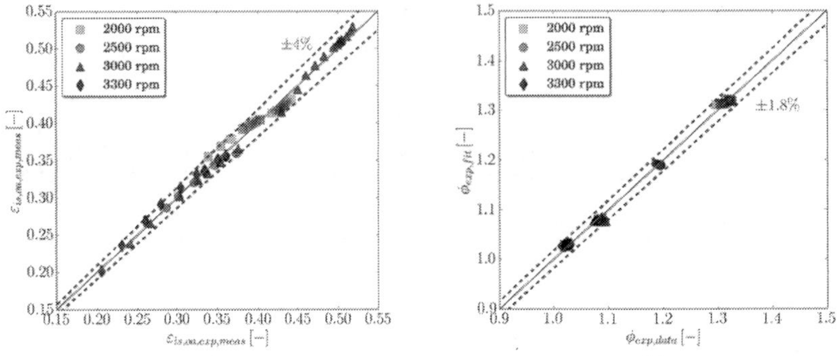

(a) (b)

Figure 4. (a) Comparison between Pacejka-based model and experimental data in terms of parity plot between measured and calculated overall isentropic effectiveness; (a) parity plot of the filling factor.

The semi-empirical model of the SSE presents a higher level of complexity in terms of number of equations to be solved and related numerical issues. Thus, 9 parameters have to be identified by minimizing an object function based on the experimental data. The objective function represents a cumulative error of the three main variables considered in the calibration process, i.e. mass flow rate, discharge temperature and power output. Its expression is given by:

$$r_{SemiEmp} = \sqrt{\sum_i \left(\frac{\dot{m}_{r,i} - \dot{m}_{r,meas,i}}{\dot{m}_{r,meas,i}}\right)^2} + \sqrt{\sum_i \left(\frac{\dot{W}_{exp,i} - \dot{W}_{exp,meas,i}}{\dot{W}_{exp,meas,i}}\right)^2} + \sqrt{\sum_i \left(\frac{T_{r,ex,,i} - T_{r,ex,meas,i}}{T_{r,ex,meas,max} - T_{r,ex,meas,min}}\right)^2} \quad (8)$$

To be noted that the term accounting for the difference between measured and calculated discharge temperature has been given a higher weight. The parameter identification is done by using a genetic algorithm with proper bounds for each of the parameters listed in Tab. 3.

6. Results and discussion

The described models have been implemented in the Python language. The minimization process of the objective function for the Pacejka equation requires a limited computational effort to converge. However, the bounds of the SLSQP should be refined a few times in order to achieve the best fit. The set of identified parameters are listed in Tab. 4. By considering all the rotational speeds, i.e. 2000 rpm, 2500 rpm, 3000 rpm and 3300 rpm, a coefficient of determination, R^2, of 99.32% is obtained for the overall isentropic effectiveness with a maximum relative error of 4.05%, as shown in Fig. 4(a). If only the rotational speeds of 2000 rpm and 3000 rpm are

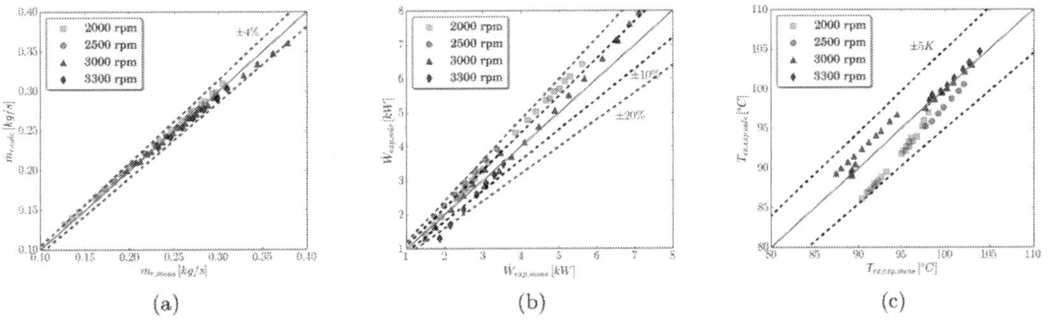

Figure 5. Parity plots with the optimized parameters: (*a*) mass flow rate; (*b*) power output; (*c*) discharge temperature.

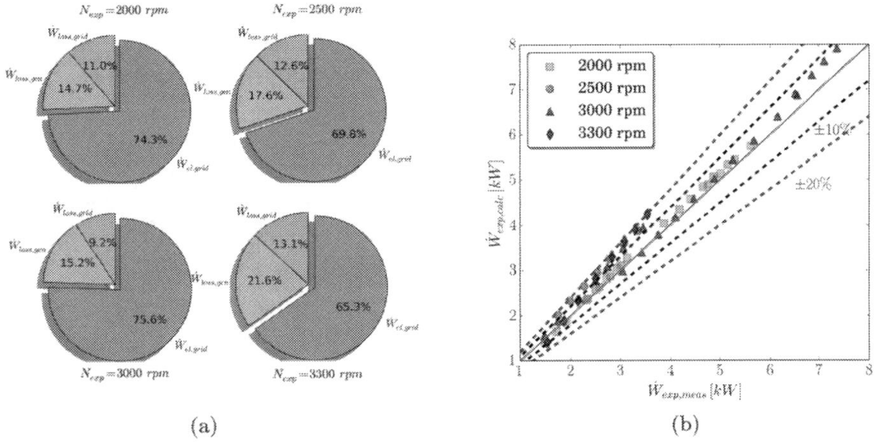

Figure 6. (*a*) Distribution of the electro-mechanical losses associated with the coupling of the expander shaft to the generator and inverter, at different rotational speed; (b) improved agreement of the electrical power output by integrating specific correlations of for generator and inverter losses.

considered because investigated during the same day of testing and throughout a wider range of conditions, the R^2 increases slightly to 99.69% and the relative error is reduced by the 30%. The filling factor data points are fitted with R^2 of 99.71% and a maximum relative error of 1.76%, as shown in Fig. 4(b).

The search of the optimized parameters is more involved and lengthy for the semi-empirical model. Although the total number of parameters of the models is comparable, the solution process of the semi-empirical model includes more iterative loops, especially for the calculation of the flow through fictitious nozzles and heat transfer, and, moreover, the parameters are coupled because it is a physical-based model. The optimization process starts by considering only the nominal speed at 3000 rpm in order to obtain a better set of guess parameters. Then, all the data points are considered in the optimization and the genetic algorithm keeps evolving by selecting new generations of parameters based on a natural selection process. The time required to obtain the optimized parameters varies a lot, depending on the selected tolerance of the converging criteria. Reasonable maximum relative errors between measured and calculated values of mass flow rate, power output and discharge temperature have been selected as the upper limit of the results presented in [9]. In particular, the optimization process has been considered successful if the maximum relative errors between measured and estimated values were 5%, 10% and 5 K for the mass flow rate, power output and discharge temperature, respectively. The optimized parameters are listed in Tab. 3. The parity plots are shown in Fig 5. As a result, the semi-empirical model was able to capture the trend of the experimental data. The mass flow rate is estimated within 4% at all the rotational speeds which means that the geometric parameters and the leakage model have been able to match the real behavior of the machine, as shown in Fig. 5(a). The discharge temperature is predicted within 3 K only at 2500 rpm, 3000 rpm and 3300 rpm. At 2000 rpm, the discharge temperature is underestimated in the range of 3 K up to 5 K, as shown in Fig. 5(c). There is a certain dependency of the heat transfer rate with the rotational speed and therefore the current correlation used should be further modified. Finally, the power output has been predicted within 10% only at the nominal speed of 3000 rpm and few points at 3300 rpm. The power output is generally overestimated at all the other rotational speeds, i.e. 2000 rpm and 2500 rpm. The parity plots presented in Fig. 5 have been obtained by considering the mechanical losses to vary linearly with a mechanical loss torque as shown in Eq. 5. An additional term proportional to the rotational speed similar to the one proposed in [11] was also added to account for the electrical losses from the generator and inverter. By observing Fig. 5(b), it is clear that a different model has to be used to better describe the mechanical losses especially at lower speeds. Furthermore, in the optimization process, the value of the generator efficiency and inverter effectiveness should be considered. Therefore, the performance maps of generator and inverter have been included in the semi-empirical model in order to better estimate the shaft power. The efficiency of the generator depends on the torque and rotational speed, instead the effectiveness of the inverter is mainly related to the electrical power [7]. In Fig. 6(a), a comparison between the actual electrical power output and the electro-mechanical losses for each rotational speed is proposed. Two general considerations can be made related to the uncertainties of the semi-empirical model on the predictions of the electrical power output. First of all, the electrical losses of the inverter are almost constant throughout the power range investigate except at below 2000 W, where they decreases sharply [7]. Secondly, the mechanical efficiency of the generator is strongly related to the rotational speed and the torque applied. At low rotational speed and low torque the efficiency decreases significantly [7]. These behaviors can be found in Fig. 6(a). By recalling that for the rotational speeds of 2500 rpm and 3300 rpm, low values of the power output were achieved, the mechanical efficiency of the generator accounts for 17.6% and 21% of the total shaft power, respectively. The mechanical loss term have been modified with an additional contribution proportional to the internal power, i.e. $\alpha_{loss}\dot{W}_{in}$. As a result, the electrical power output has been predicted within 10% for all the experimental points, as shown in Fig. 6(b). Mass flow rate and discharge temperature predictions slightly improved and the maximum relative errors have been reduced to 3.5% and 4 K,respectively. To be noted is that if only the nominal rotational speed is considered, the mass flow rate, the power output and the discharge temperature are predicted within ±3%, +7% and +3 K respectively. It is important to mention that semi-empirical models such as those one proposed in this paper, are calibrated and optimized for a particular machine and for a range of operating conditions. Extrapolating results to larger/smaller machines may result in large errors. For this reason, a larger single-screw expander will be tested and the model extended to include the scale effects and generalize the model from a single machine to a class of machines. The computational time of a single run of the semi-empirical model is around 2.5 seconds. Instead the validation for the all 60 steady-state points takes around 97 seconds. As a comparison, the fastest simulation of the corresponding deterministic model takes approximately 30 seconds.

7. Conclusions

In his paper, the experimental data of a 122 mm single-screw expander for organic Rankine cycle applications have been discussed and further used to calibrate two low-order models, for instance the Pacejka equation for the overall isentropic effectiveness and a semi-empirical model based on the physical aspects of the expander. From the experimental data, the maximum power output obtained was 7.364 kW and a maximum overall isentropic effectiveness of 51.91% was achieved. The calibration of the Pacejka equation resulted in a maximum relative error of 4.05%. By means of a genetic algorithm, the parameters of the semi-empirical model have been determined and the model was able to capture the physical behavior of the SSE. Mass flow rate, power output and discharge temperature have been initially predicted within 5%, 10% and 5 K, respectively for most of the experimental points. A better characterization of the electro-mechanical losses to account for the generator and the inverter has been successively included which led to improved agreement between the measured and calculated values. The maximum isentropic efficiency obtained at the shaft was around 62%.

Acknowledgment

The authors would like to thank graduate student Nelson James from the Herrick Laboratories, Purdue University for his contribution related to the implementation of the genetic algorithm used in this paper.

References

[1] Uusitalo A, Turunen-Saaresti T, Honkatukia J, Colonna P and Larjola J 2013 *J. Eng. Gas Turbines and Power* **135** 1–9

[2] Declaye S, Quoilin S, Guillaume L and Lemort V 2013 *Energy* **55** 173–183

[3] Avadhanula V K and Lin C S 2014 *J. Eng. Gas Turbine and Power* **136** 1–8

[4] Ziviani D, Bell I, Paepe M D and van den Broek M 2015 Update on single-screw expander geometry model integrated into an open-source simulation tool *9th Int. Conf. on Compressors and their Systems, City University of London, London* 39

[5] Gusev S, Ziviani D, Bell I and M De Paepe M v d B 2014 Experimental comparison of working fluids for organic rankine cycle with single-screw expander *Proceedings of 15th Int. Refrig. Air Cond. Conf. at Purdue* 2653

[6] Bell I H, Wronski J, Quoilin S and V L 2014 *Ind. Eng. Chem. Res.* **53** 2498–2508

[7] Melotte N 2012 *Experimental study and dynamic modeling of a Waste Heat Recovery Organic Eankine Cycle* Master's thesis University of Liege

[8] Desideri A, van den Broek M, Gusev S, Lemort V and Quoilin S 2014 Experimental campaign and modeling of a low-capacity waste heat recovery system based on a single screw expander *Proceedings of 22nd Int. Compr. Eng. Conf. at Purdue* 1451

[9] Lemort V, Quoilin S, Cuevas C and Lebrun J 2009 *Applied Thermal Engineering* **29** 3094–3102

[10] Ziviani D, Bell I, De Paepe M and van den Broek M 2014 Comprehensive model of a single screw expander for orc-systems applications *22th Int.Compressor Engineering Conf. at Purdue* 1506

[11] Lemort V, Declaye S and Quoilin S 2011 *Proc. IMechE, Part A: J. Power and Energy* **226** 126–136

The Development of a Small High Speed Steam Microturbine Generator System

Adrian Alford[1], **Philip Nichol**[1] and **Ben Frisby**[2]

[1]Corac Energy Technologies, Stirling Road, Slough SL1 4ST
[2]Spirax Sarco Ltd, Runnings Road, Cheltenham, GL51 9NQ

E-Mail: Philip.Nichol@corac-cet.co.uk

Abstract. The efficient use of energy is paramount in every kind of business today. Steam is a widely used energy source. In many situations steam is generated at high pressures and then reduced in pressure through control valves before reaching point of use. An opportunity was identified to convert some of the energy at the point of pressure reduction into electricity. This can be accomplished using steam turbines driving alternators on large scale systems. To take advantage of a market identified for small scale systems, a microturbine generator was designed based on a small high speed turbo machine. This gave rise to a number of challenges which are described with the solutions adopted. The challenges included aerodynamic design of high efficiency impellers, sealing of a high speed shaft, thrust control and material selection to avoid steam erosion. The machine was packaged with a sophisticated control system to allow connection to the electricity grid. Some of the challenges in packaging the machine are also described. The Spirax Sarco TurboPower has now concluded performance and initial endurance tests which are described with a summary of the results.

1. Background

1.1 Expertise of partners

Spirax-Sarco Limited has over 70 years of experience in research and innovation in most aspects of saturated steam generation and transmission. The company provides expert advice to UK government and other organizations to formulate strategies and standards around steam systems, boiler operations and sterilization. Spirax Sarco offers an extensive portfolio of products and services, creating solutions for steam-using organisations worldwide.
Corac Energy Technologies have considerable experience in the design and development of innovative turbo-machinery for specialist applications in the oil and gas and industrial markets.
The two companies applied their combined expertise to this development project

1.2 Market need & drivers

Energy costs are of significant concern to industry worldwide, and play a significant part in the viability and profitability of many processes. According to the UK manufacturers' organization, EEF

Content from this work may be used under the terms of the Creative Commons Attribution 3.0 licence. Any further distribution of this work must maintain attribution to the author(s) and the title of the work, journal citation and DOI.
Published under licence by IOP Publishing Ltd

Executive Survey 2014 [4], gas is the dominant energy source for manufacturing and accounts for 60% of use. Yet electricity prices are of particular concern because they represent the larger cost, with 66% of utility spend on electricity. In its survey, EEF found that the cost of electricity was the most cited energy challenge for UK companies over the next two years.

Figures from the DECC Quarterly Energy Prices December 2014 report [3] highlight that the average price of electricity for the industrial sector, including Climate Change Levy, has risen by 108 per cent in real terms since 2003, and by 3.3 per cent in the latest year.

Against this backdrop of rising energy costs, an opportunity was identified to reduce the energy costs of small to medium industrial steam users by enabling local, small scale power generation.

Industrial steam boilers are typically designed to operate at higher pressures than required by the end use application. Generating and distributing steam at higher pressure offers two advantages: Firstly, the boiler's thermal storage capacity is increased, helping it to cope more efficiently with fluctuating loads and minimising the risk of producing wet and dirty steam; secondly, smaller bore steam pipes can be used for distribution.

Having distributed the steam from the boiler, its pressure needs to be reduced to that required by the application. This reduction is conventionally performed by a pressure reducing station comprising a pressure reducing valve and various items of ancillary equipment.

An alternative approach is to pass the plant steam through a microturbine installed in parallel with a pressure reducing station, which allows the energy released by the resulting pressure drop to supplement the electricity supply (Figure.1). The outlet steam is used by the downstream application as in a conventional system.

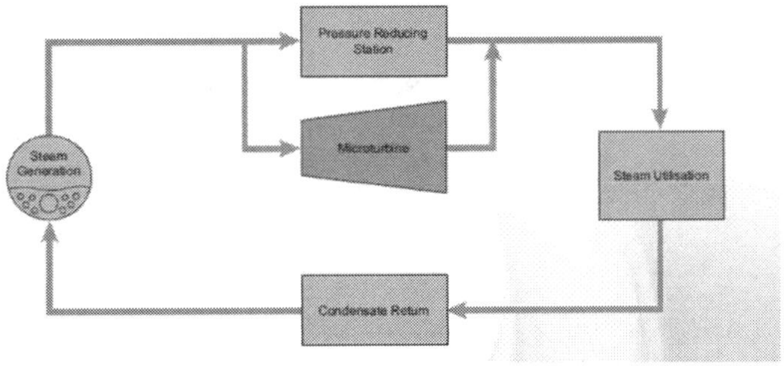

Figure 1. Proposed installation of system.

A microturbine can generate high value electrical energy for the consumer to be fed back into the grid and consumed locally or exported. While this approach is used by large steam users, the opportunity identified is for smaller steam users with boilers typically, less than 6MWth. Industries with potential for using the Spirax Sarco TurboPower include food and drink, chemicals processing, pharmaceuticals and hospitals.

2. Outline Machine Specification

2.1 General Requirements

A specification for the prototype machines was developed by the partner companies:

The use of a microturbine should provide low cost electricity generation from low steam flows. To make an attractive investment, it must have low life cycle costs. For the widest application, there should be no possibility of steam contamination.

To embrace the widest range of applications the microturbine was specified so as to have an electrical output power from the package of 50kW. By using two microturbines in one package a 100 kW unit was specified.

IOP Conf. Series: Materials Science and Engineering **90** (2015) 012062 doi:10.1088/1757-899X/90/1/012062

The microturbine should offer a solution to those who produce steam at medium pressure up to 20 BarG before reducing it through a pressure reducing valve to a lower pressure, typically between 1BarG and 4BarG for use.
The target flow range for the application was 1,000 kg/h to 4,000 kg/h.
The unit would be likely to be installed near a steam using process, driving the decision to specify the operating ambient temperature for the unit at 40°C. A self-contained air cooling system was required to maximise installation flexibility. The package should allow ducting of cooling air to the machine.
The steam quality and pressure ranges were specified to embrace the widest range of applications with the steam quality being dry saturated. Wet steam provides a significant challenge for this type of high speed machine and it was accepted that the machine should be preceded by a separator to protect the rotating parts from damage by 'slugs' of water.
Most importantly, the machine should be able to produce the maximum amount of energy from the widest range of steam supply and exhaust pressures. This was to embrace not only a wide range of applications but also large variations in steam flow rates.
A high speed machine with the turbine and generator on a single shaft was specified to provide an efficient solution with few moving parts having, inherently, low maintenance requirements and few wearing components.

2.2 Efficiency Targets
The generation efficiency of the unit must be high for the system to provide an attractive investment for a steam user. The use of the downstream steam by a process means that the outlet steam is not a loss from the system, giving more efficient electricity generation than a condensing turbine.
Generating efficiency is the electricity generated, divided by the energy input that is used to perform this generation, and figures in excess of 70% are expected from the steam microturbine. To realise this, any losses of steam from the system must be minimised as they significantly affect efficiency making effective sealing paramount. A maximum steam loss of 2% of the mass flow rate through the microturbine was specified. Losses in the conversion process through the generator and inverters must also be minimised along with loads from ancillary equipment in the package, for example cooling fans and pumps.

2.3 Physical Requirements
Steam pressure reduction usually occurs near the point of use. The Spirax Sarco TurboPower is intended to be used at pressure reducing stations rather than in the boiler house. This presents a challenge as space and access is often restricted at these locations, and it is likely that there will be people working nearby. This means that the unit has to be compact and easily transportable by forklift or pallet truck, should fit through standard size double doors and be quiet enough not to necessitate hearing protection when working nearby.

2.4 Legislation Requirements
The machine was specified to comply with European legislation with easy adaptation to global requirements.
Within the UK, the connection of generation devices to the National Grid is subject to regulation. If the current is greater than 16A per phase, the requirements of the Energy Networks Association Engineering Recommendation G59/2 "Recommendations for the Connection of Generating Plant to the Distribution Systems of Licensed Distribution Network Operators (DNOs)" [3] must be met before connecting to the network. Requirements outside of the UK vary, but it is common to find similar requirements to those of G59/2.

2.5 Consideration of Alternative Technologies
A number of technologies were considered for this application:
Axial turbines were found to be too expensive for the low flow rates involved.
A number of manufacturers are offering screw expanders in similar sizes. These tend to have lower efficiency due to the number of rotating parts and friction with higher steam losses due to the larger number of seals required.
Reciprocating engines may be used but have lower efficiencies and high costs.
The high speed radial machine was selected as it promised high efficiencies and lower costs. It should also be noted that the proposed unit has an integral generator so that there is no gearbox or drive coupling with associated lubrication and cooling systems. Gear transmissions have higher losses due to friction and higher maintenance requirements than a single shaft gear-less system.

3. Prototype Machine

3.1 Overview
The prototype machine to meet the specification was based upon an integral high speed air compressor that was described in a previous paper. [1]
The rotating unit is a radial flow turbine driving a permanent magnet generator on a single shaft. The shaft is supported on hybrid tilting pad air bearings. One end of the shaft retains the magnets for the generator and the opposite end of the shaft supports the turbine and shaft seal. The machine contains no lubricant so the risk of steam contamination is negligible. See Figure 3.

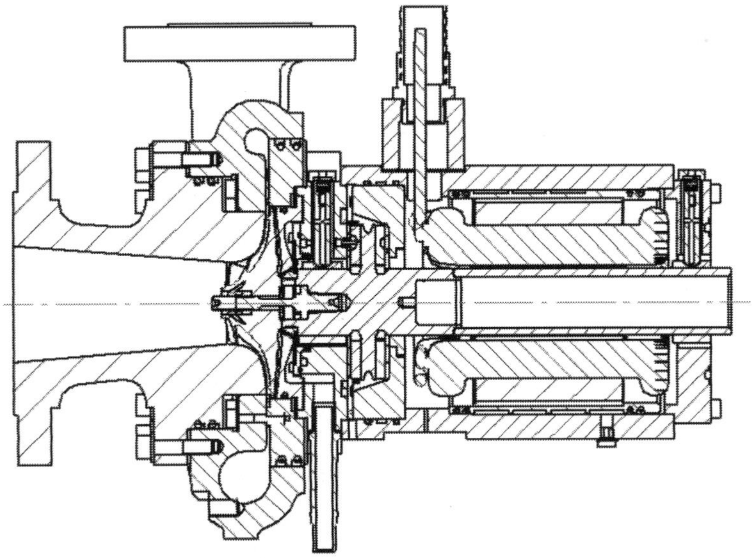

Figure 3. Section through microturbine.

3.2 Bearings
An early decision was taken to use Corac's hybrid tilting pad air bearings for this range of machines. The high speeds, expected high temperatures and economic considerations precluded the use of any other bearing technology. The system specified had been proven in previous applications.
The journal bearings consist of three tilting pads supported on hardened stainless steel supports. The supports allow the bearing pads to tilt freely in all directions. The top bearing pad is spring loaded to allow for thermal expansion and providing a pre-load that may be adjusted to improve shaft stability. Whilst the choice of materials means the bearings are capable of sustaining many 'dry' starts and stops

it was decided to use a hybrid design for these machines to ensure extended bearing life in harsh environments. Compressed air is supplied to the journal bearings via small ports in the pads to allow the shaft to float before the machine starts to rotate. Thus the bearings operate in a hydrostatic mode. When a pre-set speed has been reached the compressed air is switched off so the bearings operate hydrodynamically.

The thrust bearing consists of two sets of opposing tilting pad air bearings. Due to the high loads encountered it was necessary to maximise the capacity of the bearings within a very confined space. The disc on the shaft that provides the counter-face for the bearings is limited in diameter by the stresses imposed by rotating at such high speeds. This has the effect of limiting the bearing surface area available. The system employed uses a combination of hydrostatic and hydrodynamic operation. An external compressed supply is fed to the bearings when required under command of the machine controller.

3.3 Turbine

A radial inflow turbine was selected for this duty. Design metrics for the duty fall within normal bounds for this class of device and in the absence of condensation a high adiabatic efficiency would be expected. As the steam entering the package is dry saturated, condensation would occur within the stage, leading to a reduction in adiabatic efficiency and potentially to erosion. It was established that good design should allow these risks to be mitigated.

3.4 Generator

The generator is a two pole permanent magnet machine which is integrated on the same shaft as the turbine. The windings are formed as six turns around an electrical steel stator in an aluminium casing. A challenge with this type of machine is to limit the length of the end turns so as to keep the overall shaft length as short as possible. In this machine the end turns are formed after winding to control the shape.

The machine uses water cooling in a jacket around the stator and forced air cooling in the air gap. Performance was optimised to give the desired output with a margin within the capability of a Class H insulation system.

Figure 4. Prototype Spirax Sarco TurboPower Package.

3.5 Package

The approach to the package design was to employ best the practice used in packaging of similar machinery to produce a robust, compact and aesthetically pleasing unit. See Figure 4. It was

IOP Conf. Series: Materials Science and Engineering **90** (2015) 012062 doi:10.1088/1757-899X/90/1/012062

determined that the package should be capable of being configured as a single 50kW machine or a dual 100kW machine.

The package was designed to be lifted into position and connected directly to steam supply and outlet via external flanges. The unit was also designed to be connected directly to the electricity network having the necessary protection devices in the control panel.

The package was designed to house the expander unit with the steam control valves, an inlet steam strainer, and the pneumatic controls for the steam valves and air bearings. Integral to the package was a control panel housing the electrical system for the generator.

The specification was for an integral water to air cooling system for the generator and power electronics, provided by a fan-cooled radiator and water circulating pump within the unit. To maximise efficiency both the radiator fan and the control panel fans were thermostatically controlled.

The package was clad with sound attenuating panels designed to be easily removed for maintenance.

3.6 Controls

The entire machine was designed to be controlled by a single high speed micro-controller with be-spoke software developed for this range of machines. A touch screen HMI provides continuous running information, alerting the user to fault or maintenance conditions and logging energy efficiency data.

Steam flow is controlled by a pneumatically operated valve on the inlet of the turbine, the primary control parameter being the outlet pressure of the steam to the users' process. The secondary control parameter is the electrical power output of the machine controlled to be the maximum available from the prevailing steam conditions. See Figure 5.

An inverter and Active Front End (AFE) unit is installed in the panel to control the running of the generator and the supply of power to the grid. In simple terms, the inverter converts the high

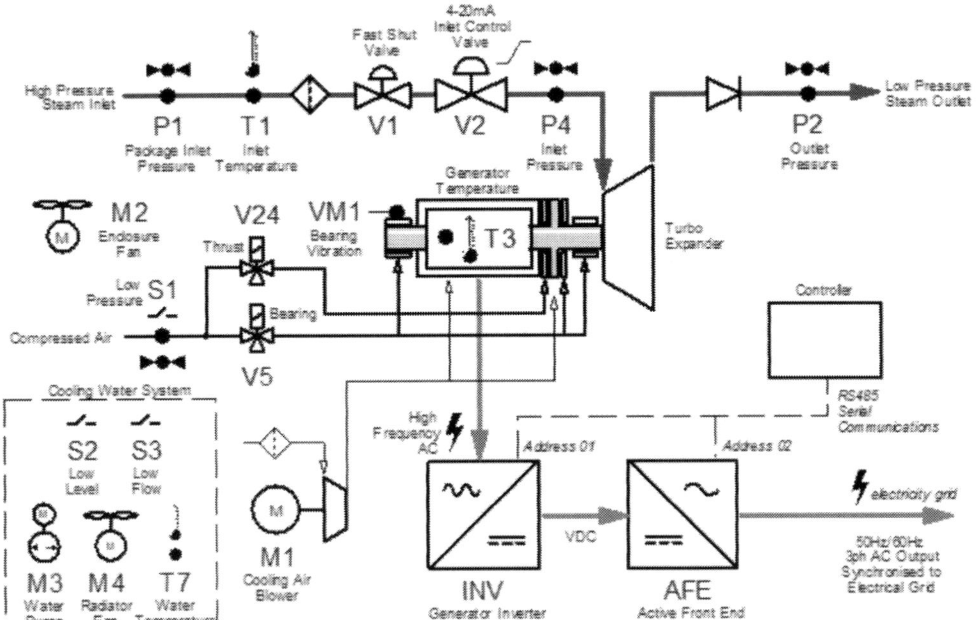

Figure 5. System Schematic Diagram.

frequency AC output of the generator into DC and the AFE converts the DC into 50Hz or 60Hz AC power to be fed into the grid.

To meet the UK regulations for supplying power to the grid it was necessary to install a compliant relay that monitors the power being exported to the grid.

The inverter selected uses a vector-less system needing no external speed sensor. Values of speed, voltage, current and power are continually calculated by the inverter and AFE and fed directly to the controller. An integral over speed avoidance system was built into the package.

4. Technical challenges & solutions

4.1 Aerodynamic System Design

Initial design work was carried out using a 1D package with a perfect gas assumption. This achieved a basic design which was optimized using a 3D CFD package (Ansys CFX v.14.0). The characteristics to be optimized centred on the reduction in impeller diameter and mass, and the ability to maintain efficient and reliable operation in a two-phase condensing steam flow. Impeller diameter had to be minimized to reduce thrust and back plate windage. To ensure the rotor had the required margin below the first bending critical speed the impeller mass had to be minimized. Both of these requirements tended to reduce stage adiabatic efficiency.

Further trade-offs against efficiency resulted from the need to keep the relative velocity at the impeller tip as low as possible to reduce droplet erosion, but high enough to maintain entrainment of these droplets and ensure their clearing from the impeller against the centrifugal field. These seemingly conflicting requirements were initially addressed using a perfect gas model due to computational speed, though it was known that condensation would occur within the stage. Impeller diameter reduction was achieved by using a non-radial blade angle at inlet which also allowed high blade loading without increasing gas relative velocities. A reduction in axial length was achieved while controlling secondary flows by adding splitter blades. This helped diameter minimization by reducing blade loading.

Once the required specification was achieved, design moved to a non-equilibrium condensing steam model within the CFD package. At this point, large secondary flows and a significant reduction in adiabatic efficiency were demonstrated by the use of this more realistic model. The model took approximately 4 times as long to converge as the perfect gas model, and a significant number of further design iterations were necessary in order to return the stage efficiency at previously modelled levels.

This model estimated the droplet size generated within the stage and tracked the droplet path and the degree to which the droplet lagged the vapour flow. This allowed the establishment of droplet clearing in the free stream. The effects of wall wetting and liquid re-entrainment were not modelled. Significant uncertainty therefore remained beyond the design stage regarding the reduction in efficiency which would be experienced in the tested machine due to these phenomena. Though a small number of radial inflow condensing turbines have been reported in literature,[2] very few provide any information on design choices to assist a designer.

4.2 Shaft sealing

A tight specification of 2% steam flow at an inlet pressure of 10 BarG was specified for the seal. A number of configurations were experimented with to establish the best configuration. Experiments were carried out on a test rig that demonstrated that the use of a single labyrinth seal would be likely to achieve a leakage rate of around 1.6 % with a seal inlet pressure of 4BarG

The key to operation of the shaft seal was to reduce the pressure reaching the seal from the inlet to the turbine. The use of vanes to create an opposing pressure behind the turbine was assessed but it was found that the windage losses would have resulted in unacceptable heating and a reduction in efficiency. The method adopted was to have an axial toothed labyrinth seal at the back of the turbine and controlling the pressure behind the wheel to maintain the inlet pressure to the atmospheric shaft seal at a reasonable level. The final geometry of shaft seal can be seen in Figure 6.

IOP Conf. Series: Materials Science and Engineering **90** (2015) 012062 doi:10.1088/1757-899X/90/1/012062

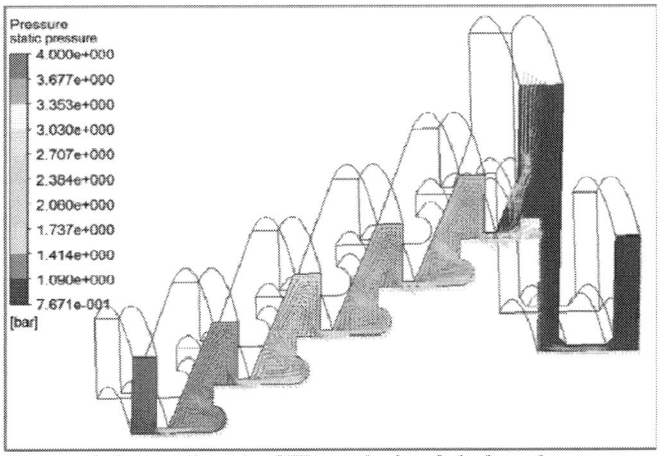

Figure 6. Streamlines in CFD analysis of shaft seal geometry.

4.3 Thrust Load Control

Maintenance of thrust loads during running within the capabilities of the gas thrust bearing was initially addressed by the minimization of impeller diameter. A further parameter which influences thrust load is the effective diameter of the shaft seal. Though a large diameter shaft seal will reduce thrust loads when running at the design point, it will increase static thrust load when starting.

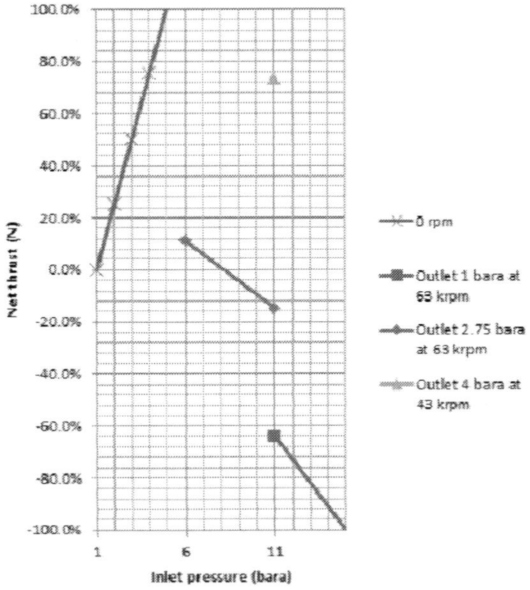

Figure 7. Graph of thrust bearing load under different operating conditions

To maintain thrust loads within the capacity of the thrust bearing during machine starting, it was necessary to use the smallest diameter seal that could be fitted to the shaft. It was also necessary to balance thrust loads by the use of a seal on the impeller back and partial venting of the impeller back-plate region.

Extensive modelling of this passive thrust control system was necessary to ensure that the loads generated were sufficiently low through the entire operational envelope with adequate margin of safety. Figure 7 shows the resultant thrust loads under various operating conditions.

Thrust loads act in both directions, depending on the operating point, necessitating a double-sided thrust bearing design. To support heavier loads acting in one direction, the corresponding side of the thrust bearing was configured to be a hydrostatic bearing with higher capacity than the hydrodynamic bearing on the opposite face.

4.4 Material choices

Using expertise within Spirax Sarco, material choices were made for the critical components in areas that would be exposed to steam. Of concern was the need to have impellers that would last for extended periods when exposed, not only to dry saturated steam but also to impact by water droplets from wet steam reaching the turbine or condensation taking place within the turbine. Some research gave assurance that titanium provides good erosion resistance in this environment. The light weight, high strength and reasonable machinability of titanium also made it a natural choice for this application.

At the conclusion of an endurance running test, the impeller, in particular, was examined for signs of erosion. After 1000 hrs there was no visible damage or measurable weight loss. Other components in contact with the steam such as the volute and nozzles were made from 316 stainless steel and have been found to give excellent service

4.5 Over speed Avoidance System.

It was necessary to design a system that would prevent the generator 'running away' in the event of grid loss or a failure of the inverter. Traditional low speed machines may be controlled by 'governor' systems that shut off the steam supply as the speed rises. In this case the low inertia of the machine giving rise to very high potential acceleration rates means that such a system would not act quickly enough. The system employed has a fast acting steam shut-off valve on the inlet to the turbine that is directly operated in the event of grid loss or other potential over speed event. This valve is pneumatically operated with a normally closed function so that in the event of loss of air pressure or power it will close. In the event of failure of this valve, the slower acting, inlet control valve was also designed to close automatically.

A secondary system was designed to act with the steam valves: In the event of grid loss, other over speed event or inverter fault the generator windings are connected via a pair of fail-safe contactors to air-cooled resistors that are sized to absorb the output of the generator until the steam valves operate. These systems are designed to operate, even in the event of a failure of the central controller. Thorough testing has proved the efficacy of this system under a number of dual failure modes.

5. Performance testing

A rig was constructed to test the microturbine package across a range of steam and environmental conditions. It was important to provide an installation that matches conditions with industrial steam users. The unit was installed in parallel with a pressure reducing valve, and the facility to vary running conditions and record test parameters as enabled.

The test rig was supplied with a 30BarG steam supply reduced to test pressures in the range of 4 to 14 BarG. The supply of saturated steam was conditioned via a steam separator. Temperature and pressure measurements were used to confirm saturated conditions.

A valve placed downstream of the microturbine provided the facility to control the load, mimicking a downstream application. A power meter was used to measure the net power output from the prototype package. An inline variable area steam flow meter was used to measure the steam mass flow rates passing into the microturbine.

The unit was tested at a wide range of conditions, with varying upstream and downstream pressures resulting in a range of steam mass flow rates passing through the device. Power output was recorded at each set of conditions. The plot shown in Figure 8 illustrates the power output recorded over a range of steam inlet and outlet pressures and different turbine speeds. Useful electrical power achieved, even at low steam flow rates providing a useful turndown range.

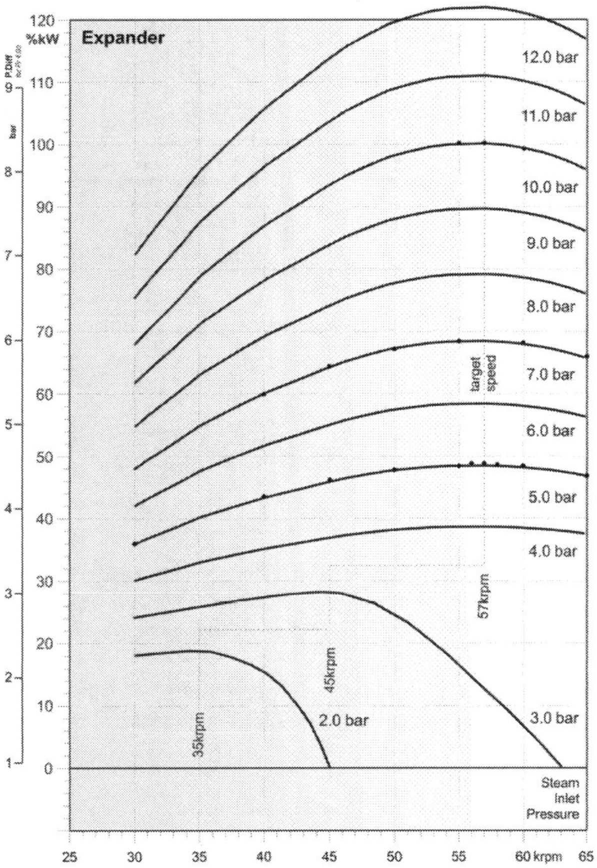

Figure 8. Performance map

6. Minimising losses

Minimising losses is essential to achieve an attractive investment for a customer. Live steam tests were conducted to measure the seal leakage rates to ensure that they were within expectations and it can be seen the table in Figure 9 that the leakage rates were below 1% of the flow rate through the turbine. The tests also validated CFD models of the labyrinth seal.

Inlet Pressure (Barg)	*Leakage (%) of Steam Mass Flow*
6.9	0.79
9.9	0.88
10.3	0.82

Figure 9. Table of steam seal leakage rate measured as a proportion of the total steam mass flow rate through the turbine.

Electrical losses were measured and are summarised in Figure 10. It should be noted that the overall value of these losses is very low and would compare very favourably with a traditional turbine driving a generator through a gear box.

IOP Conf. Series: Materials Science and Engineering **90** (2015) 012062 doi:10.1088/1757-899X/90/1/012062

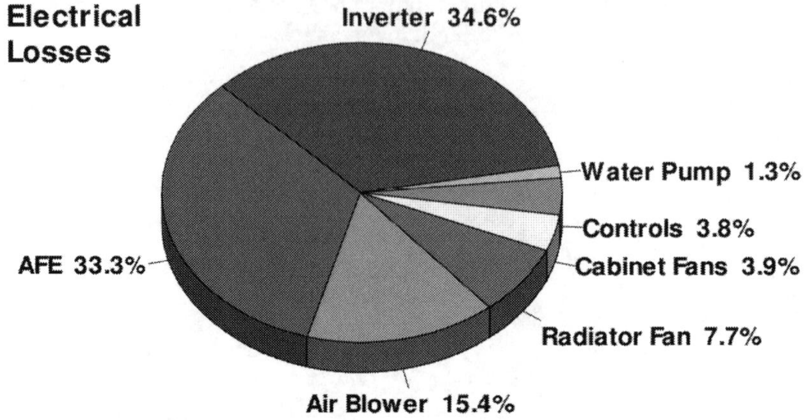

Figure 10. Breakdown of Electrical Losses on Test Package

7. Endurance test

The unit was run within the test facility for 1000 hours before being disassembled and thoroughly inspected. Aerodynamic components, bearings, steam sealing components and generator electrical performance were all assessed and no significant signs of wear or degradation were observed.

8. Conclusions and recommendations

This highly innovative machine has performed in terms of efficiency, reliability and flexibility in a way that met all the expectations.
The prototype machine is capable of being developed with relatively little effort into a competitive product that will bring energy recovery to a range of small steam users.

References

1. Successful Trials of Turbo-boosting of Positive Displacement Compressors.
A.Alford & P. Nichol. International Conference on Compressors and their Systems, London, 2011.
2. 150 kW Class Two-Stage Radial Inflow Condensing Steam Turbine System
Kazutaka Hayashi, Hiroyuki Shiraiwa, Hiroyuki Yamada, Susumu Nakano and Kuniyoshi Tsubouchi Hitachi Engineering & Services Co. Ltd., Hitachi, Ibaraki, Japan. ASME 2011 Turbo Expo: Turbine Technical Conference and Exposition. Paper No. GT2011-46192,
3. DECC 2014, Quarterly Energy Prices: December 2014 p11
4. EEF 2014, Business Productivity and Energy Efficiency p30
5. Energy Networks Association Engineering Recommendation G59/2 "Recommendations for the Connection of Generating Plant to the Distribution Systems of Licensed Distribution Network Operators (DNOs

Copyright:
The copyright to this document is the property of Corac Energy Technologies Ltd and Spirax Sarco Ltd.

The impact of component performance on the overall cycle performance of small-scale low temperature organic Rankine cycles

M White and A I Sayma
School of Mathematics, Computer Science and Engineering, City University London, UK.

Email: Martin.White.1@city.ac.uk

Abstract. Low temperature organic Rankine cycles offer a promising technology for the generation of power from low temperature heat sources. Small-scale systems (~10kW) are of significant interest, however there is a current lack of commercially viable expanders. For a potential expander to be economically viable for small-scale applications it is reasonable to assume that the same expander must have the ability to be implemented within a number of different ORC applications. It is therefore important to design and optimise the cycle considering the component performance, most notably the expander, both at different thermodynamic conditions, and using alternative organic fluids. This paper demonstrates a novel modelling methodology that combines a previously generated turbine performance map with cycle analysis to establish at what heat source conditions optimal system performance can be achieved using an existing turbine design. The results obtained show that the same turbine can be effectively utilised within a number of different ORC applications by changing the working fluid. By selecting suitable working fluids, this turbine can be used to convert pressurised hot water at temperatures between 360K and 400K, and mass flow rates between 0.45kg/s and 2.7kg/s, into useful power with outputs between 1.5kW and 27kW. This is a significant result since it allows the same turbine to be implemented into a variety of applications, improving the economy of scale. This work has also confirmed the suitability of the candidate turbine for a range of low temperature ORC applications.

1. Nomenclature

a	Speed of sound, m/s
D	Rotor diameter, m
h	Enthalpy, J/kg
$\dot{m}$	Mass flow rate, kg/s
N	Rotational speed, RPM
P	Pressure, kPa
PP	Pinch point
PR	Pressure ratio
$\dot{Q}$	Heat input, J/s
s	Entropy, J/(kg K)
T	Temperature, K
$\dot{W}$	Power, J/s
η	Isentropic efficiency, %
μ	Viscosity, Pa s
ρ	Density, kg/m^3
Δh_s	Turbine isentropic enthalpy drop, J/kg
ΔT_{sh}	Degree of superheat, K

Subscripts

1-4	ORC locations
c	Cold source
ci	Cold source inlet
cp	Cold source pinch point
e	Expander
h	Heat source
hi	Heat source inlet
hp	Heat source pinch point
n	Net work
o	ORC
p	Pump
s	After isentropic process

Content from this work may be used under the terms of the Creative Commons Attribution 3.0 licence. Any further distribution of this work must maintain attribution to the author(s) and the title of the work, journal citation and DOI.
Published under licence by IOP Publishing Ltd

IOP Conf. Series: Materials Science and Engineering **90** (2015) 012063 doi:10.1088/1757-899X/90/1/012063

2. Introduction

The organic Rankine cycle (ORC) is a promising technology for the sustainable, localised generation of power from heat sources such as solar, biomass, geothermal and waste heat. These applications cover heat source temperatures ranging from 80°C to 300°C, whilst power outputs range from few kilowatts up to few megawatts. ORC technology has been successfully commercialised for large-scale applications [1,2]. For small-scale applications the economy of scale and technical challenges such as the development of an efficient and economical expander restrict widespread implementation, although bespoke systems are available [3,4].

Typically ORCs are characterised by the heat source temperature. High temperature heat sources such as exhaust gases from internal combustion engines require very high pressure ratios, which are achieved by a turbo-expander such as those presented in [5,6]. The relatively small enthalpy drop of organic fluids permits these pressure ratios to be achieved over a single stage, but at the expense of introducing supersonic flows within turbine. Piston expanders may also be used for high temperature heat recovery applications, however their implementation with ORCs is currently limited [7]. Comparatively, low temperature applications have lower pressure ratios, permitting the use of volumetric expanders, such as screw and scroll machines. However these expanders are typically obtained by the reverse operation of existing compressors, and therefore the achievable efficiency may be limited [8,9]. Furthermore, at the target power of 10kW, screw expanders experience high leakage flows [10], whilst scrolls are yet to be proven [11]. For a 10kW ORC utilising a low temperature heat source (100-120°C), the authors believe an efficient radial turbine to be the most suitable choice.

Thermodynamic analysis and working fluid selection remain critical, highlighted by the abundance of literature on the topic [12,13]. Due to the large number of interdependencies, the implementation of a suitable optimisation algorithm is also essential [14,15]. At this point the distinction between two types of cycle design should be made. The first comprises of designing an optimal system for a particular application assuming that each component performs at its design point. This simplifies turbine modeling since the target efficiency can be specified; the turbine will then be designed to achieve this. However, this approach may lead to high cost, bespoke systems that may not be economical to implement. The second optimises a group of existing components to generate the maximum power from a specified heat source. This may be beneficial for small-scale systems since the key component, the turbine, can be manufactured on a relatively large scale but then implemented into a range of different applications. In this instance component models should be coupled with cycle analysis to account for component off-design performance, where the search for optimal cycle conditions may move the expander performance away from design conditions. Therefore it is critical to accurately predict the impact of turbine off-design performance on the cycle. At present, the assumption of constant turbine efficiency is a shortcoming within many thermodynamic papers.

Turbine performance is characterised by a performance map that plots mass flow rate and efficiency as functions of pressure ratio and rotational speed. Similitude theory is used to non-dimensionalise this map, and has been successfully validated for ideal gases [16]. With care it can also be used to predict performance following a change in working fluid, so long as the thermodynamic properties of the two fluids are not significantly different [17]. The application of similitude theory to ORC turbines has been considered by a number of authors [18-20]. However, this is often with reference to simplified non-dimensional parameters, applicable only to ideal gases. Until recent preliminary work, the suitability of applying similitude theory to real gases such as the organic fluids found within ORCs had not been investigated or validated.

Previous work has led to the design of a candidate ORC turbine [21]. The turbine performance has been evaluated using CFD and the performance map has been obtained. This paper demonstrates a modeling methodology that combines this turbine map with cycle analysis in a bid to establish the range of heat source conditions that can be effectively utilised using this candidate turbine. The aim of this paper is to confirm the suitability of implementing this existing design into a range of different ORC applications. Not only will this further validate this turbine design, but it will also show how the economy of scale of small-scale ORCs can be improved. Within this paper the turbine and ORC models are summarised before the modeling methodology is demonstrated by considering a case study. Having shown how the model can be used to obtain the optimal operating point for a given heat source temperature and working fluid, the model is run for a range of heat source temperatures and working fluids and the range of potential applications for this candidate turbine is established.

3. Turbine modelling

3.1 Similitude theory

The performance of a turbine is described by the isentropic enthalpy change Δh_s, power output $\dot{W}_e$, and isentropic efficiency η_e. These parameters are a function of the turbine diameter D, rotational speed N, mass flow rate $\dot{m}_o$, and the fluid speed of sound a_3, and density ρ_3 at the turbine inlet. Dimensional analysis reduces these parameters Equation 1. The parameters on the left are the head coefficient, isentropic efficiency and power coefficient, whilst the parameters on the right are the flow coefficient, Reynolds number and blade Mach number.

$$\left[\frac{\Delta h_s}{N^2D^2}, \eta_e, \frac{\dot{W}_e}{\rho_3 N^3 D^5}\right] = f\left(\frac{\dot{m}_o}{\rho_3 N D^3}, \frac{\rho_3 N D^2}{\mu}, \frac{ND}{a_3}\right) \quad (1)$$

Typically, for ideal gas applications Reynolds number effects are neglected [16]. A performance map then shows the flow coefficient and isentropic efficiency as a function of the head coefficient and blade Mach number. For real gases, preliminary analysis has shown that it may not be suitable to neglect Reynolds number effects. After comparing predictions made using a performance map to CFD results it was found that the predictions agreed with the CFD results to within ±2% when the Reynolds number deviation was between ±75%. At deviations higher than +75%, accurate predictions were only made at subsonic conditions. Combining the Reynolds number and blade Mach number, whilst neglecting the turbine diameter, results in an expression that relates only the thermodynamic properties (Equation 2). Provided that the blade Mach number is maintained, this expression may be used to determine whether the deviation in Reynolds number is within the specified range.

$$\left(\frac{\rho_3 a_3}{\mu_3}\right)_A = \left(\frac{\rho_3 a_3}{\mu_3}\right)_B \quad (2)$$

3.2 Candidate ORC turbine and performance map

Previous work has led to the design of a candidate ORC radial turbine [21]. The design specification is summarised in Table 1. Performance data was obtained using CFD, and these results were used to generate the turbine performance curves shown in Figures 1 and 2. Within the CFD, real gas properties were accounted for by generating gas property tables using REFPROP; this is a program consisting of state-of-the-art equations of state for a large array of working fluids [22]. In Figures 1 and 2 the non-dimensional parameters in Equation 1 have been reduced to $(\Delta h_s/a_3^2)$, $(\dot{m}_o/\rho_3 a_3)$ and (N/a_3), since the diameter of the turbine does not change. The reduced rotational speeds range from 50% to 125% of the design value. Although operation at these extreme rotational speeds may not be feasible, these speeds have been selected to extend the operating envelope for the purposes of analysis only. After completing the analysis the rotational speed should be verified to ensure it lies within a feasible range.

Table 1. Thermodynamic conditions and design specification for the developed ORC turbine.

Working Fluid	-	R245fa	[-]	Isentropic Efficiency	η_e	85.00	[%]
Total Inlet Temperature	T_3	350.00	[K]	Pressure Ratio	PR	2.50	[-]
Total Inlet Pressure	P_3	623.06	[kPa]	Rotational Speed	N	37,525	[RPM]
Mass Flow Rate	$\dot{m}_o$	0.70	[kg/s]	Rotor Diameter	D	66.68	[mm]

The range of applicability for the developed turbine map is established in Figure 3. Here $(\rho_3 a_3/\mu_3)$ is calculated over a range of turbine inlet temperatures and pressures, and the resulting value is expressed as a percentage deviation from the value obtained at the original design point. This clearly demonstrates that this turbine, operating with R245fa, is suitable for ORC applications with turbine inlet temperatures between approximately 315K and 375K.

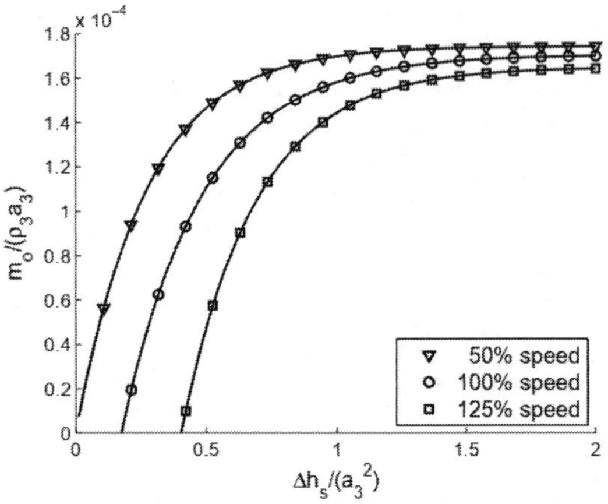

Figure 1. Variation of flow coefficient with head coefficient at 50%, 100% and 125% of the design blade Mach number.

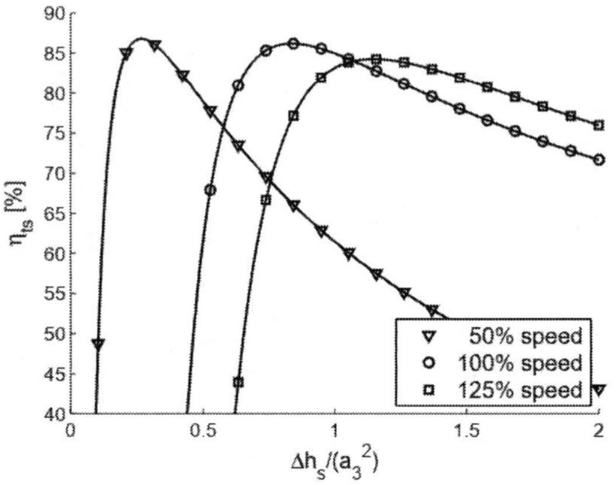

Figure 2. Variation in isentropic efficiency with head coefficient at 50%, 100% and 125% of the design blade Mach number.

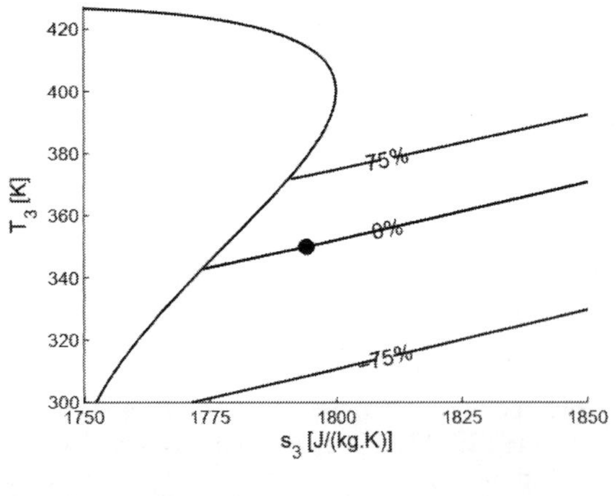

Figure 3. Deviation in Reynolds number with variations in turbine total inlet conditions for R245fa assuming that the blade Mach number has been conserved (the dot represents conditions at the turbine design point).

4. ORC modelling

A thermodynamic model has been developed that can establish the range of heat sources that can be efficiently converted into useful power using the existing turbine. Since the main focus of this work is the turbine, it is suitable to model a simple subcritical ORC without a regenerator. This simplifies the analysis, but also eases manufacture, and reduces the overall cost of the system. The thermodynamic analysis of the ORC is well covered within the literature [12,13], so the model will only be summarised here. The simple ORC model has been linked with REFPROP for real fluid properties. The T-s diagram for the ORC is shown in Figure 4, along with the notation used throughout this paper.

Assuming that the working fluid enters the pump as a saturated liquid, the input condensation temperature T_1 supplies all properties at location 1. The pump outlet conditions are then specified by the input ORC pressure ratio PR, and isentropic pump efficiency η_p (Equation 3).

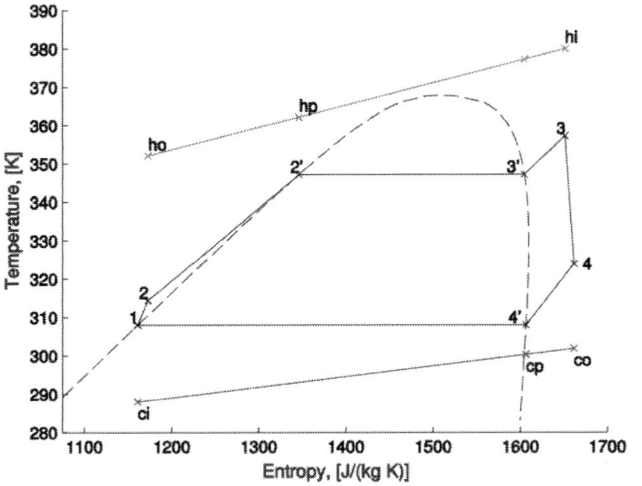

Figure 4. Typical T-s diagram for the simple subcritical low temperature ORC.

$$h_2 = h_1 + \frac{h_{2s} - h_1}{\eta_p} \tag{3}$$

Neglecting pressure drops through the evaporator the working fluid properties at locations 2' and 3' are known. The input superheat ΔT_{sh}, and evaporator pinch point PP_h, provide the ORC conditions at the location 3, and the heat source conditions at the pinch point (hp) respectively. An energy balance provides the ratio of working fluid mass flow rate $\dot{m}_o$, to heat source mass flow rate $\dot{m}_h$.

$$\frac{\dot{m}_o}{\dot{m}_h} = \frac{h_{hi} - h_{hp}}{h_3 - h_{2'}} \tag{4}$$

For the turbine the expression $(\rho_3 a_3/\mu_3)$ is first determined to ensure that it lies within the ±75% limits. The known isentropic enthalpy drop then supplies the reduced head coefficient $(\Delta h_s/a_3^2)$. Using Figure 1 this supplies the minimum and maximum flow coefficients, which correspond to 125% and 50% of the reduced rotational speed respectively. From this the limits for $\dot{m}_o$ are obtained, and an array of $\dot{m}_o$ values is constructed. For each $\dot{m}_o$ value the reduced rotational speed is obtained through interpolation of Figure 1. The reduced head coefficient and rotational speed then supplies the turbine efficiency η_e through interpolation of Figure 2. The working fluid properties at location 4 are determined using Equation 5, whilst $\dot{m}_h$ follows from Equation 4.

$$h_4 = h_3 - \eta_e(h_3 - h_{4s}) \tag{5}$$

Again, neglecting pressure drops the working fluid properties location 4' are known, whilst the condenser pinch point PP_c supplies the heat sink properties at the pinch point (cp). An energy balance then provides the required heat sink flow rate $\dot{m}_c$ (Equation 6). Having completed the cycle analysis, the performance of the cycle is assessed in terms of the net power produced $\dot{W}_n$, and the cycle efficiency η_o (Equation 7).

$$\dot{m}_c = \frac{\dot{m}_o(h_{4'} - h_1)}{(h_{ci} - h_{cp})} \tag{6}$$

$$\eta_o = \frac{\dot{W}_n}{\dot{Q}} = \frac{\dot{W}_e - \dot{W}_p}{h_3 - h_2} = \frac{(h_3 - h_4) - (h_2 - h_1)}{h_3 - h_2} \tag{7}$$

The power produced by the cycle is also compared to the maximum net power that could have be produced using the same heat source with a turbine operating at an optimal efficiency of 85%. For fixed values of T_1, ΔT_{sh}, PP_e, with a fixed heat source temperature and mass flow rate, there exists an optimal pressure ratio at which optimal power can be produced. This optimum exists because, whilst a higher pressure ratio increases cycle efficiency, it also reduces the heat source temperature drop, leading to a smaller ORC mass flow rate. Therefore a trade-off between maximising the cycle efficiency, and maximising the heat absorbed by the working fluid exists.

The top graph in Figure 5 shows this optimal pressure ratio for a range of heat sources with a fixed mass flow rate of 1.0kg/s. The bottom graph shows that for a fixed temperature, the optimal pressure ratio remains fixed, although the net power increases linearly with mass flow rate. Therefore using the input heat source temperature and the heat source flow rate calculated using Equation 4, and assuming T_1, ΔT_{sh} and PP_e remain constant, the maximum potential power can be obtained using Figure 5. A maximum turbine efficiency of 85% has been selected since this was considered to be an achievable target for design point efficiency. When comparing the actual net power to the maximum power, if the actual power is greater, this is the result of the turbine operating at a higher efficiency than 85%.

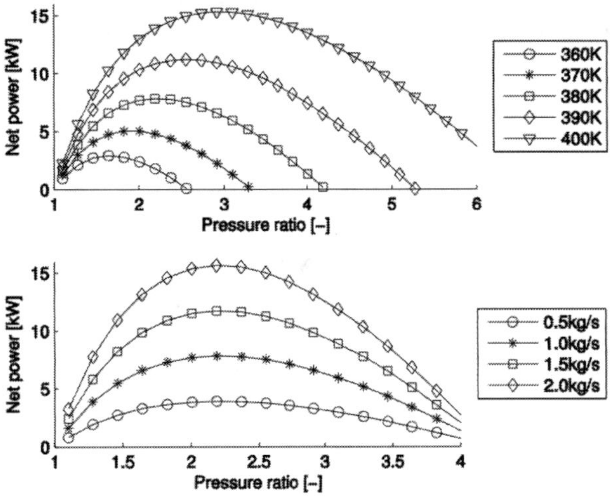

Figure 5. Variation in net power produced as a function of pressure ratio for different heat source conditions. Top: fixed heat source mass flow rate of 1.0kg/s; Bottom: fixed heat source temperature of 380K.

5. Results and discussion

To demonstrate the developed model a heat source of pressurised water at 380K and 400kPa has been considered. The ORC working fluid is R245fa. Throughout this paper T_1, PP_c, PP_h, ΔT_{sh} and η_p have all been fixed because the focus is on the turbine performance. Therefore the impact of variations in pump and heat exchanger performance, although important, has been neglected. T_1 and PP_c dictate the condenser area and the heat sink mass flow rate. The heat sink temperature is set at 288K, whilst setting $T_1 = 313$K and $PP_c = 10$K corresponds to an approximate 15K rise in the heat sink temperature through the condenser. The value for PP_h has been estimated to be 15K; both pinch points dictate the size of the heat exchangers, and represent a trade-off between performance and cost. The values selected have been found to provide a reasonable balance between these two aspects. It has been widely shown that superheating is not necessary for organic fluids, but a small superheat of 2K has been selected to ensure that the fluid is fully vaporised at the turbine inlet. Since the pump work is much smaller than the turbine work the impact of variations in pump efficiency are expected to be negligible; therefore the pump efficiency is assumed to be constant at 70%.

The ORC model was then run for a range of pressure ratios, and a range of working fluid mass flow rates was established at each pressure ratio. The resulting performance map shows how the net power, as a percentage of the maximum potential power, varies with pressure ratio and $\dot{m}_o$ (Figure 6).

The black lines indicate the heat source mass flow rate in kg/s. This figure is useful since, for a specified heat source mass flow rate at 380K, it is easy to assess the feasibility of using this turbine. For example, for a heat source mass flow rate of around 1.0kg/s, with a pressure ratio of 2.2, the turbine efficiency is high and 100% of the maximum potential net power can be achieved. The optimal operating point is shown as a black dot. As the mass flow rate changes, turbine performance deteriorates leading to a lower percentage of the maximum potential power being produced. For these heat sources an alternative turbine may offer better performance, and further analysis would be required to establish whether the improved performance would outweigh the increased costs of developing a different design. None the less, for a heat source mass flow rate between 0.7kg/s and 1.2kg/s a reasonable performance can be achieved with this existing design operating with R245fa. For these values the net power produced should remain above 90% of the maximum potential power.

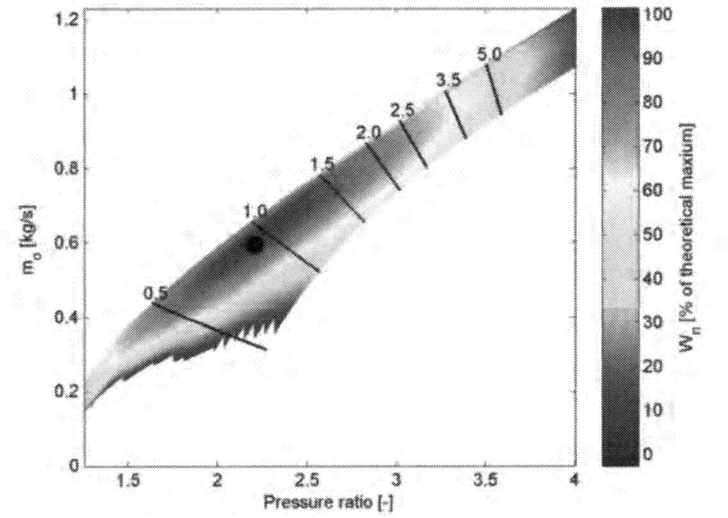

Figure 6. Contour of the net power produced by an ORC operating with the candidate turbine as a percentage of the maximum potential power. Heat source of water at 380K, and R245fa as working fluid. The black lines indicate the heat source mass flow rate in kg/s, whilst the black dot represents the point of optimal operation.

Although useful, Figure 6 only applies to one heat source, and one working fluid. The same study has therefore been repeated for a range of working fluids, at three different heat source temperatures. At each combination of heat source temperature and working fluid, the optimal heat source mass flow rate will be established, thus establishing the range of heat sources that can be effectively converted into power using this turbine. The three heat source temperatures selected for this study are 360K, 380K and 400K, and these span the range that could be effectively converted into power using this turbine. Below 360K the cycle efficiency would reduce significantly impacting the economic feasibility, whilst above 400K the pressure ratio increases, increasing the likelihood of supersonic flows. For these applications it would be more suitable to implement a supersonic turbine design.

A review of screening studies resulted in 50 potential fluids. However, after considering suitable operating conditions, the evaporation and condensation pressures were estimated. Fluids with sub-atmospheric condensation pressures or supercritical evaporation pressures were neglected. Common refrigerant mixtures were neglected as these were found to have high saturation pressures, near critical evaporator pressures or negative gradient saturated vapour domes. The values for T_1, PP_c, PP_e, ΔT_{sh} and η_p were again fixed, and the model was run over a range of pressure ratios for each heat source temperature and working fluid, and the optimal operating point was obtained.

Figure 7 summarises the results in terms of the optimal heat source mass flow rate and net power produced for each working fluid. The top-right graph in Figure 7 summarises all of the results, with each marker representing the result obtained for a particular working fluid at the respective heat source temperature. The remaining graphs expand these results, showing which working fluid each marker represents. The most striking observation is the large spread of heat source mass flow rates that be effectively utilised by this turbine. For example, for a heat source temperature of 400K the same turbine can effectively convert a heat source ranging from 0.47kg/s to 1.65kg/s by simply changing the working fluid. The resulting power output ranges from 7.03kW up to 26.16kW. Furthermore, across all of the operating points shown in Figure 7, the location of the optimal point (as previously described in Figure 6) is consistently close to 100% of the maximum potential power, thus corresponding to turbine isentropic efficiencies close to 85%. This confirms that at the corresponding heat source conditions, the ORC is operating at an optimal pressure ratio that corresponds to the optimal head coefficient for the candidate turbine. In other words, for the same heat source conditions it would be unlikely an alterative turbine would offer much improvement on the turbine, and cycle, performance.

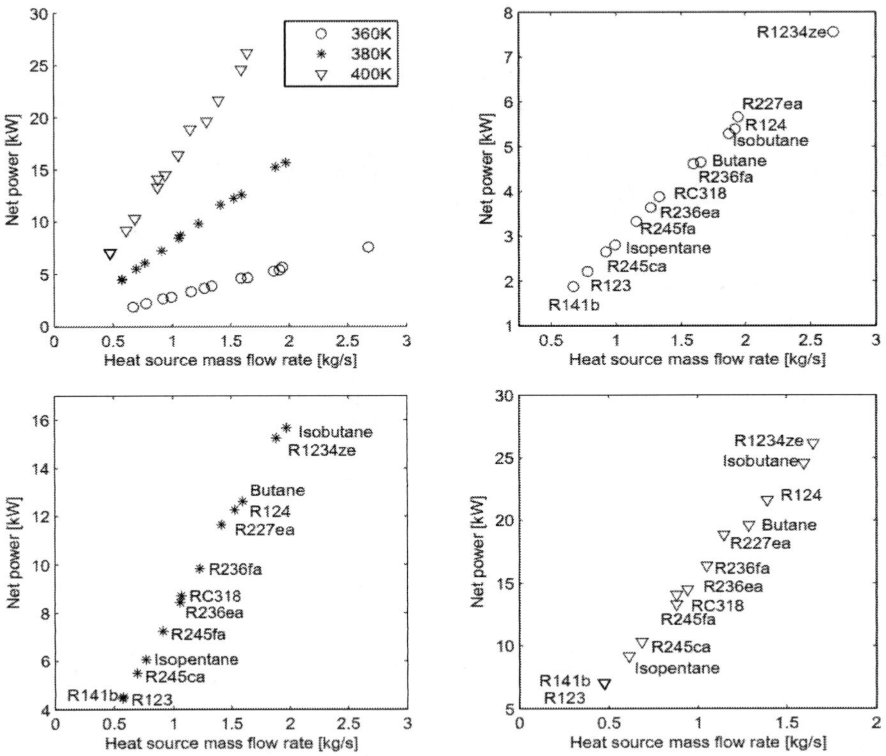

Figure 7. Cycle analysis results showing the heat source mass flow rates that can be accommodated by an ORC utilising the candidate turbine at each combination of heat source temperature and working fluid. Top left: summary of all results; top right: 360K; bottom left; 380K; bottom right; 400K.

From the results it is observed that as the heat source temperature increases, the non-dimensional speed at which optimal performance is obtained also increases. For 360K the optimal points in Figure 7 correspond to between 60% and 80% of the design non-dimensional rotational speed. For the 380K and 400K cases this increases to between 80% and 100%, and 100% and 120% of the design speed respectively. This suggests that below 360K, and above 400K, the non-dimensional speed would move outside practical limits for this turbine, confirming the heat source temperatures selected for this study.

The resulting cycle and turbine efficiencies are shown in Figure 8. As expected cycle efficiency increases with increasing heat source temperature, however there is only a small variation in cycle efficiency amongst the different working fluids. The turbine efficiency remains high for all cases; although it is clear the turbine operates most efficiently for the 380K heat source.

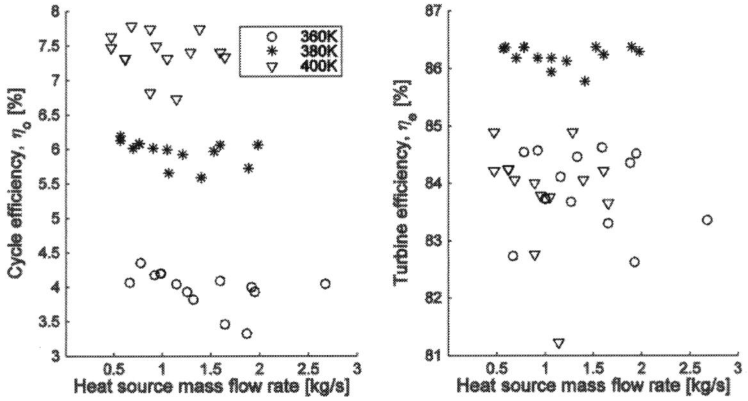

Figure 8. Cycle analysis results showing variation in cycle and turbine efficiencies at the different heat source temperatures.

Overall, Figure 7 suggests that the same turbine can be utilised within a number of different ORC applications by selecting a suitable working fluid. This allows the same turbine to be produced on a relatively large scale, improving the economic feasibility of the system. Then, when supplied with a defined heat source, Figure 7 can be used to indicate the most appropriate working fluid that will allow the turbine to operate most effectively, therefore producing the maximum power. It should be noted that additional mechanical factors, such as the bearing system and generator, are not taken into consideration during this study, and this may limit the feasibility of using the same turbine assembly across a wide range of power outputs. However in these instances, even if modifications to the mechanical design may be required, the costs associated with the aerodynamic design and manufacture of the stator and rotor assembly can be avoided.

6. Conclusion

This paper has extended work on the development of a radial turbine for low temperature ORC applications by combining a turbine's performance map with a thermodynamic analysis of the cycle. When supplied with a particular heat source temperature, this model has been used to establish the range of heat source mass flow rates that can be effectively converted into power using an ORC that incorporates this existing turbine design. It has been shown that when provided with a particular heat source temperature and working fluid there is an optimal heat source mass flow rate that can be accommodated by this design. This fact has been used to show that the same turbine can be effectively utilised within a range of different ORC applications, simply by changing the working fluid. This is a significant result, since it allows the same turbine design to be produced on a relatively large scale, but implemented into a variety of applications, improving the economy of scale. The results indicate that the examined turbine can be used for heat source temperatures ranging from 360K to 400K, with mass flow rates between 0.45kg/s and 2.7kg/s. The resulting power output ranges between 1.5kW and 27kW, depending on the working fluid selected. Further work should extend this analysis to investigate how similitude theory can be used to geometrically scale the existing turbine design, making it suitable for an even wider range of potential applications.

7. Acknowledgement

This work was funded by the UK Engineering and Physical Sciences Research Council (EPSRC).

8. References

[1] Turboden, 2014, "Turboden – Organic Rankine Cycle Turbogenerators for Clean Electric Energy Production", www.turboden.eu, [Accessed: 15/11/2015].

[2] ORMAT Technologies Inc., 2014, "ORMAT Technologies Inc.", www.ormat.com [Accessed: 15/11/2015].
[3] Infinity Turbine, 2014, "Infinity Turbine", www.infinityturbine.com, [Accessed: 09/12/2014].
[4] Electratherm, 2014, "Energy Efficient Green Machine Heat to Power Generation System", www.electratherm.com, [Accessed: 09/12/2014].
[5] Casati, E., Vitale, S., Pini, M., and Persico, P., 2014, "Centrifugal Turbines for Mini-Organic Rankine Cycle Power Systems", Journal of Engineering for Gas Turbines and Power, **136**(12).
[6] Lang, W., Colonna, P., and Almbauer, R., 2013, "Assessment of Waste Heat Recovery From a Heavy-Duty Truck Engine by Means of an ORC Turbogenerator", Journal of Engineering for Gas Turbines and Power, **134**(4).
[7] Bao, J., and Zhao, L., 2013, "A review of working fluid and expander selections for organic Rankine cycle", Renewable and Sustainable Energy Reviews, **24**, pp. 325-342.
[8] Bracco, R., Clemente, S., Micheil, D., and Reini, M., 2013, "Experimental tests and modelization of a domestic-scale ORC (Organic Rankine Cycle)", Energy, **58**, pp. 107-116.
[9] Declaye, S., Quoilin, S., Guilliaume, L., and Lemort, V., 2013, "Experimental study on an open-drive expander integrated into an ORC (Organic Rankine Cycle) system with R245fa as working fluid", Energy, **55**, pp. 173-183.
[10] Leibowitz, H., Smith, I. K., and Stosic, N., 2006, "Cost Effective Smalle Scale ORC Systems for Power Recovery from Low Grade Heat Sources", IMECE2006-14284, Proc. 2006 ASME International Mechanical Engineering Congress and Exposition, Chichago, Illinois, USA.
[11] Quoilin, S., Van Den Broek, M., Declaye, S., Dewalleg, P., and Lemort, V., 2013, "Techno-economic survey of Organic Rankine Cycle (ORC) systems", Renewable and Sustainable Energy Reviews, **22**, pp. 168-186.
[12] Saleh, B., Koglbauer, G., and Wendland., M., and Fischer., J., 2007, "Working fluids for low-temperature organic Rankine cycles", Energy, **32,** pp. 1210-1221.
[13] Tchanche, B. F., Papadakis, G., Lambrinos, G., and Frangoudakis, A., 2009, "Fluid selection for a low-temperature solar organic Rankine cycle", Applied Thermal Engineering, **29**, pp. 2468-2476.
[14] Sun, J., and Li, W., 2011, "Operation optimization of an organic rankine cycle (ORC) heat recovery power plant", Applied Thermal Engineering, **31**, 2032-2041.
[15] Wang, Z. Q., Zhou, N. J., Guo., J., and Wang, X. Y., 2012, "Fluid selection and parametric optimization of organic Rankine cycle using low temperature waste heat", Energy, **40**, 107-115.
[16] Moustapha, H., Zelesky, M. F., Baines, N. C., and Japiske, D., 2003, *Axial and Radial Turbines***,** Concepts ETA, Inc.
[17] Aungier, R. H., 2006, *Turbine Aerodynamics: Axial-Flow and Radial In-flow Turbine Design and Analysis,* ASME, United States.
[18] Manente, G., Toffolo, A., Lazzaretto, A., and Paci, M., 2013, "An Organic Rankine Cycle off-design model for the search of the optimal control strategy", Energy, **58**, pp. 97-106.
[19] Calise, F., Capuozzo, C., Carontenuto, A., and Vanoil, L., 2014, "Thermoeconomic analysis and and off-design performance of an organic Rankine cycle powered by medium temperature heat sources", Solar Energy, **103**, pp. 595-609.
[20] Li, J., Pei, G., Li, Y., Wang, D., and Ji, J., 2013, "Examination of the expander leaving loss in variable organic Rankine cycle operation", Energy Conversion and Management, **65**, pp. 66-74.
[21] White, M., and Sayma A. I., 2015, "The One-Dimensional Meanline Design of Radial Turbines for Small Scale Low Temperature Organic Rankine Cycles", Proc. GT2015 ASME Turbo Expo: Turbine Technical Conference and Exposition, Montreal, Canada.
[22] Lemmon, E. W., Huber, M. L., and McLinden, M. O., 2013, "NIST Standard Reference Database 23: Reference Fluid Thermodynamic and Transport Properties-REFPROP", Version 9.1, National Institute of Standards and Technology, Standard Data Program, Gaithersburg.

Update on single-screw expander geometry model integrated into an open-source simulation tool

D. Ziviani[1], **I. H. Bell**[2], **M. De Paepe**[1], **M. van den Broek**[1]

[1]Department of Flow Heat and Combustion Mechanics - Ghent University - UGent, Sint-Pietersnieuwstraat 41, 9000 Gent, Belgium
[2]Aerospace and Mechanical Engineering Department, University of Liège, Campus du Sart Tilman B49, B-4000 Liège, Belgium

E-mail: davide.ziviani@ugent.be, ian.h.bell@gmail.com, martijn.vandenbroek@ugent.be, michel.depaepe@ugent.be

Abstract. In this paper, a mechanistic steady-state model of a single-screw expander is described with emphasis on the geometric description. Insights into the calculation of the main parameters and the definition of the groove profile are provided. Additionally, the adopted chamber model is discussed. The model has been implemented by means of the open-source software PDSim (Positive Displacement SIMulation), written in the Python language, and the solution algorithm is described. The single-screw expander model is validated with a set of steady-state measurement points collected from a 11 kWe organic Rankine cycle test-rig with SES36 and R245fa as working fluid. The overall performance and behavior of the expander are also further analyzed.

1. Introduction

In the optimization of organic Rankine cycle (ORC), there is an open question about the choice of the appropriate expansion device for different applications and working conditions. Because of their wide power range, overall efficiency and good part-load performance, screw expanders are considered one of the most promising technologies. However, challenges arise especially at the lower capacities where other positive displacement machines , e.g. scroll, rotary vane, spool, etc., are more cost-effective. Since the early developments in the 1930s, twin-screw compressors have been widely used in the refrigeration and process gas industries. The single-screw compressor (SSC) concept was developed in the 1960s by B. Zimmern [1]. Its configuration has a number of advantages over the twin-screw such as balanced loading of the main screw, higher volumetric efficiency, long working life, low vibrations and a simplified configuration. Single-screw machines also require less demanding lubricating conditions than twin-screw machines. Due to these characteristics, ORC systems based on single screw expander (SSE) have recently gained attention. In the literature, a number of studies are dedicated to model and characterize the performance of SSCs. However, many of the results are either incomplete with respect to the theory behind the machine design or reveal inconsistencies in the data provided (e.g.[2] and[3]). On the other hand, SSEs have only been investigated experimentally [4], but the experimental data published was limited to compressed air as working fluid, not realistic for ORC applications.

Content from this work may be used under the terms of the Creative Commons Attribution 3.0 licence. Any further distribution of this work must maintain attribution to the author(s) and the title of the work, journal citation and DOI.
Published under licence by IOP Publishing Ltd

IOP Conf. Series: Materials Science and Engineering **90** (2015) 012064 doi:10.1088/1757-899X/90/1/012064

Nomenclature

C	Distance (m)
D	Diameter (m)
h	Specific Enthalpy (kJ/kg)
	Heat Transfer Coefficient (kW/m^2K)
i	Transmission Ratio (-)
L	Length (m)
m	Mass (kg)
$\dot{m}$	Mass Flow Rate (kg/s)
p	Pressure (kPa)
$\dot{Q}$	Heat Rate (kW)
T	Temperature (K)
u	Specific Internal Energy (kJ/kg)
v	Specific Volume (m^3/kg)
V	Volume (m^3)
$\dot{V}$	Volumetric Flow Rate (m^3/s)
$\mathbf{v}$	Velocity Vector (m/s)
w	Tooth width (m)
z	Number of Grooves or Teeth ()
α	Angle (rad)
β	Angle (rad)
	Expansion Ratio(-)
δ	Clearance (m)
ε	Effectiveness (-)
η	Efficiency (-)
θ	Rotation Angle (rad)
ω	Angular Speed (rad/s)
Ω	Angle (rad)

Subscript

CV	Control Volume
exp	expander
el	electric
d	discharge
eff	effective
f	fixed
g	groove
int	internal
is	isentropic
l	liquid
m	mechanical
meas	measured
nom	nominal
s	suction
sh	shaft
sr	screw rotor
sw	starwheel
th	theoretical
v	volume

The authors proposed a new mechanistic model of SSE with a preliminary validation [5] and additional experimental results were also discussed [6]. The main issues that have been pointed out were the profile generation method and the mathematical description of the SSE geometry. In this paper, the equations used for the profile generation are provided on the basis of the work published by Yang S-C [7]. Additional geometry aspects such as suction and discharge angles, sealing lines, swept volume calculation are further analyzed. The deterministic model is briefly described as well as the integration into the open-source software PDSim [8]. An 11 kWe SSE, shown in Fig. 1(a) with a 122 mm rotor is used as reference case. The expander has been tested with two working fluids, i.e. SES36 and R245fa, and the details of the experimental setup are presented in the companion paper [9].

2. Single-screw expander geometry description

The expansion process of the SSE occurs simultaneously on both sides of the rotor. The inlet flow is split in two parts to reach each of the inlet ports. During the suction process, Fig. 1(b), the groove is filled by working fluid until the left flank of the groove passes the right edge of the triangular -shaped port. At suction closure, the working fluid starts the expansion process, Fig. 1(c). When the right side of the engaged tooth begins to leave the groove, the discharge process occurs, Fig. 1(d). Reference studies about the geometry of SSE can be found in literature concerning SSC, for example [2, 10–12]. However, the formulation is not always comprehensive and many aspects are missing. In the following, a description of the principles behind the SSE design are highlighted and the symbols adopted are in accordance with the open literature in order to facilitate the reading. The geometric description of the SSE involves a large number of variables and parameters. In particular, it is possible to group the variables in three groups: independent, dependent and limit-set variables. Besides the variables, a number of empirical

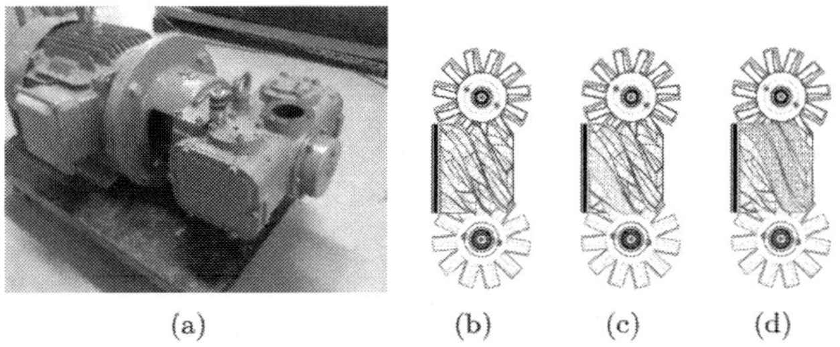

Table 1. Geometric parameters of 11 kWe single-screw expander tested.

Engaging ratio	[-]	11/6
D_{sr}	[mm]	122
D_{sw}	[mm]	132
$V_{g,max}$	[cm^3]	57.39
$r_{v,built-in}$	[-]	4.7
L_{rotor}	[m]	121

Figure 1. (*a*) 11 kWe single-screw expander considered in the present study; Graphic representation of the single-screw expansion process: (*b*) suction; (*c*) closed expansion; (*d*) discharge. To be noted that the expansion process occurs simultaneously also on the opposite side of the rotor.

design parameters have to be selected. The independent variables include the number of grooves of main rotor, z_{sr}, the number of teeth of the starwheels, z_{sw}, the number of starwheels, N_{sw}, the main rotor diameter, D_{sr}, the starwheel outer and inner diameters, D_{sw}, $D_{sw,in}$ and the tooth width. The considered SSE has the standard configuration of 6 grooves in the main rotor and 11 teeth in the starwheels, which automatically defines the correlation between rotation angle and angular speed of the meshing pair:

$$i = \frac{z_{sw}}{z_{sr}} = \frac{\theta_{sr}}{\theta_{sw}} = \frac{\omega_{sr}}{\omega_{sw}} = \frac{11}{6}. \tag{1}$$

where ω_{sr} and ω_{sw} are the angular speed of main rotor and starwheel, respectively. The number of grooves typically is a multiple of the number of starwheels in order to balance the forces acting on the main rotor. In fact, the expansion or compression process will be symmetric in both sides of the rotor since the same number of grooves are engaged. It should be noted that the number of starwheels and their positions determine the type of configuration of the single screw machine, i.e. CC, CP, PC, PP [13]. The most common one is Cylinder-Plate (CP) and the reason behind it is explained in the following section dealing with the kinematics of meshing process. The number of teeth of the starwheel should not have a common factor with the number of grooves to guarantee the uniformity of the meshing and control the wear of the teeth. Increasing or decreasing the number of teeth has also the effect of modifying the swept volume [10]. The diameters of main rotor and starwheel are related by a design parameter, $D_{sw} = \lambda_d D_{sr}$. Normally, D_{sw} is either equal to or slightly larger (up to 10%) than D_{sr}. The diameter of the starwheel and the distance between the center axes of main rotor and starwheel, $d_{sr,sw}$ (in many cases $d_{sr,sw} = 0.8\,D_{sr}$) , influence the maximum groove volume and the degree of penetration of the tooth into the groove. The higher the degree of penetration and the higher the groove volume. However, there exists a maximum around 0.45$\sim$ 0.50 of penetration after which the width of the tooth has to be reduced in order to guarantee the structural integrity of the groove. For obvious reason there also is a structural limit related to the shaft diameter, $R_{sr} - (R_{sw} - R_{sw,in}) > R_{sr,sh}$. Dependent variables are related to the calculation of engaging angles and the length of the rotor sections. By referring to Fig. 2(a), it is possible to define:

$$\begin{aligned} \beta_s &= \cos^{-1}\left(\frac{d_{sr,sw} - R_{sr}}{R_{sw}}\right), \beta^* = \sin^{-1}\left(\frac{w}{2R_{sw}}\right) \\ \beta^*_{in} &= 2\sin^{-1}\left(\frac{w}{2R_{sw,in}}\right), \beta^*_{in} < \frac{2\pi}{z_{sw}} \end{aligned} \tag{2}$$

as well as:

$$L_{rot,eff} = L_s + L_d = R_{sw} \left(\sin \beta_s + \sin \beta_d \right) \ , \ L^* = (0.1 \sim 0.15) D_{sr} \tag{3}$$

The selection of discharge angle β_d is related to the total engaging angle of the starwheel α_{sw} and typically $\beta_d = 0.7\,\beta_s$. The starwheel engaging angle and the corresponding groove wrap angle around the main rotor, α_{sr}, have an upper limit established by:

$$\alpha_{sr} = \alpha_{sw} \frac{z_{sw}}{z_{sr}} < \frac{2\pi}{N_{sw}} \tag{4}$$

Typically, $\alpha_{sw} \cong 90°$ and the explanation is given by the following relationship between the number of teeth engaged simultaneously, $n_{sw,eng}$, and the angular spacing of the teeth,γ:

$$\begin{aligned} \alpha_{sw} &= 2\beta^* n_{sw,eng} + 2.5 \left(\gamma - 2\beta^* \right) \ , \gamma = 2\pi / z_{sw} \\ \alpha_{sw} &= \beta_{open} + \beta_s + \beta^* \end{aligned} \tag{5}$$

where β_{open} represents the point of discharge opening. It is therefore a variable that defines a limit and determines how far the groove wraps around the rotor prior discharge. In other words, it influences the swept volume at the discharge angle and therefore the internal volume ratio. Finally, the tooth width is usually regarded as independent variable. The wider the tooth, the wider is the groove which results in an increased groove volume. However, there is a structural limit related to the minimum groove tip thickness, Δ. The analytic expression of the tooth width w, with the geometry and structural limits is given by:

$$\begin{aligned} &\Delta = \xi\, D_{sr}, \ \xi = 0.017 \sim 0.025 \\ &w = 2 \left(d_{sr,sw} - \frac{D_{sr}}{2} \right) \sin \frac{\gamma}{2} - \xi\, D_{sr} \cos \frac{\gamma}{2} \end{aligned} \tag{6}$$

A good approximation can also be obtained as $w \approx 0.3\, R_{sr}$. The geometric relationships can be developed on the basis of a reference system defined as in Fig. 2(a). In particular, a set of cylindrical coordinates (z, r, θ_{sr}) are positioned so that the z axis is coincident with the main rotor rotation axis and the r axis is perpendicular to the z axis and has its origin in correspondence of the perpendicular intersecting line from the starwheel rotation axis. A general point on the starwheel rotates in the z, r plane defined by the main rotor coordinates. On the starwheel an analogous reference system can be defined associated with the rotation angle θ_{sw} (see Fig. 2(a)).

3. Profile generation: direct problem of the main rotor

The geometry and kinematics of the SSE are directly linked and it is necessary to solve the groove profile generation in order to accurately describe the machine. From the fundamentals of gear kinematics, the single-screw type of meshing can be seen as a double meshing worm gear [14]. The conjugate worm gear surfaces rely on continuous contact lines that ensure the load-transmission. By transferring this concept to an expander (or compressor), the calculation of the contact lines and the precision during the manufacturing determine the sealing capabilities of the expansion (or compression) chamber. Generally, both the direct and indirect problems have to be solved [14]. In the following, only the direct problem, i.e. the cutting tool (tooth profile in this case) is given and the conjugated groove profile is calculated. The SSE considered in this study presents a single straight line meshing profile (SSLMP). From the gear kinematics theory [14], the meshing process of the screw rotor (sr) with two starwheels (sw) can be described as an envelope of two-parameter family of surfaces with a cylindrical-plate (CP) configuration. In particular, the generating surface is represented by the starwheel, i.e. plate, (direct problem). The profile

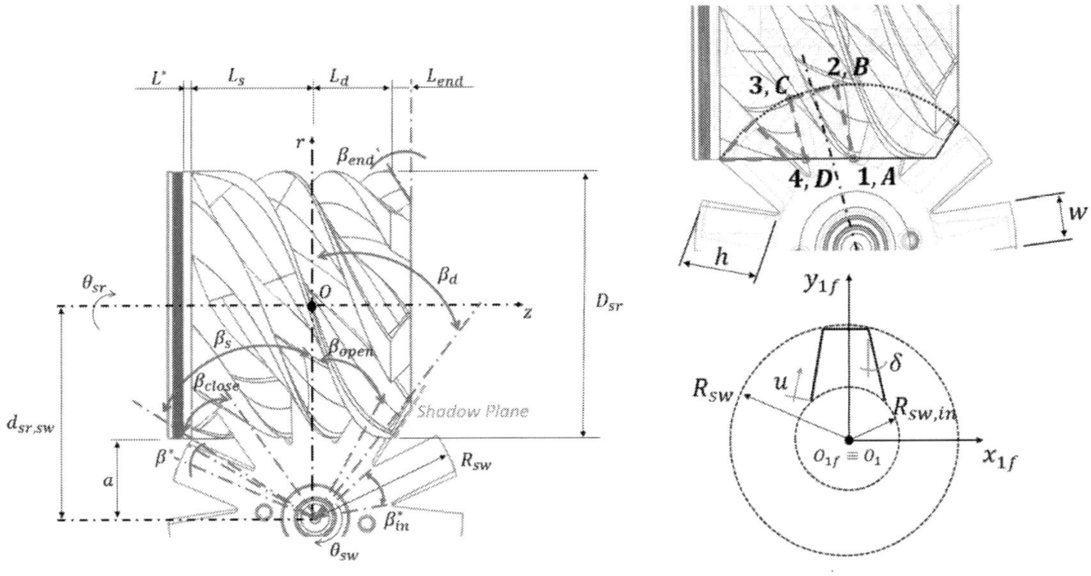

(a) (b)

Figure 2. (*a*) Schematic of the main geometric parameters of the SSE; (*b*) description of a single engaging tooth into the mating groove.

of the starwheel is obtained by applying the inverse envelope concept. A right-handed fixed, $O_{i,f}(x_{if}, y_{if}, z_{if})$, and rotating, $O_i(x_i, y_i, z_i)$, coordinate systems have to be defined for both screw rotor ($i = 1$) and starwheel ($i = 2$). Each starwheel tooth engages the mating groove with three sides, identified by segments $\overline{AB}$, $\overline{BC}$ and $\overline{CD}$. The generic representation of a starwheel tooth is proposed in Fig. 2(b). A tooth with parallel flanks is obtained by fixing $\delta = 0$. Each segment of the tooth is described by a parameter u. The second parameter is the thickness of the tooth, t. By adapting the notation introduced in [14] and by applying the procedure described in [7], the system of coordinates in the screw rotor reference frame O_1 is given by:

$$\begin{cases} x_1 = -x_2 \sin\theta_{sw} \cos\theta_{sr} - y_2 \cos\theta_{sw} \sin\theta_{sr} + z_2 \cos\theta_{sr} + d_{sr,sw} \sin\theta_{sr} \\ y_1 = -x_2 \sin\theta_{sw} \cos\theta_{sr} - y_2 \cos\theta_{sw} \cos\theta_{sr} - z_2 \sin\theta_{sr} + d_{sr,sw} \cos\theta_{sr} \\ z_1 = x_2 \cos\theta_{sw} - y_2 \sin\theta_{sw} \end{cases} \tag{7}$$

where the relationship between θ_{sw} and θ_{sr} is given in (1). Eq. (7) can be further reworked to be used to generate the meshing curve. In particular:

$$\begin{cases} x_1 = -r \sin(\theta_{sr}) \cos(\theta_{sw} - \delta_u) + d_{sr,sw} \sin\theta_{sr} \\ y_1 = -r \cos(\theta_{sr}) \cos(\theta_{sw} - \delta_u) + d_{sr,sw} \cos\theta_{sr} \\ z_1 = -r \sin(\theta_{sr}) \cos(\delta_u - \theta_{sw}) \\ \delta_u = \pm \sin\left(\frac{w}{D_{sw}}\right) \\ r = \sqrt{\left(R_{sw} \cos\left(\sin^{-1}\left(\frac{w}{D_{sw}}\right) + h_{tooth}\right)\right)^2 + \frac{w^2}{4}} \end{cases} \tag{8}$$

The rotor is obtained through 3D cut-sweep by implementing Eq. 8 in a parametric form with the main rotor angular position as the running parameter. The meshing curve and the 3D model of the rotor are shown in Fig. 3(a).

4. Unwrapped groove curves and volume ratio calculation

At this point, having defined all the geometric parameters of the main rotor, it is possible to derive the helical equation of the groove in order to be able to represent it unwrapped on a plane [15]. A local reference system (Θ, z) is defined in order to obtain the 2D groove graphics where Θ is the unwrapped curvilinear coordinate of the rotor and z is the coordinate along the main rotor axis. In the (Θ, z), the extremities of each groove line have a distance equal to the wrap angle. Mathematically:

$$\begin{cases} \Theta = a \tan \theta_{sw} \pm \frac{w}{2 \cos \theta_{sw}} \\ z = R_{sw} i \left(\theta_{sw} + \beta_d\right), \end{cases} \quad - \beta_{open} \leq \theta_{sw} \leq \alpha_r(\alpha_l)$$
$$\alpha_r(\alpha_l) = \cos^{-1}\left(\frac{d_{sr,sw} - R_{sr}}{R_{sw}}\right) \pm sin^{-1}\left(\frac{w}{D_{sw}}\right) \tag{9}$$

where the $\pm$ sign is related to the subscripts r, $(+)$, and l, $(-)$, i.e. right and left sides of the groove, β_{open} is the opening angle which is influenced by the rotor end angle, β_{end}. The unwrapped groove is plotted in Fig. 3(c). The planar representation of the groove is necessary in order to calculate the sealing lines of the groove tips. The length of the groove tip leakage line at each main rotor angular position is given by,

$$L_{leak,tip} = \sum_{i \cdot \beta_s}^{\theta_{sr}} \sqrt{(\Delta\Theta)^2 + (\Delta z)^2} \tag{10}$$

Additionally, it can be used to estimate the internal built-in volume ratio. In fact, by introducing the real geometry of the suction port and its location, a crossing-algorithm can be implemented to recognize when the groove is crossing the suction port until suction closure, as shown in Fig. 3(d) and Fig. 3(e). Since the angle at which the discharge process begins can be calculated, β_{open}, the volume ratio can be determined. The correct estimation of the volume ratio allows to analyze the matching problem of the expander with the application. The evolution of the real suction port area, shown in Fig. 4(a),with respect the main rotor rotating angle is proposed in Fig. 4(b). A similar behavior can also be found in [10]. As a consequence, it is possible to optimize the suction process by investigating the shape and the area of the suction port. To be noted is that in Fig. 3(c), that at the current position of incipient discharge opening, the groove volume is not at its maximum. This is a characteristic of the single screw. By delaying the opening (β_{open}) it is possible to adjust the volume at discharge. In general, the swept volume can be defined as a function of D_{sr}^3, [2]. The swept volume is calculated by correlating the engaging tooth area into the corresponding groove and the tooth engaging angle. Generally, two approaches can be considered to obtain the engaging area, i.e. through a numerical integration of the tooth engaging area [16] or a polygonal approach which requires the definition of four corner points of a tooth at each rotation angle [12, 17]. A comparison between the different mathematical models of the groove volume is proposed in Fig. 4(b). To be noted that an analytic solution can be found by in integrating the function described in [16]. The numerical difference between the numerical integration and the analytic solution is negligible. However, the latter one reduces the overall computational time. A slight difference, around 3%, exists with respect to the polygonal approach. The polygonal description of the tooth is the adopted method to be included in the deterministic model because it allows the accurate calculation of the sealing lines on the tooth. The theoretical volume flow rate is given by:

$$\dot{V}_{th} = 2 z_{sr} \frac{V_{g2}}{r_{v,builtin}} n. \tag{11}$$

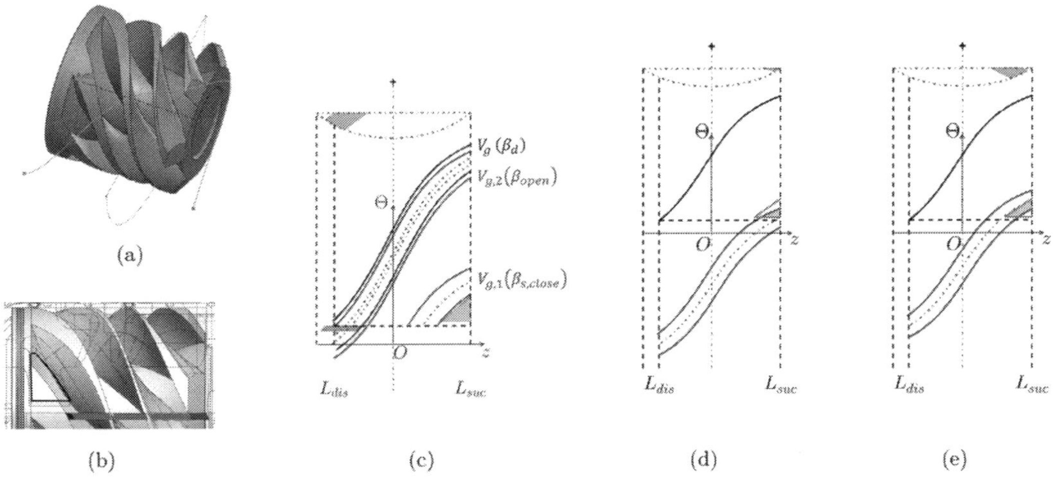

Figure 3. (*a*) Screw rotor profile generation in SolidWorks®; (*b*) 3D view of meshing condition at the beginning of the discharge, i.e. $\theta_{sw} = \beta_{open}$; (*c*) 2D representation of a groove where $V_{g,1} = V_g(\theta_{sw} = \beta_{s,closure})$,$V_{g,2} = V_g(\theta_{sw} = \beta_{open})$ and $r_{v,built-in} = V_{g,2}/V_{g,1}$; (*d*) Groove tip leading edge crossing the suction port; (*e*) groove tip trailing edge closing the suction port.

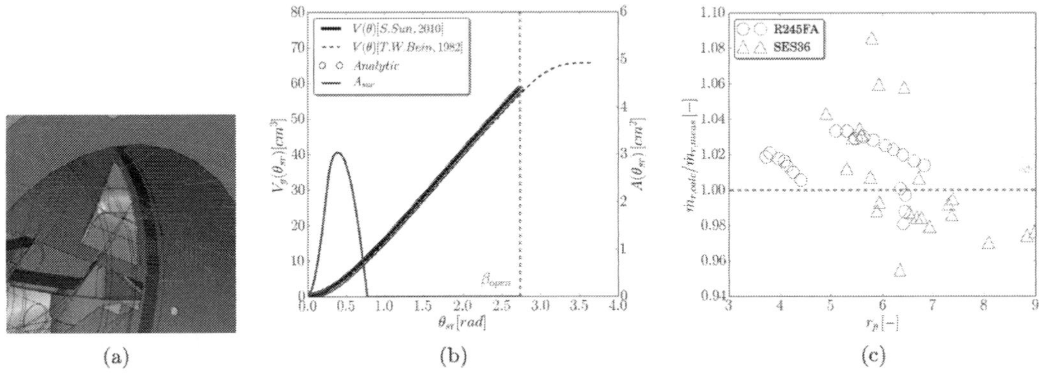

Figure 4. (*a*) 3D view of the suction port area; (*b*) Suction port area and comparison of different models for calculating the groove volume at each rotating angle of the main rotor; (*c*) Parity plot for calculated and measured values of mass flow rate for R245FA and SES36 .

5. Mechanistic model and Integration into PDSim

The mechanistic model is implemented by using a generalized framework for the simulation of positive displacement (PD) compressors and expanders, namely PDSim, developed by Bell [8]. The open-source software package[1] represents a useful tool that aims to overcome the non-availability of commercial software about mechanistic models of PD machines. PDSim has been developed in the Python programming language and its object-oriented structure allows to adapt the general core code to any PD compressor or expander. The basic idea is to integrate a geometry model specific for each machine into a general multi-chamber (or multi-control

[1] The PDSim website is at http://pdsim.sourceforge.net.

Table 2. Summary of the minimum and maximum measured values of the variables considered for both working fluids.

	Independent variables			Dependent variables			Performance indicators	
	$p_{su,exp}$ (kPa)	$p_{ex,exp}$ (kPa)	$T_{su,exp}$ (°C)	$\dot{m}$ (kg/s)	$\dot{W}_{el,exp}$ (W)	$T_{ex,exp}$ (°C)	$\eta_{is,sh,exp}$ (-)	ϕ_{FF} (-)
R245fa								
Min	566	120	106.7	0.12	1283	75.5	0.377	1.038
Max	1230	232	124.9	0.37	7364	104.6	0.624	1.331
SES36								
Min	450	90.7	91.65	0.150	903.8	73.99	0.542	0.897
Max	1028	177.9	125	0.406	7865	102.0	0.7232	1.117

volumes) approach based on a set of governing equations, standard flow models library, heat transfer and mechanical losses routines, as outlined in [5]. The analysis of a compressor or expander is based on a control volume approach which is representative of the internal working chamber. By assuming uniform temperature in the control volume, negligible kinetic energy and gravitational effects, thermal interaction through heat transfer, open control volume in terms of mass flow rate in and out, the conservation of mass and energy are imposed on the control volume. The thermo-physical properties of the working fluid are retrieved from the CoolProp library [18]. Due to the fact that the equations of state implemented are explicit in temperature and density, the governing equations have been expressed explicitly in terms of the derivative of temperature and density with respect the crack angle. For positive displacement machines, it is convenient to express the governing equations in terms of crank angle rotation rather than by time-dependent equations. The presence of lubricant oil (referred from now on as liquid) is accounted by assuming an homogeneous mixture. The additional state variable needed to fully describe a mixture of refrigerant and liquid has been chosen to be the liquid mass fraction, x_l. Therefore, the set of differential equations to be solved simultaneously for each control volume are the conservation of the total mass , the conservation of energy and the conservation of liquid mass. In particular, the conversation of energy is expressed in terms of $dT/d\theta$ and by recalling that $m_l = x_l \cdot m_{CV}$, it is possible to express the conservation of the liquid mass, m_l, in terms of the derivative of the liquid mass fraction with respect the crank angle, $dx_l/d\theta$. The complete mathematical derivation of the governing equations can be found in [19]. More details about the cycle solver algorithm can be found in [8].

6. Results and discussion

A total of 60 steady-state points have been determined from the experimental data for both R245fa and SES36. The description of the ORC test rig and the working fluid comparison have been presented in [6]. A specific analysis of the SSE with R245FA is proposed in a companion paper [9]. In Tab. 2, the minimum and maximum working conditions for both working fluids are listed. To be noted that only the electrical power output at the inverter was measured. However, by introducing the correlations for the inverter effectiveness and generator efficiency previously obtained [9], it has been possible to estimate the shaft power . Therefore,the expander isentropic efficiency at the shaft is defined as:

$$\eta_{is,sh,meas} = \frac{\dot{W}_{sh,calc}}{\dot{W}_{is,exp}} = \frac{\dot{W}_{el,meas}}{\eta_{mech,gen}\,\varepsilon_{el,inv}\left[\dot{m}_{meas}\left(h_{su,exp} - h_{ex,is,exp}\right)\right]} \tag{12}$$

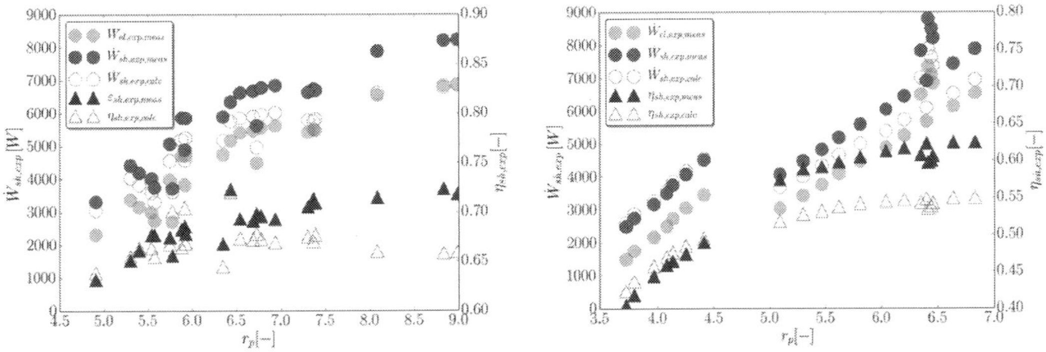

(a) (b)

Figure 5. Comparison between predicted and measured shaft power and isentropic efficiency: (*a*) SES36; (*b*) R245FA.

In the following, the analysis of the experimental data is limited to the nominal rotational speed of 3000 rpm. The combination of the water-glycol air-cooled condenser and the thermophysical properties of the fluids have an impact on the overall performance. In particular, SES36 was tested during winter which allowed to have low condensing temperature and pressure with benefits on the power output and isentropic efficiency. R245fa presents a higher condensing pressure compared to SES36, and the spring conditions were not suitable to assess the maximum performance of the SSE. The tuning of the mass flow rate is mainly related to the leakage gap areas and the uncertainties associated to the non-perfect contact areas of the tooth flanks with the corresponding groove sides. Under the strong assumption that the leakage areas are constant, the mass flow rate is predicted within 3.5% for both fluids operating conditions, as shown in Fig. 4(c). By looking at the behavior of the mass flow rate relative errors at different pressure ratio, it is clear that the gaps vary with the operating conditions, typically due to rotor deformations, presence of oil and additional clearances during the meshing process. A correlation should be introduced to take into account rotational speed, pressure ratio. The predictions of the shaft power and the isentropic efficiency are more affected by inaccuracy, as shown in Fig. 5(a) and Fig. 5(b). Two source of errors can be identified: the measured shaft power is actually calculated by introducing generator and inverter performance maps since real measurements at the shaft were not available; the mechanical losses have been assumed to vary linearly with the rotational speed. The mechanical losses can be correctly estimated only if the flow irreversibilities and heat losses have been captured by the model. As a result, the shaft power has been predicted within 17% for both fluids. Slightly better convergence is obtained for R245FA with a mean relative error for the shaft power of 9.7%.

7. Conclusions

In this paper the geometry of a single-screw expander (SSE) has been analyzed in depth with emphasis on the profile generation and calculation of the main parameters. A 11 kWe SSE has been considered as test case and the geometry model has been validated with the real dimensions. The experimental performance has also been presented for two different fluids, i.e. SES36 and R245fa. Finaly, the geometry model has been integrated into a mechanistic model by using the open-source software PDSim.

Acknowledgment
The authors would like to thank Feilong Liu for his help in the understanding of the single-screw manufacturing process and profile generation design methods.

References

[1] Zimmern B and Patel G C 1972 Design and operating characteristics of the zimmern single screw compressor *International Compressor Engineering Conference* paper 16

[2] Sun G 1988 The investigation of some basic geometric problems of the single screw co *International Compressor Engineering Conference* paper 630

[3] Jin G, Zhang S and Yu X 2006 Theoretical analysis of diameter ratio of engagement pair for singe screw compressor *International Compressor Engineering Conference* paper 1831

[4] He W, Wu Y, Peng Y, Zhang Y, Ma C and Ma G 2013 *Applied Thermal Engineering* **51** 662 – 669

[5] Ziviani D, Bell I H, De Paepe M and van den Broek M 2014 Comprehensive model of a single screw expander for orc-systems applications *22th Int.Compressor Engineering Conf. at Purdue* 1506

[6] Gusev S, Ziviani D, Bell I H, De Paepe M and van den Broek M 2014 Experimental comparison of working fluids for organic rankine cycle with single-screw expander *Proceedings of 15th Int. Refrig. Air Cond. Conf. at Purdue* 2653

[7] Yang S C 2002 *Proceedings of IMechE, Part C: J. of Mechanical Engineering Science* **216** 343–351

[8] Bell I H, Lemort V, Groll E A, Braun J E and Horton W T 2013 Development of a generalized steady-state simulation famework for positive displacement ccompressor and expanders *Proceedings of 8th Int. Conf. on Compressors and their Systems, City University of London, London* pp 717–729

[9] Ziviani D, Desideri A, Lemort V, De Paepe M and van den Broek M 2015 Low-order models of a single-screw expander for organic rankine cyle applications *9th Int. Conf. on Compressors and their Systems, City University of London, London* 38

[10] Chan C Y, Haselden G G and Hundy G 1981 *Int. J. Refrigeration* **4** 275–280

[11] Bein T W and Hamilton J F 1982 Computer modeling of an oil flooded single screw air compressor *International Compressor Engineering Conference* paper 383

[12] Hirai T, Noda S, Sagara Y and Tsuzi K 1986 Performance analysis of oif single screw compressor *International Compressor Engineering Conference* paper 520

[13] Masuda M, Ueno H, Inoue T, Hori K and Hossain M 2013 Effact of variable volume index on performance of single screw compressor *Proceedings of 8th Int. Conf. on Compressors and their Systems, City University of London, London* pp 257–264

[14] Litvin F L 1994 *Gear Geometry and Applied Theory* (Prentice-Hall, New Jersey)

[15] Lundberg A and Gianvall R 1979 *Int. J. Refrigeration* **2** 221–232

[16] Sun S, Wu W, Yu X and Feng Q 2010 Analysis of oil film force in single screw compressor *International Compressor Engineering Conference at Purdue* 2024

[17] Ignatiev K M 2012 Approach to the numeric geometry analysis of positive displacement compressors, its application to a single screw compressor simulation and verification by experiment *International Compressor Engineering Conference* 2059

[18] Bell I H, Wronski J, Quoilin S and Lemort V 2014 *Ind. Eng. Chem. Res.* **53** 2498–2508

[19] Bell I H 2011 *Theoretical and Experimental Analysis of Liquid Flooded Compression in Scroll Compressors* Ph.D. thesis Purdue Univeristy

IOP Conf. Series: Materials Science and Engineering **90** (2015) 012065 doi:10.1088/1757-899X/90/1/012065

Modelling, sizing and testing a scroll expander for a waste heat recovery application on a gasoline engine

Arnaud LEGROS[1], Ludovic GUILLAUME[1], Mouad DINY[2] and Vincent LEMORT[1]

[1] Energy Systems, University of Liège
7, chemins des chevreuils
Bâtiment B49 (P33)
4000 Liège
Belgium

[2] PSA Peugeot Citroën
18 rue des Fauvelles
92256 La Garenne-Colombes Cedex
France

Vincent.lemort@ulg.ac.be

Abstract. Waste heat recovery technologies in a mobile application emerge every time energy becomes a valuable resource. It has been the case in the 70s with oil crisis and it is starting to regain some interests now due to the continuously rising price of oil and due to the restrictive standards imposed by the different governments. This paper deals with the recovery on the exhaust gases of an internal combustion engine by using a Rankine system. The study focuses on the expander, which is one of the most important components of the system. The use of a scroll expander operating with steam is currently investigated through simulation and experimentation. This paper presents the modelling of a scroll expander. The model is a detailed model including various losses such as leakage, friction or under or over expansion. This model has been used to design and size a tailor-made scroll expander. This was necessary due to the small amount of expanders on the market and also to have a machine that fits our application. After designing the machine, a prototype has been built. It has also been tested on our prototype bench of waste heat recovery on a gasoline engine, by means of a Rankine cycle. Measured performance will be presented, analysed and compared to predictions by the model. The first results will be presented here and discussed in order to give recommendations for the design of next prototypes.

1. Introduction

After various restrictive laws on pollutants in the exhaust gases of passenger car engines, future legislations will be focused on the dioxide carbon emissions. Figure 1 shows that Europe is currently the leader on the average CO_2 emissions and its target for 2020 is 95 g of CO_2 per km. Similar goals are set by other governments, such as USA or China, within the same period of time. However, the trend is to reduce the carbon dioxide emissions.

Content from this work may be used under the terms of the Creative Commons Attribution 3.0 licence. Any further distribution of this work must maintain attribution to the author(s) and the title of the work, journal citation and DOI.
Published under licence by IOP Publishing Ltd

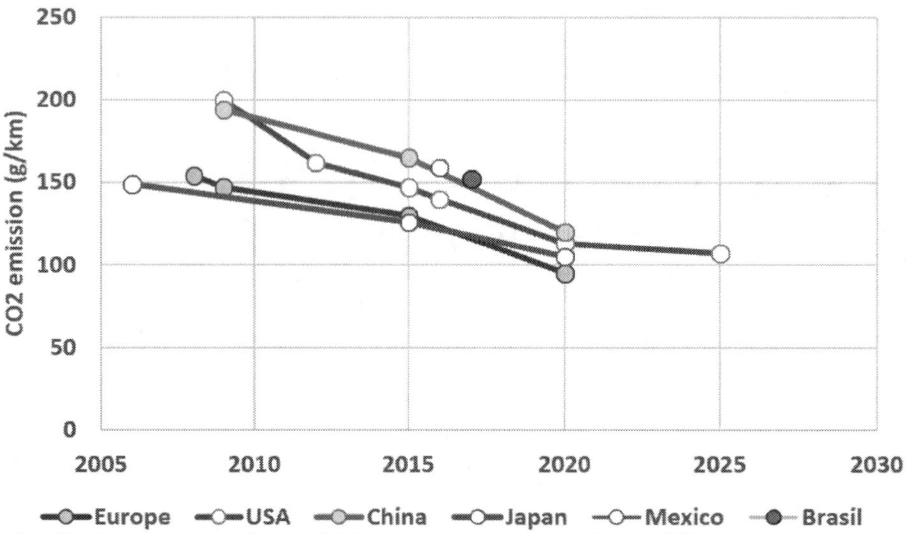

Figure 1: Carbon dioxide emissions regulations of different governments across the world [1].

In order to reduce the carbon dioxide emissions, the engine efficiency has to be improved and its fuel consumption decreased. With an efficiency order of magnitude of about 35%, passenger car gasoline engines show various sources of losses. The major losses in an engine are thermal losses and are present in the engine coolant and in the exhaust gases. Those thermal losses represent roughly 60% of the energy released during the combustion process and this percentage can vary widely when the engine is solicited at part load [2, 3].

The coolant and the exhaust gases are two losses that can be exploited. However, this paper focusses on the waste heat available in the exhaust gases only. That choice has been justified by the fact that the exhaust gases have a higher exergy content than the coolant [4].

Lots of technologies can be used to recover waste heat from the gases. Currently, almost none of them are mature and then are still in research phase. Legros et al. [1] have conducted an extensive review of those technologies and provide a comparison between them. The outcome of that comparison shows some advantages of the Rankine cycle heat engine.

The present paper will present the design steps of the expander of the Rankine cycle system used in a waste heat recovery application and introduce some preliminary experimental results.

2. Sizing the expander

Available expanders are not numerous on the market and most of them are not suitable for a mobile application such as waste heat recovery on passenger cars. Therefore, a new expander has to be designed to match more closely the project requirements.

Volumetric expanders are preferred to turbine expanders. Indeed, the latter work at higher speeds and are less suited for small power applications (ranging from 0 to 10kW). Moreover, the volumetric expanders tends to show better isentropic efficiencies [5]. Volumetric expanders are characterized by a displacement and, for most of them, a volume ratio.

Among the different technologies of volumetric expanders, the most common are scroll, screw and piston expanders [5]. Isentropic efficiency is a key parameter of the machine because the energy that has to be released to the condenser will be lower with a higher efficiency for a given amount of energy transfer to the evaporator. Scroll expanders usually show higher efficiencies than piston expanders [6, 7] and screw expander are usually for larger power range.

Scroll expander technology has been selected for the present waste heat recovery application and the working fluid is water. The design of the expander is divided in two steps. The first one aims at defining the characteristics of the expander and the second one will define the geometry that fits those characteristics. The following subsections will introduce to the driving cycles and the methodology and reasoning used to size the scroll expander.

2.1. Driving cycles

Passenger car fuel consumption are evaluated on normalized driving cycle. The current cycle in Europe is called NEDC, or New European Driving Cycle. This cycle is rather old and will soon be changed to have a cycle more representative of the actual behavior of car drivers. The future cycle is called WLTC and stands for World harmonized Light vehicles Test Cycle.

An analysis of those cycles has been conducted. A frequential distribution of the available power in the exhaust gases reveals that the power range is very widespread but it is mainly located in the first class as shown on Figure 2. The frequential distribution may slightly varies depending on the engine or the driving cycle but the trend is always the same. On a mid-size passenger car gasoline engine, more than 99% of the time the power available in the exhaust gases is less than 50kW with a peak power of 84kW.

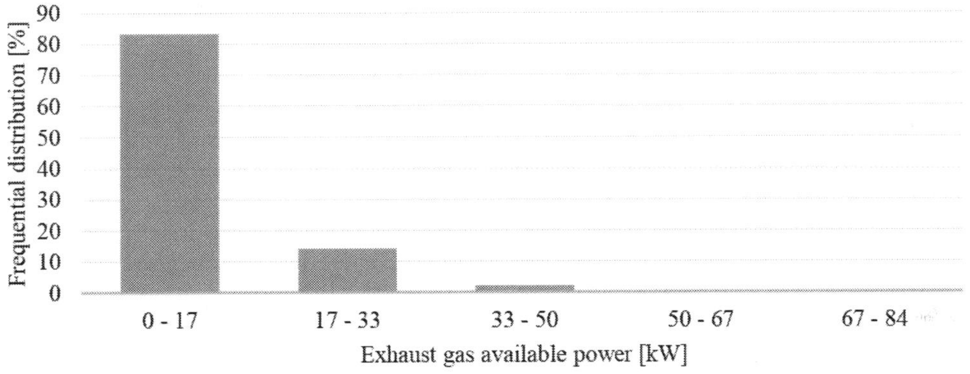

Figure 2 : Frequential distribution of a gasoline engine exhaust gases available power on a WLTC cycle

That result coupled to the highly transient behavior of an engine do not allow a design of a nominal point. Instead, an optimization has been conducted. By reducing the power range to the three first classes (0 to 50kW), an optimization of the produced power by the expander can be done.

2.2. Characteristics of the expander

The characteristics of the expander are its displacement and its volume ratio. Those characteristics were determined based on a Rankine cycle steady-state model coupled to the exhaust gases on a driving cycle. The model is similar to the one developed by Quoilin et *al.* [8].

2.2.1. Volume ratio. The volume ratio is defined by the ratio of the volume of the chamber right before the opening to the discharge chamber and the displacement. This ratio is responsible of over- or under-expansion losses. Those losses appear because the pressure at the end of the expansion is different from the exhaust pressure. In the case of over-expansion, the pressure after the expansion is under the exhaust pressure. The under-expansion leads to a pressure superior to the exhaust pressure after the expansion and if the volume ratio would have been higher, the produced power could have been increased.

Looking at existing scroll machines, their volume ratio varies roughly between 2.4 to 5 [9, 10]. It must also be noted that the higher the volume ratio, the more contact points there are in the expander. Those contact points defines the different chambers and that number of pairs of contact points should be minimised in order to reduce the internal leakage [5].
The sensitivity of the volume ratio on the produced power is also very interesting. An expression has been derived and is given in equation (1). Dividing that expression by the isentropic power gives a ratio of the residual power after a reduction of the volume ratio. Plotting this expression reveals that a significant decrease of the volume ratio can be made without reducing too much the mechanical power.

$$w_{red} = P_{su} v_{su} \left(\frac{\gamma \left(r_{v,red}^{\frac{1-\gamma}{\gamma}} - 1 \right)}{\gamma - 1} + r_{v,red} \left(r_{p,red} - \frac{1}{r_p} \right) \right) \quad (1)$$

As a conclusion, a small built-in volume ratio is preferable for two reasons. First, a smaller volume ratio will lead to a more compact machine and secondly, working with a smaller volume ratio will reduce the number of contact pairs and therefore, internal leakage.

2.2.2. Displacement. The displacement of the machine is the parameter that has been optimized to maximise the power produced by the expander. The displacement is function of several parameters such as the working fluid nature, the evaporating pressure and the mass flow rate.
In order to maximise the mechanical power, the Rankine cycle simulation code is run with inputs set as the mean value of the three power classes defined in the previous subsection. Three simulation results are then obtained and by varying the displacement and the evaporating pressure, three cartography maps can be drawn, displaying the rotational speed in function of the displacement and the evaporating pressure. One of this cartography is presented in the Figure 3 and is for the second class.

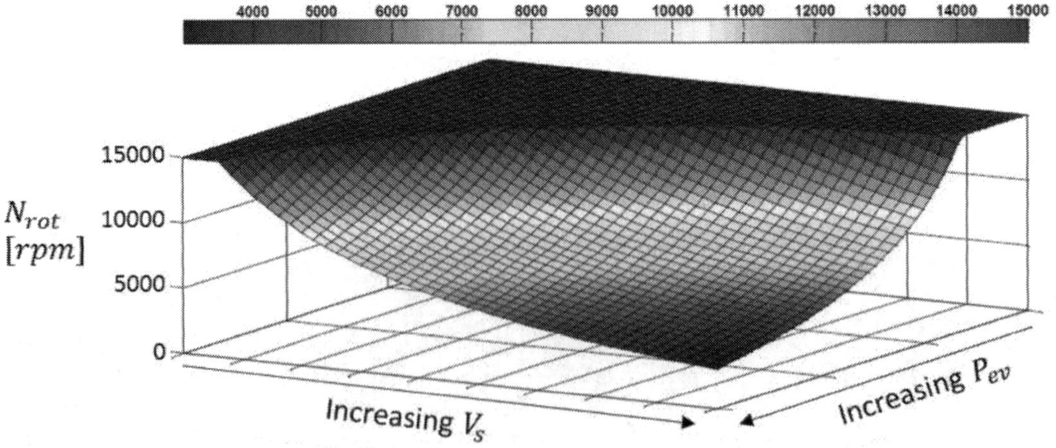

Figure 3: Cartography map of the rotational speed in function of the displacement and the evaporating pressure

To understand Figure 3, bounds on the variables must be first set. The boundary of the rotational speed has been estimated based on the observation of what is achievable technically in scroll machines. For the purpose of presentation of the map, rotational speed that would have overcome the upper bound is set to that bound.

Therefore, if the displacement of the expander is too small, the machine can not physically run sufficiently fast to be able to recover the available power in the exhaust gases. Using evaporating pressure that is too low leads to the same conclusion.
On another hand, if we look at the lowest available power class, the highest displacement/evaporating pressure would like to ridiculously low rotational speed. Low rotational speed leads to high internal leakage and therefore, reduced isentropic efficiency.
Based on that reasoning, an optimal displacement should be encountered on the overall cycle. Running the simulation code on the overall cycle for several displacement provides the Figure 4. A maximum of mechanical power appears for a given displacement. The overspeed frequency is also displayed and it can clearly be seen that lowest displacements results in high overspeed frequency. In a case of overspeed, the cycle controller would have to reduce the amount of energy injected into the cycle, decreasing then the mechanical power that can be produced.

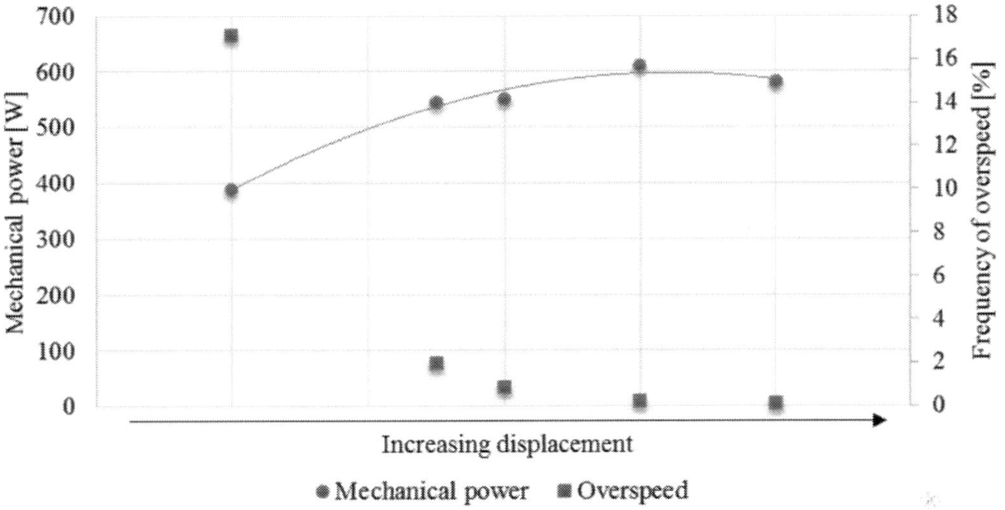

Figure 4: Mechanical power and overspeed frequency in function of the displacement for a WLTC

2.3. *Definition of the geometry*

The definition of the geometry of the scroll expander requires more than a simplified model such as the one used in the previous analysis. A detailed model of scroll expander has therefore been build but the explanation of such a model is out of the scope of this article. More information can be found in the literature [11-13].
Based on the characteristics defined in the previous subsection, a geometry can be defined. Several geometry can exist and return the same characteristics. An optimisation could have been run to define the best geometry. However, the simulation model was not calibrated and losses could not be predicted correctly. This imprecision in the model could results in different results of the optimisation and would make inefficient the use of an optimisation.
The shape of the suction port has also been studied to reduce the pressure drop losses. In scroll machine, the moving involute can pass over the admission port, leading to a reduction of admission area. This decrease leads in a pressure drop. Since the clearance volume in scroll expander is not a drawback, its minimisation was not mandatory and a non-negligible clearance volume has been implemented in order to maximize the suction port cross sectional area. Parametric study show that such an increase of the area of the suction port allows the reduction of more than 50% of the pressure drop at the highest rotational speed [11].
A 2D view of the expander geometry is provided in Figure 5. The shape of the suction port can be seen.

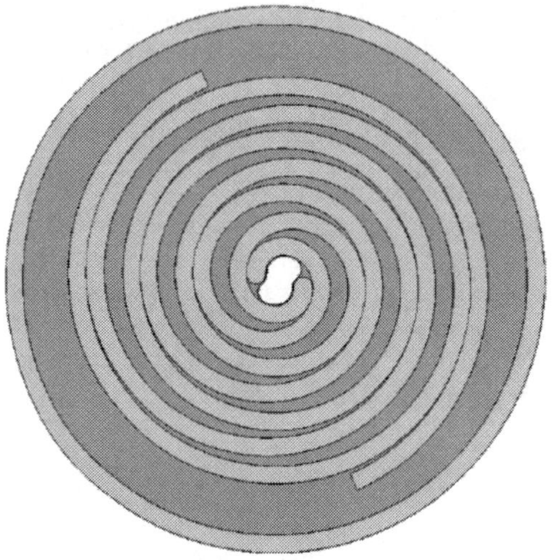

Figure 5: Geometry of the designed scroll expander adapted to a waste heat recovery application for passenger car

2.4. *High temperature investigations*

The investigated Rankine cycle and expander uses water as the working fluid. In order to achieve higher cycle efficiency, evaporating pressure needs to be increased and therefore, the temperature of the working fluid entering the expander is also high. Literature reports few scroll expanders but all of them showed a maximal operating temperature of 200°C.

In order to increase that temperature limit, investigations on the tribology of the expander have been conducted. Those investigations have been also mandatory due to the weak miscibility of water and oil. The mix of water and lubricating oil makes the design very complex by adding a lot of constraints to the system (complexity of the separation, additional pump…). Therefore, the design of the expander has been focussed on an oil-free expander. The study of the material in contact in the scroll expander has revealed autolubricating materials that work with high temperature and provide low friction coefficient, reducing at the same time the friction losses. The involutes are made of coated aluminium and the tip seal are made of an autolubricating material. The details for the study fall out of the scope of this article and more details can be found in [14].

3. Experimental results

After the design of the expander, the manufacturing of the machine has been achieved. The scroll expander has been tested in a Rankine cycle test bench. The Rankine cycle is coupled to a gasoline engine to provide exhaust gases into the evaporator. The scroll expander is coupled to an alternator, controlled by a variable resistance. The alternator consists in a synchronous electric machine with an efficiency higher than 90%.

Pressure and temperature measurements are available at the supply and exhaust of the expander. The mass flow rate of working fluid and the electrical power produced by the generator are also measured. Temperature next to the bearings are also measured. The first test campaign is presented in this article. The measurements were made by varying the rotational speed of the alternator for various mass flow rate. A diagram of the test bench and the sensors position are displayed in Figure 6. For simplicity of control, the cycle is an open loop.

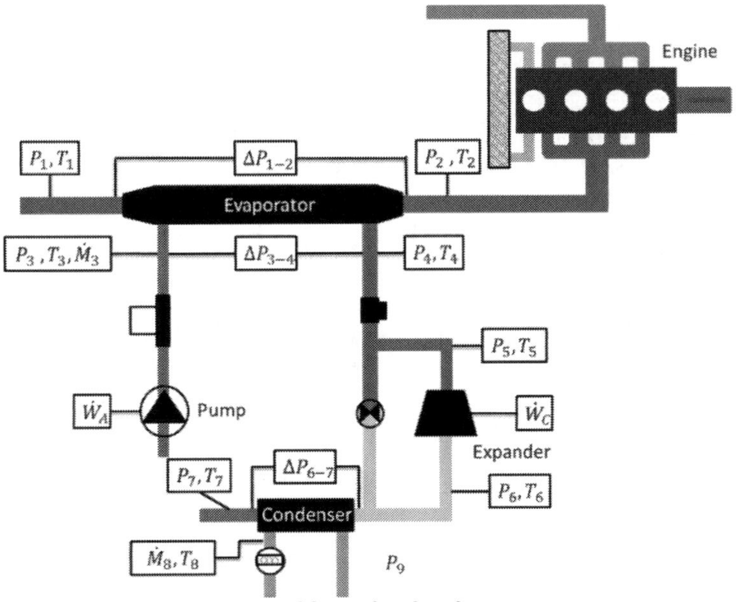

Figure 6: Diagram of the test bench and sensors position

Isentropic efficiency and filling factor are presented in Figure 7. It can directly be noted that the maximal isentropic efficiency reached is about 26%. That efficiency is relatively low in comparison of what could be observed in literature. Isentropic efficiency is defined by the equation (2).

$$\epsilon_s = \frac{\dot{W}_{el}}{\dot{M}(h_{su}-h_{ex.s})} \tag{2}$$

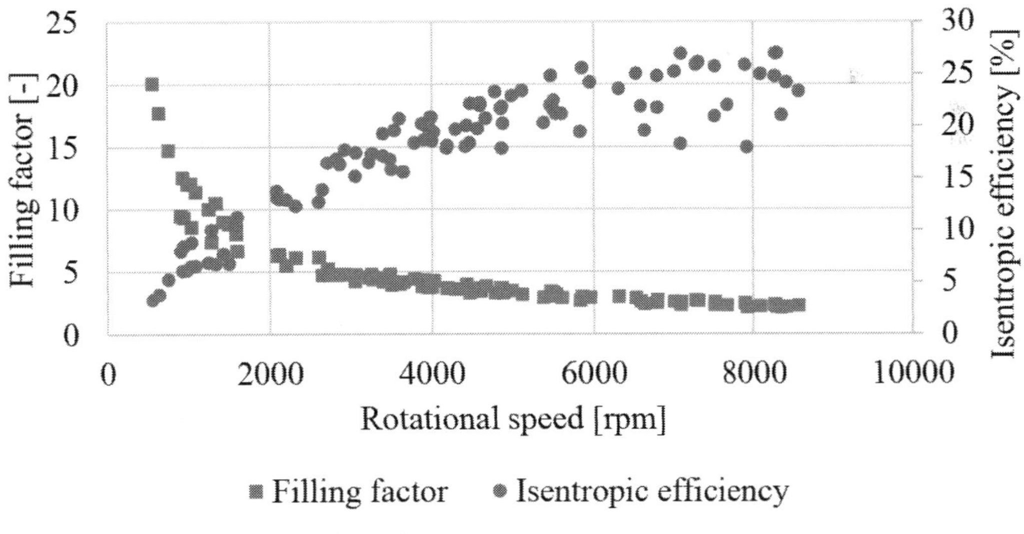

Figure 7: Performances of the designed scroll expander

In order to understand why the efficiency is so low, the filling factor can be evaluated with the equation (3). The filling factor is presented in Figure 7. For very low rotational speeds, the filling factor in very high and leads to very poor efficiency. The increase of rotational speed tends to reduce the filling factor and a minimal value of around 2 could be reached. Although the decrease of filling factor is impressive, a value of 2 is still important and an analysis of the expander revealed that flank leakages are important due to a pretty high leakage clearance.

$$\varphi = \frac{\dot{M}}{\rho_{su} V_s N_{rot}} \quad (3)$$

Looking at the temperature of the bearings, a trend with the rotational speed is shown in the Figure 8. That dependence with the rotational speed provides evidence of mechanical losses in the bearings.

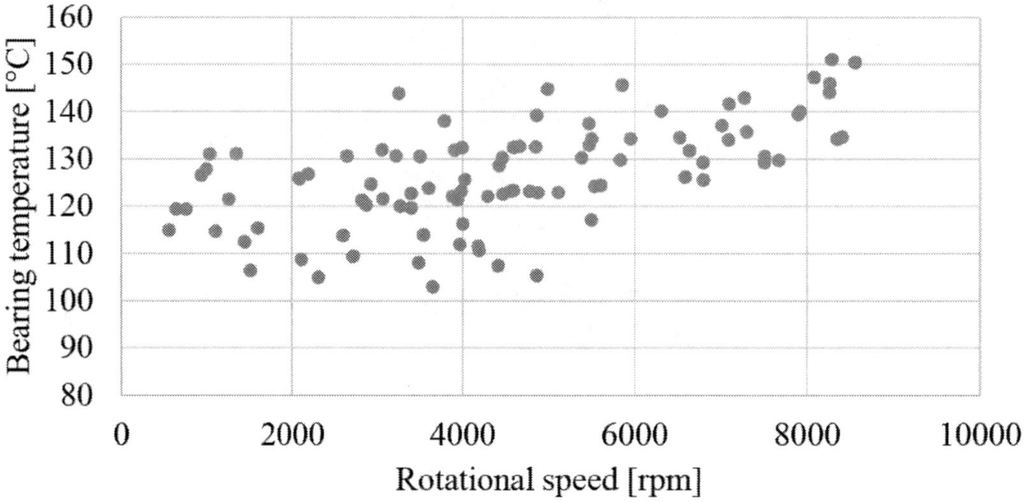

Figure 8: Bearing temperature in function of the rotational speed

4. Conclusion

A design methodology has been defined and applied to size a new oil-free scroll expander for a waste heat recovery steam Rankine cycle. This has been mandatory since the scroll expanders available on the market are not suited to a waste heat recovery application on a passenger car gasoline engine.

The design methodology has been successfully applied and the sized expander has been manufactured and tested. The preliminary tests show a rather low isentropic efficiency, mainly affected by a huge internal leakage and, secondarily, by the friction losses. Additional investigations regarding the leakage clearance and the tip seal configuration will be conducted in order to increase the isentropic efficiency.

An Oil-free machine was manufactured and tested without mechanical issues, helping in simplifying the Rankine loop for vehicle integration.

The compactness of the machine is very good and the expander is currently similar in size as a compressor for air conditioning applications in automotive and commercial vehicle applications.

The experiments showed that high temperatures could be achieved and a maximum temperature of 240°C at the inlet of the expander has been reached.

References

[1] A. Legros, L. Guillaume, M. Diny, H. Zaïdi, and V. Lemort, "Comparison and impact of waste heat recovery technologies on passenger car fuel consumption in a normalized driving cycle," *Energies,* vol. 7, pp. 5273-5290, 2014.

[2] R. E. Chammas and D. Clodic, "Combined cycle for hybrid vehicles," in *SAE World congress,* Detroit, 2005.

[3] J. C. Conklin and J. P. Szybist, "A highly efficient six-stroke internal combustion engine cycle with water injection for in-cylinder exhaust heat recovery," *Engery,* 2010.
[4] T. Endo, S. Kawajiri, Y. Kojima, K. Takahashi, T. Baba, S. Ibaraki, *et al.*, "Study on maximizing exergy in automotive engines," *SAE technical paper,* 2007.
[5] V. Lemort, L. Guillaume, A. Legros, S. Declaye, and S. Quoilin, "A comparison of piston, screw and scroll expanders for small-scale Rankine cycle systems," in *The 3rd international conference on microgeneration and related technologies*, Naples, 2013.
[6] S. Declaye, S. Quoilin, L. Guillaume, and V. Lemort, "Experimental study on an open-drive scroll expander integrated into an ORC (Organic Rankine Cycle) system with R245fa as working fluid," *Energy,* vol. 55, pp. 173 - 183, 2013.
[7] R. Daccord, J. Melis, T. Kientz, A. Darmedru, R. Pireyre, N. Brisseau, *et al.*, "Exhaust heat recovery with Rankine piston expander," in *ICE Powertrain electrification & energy recovery*, Rueil-Malmaison, France, 2013.
[8] S. Quoilin, V. Lemort, and J. Lebrun, "Experimental study and modeling of an Organic Rankine Cycle using scroll expander," *Applied Energy,* vol. 87, pp. 1260-1368, 2010.
[9] *Scroll expanders*. Available: http://airsquared.com/products/expanders
[10] D. Cretegny, "Etude & frabrication d'un expanseur scroll à haut rapport de volume," 2007.
[11] A. Legros, L. Guillaume, V. Lemort, M. Diny, I. Bell, and S. Quoilin, "Investigation on a scroll expander for waste heat recovery on internal combustion engine," presented at the International conference on compressors and their systems, London, 2013.
[12] I. Bell, "Theoretical and experimental analysis of liquid flooded compression in scroll compressors," Purdue University, 2011.
[13] V. Lemort, "Contribution to the characterization of scroll machines in compressor and expander modes," PhD thesis, University of Liège, 2008.
[14] A. Legros, H. Zaïdi, M. Diny, M. Diaby, and V. Lemort, "Etude tribologique sur les lubrifiants solides à haute température dans des machines de détente," presented at the La Tribologie : Fondamentaux et Applications Complexes, Lyon, 2013.

IOP Conf. Series: Materials Science and Engineering **90** (2015) 012066 doi:10.1088/1757-899X/90/1/012066

Experimental performance of a piston expander in a small-scale organic Rankine cycle

JF Oudkerk, R Dickes, O Dumont and V Lemort

Thermodynamic Laboratory, University of Liège, Campus du Sart Tilman B49, B-4000 Liège

E-mail: jfoudkerk@ulg.ac.be

Abstract. Volumetric expanders are suitable for more and more applications in the field of micro- and small-scale power system as waster heat recovery or solar energy. This paper present an experimental study carried out on a swatch-plate piston expander. The expander was integrated into an ORC test-bench using R245fa. The performances are evaluated in term of isentropic efficiency and filling factor. The maximum efficiency and power reached are respectively 53% and 2 kW. Inside cylinder pressure measurements allow to compute mechanical efficiency and drown P-V diagram. A semi-empirical simulation model is then proposed, calibrated and used to analyse the different sources of losses.

1. Introduction

The piston expander is among the oldest displacement machines. It has been largely used during the Industrial Revolution and up to the beginning of the 20th century. There is currently a large regain of interest for piston expanders for small-scale steam and organic Rankine cycle systems, especially used in micro-CHP and waste heat recovery on internal combustion engines ([1], [2], [3], [4]). Piston expanders actually show some advantages over other expansion machines, such as larger built-in volume ratio, high achievable operating pressures and temperatures, their ability to ingest liquid and low rotational speeds. Among technologies of piston expanders, axial machines show the additional advantage to be compact, which makes them suitable for mobile applications. However, there is a lack of scientific and technical literature on such machines.

The present paper describes an experimental investigation conducted on a swash-plate piston expander, characterized by a total cylinder volume of 195 cm^3. The lubrication is ensured by an external oil loop with injection of oil at mains friction points. As represented in Figure 1, the admission and exhaust processes are achieved by means of systems that induce symmetrical ports opening and closing timings with respect to TDC (Top Dead Center) and BDC (Bottom Dead Center). This expander was tested over several operating conditions and the measured performance is analyzed and presented in this study. A semi-empirical model (lumped parameter model) of a piston expander is then proposed, calibrated and validated based on experimental data. This model is finally used to simulate the performance and to analyze the different sources of losses.

Content from this work may be used under the terms of the Creative Commons Attribution 3.0 licence. Any further distribution of this work must maintain attribution to the author(s) and the title of the work, journal citation and DOI.
Published under licence by IOP Publishing Ltd

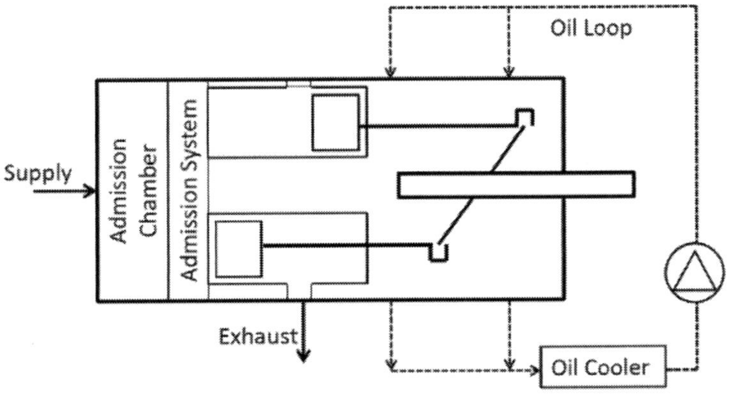

Figure 1: Schematic diagram of the swash-plate piston expander

2. Description of the test bench

The piston expander is integrated into an ORC cycle test bench using R245fa. Figure 2 represents this ORC composed of a pump, a recuperator, a boiler, a condenser and a liquid tank. The mass flow rate is controlled by adjusting the pump speed through a variable-frequency drive (VFD). The boiler is fed with thermal oil and the condenser is water-cooled.

The expander is connected to an asynchronous electrical motor. In order to start the expander and control its speed, the motor is driven by a four quadrants variable-frequency drive. As shown in Figure 2, several sensors are placed in order to measure the performance of the expander. The flow is measured at the pump outlet by a coriolis-effect mass flow meter. A torque sensor is placed between the expander and the generator in order to compute the mechanical power. Finally, the temperature and the pressure of the working fluid at the inlet and the outlet of the expander are measured with piezoresistive pressure sensors and Type-T thermocouples.

In addition of these measurements, the pressure in the cylinder is measured with the help of piezoelectric pressure sensor and the angular position of the shaft with a rotary encoder. This measurement system allows drawing P-V diagram and therefore indicated power and mechanical efficiency can be evaluated.

IOP Conf. Series: Materials Science and Engineering **90** (2015) 012066 doi:10.1088/1757-899X/90/1/012066

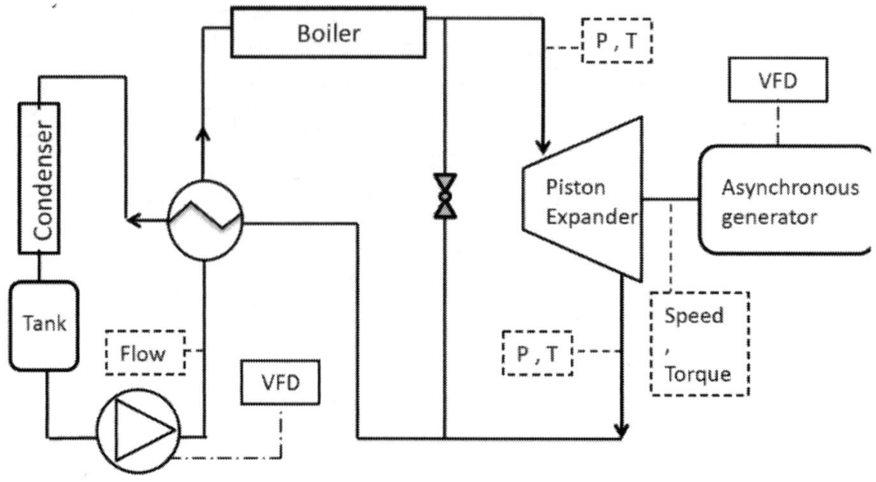

Figure 2: Schematic diagram of the test bench.

3. Experimental results

The initial goal of the experimentation was to test the expander for an exhaust pressure of 3bar, supply pressure varying from 18 to 30 bar and rotational speed from 1000 to 4000 RPM. Unfortunately, for high flow rate operating point, the pressure was higher than 3 bar due to the actual limitations of the test rig. Generally, the exhaust pressure varies along the achieved steady-state points. Table 1 shows the matrix of tests and the exhaust pressure obtained. The empty cases are operating conditions for which the expander consumes power and are not considered further.

Table 1 : Matrix test and exhaust pressure.

		Psu [bar]				
	Pex [bar]	18	21	24	27	30
RPM	1000	2.96	2.91	2.9	3.02	2.96
	2000		3.09	2.9	3.29	3.58
	2500		3.06	3,01	3,5	3.94
	3000		3.08	3.15	3.48	3.87
	3500		3.08	3.21	3.55	4.02
	4000		2.876	3.28	3.66	3.98

During the experimental campaign, a total of 60 steady-state operating points was measured. For a given rotational speed, the supply pressure is controlled by modifying the mass flow rate, which is imposed by the pump. The superheating is controlled by the oil boiler. For these points, the superheating was kept between 4 and 17 K, the mass flow rate varies from 29 to 105 g/s and the mechanical power output is comprised between 0.3 and 2 kW.

IOP Conf. Series: Materials Science and Engineering **90** (2015) 012066 doi:10.1088/1757-899X/90/1/012066

Figure 3 shows the evolution of shaft and indicated isentropic efficiencies in terms of pressure ratio. These isentropic efficiencies are respectively defined as the ratio between the measured shaft and indicated power and the isentropic power:

$$\epsilon_{s,sh} = \frac{\dot{W}_{\mathrm{sh}}}{\dot{M}.(h_{\mathrm{su,exp}} - h_{ex,exp,s})} \quad , \quad \epsilon_{s,in} = \frac{\dot{W}_{\mathrm{in}}}{\dot{M}.(h_{su,exp} - h_{ex,exp,s})} \tag{1}$$

where $\dot{M}$ is the mass flow rate, $h_{su,exp}$ the supply enthalpy and $h_{ex,exp,s}$ the isentropic exhaust enthalpy,

The mechanical power is computed with the help of the torque (T_m) and the rotational speed (N):

$$\dot{W}_{sh} = N.2.\pi.T_m \tag{2}$$

The indicated power is computed by integrating the pressure inside the cylinder (P_{cyl}) with the volume of the cylinder (V_{cyl}) over one revolution to obtain the indicated work. This work is then multiplied by the rotational speed to compute the power:

$$\dot{W}_{in} = N.\oint P_{cyl}.dV_{cyl} \tag{3}$$

The ratio between the shaft and the indicated power is the mechanical efficiency of the expander:

$$\eta_m = \frac{\dot{W}_{sh}}{\dot{W}_{in}} = \frac{\epsilon_{s,sh}}{\epsilon_{s,in}} \tag{4}$$

For low pressure ratios, the shaft isentropic efficiency sharply decreases while the indicated isentropic efficiency remains quite constant. This means a decrease of the mechanical efficiency. Indeed, by reducing the pressure ratio, the indicated power decreases and the constant mechanical losses take more and more part leading to a decrease of the mechanical efficiency. The isentropic efficiencies also vary with the rotational speed. The indicated efficiency is lower for 1000 RPM. This can be explained by, as shown below, a larger impact of internal leakages for low rotational speeds. However, these leakage losses are balanced by a better mechanical efficiency and lower supply pressure drop for low rotational speed. This leads to the conclusion that a tradeoff can be made in order to find an optimal rotational speed.

Figure 3 also shows that, as expected because of the limitation of the test rig, high pressure ratio is achieved for 1000 RPM while it is limited for higher rotational speed. Further work should test the expander under higher pressure ratios at higher rotational speeds.

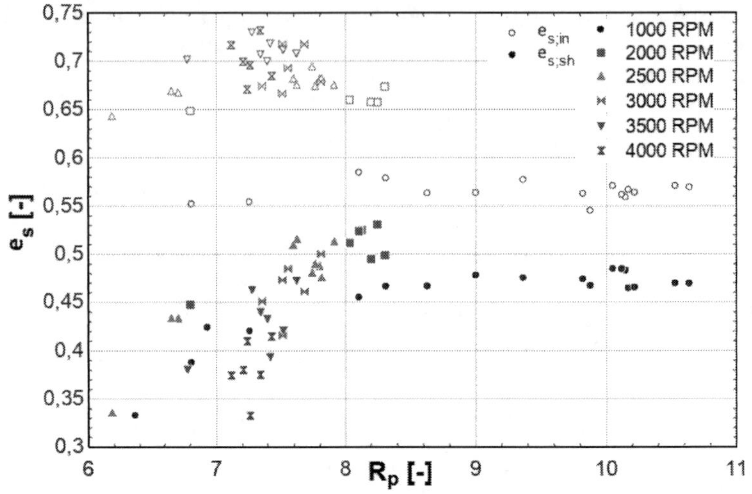

Figure 3 : Evolution of the shaft and indictaed isentropic efficiencies with the pressure ratio

The filling factor is used to evaluate the volumetric performance. The filling factor is defined as the ratio between the measured mass flow rate and a theoretical mass flow rate, computed based on the expander displacement:

$$FF = \frac{\dot{M}}{N.\rho_{su}.V_{IC}} \tag{5}$$

where N is the rotational speed, ρ_{su} the density at the inlet and V_{IC} the inlet closing volume defined as the cylinder volume when the intake port closes (see Figure 5).

As it can be seen in Figure 4, the filling factor increases with the pressure ratio and decreases with the rotational speed. The evolution with the speed is due to the increase of the leakage and the decrease of the supply pressure drop for low rotational speed. The value of the filling factor is always lower than 1 despite of the leakage. This is due to the presence of a clearance volume in the cylinder. So when the fluid is admitted in the cylinder, a part of the volume is already taken by some fluid trapped into this clearance volume. The effect of the clearance volume decreases when the pressure ratio increases because the mass flow increases and the then, the amount of fluid trapped into the clearance volume have less impact. Indeed, for a given exhaust pressure, the amount of fluid trapped into the clearance volume remain slightly constant while the mass flow increases with the supply pressure.

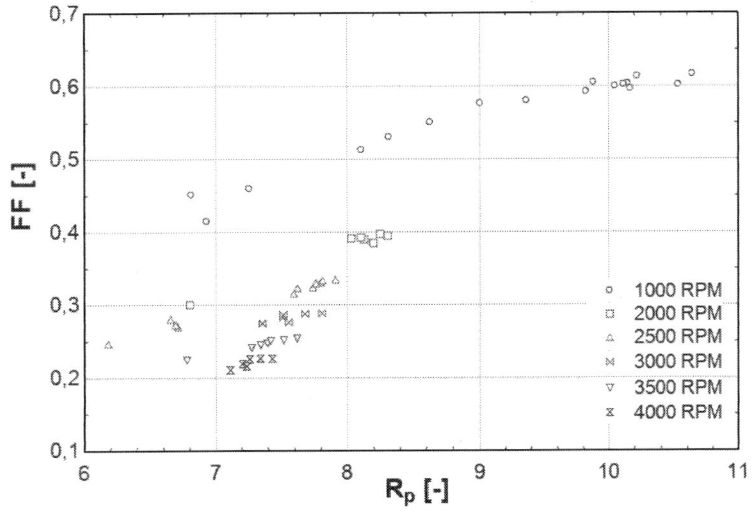

Figure 4 : Evolution of the filling factor with the pressure ratio

Figure 7 and Figure 8 show P-V diagram drawn with the measured in cylinder pressure for four operating conditions, two supply pressures and two rotational speeds. The inlet and exhaust closing volumes (V_{IC} and V_{EC}) shown on these figures are also the opening volumes (V_{IO} and V_{EO}) because of the symmetry of the admission and exhaust system. The evolution of the pressure drop with the speed and the supply pressure can be observed.

4. Piston expander semi-empirical model

The piston expander model, proposed by Glavatskaya et al. [1] and based on the semi empirical model of a scroll-expander proposed by Lemort et al. [5], has been adapted to represent the investigated swash-plate expander. The major difference between the modeling of a scroll and a piston expander lies in the presence of a clearance volume in the latter machine, which is responsible for the recompression of part of the fluid after the exhaust process. This difference yielded a modification of the scroll expander model to take into account the recompression process.

As in the scroll model, the evolution of the fluid through the expander is decomposed into six steps standing for the main physical features. The suction and the discharge processes are both considered as consecutive adiabatic pressure drop and isobaric cooling down (or heating up). After the suction process, the mass flow is split into two parts. The biggest part, the internal flow, following the expansion processes while the other one, the leakage flow, by-passes this expansion process. These two flows are mixed in an adiabatic way just before the discharge process.

Adiabatic pressure drops are computed assuming an isentropic flow through a convergent nozzle. The cross-sectional areas of the supply and exhaust fictitious nozzles are A_{su} and A_{ex}. Heat transfers are modeled as transfers between the fluid (or ambient air for ambient loses) and a fictitious wall, at uniform temperature, with overall heat transfer coefficients AU_{su}, AU_{ex} and AU_{amb}.

More information about the overall model can be found in [5]. The present paper focuses on the internal expansion process. The latter is divided in six steps as proposed by Stumpf [6] and shown in Figure 5:

- 1-2: isobaric intake
- 2-3: isentropic expansion
- 3-4: constant machine volume expansion
- 4-5: isobaric exhaust

- 5-6: isentropic compression
- 6-1: constant machine volume compression

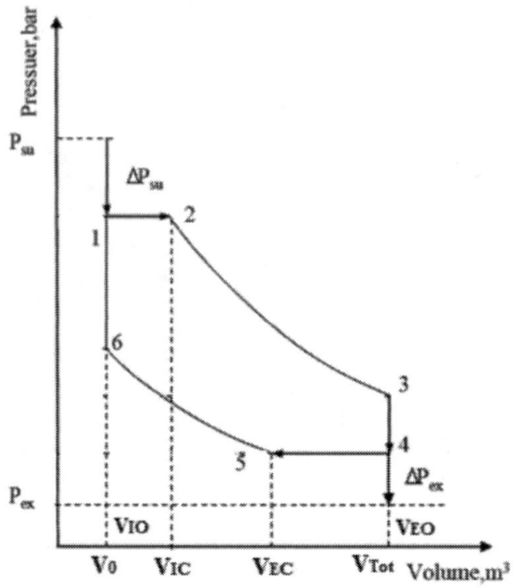

Figure 5 : Expansion process

This expansion process differs of the one of the studied expander. Indeed, as evoked above, the opening and closing volume are symmetric compared to TDC and BDC. In other words, during the compression stroke 5-6, the intake port opens at the same volume as in point 2 (see Figure 6) before the piston reaches the TDC. In the same way, during the expansion stroke 2-3, the exhaust port opens at volume in point 5 before the piston reaches the BDC.

The thermodynamic state of the fluid in point 2 is equal to that of the entering fluid (after pressure drop and heat transfer). The state at point 3 is computed with an isentropic expansion.

In the same way, the state of the fluid at point 5 is assumed to be the state just before the mixing with leakage flow. An isentropic compression allows computing the state at point 6.

The indicated work of the expansion cycle is given by the area under the PV diagram. This area can be computed by summing these four terms:

$$W_{in} = \frac{V_2}{v_2}.(h_2 - h_3) + V_{Tot}.(P_3 - P_4) - \frac{V_5}{v_5}.(h_6 - h_5) - V_0.(P_1 - P_6) \tag{6}$$

where $V_2 = V_{IC}$ is the intake closing volume, $V_5 = V_{EC}$ the exhaust closing volume, V_0 the clearance volume and v_2 and v_5 are specific volumes. V_{IC} and V_{EC} are defined with the following geometrical parameters:

$$f_{IC} = \frac{V_{IC}}{V_{Tot}} \ ; \ f_{EC} = \frac{V_{EC}}{V_{Tot}} \tag{7}$$

The enthalpy at point 5 is computed by the energy balance over the expansion process:

$$\dot{W}_{in} = \dot{M}_{in}.(h_2 - h_5) \tag{8}$$

The mass flow depends on geometric parameters and operating conditions:

$$\dot{M} = \dot{M}_{in} + \dot{M}_{leak} = N.\left(\frac{V_{IC}}{v_2} - \frac{V_{EC}}{v_5}\right) + \dot{M}_{leak} \tag{9}$$

where the leakage flow rate ($\dot{M}_{leak}$) is computed in a similar way than the pressure drops, i.e. by introducing a fictitious nozzle of throat cross sectional area A_{leak}. The internal flow rate ($\dot{M}_{in}$) is computed by subtracting the mass trapped in the clearance volume out of the mass in the cylinder at the beginning of the expansion.

This model allows computing the indicated power but for the shaft power a model of the mechanical losses is needed. As suggested by Navaro for reciprocating compressor [7], the mechanical loss power is a function of indicated power and rotational speed:

$$\dot{W}_{loss} = A.\dot{W}_{in} + B.RPM^2 \tag{10}$$

The coefficient A and B are determined with linear regression. A coefficient of determination of R^2=98.9 % was obtained with A =0.1427 [-] and B = $4.5682.10^{-5}$ $[W.min^2]$.

Finally, the model requires five inputs: the supply and exhaust pressures, the supply and ambient temperatures and the rotational speed. The main outputs are the mass flow rate, the shaft power and the exhaust temperature. The six parameters of the model, given in Table 1, are tuned to obtain the best fit with the measurements.

Table 1: Calibrated parameters of the model

A_{su} $[mm^2]$	A_{ex} $[mm^2]$	A_{leak} $[mm^2]$	AU_{su} $[W/K]$	AU_{ex} $[W/K]$	AU_{amb} $[W/K]$
2.01	78.54	0.237	$1.\left(\frac{\dot{M}}{0.013}\right)$	$2.\left(\frac{\dot{M}}{0.013}\right)$	4

With these parameters, the error is below 5% for mass flow and shaft power and 5K for exhaust temperature. The comparison between measured and calculated outputs is shown on Figure 6.

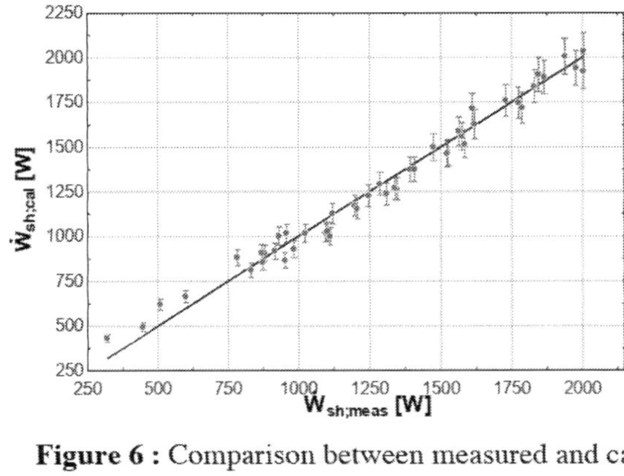

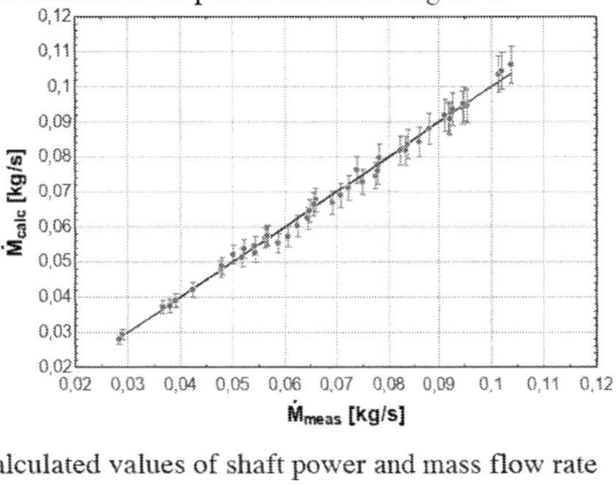

Figure 6 : Comparison between measured and calculated values of shaft power and mass flow rate

Figure 7 and Figure 8 show comparison of P-V diagrams predicted by the model and those measured. As it can be seen, the calculated diagrams are closed to the measured despite of the difference of the expansion process (different opening and closing volume).

IOP Conf. Series: Materials Science and Engineering **90** (2015) 012066 doi:10.1088/1757-899X/90/1/012066

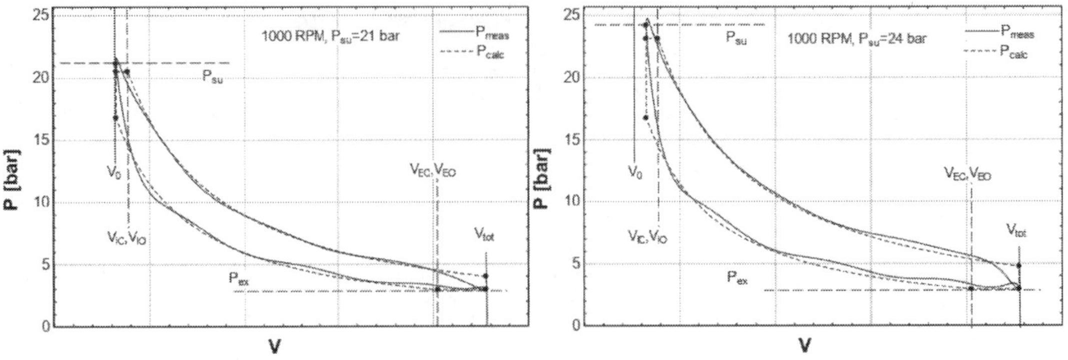

Figure 7 : P-V diagrams for RPM=1000 and for $\boldsymbol{P_{su}} = \mathbf{21}$ bar (left) and $\boldsymbol{P_{su}} = \mathbf{24}$ bar (right)

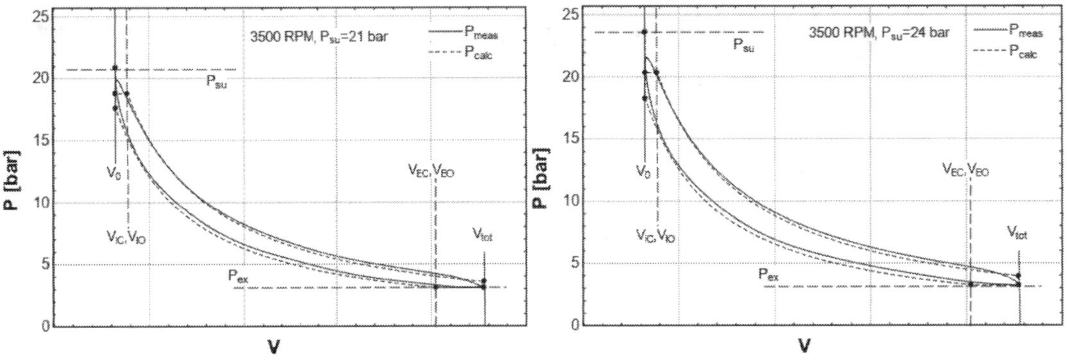

Figure 8 : P-V diagrams for RPM=3500 and for $\boldsymbol{P_{su}} = \mathbf{21}$ bar (left) and $\boldsymbol{P_{su}} = \mathbf{24}$ bar (right)

5. Model analysis

5.1. Decomposition of the losses

The model is used to analyse the impact of each source of losses on the isentropic efficiency. Figure 9 shows disaggregation of different losses. First, it can be seen that the built-in volume ratio leads to a decrease of the isentropic efficiency with the pressure ratio due to under expansion losses. For a pressure ratio of 5.8, the work to compress the fluid trapped into the clearance volume becomes equal to the work produced during the expansion and then the expander cannot produce anymore power. Then, different sources of losses are added. Heat transfers slightly decrease the performances. The leakages have a strong impact on the efficiency and flatten the curve because they have more impact at low pressure ratios. The pressure drop does not impact much more the efficiency because the power and the flow rate decrease in the same time. The pressure drop affects more the compactness of the expander. Finally, the mechanical losses decrease the efficiency and, as observed on the measurement, have more impact for low pressure ratios.

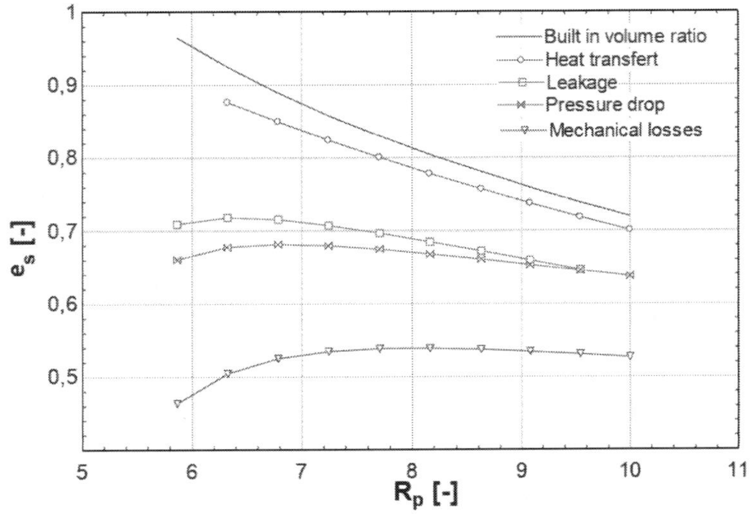

Figure 9 : Impact of each source of losses on the isentropic efficiency

5.2. Optimal rotational speed

Figure 10 shows the evolution of mechanical efficiency, leak ratio (which is the ratio between the leakage flow rate and the total mass flow rate) and isentropic efficiencies in terms of rotational speed. It can be seen that, when rotational speed increases, the leak ratio decreases leading to an increase of indicated isentropic efficiency. However, at the same time, the mechanical efficiency decreases. These two opposite effects lead to a rotational speed that maximises the isentropic efficiency. This optimal speed is 2000 RPM and doesn't change with the supply pressure. That corresponds to a mean piston speed of 2 m/s.

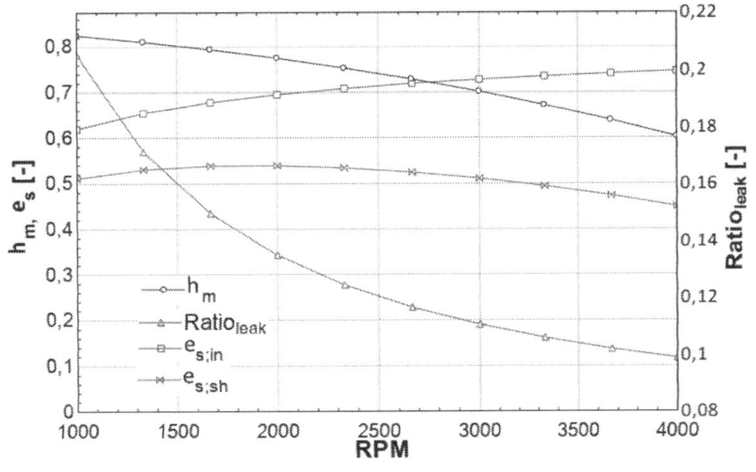

Figure 10 : Influence of rotational speed on isentropic efficeincy for $\mathbf{P_{su} = 24}$ bar and $\mathbf{P_{ex} = 3}$ bar

6. Conclusion

A swash-plate piston expander was tested in an ORC system with R245fa. Measurement of inside cylinder pressure allowed computing indicated power and thus mechanical efficiency. The maximal achieved isentropic efficiency was 53 %. The filling factor was always below 1 despite of leakage because of the clearance volume. The mechanical efficiency is comprised between 50 and 85% and is strongly related to the shaft torque. Unfortunately the test rig did not allow reaching high pressure ratio for high RPM.

A semi-empirical model of a scroll expander was adapted to describe the piston expander. The model is able to predict measured mass flow rate and mechanical power within 5%. This model was used to analyse the different source of losses and determine the optimal rotational speed.

References

[1] Glavatskaya Y, Podevin P, Lemort V, Shonda O, Descombes G 2012 Reciprocating Expander for an Exhaust Heat Recovery Rankine Cycle for a Passenger Car Applicatio *Energies* **5**(6) 1751-1765

[2] Seher D, Lengenfelder T, Gerhardt J, Eisenmenger N, Hackner M, Krinn I 2012 Waste Heat Recovery for Commercial Vehicles with a Rankine Process, 21st Aachen Colloquium Automobile and Engine Technology

[3] Daccord R, Melis J, Kientz T, Darmedru A, Pireyre R, Brisseau N, Fonteneau E 2013 Exhaust heat recovery with rankine piston expander In Proceedings of ICE Powertrain Electrification & Energy Recovery, Rueil-Malmaison, France, 28 May 2013

[4] Wronski J, Skovrup M J, Elmegaard B, Risla H N, Haglind F 2012 Design and Modelling of a Novel Compact Power Cycle for Low Temperature Heat Sources, In Proceedings of ECOS 2012, The 25th International Conference on Efficiency, Cost Optimization, Simulation and Environmental Impact of Energy Systems, June 26-29, 2912, Perugia, Italy

[5] Lemort V, Quoilin S, Cuevas C, Lebrun J 2009 Testing and modeling a scroll expander integrated into an Organic Rankine Cycle *Applied Thermal Engineering* **29** 3094-3102

[6] Stumpf J. 1922 *The una-flow steam-engine*, Second Edition

[7] Navarro E, Granryd E, Urchueguia JF, Corberan JM 2007 A phenomenological model for analyzing reciprocating compresors *International Journal of Refrigeration* **30** 1254-1265

IOP Conf. Series: Materials Science and Engineering **90** (2015) 012067 doi:10.1088/1757-899X/90/1/012067

The development of a control system for a small high speed steam microturbine generator system

A Alford[1], **P Nichol**[1], **M Saunders**[1] and **B Frisby**[2]

[1] Corac Energy Technologies Ltd., Sterling Road, Slough, Berkshire, SL1 4ST, UK
[2] Spirax Sarco Ltd., Runnings Road, Cheltenham, GL51 9NQ, UK

Abstract: Steam is a widely used energy source. In many situations steam is generated at high pressures and then reduced in pressure through control valves before reaching point of use. An opportunity was identified to convert some of the energy at the point of pressure reduction into electricity. To take advantage of a market identified for small scale systems, a microturbine generator was designed based on a small high speed turbo machine. This machine was packaged with the necessary control valves and systems to allow connection of the machine to the grid. Traditional machines vary the speed of the generator to match the grid frequency. This was not possible due to the high speed of this machine. The characteristics of the rotating unit had to be understood to allow a control that allowed export of energy at the right frequency to the grid under the widest possible range of steam conditions. A further goal of the control system was to maximise the efficiency of generation under all conditions. A further complication was to provide adequate protection for the rotating unit in the event of the loss of connection to the grid. The system to meet these challenges is outlined with the solutions employed and tested for this application.

1. Background

This machine was developed in a joint project between Spirax-Sarco Ltd and Corac Energy Technologies Ltd.

Spirax-Sarco Limited has over 70 years of innovation and research in almost every aspect of saturated steam generation and transmission. The company frequently provides expert advice and consultations to UK government and other autonomous organizations to formulate strategies and standards around steam systems, boiler operations and sterilization.

Corac Energy Technologies Ltd has considerable experience in the design and development of innovative turbo-machinery for specialist applications in the oil and gas and industrial markets.

Content from this work may be used under the terms of the Creative Commons Attribution 3.0 licence. Any further distribution of this work must maintain attribution to the author(s) and the title of the work, journal citation and DOI.
Published under licence by IOP Publishing Ltd

2. Introduction

Steam is usually generated in boilers at high pressures to simplify transmission. Generating and distributing steam at higher pressure offers two advantages. Firstly, the boiler's thermal storage capacity is increased, helping it to cope more efficiently with fluctuating loads. Secondly, smaller bore steam pipes can be used for distribution. On leaving the distribution system the pressure needs to be reduced to that required by the application. This pressure reduction is conventionally performed by a pressure reducing station comprising a pressure reducing valve and ancillary equipment (Figure1a).

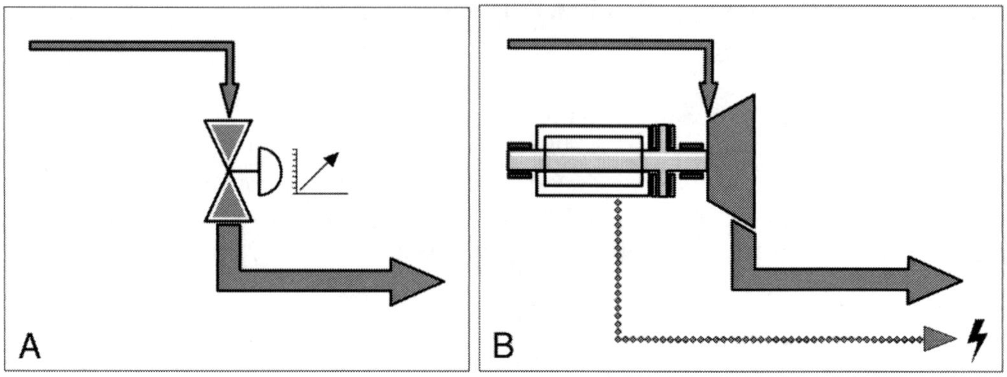

Figure 1. Conventional pressure reducing valve (a); Microturbine expander generator (b)

An alternative to this approach is to pass the high pressure steam through a microturbine, typically installed in parallel with a conventional pressure reducing station, enabling operators to use the energy released by the pressure drop to supplement their electricity supply (Figure.1b). The outlet steam is used by the downstream application as in a conventional system.

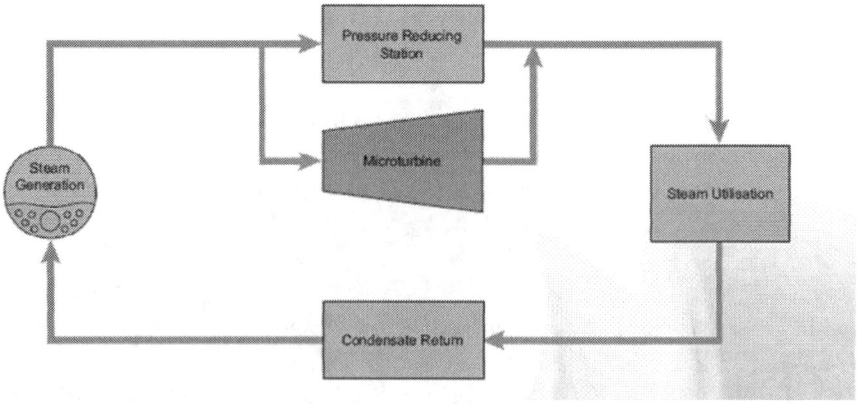

Figure 2. Steam generation and utilisation cycle

A microturbine generator in this application can produce high value electrical energy for the consumer which will be fed back into the local grid. See figure 2. This approach is not uncommon for large steam users, such as oil refineries, the opportunity identified is for smaller steam users with boilers sized under 6MWth. Industries with particular potential for microturbine generators include food and drink, chemicals processing, pharmaceuticals and hospitals.

Traditional, grid-scale fixed speed steam generation turbines are supplied with steam at reasonably constant pressure and exhaust to a condenser so that the steam inlet and outlet pressures are fairly constant. The turbine reacts to changes in electrical load by throttling the steam inlet valve to maintain the generator at constant output frequency within tight limits, as this dictates the frequency of the electricity supplied. Micro-turbines with directly grid connected induction generators will not influence grid frequency and can therefore be operated over a wide range of conditions dictated by process conditions only. The downside to this is that the expander cannot have its speed adjusted to enable it to operate at peak efficiency over a wide range of inlet and outlet conditions. The system under consideration has to respond to changes of steam inlet pressure and flow rate to ensure that the downstream process is not adversely affected. The variable speed capability of a system with inverter control can allow the expander to be operated at peak efficiency, taking into account turbine efficiency, generator efficiency and frictional losses amongst others. From an electrical demand point of view it was assumed that the grid has an infinite ability to accept power.

3. The machine

The machine consists of a microturbine driving a high speed generator on a single shaft running on air bearings. The machine is installed in a package with controls, steam valves and a power conversion system. The prototype unit is shown in Figure 3 ready for testing.

Figure 3. Prototype Spirax Sarco TurboPower package

4. Control objectives

The control objectives for this system were dictated by the needs of the typical applications identified during market research. As shown in Figure 2 the system must produce the maximum electrical power whilst maintaining the steam supply needs of the downstream process. This is summarised as:

- Maintain downstream pressure to a predetermined target
- Generate optimal electrical output power under all operational conditions
- Maintain the safety and integrity of the system.

IOP Conf. Series: Materials Science and Engineering **90** (2015) 012067 doi:10.1088/1757-899X/90/1/012067

Downstream steam pressure must be maintained to a pre-determined level, the production of electricity may be compromised to allow this.

5. Electrical System design

The system to extract energy from the steam and convert it to electricity which is then fed into the grid is as follows:

With reference to Figure 4 the microturbine drives a permanent magnet synchronous generator on a common shaft; high frequency AC current is fed into an inverter; the inverter converts the AC current into DC current; the DC current is then fed into an Active Front End (AFE) which converts the DC current into AC current at the grid frequency. The inverter is also used to control the speed of the generator.

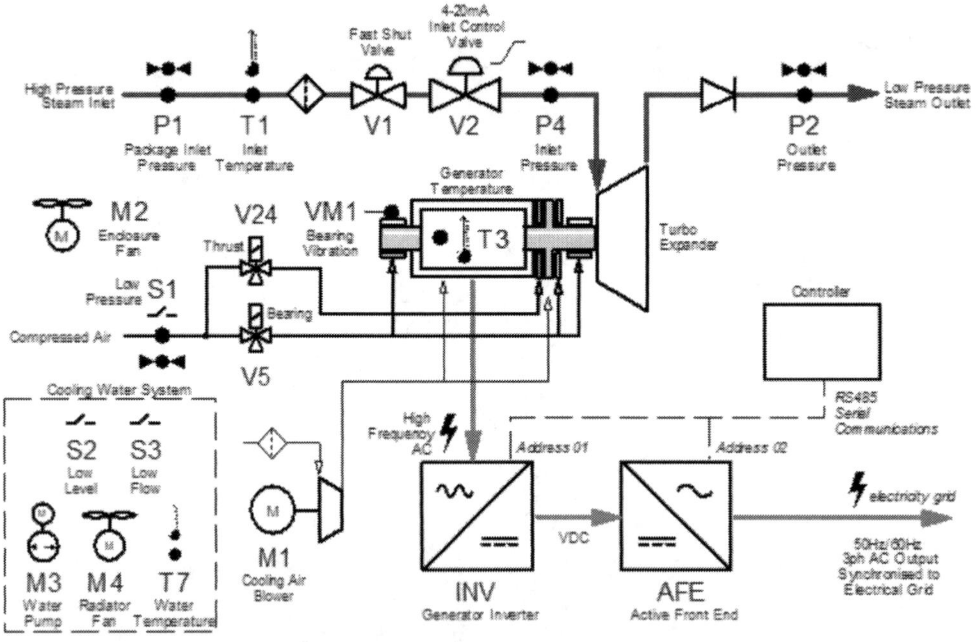

Figure 4. Expander generator package schematic diagram

Key components of the power converter system are line filters, inverter, DC-link capacitors, AFE or grid-tie inverter and electrical braking system.

6.1 Inverter drive

An inverter operated in power generation mode offers a readily available solution for generator power conversion, monitoring and control.

A generator is driven by an external rotational force which is high pressure steam applied to the turbine. The inverter, which is connected to the generator, provides variable voltage and frequency in active rectification mode during normal operation when the machine is generating, and variable voltage and frequency inverting mode when the machine is motoring. Using power from the grid to motor the generator is necessary for the start-up sequence, which is described later. The D.C. link voltage is kept constant independent from the mains voltage, whilst current varies according to demand.

6.2 *Active front end*

An active front end (AFE), or grid-tie inverter, converts DC power, in this instance generated as output by the inverter, to AC power matched and synchronised to the electricity supply grid. The AFE acts autonomously, monitoring available DC input power and regulating power output in proportion. This allows the control of power factor and transfer of power from the D.C. bus to the grid and vice versa. The AFE also acts as a commutating reactor, minimising harmonics at current switching between power switches and free-wheeling diodes within the inverter. A benefit of this approach is that the output frequency and voltage of the system can be changed at will merely by adjusting inverter parameters.

To supply electricity into the grid in the UK it is necessary to comply with a set of regulations known as G59. The AFE provides the necessary control and monitoring to meet the limits set by these regulations.

6. Controller selection

A controller was specified to provide control, protection and sequencing to the machine. The controller had to provide an HMI, control by communications connection to the inverter and provide connections for all sensors and valves.

Figure 5. Controller, concept

The sensor input and control output requirements were assessed to be within the scope of either a typical PLC (Programmable Logic Controller) or a customised embedded controller. The concept for the chosen controller is shown in Figure 5.

After assessing the specification requirements, the selection criteria were driven by three key factors:

- Volume manufacturing cost
- Human interface specification
- Control response speed

IOP Conf. Series: Materials Science and Engineering **90** (2015) 012067 doi:10.1088/1757-899X/90/1/012067

9.1 Controller cost considerations

The client required a solution that could be readily implemented, without modification, on a production unit. Projected volumes were not high but sufficiently large to question the cost of a standard PLC in comparison to a standard or custom embedded controller. A typical embedded industrial controller offers a single unit solution incorporating processing, memory, I/O, data connectivity, power supply and human interface. In general this configuration exhibits a lower cost than the modular generic configurations offered by a PLC. A potential draw back when considering an embedded controller is coping with changes in requirement.

9.2 Human interface requirements

The client's human machine interface (HMI) requirement was for a high resolution, full colour, touch screen with high visibility illuminated indicators for power, status and fault indication.. High resolution touch screen HMI options are readily available for PLCs. Industrial machine embedded controllers usually offer both graphic screen and LED indicators as standard.

9.3 Controller Connectivity

Part of the client requirement was for the controller to be able to communicate with other control and monitoring systems either by digital signals or by use of standard communication protocols. For the future having a controller that may be accessed via the Internet was deemed to be desirable. The embedded controllers available offer these features.

9.4 Controller speed of response

The required speed of response of the controller is critical in this application. Certain functions: high vibration and over speed avoidance, for example, demanded a very fast response. The required application program cycle needed to be in excess of ten times a second. Although PLC s exist that happily satisfy this requirement the total program cycle times of PLCs in the project cost range tended to be in the order of 150-250ms, 4-6 times a second. A typical industrial embedded controller has response times in the order of 50ms, 20 times a second. The cycle time requirement was the most important factor leading to the choice of the embedded controller.

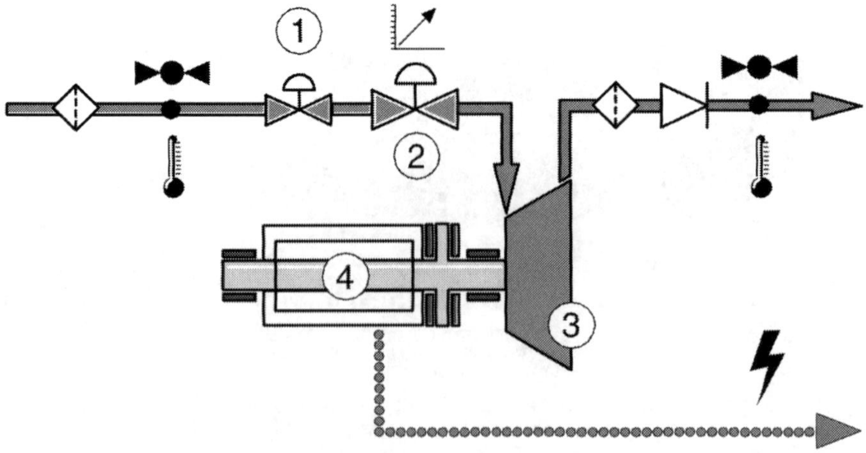

Figure 6. Steam control schematic

7. System configuration

Figure 6 shows the basic schematic diagram for the machine: Steam flow to the generator turbine (3) is regulated by an inlet control valve (2). Emergency shutdown, over speed protection and steam isolation are provided by a fast acting shut-off valve (1). The generator (4) is connected to an inverter drive, operated in generation mode, providing output power rectification, monitoring functions and speed control.

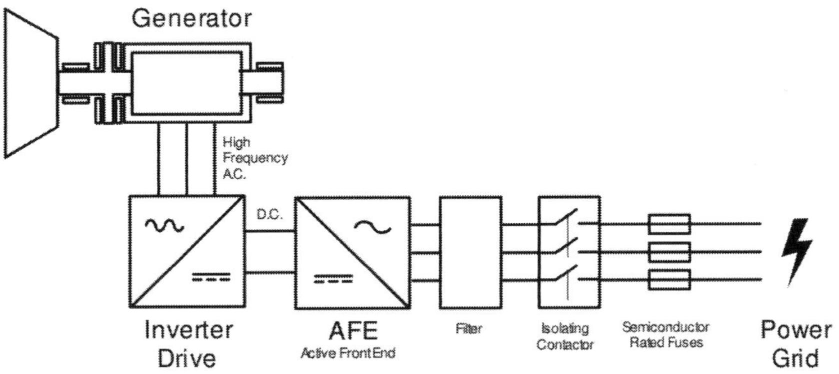

Figure 7. Electrical power generation schematic

With reference to Figure 7: The rectified output of the inverter is connected to an active front end inverter (AFE) which has the sole purpose of inverting available D.C. power from the inverter to an output matched and synchronised to the electricity grid. The AFE inverter also offered functions to export electricity into the grid. The package was designed for connection to a non-island power grid system; that is a system that is connected to the national or local electricity grid.

8. Control of the microturbine

This section describes some of the special control routines that were developed to meet some of the difficulties unique to a steam machine.

Unpredictable behaviour of the machine due to thermal effects of steam was something that had to be understood during the development process. Of particular concern were the effects experienced when expansion of steam through a control valve tends towards superheating and the presence of, even small, amounts of condensate holds the steam in a saturated condition at saturation temperature. Over time, condensate will evaporate and temperatures unexpectedly rise in association with density changes. Any re-introduction of condensate will cause temperatures to fall to saturated levels again. Even with the most effective condensate management practises in place, condensate can appear periodically. The systems control engineers on this project had to take the potential variability of this parameter into account.

9.1 Start-up routine

It was found that Condensate tends to accumulate, particularly in lower voids within control valves and similar components, after a short period of cooling. Introduction of steam into a cool system obviously generates condensate.

During a cold start sequence, steam introduced to the turbine is likely to be two phase with a degree of suspended condensate. The effects of this were alleviated by designing a cycle to pre-heat the system prior to operation. A small flow of steam was allowed to exert a low pressure within the stationary turbine system which pre-heated the components; remaining condensate was drained through steam traps. During this cycle it was necessary to prevent unintentional rotation of the turbine: the inverter was used to energise the generator to hold the turbine shaft stationary during this process. A small amount of power is consumed for a short period during this cycle. A greater challenge was the discharge of pockets of condensate collected within components on the inlet line to the turbine: As flow increased there was a tendency for the condensate to be carried into the turbine. On occasions this lead to 'water hammer' effects sufficiently severe to disrupt turbine stability.

Mitigation for this was challenging: Experimentation demonstrated condensate pick-up was associated with a particular steam flow rate. Knowing this a turbine speed was calculated that matched turbine injection angle velocity to impeller blade velocity with the goal of allowing condensate to pass through the impeller with minimal blade impact. To achieve this, the turbine had to be driven. 'Motoring' the turbine at this calculated speed while steam flow was slowly increased through the condensate pick-up flow range allowed the condensate to pass through the turbine with minimal stability disruption. Again the trade-off for this process is a small consumption of electrical power.

9.2 Acceleration, 'winning back lost power'

The pursuit of maximum efficiency identified an opportunity to recover power consumed during the warm-up and condensate management cycle. This was implemented by using carefully controlled steam flow to accelerate the turbine rather than electrically driving the turbine to operational speed, this resulted in the acceleration cycle power consumption being eliminated. This function was found to be simpler than originally anticipated, relying predominantly on timing, with sufficient margin to accommodate varying circumstances. The key constraint was turbine bearing stability, where care had to be taken to accelerate sufficiently fast to avoid any power consuming assistance from the inverter drive but slow enough to avoid affecting the shaft stability with excessive changes of torque. The end result of tuning this function was generation of power throughout the acceleration cycle with power output increasing as operational speed was approached. Once at operational speed, flow could be proportionally increased to provide optimum power generation for the given steam conditions using measurements of power and generator speed provided by the inverter.

9.3 Turbine shaft bearing stability

The low friction, gas bearings utilised in the micro turbine design could experience instability, indicated by transient shaft orbit measurement deviations, in response to rapid, transient rates of change of turbine loading. The stability boundaries turned out to be roughly analogous to the percentage of the change referenced to the mass flow rate: At lower steam flow rates a small rate of change had a similar impact to a higher rate of change at a higher flow rate. At lower valve openings the rate of change of position of the inlet valve control needed to be constrained to a greater degree, being proportionally relaxed as the valve opened further and steam flow rates increased. Although this constraint impacted heavily on control response a compromise was achieved with acceptable limitations.

9.4 Control of the steam inlet valve

Initial models suggested steam flow control, from which outlet pressure control is derived, would be influenced by turbine speed. This concept relied on the principle that higher impeller speeds would restrict flow due to the generation of a higher ΔP opposing flow. In practise, the degree of variation of steam flow across the speed range (defined by the torque limit) was measured to be less than 5%, the load control tasks were therefore required to be dealt with by control of the inlet steam valve.

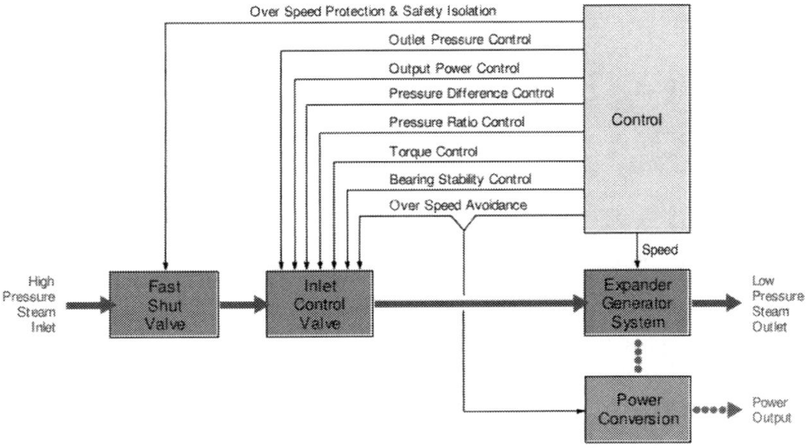

Figure 8. Control overview

This presented a significant challenge with the inlet control valve being simultaneously tasked with a number of key control functions as can be seen in Figure 8. Attention was given to understanding the interaction of control functions with strategies implemented to mitigate undesirable effects from conflicting requirements whilst meeting the original control system specification.

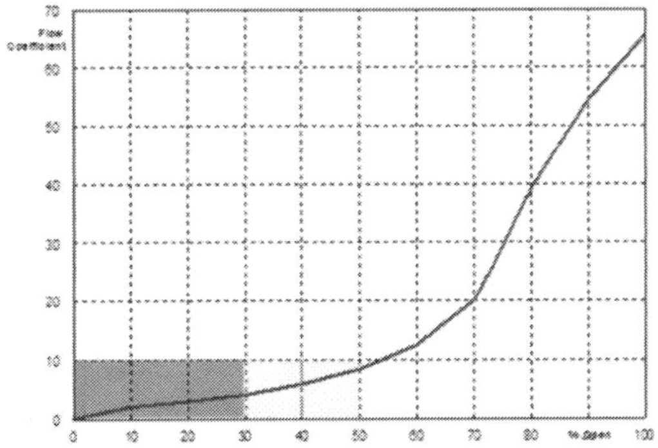

Figure 9. Inlet control valve characteristic

To be able to successfully implement proportional inlet valve control, with a differential rate of mass flow change curve, could have required the expense of additional monitoring and modified valve control or the requirement to map and implement logic for all conditions.

These requirements were avoided by using a readily available valve with an inverse flow coefficient characteristic. The inverse curve provided lower rates of mass flow change at lower percentage open values. The characteristics of the valve used are shown in Figure 9.

9. Generator control

9.1 Control Concept

In this system the output frequency is controlled by the inverter and the steam inlet valve is used to control the pressure ratio across the microturbine. The generator speed was adjusted to provide maximum output power for a given set of steam conditions.

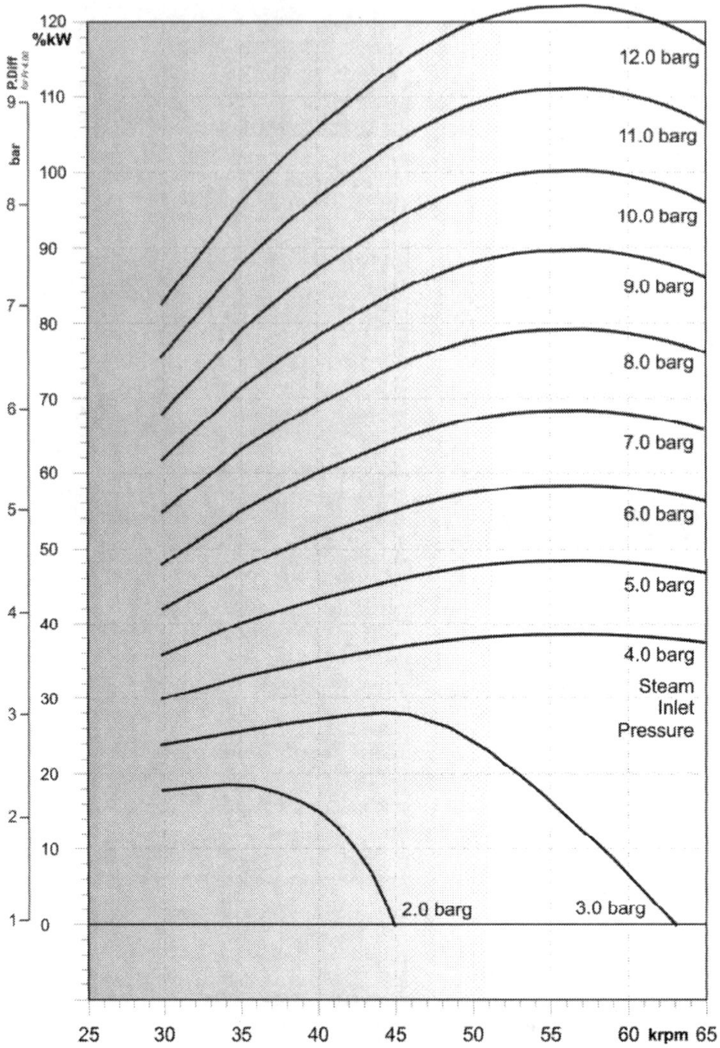

Figure 10. Performance curves

Figure 10 shows the results of a series of characterization tests for the microturbine and generator. In these tests steam inlet and outlet conditions were established and the generator was run at a range of speeds with output power and other data being recorded. As can be seen the maximum output power for given steam inlet and outlet pressures occurred within a relatively narrow speed range. Across this range, steam mass flow and expander efficiency varied; it is therefore practical to vary the machine rotational speed to achieve maximum efficiency and therefore maximum power output for a given steam demand and required downstream steam pressure.

9.2 Transient steam fluctuations

Abrupt increases in mass flow resulting from rapid inlet pressure or outlet pressure changes were found to exceed the operational torque limit of the inverter and generator system. Loss of inverter synchronisation with the generator rotor resulted in attempts by the inverter to regain synchronisation, leading to excessive currents causing high temperatures in power switching components. When synchronisation was completely lost, over speed events occurred. To mitigate these effects, the decision was taken to temporarily increase the generator speed to reduce torque during the duration of excessive mass flow events while the inlet control valve opening was reduced to limit the mass flow. Failure of this method to maintain torque at manageable levels causing loss of synchronisation triggers the over speed protection system.

9.3 Over speed avoidance and protection

The turbine impeller and shaft assembly has low mass and low inertia. There is a potential for uncontrolled acceleration at a high rate with a risk of rotating unit damage due to over speeding.

Turbine speed is regulated and maintained by the inverter predominantly by varying the electrical loading of the generator. The higher the electrical loading, the greater the braking forces. In the event of a sudden removal of electrical loading due to grid loss or an inverter failure, the generator would accelerate in an uncontrolled manner. To prevent the speed from exceeding safe limits a fast acting, fail-safe shut-off valve was fitted to the steam inlet. Although the valve can isolate the steam supply within 1.5 seconds this is not fast enough to prevent over speeding in all situations. Without any loading it was estimated the turbine could accelerate at circa 20krpm/second. Extremely fast acting valve types were reviewed but were found to be prohibitively expensive. An alternative strategy was introduced: a fixed resistive electrical load was employed, that was switched across the output of the generator in the event of over speed or grid loss. This system, which was independent from the controller, offered a simple, reliable and low cost method of arresting acceleration while the fast shut-off valve was closing. With inlet steam flow and pressure at full output power level, tests demonstrated this method prevented acceleration then slowed the turbine once the shut-off valve had closed. The combination of a readily available fast shut-off valve and resistive braking proved to be highly successful under all testing regimes.

With the turbine driven by steam supplied remotely from a source beyond the control boundary of the package, the potential for abrupt changes of pressure without warning also had to be considered. The initial defence against an abrupt increase in steam pressure is the inlet control valve. The rate of closing of the inlet control valve had to be restricted in order to avoid causing shaft instability issues. In the event of the inlet control valve being unable to arrest acceleration sufficiently the emergency over speed avoidance mechanism detailed above was activated using the speed signal from the inverter.

9.4 Linking to the grid

Within the UK, the connection of any form of electricity generation device to run synchronised with the National Grid is subject to regulation. If the power generation is greater than 16A per phase, the requirements of the Energy Networks Association Engineering Recommendation G59/2 "Recommendations for the Connection of Generating Plant to the Distribution Systems of Licensed Distribution Network Operators (DNOs)" must be met before connected to the network.

The regulation ensures a process is followed within which a proposal is submitted to the DNO, the connection is detailed, and a connection agreement is made between the generator and the DNO. Testing and commissioning is also completed in conjunction with the DNO.

To meet this regulation, a G59 compliant mains protection relay must be installed between the generator and the grid connection. This is an electronic monitoring device which examines the quality and stability of the electricity being exported. Pre-programmed limits of certain parameters such as voltage, frequency, phase angle etc. as dictated by the DNO are held within the device. Should any of these parameters exceed limits; the relay will open a protective circuit breaker, disconnecting the generator from the grid.

IOP Conf. Series: Materials Science and Engineering **90** (2015) 012067 doi:10.1088/1757-899X/90/1/012067

In theory the control systems for this machine met these requirements. In practice it was found necessary to fit an approved grid connection protection relay that had been approved to these regulations.

Requirements outside of the UK vary from country to country, but it is common to find similar requirements to those of G59/2.

The G59 approval process involves demonstrating that each and every criterion of the standard is met by test and measurement. This process took around four hours and placed a heavy reliance on the power generation package to be able to tolerate, and immediately recover without repair, from each test. The G59 approval process focuses totally on the grid interface and interfacing variables, paying little attention to the generating equipment itself. The steam microturbine generator system survived this process without issue and the installation was approved.

10. Conclusions

The control system deployed achieved all of the objectives that were identified in the specification and has proved to be inherently reliable. The use of standard control equipment such as inverters has ensured short development times and reasonable costs.

The problems identified with condensation and wet steam in the supply are believed to be realistic problems in most industrial installations and the design of the control system has given a high degree of robustness to what might be viewed as a sensitive high speed machine.

A highly robust and cost effective method of preventing over speed and damage due to loss of grid connection was developed and successfully implemented.

The high speed generator with an electronic power conversion system offers immense flexibility and consistency of control compared to fixed speed gear driven machines.

Copyright:
The copyright to this document is the property of Corac Energy Technologies Ltd and Spirax Sarco Ltd.

A thermal model for analysis of hermetic reciprocating compressors under the on-off cycling operating condition

S K Lohn, M C Diniz and C J Deschamps[1]
POLO Research Laboratories
Federal University of Santa Catarina, 88040900, Florianopolis, SC, Brazil

E-mail: deschamps@polo.ufsc.br

Abstract. The on-off cycling operating condition of compressors is very common in low capacity refrigeration systems, being characterized by alternate periods in which the compressor is either operating (on) or idle (off). Thermal interactions between the compressor components affect its performance during the operating period and establish the initial condition for the compressor start up from idle condition. This paper presents a numerical model to predict the temperature field of hermetic reciprocating compressors under on-off cycling conditions. The model adopts a lumped formulation for control volumes formed in the fluid solution domain and the finite volume method to solve heat conduction in the solid components. Some required heat transfer coefficients were experimentally adjusted. Predictions for temperature were compared to measurements and good agreement was observed, especially for the thermal transient during the period in which the compressor is off.

1. Introduction

Reciprocating compressors are widely applied in low capacity refrigeration applications, commonly associated with an on-off cycling operating condition characterized by alternate periods in which the compressor is either operating (on) or idle (off).

Due to economic and environmental issues, the refrigeration industry is oriented towards performance improvement, thus requiring the analysis of aspects related to the transient operation of the system. The current literature related to transient system conditions is focused on the heat exchangers and expansion device [1]. Therefore, it is necessary to improve the knowledge on the compressor performance during the on-off cycling condition in order to increase the system efficiency. Thermal interactions between internal components of hermetic reciprocating compressors strongly influence their thermodynamic efficiency, as well their reliability. On the other hand, thermal interactions during the off period define the initial condition for the compressor start-up.

Heat transfer inside hermetic compressors has been studied both experimentally [2] and numerically [3]. Despite the good insight provided by experimentation, the use of numerical approaches presents significant benefits, such as versatility and fast results. Heat transfer in hermetic compressors is modelled employing different degrees of complexity, and a wide variety of steady state models is available in the literature [4]. Simple semi-empirical models, such as the one developed by Todescat *et al.* [5], provide very fast predictions of the temperature field. Other models, such as the totally differential model of Raja *et al.* [6] are characterized by high computational cost, although

[1] To whom any correspondence should be addressed.

Content from this work may be used under the terms of the Creative Commons Attribution 3.0 licence. Any further distribution of this work must maintain attribution to the author(s) and the title of the work, journal citation and DOI.
Published under licence by IOP Publishing Ltd

allowing deeper understanding of the phenomena. As a compromise, Ribas Jr. [7] developed a hybrid model that combines a differential approach for the solid domains and a lumped formulation for the gas chambers.

This paper presents a numerical model to predict the temperature field of a hermetic reciprocating compressor under on-off cycling conditions, based on the hybrid approach [7]. Predictions for temperature in different regions inside the compressor under the transient condition were validated through comparisons with measurements.

2. Compression cycle model

The thermal model requires input data related to the compression cycle. The modelling of the compression cycle employed in this work is based on four groups of equations [5], following a transient lumped formulation. The time scale of the compression cycle (milliseconds) is considerably smaller than that of the transient temperature field of solid components (minutes). Hence, the results obtained through the compression cycle model are integrated over a representative cycle to be used as input data in the thermal model.

The first group of equations calculates the instantaneous volume of gas inside the cylinder with respect to the compressor geometrical parameters. The second group of equations are the conservation equations (mass and energy) applied to the gas inside the cylinder. With the instantaneous cylinder volume and the mass of gas inside the cylinder, obtained through equation (1), it is possible to calculate the gas density.

$$\frac{\partial m_G}{\partial t} = \dot{m}_{suc} - \dot{m}_{suc,b} - \dot{m}_{dis} + \dot{m}_{dis,b} - \dot{m}_{leak}\,. \tag{1}$$

The manipulation of the lumped energy balance originates the following equation to predict the gas temperature during the compression cycle:

$$\begin{aligned}\frac{dT_G}{dt} &= \frac{1}{m_G c_{v,G}}\left[H_G A_w T_w - h_G\frac{dm_G}{dt} + \sum_{in}\dot{m}h - \sum_{out}\dot{m}h\right] \\ &- \frac{T_G}{m_G c_{v,G}}\left[H_G A_w + \left.\frac{\partial p_G}{\partial T_G}\right|_v \frac{d\forall_G}{dt} - \left.\frac{\partial p_G}{\partial T_G}\right|_v v_G\frac{dm_G}{dt}\right]\end{aligned}\,, \tag{2}$$

where H_G is the convective heat transfer coefficient estimated with the correlation of Annand [8]. Two temperatures are very important in equation (2): the refrigerant temperature in the suction chamber (T_{sc}) and the temperature of the cylinder wall (T_w). These temperatures can either be specified from experimental data or calculated with a thermal model, as in this work. With the temperature estimated through equation (2) and the density obtained via equation (1), the instantaneous pressure is calculated using an equation of state. The equation of state and all other thermodynamic properties are obtained using the Refprop library [9].

The third group of equations represents the valve dynamics. A one-degree of freedom mass-spring model is used, with effective force areas being used to characterize the force acting on the valve [10]. Finally, the fourth group of equations predicts the mass flow rates through the valves and the gas leakage through the piston-cylinder gap. Theoretical estimates of mass flow rate in valves are obtained from the isentropic flow through a convergent nozzle and then corrected by using the concept of effective flow area to account for viscous effects. Leakage through the piston-cylinder gap was modelled as a fully developed laminar Couette-Poiseuille flow [11].

Predictions from the compression cycle model that are used as input data in the thermal model are given in Table 1. The amount of heat generated due to friction in the piston-cylinder gap, $\dot{Q}_{pc}$, is calculated with the Couette-Poiseuille velocity profile, whereas the electrical efficiency, η_{ele}, is obtained experimentally.

Table 1. Input data from compression cycle model to thermal model.

Variable	Description	Calculation
$\dot{m}$	Mass flow rate	$f\int \dot{m}_{dis}dt$
$\dot{Q}_W$	Heat transfer rate inside cylinder	$\dot{Q}_{pc} + A_W f\int H_G(T_W - T_G)dt$
$\dot{W}_{ind}$	Indicated power	$f\int p_G d\forall_G$
$\dot{Q}_{bea}$	Mechanical losses	Obtained from experimental data
$\dot{Q}_{mot}$	Electrical losses	$(\dot{W}_{ind} + \dot{Q}_{bea} + \dot{Q}_{pc})(1/\eta_{ele} - 1)$

3. Thermal model

Figure 1 presents a schematic view of the compressor simulated in this work, identifying the solution domains of each model employed. The hybrid thermal model solves the three-dimensional heat conduction in solid components with a differential formulation and the control volumes with gas through a lumped modelling approach.

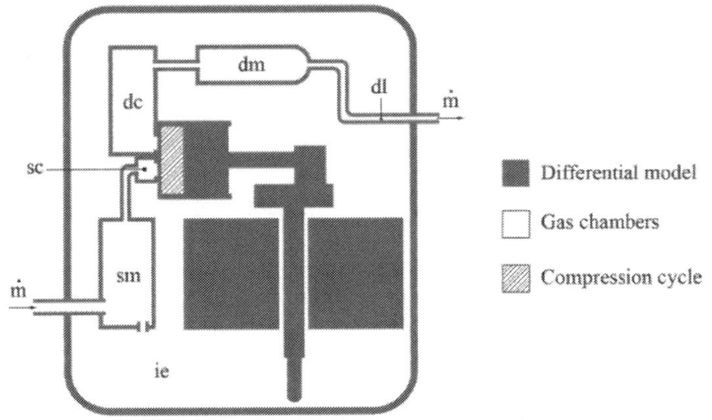

Figure 1. Domains each sub models in the thermal model.

3.1. Solid components

The differential model is solved, with suitable boundary conditions, through a commercial code [12] that uses the finite volume method. On the external surface of the compressor shell, convective heat transfer is established with the ambient. For the internal solid components in contact with refrigerant fluid (motor, cylinder, crankcase, etc.) a convective boundary condition was also specified:

$$H_{ch}\left(T_{sol} - T_{ch}\right) = -k\frac{dT}{d\hat{n}}, \tag{3}$$

where T_{sol} is the temperature of a generic solid surface, T_{ch} is a temperature of a generic gas control volume and H_{ch} is the convective heat transfer coefficient. There are also regions inside the compressor where internal heat generation must be specified. For instance, heat generation due to electrical motor inefficiencies is specified in the finite volumes that discretize the electrical motor, i.e.:

$$\dot{Q}'''_{mot} = \frac{\dot{Q}_{mot}}{\forall_{mot}} . \tag{4}$$

Heat generation due to mechanical losses is characterized as a source term in a virtually defined thin wall between the shaft and the bearing [12]. Therefore, similarly to the heat generation in the motor:

$$\dot{Q}'''_{bea} = \frac{\dot{Q}_{bea}}{\forall_{sur}} . \tag{5}$$

Finally, the total heat transfer at the cylinder wall is also used as boundary condition in the differential model:

$$\frac{\dot{Q}_W}{A_W} = -k\frac{dT}{d\hat{n}} . \tag{6}$$

3.2. Gas control volumes

The solution domain of gas inside the compressor is divided into several control volumes (internal environment, suction muffler, discharge chamber, discharge muffler and discharge line) and the temperature in each one of them is calculated using lumped energy balances. The transient term on the energy balances was only employed in internal environment control volume and the time step depends on the condition (on/off). Therefore, the implementation of the models for the on and off periods of the compressor will be addressed separately.

3.2.1. Off period. During the period in which the compressor is off, the compression cycle is not simulated. Furthermore, it is not necessary to simulate all the gas control volumes of the compressor immediately after the shutdown, as the higher heat capacity of the solid components forces thermal equilibrium between the solid walls and the adjacent gas in almost all the gas control volumes. The exception is the internal environment (ie) and, for this reason, only this control volume was simulated.

Immediately after the compressor is turned off, pressure in all control volumes defined for the gas domain is considered equal to the equalization pressure. The energy balance for the compressor internal environment is given by:

$$H_{ch}(T_{sol} - T_{ie})\,\Delta t = m c_P (T_{ie}^{t+\Delta t} - T_{ie}^{t}) . \tag{7}$$

During the off period, heat transfer is mainly caused by natural convection, with the convective heat transfer coefficients being estimated using a modified version of the Catton equation for a rectangular cavity [13]:

$$H_{ch} = C\left[\left(\frac{\Pr}{0.2+\Pr}\frac{\beta}{\alpha\nu}\right)^{0.28} k\right]\left[T_{sol} - T_{ie}\right]^{0.28} , \tag{8}$$

where C is an experimentally calibrated coefficient required to adjust the Catton equation to the complex geometries of the compressor components. Four heat transfer coefficients are necessary to represent the convective heat transfer in the compressor internal environment, and a different constant C is required for each one. A new temperature for the internal environment chamber is calculated using equation (7).

3.2.2. On period. The simulation of the compression cycle is carried out when the compressor is switched on. The transient term in the energy balances on the control volumes is only important in the equation for the internal environment, as the thermal capacities of the remaining volumes are negligible. This energy balance remains in the same form shown in equation (7). The heat transfer coefficients for the internal environment are calculated using a modified version of the correlation for turbulent heat transfer over flat plates:

$$H_{ch} = C\left(\dot{m}_{oil}^{0.8}\right)\left(\mathrm{Pr}^{0.33}\, k\mu^{-0.8}\right), \quad (9)$$

where the mass flow rate, $\dot{m}_{oil}$, is calculated using the procedure presented by Drost [14]. The other control volumes for the gas (suction and discharge paths) are simulated using steady state energy balances given by

$$H_{ch}\left(T_{sol} - T_{ch}\right) = \dot{m}\left(h_{out} - h_{in}\right), \quad (10)$$

where the mass flow rate, $\dot{m}$, is calculated in the compression cycle model. In this case, the convective heat transfer coefficients were estimated using modified equations for turbulent flow inside ducts, given by:

$$H_{ch} = C\left(\dot{m}^{0.8}\right)\left(\mathrm{Pr}^{0.33}\, k\mu^{-0.8}\right), \quad (11)$$

where C is a constant that is experimentally adjusted, since it is affected by the compressor geometry. In total, there are eight heat transfer coefficients to be determined for the on period.

3.3. Coupled solution

The thermal model is simulated in a coupled manner with the compression cycle model explained in Section 2. The coupled solution is represented in the flowchart of Figure 2, where the inputs and results of each sub model can be observed. The inputs in dark grey are given by experimental data, while the ones in light grey are updated by the different models. Every time the compression cycle model is called, five cycles are simulated in order to achieve a fully established cyclic condition for the input conditions. After receiving data from the compression cycle model, the thermal model runs a certain number of time steps while maintaining constant the input data received from the compression cycle model. During these iterations, the differential model and the lumped models for the control volumes exchange information through heat transfer between components. After the specified number of iterations, the thermal model provides the temperatures to be used as input in the simulation of the compression cycle. The procedure is repeated along the time defined for the on period.

4. Results

As mentioned in the previous section, it is necessary to adjust the constants C for the heat transfer coefficients of the thermal model based experimental data. To this extent, a R134a compressor with a displacement volume of 7.55 cm^3 was tested in a hot cycle test bench, being submitted to the operating condition characterized by T_{evap} = -23.3°C (p_{evap} = 0.115 MPa) and T_{cond} = 40.5°C (p_{cond} = 1.03 MPa). Both the gas inlet temperature and the external environment temperature were set to 32°C. The compressor was instrumented with thermocouples in locations in the gas and solid: (a) gas: suction line, suction muffler inlet, suction muffler outlet (suction chamber), discharge chamber, discharge muffler, discharge line, internal environment; (b) solid: cylinder, shell and motor.

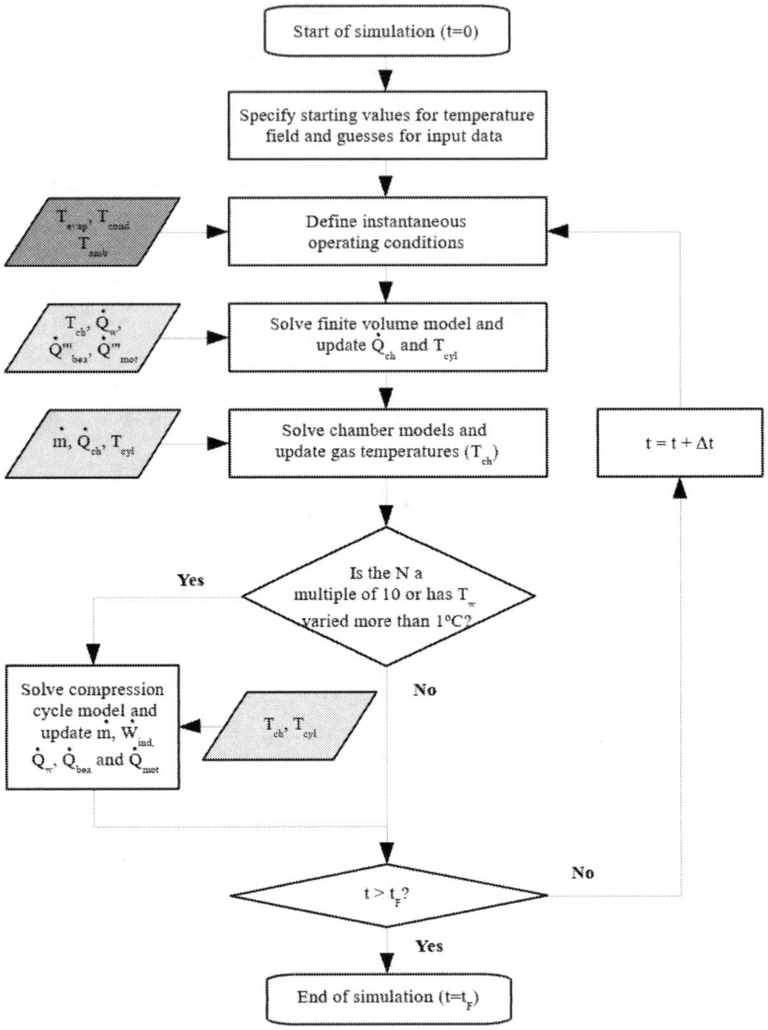

Figure 2. Flowchart of coupled simulation.

The temperature distribution obtained after the compressor reached thermal equilibrium was used to adjust the constants C in the heat transfer correlations for the on period. For the suction region, the heat transfer coefficients were within 30 and 60 W/m^2K during the on period. In the discharge region, the values ranged from 100 to 250 W/m^2K. For the off period, an experimental temperature distribution obtained 25 minutes after the compressor was turned off was adopted to adjust the constants C. For the off period the heat transfer coefficients are lower since heat transfer occurs by natural convection.

The mesh used to discretize the solid components was generated for the numerical model after introducing simplifications in the original geometry. In regions of the fluid flow where heat flux and exchange of momentum are intense, special care was taken to ensure that the gradients were correctly represented. The mesh generated had a total of 1.3 million volumes to guarantee the truncation errors were negligible. Due to the very different heat capacities of the gas and solid parts, the time step in the solution of the problem is limited by the thermal transient of the gas inside the chambers. The values used for the time step were $\Delta t = 7s$ for the off period and $\Delta t = 2s$ for the on period.

IOP Conf. Series: Materials Science and Engineering **90** (2015) 012068 doi:10.1088/1757-899X/90/1/012068

For the off period, the simulation model was validated by comparing numerical and experimental results for temperature distribution of the compressor, during the 210 minutes after the compressor was shut down from a thermally stabilized condition. Figures 3 and 4 show very good agreement between numerical results and experimental data. During the off period, there is no gas compression, heat generation in the motor and bearings. Hence, the compressor average temperature is reduced due to the heat transfer to the external environment. As can be observed, the decay of temperature is similar for the solid and gas regions. This indicates that the temperature distribution during the off period is governed mainly by the heat capacity of the compressor solid parts.

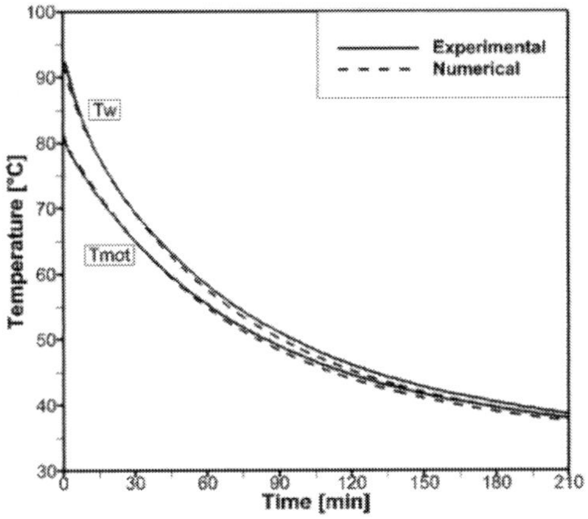

Figure 3. Numerical vs. experimental results of T_w and T_{mot} for the off period.

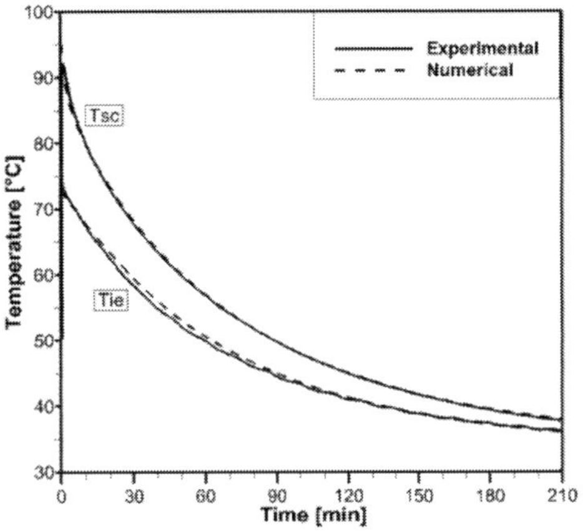

Figure 4. Numerical vs. experimental results of T_{sc} and T_{ie} for the off period.

Figures 5 and 6 show the comparison between numerical and experimental results for a period after the compressor is switched on. The agreement is still good, despite the presence of some deviations in the first 90 minutes of operation. In fact, better agreement is observed as the thermal equilibrium condition as reached ($t > 180$ min), because the constants C for the heat transfer coefficients were adjusted using the temperature distribution for this condition.

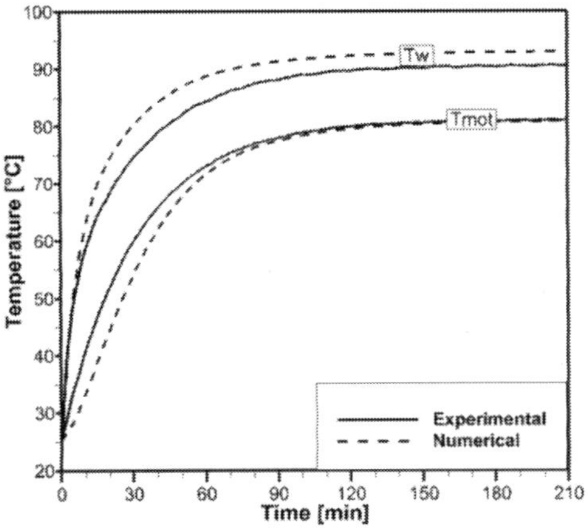

Figure 5. Numerical vs. experimental results for T_w and T_{mot} for the on period.

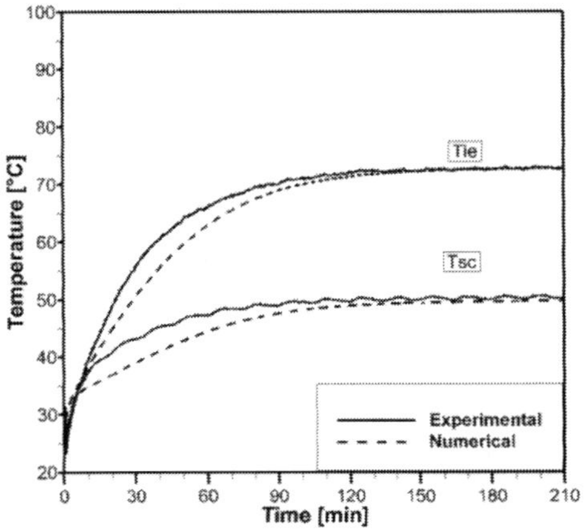

Figure 6. Numerical vs. experimental results for T_{ie} and T_{sc} for the on period.

After being validated, the model can be used to evaluate the compressor temperature distribution for cycling operating conditions. One test was defined by maintaining the compressor on for 30 minutes and then off during the next 30 minutes. Results for temperature variation at the cylinder wall and electrical motor are shown in Figures 7 to 10. Comparing Figures 7 and 8, it can be seen that the variation of the cylinder temperature is greater than that of the motor during the on-off cycling. Figure 9 shows an interesting behaviour for the suction chamber temperature. While during the on condition T_{sc} is lower than the T_{ie} (Figure 10), immediately after the compressor is turned off the temperature of the suction chamber rises due to heat transfer with discharge chamber and cylinder.

Figure 11 illustrates the temperature field in the solid components during one cycle of the on-off transient operating condition. Figure 11 (a) shows the temperature of the solid parts immediately after the compressor is turned on, i.e., after 30 minutes of off period. Figures 11 (a) to 11 (d) show the increase in compressor temperature that occurs during the on period. This aspect is more pronounced for solid components near the discharge system. The highest temperatures can be seen in Figure 11 (d), which is the moment when the compressor is switched off. The cooling down process after the compressor is switched off is illustrated by Figures 11 (e) and (f).

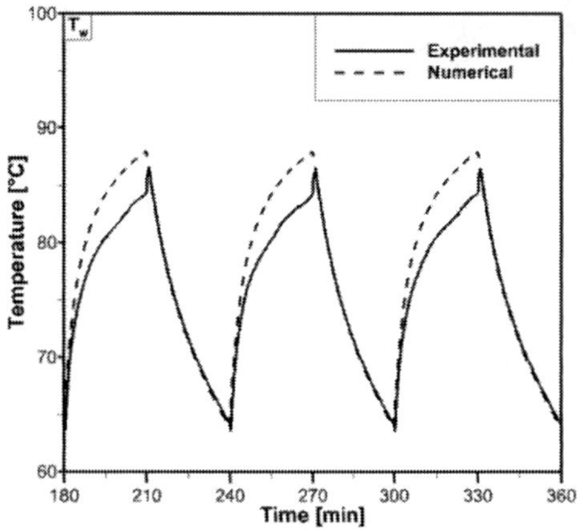

Figure 7. Cycling results for T_w.

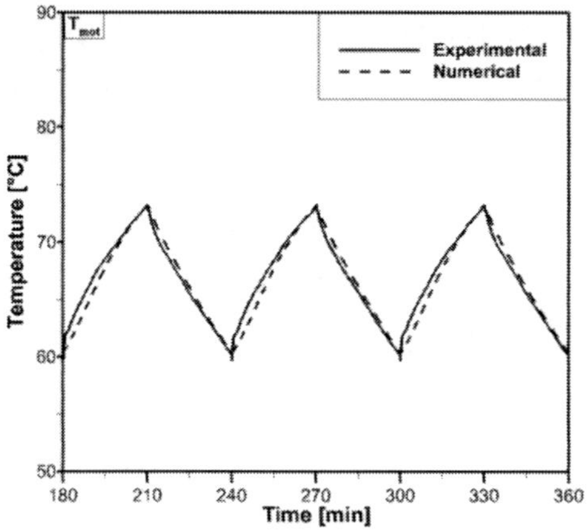

Figure 8. Cycling results for T_{mot}.

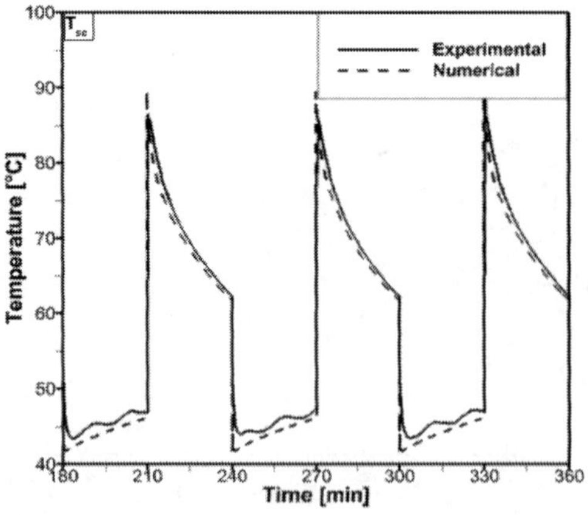

Figure 9. Cycling results for T_{sc}.

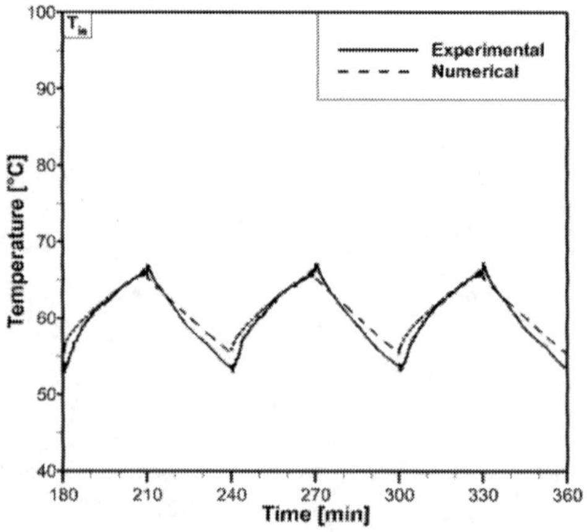

Figure 10. Cycling results for T_{ie}.

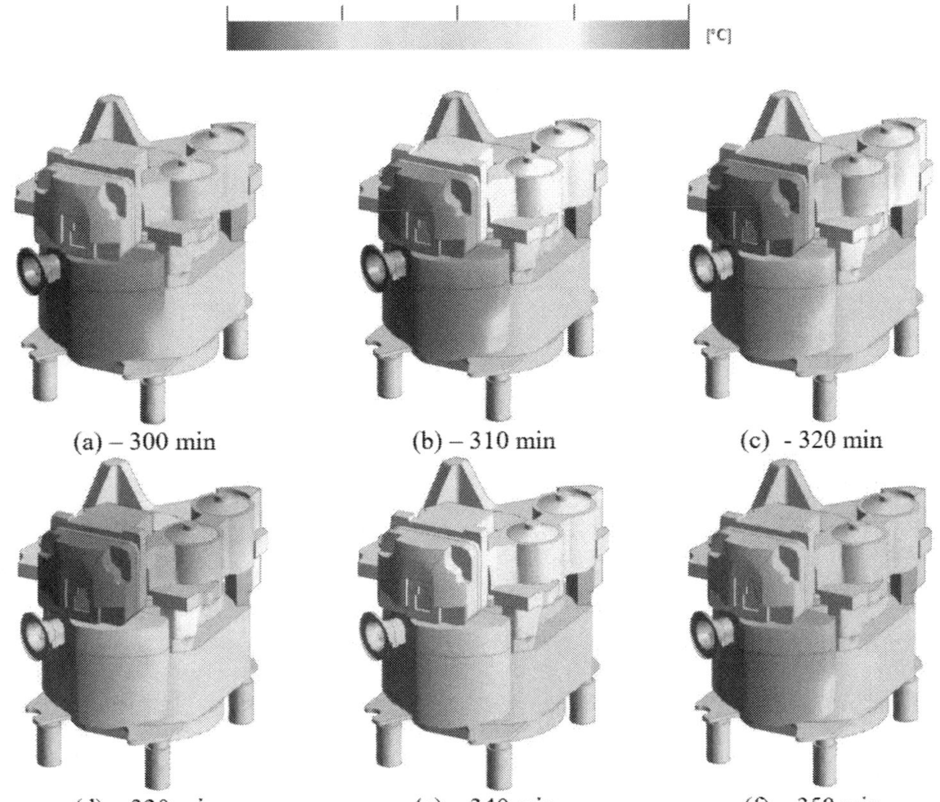

Figure 11. Transient temperature distribution of the solid parts.

5. Conclusions

This paper presented a thermal model developed to predict the temperature distribution of a small reciprocating compressor under on-off cycling condition. The simulation approach is based on a coupled solution procedure of three models: (i) a transient lumped model for the compression cycle, (ii) lumped energy balances for the remaining gas inside the compressor and (iii) finite volume model to solve heat conduction in the solid components. Some correlations of heat transfer coefficients were adjusted by using experimental data of temperature distribution. The model was validated through comparisons with temperature measurements in a period after the compressor was shut down (off period) and in a period after the compressor was switched on (on period). Good agreement was verified in both conditions, especially in the period when the compressor was off. The model was also applied in cycling operating conditions; again showing good agreement with measurements.

Acknowledgments

The present study was developed as part of a technical-scientific cooperation program between the Federal University of Santa Catarina and EMBRACO. The authors also acknowledge the support provided by EMBRACO and CNPq (Brazilian Research Council) through Grant No. 573581/2008-8 (National Institute of Science and Technology in Refrigeration and Thermophysics) and CAPES (Coordination for the Improvement of High Level Personnel).

General and Greek symbols					
A	Heat transfer area	(m^2)	p	Pressure	(Pa)
C	Constant	(-)	Pr	Prandtl number	(-)
c	Specific heat	(J/kg.K)	$\dot{Q}$	Heat transfer rate	(W)
f	Frequency of operation	(Hz)	T	Temperature	(K)
h	Enthalpy	(J/kg)	$\forall$	Gas volume	(m^3)
H	Heat transfer coefficient	(W/m^2K)	$\dot{W}_{ind}$	Indicated Power	(W)
k	Thermal conductivity	(W/m.K)	α	Thermal diffusivity	(m^2/s)
m	Mass	(kg)	β	Coefficient of thermal expansion	(1/K)
$\dot{m}$	Mass flow rate	(kg/s)	v	Specific volume	(m^3/kg)
N	Time step number	(-)	μ	Dynamic viscosity	(Pa.s)

Subscripts					
amb	External ambient	ele	Electrical	pc	Piston cylinder gap
b	Backflow	G	Gas inside cylinder	sol	General solid surface
bea	Bearings	ie	Internal environment	sc	Suction chamber
ch	General gas chamber	in	Inlet	sm	Suction muffler
dis	Discharge	leak	Leakage	suc	Suction
dc	Discharge chamber	mot	Motor	w	Cylinder wall
dm	Discharge muffler	oil	Oil		
dl	Discharge line	out	Outlet		

References

[1] Hermes C J L and Melo C 2008 A first-principles simulation model for the start-up and cycling transients of household refrigerators *Int. J. Refrig.* **31** 1341-57

[2] Dutra T and Deschamps C J 2013 Experimental characterization of heat transfer in the components of a small hermetic reciprocating compressor *Appl. Therm. Eng.* **58** 499-510

[3] Ooi K T 2003 Heat transfer study of a hermetic refrigeration compressor *Appl. Therm. Eng.* **23** 1931-1945

[4] Ribas Jr F A, Deschamps C J, Fagotti F, Morriesen A and Dutra T 2008 Thermal analysis in reciprocating compressors – a critical review *Proc. Int. Compressor Eng. Conf. (Purdue)* Paper 1907

[5] Todescat M L, Fagotti F, Prata A T and Ferreira R T S 1992 Thermal energy analysis in reciprocating hermetic compressor *Proc. Int. Compressor Eng. Conf. (Purdue)* 1419-28

[6] Raja B, Seckar S J, Lal D M and Kalanidhi A 2003 A numerical model for thermal mapping in hermetically sealed reciprocating compressor *Int. J. Refrig.* **26** 652-58

[7] Ribas Jr F A 2007 Thermal analysis of reciprocating compressors *Proc. Int. Conf. on Compressors and their Syst. (London)* 277-87

[8] Annand W J D 1963 Heat transfer in the cylinders of reciprocating internal combustion engines *Proc. I. Mech. Eng.* **117** 973-96

[9] NIST 2007 *Refprop – Reference fluid thermodynamic and transport properties* (NIST)

[10] Deschamps C J, Prata A T and Ferreira R T S 2000 Modelling of Turbulent Flow Through Radial Diffusers *J. of Brazilian Society of Mechanical Sciences* n 22 pp 31–41

[11] Ferreira R T S and Lilie D E B 1984 Evaluation of the Leakage Through the Clearance Between Piston and Cylinder in Hermetic Compressor *Proc. Int. Compressor Eng. Conf. (Purdue)* pp 1–6

[12] ANSYS Inc 2011 *User's guide v.12.0.0.0* (ANSYS)

[13] Catton I 1978 Natural convection in enclosures *Proc. Int. Heat Transfer Conf. (Toronto)* **6** 13-31

[14] Drost R T and Quessada J F 1992 Analytical and experimental investigation of a scroll compressor lubricating system *Proc. Int. Compressor Eng. Conf. (Purdue)* 551-60

IOP Conf. Series: Materials Science and Engineering **90** (2015) 012069 doi:10.1088/1757-899X/90/1/012069

Compressor-fan unitary structure for air conditioning system

N Dreiman
Consultant, P.O.Box 144, Tipton, MI 49287 USA

E-mail: ndreiman@frontier.com

Abstract. An extremely compact, therefore space saving unitary structure of short axial length is produced by radial integration of a revolving piston rotary compressor and an impeller of a centrifugal fan. The unitary structure employs single motor to run as the compressor so the airflow fan and eliminates duality of motors, related power supply and control elements. Novel revolving piston rotary compressor which provides possibility for such integration comprises the following: a suction gas delivery system which provides cooling of the motor and supplies refrigerant into the suction chamber under higher pressure (supercharged); a modified discharge system and lubricating oil supply system. Axial passages formed in the stationary crankshaft are used to supply discharge gas to a condenser, to return vaporized cooling agent from the evaporator to the suction cavity of the compressor, to pass a lubricant and to accommodate wiring supplying power to the unitary structure driver –external rotor electric motor.

1. Introduction

A heating, ventilation, and air conditioning system generally includes a closed loop refrigeration system with at least one evaporator, at least one condenser, at least one compressor and at least one airflow fan driven by an electric motor. The compressor motor and the airflow fan motor have, as usual, separate power wiring, running capacitors, start capacitors and another operation control elements. A combined rotary compressor - airflow fan unitary structure for air conditioning system has a single electric motor which drives both, the compressor and the fan. The unitary structure consists of a revolving piston rotary compressor with the stationary crankshaft and an impeller wheel (or plurality of impeller wheels) attached externally to the revolving piston compressor [1]. The mechanical structure and operating principle of the revolving piston rotary compressor with stationary crankshaft which provides possibility for such assembly are described below [2].

2. Motion analysis

2.1. *Kinematics of the revolving piston compressor*. A rotor block and a revolving piston of novel rotary compressor are considered, consequently, as internal cylinder (driver) and external cylinder

Content from this work may be used under the terms of the Creative Commons Attribution 3.0 licence. Any further distribution of this work must maintain attribution to the author(s) and the title of the work, journal citation and DOI.
Published under licence by IOP Publishing Ltd

(follower) having rolling contact at point P with transmission of the motion from the driver to the follower through the rigid link C (coupler) and a line of contact. According to the compressor structure, the vane presented in figure 1A, 1B as link C, separates suction and compression chambers and extends radially inward toward rotor block center of rotation O_R through the center of the bushing mounted in the rotor block. The line connected piston center $\mathbf{O_P}$ and point $\mathbf{P_R}$ is considered as the part of rotational link with $\mathbf{O_P P_R = d}$, $\mathbf{O_R}$ and the piston center $\mathbf{O_P}$ are individually taken as the two immovable points of the mechanism with $\mathbf{O_R O_P = e}$ (fixed link). The line $\mathbf{O_P P_P = f}$ considered as a driver. The equation of motion of the compressor elements such as integral revolving piston – vane, rotor block with vane bushing can be derived by choosing the orthogonal coordinate system fixed to the stationary crankshaft with origin in the rotor block center $\mathbf{O_R}$, the **X** axis lie along common line, namely the line of centers directed to the point of contact **P** and the **Z** axis coincided with the stationary crankshaft axis. For plane triangle $\nabla \mathbf{O_P P_R O_R}$ (see figure 2A), as per the Law of Sines and the Law of Cosines

$$\mathbf{e \sin\theta = f \sin\phi \; ; \; d = e \cos\theta + \mathit{f} \cos\phi} \tag{1}$$

Due to the fact that the revolving piston-vane is integral part and rotates as one solid body around **Z**-axis, there is no reciprocating movement of the vane. The flats of the bushing implanted in the rotor block slide along the suction and discharge sides of the vane with change of the turning angle θ. The part of the vane **L (θ)** exposed to the suction and discharge pressure inside of the working chamber will be

$$\mathbf{L(\theta) = P_R P_P = R_P - d}, \tag{2}$$

where radius of the piston $\mathbf{R_P = R_R + e}$, and radius of the rotor block $\mathbf{R_R = f}$. After substitution and transformation, exposed length of the vane can be expressed as

$$\mathbf{L(\theta) = e(1 - \cos\theta) + f\{1 - [1 - (\lambda \sin\theta)^2]^{1/2}\}}, \tag{3}$$

where $\lambda = \mathbf{e / f}$. We can use binominal series shown below for the last term of the Eq.3

$$\mathbf{(a + b)^n = a^n + n\, a^{n-1} b + n(n-1)/2!\, [a^{n-2} b^2] + n(n-1)(n-2)/3!\, [a^{n-3} b^3] \ldots} \tag{4}$$

with $\mathbf{a} = 1$, $\mathbf{b} = -(\lambda\sin\theta)^2$ and $\mathbf{n} = ½$. After expanding the root into a binominal expansion with inclusion of the first two terms as the series rapidly converges, we obtain

$$\mathbf{L(\theta) = e(\mathit{1} - \cos\theta + \lambda^2/2 \sin^2\theta - \lambda^3 \sin^4\theta / 8 + \ldots\ldots)} \tag{5}$$

Thus the length of the exposed part of the vane very approximately (error 0.15%) will be:

$$\mathbf{L(\theta) = e(\mathit{1} - \cos\theta + \lambda^2/2 \sin^2\theta)} \tag{6}$$

The bushings inside of the rotor block cavity slides along the vane sides and turns to accommodate vane position with the change of the piston-rotor block position (see figure 1B). Differentiating Eq.(5) with respect to time and denoting $\mathbf{d\theta / dt = \omega}$ we obtain the sliding speed $\mathbf{V_{BV}}$ between the bushing and vane with change of the turning angle.

$$V_{BV} = e\,\omega(\sin\theta + \lambda^2/2\,\sin 2\theta) \quad (7)$$

where ω is angular velocity of the rotor block.

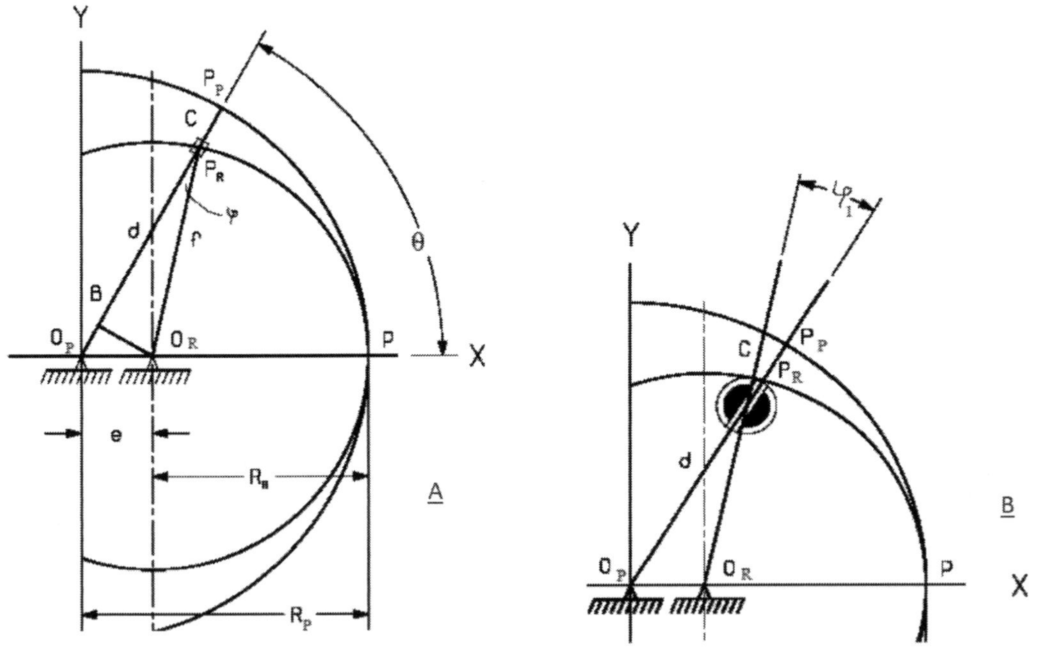

Figure 1. A schematic view of the revolving piston compressor mechanism

The rotor block bushing turning angle ϕ_I (see figure.1B) with change of the vane rotational angle θ will be

$$\phi_I = \arcsin\,[\,e\,\sin\theta\,/(\,R_R - R_B\,)\,] \quad (8)$$

where R_B is the radius of the roller bushing. The values of the bushing turning angle ϕ_I with change of the rotational angle θ in the limits 0-2π (single revolution) are shown in figure.2.

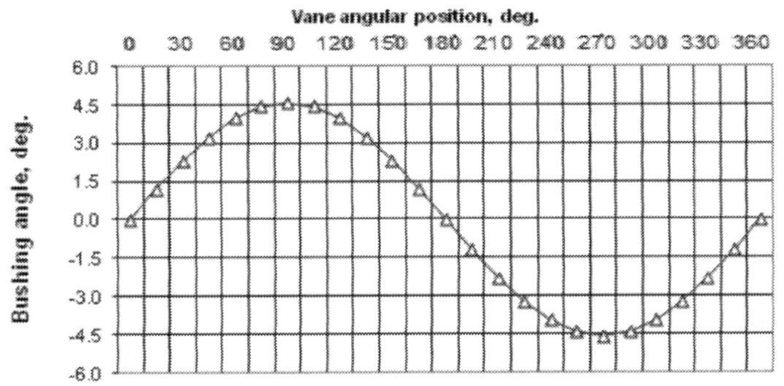

Figure 2. Turning angle of the bushing with a change of the vane angular position.

2.2. Velocity. Motion of the internal cylinder is transmitted to the external cylinder through the driving link **C.** If driving link **C** rotates counterclockwise with constant angular velocity, the follower will revolve in the same direction at a varying (accelerating and decelerating) speed. The relative velocity between the cylinders is a function of the driver angle θ, radii of the contact cylinders, centers distance. The linear velocities of the piston v_P and the rotor block v_R are shown below:

$$\mathbf{V_P = dS_P / dt = \omega\, dS_P / d\theta = \omega\, R_P} \tag{9}$$

$$\int_0^{\theta} v_R = \mathbf{dS_R / dt = \omega}\, d\, [\, \textstyle\int \mathbf{dx\, d\theta}\,] / \mathbf{d\theta = \omega\, d} \tag{10}$$

0

where $\mathbf{S_P}$ and $\mathbf{S_R}$ are revolving arcs of the piston and rotor block. The relative velocity of the rotor block to the revolving piston

$$\Delta v = \omega\, (\mathbf{R_P} - \mathbf{d}) \tag{11}$$

Taking derivative we find that

$$\mathbf{d}v/\mathbf{dt} = \omega\, [\, (\, \mathbf{d}d / \mathbf{d}\theta\,)(\mathbf{d}\theta / \mathbf{dt}\,) = \omega^2 (\mathbf{d}d / \mathbf{dt}) = -\omega^2 \lambda\, \mathbf{R_R} \sin\theta\, (1 + \lambda/2 \cos\theta\,) \tag{12}$$

The extreme values of the roller relative velocity are shown below:

$$\Delta v_{max} = 2\omega\, (\mathbf{R_P} - \mathbf{R_R}) \quad \text{and} \quad \Delta v_{min} = \mathbf{0} \tag{13}$$

Effect of the $\mathbf{R_P} / \mathbf{R_R}$ ratio on the values of the relative velocity between the rotor block and the revolving piston at different vane angular positions is shown in figure.3. It can be seen that rather smaller ratio gives us lower relative velocity between rotating in one direction rotor and piston .

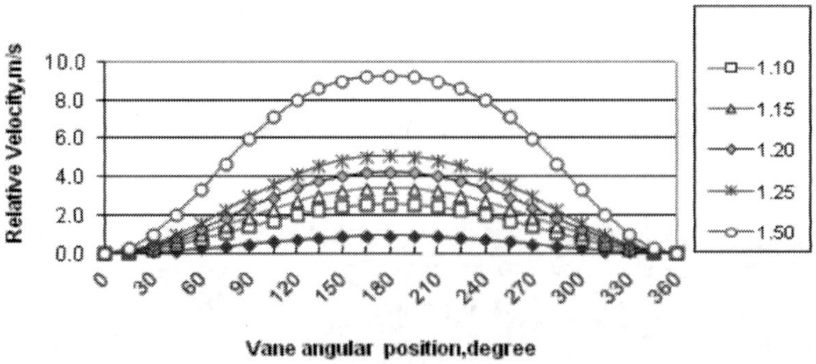

Figure 3. Effect of the radii ratio on the relative velocity of the cylinders.

3. Structure of the revolving piston rotary compressor

A construction of a compressor-fan unitary structure combines a revolving piston rotary compressor having a stationary crankshaft with an impeller wheel or plurality of wheels used in radial flow (centrifugal) fans. Developed for such unitary structure novel revolving piston rotary compressor contains (see figure 4) a rotor 1 of the driver -external rotor electric motor 2,which is integrated with concentrically situated pump rotatably assembled on the stationary crankshaft, [3] . A mounting plate 3 is rigidly fixed to the upper end of the stationary crankshaft which supports the assembled on it motor stator 4, spinning around it rotor block 5, a revolving piston 6, and an impeller wheel 61. The revolving piston is disposed eccentrically outside of the rotor block with the direct (no operating clearance) contact between the inner peripheral surface of the revolving piston and external peripheral surface of the rotor block. The direct contact of the revolving piston inner surface with the rotor block lies in the plane passing through the lines of centers of their rotation (see figure 1, point **P**). In such kinematic pair (an external and internal cylinder) a motion of the rotor block (driver) will be transmitted to the revolving piston (follower), and both cylindrical parts rotate in the same direction due to a force developed at the contact line. At the point of contact **P** the tangential velocities of the rotor block and the revolving piston are unidirectional and equal. It means that there is no sliding friction and frictional losses at the contact line of both parts are minimal due to the relatively low value of the rolling friction. In addition, there no high-low side leak and the angular momentum developed through such contact will be supplemental to the momentum transferred from the rotor block to the revolving piston through the coupler – vane which is rigidly fixed to the piston.

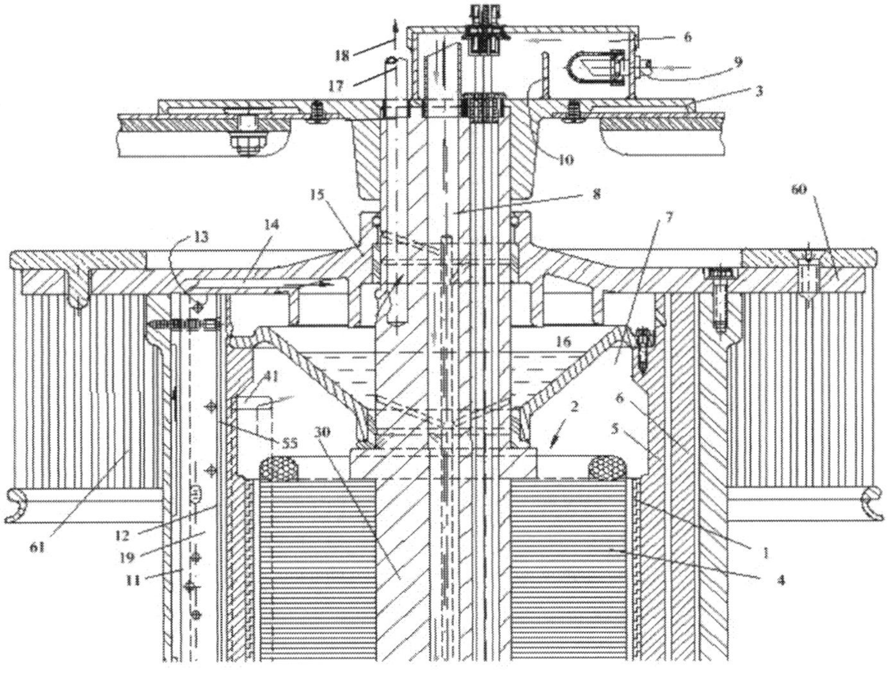

Figure 4. A side sectional view of a revolving piston rotary compressor-impeller wheel unitary structure.

3.1 Suction System .The suction system of the novel rotary compressor consists of an input suction cavity 6 which is disposed on the mounting plate, an electric motor compartment 7 equipped inside with an impeller (not shown) which has been rigidly fixed to the revolving rotor block head below the stator, and a variable volume suction chamber 40 (see figure 5). The motor compartment 7 is in fluid communication as with the input suction cavity through a channel 8 formed inside of the crankshaft, so with the suction chamber through a vertical channels 41 formed in the wall of rotor block 5 and a suction port 42, as shown in figure 5. A suction inlet 9 directs a vapor-liquid mixture of refrigerant and lubricating oil through a screen (to filter an impurities) into the inner volume of the input suction cavity 6, where gas flows to the top and the liquid collects behinds the separating wall 10 having metering oil delivery hole at the bottom. The refrigerant drawn from the suction input cavity 6 will be delivered to the part of the electric motor compartment spaced below the stator where centrifugal force triggered by rotation of the rotor block and rigidly fixed to it impeller will forcibly guide oil to the mitering (bleeding) hole formed in the side wall of the rotor block. The vapor portion of the refrigerant, pressure of which has been increased by the impeller, will be supercharged through the suction port to the suction chamber. Such arrangement of the suction system prevents direct delivery of gas-liquid-oil mixture to the suction chamber by double separation of liquid and vapor, increases the liquid refrigerant storage capacity and provides cooling of the compressor motor during and after its duty cycle.

3.2 Revolving Piston - Vane Assembly. The vane 19 of the compressor separates suction chamber 40 and compression chamber 44, as shown in figures 4, 5, and has an axial edge 11 rigidly fixed without operating clearance to the inner periphery of the revolving piston 6 with opposite axial edge 12 fitted in between sliding segmental bushings 43 mounted in the rotor block. The vane, however, not only serves to separate the working cavity between rotor block and revolving piston into suction chamber and compression chamber, but it forms also a mechanical connection (coupler), between the driver (rotor block) and follower (revolving piston) so, that the external rotor motor revolves simultaneously the rotor block 5 and the revolving piston 6. Such coupling arrangement also prevents slippage in between cylindrical surfaces at the line of the cylinders contact. The vane radial edges 13 are axially rigidly fitted in between the revolving piston heads without operating clearances. It eliminates frictional losses and also blocks high-low side leaks. Integration of the vane with the revolving piston excludes frictional and leakage losses observed in contemporary rotary compressors where a vane and roller slide against facing surfaces of the stationary cylinder block and stationary heads. There is also no" grinding" interference between vane and roller (roller and spring are eliminated) and, in addition, nose of the vane (axial edge 12 does not required any special treatment, such as precision machining, TiN coating, special vane tip seal, etc., recommended and used in prior art rotary compressors.

3.3. Discharge System. Referring now to figures 4, 5 and 6, the discharge system of the compressor comprises a discharge valve system arranged in the vane 10, a horizontal passage 14 formed in the revolving piston head 15, cone-shaped gas expansion cavity 16 which is in fluid communication with the discharge line 18 through a discharge gas exhaust channel 17 formed axially in the crankshaft.

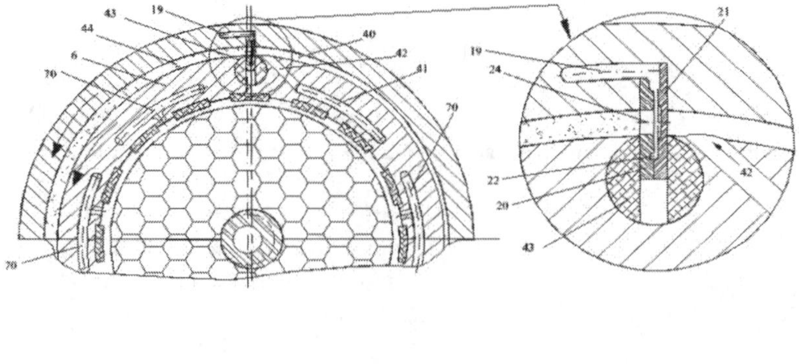

Figure 5. A partial horizontal cross-sectional view of the revolving piston rotary compressor.

The vane consists of a discharge side plate 20 (see figure 6) and suction side plate 21which are held together (by welding, riveting, plurality of screws, etc.), and a discharge valve cavity 22. At least one reed type valve 23 is clamped to the back of the discharge side plate against discharge port opening 24 facing compression chamber 44. The suction side plate 21 is formed with a valve stop which prevents excessive movement of the reed valve. The discharge valve cavity 22 is in fluid communication with

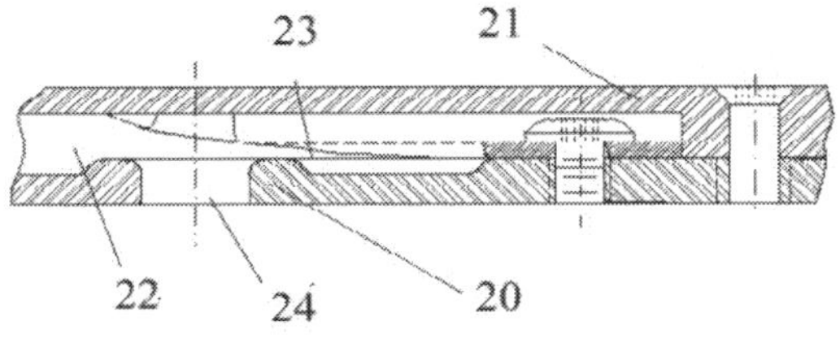

Figure 6. A partial sectional view of the vane-discharge valve module.

the horizontal passage 14 through longitudinal channel 19 formed in the wall of the revolving piston.

3.4 Lubrication The oil accumulation and distribution passageways (see figures 4, 5,7) comprise an oil pump 45 having an oil input cavity 46, an oil discharge cavity 47, a cone shaped oil retaining reservoir 48 spaced above the electric motor compartment, plurality of vertical channels 70 formed in the wall of rotor block. The oil pump discharge side is in fluid communication with formed in the stationary crankshaft and axially extending channels through which oil delivered to plurality of oil passages to lubricate the compressor bearings 53, and a channel 56 which supplies oil to the reservoir. The rotor block vertical channels 70 are in fluid communication as with the reservoir so with the input side of the oil pump through the orifices 54.

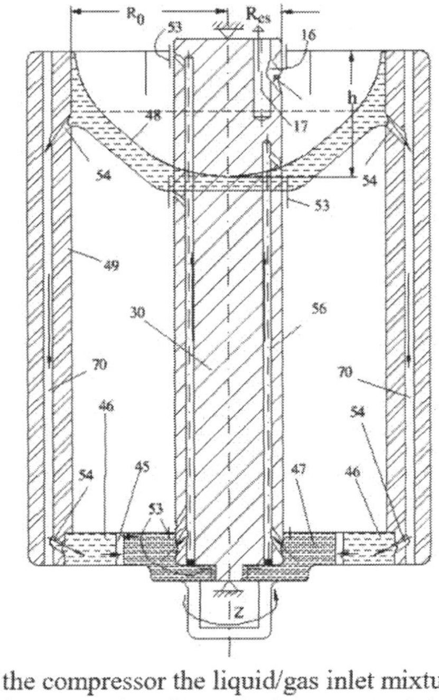

Figure 7. An oil circulation pattern of the revolving piston rotary compressor.

During operation of the compressor the liquid/gas inlet mixture is forced under pumping pressure into the reservoir 48 are thrown against the wall of the reservoir by action of centrifugal force. This cause a boundary layer to be formed between the oil and relatively lighter gas in axial region of turbulence chamber (reservoir).The layer in most instances takes shape of the funnel which in longitudinal section approximates a cone-shape or parabolic form. Figure 7 depicts a fluid contained in the reservoir that rotates with a constant angular velocity ω about the vertical *Z* axis. The gravity vector acts in the negative *Z* direction. For steady-state rotation, the fluid rotates with the reservoir as a rigid body. The fluid volume underneath the paraboloid of revolution is the original fluid volume and the shape of the paraboloid of revolution depends only upon the angular velocity ω. The volume of the vapor separated from oil

$$v = \int_0^h \frac{R_0^2}{h} x dx - \pi h R_{cs}^2 = \pi h / 2 (R_0^2 - R_{cs}^2) \quad (14)$$

where

$$h = \omega^2 R_0^2 / 2g \quad (15)$$

Here **h** is the depth on the axis Z, $\mathbf{R_0}$ is the radius of the rotor block, $\mathbf{R_{cs}}$ is the radius of the crankshaft, ***g*** is the acceleration of gravity. The reservoir has an axial outlet 17 formed at the top portion of the stationary crankshaft for discharge of the vapor collected above the parabaloid surface and plurality of oil outlets 54connected to plurality of vertical channels 70 through which oil flows downwardly under gravity into an intake cavity of the oil pump.
The reservoir has cone-like shape which diverges outwardly from the lower portion to a circular lip and spinning coaxially with the rotor block. By providing such structure, the whirling oil-vapor mixture formed in such turbulence chamber is prevented from uncontrollable movement and possible extension of the vertex toward the bottom of the reservoir.
During operation of the compressor the lubricant delivered into the reservoir 48 through an axial slot 56 is thrown outward by centrifugal force and forms also an annular seal of liquid at the periphery of chamber and along the axial edge 12 of the vane. Any leakage that may take place past the sides of the vane or past the ends 13 of the rotor block facing piston heads will be leakage of the lubricant inward into the crescent-shaped space, as this lubricant is under the delivery pressure (discharge + pumping) in addition to the pressure due to centrifugal force.

4. Arrangement of the compressor-fan structure in an air conditioning system

Figure 8 illustrates the operation of some typical cooling units equipped with the compressor-fan unitary structure, where figure 8A shows the airflow in a window / wall mounted unit and figure 8B shows the air circulation in a condensing unit.

Referring now to the figure 8, it will be seen that a combined revolving piston compressor-fan unitary structure is compact and integrates an electric motor, rotor block, revolving piston, and an impeller wheel of a centrifugal fan on short axial length of a stationary crankshaft. Disposed around and extending axially along the compressor-fun structure is a scroll shaped housing, which is fixed to the frame of an air conditioning unit (not shown) and which performs also as a safety screen surrounding a spinning assembly. The construction of the revolving piston rotary compressor makes it possible to attach an impeller wheel (or plurality of wheels) externally to the extended flange of upper or/and lower piston head or fix it to the circumference of the revolving piston. A magnetic clutch controlled by a thermostat can be utilized to engage or disengage of the impeller wheel. The compressor-fan structure employs single driver- external rotor electric motor, which comprises a rotor installed outside of the stator. Such motors have much higher torque, wider speed range, higher operating efficiency even at low rotational speed, outstanding power density, low starting current, excellent dynamic characteristics, more compact design, reduced noise, vibration and lower fabrication and assembly cost than comparable internal rotor electric motors. A suction gas cools the stator spaced inside of the motor compartment of the revolving piston rotary compressor which additionally cooled also externally during operation by an air flowing through an annular space between the compressor exterior and the impeller wheel. The driver-external rotor motor of the formed unitary structure can run at 3000, 1500, 1000, 750 min^{-1} (50Hz) and is suitable for operation on frequency converters with all-pole sinusoidal filters.

The revolving piston rotary compressor-fan unitary structure eliminates duality of motors, related electrical power transmission and control elements.

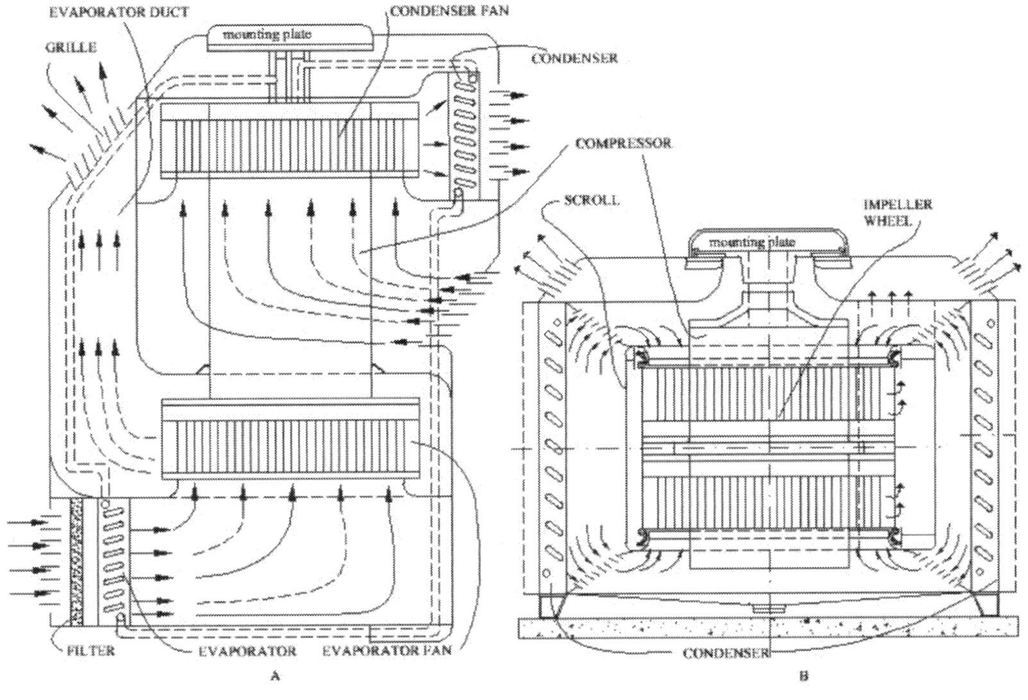

Figure 8. A schematic vertical cross section of wall / window (A) and air-cooled condensing unit (B) utilizing compressor-fun unitary structure

Use of the unitary structure in the unit with radial flow belt-driven fan may exclude also such mechanical parts as belts, motor and fan pulley, belts tension adjusters, etc., and improve their performance.

The stationary crankshaft supporting the unitary structure is mounted centrally and has only one extension projecting outwardly from upper part of the structure with refrigerant inlets/outlets and power input spaced at the top of the stationary crankshaft. Such a design item made it possible to form storage space above the mounting plate and locate in this space suction and liquid line connection couplings, service valves, and a terminal box which may housing an electrical terminal, start and run capacitors, a solid state relay, a thermally triggered overload protector, etc.. It simplifies assembly, service and inspection of an air conditioning unit. The unitary structure can be readily removed for exchange, repair, when it requires servicing. Due to symmetrical design of all parts, rotating around their centers of gravity (stationary crankshaft with rigidly fixed to it stator), only a relative rotation around different centers will happen and therefore no counterweight and practically no balancing is needed and if any imbalance is found it may be removed externally.

Conclusions.

1. A compressor-fan unitary structure is composed of a revolving piston rotary compressor with a fan. The unitary compressor-fan structure utilizes a single external rotor electric motor to drive compressor and cooling fan used in an air conditioning system.

2. Design of novel revolving piston rotary compressor which provides possibility for such structure comprises a stator of the external rotor motor fixed on the stationary crankshaft, rotor block concentrically spaced to the stator and driving a piston which is surrounded and eccentrically set to the rotor block. The piston is equipped with a rigidly fixed a vane- discharge valve module spaced inside of the piston cavity and bolting lugs spaced externally.

3. There are no operating axial clearances between the vane module and the piston heads, and radial operating clearances between the unidirectional spinning rotor block and the revolving piston. The frictional and leakage losses between the radial ends surfaces of the rotor block and facing surfaces of the piston heads will be minimal because of relatively low rubbing speed between synchronously revolving in one direction components and formation of high pressure oil seal film formed along the contacts due to a centrifugation.

4. A suction side delivery scheme has two- stage process of liquid-vapor separation: first- in the input cavity due to gravity and second-in the motor compartment due to a centrifugation. The vapor is supercharged in the suction chamber and simultaneously cools motor. A direct delivery of a suction gas to the suction chamber is eliminated. The compressor cooled also by an air movement triggered by the fan which is fixed externally to the pistons bolting lugs.

5. The unitary structure is arranged on a stationary crankshaft of short axial length and rotates around the center of gravity. The forces triggered by spinning of the unitary structure are symmetrically load bearings spaced below and above the stator which is fixed to the stationary crankshaft.

6. An extremely compact, therefore space saving unitary structure provides a cooling air movement and operation of the compression cycle by utilizing single driver-external rotor electric motor. It reduces, cost, assembly and service time of the air conditioning system due to lower number of components (single motor and one set of control elements) and related wiring procedures.

References.

[1] Dreiman N 2014 Combined revolving piston rotary compressor-fan unitary structure *United State Patent Office* patent pending.

[2] Dreiman N 2014 Rotary compressor with the stationary crankshaft *Int. Compressor Conf. at Purdue, USA 14-17 July, 2014*, pp 144-154

[3] Dreiman N 2014 Revolving piston rotary compressor with stationary crankshaft *United State Patent Office 14/450,215 class418*

Observation of Oil Flow Characteristics in Rolling Piston Rotary Compressor for Reducing Oil Circulation Rate

S j Song[1], K Y Noh[1], B C Min[1], J S Yang[2], G M Choi[3], D J Kim[3]

[1] Graduate School of Mechanical Engineering, Pusan National University, San 30, Jangjeon-dong, Geumjeong-gu, Busan 609-735, Korea
[2] Rolls-Royce University Technology Centre, Pusan National University, San 30, Jangjeon-dong, Geumjeong-gu, Busan 609-735, Korea
[3] Department of Mechanical Engineering, Pusan National University, San 30, Jangjeon-dong, Geumjeong-gu, Busan 609-735, Korea

E-mail : Choigm@pusan.ac.kr

Abstract. The oil circulation rate (OCR) of the rolling piston rotary compressor is a significant factor which affects the performance of refrigeration system. The increase of oil discharge causes decreasing of the heat transfer efficiency in the heat exchanger, pressure drop and lack of oil in lubricate part in compressor. In this study, the internal flow of compressor was visualized to figure out the oil droplet flow characteristics. The experiments and Computational Fluid Dynamics (CFD) simulations were conducted in various frequency of compressor to observe the effect of operation frequency on oil droplet flow characteristics for reducing OCR. In situ, measurement of oil droplet diameter and velocity were conducted by using high speed image visualization and Particle Image Velocimetry (PIV). The flow paths were dominated by copper wire parts driving the motor which was inserted in compressor. In order to verify the reliability of CFD simulation, the tendency of oil flow characteristics in each flow path and the compressor operating conditions were applied in CFD simulation. For reducing OCR, the structure such as vane, disk and ring is installed in the compressor to restrict the main flow path of oil particle. The effect of additional structure for reducing OCR was evaluated using CFD simulation and the results were discussed in detail.

1. Introduction

In rolling piston rotary compressor the oil is essential for lubricating the sliding parts and filled in the sealed parts to reduce refrigerant leakage. The oil circulation rate called OCR is a significant factor which affects the performance of refrigeration system. Increase of the amount of oil discharge causes decrease of the heat transfer efficiency in the heat exchanger, pressure drop and lack of oil in lubrication part in compressor. Therefore, OCR reduction is necessary to improve the performance of refrigeration system. The rolling piston rotary compressor is divided in two types such as constant velocity type and inverter type. The inverter type rolling piston rotary compressor is using Brushless DC (BLDC) motor and it has high efficiency compared with constant velocity type. Moreover, by using the BLDC motor, the flow paths are formed inside of the compressor and these are changed by different muffler type.

Kamal shrma et al. (2010) carried out the CFD simulation of rolling piston rotary compressor for reducing OCR. This paper represents the importance of the gas velocities and turbulence in the rolling

[1] Corresponding author. Tel.: +82 51 510 2476; E-mail address: choigm@pusan.ac.kr

Content from this work may be used under the terms of the Creative Commons Attribution 3.0 licence. Any further distribution of this work must maintain attribution to the author(s) and the title of the work, journal citation and DOI.
Published under licence by IOP Publishing Ltd

piston rotary compressor for the reducing OCR. Yong Jae Kim et al. (2004) conducted the visualization inside the rotary compressor to observe the oil movement when the compressor was operating. Liyang Deng et al. (2012) studied the oil distribution in the rotary compressor and compared the influence of oil discharge rate in different operation condition.

In this study, to find the way to reduce OCR, the oil characteristics were observed by experiments and the experimental results were applied to CFD simulation. The rolling piston rotary compressor was visualized by using a high speed camera and particle image velocimetry(PIV) to investigate the flow characteristics. Also, for reducing OCR, the research on the effect of the structure installed in the compressor was carried out using CFD simulation.

2. Experimental Setup

To visualize the oil behavior inside the rolling piston rotary compressor, the sight glasses were installed at the top and side of the compressor. Figure 1 shows the schematic diagram of the experimental setup and continuous image capturing system. In this case, the laser beam was inserted vertically to the sight glass at the top of the compressor by mirror which was placed above the compressor. The optical setup was adjusted to visualize the flow path in the compressor. In addition, to obtain the position of main flow path at the top of the compressor, the install position of high speed camera and laser were changed. The specification of the high speed camera and laser is shown in Table 1. Figure 1(b) shows the schematic diagram of PIV system. The specification of PIV is in Table 2.

For a test model, the experimental condition was P_s : 10.15kgf/cm^{2}A, P_d : 23.37kgf/cm^{2}A, T_s : 18.3°C, T_{cond} : 37.8 °C, T_{eva} : 7.2 °C and operating frequency was 40, 50, 60 Hz. Where, P_s is the suction pressure measured at the suction port at the compressor, P_d is the discharge pressure measured at the discharge port at the compressor, T_s is the suction temperature measured at the suction port at the compressor, T_{cond} is the condensing temperature and T_{eva} is the evaporating temperature.

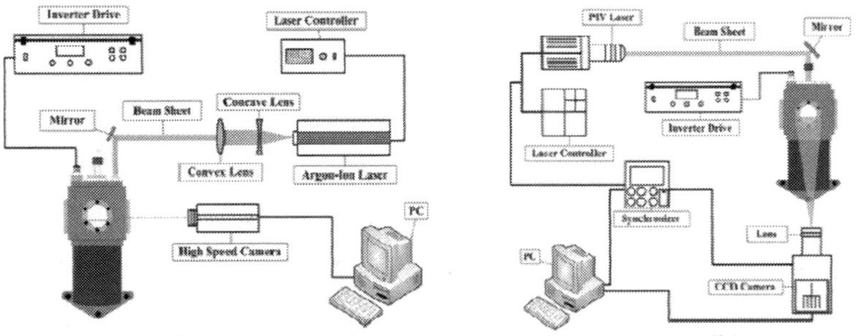

(a) (b)

Figure 1. Schematic of experimental setup : (a) high speed camera (b) PIV

Table 1. Specification of high speed camera and laser

High speed camera

Resolution	1024X1024, 128(W)X16(H)
Frame rate	1000fps@1024X1024
Shutter	1.5㎲ (Min.)
Exposure time	940 μs
Lens	50mm , f1.4

Laser

Output Power	Line thickness
0.8~5W	60 ㎛

Table 2. Specification of PIV and laser

PIV

Resolution	Frame rate	Lens
2048 x 2048	over 15frame/second with no binning	AF MICRO 105mm 1:2.8 D

Laser

2 x 120mJ@ 532nm	15Hz Repetition Rate

3. Numerical Analysis

To improve the reliability of the results, the CFD simulation was divided in two parts. The first case is the lower part of the motor and stator with muffler and the other is the whole parts which are placed above the muffler. Figure 2(a) and 2(b) show the mesh model of the muffler and flow field, and the mesh model of the whole part except muffler, respectively. For the boundary condition, the pressure inlet and pressure outlet were used, and the oil distribution at the inlet is almost uniform because the mesh is uniform at the inlet boundary condition. Numerical simulations were calculated by the discharge pressure when the valve is maximum opened.

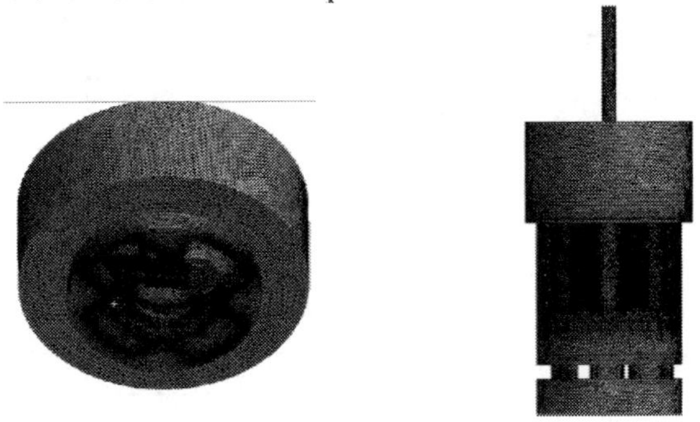

(a) (b)
Figure 2. Mesh model : (a) Muffler mesh model (b) Whole mesh model

In rolling piston rotary compressor, the flow paths are formed to A, B, C and D zones in the shell as shown figure 3(a). The position A is called air cut and B, C, D is copper cut, air gap, vent hole respectively. Also the location of holes at the muffler affects the refrigerant and oil flows in the compressor. In test model, the outlets at the muffler were placed as shown in figure 3(a). Also to observe the effect of the flow characteristics on the location of muffler outlet, the muffler holes were shifted as shown in figure 3(b). In this paper, the CFD simulation results of figure 3(a) and 3(b) is called Case 1 and Case 2, respectively.

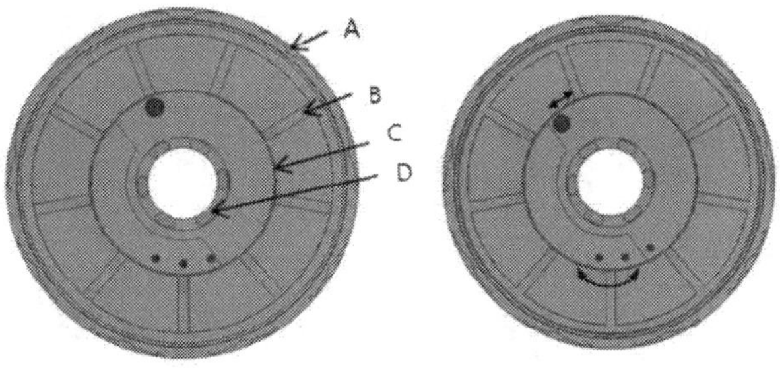

(a) (b)
Figure 3. Outlet position at muffler: (a) Parallel with copper cut (b) Shifted with copper cut

To observe the effect of additional structure on OCR, the ring and the disk were installed as shown in figure 4(a) and 4(b), respectively.

The CFD simulation was carried out at the compressor frequency of 60Hz and validated with the experimental results, the particle size and flow characteristics. In this simulation results, the steady state analysis was conducted and the operating fluid was single phase R410a. In addition, to simulate the rotation inside compressor, the Multiple Reference Frame (MRF) model was used. The specification of the simulation conditions are in Table 3. The commercial CFD tool ANSYS Fluent 14v was used for simulation.

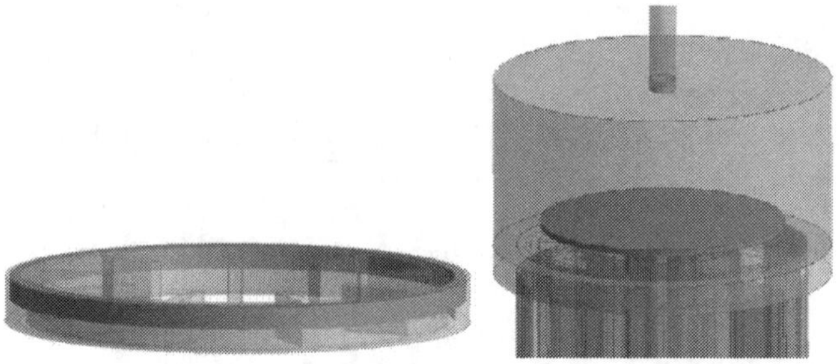

(a) (b)

Figure 4. Two kinds of plate : (a) Ring type (b) Disk type

Table 3. Specification of numerical analysis condition

Turbulence Model	Realizable K-epsilon model
Fluid	R410A, PVE oil (μ : 0.0029kg/m•s)
Boundary condition	Pressure inlet(2385955pa)
	Pressure outlet(2291814pa)
Applied Model	DPM model
	MRF model
Frequency	60Hz

4. Results and Discussion

4.1 Experimental Results

Figure 5(a) shows the experimental test model and Figure 5(b) and (c) show the top view of the high speed images at 40Hz and 60Hz respectively. As shown in Figure 5(b) and 5(c), the left side and right side are the copper wire and the bottom side is the rotor. Most of oil was discharged at the copper cut from the results of figure 5(b) and (c). In addition, comparing the results of 40Hz and 60Hz, the oil discharge amount of 60Hz was larger than that of 40Hz.

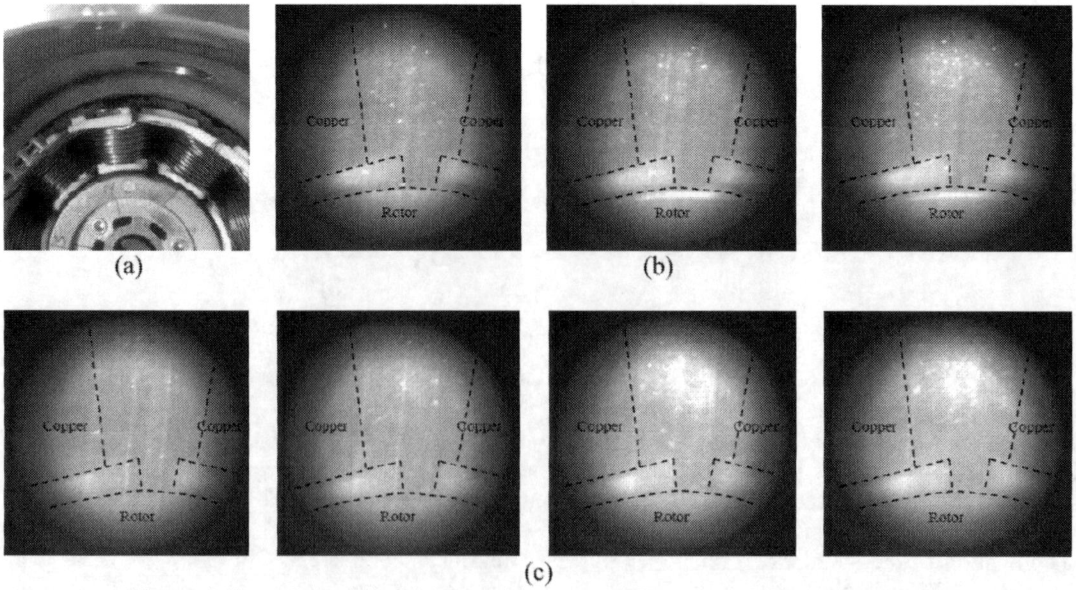

Figure 5. Top view of the high speed image : (a) Test model (b) 40Hz (c) 60Hz

Figure 6 shows the velocity field obtained by PIV. As shown in figure 6 the oil flow direction is the upper right corner of the diagonal direction. Also the average velocity was about 3m/s .

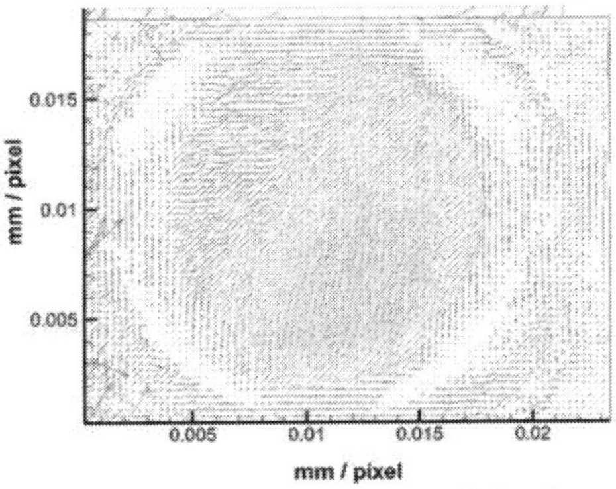

Figure 6. Vector field of oil droplet

4.2 Analysis of Numerical Simulation

4.2.1 The Flow Characteristic at Muffler

Figure 7 shows the vector field of the lower cavity and inside of the muffler. When the refrigerant and oil are discharged from muffler, the direction is not vertically to the outlet surface. To apply the flow and particle direction to whole simulation condition, the part of muffler was simulated. At muffler, the 4 discharge holes were placed. As shown in figure 7, the refrigerant was discharged from muffler to outer region of the compressor.

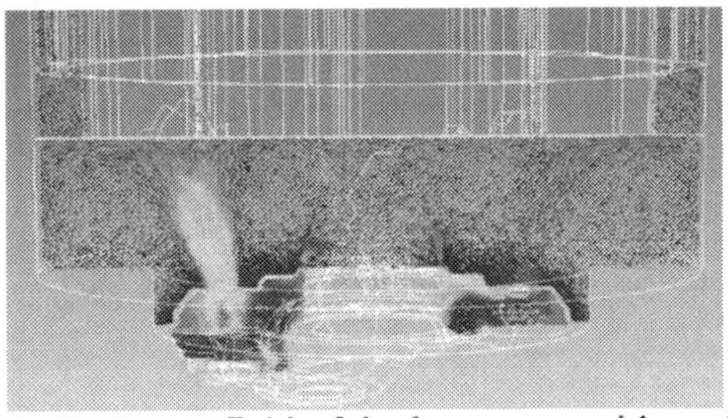
Figure 7. Vector field of the lower part with muffler

4.2.2 The Flow Characteristic with Inlet position Change

Figure 8 (a) and (b) show the vector field of case 1 and case 2, respectively. The flow direction of simulation result is similar with the experimental result. In addition, the average velocity at the copper cut has the same order as the average velocity of experimental result. In Figure 8 (a), the most of refrigerant flow passes through the copper cut. However, in Figure 8 (b), the main path of flow is different from case 1, because of the location change of muffler outlet. The reason of this result is because the refrigerant flow run into the weight balance of lower cavity and the direction of flow is changed to outer region of the lower cavity.

Figure 9 shows the particle tracks of case 1 and case 2. Figure 9 (a) and (b) are side view and figure 9 (c) and (d) are top view of the particle track. In case 1, the most of particle passes through the copper cut as same result with the refrigerant flow. The particles passed through the copper cut, rotate at the center of the upper cavity as shown in Figure 9(c) and these were discharged to outlet. However, in case 2, the particles were run into the weight balance as same of refrigerant and these are passed through the both air cut and copper cut. The particles which were passed to air cut, rotate outer of the shell as shown in Figure 9(d) and it affects the amount of discharged particle through outlet reduce.

(a) (b)

Figure 8. Vector field : (a) Case 1 (b) Case 2

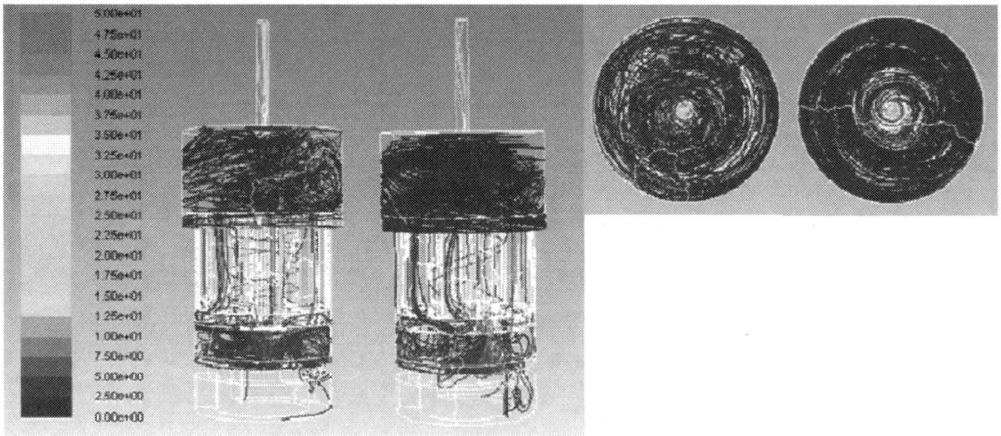

(a) (b) (c) (d)

Figure 9. Particle track : (a) Side view of case 1 (b) Side view of case 2
(c) Top view of case 1 (d) Top view of case 2

4.2.3 The Flow Characteristics with Additional Structure Installed

Figure 10 shows the vector field when the structures, such as the ring and the disk, were installed. To observe the flow characteristic change more efficiently when the structure was installed, the case 1 with disk and case 2 with ring were compared. As shown in figure 10 (a), the flow is bumped into the disk and it makes the recirculation zone near the disk. Figure 10 (b) and (c) show the model with the ring which was simulated at the case 2. As shown in figure (b) and (c) the flow which passes through the air cut is interrupted by the ring and it circulates at the bottom of the ring.

Figure 11 shows the particle track of case 1 and case 2 with structure. Figure11 (a) and (b) are side view and figure11 (c) and (d) are top view of the particle track. Comparing the particle track of case 1 and case 1 with disk, the particle of case 1 with disk rotated outer of the upper cavity because the disk is placed at the center of the upper cavity and the particles pass through the outer of the disk. Also comparing the particle track of case 2 and case 2 with ring, the number of particles which is rotated at upper cavity is reduced because the particle is bumped into the ring and the direction is changed to downward.

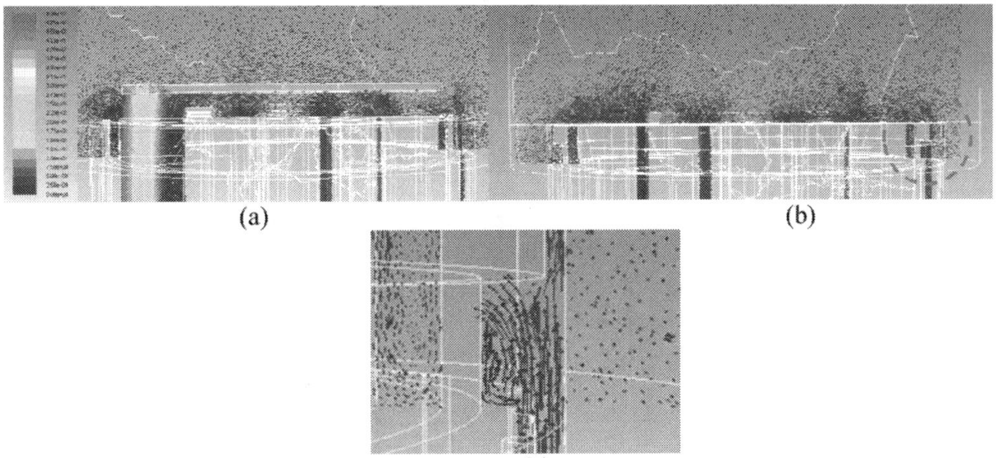

(a) (b) (c)

Figure 10. Vector field of additional structure case : (a) Disk with case 1 (b) Ring with case 2 (c) Detailed view of figure 10 (b)

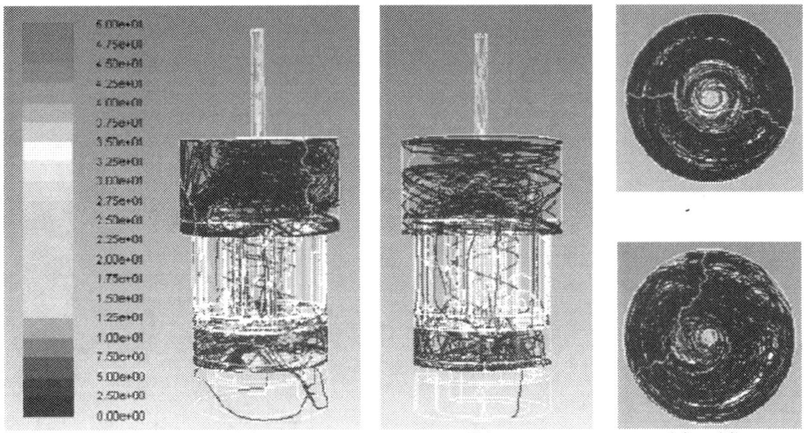

(a) (b) (c) (d)

Figure 11. Particle track of additional structure case : (a) Side view of disk with case 1 (b) Side view of ring with case 2 (c) Top view of disk with case 1 (d) Top view of ring with case 2

4.2.4 Particle and Mass Flow Rate

In this study, the mass of particle is unconsidered and the discharged particle rate in each case was considered as shown in Figure 12. Comparison of the case 1 and disk with case 1, the discharged particle rate reduces about 4% because the disk interrupts the particle discharge to the center of the upper cavity. In addition, comparison of the case 2 and ring with case 2, the discharged particle rate reduces about 2%.

Figure 13(a) shows the mass flow rate in each case. The positions to measure the average mass flow rate are shown in Figure 13(b). Each position has the distance of 10mm from position 1 to upper cavity in figure 13(b). As shown in figure 13, the mass flow rate of all cases is similar. It means that the installation of disk and ring is influenced a little in COP. In figure 12 and 13, the mass flow rate of refrigerant is similar but the oil reduces when the disk and ring was installed. The OCR is the function of the refrigerant and oil. Therefore, the installation of the ring and disk is effective for reducing OCR.

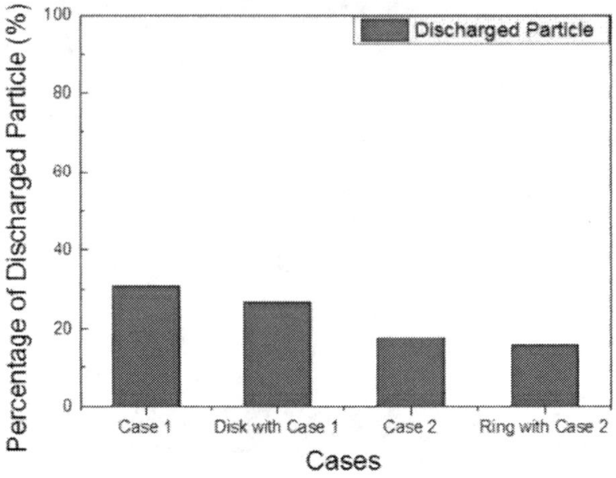

Figure 12. Percentage of discharged particle with variable cases

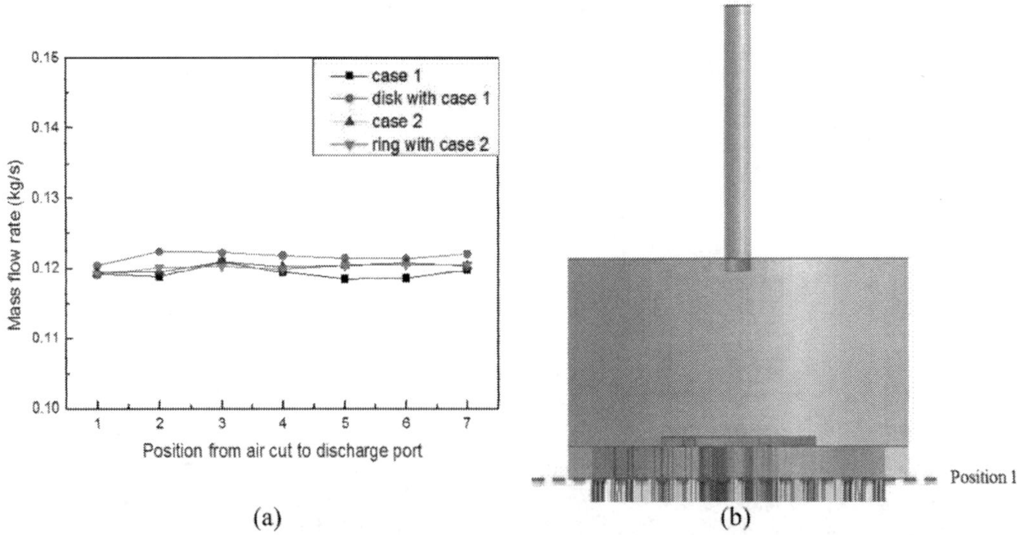

Figure 13. : (a) Mass flow rate in each position (b) Position measuring at the upper cavity

5. Conclusion

In this paper, the oil paths in the rolling piston rotary compressor were visualized and simulated using high speed camera and PIV to observe the effect of additional structure in compressor. The following conclusions are obtained from the experiment and CFD results.

(1) Most of oil passes through the copper cut and the amount of discharged oil was higher with increase of the compressor frequency from the experimental results.

(2) The position of muffler outlet is influenced to the main refrigerant and oil paths which were formed by stator and rotor.

(3) When the oil rotates outer region of the upper cavity shell, the amount of discharged oil reduces to compare when the oil rotates center of the upper cavity.

(4) The installation of structures such as ring and disk is significantly influenced in reduction of OCR.

Reference

[1] Toyama T, Matsuura H, and Yoshida Y 2006 Visual Techniques to Quantify Behavior of Oil Droplets in a Scroll Compressor *International Compressor Engineering Conference* p 1788.
[2] Deng L, Liang S, Liu Q, Wu J and Xu J 2012 CFD Analysis of Oil Discharge Rate in Rotary Compressor *International Compressor Engineering Conference* p 2067.
[3] Sharma K, Rao K, V Kiran M. N. S and Gopinathan A 2010 CFD Analysis of Discharge Gas flow in Rotary Compressor for OCR reduction *International Compressor Engineering Conference* p1965.
[4] Kim Y J, Lee S K, Youn Y and Ko H S 2004 Experimental Study of Refrigerant-Oil Behavior Inside Rotary Compressor *International Compressor Engineering Conference* p 1698.
[5] Min K and Hwang, I 2000 Oil Circulation Rate in Rotary Compressor: Its Measurement and Factors Affecting the Rate *International Compressor Engineering Conference*. p 1395.
[6] ANSYS Fluent manual

IOP Conf. Series: Materials Science and Engineering **90** (2015) 012072 doi:10.1088/1757-899X/90/1/012072

Performance Testing of Jefferson Lab 12 GeV Helium Screw Compressors

P Knudsen, V Ganni, K Dixon, R Norton, J Creel

Cryogenics Group, Engineering Division, Thomas Jefferson National Accelerator Facility, 12000 Jefferson Ave., Newport News, VA 23606, USA

E-mail: knudsen@jlab.org

Abstract. Oil injected screw compressors have essentially superseded all other types of compressors in modern helium refrigeration systems due to their large displacement capacity, reliability, minimal vibration, and capability of handling helium's high heat of compression. At the present state of compressor system designs for helium refrigeration systems, typically two-thirds of the lost input power is due to the compression system. It is important to understand the isothermal and volumetric efficiencies of these machines to help properly design the compression system to match the refrigeration process. It is also important to identify those primary compressor skid exergetic loss mechanisms which may be reduced, thereby offering the possibility of significantly reducing the input power to helium refrigeration processes which are extremely energy intensive. This paper summarizes the results collected during the commissioning of the new compressor system for Jefferson Lab's (JLab's) 12 GeV upgrade. The compressor skid packages were designed by JLab and built to print by industry. They incorporate a number of modifications not typical of helium screw compressor packages and most importantly allow a very wide range of operation so that JLab's patented Floating Pressure Process can be fully utilized. This paper also summarizes key features of the skid design that allow this process and facilitate the maintenance and reliability of these helium compressor systems.

1. Introduction

Helium refrigeration systems are extremely energy intensive processes [1], especially 2-K helium refrigeration systems [2]. Even for large systems [1,3], which are typically more efficient [4], it is achievable to obtain 1 W of cooling at 4.5-K for ~250 W of input power; or, 1 W of cooling at 2-K (~30 mbar) for ~750 W of input power. The cold box, which is comprised of heat exchangers, turbines and process control valves, depends solely upon the availability provided by the helium compression system. Often, great effort is expended in the cold box design (i.e., how the supplied availability is utilized), but relatively little effort may be given to the helium compression system design, even though two-thirds of the input power losses are attributable to the compression system [5,6]. The reliability and efficiency of the entire helium system is largely governed by the reliability and efficiency of the helium compression system. Oil flooded screw compressors have improved the former without sacrificing the later [7]. A well matched compression system and cold box system is essential for efficient operation of helium refrigeration systems, especially since there is typically a considerable margin (usually, at least 50%) involved in estimating the loads on helium a refrigeration system. Additionally, these systems are usually commissioned in phases, requiring the refrigeration

Content from this work may be used under the terms of the Creative Commons Attribution 3.0 licence. Any further distribution of this work must maintain attribution to the author(s) and the title of the work, journal citation and DOI.
Published under licence by IOP Publishing Ltd

system to support partial loads. And, often the load requirements can evolve considerably due to changes in the physics being supported and issues encountered during the project with accelerator and/or major beam end station components (e.g., magnets, targets, etc.). These factors, in addition to decreasing operating budgets, can require a considerable variability in the actual load being supported, even during a fiscal year.

2. Background

The Cryogenics Group at Thomas Jefferson National Accelerator Facility (JLab) has operated screw compressors, as part of helium refrigerators, from many manufacturers for 25 years. These compressors range from 100 to ~2000 kW and are from a wide range of compressor manufacturers and skid packagers. Some staff members have also worked in industry, supplying many of the helium refrigerator systems used at government laboratories and in industry, which includes the compressor and cold box systems. The limitations imposed by these compressor skid designs has become quite apparent to those staff members with a background of having specified their design in industry and then exposed to their operation and maintenance.

As mentioned previously, it is not common for helium refrigerators to run at a single operating condition for the majority of their life and many times these conditions may not often be the maximum capacity. A single operating condition can be more common in industry where the objective is the maximum capacity, except during maintenance shut-down periods. The realization of the now patented Ganni – Floating Pressure Process Cycle [8] during the design of the cryogenic system for the Superconducting Super-Collider Laboratory (SSCL) became the primary motivation to design a compressor skid capable of a wide range of operation of suction and discharge pressures.

It was fortuitous that the request by NASA Johnson Space Center (JSC) to the JLab Cryogenics Group to design and specify a 12 kW 20-K helium refrigerator [9] for the James Webb program provided an opportunity to develop a wide range helium compressor skid for that refrigerator which would use the Floating Pressure Process. Such a deviation from the 'norm' would not have been possible without the support of NASA-JSC project team. This development effort for NASA-JSC afforded an acceptable level of risk for the JLab 12 GeV project. Consequently, by the summer of 2013, the 12 GeV upgrade helium compression system was successfully commissioned, composed of compressor skids completely design by JLab and built-to-print by industry, demonstrating a very wide range of operation to allow full implementation of the Floating Pressure Process [3,10].

These compressor skid design packages are being used for the helium refrigerator system supporting the Michigan State University (MSU) – Facility for Rare Isotope Beams (FRIB) [11], having incorporated minor improvements realized during JLab 12 GeV commissioning. Salient 12 GeV compressor skid specifications are shown in table 1 and the high pressure (HP) stage skid is shown in figure 1.

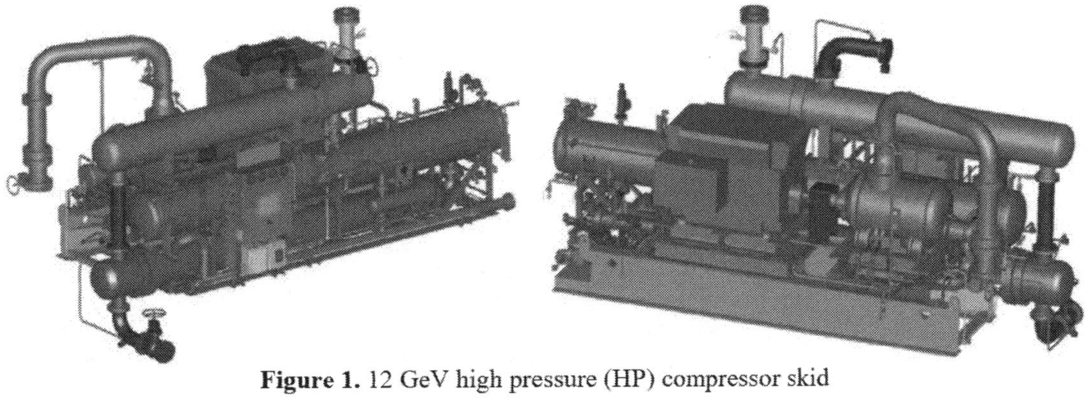

Figure 1. 12 GeV high pressure (HP) compressor skid

Table 1. 12 GeV compressor skid specifications (selected)

Stage		Units	HP	MP	LP	*Notes*
No. Units			1	1	3	
Compressor			Howden WLViH 321/193 (Variable Volume)	Howden WLVi 321/165 (Variable Volume)	Howden WLVi 321/193 (Variable Volume)	
Displacement		[m^3/s]	1.774	1.577	1.774	*(a)*
Motor Rating		[kW]	1864	671	671	*(b)*
Motor Service Factor			1.15	1.15	1.15	
Weight (Est.)		[t]	30	22	20	*(c)*
Oil Charge (Est.)		[ℓ]	1400	1100	1000	
Water Flow (Est.)		[ℓ/s]	61	23	21	*(d)*
Oil Cooler	Type		AEU: Shell (Oil, 1 pass) & Tube (Water, 2 pass)			
	Duty	[kW]	1671	617	617	
	(UA)	[kW/K]	77.6	29.4	31.2	*(e)*
Helium After-Cooler	Type		AEU: Shell (Water, 1 pass) & Tube (Helium, 2 pass)			
	Duty	[kW]	448	184	105	
	(UA)	[kW/K]	20.4	8.44	4.66	*(e)*
Oil Pump Motor		[kW]	5.59	3.73	3.73	
Rotor Oil Injection Filter		[mm abs.]	8	8	8	*(f)*
Bearing Oil Injection Filter		[mm abs.]	8	8	8	*(f)*
Helium Pressure Rating		[barg]	22.4	12.1	12.1	
Oil Pressure Rating		[barg]	26.5	15.5	15.5	

Notes: *(a)* at 59.17 Hz; *(b)* Westinghouse 4160 V; *(c)* Metric tonnes; *(d)* At 8.3 K temperature difference; *(e)* (UA) is the net thermal rating; *(f)* 98% efficiency

3. Key design features

The JLab 12 GeV (and MSU-FRIB) compressor skids incorporate well proven components and have several key design features; a wide range of operation, good efficiency, considerations for maintenance technician and environmental safety, improved reliability, improved maintainability and lower cost.

3.1. Wide range of operation

Conventional oil-flooded rotary screw helium compressor skid packages can be rather limited in the range of suction and discharge pressure ranges that they can operate; especially in regards to their bulk oil separation (BOS). The helium cold box can function to serve as an ultimate purifier, since there are no known contaminates the will escape being condensed and/or solidified at 4.5-K. So, although dry rotary screw compressors have their appeal in this regard for helium refrigeration systems, they inherently require tighter tolerances and experience higher compression temperatures, especially when helium is the 'refrigerant'. And, notwithstanding even these considerations, if oil is used to lubricate the bearings and timing gears, a comment attributable to Sam Collins (Professor Emeritus of the Massachusetts Institute of Technology, internationally known as the father of practical helium liquefiers) would seem to seal their fate; 'It makes little difference whether it is one drop or one gallon of oil in the helium it still needs to be separated for helium liquefiers.'

The BOS's for the 12 GeV (and MSU-FRIB) compressors are long horizontal vessels without flanges, composed of a short cyclonic separation section followed by and inertial separation section. The cyclonic and inertial sections are separated by a weir and the oil forms the 'seal' between the two sections. The entire vessel is sloped 'downhill' ~25 mm (end-to-end). The inertial section has two supply headers that direct the helium-oil mixture at a 15 degree angle down from the horizontal plane

IOP Conf. Series: Materials Science and Engineering **90** (2015) 012072 doi:10.1088/1757-899X/90/1/012072

into the vessel volume. A return header toward the top center of the vessel collects the helium before going to the after-cooler. There are no glass fiber coalescing elements in this vessel and unlike typical BOS's is has a considerably smaller diameter, but allows a reasonable transit time as the oil flows to the outlet, then onto the oil cooler below. Table 2 summarizes a number of the BOS design parameters (LP – low pressure stage, MP- medium pressure stage, HP – high pressure / swing stage). Figure 2 depicts the BOS cross-sections for the HP stage and a cross-section of the helium gas velocity distribution of the HP stage (calculated; velocity scale is logarithmic and ranges from 1 mm/s to 1 m/s with the mid-section ~20 cm/s). Table 3 shows the range of conditions tested during commissioning for the BOS's and table 4 shows the range of conditions that they have been operated at since being commissioned (the built-in-volume, BVR, setting during operation is: 2.4 for LP stages and 2.2 for MP and HP stages).

Table 2. Bulk oil separator (BOS) design parameters (selected)

		HP	**MP**	**LP**	***Notes***
Outside Diameter	[m]	0.762	0.762	0.610	
Overall Length (Head-to-Head)	[m]	6.78	6.28	6.23	
Minimum Oil Volume	[ℓ]	276	254	225	
Nominal Oil Volume	[ℓ]	407	374	330	*(a)*
Oil-Helium Surface Area Per Volume	[m^2/m^3]	10.9	10.6	10.5	*(b)*
Oil Transit Time	[s]	22	81	55	*(b)*
Max. Carry-Over Droplet Size	[μm]	<100	<100	<100	*(c)*
Total Oil Carry-Over	[ppmw]	500	500	250	*(c), (d)*

Notes: *(a)* 15.2 cm liquid level height; *(b)* At nominal volume; *(c)* To helium after-cooler; *(d)* Parts per million with respect to helium mass flow

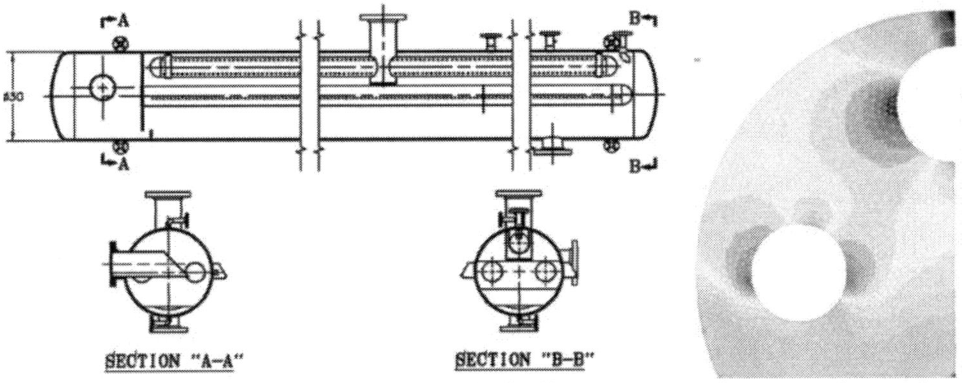

Figure 2. 12 GeV high pressure (HP) stage bulk oil separator (BOS) cross-sections and gas velocity distribution (calculated)

As there will always be oil that the BOS does not separate, the after-cooler heat transfer surface area was specified assuming 0.2% oil (by weight of helium; which marginally increased the surface area by ~10%) and designed so that coalesced oil can drain back to the compressor suction. After the helium leaves the after-cooler and the skid, there is an external coalescer dedicated to that skid. Referring to table 5, these coalescer vessels are substantial smaller in diameter than on other skids and are not subject to the high helium discharge temperatures (or high gas velocities). The compressor oil removal was found to perform adequately for all test and operational conditions listed. Additionally, except for the MP stage at high suction pressures (~3 bar), the external coalescer collected a minimal amount of oil.

IOP Conf. Series: Materials Science and Engineering **90** (2015) 012072 doi:10.1088/1757-899X/90/1/012072

Table 3. Range of conditions tested during commissioning of 12 GeV compressors

Compressor Stage	Suction Pressure [bar]	Discharge Pressure [bar]	Pressure Ratio	BVR (†)
LP Stage	1.06	3.55 to 6.08	3.33 to 5.71	2.2 to 3.2
MP Stage	1.06 to 3.04	3.55 to 6.08	2.00 to 3.33	2.2 to 3.2
HP Stage (§)	1.07 to 5.47	13.37 to 18.75	3.00 to 16.98	2.2 to 5.0 (‡)

(†) BVR increments were 0.2

(‡) Not every BVR setting was used for each test point due to manufacturer maximum suction pressure limitations (for a given BVR)

(§) HP stage can be operated as a (swing) LP or MP stage; under these conditions the suction pressure can be as low as ~1.03 atm, the discharge pressure as high as ~18.5 atm, and the pressure ratio ~18

Table 4. Range of conditions that 12 GeV compressors have been operated at since being commissioned

Compressor Stage	Suction Pressure [bar]	Discharge Pressure [bar]	Pressure Ratio
LP Stage	1.04 to 1.22	3.40 to 7.03	3.17 to 6.55
MP Stage	1.12 to 2.26	3.37 to 7.07	2.48 to 5.58
HP Stage (§)	3.19 to 5.78	7.25 to 20.04	2.27 to 3.70

(§) Same note as in table 3

Table 5. Size of selected BOS/coalescer vessels

Plant	Stage	Diameter [m]	Length [m]
CHL-1	LP	1.219	4.48
CHL-1	HP	1.219	5.40
SNS	LP	1.270	3.23
SNS	HP	1.067	3.37
CHL-2	LP	0.610	6.23
CHL-2	MP	0.762	6.28
CHL-2	HP	0.762	6.28

Note: CHL-1 is the old helium refrigerator installed at JLab in the 1990's for a 4 GeV program and used through the 6 GeV era. SNS is the helium refrigerator at Oak Ridge, TN for the Spallation Neutron Source which used compressor bodies essentially identical to CHL-1 except that they have a variable BVR; the skids are not identical to CHL-1. For CHL-1 and SNS, the BOS and coalescers are integral to each other. CHL-2 is the new helium refrigerator for the JLab 12 GeV upgrade. The BOS is part of the compressor skid, but the coalescer is external to the skid and is the same size for all stages; 0.914 m in diameter, 3.86 m long.

3.2. *Efficiency*

It has been found in practice that there is an optimum helium discharge temperature. Isothermal compression is the ideal compression process for helium refrigeration systems that use a Collins type process [12]. Increasing injected oil flow would tend to make the compression process more isothermal, assuming it mixes well with the helium during compression. However, oil viscosity increases exponentially with decreasing temperature. So, it has been found that allowing the helium discharge temperature to be as high as practically possible results in a minimum input power. To ensure good reliability of seals and bearings, the discharge temperature should not exceed 372 to 378 K, and normally 370 K is used for continuous operation. For the 12 GeV compressors, the oil injected into the rotors for cooling is not pumped, rather the differential pressure between discharge and the injection point supplies the potential, and is regulated using a control valve that seeks to maintain a set discharge temperature. The oil used to cool the bearings, which is around 20 to 30% of the total oil flow, is supplied to the bearings using a positive displacement pump equipped with a variable frequency drive (VFD) that seeks to maintain a set differential pressure (between pump discharge and compressor discharge). As such, a considerably smaller oil pump is needed than if the full oil flow was being pumped and there is a minimum of excess oil flow handled by the pump that is bypassed back to the BOS, since a VFD is employed. Also, since coalescing elements are not incorporated into the BOS, there is a minimum amount of pressure loss in the bulk oil separation process.

During commissioning, the compressors were operated over the ranges indicated in table 3, and their performance was measured. Input power to the motor, helium mass flow, as well as, helium suction, helium/oil discharge and oil injection process conditions (i.e., pressure and temperature) were measured. Total oil flow was not measured, but can be calculated using an energy balance. The manufacturer's motor efficiency (vs. load fraction) was assumed to be correct, helium and oil were assumed to be at the same temperature upon discharge from the compressor and it was assumed that there was no helium dissolved in the oil injected. The motor speed was measured in some cases (using a stroboscope) and was found to be less than 1% greater than the motor nameplate data. The calculated volumetric and isothermal efficiencies for the LP, MP and HP stages obtained from the 12 GeV compressor commissioning data are shown in figures 3, 4 and 5, respectively. The actual (measured) mass flow is used for the isothermal efficiency. These efficiencies are applicable to the compressor proper (not the overall skid). Volumetric efficiency for the MP stage is ~5% too high due to calibration issues with the suction transducer and the HP stage isothermal efficiency is ~5 to 8% too high due to the calibration of the motor power monitoring or the turn-down assumed for the current or voltage transformers. This data shows similarities to previous work at the SSCL [13].

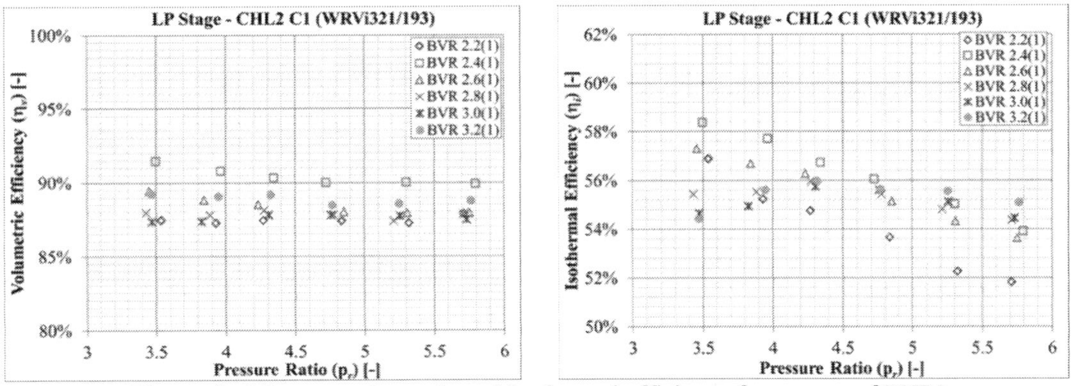

Figure 3. LP stage volumetric and isothermal efficiency for a range of BVR's

3.3. Safety, reliability, maintainability and cost

There are a number of previously mentioned factors that contribute to improved safety, reliability, maintainability and lower cost. The design of the skid is sized and arranged to allow the oil to drain first to the oil cooler and then up to the BOS. Except for a short "P" trap section of pipe between the compressor discharge flange and the BOS inlet, there are no components or sections of pipe that will hold the oil upon shut-down. This allows for ease in start-up and shut-down and the ability to verify the oil inventory, which is not the case in a number of typical designs.

Also, the skids designs allow a minimal oil charge (see table 1), as compared to typical helium compressor skid packages. This factor can be appreciated for those who have operated these machines and had to deal with the oil sprayed externally due to unexpected conditions that caused a relief valve to discharge or a line that developed a crack (such as a flex-hose or tube). The oil spray is not just messy, it can ruin the electric motors if ingested and can cause personnel health and environmental concerns.

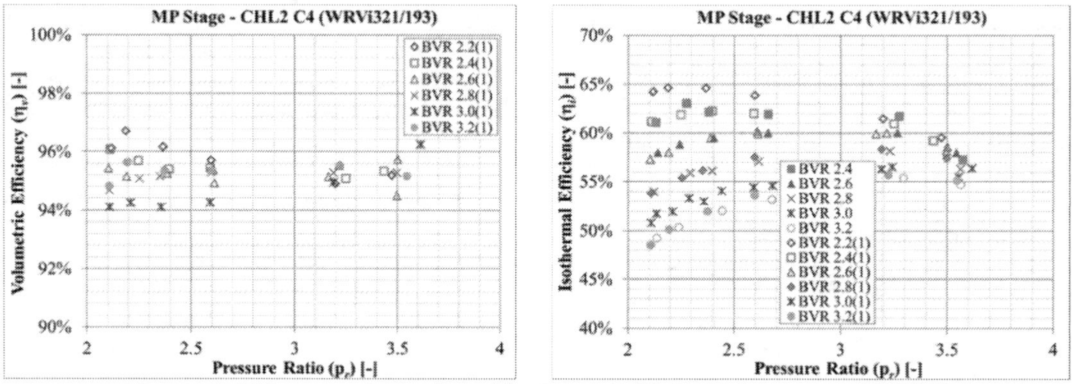

Figure 4. MP stage volumetric and isothermal efficiency for a range of BVR's

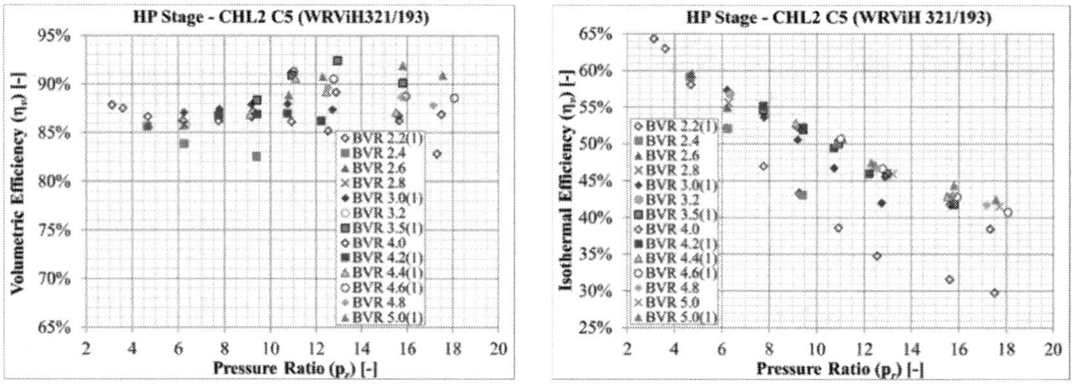

Figure 5. HP stage volumetric and isothermal efficiency for a range of BVR's

As mentioned, there are no coalescers in the BOS, and the external coalescer is not subject to high temperatures or high velocities. In typical designs, if operated outside their limited design range, not only is excessive oil carry-over an issue, these elements have been known to tear.

The BOS design in typical packages is quite large and usually flanged (see table 5). However, the BOS for the 12 GeV compressors is much smaller and has no flanges. This can tend to lower the cost of this component.

The oil pump for the 12 GeV compressor skids is downstream of the oil-cooler. In several typical designs the oil pumps are upstream of the oil-cooler (and are handling the total oil flow), which would seem better for minimizing specific pump input power (handling less viscous oil). However, if they are handling the total oil flow at high temperature, they need to be much larger (than if handling only the bearing injection flow) and they can tend to develop shaft seal leaks quite easily. The 12 GeV compressor pumps run cooler and have not developed shaft seal leaks. And, the fact that only 20 to 30% of the total oil flow is pumped allows for a much smaller pump and motor.

The entire 12 GeV compressor skid is design with minimal elevation for ease of access during maintenance and as mentioned the separate sub-skids isolating the rotating and non-rotating components minimizes the effect of vibration on a majority of the components and controls.

For those who have operated these machines, alignment of compressor and motor after maintenance can be an interesting challenge, which can be complicated by difficult access and thermal expansion effects. These can also affect their alignment over time while operating. For the 12 GeV compressors, both compressor and motor are mounted on frames with thick-plates (see figure 1). These, in turn, are mounted to the main structure below which is anchored to the floor. Additionally, universal adjustable steel chocks and substantial tabs with adjustment bolts are incorporated to ease the alignment and help prevent a 'soft foot'.

4. Conclusion

The development of the 12 GeV compressor skids has been part of an ongoing research and development effort at JLab. The present development has allowed a very wide range of operation and a full realization of the Floating Pressure Process [8]; and, after ~11,000 hours of operation (by the end of 2014 for each skid) they have proven to be efficient, reliable and easily maintained. However, there is still considerable research to be done to develop alternate BOS designs and investigate methods and processes to improve efficiency; especially in regards to (1) oil choice, oil injection temperature and oil injection method on the isothermal efficiency and (2) the behavior of the helium-oil mix in the compression process and the effect of dissolved helium in the bulk oil flow.

5. Acknowledgements

The authors would like to express their appreciation and thanks to the TJNAF management and to their colleagues for their support. This work was supported by the U.S. Department of Energy under contract no. DE-AC05-06OR23177.

References

[1] Gruehagen J, Wagner U 2004 Measured performance of four new 18 kW at 2.5 K helium refrigerators for the LHC cryogenic system *Proc. 20th International Cryogenic Engineering Conference (Beijing, China, 11-14 May 2004)*

[2] Ganni V, Knudsen P 2014 Helium refrigeration considerations for cryomodule design *Adv. Cryo. Eng.* **59** (New York: American Institute of Physics) pp 1814-1821

[3] Ganni V, Knudsen P, Dixon K, Norton R, Creel J, Arenius D 2014 Commissioning of helium refrigeration system at JLab for the 12 GeV upgrade *Adv. Cryo. Eng.* **59** (New York: American Institute of Physics) pp 944-948

[4] Knudsen P, Ganni V 2012 Process options for nominal 2-K helium refrigeration system design *Adv. Cryo. Eng.* **57** (New York: American Institute of Physics) pp 800-813

[5] Ziegler B 1985 Second law analysis of the helium refrigerators for the HERA proton magnet ring *Adv. Cryo. Eng.* **31** (New York: Plenum) pp 699-707

[6] Knudsen P 2011 JLab's 12 GeV Compressor System and Development Work *Jefferson Lab's Engineering Division Professional Development Program Cryogenics Seminars (Newport News, VA, 4 August 2011)* https://www.jlab.org/eng/JLab%20Summer%202011%20Seminar%20-%20Compr%20Work%20Dev%20Pres.pdf

[7] Toscano W M, Hoagland L C 1977 The thermodynamic performance of oil-flooded helium screw compressors *Trans. ASME J. Eng. For Industry* p 749-753

[8] Ganni V, Knudsen P 2010 Optimal design and operation of helium refrigeration systems using the Ganni cycle *Adv. Cryo. Eng.* **55** (New York: American Institute of Physics) pp 1057-1071

[9] Homan J, Redman R, Ganni V, Sidi-Yekhlef A, Knudsen P, Norton R, Lauterbach J, Linza R, Vargas G 2014 Commissioning of a 20 K helium refrigeration system for NASA-JSC chamber-A *Adv. Cryo. Eng.* **59** (New York: American Institute of Physics) pp 307-314

[10] Ganni V, Dixon K, Creel J, Knudsen P, Arenius D 2010 Compressor system for the 12 GeV upgrade at Thomas Jefferson National Accelerator Facility *Proc. of the 23rd International Cryogenic Engineering Conference (Wroclaw, Poland, 19-23 July 2010)* pp 859-863

[11] V. Ganni, Knudsen P, Arenius D, Casagrande F 2014 Application of JLab 12 GeV helium refrigeration system for the FRIB accelerator at MSU *Adv. Cryo. Eng.* **59** (New York: American Institute of Physics) pp 323-328

[12] Knudsen P, Ganni V 2006 Simplified helium refrigeration cycle analysis using the 'Carnot

Step' *Adv. Cryo. Eng.* **51** (New York: American Institute of Physics) pp 1977-1986
[13] Ganni V, Knudsen P, Creel J, Arenius D, Casagrande F, Howell M 2008 Screw compressor characteristics for helium refrigeration systems *Adv. Cryo. Eng.* **53** (New York: American Institute of Physics) pp 309-315

The Benefits of Installing Remote IO to a Two Stage Oil Free Screw Compressor Package in an Upstream Oil and Gas Application

Dr. William James MILLIGAN,
Howden Process Compressors Ltd, Screw Package Engineering Dept, Renfrew, UK
+44(0)141 885 7912, Billy.Milligan@Howden.com
Prof. David HARRISON
Glasgow Caledonian University,
Glasgow, UK
+44(0)141 331 3000,
D.K.Harrison@gcal.ac.uk

Abstract

A two stage oil free screw compressor package was recently delivered to a major blue chip oil and gas giant for use on one of their oil platforms located in the North Atlantic. During initial design it became clear that the amount of instrumentation required on the package would be substantial. As a result of this an elegant solution was required that would move non-critical inputs and outs (IO) from the control panel to the package itself.

This paper first describes the problems caused by the substantial amount of instrumentation required on the package. This description includes how the instrumentation is classed as critical and non-critical and what signals must still be sent over a hard wire and why. As the final location of the package is in an extremely hazardous environment, a highly corrosive marine environment, and the compressor itself is compressing highly combustible hydrocarbon gas, the protection methods employed by the instruments to prevent explosion is then discussed. The same protection methods must also be employed by the control systems IO that have been moved from the safe area to the hazardous area and this is all detailed within the paper.

The multiple communications networks that allow the remote IO system to send its values to the package control system and then onto the client's distributed control system are also discussed and how redundancy is built into this network. The paper concludes with how the changes made to this package have resulted in cost savings to the order of tens of thousands of Dollars.

Keywords: Instrumentation, Control, PLC, Remote IO,

INTRODUCTION

This paper concerns the design of the Howden low pressure compressor package being supplied for the Hebron Topsides project, off-shore Canada. This compressor package consists of two oil-free screw compressors; associated process piping, exchangers and vessels; lube oil system; main drive motor; gearbox; seal system and safe area Unit Control Panel (UCP). Under normal operating conditions, the compressors are configured to operate and maintain a set point suction pressure.

The instrumentation deployed, both in terms of quantity and level of specification, led to the delivery of the most instrumented package in Howden Process Compressors history. All of these instruments measured values had to be sent back to the package control system located in a control room in a remote

Content from this work may be used under the terms of the Creative Commons Attribution 3.0 licence. Any further distribution of this work must maintain attribution to the author(s) and the title of the work, journal citation and DOI.
Published under licence by IOP Publishing Ltd

IOP Conf. Series: Materials Science and Engineering **90** (2015) 012073 doi:10.1088/1757-899X/90/1/012073

location. A safe, reliable and easy to install method of sending these values back to the control system was required in order to reduce the cabling cost to the end user.

HAZARDOUS AREA

The compressor package is designed for use in a Zone 2 IIB T3 hazardous area according to IEC 60079-10-1, the unit control panel is designed for use in a safe area. The IEC 60079 series of standards is the international standard on the subject of potentially explosive environments caused by gases and dust. The regulations deal with how to classify areas into different zones (0, 1 and 2 for gas) depending on their likelihood of an explosion with regards to the plant and process, the substance involved and the various temperatures (ambient and minimum ignition for example). They then go on to specify different protection methods (Ex ratings) to protect against explosion within the different zones. The end user has classified the zone as a zone 2 environment which is a place in which an explosive atmosphere consisting of a mixture with air of flammable substances in the form of gas, vapour or mist is not likely to occur in normal operation but, if it does occur, will persist for a short period only [1]. Zone 2 is the least onerous of the hazardous area classifications with Zone 1meaning a potentially explosive situation more likely to occur and for a longer period and Zone 0 more serious still. Certain protection methods utilised on electrical and instrumentation, and on some mechanical equipment, can only be used in the particular Zones. For example Ex n, non-sparking apparatus, protection methods can only be deployed in Zone 2 areas but Ex ia, intrinsically safe apparatus, protected equipment can be used in Zones 0, 1 and 2. The unit control panel is designed for use in a safe area.

CONTROL SYSTEM ARCHITECTURE

There are three locations for starting and stopping the package: from the package HMI (Human Machine Interface), the control panel HMI and from the client's Process Control System (PCS). This allows the operator to initiate compressor pre-start or compressor start and stop sequences via the HMI buttons or from the client's control system. By using the Local/Remote selector switch mounted on the UCP the operator selects between Local HMI control and Remote client's control system control. The Local/Remote selector switch status is sent to the client's PCS. Additionally, maintenance functions can be selected from the HMI which enable the operation of the low voltage motors; oil heaters and stroke valves as required during plant maintenance checks.

An Allen-Bradley ControlLogix series PLC and two FactoryTalk View powered Hazardous Area HMI are supplied with the control panel to provide automatic package control and operator interface, and to provide a communication interface point with the client's Integrated Control and Safety System (ICSS). The client had two separate control systems that required an interface with the Howden UCP: the client's PCS controlled non-critical applications (indication and alarm functions) whereas the client's ICSS controlled the critical trip functions. Also within the Howden UCP a Bently Nevada 3500 Machine Condition Monitoring (MCM) system monitors the package main drive motor, gearbox and compressor temperature and vibration parameters and initiates a trip of the package if any of these parameters exceed the shutdown limits. The machine monitoring system values, alarms and trips are all transmitted to the PLC over an industrial Ethernet link using the Modbus protocol for display on the HMI and re-transmission to the platform ICSS. A simplified control system architecture schematic (reduced from nine sheets to one) showing how the package instrumentation signals are fed from the package to their end destination is shown in Figure 1.

Under normal operating conditions, the compressor control system hardware listed above is configured to operate to maintain a set suction pressure across each stage after having first performed the prestart sequence by getting the oil system up to the correct pressure and temperature and the seal system up to the correct pressure. To achieve the desired suction pressures on both stages the compressor control system first measures a process variable which in this case is the suction pressure transmitter located on each of the individual stage suction pipework. Depending on the pressure value measured the control system then has to operate the particular stages recycle control valve.

If the process variable is below the setpoint the recycle valve will open to allow the discharge gas back to the suction pipeline. If the process variable is above the setpoint the recycle valve will close. The opening and closing of this valve is controlled by a PLC proportional, integral and derivative (PID) block whose settings can be adjusted via the HMI.

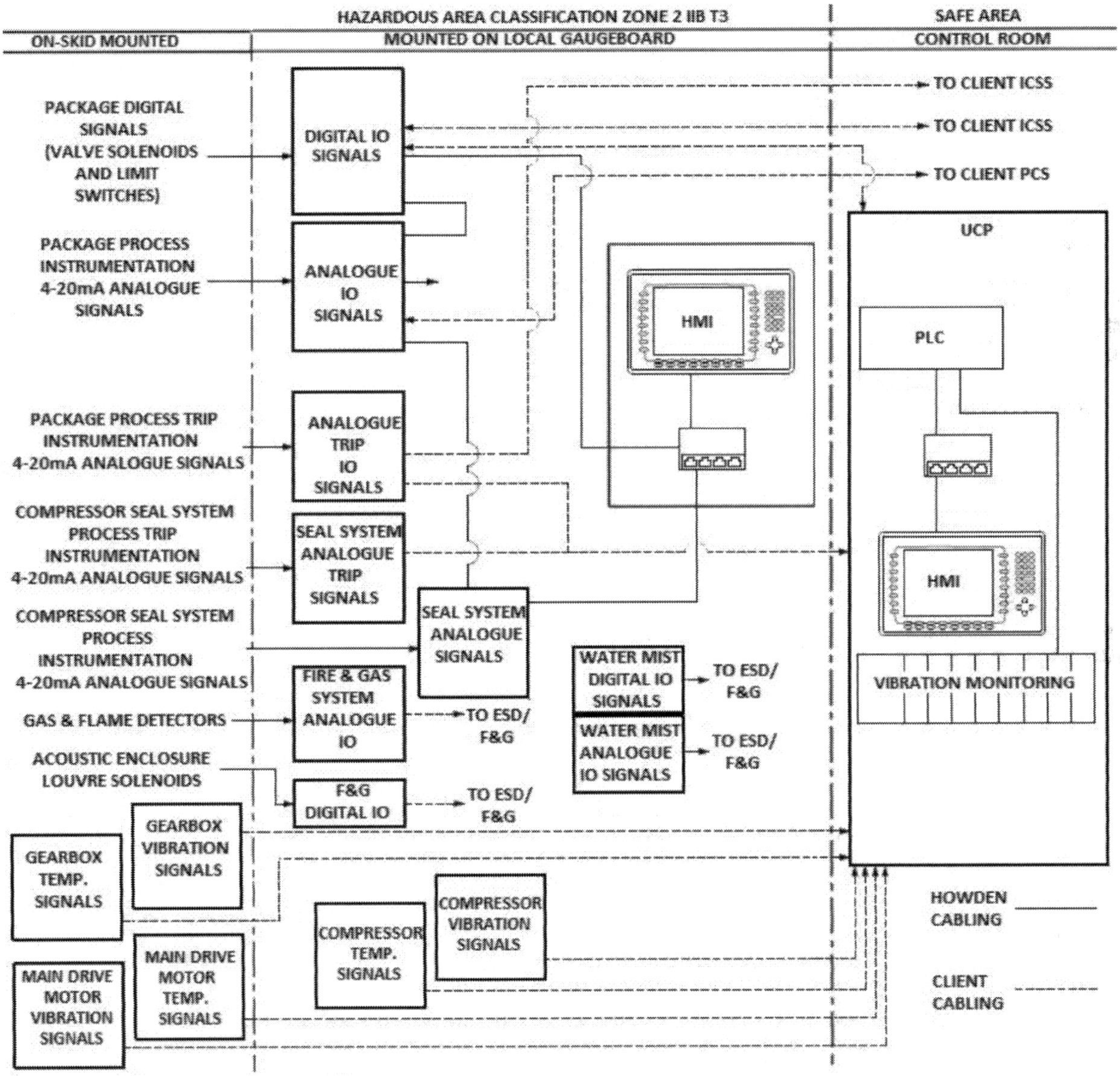

Figure 1: A simplified control system architecture drawing

On the second stage the process variable is taken from the second stage suction pipework and is compared to the setpoint inputted from the HMI. If the process variable is below the setpoint the recycle valve will open to allow the discharge gas back to the suction pipeline. If the process variable is above the setpoint the recycle valve will close. The opening and closing of this valve is again controlled by a PLC PID block whose settings can be adjusted via the HMI.

The re-cycle control valve functions as a trimming valve under PID loop control during compressor running and as an equalising valve during any of the stopping sequences. Each of the recycle valves were fitted with additional solenoids so they could be crashed fully open during a stop, trip or emergency stop sequence to equalise the suction and discharge pipe work and prevent reverse rotation.

The suction pressure set-point along with the PID gains can all be inputted to the PLC via either of the HMI systems supplied with the package. To test the operation of the recycle control valve, the HMI system operator can put the valve into manual mode and manually input a valve position set-point in %. An operator in the field can then visually verify if the valve moved to the desired position before putting the valve back into automatic control.

After the pre-start sequence is complete and the compressor main drive motor has been running for a set period of time, the recycle control valves are released to automatic control. A fifteen second time delay is put in between the enabling the automatic control of the 1st stage and 2nd stage recycle controllers. This is to prevent the controllers hunting during start-up as the system establishes healthy pressures. It is expected that the 2nd stage recycle valve will, to a certain extent, follow the position of the 1st stage recycle control valve even though they are controlled independently. As the 1st stage suction pressure falls, the 1st stage recycle valve will open sending less process gas forward, which will impact on the 2nd stage suction pressure causing the 2nd stage recycle valve to open. As the suction pressure increases and the 1st stage recycle valve closes, this will send more gas forward into the 2nd stage suction pipework, causing the 2nd stage recycle valve to close.

The control system deployed on this package had to carry out many different functions. On the gas system there was pressure and temperature control across two different stages, oil pressure and temperature control, seal system pressure and temperature control, drive train monitoring and the inclusion for fire and gas detection equipment. During the initial design phase each of the instruments and valves were segregated into different junction boxes first between analogue and digital signals and then further segregated between critical and non-critical IO. All of these systems coupled with the end user specifications that detail separate instruments for alarm and trip signals and then add in that the package is compressing hydrocarbon gas in a potentially explosive environment means Canadian National and International Hazardous Area regulations must also be applied lead to a package bristling with instrumentation. The design was peer reviewed during a hazards and operability study (HAZOP) with the topsides main contractor and end user who understand their process best. During the HAZOP additional transmitters were included and it became more evident that this package would contain a huge amount of instrumentation.

INSTRUMENTATION

This leads to a compressor package featuring 24 pressure transmitters, 17 differential pressure transmitters, 21 temperature transmitters, 7 level transmitters, 19 flow transmitters, 27 RTDs, 8 accelerometers, 25 vibration probes, 4 Flame and Gas detectors, 7 modulating control valves and 11 solenoids valves. All of the process or machine condition monitoring measurements gathered by this instrumentation must to be sent back to the relevant control system in order for it to control the package efficiently and safely.

Before discussing how this instrumentation is cabled back to the control system it is important to understand the robust nature of the design of the instrumentation systems due to the environment it is deployed in. Instruments, gauges and detectors are provided with stainless steel 316 cases, housings, electronics enclosures and fixings. This protects them from corrosion in the marine environment. The

Degree of Protection of control stations, junction boxes and enclosures are IP 66. This means all enclosures are completely protected against dust and protected against powerful jets of water from all directions [2]. All cabinets and panels have bottom cable entries only to assist in the prevention of water ingress.

All process transmitters are fitted with local LCD indicators to aid the commissioning engineers and operators in fault finding. In addition to the IP rating all process transmitters exposed to the elements are installed into a weatherproof heated enclosure complete with window to the operators can still read the LCD display. All metallurgy of the wetted parts of process instruments, such as control valves, pressure transmitters and differential pressure transmitters match or exceed that of the associated piping classification to prevent corrosion due to dissimilar metals.

All process temperature transmitters RTDs are provided with dual sensing elements within the thermowells so if one of the RTDs fails then the transmitter can use the other. Thermowells have been provided for all temperature measuring devices used in flammable, toxic, or otherwise hazardous, pressurised, or vacuum systems. Thermowells were not required for temperature measurement of machinery bearings, or motor windings where there is no risk to personnel from the process fluid during removal of the measuring element. All thermowells provided in dynamic lines have had wake frequency calculations performed to ensure the stress limits of the thermowell materials and its design are not exceed and will not fail in service.

CABLING

In addition to the high specification instrumentation required on the package the infrastructure taking the instruments signals to and from the package was also of a very high level due to the harsh environment and the nature of the process. There were two main different types of cabling used on the package. One type for digital IO signals and another for the analogue IO signals. The main difference between the two cabling types was the screens. The analogue cabling had a screen wire for every analogue signal whereas the digital cabling only had one overall screen wire as noise or interference affects the digital signals to a lesser extent. In addition to the different types of cable used on the package, all of the home run cables (cables ran from the skid edge to their required destination off-skid) in critical service were also fire resistant. Critical services on a top sides platform include: emergency power generation and distribution systems; UPS and battery charger systems; emergency and escape lighting systems; fire and gas systems; fixed fire fighting and firewater pump systems; emergency shutdown systems; critical instrument signals and alarms; lifesaving systems; safety telecommunications systems and PA/GA; navigation and obstruction signals, helicopter landing and operations systems; and critical platform make-safe systems. Of these systems the screw compressor package included fire and gas systems, fire fighting systems, emergency shutdown signals and critical instrument and alarm signals.

All cables installed on the package and all home run cables were armoured with a tinned annealed copper metal braid and certified for use in a hazardous location. The outer sheath was made from a low smoke zero halogen, flame retardant, UV and ozone resistant, extruded thermosetting compound. This is quite typical for offshore applications as was the cold bend radius testing carried out to ensure the cable would not become brittle and break at extremely low temperatures found in the North Atlantic Ocean. All cabling supplied for the package was suitable for ambient temperatures from -17 to +40 degrees Celsius. The cable outer sheath was grey or blue depending on the hazardous area method of protection applied to the instrument it was connected to. The inner conductor colours were the same for all types of cabling: black as a positive and white as the negative.

All armoured cables were terminated to the instrument and to the junction box with cable glands. All cable glands supplied were nickel-plated bronze with metric threads with earth tags to provide additional equi-potential bonding. All cable glands were suitably rated and certified for use in a hazardous environment and any stopping plugs fitted into unused drilled holes in certified equipment were also suitably certified.

It is important to note for the purposes of this paper that the client specified that all digital communication cables should be treated as a signal cable and given the same protection from damage. This will become relevant during the benefits discussion.

Power and signal wiring systems of different groups were to maintain minimum parallel separation using separate installation methods. For example, all power and signal cable groups were run in separate metal enclosed trays. Further to this all intrinsically safe cabling (blue sheathed cables rather than grey) were run in another separate tray so at times three different 316 marine grade stainless steel trays, with covers, were run: one carrying power to motors, one for signals to flame proof instruments (grey cabling) and one to intrinsically safe instruments (blue cabling). It was by all means a very robust design.

Another design feature of the wiring system was the spare allocation designed into the package to allow for future expansion. All home run multi-conductor cables had to have at least 20% spare capacity which also impacted on the field junction boxes as they then too also had to have the capacity to terminate the additional field instrumentation and the 20% spare capacity of the multi-conductor cables. The same home run multi-conductor cabling also had to be terminated at its off-skid destination which meant a minimum 20% spare allocation in the UCP as well as the junction boxes.

The substantial amount of instrumentation and now the high specification of cabling and cabling methods required meant the cost of cabling these signals back to the safe area control panel or the clients control system would be extremely costly. They were looking for an elegant, and less costly, solution. A decision to terminate the non-critical signals into remote IO terminals was made during the initial design phase to try and reduce the amount of cabling at site. This meant effectively moving the IO terminations from these signals from the PLC located in the safe area control panel out to the package and send the signal back via a communications link. As the cable route distance was 250m from the package to the control panel, a fibre optic physical layer with TCP/IP protocol would be used (client specified) to send the data back.

Previously mentioned in the text is the zone classification, Zone 2 IIB T3, of the area the package is deployed within. As the package contains many instruments and control equipment a few protection concepts were employed in order to prevent a catastrophe. Firstly, all of the process instruments such as the pressure, level and temperature instruments were all certified Ex d. The lubrication oil pumps and the oil heater were also certified Ex d. A flameproof enclosure (Ex d) is defined as an enclosure in which the parts which can ignite an explosive gas atmosphere are placed and which can withstand the pressure developed during an internal explosion of an explosive mixture, and which prevents the transmission of the explosion to the explosive gas atmosphere surrounding the enclosure [3]. Ex d rated equipment can be placed in Zones 1 and 2 only.

The machine condition monitoring instruments were certified Ex ia. Intrinsically safe protection type (Ex ia) is defined as a type of protection based on the restriction of electrical energy within the apparatus and of interconnecting wiring exposed to the potentially explosive atmosphere to a level below that which can cause ignition by either sparking or heating effects [4].

The Allen-Bradley Flex IO modules installed within the junction boxes on the package were certified Ex n, which means they contained non-incentive and or normally no sparking circuits [5].

COMMUNICATIONS

An extensive communication network is installed within the UCP to allow all the information to get to where it needs to go. Firstly, a Device Level Ring (DLR) was designed for the entire package network comprising of the three PLC racks; the HMI and the two fibre optic media convertors in the safe area UCP. These were linked in the same ring to two other fibre media convertors; the package HMI and the four Flex IO nodes installed on the package. The DLR was chosen so that if any one port on any one of the pieces of kit failed then the entire system could still communicate with each other (the network changes from a ring to a bus topology) [6].

A redundant communication connection between the PLC and the clients PCS is installed. This comprised of two RS485 physical connections transmitting information between the PLC and the PCS using Modbus RTU protocol. The RS485 standard is a key component in transferring digital information between control system terminals [7]. Again, if one of the communications links between the PLC and the DCS was removed or damaged, the remaining link will continue to operate. The DCS is configured as the Master, whereby it schedules the messages backwards and forwards between the two nodes, and the PLC is the Slave.

RESULTS AND CONCLUSIONS

When calculating the cost benefits to the end user the hardware costs of the Flex IO modules must be accounted for. This comprised of FLEX power modules, Ethernet communications modules, analogue input modules, analogue output modules, digital input modules, digital output modules, module DIN rail mounts and fibre media convertors. All of this hardware was split across four junction boxes on the package. The hardware costs for the remote IO system was $15,500. As the cable for this project was sourced in North America and for the ease of comparison all prices will be given in USD. An armoured Ethernet cable was used to communicate between the four junction boxes but a fibre optic cable was required to communicate with the UCP. 250m of this fibre optic cable cost $4,375 (courtesy of Ken O'Brien, Source Energy Atlantic Inc.) so we have a combined total hardware costs of $19,875 for the remote IO system.

Figure 2: The compressor package on test with the author shown for scale reference

The above replaced 2 x 24 pair cable for the conventional process analogue instruments, 1 x 24 pair cable for the package digital signals and a further 2 x 24 pair analogue instrumentation cable for the seal system analogues. A 250m run of the 24 pair cable is $8750, and the remote IO system replaced five of these cables therefore the total amount of hardware replaced is $43,750. To conclude, by taking the Flex IO hardware and fibre cabling costs from the replaced copper cabling costs, a saving of $23,875 on the hardware alone and the end user is still left with a safe, robust and easy to install method of sending the instrumentation values back to the control system.

What has not been taken into account is the installation time required, or the installation materials required, for the installation teams in Korea who are building the platform and responsible for installing the cables running from the package to the control room. This would lead to an even larger saving but requiring subtracted from this saving would be the time taken to install the fibre cable, the time taken to install the armoured Ethernet cable on skid by Howden, and the time taken by the author to first design the system. Perhaps what counts but cannot be counted is the customer asking to reduce cabling costs and installation time and an elegant solution being delivered.

Future work will see the removal of process transmitter package wiring altogether for non-critical applications in the oil and gas or harsh environment applications with the advent of wireless pressure and temperature transmitters. The design of these packages are client specification driven so until wireless transmitters are specified then they will not be installed and as blue chip oil and gas companies are quite conservative at adopting new technologies the author does not expect to see them specified for another ten years yet.

REFERENCES

[1] International Electrotechnical Commission IEC 60079-10 (2008) Explosive atmospheres Part 10-1: Classification of areas – Explosive gas atmospheres, 1st ed, Geneva: IEC, p.8, ISBN 2-8318-1019-3.

[2] International Electrotechnical Commission IEC 60529 (2001) Degrees of protection provided by enclosures (IP Code), 2nd ed, Geneva: IEC, pp.14-16.

[3] British Standards Institution. (2007) Explosive atmospheres – Part 1: Equipment protection by flameproof enclosures “d”, London: BSI, p.8, ISBN 978 0 580 54669 3.

[4] British Standards Institution. (2007) Explosive atmospheres – Part 11: Equipment protection by intrinsic safety “i”, London: BSI, p.10, ISBN 978 0 580 50199 9.

[5] British Standards Institution. (2007) Explosive atmospheres – Part 15: Equipment protection by type of protection "n", London: BSI, p.10, ISBN 978 0 580 59523 3.

[6] Nance B. (1997) Introduction to Networking, 4th Ed, Indianapolis: Que Corporation, pp.13-15, ISBN 0 7897 1158 3.

[7] Bailey D. and Wright E. (2003) Practical SCADA for Industry, 1st ed, Oxford: Elsevier, p.182, ISBN 07506 5805 3.

Inlet Diameter and Flow Volume Effects on Separation and Energy Efficiency of Hydrocyclones

Ş Erikli[1, 2], A B Olcay[2]
[1] Dalgakiran Makine A.Ş., Istanbul, Turkey;
[2] Yeditepe University, Mechanical Engineering Department, Atasehir, Istanbul, 34755, Turkey

sukruerikli@dalgakiran.com; bahadir.olcay@yeditepe.edu.tr

Abstract. This study investigates hydrocyclone performance of an oil injected screw compressor. Especially, the oil separation efficiency of a screw compressor plays a significant role for air quality and non-stop working hour of compressors has become an important issue when the efficiency in energy is considered. In this study, two separation efficiency parameters were selected to be hydrocyclone inlet diameter and flow volume height between oil reservoir surface and top of the hydrocyclone. Nine different cases were studied in which cyclone inlet diameter and flow volume height between oil reservoir surface and top were investigated in regards to separation and energy performance aspects and the effect of the parameters on the general performance appears to be causing powerful influence. Flow inside the hydrocyclone geometry was modelled by Reynolds Stress Model (RSM) and hydro particles were tracked by Discrete Phase Model (DPM). Besides, particle break up was modelled by the Taylor Analogy Breakup (TAB) model. The reversed vortex generation was observed at different planes. The upper limit of the inlet diameter of the cyclone yields the centrifugal force on particles to decrease while the flow becomes slower; and the larger diameter implies slower flow. On the contrary, the lower limit is increment in speed causes breakup problems that the particle diameters become smaller; consequently, it is harder to separate them from gas.

1. Introduction

Cyclones known as separators are commonly employed to extract particles or droplets from the system with the help of centrifugal forces. Besides, cyclone structures can be easily manufactured and they have low maintenance and investment costs. While these advantages make the cyclone systems quite popular in the industry, their complex flow characteristics with swirls cannot be solved with mathematical models. Therefore, experimental or numerical approaches have been used to determine the performance of the cyclones. There are many experimental studies regarding to the design of cyclone systems [1] [2] [3]. However, experimental approaches have its own difficulties to identify the flow inside the cyclone since visualization of entire fluid flow inside a closed geometry is very challenging. Hence, observations from inlet and outlet measurements have been typically used to quantify performance parameters of these systems while numerical studies can help to understand the flow characteristics in cyclone systems locally. As a result of improvements in computer processors and solver programs, the time needed to complete a numerical solution has been decreasing; therefore, nowadays many numerical studies [4] [5] [6] related to fluid flow in cyclone systems can be found.

Content from this work may be used under the terms of the Creative Commons Attribution 3.0 licence. Any further distribution of this work must maintain attribution to the author(s) and the title of the work, journal citation and DOI.
Published under licence by IOP Publishing Ltd

As the fluid flow inside cyclone systems is inherently turbulent, the studies about the turbulence model for swirling flows [6] are investigated. It was realized that the Reynolds Stress Model (RSM) has been widely accepted for swirling flows in these studies while Reynolds Averaged Navier-Stokes (RANS) have been generally employed for fluid flow simulations since this method gives results with reasonable error [7] compared to Large Eddy Simulation (LES) solutions.

In addition to fluid flow simulations in cyclone systems, the particles can be essentially observed with two different methods, namely volume of fluid (VOF) and discrete phase method (DPM). The oil reservoir at the bottom of the oil tank and desired number of particles can be modelled in VOF method with certain restrictions. The limitations in VOF model can be mainly summarized as computational time being longer in VOF than DPM method and VOF method is being unable to model the surface tension or velocity on the droplet surface. On the other hand, DPM is a simulation technique widely used for particle flows and in DPM method particles can be traced and monitored. Furthermore, different particle models such as collision of particles or particle break-ups can be employed. Studies [8] show that the DPM results in cyclone modelling become more accurate with Euler-Lagrangian approach.

When flow characteristics in cyclones and hydrocyclones are considered, gas flow characteristics are observed to be the same for both systems while the particle characteristics show some differences. Particularly, dust cyclones are used for separate solid particles such as dusts while hydrocyclones are utilized for separation of liquids from gas flow. When solid particles flow in the gas phase, there is no change in the shape of solid particles whereas shape of the liquid particle in gas flow can change due to speed, pressure or specification of a gas [9]. However, there is a common challenge for cyclones and hydrocyclones that small particles are typically hard to separate from large particles. There are many studies regarding to dust cyclone geometries and turbulence models used in these studies can be employed for hydrocyclones; however, the studies about particles are not applicable for hydrocyclones. When the fluid flow speed becomes large enough, the liquid droplets break-up and the droplets start to flow as small droplets; as a result, the high speed in hydrocyclones becomes an undesirable situation. On the other hand, collision of solid particles can be interpreted as elastic collisions while the liquid particle collision causes both coalescence and elastic collision. A particle diameter increases after a collision and if the particle diameter passes the already set critical diameter value, the particle may break up to smaller particles. Therefore, the break up and coalescence effects must be considered for hydrocyclones [10]. The breakup effect on separation performance was studied by Gao et al. [11] extensively.

The cyclone geometry was investigated by different researchers. The famous experimental studies based on dust cyclones are examined again with different methods or simulations [5]. Noroozil et al. [12] not only investigated the cyclone inlet diameter effects on cyclone performance, but also the angle of the inlet port of the cyclone. The cyclone main body diameter performance effect on its performance was studied by Yetilmezsoy, 2006 [13].

In this study the inlet diameter and flow volume effects on separation and energy efficiency of hydro cyclones were simulated and investigated. Simulation was based on RANS method. The particles were traced with DPM method and the coalescence and breakup effects were considered. As a result decreasing inlet diameter increases breakup of droplets and the tiny diameters cannot separate from flow and decreasing the diameter of the main [14] body of the cyclone increases the pressure drop. The optimum dimension rate between main body and inlet diameter is tried to be found and discussed.

2. Numerical Model

One of the methods to compress air in industry is to use oil injected screw compressors. In this type of compression, oil is injected into airend to ensure lubrication for rotating screws and provide seal for air leakage (i.e., preventing back flow of air). Once the compression is achieved, oil and air mixture needs to be sent to the hydrocyclones where air is separated from oil. In this study, a simplified numerical model was generated to study performance of separation and energy efficiencies of hydrocyclones. This model was also described and used by Gao et al. (2014) to investigate effects of the central channel on the flow field in an oil-gas hydrocyclones. In this study, three dimensional hydrocyclone with rectangular inlet has been employed and shown in Figure 1. Parameters to describe the hydrocyclone are given in Table 1.

Table 1 Hydrocyclone geometric parameters

Swirl volume height (H)		2D			2.5D			3D		
Hydrocyclone inlet duct	Height (a)	0.3	0.4	0.5	0.3	0.4	0.5	0.3	0.4	0.5
	Width (b)	0.15	0.15	0.15	0.15	0.15	0.15	0.15	0.15	0.15
	Area (a*b)	0.045	0.06	0.075	0.045	0.06	0.075	0.045	0.06	0.075
Hydrocyclone central duct	Diameter (d)	0.5D			0.5D			0.5D		
	Height (h)	D			D			D		
	Inlet length (L)	D			D			D		
Hydrocyclone diameter (D)		D			D			D		

The numerical model illustrated in Figure 1 was generated with SolidWorks based on the parameters provided in Table 1. Three-dimensional hydrocyclone geometry was imported to ICEM-CFD for meshing (Figure 2). Number of elements were different for each simulated case because geometry was changing from one case to another. Table 2 below provides mesh information of the studied nine different cases. In these models, grid independency is ensured by using Gao et al. (2014) [10]. Briefly, three different swirl volume height (H = 2D, 2.5D, and 3D), with three different hydrocyclone inlet duct sizes (0.15 x 0.3, 0.15 x 0.4 and 0.15 x 0.5) were used to study the effect of separation and energy efficiency of hydrocyclones. In order to observe the effects of inlet diameter and swirl volume height on hydrocyclone performance, nine different combinations from diameter and height sets were created and listed in Table 1. Figure 2 shows the computational grid of the numerical model for swirl volume height is being 3D and height of hydrocyclone inlet duct is being 0.5. Total number of elements is 258,064.

Table 2 Studied cases (total of nine different models was generated for the study).

Swirl volume height (H)	2D			2.5D			3D		
Height of hydrocyclone inlet duct (a)	0.3	0.4	0.5	0.3	0.4	0.5	0.3	0.4	0.5
Number of Cells	203269	206473	203353	226227	229088	245907	255255	257837	258064

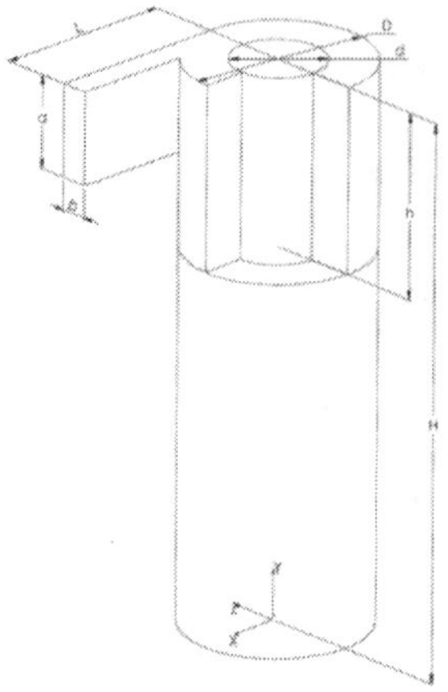

Figure 1. Numerical model with parameters

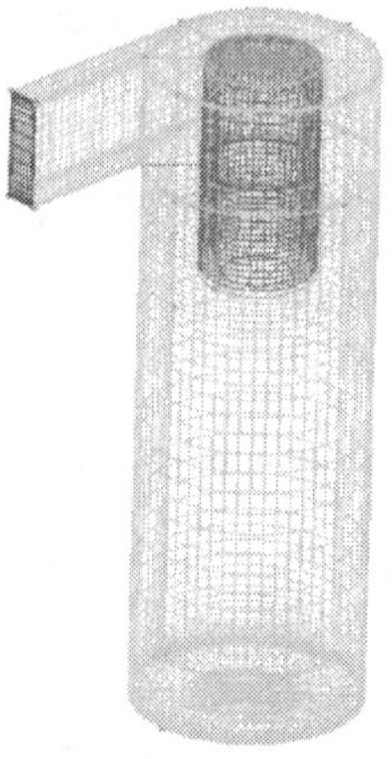

Figure 2. Numerical model with computational grid

Once the meshing was completed, boundary conditions were defined. Pressure and velocity were coupled in equations of motion for incompressible flows. As an outlet boundary condition a static pressure value of pressure outlet was specified while velocity inlet boundary condition was assigned for the inlet. Thus, the velocity at the outlet surface must adjust itself to match the rest of the flow field, similarly at the inlet surface the pressure needs to adjust itself so that the inlet and outlet boundary conditions were not over specified mathematically. DPM particle injection surface was also employed to the inlet surface. In DPM modeling settings, inlet and outlet surfaces were assumed to be an escape boundary condition. All the walls were modeled with no slip boundary wall condition except for the bottom wall. The bottom wall was set to no slip boundary wall condition for turbulence model and trap model for DPM model. Gravity was also defined along -y direction for all simulations. Air and oil material properties were defined for computational grid. Specifically, the air density and viscosity were taken as 3.5 kg/m³ and 2.09×10^{-6} kg/m.s, respectively at 0.2 MPa pressure and the oil density and viscosity were taken as 830 kg/m³ and 5.08 kg/m.s, respectively at 95°C.

There were mainly three assumptions made during this study. First assumption was about the inlet of the cyclone. In this study, a rectangular duct was used instead of a circular pipe. Second assumption was regarding to the oil reservoir. The reservoir oil below the tank was assumed to be a wall with trap condition so that oil surface flow of oil reservoir remained the same. Therefore, the bubble and the returning cached particles would be eliminated. Third assumption was about the shape of the particles. The particles in the study were proposed to be a spherical ball and no change in their shape was allowed during simulations.

In the numerical model the RSM is preferred for the viscous turbulence model. For tracing droplets DPM model was used with sub modules that are Saffman Lift force, Discrete Random Walk (DRW), Breakup, Stochastic Collision and Coalescence. There are a lot of combinations of turbulence solver models for solving different flow problems moderately or decreasing computational time in Fluent. Kaya and Karagöz [6] discusses the best combination for solver models. The most convenient combination was found to be PRESTO for pressure discretization, SIMPLEC for pressure velocity coupling, QUICK for momentum discretization, second-order upwind for turbulent kinetic energy and turbulent dissipation rate and lastly first order upwind for Reynolds stress based on their study. In turbulence modelling, the Reynolds Stress Model (RSM) was utilized to solve Reynolds-averaged Navier-Stokes equations. Although use of RSM increases the computation time of the simulation, this model is typically recommended for cyclone, highly swirling flows in combustors, rotating flow

IOP Conf. Series: Materials Science and Engineering **90** (2015) 012074 doi:10.1088/1757-899X/90/1/012074

passages, and the stress-induced secondary flow in ducts. Discrete phase model (DPM), utilized in this study, is widely used for the particle flows. In this model particles were added to the flow to track the liquid droplets. The volume fraction of oil particles was generally lower than 10-12% in the flow so that they did not influence the main phase flow of inside the oil tank. The velocity of the particle was dependent on shear force in the particle, main phase velocity and particle. Another additional parameter was about changing velocity of particle is acceleration "a" due to other forces on the particle.

$$\frac{u_p}{dt} = \frac{1}{\tau_p}\left(u - u_p\right) + a \quad (1)$$

After some analytical integration, the particle velocity at new location n+1 was formulated as:

$$u_p^{n+1} = \frac{u_p^n + \Delta t\left(a + \frac{u^n}{\tau_p}\right)}{1 + \frac{\Delta t}{\tau_p}} \quad (2)$$

And the new location was computed with basic location formula of physics.

$$x_p^{n+1} = x_p^n + \frac{1}{2}\Delta t\left(u_p^n + u_p^{n+1}\right) \quad (3)$$

Forces acting on a particle, defines the acceleration of a particle, were formulated with Lagrangian model in Fluent. The acceleration of a particle was basically dependent on a drag force, density difference between main flow and particle, and an additional acceleration force (F_x) where this force played an important role when the density of the main phase was greater than the density of the particle.

$$\frac{u_p}{dt} = F_D\left(u - u_p\right) + \frac{g_x(\rho_p - \rho)}{\rho_p} + F_x \quad (4)$$

$$F_D = \frac{18\mu}{\rho_p\ _p^2}\frac{C_D Re}{24} \quad (5)$$

$$Re \equiv \frac{\rho d_p |u_p - u|}{\mu} \quad (6)$$

The oil particles have greater density than air; therefore, F_x becomes negligible. The Saffman lift constant is greater than other additional acceleration force terms, so only Saffman lift force was used in the current study.

$$F_x = \vec{F} = \frac{2K v^{1/2} \rho d_{ij}}{\rho_p\ _p\left(\ _{lk} d_{kl}\right)^{1/4}}\left(\vec{v} - \vec{v}_p\right) \quad (7)$$

Instantaneous oil droplet velocity to particle observation in turbulent flow was included when DRW model was activated. The DRW model, uses the fluctuating velocity components, is a discrete piecewise constant functions of time:

$$u = \bar{u} + u' \quad (8)$$

As a result of collision of particle, the particle diameter increases. When the particle diameter and its velocity reach to a certain value, the particle may breakup more tiny particles before it is collided again. The collision possibility of a particle with N-1 particles is approximately (1/2) N^2. When two particles collided, they may be coalesced or bounced. The reaction of the collision of two particles can be calculated by a collisional weber number.

$$We_c = \frac{\rho U_{rel}^2 \bar{D}}{\sigma} \quad (9)$$

When two particles collided, the event results in coalescence or elastic collision. There is another type of coalescence that while two particle flows to same direction if they get closer each other below the critical distance they coalescence. The critical offset is dependent to the Weber number and relative radius between particles. The critical offset formulated by O'Rouke [15] on below.

$$b_{crit} = (r_1 + r_2)\sqrt{min(1.0, \frac{2.4f}{We})} \quad (10)$$

The actual collision parameter b equals to$(r_1 + r_2)\sqrt{Y}$. Y is a random number between 0 and 1. If the b_{crit} is greater than the actual b, particules are collided with coalescence. The term f is a related the ratio between particle radius.

$$f\left(\frac{r_1}{r_2}\right) = \left(\frac{r_1}{r_2}\right)^3 - 2.4\left(\frac{r_1}{r_2}\right)^2 + 2.7\left(\frac{r_1}{r_2}\right) \quad (11)$$

O'Rouke [17] claims that the coalescence causes energy loss; therefore, the coalescent particle velocity cannot be calculated because momentum equation but itself would not be enough. The derived expression for the coalescent particle velocity can be computed from:

$$v_1' = \frac{m_1v_1+m_2v_2+m_2(v_1+v_2)}{m_1+m_2}\left(\frac{b-b_{crit}}{r_1+r_2-b_{crit}}\right) \quad (12)$$

In oil droplet breakup, Taylor analogy method is used for low-Weber-number injections. As a result weber number less than 100 the TAB model can be applicable. If one of the particle distortion equals the half of the droplet diameter, droplet assumed to be broken up. The break up occurs when y gets higher than 1 and y can be calculated by formula below.

$$y(t) = We_c + e^{-(t/t_d)}\left[(y_0 - We_c)\cos(\omega t) + \frac{1}{\omega}\left(\frac{dy_0}{dt} + \frac{y_0 - We_c}{t_d}\right)\sin(\omega t)\right] \quad (13)$$

3. Results and Discussions

Gao et al. (2014) discusses their simulation and experimental results, so results of the simulation can be compared with the experimental results. The deviation among the current study, experimental and simulation results by Gao's team is in acceptable level. The results provided in Table 3 are also plotted in Figure 2. For the flow with 14.4 m/s velocity the Low-Re Corrections option is enabled.

Table 3 Pressure loss variations with inlet velocity for Gao's experimental and numerical studies and current study

Inlet Velocity (in m/s)	18.10	17.29	16.89	16.30	15.20	14.40
Pressure drop comparison (in Pa)						
Experimental results	4495.3	3711.46	2808.99	2235.87	1861.66	1468.11
Simulation results by Xiang	4299.10	3487.84	2534.73	1913.09	1473.48	1320.43
Simulation results of current study	4239	3382	2645	2144	1529	1510
Deviations in simulation results						

Simulation results by Xiang	4%	6%	10%	14%	21%	10%
Simulation results of current study	6%	9%	6%	4%	18%	-3%

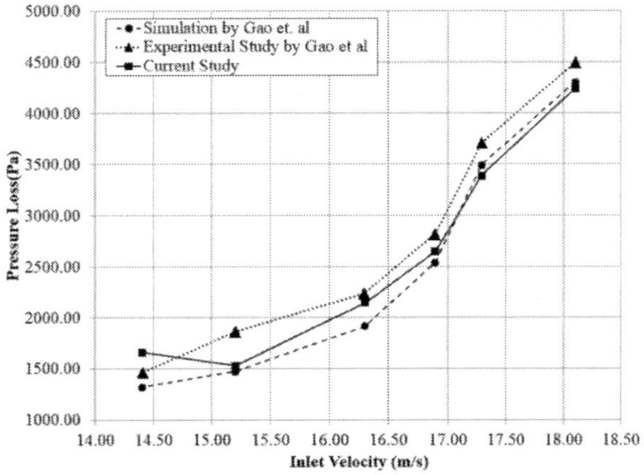

Figure 3. Pressure loss variations with inlet velocity for Gao's experimental and numerical studies and current study

Tangential velocities on the line parallel to the Z axis at the 75 mm offset from the bottom surface illustrated in Figure 3 show us the central reverse flow swirl channel shape. Besides, it is compared with the simulation results by Gao et al. [10] and they are in acceptable range.

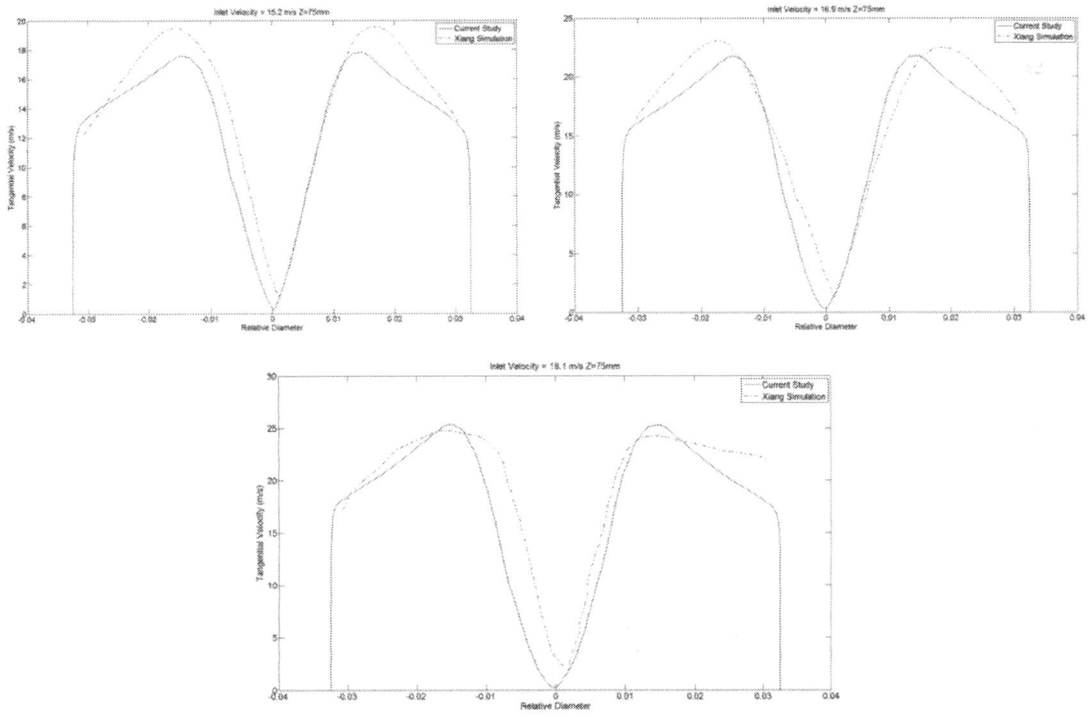

Figure 4. Tangential velocities on the line parallel to the Z axis at the 75 mm offset from y = 0

As a result of high swirl flow velocity the lowest height cyclone has highest pressure lost as shown in Figure 4. Furthermore, for the same volumetric flow rate while inlet diameter is decreasing, flow speed increases therefore it is obvious that the pressure loss increase is inversely proportionally to the inlet diameter.

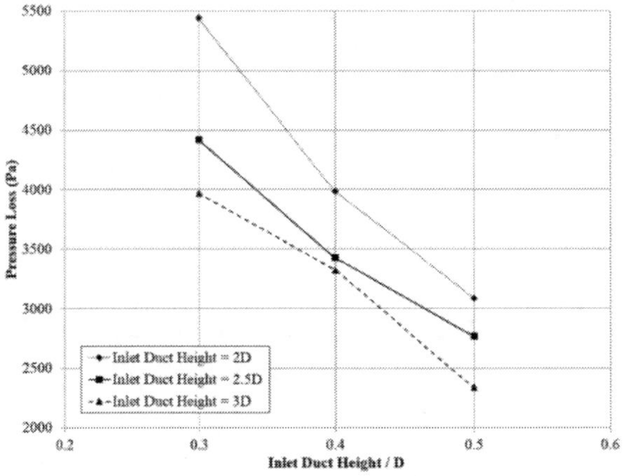

Figure 5. Pressure loss variations with different inlet duct height and swirl volume height

Inlet velocity of the cyclone is known to be effective for the separation performance. The higher inlet velocity causes to increase breakup rate of the particles and the lower inlet velocity slowdowns swirling flow yielding decrease in centrifugal force effect on the particles and decrease in the separation performance. Cyclone height is another effective parameter for the separation; it basically defines the end plane of the swirling flow. The particles flow through the swirling flow and the particles crashes the oil surface at the end of the swirl; therefore, they would be trapped. If the cyclone height is too long, the swirl slows down and more particles would start to discharge from the port. If the cyclone height is too low, there would not be enough time for particles to reach the cyclone surface because particles must flow with swirl. The separation efficiency is calculated from equation 14 below which is a function of total mass of the particles at the inlet and outlet. As a result of breakup and coalescence, the total number of particles may change in flow volume.

$$eff\% = \frac{m_{in}-m_{out}}{m_{in}} \tag{14}$$

Figure 5 shows change of separation efficiency along with inlet duct height. It is noticed that while the cyclone height is increasing from 3D to 2D, separation performance seems to be increasing. On the other hand, for each swirl volume height separation performance shows a decline when inlet duct height is increased. In other words, Figure 5 illustrates that when the inlet diameter decreases, the separation performance shows an increase. Therefore, the separation efficiency increase is inversely proportional with decreasing inlet area.

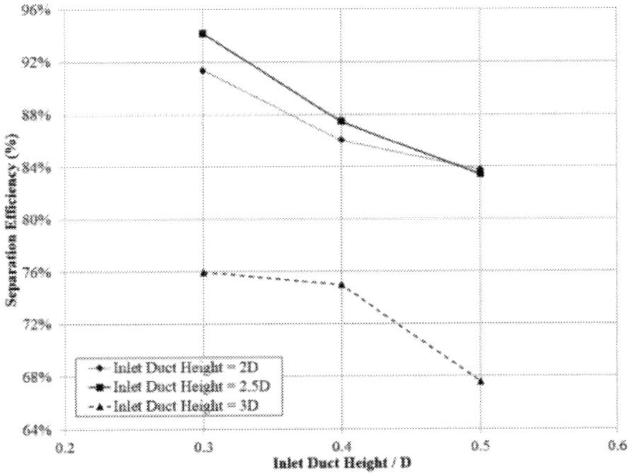

Figure 6. Separation efficiency with different inlet duct height and swirl volume height

4. Conclusions

The optimum point of the separation performance is found to be related with velocities and the flow angles in the cyclone. There are many parameters that can change flow velocity and flow angles. Therefore, prediction of the relation between cyclone geometry and performance is more reliable with performing experiments or simulations.

The height of the cyclone seems to be a significant parameter for separation efficiency and energy consumption. It was noted that while the height is decreasing, the rise of pressure loss is also increased. Another important point is that if the cyclone height is too short, the oil below the tank may be suctioned to the outlet port as a result of low pressure area at the middle of the cyclone. This effect cannot be observed in this study because the oil surface below the tank was assumed to be a surface.

Change in the inlet port diameter seems to be affecting separation efficiency. Specifically, the influence was more for short cyclones than long ones. In short cyclone geometry, the particles are trapped at the bottom surface because swirl changes direction from downward to the upward more sharply in close of bottom trap condition surface. The particles cannot hang on the multiphase flow due to sharp turns and separates. When the bottom surface is closer to the inlet port, the differences of velocities between downward and upward are more from the inlet velocity near the bottom surface, thus increases sharp turn effect.

While the inlet diameter is getting smaller, the separation performance increases the increment rate of the efficiency curve inclination decreasing. Because the high velocity inlet results in more efficient separation, the inlet velocity shows an increase and the breakup possibility of the particles also increases yielding separate particles with 10^{-5} mm diameter.

5. References

[1] J. C. J. F. X. P. Xiang Gao, "Numerical investigation of the effects of the central channel on the flow field in an oil–gas cyclone separator," *Internal Journal of Refrigeration,* no. 36, p. 10, 2013.

[2] Y.-R. C. C.-H. W. R.-M. W. Eldin Wee Chuan Lim, "Experimental and computational studies of multiphase hydrodynamics in a hydrocyclone separator system," *Chemical Engineering Science 65,* 2010.

[3] A.J.Hoekstra, Gas flow field and collection efficiency of cyclone separators, 2000.

[4] J. Gimbun, "Cfd Simulation of Aerocyclone Hydrodynamics and Performance at Extreme Temperature," *Engineering Applications of Computational Fluid Mechanics* , vol. Vol. 2, p. Pp. 22–29, 2008.

[5] D. W. T. N.-M. K. Nageswararao, "Two empirical hydrocyclone models revisited," *Minerals Engineering,* vol. 17, pp. 671-687, 2004.

[6] T. S. Y. C. T. G. C. a. A. F.-R. Jolius Gimbun, "CFD Study on the Prediction of Cyclone Collection Efficiency".

[7] I. K. F.Kaya, "Performance analysis of numerical schemes in highly swirling turbulent flows in cyclones," *Current Science* , vol. 94, no. 10, 2008.

[8] J. D. G. Gronald, "Simulating turbulent swirling flow in a gas cyclone: A comparison of various," *Powder Technology,* vol. 205, pp. 160-171, 2011.

[9] S. R. M. P. B. K. Kozic, "Comparison Of Euler-Euler And Euler-Lagrange Approach In Numerical Simulation Of Multiphase Flow In Ventilation Mill-Air Mixing Duct," *Third Serbian (28th Yu) Congress On Theoretical And Applied Mechanics Vlasina Lake, Serbia, ,* no. B-08, 5-8 July 2011.

[10] R. D. N. T. M. L. Q. E. G. G. M. a. P. V. Utikar, Computational Fluid Dynamics, Hydrodynamic Simulation of Cyclone Separators, H. W. Oh, Ed., InTech, 2010.

[11] "Numerical modelling of droplet break-up for gas atomization," *Computational Materials Science,* vol. 38, p. 282–292, 2006.

[12] Y. Z. X. Y. Y. C. X. P. Xiang Gao, "The Research On The Performance Of Oil-gas Cyclone Separators In Oil Injected Compressor Systems With Considering The Collision And Breakup Of Oil Droplets," *Purdue University Purdue e-Pubs International Compressor Engineering Conference,* no. 2119, p. 10, 2012.

[13] S. H. H. Sooran Noroozi, "CFD Simulation of Inlet Design Effect on Deoiling Hydrocyclone Separation Efficiency," *Chem. Eng. Technol.,* no. No. 12, p. 1885–1893, 2009.

[14] K. Yetilmezsoy, "Determination of Optimum Body Diameter of Air Cyclones Using a New Empirical Model and a Neural Network Approach," *ENVIRONMENTAL ENGINEERING SCIENCE,* vol. 23, no. 4, 2006.

[15] K. N. a. A. L. T.C. RAO, "INFLUENCE OF FEED INLET DIAMETER ON THE HYDROCYCLONE BEHAVIOUR," *International Journal of Mineral Processing,* vol. 3, pp. 357-363, 1976.

[16] J. C. J. F. X. P. Xiang Gao, "umerical and experimental investigations of the effects of the breakup of oil droplets on the performance of oilegas cyclone separators in oil-injected compressor systems," *Computer&Fluids 92,* no. 45-55, p. 10, 2014.

[17] P. B. F. O'Rourke, "Modeling of Drop Interactions in Thick Sprays and a Comparison," *Proceedings of the Institution of Mechanical Engineers,* pp. 101-106, 1980.

Research on the two-phase flow and separation mechanism in the oil–gas cyclone separator

L Z Wang[1], X Gao[1], J M Feng[1, 3] and X Y Peng[1, 2]

[1] School of Energy and Power Engineering, Xi'an Jiaotong University, Xi'an 710049, China
[2] State Key Laboratory of Multiphase Flow in Power Engineering, Xi'an Jiaotong University,Xi'an 710049,China

E-mail: jmfeng@mail.xjtu.edu.cn

Abstract. The cyclone separator has attracted increasing attention due to its small size, rapid construction and high separation efficiency. This study investigated its gas–liquid two-phase flow and separation characteristics experimentally and numerically. A numerical model of two-phase flow in the cyclone separator was proposed using the Euler–Lagrange method. The distribution of pressure, tangential and axial velocity in the gas-phase flow field was obtained, and the oil droplet movement was traced. Separation efficiency was also studied experimentally, and the diameter distributions of oil droplets at the inlet and the outlet of the separator were measured by a Malvern laser particle size analyser to verify the simulation model. Based on high-speed photography technology, the oil film distribution and flow pattern on the wall of the cyclone separator were visualised. The variation of oil–gas two-phase flow in the cyclone separator was compared under various inlet flow rates. Based on the results, an improved structure was proposed, and the performance of the improved separator was investigated experimentally.

1. Introduction

In oil-injected compressor units, the compressed gas should be separated from the oil to ensure circulation of the lubricating oil and to keep the exhaust gas clean. Improving efficiency at the first separation level effectively reduces the filter load of the second separation, thereby extending the filter's service life and reducing costs [1]. Improving the efficiency and the structure of the oil–gas separator was therefore the focus of this study [2].

A gravity sedimentation separator can only be used to separate larger oil droplets, those with a diameter of at least 100 um [3]. Inertia collision separators perform poorly when separating oil droplets with a diameter of less than 25 um, so are not used in more demanding environments [4]. In contrast, the oil–gas cyclone has very good performance prospects because of its rapid construction, small size, high efficiency and small resistance loss. Nonetheless, if designed or operated improperly, it will perform poorly. At present, the design and practicality of the oil–gas cyclone separator mainly depends on experience. Given the current lack of a theoretical basis, the two-phase flow process in the separator was studied experimentally and numerically; the separator's performance under actual operator conditions, and the particle size distribution of import oil droplets, were tested at the same time.

Content from this work may be used under the terms of the Creative Commons Attribution 3.0 licence. Any further distribution of this work must maintain attribution to the author(s) and the title of the work, journal citation and DOI.
Published under licence by IOP Publishing Ltd

2. Numerical simulation study

2.1. Physical model and simulation method

2.1.1. Physical model. The investigation was carried out in the oil–gas cyclone separator shown in Figure 1, and the geometrical details of the separator are presented in Table 1. The oil–gas mixture entered tangentially through the inlet, and rotated through the inner and outer cylinders. Due to the centrifugal force, droplets moving to the wall were separated, and the remaining droplets followed the discharge gas to escape through the outlet.

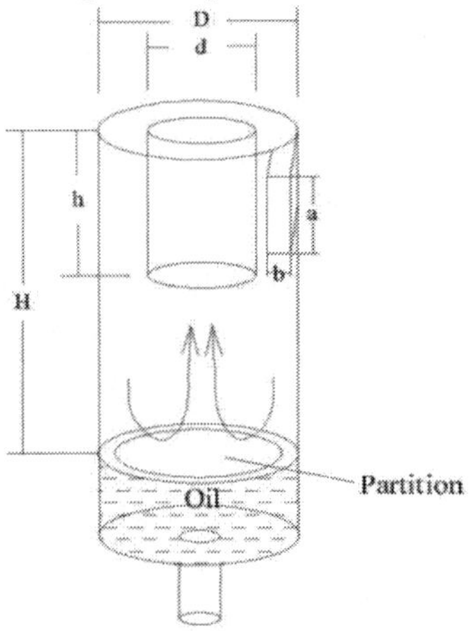

Figure 1. Structure of the oil–gas cyclone separator

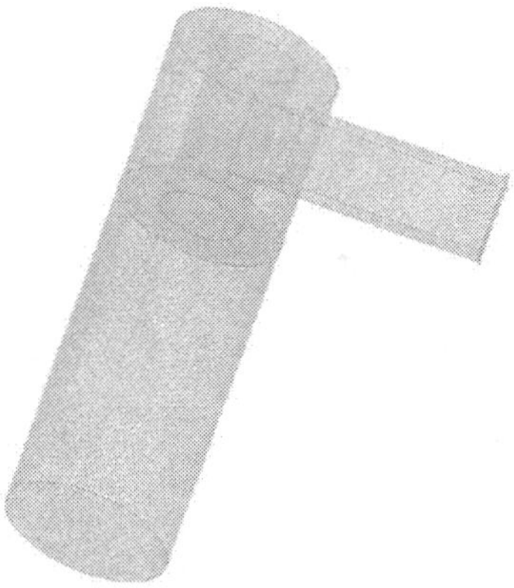

Figure 2. Mesh of the oil–gas cyclone separator

Table 1. The structural parameters

Dimension	Cyclone height (H)	Central channel height (h)	Cyclone diameter (D)	Central channel diameter (d)	Inlet pipe height (a)	Inlet pipe width (b)	Inlet pipe length (L)
Value (mm)	600	200	200	100	100	300	600

2.1.2. Turbulence model. The gas flow field requires a complex three-dimensional turbulent flow in the separator, involving significant anisotropy, and so the seven-equation RSM (Reynolds Stress Model) was used for the gas-phase flow field turbulence model. The Reynolds number of gas flow field is in the range of 250000 to 350000, so it is turbulent flow completely. Turbulence intensity I and hydrodynamic diameter D_H were used to describe turbulence, and they are given by [5]

$$I \cong 0.16\left(\mathrm{Re}_{D_H}\right)^{-1/8} \tag{1}$$

$$D_H = \frac{4A}{P} \tag{2}$$

where A is the flow area/m^2, P is the wetted perimeter/m, Re_{D_H} is the Reynolds number under the hydrodynamic diameter and $\mathrm{Re}_{D_H} = \frac{\rho u D_H}{\mu}$, u is the flow velocity.

For an incompressible fluid, the governing equations for continuity and balance of momentum are given as [6]

$$\frac{\partial \bar{u}_i}{\partial x_i} = 0 \tag{3}$$

$$\frac{\partial \bar{u}_i}{\partial t} + \bar{u}_j \frac{\partial \bar{u}_i}{\partial \bar{x}_j} = -\frac{1}{\rho}\frac{\partial \bar{P}}{\partial x_i} + v\frac{\partial^2 \bar{u}_i}{\partial x_j \partial x_j} - \frac{\partial}{\partial x_j} R_{ij} \tag{4}$$

where $\bar{u}_i$ is the mean velocity, $\bar{x}_i$ is the position, $\bar{P}$ is the mean pressure, ρ is the gas density, v is the gas kinematic velocity and $R_{ij} = \overline{u_i' u_j'}$ is the Reynolds stress tensor. Here, $u_i' = u_i - \bar{u}_i$ is the i th fluctuating velocity component. The RSM turbulence model provides differential transport equations for evaluating the turbulence stress components:

$$\frac{\partial}{\partial t} R_{ij} + \bar{u}_k \frac{\partial}{\partial x_k} R_{ij} = \frac{\partial}{\partial x_k}\left(\frac{v_t}{\sigma^k}\frac{\partial}{\partial x_k} R_{ij}\right) - \left[R_{ik}\frac{\partial \bar{u}_j}{\partial x_k} + R_{jk}\frac{\partial \bar{u}_i}{\partial x_k}\right]$$
$$- C_1 \frac{\varepsilon}{k}\left[R_{ij} - \frac{2}{3}\delta_{ij} K\right] - C_2\left[P_{ij} - \frac{2}{3}\delta_{ij} P\right] - \frac{2}{3}\delta_{ij}\varepsilon \tag{5}$$

The turbulence production terms P_{ij} are defined as

$$P_{ij} = -\left[R_{ik}\frac{\partial \bar{u}_j}{\partial x_k} + R_{jk}\frac{\partial \bar{u}_i}{\partial x_k}\right], \quad P = \frac{1}{2} P_{ij} \tag{6}$$

where P is the fluctuating kinetic energy production. v_t is the turbulence(eddy) kinematic viscosity and $\delta^k = 1$, $C_1 = 1.8$, $C_2 = 0.6$ are empirical constants. The transport equation for the turbulence dissipation rate, ε, is given as

$$\frac{\partial \varepsilon}{\partial t} + \bar{u}_j \frac{\partial \varepsilon}{\partial x_j} = \frac{\partial}{\partial x_j}\left[\left(v + \frac{v_t}{\sigma_\varepsilon}\right)\frac{\partial \varepsilon}{\partial x_j}\right] - C^{\varepsilon 1}\frac{\varepsilon}{K} R_{ij}\frac{\partial \bar{u}_i}{\partial \bar{x}_j} - C^{\varepsilon 2}\frac{\varepsilon^2}{K} \tag{7}$$

where $K = \frac{1}{2}\overline{u_i' u_j'}$ is the fluctuating kinetic energy. The values of the constants are $\sigma^{\varepsilon} = 1.3$, $\sigma^{\varepsilon 1} = 1.44$, $\sigma^{\varepsilon 2} = 1.92$.

For the oil droplets, the DPM (Discrete Phase Model) was used to track and simulate the oil droplets, and the uncoupled mode was chosen for the calculation. Because the volume fraction of the oil droplets was less than 10%, so the effect of the oil–gas interaction was ignored. When running the simulation, the gas-phase flow field was calculated first, and then the discrete phase load to simulate the oil droplets trajectories. The oil droplets diameters could be defined directly and they are in the range of 10um to 100um.

2.1.3. Numerical schemes. The numerical schemes for the gas flow field in this study were drawn from Kaya and Karagoz [7]. They investigated the performance of different discretisation schemes and the suitability of various numerical schemes in highly complex swirling flows in the simulation of cyclone separators. The best settings are summarised in Table 2.

Table 2. Numerical schemes for the current simulations

Numerical setting	Discrete scheme
Pressure discretisation	PRESTO
Momentum discretisation	QUICK
Turbulent kinetic energy	Second-order upwind
Turbulent dissipation rate	Second-order upwind
Reynolds stress	First-order upwind

2.1.4. Boundary condition. For the gas-phase flow field, the constant velocity inlet and pressure outlet boundary were used, and the wall was regarded as a no-slip boundary. For the oil droplets phase, a

IOP Conf. Series: Materials Science and Engineering **90** (2015) 012075 doi:10.1088/1757-899X/90/1/012075

constant inlet velocity at the inlet boundary was also used, and the value was kept the same as that of the gas-phase field. The outlet was defined as 'escape', meaning that once the oil droplets came to the outlet, they were free; the wall was defined as a trap boundary, meaning that once the oil droplets came to the wall they were separated from the inlet mixture. The physical properties of the oil in the simulation were the same as those in the lubricating oil in the experiment: the density was 830 $kg{\cdot}m^{-3}$; the dynamic viscosity was 3.32×10^{-3} $kg{\cdot}m^{-1}{\cdot}s^{-1}$; and the surface tension coefficient was $0.025N{\cdot}m^{-1}$. The Weber number range is 65~650, far more than 1, so the effect of surface tension of the oil droplets can be ignored during calculation.

2.1.5. Grid. The oil gas cyclone separator was divided into the entrance, the upper cylinder, the lower cylinder and the outlet pipe, and then the structured grids were generated and meshed separately, as shown in Figure 2. The tangential velocity, static pressure and total pressure loss were compared when the grid quantities were 101944, 217292 and 397119. The results showed that the difference under the maximum grid quantity and the minimum grid quantity was less than 8%. Considering the complexity of the three-dimensional flow, we chose the grid of 217292 for our calculations.

2.2. Numerical simulation results

The following method was used to simulate the high-efficiency Stairmand cyclone, as the tangential and axial velocity distributions of this system had been measured in a previous study [8]. Figure 3 illustrates the comparison of the simulated and tested tangential and axial velocity distributions. Given the complexity of the turbulence swirling flow in the cyclones, the agreement between the simulation and the experimental data was considered to be acceptable.

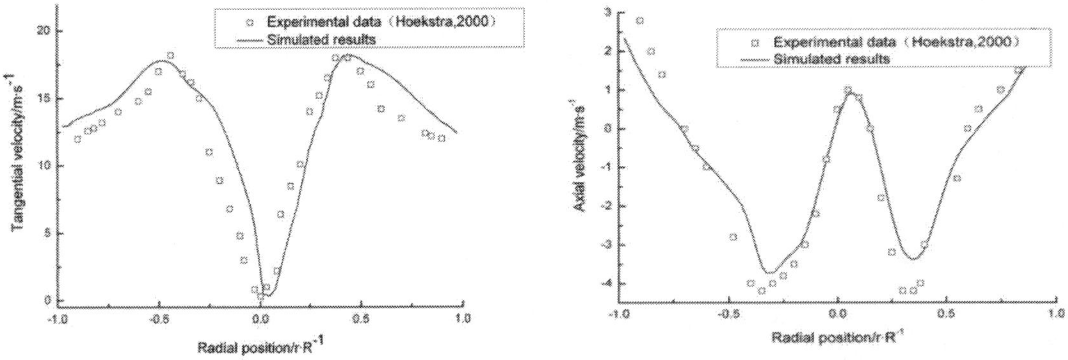

(a) Tangential velocity (b) Axial velocity

Figure 3. Comparison between the simulation results and the experimental results

Figure 4 shows the pressure field distribution on the cross-section of Y=0 at the discharge velocity of $15m{\cdot}s^{-1}$ and discharge pressure of 1atm. It can been seen that both the total pressure field and the static pressure field were symmetrical along the height direction of the separator, and that the symmetry of the static pressure field was better than that of the total pressure. The pressure decreased gradually from the wall to the central axis; both the total pressure and the static pressure were lowest in the central axis, and had their highest values near the wall. The cross-section total rate in the cyclone separator on the cross-section of Y=0 was obtained, and the distribution of tangential velocity and axial velocity are shown in Figure 5 at a discharge pressure of 1atm.

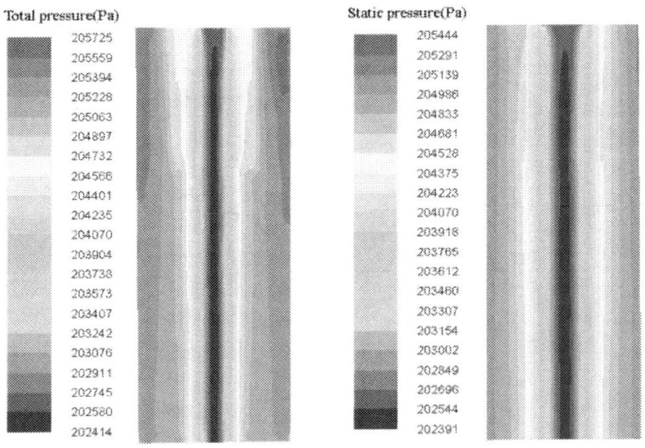

Figure 4. The pressure field distribution (v_{in} =15 m·s^{-1}, Y=0)

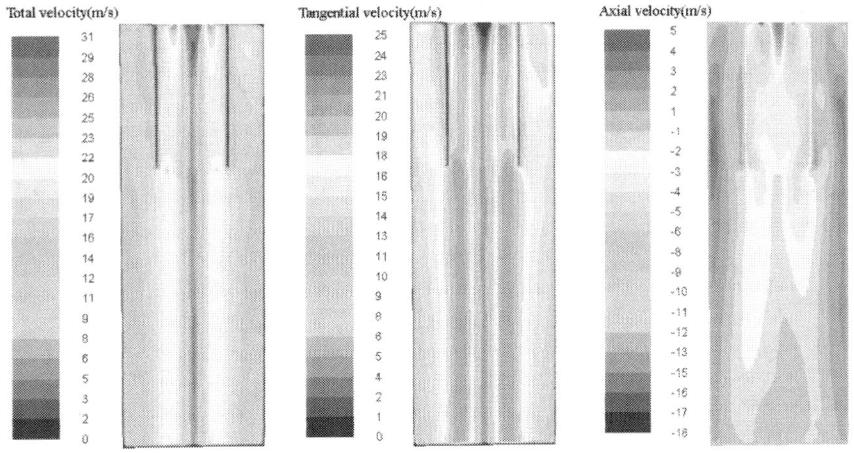

Figure 5. The velocity field distribution (v_{in} =15m·s^{-1}, Y=0mm)

The total velocity and the tangential velocity increased gradually from the wall; after reaching the maximum, they decreased and reached the lowest values at the central axis. Judging from the pressure field distribution in Figure 4, we know that the potential energy of gas continuously converts to kinetic energy. Regarding velocity, the total velocity and the tangential velocity changed very little along the height direction of the separator, and the axial velocity dropped to 2m·s^{-1} at the bottom of the separator; at the entrance of the internal cylinder, the velocity increased significantly, reaching 8m·s^{-1}. These phenomena show that changes in the entrance area produced stronger airflow in the local vortex.

Figure 6 compares the tangential velocity and the axial velocity on cross-section Y=0 and position Z=0.5H mm in the oil–gas cyclone separator at inlet velocities of 14 m·s^{-1}, 16 m·s^{-1} and 18m·s^{-1}, respectively. The tangential velocity distribution was M-shaped, while the axial velocity distribution was W-shaped. After further analysis, increasing the inlet velocity was found not to change the distribution of the velocity field inside the separator, but it significantly increased the tangential velocity, while the axial velocity remained essentially unchanged.

IOP Conf. Series: Materials Science and Engineering **90** (2015) 012075 doi:10.1088/1757-899X/90/1/012075

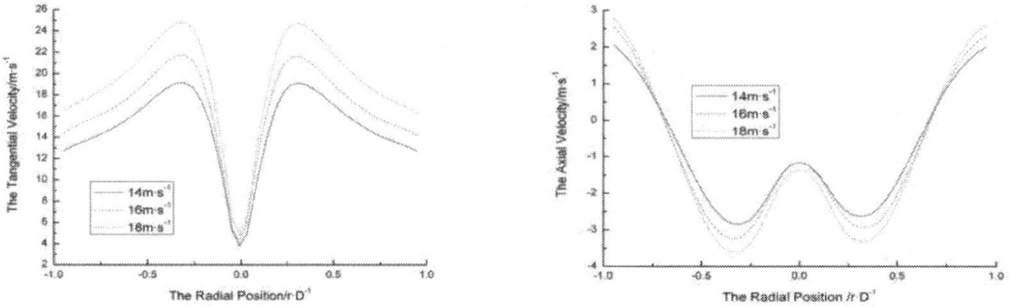

(a) Tangential velocity (b) Axial velocity

Figure 6. The tangential and axial velocities under different inlet velocities (Z=0.5 H mm)

3. Experimental study

3.1. Experimental system

Figure 7 shows the experimental system. The diameter distribution of the oil droplets was measured using a similar method to that introduced in the literature [9]. A four-way observation tube was installed at the inlet and the outlet of the separator, and high-purity nitrogen was injected between the optical glass and the main flow during the test to ensure the purity of the optical glass, thus ensuring that the Malvern laser particle size analyser was functioning normally.

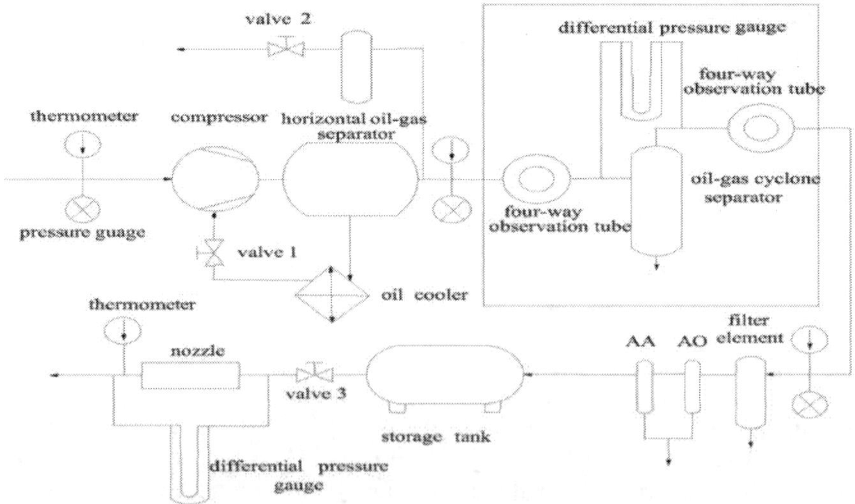

Figure 7. Flow chart of the experimental system

3.2. Experimental results

The experiment focused on the distribution and the flow patterns of the oil film on the inner surface of the oil–gas cyclone separator, and the results are shown in Figure 8, when the inlet velocity was 15m·$s^{-1}$and the discharge pressure was 1 atm. Figure 9 shows the division of the inner surface according to the characteristics of the oil film flows.

Figure 8. Partition of oil film at the inlet

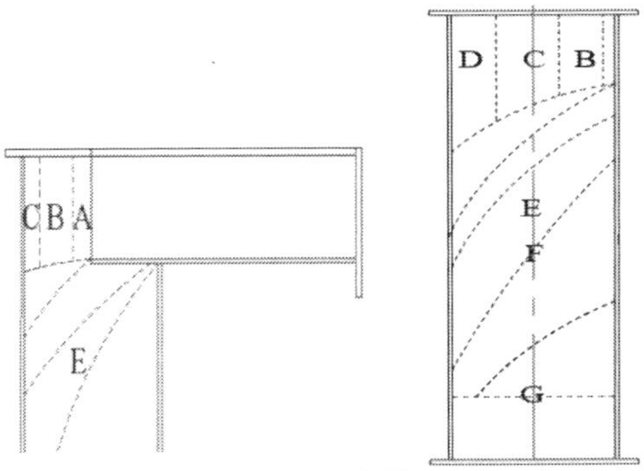

Figure 9. Partition of oil film along the height

Different parts of the oil film flow pattern were observed by high-speed camera, and Figures 10(a)–(g) show the flow pattern and the seven corresponding areas of A–G, respectively, when the inlet velocity was 15 $m \cdot s^{-1}$ and the discharge pressure was 1 atm. In region A, the oil film developed gradually to a spreading planar flow from a small stocks flow, meaning that at the inlet of the separator, the oil film increased because the oil droplets constantly bumped into the wall. In region B, groups of intermittent oil droplets moved in with the airflow, with oil droplets from the incoming flow colliding frequently with the oil film on the wall. The oil film in region C was a corrugated flow, reflecting the formation, development and disappearance of the 'pit' as oil droplets impacted on the oil film. In region D, the oil film developed gradually into a declining stock flow from the corrugated flow, as the oil in the oil film could not completely cover the entire tube wall as the spreading surface of the oil film continued to expand. In region E, we see that the oil film took on a sloped 'scales' flow, with 'scales' that were very thick, and corrugated on surface with small bubbles inside, and obviously thicker than the oil films in other areas. Region F featured a small flow of oil film, where oil droplets collided with a dry wall area without oil film flow, meaning transportation of oil in this area was very poor, with no obvious flow channel. In region G, the flow pattern of oil film at the interface of the oil and gas was extremely complex, with strong disturbance airflow, making the interface very unstable, and the lubricating oil which had been separated contained a large amount of bubbles. On the inner wall of the separator, the lubricating oil contained oil that flowed to the bottom along the surface of cylinder body, and oil that sprayed up under the effect of airflow.

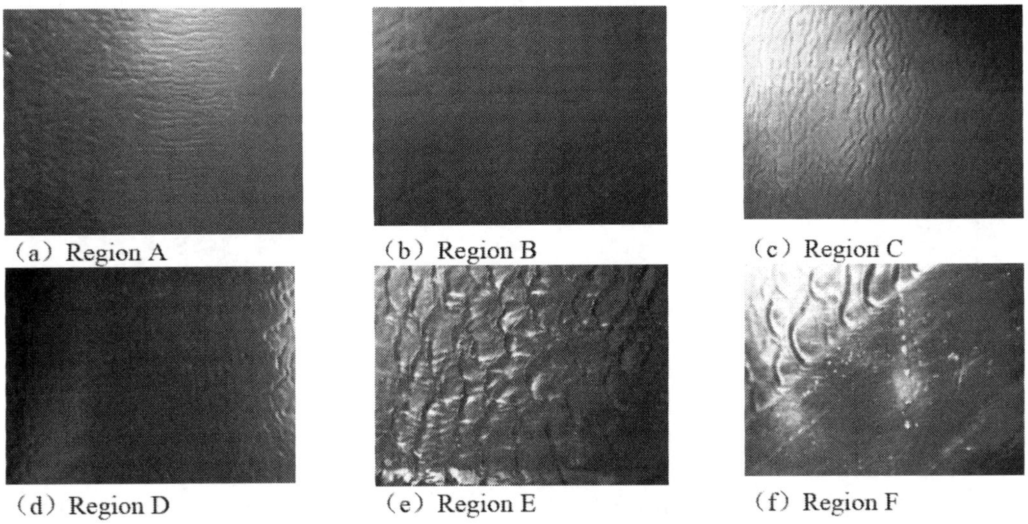

(a) Region A (b) Region B (c) Region C
(d) Region D (e) Region E (f) Region F

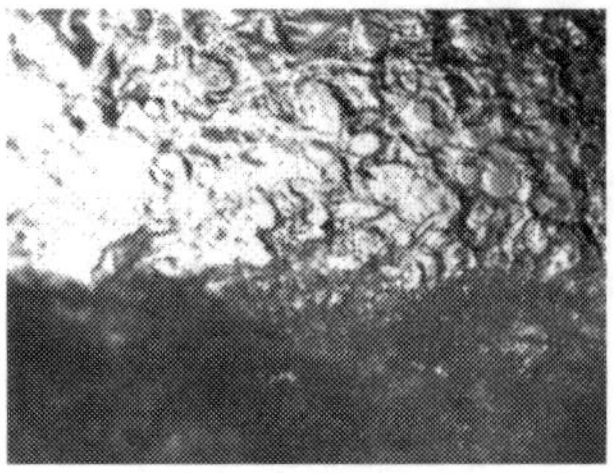

(g) Region G

Figure 10. The oil film flow patterns of regions A–G

Figure 11 shows the separation efficiency of 70–80% when the inlet velocity was 14.4-18.1m·s^{-1} and the discharge pressure was 0.2 MPa. When the inlet velocity was less than 17m·s^{-1}, the separation efficiency increased with the increase of the inlet flow velocity, and the separation efficiency decreased with the increase of the inlet flow velocity if the inlet velocity was greater than 17m·s^{-1}. For the simulation results, the separation efficiency at 17 m·s^{-1} was also slightly higher than that at other velocities. This suggests that there is a critical maximum value for separation efficiency with the change of the inlet velocity, which determines the relationship between the inlet velocity and the separation efficiency. This critical value is the key in the design of the separation, and for the separator studied in this paper, when the inlet velocity is 17 m·s^{-1}. In addition, the measured separation efficiency was lower than the simulated results. This can be explained by the fact that in the numerical simulation the oil level on the bottom was treated as the wall boundary and the trap boundary condition was applied, so that the oil droplets were assumed to be separated if they reached this surface. Due to the effect of the high-speed swirling flow, the oil separated on the baffle might be blown back into the flow field and finally escape through the outlet. Moreover, the effect of the 'dust ring' [10] was ignored in the simulation.

When comparing the droplets' distribution between the inlet and the outlet, Figure 12 shows that the number of droplets with a diameter greater than 45 um reduced significantly. The increase in the percentage of oil droplets smaller than 20um at the outlet implied that droplet breakup happened during the separation process. It was also evident that the percentage of smaller-diameter oil droplets increased at the outlet as the inlet velocity increased. This meant that a higher velocity inside the separator led to easier droplets breakup. The percentage of 10 um droplets increased in volume from 3% at the inlet to 7.5% at the outlet when the inlet velocity was 15.2 m·s^{-1}. If the inlet velocity reached 18.1 m·s^{-1}, the percentage of 10 um-diameter droplets would increase to 10% [11]. We conclude that when the inlet velocity reaches a certain critical value, more small oil droplets will be produced in the separator, thereby decreasing separation efficiency. This conclusion is consistent with the 'the critical velocity' which is analysed in Figure 11.

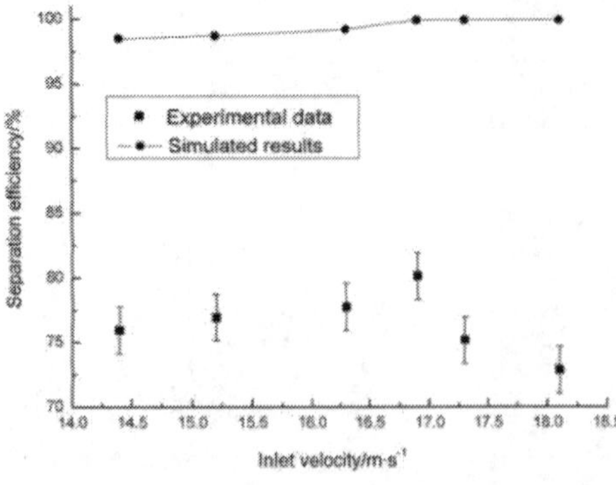

Figure 11. Separation efficiencies under different inlet velocities

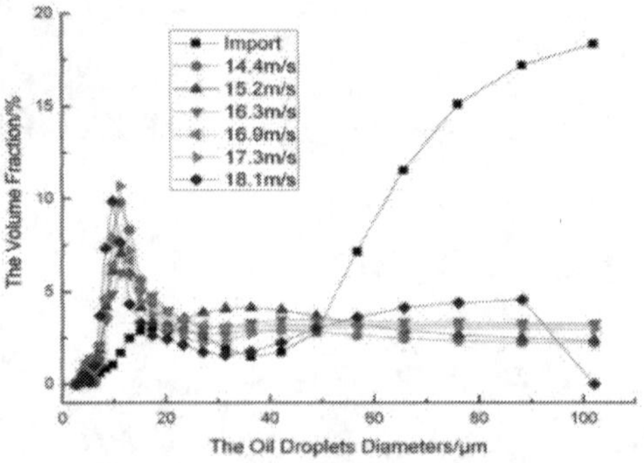

Figure 12. Oil droplet distribution at the inlet and outlet

4. Improvement of the separator

According to the above findings, two problems exist in conventional oil–gas cyclone separators. First, spraying the separated oil at the bottom of the separator into the gas-phase flow field unfavourably affects separation, and, with the increase of the inlet velocity, the amount of the oil going into the gas-phase flow field will increase significantly. Second, the numerical simulation and experimental results both prove that with an increase of the oil film thickness, more small oil droplets will splash from the oil film because of the impact of the gas flow, thereby decreasing separation efficiency.

Figure 13 shows a proposed solution, with a porous cylinder added to the separator with a diameter 4mm less than that of the original cylinder, providing a 5 mm sandwich between the external separator and the porous cylinder. This layer prevents the oil droplets from returning to the mainstream through the round hole into the interlayer; because it is difficult for an oil film to form on the porous surface, it can avoid the oil droplets adhering to the oil film surface, and reduce the possibility of splash. We can see from Figure 13 that the oil film turned to femoral and flowed downward vertically; this means that there is no gas flow in the interlayer space, so the lubricating oil flows naturally under the action of gravity. Looking at the comparison of static pressure loss shown in Figures 14, we can see that there is less pressure loss in the improved structure. So we add a porous cylinder to a real oil–gas cyclone separator and applied it to an oil- injected compressor system for further study. Restricted to the processing conditions, the distance between porous cylinder and external separator is 10 mm, The improved separator is shown in Figure 15 and the geometrical details are presented in Table 3. The Figure 16 shows the comparison results of the separation efficiency, from which can we see that the improved separator is more efficient.

Figure 13. The distribution of the oil film in the improved separator

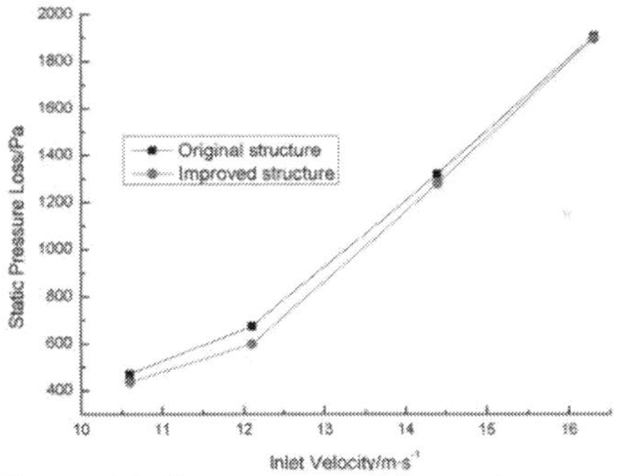

Figure 14. Comparison of the static pressure loss of the original and improved structures

Figure 15. The improved oil-gas cyclone separator

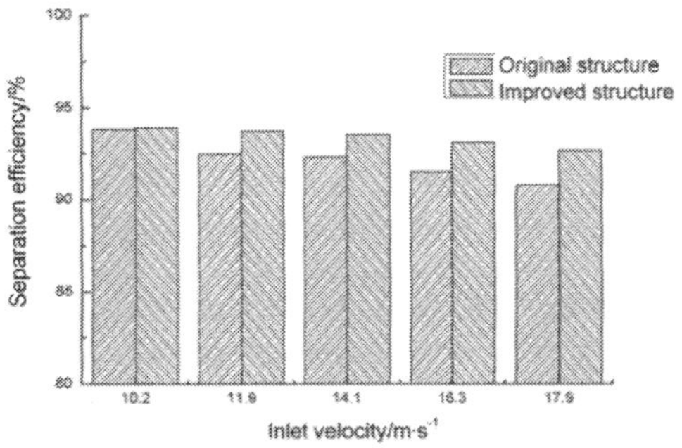

Figure 16. Comparison of the separation efficiency of the original and improved structures

Table 3. The structural parameters

Dimension	Cyclone height (H)	Central channel height (h)	Cyclone diameter (D)	Central channel diameter (d)	Inlet pipe diameter (a)	Inlet pipe length (L)
Value (mm)	550	240	240	120	25	100

5. Conclusion

(1) The two-phase flow simulation model in the oil–gas cyclone separator was built using the Euler–Lagrange method. The distribution of the gas flow field and oil droplets trajectories were obtained, and the simulation results were verified by experiments.
(2) The distribution of the oil film on the outside wall of the separator was visualised using high-speed photography technology. According to the oil film distribution and the flow pattern, the wall was divided into seven different regions, A–G, and each of them was described in detail. The fog type annular two-phase flow at the separator inlet was identified as the root cause of the distribution of the oil film and flow pattern.
(3) The results showed that static pressure loss increased with the increase of inlet velocity, while separation efficiency initially increased, and then decreased, with the increase of inlet velocity. There exists a maximum value of separation efficiency affected by inlet velocity; and avoiding the generation of small oil droplets is an effective way to improve separation efficiency.
(4) An improved structure was proposed, which took into account the two main factors that affect the separation efficiency. The experimental results showed that the new structure is feasible and efficient.

References

[1] Hammerl K, Tinder L and Frank M 2000. Modification in the design of the oil injected system for crew compressor. *Purdue Compressor Technology Conference.* 987-994.
[2] Wiencke B 2011. Fundamental principles for sizing and design of gravity separators for industrial refrigeration. *International Journal of Refrigeration.* **34** 2092-2108.
[3] Liu C S 1994. Development and field test of separator in prying loading type gas collecting device. *Natural Gas and Petroleum.* **12** 36-40.
[4] Wang S Z 2007. Applied exploration of gas-liquid separator and high efficiency separator. *Small Nitrogenous Fertilizer.* **35** 25.
[5] Jing S R and Zhang M Y 2001. *Fluid Mechanics* (Xi'an: Xi'an Jiaotong University Press) p189.
[6] Elsayed K and Lacor C 2011. Numerical modeling of the flow field and performance in cyclones of different cone-tip diameters. *Computers and Fluids.* **51** 48-59.
[7] Kaya F and Karagoz I 2008. Performance analysis of numerical schemes in highly swirling turbulence flow in cyclone. *Current Science.* **94** 1273-1278.
[8] Hoekstra A J 2000. *Gas flow field and collection efficiency of cyclone separators.* (PhD thesis) (Technical University Delft, Stevinweg, Netherlands).
[9] Feng J M, Chang Y F and Peng X Y 2008. Investigation of the oil–gas separation in a horizontal separator for oil-injected compressor units. *Journal of Power and Energy.* **222** 403-411.
[10] Wang B, Xu D L, Chu K W and Yu A B 2006. Numerical study of gas–solid flow in a cyclone separator. *Applied MathematicalModelling.* **30** 1326-1342.
[11] Gao X, Chen J F, Feng J M and Peng X Y 2013. Numerical and experimental investigations of the effects of the breakup of oil droplets on the performance of oil–gas cyclone separators in oil-injected compressor systems. *Refrigeration.* **36** 1894-1904

IOP Conf. Series: Materials Science and Engineering **90** (2015) 012076 doi:10.1088/1757-899X/90/1/012076

Experimental performance evaluation of heat pump-based steam supply system

T Kaida, I Sakuraba, K Hashimoto, H Hasegawa

Central Research Institute of Electric Power Industry (CRIEPI), Japan

E-mail: kaida@criepi.denken.or.jp

Abstract. Heat pumps have become increasingly important as a technology to reduce primary energy consumption and greenhouse effect gas emission. They are presently used mainly on residential air-conditioning and domestic hot water and are expected to spread to industrial heating processes. In 2011, Kobe Steel, Ltd. developed and commercialized two heat pump-based steam supply systems; the high efficiency steam supply system with a steam temperature of 120°C (SGH120) and the system which enables a steam temperature of 165°C (SGH165). For promoting the spread of these industrial heat pumps and enhancing the reliability of them, we investigate experimentally steam generation rate, energy efficiency and controlled performance of the SGH165 under various operating conditions on the assumption of actual different industrial processes, and evaluate technical possibilities for better performance.

1. Introduction

Heat pumps contribute to primary energy saving, to greenhouse effect gas abating and to increasing renewable energy use, so they are expected to become more popular and common. In Japanese residential and commercial sector, heat pumps have already penetrated in the form of air-conditioners. In addition, a heat pump water heater (Eco Cute), using carbon dioxide as a refrigerant, was developed and commercialized in 2001 (see [1]), and its cumulative shipments exceeded four million by the end of 2013.

In contrast, in the industrial sector, heat pumps for the heating process are uncommon, for it is difficult to achieve higher output temperature with high efficiency, low initial cost and high reliability. In recent years, Japanese manufactures overcome the difficulty of commercializing heat pumps for industrial use, and installation examples have been reported. Among them, the heat pump system capable of supplying steam with temperatures at 120°C and above is only the Steam Grow Heat pump (SGH), which is commercialized in 2011 (see [2]), by co-development of Kobe Steel, Ltd. and electric companies in Japan. This includes the following characteristics; efficient use of unused thermal energy by waste heat recovery and upgrading, reduction of dissipation heat loss by set close to processes in distributed arrangement. It is hoped that this system will contribute to better energy saving on industrial heat processes, for example distilling, condensation, disinfect and drying, compared to existing inefficient steam supply systems which are usually arranged in the center boiler room and send to each building with dissipating heat from steam piping lines.

In order to promote the spread of industrial heat pumps, it is important to enhance the reliability of them not to mention the high efficiency and the low initial cost. In addition, the performance data

Content from this work may be used under the terms of the Creative Commons Attribution 3.0 licence. Any further distribution of this work must maintain attribution to the author(s) and the title of the work, journal citation and DOI.
Published under licence by IOP Publishing Ltd

under various operating conditions are required by industrial customers. Moreover, performance evaluations are also important for continuous improvement.

The authors investigate experimentally the energy efficiency and controlled performance of the SGH, and provide the highly reliable evaluation results adapted to the actual state of use. In addition we evaluate technical possibilities for better performance.

2. Outline of Steam Glow Heat Pumps (SGH120 and SGH165)

Figure 1 shows the external appearances of the SGH120 and the SGH165.

The SGH120 is composed of a heat pump unit and a flash tank. The heat pump unit lifts the heat from the heat source water (35-70°C) corresponding to the warm effluent and sends the heat to the pressurized circulating water. In the flash tank, the pressurized water is decompressed and evaporated. Consequently, the flash steam (up to 120°C, 0.1MPaG) is supplied to each process, and the remaining saturated water is back to the heat pump unit.

The SGH165, which is the objective of this study, has a steam compressor in addition to the above system. The steam compressor compresses the saturated steam generated in the flash tank. For preventing the superheat of the discharge steam, water is injected to the compressor. The discharge steam (up to 175°C, 0.8MPaG) is separated from the mist at the drain separator and supplied to each process.

The refrigerant is R245fa for SGH120, and a mixture of R134a and R245fa for SGH165 to achieve a good performance. Because of operating under higher pressure ratio and higher temperature compared to existing heat pumps, a newly-developed screw compressor is equipped and refrigerant mist is sprayed into a motor for cooling down.

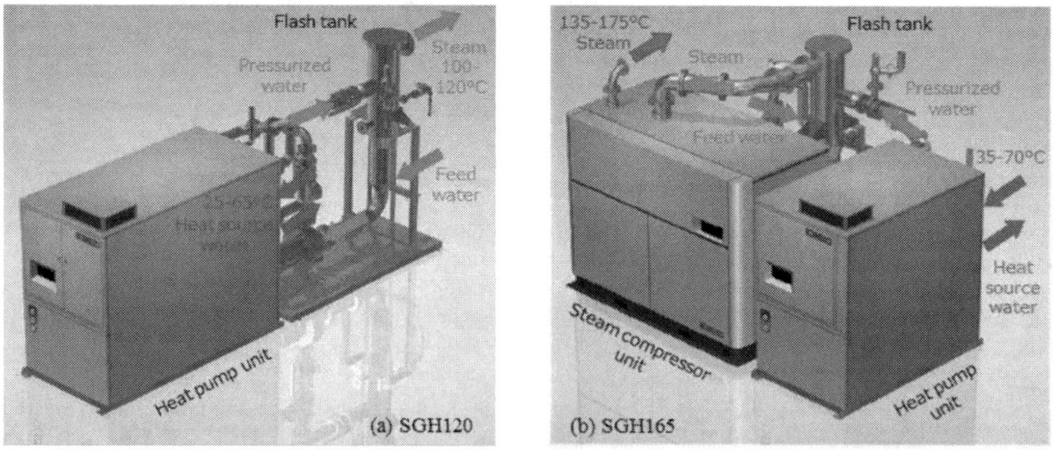

Figure 1. Steam Grow Heat pumps

3. Experimental apparatus and method

3.1. Experimental apparatus

The experimental apparatus, which was previously constructed (see [3]), is shown in figure 2. We can evaluate steam supply heat pump having a rated power output of less than 600kW (up to 200°C and 1 ton/h). The apparatus is composed of four facilities.

The first "brine cooling facility" includes air-cooled brine chillers, a brine tank, brine pumps, three plate-type heat exchangers. The facility is the cold source of the apparatus and supplies brine at 0°C to the following three facilities.

The second "heat source water facility" includes a water tank, a brine cooler, electric heaters, water pumps, and flow-control valves. The heat source water temperature is adjusted by the brine cooler and electric heaters at range of 10-90°C, and the flow rate is adjusted by the valves at range of 2-400m^3/h.

The third "cooling water facility" includes a water tank, a brine cooler, electric heaters, water pumps, flow-control valves and a cooling tower. The temperature and flow rate control methods are the same as for the heat source water facility.

The last "steam cooling facility" includes an pressure-control valve, an air cooler, a water cooler (condenser), two water tank, a brine cooler, an electric heater, a water pump, flow-control valves. The facility is a closed system: the steam generated from the machine to be tested is cooled and condensed, and the water is returned to the machine. The temperature of the feed water is adjusted by the brine cooler and the electric heater at range of 10-95°C. The discharge steam pressure is controlled by the regulation valve, and the pressure and flow rate are adjusted at the range of 0-1.0MPaG and 2-1000kg/h, respectively.

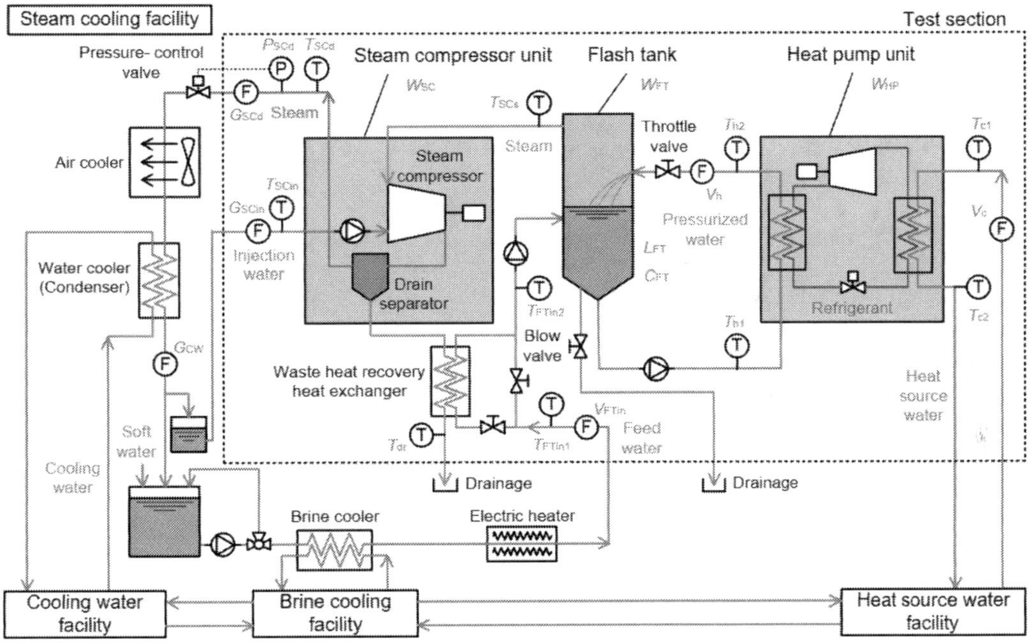

Figure 2. Schematic diagram of experimental apparatus and test section

3.2. Measure points and data acquisition systems

SGH165 needs heat source water, feed water, injection water and electric power, and can supply steam with drainage. As shown in figure 2, we measured each of the inlet-outlet temperatures, the flow rates, the discharge steam pressure, and the consumption powers.

Measuring the temperatures, four-line type high-precision resistance thermometers was used. The sensor was connected to a high-precision temperature logger and has high-accuracy of ±0.005K. The flow rates of the heat source water, pressurized water and feed water were measured by electromagnetic flowmeters with ±0.5% accuracy of reading. The flow rates of injection water and discharge steam were measured by a Coriolis flowmeter with ±0.24% accuracy of reading and a vortex flowmeter with ±1% accuracy of reading, respectively. As well, condensed water was measured by a Coriolis flowmeter with ±0.1% accuracy of reading in order to check validation each other. The electric power consumptions were measured by an electric energy meter with accuracy ±0.06%.

All sensor measurements were collected using a dedicated PC via convenient data acquisition software.

3.3. Performance evaluation indexes

The coefficient of performances (COP) of the heat pump and the system are evaluated using (1) and (2), respectively.

$$COP_{\mathrm{HP}} = \frac{Q_{\mathrm{HP\,h}}}{W_{\mathrm{HP}}} = \frac{\rho_{\mathrm{h2}} c_{p\,\mathrm{h2}} V_{\mathrm{h2}} \left(T_{\mathrm{h2}} - T_{\mathrm{h1}}\right)}{W_{\mathrm{HP}}} \tag{1}$$

$$COP_{\mathrm{Sys}} = \frac{Q_{\mathrm{Sys}}}{W_{\mathrm{HP}} + W_{\mathrm{SC}}} = \frac{G_{\mathrm{SCd}} \left(h_{\mathrm{SCd}} - h_{\mathrm{FTin}}\right)}{W_{\mathrm{HP}} + W_{\mathrm{SC}}} \tag{2}$$

Note that the feed water pump and the circulation pump input powers are considered out of the system. The uncertainties of the heat pump COP and the system COP are up to ±1.0% and ±1.5%, respectively.

4. Experimental results and discussion

4.1. Operating characteristics

Figure 3 shows the operating characteristics of the SGH165 as time-series variations. The heat pump unit starts (see W_{HP}) and heats up the circulating water (see T_{h2}). T_{h2} reaches 100°C and then, the flash steam is generated and supplied to the steam compressor unit (see T_{SCs}). After the suction pressure of the steam compressor increases, the compressor runs to make the pressure constant (see W_{SC}) and generates the steam (see P_{SCd}, T_{SCd} and G_{SCd}). For keeping the flash tank water level constant, the feed water is supplied to the tank (see L_{FT} and G_{FTin}). The water level is measured by the equipped sensor of differential pressure type in the flash tank. In this test condition, the heat exchanger for the waste heat recovery from the steam compressor drainage is used, so the flash tank inlet temperature is increased (see T_{FTin1} and T_{FTin2}). When we assume that the temperature of the drainage into the heat exchanger is the saturated temperature at the steam discharge pressure, the drainage mass flow rate is calculated by the heat balance of the heat exchanger and illustrated (see G_{dr}). About half of the injection water is drained outside the system.

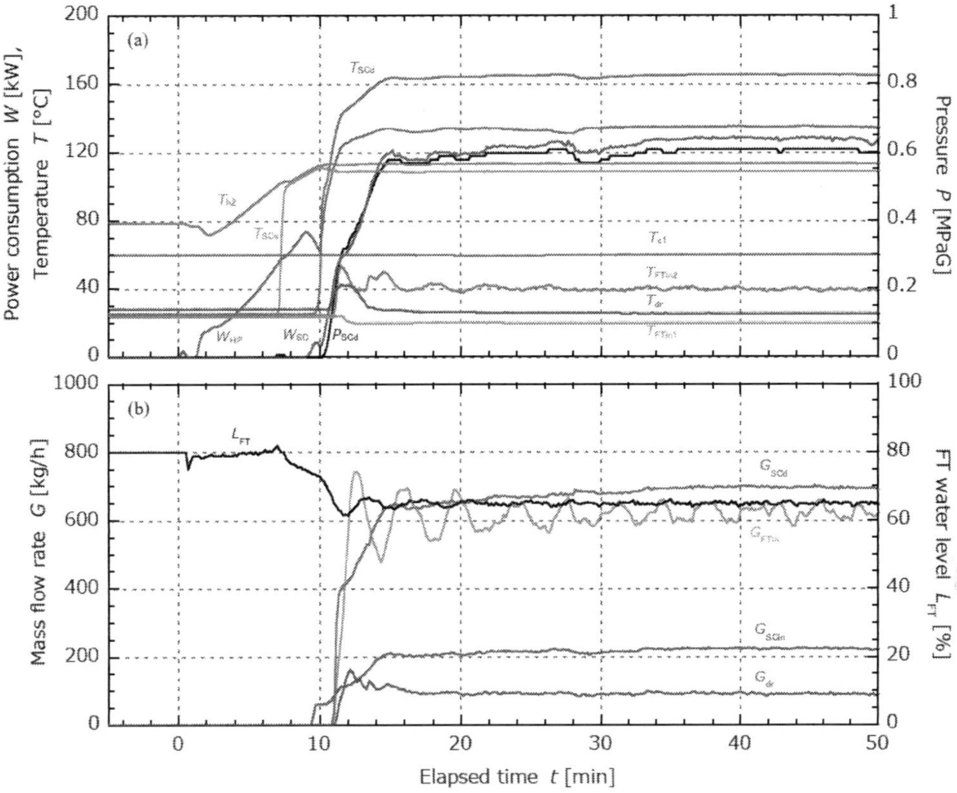

Figure 3. Operating characteristics of the SGH165

4.2. *Steady-state characteristics*

On the assumption of actual different industrial process temperatures, the steady-state data were obtained under various heat source temperatures, feed water temperatures and steam discharge pressures. This steady-state means meeting the following conditions;

- The system mass balance (MB) and the system heat balance (HB) are kept at or near 1.0. Each balance is calculated by using (3) and (4), respectively, where G_{loss} is the loss from the shaft seal part of the steam compressor and the value is about 30kg/h obtained by a preliminary test.

$$MB_{Sys} = \frac{G_{SCd} + G_{dr} + G_{loss}}{G_{FTin} + G_{SCin}} \tag{3}$$

$$HB_{Sys} = \frac{Q_{Sys}}{Q_{HP\,c} + W_{HP} + W_{FT} + W_{SC}} \tag{4}$$

- The heat source water inlet temperature and the feed water temperature are kept constant.
- The discharge steam pressure and flow rate are kept constant.

As shown in figure 4, the steam discharge mass flow rate and the system COP increase with the heat source inlet temperature under the condition that the steam discharge pressure and the feed water are constant. Especially, the influence of heat source temperature on the steam discharge rate is marked. Figure 5 shows the results related to steam discharge pressures. The steam discharge rate increases with the pressure, but the system COP decreases. Figure 6 shows the results related to feed water temperatures. The steam discharge rate increases with the temperature, but the system COP slightly decreases. This figure also illustrates the case without the waste heat recovery heat exchanger

(HRHX). The steam discharge rate slightly increases in the case of having the exchanger, especially when the feed water temperature is low, but the change of the system COP is negligible little.

We can conclude the following. It is most important to secure the higher heat source temperature, which depends on the warm effluent temperature. The waste heat recovery heat exchanger is valid for slightly increasing of the steam discharge rate when the feed water temperature is low, but there is little change in the system COP.

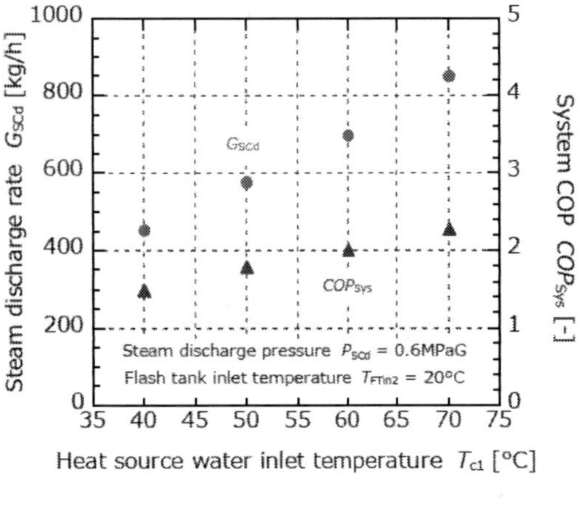

Figure 4. Influence of heat source temperature on steam discharge rate and system COP

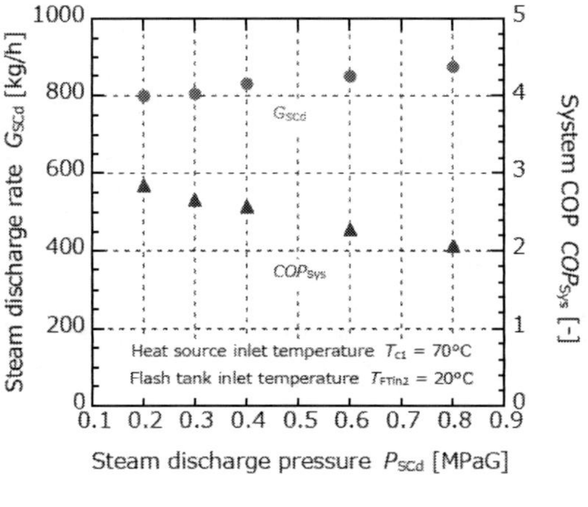

Figure 5. Influence of steam discharge pressure on steam discharge rate and system COP

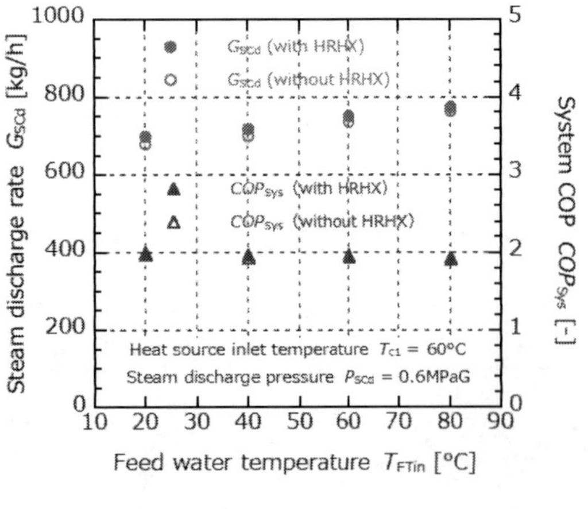

Figure 6. Influence of feed water temperature and effect of waste heat recovery

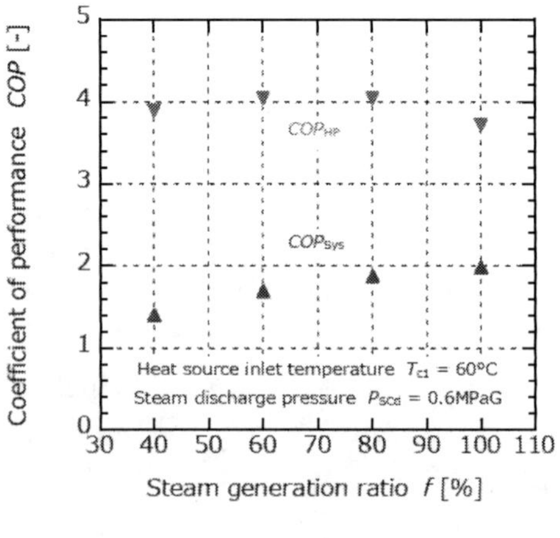

Figure 7. Part-load performance

4.3. Part-load performance

Figure 7 shows the part-load performance. The data were obtained by narrowing the steam pressure-control valve. The minimum of the steam generation rate was restricted by the refrigerant compressor rotating speed of 1500rpm. The system COP decreases with decreasing the steam generation rate. On the other hand, the heat pump COP has maximum at the range of the steam generation rate of 60-80%.

We can conclude that the part-load performance has a large influence on the system performance, so it is recommended to operate the system at the rated power.

4.4. Influence of condensed water blow

As generating the steam, the water in the flash tank becomes condensed and the electric conductivity becomes more. For continuous operation, it is required to blow the condensed water.

Figure 8 shows the time-series change. In this condition, the bowing is operated with a period of 24 minutes and the blowing time is 6 minutes. During the blowing, the net steam discharge rate is decreased by about 5%. In contrast, as shown in figure 9, the change of the system COP is negligible little, so the symbol of the COP (without blowing) is covered by that of the COP (with blowing).

We can conclude that the condensed water blow has a small influence on the system performance.

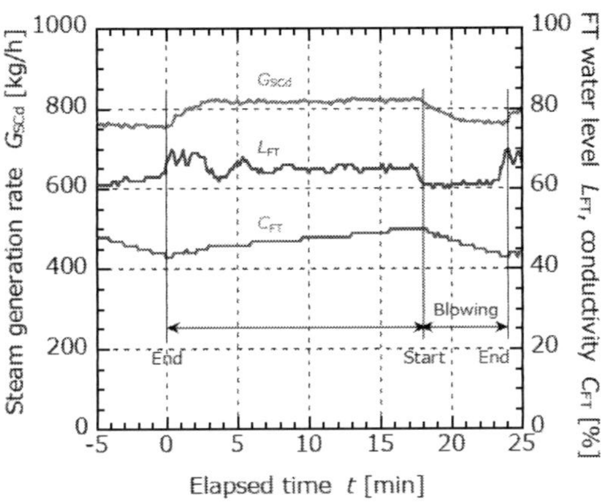

Figure 8. Time-series of condensed water blow

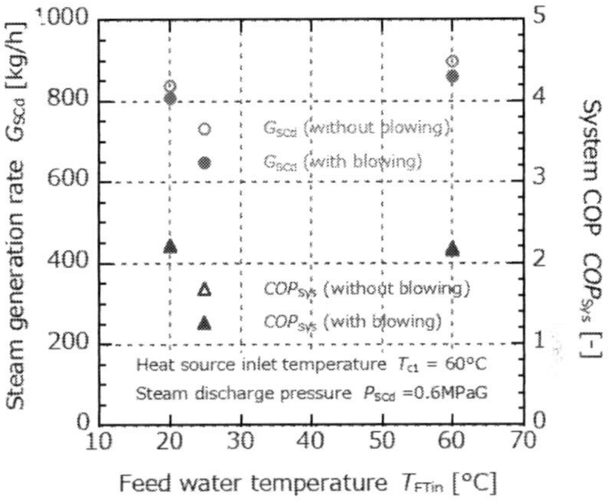

Figure 9. Influence of condensed water blow

4.5. Load-following capability

In actual plants, steam demand fluctuates, so the steam supply system is required to follow the fluctuation.

Figure 10 shows the load-following capability when the steam supply rapidly decreased by 20%. As soon as the steam discharge rate decreases, the steam compressor instantly loads down (see W_{SC}) and makes it possible to remain the discharge pressure constant (see P_{SCd}). On the other hand, the heat pump loads up (see W_{HP}) and the pressurized water temperature increases (see T_{h2}). When T_{h2} reaches a temperature, the heat pump loads down, and the steam compressor loads up. The result is that the steam pressure and flow rate are kept constant while the heat pump and steam compressor loads slowly fluctuate with opposite-phase.

Figure 11 shows the load-following capability when the steam supply rapidly decreased by 40%. As soon as the steam discharge rate decreases, the steam compressor instantly loads down (see W_{SC}), but the steam discharge pressure goes disturbed in some degree (see P_{SCd}).

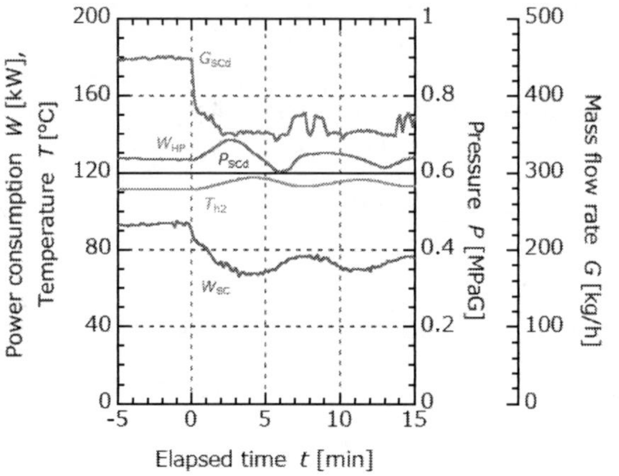

Figure 10. Load following capability (steam supply rapidly decreased by 20%)

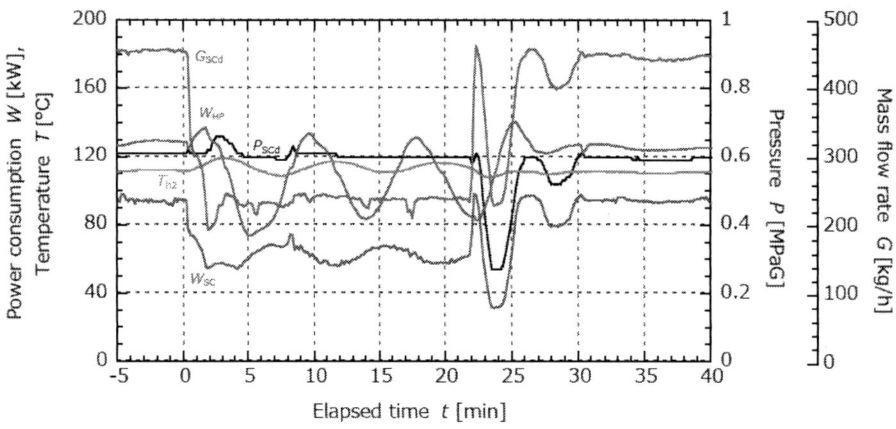

Figure 11. Load following capability (steam supply rapidly decreased by 40%)

4.6. Discussion for better performance of heat pump-based steam supply system

The current state of the system COP is insufficient for use in wide range of heat source water temperatures. As shown in section 4.2, the system COP is influenced by heat source water temperature more than by feed water temperature. As we estimate effects of exchanging the heat from the steam compressor drainage to the heat source water not to the feed water, the heat source water inlet temperature will increase by about 0.2K, so there will be little hope of the system COP enhancement. For enhancement, it is necessary to drastically enhance the heat pump COP. It is desirable to develop the refrigerant suited to high temperature output as an alternative to R245fa.

Regarding load-following capability, the responsivity of the steam compressor is so good, and the system load-following capability will be expected to be better by tuning up the PID controller. For enhancing the reliability, it is necessary to confirm the load-following capabilities related to heat source temperatures and flow rates. This will be the subject of future investigations.

5. Conclusion

For promoting the spread of these industrial heat pumps and enhancing the reliability of them, we investigate experimentally steam generation rate, energy efficiency and controlled performance of the

SGH165 under various operating conditions on the assumption of actual different industrial processes, and then we can conclude the following.

For high efficiency, it is most important to secure the heat source temperature as highly as possible, although the temperature depends on the warm effluent temperature and has a limit. The waste heat recovery heat exchanger is valid for slightly increasing of the steam discharge rate when the feed water temperature is low, but there is little change in the system COP. The condensed water blow has a small influence on the system performance. The SGH165 can uneventfully follow the load down when the steam supply rapidly decreased by 20%.

We will examine the possibility of enhancing the heat pump COP and will confirm the load-following capabilities related to heat source temperatures and flow rates.

References

[1] Saikawa M, Hashimoto K, Kobayakawa T, Kusakari K, Ito M and Sakakibara H 2000 Development of prototype of CO_2 heat pump water heater for residential use *Proc. of the 4th IIR Gustav-Lorentzen Conference on Natural Working Fluids (West Lafayette, USA, 25-28 July 2000)* pp 51-57

[2] Iizuka K and Maeda M 2011 Development of high efficiency steam supply systems "Steam Grow Heat pump" *Mechanical Engineering Congress (Tokyo, Japan, 11-14 September 2011)* W091008 (in Japanese)

[3] Hashimoto K, Kaida T, Hasegawa H, Sakuraba I, Fujinawa T and Watanabe C 2014 Development of large capacity heat pump performance evaluating apparatus *Proc. of the 11th IEA Heat Pump Conference (Montreal, Canada, 12-16 May 2014)* Poster P.4.7

IOP Conf. Series: Materials Science and Engineering **90** (2015) 012077 doi:10.1088/1757-899X/90/1/012077

Reciprocating and Screw Compressor semi-empirical models for establishing minimum energy performance standards

Hassan Javed, Peter Armstrong
Masdar Institute of Science and Technology, Abu Dhabi, UAE

E-mail: hjaved@masdar.ac.ae ; parmstrong@masdar.ac.ae

Abstract. The efficiency bar for a Minimum Equipment Performance Standard (MEPS) generally aims to minimize energy consumption and life cycle cost of a given chiller type and size category serving a typical load profile. Compressor type has a significant chiller performance impact. Performance of screw and reciprocating compressors is expressed in terms of pressure ratio and speed for a given refrigerant and suction density. Isentropic efficiency for a screw compressor is strongly affected by under- and over-compression (UOC) processes. The theoretical simple physical UOC model involves a compressor-specific (but sometimes unknown) volume index parameter and the real gas properties of the refrigerant used. Isentropic efficiency is estimated by the UOC model and a bi-cubic, used to account for flow, friction and electrical losses. The unknown volume index, a smoothing parameter (to flatten the UOC model peak) and bi-cubic coefficients are identified by curve fitting to minimize an appropriate residual norm. Chiller performance maps are produced for each compressor type by selecting optimized sub-cooling and condenser fan speed options in a generic component-based chiller model. SEER is the sum of hourly load (from a typical building in the climate of interest) and specific power for the same hourly conditions. An empirical UAE cooling load model, scalable to any equipment capacity, is used to establish proposed UAE MEPS. Annual electricity use and cost, determined from SEER and annual cooling load, and chiller component cost data are used to find optimal chiller designs and perform life-cycle cost comparison between screw and reciprocating compressor-based chillers. This process may be applied to any climate/load model in order to establish optimized MEPS for any country and/or region.

1. Introduction

Energy efficiency in buildings is a widely researched topic. Demand side management outperforms other energy, cost and emission reduction measures [6], but is it really true in developing countries where no MEPS and rating systems are practiced? United Arab Emirates (UAE) has hot and humid weather conditions for most of the year, which translates to cooling requirement of 40-55% of annual electricity consumption [3, 10]. Executive Affairs Authority in Abu-Dhabi is motivated to establish minimum energy performance standards by exploring various chiller design options appropriate to UAE conditions, development of various conventional and innovative design configurations are tested. Flexible heat pump models with component based sub-models for evaporator, compressor and condensers are developed so various design configurations can be extensively tested and compared for given boundary conditions. Compressors are the most intensive energy consuming components in chiller systems, hence evaluation of various compressor types is considered as a potential design option. This paper compares screw and reciprocating compressor semi-empirical models and their use in establishing MEPS for chillers.

Literature related to heat pump and components models for chiller systems is extensively reviewed by Armstrong et al., Zakula et al., Qureshi et al., Javed et al, Tauha et al. [1, 2, 5, 13, 19, 22], so literature review for this paper will specially focus on screw and reciprocating compressor modeling and applications. Compressors can be broadly classified into positive displacement and dynamic compressor types [8]. Positive displacement compressors, where compression is achieved through reduction of compression chamber volume, types include reciprocating, screw, scroll and rotary

Content from this work may be used under the terms of the Creative Commons Attribution 3.0 licence. Any further distribution of this work must maintain attribution to the author(s) and the title of the work, journal citation and DOI.
Published under licence by IOP Publishing Ltd

compressors. Dynamic compression is achieved through kinetic energy transfer; centrifugal compressor is the most common type [14]. Development of Compressor maps is central to compressor modelling, such maps may involve physics based, empirical or semi-empirical models [1, 18].

Sjoholm presents some of the earlier work related to variable volume ratio and capacity control in screw compressors [20,9]. Le at al. developed screw chiller compressor model based on simple efficiency loss calculation [16, 17]. Fujiwara et al. developed empirical relation for flow and heat transfer characteristics of screw compressor; furthermore it models the effects of oil on gas leakage and cooling [11, 12]. Stosic et al. developed a thermodynamic performance simulation model which accounts for detailed modelling of oil injection in a screw compressor [21]. Reciprocating compressors are also extensively described in literature, both simple and detailed models. The Air Conditioning and Refrigeration Institute developed a bi quadratic empirical model to represent mass flow rate and compressor power for fixed speed compressors, but any interpolation with this model leads to erroneous results and the model cannot be used for variable speed applications [7]. Models proposed by Popovic et al., Jahnig et al. and Armstrong et al. are valid over wide range of conditions and are based on volumetric efficiency which takes into account actual thermodynamic process for compression [18, 15, 1].

Literature reviewed readily describes under and over-compression for screw compressors in terms of isentropic work (power law) functions, but not well described, in terms of a polynomial function of reasonably low order. This paper demonstrates a semi-empirical screw compressor model adjustment for under-over compression losses. The adjusted model represent screw compressor performance with six bi-quadratic parameters, and two new parameters volume index and smoothing band. The Physical UOC model is derived, which evaluates effective volume index of the compressor under investigation. Holistic objective of the research work is to demonstrate the importance of accurate compressor models in establishing MEPS, air and evaporatively cooled chillers are modeled and compared for screw and reciprocating compressor designs.

2. Screw and Reciprocating Compressor Model

A complete representation of compressor performance requires two sub-models to compute refrigerant flow rate and input power given compressor speed and pressure ratio. The models are based on the fact that flow losses and internal power dissipation are more-or-less proportional to refrigerant flow rate and compressor work with an additional loss component proportional to departure from the built-in volume index. Thus we can expect to get by with low-order models to empirically represent deviations from the idealized models of (1) flow proportional to speed and (2) constant isentropic efficiency. Model coefficients are estimated from measured performance observed over the range of pressure ratio and mass flow rate encountered in chiller operation.

2.1. Screw Compressor Model without Over-Under Compression

Capacity and isentropic efficiency models are extensively described in Javed et al., 2014. Data required for such model training and analysis may be obtained from compressor selection tools[1]. A common isentropic model formulates compressor isentropic efficiency as a bi-cubic relation [19] involving pressure ratio(π), and compressor speed(f) shown in figure 1, coefficients in Table 1. Bi-cubic regression models are of the form

$$y = \sum_i C_i g_i\left(\bar{X}\right) + \varepsilon$$

[1] www.bitzer.de/websoftware

where, $\boldsymbol{y}$ is the dependent variable (η_{isen}), $\overline{\boldsymbol{X}} = \overline{\mathrm{X}}(f,\pi)$ the vector of independent variables, $\boldsymbol{g}_i(\overline{\boldsymbol{X}})$ is the ith bi-cubic term, $\boldsymbol{C}_i$ is the regression coefficient associated with $g_i(X)$ and ε is the error or residual.

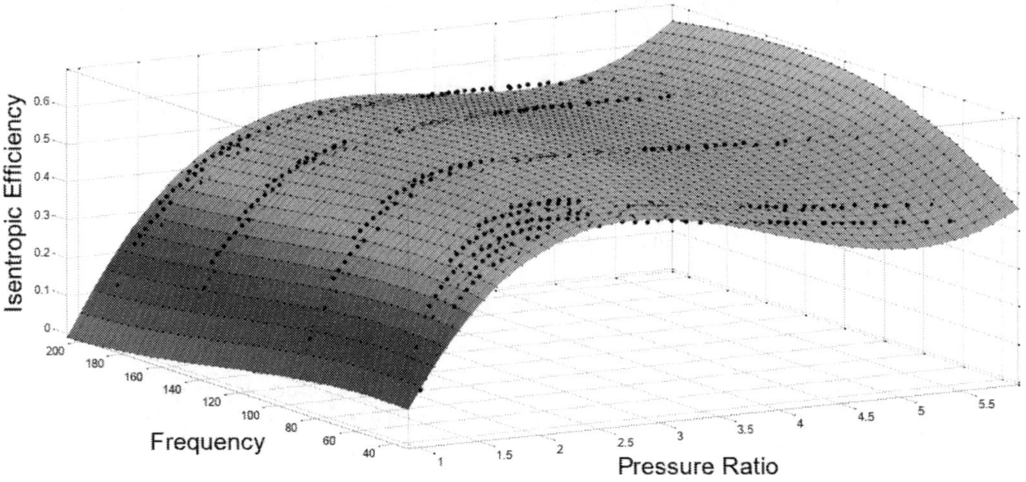

Figure 1 : Isentropic Efficiency Map for range of Pressure ratios and compressor speed

Table 1 : Bi-cubic relation for η_{isen} producing RMS=0.02431 for points shown in Figure 1

Coefficient	Term	Value	t-stat
C1	f^3	4.74E-08	4.4284
C2	^{3}	2.32E-02	37.489
C3	f^2	-1.59E-05	-3.9376
C4	$_2$	-2.75E-01	-43.762
C5	$\pi.f^2$	-2.79E-06	-7.6101
C6	$\pi^2.f$	5.57E-06	0.4527
C7	$\pi.f$	9.76E-04	8.3714
C8	f	4.96E-04	0.9961
C9		9.64E-01	45.707
C10	Constant	-5.96E-01	-22.130

It is apparent that even with an increase of model order from bi-quadratic to bi-cubic the general polynomial model does not well represent isentropic efficiency of a screw compressor

2.2. Screw Compressor Over-Under compression Model

Under- and over-compression (UOC) processes affect the isentropic efficiency of a screw compressor. The theoretical UOC model involves a compressor-specific (but sometimes unknown) volume index parameter and the real gas properties of the refrigerant used. Isentropic efficiency is estimated by a quotient of the UOC model and a bi-cubic, used to account for flow, friction and electrical losses. A smoothing parameter is used to flatten the UOC model peak. The unknown volume index and bi-cubic coefficients may be identified by curve fitting to minimize root mean square error (RMSE) and coefficient of variance (COV).

Isentropic work done before the compression pocket reaches the discharge port is given by

$$\dot{W}_{rev} = \dot{m}P_{suc}v_{suc}\left[(\pi_{ext})^{\left(\frac{n-1}{n}\right)} - 1\right]$$

$\dot{m}, P_{suc}, v_{suc}$ =Mass flow rate of refrigerant; Suction Pressure and suction specific volume
$_{ext}$ = Actual pressure ratio, without UOC
The exponent n is given in Armstrong et al. [1] (real gas properties), where $v_{i,o}$ is inlet and outlet specific volume.

$$n = \frac{P_{suc}v_{suc}}{RT_{SUC}}$$

First we consider an ideal compressor where the only compressor losses were from under or over compression. The work done when $\pi_{vi} >$ $_{ext}$ (over-compression) would be

$$\dot{W}_{oc} = \dot{m}P_{suc}v_{suc}\left[(\pi_{vi})^{\left(\frac{n-1}{n}\right)} - 1\right]$$

Conversely the work done when $\pi_{vi} <$ $_{ext}$ (under compression) would be

$$\dot{W}_{uc} = \dot{m}P_{suc}v_{suc}\left[\left((\ \ _{ext})^{\left(\frac{n-1}{n}\right)} - 1\right) + \left[\left(\left(\frac{\pi_{ext}}{\pi_{vi}}\right)^{\left(\frac{1}{n}\right)} - 1\right)\left(\left(\frac{\pi_{ext}}{_{vi}}\right)^{\left(\frac{n-1}{n}\right)} - 1\right)\right]\pi_{vi}^{\left(\frac{n-1}{n}\right)}\right]$$

Thus the isentropic efficiency of a hypothetical compressor that suffers only UOC losses is

$$\eta_{UOC} = \frac{\dot{W}_{rev}}{\dot{W}_{UOC}} \quad \text{where } \dot{W}_{UOC} = W_{OC} \quad _{vi} > \pi_{ext} \ ; \dot{W}_{UOC} = W_{UC} \quad _{vi} < \pi_{ext}$$

The theoretical isentropic efficiency incorporating under and over compression is shown in figure 2. Design pressure ratio (π_{vi}) if unknown, can be obtained empirically by minimizing RMSE {ηisen-η_{UOC}(π_{vi},f)*bi-quad(π_{ext},f)} 0.0177 and COV. Design pressure ratio of 1.7 is obtained. The Matlab Smoothing function (moving average) is applied to the UOC model with a span value selected to minimize RMSE and COV.

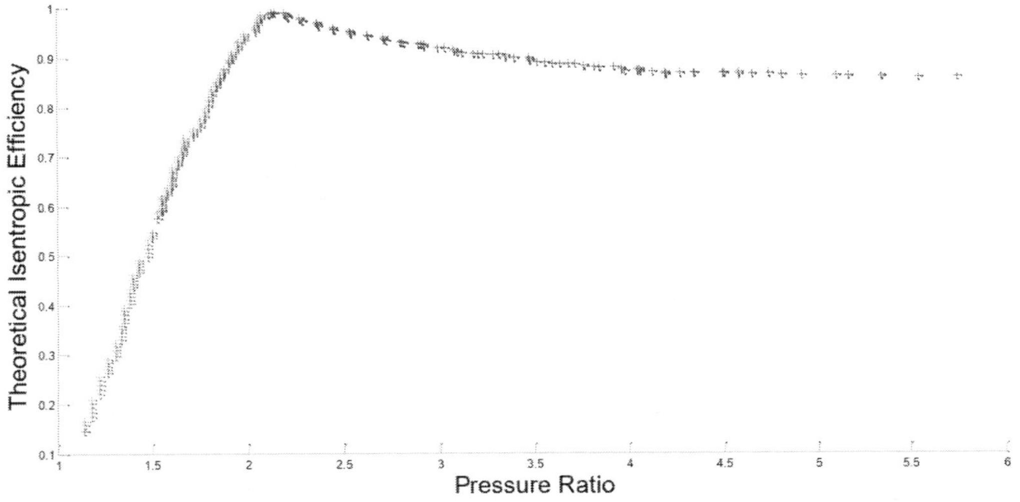

Figure 2 : Theoretical Isentropic Efficiency (η_{UOC}) Map for Screw compressor, accounting only for UOC losses

2.3. Screw Compressor Model Adjusted for Over-Under compression

A bi-Cubic relation in compressor speed (f) and external pressure ratio (π_{ext}) is developed by normalizing UOC curve to isentropic efficiency of the compressor with its UOC losses hypothetically removed (Figure 1). The model for isentropic efficiency considering all the other losses is therefore of the following form, shown in Figure 3, coefficients in table 2.

$$y = \frac{\eta_{isen}}{\eta_{UOC}} = \eta_{other} + \varepsilon$$

where η_{isen} is the test data

$\eta_{other} = f(\pi_{ext}, f)$ and ε is the error or residual.

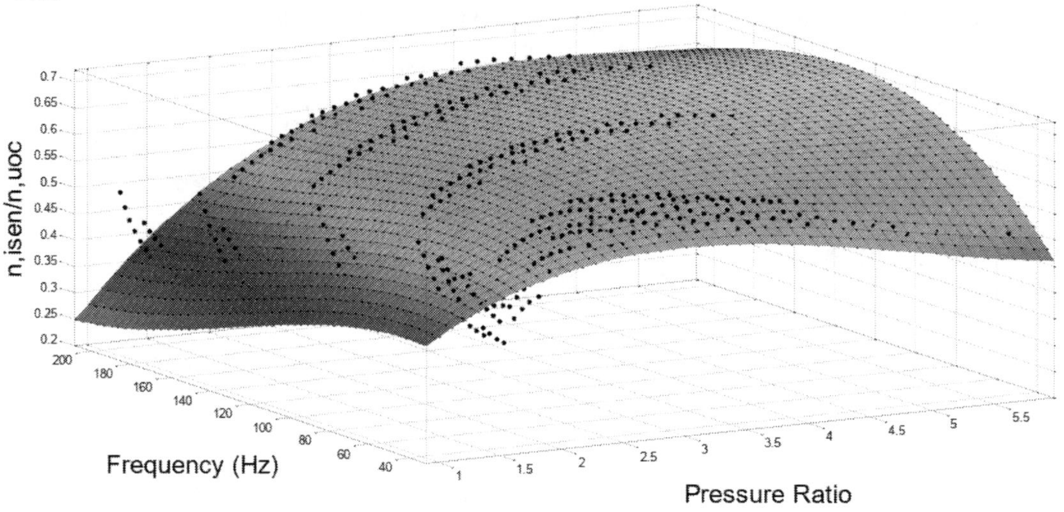

Figure 3 : Adjusted Isentropic Efficiency Map with UOC losses for Screw Compressor

Table 2 : Bi-cubic relation for η_{isen}/η_{UOC} producing RMS=0.01912 for points shown in Figure 3

Coefficient	Term	Value	t-stat
C1	f^3	6.27E-06	7.1979
C2	π^3	-3.98E-03	13.479
C3	f^2	-2.11E-05	-6.3965
C4	π^2	5.67E-02	-15.433
C5	$\pi.f^2$	-3.41E-06	-12.083
C6	$\pi^2.f$	-5.93E-05	-6.7629
C7	$\pi.f$	1.62E-03	17.448
C8	f	2.24E-04	0.5460
C9	π	1.95E-01	13.173
C10	Constant	2.58E-01	12.362

2.4. Reciprocating compressor Model

Data used for capacity and isentropic model of a representative reciprocating compressor were generated from the compressor selection tool [2], used by Armstrong et al., 2009(a). To characterize the

[2] http://www.carlylecompressor.com/

compressor performance evaluation is done at 500 points, by generating the following data set, Table 3. The sizing tool output includes compressor input power, mass flow rate. Compressor inlet state is known, and compressor outlet enthalpy is calculated using model presented in Armstrong et al., 2009a. Performance is evaluated over four girds of operating conditions resulting in 5x5x5x4=500 unique conditions.

Table 3 : Grid values of the four variables used to span the full range of operating conditions

Shaft Speed (RPM)	Condensing Temp (°F)	Evaporating Temp (°F)	Evaporator Superheat (°R)
900	80	30	0
1100	90	35	5
1300	100	40	10
1525	110	45	20
1750	130	50	

The compressor isentropic model is a bi-quadratic relation (Qureshi 2013) involving pressure ratio(π), and compressor speed(f) shown in figure 4, (coefficients in table 4).

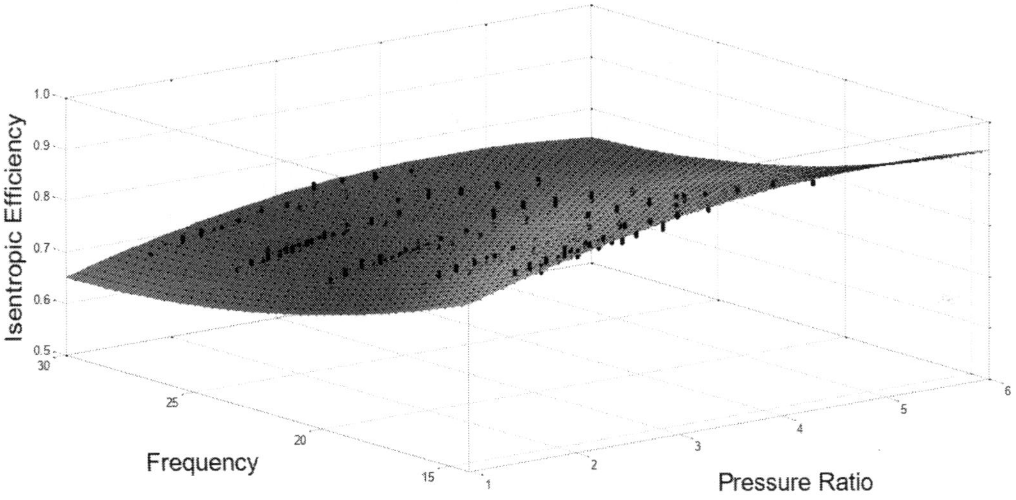

Figure 4: Reciprocating Compressor Isentropic Efficiency Map for range of Pressure ratios and compressor speed

All 500 selection tool results are shown based on the input value shown in Table3.

Table 4: Bi-quadratic relation for η_{isen} producing RMS = 0.01042 for points shown in Figure 6

Coefficient	Term	Value	t-stat
C_1	Constant	1.204E 00	81.13
C_2		8.664E-02	15.85
C_3	f	-4.166E-02	-38.64
C_4	^{2}	-8.195E-03	-10.50
C_5	$\pi.f$	3.595E-04	-2.77
C_6	f^2	6.993E-04	30.40

3. Application to Establishing Chillers MEPS

Heat pump products, like other energy-intensive capital equipment, compete in the market on both first cost (price) and operating cost. Although life-cycle cost (LCC) may be estimated, the average buyer is influenced more by first cost. To reduce the impact of this irrational behavior and simultaneously transform markets toward higher fleet efficiency, many countries adopt minimum equipment performance standards (MEPS) for heat pump and A/C equipment. The efficiency bar generally aims to minimize LCC of a given machine type and size category serving a typical load profile. Measures that could be cost-effective in UAE have not been thoroughly researched. Moreover available high performance chillers are rarely used because of subsidized electricity rates and weak price elasticity. One possible solution is to establish minimum energy performance standards by exploring various heat pump design options appropriate to UAE conditions. Trying out different compressors types is proposed as one such design option. Each heat pump component is modeled from first principles and/or empirical relations known to be valid over a wide range of conditions, capacity fractions and loads. Optimized air-cooled heat pump designs were evaluated, from an extension of previous work [13]. The component-based models are used to produce tabular performance maps. An empirical UAE cooling load model was adapted to produce hourly loads, scalable to any building, from TMY weather data. Together with a given chiller performance map, the cooling load model can be used to estimate annual electricity use.

Decoupling the evaporator from the rest of the system leaves two refrigerant side boundary conditions i.e. capacity and evaporating temperature and dry bulb/wet bulb temperature as the third boundary condition for air cooled /evaporatively-cooled heat pump respectively. In this paper, a comparison is made between screw compressor (air-cooled and evaporatively-cooled) and reciprocating compressor (air-cooled and evaporatively-cooled) heat pump designs [5].

The compressor and condenser represent a sub-system with capacity Q_e, evaporating temperature, T_e, and outdoor wet bulb temperature, T_x, as boundary conditions. Figure 5 and 6 map the performance of (a) air-cooled chiller (b) evaporatively-cooled chiller for screw and reciprocating compressor respectively. Figure 5 (a) shows that specific power has minimum values between 0.5-0.6 Capacity Fraction (CF). Hence the chiller is cycled for low Capacity Fraction (CF) over annual energy runs.

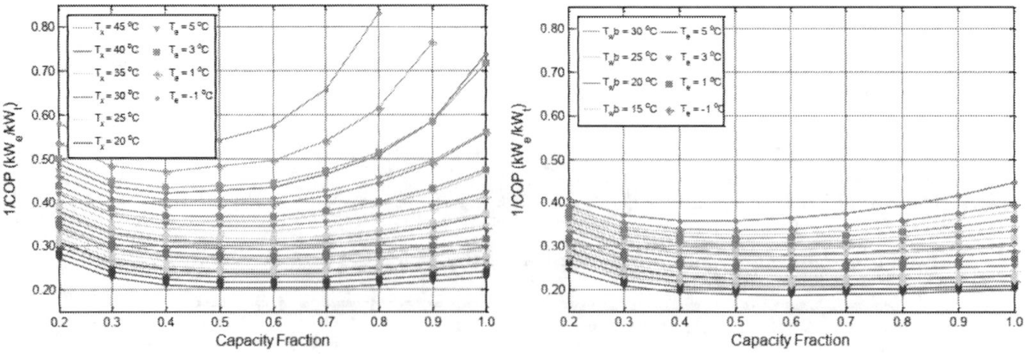

(a) **(b)**

Figure 5: Comparison (a) Air-cooled and (b) Evaporatively-cooled chiller performance map--(Screw Compressor)

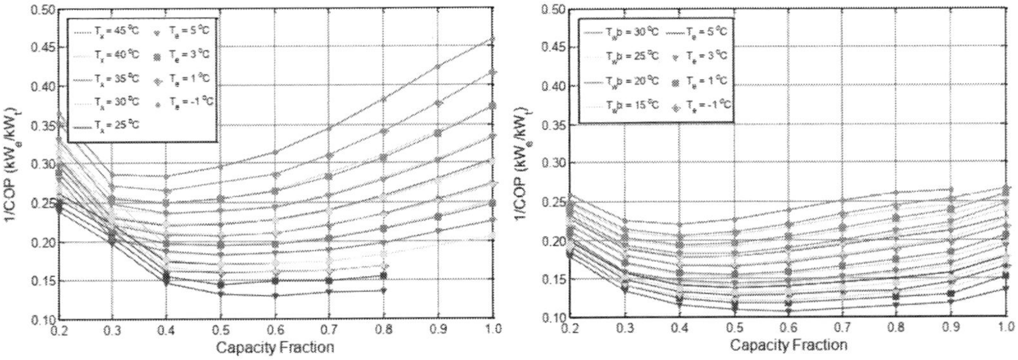

(a) (b)

Figure 6: Comparison (a) Air-cooled and (b) Evaporatively-cooled chiller performance map--(Reciprocating Compressor)

The SEER comparison for air-cooled and evaporatively-cooled screw and reciprocating chillers is presented in table 5. In case of air cooled chillers percentage savings of 36.4% is reported with reciprocating compressors, whereas in case of evaporatively-cooled chillers savings are over 59.7% compared to respective base case.

Table 5: SEER Comparison for Screw and Reciprocating Compressor

	Air Cooled chiller	**Evaporatively-Cooled Chiller**
Screw Compressor (SEER)	3.3506	3.9632
Reciprocating Compressor (SEER)	4.5727	6.3309
SEER Percent (%) Increase	**36.4**	**59.7**

4. Conclusion

The isentropic efficiency of positive displacement compressors with outlet check valves (reciprocating and rolling-piston compressors) is typically represented by a low order polynomial in external pressure ratio and shaft speed. But the isentropic efficiency of a fixed-port compressor (e.g. screw or scroll compressor) is not well modelled by a polynomial in pressure ratio and speed even with higher polynomial order. The reason for this difficulty is the under- and over-compression losses which, although readily described in terms of isentropic work (power law) functions, are not well described, even for a fixed speed compressor, in terms of a polynomial function of reasonably low order.

A model that combines the isentropic efficiency of a hypothetical compressor with no UOC losses, and the isentropic efficiency of a hypothetical compressor with only UOC losses, is found to represent screw compressor performance with only two new parameters, volume index and smoothing band, in addition to the ten bi-cubic parameters.

The expressions for under- and over-compression specific work are functions of external pressure ratio and both reduce to isentropic work when external pressure ratio equals the design pressure ratio, π_{vi}, that corresponds to the effective volume index of the machine in question. A moving average of the two isentropic efficiency functions reflects the fact that discharge ($\pi_{vi} > \pi_{ext}$) or back flow ($\pi_{vi} < \pi_{ext}$) at the discharge port is not instantaneous. The UOC model therefore incorporates a moving average window width as well as the volume-index parameter, π_{vi}.

The annual energy use of identical chillers, one with a typical $\pi_{vi} = 1.7$ screw compressor and the other with a reciprocating compressor of the same capacity, is simulated using the aggregate cooling load model of Abu Dhabi [3]. The reciprocating chiller is found to use from 35-60% less energy than the screw chiller.

5. Acknowledgments

We thank Executive Affairs Authority (EAA) and Masdar Institute of Science and Technology (MIST) for supporting this research. We also thank Mr Burhan of SKM (chiller manufacturer) for useful discussion and encouragement.

6. References

[1] Armstrong, P., W. Jiang, D.W. Winiarski, S. Katipamula, L.K. Norford, and R.Willingham. (2009). Efficient low-lift cooling with radiant distribution, thermal storage and variable-speed chiller controls—Part I: Component and subsystem models. *HVAC&R Research*, 15(2):367–401.

[2] Armstrong, P., W. Jiang, D.W. Winiarski, S. Katipamula, L.K. Norford, and R.Willingham. (2009). Efficient low-lift cooling with radiant distribution, thermal storage and variable-speed chiller controls—Part II: Annual Energy Use and Savings. *HVAC&R Research*, 15(2):403–433.

[3] Ali, M.T., M.M. Mokhtar, M. Chiesa, P.R. Armstrong, (2011). A cooling change-point model of community-aggregate electrical load, Energy and Buildings, 43(1):28-37

[4] Ali, M.T, P.R. Armstrong, (2015). First-principles evaporator model for fast solution when coupled to an outdoor unit performance map, ASHRAE, Atlanta.

[5] Ali, M.T., H. Javed, P.R. Armstrong (2014). Optimized heat pump design for GCC countries, Part II: Baseline chiller instrumentation and best practice guidance, Proc. 11th IEA Heat Pump Conference

[6] Afshari,.A C.Nikolopoulou, M.Martin, (2014). Life-Cycle analysis of building retrofits at the urban scale-A case study in UAE, *Sustainability*, 6,453-473

[7] ARI. 1999. ANSI/ARI 540-1999(R2002), *Positive Displacement Refrigerant Compressors and Compressor Units*. Air Conditioning and Refrigeration Institute.

[8] ASHRAE. 2008. Compressors. *ASHRAE Handbook: HVAC Systems and Equipment*, Ch. 37.

[9] Brasz, J. (2010). A proposed centrifugal refrigeration compressor rating method, ICEC at Purdue.

[10] Friedrich, L.A. (2013). An Urban Energy Baseline Model for Measurement & Verification of Building Energy Efficiency Retrofits in Abu-Dhabi, MSESM Thesis, Masdar Institute of Science and Technology.

[11] Fujiwara M, Kasuya K, Matsunaga T, Watanabe M. Computer modeling for performance analysis of rotary screw compressor, Proceedings of the International Compressor Engineering Conference at Purdue, USA, pp. 536–43, 1984.

[12] Fujiwara M, Osada Y. Performance analysis of an oil injected screw compressor and its application. *International Journal of Refrigeration* 1995; 18(4):220–7.

[13] Javed,.H Ali, M.T., P.R. Armstrong (2014). Optimized heat pump design for GCC countries, Part I: Alogritham, Proc. 11th IEA Heat Pump Conference, Montreal

[14] Bloch. H.P. *A Practical Guide to Compressor Technology*, Edition 2, ISBN-13: 978-0471727934

[15] Jahnig, D.I., D.T. Reindl, and S.A. Klein. 2000. A semi-empirical method for representing domestic refrigerator/ freezer compressor calorimeter test data. *ASHRAE Transactions* 106(2).

[16] Le, C.V., Bansal, P.K., Tedford, J.D., 2004. Three-zone system simulation model of a multiple-chiller plant. Applied Thermal Engineering 24, 1995–2015.
[17] Le, C.V., Bansal, P.K., Tedford, J.D., 2005, Simulation model of a screw liquid chiller for process industries using local heat transfer integration approach. Proc. IMechE 219 Part E J. Process Mech. Eng., 95–107.
[18] Popovic, P., and H.N. Shapiro. 1995. A semi-empirical method for modeling a reciprocating compressor in refrigeration system. *ASHRAE Transactions* 101(2):367-382.
[19] Qureshi, O.A, H.Javed, A.Afshari, P.R. Armstrong, (2013). Optimal model-based control of chiller tower fan and cooling water pump, ASHRAE, Seattle.
[20] Sjoholm, L., Variable volume-ratio and capacity control in twin-screw compressors (1986). International Compressor Engineering Conference. Paper 548.
[21] Stosic N, Milutinovic L, Hanjalic K, et al. Investigation of the influence of oil injection upon the screw compressor working process. *International Journal of Refrigeration* 1992; 15(4):206–20.
[22] Zakula, T., N.T. Gayeski, P.R. Armstrong and L.K. Norford (2011). Variable-speed heat pump model for a wide range of cooling conditions and loads, *HVAC&R Research*, 17(5), 670-691.

IOP Conf. Series: Materials Science and Engineering **90** (2015) 012078 doi:10.1088/1757-899X/90/1/012078

Prediction of uncertainty of 10-coefficient compressor maps for extreme operating conditions

Howard Cheung

Postdoctoral Research Fellow, Ray W. Herrick Laboratories, School of Mechanical Engineering, Purdue University, 177 S. Russell St., West Lafayette, IN 47907-2031, US

E-mail: cheung@purdue.edu

Christian K. Bach

Assistant Professor, Mechanical and Aerospace Engineering, Oklahoma State University, 218 Engineering North, Stillwater, OK 74078-5016

E-mail: cbach@okstate.edu

Abstract. Empirical compressor maps are a simple and reliable approach for heating and cooling system designers to estimate compressor refrigerant mass flow rate and power consumption quickly. These maps were used for a long time since most compressor manufacturers build the maps with extensive test matrices, leading to good accuracy. However, the situation changes when engineers extrapolate the maps to investigate the compressor's performance under extreme operating conditions such as for cold climate heat pump applications or under conditions with system faults. Engineers are not confident on the exact uncertainty of the extrapolation, and often claim that the inaccuracy of their studies is a result of high extrapolation uncertainty. This paper presents a method to estimate the extrapolation uncertainty due to the structure of the test matrix that trains the manufacturer maps and helps the investigators to understand if the extrapolation is the main cause of their inaccuracy. To verify that the method can estimate the uncertainty due to extrapolation, the study builds 10-coefficient compressor maps trained by different test matrices of the same size and different operating points. The maps are used to estimate the compressor performance under different operating points and their estimation uncertainties are compared. The results show that the component of the uncertainty that depends on the structure of the test matrix is small at operating conditions within the test matrix but grows significantly as the map is used to estimate outputs further away from the operating conditions within the test matrices.

1. Introduction

Empirical compressor maps for positive displacement compressors, such as the 10-coefficient polynomials outlined in ANSI/AHRI Standard 540 [1] and European Standard EN 12900 [2] are widely used in academia and industry. These maps are trained with experimental data and then used to calculate power input, mass flow rate, current, and compressor efficiency. However, the maps contain squared and cubic terms of the inputs (suction and discharge dew point temperature) and are therefore potentially inaccurate when used for extrapolation outside of the training data range. Manufacturers often state an accuracy of $\pm5\%$ for tabulated performance data (e.g. [3] or [4]). This paper shows a method that estimates the resulting uncertainty for

Content from this work may be used under the terms of the Creative Commons Attribution 3.0 licence. Any further distribution of this work must maintain attribution to the author(s) and the title of the work, journal citation and DOI.
Published under licence by IOP Publishing Ltd

the 10 coefficient polynomial at a certain point other than the training data set.
This paper does not address interpolation between different speeds of variable speed compressors. However, the shown methodology can be extended to predict the uncertainty of the outputs for interpolating between fixed speeds rather than using different fixed speed compressor maps as done by e.g. [5] and [6].

2. Linear regression

Finding the coefficients of the 10 coefficient maps can be treated as a linear regression problem, with the evaporating and condensing temperature and their linear, squared and cubed combinations being the independent variables $\vec{x}$ and the power consumption of the compressor being the estimated dependent variable y. To better understand the uncertainty sources of the model, a brief review of linear regression along with how this applies to the polynomial compressor map is shown subsequently.

2.1. Review

Linear regression is used to estimate the parameters of a linear model for a system that might have slightly nonlinear behavior. If we had the parameters $\vec{\beta}_{true}$ for the true model that is the best linear model to describe the actual system, we could use the independent inputs to our model ($\vec{x}^T$) to calculate the true output y_{true} up to an error ε caused by the difference between model and system, e.g.

$$y_{true} = \vec{x}^T \vec{\beta}_{true} + \varepsilon. \tag{1}$$

Unfortunately it is not possible to obtain the true model, since it is not possible to obtain the infinite number of input/output data points from the system. Therefore, we need to estimate the model parameters to calculate an estimate of the output. To emphasize the difference between the two models, a hat ($\hat{}$) is used for the estimated values (estimated model parameters $\hat{\vec{\beta}}$ and estimated model output $\hat{y}$), and the estimated linear model is

$$\hat{y} = \vec{x}^T \hat{\vec{\beta}}. \tag{2}$$

The parameters for the model can be estimated using measurement data. For non-weighted reduction of the squares of the errors, it can be shown that the parameter vector $\hat{\vec{\beta}}$ can be calculated as

$$\hat{\vec{\beta}} = (\mathbf{X}_{train}^T \mathbf{X}_{train})^{-1} \mathbf{X}_{train}^T \vec{y}_{train}, \tag{3}$$

where the subscript *train* is used for training data. The training data input matrix $\mathbf{X}_{train}$, composed of input data vectors $\vec{x}_{train,i}$ for each data point i is constructed as

$$\mathbf{X}_{train} = \left[\begin{array}{ccccc} \vec{x}_{train,1} & \cdots & \vec{x}_{train,i} & \cdots & \vec{x}_{train,n} \end{array} \right]^T, \tag{4}$$

where n is the total number of data points. The output training data vector is composed of output values $y_{train,i}$ for each data point as

$$\vec{y}_{train} = \left[\begin{array}{ccccc} y_{train,1} & \cdots & y_{train,i} & \cdots & y_{train,n} \end{array} \right]^T. \tag{5}$$

The accuracy of the model, measured by the mean sum of square of errors σ, is

$$\sigma = \sqrt{\frac{\Sigma_i (y_{train,i} - \vec{x}^T \hat{\vec{\beta}})^2}{n-1}}. \tag{6}$$

2.2. Polynomial compressor map as linear regression problem

ANSI/AHRI Standard 540 [1] estimates the compressor power consumption $\hat{\dot{W}}$ as

$$\hat{\dot{W}} = \hat{\beta}_1 + \hat{\beta}_2 T_{evap} + \hat{\beta}_3 T_{cond} + \hat{\beta}_4 T_{evap}^2 + \hat{\beta}_5 T_{evap} T_{cond} + \hat{\beta}_6 T_{cond}^2 + \hat{\beta}_7 T_{evap}^3 + \hat{\beta}_8 T_{evap}^2 T_{cond} + \hat{\beta}_9 T_{evap} T_{cond}^2 + \hat{\beta}_{10} T_{cond}^3, \quad (7)$$

where $\hat{\beta}_i$ are the estimated model coefficients and T_{evap}, and T_{cond} are the evaporation and condensing dew point temperatures. Referring to Eqn. (2), $\hat{\dot{W}}$ takes the place of $\hat{y}$, the estimated parameter vector $\hat{\vec{\beta}}$ is composed of the $\hat{\beta}_i$, and the vector of independent variables that contains the saturation temperature variables, $\vec{x}$, is constructed as

$$\vec{x} = \begin{bmatrix} 1 & T_{evap} & T_{cond} & T_{evap}^2 & T_{evap}T_{cond} & T_{cond}^2 & T_{evap}^3 & T_{evap}^2 T_{cond} & T_{evap} T_{cond} & T_{cond}^3 \end{bmatrix}^T. \quad (8)$$

3. Uncertainty of steady state measurements

The training data for compressor maps is obtained from steady state time series data. Steady state time series data do not show any significant monotonic trends with time but rather shows periodic and random fluctuations (noise) around an average value. The mean value a of the measured variable can be defined as

$$a = \Sigma_{i=1}^{N} \frac{a_{mea}(t_i)}{N}, \quad (9)$$

where $a_{mea}(t_i)$ is a discrete measurement of the variable at time step i, and N is the number of measurements in a time series. Measurement devices have a time-independent uncertainty often called zero-order uncertainty. The limited number of samples in combination with the fluctuations of the variable due to environmental noise leads to the first-order uncertainty. Zero-order uncertainty is typically provided by the manufacturer of the measurement device, such as $\pm 0.5K$ for T-type thermocouples, or 0.5% of the measured flow rate for Coriolis mass flow meters. First-order uncertainty is not known in advance but rather needs to be approximated by statistically analyzing the time series data of the steady state measurement. Taylor and Kuyatt (1994)[7] give the overall measurement uncertainty as

$$\Delta a = \sqrt{\Sigma_{i=1}^{N}\left(\frac{\Delta a_{mea}(t_i)}{N}\right)^2 + \left(\frac{t_{N-1,1-\alpha/2}}{N}\right)^2 \frac{\Sigma_{i=1}^{N}(a_{mea}(t_i) - a)^2}{N-1}} \quad (10)$$

where $t_{N-1,1-\alpha/2}$ is the statistic of the Student's t-distribution with a degree of freedom of N-1 and level at $1-\alpha/2$. The first part of the uncertainty is the sum of squares of the zero-order uncertainty of each observation within the steady state measurement. The second part of the uncertainty is given as the confidence interval of the average value at the level of $1-\alpha/2$. In practice, α is taken as 0.1, and a 95% confidence interval is usually used in the approximation.

4. Generating experimental results of steady state data with uncertainties from compressor map data

In this paper, no real measurement data are taken for training data, and only ideal compressor performance data from the manufacturers without uncertainty is available. In order to test the uncertainty calculation of the compressor map output, a set of training data with uncertainties is approximated from the ideal compressor performance data. If the ideal compressor performance data are the true values of the measurement, its measurement value will be different because it is

IOP Conf. Series: Materials Science and Engineering **90** (2015) 012078 doi:10.1088/1757-899X/90/1/012078

subjected to the zero-order uncertainty of the measurement device and the first-order uncertainty of the noise of the measurement environment. Assuming that the uncertainty is the confidence interval of the measurement value at a level of $1 - \alpha/2$ and the measurement value follows a normal distribution with a mean value around the true value of the measurement, the standard deviation of the normal distribution will be given by

$$\sigma_{mea} = \frac{\sqrt{(\Delta a_{zero-order}(a = a_{true}))^2 + (\Delta a_{first-order}(a = a_{true}))^2}}{z_{1-\alpha/2}} \tag{11}$$

where $z_{1-\alpha/2}$ is the z-normal distribution statistics given at a level of $1 - \alpha/2$. The observations within steady state can then be approximated by running a random number generator following a normal distribution with mean at the true value of a and standard deviation at σ_{mea} multiple times. These values can then be analyzed with the method listed in the section 3 to calculate the value and uncertainty of steady state measurement.

5. Sources of uncertainty

Uncertainty of the compressor map output is the range where the true value of the output may be relative to the map output. It consists of multiple components and can be grouped as follows:

- Uncertainty due to inputs
- Uncertainty due to training data
- Uncertainty due ot model random error
- Uncertainty due to outputs

5.1. Uncertainty due to inputs

Uncertainty due to inputs is the uncertainty propagated to the map output due to the uncertainty in the inputs to the maps. The inputs to the map (evaporating and condenser temperature) are usually obtained by converting pressure measurements to saturation temperatures with refrigerant equations of state. Therefore the estimated saturation temperature contains uncertainty from both the equation of state and the pressure measurement. The equation of state of R22 estimates saturation pressure at an uncertainty of 0.2% [7]. When the equation of state estimates an saturation temperature at a given pressure, this uncertainty is transformed into a component of the uncertainty of the saturation temperature as shown in

$$\frac{\Delta P_{sat,EOS}}{P_{sat}(T)} = 0.2\% = 0.002 \tag{12}$$

and

$$\Delta T_{sat,EOS} = \left| \frac{\partial T_{sat}(P)}{\partial P} \right| \Delta P_{sat,EOS}, \tag{13}$$

where $\Delta P_{sat,EOS}$ is the uncertainty of saturation pressure as a result of uncertainty of the equation of state, $P_{sat}(T)$ is the saturation pressure from temperature T, $\Delta T_{sat,EOS}$ is the uncertainty of saturation temperature as a result of uncertainty of the equation of state and $T_{sat}(P)$ is the saturation temperature at pressure P.

The component of the saturation temperature uncertainty due to pressure measurement is calculated by

$$\Delta T_{sat,mea} = \left| \frac{\partial T_{sat}(P)}{\partial P} \right| \Delta P_{sat,mea}, \tag{14}$$

where $\Delta T_{sat,mea}$ is the uncertainty of saturation temperature as a result of pressure measurement and $\Delta P_{sat,mea}$ is the uncertainty of pressure measurement. The overall uncertainty of the saturation temperature is given by

$$\Delta T_{sat} = \sqrt{(\Delta T_{sat,EOS})^2 + (\Delta T_{sat,mea})^2}. \tag{15}$$

The uncertainty of the map output propagated from the inputs of condensing temperature and evaporating temperature is calculated by

$$\Delta \hat{\dot{W}}_{input} = \sqrt{\left(\frac{\partial \hat{\dot{W}}}{\partial T_{evap}} \Delta T_{evap}\right)^2 + \left(\frac{\partial \hat{\dot{W}}}{\partial T_{cond}} \Delta T_{cond}\right)^2}, \tag{16}$$

where $\Delta \hat{\dot{W}}_{input}$ is the uncertainty due to inputs at the map output, ΔT_{evap} is the uncertainty of evaporating temperature at input and ΔT_{cond} is the uncertainty of condensing temperature at input.

5.2. Uncertainty due to training data

Uncertainty due to training data is the uncertainty propagated to the map output from the training data through the map coefficients. This can be understood by considering the estimation of the map coefficients as a function of the training data as

$$\hat{\vec{\beta}} = g(T_{evap,train,1}, ..., T_{evap,train,n}, T_{cond,train,1}, ..., T_{cond,train,n}, \dot{W}_{train,1}, ..., \dot{W}_{train,n}), \tag{17}$$

where n is the number of training data points. Through function g and $\hat{\vec{\beta}}$ in Eqn. (17), as suggested in [8], the uncertainty component due to training data is calculated

$$\Delta \hat{\dot{W}}_{train} = \sqrt{\begin{array}{l} \Sigma_{j=1}^{n}\left(\Sigma_{i=1}^{m}\left(\frac{\partial \hat{\dot{W}}}{\partial \beta_i}\frac{\partial \beta_i}{\partial T_{evap,train,j}}\right)\Delta T_{evap,train,j}\right)^2 \\ +\Sigma_{j=1}^{n}\left(\Sigma_{i=1}^{m}\left(\frac{\partial \hat{\dot{W}}}{\partial \beta_i}\frac{\partial \beta_i}{\partial T_{cond,train,j}}\right)\Delta T_{cond,train,j}\right)^2 \\ +\Sigma_{j=1}^{n}\left(\Sigma_{i=1}^{m}\left(\frac{\partial \hat{\dot{W}}}{\partial \beta_i}\frac{\partial \beta_i}{\partial \dot{W}_{,train,j}}\right)\Delta \dot{W}_{train,j}\right)^2 \end{array}}, \tag{18}$$

where $\Delta \hat{\dot{W}}_{train}$ is the uncertainty due to training data at the map output, $\Delta T_{evap,train,j}$ is the uncertainty of the evaporating temperature at the jth data point, $\Delta T_{cond,train,j}$ is the uncertainty of the condensing temperature at the jth data point, $\Delta \dot{W}_{train,j}$ is the uncertainty of the power consumption at the jth data point, and m is the number of coefficients in $\hat{\vec{\beta}}$.

5.3. Uncertainty due to model random error

In linear regression, the random error between true model and actual system, ε in Eqn. (1) is assumed to be normally distributed around zero with some finite variance. This variance becomes part of the uncertainty of the uncertainty of the map output and can be presented in the form of confidence intervals. Statistic textbooks [9, 10] illustrated that the confidence interval of the variance can be calculated as

$$\Delta\hat{\dot{W}}_{model} = t_{n-m,1-\alpha/2}\sigma\sqrt{1+\vec{x}^T\left(\mathbf{X}_{train}^T\mathbf{X}_{train}\right)^{-1}\vec{x}} \tag{19}$$

where $\Delta\hat{\dot{W}}_{model}$ is the uncertainty due to model random error at the map output.

One significant term in Eqn. (19) is $\vec{x}^T(\mathbf{X}_{train}^T\mathbf{X}_{train})^{-1}\vec{x}$ which is the leverage of a regression model [11]. It estimates how deviated the current input vector to the map is relative to the training data of the regression model, and the magnitude of the leverage grows with the deviation. This helps to understand if the estimation is related to the training data and if (or "how much") the model is applicable at the current situation described by the input vector.

5.4. Uncertainty due to output

Uncertainty of the map output should represent the probable range where the true value lies relative to the map output. However, since the map is built from data obtained from measured power consumption instead of true values of the power consumption, the uncertainty components of the map output propagated from other sources only estimate the uncertainty of the estimation with the measured power consumption. Another component of uncertainty must be introduced so that the map output uncertainty is the uncertainty to the true map output. This uncertainty component can be approximated with the uncertainty of the measured power consumption with its true values by

$$\Delta\hat{\dot{W}}_{output} = \frac{1}{n}\Sigma_{i=1}^{n}\frac{\Delta\dot{W}_{train,i}}{\dot{W}_{train,i}}. \tag{20}$$

5.5. Overall uncertainty

The overall uncertainty of the map output is given by the square of the sum of all uncertainty components as

$$\Delta\hat{\dot{W}} = \sqrt{\left(\Delta\hat{\dot{W}}_{input}\right)^2 + \left(\Delta\hat{\dot{W}}_{train}\right)^2 + \left(\Delta\hat{\dot{W}}_{model}\right)^2 + \left(\Delta\hat{\dot{W}}_{output}\right)^2}. \tag{21}$$

6. Results and discussion

To study how this calculation method of compressor map output uncertainty helps to determine extrapolation uncertainty, the performance data of a $4.5kW$ R22 reciprocating compressor is used. The tabulated performance data shows the compressor power consumption at different evaporating and condensing temperature with a constant suction superheat 11 K within its operating range as shown in Figure 1.

Since the manufacturer data does not include experimental uncertainty, it is assumed that the data in the specification are the true values of the performance, and they are used to approximate the observations during steady state experiments with the assumptions in Table 1.

Table 1. Assumptions to approximate results in experiments of steady state compressor operation

Variable	Zero-order uncertainty	First-order uncertainty
Power consumption	0.50%	3%
Evaporating pressure	0.80%	0.9kPa
Condensing pressure	0.80%	0.4kPa

IOP Conf. Series: Materials Science and Engineering **90** (2015) 012078 doi:10.1088/1757-899X/90/1/012078

The steady state period is assumed to be 10 minutes long with data acquired at 0.1Hz in the experiment. The experimental observations at all data points in Figure 1 are generated by the method listed in Section 4.

6.1. Study cases and comparison of maps

To examine if the uncertainty of compressor map changes as the map extrapolates, various compressor maps trained from different ranges of data are compared. Six maps with different ranges of training data are designed for the study and the ranges of the training data in these maps are shown in Figure 2.

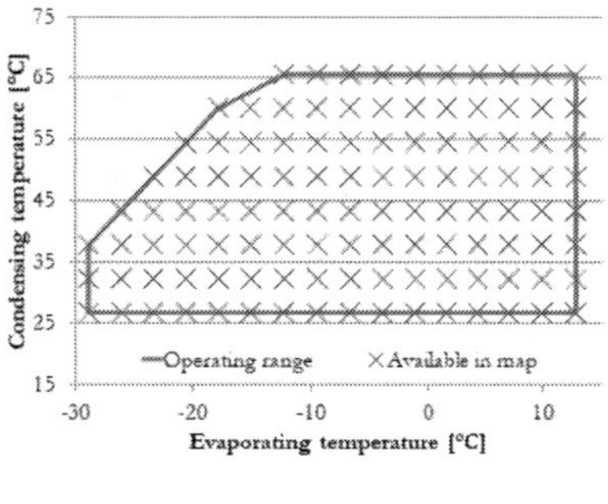

Figure 1. Operating envelope of compressor chosen for this study and available data points from manufacturer map.

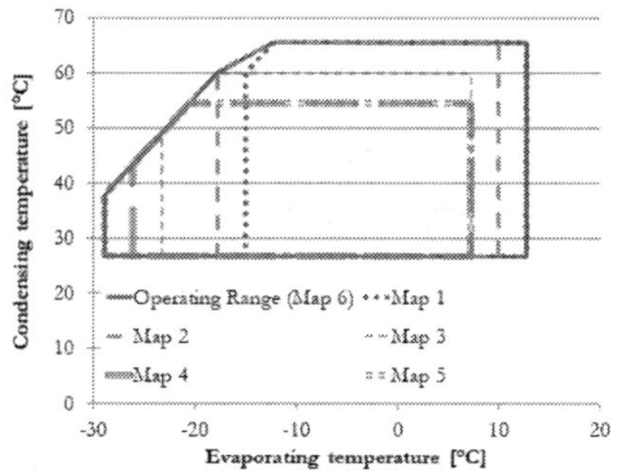

Figure 2. Range of training data for different maps.

Figure 2 shows that Maps 1 to 5 have their ranges of training data shifting from the top right-handed corner of the operating range of the compressor to its bottom left-handed corner, while Map 6 covers the entire operating range. Since it is unfair to compare maps that are generated by different number of data points, the training data points in each map are arranged such that each map contains 70 data points.

6.2. Effect of range of training data on map output

To study how the range of training data affects the extrapolation uncertainty and accuracy, the difference between the map output and approximated measurement and the map output uncertainty for all data points in Figure 1 are plotted in Figure 3, and a similar figure with their uncertainty components is plotted in Figure 4.

Figure 3 shows that inaccurate map outputs are associated with higher uncertainty and the uncertainty calculation method is a good indicator of the accuracy of the map output. Figure 4 shows that the high uncertainty of the inaccurate data points in Figure 3 are primarily a result of high uncertainty from model random error and training data.

The increase of uncertainty from model random error with inaccuracy can be explained by the leverage term $\vec{x}^T(\mathbf{X}_{train}^T\mathbf{X}_{train})^{-1}\vec{x}$ in Eqn. (19) which increases as the map extrapolates, and map extrapolation results in lower accuracy. Hence the uncertainty from model random error increase with a decrease of the map accuracy and applicability.

To explain the increase of uncertainty from training data with a decrease of map accuracy in Figure 4, the squared terms in Eqn. (18) are labeled as uncertainty from training data per measurement, and their values from two map outputs of Map 1 are plotted as Figures 5 and 6.

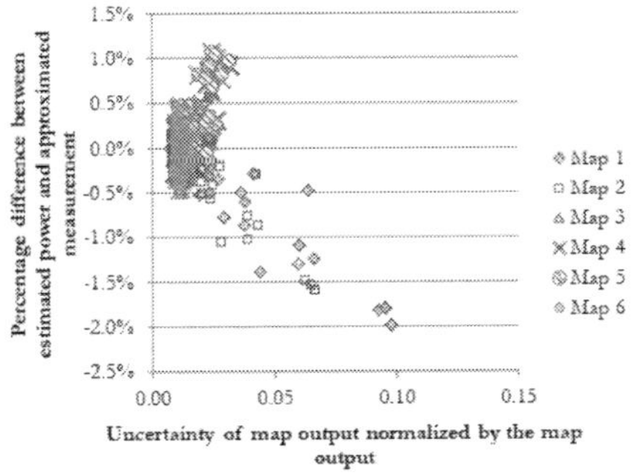

Figure 3. Change of accuracy of maps with output uncertainty in different maps.

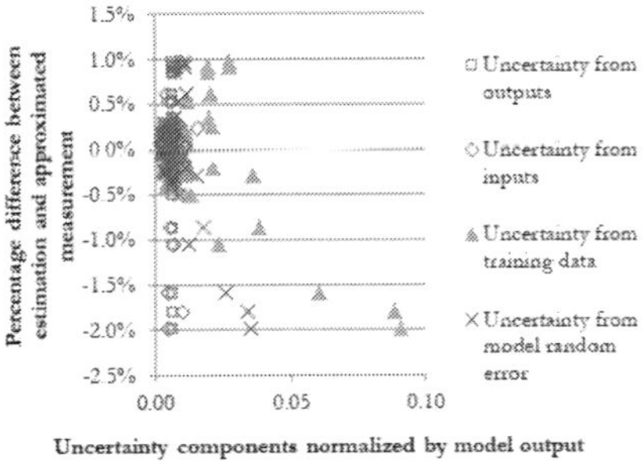

Figure 4. Change of accuracy of maps with different uncertainty components.

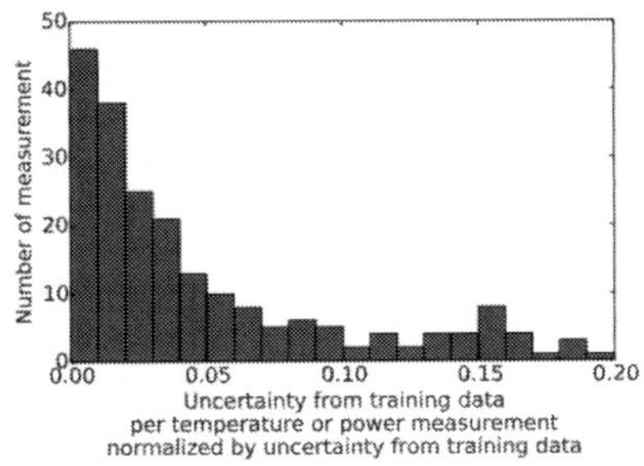

Figure 5. Histogram of uncertainty terms in Eqn. (18) for $T_{evap} = -1.1°C$ and $T_{cond} = 60.0°C$.

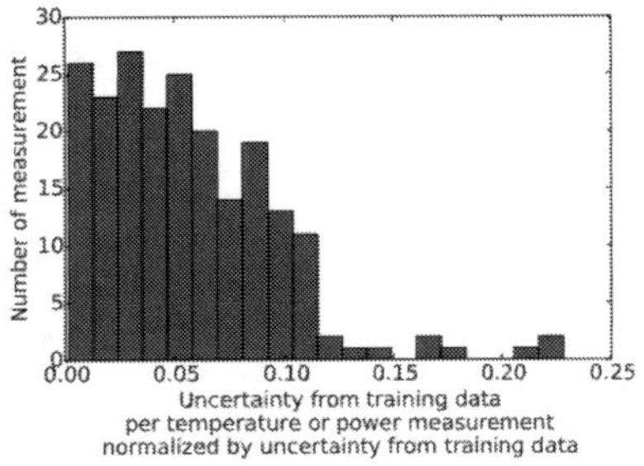

Figure 6. Histogram of uncertainty terms in Eqn. (18) for $T_{evap} = -28.9°C$ and $T_{cond} = 26.7°C$.

Figure 5 shows a histogram that is more left-skewed than Figure 6. This is because Figure 5 is obtained from a data point inside the training data range of Map 1 (see Figure 2), and the map only needs information from a few data points around $T_{evap} = -1.1°C$ and $T_{cond} = 60.0°C$ to estimate its map output. Hence the estimation does not depend on most of the data points, and their training data uncertainties do not propagate to the map output as shown by Figure 5.

However, Map 1 extrapolates to the lower left handed corner in Figure 2 for the condition in Figure 6, and the estimation is significantly affected by multiple data points in the training data. If any of these significant training data points change, the estimation result at this condition will be changed significantly. Hence much more training data points propagate their uncertainty to the map output under the condition in Figure 6 than that in Figure 5. This also explains that the uncertainty of training data is a significant component in the extrapolation uncertainty of the map output, and the uncertainty increases as the map applicability and accuracy decrease.

6.3. Effect of number of training data points on map output

To understand how the number of training data points affect the map output, another map (Map 7) was created with the same range of training data as Map 1 and 24 data points only, as shown in Figure 7. The accuracy of Map 7, rated by coefficient of variation *cov* as

IOP Conf. Series: Materials Science and Engineering **90** (2015) 012078 doi:10.1088/1757-899X/90/1/012078

$$cov = \frac{\text{Mean of the sum of squares of errors}}{\text{Mean value of estimated values}} = \frac{\sigma}{1/n \cdot \Sigma_1^n \hat{y}_i} \quad (22)$$

is compared with that of Map 1 in Table 2.

Table 2. Accuracy of Maps 1 and 7.

Map	Coefficient of variation based on training data poinst only (cov_{train})	Coefficient of variation based on all data points in Figure 1 (cov_{all})
Map 1	0.18%	0.31%
Map 7	0.17%	1.10%

Table 2 shows that the accuracy of Map 7 is similar to that of Map 1 at the training data points, but the analysis with all available data points shows that Map 7 is less accurate than Map 1. This shows that using fewer training data points reduces map accuracy. To understand if it is related to extrapolation, the difference between the power consumption estimation and their approximated measurement with the distance, e.g. $\sqrt{\Delta T_{cond}^2 + \Delta T_{evap}^2}$, from the nearest training data point as shown in Figures 8 and 7.

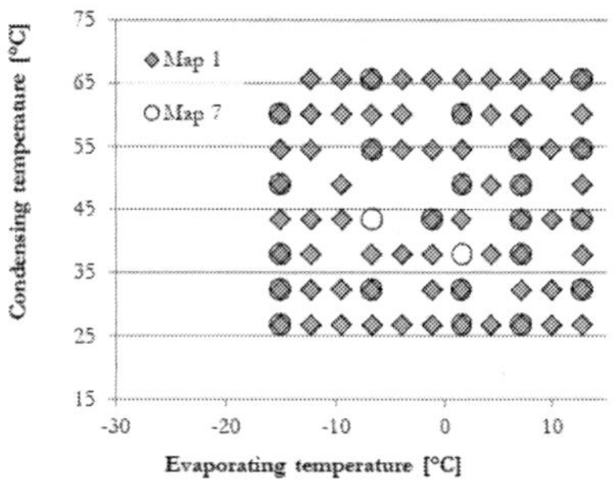

Figure 7. Training data points for Maps 1 and 7. Note the absence of low temperature training data.

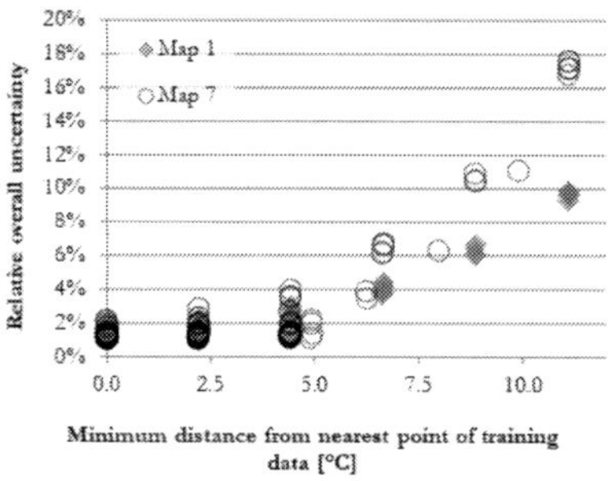

Figure 8. Change of accuracy with distance to nearest training data point for maps 1 and 7.

As shown in Figures 8 and 7, although the range of training data points in both maps is the same, the effect of the reduced number of points in the training data is pronounced for large distances (e.g. extrapolated outside training data range of map) of 10°C or larger and can increase the uncertainty from 10% (Map 1) to around 17% (Map 7) in the most extreme case.

This shows that the inaccuracy due to the reduction of training data can be shown by uncertainty. To understand why uncertainty can show this, Figures 9 and 10 are plotted.

Figures 9 and 10 show that both the accuracy of Maps 1 and 7 decreases as their uncertainty from model random error and the uncertainty from training data increase, and the reduction of Map 1 accuracy in both figures are more rapid than Map 7. This shows that Map 7, despite having the same training data range as Map 1, is less applicable and accurate as Map 1. This shows that extrapolation uncertainty and accuracy are not only affected by the range of training data but also the number of training data points.

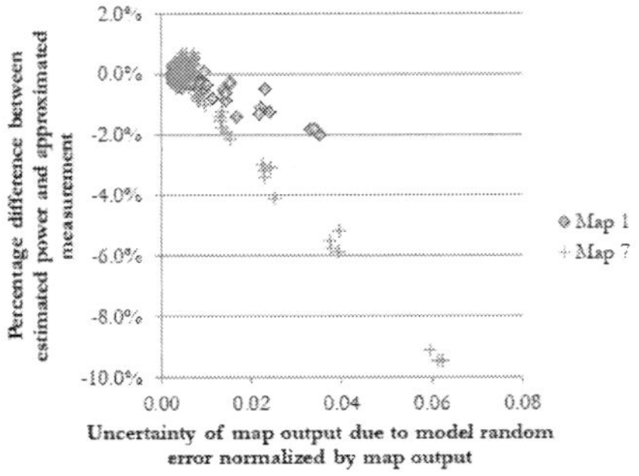

Figure 9. Change of accuracy of maps with uncertainty from model random error in Maps 1 and 7.

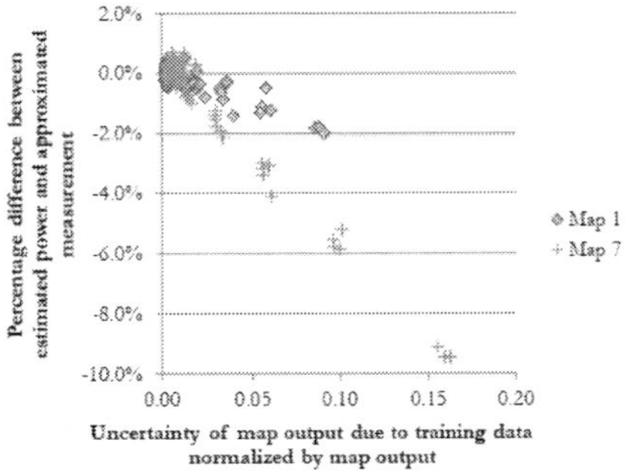

Figure 10. Change of accuracy of maps with uncertainty from training data in Maps 1 and 7.

7. Conclusions

To conclude, this paper demonstrates how uncertainty of 10-coefficient compressor maps can be calculated based on the uncertainty due to inputs, the uncertainty due to model random error, the uncertainty due to training data and the uncertainty from outputs. It shows a method to assess the acuracy and applicability of the map output under extrapolation without measuring the output variable. It also shows that the range of training data and the number of training data points affect the accuracy and applicability of the map.

8. Future work

One of the basic assumptions in this paper is that individual measurement values are normally distributed around their average value. However, this might not be the case for actual measurement data. Therefore, experiments with actual compressor measurements should be conducted to check whether this assumption is appropriate.

References

[1] ANSI/AHRI standard 540: 2004 standard for performance rating of positive displacement refrigerant compressors and compressor units.
[2] EN 12900:2013. Refrigerant compressors - rating conditions, tolerances and presentation of manufacturer's performance data.
[3] Emerson Climate Technologies 2006 Copeland CF and CS compressors 50 hertz - the final solution for HFC performance. Tech. rep. Sidney, Ohio
[4] Bristol Compressors International Inc 2015 Performance table for H23A463DBL, revision 5. Tech. rep.
[5] B Shen, O Abdelaziz, and C K Rice 2014 Compressor selection and equipment sizing for cold climate heat pumps. *2014 IEA HP conference*
[6] S L Caskey, D Kultgen, E A Groll, W Hutzel, and T Menzi 2012 Simulation of an air-source heat pump with two-stage compression and economizing for cold climates.
[7] B N Taylor and C E Kuyatt 1994 *NIST Technical Note 1297, 1994 Edition: Guidelines for Evaluating and Expressing the Uncertainty of NIST Measurement Results* (Gaithersburg: National Institute of Standards and Technology)
[8] L Song, G Wang, and M R Brambley 2013 *HVAC&R Research* **19** 335 – 345
[9] D C Montgomery 2005 *Design and Analysis of Experiments.* 6th ed (Jon Wiley & Sons)
[10] F A Graybill and H K Iyer 1994 *Regression Analysis: Concepts and Applications.* (Duxbury Pr)
[11] A C Atkinson 1987 *Transformation and Regression.* (Oxford Statistical Science Series)

The origin of gas pulsations in rotary PD compressors: ΔP or ΔU?

P X Huang, S Yonkers
Hi-Bar MC Technologies, LLC, USA

D Hokey
GE Oil & Gas, USA

ABSTRACT

This paper is a continuing work from the same authors on the same topic by analogizing the transient shock tube process to the gas pulsation phenomena [1, 2]. It has been demonstrated in the previous theoretical investigation that gas pulsations can be generated either by a sudden velocity change (ΔU) say from a piston or lobe movement, or by the sudden opening of a pressure difference (Δp_{41}) such as during the discharge phase of a rotary PD (Positive Displacement) compressor (screw, scroll or Roots) under UC (Under-Compression) or OC (Over-Compression) conditions. It is further reasoned that the more dominant pulsation source should be from the Δp_{41} induced source, but without backing from experiments.

This paper is aimed at validating the theoretical predictions through experimental investigation conducted on a Roots blower, a special 100% UC case. This further clarification is important not only to prove the theory, but would pave the way for potential gas pulsation control methods if exact origin and nature of gas pulsations can be located and understood. Two schemes of tests under different load and speed conditions were carried out and results confirm the previous analytical conclusion that the dominant source of the gas pulsations for the 100% UC case is mainly from the sudden release of a pressure difference (Δp_{41}) while the non-uniform rotor movement (ΔU) induced pulsation is about one order of magnitude lower. More experimental work on other types of rotary compressors such as screw or scroll is desirable in the future.

NOMENCLATURE

CW	compression waves pulsation component in PD compressor, shock wave in shock tube
EW	expansion waves pulsation component in PD compressor and shock tube
IFF	induced fluid flow pulsation component
OC	over compression
P, p	absolute gas pressure
T	absolute temperature
UC	under compression
ΔU	piston or contact surface velocity in shock tube, IFF velocity in PD compressor
W	shockwave velocity in shock tube, CW velocity in PD compressor
ρ	gas density

Subscripts

1	initial low pressure in shock tube, or PD compressor inlet, in cavity
2	pressure after shock wave in shock tube
4	initial high pressure in shock tube or at compressor outlet or jet port
jet	Roots blower jet port
inlet	Roots blower inlet port
outlet	Roots blower outlet port

1. INTRODUCTION

Gas pulsations are a group of phenomena characterized by the simultaneous flow rate and pressure fluctuations with relatively low frequency and high amplitude. They commonly exist in HVACR, energy and other processing

Content from this work may be used under the terms of the Creative Commons Attribution 3.0 licence. Any further distribution of this work must maintain attribution to the author(s) and the title of the work, journal citation and DOI.
Published under licence by IOP Publishing Ltd

industries, and are widely accepted to be inherently associated with PD type gas machinery such as reciprocating or rotary compressors, expanders and Roots type blowers. Moreover, they are believed to be one of the primary sources responsible for exciting or inducing system vibrations, noises and fatigue failures.

The conventional methods of modeling gas pulsations are mostly acoustic models based on small perturbation method or using various CFD schemes aimed at solving nonlinear differential equations for unsteady flows. However, as demonstrated in Table 1 [1], the magnitude of gas pulsations generated by industrial gas machinery is typically ranging from 0.02 – 2 bar (0.3 - 30 psi) or equivalent to 160-200 dB. By comparison, the pressure fluctuations of acoustic waves are typically less than 0.0002 bar (0.003 psi) or equivalent to the pressure level of 120 dB as the limit for using the linear acoustic equations according to Beranek [3].

Table 1: Magnitude of acoustic waves in comparison with industrial gas pulsations from Huang [1]

	Acoustic Waves ⟸⟹			Gas Pulsations			
Pressure Pulsation, bar	0.000002	0.00002	0.0002	0.002	0.02	0.2	2
Pressure Pulsation, psi	0.00003	0.0003	0.003	0.03	0.3	3	30
Sound Pres. level, L_p, dB	80	100	120	140	160	180	200

In order to get a glimpse of a more realistic physical picture of the gas pulsation phenomena while avoiding solving the full set of non-linear differential equations, the present author has proposed an alternative shock tube theory [1, 2] by analogizing the transient shock tube process to the gas pulsation phenomena as occurred typically in the discharge phase of rotary PD compressors such as screw, scroll or Roots. It has been demonstrated that gas pulsations can be generated either by a sudden velocity change (ΔU) say from a lobe or valve movement, or by the sudden opening of a pressure difference ($\Delta p_{41} = p_4 - p_1$) such as during the discharge phase of a rotary compressor (screw, scroll or Roots) under an UC (Under-Compression) or OC (Over-Compression) condition. It is further argued that the more dominant pulsation source should be from the later case (Δp_{41} induced), but without giving experimental proof.

This paper is aimed at validating the above theoretical predictions by the shock tube theory through experimental work. This would pave the way for devising new gas pulsation control strategy if the exact source and nature of gas pulsations can be located and understood. A brief review of the shock tube theory will be conducted first and followed by experimental investigation. The results are then compared between theory and test and conclusions are drawn.

2. GAS PULSATION SOURCES: A SHOCK TUBE THEORY

2.1 Brief Review of the Shock Tube Theory and Results

Refer to Huang [1, 2] for more detailed descriptions of the shock tube analogy with gas pulsations of PD compressors. However, highlights are summarized below in order to provide necessary bases for experimental comparison later. In general, two types of primary sources of gas pulsations [4] can be correlated to two types of shock tubes shown in Figures 1 & 2. The first type is Δp induced gas pulsation simulating the suddenly burst diaphragm shock tube and the second type is ΔU induced gas pulsation simulating a suddenly accelerated piston shock tube. An analogy is established between the shock tube at the moment just before and after the sudden diaphragm bursting and the discharge process of a rotary screw compressor under off-design conditions of an UC or OC, and the analogous equivalents are correlated in Table 2 and Figure 3.

Table 2: Analogy – diaphragm shock tube vs. rotary screw discharge under UC or OC

Device: shock tube	Device: rotary screw outlet
moment: before ($t = 0^-$) and after ($t = 0^+$) diaphragm opening	moment: before ($t = 0^-$) and after ($t = 0^+$) lobe opening
location: near front and back of the diaphragm with Δp_{41}	location: opening lobe experiencing Δp_{41}
high pressure region: p_4	outlet side pressure (UC): p_4
low pressure region: p_1	cavity side pressure (UC): p_1

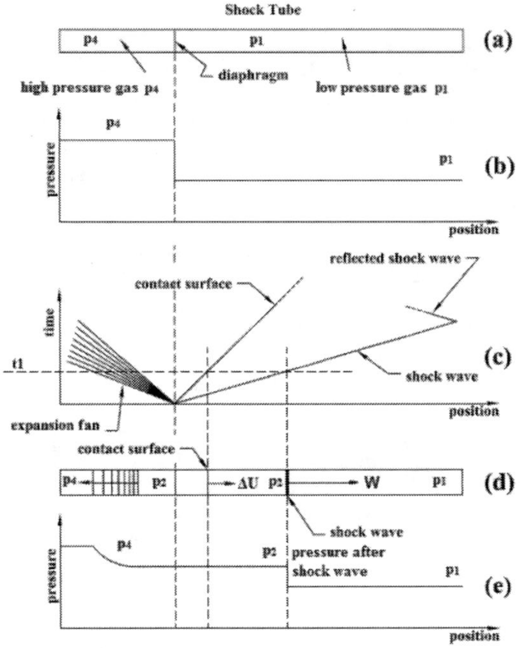

Figure 1(a-e): Diaphragm-triggered shock tube wave diagram from Anderson [4]

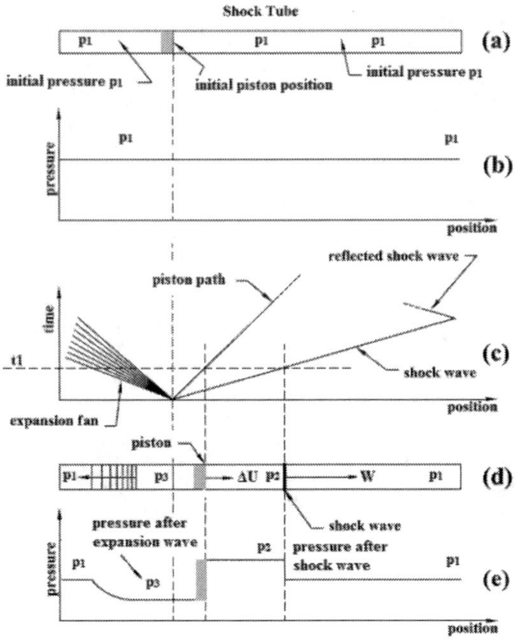

Figure 2(a-e): Piston-triggered shock tube wave diagram from Anderson [4]

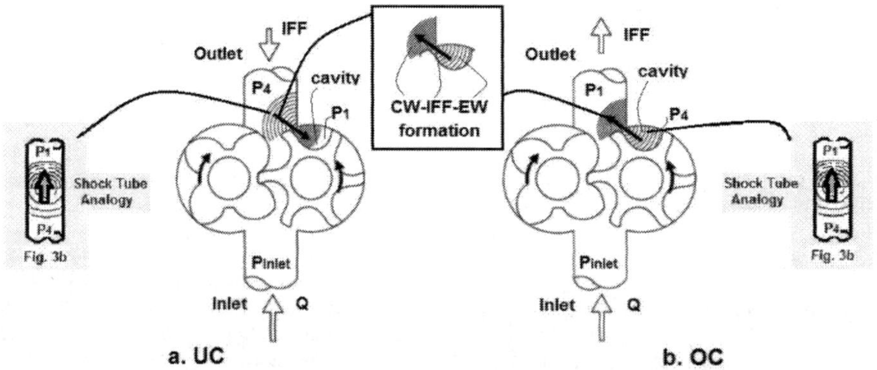

Figure 3(a-b): Gas pulsation sources of a screw compressor according to the shock tube theory from [2]

The results of the diaphragm shock tube analogy can be best summarized by the Gas Pulsation Rules [1, 2] as quoted below. In theory, these rules are applicable to different gases and for gas pulsations generated by different types of PD gas machinery, such as engines, expanders, pressure compressors and vacuum pumps, reciprocating or rotary.

1. Rule I: For any two divided compartments, either moving or stationery, with different gas pressures p_1 and p_4, there will be no or little gas pulsations generated if the two compartments stay divided (isolated from each other).

2. Rule II: If, at an instant, the divider between the high pressure gas p_4 and the low pressure gas p_1 is suddenly removed in the direction of divider surface, gas pulsations are instantaneously generated, at the location of the divider and at the instant of the removal, as a composition of a fan of Compression Waves (CW) or a quasi-shockwave, a fan of Expansion Waves (EW) and an Induced Fluid Flow (IFF) with magnitudes as follows:

$$CW = p_2 - p_1 = p_1 [(p_4/p_1)^{1/2} - 1] = (p_4 \times p_1)^{1/2} - p_1 \quad (1)$$

$$EW = p_4 - p_2 = CW * (p_4/p_1)^{1/2} = p_4 - (p_4 \times p_1)^{1/2} \quad (2)$$

$$\Delta U = (p_2 - p_1)/ (\rho_1 \times W) = CW/ (\rho_1 \times W) \quad (3)$$

Where ρ_1 is the gas density at low pressure region, W is the speed of the lead compression wave, ΔU is the velocity of Induced Fluid Flow (IFF);

3. Rule III: Pulsation component CW is the action by the high pressure (p_4) gas to the low pressure (p_1) gas while pulsation component EW is the reaction by low pressure (p_1) gas to high pressure (p_4) gas in the opposite direction, and their magnitudes are such that they approximately divide the pre-open pressure ratio p_4/p_1, that is, $p_2/p_1 = p_4/p_2 = (p_4/p_1)^{1/2}$. At the same time, CW and EW pair together to induce the third pulsation component, a unidirectional fluid flow IFF in a fixed formation of CW-IFF-EW.

In general, Gas Pulsation Rules explain the relationship between an UC (or OC) and gas pulsations as a cause (pre-opening pressure difference $p_4 - p_1$) and the effect (post-opening results) that are the two aspects of the same phenomena. Moreover, Rules I & II give the two sufficient conditions that link the UC (or OC) and gas pulsation events:

a) The existence of a pressure difference $p_4 - p_1$ from either an UC or OC;
b) The sudden opening of the divider separating the pressure difference $p_4 - p_1$.

Based on these two conditions, it can be determined that the location and moment that trigger the under compression action and gas pulsation generation are at the discharge and at the instant when the discharge port suddenly opens. Because all PD compressors or expanders convert energy between shaft and gas by dividing incoming continuous gas stream into parcels of cavity size and then discharges each cavity separately at the end of each cycle, there always exists a "sudden" opening at discharge phase to return the discrete gas parcels back to a continuous gas stream again. Therefore both sufficient conditions are satisfied at the moment of the discharge opening if compressor operates at off-design conditions such as an UC or OC.

Rule I also implies that there would be no or little gas pulsations during the suction, transfer and internal compression phases of a cycle because of the absence of either a pressure difference ($p_4 - p_1$) or a sudden opening. The focus instead should be placed upon the discharge phase, especially at the moment when the discharge port is suddenly opened and under off-design conditions like either an UC or OC.

Rule II also reveals the nature and composition of gas pulsations as a combination of large amplitude Compression Waves (CW) or a quasi-shockwave, a fan of Expansion Waves (EW) and an Induced Fluid Flow (IFF). These waves are non-linear waves with ever changing wave front during propagation. This is in direct contrast to the acoustic waves that are linear in nature and wave fronts stay the same and do not induce a mean through flow. It is also noted that the three different components (CW, EW and IFF) are generated as a homologous, inseparable whole simultaneously and in a fixed formation CW-IFF-EW. It is believed that this formation reflects the dynamics of the transient UC (or OC) and pulsation events with the wave fronts CW and EW as the moving forces driving the induced fluid flow IFF in between. In turn, the source of CW and EW is simply a re-distribution of the pre-opening UC (or OC) pressure difference Δp_{41} that is now being suddenly released and turned into a moving force pushing the flow (IFF) at front by CW and pulling the flow (IFF) from behind by EW at the same time. This new physical picture implies that gas pulsations would be difficult to control because it's not one component (just IFF as suggested by the conventional theory) but all three components have to be dealt with as a whole.

Rule III shows further that the interactions between two gases at different pressure are mutual so that for every CW pulsation, there is always an equal but opposite EW pulsation in terms of pressure ratio ($p_2/p_1 = p_4/p_2$). Team together, they induce a unidirectional fluid flow pulsation (IFF) in the same direction as the compression waves (CW).

2.2 Quantitative Comparison of Pulsation Strength of ΔP Induced with ΔU Induced

To get a quantitative idea for the two primary pulsation sources, Δp_{41} (=p_4- p_1) and ΔU, Table 3 lists some typical values of industrial strength pulsation components: CW, EW and IFF, corresponding to Δp_{21} (=p_2-p_1), Δp_{42} (=p_4-p_2) and ΔU respectively from Equations (1)-(3). According to Equation (3), ΔU value in Table 3 can also be interpreted as the velocity of a piston-triggered shock tube that generates a CW gas pulsation with magnitude of Δp_{21} (=p_2-p_1) on the pressure side and an EW gas pulsation with magnitude of Δp_{42} (=p_4-p_2) on the vacuum side.

Table 3: Comparison of gas pulsation strengths: ΔP_{41} vs. ΔU

W in Mach Number	1.03	1.06	1.09	1.12	1.15	1.18	1.21
p_4/p_1	1.146	1.306	1.480	1.666	1.870	2.080	2.312
Δp_{41} (=p_4 - p_1), bar	0.146	0.306	0.480	0.666	0.867	1.083	1.310
EW, Δp_{42} (=$p_4 - p_2$), bar	0.071	0.162	0.259	0.369	0.491	0.625	0.770
CW, Δp_{21} (=p_2 - p_1), bar	0.075	0.144	0.220	0.297	0.376	0.458	0.542
IFF, ΔU, m/sec	17.1	33.7	50.0	65.7	81.1	96.2	111

From Table 3, it can be seen that a sudden release of a pressure difference Δp_{41} (=2.2 psi=0.146 bar) would generate an induced flow with velocity of ΔU (=17.1 m/s). On the other hand, it can be interpreted that an abrupt velocity change ΔU (=17.1 m/s) from a piston or lobe or valve movement would generate a CW pulsation front Δp_{21} (=1.1 psi=0.075 bar) and an EW pulsation front Δp_{42} (=1.0 psi=0.071 bar). But for rotary type PD compressors such as screw or scroll or Roots, it was argued in [1] that it is common to see Δp_{41} range from a few psi up to tens of psi but rarely see an abrupt rotary lobe velocity change (acceleration or deceleration) more than 17 m/s. This is due to the nature of smoother kinematics of rotary movement in contrast to reciprocating type.

From Gas Pulsation Rules I & II, Δp_{41} induced gas pulsations take place only at the moment of the discharge phase when the compression cavity is suddenly opened to an outlet pressure difference, a transient (point) source. On the other hand, ΔU induced gas pulsations occur all the time due to the constant interactions between gas and positive displacement driver surrounding the compressor cavity. Figure 4 plots both Δp and ΔU induced gas pulsations in terms of pressure ratios (p_2/p_1) or (p_4/p_2) as a function of UC or OC pressure ratio (p_4/p_1) of a compressor on logarithm scale so that an equal EW and CW relationship exists according to Pulsation Rule III.

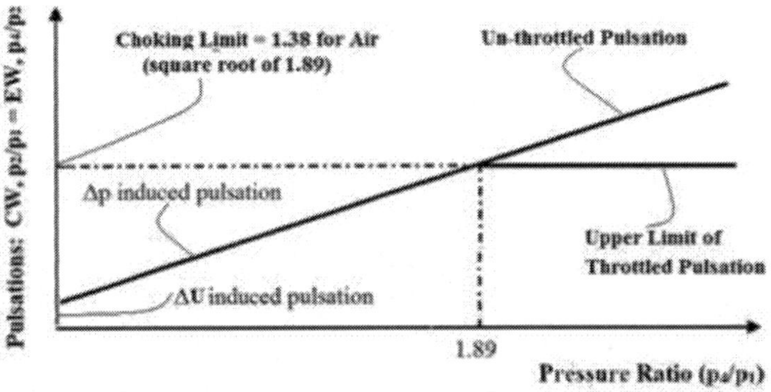

Figure 4: Gas pulsations CW, IFF and EW as a function of pressure ratio of UC or OC

It can be seen from Figure 4 that ΔU is predicted to be small when compared with Δp_{41} for rotary type PD compressors such as screw, scroll or Roots. However, this prediction is based on pure theoretical speculation. The challenging question this paper tries to answer is: will experiments support this speculation?

3. GAS PULSATION SOURCES: EXPERIMENTAL INVESTIGATION

3.1 Test Objective and Test Machine Selection

The objective of experimental investigation is to find out which source, ΔU or Δp_{41}, is a dominant source either at the inlet or outlet of a rotary PD compressor that could influence upstream and downstream flow. In practice, it is difficult to distinguish the effect of ΔU from Δp_{41} due to the fact that two are mingled together. It was reasoned above that unlike the Δp_{41} induced gas pulsations that take place only at the moment when the compression cavity is suddenly opened to an outlet pressure difference, ΔU induced pulsations occur all the time because of the inherent nature of internal compression for most of the positive displacement type compressors such as screw or scroll. However, Roots type compression is an exception, the only one that possesses NO internal compression (or 100% pure UC), while has a rotary kinematic motion and rotary discharge valve that are similar to those of a screw or scroll.

For clarification, let's exam a complete cycle of a classical Roots blower as illustrated from Figures 5a to 5d by following one flow cell in a typical 3-lobe configuration. In Figure 5a, low pressure gas first enters the space between lobes of a pair of rotors as they are open to inlet during their outward rotation from inlet port to outlet port. At the lobe position shown in Figure 5b, the gas becomes trapped between the two neighboring lobes and casing forming a cavity as it is transported from the inlet to the outlet. Then the trapped gas in cavity is suddenly opened to higher pressure outlet as shown in Figure 5c. According to the shock tube theory for UC [5], the lobe opening phase in Figure 5c resembling the diaphragm bursting of a shock tube as shown in Figure 1d would generate a series of compression waves that sweep through the low pressure gas inside the cavity and compresses it at the same time at the speed of wave. After the compression, the rotors continue to move until lobes from two rotors meet again, mesh out the compressed gas from the cavity to outlet port and return to inlet suction position to start the next cycle, as shown in Figure 5d.

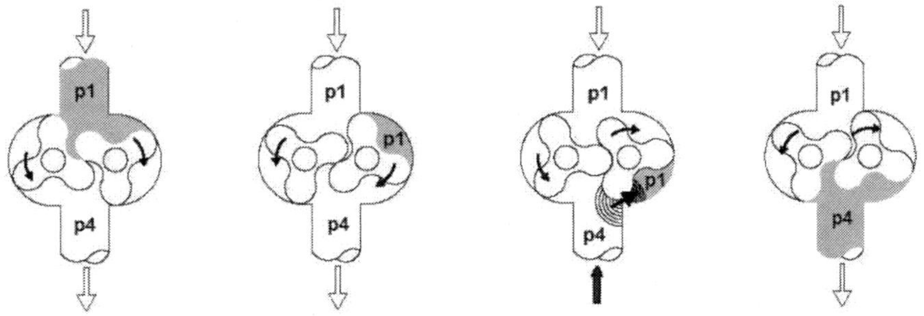

a. Suction **b. Transfer** **c. Compression** **d. Discharge**

Figure 5: Compression cycle of a classical Roots according to the shock tube theory

The Roots cycle is unique because there is NO cavity volume variation during the transport and compression phases. That is, there is NO internal compression like the constant meshing of twin screw rotors or scroll disks. The meshing lobes of Roots only interact together with gas during the suction and discharge phases while during the transfer and compression phases, the gas stays inside the fixed volume cavity formed by the two neighboring lobes and casing.

According to Gas Pulsations Rules I & II, the Roots transfer phase should have NO pulsations because of the absence of any pressure difference or lobe meshing. Lobe interaction only takes place during the Roots suction and discharge phases that could induce ΔU type pulsations. Moreover, there should exist strong Δp_{41} type pulsations during the Roots discharge phase because it is a case of 100% UC. It can be further anticipated that the larger the pressure rise from Roots inlet to outlet, the higher Δp_{41} pulsations would be induced according to Rule II. The resulting gas pulsations are a composition of pressure waves (CW, EW) and induced backflow (IFF) as shown in Figure 5c. Because of its unique features, the Roots compression is selected as the test bed and a 75HP Roots blower is prototyped first for this purpose.

3.2 Test Schemes for Source Differentiation

Two test schemes are devised to differentiate the contribution between ΔU and Δp_{41}. Scheme A is to measure directly the gas pulsations at the conventional blower inlet and outlet as shown in Figure 6a under different speed (relate to ΔU effect) and pressure rise (relates to Δp_{41} effect) conditions. The strategy is, since it is expected that the inlet pulsation has dominant contribution from ΔU (little Δp effect) while outlet pulsations has dominant contribution from Δp_{41}, the comparison of the pulsation magnitude of the inlet and outlet pulsations would tell which effect is dominant at the inlet or outlet. Note that the case for No load condition (when there is no pressure rise from inlet to outlet, or $\Delta p_{41} = 0$) is special because it can reveal the full effect of ΔU pulsation without any contribution from Δp_{41}.

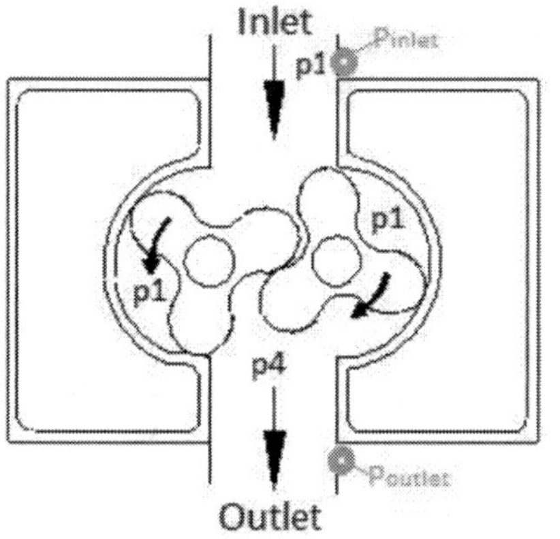

Figure 6a: Test scheme-A setup

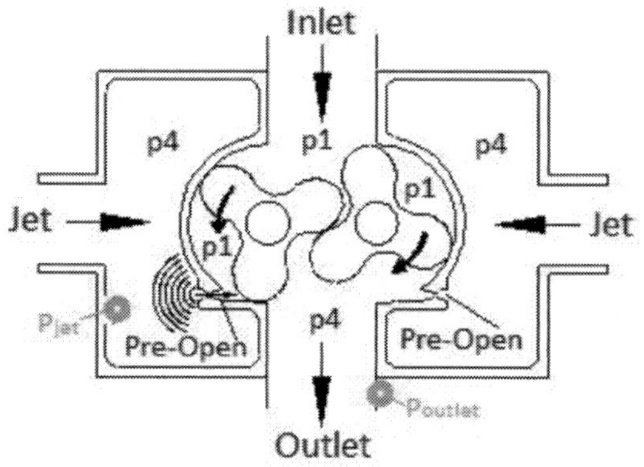

Figure 6b: Test scheme-B setup

An alternative way or scheme B is devised to totally separate Δp_{41} effect from ΔU effect by measuring gas pulsations at a non-conventional jet port connected to blower cavity through a pre-opening as shown in Figure 6b. The position of the pre-opening is such that it is at least 120 degrees away (for 3-lobe) or 180 degrees away (for 2-lobe) from the inlet opening, but long before the discharge opening. Remember that during the transfer phase of Roots cycle, there is neither rotor meshing effect nor cavity volume variation. Moreover, lobe is rotated at a constant speed with no acceleration or deceleration until it rotates through the pre-opening and the cavity is suddenly opened to the pressure difference Δp_{41}. In another word, this location is expected to "feel" the full Δp_{41} effect but little ΔU effect. To physically isolate Δp_{41} effect measured at the jet port from ΔU effect measured at the outlet, the test is setup as an exhauster so that the inlet is subjected to a vacuum load (Δp_{41}) while both jet and outlet ports are separately open to the same atmospheric pressure.

3.3 Test Results and Discussion

The measured gas pulsations at different speeds and loads are shown as colored points in Figures 7a and 7b respectively for schemes A and B. For consistency, both tests use the same dynamic pressure probes and recording instruments, and the test piping setup follows ASME PTC-9 specification (not shown in Figures 6a & 6b).

Test results of scheme A in Figure 7a show that the inlet pulsation is low and almost at a constant level around 0.2-0.4 psi, independent of either the rotor speed (range from 2400-3600 RPM, nominal) or pressure rise Δp_{41} (range from 1-16 psi). While the outlet pulsation is much higher, ranging from 0.4–2.4 psi and approximately proportional to the pressure rise from inlet to outlet. The blower speed does not show much effect at the outlet either as at the inlet. It is also interesting to note that an extended line from the average of the outlet pulsation data almost intersects with the extension line from the average of the inlet pulsation data at the point where $\Delta p_{41} = 0$. This suggests that the ΔU effect from the outlet side is almost the same as the inlet side, agreeing with predictions made from Pulsation Rule II that there would be no or little gas pulsations at both the suction (inlet) and discharge (outlet) under the no load condition ($\Delta p_{41} = 0$). The trend and order of magnitude of outlet gas pulsation shown in Figure 7a are also consistent with theoretical predictions in Figure 4 that stronger gas pulsation takes place during the compression and discharge phase when lower pressure cavity is suddenly opened to higher pressure outlet.

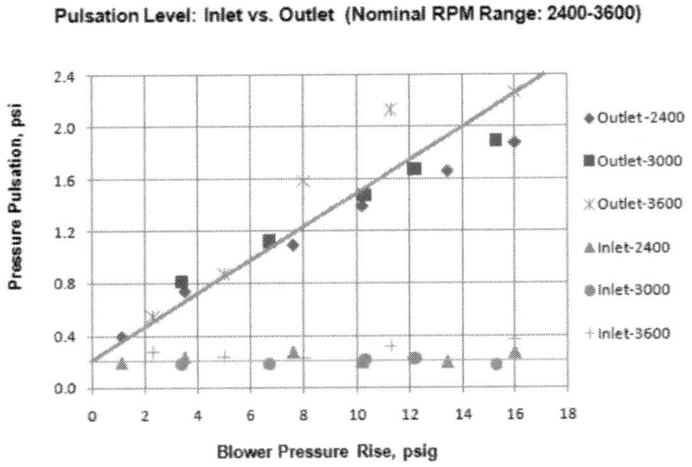

Figure 7a: Results of test scheme-A

Test results of scheme B in Figure 7b show that gas pulsations can be triggered to take place before the outlet upon the sudden pre-opening of the low pressure cavity to higher pressure jet port shown in Figure 6b. Because test is conducted under an exhauster mode so that outlet and jet ports can see the same atmospheric pressure, the level of absolute pressure rise from inlet to outlet is smaller than the blower mode as shown for scheme A with the same pressure ratio. The result in Figure 7b shows that pulsation at the jet port is high ranging from 0.2 to 0.9 psi while varying proportionally with pressure rise (Δp_{41}), the same trend as the outlet pulsation in Figure 7a. However, the outlet pulsation is now low, less than 0.15 psi and is almost a constant, which is the same trend as the inlet pulsation in Figure 7a and is also independent from the exhauster pressure rise (Δp_{41}) and speed. It is interesting again to note that an extended line from the average of the scattered jet pulsation data intersects with the zero point of both pulsation and pressure rise where $\Delta p_{41} = 0$. This suggests that the jet port does not see much of the ΔU effect, unlike the outlet gas pulsations that have both Δp_{41} and ΔU effect mingled as shown in Figure 7a. This observation is consistent with the theoretical prediction discussed in paragraph of "3.2 Test Schemes for Source Differentiation" that the jet port is the transient source "point" expected to "feel" 100% Δp_{41} effect so that outlet has only 100% ΔU effect left for test setup shown in Figure 6b.

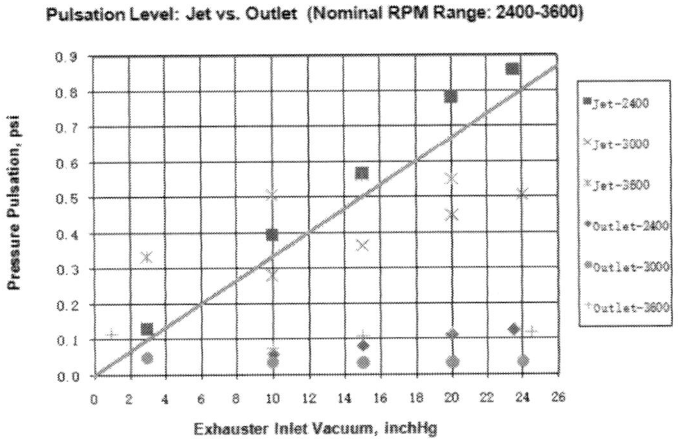

Figure 7b: Results of test scheme-B

IOP Conf. Series: Materials Science and Engineering **90** (2015) 012079 doi:10.1088/1757-899X/90/1/012079

More important is the implication from the fact shown by scheme B and Figure 7b that the dominant Δp_{41} source can be actually separated from the ΔU source and directed away from the outlet to a different location that could be eventually treated and controlled. In fact, this finding or insight is the basis for a new pulsation control method, called shunt pulsation trap that is beyond the scope of this paper. However, interested readers can refer to references [2, 6] for more details.

4. CONCLUSIONS

Pressure pulsations of a Roots blower (100% Under Compression) are measured at inlet/outlet for scheme A and at outlet/jet port as scheme B under different speed and load conditions. Though there are some scattering among the dynamic pressure data due to the wave nature of pressure pulsations, conclusions can be drawn based on the average behaviors from these tests as follows:

1. The ΔU gas pulsation mainly takes place at the inlet and outlet of the blower where two rotors are engaged in constant meshing interaction. However, the pulsation level due to ΔU effect is low (about an order of magnitude lower) compared with Δp_{41} gas pulsation and is independent of rotor speed and blower loads (pressure rises Δp_{41});
2. The Δp_{41} gas pulsation mainly takes place at the outlet of the blower triggered by the sudden opening of the low pressure cavity with $p_1$to the higher pressure outlet with p_4. Moreover, the pulsation level due to Δp_{41} ($=p_4-p_1$) is almost one order of magnitude higher than the ΔU effect, and is approximately proportional to the magnitude of blower load Δp_{41};
3. The Δp_{41} gas pulsation (a point source) can be separated from the ΔU gas pulsation (a constant source at inlet and outlet from rotor pair and gas interacting) by a pre-opening or jet port.

Conclusions 1-2 validate experimentally some of the theoretical predictions from the previous papers [1, 2], that is, ΔU pulsation is taking place throughout the rotor intermeshing process at the inlet and outlet while Δp_{41} pulsation takes place at a single point and moment when and where the Δp_{41} is suddenly exposed. This insight leads to Conclusion 3 that is the foundation for a new pulsation control method, called Shunt Pulsation Trap, as detailed in previous papers [2, 6].

Though the tests conducted in this work is on a Roots type blower that is operating under 100% under-compression (UC), the general principle should also apply to other types of rotary PD compressors such as screw or scroll with a partial UC or OC. However, direct experiments are more desirable for rotary PD compressors that possess internal compression as well as UC and OC. With a better understanding of the nature of gas pulsations, it is anticipated that future generation of rotary PD compressors can be designed to be even simpler in structure, smaller in size and smoother running than those used today.

REFERENCES

1. Huang, P., *Gas Pulsations: A Shock Tube Mechanism.* The 2012 International Compressor Engineering Conference at Perdue, 2012a.
2. Huang, P., Yonkers, S., Hokey, D., *Screw Pulsation Generation and Control: A Shock Tube Mechanism.* The 2013 International Compressor Engineering Conference at City University, 2013.
3. Beranek, L., *Noise & Vibration Control*, Inst. of Noise Control Engineer, New York, p. 396, 1988.
4. Anderson, J., *Modern Compressible Flow*, McGraw-Hill Book Company, New York, p. 172–205, 1982.
5. Huang, P., *Under Compression: An Isochoric or Adiabatic Process?* The 2012 International Compressor Engineering Conference at Perdue, 2012.
6. Huang, P., Yonkers, S., Hokey, D., *Gas Pulsation Control Using a Shunt Pulsation Trap.* The 2014 International Compressor Engineering Conference at Perdue, 2014.

TCO optimization during design phase – assessment of bearing concepts by calculation and simulation

Dipl.-Wirtsch.-Ing. Mark Dudziak

Schaeffler Trading (Shanghai) Co. Ltd., Shanghai, China

mark.dudziak@schaeffler.com

Dipl.-Ing. (TH) Andreas Krome

Schaeffler Technologies GmbH & Co. KG; Herzogenaurach, Germany

andreas.krome@schaeffler.com

Abstract. Energy efficiency has been a key issue for screw machines for many years. This article describes a new analytical procedure for determining rolling bearing friction and the use of the procedure for designing compressor bearing supports. Further analytical options for a calculative assessment of the overall efficiency of bearing support alternatives are presented in the design phase based on parameter analyses performed by the BearinX® calculation software, which includes the new friction calculation by means of a mechanical, tribological model.

1. State of the Art, Determining Bearing Friction Empirically

Energy efficiency has been a key issue for screw machines for many years. Operating and energy costs make up a considerable percentage of the total cost of ownership (TCO) and therefore are increasingly coming to the fore for operators.
In addition to intelligent control and drive technology, many compressor manufacturers are working intensely to increase the efficiency of the screw group itself. An optimized rotor and gearbox bearing support offers good savings potential.
Primarily rolling bearings in a great variety of bearing designs and arrangements are being used to provide bearing support in screw machines.

Previously, the selection of bearings was mainly determined according to the following aspects:

- Reliability and long operating life
- Available installation space, bearing rigidity
- Ease of assembly
- Bearing cost and availability
- Experience from existing solutions

Established and internationally standardized calculation methods for bearing operating life (ISO 281 [1]) are on hand for optimizations designed to increase productivity.
To date, it has not been possible to sufficiently assess friction and power loss due to relatively imprecise calculation formulas.

Content from this work may be used under the terms of the Creative Commons Attribution 3.0 licence. Any further distribution of this work must maintain attribution to the author(s) and the title of the work, journal citation and DOI.
Published under licence by IOP Publishing Ltd

IOP Conf. Series: Materials Science and Engineering **90** (2015) 012080 doi:10.1088/1757-899X/90/1/012080

Bearing manufacturers offer empirical calculation models that enable calculative estimations based on the load, speed, and operating viscosity of the lubricant and bearing design.
The formulas are based on research by Palmgren in 1957 (Figure 1) leading to the ISO 15312 [2]. They were derived from measuring results in tests with laboratory test stands. However, they provide approximations of the measuring results and are only sufficiently precise in narrow ranges. The friction calculation of individual bearing arrangements for specific operating conditions is therefore only possible on a very rough scale.

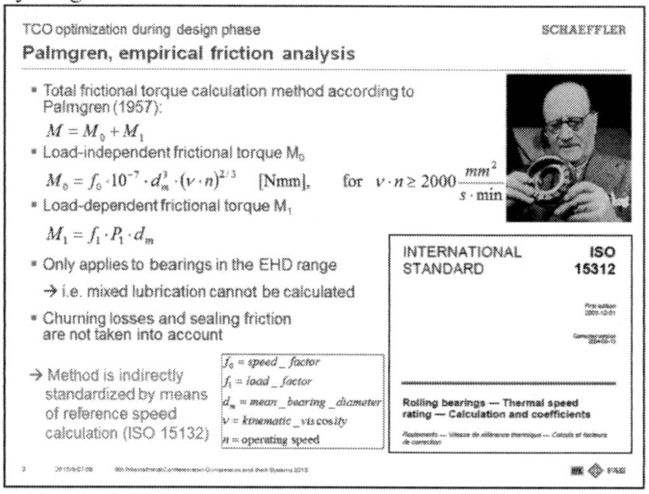

Figure 1: Friction types in rolling bearings

2. Determining Bearing Support Friction Analytically

In contrast to the empirical determining process, an analytical calculation begins with a mechanical, tribological model of a rolling bearing.
The mechanical model serves to precisely map force application, load distribution in the bearing, shaft deflection and skew position, operating clearance, the internal bearing structure, and more. The linked tribological model describes the behavior of the different tribological phenomena.
Figure 2 shows the most important friction components of a rolling bearing.

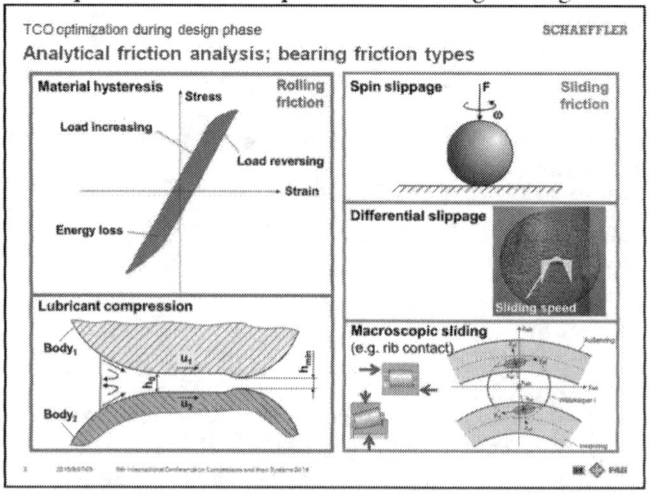

Figure. 2: Friction types in rolling bearings

Unlike with the empirical method, factoring in these friction components results in the addition of fundamental parameters such as those shown in Figure 3.

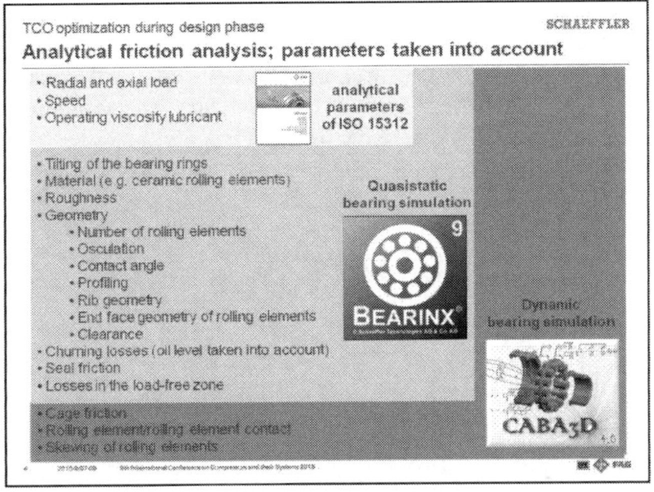

Figure 3: Rolling bearing friction, influencing parameters

The BearinX® calculation software uses this kind of mechanical, tribological model to calculate the friction of rolling bearing supports. While BearinX® is a quasi-static model that offers fast calculation times and high accuracy, a dynamic simulation with the CABA3D calculation software provides enhanced analysis possibilities with additional consideration of cage friction, rolling element contact, and rolling element skewing. A comparison of the calculated friction coefficients and those measured in the test confirms a high degree of conformity and thus the quality of the procedure developed (Figure 4).

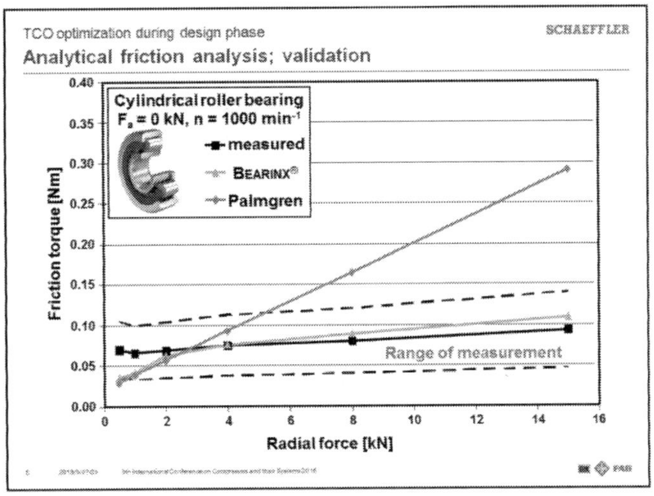

Figure 4: Comparison of the measured and calculated friction torque force

Figure 4 clearly shows that the empirical method according to Palmgren only provides realistic friction coefficients in a narrow range (approx. 1 to 5kN here). However, the friction calculated using BearinX® is so precise that it lies within the measuring value spread established in the test (0.5 to 15 kN).

3. Extended Application Analyses

IOP Conf. Series: Materials Science and Engineering **90** (2015) 012080 doi:10.1088/1757-899X/90/1/012080

The calculation procedure integrated into the BearinX® bearing design program offers a powerful tool for the individual modeling and analysis of bearing supports.

BearinX® makes it possible to calculate individual bearings and also to model shaft systems and gearboxes while allowing for elastic shaft systems. For instance, in the process, it is possible to investigate details of the individual rolling contact, lubricant film formation, bearing rigidity, shifting, rigidity in the operating points, and even the dynamic analysis of natural vibration behavior (Figure 5).

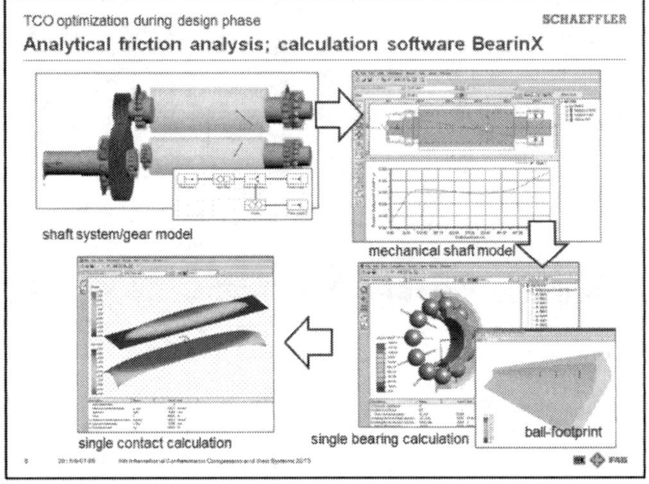

Figure 5: Modular calculation with BearinX®

So-called parameter analyses can be compiled for calculation models. To do this, a user-defined number of model input parameters is given with an initial value and a final value. The program then calculates independently specified output parameters and graphs their change as a function of the input parameters.
For example, a typical analysis would be the operating clearance of a bearing as a function of component temperature with the specified shaft and housing materials/seats and bearing play class.
In the design phase, these parameter analyses are used to investigate the parameters' effects on the command variables being analyzed.
It becomes clear in the process which parameters have the greatest effect. They can then be systematically altered so that they change the overall system behavior in the desired direction.
This frequently makes it possible to avoid or at least reduce complicated, expensive, and time-consuming tests.

3.1. Bearing Friction Analysis:

At the outset of friction analysis, it is interesting to see what kinds of friction the various types of bearings produce.

For instance, the bearing friction analysis for various types of bearings as a function of the induced force as shown in Figure 6 is helpful in this respect. Instead of individual bearings, bearing combinations were selected for this, of the kind frequently used on the pressure side in screw compressors.
The profiles confirm the known friction behavior of the various bearing types. Here, tapered roller bearings generate the greatest friction forces, whereas angular contact ball bearings and four point contact bearings exhibit considerably lower levels of friction.

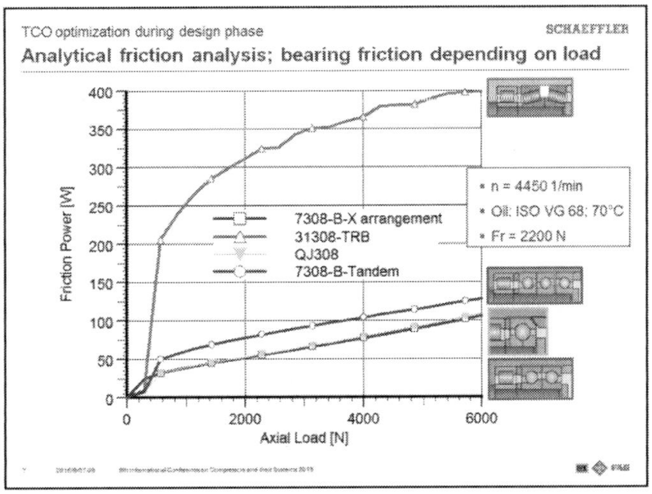

Figure 6: Bearing friction depending on the load

3.2. Bearing Rigidity:

The radial and axial bearing rigidity also represent important parameters.

The following figure shows the axial shift of the various alternatives as a function of the axial bearing load.

The greatest axial rigidity levels for the tapered roller bearing support are given here. The ball bearing solutions exhibit lower levels of rigidity.

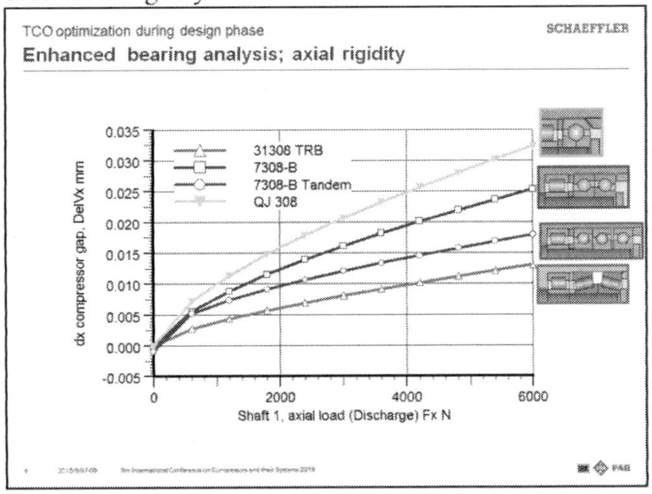

Figure 7: Parameter analysis of the axial rigidity

The radial rigidity depends on the type and size of the bearing as well as on the radial bearing play. Figure 8 indicates the radial rigidity of an NU308-JP cylindrical roller bearing.

The radial bearing play has been reduced here between CN, C3, and C2 in stages. The graph shows that the bearing spring deflection can be reduced by restricting the operating clearance. It is necessary to conduct a temperature analysis, particularly in the case of restricted operating clearance, to prevent inadmissibly high bearing preload levels due to thermal expansion.

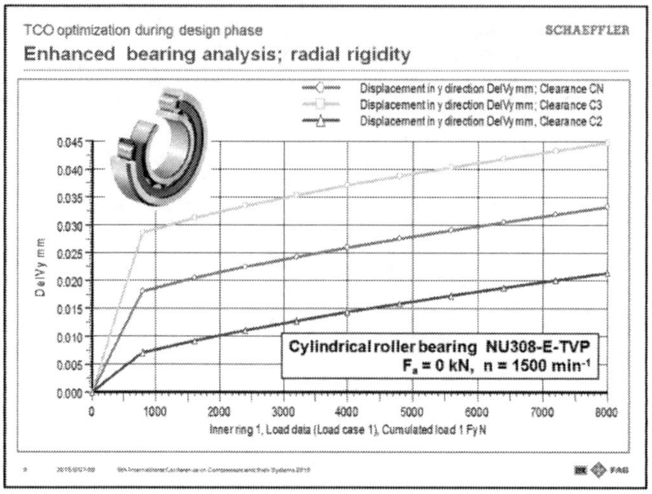

Figure 8: Parameter analysis of the radial rigidity

3.3. *Bearing Kinematics at High Speeds*

For high-speed screw machines such as oil-free compressors, it is essential to factor in the bearing kinematics in addition to questions of clearance and operating life.
The speed coefficients (nxdm) typically range between 600,000 mm/min. and 1.5 million mm/min.
Centrifugal forces result in unfavorable spin/roll conditions that produce large sliding movements. In favorable lubrication conditions, these sliding movements lead to increased frictional heat or, in mixed friction conditions, to increased fretting. The frictional heat produced must be conveyed away from the contact through respectively high quantities of oil. These oil quantities, in turn, lead to increased churning losses, which once again reduce efficiency. Suitable measures must be adopted to guarantee a minimum bearing load for all operating conditions to avoid negative circumstances and ensure a sufficient operating life.

Parameter analyses of the specific model make it possible to very accurately predict the conditions at which the bearing kinematics will begin to be adversely affected.

Figure 9 below shows that the current bearing configuration exhibits considerable spin/roll effects as of approx. 10,000 rpm, corresponding to a speed coefficient (nxdm) of approx. 435,000 mm/min.
These effects may be reduced to admissible levels by increasing the axial preload accordingly or by using ceramic balls, for example.

The use of range springs in the calculation model serves to identify shifting at any position in the model.
For instance, this makes it possible to analyze in detail the effect of centrifugal force on the axial skew of the face surfaces with angular contact ball bearings, as shown in Figure 10. Not taking this into due consideration can lead to rotor/housing contact at high speeds.

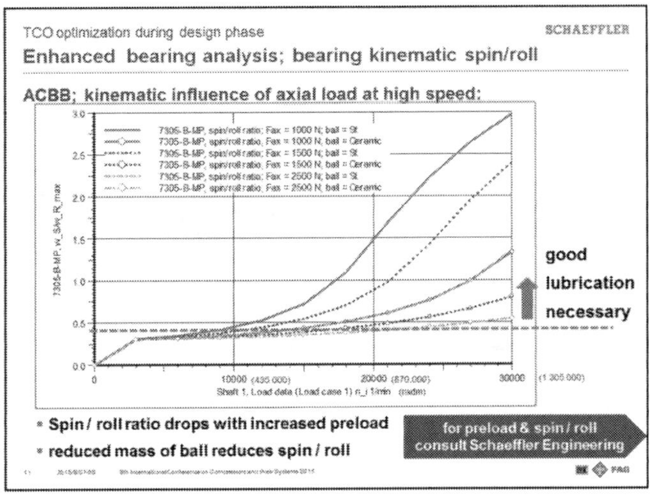

Figure 9: Parameter analysis of the spin/roll ratio

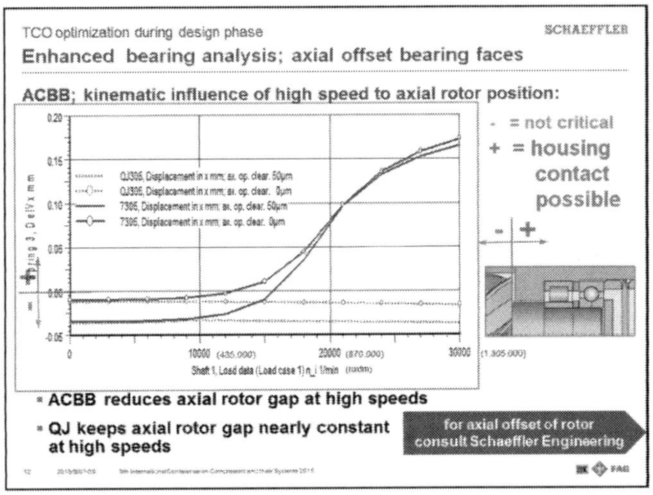

Figure 10: Parameter analysis of axial rotor position depending on the rotor speed

4. Example of a bearing optimization of a twin screw compressor HP side

Target is to optimize the bearing arrangement of a twin screw compressor on the High Pressure (HP) side regarding friction and life rating.

Initially the radial loads are carried by a radial Cylinder Roller Bearing (CRB) and the axial loads are carried by a Taper Roller Bearing (TRB). Basic compressor data and the BearinX model of the bearing set up is shown in the following slide (Figure 11).

TRB offer high axial load capacity. Due to the internal design of TRBs an axial load to the tapered rollers is generated by axial or radial loads to the bearing. This axial force is transferred from roller to rib of the related bearing ring producing sliding friction. Bearings without this kind of sliding friction (e.g. Angular Contac Ball Bearings (ACBB)) create smaller friction.

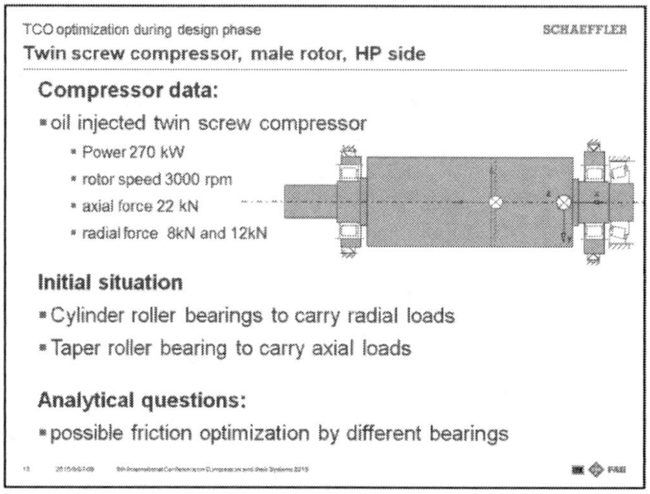

Figure 11: Basic compressor parameters and BearinX model

Figure 12 shows the cumulated calculated friction for this specific arrangement depending on the shaft speed.

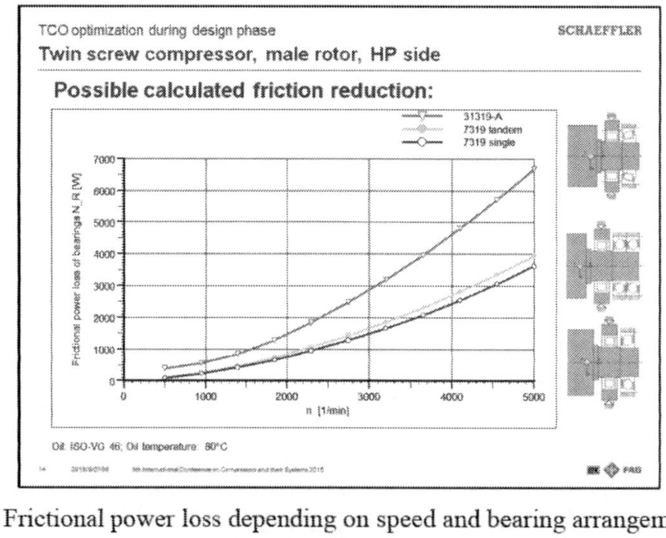

Figure 12: Frictional power loss depending on speed and bearing arrangement

The TRB arrangement shows significant higher friction than ACBB arrangements, which even grows with higher speeds. But aside reduced friction the alternative bearing arrangement needs to reach comparable and sufficient life times as the initial bearing solution. Here the modified reference rating life Lhmr offers the best calculable life time to judge load, contamination and lubrication.

The table in figure 13 gives an overview regarding life times, friction and contact pressure for each bearing alternative at nominal conditions of the compressor.

With pressures below 1000 N/mm² the initial solution shows low bearing loading. This results in very high calculated bearing life ratings with values > 650.000 h. With life time target of 40.000 h at failure rates < 1% the calculated life ratings should show values above 160.000h.

For the CRBs values like this are reached by using bearings with the same shaft and housing diameter, but reduced width.

For the first alternative the TRBs are replaced by ACBBs size 7319-B in tandem arrangement. The life time and pressure values reveal that even a single bearing arrangement is possible. A single 7319-B still reaches life ratings > 570.000 h. So a friction reduction of 1200W which is about 48% of the initial value is possible.

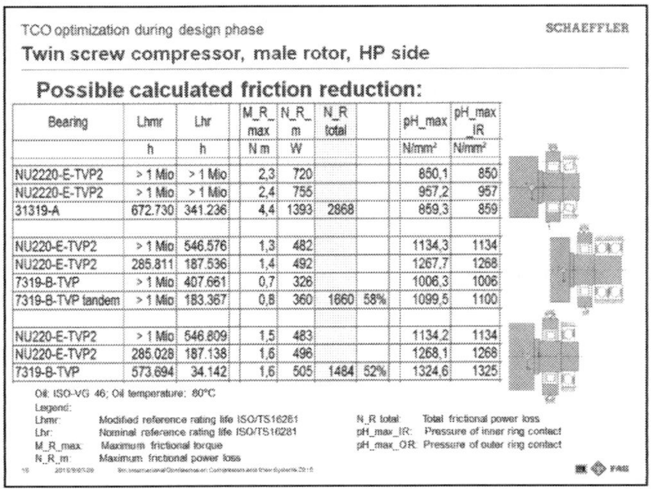

Bearing	Lhmr	Lhr	M_R_max	N_R_m	N_R total		pH_max	pH_max_IR
	h	h	N m	W			N/mm²	N/mm²
NU2220-E-TVP2	> 1 Mio	> 1 Mio	2,3	720			850,1	850
NU2220-E-TVP2	> 1 Mio	> 1 Mio	2,4	755			957,2	957
31319-A	672.730	341.236	4,4	1393	2868		859,3	859
NU220-E-TVP2	> 1 Mio	546.576	1,3	482			1134,3	1134
NU220-E-TVP2	285.811	187.536	1,4	492			1267,7	1268
7319-B-TVP	> 1 Mio	407.661	0,7	326			1006,3	1006
7319-B-TVP tandem	> 1 Mio	183.367	0,8	360	1660	58%	1099,5	1100
NU220-E-TVP2	> 1 Mio	546.809	1,5	483			1134,2	1134
NU220-E-TVP2	285.028	187.138	1,6	496			1268,1	1268
7319-B-TVP	573.694	34.142	1,6	505	1484	52%	1324,6	1325

Figure 13: rating life, frictional power loss and contact pressure at nominal condition

In comparison to the TRB 31319 the ACBB 7319-B reaches smaller axial stiffness. So the shaft displacement will be bigger at nominal loads which results in a bigger axial clearance between rotor and housing. The calculated displacements are shown in figure 14. The 7319-B tandem arrangement nearly reaches the TRB displacement, but the single 7319-B displacements are increased by ~20 μm. Here the compressor manufacturer needs to judge the efficiency reduction based on this.

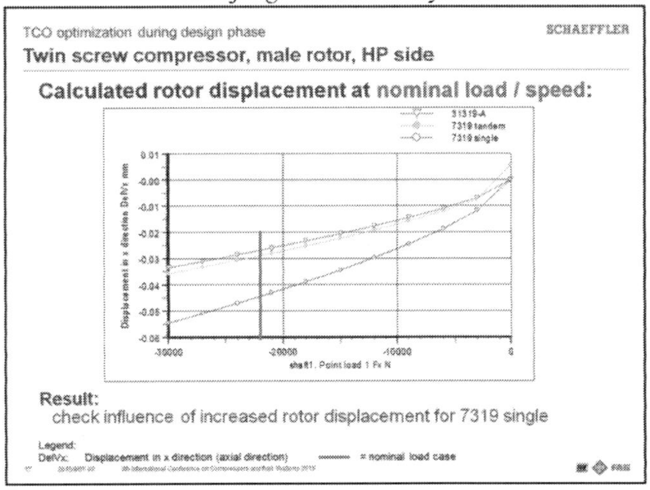

Figure 14: axial displacement at nominal condition.

Due to centrifugal forces to the balls single ACBB or ACBB in tandem without counter ACBB produce a shaft displacement depending on axial load and speed. This shaft displacement reduces the axial clearance between rotor and housing.

For high axial load conditions the displacement is small. For low load conditions the displacements may reach values where the rotor gets in contact with the housing.

Figure 15 shows the behavior at idle mode, where just 1500 N axial load is generated.
According to the BearinX parameter analysis the displacement at 3000 rpm is about 0,25 mm for a single 7319-B. The tandem arrangement even produces bigger displacements in the range of 0,35 mm. So it is necessary to carry out a test at idle mode to check whether the oil lubrication is sufficient to separate the axial contact surfaces rotor/housing while idle operation.
Alternatives to prevent this rotor displacement are to use an additional ACBB in X arrangement or to use a 4 point contact bearing (QJ series).

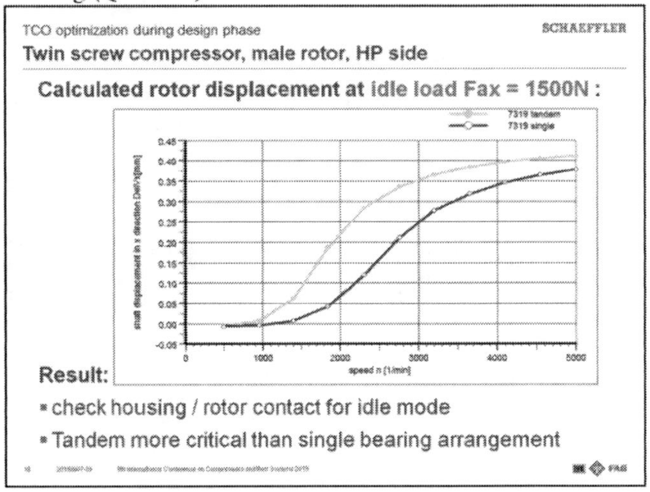

Figure 15: axial displacement at idle condition depending on shaft speed.

As final result the analysis reveals possible friction reduction of 1200W which offers savings of energy costs about 3800,- € .

These savings are based on energy costs of 0,1 €/kWh and 40.000 h operating time by taking 20% idle mode into account.

5. Summary / Conclusion:

In order to assess the overall efficiency of a compressor, aspects besides friction, such as rigidity, temperature behavior, and bearing kinematics, need to be considered.
Calculation software with a mechanical, tribological model enable calculative, comprehensive preliminary investigations during the design phase, in this way presenting decisive optimization options already at this stage and leading compressor manufacturers to the sought-after solution more quickly.

References

[1] ISO 281 Rolling bearings — Dynamic load ratings and rating life
[2] ISO 15312 Rolling bearings — Thermal speed rating – calculation and coefficients

IOP Conf. Series: Materials Science and Engineering **90** (2015) 012081 doi:10.1088/1757-899X/90/1/012081

A Fault Alarm and Diagnosis Method Based on Sensitive Parameters and Support Vector Machine

Jinjie Zhang[1] Ziyun Yao[2] Zhiquan Lv[3] Qunxiong Zhu[1] Fengtian Xu[4] Zhinong Jiang[4]

1. College of Information Science and Technology, Beijing University of Chemical Technology, Beijing, 100029, P. R. China.
2. PetroChina Beijing Gas Pipeline Co., Ltd., Beijing, 100101, P. R. China.
3. Aromatic plant of PetroChina Liaoyang Petrochemical Company, Liaoyang, 111003, P. R. China.
4. Diagnosis and Self-Recovery Engineering Research Center, Beijing University of Chemical Technology, Beijing, 100029, P. R. China.

E-mail: Jiangzhinong@263.net

Abstract. Study on the extraction of fault feature and the diagnostic technique of reciprocating compressor is one of the hot research topics in the field of reciprocating machinery fault diagnosis at present. A large number of feature extraction and classification methods have been widely applied in the related research, but the practical fault alarm and the accuracy of diagnosis have not been effectively improved. Developing feature extraction and classification methods to meet the requirements of typical fault alarm and automatic diagnosis in practical engineering is urgent task. The typical mechanical faults of reciprocating compressor are presented in the paper, and the existing data of online monitoring system is used to extract fault feature parameters within 15 types in total; the inner sensitive connection between faults and the feature parameters has been made clear by using the distance evaluation technique, also sensitive characteristic parameters of different faults have been obtained. On this basis, a method based on fault feature parameters and support vector machine(SVM) is developed, which will be applied to practical fault diagnosis. A better ability of early fault warning has been proved by the experiment and the practical fault cases. Automatic classification by using the SVM to the data of fault alarm has obtained better diagnostic accuracy.

1. Introduction

Reciprocating compressor, piston gas engine and reciprocating pump are widely applied in the industry enterprise, but the faults of these machines occurred frequently because of the obsolete fault warning and diagnosis ability. The fault diagnosis of reciprocating machinery represented by reciprocating compressor is the hard task in the field of mechanical fault diagnosis[1]. The fault warning and diagnosis level has been keeping in a relatively weak stage. The accidents of reciprocating compressor would cause serious environment pollution and financial losses.

Study on reciprocating compressor fault diagnosis mainly focus on valve and piping failure. Typical fault diagnosis researches include: the faults of reciprocating compressor valve and driving belt diagnosed by M. Ahmed with the tool of neural networks and support vector machines[2]; the fault diagnosis of valve and bearing studied by L. Jinru with the tool of Wavelet Neural Network and Genetic Algorithm[3]; research on fault diagnosis of reciprocating compressor indicating diagram by

Content from this work may be used under the terms of the Creative Commons Attribution 3.0 licence. Any further distribution of this work must maintain attribution to the author(s) and the title of the work, journal citation and DOI.
Published under licence by IOP Publishing Ltd

Kun Feng with SVM and PCA[4]. In terms of working principle and efficiency of reciprocating compressor, the efficiency, thermodynamic models and dynamic models have been studied by D. Ndiaye, P. Stouffs, C. Pérez-Segarra and M. Elhaj with theoretical simulation and experimental study[5-8].

How to applied the online monitoring system in fault diagnosis of reciprocating compressor has always been a hot research spot. The monitoring signals of the online monitoring system of reciprocating compressor have been analyzed and summarized by S.M.Schultheis[9]; the optimized location of sensor installation has been studied by N.K.Verma[10]. In addition, a warning method of online monitoring system has been studied by W.Zhong-Qing[11]; the construction of reciprocating compressor automatic diagnosis expert system has been studied by Zhinong Jiang and has made certain progress[12].

But the actual level of warning and fault diagnosis of reciprocating compressor has not been improved. How to effectively improve the accuracy of fault warning and realize automatic fault diagnosis is a significant practical and valuable work. Represented by the reciprocating compressor, the criterion of the monitoring warning parameters of reciprocating machinery has not been built. Related standards, such as API and ISO only require the warning parameters in the way of vibration effective value while other monitoring parameters are limited[13-14]. These all lead to low fault sensitivity of warning parameters and low efficiency of online monitoring diagnosis system.

This paper propose a reciprocating compressor fault warning and diagnosis method based on sensitive parameters extracted by scatter matrix method. Experimental and practical fault cases demonstrate that the accuracy and the sensitivity of the new method is superior to the traditional methods. The intelligent fault diagnosis method based on support vector machine achieved the accurate diagnosis result of reciprocating compressor common faults. This new method will improve the practical effect of online monitoring and diagnosis system.

2. Sensitive parameters extraction of reciprocating compressor

2.1. Online monitoring system of reciprocating compressor

At present, the online monitoring system of reciprocating compressor is widely applied. The monitoring signals are composed of key phase, piston rod displacement, crankcase vibration, crosshead acceleration, valve temperature and cylinder dynamic pressure. A set of typical reciprocating compressor online monitoring system is shown in Figure 1.

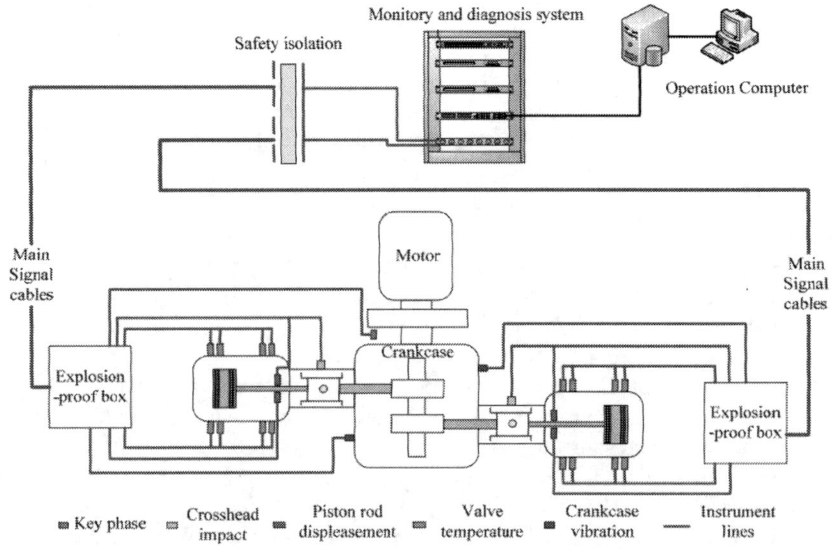

Figure 1 The compose of reciprocating compressor online monitoring system

The new monitoring and alarm parameters with different warning modes of reciprocating compressor are proposed in Table 1. These parameters overcome the weaknesses of original alarm parameters in alarm mode and alarm lines settings. Deteriorate features and relative variation trends of different faults are considered by the warning parameters in Table 1 with using the warning method of absolute value and relative value. It is significant for catching the relative signal changes of early failure. Meanwhile, most of the warning line setup modes are changed to band pass in consideration of the instrument fault signal features in the monitoring system.

Table 1 Improved warning parameters and methods

Warning methods	Monitoring signal	Warning parameters	Set of warning lines
Warning methods of absolute value	Displacement of piton rod	Peak-to-peak value of displacement waveform	Band pass
		Phase position of displacement waveform	Section alarm
		Volatility of displacement average value	Low pass
	Acceleration of crosshead	Peak value of acceleration waveform	Band pass
		Phase position of acceleration waveform	Section alarm
		DC value of acceleration waveform	Band pass
	Vibration speed of crankcase	Effective value of velocity waveform	Band pass
		DC value of velocity waveform	Band pass
		Volatility of velocity waveform effective value	Low pass
	Valve temperature	Absolute value of temperature	Band pass
		Rate of temperature change	Low pass
Warning methods of relative value	Displacement of piston rod	Slow changing rate of displacement average value	Band pass
		Fast changing rate of displacement average value	Band pass
	Valve temperature	Relative temperature value (to other air valve)	Band pass
		Relative temperature value (to itself)	Band pass

2.2. Sensitive parameters extraction algorithm

The multi-source signal fusion is an effective fault diagnosis method due to that the vibration signal characteristics are not obvious in the reciprocating machinery fault diagnosis. Based on practical experience, fault characteristics are summarized to establish the warning parameters which is shown in Table 1 instead of using algorithm directly such as wavelet transform, EMD or PCA. Then the corresponding relations between different failures and corresponding features are studied from fault data based on the distance evaluation method to extract the fault sensitive parameters.

2.2.1 Characteristic assessment technique

The distance assessment technique is relatively improved and simple which evaluates the sensitivity of different features based on the distance between inter-class and intra-class of different characteristics. The sensitive features can be determined when inter-class feature distance of the same patterns is the shortest and intra-class of the different patterns is the longest. The shorter the inter-class distance in same patterns and the longer the intra-class distance in different patterns, the more sensitive the feature will be[15].

A dimensionless processing method of fault feature is proposed in this paper. Normalization processing is the typical method based on the mean value and the estimated variance value of the data. This method is effective to historical data but unusable when online analyzing the real time data in practical fault diagnosis. The process method is:

$$V(i,j)=\frac{f(i,j)-N(i,j)}{F(i,j)-N(i,j)} \tag{1}$$

In the equation:

$V(i,j)$: the dimensionless value of j feature of number i fault;

$f(i,j)$: the current value of j feature of number i fault;

$N(i,j)$: the normal value of j feature of number i fault;

IOP Conf. Series: Materials Science and Engineering **90** (2015) 012081 doi:10.1088/1757-899X/90/1/012081

$F(i,j)$: the warning value of j feature of number i fault.

The dimensionless value $V(i,j)$ defined above ranges in [0, 1]. Considering the relationship between current parameter value, the historical normal value and warning value of various fault features, the boundary conditions of the mode are:

$$f(i,j)-N(i,j)\leq 0,\ V(i,j)=0; \tag{2}$$

$$f(i,j)-F(i,j)\geq 0,\ V(i,j)=1. \tag{3}$$

The distance of the normalized data is calculated when distance assessment is applied to feature selection, so after introducing the warning value, the normalized value can greatly reflect the sensibility of each fault parameters.

The scatter matrix measures the detachable degree of characteristics and patterns by calculating the variance ratio of inter-class and intra-class. The steps of scatter matrix method are shown as follows.

1) Receive all the feature parameters of normal data and the m-th fault data, and the k-th fault feature parameter data is recorded as D_k;

2) Calculate the intra-class scatter matrix of the k-th fault feature parameter:

$$S_{wk}=\sum_{i=1}^{M}P_i\sum\nolimits_i . \tag{4}$$

Where M is the number of data categories and is 2, reflect the normal class and fault class. P_i is the prior probability of class i, $P_i=n_i/N$. n_i is the sample number of class i in the k-th fault feature parameters. N is the total sample number of all categories in the k-th fault feature parameter. $\sum_i$ is the intra-class covariance matrix of class i in the k-th fault feature parameters and its algorithmic method is:

$$\sum\nolimits_i=\frac{1}{n_i}\sum(x_i-\mu_i)(x_i-\mu_i)^T=E[(x_i-\mu_i)(x_i-\mu_i)^T]. \tag{5}$$

Where x_i is the value of class i of the k-th fault feature parameter, and it is a column vector composed of all values. μ_i is the intra-class mean value of class i in the k-th fault feature parameters, and its algorithmic method is: $\mu_i=\frac{1}{n_i}\sum x_i=E(x_i)$. $tr\{S_{wk}\}$ is the trace of S_{wk}, which is the mean measure of all categories characteristic variance in the k-th fault feature parameters.

3) Calculate the inter-class scatter matrix of the k-th fault feature parameters:

$$S_{bk}=\sum_{i=1}^{M}P_i(\mu_i-\mu_0)(\mu_i-\mu_0)^T. \tag{6}$$

Where μ_0 is the intra-class mean value of class i in the k-th fault feature parameters: $\mu_0=\sum P_i\mu_i$. Obviously, $tr\{S_{bk}\}$, which is the trace of S_{bk}, is a measurement of the mean distance between mean value of each class and the global mean.

4) Calculate the hybrid between-class scatter matrix of the k-th fault feature parameters:

$$S_{mk}=E[(x-\mu_0)(x-\mu_0)^T]=S_{wk}+S_{bk}. \tag{7}$$

$tr\{S_{mk}\}$,which is the trace of S_{mk}, is the variance sum of characteristic values on the global mean.

5) Calculate the guide coefficient *Jk* for distance estimation of the k-th fault feature parameters:

$$J_k=tr\{S_{mk}\}/tr\{S_{wk}\}, \tag{8}$$

or

$$J_k=|S_{mk}|/|S_{wk}|=|S_{wk}^{-1}S_{mk}|, \tag{9}$$

or

$$J_k=tr\{S_{wk}^{-1}S_{mk}\}. \tag{10}$$

6) For all fault feature parameters, repeat step 1) to 5), then the guide coefficient vector J for distance estimation will be obtained. The vector needs normalization processing to get vector MJ. MJk is the sensitivity coefficient of the k-th feature parameter after calculated by the scatter matrix. And the calculate process is as follows:

$$MJk = (Jk\text{-}min(J))/(max(J)\text{-}min(J)). \tag{11}$$

Where *max* represents the maximum value, and *min* represents the minimum value. If the element of vector *MJ* is greater than 0.6, its corresponding fault feature parameter is the sensitive parameter of the m-th fault.

3. Experiment and the actual faults test

3.1. Fault experiment

The typical faults of reciprocating compressor include valve fault, moving elements faults and seal assembly faults. The diagnosis method of valve fault is more mature and these faults can be diagnosed in advance by temperature and vibration signals. Due to the weaker damage of the valve fault, few attention is paid to it. So that the fault simulating experiments are composed of three typical faults: the cylinder abrasion, the support ring abrasion and the fracture of piston rod. The cylinder abrasion is difficult to be monitored in actual, due to it is difficult to judge the wear of the cylinder through manual inspection. The fracture of piston rod belongs to the dangerous faults of reciprocating compressor, is hard and dangerous to be simulated. In the experiment, using the vibration protection system to realize the automatic protection and prevent malignant accidents. The Figure 2 and Figure 3 show the experiment process.

Figure 4 shows the signals differences between the faults and common condition. From the crosshead vibration data, different characteristics corresponding to the fracture of piston rod and cylinder abrasion, including acceleration peak value, peak value phase, average peak value. The change trend of some characteristics after normalization are shown in Table 2. In feature extraction, characteristics of crosshead vibration, piston rod displacement, crankcase vibration and temperature were all extracted. Abscissas of Figure 4 represent sampling points, different points corresponding to different rotate speed of the compressor.

Figure 2 Damaging the surface of supporting ring and piston ring

Figure 3 Removing the piston rings and support rings of piston

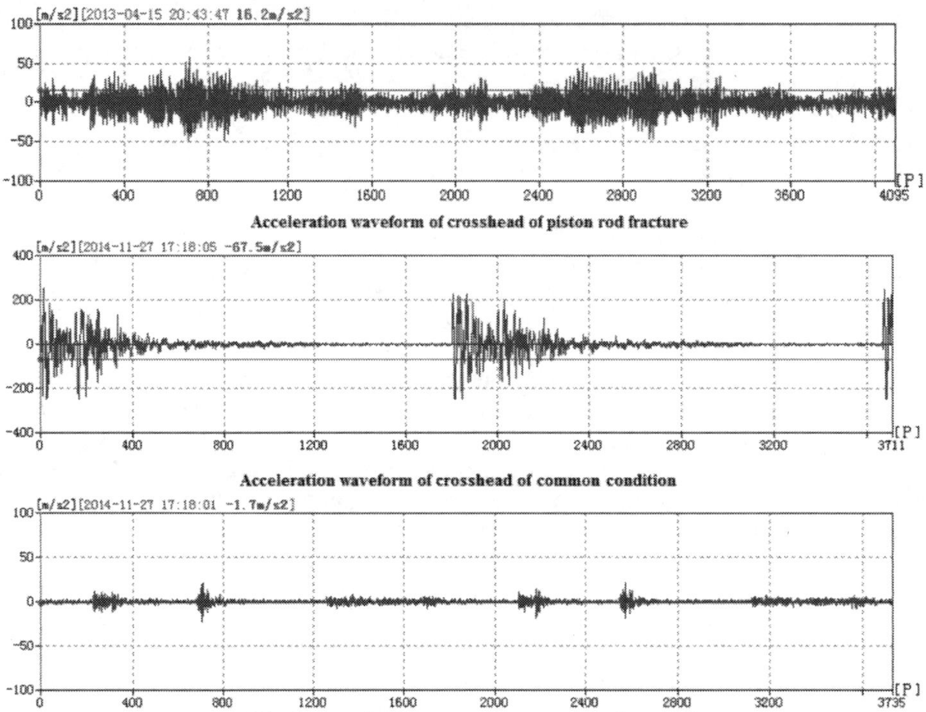

Figure 4 Data of fault simulation experiment of reciprocating compressor

Table 2 The value of parts of features

Class	Normal						Fault					
time	time1	time2	time3	time4	time5	time6	time1	time2	time3	time4	time5	time6
Displacement peak-to-peak	0.14	0.12	0.18	0.18	0.17	0.23	0.23	0.23	0.24	0.27	0.31	0.32
Acceleration peak	0.08	0.03	0.08	0.09	0.11	0.13	0.34	0.43	0.34	0.40	0.53	0.47
RMS	0.21	0.22	0.31	0.25	0.25	0.35	0.79	0.83	0.82	0.82	0.86	0.76

3.2. Sensitive parameters extraction and analysis

In the experiments, aiming at 100 fault samples and 100 common samples taken from the three typical faults, three methods: scatter matrix method, ReliefF method and entropy evaluating method, are applied to extract and comprehensively evaluate the sensitive feature parameters. While the results of ReliefF and entropy evaluating method are used for comparison. The results of all these three methods of sensitive parameters extraction are shown in Figure 5 to Figure 7.

In Figure 5 to Figure 7, the abscissas represent the number of parameters extracted and the ordinate represents the sensitive feature index which is in 0 to 1. Because numbers of measuring points are different, the parameters extracted of faults are also different. It can be seen that the results of ReliefF method are relatively poor, we focus on the scatter matrix method and entropy evaluating method to analysis the results of sensitive parameters extraction.

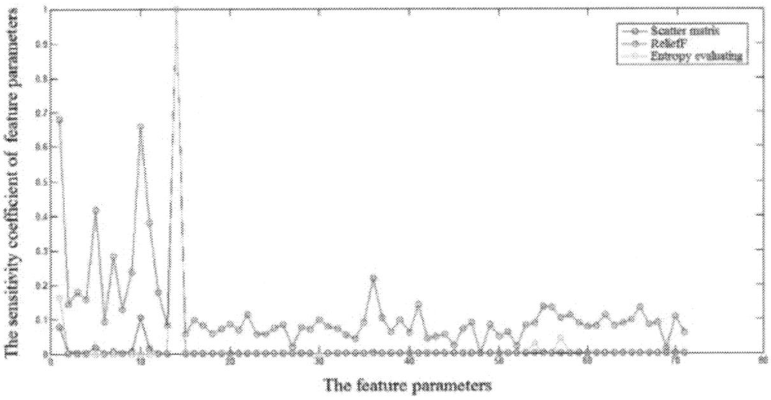

Figure 5 Schematic evaluation results of the support ring abrasion

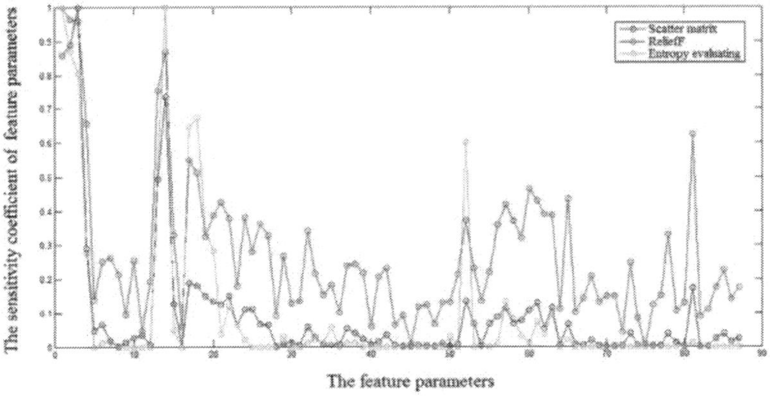

Figure 6 Schematic evaluation results of the cylinder abrasion

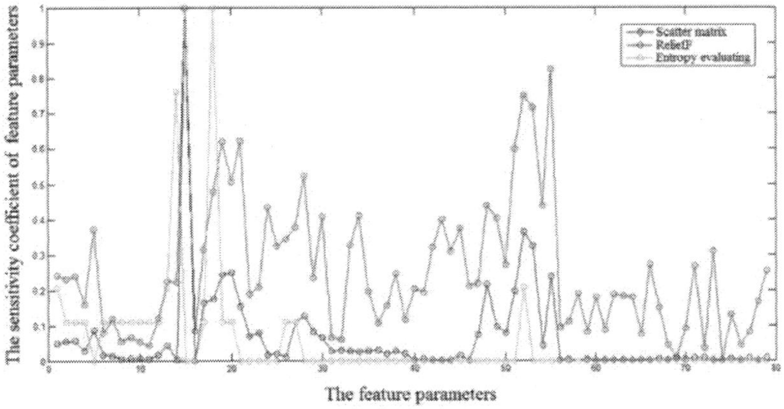

Figure 7 Schematic evaluation results of the fracture of piston rod

From Table 3, the scatter matrix method has good performance to reflect the sensitivity of characteristic parameters and the result is in consistence with practical experience of fault diagnosis. The extraction result is the most refined, but there are also other sensitive features may be lost. However, this method has the ability of self optimization. With the increase of the actual faults data, the different characteristics of the data will be exhumed and the sensitive parameters extraction will continue to be improved.

Table 3 The extraction results of the Scatter matrix method and Entropy evaluating method

Sensitive parameters extraction		
Fault	**Scatter matrix method**	**Entropy evaluating method**
Support ring abrasion	**14**: the relative variation of the piston rod displacement	**14**: the relative variation of the piston rod displacement
Cylinder abrasion	**1**: the peak-peak value of the piston rod displacement, **2, 3**: the waveform phase characteristics of piston rod displacement **14**: the relative variation of piston rod displacement	**1**: the peak-peak value of the piston rod displacement **2, 3**: the waveform phase characteristics of piston rod displacement **14**: the relative variation of piston rod displacement **17, 18**: the phase characteristics of cylinder vibration waveform **52**: RMS of crankcase box vibration
Fracture of piston rod	**15**: the peak value of cylinder vibration	**14**: the relative variation of piston rod displacement **18**: the phase characteristics of cylinder vibration waveform

3.3. Fault automatic diagnosis

The artificial neural network always be used to finish fault automatic diagnosis. Based on the artificial neural network, the inner node weight of fault classifier can be self-amended by the study of case samples. Then it will be equipped with the ability of self-study and self-optimization. The classifier of the artificial neural network will be improved. Considering the different performance of different transducers of each fault, the classifier can be designed as multi-fault classifier and two-fault classifier.

(1) Two-fault classifier

Taking advantage of the sample data of supporting ring abrasion and the cylinder abrasion, each fault sensitive parameters are extracted as above. Using the sample data shown above in this procedure, each fault classifier is trained by each fault sensitive parameters and composed two-fault classifier unit. Then 100 groups of testing data are selected to verify the classified performance, as shown in Table 4.

Table 4 Classification results of two-fault classifier

Testing data	the cylinder abrasion	supporting ring abrasion	Piston rod fracture
Sensitive feature extraction	99%	97%	98%
Without extraction	86%	81%	88%

The character of the classifier is that all the faults data should be classified by the corresponding specific classifier. For example, cylinder classifier can classify the “abrasion” fault and “non-abrasion”.

(2) Multi-fault classifier

The type of faults in practical diagnosis contains more than two types, so multi-fault classifier is constructed in this paper as well. The union set of sensitive parameters of three faults are collected to train the multi-fault classifier with the same sample data stated above. Then 100 groups of testing data of three faults are tested to show the performance of classifier. The result of the classifier is shown in Table 5, and it illustrates that fault classifier with multi-classifier construction can also process the classification correctly and effectively.

Table 5 Classification results of multi-fault classifier

Testing data	the cylinder abrasion	supporting ring abrasion	Piston rod fracture
Sensitive feature extraction	95%	97%	100%
Without extraction	81%	83%	91%

The data in Table 4 and Table 5 reflect that the fault diagnosis and classification results with sensitive feature extraction in this paper are better than those without sensitive feature extraction. However the faults data is limited, fault classification of reciprocating compressors operating in realistic may not be good enough. The working environment, unit structure and fault characteristics of these reciprocating compressors will be different. As a result, the next work is using sensitive feature extraction method to the online monitoring and diagnosis work of reciprocating compressors. Based on a large number of data, the method will be further tested. Also, the sensitive feature extraction method and classification result will be improved. The improvement flow chart is shown as follows.

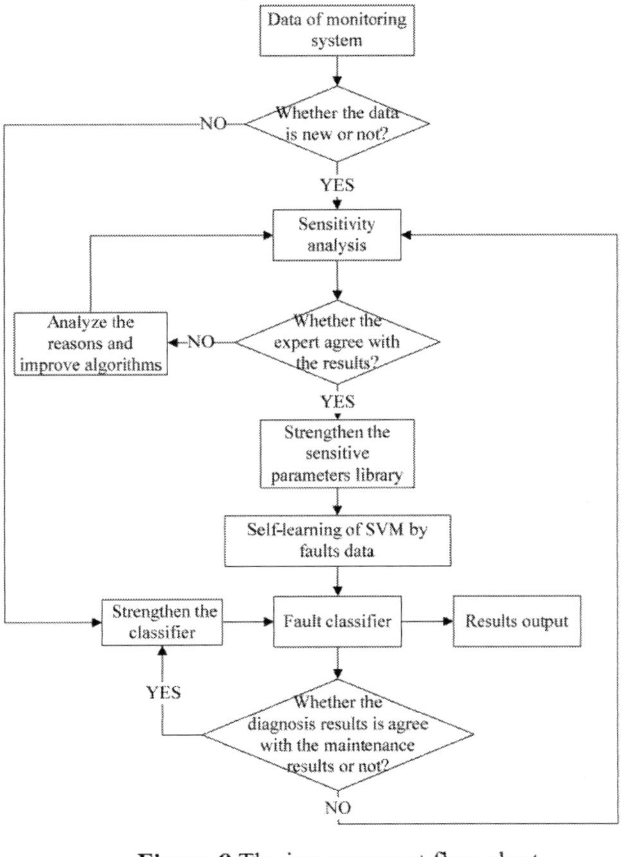

Figure 8 The improvement flow chart

4. Conclusions

Aiming at the weak sensitivity of fault parameters and difficulty of fault automatic diagnosis in reciprocating compressor fault diagnosis, this paper extracts the fault parameters based on the data of online monitoring system. By using the scatter matrix method, the sensitive parameters are extracted from the original features to illustrate the relationship between fault types and fault characteristics. It greatly reduces the process dimensionality of data. The method can also be used to improve the pertinence and effectiveness of warning parameters in online monitoring system.

Based on the fault sensitive parameters, a kind of fault diagnosis framework is built with artificial neural network, and fault typical classifier is trained by experiment data and real fault case data to test the accuracy. The accuracy rate is increased by the sensitive parameters extraction in this paper.

The research proposed in this paper is closely related to the requirements of realistic fault diagnosis of reciprocating compressor, and it has good practical application value. Next, the diagnosis framework proposed in this paper need to be integrated into the online monitoring system to improve the training of new faults data. The fault warning and diagnosis should be improved to increase the practicability of the online monitoring system.

Acknowledge
This work was supported by the National Basic Research Program of China (973 Program) under Grant No. 2012CB026000 and the National High Technology Research and Development Program of China (863 Program) under Grant No.2014AA041806.

Corresponding author
Zhinong Jiang, Email: jiangzhinong@263.net

References

[1] Zhang Jinjie, Jiang Zhinong, Wang Yao, Xu Fengtian 2014 Numerical simulation and experimental study on plate valve transient motion and fatigue fracture principles, *International Journal of COMADEM*, **17** 17
[2] M. Ahmed, S. Abdusslam, M. Baqqar, F. Gu, 2011 A. Ball, Fault classification of reciprocating compressor based on Neural Networks and Support Vector Machines, *in: Proceedings of the 17th International Conference on Automation & Computing, IEEE, UK*, 213
[3] L. Jinru, L. Yibing, Y. Keguo, 2008 Fault diagnosis of piston compressor based on Wavelet Neural Network and Genetic Algorithm, *in: Proceedings of the 7th World Congress on Intelligent Control and Automation, IEEE, Chongqing, China*, 6006
[4] K. Feng, Z. Jiang, W. He, B. Ma, 2011 A recognition and novelty detection approach based on Curvelet transform, nonlinear PCA and SVM with application to indicator diagram diagnosis, *Expert Systems with Applications* **38** 12721
[5] C. Pérez-Segarra, J. Rigola, M. Soria, A. Oliva. 2005 Detailed thermodynamic characterization of hermetic reciprocating compressors. *International journal of refrigeration*. **28** 579
[6] D. Ndiaye,, M. Bernier. 2010 Dynamic model of a hermetic reciprocating compressor in on–off cycling operation (Abbreviation: Compressor dynamic model). *Applied Thermal Engineering* **30** 792
[7] P. Stouffs, M. Tazerout, P. Wauters. 2001 Thermodynamic analysis of reciprocating compressors[J]. *International journal of thermal sciences* **40** 52
[8] M. Elhaj, F. Gu, A. Ball, A. Albarbar, M. Al-Qattan, A. Naid 2008 Numerical simulation and experimental study of a two-stage reciprocating compressor for condition monitoring, *Mechanical Systems and Signal Processing* **22** 374
[9] S.M.Schultheis, C.A.Lickteig, R.Parchewsky. 2007 Reciprocating Compressor Condition Monitoring. *Proceedings of The Thirty-Sixth Turbomachinery Symposium* 107
[10] N.K. Verma, J. Kadambari, B. Abhijit, S. Tanu, T. Subramaniam. 2011 Statistical approach for finding sensitive positions for condition based monitoring of reciprocating air compressors. 2011 *IEEE International Conference on System Engineering and Technology (ICSET)* 10
[11] ZhongQing Wei, ZhiNong Jiang, Bo Ma, Xin Zhong 2010 The Slow-Changing Alarm system of condition monitoring for rotating machinery, *WSEAS transactions on system* **9** 52
[12] Z. JIANG, J. ZHANG, M. JIN, B. MA, 2013 An expert system based on multi-source signal integration for reciprocating compressor, *WSEAS transactions on system* **12** 266
[13] API standard 618 2007 Reciprocating Compressors for Petroleum, Chemical, and Gas Industry Services
[14] ISO 10816 1995 Mechanical vibration - Evaluation of machine vibration by measurements on non-rotating parts -Part 6: Reciprocating machines with power ratings above 100 kW
[15] Lei Yaguo, He Zhengjia, Zi Yanyang 2008 Fault diagnosis based on novel hybrid intelligent model. *Chinese journal of mechanical engineering* **44** 112

Investigation of three lobes roots blower with special ejector

S.V. Vizgalov, G.N. Chekushkin, and M.V. Volkov

Kazan National Research Technological University, 68, K. Marx Str., Kazan, Russian Federation

E-mail: sv_kstu@rambler.ru

Abstract. A volumetric Roots type blower with spur three or two lobes rotors has a high performance at a small unit size and is used mainly at the pressure ratio in the range of 1,4 – 1,6. A significant limitation of this parameter is due to the nature of the compressor working process, in particular a rapid, almost instantaneous increase in the pressure in the working chamber because of back flow. This leads to a decrease in performance and efficiency, and increase of the compressor noise level. One of the ways to overcome this drawback is the use of bypassing of part of compressed and cooled gas at injection through the ejector into the working chamber that is simultaneously disconnected from both discharge and suction. Due to this, the pressure increase process becomes more gradual as related to the rotor angle of rotation, and additional suction of portions of gas from the suction line by ejection effect becomes possible, which increases the efficiency and performance of the compressor. Experimental data confirm the efficacy of this method. Research of the Roots blower combined action with an ejector attachments based on the developed mathematical model has been presented.

1. Introduction

The Roots blower belongs to positive displacement machines and has a feature of gas compression workflow associated with the working chamber volume stability transferred from the gas suction side to the gas discharge side of gas and great influence of gas backflow. As a result of this the compression process is almost isochoric, which leads to a drastic reduction in performance, volumetric and adiabatic efficiencies, pressure pulsation and noise level increase with increasing rotors speed and pressure in the discharge line. There are several well-known methods, which were adapted to the Roots blower with three or more lobes rotors. In keeping with those methods the high pressure gas is bypassed to the working chamber with the aim of the gas pulsation decrease in the discharge line.This method is not based on physical chamber volume reduction, but it helps to reduce the gas discharge pressure pulsation. It is implemented in the simplest way by executing the channels [1] on the internal surface of the housing and by gas bypassing without cooling it. Other studies provide compressed gas cooling in an integrated heat exchanger before the gas is bypassed through external piping connected to the compressor housing [2, 3]. In this case, the blower can be operated at higher pressure ratio of up to 1,8.

The more appropriate method is bypassing the compressed gas into the working camber of the blower through the ejector. This design of Roots blower with a discharge heat exchanger for cooling

Content from this work may be used under the terms of the Creative Commons Attribution 3.0 licence. Any further distribution of this work must maintain attribution to the author(s) and the title of the work, journal citation and DOI.
Published under licence by IOP Publishing Ltd

gas and two ejectors is shown in Fig. 1. A part of gas in the amount of m_{byps} is taken off through the pipelines 5 from the flow of gas compressed and cooled in the heat exchanger 4 and fed into the active ejector nozzle 6 wherein the expansion and acceleration of the gas flow and an additional fresh low pressure gas portion leak-in through the passive nozzle 8 into the mixing chamber 7 in the amount of m_{add} take place. The total flow is supplied through the diffuser 9 and the rotary union 10 via the additional slit ports 3 located on the housing into the working cavity. The ports 3 are arranged along the entire length of the housing such that they are overlapped by the rotor when connecting the working cavity with the discharge port. Thus, we achieve both additional suction of fresh gas through the ejector and a more gradual increase in pressure within the cavity during the transfer thereof from suction to discharge during rotate of rotor. Adiabatic and volumetric efficiency of this supercharger layout depends on several factors: the location of the slit ports and the ejector nozzle geometric dimensions, depending on the pressure ratio of the supercharger.

The main objective of the study is to determine the effectiveness of this blower layout when operated at different discharge pressure conditions and a fixed position slit ports located on the housing blower. Besides, since the pressure ratio in the supercharger does not exceed 1,9 and the working gas is air, the ejector operating condition is subcritical and the nozzle has a conical shape with a decreasing cross section.

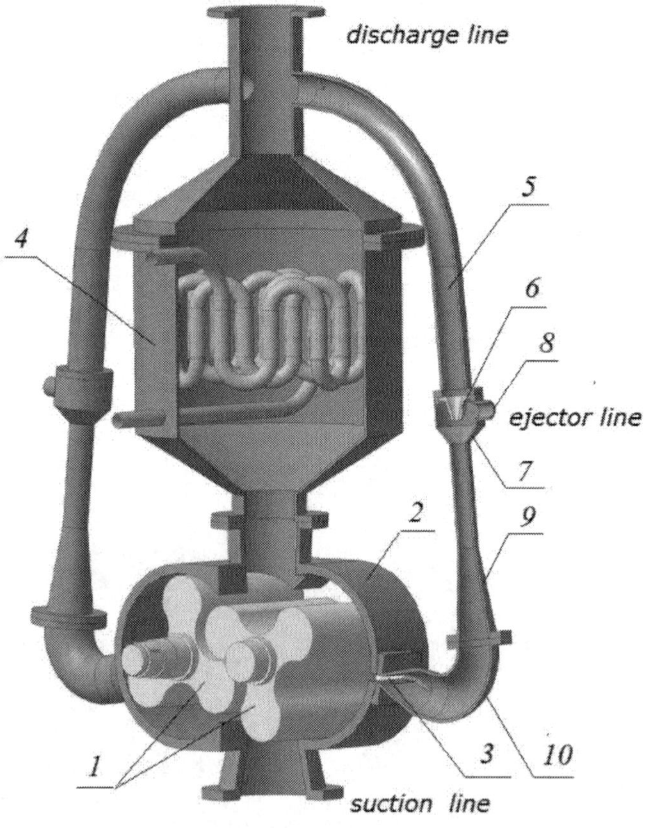

Figure 1: Roots blower with ejectors

2. Mathematical model

The simulated Roots blower has three lobes rotors, with a cycloid-circular rotor profile, theoretical capacity of 4,85 m^3/min, the rotor diameter of 113,7 mm, length of 150 mm, center distance between rotors is 75 mm. The working gas is air.

Design scheme of the supercharger is shown in Fig. 2. There are simultaneously 5 chambers in the machine at different stages of the working process designated as:

I – gas suction process into the working cavity of expanding volume $V=f(\varphi)$, where the rotor rotation angle is $\varphi= -\gamma_{inlet}$ - $\pi/3-\gamma_{inlet}$ (Fig. 2);

II, III – transfer of the constant volume cavity to the discharge side V=const; the cavity transfer angle is $\Delta\varphi=\pi/3+\gamma_{inlet}+\gamma_{out}$;

IV – reverse gas leakage process (back flow) that starts at an angle of $\varphi=2\pi/3+\gamma_{inlet}+\gamma_{out}$ and ends at an angle of φ_{bf_end} provided pressure equalization in the adjacent cavities IV and V;

V – process discharge of gas from the decreasing volume cavity from the IV and V cavities unification movement until the rotation angle of $\varphi=\pi/3+\varphi_{bf\_end}$.

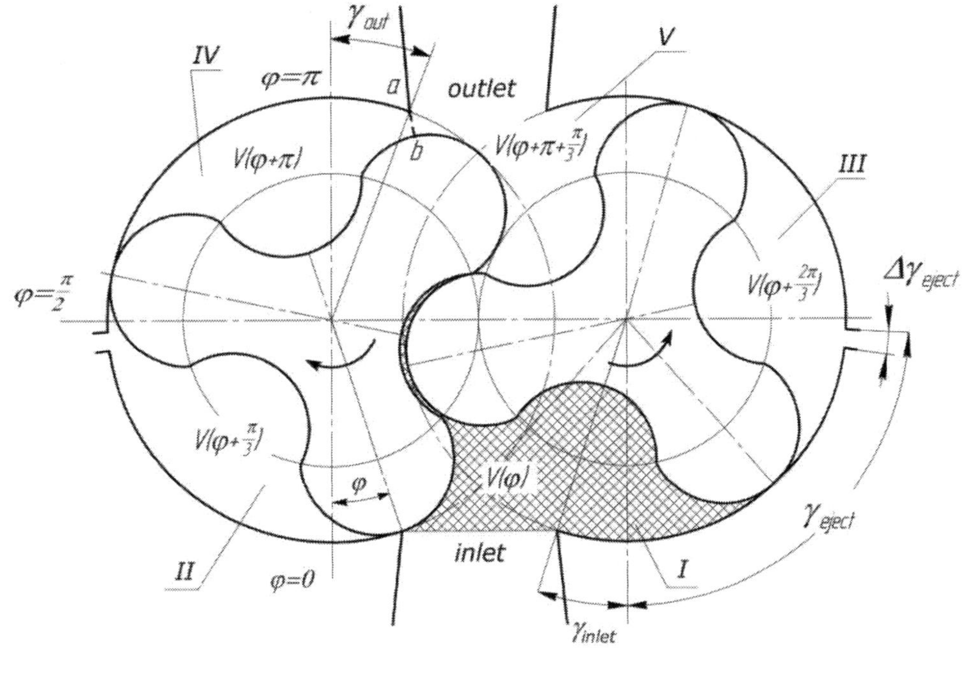

Figure 2: Roots blower design diagram

The dependence of the volume of each of the cavities on the angle of rotor rotation was determined by numerical integration, and are not given here.

At any position of the rotors determined by the angle φ, the cavities are displaced relative to each other to an angle of $\pi/3$. Gas leakage through the gaps occur between the chambers. Also, the chambers intercommunicate for some time with the inlet port - chamber I, gas discharge port - chamber V, as well as through the special slit ports with an ejector attachment - chambers II and III, through which the additional leakage of both fresh gas mass of m_{add} and discharge line bypassed flow mass of m_{byps} into the chamber takes place; the chamber IV is connected to the discharge line through the port a-b, the size of which is determined by the normal drawn from point 'a' onto the rotor profile; reverse gas leakage occurs through this port; this flow rate is determined by the pressure difference upstream and downstream of the port. Location of additional ports on the cylindrical part of the housing for ejector connection is symmetric and determined by the angle of γ_{eject}. The condition of absence of gas ejection through these ports is overlapping thereof by the rotor blade apex before the sharp rise in pressure in the reverse flow process; thus, the angle of $\gamma_{eject} \leq \pi/3 + \gamma_{out}$. The length of the port with respect to the angular coordinate is given by the value of $\Delta\gamma_{eject}$.

Simulated blower has the following design parameters:

$$\gamma_{inlet} = 18{,}093 \text{ deg}, \ \gamma_{out} = 26{,}058 \text{ deg};$$
$$\gamma_{eject} = 85 \text{ deg}; \quad \Delta\gamma_{eject} = 4{,}5 \text{ deg}.$$

The mathematical model of the Roots type compressor was developed on the basis of energy equilibrium equations in the working chamber and the ideal gas law. It takes into account the chamber volume change according to the rotor rotation angle V=f(φ), the gas leak-in during the suction period, the gas mass transfer through the clearances between the adjacent chambers having various pressures, the gas heat exchange with the chamber walls, and the gas back-streaming into the working chamber at the discharge side. In this case the gas thermodynamic parameters in the volume of the working chamber are homogeneous at any period of time. The pressure and temperature gradients are zero in the chamber.

The equations of the model in the form of differential pressure and temperature dependences on the rotor rotation angle φ are presented in the research works by authors Ibraev, Vizgalov [4, 5, 6]:

$$\begin{cases} \dfrac{dp}{d\varphi} = \dfrac{k-1}{\omega \cdot V} \cdot \left(\dfrac{\omega \cdot dQ}{d\varphi} + \sum \dot{m}_{inf} \cdot h_{inf} - \sum \dot{m}_{lks} h - \dfrac{k}{k-1} \cdot \omega \cdot p \dfrac{dV}{d\varphi} \right) \\ \dfrac{dT}{d\varphi} = \dfrac{(k-1) \cdot T}{P \cdot V \cdot \omega} \cdot \left(\dfrac{\omega \cdot dQ}{d\varphi} + \dfrac{k-1}{k} \cdot \left(\sum \dot{m}_{inf} - \sum \dot{m}_{lks} \right) \cdot h + \sum \dot{m}_{inf} \cdot h_{inf} - \sum \dot{m}_{inf} \cdot h - p \dfrac{\omega \cdot dV}{d\varphi} \right) \end{cases} \quad (1)$$

where *p, T* are pressure and temperature of gas in the working chamber respectively, *V*- volume of chamber; *dQ* is the elementary heat flux on the working chamber; *k* is the adiabatic index (*k*=1,4 for air); ω is the rotor angular velocity, rad·s^{-1}; *m* is the mass flow through the ports and gaps. Indices: "*inf*" relates to the gas inflows into the chamber, "*lks*" relates to the leaks of gas from the chamber. The gas enthalpy in this case can be defined as for the ideal gas $h=C_p \cdot T$, where C_p is the isobaric heat capacity of gas.

The elementary heat flux in the equation (1), taking into account the convective heat transfer between the gas and the walls of the chamber was determined by the equation

$$\frac{dQ}{d\tau} = \omega \cdot \frac{dQ}{d\varphi} = \bar{\alpha}(\varphi) \cdot f_w \cdot (T_w - T), \quad (2)$$

where the heat transfer coefficient $\bar{\alpha}$ averaged over the surface of the walls f_w was determined according to [7], where the experimental data were summarized by the equation of Nusselt for the periods of suction, transfer of the isolated chamber, and discharge; T_w – temperature of wall (experimental).

Gas flow through the narrow clearances between the rotors and the housing compressor, between profiles of rotors were determined by the method proposed S.E. Zakharenko [8] by the following equation:

$$\dot{m} = \sqrt{\frac{\rho_1 \cdot p_1 \cdot \ell^2 \cdot \delta^2 \left[\left(\dfrac{p_1}{p_2} \right)^2 - 1 \right]}{\ln(p_1 / p_2) + \xi + \lambda \cdot \Delta}}, \text{ kg/s} \quad (3)$$

where ℓ, δ - clearance sizes, the front length and the minimum height of clearance; ξ - coefficient of local resistances at the inlet and the outlet of the clearance due to a sudden contraction and expansion;

λ - friction coefficient of the gas in the clearance as a function of the Reynolds number, Δ - shape factor of the clearance, were determined by work [8].

The blower operating area with respect to the pressure ratio value of $\Pi=p_{disch}/p_{suc}$ is that the expansion ratio in the ejector nozzle is below the critical pressure ratio by using part of the compressed and cooled gas as the motive stream

$$\frac{p}{p_{disch}} \leq \left(\frac{2}{k+1}\right)^{\frac{k}{k-1}} = 0,529, \tag{4}$$

where p is air pressure in the suction line, $p= p_{suc} =$ 98,066 кPa; p_{disch} - pressure in discharge line (the maximum value is 187,0 kPa).

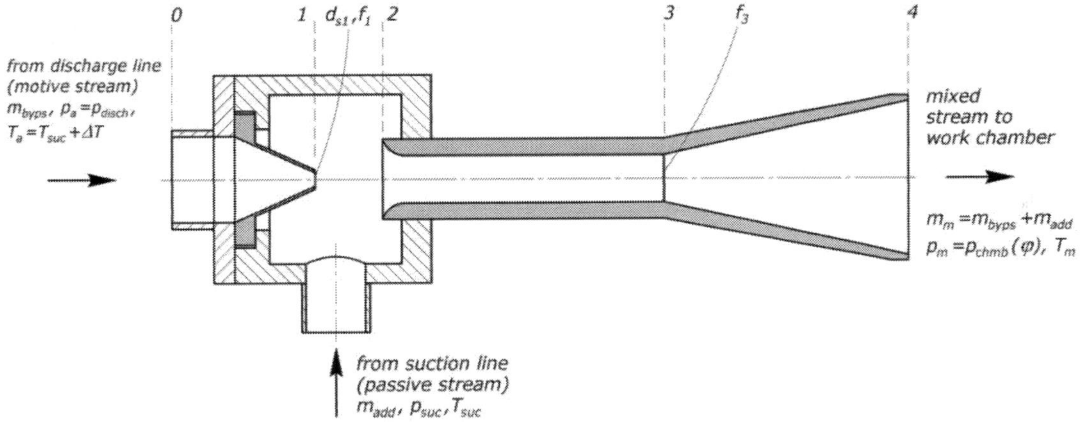

Figure 3: Ejector design schematic

The design schematic of the proposed injector is shown in Fig. 3. To determine the additional gas mass m_{add} fed into the working cavity, it is necessary to know the characteristic of the ejector in the form of the dependence of the ejection coefficient u depending on the ratio of $\Delta p_{chamb}/p_{suc}$, as well as active flow rate in the nozzle of m_{byps}.

This characteristic was determined by calculation based on the law of conservation of momentum in the flow in the ejector and the equations obtained in the research work [9]. The value of the ejection coefficient u was set in the range of 0,01 to 15,0. The basic equation of the characteristic of the ejector is as follows

$$\frac{\Delta p_{chamb}}{p_{suc}} = \frac{k}{2(k+1)}\frac{1}{\Pi_{a.p}}\frac{\psi_1^2\psi_2^2\lambda_{a.p}^2}{\left[\left(\frac{1}{\psi_3}-0,5\right)\frac{v_m}{v_a}(1+u)^2-(\psi_2\psi_4-0,5)\frac{v_n}{v_a}n\cdot u^2\right]}, \tag{5}$$

where $\Delta p_{chamb}= p(\varphi)-p_{suc}$ – is the difference of working chamber pressure and suction pressure; ψ_1, ψ_2, ψ_3, ψ_4, are the velocity coefficients, recommended values based on the experimental results are $\psi_1=0,95$, $\psi_2=0,975$, $\psi_3=0,9$, $\psi_4=0,925$ [9]; the parameter $\Pi_{a.p.}$ is a gas-dynamic pressure function and is defined as

$$\Pi_{a.p} = \frac{p_{suc}}{p_{disch}} = \frac{1}{\Pi}. \tag{6}$$

The gas-dynamic velocity function of

$$\lambda_{a.p} = \sqrt{\frac{k+1}{k-1}(1-\Pi_{a.p}^{\frac{k-1}{k}})}\ , \tag{7}$$

where ($\lambda = w_{isentr}/w_{sound}$) is determined by the value $\Pi_{a.p}$; the specific volume ratio is equal to

$$\frac{v_m}{v_a} = \frac{p_{disch} T_m}{p(\varphi)\ T_a},\quad \frac{v_n}{v_a} = \frac{p_{disch} T_{suc}}{p_{suc} T_a}; \tag{(8), (9)}$$

mixture temperature in the ejector is

$$T_m = \frac{T_a + u \cdot T_{suc}}{1+u}; \tag{10}$$

temperature of the motive stream fed to the nozzle is $T_a = T_{suc} + \Delta T$; ΔT is undercooling of gas in the heat exchanger.

The n parameter in the equation (5) is defined as

$$n = \frac{\left(\frac{f_3}{f_a^*}\right)_{opt}}{\left(\frac{f_3}{f_a^*}\right)_{opt} - \frac{1}{q_{a.p}}}, \tag{11}$$

where the gas-dynamic function $q_{a.p}$ ($q = \rho \cdot w_{isentr}/\rho_{s1} \cdot w_{s1}$) is defined as

$$q_{a.p} = \left(\frac{k+1}{2}\right)^{\frac{1}{k-1}} \lambda_{a.p} \left(1 - \frac{\lambda_{a.p}{}^2 (k-1)}{k+1}\right)^{\frac{1}{k-1}}. \tag{12}$$

The parameter equal to the optimal ratio of the ejector mixing chamber area to the nozzle area is defined by the following equations

$$\left(\frac{f_3}{f_a^*}\right)_{opt} = \frac{\sqrt{b^2 - 4a \cdot c}}{2a} - \frac{b}{2a}, \tag{13}$$

where coefficients are $a = \psi_1 \cdot \psi_2 \cdot q_{a.p}$,

$$b = -\left[\psi_1 \psi_2 + 2\varepsilon_{a.p}\left((\frac{1}{\psi_3} - 0,5)\frac{v_m}{v_a}(1+u)^2 - (\psi_2 \psi_4 - 0,5)\frac{v_n}{v_a} u^2\right)\right],$$

$$c = \frac{2\varepsilon_{a.p}}{q_{a.p}}(\frac{1}{\psi_3} - 0,5)\frac{v_m}{v_a}(1+u)^2,$$

$$\varepsilon_{a.p} = \left(1 - \frac{k-1}{k+1}\lambda_{a.p}^2\right)^{\frac{1}{k-1}},$$

where $\varepsilon_{a.p}$ is the gas relative density ($\varepsilon = \rho_{isentr}/\rho^*$).

The discharge bypass active flow consumption was defined as

$$m_{byps} = \frac{k \cdot \Pi_a \cdot p_{disch} \cdot q_{a.p}}{a_a} F_{s1}, \tag{14}$$

where F_{s1} is the nozzle exit area (d_{s1} is the determining size), a_a is the motive flow critical speed in the nozzle defined as

$$a_a = \sqrt{\frac{2k}{k+1} R T_a}\ , \quad \Pi_a = \left(\frac{2}{k+1}\right)^{\frac{k}{k-1}}, \tag{(15), (16)}$$

where R is a gas constant.

Consumption of the additional passive nozzle inflow

$$m_{add} = u \cdot m_{byps} . \tag{17}$$

Thus, having the dependence

$$\frac{\Delta p_{chamb}}{p_{suc}} = f(u, d_{s1}) \tag{18}$$

and knowing the current pressure in the chamber $p(\varphi)$, the ejection coefficient value and the flow m_{add} are defined.

The system of equations of the mathematical model was solved numerically using the method of Runge - Kutt - Feldberg [10]. Diagrams of pressure and temperature according to the angle of the rotor rotation resulted from solving the equations of mathematical model. Indicated power and indicated adiabatic efficiency of the blower were determined by integrating the curve $p = f(\varphi)$

$$N_{ind} = 6\frac{\omega}{2\pi}\int_{\varphi_0}^{\varphi_k} p(\varphi)d\varphi , \quad \eta_{ad} = N_{ad}/N_{ind} , \tag{19), (20}$$

where N_{ad} is the adiabatic compressor power.

Volumetric efficiency was determined as:

$$\eta_v = \frac{\pi \cdot (\sum m_{disch}(\varphi) - m_{byps})}{3\omega \cdot m_{is.cham}} , \tag{21}$$

where $\sum m_{disch}(\varphi)$ is the total value of the mass flow rate of gas through the outlet port to discharge line in during one revolution of the rotor, $m_{is.cham}$ is the mass of gas in isolated working chamber separated from the suction chamber at the angle φ=(60°-γ_{inlet}).

3. Result and conclusions

Calculations of integral characteristics of Roots blowers was based on integration of mathematical model differential equations and finally are presented as dependence of volumetric efficiency η_v, adiabatic efficiency η_{ad} and temperature increasing on pressure ratio Π (The range of pressure ratio of 1,25 to 1,87). The points of gas injection on the housing (one of each side of the housing) of the Root's blower (angle γ_{eject}) and geometric parameters of the ejector were fixed during the all calculations (ejector's nozzle outlet diameter d_{s1}=5 mm).

Calculated pressure-shaft angle and temperature-shaft angle diagrams of Roots blower were obtained using the developed mathematical model. The calculation results in the case of a pressure ratio of 1,6 and a rotating speed of 3000 rpm are presented in Figure 4. They include the Roots blower characteristics without the ejector (solid line), with the ejector and without bypass cooling (dashed line) and with the ejector and bypass cooling (dash-dotted line). In other working conditions the calculated characteristic curves have similar evolutions. The specific zones depending on the shaft angle and according to the schematic, which is presented in Figure 2, were denoted as I-V. There were two variants using the ejector: with and without cooling of the motive stream. In the case of the motive stream cooling its temperature was assumed 5 degree Celsius above the suction gas temperature (undercooling ΔT=5K). It can be seen from the graph that in the case of using ejector the gas pressure starts increasing before the disconnection of the working chamber from the blower suction port (shaft angle of 20,8 deg.). It occurs due to the stream starting flowing through the ejector. For the following transporting process (zones II and III) there is the significant increase in the gas pressure, but at the same time the stream of the induced air falls rapidly. The rates of the temperature curves changes are similar to the rates of the pressure curves changes. However, influence of the motive stream cooling on the temperature curves is more obvious. On the whole it should be noted that the motive stream cooling has a marginally impact on the pressure and temperature curves.

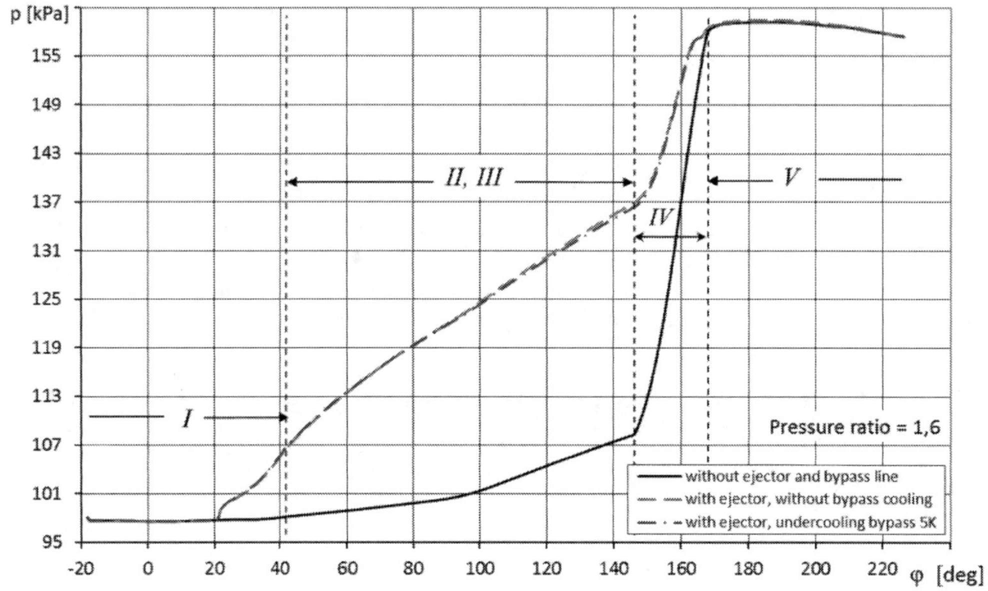

a)

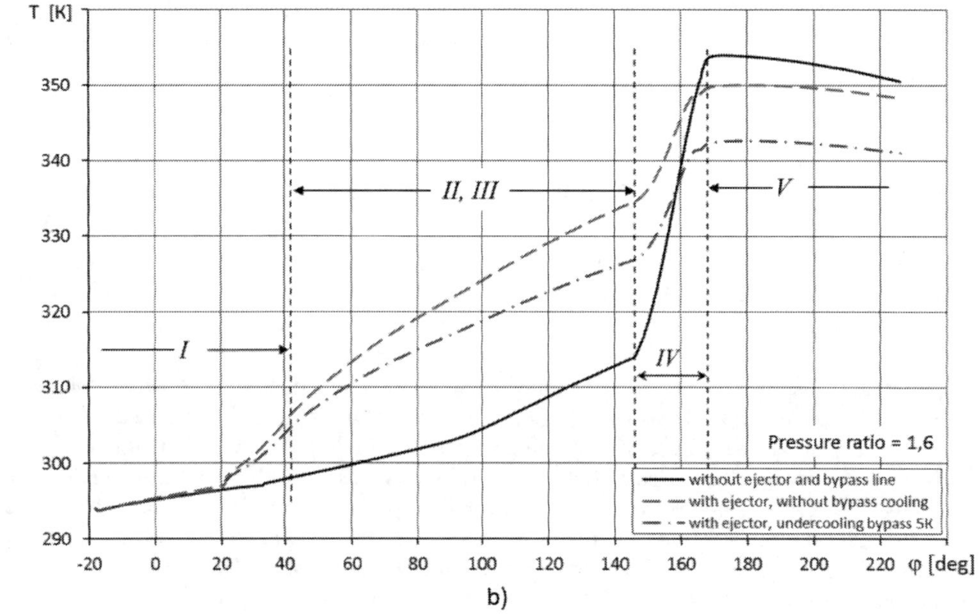

b)

Figure 4: Influence of ejector and cooling bypass on P-φ and T-φ diagrams (Π=1,6)

Obtained integral parameters are presented in Fig. 5 (a, b), 6 where the curve 1 shows the characteristics of Root's blower without ejector. In this case it should be noted that the volumetric efficiency and adiabatic efficiency decrease smoothly, but at the same time the discharge temperature increases significant. The volumetric efficiency has a parabolic behavior. The adiabatic efficiency has a quasi-linear behavior. The results of calculation of Root's blower characteristics with ejector and without a cooling heat exchanger are presented by the curve 2. In this case the discharge temperature are reduced, which can be explained by an injection of low temperature gas from suction. At the same time the η_v and η_{ad} are increased and their decreasing rate depending on pressure ratio are reduced. It should be noted that the η_v depending on pressure ratio has been increasing since $\Pi \approx 1{,}75$. In this case ejectors parameters are approaching to optimal and, hence, mass flow of injected gas from a suction are increased.

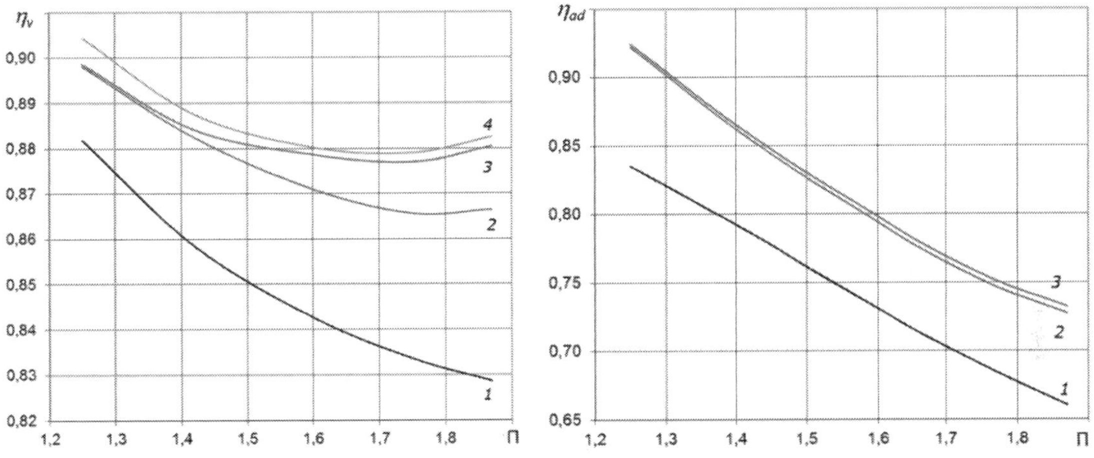

(a) *(b)*

Figure 5: Roots blower efficiency at pressure ratio
(1- Roots blower without ejector, 2- with ejector and without cooling, 3 - with cooling ΔT=20K, 4- with cooling ΔT=5K)

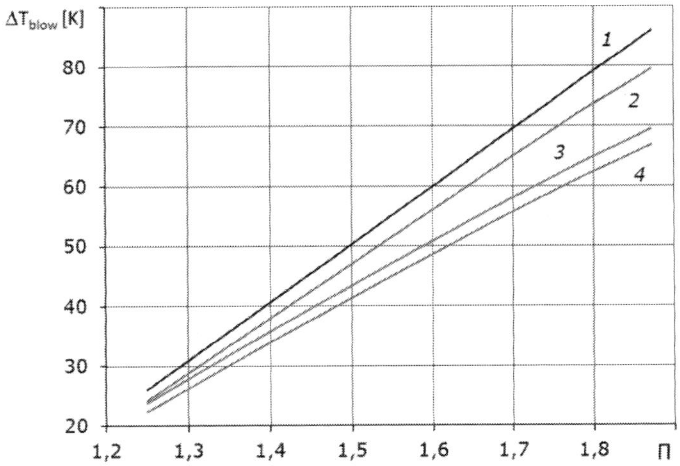

Figure 6: The temperature difference $\Delta T_{blow}=T_{disch}-T_{suc}$ at pressure ratio
(1- Roots blower without ejector, 2- with ejector and without cooling, 3 - with cooling ΔT=20K, 4- with cooling ΔT=5K)

Influences of motive stream temperature are presented by curves 3 and 4. The temperatures of motive stream have been taken as a function of suction temperature $T_a=T_{suc}+\Delta T$, where ΔT has been taken 20K for the curve 3 and 5K for the curve 4 respectively. In this case the η_v are increased by 0,5% on low pressure ratio side and by 1,65% on pressure ratio of Π=1,8 respectively. At the same time the increasing of η_{ad} is absolutely insignificantly. It can be explained by increasing density of motive stream's gas. Hence, there are no reasons to significant cooling motive stream.

The presented investigation shows advisability of using ejector for the Root blowers. But this investigation should be continued to determine the optimal ejector parameters depending on construction of Root's blower and its nominal mode.

References

[1] US Pat. 4,215,977 Pulse-Free blower, 1980.

[2] PST/US88/04066 Hi-ratio reciprocation gas compressor, 1989.

[3] US Pat. 5,439,358 Recirculating rotary gas compressor, 1995.

[4] Ibraev A M, Vizgalov S V, Khisameev I G. 2013 Analysis of the basic geometrical parameters influence on the efficiency of the Roots-type compressor on the basis of thermodynamic processes simulation. *8th Int. Conf. on Compressors and their Systems* (London: City University) 739.

[5] Vizgalov, S V, 2004, Vliyanie vnutrennego okhlazhdeniya na effektivnost rabochego protsessa shesterenchatogo kompressora (Influence of internal cooling to efficiency of working process of Root's compressor(in Russian)). Dissertation, Kazan national research technology university. 212p.

[6] Ibraev, A.M., 1987. Povyshenie effektivnosti raboty rotornykh nagnetatelei vneshnego szhatiya na osnove analiza vliyaniya geometricheskikh parametrov na ikh kharakteristiki (Higher efficiency of rotary blower by analysis of geometry parameters to them characteristics (in Russian)). Dissertation, St.Petersburg State Polytechnical University. 208 p.

[7] Sharapov I I, Saifetdinov A G, Ibraev A M, Khamidullin M S, Khisameev I G 2013 Investigation of heat exchange in the working chamber of rotary compressor. *8th Int. Conf. on Compressors and their Systems* (London: City University) 227.

[8] Zakharenko S. E. K voprosu o protechkakh gaza cherez shcheli (On the question of gas leakages through the gaps ((in Russian)) Trudy LPI. - №2. - M.-L., Mashgiz, 1953. - 144-160 p.

[9] Sokolov, E., Zinger, N.M. Struinie apparati (Jet device (in Russian)), 1989. Moscow.: Energoatomizdat, 1989. – 352 p.

[10] Forsythe G., Malcolm, M., Moler C., 1977. Computer methods for mathematical computations. Prentice-Hall, Inc., Englewood cliffs, 1977.

IOP Conf. Series: Materials Science and Engineering **90** (2015) 012083 doi:10.1088/1757-899X/90/1/012083

Design and validation of new cavity profiles for diaphragm stress reduction in a diaphragm compressor

Chen Jiahao; Jia Xiaohan; Xue Chuang; Peng Xueyuan

No.28 Xianning West Road, Xi'an, China, People's Republic of

porcupie@126.com

Abstract. The application of diaphragm compressors is severely limited by the low reliability and short life of diaphragms. The cavity profile has considerable influence on a diaphragm's stress level and therefore on its fatigue life. This experimental and theoretical analysis of the traditional cavity profile (as defined by a single-parameter generatrix) indicated that the short life of diaphragms may be caused by the superposition of peak radial stress and additional stress due to the discharge holes in the central region that occur when the diaphragm clings to the cavity surface. Therefore, two new kinds of cavity profiles are proposed to decrease the diaphragm's radial and total stress. The new cavity profiles based on different generatrices were designed in an attempt to reduce the peak radial stress in the central region. These cavities were manufactured to experimentally validate the designs. Experiments were carried out under various design conditions to investigate the radial stress distribution in the diaphragms. The reductions of radial stress on the diaphragms in the new cavities were compared with stress levels found in the traditional design. Although their cavity volumes and radii were all the same, the radial stress levels on diaphragms in the cavities designed by the new generatrices were less than those found in the traditionally designed cavity. The greatest decline in radial stress on the diaphragms in the cavities based on new generatrices was 13.8%. With the additional stress taken into account, the decline in radial stress in the diaphragm's central region reached as high as 19.6%.

1. Introduction

A diaphragm compressor is a type of compressor that guarantees gas purity without leakage. This device is commonly applied in dealing with rare, flammable, poisonous or corrosive gases. Diaphragm compressors normally vary in discharge pressure between 0.1 MPa and 300 MPa. [1] Therefore, these compressors are generally used in the petrochemical, pharmaceutical and food industries. [2]

Failure of the diaphragm is the most common type of malfunction for diaphragm compressors. The stress caused by deflection of the diaphragm as it comes into contact with the cavity surface affects its

Content from this work may be used under the terms of the Creative Commons Attribution 3.0 licence. Any further distribution of this work must maintain attribution to the author(s) and the title of the work, journal citation and DOI.
Published under licence by IOP Publishing Ltd

fatigue life. The cavity profile also has a considerable influence on the diaphragm's stress level and therefore its fatigue life. The symmetric cavity profile is shaped by the rotation of a generatrix. The traditional generatrix equation for a cavity profile is a single parameter polynomial.

Wu et al. studied the stress distribution on a diaphragm as it clung to the cavity surface and proposed an optimisation algorithm for the traditional generatrix equation of the cavity profile. [3] Ban analysed stress variations in diaphragms during compression using an iteration method. [4] Wang and Feng verified the reliability of thin plate large deflection theory and designed a new generatrix composed of two polynomials. [5] Kalshnikov [6] et al. researched the fatigue life of diaphragms under variable loads and provided two models for calculating diaphragm reliability. Antonova et al. proposed a method to evaluate the reliability of a diaphragm compressor. Altunkhov [7] et al. found that the diameter of the discharge holes and the oil pressure were the two main factors that influenced a diaphragm's fatigue life. In recent years, however, little research has been conducted on the generatrices of the cavity profiles for diaphragm compressors.

Previous studies and cases of diaphragm failures suggest that excessive radial stress is the root cause of such failures. One of the peaks in radial stress occurs in the central region of the diaphragm's gas side, as this area endures additional stress caused by the discharge holes. Thus, the superposition of these two kinds of stress results in stress concentration, which has a significant adverse effect on diaphragm performance. This paper presents two new kinds of generatrices of the cavity profiles for diaphragm compressor and indicates their optimisation algorithms. Experimental and mathematical methods were applied to verify the diaphragm stress distributions under the proposed design theories.

2. Design theory for the cavity profile

2.1 Diaphragm deformation theory

The primary structure of a diaphragm compressor is shown in Figure (1).

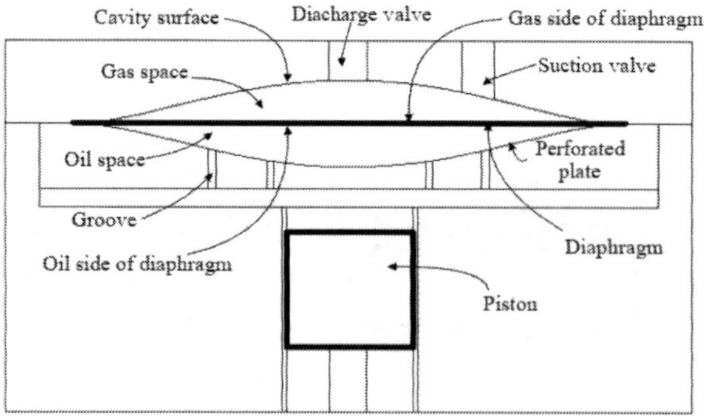

Figure 1.

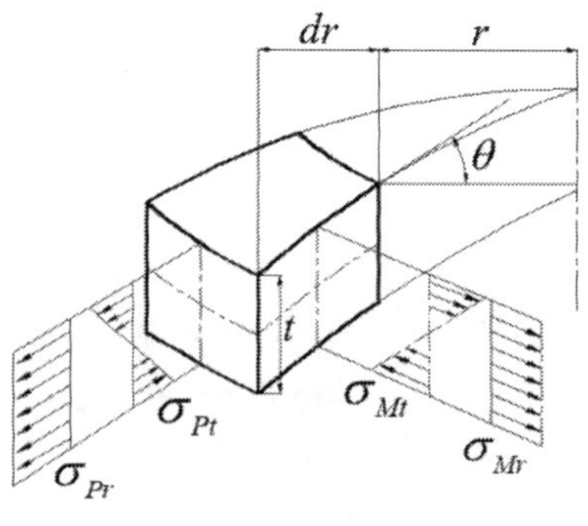

Figure 2.

The circular and symmetric deformation of the diaphragm in the compressor can be viewed as a combined effect of stretching and bending, such that the deflections and stresses are determined by the radius r. The radial tensile stress σ_{Pr}, radial bending stress σ_{Mr}, circumferential tensile stress σ_{Pt} and circumferential bending stress σ_{Mt} are shown in Figure (2).

The stress levels involved were calculated as follows, based on Von Karman's equation [8]:

$$
\begin{cases}
\sigma_{P_r} = \dfrac{1}{\mu-1}\left[\dfrac{d\left(\frac{1}{r}\int r\int \frac{-E(\frac{dH}{dr})^2}{2r}dr\right)}{dr}\Bigg|_{r=R} - \dfrac{\mu\int r\int \frac{-E(\frac{dH}{dr})^2}{2r}dr}{r^2}\Bigg|_{r=R}\right] + \\
\dfrac{1}{r^2}\int r\int \dfrac{-E(\frac{dH}{dr})^2}{2r}dr \\
\sigma_{M_r} = \dfrac{tE}{2(1-\mu^2)}\left(\dfrac{d^2H}{dr^2} + \dfrac{\mu}{r}\dfrac{dH}{dr}\right) \\
\sigma_{P_t} = \dfrac{d\left(\frac{1}{r}\int r\int \frac{-E(\frac{dH}{dr})^2}{2r}dr\right)}{dr} + \\
\dfrac{1}{\mu-1}\left[\dfrac{d\left(\frac{1}{r}\int r\int \frac{-E(\frac{dH}{dr})^2}{2r}dr\right)}{dr}\Bigg|_{r=R} - \dfrac{\mu\int r\int \frac{-E(\frac{dH}{dr})^2}{2r}dr}{r^2}\Bigg|_{r=R}\right] \\
\sigma_{M_t} = \dfrac{tE}{2(1-\mu^2)}\left(\dfrac{1}{r}\left(\dfrac{dH}{dr}\right) + \mu\dfrac{d^2H}{dr^2}\right)
\end{cases},
$$

Equation 1

where t is the thickness of the diaphragm, E is the Young modulus, μ is Poisson's ratio and H is the deflection of the diaphragm.

The tensile stress was uniform in the thickness direction, but the bending stress was linearly distributed. Therefore, the maximal radial and circumferential stress existed on the sides of the diaphragm. The radial stress on the gas side σ_{Gr}, the radial stress on the oil side σ_{Or}, the circumferential stress on the gas side σ_{Gt} and the circumferential stress on the oil side σ_{Ot} were calculated as having the following relations:

$$
\begin{cases}
\sigma_{Gr} = \sigma_{Mr} - \sigma_{Pr} \\
\sigma_{Or} = \sigma_{Mr} + \sigma_{\mathrm{Pr}} \\
\sigma_{Gt} = \sigma_{Mt} - \sigma_{Pt} \\
\sigma_{Ot} = \sigma_{Mt} + \sigma_{Pt}
\end{cases}.
\qquad
\begin{cases}
\dfrac{dH}{dr}\Big|_{r=0} = \theta\Big|_{r=0} = 0 \\
\dfrac{d^2H}{dr^2}\Big|_{r=0} = \theta'\Big|_{r=0} \le 0 \\
\dfrac{dH}{dr}\Big|_{r=R} = \theta\Big|_{r=R} = 0
\end{cases}
$$

Equation 2. **Equation 3.**

The pressure on the oil side compelled the diaphragm to cling tightly to the cavity surface in the last period of the discharge process. Thus, the maximal deformation of the diaphragm was determined by the shape of the cavity surface, which was determined by the generatrix of the cavity profile. At this point in the process, the stress level of the diaphragm reached its maximum, which should be treated as the design condition for the cavity profile when evaluating the fatigue life of the diaphragm. In this paper, therefore, references to the stress of the diaphragm refer to the stress level as the diaphragm clings tightly to the cavity surface.

2.2 Design condition of the cavity profile

To avoid stress concentration, the following conditions had to be met. First, the generatrix had to be smooth in the central region of the cavity. Second, the cavity centre had to be convex. Third, the deflection angle θ had to be zero at the cavity edge. These conditions were expressed as the equation 3.

With the traditional generatrix, most diaphragmatic fractures occurred in three positions: the diaphragm's edge, its centre and its region in contact with the outer groove of the perforated plate (see Figure 3). At the edge of the diaphragm, the crack shown in Figure 3(a) indicates that excessive radial stress (which is perpendicular to the crack) was the main reason for fractures in this region. In the centre of the diaphragm, where the discharge holes were settled (as shown in Figure 3(b)) fractures resulted from the superposition of the peak radial stress and the additional stress that occurred as the diaphragm clung tightly to the cavity surface.

These considerations indicate the paramount importance of radial stress control for effective cavity profile design. Moreover, as Moskalev suggested, reducing radial stress should be a primary design principle [9]. In view of the additional stress caused by the discharge holes, the allowable radial stress in the centre, σ_c, was lower than that for the other parts of the diaphragm, σ_a. These stress conditions were expressed as follows:

$$\begin{cases} R > r \geq R_1, \sigma_r < [\sigma_a] \\ R_1 > r \geq 0, \sigma_r < [\sigma_c] \end{cases},$$

Equation 4

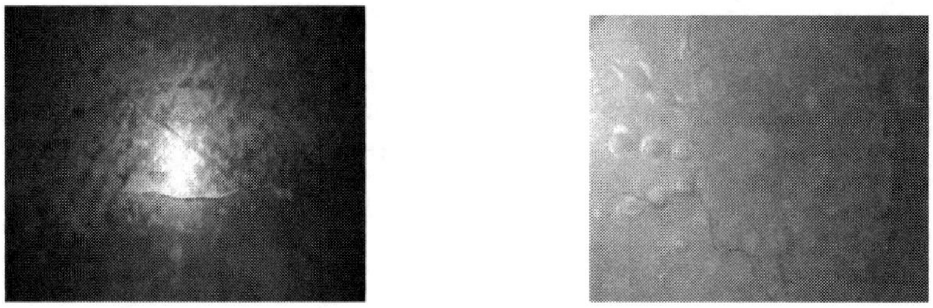

(a) (b)

Figure 3

where R_1 is the radius of the discharge region's edge (in which the discharge holes are settled).

The traditional generatrix of the cavity profile was determined by Equation (5):

$$H = H_0 \frac{1}{Z-1}[2(\frac{r}{R})^{Z+1} - (Z+1)(\frac{r}{R})^2 + (H_0 - 1)] \cdot$$

Equation 5

Based on the design parameters given in Table (1), the parameters Z and H_0 of the traditional generatrix were determined by a traversal method. When $Z = 8.8$ and $H_0 = 8.3$, the cavity volume reached a maximal value of $974\ cm^3$. Under this condition, the stress distribution of the diaphragm in the cavity was as shown in Figure (4).

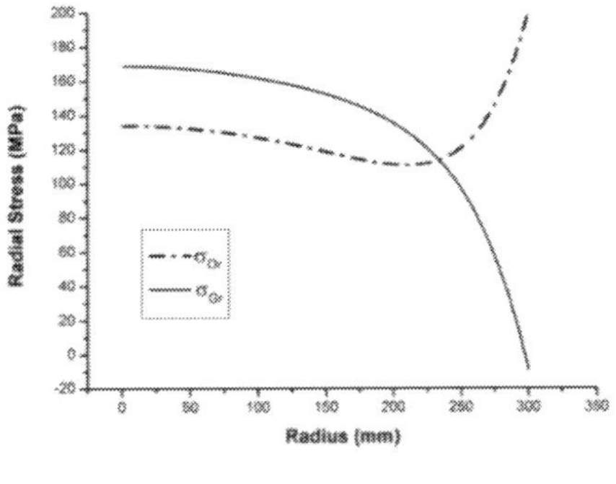

Figure 4.

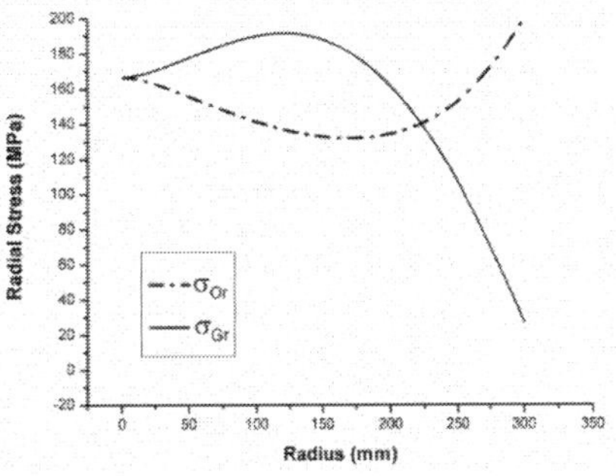

Figure 5.

Parameters	Values
R/mm	300
E/MPa	210000
μ	0.3
$[\sigma_a](20 < \mathrm{r} \leq \mathrm{R})$/MPa	200
$[\sigma_c](r \leq 20)$/MPa	170
t/mm	0.5

Table 1

However, as defined by the traditional generatrix, the radial stress distribution of the diaphragm in the cavity was non-uniform, as the radial stress at the edge and in the centre was much larger. Therefore, the peak of radial stress at the centre, which was subjected to additional stress caused by the discharge holes, seriously limited the performance of the diaphragm. The radial bending stress of the diaphragm in the cavity as defined by the traditional generatrix was expressed as follows:

$$\sigma_{Mr} = \frac{tEH_0}{2(Z-1)(1-\mu^2)}\left[\frac{2(Z+\mu)(Z+1)(r^{Z-1})}{R^{Z+1}} - \frac{2(1+\mu)(Z+1)}{R^2}\right].$$

Equation 6

In light of the above considerations, two new kinds of generatrices were designed.

2.4 New generatrices of the cavity profile

2.4.1 Two-parameter generatrix

To prevent the peak value of bending stress from appearing in the centre of the diaphragm, a two-parameter generatrix was designed. The deflection angle of this generatrix was expressed by n and m, two independent parameters. In considering the boundary condition shown in Equation (3), the equation for the two-parameter generatrix was expressed using H_0 as the maximal deflection in the cavity profile. Under the design conditions shown in Table (1), the optimal parameters were found by a traversal method. The cavity volume achieved a maximum value of $1038cm^3$ when $H_0 = 8.2, n = 3.1$ and $m = 3$. The radial stress of the diaphragm was calculated with Equation (1) and the distribution is shown in Figure (5).

2.4.2 Three-parameter generatrix

As with the two-parameter generatrix described above, the deflection angle of the three-parameter generatrix was expressed with n, m and k as three independent parameters, and with a as a control factor. In consideration of the boundary condition set by Equation (3), we generated the expression as follows:

$$H(r)=\frac{H_0}{a(\frac{(n-k)r}{(k+1)(n+1)})+\frac{(m-k)r}{(k+1)(m+1)}}\left[\frac{ar}{n+1}(\frac{r}{R})^{n+1}+\frac{r}{m+1}(\frac{r}{R})^{m+1}-\frac{(a+1)r}{k+1}(\frac{r}{R})^{k+1}\right]+H_0\cdot$$

Equation 7

The five parameters H_0, n, m, k and a had to be determined first. However, these parameters could not be obtained by the common traversal method due to the inexplicit constraints. Thus, a new optimisation algorithm was designed to find the optimum parameters of the generatrix. The boundaries of the parameters had to be determined first. According to engineering experience, H_0 ranged from 4 to 20. The control factor a can be derived from Equations (1), (2) and (7), and was calculated as follows:

$$a=\frac{[(m-k)-\frac{2\sigma_{MR}R^2(1-\mu^2)(m-k)}{EtH_0(k+1)(m+1)}]}{\frac{2\sigma_{MR}R^2(1-\mu^2)(n-k)}{(k+1)(n+1)}-(n-k)},$$

Equation 8

where σ_{MR} is the bending stress of the diaphragm on the edge $(r=R)$. Under the design condition shown in Table (1), σ_{MR} ranged from 60 MPa to 90 MPa.

According to the design conditions given in Table (1), when $a=-0.566$, $H_0=7.9$, $n=8.9$, $m=7.8$ and $k=2.7$, the cavity volume achieved a maximum value of $1073\ cm^3$. The stress distribution on the diaphragm in this condition is shown in Figure (6).

3 Verification of the new generatrices

3.1 Finite element method

To verify this design theory, the finite element method was used to calculate the radial stress. In performing this calculation, the cavity was defined by the generatrix mentioned in Section 2.4.1. The model for a quarter of the diaphragm and cavity was established. The results, which matched well with the theoretical estimation, are shown in Figures (7) and (8).

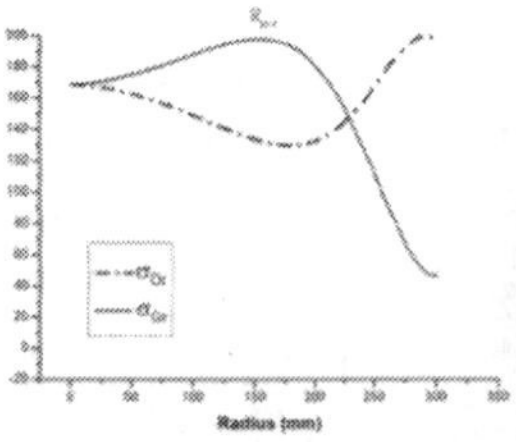

Figure 6.

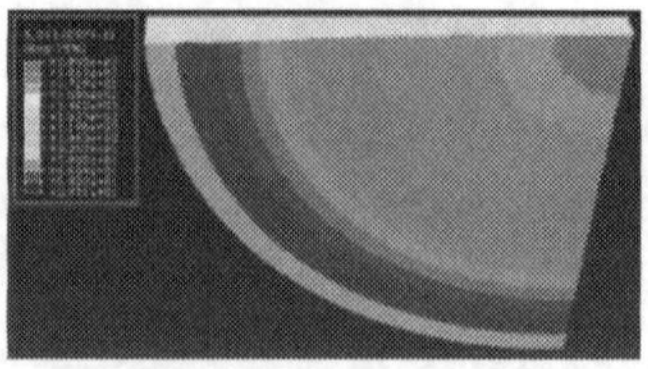

Figure 7.

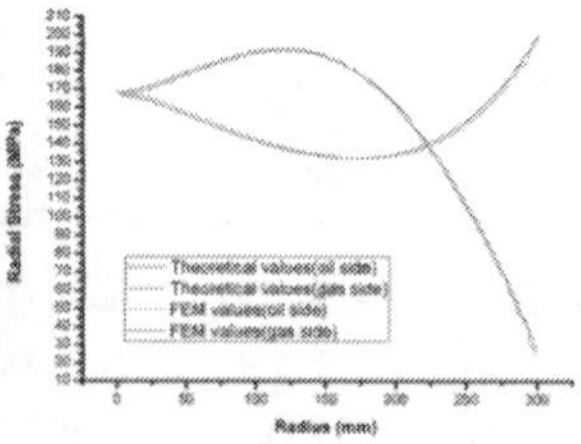

Figure 8.

3.2 Experimental method

3.2.1 Experiment on the cavity defined by the two-parameter generatrix

The relation between stress and strain in the diaphragm can be expressed as follows. Variables σ_r and σ_t are the radial and the circumferential stresses, respectively, and ε_r and ε_r are the radial and the circumferential strains, respectively. Accordingly, a test rig was built using a V-type diaphragm compressor. The two cavity profiles were separately designed with a traditional generatrix and a two-parameter generatrix.

Strain gauges were applied for the strain measurements. One displacement sensor was set near the discharge valve on each cavity to monitor the displacement of the diaphragm. A schematic diagram of the test rig is shown in Figure (9).

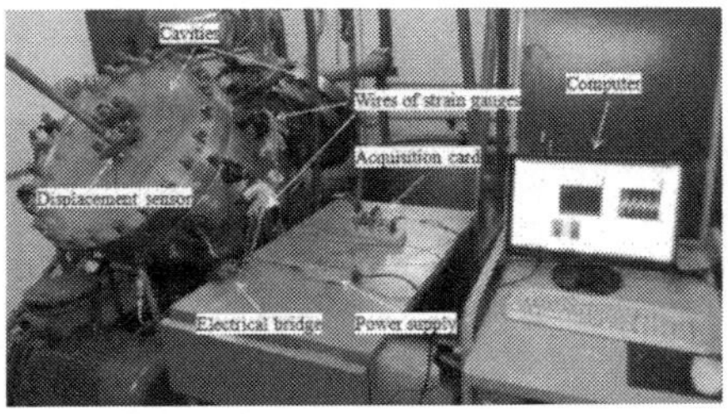

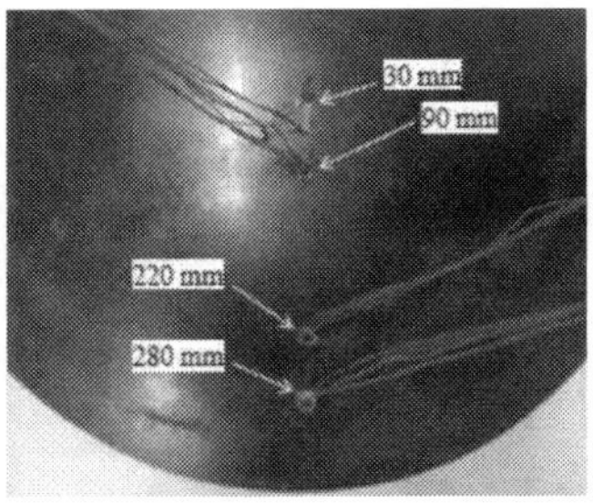

Figure 9. **Figure 10.**

The strain gauges were set equispaced along the radius for determining the radial stress distribution. [10-11] As shown in Figure (10), the strain gauges were set in both the radial and circumferential directions, at radii of 30 mm, 70 mm, 135 mm, 180 mm, 225 mm and 300 mm.

To avoid damage to the perforated plate during the working process, the strain gauges and wires were set at the radii of 30 mm, 70 mm, 135 mm, 180 mm, 225 mm and 300 mm upon the circular and radial grooves, as shown in Figure (11). Red spots indicate the positions of the strain gauges.

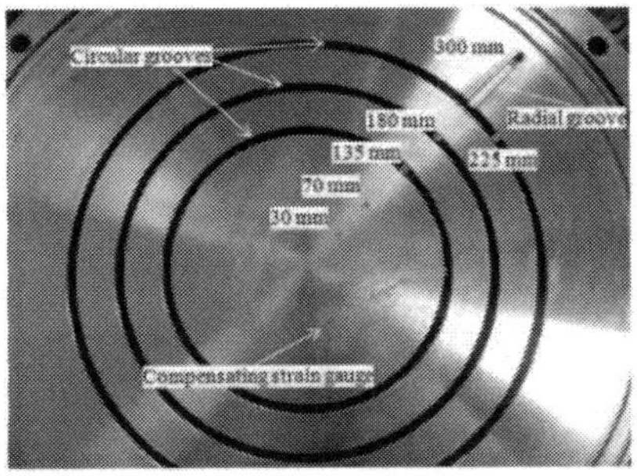

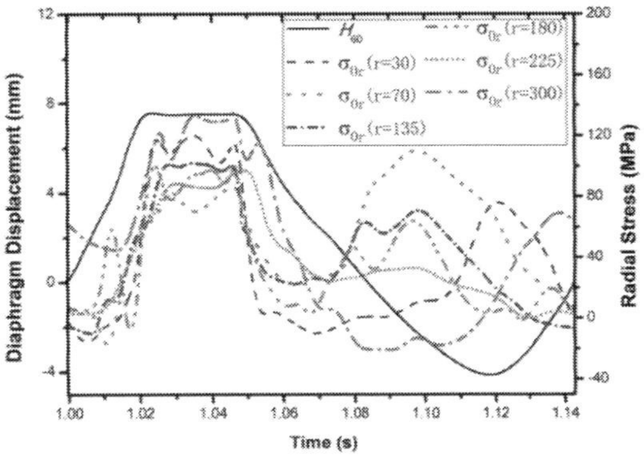

Figure 11. **Figure 12.**

Figure (12) shows the radial stress and the displacement in one operation cycle, as described above. Figure (13) shows the theoretical and experimental radial stresses of the diaphragm as it clung tightly

IOP Conf. Series: Materials Science and Engineering **90** (2015) 012083 doi:10.1088/1757-899X/90/1/012083

to the cavity surface. The trend of these theoretical stresses agreed well with the experimental stresses. However, the majority of the theoretical values were higher and the relative error was 12.3%.

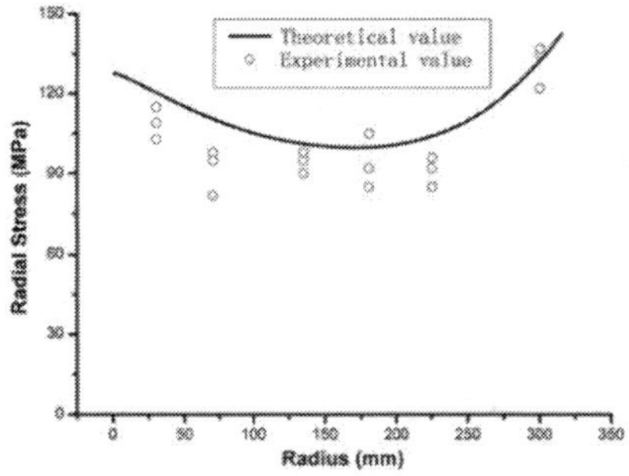

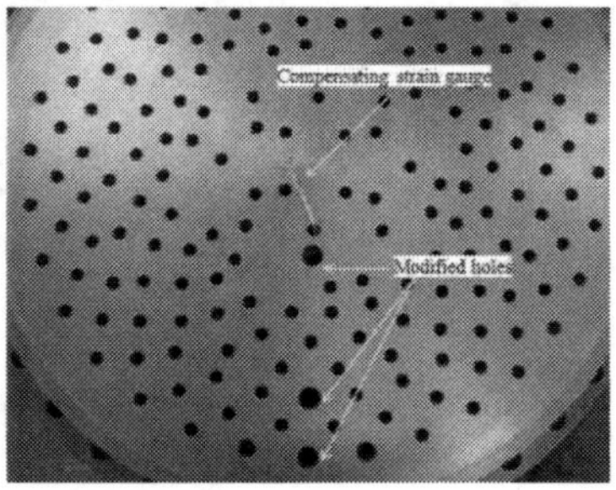

Figure 13. **Figure 14.**

3.2.2 Experiment on the cavity as defined by the traditional generatrix

The traditional generatrix of the cavity used in this experiment can be expressed as follows:

$$H(r) = 8.3\frac{1}{8.8-1}[2(\frac{r}{300})^{8.8+1} - (8.8+1)(\frac{r}{300})^2 + (8.8-1)].$$

Equation 9

Figure (15) shows the results of the experiment, including the radial stresses at different radii and the displacement of the diaphragm. The comparison of the theoretical and experimental radial stresses are shown in Figure (16). The average relative error was 11.2%.

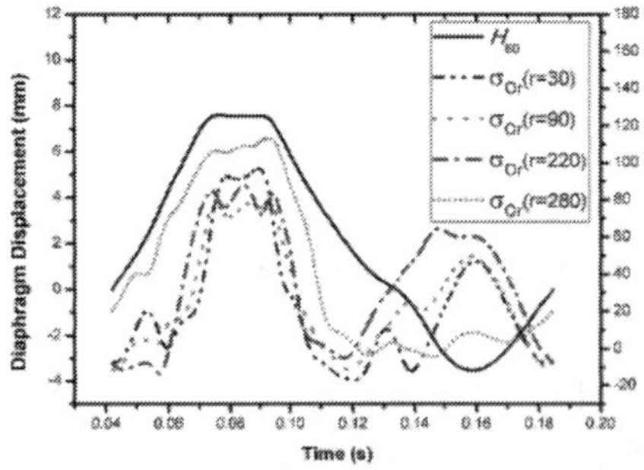

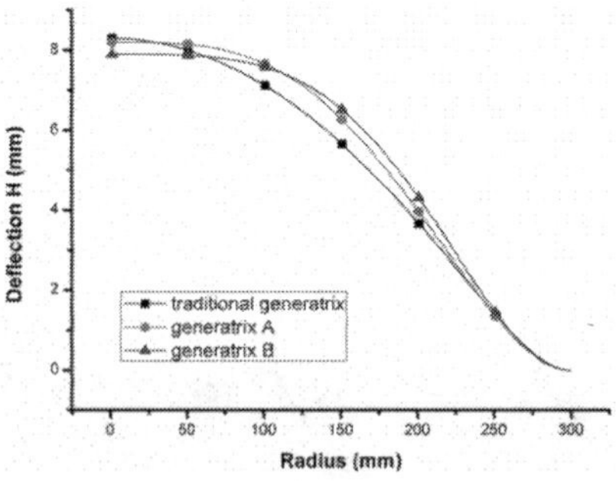

Figure 15. **Figure 16.**

4 Results and discussion

The proposed new generatrices design theory was validated by both experimental and finite element methods. The characteristics of the cavity profiles defined by the new and traditional generatrices were compared. The radial stress distributions of the diaphragms were investigated under the same set of design conditions.

The shapes of the generatrices are shown in Figure (16).

The radial stress level of the diaphragm is one of the most important evaluation criteria for a diaphragm compressor. Thus, a comparison analysis on the radial stresses of the diaphragms in the cavities as defined by the three generatrices was conducted, with the conditions of cavity radius, cavity volume and diaphragm properties held constant.

First, the cavity radii were set at 250 mm, 300 mm, 350 mm and 400 mm. The cavity volumes defined by the two generatrices were the same at each radius and were set as 542 cm^3, 974 cm^3, 1592 cm^3 and 2417 cm^3, respectively. Figure (18) shows the radial stress distributions of the diaphragms in the cavities as defined by different generatrices at various radii. For all of the cavity radii, the peak radial stress on the diaphragm in the cavities defined by the two generatrices occurred on the edges of the diaphragms and on the oil side. Therefore, the maximum radial stress on the diaphragm in the cavities defined by the new generatrix was always lower at each radius. In the centre of the diaphragm, the radial stress of the diaphragm in the cavity defined by the new generatrix was also lower, if we consider the additional stress caused by the discharge holes. The rates of decline in radial stress on the diaphragm in the cavity defined by the new generatrix were compared with those for the traditional generatrix, as shown in Figure (17). The rates of decline in radial stress on the diaphragm in the cavity defined by the new generatrix varied slightly with the increase in cavity radius. The radial stress on the diaphragm in the cavity defined by the new generatrix also showed a decline of 10.3% in peak radial stress. The radial stress in the central region of the diaphragm showed a decline from 14.0% to 10.1% when the cavity radius changed from 250 mm to 400 mm.

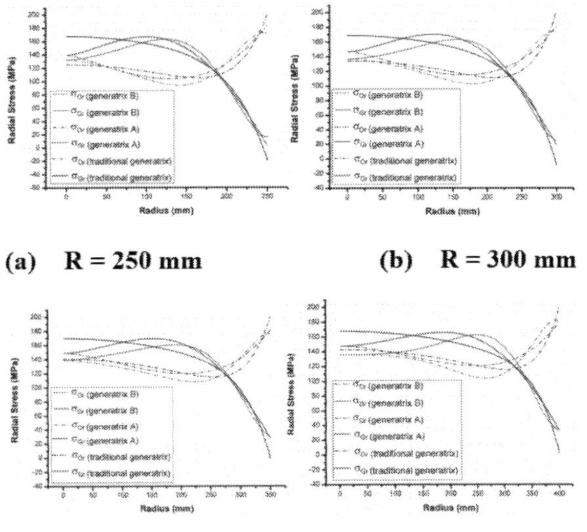

(a) R = 250 mm (b) R = 300 mm

(c) R = 350 mm (d) R = 400 mm

Figure 17.

5 Conclusions

Two new kinds of generatrices for the cavity profile of a diaphragm compressor were designed to decrease the radial stress level of the diaphragm. To find the optimum parameters for the new generatrices, optimisation algorithms were developed. The design theory for the diaphragm's radial stress was verified by experimental and mathematical methods. The radial stress on the diaphragm was investigated under various design conditions. The following conclusions can be drawn.

(1) Radial stress should be the key principle in the design of the cavity profile for a diaphragm compressor. The peak radial stress should be shifted away from the central region of the diaphragm to avoid the superposition of additional stress caused by the discharge holes.

(2) The levels of radial stress on the oil side of the diaphragm were tested and the experimental results agreed with the theoretical estimation. The radial stress of the diaphragm in the cavity defined by the new generatrix was calculated by a mathematical method and showed good agreement with the theoretical estimation.

(3) Compared with the radial stress on the diaphragm in the cavity defined by the traditional generatrix, the radial stress on the diaphragm in the cavity defined by the new generatrices declined by up to 19.6%. The peaks of radial stress changed from 10.0% to 14.0% when the cavity radius shifted from 400 mm to 250 mm. When the diaphragm thickness changed from 0.3 mm to 0.5 mm, the peak of radial stress showed a decline of between 10.0% and 13.0%, and the decrease in the rate of radial stress in the central region varied between 14.1% and 18.3%.

References:

[1] Li J, Liang L, Jia X, et al. A new generatrix of the cavity profile of a diaphragm compressor. *Proceedings of the Institution of Mechanical Engineers, Part C: Journal of Mechanical Engineering Science* 2013: 0954406213511077.

[2] Altukhov SM. Diaphragm compressors. *Chem Pet Eng* 1967; 3(8): 619-623. Wu B, Wu LZ, Wang XG, et al. Analysis and Study of the Diaphragm Cavity Profile of Diaphragm Compressors. *Compressor Technology* 2008; 4(1): 1-4 (in Chinese).

[3] Wu B, Wu LZ, Wang XG, et al. Analysis and Study of the Diaphragm Cavity Profile of Diaphragm Compressors. *Compressor Technology* 2008; 4(1): 1-4 (in Chinese).

[4] Ban L. *Optimization Research on Concave Surface of Diaphragm Compressor*. MsD Thesis, Institute of Mechanics, Chinese Academy of Sciences 1998 (in Chinese).

[5] Wang DS and Feng HP. Design of Diaphragm Chamber Curve Surface Shape with Double Function of a Diaphragm Compressor. *Journal of Xi'an Jiaotong University* 1989; 23(3): 25-32 (in Chinese).

[6] Kalshnikov VF, Kuznetsov VD and Antonova LI. Calculation of the Reliability of A Compressor Diaphragm Under Variable Loading. *Chem Pet Eng* 1987; 23(4): 168-170.

[7] Antonova LI, Kalashnikov VF and Degtyareva TS. Calculation of the Reliability of the Parametric and Functional Elements of Diaphragm Compressors. *Chem Pet Eng* 1987; 23(4): 153-155.

[8] Turvey GJ. A study of the behaviour of square plates at large deflections according to the theories of Foeppl and Von Karman. *J Strain Anal Eng* 1978; 13(1): 11-16.

[9] Yu YZ and Liu Y. *Special Compressors*. Beijing: China Machine Press 1989.

[10] Lancaster E R and Palmer S C. Strain concentrations in pressurized dented pipes. Proceedings of the Institution of Mechanical Engineers. *Proc ImechE, Part E: J Process Mech Eng* 1996; 210(1): 29-38.

[11] Leach J, Soden P D W and Kitching R. Strain distributions in hemispherical GRP vessels with radial branch connections subject to internal pressure and radial load. *Proc ImechE, Part E: J Process Mech Eng* 1994; 208(1): 45-58.

Keeping Your Compressor Healthy: Developing the Right Lubricant Formulation is the Key

Joseph A. Karnaz[1] **and Derek W. Kultgen**[2]
[1]Global Technology Leader, CPI Fluid Engineering, Midland, MI, USA
[2]Development Lab Manager, CPI Fluid Engineering, Midland, MI, USA

E-mail: jska@cpifluildeng.com
E-mail: deku@cpifluideng.com

Abstract. Selecting the correct compressor lubricant is crucial to the duration of the compressor and the refrigerant systems' useful life. However, developing an optimized lubricant for a refrigeration system requires a multitude of screenings and tests. The compatibility and stability of the lubricant with the refrigerant and compressor components needs to be examined at various accelerated conditions. The lubricant and refrigerant working viscosity must be determined at various refrigerant concentrations, temperatures and pressures as the diluted refrigerant in the lubricant has a significant effect on the viscosity. The correct lubricant formulation needs to be investigated for optimal performance. A compressor lubricant can provide many benefits to a refrigeration system such as bearing durability, sealing, and increased efficiency. Sometimes it is necessary to formulate the lubricant in order to optimize system performance. Specifically, this study investigated anti-wear properties of different oil additives to create a more robust refrigeration system. Many different additives and concentrations were considered and screened. Pending a successful screen test; these different additives' anti-wear properties were analyzed using bench top tribology tests. To reduce uncertainty and provide more in-situ results the different additives were operated in a refrigerant compressor on a gas-loop testing apparatus. Oil samples were taken periodically during the test duration for analysis. Lastly, upon test completion the compressors were dismantled and the parts were examined to determine the effectiveness of the anti-wear additives.

1. Introduction

Charging a refrigerant compressor with the incorrect or subpar lubricant can have devastating effects on the compressor and risks the contamination of the entire refrigeration system. This is why great emphasis is put on not only testing a compressor in a system but also selecting the optimal lubricant for different refrigerants, compressors, and systems. The significance of utilizing the appropriate lubricant is amplified by the fact refrigeration systems are expected to operate for many years without major repairs. Additionally, there is extreme limitation on the ability for the system to undergo regular lubricant changing intervals. The above factors are the motivation for formulating lubricants that improve refrigeration systems efficiency, working fluid purity, and reliability [1]. Specifically, this study aimed at beginning the process of discovering a new additive that when combined with the appropriate base lubricant would provide anti-wear characteristics to ensure compressor longevity.

Content from this work may be used under the terms of the Creative Commons Attribution 3.0 licence. Any further distribution of this work must maintain attribution to the author(s) and the title of the work, journal citation and DOI.
Published under licence by IOP Publishing Ltd

However, there are many obstacles to overcome before a new additive can go into production. Firstly, an appropriate base fluid lubricant to be additized must be determined. Next, the compatibility of the formulation (lubricant + additive) with the working fluid (refrigerant) and system materials needs to undergo investigation. This investigation is very important since additive chemistry types used in refrigeration applications are limited due to the design and operation of a refrigeration system along with potential environmental, health and safety regulations that need to be met. Pending successful compatibility, the ability of the additive to actually provide anti-wear performance is examined by utilizing bench top tribology tests. Ultimately, the formulation is operated in an actual refrigeration compressor to provide in-situ results. Lastly, the compressor parts undergo a surface finish analysis to confirm the additive's effect or lack thereof. A design of experiment (DOE) is an effective way of evaluating numerous additive candidates.

2. Screen Tests

Generally, the first stage to develop a new lubricant or formulation is to perform screen testing. Screen tests produce timely results to assist in the elimination of unsuitable lubricants or additives. Additionally, most screen tests do not require a lot of resources. Screen testing performed prior to this study included miscibility, sealed tube stability, and bench top tribology. However, before newly developed additives are screened, the base lubricant should be determined. Base lubricant selection is based off of the refrigerant utilized, compressor design, and the system application. One of the most sophisticated methods for obtaining data to make a base fluid selection is to operate different base fluids in a pressure, viscosity, and temperature (PVT) apparatus.

2.1. Pressure, Viscosity, and Temperature (PVT) Apparatus

The PVT apparatus measures the temperature, pressure, density, and dynamic viscosity of the lubricant or lubricant/refrigerant combination [2]. Charging the system with a known mass of lubricant and refrigerant in addition with a vapor space correction computes the concentration of refrigerant to oil by mass. Ultimately, the PVT outputs essential data such as solubility and Daniel plots. An example of a Daniel plot can be seen in Figure 1.

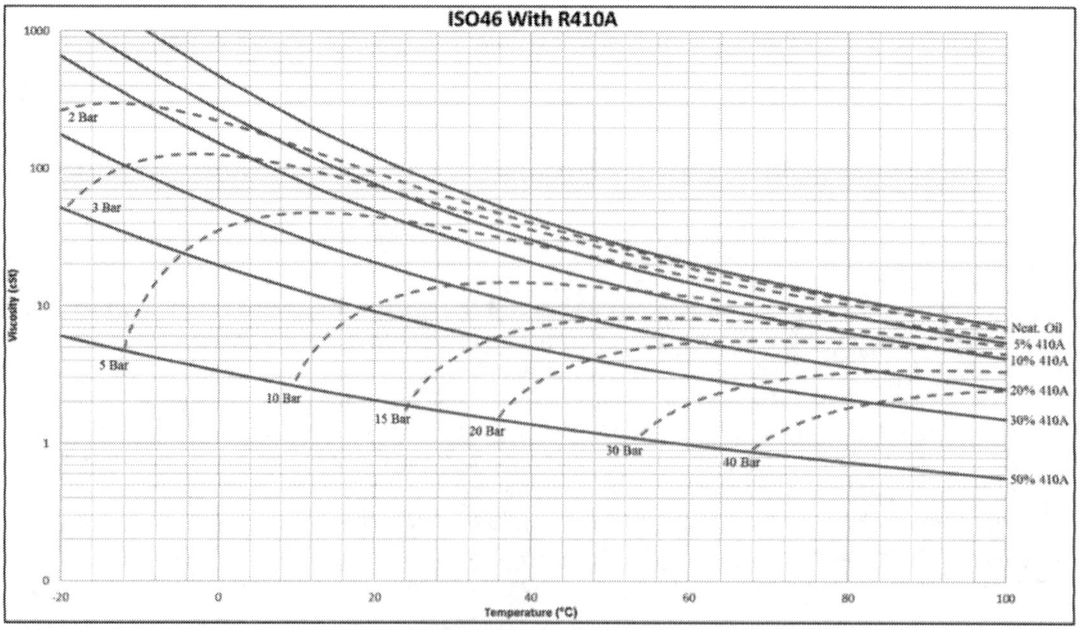

Figure 1. Daniel Plot

IOP Conf. Series: Materials Science and Engineering **90** (2015) 012084 doi:10.1088/1757-899X/90/1/012084

The Daniel plot is an exceptional example of the effect the refrigerant has on the lubricant's kinematic viscosity. The illustration clearly shows a drastic decrease in working viscosity as more refrigerant is dissolved into the lubricant. Essentially, without performing a PVT analysis on the base fluid the effectiveness of an anti-wear additive could translate into indiscernible results. The design of this experiment is to evaluate the additive and not the base fluid lubricating properties. For this reason the base fluid working viscosity needs to be such that it is not too low resulting in no additive being effective or too high where the base lubricant carries the load. Utilizing PVT analysis assists in selecting a base fluid with the optimal viscosity for the screening test.

2.2. Sealed Tube Stability and Miscibility

Lubricants with different additives and concentrations were measured for miscibility and evaluated for system stability. One method to perform stability tests is to seal a formulated lubricant and refrigerant with material catalysts (in this case aluminum, copper, and iron) in a glass tube [3]. A vacuum was pulled on the tube and then refrigerant was added. Once, the desired concentration of refrigerant was reached the glass tube was sealed shut with a torch. These samples were then heat aged at 175°C (347°F) for fourteen days. Subsequently, solutions were examined for changes in color, acid value, metal content in solution, and catalysts were examined for stains and tarnish. Stability tests assure that the formulations will be compatible with materials found in refrigeration systems.

Miscibility screening provides information regarding the ability of the lubricant and refrigerant to remain as a single phase over a range of operating conditions [4]. Conditions that allow for the change from single phase liquid to two phase liquid can have various detrimental effects on system performance. Miscibility testing focuses on the temperature at which the oil and refrigerant separate and a two phase condition exists. Miscibility can be evaluated over a wide temperature range and lubricant concentrations, for this particular test the focus was on lower temperature and concentration miscibility. Though the same base fluid was used for each additive response study, addition of certain additive chemistries can change the miscibility, requiring the need for this investigation. Results from miscibility testing can be seen in Table 1.

Table 1. Miscibility Results

Sample Set 1	Immiscible Temperature °C/(°F)	Sample Set 2	Immiscible Temperature °C/(°F)	Sample Set 3	Immiscible Temperature °C/(°F)
EXP-4957	-30 / (-22)	EXP-4969	-25 / (-13)	EXP-5031	-25 / (-13)
EXP-4958	-10 / (14)	EXP-4970	-25 / (-13)	EXP-5032	Immiscible
EXP-4960	-25 / (-13)	EXP-5020	-30 / (-22)	EXP-5033	0 / (32)
EXP-4963	-5 / (23)	EXP-5021	15 / (59)	Baseline	-30 / (-22)
EXP-4964	-30 / (22)	EXP-5024	-30 / (-22)	State-of-the-Art	-30 / (-22)
EXP-4966	-5 / (23)	EXP-5029	-10 / (14)		
EXP-4968	-10 / (14)	EXP-5030	-25 / (-13)		

The tests began at a temperature of 20°C (68°F) and then were placed in a cooling bath. The bath temperature was decreased by 5°C (9°F) intervals and the tube was inspected at each interval. The solution was immiscible below the corresponding temperature listed in Table 1. The importance of miscibility testing is to ensure the oil doesn't separate from the refrigerant in areas of the system that can have negative effects to system operation, efficiency and compressor lubrication. Miscibility and stability testing results are used for making final decisions on potential additive candidates. Succeeding successful miscibility and stability testing the additives actual anti-wear performance was screened by using a bench top tribology apparatus.

2.3. Bench Top Tribology

Bench top tribology tests are extremely useful for testing lubricants [5]. Tribology testing gives the ability to accelerate the wear between two or more surfaces. These surfaces are then inspected to determine the lubricants effectiveness. The bench top tribology test used was a four-ball machine. The four-ball machine operates by stacking four stainless steel spheres in a pyramidal shape with a lubricant film between them. A load is then applied to the top sphere while it spins and the three lower balls remain stationary. Concluding the operation of the four-ball machine, the scars on all of the spheres are measured and averaged. The average scar diameter and coefficient of friction of each proposed formulation is listed in Table 2. Please note that samples with poor miscibility are not included in the four-ball results as they were eliminated from the test matrix.

Table 2. Four-Ball Results

Sample Set 1	Average Scar Diameter (mm)	Coefficient of Friction	Sample Set 2	Average Scar Diameter (mm)	Coefficient of Friction
EXP-4957	0.231	0.042	EXP-5030	0.237	0.063
EXP-4964	0.24	0.057	EXP-5031	0.247	0.053
EXP-4969	0.261	0.053	Baseline	0.634	0.072
EXP-4970	0.264	0.059	State-of-the-Art	0.241	0.059
EXP-5020	0.473	0.092			
EXP-5024	0.286	0.055			

It is clear that according to the four-ball test some formulations are equal or better than the state-of-the-art anti-wear additive. Additionally, the worst performing sample was the baseline (un-additized base fluid). However, it must be noted that all of the results from the screening tests (miscibility, stability, tribology) need to be analyzed adjacent to each other. Essentially, if a formulation fails the stability or miscibility test; its' anti-wear performance is irrelevant to a refrigeration system.

3. Compressor Testing

As previously stated bench top tribology tests are a useful tool to simulate a wearing condition and help expedite the additive elimination process. However, most bench top tribology tests neglect significant variables. Specifically, the four-ball test ignores the effect of the refrigerant, pressure, and multiple catalytic materials. Therefore, to obtain real world results, the finalists in the additive matrix were tested in an actual refrigerant compressor for endurance testing [6].

3.1. Testing Rig

The rig utilized for compressor endurance testing was a gas-loop apparatus. A picture of the rig alongside of its' schematic can be seen in Figures 2a and 2b. The compressor endurance test rig consisted of a fan (F1) driven air-cooled tube and fin heat exchanger (HX), a receiver (REC), two metering valves (MV1 & MV2), solenoid valve (SV), fan (F2) cooled compressor with accumulator (COMP), two pressure transducers (P), and three thermocouples (T). The suction and discharge of the compressor had one thermocouple and pressure transducer each. The third thermocouple was placed on the shell of the compressor. When the shell temperature exceeded its' set point the lower fan (F2) would operate to maintain the desired temperature. The top fan was variable speed and was used to control the discharge pressure when running continuously. Both metering valves opening were manually adjusted to set the differential pressure.

Figure 2a. Picture of Testing Rig

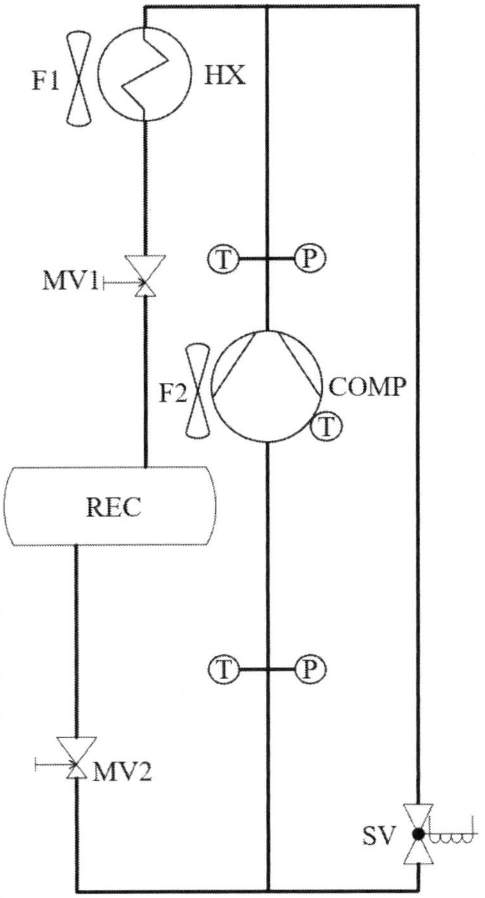

Figure 2b. Schematic of Testing Rig

The compressors utilized were a 2460W (8400 BTU/HR) hermetic rolling piston/vane rotary. The compressors were received dry of oil and a port was brazed into the lower side of the shell for oil charging and periodic sampling. The solenoid valve allowed the equalization of the high and low side. This allowed the pressures to rapidly equalize and increase the frequency at which the compressor could cycle on and off. The receiver provided additional volume for an intermediate pressure and the compressor was equipped with an accumulator to help prevent liquid refrigerant from entering the suction of the compressor.

3.2. Methodology

The endurance testing involved operating numerous rotary compressors with most of the compressors utilizing a different experimental anti-wear additive and R410A refrigerant. Each compressor underwent a relaxed break-in period and a severe condition period. The compressor operated continuously for twenty-four hours during the break-in period with the low side set to 6.9 bar (100 psig) and a high side of 27.6 bar (400 psig). The shell temperature of the compressor was set to not exceed 60°C (140°F) during the break-in. After the break-in procedure, the compressor was shut down so that an oil sample could be taken.

Next, the compressor was set to a higher load of 6.9 bar (100 psig) on the low side and 48.3 bar (700 psig) on the high side with a shell temperature of 90°C (194°F). During these severe conditions the compressor operated cyclically for 500 hours. The frequency was set for ten seconds of operating and seven seconds of idling during the entirety of the severe condition testing. This resulted in the total number of cycles being roughly 100,000. In addition to counting the number of cycles; the suction, discharge, ambient, and shell conditions were recorded for the duration of the test. The data acquisition frequency was 2.5 seconds during the operation portion of the cycle.

4. Initial Results

The investigation of the compressor test results began with using inductively coupled plasma atomic emission spectroscopy (ICP-AES) instrumentation on all of the oil samples [7]. The ICP-AES operation involves injecting the oil sample into a high temperature plasma source which moves electrons in an atom to an excited state. When the electrons fall to a lower energy state (ground state) photons are emitted. Since most atoms will emit light at different wavelengths a detector is used to measure both the wavelength of the emission and the intensity. This enables the quantification in parts per million (ppm) of various metals species present in the oil from the endurance test. A supplemental investigation involved visually inspecting the compressor parts.

The compressors operated during the test were hermetically sealed. Therefore, the shell of the compressor had to be carefully cut so the internal working components could be inspected. Once, the shell was open the internal mechanism was dismantled to inspect the bearing surfaces.

4.1. Oil Draws

As previously mentioned, a small volume of oil was withdrawn from the compressor periodically throughout the test. These small samples were analyzed for dissolved metal content by ICP. A sampling of the ICP results for dissolved iron is outlined in Table 3.

Table 3. Oil Sample ICP Results

Sample	Element Content (ppm)	
	After 24 Hour Break-In	After 500 Hours
	Fe	Fe
EXP-4957	0	0
EXP-4964	0	1.7
EXP-4969	0	1.2
EXP-4970	0.1	0.5
EXP-5020	2.0	23.8
EXP-5024	19.5	205.8
EXP-5030	0	15.5
EXP-5031	6.6	228.3
Baseline	5.4	224.9
State-of-the-Art	0	0

Levels of dissolved metals in the lubricant, specifically iron, are a potential indication for the lack of proper lubrication in a compressor. When investigating rotary compressor wear performance, the measure of iron levels provides exceptional evidence of the lubricant formulations effectiveness. Other metals were measured but aren't listed as they may indicate compositional additive content or other non-lubrication related reactions.

4.2. Compressor Components

Upon dismantling the compressor, a subjective evaluation of the compressor components was performed by visually inspecting them and rating their condition. The inspection was broken down into six categories which included polish, wear, copper plating, residue, chemical staining, and scratches.

Polish describes a condition where traditional wear isn't present but, it is clear that the compressor operated. Wear describes the condition of parts showing signs of excess friction such as dark regions. Copper plating is very obvious due to the reddish-orange color of the parts. Residue describes debris or actual solids that were clinging to the parts. Chemical staining results from a reaction occurring inside the compressor during operation that tarnished the parts a different color. Lastly, a scratch was considered an easily identifiable groove or cavitation found on the part. Scratches aren't necessarily considered wear as it is typically the result of foreign debris getting lodged between two moving parts.

Table 4. Key for Table 5

Description	Polish	Wear	Copper Plating	Residue	Chemical Staining	Scratch
None	0	-	-	-	-	-
Light (Small)	1	4	A	L	X	E
Medium	2	5	B	M	Y	
Heavy (Large)	3	6	C	N	Z	F

Table 5. Visual Results of Compressor Parts

	Oil Sample									
Part	EXP-4957	EXP-4963	EXP-4966	EXP-4970	EXP-5020	EXP-5024	EXP-5030	EXP-5031	Base line	State-of-the-Art
Roller	2 -	1 -	2 X	3 -	3 A	1 F	1 F	1 -	4 A	1 -
Vane	2 L	2 L	2 XL	2 -	3 -	2 -	2 F	1 -	4 -	1 -
Crankshaft	1 -	1 Y	2 X	2 -	1 A	1 XL	1 B	2 A	3 X	1 -
Outboard	2 -	1 Z	4 Z	2 -	2 A	1 AX	3 B	2 A	3 -	2 -
Main	1 -	2 Y	2 X	2 A	2 A	1 -	1 A	1 -	2 A	1 -

5. Validation

Since, a four-ball test neglected important variables, metals analysis data doesn't precisely measure wear, and visual inspection is subjective; due to this, methods to validate the above were performed. The validation tests consisted of surface finish analysis and Auger. Another advantage to conducting the validations tests was the ability to compare new and unused compressor parts to those found in the experimentation.

5.1. Surface Finish

The instruments used to quantify the surface finish performed Stylus Profilometry and Optical Profilometry. Both methods measure the indentations in the part due to wear [8]. The Stylus method provides a 2-dimensional representation of wear whereas; the Optical method utilizes a color gradient to illustrate depth in a 3-dimensional wear profile. Figure 3 shows pictures of the compressor's vane tips, two after the 500 hour endurance test and one of a new vane tip.

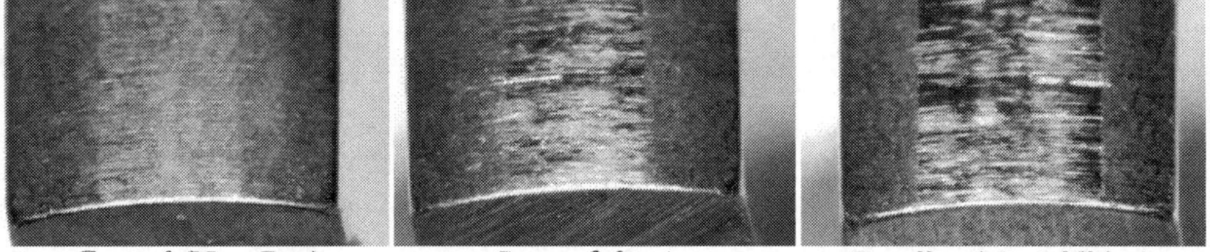

Control (New Part) State-of-the-Art Baseline (No Additive)

Figure 3: Photographs of Vane Tip from Compressors Parts

5.1.1. Stylus Profilometry. Figure 4 shows the surface finish results of the vane tip surface from the baseline and state-of-the-art. The compressor tested without an anti-wear additive shows a significant surface finish change (increased surface roughness) compared to the state-of-the-art additive formulated lubricant compressor test. The state-of-the-art indicates only slight changes to surface finish when compared to the new part.

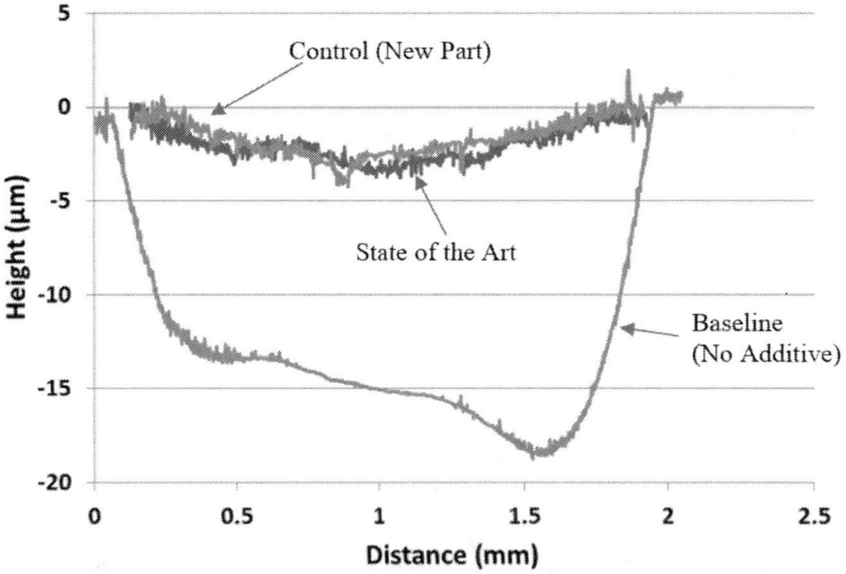

Figure 4: Surface Finish Measurements of Compressor Parts - Vane Tip

5.1.2. Optical Profilometry. Optical Profilometry is another way to evaluate the surface of a bearing. It is easy to see the surface roughness difference between the three vane tip surfaces in Figure 5a. Figure 5b shows a calculation of the surface profile curvature differences of the vane tip. The lower curve indicates a more worn surface of the baseline sample while, the similar profiles between the new compressor part and the state-of-the-art compressor test illustrate little to no surface change. These results correspond well with the stylus profilometry results and both results show the leading and trailing edges of the wear pattern.

Figure 5a. Optical Measurements

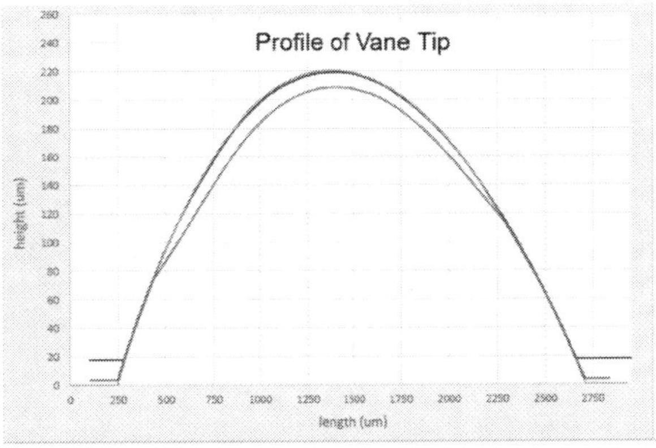

Figure 5b. Surface Profile Calculations

5.2. Auger

When differentiating the effectiveness of additives, it is useful to understand how the additive interacts with the bearing surface. Most anti-wear additives function or activate when there is failure of the base fluid to provide the appropriate bearing film thickness. Auger analysis looks at the surface of a material at a micro level (to a depth of approximately 5 nanometers) and each formulated lubricant is composed of elements that can be detected via Auger. However, this type of analysis does have some limitation. An Auger bearing surface analysis example is shown below in Figure 6. Specifically, these results were obtained from performing Auger chemical analysis on the vane tip from the state-of-the-art compressor test.

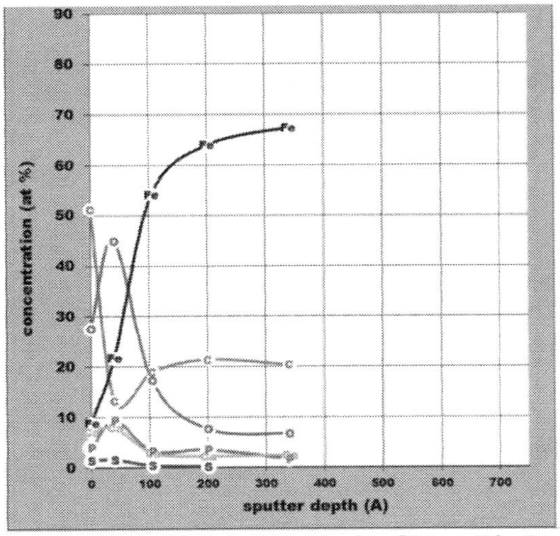

Figure 6: Auger Chemical Analysis of Vane Tip Surface

Examination of the Auger results above show at no depth into the surface metal; the primary elements present are carbon, oxygen, and some phosphorus. These elements are most likely from the formulated

IOP Conf. Series: Materials Science and Engineering **90** (2015) 012084 doi:10.1088/1757-899X/90/1/012084

lubricant as well as, a combination of residual lubricant and reactive chemistry of the additives on the surface. When the analysis moved deeper into the surface the concentration of the lubricant elements decreased and bearing surface became more prevalent as indicated by the increase in iron. This phenomenon gives the notion that the additive response to the surface was not significant. This could be a function of initial additive level or additive surface activeness at these conditions. Evaluation of other bearing surfaces or tests with different additives might show a different additive response via this method.

6. Conclusion

Changes in the HVAC&R industry have always driven the need to find solutions that maintain reliable and energy efficient products. Today these changes continue not only in regards to refrigerant but also other aspects of the chemical industry. System chemistry and lubrication have always been critical parameters that need to be investigated and bench top qualification is helpful to screen products.

This paper demonstrated methods to evaluate various anti-wear additives designed to replace or be substitutes for current state-of-the-art chemicals. Lubricant additive formulation criterion is not limited to compressor performance alone. Today restrictions reside in chemical registration and environmental acceptance that can limit additive options and concentrations. Methods such as miscibility, stability, wear testing and compressor testing were used to screen the acceptability of various additive candidates. Visual observation along with more sophisticated methods of evaluating the effect of additive response on the compressor bearing surface during operation are the qualifying steps to approving a formulated lubricant acceptable for OEM system and field trial testing.

The beginning stages of evaluation showed some additive chemistries were too reactive or incompatible for effective use in HVAC&R applications. Compressor testing and bearing surface finish evaluation are still in the early stages of investigation but some candidates appear to have the potential to be acceptable replacements or prospective anti-wear additives to provide performance improvements. Additionally, work is ongoing to identify the appropriate anti-wear candidates that not only meet performance requirements but also conforms to chemical sustainability.

References

[1] J. Karnaz. K. Liu. "Lubricant and Refrigerant Properties – The Need for Lubricant Optimization with Various Types of Alternative Refrigerants". The 5th International Conference on Cryogenics and Refrigerants. Hangzhou (2013).

[2] C. Seeton. "Solubility, Miscibility and Liquid Viscosity of Lubricants with CO_2 and Propane". ASHRAE Winter Meeting. Chicago (2005).

[3] ASHRAE Standard 97. "Sealed Glass Tube Method to Test the Chemical Stability of Materials For Use Within Refrigeration Systems. (2007).

[4] J. Karnaz. "Compressor and System Energy Efficiency Improvements through Lubricant Optimization". 8th International Conference on Compressors and Their Systems. London (2013).

[5] A. Suh, A. Polycarpou, T. Conry. "Fundamental Investigation on the Tribological Failure Mechanism of Compressor Surfaces, Scuffing: Detailed Rouoghness Analysis of Al390-T6". ACRC Project #127 (2003).

[6] J. Karnaz. "Utilizing Lubricant Refrigerant Interaction Data for Compressor and System Design. Advanced Compressor Modeling (Compressor 102)". Purdue University Short Course (2012).

[7] D. Skoog, F. Holler, S. Crouch. "Principles of Instrumental Analysis". Thomson Brooks/Cole. Belmont, CA. 2007, 6th ed.

[8] R. Tuomas. "Properties of Oil and Refrigerant Mixtures – Lubrication of Ball Bearings in Refrigeration Compressors". Lulea University of Technology 2006.

IOP Conf. Series: Materials Science and Engineering **90** (2015) 012085 doi:10.1088/1757-899X/90/1/012085

Carbon and energy saving markets in compressed air

R Cipollone, PhD
Professor
University of L'Aquila, Department of Industrial and Information Engineering and Economy, Via Gronchi 18, 67100 L'Aquila, Italy
e-mail: roberto.cipollone@univaq.it

Abstract. CO_2 reduction and fossil fuel saving represent two of the cornerstones of the environmental commitments of all the countries of the world. The first engagement is of a medium to long term type, and unequivocally calls for a new energetic era. The second delays in time the fossil fuel technologies to favour an energetic transition. In order to sustain the two efforts, new immaterial markets have been established in almost all the countries of the world, whose exchanges (purchases and sales) concern CO_2 emissions and equivalent fossil fuels that have not been emitted or burned. This paper goes deep inside two aspects not yet exploited: specific CO_2 emissions and equivalent fossil fuel burned, as a function of compressed air produced. Reference is made to the current compressor technology, carefully analysing CAGI's (Compressed Air Gas Institute) data and integrating it with the PNUEROP (European Association of manufacturers of compressors, vacuum pumps, pneumatic tools and allied equipment) contribution on the compressor European market. On the base of energy saving estimates that could be put in place, this article also estimates the financial value of the CO_2 emissions and fossil fuels avoided.

1. Introduction

Fossil fuel shortage and carbon increase in the atmosphere are amongst the most important current concerns of the political actions. Carbon content increase in the atmosphere calls for a new energetic economy which can only be based on renewables: it is a mean to long term problem. Fossil fuel shortage is a short to mean term problem, dominated by public and private financial and economic interests.

Fossil fuels are responsible of almost 90 % of the final energy uses and the absolute increase of the energy produced by renewables is completely nullified, in percentage terms, by the continuous increase of the energy request. Most recent data, [1,2,3] report that global primary energy consumption accelerated in 2013 despite stagnant economic growth: it increases by 2.3 %, greater than it was in 2012 (+1.8 %). In terms of CO_2, the result of this fossil economy produces a significant contribution to the continuous carbon content increase into the atmosphere.

The scientific community agrees about a positive feedback between climate change and carbon cycle. The magnitude of this feedback is at the centre of the scientific debate. Reference Institution is IPCC, [4] and, based upon current understanding, the stabilization of CO_2 at 450 ppm will likely result in a global equilibrium warming of 1.4°C to 3.1°C, with a best guess of about 2.1°C. This would require a reduction of current annual greenhouse gas emissions by 70-80 % by 2100. In spite of this awareness, CO_2 concentration ranked in 2013 at an unfortunate level of 400 ppm, so an urgent intervention should be in the agenda of policy makers. CO_2 emissions grew by 2.3 % in 2013 with respect to 2012, reaching the absolute value of 35.3 Gt CO_2/year, [3].

Content from this work may be used under the terms of the Creative Commons Attribution 3.0 licence. Any further distribution of this work must maintain attribution to the author(s) and the title of the work, journal citation and DOI.
Published under licence by IOP Publishing Ltd

Concerning fuel shortage, at present, fossil fuel reserves (P1 type) have a horizon of exhaustibility of 53 years, 55 years and 113 years for the oil, gas and coal respectively, [1,3]. New production technologies and new unconventional oil and gas sources have increased the reserve levels well above conventional availability. Almost in all the countries in the world, CO_2 emissions, fossil fuel saving (and energy production from renewable sources) have been committed. Target short-term year is 2020 and all the countries are very aligned to the same reduction percentages, also in those countries which were in the past, far enough to a common awareness (China, India, USA, etc...).
In order to favour the reaching of the reduction goals, some environmental new markets have been established. In many countries, under a strict public regulation and control in terms of working rules and quantities to be exchanged, CO_2 emission and fossil fuel saving trading markets have been launched. Similarly, Carbon taxes have been established as well as financial incentives recognized when high efficiency components are chosen. As in all the "tangible" markets a specific good is traded, on the CO_2 and fossil fuel saving markets, CO_2 and fossil fuels that have not been emitted or burned are traded. Subjects which have to reach CO_2 emissions or fossil fuel saving targets can buy quotas sold by subjects which sell "certified quantities" of CO_2 not emitted or fossil fuel not burned. The experience made till now demonstrates how far these markets are from being "ideal" (price established by demand and supply balance), but, nevertheless, CO_2 avoided as well fossil fuels not burned today have a "price" and they represent opportunities to sale or purchase. The compressed air sector cannot neglect this opportunity.
Compressed air makes use of electrical energy; 10 % of the industrial electricity is consumed to produce compressed air, [5-11]. This figure grows up to 20 % if other, non-industrial compressor needs are considered, like those in the residential sector, transportation, and agriculture.
Moreover, the energy saving potential of the compressed air sector is great and many studies are aligned to consider that a 40-50 % of the present consumption could be saved, [5,11,13,14]. Electricity saved means CO_2 not emitted and fossil fuels avoided for the respective quota.
In this paper the carbon dimension of the compressed air and the related fossil fuel saving are discussed.
By a proper elaboration of CAGI's data [12], Carbon emission by the present compressor technology has been evaluated, as a function of pressure delivered and flow rate (machine size). Thousands of data have been re-elaborated from original datasheets in order to give consistency. The analysis made use of a wider compressor specific consumption analysis, [13] and a quantitative estimation of the potential saving in existing compressors, [14].
Based on a recent study, [7], which confirmed previous knowledge in the matter, the potential energy saving has been returned in terms of CO_2 and fossil fuel credits which are tradable. Making reference to the European context, in which market prices are more established, a financial value of this trading has been evaluated and the need of a common synergic action of the sector is fully justified.

2. Compressed air electricity consumptions

The electricity consumption in 2012 was 18312 TWh [15] whose fossil fuel share is close to 70 %: this means almost 12820 TWh are produced by burning oil, gas and coal. The main use of electricity is for industry, transport and building. When a projection for future consumption is made, electricity consumption will grow: BRIC countries in 2020 will have a dominant share (40 %) and China will become the most important electricity user (greater than USA and Europe and close to the sum of these two). Compressed air production in developed countries accounts for a mean value around 10 % of the overall electricity consumption in the industrial sector. This share is referred to the industrial needs of compressed air. Adding all the other "compression needs" i.e. the commercial and residential markets (portable tools, air pumps, pneumatic heating, ventilation, air conditioning, personal uses, etc...), the consumption grows to 20 % of the industrial electricity needs.
An important issue is to subdivide electric energy consumption of the compressed air use into different causes: a useful concept is to consider actions external to the compressors or related to them. Required flow rate and line pressure fall into the first class; compressor energy specific consumption (kWh_{el}/m^3)

into the other class. Energy consumed for compressed air (in a specific production site) is given by the flow rate delivered multiplied by the specific energy consumption of compressors. Great leakages and lack of pressure level optimization are usually found in many compressed distribution lines. The first contribution is quantified by means of a percentage of the air wasted which ranks at the unbelievable and embarrassing level of 20-40 %. In absence of this awareness, the interventions on the compressor unit side are certainly beneficial but mostly useless.

An additional significant energy saving is related to pressure delivery optimization reachable by a proper compressed air piping design and filter replacement. Reducing pressure level represents a very effective action: the consumption of a compressor depends as first approximation of the fourth square of the pressure ratio. If pressure is decreased of one bar, it is easy to verify that reducing, for instance, from 8 to 7 pressure ratio, a saving of 10 % is reached. At higher operating pressures this benefit is lower, for example decreasing line pressure from 15 to 14 bar, the beneficial effect on the compressor consumption is reduced to 4%.

Concerning the compressor technology, the potential saving on the specific energy consumption is significant, too. A useful reference could be found in [13] in which real existing machines have been compared using the CAGI's datasheets and re-elaborated [14] in order to evaluate the maximum potential saving in existing compressors and highlight the importance of machine replacement.

3. Carbon dimension of the compressed air

CO_2 emission due to electricity production is extremely variable: it depends mainly on the share of fossil fuel used and on the global efficiency of the power plants. European emission is close to 0.325 $kgCO_2/kWh$ but the spread among different countries is great. Norway has the lowest emission factor (0.01 $kgCO_2/kWh$), ninety times lower than Greece. Switzerland, Sweden, France, Austria, Finland, Belgium, Denmark have values below the European datum, while others countries stay above (Luxemburg, Italy, Spain, Netherlands, UK, Germany, Portugal, Ireland). Important differences are still present when comparing Europe with others territorial contexts. BRICS countries have CO_2 emission values well above those in Europe: 2.25 times in China, 3 times in India and South Africa. This applies also to USA, which has a specific CO_2 emission 1.75 greater than in Europe.

Compressed air considered as a commodity is an electricity user: it has different impacts on CO_2 emission in different countries. So, in the countries with higher emission factors, the energy saving potential will have a greater importance. In an extensive work [13], datasheets from CAGI were elaborated in order to compare specific energy consumption of different compressors from many manufacturers. Thousands of data points have been processed in order to compare different machines whose testing conditions were not referred to exactly the same operating conditions, [14]. The result is reported in Figure 1. An extensive analysis is available in [14]: the most interesting datum is the presence in the market of significant scatter among different machines that demonstrates the use of compressors of different design generations. Air cooled compressors, operating at 8 bar (a), at 10 m^3/min have a mean specific consumption equal to 6.8 $kW/m^3/min$ with a scatter equal to 1.5 $kW/m^3/min$. At 30 m^3/min, mean specific consumption decrease to 6.5 $kW/m^3/min$ with a scatter of 1 $kW/m^3/min$. Bigger machines ($>50m^3/min$) have a specific consumption which approaches 6.2 $kW/m^3/min$. The wide scatter invites to seriously consider machine replacement. Recently, PNEUROP [7] explored the compressor efficiency as function of flow rate in the European market. Recalculating specific energy consumption produces scatters very close to those found from CAGI's data. From a quantitative point of view, among different machines on the market, a difference close to 20-40 % can be reached. A useful rule of thumb can be set up for the overall energy saving in compressed air use. Considering that, simple leakage interventions could account for a reduction in flow rate of about (20-40) % and pressure refining and machine replacement could account for a specific consumption reduction close to 10% and (20-40) %, when a datum is known for the overall machine consumption reduction, the overall saving can be easily estimated by doubling this value.

Figures 2a) -2d) show the specific CO_2 emissions evaluated making reference to the European context (screw and sliding vane rotary types) (*ccm*: carbon per cubic meter of compressed air.

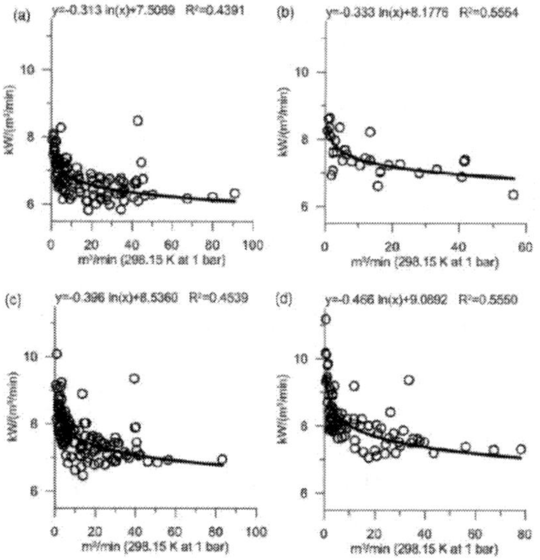

Figure 1a) Specific energy, rated flow–air cooled. Pressure delivered: (7, 8, 9, 10) bar (a)

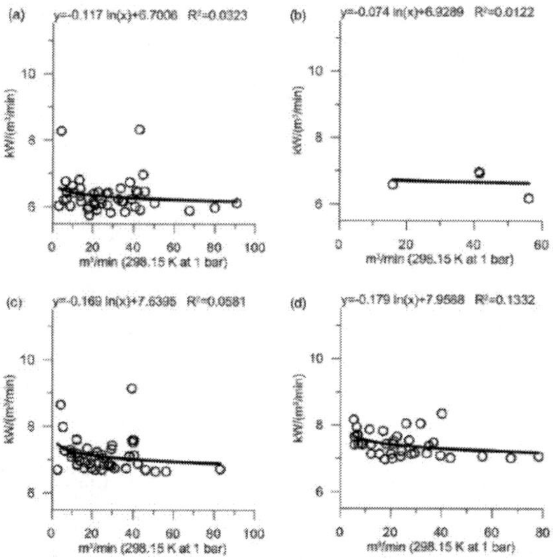

Figure 1b) Specific energy, rated flow–water cooled. Pressure delivered: (7,8,9,10) bar (a).

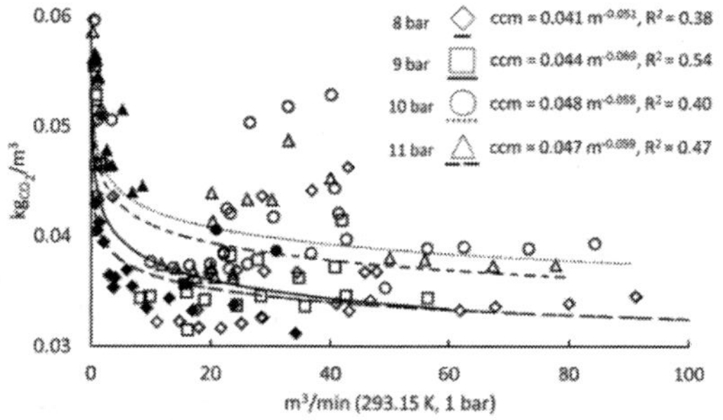

Figure 2a. Specific CO_2 emissions. Compressors operates at fixed speed, air cooled (black symbols refer to sliding vane rotary machines)

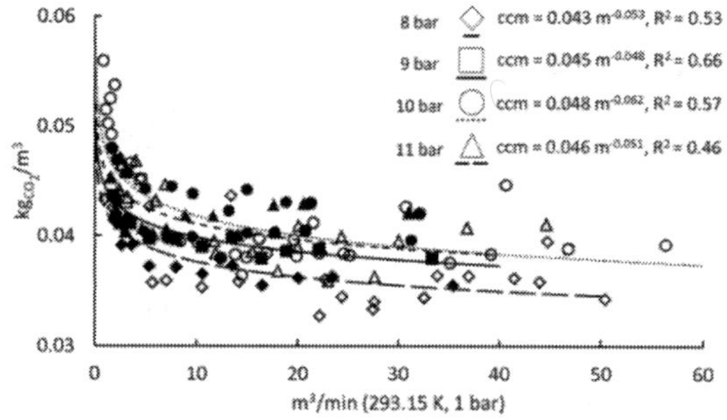

Figure 2b. Specific CO_2 emissions. Compressors operates at variable speed, air cooled (black symbols refer to sliding vane rotary machines).

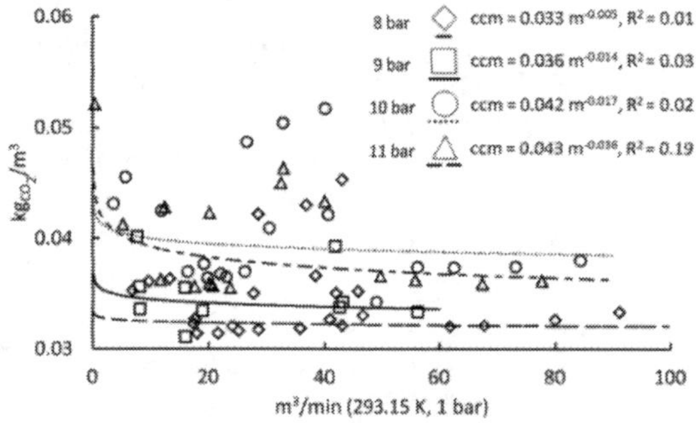

Figure 2c. Specific CO_2 emissions per unit mass of compressed air. Compressors operates at fixed speed, water cooled.

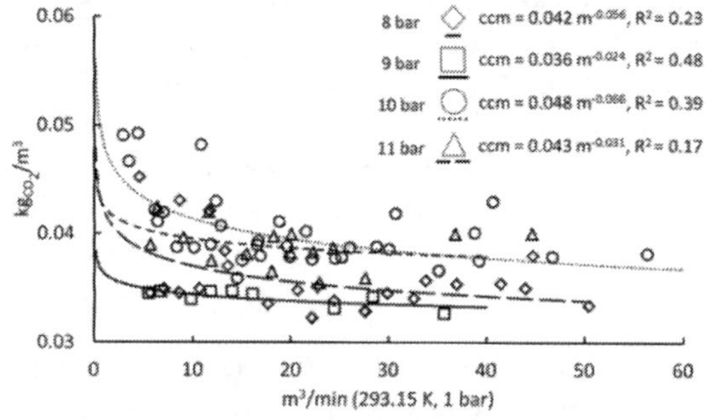

Figure 2d. Specific CO_2 emissions per unit mass of compressed. Compressors operates at variable speed, water cooled

The following considerations apply:

a) *ccm* decreases in general with respect to flow rate. So, bigger machines are less polluting: for air cooled machines operating at 8 bar (a), at 5 m^3/min, 20 m^3/min and 50 m^3/min, mean *ccm* is 0.037 $kgCO_2/m^3$, 0.035 $kgCO_2/m^3$ and 0.033 $kgCO_2/m^3$, respectively. A very high scatter among machines is present, meaning the coexistence on the market of many machines conceived in the past. The scatter doesn't allow to estimate the benefit of water cooled machines (*ccm* of about 0.039 $kgCO_2/m^3$, 0.036 $kgCO_2/m^3$ and 0.034 $kgCO_2/m^3$ at the same flow rates);

b) when flow rate increases, *ccm* shows an asymptotic trend, with lower values for both mean *ccm* and scatter (for a flow rate of 50 m^3/min, the mean *ccm* value is about 0.033 $kgCO_2/m^3$ and the scatter is 0.027 $kgCO_2/m^3$);

c) for higher delivered pressures, *ccm* differences among air and water cooled increase, inviting to water cooled version. For a delivered pressure of 11 bar (a), the scatter further reduces, leading to a compressors population that is tightly distributed around a common value: for a flow rate of 30 m^3/min, the mean *ccm* value is around 0.035 $kgCO_2/m^3$, with a scatter of about 0.002 $kgCO_2/m^3$.

4. Environmental related markets

4.1. Carbon Market

The first international attempt to reduce CO_2 growth in the atmosphere was introduced through the Kyoto Protocol in 1997. Reduction commitments were differentiated by country. A mean emission reduction accepted by the countries signing the Protocol was by 5% compared to 1990 level, over the period from 2008-2012. One of the ways to reach this target was the assignment of a value to Carbon not emitted.

On January 1st 2005 the European Union Emission Trading Scheme (EU-ETS) was launched. It was governed by the EU ETS Directive (2003/87/EC) and revised in 2009 (2009/29/EC). Provisions of the Directive had to be transposed into national law by 31 December 2012. The EU ETS actually covers around 11 000 large greenhouse gas emitting installations in the energy and industry sectors.

Sellers and buyers, according to defined rules, could operate on this market, so giving to sellers the possibility to make a profit and to buyers to have an additional measure to reach the emission reduction targets. The unit exchanged on this market was the ton of CO_2 (tCO_2) avoided and it remained as unit for the others carbon-pricing instruments. Presently, EU-ETS is covers 45% of the EU's greenhouse gas emissions. On the worldwide base, EU emissions accounts for a 10-12 % of the whole emission scenario.

A "quantity-based" instrument was applied in order to define Carbon trades: policy issues a pre-determined quantity of emissions allowances and require that covered installations surrender an allowance for each ton of CO_2 they emit during the year. The availability of allowances (scarcity or abundance), combined with their tradability, creates a market price for allowances. This carbon price motivates committed emitters to choose between abatement options (investments) and purchasing on the market.

Carbon price on the EU-ETS market had a strong volatility, due to many aspects (political, social, related to the accountability, etc...) which have drugged the market. Today's situation is very unfortunate: 6-7 €/tCO_2 is the reference price due to the big oversupply of credits mainly generated by the deeply depressed economy, significant uncertainties about the allowance scarcity post-2020 and overlapping policies.

The potentiality of carbon initiatives spread all around the world: about 40 national and over 20 sub-national jurisdictions are putting a price on carbon, [16,17]. The geographical distribution of these carbon markets incorporates carbon taxes, emission trading schemes, offset and result based financing and are widespread in all the countries in the world. An important result is that the world's two largest emitters (China and USA) are also institutionalizing a carbon pricing.

IOP Conf. Series: Materials Science and Engineering **90** (2015) 012085 doi:10.1088/1757-899X/90/1/012085

Data shows carbon prices from under 0.8 €/tCO_2 to 150 €/ tCO_2 and differences do not reflect the economic fundamentals of a mature market. Such a large variation of the price is related to the different measures adopted: Carbon taxes (which are fixed) or national and subnational ETS (whose price is the consequence of a traded market).

One important issue for the role of compressed air in carbon markets is the prediction of carbon price, which for sake of simplicity, here is discussed concerning EU-ETS.

The main driving factors of CO_2 allowance prices can be divided into two macro categories, [18]: (i) policy and regulatory issues, and (ii) market fundamentals that directly concern the production of CO_2 and thus demand and supply of CO_2 allowances. Policy and regulatory issues have a long-term impact on prices: political decisions, regulatory and operating guidelines, market events related to allowances verification, length of trading periods, uncertainty about future political decisions are among the most important causes. Market fundamentals are related to the balance between demand and supply. Demand for permits is affected by economic growth and weather; supply depends on commodity prices (coal, oil, and gas), technology advancement, renewable energy production and the importance of other existing instruments (CER, ERU, CDM, IJ). The degree of freedom accepted on allowances, banking and borrowing are additional issues which modify prices.

The copious literature on the subject is oriented to predict a price close to 12 €/tCO_2 at most [19], 14 €/tCO_2 in 2020 [20], 11.34 €/tCO_2 [21], 16 €/tCO_2 in the short term and 30 €/tCO_2 in the medium term [22], around 8 €/tCO_2 in the short term and 35 €/tCO_2 in the medium term [23], 15 €/tCO_2 in the short term and 22 €/tCO_2 in the medium term according to the 550 ppm scenario or 50 €/tCO_2 in the short term and 70 €/tCO_2 in the medium term according to the 450 ppm scenario [18]. Bloomberg New Energy Finance (2013) estimates a price of 30 €/tCO_2 without back-loading in the mean term.

For what concerns this analysis, an average value of 15 €/tCO_2 by 2010 has been assumed. In the eventuality of a more stringent CO_2 commitments, like for instance the 450 ppm scenario by IEA [3], the 30-50 €/tCO_2 range could be reached.

4.2. Fossil fuel saving market

As for CO_2, fuel saving also is exchanged in its own market where demands of saving (requested, and therefore bought by subjects which are obliged by law to reach certain targets) is balanced by supply (offered and therefore sold, by subjects which have reached certified savings). The units exchanged are called Energy Saving Certificates (ESC) and they could be converted into equivalent fossil fuel.

ESCs have been established in several countries as a mechanism through which third parties, such as commercial and industrial companies, can help the other parties comply with the savings targets imposed on them by the local legislation. ESCs are tradable, and are associated to a unit of energy saved in electrical form or, directly, a ton of oil equivalent, TOE. This last definition allows to consider wider savings, i.e., of a thermal type.

When electrical energy saving has to be transformed into TOE, the efficiency of the electric generation must be assumed. It is quite different in different countries (30 % in France and Sweden, to 46 % in Luxembourg and Netherlands). Mean European datum is 37.5 % and, when referring it to the worldwide scenario, most part of the countries has much lower efficiency with respect to the mean averaged European level. These efficiency variations suggest a common equivalence rule in a scenario in which the ESCs markets would be homogenized and traded in all the world in order to avoid financial speculations. Same consideration applies to the carbon markets, too.

When thermal energy saving has to be transformed into TOE, the determination is much easier, being the equivalence coefficient equal to the baseline boiler or to the steam generator efficiency.

The UK was the first country to set up an Energy Savings Certificates scheme, starting in 2002. Having achieved good results during the first period (2002-2005), the mandatory savings requirement for energy suppliers was doubled, to a total of 130 TWh cumulative and actualized. The actualization takes into account the obsolescence of an energy saving device during time, from its first installation, [24].

A fossil fuel savings trading scheme has also been tested in Italy since 2005. These schemes (also implemented in Denmark and in the Flemish region of Belgium) have been recognized to be effective at the European level, as directive 2012-27-UE on energy efficiency, adopted in late 2012, requires Member States to enact similar schemes by 2014.
Many other countries have expressed an interest in this type of mechanism and are studying its applications. In USA currently, 21 states have energy efficiency targets, mandatory or voluntary; a strong diversification among them occurs, [25]. New South Wales's (Australia) ETS program is ready to be launched. Recently, India released a Climate Action Plan that encouraged the creation of a national market for ESCs and China also is debating on the scheme that should be more suitable, [26].
In general, the process of measuring the energy saved requires establishing a baseline for energy use and/or demand before and after the implementation of a project. Specific market design rules are needed in order to evaluate the effectiveness of the saving, its consistence with respect to standards, the equivalence between thermal, mechanical and electrical savings. Many compliance markets require third party verification using a measurement and verification protocol.
At the present time, the spread of these market-based tools is wide in the world. The energy saving commitments are, in fact, strongly country based, answering to specific national commitments.
One important aspect, which increases the difficulty in comparing ESCs schemes, is related to the period in which the certificates are generated and, therefore, spent.
Most established ESCs markets are certainly in Europe, in UK, France (FR), Denmark (DK) and Italy (IT): these report a main difference which is worthy of attention.
In UK, FR and DK the saving is rewarded in the year the energy saving technology takes place – all savings are calculated and contributed to the first year. The number of certificates NESC are calculated according to the energy saving (with respect to an energy baseline consumption), to the life span (LSPAN, technical lifetime) of the activity and to a coefficient which represents the discount coefficient which considers that the original saving is diminishing during time due to the obsolescence. LSPAN is defined for different technologies and it doesn't match in different countries. These certificates are recognized the same time as the investment for the saving takes place (one at a time), supposing that the saving will take place over years thereafter.
In Italy, a financial lifespan is fixed for different energy saving actions and (5 or 8 years) and during this period certificates are paid, when the saving is in reality put in place. The value of these certificates depends on the current values on the market: so, uncertainties can occur about the future revenues. Certificates are eligible effectively when energy is saved, [27], avoiding operative risks. ESCs are in expressed in TOE (Tons of oil equivalent) and a specific coefficient is introduced which transforms energy saved (thermal or electrical) into TOE. ESCs are evaluated according to a discounting procedure of the saving during the period between technological life span (LSPAN) and overall ESCs are redistributed during the overall lifespan period. A discount coefficient is defined. A comprehensive description is reported in [28, 29].
Generally speaking, price comparison among ESCs in UK, FR, DK and IT is not straightforward and advantages and drawbacks are a matter of discussion: a comprehensive analysis can be found in [30-33].
In spite of the differences, when reference is made for "conventional" saving measures (lamp, boiler replacement, insulation of building, etc…), the overall discounted financial value gets close to a common value, [24,25]. Considering the scope of the ESCs (primary energy saving in fossil fuel form), one could say that the Italian market appears be more mature and comprehensive.

5. Financial value of the compressed air energy saving on the environmental markets

Energy consumption at European level due to compressed air has been updated by PNEUROP recently, [7]. Results of energy consumption for fixed speed compressors (screw and vanes) was 36.58 TWh, while for variable speed the figure decreases to 19.17 TWh. An overall consumption close to 56 TWh was estimated. This datum is 30 % less than a reference estimate presented in [5], equal to 80

TWh. The older estimate was produced in a period in which a weak energy saving awareness was present in the compressed air industry. Apart from the unavoidable uncertainties, the difference could be considered as the energetic improvement of the sector in the past 15 years but also as a result of the recent, still ongoing, economic crisis. Intrinsic uncertainties are very frequent when dealing with this analysis.

The potential energy saving associated to the compressor replacement in the actual European market depends on the target of the efficiency, or of the specific energy, below which a compressor can't be operated. A threshold must be defined for the compressor specific consumption dependent on flow rate and pressure delivered, or for the compressor efficiency only dependent by flow rate.

Some attempts to define an energy baseline have been done: in [34] specific energy consumption at 7 bar per Nm^3 (0°C@1 bar) is defined equal to 130 Wh/Nm3 when premium machines are set at 85 Wh/Nm^3. Nevertheless, real compressors operate at different pressures and the baseline energy datum (130 Wh/Nm^3) can't be used. PNEUROP, [7], made a wider reference, introducing the adiabatic efficiency of the compressor: a wide market characterization is presented in terms of the adiabatic efficiency reported as a function of flow rate, for fixed or variable speed compressor technology. Observing the distribution of the data of different machines on the compressor adiabatic efficiency vs flow rate, some replacement scenarios have been done setting a cutoff curve for the efficiency. According to PNEUROP, an energy saving in the range 0.7-2.1 TWh is reachable for fixed speed compressors, thanks to different replacement scenarios. With the same assumptions, a saving in the range 0.3-1.2 TWh is achievable for variable speed compressors. For all compressors, 1-3.3 TWh is the potential range of achievable energy improvement.

Considering the rule of thumb, a double saving can be assumed. Therefore, 2-7 TWh can be "put on the energy saving market" if the whole compressed air chain is improved (leakages & pressure losses optimization and compressor replacement).

The quantity of CO_2 that could be traded is equal to 1-2.5 10^6 tCO_2, whose value is 15-35 M€ on the ETS-EU market, at 15 €/tCO_2. This result has been calculated according to the European efficiency electricity generation (0.325 $kgCO_2$/kWh). Finally, compressed air energy saving, considering the CO_2 credits and ESCs, has a value in the range 165-385 M€.

In a scenario in which both energy saving and CO_2 reduction become more severe and stricter commitments will take place, the overall near future values could approach 200-500 M€, considering a highly plausible CO_2 price at 50 €/tCO_2.

This result could be achieved if the compressed air sector would perform a lobby action at the European level according to which efforts done on energy efficiency by compressed air users would be recognized as tradable in the EU-ETS market.

The electrical saving of 2-7 TWh can be transformed into TOE saved, when a national average grid efficiency is defined. With an average electrical efficiency equal to 0,187 10^{-3} TOE/kWh (Italy), the compressed air saving at European level has a value of 0.56-1.3 10^6 TOE. Adopting as it is allowed, a technological life span equal to 15 years and financial lifespan equal to 5 years, and an energy discounting coefficient equal to 0.02, the overall saving in the technological lifespan period is 2.65 times the saving evaluated at the first year. The overall fossil fuel saved fits in the range 1.5-3.5 10^6 TOE. This saving is equally redistributed in a 5 year period (financial lifespan), one-fifth of the total for each year. The corresponding financial value would be 150-350 M€ considering an average price on the market close to 100 €/TOE.

A useful number can be found referring the overall mean financial saving (assumed equal to 350 M€) to the overall electricity saving (a mean value equal to 4.5 TWh). A specific value related to the compressed air chain is equal to 8 10^{-2} €/kWh which is really noteworthy.

A common synergic activity at the compressed air sector level should be put in place.

6. Conclusions

Compressed air in industry is fully based on the use of electrical energy, most part of which is produced using fossil fuels. The saving potential is huge, in a range of 25-50 % of the present consumption if leakage control, pressure readjustment, and compressor replacement are considered. This last contribution represents 50 % of the overall saving. CO_2 and fossil fuel savings are committed in almost all the countries in the world. Two new immaterial markets have been established: carbon and fossil fuel avoided.

Making reference to CAGI's data and to the more recent PNEUROP analysis, the paper describes (a) the carbon dimension of the actual compressor technology, (b) the equivalent fossil fuel consumption associated to the unit of compressed air.

Concerning the carbon content of compressed air in the European context, at 8 bar (a), the CO_2 emission is in the range of 37-33 gCO_2/m^3 when flow rate is in the range 5-50 m^3/min respectively. This datum should be multiplied by a factor of 2.25 in China, 3 in India and South Africa and1.75 in USA.

Specific energy consumption of different compressors reveals a significant scatter among different machines. Air cooled compressors, operating at 8 bar (a), at 10 m^3/min have a mean specific consumption equal to 6.8 kW/m^3/min with a scatter equal to 1.5 kW/m^3/min. At 30 m^3/min, mean specific consumption decreases at 6.5 kW/m^3/min with a scatter of 1 kW/m^3/min. Larger machines (> 50 m^3/min) have a specific consumption which approaches 6.2 kW/m^3/min. The scatter could be considered as the energy gain when machines are replaced.

PNEUROP's assumptions on the energy saving referred to machine replacement in the European context is 1-3.5 TWh. An overall energy saving considering leakage control and pressure readjustment can be easily reachable. The "doubling" rule of thumb allows to estimate the overall energy saving, equal to 2-7 TWh, including a "mean" leakage control and pressure readjustment.

This result corresponds to a tradable quantity of carbon credits equal to 1-2.5 10^6 tCO_2 and to a tradable saving of fossil fuel equal to 1.5-3.5 10^6 TOE, considering a technological life span equal to 15 years and a financial life span equal to 5 years.

Considering a mean future carbon credit exchanged at 50 €/CO_2 and a mean future fossil fuel saving exchanged at 100 €/TOE, a specific market value (considering the CO_2 not emitted and the fuel saved) of the compressed air chain is close to 8 10^{-2} €/kWh. This datum is close to the net electricity cost in many countries and it represents a financial potential for compressed air users which can't be neglected. A positive lobbying action of compressed air users could be done at European level.

Acknowledgments

The Author is grateful to Ing. E. Mattei S.p.A. for continuous research funding and support. Dr. Giulio Contaldi, CEO of the Company, is particularly acknowledged.

References

[1] BP Energy Outlook 2035, January 2014 available at www.bp.com/energyoutlook#BPstats
[2] Bergen @ Energi AS, Energy Outlook, 2014, available at https://www.bergen-energi.com/
[3] Energy Outlook 2014 (WEO-2014), available at www.worldenergyoutlook.org/
[4] IPCC Fifth Assessment Report, Climate Change 2013: The Physical Science Basis" (AR5) available at: https://www.ipcc.ch/publications_and_ data/
[5] P.Radgen, E. Blaustein (Eds), Compressed Air systems in the European Union. Energy, Emissions, Saving, Potential and Policy actions, LOG_X, ISBN 3-932298-16-0
[6] R. Saidur et al. A review on compressed air energy use and energy saving, Journal of Renewable and sustainable energy Reviews, 14 (2010), 1135-1153
[7] PNEUROP, European Association of manufacturers of compressors, vacuum pumps, pneumatic – See documents on: http://www.eco-compressors.eu/documents.htm).
[8] Compressed air energy input and useful energy (Adapted from North West Energy Alliance), Copyright 2007 CEA Tech. Inc

[9] Compressed air: opportunities for the business – © The Carbon Trust 2012, January 2012.
[10] GPG 385 Energy efficient compressed air systems – Carbon Trust by British Compressed Air Society Ltd, February 2005
[11] F. Da Cunha, Compressed air: energy efficient reference guide, CEA Technology Inc. (CEATI) Customer Energy Solution Interest Group (CESIG) , 2007
[12] CAGI, Compressed Air Gas Institute datasheets, available at www.cagi.org
[13] R. Cipollone, Sliding vane rotary compressor technology and energy saving, Proceeding of the IMECHE, Part E: Journal of Process Mechanical engineering, 2014, available at http://pie.sagepub.com/contents/early /2014/08/13 /09544089 14546146
[14] R. Cipollone, D. Vittorini, Energy saving potential in existing compressors, Purdue University, International Compressor Engineering Conference, Paper 2379, available at http://docs.lib.purdue.edu/icec
[15] Energy Outlook 2013 (WEO-2013), available at www.worldenergyoutlook.org/
[16] World Bank Group, Climate Change, State and trends of carbon prizing, Report n. 88284 Washington DC, May 2014
[17] McKynsey&Company - Pathways to a low carbon economy – http://www.mckinsey.com/global GHGcostcurve, 2010
[18] C. Carraro, A. Favero- The Economic and Financial Determinants of Carbon Prices, Czech Journal of Economics and Finance, 59, 2009, no. 5, pag. 396-409
[19] Sander de Bruyn, Dagmar Nelissen, Marnix Koopman, Carbon leakage and the future of the EU ETS market - Impact of recent developments in the EU ETS on the list of sectors deemed to be exposed to carbon leakage, Delft, CE Delft, April 2013 (available from www.cedelft.eu)
[20] A.Öko-Institut, Strengthening the European Union Emissions Trading Scheme and raising climate ambition, Berlin : Öko-Institut, 2012
[21] B.Point Carbon, Nina Chestney, Poll – Analysts cut EU, UN carbon forecasts, Point Carbon website, Reuters News, accessed 18 December 2012, at www.pointcarbon.com/news/ reutersnews/1.2087144
[22] F. Roches, Scenarios for the ETS Carbon Prize, Euroelectric Conference "The Future role of EU-ETS" Brussels October 5th, 2102
[23] A. von Schemde, B. Tennbakk - TCG-Insight 5-2014: EU ETS reform is set to increase the cost of emitting CO_2,Thema Consulting Group, available at http://www.t-cg.no/
[24] ADEME Energy Saving Certificates 2011-2013, Knowledge for Action – 2012, www.ademe.fr
[25] Robyn Liska and Devin Canavan, Bottom Line on Energy Savings Certificates, October 2008 available at http://www.wri.org/publication/bottom-line-energy-savings-certificates
[26] W. Xiaodong - White certificates trading, green certificates trading, emission trading: Which one to choose ? EASCS, September 11th, 2013, available at: https://www.thepmr.org/system/files/ documents/WB_Beijing %20 Office%20coordination%20of%20tra ding%20schemes%20September%202013.pdf
[27] M. Togeby, K. D. Mikkelsen, E. James-Smith - Design of White Certificates Comparing UK, Italy, France and Denmark, Ea Energy Analyses, November 2007, available at: http://www.ea-energianalyse.dk /reports/710_White_certificates_ report_19_Nov_07.pdf
[28] ENEA Working Group on white certificates I titoli di efficienza energetica. Cosa sono e come si ottengono i "certificati bianchi" alla luce della Delibera EEN 9/11 GUIDA OPERATIVA/2, 2012 ENEA Agenzia nazionale per le nuove tecnologie, l'energia e lo sviluppo economico sostenibile, available at http://www.enea.it/it/ produzione- scientifica/ pdf-volumi/v2012-guida-cb2.pdf
[29] L. De Sanctis, D. Ranieri - Il Meccanismo dei certificati bianchi, Naples Energy-Med, April 2013 available at: http://www.enea.it/it/comunicazione /events/energy-med-2/Napoli12aprile2013bis.pdf
[30] S. Rezessy, P. Bertoldi – Energy Supplier obligations and white certificate Schemes: Comparative analysis of results in the European Union, Energy Policy V38:3, March 2010
[31] L. G. Giraudet, D. Finon – On the road to a unified market for energy efficiency: the contribution of white certificates schemes – Fondazione ENI E. Mattei, Sustainable development Series, nota di Lavoro 132.2010, Editor. Carlo Carraro available at: http://www.feem.it/userfiles/attach/ 201010191056554 NDL2010-132.pdf
[32] R. Cowart- The Regulatory Assistance Project-Energy Savings Obligations and White Certificates- Energy Regulators Regional Association– http://www.raponline.org
[33] M. Togeby, et al. - Design of White Certificates Comparing UK, Italy, France and Denmark, Ea Energy Analyses, November 2007 - http://www.ea-energianalyse.dk/reports/710 White certificates report 19 Nov 07.pdf
[34] European Commission. Reference Document on Best Available Techniques for Energy Efficiency, February 2009 available at: http://eippcb.jrc.ec.europa.eu/reference/BREF/ENE_Adopted_02-2009.pdf